# Diagnosis and Treatment of Spinal Cord Injury

# Diagnosis and Treatment of Spinal Cord Injury

## The Neuroscience of Spinal Cord Injury


Edited by

**Rajkumar Rajendram**
Department of Medicine, King Abdulaziz Medical City, King Abdulaziz International Medical Research Center, Ministry of National Guard - Health Affairs, Riyadh, Saudi Arabia
College of Medicine, King Saud bin Abdulaziz University of Health Sciences, Riyadh, Saudi Arabia

**Victor R. Preedy**
King's College London, London, United Kingdom

**Colin R. Martin**
Institute for Health and Wellbeing, University of Suffolk, United Kingdom



ELSEVIER

ACADEMIC PRESS
An imprint of Elsevier

Academic Press is an imprint of Elsevier
125 London Wall, London EC2Y 5AS, United Kingdom
525 B Street, Suite 1650, San Diego, CA 92101, United States
50 Hampshire Street, 5th Floor, Cambridge, MA 02139, United States
The Boulevard, Langford Lane, Kidlington, Oxford OX5 1GB, United Kingdom



ISBN 978-0-12-822498-4
SET ISBN 978-0-323-99198-8

For information on all Academic Press publications
visit our website at https://www.elsevier.com/books-and-journals

*Publisher:* Nikki Levy
*Acquisitions Editor:* Natalie Farra
*Editorial Project Manager:* Timothy Bennett
*Production Project Manager:* Paul Prasad Chandramohan
*Cover Designer:* Mark Rogers

Typeset by STRAIVE, India

# Dedication

I would like to dedicate this book to my wonderful daughter, Dr. Caragh Brien, of whom I am so incredibly proud.

Colin R. Martin

# Contents

## 29. Exploring the exogenous and endogenous effects of melatonin on spinal cord injury

*Dewan Md. Sumsuzzman, Kanghui Park, and Yonggeun Hong*



## 30. High-intensity interval training in individuals with spinal cord injury

*Byron Lai, Jereme Wilroy, and Gordon Fisher*



## 31. Stem cells and chronic spinal cord injury: Overview

*Leonardino A. Digma, Pavan S. Upadhyayula, Joel R. Martin, and Joseph D. Ciacci*

# Contributors

*Numbers in parentheses indicate the pages on which the authors' contributions begin.*

**L. Abou** (117), Department of Kinesiology and Community Health, University of Illinois at Urbana-Champaign, Urbana, IL, United States

**Monzurul Alam** (245), Department of Biomedical Engineering, The Hong Kong Polytechnic University, Hong Kong

**George A. Alexiou** (191), Department of Neurosurgery, University Hospital of Ioannina, Ioannina, Greece

**Melchor Alvarez de Mon** (131), Hospital Universitario Príncipe de Asturias, Madrid; School of Department of Medicine and Medical Specialties, School of Medicine, Universidad de Alcalá (UAH), Alcalá de Henares, Spain

**Ángel Arévalo-Martín** (131), Laboratory of Neuroinflammation, Hospital Nacional de Parapléjicos, Toledo, Spain

**Stefano Artiaco** (299), Department of Orthopaedics and Traumatology, CTO Hospital, University of Turin, Turin, Italy

**Sevda C. Aslan** (229), Department of Neurological Surgery, Kentucky Spinal Cord Injury Research Center, University of Louisville, Louisville, KY, United States

**Manuel Astray-Lopaz** (173), Unidad de Lesionados Medulares. Hospital de A Coruña, A Coruña, Spain

**Linda Barclay** (501), Department of Occupational Therapy, Monash University, Frankston, VIC, Australia

**Bruno Battiston** (299), Department of Orthopaedics and Traumatology, Hand and Upper limb Surgery Unit, CTO Hospital, Turin, Italy

**Bruce E. Becker** (491), University of Washington School of Medicine, Seattle, WA, United States

**Sophia Berg** (453), Department of Health Sciences, College of Education and Professional Studies, Central Washington University, Ellensburg, WA, United States

**David J. Berlowitz** (147), Department of Physiotherapy, The University of Melbourne; Institute for Breathing and Sleep, Austin Health, Heidelberg, Melbourne, VIC, Australia

**Ángela Bernabéu-Sanz** (95), Magnetic Resonance Department, Inscanner SL, Alicante; Anatomy and Histology Department, Institute of Bioengineering, University Miguel Hernández, Elche, Spain

**Étienne Bourassa-Moreau** (15), Department of Spinal Surgery, Sacré-Coeur Hospital of Montreal, Université de Montreal, Montréal, QC, Canada

**James J. Bresnahan** (41), Department of Rehabilitation Medicine, Thomas Jefferson University Hospital, Philadelphia, PA, United States

**Denisse Calderón-Vallejo** (475), Department of Physiology and Pharmacology; Department of Morphology, Center for Basic Sciences, Autonomous University of Aguascalientes, Aguascalientes, Ags, Mexico

**S. Canavero** (313), Global Initiative to Cure Paralysis (GICUP), Department of Orthopedics, Ruikang Hospital, Nanning, China; Heaven/Gemini Global initiative, Turin, Italy

**Ana Paula Silva Champs** (83), SARAH Network Rehabilitation Hospitals, Belo Horizonte, Minas Gerais, Brazil

**Anthony Charles** (3), Department of Surgery, University of North Carolina, Chapel Hill, NC, United States

**K. Chhatwani** (117), Department of Kinesiology and Community Health, University of Illinois at Urbana-Champaign, Urbana, IL, United States

**Joseph D. Ciacci** (397), Department of Neurological Surgery, University of California San Diego, San Diego, CA, United States

**Connor A. Clemett** (411), Department of Anatomy and Medical Imaging, School of Medical Sciences, Faculty of Medical and Health Sciences, University of Auckland, Auckland, New Zealand

**Angela Coniglio** (217), Spine Surgery Unit, Department of Orthopaedics and Traumatology, CTO Hospital, Turin, Italy

**Joshua D. Crossman** (359), Department of Pharmacology and Toxicology, College of Pharmacy, University of Arizona; Southern Arizona VA Health Care System, Tucson, AZ, United States

**Antonio Lopes da Cunha Júnior** (83), SARAH Network Rehabilitation Hospitals, Belo Horizonte, Minas Gerais, Brazil

**Jehane H. Dagher** (55), Montreal Rehabilitation Institute, Gingras Lindsay Pavillon; Physical Medicine and Rehabilitation, McGill University Health Centre, Montreal General Hospital Site, Montreal, QC, Canada

**Gustavo Correa Netto de Melo** (83), SARAH Network Rehabilitation Hospitals, Belo Horizonte, Minas Gerais, Brazil

**Camila Quel de Oliveira** (525), Discipline of Physiotherapy, Graduate School of Health, University of Technology Sydney, Ultimo, NSW, Australia

**María del Carmen Díaz-Galindo** (475), Department of Physiology and Pharmacology, Center for Basic Sciences, Autonomous University of Aguascalientes, Aguascalientes, Ags, Mexico

**Mª Eugenia Díaz-Recarey** (259), Unidad de Lesionados Medulares, Complejo Hospitalario Universitario A Coruña, A Coruña, Spain

**Leonardino A. Digma** (397), Department of Neurological Surgery, University of California San Diego, San Diego, CA, United States

**Dennis G. Dyck** (491), Psychology Department, Washington State University Spokane, Spokane, WA, United States

**Guillermo Asín-Prieto Eng** (563), Neurorehabilitation and Biomechanics Unit (HNP-SESCAM), Associate Unit CSIC, Biomechanics and Technical Aids Department, National Hospital for Paraplegics, SESCAM, Toledo; Gogoa Mobility Robots, S.L., Abadiño, Vizcaya, Spain

**Eduardo Fernández-Jover** (95), Anatomy and Histology Department, Institute of Bioengineering, University Miguel Hernández, Elche; Centro de Investigación Biomédica en Red (CIBER BBN), Madrid, Spain

**M. Elena Ferreiro-Velasco** (173), Unidad de Lesionados Medulares. Hospital de A Coruña, A Coruña, Spain

**María Elena Ferreiro-Velasco** (259), Unidad de Lesionados Medulares, Complejo Hospitalario Universitario A Coruña, A Coruña, Spain

**Gordon Fisher** (385), Department of Human Studies, Nutrition Obesity Research Center, Center for Exercise Medicine, University of Alabama at Birmingham, Birmingham, AL, United States

**Salomón Flórez-Jiménez** (15), Department of Spinal Surgery, Sacré-Coeur Hospital of Montreal, Université de Montreal, Montréal, QC, Canada

**Federico Fusini** (217, 299), Department of Orthopaedics and Traumatology, ASL CN1, Regina Montis Regalis Hospital, Mondovì, Cuneo; Department of Orthopaedics and Traumatology, CTO Hospital, University of Turin, Turin, Italy

**Angel Gil-Agudo** (563), Neurorehabilitation and Biomechanics Unit (HNP-SESCAM), Associate Unit CSIC, Physical Medicine and Rehabilitation Department; Neurorehabilitation and Biomechanics Unit (HNP-SESCAM), Associate Unit CSIC, Biomechanics and Technical Aids Department, National Hospital for Paraplegics, SESCAM, Toledo, Spain

**Ibone Gimenez-Jiménez** (173), Unidad de Lesionados Medulares. Hospital de A Coruña, A Coruña, Spain

**Massimo Girardo** (217), Spine Surgery Unit, Department of Orthopaedics and Traumatology, CTO Hospital, Turin, Italy

**Gabrielle Gour-Provençal** (55), Montreal Rehabilitation Institute, Gingras Lindsay Pavillon, Montreal, QC, Canada

**Marnie Graco** (147), Department of Physiotherapy, The University of Melbourne; Institute for Breathing and Sleep, Austin Health, Heidelberg, Melbourne, VIC, Australia

**Francisco Gutiérrez-Henares** (131), Rehabilitation Department, Hospital Nacional de Parapléjicos, Toledo, Spain

**Susan J. Harkema** (229), Department of Neurological Surgery, Kentucky Spinal Cord Injury Research Center, University of Louisville, Louisville, KY, United States

**L. Hawari** (117), Department of Kinesiology and Community Health, University of Illinois at Urbana-Champaign, Urbana, IL, United States

**P. Henderson** (313), Microsurgical Research Society, Icahn School of Medicine at Mount Sinai, New York, NY, United States

**April N. Herrity** (229), Department of Neurological Surgery, Kentucky Spinal Cord Injury Research Center; Department of Physiology, University of Louisville, Louisville, KY, United States

**Gillean Hilton** (501), Department of Occupational Therapy, Monash University, Frankston, VIC, Australia

**Yonggeun Hong** (373), Department of Rehabilitation Science, Graduate School of Inje University; Department of Physical Therapy, College of Medical Healthcare & Engineering, Inje University, Gimhae, Gyeongsangnam-do, Republic of Korea

**Charles H. Hubscher** (229), Department of Anatomical Sciences and Neurobiology, Kentucky Spinal Cord Injury Research Center, University of Louisville, Louisville, KY, United States

**Takahide Itokazu** (347), Department of Molecular Neurosciences; Department of Neuro-Medical Science, Graduate School of Medicine, Osaka University, Suita, Japan

**Francisco Jaramillo-González** (475), Department of Morphology, Center for Basic Sciences, Autonomous University of Aguascalientes, Aguascalientes, Ags, Mexico

**Lavina Jethani** (337), Physical Medicine and Rehabilitation, University of Texas Health Science Centre, Houston, TX, United States

**Georgios Kafritsas** (191), Department of Neurosurgery, University Hospital of Ioannina, Ioannina, Greece

**Ali Reza Khalatbary** (437), Molecular and Cell Biology Research Center, Faculty of Medicine, Mazandaran University of Medical Sciences, Sari, Iran

**Khosrow Khalifeh** (325), Department of Biology, Faculty of Sciences; Department of Biotechnology, Research Institute of Modern Biological Techniques, University of Zanjan, Zanjan, Iran

**C.Y. Kim** (313), Global Initiative to Cure Paralysis (GICUP), Department of Orthopedics, Ruikang Hospital, Nanning, China; Heaven/Gemini Global initiative, Turin, Italy; Konkuk University School of Medicine, Seoul, South Korea

**Kyoung-Tae Kim** (425), Department of Neurosurgery, School of Medicine, Kyungpook National University, Kyungpook National University Hospital, Daegu, Republic of Korea

**Philip J. Koehler, III** (41), Department of Rehabilitation Medicine, Thomas Jefferson University Hospital, Philadelphia, PA, United States

**Radha Korupolu** (337), Physical Medicine and Rehabilitation, University of Texas Health Science Centre; TIRR Memorial Hermann Hospital, Houston, TX, United States

**Elisa Lacroce** (549), Department of Chemistry, Materials and Chemical Engineering "Giulio Natta", Politecnico di Milano, Milan, Italy

**Byron Lai** (385), Department of Pediatrics, Division of Pediatric Rehabilitation Medicine, University of Alabama at Birmingham, Birmingham, AL, United States

**Rodrigo Lanna de Almeida** (69), Department of Spinal Cord Injury, Heidelberg University, Heidelberg, Germany

**Subum Lee** (425), Department of Neurosurgery, Korea University Anam Hospital, Seoul, Republic of Korea

**Jean-François Lemay** (199), School of Rehabilitation, Faculty of Medicine, University of Montreal, Montreal, QC, Canada

**Annie Levasseur** (287), Department of Biomedical Sciences, University of Montreal, Research Center CIUSSS Nord-de-l'Île-de-Montreal, Montreal, QC, Canada

**Yuanyuan Li** (461), Department of Neurosciences, University of California San Diego, La Jolla, CA, United States

**Elisa López-Dolado** (131), Rehabilitation Department, Laboratory of Interfaces for Neural Repair (LINER), Hospital Nacional de Parapléjicos, Toledo; Department of Medicine and Medical Specialties, School of Medicine, Universidad de Alcalá (UAH), Alcalá de Henares, Spain

**Paul Lu** (461), Department of Neurosciences, University of California San Diego, La Jolla; Veterans Administration San Diego Healthcare System, San Diego, CA, United States

**Megan K. MacGillivray** (539), Human Kinetics, St. Francis Xavier University, Antigonish, NS, Canada

**Jean-Marc Mac-Thiong** (15, 287), Department of Spinal Surgery, Sacré-Coeur Hospital of Montreal, Université de Montreal, Montréal; Department of Surgery, Faculty of Medicine, University of Montreal, Hôpital du Sacre-Coeur de Montreal, CIUSSS Nord-de-l'Île-de -Montreal, Montreal, QC, Canada

**Raquel Madroñero-Mariscal** (131), Rehabilitation Department, Fundación del Lesionado Medular and Hospital Universitario Infanta Leonor, Madrid; Laboratory of Interfaces for Neural Repair, Hospital Nacional de Parapléjicos, Toledo, Spain

**Gabriela Afonso Galante Maia** (83), SARAH Network Rehabilitation Hospitals, Belo Horizonte, Minas Gerais, Brazil

**Joel R. Martin** (397), Department of Neurological Surgery, University of California San Diego, San Diego, CA, United States

**Alessandro Massè** (217, 299), Department of Orthopaedics and Traumatology, CTO Hospital, University of Turin, Turin, Italy

**Gilles Maurais** (15), Department of Spinal Surgery, Sacré-Coeur Hospital of Montreal, Université de Montreal, Montréal, QC, Canada

**Antonio Montoto-Marqués** (259), Unidad de Lesionados Medulares, Complejo Hospitalario Universitario A Coruña, A Coruña, Spain; Facultad de Ciencias de la Salud, Universidad de A Coruña, A Coruña, España

**Kerri A. Morgan** (513), Washington University School of Medicine, St. Louis, MO, United States

**W. Ben Mortenson** (539), Rehabilitation Research Program, G.F. Strong Rehabilitation Centre; Department of Occupational Science and Occupational Therapy, University of British Columbia; International Collaboration on Repair Discoveries, Blusson Spinal Cord Centre (3rd Floor), Vancouver, BC, Canada

**Kristin E. Musselman** (199), Department of Physical Therapy, Temerty Faculty of Medicine, University of Toronto, SCI Mobility Lab, Lyndhurst Centre, Toronto Rehabilitation Institute-University Health Network, Toronto, ON, Canada

**Simon J. O'Carroll** (411), Department of Anatomy and Medical Imaging, School of Medical Sciences, Faculty of Medical and Health Sciences, University of Auckland, Auckland, New Zealand

**Fabiana Gonçalves Oliveira** (83), SARAH Network Rehabilitation Hospitals, Belo Horizonte, Minas Gerais, Brazil

**Patricia Orme** (41), Department of Rehabilitation Medicine, Thomas Jefferson University Hospital, Philadelphia, PA, United States

**Natalia Padula** (525), Acreditando Center of Neuromotor Recovery, Health and Wellness, Sao Paulo, Brazil

**Ramiro Palazón-García** (107), Physical Medicine and Rehabilitation Department, Hospital Nacional de Parapléjicos, Toledo, Spain

**Kanghui Park** (373), Department of Physical Therapy, Dong-ju College, Busan, Republic of Korea

**Vinood B. Patel** (577), University of Westminster, School of Life Sciences, Department of Biomedical Science, London, United Kingdom

**James C. Pendleton** (41), Department of Rehabilitation Medicine, Thomas Jefferson University Hospital, Philadelphia, PA, United States

**Giuseppe Perale** (549), Faculty of Biomedical Sciences, University of Southern Switzerland (USI), Lugano, Switzerland; Ludwig Boltzmann Institute for Experimental and Clinical Traumatology, Vienna, Austria

**Victor R. Preedy** (577), Diabetes and Nutritional Sciences Research Division, Faculty of Life Science and Medicine, King's College London, London, United Kingdom

**Kelly Pritchett** (453), Department of Health Sciences, College of Education and Professional Studies, Central Washington University, Ellensburg, WA, United States

**Laura N. Purcell** (3), Department of Surgery, University of North Carolina, Chapel Hill, NC, United States

**J. Luis Quintanar** (475), Department of Physiology and Pharmacology, Center for Basic Sciences, Autonomous University of Aguascalientes, Aguascalientes, Ags, Mexico

**Rajkumar Rajendram** (577), College of Medicine, King Saud bin Abdulaziz University for Health Sciences; Department of Medicine, King Abdulaziz Medical City, King Abdullah International Medical Research Center, Ministry of National Guard Health Affairs, Riyadh, Saudi Arabia; Stoke Mandeville Hospital, Buckinghamshire Healthcare NHS Trust, Aylesbury, United Kingdom

**Alessandro Rava** (217), Department of Orthopaedic and Traumatology, CTO Hospital, University of Turin, Turin, Italy

**X.P. Ren** (313), Global Initiative to Cure Paralysis (GICUP), Department of Orthopedics, Ruikang Hospital, Nanning, China; Heaven/Gemini Global initiative, Turin, Italy

**Brianna Rice** (453), Department of Health Sciences, College of Education and Professional Studies, Central Washington University, Ellensburg, WA, United States

**L.A Rice** (117), Department of Kinesiology and Community Health, University of Illinois at Urbana-Champaign, Urbana, IL, United States

**Andréane Richard-Denis** (287), Department of Medicine, Faculty of Medicine, University of Montreal, Hôpital du Sacré-Coeur de Montréal, CIUSSS Nord-de-l'Île-de-Montréal, Montreal, QC, Canada

**Salvatore Risitano** (217, 299), Department of Orthopaedics and Traumatology, CTO Hospital, University of Turin, Turin; Department of Orthopaedics and Traumatology, Maggiore Hospital of Chieri, Chieri, Italy

**Leonardo Fonseca Rodrigues** (161), Hospital São Vicente de Paulo, Rio de Janeiro, RJ, Brazil

**Miguel Rodríguez-Cola** (131), Internal Medicine Department, Hospital Nacional de Parapléjicos, Toledo, Spain

**Antonio Rodríguez-Sotillo** (173, 259), Unidad de Lesionados Medulares. Hospital de A Coruña; Unidad de Lesionados Medulares, Complejo Hospitalario Universitario A Coruña, A Coruña, Spain; Facultad de Ciencias de la Salud, Universidad de A Corunã, A Corunã, Espanã

**Filippo Rossi** (549), Department of Chemistry, Materials and Chemical Engineering "Giulio Natta", Politecnico di Milano, Milan, Italy

**Sebastian Salvador-De La Barrera** (173, 259), Unidad de Lesionados Medulares. Hospital de A Coruña; Unidad de Lesionados Medulares, Complejo Hospitalario Universitario A Coruña, A Coruña, Spain

**Bonita J. Sawatzky** (539), International Collaboration on Repair Discoveries, Blusson Spinal Cord Centre (3rd Floor); Department of Orthopaedics, University of British Columbia, Vancouver, BC, Canada

**Michele Scandola** (27), NPSY-Lab.VR & SCI-Research Group, Department of Human Sciences, University of Verona, Verona, Italy

**Rick G. Schnellmann** (359), Department of Pharmacology and Toxicology, College of Pharmacy, University of Arizona; Department of Neurosciences, College of Medicine, University of Arizona; Southern Arizona VA Health Care System; College of Pharmacy, University of Arizona; Southwest Environmental Health Science Center, University of Arizona; Center for Innovation in Brain Science, University of Arizona, Tucson, AZ, United States

**Natalie E. Scholpa** (359), Department of Pharmacology and Toxicology, College of Pharmacy, University of Arizona; Southern Arizona VA Health Care System, Tucson, AZ, United States

**Benjamin R. Scoblionko** (41), Department of Rehabilitation Medicine, Thomas Jefferson University Hospital, Philadelphia, PA, United States

**Noritoshi Sekido** (273), Department of Urology, Toho University Ohashi Medical Center, Meguro-City, Tokyo, Japan

**Akram Shirdel** (325), Department of Biochemistry, Faculty of Biological Sciences, Tarbiat Modares University, Tehran, Iran

**Epiphani C. Simmons** (359), Department of Pharmacology and Toxicology, College of Pharmacy; Department of Neurosciences, College of Medicine, University of Arizona, Tucson, AZ, United States

**Ethan Simpson** (539), Rehabilitation Research Program, G.F. Strong Rehabilitation Centre; Department of Occupational Science and Occupational Therapy, University of British Columbia; International Collaboration on Repair Discoveries, Blusson Spinal Cord Centre (3rd Floor), Vancouver, BC, Canada

**Gurkaran Singh** (539), Rehabilitation Research Program, G.F. Strong Rehabilitation Centre; Department of Occupational Science and Occupational Therapy, University of British Columbia; International Collaboration on Repair Discoveries, Blusson Spinal Cord Centre (3rd Floor), Vancouver, BC, Canada

**Crystal Lederhos Smith** (491), Elson S. Floyd College of Medicine, Washington State University, Spokane, WA, United States

**Maria Marta Sarquis Soares** (83), Internal Medicine Department, School of Medicine, Universidade Federal de Minas Gerais, Belo Horizonte, Minas Gerais, Brazil

**Tania Cristina Leite de Sampaio e Spohr** (161), Instituto Estadual do Cérebro Paulo Niemeyer, Rio de Janeiro, RJ, Brazil; Centogene GmbH, Rostock, Germany

**Dewan Md. Sumsuzzman** (373), Department of Rehabilitation Science, Graduate School of Inje University; Department of Physical Therapy, College of Medical Healthcare & Engineering, Inje University, Gimhae, Gyeongsangnam-do, Republic of Korea

**Kelly L. Taylor** (513), Murray State University, Paducah, KY, United States

**Paolo Titolo** (299), Department of Orthopaedics and Traumatology, Hand and Upper limb Surgery Unit, CTO Hospital, Turin, Italy

**Susan Tucker** (513), Washington University School of Medicine, St. Louis, MO, United States

**Janelle Unger** (199), School of Physical Therapy, Western University, London, ON, Canada

**Pavan S. Upadhyayula** (397), Department of Neurological Surgery, University of California San Diego, San Diego, CA, United States

**Renato Vilas-Boas** (259), Unidad de Lesionados Medulares, Complejo Hospitalario Universitario A Coruña, A Coruña, Spain

**Spyridon Voulgaris** (191), Department of Neurosurgery, University Hospital of Ioannina, Ioannina, Greece

**Jereme Wilroy** (385), Department of Physical Medicine and Rehabilitation, University of Alabama at Birmingham, Rehabilitation Center, Birmingham, AL, United States

**E. Wong** (117), Department of Kinesiology and Community Health, University of Illinois at Urbana-Champaign, Urbana, IL, United States

**Toshihide Yamashita** (347), Department of Molecular Neurosciences; Department of Neuro-Medical Science, Graduate School of Medicine, Osaka University, Suita, Japan

**R. Yarnot** (117), Department of Kinesiology and Community Health, University of Illinois at Urbana-Champaign, Urbana, IL, United States

**Avital Yohann** (3), Department of Surgery, University of North Carolina, Chapel Hill, NC, United States

**Hengze You** (245), Department of Biomedical Engineering, The Hong Kong Polytechnic University, Hong Kong

**Yong-Ping Zheng** (245), Department of Biomedical Engineering, The Hong Kong Polytechnic University, Hong Kong

**Andreas Zigouris** (191), Department of Neurosurgery, University Hospital of Ioannina, Ioannina, Greece

# Preface

Spinal injury affects about 10 million people annually worldwide. Many of these injuries are preventable and occur due to falls, violence, and road traffic accidents. However, spinal cord damage may also result from nontraumatic injury, for example, due to toxins, infection, cancer, intervertebral disc damage, and vascular disease.

Spinal injuries may cause significant, lifelong disabilities and thereby impact the family unit. Symptoms are varied and include paresthesia, spasticity, and loss of motor control. In extreme cases, affected individuals are chair- or bedbound for life. These debilitating injuries may also be accompanied by severe pain. Loss of bladder and bowel control increases dependency and further reduces quality of life.

Diagnosis is generally based on symptoms and various imaging techniques in the clinical context of a traumatic injury to the spinal cord. Various syndromes are associated with spinal cord injury (e.g., central, anterior, and posterior cord syndromes). However, a variety of pathological processes are involved in the initiation of primary and secondary damage to the spinal cord.

Physical trauma may be accompanied by multiorgan cellular and biochemical injury. These injuries include neurodegeneration, free radical damage, changes in gene expression, and physiological changes such as loss of the regulation of blood pressure. So the manifestation of symptoms and signs from any given injury may be highly variable.

The patient's functional outcome predominantly depends on the level of spinal column injury (i.e., cervical, thoracic, lumbar, or sacral), the location of the damage within the spinal cord (e.g., central, anterior, posterior), and the severity. This functional outcome may be modified to some extent by medical therapies and rehabilitation.

To fully comprehend and positively influence the trajectory of patients' outcomes, it is necessary to understand the fundamental principles of the conditions that arise as a result of spinal cord injury. Presently, the availability of much of this information is sporadic, in different scientific domains, and designed for different scientific specialties. Spinal cord injuries are diverse, so a multidisciplinary approach is needed. This is addressed in the two-volume set *The Neuroscience of Spinal Cord Injury* comprising the 2 books:

*Diagnosis and Treatment of Spinal Cord Injury*
*Cellular, Molecular, Physiological, and Behavioral Aspects of Spinal Cord Injury*

This book, *Diagnosis and Treatment of Spinal Cord Injury*, has the following six sections:

- *Setting the scene: introductory chapters*
- *Clinical features of spinal injury*
- *Diagnosis and evaluation*
- *Treatments: experimental and clinical*
- *Rehabilitation in spinal injury*
- *Resources*

Each chapter has the following sections:

- *An abstract (published online)*
- *Key facts*
- *Mini-dictionary of terms*
- *Applications to other areas of neuroscience*
- *Summary points*

The sections *Key facts, Mini-dictionary of terms,* and *Summary points* enable the reader to cross the transintellectual and transdisciplinary divides. The section *Applications to other areas of neuroscience* presents the translational aspects of the chapter and the applicability of the information.

*The Neuroscience of Spinal Cord Injury* is designed for research and teaching purposes. It is suitable for neurologists, surgeons, trauma specialists, psychologists, health scientists, public health workers, doctors, pharmacologists, and research scientists. It is valuable as a personal reference book and also for academic libraries that cover the domains of trauma and neurology. Contributions are leading national and international experts including those from world-renowned institutions. It is suitable for undergraduate and postgraduate students as well as lecturers and academic professors.

**Rajkumar Rajendram, Victor R. Preedy, and Colin R. Martin (Editors)**

Section A

# Setting the scene: Introductory chapters

Chapter 1

# Traumatic spinal cord injury and outcomes in low-resource settings

Avital Yohann, Laura N. Purcell, and Anthony Charles
*Department of Surgery, University of North Carolina, Chapel Hill, NC, United States*

## Abbreviations

| | |
|---|---|
| **ASIA** | American Spinal Injury Association |
| **ATLS** | advanced trauma life support |
| **CT** | computerized tomography |
| **GSW** | gunshot wound |
| **HIC** | high-income country |
| **ICU** | intensive care unit |
| **LMIC** | low- and middle-income country |
| **MRI** | magnetic resonance imaging |
| **RTI** | road traffic injury |
| **TSCI** | traumatic spinal cord injury |
| **YLD** | years lived with disability |

## Introduction

Traumatic spinal cord injuries (TSCIs) are devastating under any circumstance. However, patients in low-resource settings face unique challenges as they experience delays to care, a lack of prehospital care, and limited proper immobilization during transport, resulting in significantly higher in-hospital mortality than patients in high-income countries (HICs) (Bickenback, 2013; Eaton, Mukuzunga, Grudziak, & Charles, 2019; Lakhey, Jha, Shrestha, & Niraula, 2005). While long-term mortality of TSCI in HICs is now primarily due to chronic diseases, TSCI patients in low-income countries continue to die of preventable secondary causes (Bickenback, 2013). Those who survive struggle with access to proper assistive devices, rehabilitation services, follow up and health maintenance, and community reintegration (Bickenback, 2013). Moreover, while TSCI incidence is stable or decreasing in HICs, rates in low- and middle-income countries (LMICs) continue to rise (Jazayeri, Beygi, Shokraneh, Hagen, & Rahimi-Movaghar, 2015). Though most of the available evidence regarding the epidemiology and outcomes of TSCIs comes from HICs, the limited data from LMICs provide essential insights into improving TSCI prevention and management. This chapter will discuss the epidemiology, etiology, treatments, and outcomes of TSCIs in different LMICs globally.

## Epidemiology

The worldwide incidence of traumatic spinal cord injury (TSCI) is between a quarter and 1 million people annually (Bickenback, 2013). The Global Burden of Disease Study reported a global age-standardized incidence rate of 13 per 100,000 and a prevalence of 368 per 100,000 in 2016. The regions with the highest reported incidence of TSCIs are central and eastern Europe and central Asia. Because of armed conflict, Syria, Yemen, Iraq, and Afghanistan have the highest incidence (James et al., 2019) (Fig. 1). The incidence of TSCI appears to be stable or decreasing in HICs, but increasing in LMICs (Jazayeri et al., 2015). In 2016, TSCIs were responsible for an estimated 9.5 million years lived with disability (YLDs) (James et al., 2019). Available evidence on the epidemiology of TSCIs comes from HICs, but the limited data from LMICs provide essential insight into improving TSCI prevention and management strategies.

**Diagnosis and Treatment of Spinal Cord Injury. https://doi.org/10.1016/B978-0-12-822498-4.00001-4**

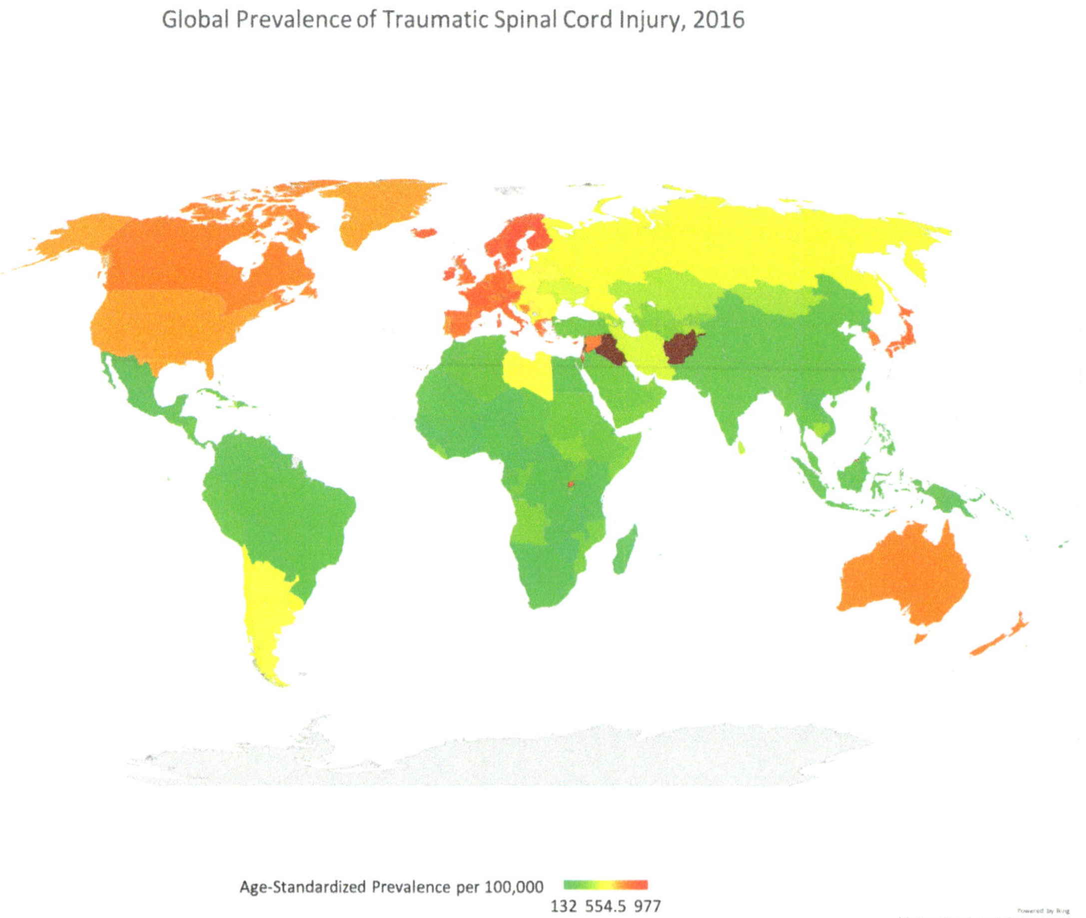


FIG. 1 Global prevalence of traumatic spinal cord injury. The figure illustrates the age-standardized prevalence of TSCI per 100,000 people by country in 2016 ( James et al., 2019). The countries with prevalence over 1300 per 100,000, shown in *dark red*, are Lebanon, Iraq, and Afghanistan.

### Sex and age

Males are disproportionately affected by TSCIs (Kang et al., 2017), with some countries reporting a male preponderance of over 90% (Deconinck, 2003; Gosselin & Coppotelli, 2005). Males are more likely to be in situations that place them at risk for TSCI or participate in high-risk activities (Kang et al., 2017) (Fig. 2).

In HICs, TSCIs follow a bimodal age distribution, with one peak in the second to third decade and another peak after age 65 (Kang et al., 2017). In LMICs, TSCIs predominantly affect younger patients. In the Eastern Mediterranean (specifically Iran, Pakistan, and Afghanistan), the mean age range is 26–38 years (Chabok et al., 2010; Deconinck, 2003). Similarly, in Southeast Asia, Sub-Saharan Africa, and Latin America, the mean age is 30–44 years (Barbetta et al., 2018; Bellet, Rashid, Jusabani, Dekker, & Temu, 2019). Studies in China and India report a mean age range of 45–51 years (Aleem et al., 2017; Zhou et al., 2016).

## Etiology

The most common cause of TSCIs in most regions is falls, followed by road traffic injuries (RTIs) ( James et al., 2019).

### Falls

Falls are the most common cause of TSCIs in Southeast Asia, parts of the Eastern Mediterranean, Sub-Saharan Africa, and the Western Pacific. In Southeast Asia, falls are responsible for 53%–84% of TSCIs (Bellet et al., 2019; Dhakal, Paudel, Dhungana, Gurung, & Kawaguchi, 2018). Falls in these regions are primarily occupation-related (Agarwal, Sud, &

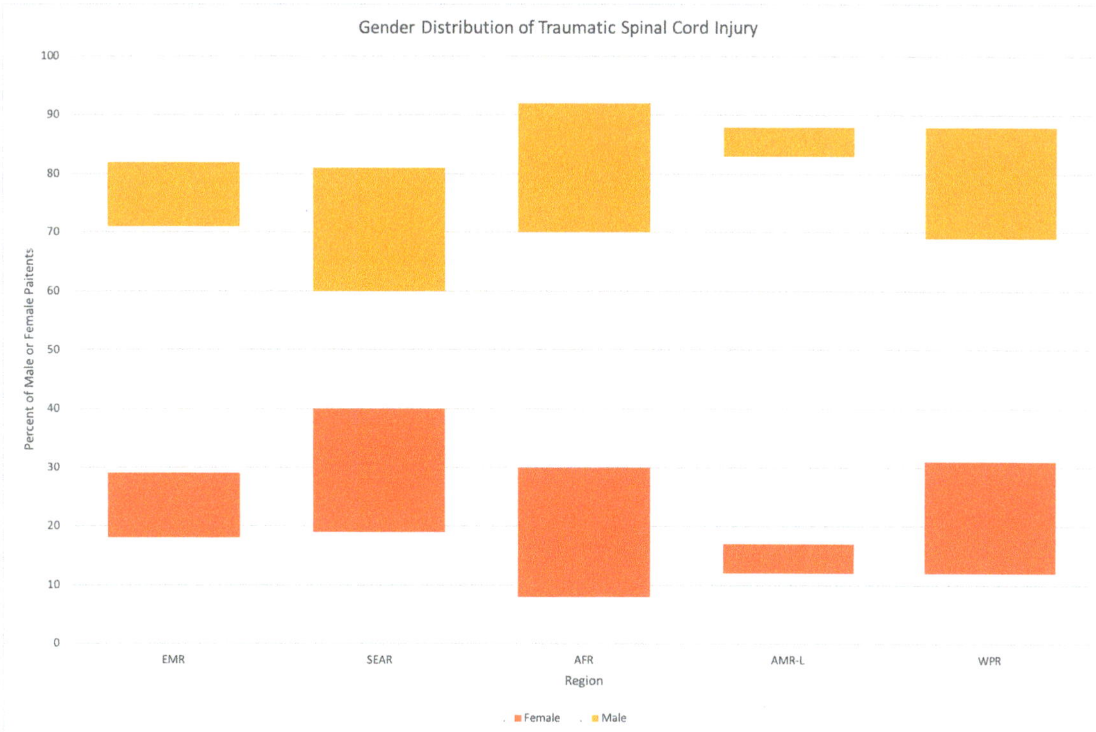


FIG. 2 Gender distribution of traumatic spinal cord injury by region. The range of proportions of spinal cord injured patients by sex is shown by region. *EMR*, Eastern Mediterranean; *SEAR*, Southeast Asia; *AFR*, Africa; *AMR-L*, Latin America, *WPR*, Western Pacific.

Shishehbor, 2016; Dhakal et al., 2018). In Iran and Pakistan, construction workers and well diggers are at high risk for falls due to unsafe work environments (Derakhshanrad, Yekaninejad, Vosoughi, Sadeghi Fazel, & Saberi, 2016; Rathore, Hanif, Farooq, Ahmad, & Mansoor, 2008). In Sub-Saharan Africa, falls are common from roofs, scaffolds, trees, or electric poles (Leidinger et al., 2019; Löfvenmark et al., 2015). In Cambodia, approximately 50% of TSCIs are secondary to falls from trees or houses (Choi et al., 2017; Makhni et al., 2019). Malaysia's high rates of falling TSCIs are attributed to migrant workers in high-risk conditions with limited safety regulations (Ibrahim et al., 2013). In China, falls are responsible for over 60% of TSCIs due to increased construction of buildings and subways (Liu et al., 2012; Liu, Liu, Gao, Li, & Li, 2020).

Falls from standing are also common. LMIC's growing aging population has resulted in more individuals vulnerable to TSCIs from ground-level falls (Yuan, Shi, Cao, Li, & Feng, 2018). Also, ground-level falls while carrying heavy objects atop one's head are common, particularly for females in Sub-Saharan Africa (Moshi, Sundelin, Sahlen, & Sörlin, 2017).

## Road traffic injuries

RTIs are the most common cause of TSCIs in Sub-Saharan Africa, Latin America, and parts of the Western Pacific. Studies from Ghana, Nigeria, and Botswana report up to 80% of TSCIs are due to RTIs (Ametefe et al., 2016; Löfvenmark et al., 2015). While in Malawi, Tanzania, and Sierra Leone, RTIs and falls are evenly distributed (Eaton, Hanif, Grudziak, & Charles, 2017; Gosselin & Coppotelli, 2005). In Latin America, 39%–44% of TSCIs are due to RTIs (Bellucci et al., 2015). Within the Western Pacific, RTIs are the most common cause of TSCI in Malaysia, responsible for 66% of cases (Ibrahim et al., 2013). One-third of TSCIs in Cambodia are due to RTIs, most of which can be attributed to motorcycle collisions (Choi et al., 2017; Makhni et al., 2019).

RTIs have surpassed falls as the leading cause of TSCIs in locations with rapid increases in motorization. In Brazil, RTIs are the leading etiology of TSCI in 2003, as the number of motorcycles in Brazil has increased by 491% (Bellucci et al., 2015). In high population centers in China, such as Beijing, RTIs have surpassed falls as the most common cause of TSCIs, as the number of private vehicles has increased from 1.2 million in 2002 to 4.8 million in 2010 (Chen et al., 2017; Hua et al., 2013; Yuan et al., 2018). Factors contributing to high rates of RTIs in Sub-Saharan Africa and other LMICs include lack of road maintenance, unsafe driving, road unworthy vehicles, and poor enforcement of traffic laws (Ametefe et al., 2016; Chen et al., 2017; Obalum, Giwa, Adekoya-Cole, & Enweluzo, 2009).

### Violence and other causes

Violence has generally become a less common cause of TSCIs worldwide. However, in areas with active conflict and terrorism, such as in North Africa and the Middle East (specifically Syria, Yemen, Iraq, and Afghanistan), it is the most common cause of TSCIs ( James et al., 2019). A study in Afghanistan found 59% of TSCI cases were due to bullets or shrapnel (shelling, gunshots, and mine explosions) (Deconinck, 2003). In Brazil, gun violence is responsible for nearly 30% of TSCIs (Barbetta et al., 2018; Bellucci et al., 2015). In South Africa, violence is a significant contributor, as up to 60% of TSCIs are attributed to gun violence, stabbing, and blunt assault mechanisms (Eaton et al., 2019; Löfvenmark et al., 2015). Assault is a less commonly reported cause of TSCIs in the rest of Sub-Saharan Africa, accounting for 2%–10% (Eaton et al., 2019; Löfvenmark et al., 2015).

Finally, sports-related activities compose a small proportion of TSCIs globally, such as diving into shallow water (6%–12% of Brazil's TSCIs), winter sports ($\leq$5% of TSCIs in China) (Bellucci et al., 2015; Chen et al., 2017; Liu et al., 2020). In Sub-Saharan Africa, specifically Nigeria and Botswana, sports-related activity is responsible for an estimated 2% of TSCIs (Löfvenmark et al., 2015; Obalum et al., 2009). A study from Iran reported a relationship between education level and TSCI etiology. Sports become a more common cause of TSCI with increasing years of education (primarily diving, gymnastics, wrestling), while violence and being struck by falling objects became less common (Derakhshanrad et al., 2016).

### Spinal cord level and severity

Overall complete cervical injuries are the most common, however significant TSCI level and severity variations exist globally (Kang et al., 2017). For example, the thoracic spine is the most commonly injured spinal level in Pakistan, Iran, and Afghanistan, with cervical injuries composing 30% of TSCIs (Deconinck, 2003; Derakhshanrad et al., 2016; Rathore et al., 2008). However, most Afghani patients have complete thoracic or lumbar injuries, and only 7% have cervical injuries, which is reflected in improved survival (Deconinck, 2003).

In Sub-Saharan Africa, the cervical spine is the most common injured site (30%–68% of cases), followed by the thoracic (10%–39%) and lumbar spine (4%–59%) (Ametefe et al., 2016; Bellet et al., 2019). Injury severity, however, varies significantly. In South Africa, Malawi, Tanzania, and Ghana, complete injuries range from 23% to 48% of patients, whereas up to 92% of TSCIs are complete in Nigeria and Sierra Leone (Ametefe et al., 2016; Bellet et al., 2019; Gosselin & Coppotelli, 2005; Obalum et al., 2009).

In Brazil, the most common location of TSCI is the thoracic spine (46%–61%), with approximately 60% being complete (Barbetta et al., 2018; Bellucci et al., 2015). Cervical injuries in Brazil were more common in automobile collisions, whereas thoracic injuries were more associated with motorcycle collisions and GSWs (Barbetta et al., 2018).

In China, 47%–76% of TSCIs are reported to be cervical, and 14%–55% of injuries are complete (Chen et al., 2017). Beijing has higher tetraplegia rates and complete injuries than other regions in China (Yuan et al., 2018). The cervical spine is generally the most common location of injury in Southeast Asia and the Western Pacific. In both regions, falls are associated with lumbar injuries and motorcycle and automobile collisions with cervical injuries (Agarwal, Upadhyay, & Raja, 2007; Choi et al., 2017). In Cambodia, lumbar injuries are more common than cervical injuries (Fig. 3) (Choi et al., 2017; Makhni et al., 2019).

## Diagnosis

Diagnosis of TSCI relies on a combination of history and clinical findings with the use of radiographic adjuncts. Clinical findings include flaccid paralysis and lack of response to painful stimuli below the level of injury, as well as bradycardia, priapism, and paradoxical breathing (Kebaish & Harris, 2020).

A complete TSCI produces total loss of all motor and sensory function below the level of injury. Nearly 50% of all TSCIs are complete. Even with a complete TSCI, the spinal cord is rarely completely transected. More commonly, loss of function is caused by a contusion to the spinal cord or ischemia of the spinal cord. In an incomplete TSCI, some function remains below the primary level of the injury.

Clinicians commonly utilize the American Spinal Injury Association (ASIA) grading scale to describe the injury severity (Kirshblum et al., 2011).

- ASIA A: a complete spinal cord injury with no sensory or motor function preserved
- ASIA B: an incomplete sensory injury with complete motor function loss

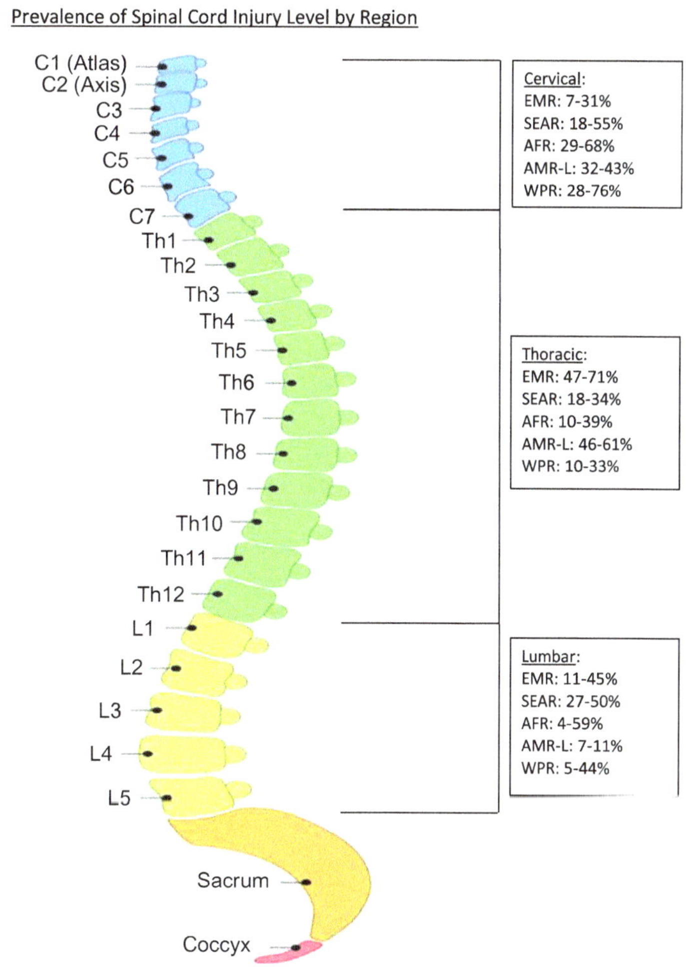


FIG. 3 Distribution of traumatic spinal cord injury level by region. The range of prevalence of injuries at different spinal levels across regions is reported from the available literature. *EMR*, Eastern Mediterranean; *SEAR*, Southeast Asia; *AFR*, Africa; *AMR-L*, Latin America; *WPR*, Western Pacific.

- ASIA C: an incomplete motor injury, where there is some movement, but less than half the muscle groups are anti-gravity (can lift against the force of gravity with a full range of motion)
- ASIA D: an incomplete motor injury with more than half of the muscle groups are anti-gravity
- ASIA E: normal

The more severe the injury, the less likely recovery will occur. Historically, the radiological diagnosis of TSCI started with plain spine radiography. However, with the technological advancements and availability, the entire spine may be imaged with computerized tomography (CT) as an initial screen to identify fractures and other bony abnormalities. For patients with known or suspected injuries, magnetic resonance imaging (MRI) helps delineate spinal cord anatomy and detect any blood clots, herniated disks, or masses that may be compressing the spinal cord. MRI is recommended if feasible and is especially useful if CT findings are discordant with symptoms and for surgical planning (Kebaish & Harris, 2020).

The majority of data regarding the imaging modalities used in diagnosing TSCI in LMICs come from Sub-Saharan Africa. Most patients received plain radiographs in those studies that reported diagnostic methods (71%–100%) (Bellet et al., 2019; Eaton et al., 2019). CT scans and MRIs are less commonly utilized. Reasons included lack of availability in the treatment facility or patient inability to pay for the diagnostic test. Between 7% and 59% of patients had a CT scan, and 41%–72% had an MRI (Ametefe et al., 2016; Leidinger et al., 2019). For reference, a 2019 survey of emergency medicine providers found that 99% and 97% of respondents in HICs reported access to in-hospital plain radiography and CT, respectively. Of respondents in LMICs, 91% reported access to in-hospital plain radiography, and 65% reported access to in-hospital CT (Alibhai, Hendrikse, & Bruijns, 2019).

## Prehospital and acute management

Care provided during the first 24–48 h following a traumatic TSCI is critical and can significantly influence TSCI outcomes (Gélis et al., 2011). Prehospital management requires rapid evaluation, including measurement of vital signs and level of consciousness; initiation of injury management, spinal immobilization to preserve neurological function until long-term spinal stability is established, and control of bleeding, body temperature, and pain; and prompt and safe access to the healthcare system.

Prehospital care is underdeveloped in low-income countries. Emergency medical services are frequently inadequate or nonexistent. For example, in Tehran, Iran, only one-third of patients with spinal injuries are transported by emergency medical services (Sharif-Alhoseini & Rahimi-Movaghar, 2014). Similarly, in Botswana, only 20% of patients are transported by ambulance (Löfvenmark et al., 2015). Difficult terrain and transportation costs delay care, and patients may not arrive at a hospital for several days after their injury. Studies from Tanzania and Nepal reported approximately one-third of patients took greater than 2 days from injury to reach a hospital (Bellet et al., 2019; Dhakal et al., 2018; Lakhey et al., 2005). In Afghanistan, the average time to a hospital after sustaining TSCI was 3.4 days (Deconinck, 2003).

Due to cost and resource limitations in LMICs, spinal precautions are infrequently practiced in the prehospital setting. Among patients with cervical spine injuries in Nepal, only 36% arrived at the hospital with a cervical collar (Munakomi, Bhattarai, & Cherian, 2017). In India, up to 98% of TSCI patients are not adequately immobilized (Aleem et al., 2017; Dhamangaonkar, Joshi, Kumar, & Goregaonkar, 2013). A study from Malawi reported that 37% of patients with TSCI had cervical collars placed at some point during their hospitalization and that no other spinal braces were utilized (Eaton et al., 2019).

## Treatment strategies

Management of spinal cord injuries is divided into phases. Acute phase management is based on Advanced Trauma Life Support (ATLS) principles, including prioritizing and treating life-threatening injuries to maximize survival, treating disabling injuries to minimize impairment, and limiting pain and suffering. Assessment should begin immediately upon arrival at the hospital and include: a medical history, signs and symptoms, neurological examination, radiological imaging, and laboratory testing (Bickenback, 2013). Signs and symptoms of spinal cord injury include weakness, sensory and motor deficits, bowel and bladder dysfunction, anatomical deformity, and localized tenderness.

Decisions about the best management strategy depend on the environment, available infrastructure, and surgical expertise. For TSCIs, there are risks and benefits to both nonoperative and surgical management. Many factors must be considered to determine the most appropriate management approach, including the level of injury, type of fracture, degree of instability, presence of neural compression, and impact of other concurrent injuries. Nonoperative management involves measures to immobilize the spine and to reduce dislocation with bed rest, traction of the spine, or the wearing of orthoses for 2–3 months. Indications for surgical management include spinal column stabilization and spinal cord decompression by reducing a dislocation or removing fracture fragments that cause neural compression.

Operative management of TSCI, within the first 48 h, has been associated with improved neurologic outcomes (Furlan, Noonan, Cadotte, & Fehlings, 2011). Nonoperative management of these injuries is associated with a fourfold increase in mortality than those with similar injuries managed operatively (Lessing, Lazaro, et al., 2020). Also, operative intervention is associated with early mobilization, better neurologic improvement, and higher return to wage-earning jobs (Lessing, Zuckerman, et al., 2020).

In LMICs, a high proportion of patients presenting with spinal cord injury do not undergo operative intervention. Studies have shown 0%, 30%, and 60% of TSCI patients undergo surgical intervention in Malawi, Nigeria, and Ghana, respectively (Eaton et al., 2019; Nwankwo & Uche, 2013). The primary reasons expressed in studies for nonoperative management or delays in surgical intervention are a paucity of neurosurgical workforce, lack of patient finances, delays in imaging access, and unavailability of operating room access (Nwankwo & Uche, 2013).

Globally, approximately 22.6 million new cases need neurosurgery consults annually, with 13.8 million persons requiring surgical intervention (Dewan Rattani, Baticulon, et al., 2019; Kanmounye et al., 2020). However, approximately 5.2 million neurosurgical cases go untreated, primarily due to lack of neurosurgical services and inadequate surgical workforce, with Africa bearing the lowest surgeon densities (Dewan, Rattani, Fieggen, et al., 2019). In 2018, the median workforce densities (neurosurgeons per 100,000 population) by WHO region were Americas (0.54), Africa (0.05), Eastern Mediterranean (0.34), Europe (0.99), South-East Asia (0.13), and Western Pacific (0.11) (Fig. 4) (Kanmounye et al., 2020). Approximately one neurosurgeon per 212,000 people is required to meet the healthcare demands from neurotrauma

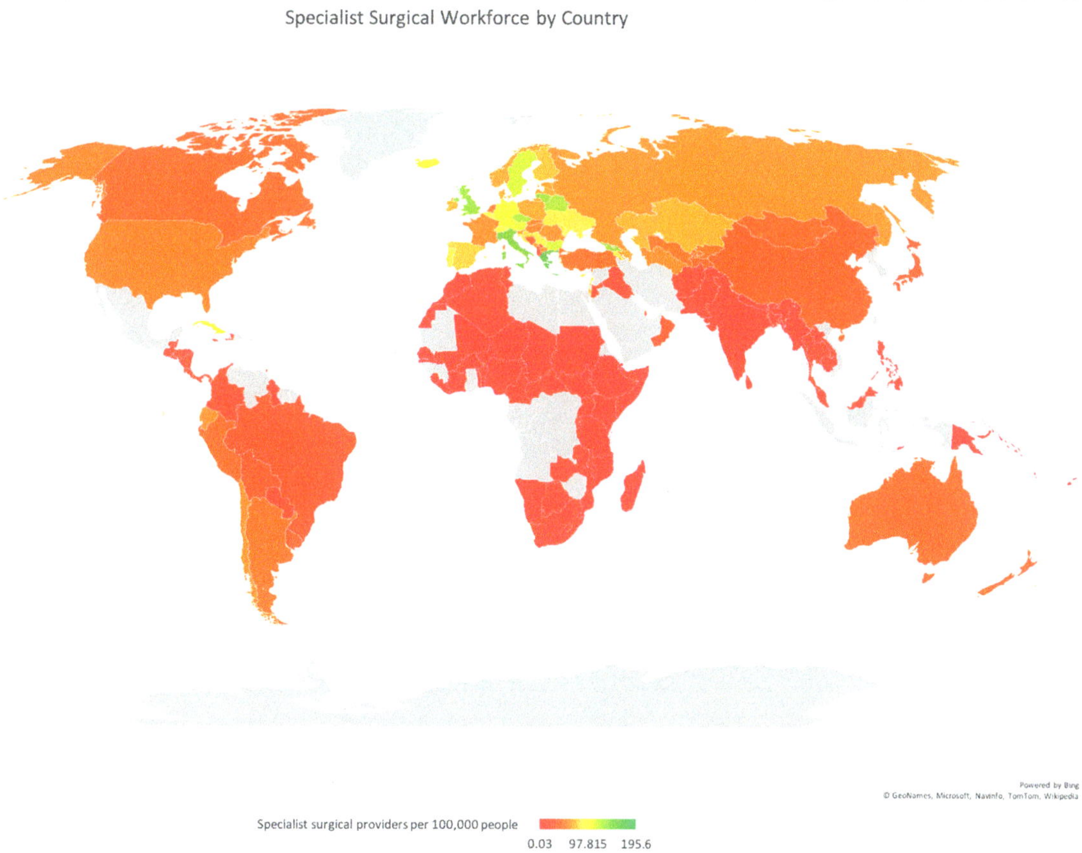


FIG. 4 Specialist surgical workforce by country. The figure displays the number of specialist surgical providers per 100,000 people (World Bank, 2018).

adequately, and an estimated 23,300 additional neurosurgeons globally are needed to address this significant deficit (Corley, Lepard, Barthélemy, Ashby, & Park, 2019).

Surgical care to patients and their families is often cost-prohibitive (James et al., 2019; Kumar et al., 2018). It is expected that patients and families must gather the funds to pay for the cost of surgical supplies, implants, or diagnostic radiology. High implant costs for spinal cord operative management can often lead to delay or deferment of operative treatment (James et al., 2019; Kumar et al., 2018). A Tanzanian study determined nonoperative management costs 212 USD per patient compared to 731 USD per patient for operative management (Lessing, Zuckerman, et al., 2020). However, operative management was determined to be cost-effective as the incremental cost per disability-adjusted life years averted was 112–158, 47–67, 50–71, 41–58 USD for quadriplegic, paraplegic, neurologic improvement, and neurologically intact patients, respectively (Lessing, Zuckerman, et al., 2020).

Ancillary support is also challenging in low-resource environments for patients needing treatment for spinal cord injuries. There is a lack of intensive care unit (ICU) beds and respiratory, physical, and occupational therapy. In the United States, the ICU bed density is approximately 20–32 beds per 100,000 population, compared to 5 beds per 100,000 in the United Kingdom, 2.8–4.6 beds per 100,000 in China, and 0.34–1.57 beds per 100,000 population in India (Gooch & Kahn, 2014; Murthy & Wunsch, 2012). There is a lack of critical care data in Sub-Saharan Africa. The estimated ICU bed density in Malawi is 0.1 beds per 100,000 population (Fig. 5).

## Post-acute medical care and rehabilitation

Appropriate medical care and rehabilitation can prevent TSCI complications and help patients achieve a meaningful life. Rehabilitation, defined as a "set of measures that assist individuals in achieving and obtaining optimal functioning in the interaction with their environments" (World Health Organization, 2011), should commence in the acute phase for people

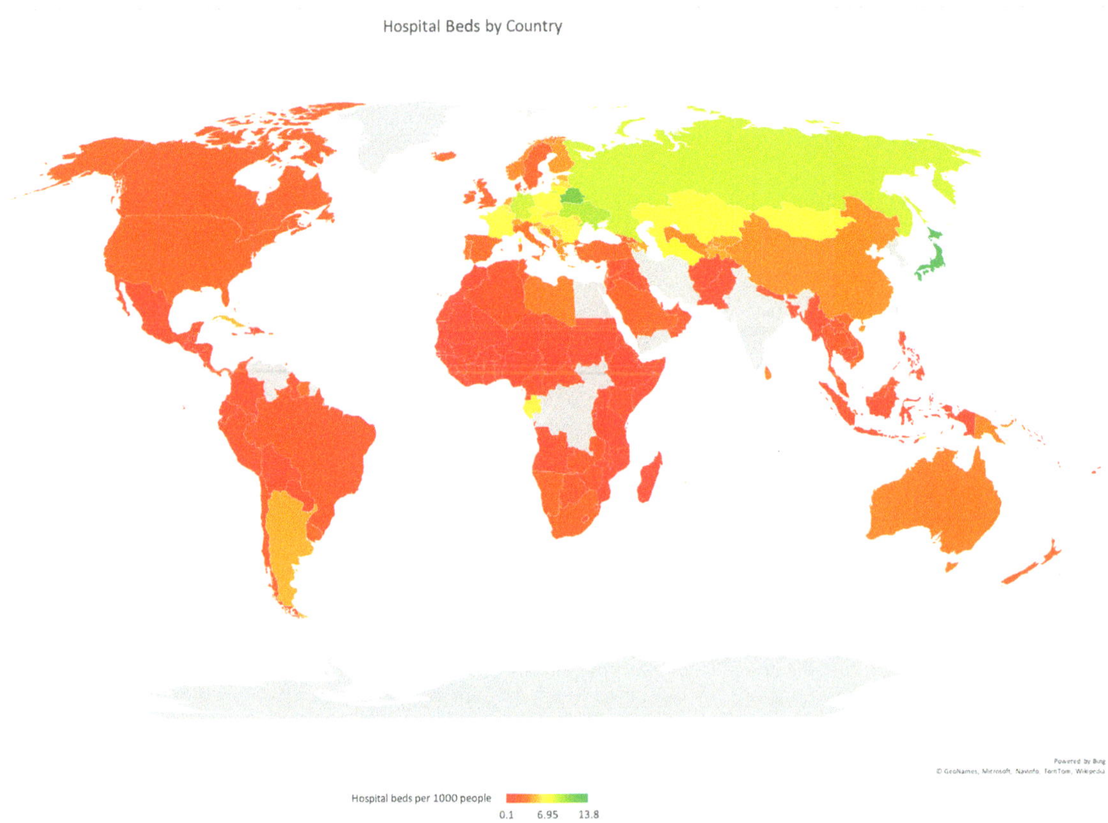


FIG. 5 Number of hospital beds by country. This figure shows the number of hospital beds per 1000 people by country (World Bank, 2015).

with TSCI, continue to be available to promote functioning, and be available in a range of different settings from the hospital through to the home and community environments.

Loss of normal bladder function is one of the most significant consequences for people who have sustained TSCIs. Inadequate management of bladder function leads to secondary complications such as UTIs, urinary retention, incontinence, stones in the kidneys and urinary tract, and reflux of urine (Consortium for Spinal Cord Medicine, 2006). These urinary complications can result in renal failure (Wolfe et al., 2010).

Neurogenic bowel is a common condition following TSCI. It is associated with many gastrointestinal problems, including poor colonic motility, prolonged bowel transit time, chronic constipation, abdominal distension, and fecal incontinence (Correa & Rotter, 2000). People with TSCI who experience a neurogenic bowel are often afraid of possible bowel incontinence, which can significantly impact an individual's ability to return to former social roles and activities (Krassioukov, Eng, Claxton, Sakakibara, & Shum, 2010; World Health Organization, 2011). Adequate management of bowel function can be challenging where resources are limited (Yasmeen, Rathore, Ashraf, & Butt, 2010).

Rehabilitation should aim to help people overcome these limitations by improving trunk and limb function, modifying the person's immediate environment, and providing assistive devices and other reasonable accommodations to enable individuals to continue family and work roles. The reality in most LMIC is that access to rehabilitation is somewhat limited, and most patients are ultimately cared for at home in the community. In a study in rural Uganda, impaired sexual (90%) and bowel/bladder function (80%), depression and chronic pain (85%), skin break-down, and urinary infections (57%) recognized complications of long-term TSCI (Stothers, Macnab, Mukisa, Mutabazi, & Bajunirwe, 2017). Major problems articulated were lack of care for patients living at home, lack of support for family caregivers, and limited aids to living, including wheel-chairs and crutches. Locally made walking aids are prevalent, but most patients rely on family members for their mobility.

In LMICs, infections and septicemia caused by urinary tract complications and pressure sores are still the leading cause of death for people with TSCI (Gosselin & Coppotelli, 2005; Hoque, Grangeon, & Reed, 1999). In Zimbabwe, 7% of individuals with TSCI died from septicemia due to pressure sores while being hospitalized (Correa & Rotter, 2000). With regular turning to relieve pressure points and proper nursing care, pressure sores could be prevented.

Life expectancy for individuals with TSCI in less-resourced settings is shorter than for the average population and persons with TSCI in HICs. Septicemia resulting from urinary tract infections and pressure sores was still the leading cause of death and had the most significant impact on life expectancy and daily life among individuals with TSCI. The findings confirm that individuals with TSCI and their families in low-income settings face substantial challenges in their everyday life because of limited financial, medical, social, and technical support. Poverty has an even more significant impact on people with TSCI in low-income countries, contributing to individual suffering and a high mortality rate due to limited access to rehabilitation services, including appropriate wheelchairs and medical treatment for complications, such as infections, pressure sores, and respiratory problems, common to people with TSCI.

Appropriate and timely medical care and rehabilitation can significantly reduce mortality, morbidity, and disability in TSCI patients. Access to specialized and health care can lead to better outcomes and a productive and enjoyable life for people with TSCI.

## Applications to other areas of neuroscience

The two most common causes of TSCI falls, and road traffic injuries (RTIs) are also the two most common causes of traumatic brain injury (TBI) ( James et al., 2019). Co-occurrence of TBI and TSCI is common, especially in cervical TSCI (Macciocchi, Seel, Thompson, Byams, & Bowman, 2008). Studies that have specifically evaluated this co-occurrence found that patients who sustain TSCI have a 16%–60% chance of sustaining a TBI (Macciocchi et al., 2008). Moreover, there is evidence that TBI in TSCI patients is significantly underdiagnosed in up to 60% of cases, which may be even higher in LMICs that have limited access to resources such as imaging and neuropsychological testing (Sharma, Bradbury, Mikulis, & Green, 2014). Not surprisingly, patients with dual diagnoses experience worse functional outcomes than patients with TSCI alone (Garlanger, Beck, & Cheville, 2018).

Recognition of this relationship has several important implications. First, efforts to reduce traumatic mechanisms can be expected to minimize TSCIs and TBIs by improving road safety and working conditions. Second, a high index of suspicion is essential for patients who are at risk for TBI, especially in low-resource settings with limited access to diagnostic adjuncts. Lastly, patients with dual diagnoses require special consideration when planning for discharge from a hospital or rehabilitation center. TSCI alone presents a massive challenge in low-resource settings. A concurrent TBI will severely impact functional independence and community reintegration.

## Key facts of global traumatic spinal cord injury epidemiology and etiology

- The global incidence is between 250,000 and 1 million annually.
- Males are more commonly affected than females, and at least two-thirds of patients are males in most countries.
- In high-income countries (HICs), TSCIs occur in a bimodal age distribution. In low- and middle-income countries (LMICs), however, TSCIs predominantly affect younger patients.
- The two most common causes of TSCIs worldwide are falls and road traffic injuries (RTIs).
- In areas where active armed conflict is common, such as the Middle East, violence is the leading cause of TSCIs.

## Summary points

- The global incidence of TSCI is between 250,000 and 1 million annually.
- Males are more commonly affected than females worldwide.
- The two most common causes of TSCI are falls and road traffic injuries (RTIs).
- There is significant global variation in the distribution of spinal cord injury level and severity.
- Patients in LMICs experience significant delays to care following TSCI.
- Due to the lack of resources in LMICs, a high proportion of patients with TSCI who would benefit from surgical treatment do not undergo operative management.
- Rehabilitation services in LMICs are limited, and patients in these settings have difficulty gaining independence and preventing long-term complications of TSCIs.
- In LMICs, infections and septicemia caused by urinary tract infections and pressure sores are the leading cause of death for people with TSCIs.
- Life expectancy for individuals with TSCIs in LMICs is lower than that of the average population and lower than that of individuals with TSCI in HICs.

## References


Agarwal, S., Sud, K., & Shishehbor, M. H. (2016). Nationwide trends of hospital admission and outcomes among critical limb ischemia patients from 2003–2011. *Journal of the American College of Cardiology*, *67*(16), 1901–1913. https://doi.org/10.1016/j.jacc.2016.02.040.

Agarwal, P., Upadhyay, P., & Raja, K. (2007). A demographic profile of traumatic and non-traumatic spinal injury cases: A hospital-based study from India. *Spinal Cord*, *45*(9), 597–602. https://doi.org/10.1038/sj.sc.3102005.

Aleem, I. S., DeMarco, D., Drew, B., Sancheti, P., Shetty, V., Dhillon, M., … Bhandari, M. (2017). The burden of spine fractures in India: A prospective multicenter study. *Global Spine Journal*, *7*(4), 325–333. https://doi.org/10.1177/2192568217694362.

Alibhai, A., Hendrikse, C., & Bruijns, S. R. (2019). Poor access to acute care resources to treat major trauma in low- and middle-income settings: A self-reported survey of acute care providers. *African Journal of Emergency Medicine*, *9*(January), S38–S42. https://doi.org/10.1016/j.afjem.2019.01.004.

Ametefe, M. K., Bankah, P. E., Yankey, K. P., Akoto, H., Janney, D., & Dakurah, T. K. (2016). Spinal cord and spine trauma in a large teaching hospital in Ghana. *Spinal Cord*, *54*(12), 1164–1168. https://doi.org/10.1038/sc.2016.57.

Barbetta, D., Smanioto, T., Poletto, M., Ferreira, R., Lopes, A., Casaro, F., … Ogashawara, T. (2018). Spinal cord injury epidemiological profile in the Sarah Network of Rehabilitation Hospitals—A Brazilian population sample. *Spinal Cord Series and Cases*, *4*(1), 2–7. https://doi.org/10.1038/s41394-018-0049-8.

Bellet, F. D., Rashid, S. M., Jusabani, M. A., Dekker, M. C. J., & Temu, R. J. (2019). The characteristics of cervical spinal cord trauma at a North Tanzanian referral hospital: A retrospective hospital-based study. *Pan African Medical Journal*, *33*, 1–7. https://doi.org/10.11604/pamj.2019.33.82.18353.

Bellucci, C. H. S., De Castro Filho, J. E., Gomes, C. M., De Bessa, J., Battistella, L. R., Rubio De Souza, D., … Barros Filho, T. E. P. (2015). Contemporary trends in the epidemiology of traumatic spinal cord injury: Changes in age and etiology. *Neuroepidemiology*, *44*(2), 85–90. https://doi.org/10.1159/000371519.

Bickenback, J. (2013). *International perspectives on spinal cord injury*. Retrieved from http://apps.who.int/iris/bitstream/10665/94190/1/9789241564663_eng.pdf.

Chabok, S. Y., Safaee, M., Alizadeh, A., Dafchahi, M. A., Taghinnejadi, O., & Koochakinejad, L. (2010). Epidemiology of traumatic spinal injury: A descriptive study. *Acta Medica Iranica*, *48*(5), 308–311.

Chen, R., Liu, X., Han, S., Dong, D., Wang, Y., Zhang, H., … Yao, M. (2017). Current epidemiological profile and features of traumatic spinal cord injury in Heilongjiang province, Northeast China: Implications for monitoring and control. *Spinal Cord*, *55*(4), 399–404. https://doi.org/10.1038/sc.2016.92.

Choi, J. H., Park, P. J., Din, V., Sam, N., Iv, V., & Park, K. B. (2017). Epidemiology and clinical management of traumatic spine injuries at a major government hospital in Cambodia. *Asian Spine Journal*, *11*(6), 908–916. https://doi.org/10.4184/asj.2017.11.6.908.

Consortium for Spinal Cord Medicine. (2006). Bladder management for adults with spinal cord injury: A clinical practice guideline for healthcare providers. *Journal of Spinal Cord Medicine*, *29*(5), 527–573.

Corley, J., Lepard, J., Barthélemy, E., Ashby, J. L., & Park, K. B. (2019). Essential neurosurgical workforce needed to address neurotrauma in low- and middle-income countries. *World Neurosurgery*, *123*, 295–299. https://doi.org/10.1016/j.wneu.2018.12.042.

Correa, G. I., & Rotter, K. P. (2000). Clinical evaluation and management of neurogenic bowel after spinal cord injury. *Spinal Cord*, *38*(5), 301–308. https://doi.org/10.1038/sj.sc.3100851.

Deconinck, H. (2003). The health condition of spinal cord injuries in two Afghan towns. *Spinal Cord*, *41*(5), 303–309. https://doi.org/10.1038/sj.sc.3101443.

Derakhshanrad, N., Yekaninejad, M. S., Vosoughi, F., Sadeghi Fazel, F., & Saberi, H. (2016). Epidemiological study of traumatic spinal cord injuries: Experience from a specialized spine center in Iran. *Spinal Cord*, *54*(10), 901–907. https://doi.org/10.1038/sc.2016.10.

Dewan, M. C., Rattani, A., Baticulon, R. E., Faruque, S., Johnson, W. D., Dempsey, R. J., … Shrime, M. G. (2019). Operative and consultative proportions of neurosurgical disease worldwide: Estimation from the surgeon perspective. *Journal of Neurosurgery*, *130*(April), 1098–1106. https://doi.org/10.3171/2017.10.JNS17347.1098.

Dewan, M. C., Rattani, A., Fieggen, G., Arraez, M. A., Servadei, F., Boop, F. A., … Park, K. B. (2019). Global neurosurgery: The current capacity and deficit in the provision of essential neurosurgical care. Executive summary of the global neurosurgery initiative at the program in global surgery and social change. *Journal of Neurosurgery*, *130*(4), 1055–1064. https://doi.org/10.3171/2017.11.JNS171500.

Dhakal, G. R., Paudel, S., Dhungana, S., Gurung, G., & Kawaguchi, Y. (2018). Epidemiological characteristics of dorsal and lumbar spine trauma presenting to a trauma hospital in Kathmandu, Nepal: Formulation of a national spine policy. *Spine Surgery and Related Research*, *2*(4), 249–252. https://doi.org/10.22603/ssrr.2017-0087.

Dhamangaonkar, A. C., Joshi, D., Kumar, R., & Goregaonkar, A. B. (2013). Neurological deficits presenting to an urban tertiary care centre in Mumbai: An epidemiological study. *Malaysian Orthopaedic Journal*, *7*(1), 36–40. https://doi.org/10.5704/moj.1303.014.

Eaton, J., Hanif, A. B., Grudziak, J., & Charles, A. (2017). Epidemiology, management, and functional outcomes of traumatic brain injury in sub-Saharan Africa. *World Neurosurgery*, *108*, 650–655. https://doi.org/10.1016/j.wneu.2017.09.084.

Eaton, J., Mukuzunga, C., Grudziak, J., & Charles, A. (2019). Characteristics and outcomes of traumatic spinal cord injury in a low-resource setting. *Tropical Doctor*, *49*(1), 62–64. https://doi.org/10.1177/0049475518808969.

Furlan, J. C., Noonan, V., Cadotte, D. W., & Fehlings, M. G. (2011). Timing of decompressive surgery of spinal cord after traumatic spinal cord injury: An evidence-based examination of pre-clinical and clinical studies. *Journal of Neurotrauma*, *28*(8), 1371–1399. https://doi.org/10.1089/neu.2009.1147.

Garlanger, K. L., Beck, L. A., & Cheville, A. L. (2018). Functional outcomes in patients with co-occurring traumatic brain injury and spinal cord injury from an inpatient rehabilitation facility's perspective. *Journal of Spinal Cord Medicine*, *41*(6), 718–730. https://doi.org/10.1080/10790268.2018.1465744.

Gélis, A., Stéfan, A., Colin, D., Albert, T., Gault, D., Goossens, D., … Coudeyre, E. (2011). Therapeutic education in persons with spinal cord injury: A review of the literature. *Annals of Physical and Rehabilitation Medicine*, *54*(3), 189–210. https://doi.org/10.1016/j.rehab.2011.03.004.

Gooch, R. A., & Kahn, J. M. (2014). ICU bed supply, utilization, and health care spending: An example of demand elasticity. *Journal of the American Medical Association*, *311*(6), 567–568. https://doi.org/10.1016/S0965-8564(99)00047-6.

Gosselin, R. A., & Coppotelli, C. (2005). A follow-up study of patients with spinal cord injury in Sierra Leone. *International Orthopaedics*, *29*(5), 330–332. https://doi.org/10.1007/s00264-005-0665-3.

Hoque, F., Grangeon, C., & Reed, K. (1999). Spinal cord lesions in Bangladesh: An epidemiological study 1994–1995. *Spinal Cord*, *37*(12), 858–861. https://doi.org/10.1038/sj.sc.3100938.

Hua, R., Shi, J., Wang, X., Yang, J., Zheng, P., Cheng, H., … An, Y. (2013). Analysis of the causes and types of traumatic spinal cord injury based on 561 cases in China from 2001 to 2010. *Spinal Cord*, *51*(3), 218–221. https://doi.org/10.1038/sc.2012.133.

Ibrahim, A., Lee, K. Y., Kanoo, L. L., Tan, C. H., Hamid, M. A., Hamedon, N. M., & Haniff, J. (2013). Epidemiology of spinal cord injury in hospital Kuala Lumpur. *Spine*, *38*(5), 419–424. https://doi.org/10.1097/BRS.0b013e31826ef594.

James, S. L., Bannick, M. S., Montjoy-Venning, W. C., Lucchesi, L. R., Dandona, L., Dandona, R., … Zaman, S. B. (2019). Global, regional, and national burden of traumatic brain injury and spinal cord injury, 1990–2016: A systematic analysis for the Global Burden of Disease Study 2016. *The Lancet Neurology*, *18*(1), 56–87. https://doi.org/10.1016/S1474-4422(18)30415-0.

Jazayeri, S. B., Beygi, S., Shokraneh, F., Hagen, E. M., & Rahimi-Movaghar, V. (2015). Incidence of traumatic spinal cord injury worldwide: A systematic review. *European Spine Journal*, *24*(5), 905–918. https://doi.org/10.1007/s00586-014-3424-6.

Kang, Y., Ding, H., Zhou, H., Wei, Z., Liu, L., Pan, D., & Feng, S. (2017). Epidemiology of worldwide spinal cord injury: A literature review. *Journal of Neurorestoratology*, *6*, 1–9. https://doi.org/10.2147/jn.s143236.

Kanmounye, U. S., Lartigue, J. W., Sadler, S., Yuki Ip, H. K., Corley, J., Arraez, M. A., & Park, K. (2020). Emerging trends in the neurosurgical workforce of low- and middle-income countries: A cross-sectional study. *World Neurosurgery*, *142*, e420–e433. https://doi.org/10.1016/j.wneu.2020.07.067.

Kebaish, K., & Harris, A. (2020). Spine and spinal cord injuries. In J. Cameron, & A. Cameron (Eds.), *Current surgical therapy* (13th ed., pp. 1221–1223). Philadelphia: Elsevier.

Kirshblum, S. C., Waring, W., Biering-Sorensen, F., Burns, S. P., Johansen, M., Schmidt-Read, M., … Krassioukov, A. (2011). Reference for the 2011 revision of the International Standards for Neurological Classification of Spinal Cord Injury. *Journal of Spinal Cord Medicine*, *34*(6), 547–554. https://doi.org/10.1179/107902611X13186000420242.

Krassioukov, A., Eng, J. J., Claxton, G., Sakakibara, B. M., & Shum, S. (2010). Neurogenic bowel management after spinal cord injury: A systematic review of the evidence. *Spinal Cord*, *48*(10), 718–733. https://doi.org/10.1038/sc.2010.14.

Kumar, R., Lim, J., Mekary, R. A., Rattani, A., Dewan, M. C., Sharif, S. Y., … Park, K. B. (2018). Traumatic spinal injury: Global epidemiology and worldwide volume. *World Neurosurgery*, *113*, e345–e363. https://doi.org/10.1016/j.wneu.2018.02.033.

Lakhey, S., Jha, N., Shrestha, B. P., & Niraula, S. (2005). Aetioepidemiological profile of spinal injury patients in Eastern Nepal. *Tropical Doctor*, *35*(4), 231–233. https://doi.org/10.1258/004947505774938756.

Leidinger, A., Kim, E. E., Navarro-Ramirez, R., Rutabasibwa, N., Msuya, S. R., Askin, G., … Härtl, R. (2019). Spinal trauma in Tanzania: Current management and outcomes. *Journal of Neurosurgery: Spine*, *31*(1), 103–111. https://doi.org/10.3171/2018.12.SPINE18635.

Lessing, N. L., Lazaro, A., Zuckerman, S. L., Leidinger, A., Rutabasibwa, N., Shabani, H. K., & Härtl, R. (2020). Nonoperative treatment of traumatic spinal injuries in Tanzania: Who is not undergoing surgery and why? *Spinal Cord*. https://doi.org/10.1038/s41393-020-0474-y.

Lessing, N. L., Zuckerman, S. L., Lazaro, A., Leech, A. A., Leidinger, A., Rutabasibwa, N., … Härtl, R. (2020). Cost-effectiveness of operating on traumatic spinal injuries in low-middle income countries: A preliminary report from a major East African referral center. *Global Spine Journal*. https://doi.org/10.1177/2192568220944888.

Liu, J., Liu, H.-W., Gao, F., Li, J., & Li, J.-J. (2020). Epidemiological features of traumatic spinal cord injury in Beijing, China. *Journal of Spinal Cord Medicine*, 1–7. https://doi.org/10.1080/10790268.2019.1654190.

Liu, P., Yao, Y., Liu, M. Y., Fan, W. L., Chao, R., Wang, Z. G., … Zhao, J. H. (2012). Spinal trauma in mainland China from 2001 to 2007: An epidemiological study based on a nationwide database. *Spine*, *37*(15), 1310–1315. https://doi.org/10.1097/BRS.0b013e3182474d8b.

Löfvenmark, I., Norrbrink, C., Nilsson-Wikmar, L., Hultling, C., Chakandinakira, S., & Hasselberg, M. (2015). Traumatic spinal cord injury in Botswana: Characteristics, aetiology, and mortality. *Spinal Cord*, *53*(2), 150–154. https://doi.org/10.1038/sc.2014.203.

Macciocchi, S., Seel, R. T., Thompson, N., Byams, R., & Bowman, B. (2008). Spinal cord injury and co-occurring traumatic brain injury: Assessment and incidence. *Archives of Physical Medicine and Rehabilitation*, *89*(7), 1350–1357. https://doi.org/10.1016/j.apmr.2007.11.055.

Makhni, M. C., Vatan, N. M., Park, P. J., Kabeer, Z. A., Cerpa, M., Choi, J.-H., … Park, K. B. (2019). Path to prevention of spinal trauma in a low- to middle-income country. *JAAOS: Global Research and Reviews*, *3*(10). https://doi.org/10.5435/jaaosglobal-d-19-00080, e1900080.

Moshi, H., Sundelin, G., Sahlen, K. G., & Sörlin, A. (2017). Traumatic spinal cord injury in the north-east Tanzania – Describing incidence, etiology, and clinical outcomes retrospectively. *Global Health Action*, *10*(1). https://doi.org/10.1080/16549716.2017.1355604.

Munakomi, S., Bhattarai, B., & Cherian, I. (2017). Prospective observational research on the clinical profile and outcome analysis among a cohort of patients sustaining traumatic cervical spine and cord injury in a peripheral tertiary spine care centre in Nepal. *F1000Research*, *6*, 1–16. https://doi.org/10.12688/f1000research.12911.1.

Murthy, S., & Wunsch, H. (2012). Clinical review: International comparisons in critical care—Lessons learned. *Critical Care*, *16*(2). https://doi.org/10.1186/cc11140.

Nwankwo, O. E., & Uche, E. O. (2013). Epidemiological and treatment profiles of spinal cord injury in southeast Nigeria. *Spinal Cord*, *51*(6), 448–452. https://doi.org/10.1038/sc.2013.10.

Obalum, D. C., Giwa, S. O., Adekoya-Cole, T. O., & Enweluzo, G. O. (2009). Profile of spinal injuries in Lagos, Nigeria. *Spinal Cord*, *47*(2), 134–137. https://doi.org/10.1038/sc.2008.93.

Rathore, M. F. A., Hanif, S., Farooq, F., Ahmad, N., & Mansoor, S. N. (2008). Traumatic spinal cord injuries at a tertiary care rehabilitation institute in Pakistan. *Journal of the Pakistan Medical Association*, *58*(2), 53–57.

Sharif-Alhoseini, M., & Rahimi-Movaghar, V. (2014). Hospital-based incidence of traumatic spinal cord injury in Tehran, Iran. *Iranian Journal of Public Health*, *43*(3), 331–341.

Sharma, B., Bradbury, C., Mikulis, D., & Green, R. E. (2014). Missed diagnosis of traumatic brain injury in patients with traumatic spinal cord injury. *Journal of Rehabilitation Medicine*, *46*(4), 370–373. https://doi.org/10.2340/16501977-1261.

Stothers, L., Macnab, A. J., Mukisa, R., Mutabazi, S., & Bajunirwe, F. (2017). Traumatic spinal cord injury in Uganda: A prevention strategy and mechanism to improve home care. *International Journal of Epidemiology*, *46*(4), 1086–1090. https://doi.org/10.1093/ije/dyx058.

Wolfe, D. L., Ethans, K., Hill, D., Hsieh, J. T. C., Mehta, S., Teasell, R. W., & Askes, H. (2010). Bladder health and function following spinal cord injury. In J. J. Eng, R. W. Teasell, W. C. Miller, D. L. Wolfe, A. F. Townson, J. T. C. Hsieh, … B. M. Sakakibara (Eds.), *Spinal Cord Injury Rehabilitation Evidence. Version 3.0* (pp. 1–19). Vancouver.

World Bank. (2015). *Hospital beds per 1000 people*. Retrieved November 20, 2020, from https://data.worldbank.org/indicator/SH.MED.BEDS.ZS.

World Bank. (2018). *Specialist surgical workforce*. World Bank.

World Health Organization. (2011). *Summary world report on disability* (pp. 1–24). World Health. Retrieved from www.who.int/about/licensing/copyright_form/en/index.html%0Ahttp://www.larchetoronto.org/wordpress/wp-content/uploads/2012/01/launch-of-World-Report-on-Disability-Jan-27-121.pdf.

Yasmeen, R., Rathore, F. A., Ashraf, K., & Butt, A. W. (2010). How do patients with chronic spinal injury in Pakistan manage their bowels? A cross-sectional survey of 50 patients. *Spinal Cord*, *48*(12), 872–875. https://doi.org/10.1038/sc.2010.46.

Yuan, S., Shi, Z., Cao, F., Li, J., & Feng, S. (2018). Epidemiological features of spinal cord injury in China: A systematic review. *Frontiers in Neurology*, *9*(AUG), 1–12. https://doi.org/10.3389/fneur.2018.00683.

Zhou, Y., Wang, X. B., Kan, S. L., Ning, G. Z., Li, Y. L., Yang, B., … Feng, S. Q. (2016). Traumatic spinal cord injury in Tianjin, China: A single-center report of 354 cases. *Spinal Cord*, *54*(9), 670–674. https://doi.org/10.1038/sc.2015.173.

Chapter 2

# Biomechanics and patterns of spine injuries associated with spinal cord injury

Salomón Flórez-Jiménez, Étienne Bourassa-Moreau, Jean-Marc Mac-Thiong, and Gilles Maurais

*Department of Spinal Surgery, Sacré-Coeur Hospital of Montreal, Université de Montreal, Montréal, QC, Canada*

## List of abbreviations

**CCS** Central cord syndrome
**PLC** Posterior ligamentous complex
**TSCI** Traumatic spinal cord injury

## Introduction

Worldwide, approximately 30 million individuals live with a traumatic spinal cord injury (TSCI), and about 1 million new cases of TSCI occur annually, and these numbers are expected to increase in view of population growth. Annually, 150,000 spine fractures are treated in the United States. Among those, close to 20% are associated with spinal cord injury (SCI) (Buchowski, Kuhns, Bridwell, & Lenke, 2008). A spinal fracture and/or dislocation occurs in about 65% of all TSCI.

TSCI contributes to a considerable burden on patient's lives because of lifelong functional disabilities and major psychosocial challenges. Accordingly, the severity of the spine injuries is usually associated with the severity of the SCI, in line with increased burden of healthcare resource utilization. Although management of the underlying spine injuries further contributes to the economic burden of TSCI, it is strongly influencing the clinical outcomes of patients.

TSCIs are most often due to a traumatic event, mainly falls and motor vehicle accidents. While motor vehicle accidents remain the most common cause of TSCI, there is an increasing incidence of TSCI due to low-energy trauma such as falls (Chen, He, & DeVivo, 2016; Pickett, Campos-Benitez, Keller, & Duggal, 2006).

Because of physiological changes due to aging, elderly individuals are particularly at increased risk of sustaining a TSCI from low-energy trauma such as a fall from their height.

In keeping with the aging of the general population, a demographic shift is observed toward older TSCI patient's population (Ahn et al., 2017) with pre-existing spondylotic spinal stenosis and without significant traumatic spinal instability.

Understanding the biomechanics and patterns of the common types of spine injuries associated with SCI is central to the management of individuals with TSCI, and will be featured in this chapter.

## Pathogenesis of TSCI: The critical role of spine biomechanics

Several studies (Al-Habib et al., 2011; Wagnac et al., 2019; Wilson, Cadotte, & Fehlings, 2012) demonstrated the importance of the mechanism and level of energy of the traumatic event as predictors of the severity of TSCI. When compared with low-velocity trauma (e.g., fall from standing height), high-velocity mechanisms of trauma (e.g., motor-vehicle accidents) more often result in increased spinal cord damage, particularly in the white matter (Mattucci et al., 2019). The mechanism and level of energy of the trauma are only moderately associated with the outcomes, because they only partially reflect the level of energy that was actually transferred to the spinal cord during the trauma. During the trauma, the primary injury to the spinal cord occurs due to the physical insult and local energy transferred to the spinal cord, thereby causing direct damage to neural tissues and its blood vessels. It is believed that the morphological pattern of the spine injury is a better reflection of the underlying dynamics and local energy sustained by the spinal cord during the trauma than the mechanism of trauma and global level of energy (Goulet, Richard-Denis, Petit, Diotalevi, & Mac-Thiong, 2020). During the trauma, energy is transferred to the spinal cord by direct or indirect mechanisms originating from the spine injury.

Diagnosis and Treatment of Spinal Cord Injury. https://doi.org/10.1016/B978-0-12-822498-4.00002-6

Direct injury occurs from contusion of the spinal cord by vertebral bone and soft tissues (intervertebral disc and/or ligaments) in direct contact with the spinal cord.

Indirect injury to the spinal cord results from the motion (e.g., flexion, hyperextension, axial compression, distraction, dislocation, etc.) between spinal elements during the trauma, causing excessive tension, shear, torsion, compression, and/or bending loads on the spinal cord.

The type of loads sustained by the spinal cord will influence the extent of tissue damage (Chen et al., 2016), contributing to the heterogeneity of the clinical outcomes observed following a TSCI (Mattucci et al., 2019). Contusion causes a relatively localized damage, as opposed to dislocation (causing mainly shear stress) and distraction which result in more asymmetrical damage that can extend further rostrally and caudally (Chen et al., 2016).

The acute surgical management is critical in TSCI patients and serves two purposes: neurological decompression and spinal stabilization. Neurological decompression entails removal of bone, blood, and disco-ligamentous structures that impinges the spinal cord, whereas spinal stabilization is achieved by spinal instrumentation rigidly connecting the unstable segments of the spine. Acute surgery ultimately aims at limiting the secondary injury to the spinal cord. The pattern of spine injury contributes to the secondary injury of the spinal cord through direct and indirect mechanisms. The residual compression of the spinal cord for example from the typical vertebral bone fragment retropulsed in the spinal canal with thoracolumbar burst fractures directly exerts loads on the spinal cord. In parallel, residual misalignment or instability between spinal elements can also exert undue loads on the spinal cord through an indirect mechanism, further contributing to the secondary injury of the spinal cord if it has not been addressed properly. Careful and prompt assessment of the clinical stability is therefore key after a TSCI.

In line with the definition by White 3rd, Johnson, Panjabi, and Southwick (1975), we define traumatic spinal instability to be the loss of the ability of the spine to maintain normal alignment and motion between spinal segments in such a way that a secondary injury to the spinal cord will not occur.

## Patterns of spine injuries associated with spinal cord injuries

TSCI can be associated with a stable or unstable spine, and we suggest that TSCI should be categorized accordingly to guide its management.

### Traumatic spinal cord injury in a stable spine

Different terminologies have been used previously to describe TSCI not associated with overt spinal instability. The term SCIWORA (Spinal Cord Injury Without Radiographic Abnormality) was proposed by Pang and Wilberger Jr. (1982) in 1982 to describe TSCI occurring in children without evidence of spine injuries such as fracture or dislocation on plain radiographs or tomography. Given the inherent elasticity of the vertebral column in children, they suggest that the vertebral column can deform in flexion, hyperextension or distraction in such a way that the spinal cord can be injured without concomitant damage of the vertebral column. SCIWORA more frequently involves the cervical spinal cord because spinal mobility is greater in the cervical than the thoracic and lumbar spine. In the adult population, SCIWORA has also been used to describe TSCI associated with underlying spondylotic changes, spinal stenosis, ossification of posterior longitudinal ligament, ligamentous abnormalities, and/or traumatic disc herniation (Asan, 2018; Boese et al., 2016; Boese & Lechler, 2013; Gupta et al., 1999), in the absence of fracture or dislocation. Many authors (Como et al., 2012; Dreizin et al., 2015) warn about the misapplication of the term SCIWORA in adults considering that these patients will commonly have underlying canal stenosis and significant degenerative changes—which are radiographic abnormalities per se—and instead propose the use of the term SCIWOCTET (Spinal Cord Injury Without Computed Tomography Evidence of Trauma) or SCIWORET (Spinal Cord Injury Without Radiographic Evidence of Trauma) in these patients.

For the cervical spine, proper identification of TSCI with a stable spine is important because they are typically associated with improved recovery (Paquet et al., 2018). While the presence or absence of radiographic/tomographic findings can help to understand the pathomechanism of the injury and guide treatment, the central aspect to SCIWORA, SCIWORET, and SCIWOCTET is that they usually are not associated with overt spinal instability. For these conditions, the spinal cord is at higher risk of injury either due to the altered biomechanics of the spine or to the decreased space available for the spinal cord. The current authors believe that the overarching concept for these aforementioned terms is the absence of spinal instability, and that these terms should be used with caution because they underestimate the importance of spinal trauma and biomechanics in the context of TSCI. Radiological evidence of trauma can include a wide range of findings, and should not be interpreted without consideration of the spinal stability. The authors instead propose to use the term TSCI with a stable spine in order to account for the biomechanical and surgical aspects involved in the management of TSCI. Indeed, TSCI with a stable spine does not preclude the presence of minor spinal injuries such as spinous process fractures that do not impair the clinical stability of the spine and do not require surgical stabilization.

In adults, underlying degenerative changes are usually observed, but the literature does not provide specific outcome predictors based on the evaluation of the underlying spondylosis.

Animal models repeatedly showed a strong correlation between the amount of spinal cord compression and severity of the neurological impairment. However, in adult trauma patients the correlation between cord compression observed on MRI and the neurological impairment remains unclear. Interestingly, normal aging of the spine commonly implies degenerative changes that cause reduction of the diameter of the spinal canal called spinal stenosis, especially at the cervical level. Literature does not provide specific outcome predictors based on the evaluation of the underlying spondylosis. In a population of adult low-velocity cervical trauma—CCS—Aarabi et al., 2011 found a somewhat counterintuitive association between maximal cord compression on MRI and neurological recovery.

The CCS is a common clinical syndrome observed after a cervical TSCI and is typically observed in the presence of a stable spine. It has been described initially by Schneider, Cherry, and Pantek (1954) and is characterized by a disproportionate greater motor impairment in the upper extremities and varying degree of bladder dysfunction and sensory loss. The central cord syndrome is mainly prevalent in the elderly population with pre-existing spondylotic cervical stenosis (McKinley, Santos, Meade, & Brooke, 2007; Molliqaj, Payer, Schaller, & Tessitore, 2014). The mechanism of trauma typically involves a fall from someone's height with cervical hyperextension. During hyperextension of the cervical spine, spinal cord damage associated with the central cord syndrome results from posterior compression from inward bulging of the ligamentum flavum and from anterior compression from marginal osteophytes and/or bulging discs (Schneider et al., 1954). The greater impairment in the upper extremities results from the injury to the spinal gray matter and lateral corticospinal tracts (Collignon, Martin, Lenelle, & Stevenaert, 2002). More specifically, it is believed that the predominant involvement of the upper extremities is due to the somatotopic arrangement of the lateral corticospinal tracts, such that the medial (more central) portion of the corticospinal tract controlling upper extremity motor function sustains a greater injury than the lateral portion controlling lower extremity motor function (Schneider, Thompson, & Bebin, 1958). However, this hypothesis has been challenged by different authors who observed a uniform damage to the lateral corticospinal tract, suggesting that motor fibers descending to the upper and lower extremities are interlaced (Nathan & Smith, 1955; Quencer et al., 1992).

In a finite-element analysis, Bailly et al. suggest that the presence of hypertrophic ligamentum flavum induced specific stress and strain distributions consistent with previous histologic findings related to central cord syndrome. The presence of a disc bulging alone was not associated with such a pattern, but when combined with hypertrophic ligamentum flavum, increased the overall level of stress and strain in the lateral corticospinal tract (Fig. 1)

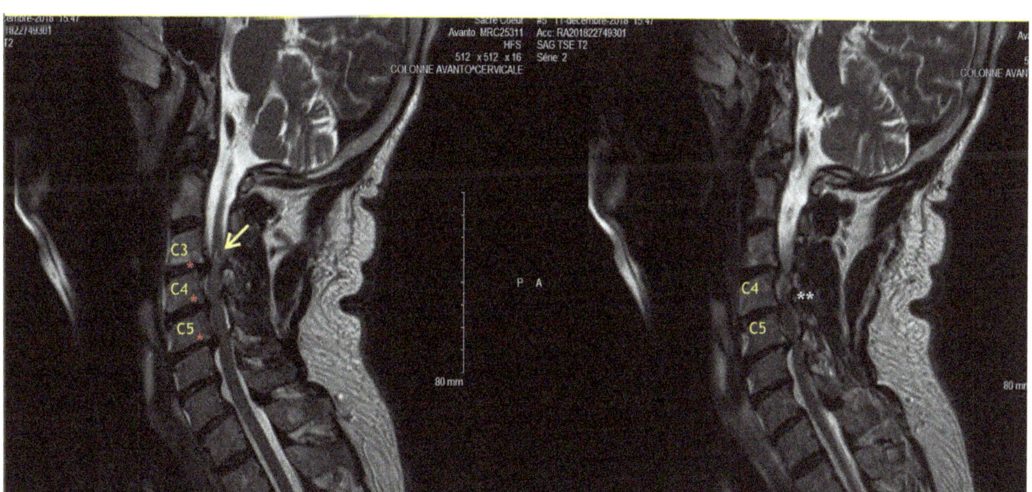


FIG. 1 Hyperextension cervical trauma in a stable spine with degenerative changes. CCS. (*) red asterisks, Bulging disc C3–C4, C4–C5, C5–C6. (**) White asterisks, thickened ligamentum flavum multilevel mainly C4–C5. Yellow arrow showing spinal cord edema and hypersignal in C3 level.

## Traumatic spinal cord injury in an unstable spine

While TSCI in a stable spine occurs predominantly at the cervical level due to the increased range of motion and frequent preexisting canal stenosis, unstable injuries occur commonly at all spinal levels. The neurological level of injury strongly relates to the neurological recovery, and it is common to distinguish between high cervical (C1–C4), low cervical (C5–C8), thoracic (T1–T10), and thoracolumbar (T11–L2) TSCI (Dvorak et al., 2014). These categories account for the functional activity of the spinal cord, including breathing, upper versus lower extremity function, trunk control, and bowel/bladder function. However, other anatomical and morphological criteria should also be considered to further delineate the level of injury and characterize the trauma biomechanics when classifying and managing the unstable spine associated with TSCI.

There are three basic categories or types used in a similar manner to the AO thoracolumbar fracture classification system to describe primary injury morphology. The "Type A" injuries are fractures that result in compression of the vertebra with intact tension band. "Type B" injuries include failure of the posterior or anterior tension band through distraction with physical separation of the subaxial spinal elements while maintaining continuity of the alignment of the spinal axis without translation or dislocation. "Type C" includes those injuries with displacement or translation of one vertebral body relative to another in any direction; anterior, posterior, lateral translation, or vertical distraction. Injuries are first classified by their level and either C, B, or A in this order (Vaccaro et al., 2016).

### *Cervical spine injuries (C0–C7)*

TSCI with an unstable spine are highly prevalent in the cervical spine for different reasons. The cervical spine is vulnerable to several types of trauma mechanisms, including motor vehicle accidents, diving, sports injuries, and falls. When compared to the thoracic and lumbar spine, the bony elements in the cervical spine are smaller and the disco-ligamentous structures are weaker so that they are more exposed to injuries.

In line with the demographic shift to an older population, the increasing prevalence of cervical degenerative changes and pre-existing spondylotic stenosis increases the likelihood to sustain a TSCI in association with a cervical trauma.

#### Upper cervical spine injuries (C0–C2)

TSCI following traumatic spine injuries occurring at the occipito-cervical or atlanto-axial level are highly distinct from subaxial injuries due to the increased risk of mortality and poor recovery. It is thought that between 8% and 19% of fatal cervical spine injuries result from occipito-cervical dislocation (Bucholz, Burkhead, Graham, & Petty, 1979). In addition, occipito-cervical and atlanto-axial levels have a highly distinct bony/soft tissue anatomy and biomechanical behavior, which require particular attention for evaluation and management. The spinal canal is also particularly large at these levels; it is widest between the C1 and C3 levels (AP diameter 16–30 mm) and progressively narrows caudally (14–23 mm).

Occipito-cervical dislocation is associated with high mortality rates because a high-energy trauma—and high level of energy transferred to the spinal cord—is required to damage the strong ligamentous structures at the occipito-cervical junction, which comprise the tectorial membrane and alar ligament (Alker et al., 1975).

TSCI is seen infrequently with isolated C1 fracture (Lee & Woodring, 1991). A TSCI occurs when the transverse ligament is torn and atlanto-axial instability occurs. The mechanism of SCI is similar to that occurring with isolated rupture of the transverse ligament or odontoid fracture: the spinal cord will be compressed between the posteriorly displaced odontoid and the posterior arch of C1.

#### Subaxial cervical spine injuries (C3–C7)

The Spinal Trauma Study Group (Vaccaro et al., 2007) provided new insight on the morphological features of subaxial cervical trauma that need to be considered in TSCI, through the proposal of a novel classification system. In line with previous studies (Chen et al., 2016), they consider that the severity and outcome of TSCI depend on the morphology of the lesion, which results in different loading conditions in the spinal cord. Based on their concept, we propose that TSCI with unstable subaxial cervical spine be classified following this increasing order of severity:

(1) Compression injury without involvement of the posterior vertebral wall.
(2) Burst injury with involvement of the posterior vertebral wall.
(3) Distraction injury such as a hyperextension injury involving rupture of anterior longitudinal ligament and intervertebral disc or a flexion-distraction injury involving disruption of posterior ligaments and facet capsules.
(4) Translation/rotation injury which is typical of bilateral facet dislocation.

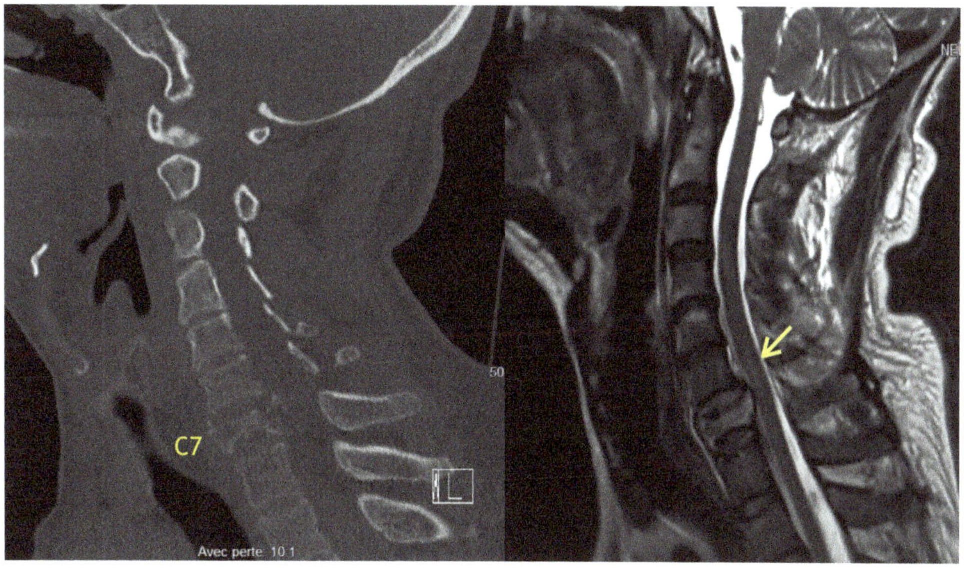


FIG. 2 Burst fracture of C7 and significant energy transferred to the spinal cord from retropulsion of the posterior wall into the spinal canal causing spinal cord edema (yellow arrow).

TSCI is unlikely with a compression injury sparing the posterior vertebral wall, unless significant kyphosis occurs (Yokota et al., 2019). A TSCI is more likely with burst fractures when significant energy is transferred to the spinal cord from retropulsion of the posterior wall into the spinal canal (Fig. 2)

Hyperextension injury causes distraction of the spinal cord that can lead to neural damage. In addition, spinal cord damage can result from inward bulging of the ligamentum flavum in the presence of a spondylotic spine, by a mechanism similar to that described above for cervical TSCI in a stable spine. A hyperextension injury—and less frequently a translation injury—is typically observed with a cervical TSCI in patients with a rigid spine (e.g., ankylosing spondylitis, diffuse idiopathic skeletal hyperostosis or ossification of the posterior longitudinal ligament). For these patients, the presence of an epidural hematoma causing the SCI should also be ruled out (Fig. 3)

Facet dislocations are most common in the subaxial cervical spine because a lesser magnitude of force is required to produce the dislocation, as compared to the thoracic and lumbar spine (Holdsworth, 1970). This is mainly due to the

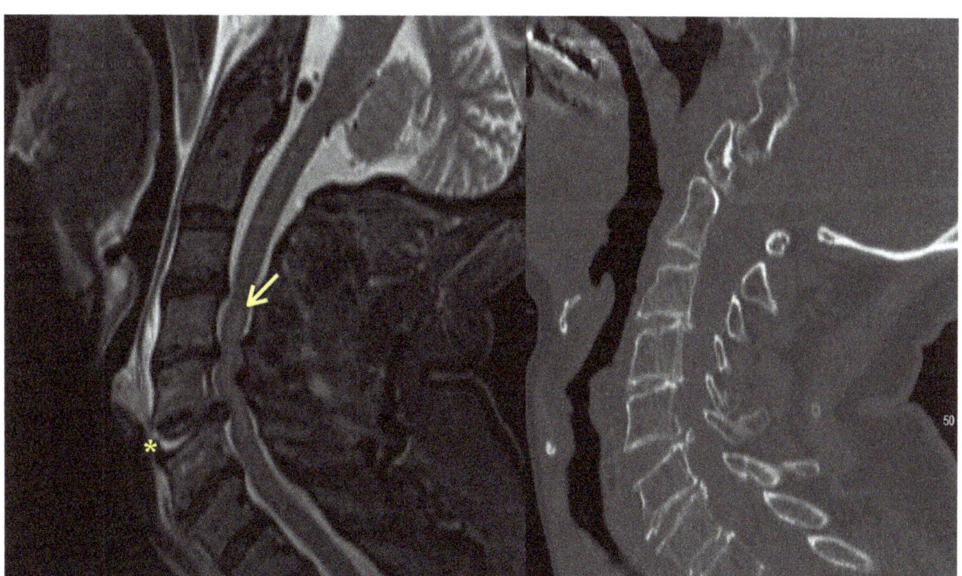

FIG. 3 Hyperextension injury in a mobile spine ct and irm, showing disc rupture (yellow asterisk) and spinal cord edema (arrow).

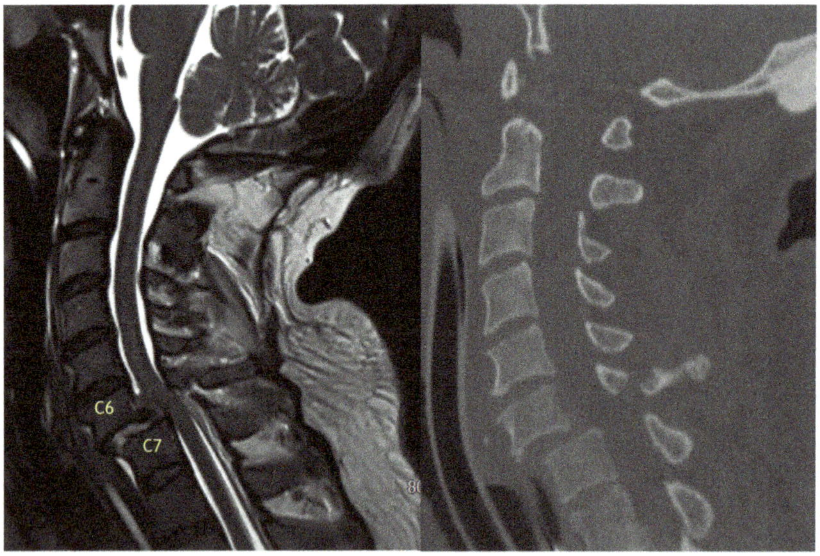


FIG. 4 Cervical dislocation C6–C7.

increased mobility, distinctive orientation of the superior articular facets (Ebraheim, Patil, Liu, Haman, & Yeasting, 2008), and smaller osseous and ligamentous structures of the cervical spine. Facet dislocation typically occurs as a result of a hyperflexion injury and is commonly associated with SCI (Sekhon & Fehlings, 2001; Tator, 1983).

During the trauma, the primary injury to the spinal cord results from the distraction of the spinal cord posteriorly as well as the resulting translation and/or rotation between dislocated vertebrae causing shear stresses on the spinal cord. After the trauma, a secondary injury to the spinal cord is accentuated by the shear and compression loading from residual translation and/or rotation between vertebrae, which needs to be managed by surgery.

Wilson et al. showed that cervical dislocation is more frequently associated with severe neurological deficit and high-energy trauma mechanisms, and results in a diminished potential for recovery (Wilson et al., 2013) (Fig. 4)

### *Thoracic spine injuries (T1–T10)*

The pathomechanism and outcome of thoracic injuries significantly differ from cervical injuries because of the inherent stability provided by the thoracic cage (i.e., ribs and sternum). Each thoracic vertebra has a diarthrodial articulation with a rib on each side, thereby increasing the stability of the thoracic spine and potentially decreasing the incidence of degenerative changes since reduced mobility.

The thoracic facet joints are aligned with the coronal plane to allow for axial rotation. The ratio of the canal size to the spinal cord size is relatively small in the thoracic spine, which increases the risk of neurologic injury even with small amounts of canal intrusion.

The tenuous blood supply (critical watershed area) in the upper thoracic region also exposes the spinal cord between T4 and T10 at increased risk of SCI with a spine trauma.

In aging individuals with osteoporosis, a minor trauma can cause a compression fracture in the thoracic spine, usually below T4 and most often at the thoracolumbar region. However, a TSCI is rarely seen with osteoporotic compression fractures, even in the presence of a marked kyphotic deformity. When a TSCI occurs after minor trauma in an osteoporotic patient, it is usually secondary to a burst fracture with a retropulsed fragment into the spinal canal, and the onset of neurological deficit is typically gradual rather than acute. For patients with a rigid spine (e.g., ankylosing spondylitis and diffuse idiopathic skeletal hyperostosis), a hyperextension or translation spine injury should be suspected in the presence of a TSCI at the thoracic level, even if a minor trauma has occurred.

More often, thoracic TSCI results from high-energy trauma mechanisms. Thorough clinical and radiological evaluation of the thoracic cage should be performed to assess the stability of the spine injury because the great majority of thoracic TSCI will involve an associated disruption of the thoracic cage.

Associated fractures of the sternum and/or ribs should raise the suspicion for severe spinal instability. Because most thoracic TSCI result from high-energy trauma, the initial neurological deficit is usually severe, and the prognosis is often poor (Bourassa-Moreau et al., 2016; Zariffa et al., 2011) (Fig. 5).

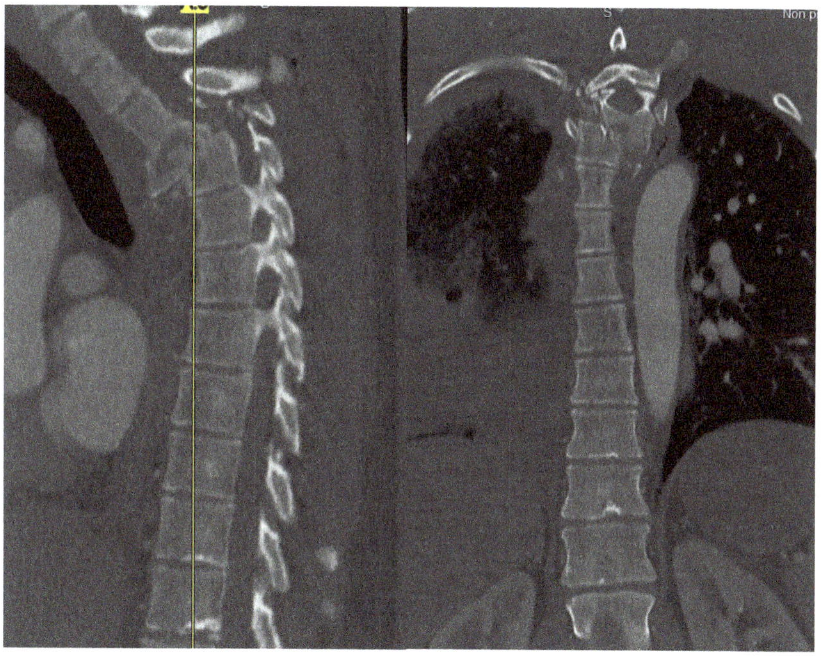

**FIG. 5** Complete injury SCI, dislocation, and chest trauma. Complex comminuted fracture with lateral dislocation of the D3–D4 vertebrae and discontinuity of the spinal canal. Multiple bilateral rib fractures, Right hemothorax. Extensive right subpleural pulmonary contusion and minimal right apical pleural detachment.

## *Thoracolumbar spine injuries (T11–L2)*

The thoracolumbar junction between T11 and L2 is the most frequent site for TSCI in an unstable spine for both osteoporotic and nonosteoporotic individuals.

There is an increased stress concentration in this region because it is a transition zone from the stiff thoracic spine/cage and the mobile lumbar spine. The straight alignment of the thoracolumbar region between the supra-adjacent thoracic kyphosis and subjacent lumbar lordosis also increases the transfer of axial compression loads through this region and resulting risk of compression and burst fractures.

While compression fractures are usually not associated with TSCI, thoracolumbar burst fractures can be associated with a TSCI due to a bony fragment retropulsed into the spinal canal and compressing the spinal cord.

In the presence of a TSCI, determining the morphological pattern of the thoracolumbar fracture seems particularly useful to characterize the severity of the injury and the prognosis for neurological recovery. Different authors (Keynan et al., 2006; Martineau, Goulet, Richard-Denis, & Mac-Thiong, 2019) propose that the energy directly transferred to the spinal cord during the trauma is strongly related to the local morphological pattern of thoracolumbar fractures, as opposed to the global level of energy or mechanism of the trauma.

The morphological pattern of thoracolumbar injury also provides insight into the dynamics of the SCI during the trauma, more so than the residual compression of the spinal cord after the trauma from the retropulsed fragment.

It is well recognized that the dynamic spinal canal encroachment during the trauma is significantly greater than the static canal encroachment seen on posttrauma imaging. Accordingly, the degree of canal occlusion observed after the trauma is not related to the neurological outcome (Boerger, Limb, & Dickson, 2000; Herndon & Galloway, 1988; Mohanty & Venkatram, 2002).

Goulet et al. have shown that posterior translation of the posteroinferior wall of the vertebral body, presence of comminution of the bony fragment retropulsed into the spinal canal and complete lamina fracture were strongly associated with neurological recovery.

These findings suggest that these morphological features—characterizing damaged bony structures directly in contact with the spinal cord—should be assessed routinely in TSCI patients to estimate the local level of energy transferred to the spinal cord and the prognosis for recovery. Conversely, vertebral body comminution, local/segmental kyphosis, and the degree of canal compromise were not associated with the neurological outcome. In their finite-element computational simulation of TSCI for thoracolumbar burst fractures, Diotalevi et al. tested 16 typical clinical scenarios with varying degree of comminution, velocity, and displacement of the bony fragment retropulsed into the spinal canal, as well as the presence or absence of a fracture of the posteroinferior vertebral body wall.

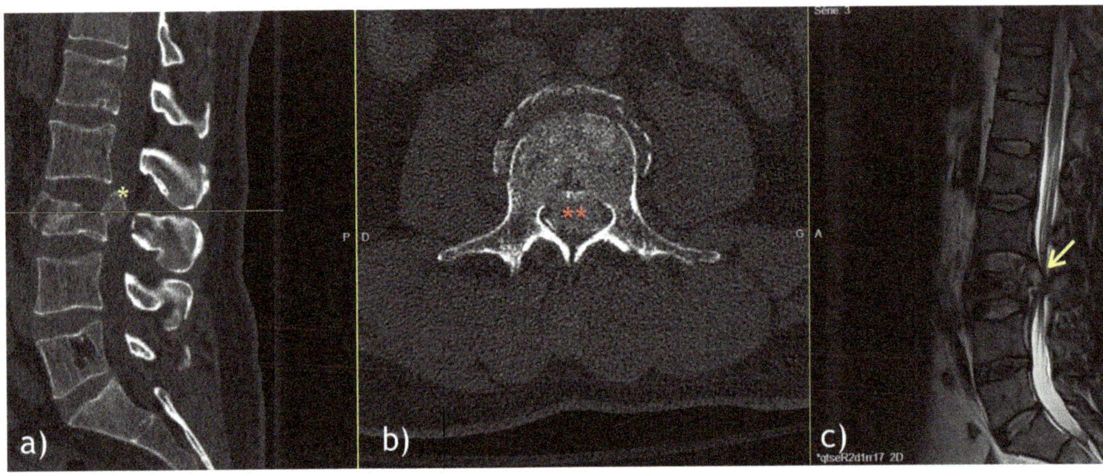


FIG. 6 Burst T3. (A) Translation posterior wall. [yellow asterisk (*)] (B) comminution of the fragment and lamina fracture [red asterisks (**)]. (C) MRI showing compression and cord edema (Yellow arrow).

They observed that the velocity of the retropulsed fragment was the most important factor influencing the stresses and strains on the spinal cord, followed by a fracture of the posteroinferior vertebral body wall and sagittal rotation of the retropulsed fragment (Diotalevi et al., 2020) (Fig. 6)

The Spinal trauma Study Group put a strong emphasis on the assessment of the posterior ligamentous complex (PLC) (supraspinous ligament, interspinous ligament, ligamentum flavum, and facet joint capsules) to further assess the stability of thoracolumbar spine injuries (Vaccaro et al., 2006). Rupture of the PLC entails a poor prognosis for healing because, unlike bone fracture, ligaments do not recover its pretrauma tensile strength after complete rupture.

Although the loss of integrity of PLC is a crucial element for surgeons to determine indication for surgery in thoracolumbar trauma, it remains to be defined whether injury to the PLC is associated with the occurrence a severity of TSCI.

Flexion-distraction spine injuries most often result from falls, car accident (seat belt injury), and unrestrained vehicle accidents (Wagnac et al., 2019). It is estimated that a small number (about 10%–15%) of flexion-distraction injuries in the thoracolumbar spine will lead to a SCI through a mechanism of distraction of the spinal cord. On the opposite, translation/rotation spinal injuries in the thoracolumbar spine typically result from a high-energy trauma mechanism and lead to severe TSCI due to the high shear stresses causing primary and secondary damage to the spinal cord. In the AO classification the type C injuries are translational injuries in any axis. No further subdivision is necessary since all Type C injuries are highly unstable due to separation, displacement, or translation of 1 vertebral body (or elements of it) relative to another in any direction.

## Applications to other areas of neuroscience

Here we reviewed the basic biomechanical concepts of traumatic spinal cord injuries associated with spine injuries. We have also reviewed the local mechanisms by which the spinal cord is damaged during traumatic events, and we describe the interactions between spinal osteoligamentous elements and the spinal cord during and after the trauma.

The knowledge transferred in this chapter will help neuroscientists to better understand and study the primary and secondary spinal cord injuries occurring after a trauma. This work also underscores the importance of combining clinical and biomechanical knowledge for assessing and treating individuals following a traumatic SCI. For example, we have highlighted the relevance of using computational models for simulating and investigating traumatic spinal cord injuries. The concepts presented in this chapter could be applied for designing new preclinical models of traumatic SCI, or for planning on innovating research protocols. It is becoming clear that combining expertise in neurosciences and biomechanical engineering is key in the development of a more comprehensive understanding of traumatic spinal cord injuries in the future.

## Mini-dictionary of terms

**Traumatic spinal cord injury in a stable spine:** SCI that causes neurological compromise and in the images there is no fracture or unstable injury, generally called CCS or SCIWORA.

**Spinal cord distraction:** Structural alteration in the white matter, sparing the most myelinated axons and causing extensive lesion cavity.

**Hypertrophic ligamentum flavum:** Thickening of the ligamentum flavum for the aging process causing central canal stenosis and associated to TSCI in a stable spine combined with trauma in hyperextension.

**Finite-element analysis:** Physical phenomenon simulation using a numerical technique called finite element method (FEM) to reproduce some situations and reduce the number of physical prototypes to optimize study designs.

**Dynamic spinal canal encroachment:** Invasion of the canal that occurs with movement or by gravity It is well recognized that the dynamic spinal canal encroachment during the trauma is significantly greater than the static canal encroachment seen on post-trauma imaging (Panjabi et al., 1995; Wilcox et al., 2003).

## Key facts

Key facts of central cord syndrome (CCS) or spinal cord injury (SCI) in a stable spine.

- Not all the SCIs happen in an unstable spine. In a spine with limited capacity of the vertebral canal, extra physiologically motion can result in SCI.
- Due to aging of population there is a growing number of patients with cervical SCI without instability associated with degenerative changes of the cervical spine.
- Traumatic central cord syndrome (TCCS) is a clinical entity that implies predominant impairment of upper extremities compared to relatively spared lower extremities in the setting of an incomplete SCI.
- TCCS is often found in elderly patients with degenerative cervical spine and no over spinal instability, despite being the most common incomplete SCI it lacks well defined diagnostic criteria.

Key facts of morphology of the fractures and biomechanical factors of SCI.

- Between variables such as low- and high-energy trauma and fracture morphological patterns, are these last ones which explain in a better way the level of energy transferred to the spinal cord.
- Cadaveric and animal models are used to study biomechanics of fractures, finding burst fractures and dislocations as the most common injury patterns, also to understand the biology of spinal cord injury and the effects of other relevant factors as injury mechanism, injury velocity, and residual compression.
- The weight-drop animal injury model was developed in 1911 and has evolved significantly, and it can classify the SCI mechanism, such as contusion, compression, dislocation, distraction, and transection.

## Summary points

- Indirect injury to the spinal cord results from the motion (e.g., flexion, hyperextension, axial compression, distraction, dislocation, etc.) between spinal elements during the trauma, causing excessive tension, shear, torsion, compression and/or bending loads on the spinal cord.
- In line with the definition by White and Panjabi, we define traumatic spinal instability to be the loss of the ability of the spine to maintain normal alignment and motion between spinal segments in such a way that a secondary injury to the spinal cord will not occur.
- Wilson et al. showed that cervical dislocation is more frequently associated with severe neurological deficit and high-energy trauma mechanisms, and results in a diminished potential for recovery.
- More often, thoracic TSCI results from high-energy trauma mechanisms. Thorough clinical and radiological evaluation of the thoracic cage should be performed to assess the stability of the spine injury because the great majority of thoracic TSCI will involve an associated disruption of the thoracic cage.
- The velocity of a retropulsed fragment was the most important factor influencing the stresses and strains on the spinal cord, followed by a fracture of the postero-inferior vertebral body wall and sagittal rotation of the retropulsed fragment.
- Although the loss of integrity of PLC is a crucial element for surgeons to determine indication for surgery in thoraco-lumbar trauma, it remains to be defined whether injury to the PLC is associated with the occurrence and severity of TSCI.

## References

Aarabi, B., Alexander, M., Mirvis, S. E., Shanmuganathan, K., Chesler, D., Maulucci, C., et al. (2011). Predictors of outcome in acute traumatic central cord syndrome due to spinal stenosis. *Journal of Neurosurgery Spine*, *14*(1), 122–130. https://doi.org/10.3171/2010.9.SPINE09922.

Ahn, H., Lewis, R., Santos, A., Cheng, C. L., Noonan, V. K., Dvorak, M. F., et al. (2017). Forecasting financial resources for future traumatic spinal cord injury care using simulation modeling. *Journal of Neurotrauma*, *34*(20), 2917–2923. Epub 2017 Aug 24. PMID: 28594315; PMCID: PMC5653147 https://doi.org/10.1089/neu.2016.4936.

Al-Habib, A. F., Attabib, N., Ball, J., Bajammal, S., Casha, S., & Hurlbert, R. J. (2011). Clinical predictors of recovery after blunt spinal cord trauma: Systematic review. *Journal of Neurotrauma*, *28*(8), 1431–1443. https://doi.org/10.1089/neu.2009.1157.

Alker, G. J., Oh, Y. S., Leslie, E. V., Lehotay, J., Panaro, V. A., & Eschner, E. G. (1975). Postmortem radiology of head neck injuries in fatal traffic accidents. *Radiology*, *114*(3), 611–617. https://doi.org/10.1148/114.3.611.

Asan, Z. (2018). Spinal cord injury without radiological abnormality in adults: Clinical and radiological discordance. *World Neurosurgery*, *114*, e1147–e1151. https://doi.org/10.1016/j.wneu.2018.03.162.

Boerger, T. O., Limb, D., & Dickson, R. A. (2000). Does 'canal clearance' affect neurological outcome after thoracolumbar burst fractures? *Journal of Bone and Joint Surgery. British Volume (London)*, *82*(5), 629–635. https://doi.org/10.1302/0301-620x.82b5.11321.

Boese, C. K., & Lechler, P. (2013). Spinal cord injury without radiologic abnormalities in adults: a systematic review. *Journal of Trauma and Acute Care Surgery*, *75*(2), 320–330. https://doi.org/10.1097/TA.0b013e31829243c9. 23702634.

Boese, C. K., Muller, D., Broer, R., Eysel, P., Krischek, B., Lehmann, H. C., et al. (2016). Spinal cord injury without radiographic abnormality (SCIWORA) in adults: MRI type predicts early neurologic outcome. *Spinal Cord*, *54*(10), 878–883. https://doi.org/10.1038/sc.2016.13.

Bourassa-Moreau, E., Mac-Thiong, J. M., Li, A., Ehrmann Feldman, D., Gagnon, D. H., Thompson, C., et al. (2016). Do patients with complete spinal cord injury benefit from early surgical decompression? Analysis of neurological improvement in a prospective cohort study. *Journal of Neurotrauma*, *33* (3), 301–306. https://doi.org/10.1089/neu.2015.3957.

Bucholz, R. W., Burkhead, W. Z., Graham, W., & Petty, C. (1979). Occult cervical spine injuries in fatal traffic accidents. *The Journal of Trauma*, *19*(10), 768–771. https://doi.org/10.1097/00005373-197910000-00009.

Buchowski, J. M., Kuhns, C. A., Bridwell, K. H., & Lenke, L. G. (2008). Surgical management of posttraumatic thoracolumbar kyphosis. *The Spine Journal*, *8*(4), 666–677. https://doi.org/10.1016/j.spinee.2007.03.006.

Chen, Y., He, Y., & DeVivo, M. J. (2016). Changing demographics and injury profile of new traumatic spinal cord injuries in the United States, 1972–2014. *Archives of Physical Medicine and Rehabilitation*, *97*(10), 1610–1619. https://doi.org/10.1016/j.apmr.2016.03.017.

Collignon, F., Martin, D., Lenelle, J., & Stevenaert, A. (2002). Acute traumatic central cord syndrome: Magnetic resonance imaging and clinical observations. *Journal of Neurosurgery*, *96*(1 Suppl), 29–33. https://doi.org/10.3171/spi.2002.96.1.0029.

Como, J. J., Samia, H., Nemunaitis, G. A., Jain, V., Anderson, J. S., Malangoni, M. A., et al. (2012). The misapplication of the term spinal cord injury without radiographic abnormality (SCIWORA) in adults. *Journal of Trauma and Acute Care Surgery*, *73*(5), 1261–1266. https://doi.org/10.1097/TA.0b013e318265cd8c.

Diotalevi, L., Bailly, N., Wagnac, E., Mac-Thiong, J. M., Goulet, J., & Petit, Y. (2020). Dynamics of spinal cord compression with different patterns of thoracolumbar burst fractures: Numerical simulations using finite element modelling. *Clinical Biomechanics (Bristol, Avon)*, *72*, 186–194. https://doi.org/10.1016/j.clinbiomech.2019.12.023.

Dreizin, D., Kim, W., Kim, J. S., Boscak, A. R., Bodanapally, U. K., Munera, F., et al. (2015). Will the real SCIWORA please stand up? Exploring clinicoradiologic mismatch in closed spinal cord injuries. *AJR. American Journal of Roentgenology*, *205*(4), 853–860. https://doi.org/10.2214/AJR.14.13374.

Dvorak, M. F., Noonan, V. K., Fallah, N., Fisher, C. G., Rivers, C. S., Ahn, H., et al. (2014). Minimizing errors in acute traumatic spinal cord injury trials by acknowledging the heterogeneity of spinal cord anatomy and injury severity: An observational Canadian cohort analysis. *Journal of Neurotrauma*, *31* (18), 1540–1547. https://doi.org/10.1089/neu.2013.3278.

Ebraheim, N. A., Patil, V., Liu, J., Haman, S. P., & Yeasting, R. A. (2008). Morphometric analyses of the cervical superior facets and implications for facet dislocation. *International Orthopaedics*, *32*(1), 97–101. https://doi.org/10.1007/s00264-006-0286-5.

Goulet, J., Richard-Denis, A., Petit, Y., Diotalevi, L., & Mac-Thiong, J. M. (2020). Morphological features of thoracolumbar burst fractures associated with neurological outcome in thoracolumbar traumatic spinal cord injury. *European Spine Journal*, *29*(10), 2505–2512. https://doi.org/10.1007/s00586-020-06420-9. Epub 2020 May 19. PMID: 32424639.

Gupta, S. K., Rajeev, K., Khosla, V. K., Sharma, B. S., Paramjit, Mathuriya, S. N., et al. (1999). Spinal cord injury without radiographic abnormality in adults. *Spinal Cord*, *37*(10), 726–729. https://doi.org/10.1038/sj.sc.3100900.

Herndon, W. A., & Galloway, D. (1988). Neurologic return versus cross-sectional canal area in incomplete thoracolumbar spinal cord injuries. *The Journal of Trauma*, *28*(5), 680–683. https://doi.org/10.1097/00005373-198805000-00022.

Holdsworth, F. (1970). Fractures, dislocations, and fracture-dislocations of the spine. *The Journal of Bone and Joint Surgery American Volume*, *52*(8), 1534–1551. https://www.ncbi.nlm.nih.gov/pubmed/5483077.

Keynan, O., Fisher, C. G., Vaccaro, A., Fehlings, M. G., Oner, F. C., Dietz, J., et al. (2006). Radiographic measurement parameters in thoracolumbar fractures: A systematic review and consensus statement of the spine trauma study group. *Spine (Phila Pa 1976)*, *31*(5), E156–E165. https://doi.org/10.1097/01.brs.0000201261.94907.0d.

Lee, C., & Woodring, J. H. (1991). Unstable Jefferson variant atlas fractures: An unrecognized cervical injury. *AJNR—American Journal of Neuroradiology*, *12*(6), 1105–1110. https://www.ncbi.nlm.nih.gov/pubmed/1763734.

Martineau, J., Goulet, J., Richard-Denis, A., & Mac-Thiong, J.-M. (2019). The relevance of MRI for predicting neurological recovery following cervical traumatic spinal cord injury. *Spinal Cord*, *57*(10), 866–873. https://doi.org/10.1038/s41393-019-0295-z.

Mattucci, S., Speidel, J., Liu, J., Kwon, B. K., Tetzlaff, W., & Oxland, T. R. (2019). Basic biomechanics of spinal cord injury—How injuries happen in people and how animal models have informed our understanding. *Clinical Biomechanics (Bristol, Avon)*, *64*, 58–68. https://doi.org/10.1016/j.clinbiomech.2018.03.020.

McKinley, W., Santos, K., Meade, M., & Brooke, K. (2007). Incidence and outcomes of spinal cord injury clinical syndromes. *The Journal of Spinal Cord Medicine*, *30*(3), 215–224. https://doi.org/10.1080/10790268.2007.11753929.

Mohanty, S. P., & Venkatram, N. (2002). Does neurological recovery in thoracolumbar and lumbar burst fractures depend on the extent of canal compromise? *Spinal Cord*, *40*(6), 295–299. https://doi.org/10.1038/sj.sc.3101283.

Molliqaj, G., Payer, M., Schaller, K., & Tessitore, E. (2014). Acute traumatic central cord syndrome: A comprehensive review. *Neurochirurgie*, *60*(1–2), 5–11. https://doi.org/10.1016/j.neuchi.2013.12.002.

Nathan, P. W., & Smith, M. C. (1955). Long descending tracts in man. I. Review of present knowledge. *Brain*, *78*(2), 248–303. https://doi.org/10.1093/brain/78.2.248.

Pang, D., & Wilberger, J. E., Jr. (1982). Spinal cord injury without radiographic abnormalities in children. *Journal of Neurosurgery*, *57*(1), 114–129. https://doi.org/10.3171/jns.1982.57.1.0114. PMID: 7086488.

Panjabi, M. M., Kifune, M., Wen, L., Arand, M., Oxland, T. R., Lin, R. M., et al. (1995). Dynamic canal encroachment during thoracolumbar burst fractures. *Journal of Spinal Disorders*, *8*(1), 39–48. https://www.ncbi.nlm.nih.gov/pubmed/7711368.

Paquet, J., Rivers, C. S., Kurban, D., Finkelstein, J., Tee, J. W., Noonan, V. K., et al. (2018). The impact of spine stability on cervical spinal cord injury with respect to demographics, management, and outcome: A prospective cohort from a national spinal cord injury registry. *The Spine Journal*, *18*(1), 88–98. https://doi.org/10.1016/j.spinee.2017.06.032.

Pickett, G., Campos-Benitez, M., Keller, J., & Duggal, N. (2006). Epidemiology of traumatic spinal cord injury in Canada. *Spine*, *31*(7), 799–805.

Quencer, R. M., Bunge, R. P., Egnor, M., Green, B. A., Puckett, W., Naidich, T. P., et al. (1992). Acute traumatic central cord syndrome: MRI-pathological correlations. *Neuroradiology*, *34*(2), 85–94. https://doi.org/10.1007/BF00588148.

Schneider, R. C., Cherry, G., & Pantek, H. (1954). The syndrome of acute central cervical spinal cord injury; with special reference to the mechanisms involved in hyperextension injuries of cervical spine. *Journal of Neurosurgery*, *11*(6), 546–577. https://doi.org/10.3171/jns.1954.11.6.0546.

Schneider, R. C., Thompson, J. M., & Bebin, J. (1958). The syndrome of acute central cervical spinal cord injury. *Journal of Neurology, Neurosurgery, and Psychiatry*, *21*(3), 216–227. https://doi.org/10.1136/jnnp.21.3.216.

Sekhon, L. H., & Fehlings, M. G. (2001). Epidemiology, demographics, and pathophysiology of acute spinal cord injury. *Spine (Phila Pa 1976)*, *26*(24 Suppl), S2–12. https://doi.org/10.1097/00007632-200112151-00002.

Tator, C. H. (1983). Spine-spinal cord relationships in spinal cord trauma. *Clinical Neurosurgery*, *30*, 479–494. https://doi.org/10.1093/neurosurgery/30.cn_suppl_1.479.

Vaccaro, A. R., Hulbert, R. J., Patel, A. A., Fisher, C., Dvorak, M., Lehman, R. A., Jr., et al. (2007). The subaxial cervical spine injury classification system: A novel approach to recognize the importance of morphology, neurology, and integrity of the disco-ligamentous complex. *Spine (Phila Pa 1976)*, *32*(21), 2365–2374. https://doi.org/10.1097/BRS.0b013e3181557b92.

Vaccaro, A. R., Koerner, J. D., Radcliff, K. E., Oner, F. C., Reinhold, M., Schnake, K. J., et al. (2016). AOSpine subaxial cervical spine injury classification system. *European Spine Journal*, *25*(7), 2173–2184. https://doi.org/10.1007/s00586-015-3831-3.

Vaccaro, A. R., Lee, J. Y., Schweitzer, K. M., Jr., Lim, M. R., Baron, E. M., Oner, F. C., et al. (2006). Assessment of injury to the posterior ligamentous complex in thoracolumbar spine trauma. *Spine J.*, *6*(5), 524–528. https://doi.org/10.1016/j.spinee.2006.01.017. Epub 2006 Jul 11 16934721.

Wagnac, E., Mac-Thiong, J. M., Arnoux, P. J., Desrosiers, J. M., Menard, A. L., & Petit, Y. (2019). Traumatic spinal cord injuries with fractures in a Quebec level I trauma center. *The Canadian Journal of Neurological Sciences*, *46*(6), 727–734. https://doi.org/10.1017/cjn.2019.252.

White, A. A., 3rd, Johnson, R. M., Panjabi, M. M., & Southwick, W. O. (1975). Biomechanical analysis of clinical stability in the cervical spine. *Clinical Orthopaedics and Related Research*, *109*, 85–96. https://doi.org/10.1097/00003086-197506000-00011. PMID: 1132209.

Wilcox, R. K., Boerger, T. O., Allen, D. J., Barton, D. C., Limb, D., Dickson, R. A., et al. (2003 Nov). A dynamic study of thoracolumbar burst fractures, *The Journal of Bone and Joint Surgery. American Volume*, *85*(11), 2184–2189. https://doi.org/10.2106/00004623-200311000-00020. 14630851.

Wilson, J. R., Cadotte, D. W., & Fehlings, M. G. (2012). Clinical predictors of neurological outcome, functional status, and survival after traumatic spinal cord injury: A systematic review. *Journal of Neurosurgery. Spine*, *17*(1 Suppl), 11–26. https://doi.org/10.3171/2012.4.AOSPINE1245.

Wilson, J. R., Vaccaro, A., Harrop, J. S., Aarabi, B., Shaffrey, C., Dvorak, M., et al. (2013). The impact of facet dislocation on clinical outcomes after cervical spinal cord injury: Results of a multicenter North American prospective cohort study. *Spine (Phila Pa 1976)*, *38*(2), 97–103. https://doi.org/10.1097/BRS.0b013e31826e2b91.

Yokota, K., Maeda, T., Kawano, O., Mori, E., Takao, T., Sakai, H., et al. (2019). Progression of local kyphosis after conservative treatment for compressive cervical spine fracture with spinal cord injury. *Journal of Orthopaedic Surgery and Research*, *14*(1), 98. https://doi.org/10.1186/s13018-019-1115-z.

Zariffa, J., Kramer, J. L., Fawcett, J. W., Lammertse, D. P., Blight, A. R., Guest, J., et al. (2011). Characterization of neurological recovery following traumatic sensorimotor complete thoracic spinal cord injury. *Spinal Cord*, *49*(3), 463–471. https://doi.org/10.1038/sc.2010.140.

Chapter 3

# Body, action, and space representations in people affected by spinal cord injuries

Michele Scandola
*NPSY-Lab.VR & SCI-Research Group, Department of Human Sciences, University of Verona, Verona, Italy*

## List of abbreviations

**BR** body representation
**PPS** peripersonal space
**RT** response times
**SCI** spinal cord injury

## Introduction

Anecdotical reports concerning modifications of the perception of the body, space, and action following a spinal cord injury (SCI) are well known to professionals that have experienced with these patients as they can report body misperceptions (e.g., phantom limb or body loss), misrepresentations of space (e.g., body or limbs position, distance lengthening), changes in action representation (e.g., illusionary and phantom limb movement leading to muscular fatigue) (Cole, 2004; Conomy, 1973; Curt, Yengue, Hilti, & Brugger, 2011; Scandola et al., 2017). These sensations, linked to pain, the level of the lesion and the time passed since the lesion onset (Scandola, Aglioti, Avesani, et al., 2017), are symptoms of neuroplastic changes in areas related to the body, space, and action representation.

A spinal cord interruption, especially when this is not developmental but acquired, abruptly changes the brain–body communication. This modification alters the balance between the body sensorial inputs, the signals sent toward it, and the brain activity, leading to abnormalities in cortical and subcortical activity. Consequently, the brain neuroplastically changes itself (Nardone et al., 2013) into brain networks that are involved in body, space, and action representations. Therefore, modifications in these cognitive functions are likely to occur.

This chapter is organized as: a first section reviewing the main literature concerning neuroplastic modifications in SCI individuals, underlying those related to the body, space, or action representations, the following three sections will show the experimental outcomes from studies concerning these representations in SCI individuals, and a fourth section will highlight the experimental results showing the effects of rehabilitative trainings on these representations.

Finally, a concluding section will present the main findings, interpretations, and connections.

## The neuroplasticity following SCI and the networks involved in the body, space, and action representations

Neuroplasticity mechanisms have been recently investigated in SCI. Different levels of activation in the sensory and motor cortices have shown, particularly a reduction of activity in areas corresponding to paralyzed limbs ( Jurkiewicz, Mikulis, Fehlings, & Verrier, 2010) along with increased activation in areas corresponding to spared body parts (Aguilar et al., 2010; Curt, Bruehlmeier, Leenders, Roelcke, & Dietz, 2002).

Other studies (Henderson, Gustin, Macey, Wrigley, & Siddall, 2011) showed the expansion of the cortical hand area toward the leg area during upper limbs movements. An enlargement of the face representation in the surrounding areas in monkeys after SCI was observed in the primary and nonprimary somatosensory cortices (Tandon, Kambi, Lazar, Mohammed, & Jain, 2009).

Other works found a general reorganization of somatotopic areas (Ghosh et al., 2012; Lundell et al., 2011) and their degeneration over time (Freund et al., 2013).

Diagnosis and Treatment of Spinal Cord Injury. https://doi.org/10.1016/B978-0-12-822498-4.00003-8

All these neuroplastic changes relate to the body, space, and action representations; indeed, all the somatotopic areas are involved in BR (Berlucchi & Aglioti, 2010). Moreover, in BR, the posterior parietal cortex, integrating multisensory inputs and outputs has a fundamental role (Bonda, Petrides, Frey, & Evans, 1995) as this brain area is directly involved in the neuroplasticity following SCI, because of the sensorial modifications.

The understanding of others' actions and space representations rely also on somatosensory cortices. In fact, the fronto-parietal "mirror" network (Rizzolatti, Cattaneo, Fabbri-Destro, & Rozzi, 2014) includes the ventral premotor cortex and the inferior frontal gyrus. Space perception involves the intraparietal sulcus, the lateral occipital complex, and the premotor cortex (Makin, Holmes, & Zohary, 2007; Sereno & Huang, 2006) in the peripersonal space representation, while the extra-personal space perception is linked to an occipito-parietal circuit, a parieto-prefrontal pathway, and a parieto-premotor one (Kravitz, Saleem, Baker, & Mishkin, 2011).

The unbalance between the somatosensory and other sensory feedbacks, causing fundamental changes in the neurological system after SCI, might cause of maladaptive neuroplasticity in these cognitive representations, causing modifications that may be subtle, or evident like phantom sensations (Moore, Stern, Dunbar, Kostyk, & Gehi, & Corkin, 2000) and neuropathic pain (Wrigley et al., 2009).

## The body representations

Body ownership is an important aspect of our body representations (BR) and can be defined as the sensation that a bodily part is part of our body. Relevant disturbances can lead to Asomatognosia ( Jenkinson, Moro, & Fotopoulou, 2018), namely abnormalities in various facets of body ownership such as the experienced existence, visual self-recognition, and sense of belonging to contralesional body parts.

This facet of BR, investigated by the rubber-hand paradigm, appears to be modified according to neuroplastic cortical modifications. In the commonest version of the rubber-hand illusion (see Fig. 1, Botvinick & Cohen, 1998), participants

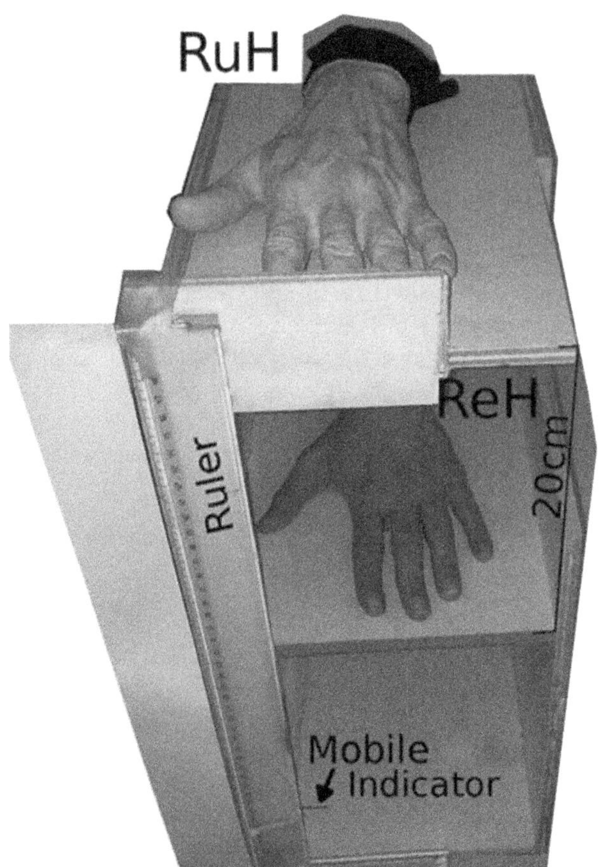


FIG. 1 Rubber hand experimental set-up used in Scandola et al. (2014): Experimental wooden box (RuH, Rubber Hand; ReH, Real Hand), the mobile indicator was always visible to the participant laterally to the box.

observe a rubber-hand, i*n a congruent position with* their own body, synchronously touched with their hand, hidden from sight (Botvinick & Cohen, 1998). An asynchronous rubber-hand/real-hand touch should not elicit a body ownership sensation and, therefore, is used as control condition (Botvinick & Cohen, 1998).

In hand representation-deprived tetraplegics, the ownership sensation toward a rubber hand can be elicited by the synchronous tactile stimulation of the left cheek and the rubber hand (see Fig. 2; Scandola et al., 2014). This is probably caused by the overlapping of the cheek somatic and motor representations over the representation of the hand that are normally contiguously represented (Ramachandran, Stewart, & Rogers-Ramachandran, 1992). A similar result with synchronous visuo-tactile stimulation of the upper back and rubber legs in SCI individuals with complete paraplegia was found (Pozeg et al., 2017). Moreover, the subjective verbal report of the sensation of body ownership toward a body part seems to be captured by the mere vision of the body part, especially when there is a total lack of tactile sensations (lower limbs in people affected by complete paraplegia: Scandola et al., 2020; rubber hand in a patient affected by tetraplegia: Tidoni, Grisoni, Liuzza, & Aglioti, 2014).

A further, implicit paradigm to study body ownership relies on the body-view enhancement effect. This task exploits the effect of visual stimuli appearing on our body that evoke faster reactions than visual stimuli appearing on the same space location, but on a neutral surface (Kao & Goodale, 2009). By a modified version of this paradigm, it was observed that individuals affected by complete paraplegia feel body ownership for the parts of the body that have spared sensory and motor functions, but also the same level of body ownership for their wheelchair, while the lower limbs show no body ownership (Scandola et al., 2019). Interestingly, the same task executed with the SCI participant seated in a different, never used before, wheelchair shows that this redefines their whole BR: there is *no* body ownership toward the new wheelchair, the lower limbs, nor for the part of the body that has spared motor and sensory functions (Scandola, Togni, et al., 2019).

The notion of body schema dates back to the seminal work of Head and Holmes (1911) and can be nowadays defined as the dynamic representation of body parts and their movements, which derives from multiple sensory and motor inputs (e.g., proprioceptive, vestibular, tactile, visual, efference copy). Patients with an impaired body schema are not able to distinguish the position of their arm when their eyes are closed.

A popular experimental paradigm investigating this aspect is the Mental Body Rotation paradigm (Parsons, 1987). According to this, individuals are asked to identify the laterality of a hand or a foot presented on screen at different rotations. The underlying mechanisms of decision imply that participants rotate the bodily stimuli congruent with their body and, therefore, response times (RTs), required to align series of gradually more rotated images to the vertical congruent with the body position, are slower. This leads to a typical RTs pattern, characteristic of a normal body schema representation, called bio-mechanical effect (Parsons, 1987).

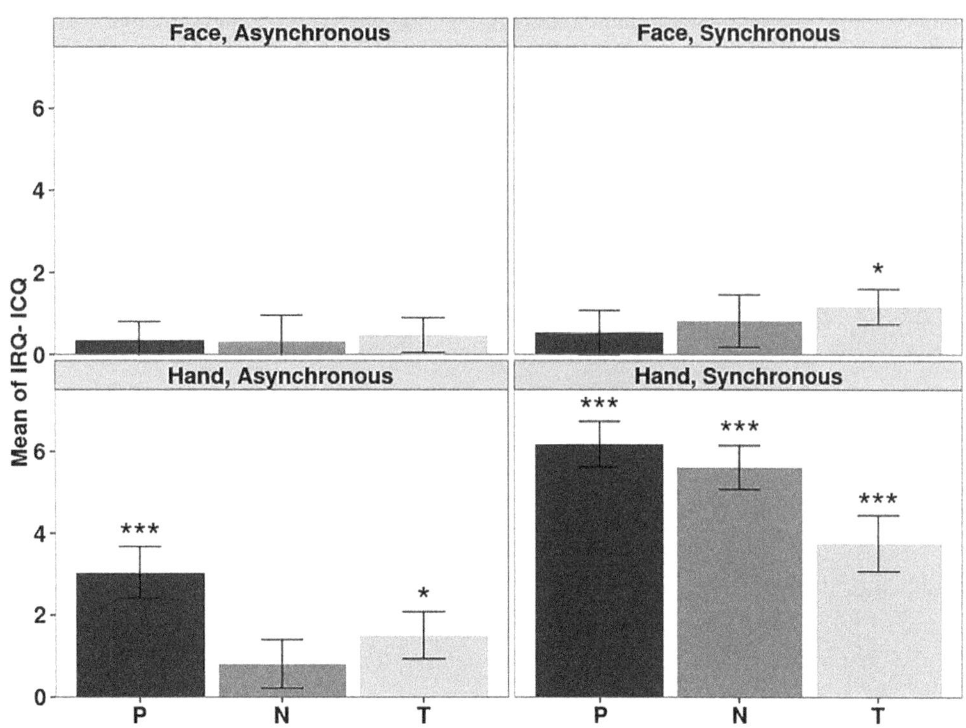


FIG. 2 Results from Scandola et al. (2014): Presence of body ownership in the paraplegics (P), tetraplegics (T), and control healthy (N) groups.

Mental Body Rotation paradigm confirms that complete SCI disrupts the influence of postural changes on the representation of the deafferented body parts (feet, but not hands) and, regardless the posture, whole-BR progressively deteriorates proportionally to SCI completeness (Ionta et al., 2016), as a direct effect of the somato-sensation lack in the affected body-parts in SCI individuals.

Indeed, in SCI participants, the misproprioception of the paralyzed body parts is not uncommon (77% of cases) and it is present in all people affected by complete tetraplegia (Scandola, Aglioti, Avesani, et al., 2017).

Also Disownership-like and Somatoparaphrenia-like sensations (sensations of not owning their own limbs, even attributing them to someone else, as in the latter case) are often reported (48% and 46%, respectively), altogether to illusory motion of paralyzed body parts (51%).

In all these cases, these bodily illusions are connected to the time since the lesion onset, the level of injury, and the presence of pain (Scandola, Aglioti, Avesani, et al., 2017).

Interestingly, these bodily illusions are less evident in presence of neuropathic and visceral pain. This apparently counterintuitive result addresses toward the fact that, in a deafferented/deefferented body, a pain sensation coming from the impaired body part is used by our cognitive system as a surrogate BR.

## The action representation

The representation of actions is fundamental in our everyday life. It allows us to plan our gestures, interact with the environment and other people. The representation of actions is based on the fronto-parietal "mirror" network (Aglioti, Cesari, Romani, & Urgesi, 2008; Avenanti, Candidi, & Urgesi, 2013; Calvo-Merino, Glaser, Grèzes, Passingham, & Haggard, 2004; Panasiti, Pavone, & Aglioti, 2016).

In matching-to-sample experiments, participants are presented with a single stimulus called the sample and then with two choice stimuli called the comparisons. The participants have to select the comparison that matches the sample. A version of this paradigm that used images of bodies that could be different for body action or body form, for the whole body, in its upper part, or in the lower part, was applied to a sample of complete paraplegics and healthy controls (Pernigo et al., 2012). Results showed that complete paraplegics participants had slower reaction times and greater errors incidence, in body images that were different for the lower part of the body, than the control group of healthy participants.

These results were confirmed and widened by using a Temporal Occlusion Paradigm (Scandola, Aglioti, et al., 2019) where participants observed two sets of videos depicting an actor who attempted to climb onto a platform using a wheelchair or rollerblades. Each video could end in three different ways: (a) successfully (the actor went up the step), (b) failing (the actor stopped before the step without going up), or (c) falling (the actor fell without going up). Each video was cut in five different durations at 600, 1200, 1800, 2400, and 3000 ms. In the shortest (600 ms) clip, only preparatory body movements were shown, while in the longest (3000 ms), the complete action (see Fig. 3).

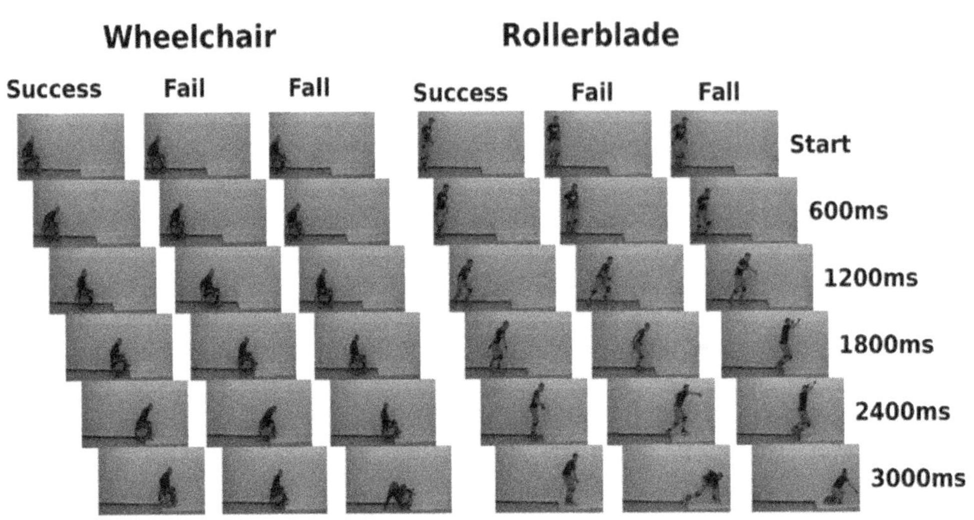


FIG. 3 Stimuli used in Scandola, Aglioti, et al. (2019): Wheelchair and rollerblade videos at different durations (Start, 600, 1200, 1800, 2400, 3000 ms) for the three endings.

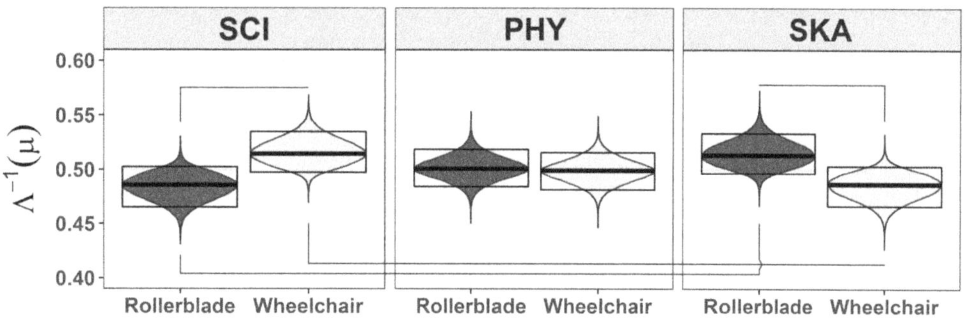


FIG. 4 Results from (Scandola, Aglioti, et al., 2019): a Temporal Occlusion Paradigm in anticipating the ending of wheelchair or rollerblade videos in *SCI* = spinal cord injured people; *PHY* = physiotherapists; *SKA* = skaters. Higher values mean better performance.

Participants had to report the outcome of the video choosing among success, fail, or fall. Participants were divided in three groups: the complete paraplegics group, the expert skaters group, and a group composed by physiotherapists that had at least 1 year of experience in a spinal-unit, but no experience in rollerblades or other skating tools.

Results from this study showed that complete paraplegics participants had a worse performance in rollerblades videos compared to the Skaters group, but also compared to the Physiotherapists group, inexperienced in skating (see Fig. 4), suggesting that abnormalities in the mirror system for the lower limbs affect the recognition of lower limbs' actions in a subtle, but still worsening way compared to a group that had no direct experience in the action.

A further indication regarding modifications in the action representation system derives from the tendon-vibration illusion, able to induce illusory perception of movement (Tidoni et al., 2015). The application of this illusion to individuals affected by SCI induced movement illusions that were qualitatively different from the healthy control sample (e.g., as if the arm wanted to extend itself, or a sensation of pushing against something) that may reflect different reorganization processes following SCI (Fusco, Tidoni, Barone, Pilati, & Aglioti, 2016).

The modifications in the action representations could also affect the learning of new motor sequences. In fact, according to Bloch and colleagues (Bloch, Tamir, Vakil, & Zeilig, 2016), it exists a specific deficit in implicit learning of motor sequences in SCI people, and it may be extended even in the motor learning abilities consequently a SCI lesion.

However, complete SCI participants demonstrated a better performance in forecasting the outcome of wheelchair action in a Temporal Occlusion Paradigm than Physiotherapists working in a Spinal Cord Unit and a group of expert skaters (Scandola, Aglioti, et al., 2019). Interestingly, experimental results confirmed that agonistic wheelchair-basket players showed a better performance with images changing only in the upper-part of the body than control healthy agonistic basketball players (Pernigo et al., 2012); moreover, a greater EEG observational contingent negative variation waveform, considered a marker of action effect prediction, during the evaluation of free throws videos, than control participants (Özkan, Pezzetta, Moreau, Abreu, & Aglioti, 2019). Finally, Bloch, Shaham, Vakil, Ashkenazi, and Zeilig (2020) with a different experimental paradigm could not replicate the impairment of implicit motor learning previously reported. Therefore, the motor learning impairment in SCI is probably not present, or it is very specific for some motor nonecological tasks.

Another aspect of action representation is the motor imagery. This can be defined as the act to internally represent an action, without its execution ( Jackson, Lafleur, Malouin, Richards, & Doyon, 2001). Nonetheless, motor imagery involves the activation of brain structures involved in action performance, such as premotor areas and the left intraparietal sulcus, even *if* at a minor level than actual action performance (Bonda et al., 1995; Decety, 1996; Nikulin, Hohlefeld, Jacobs, & Curio, 2008).

While in some studies, motor imagery seems to be spared after SCI (Decety & Boisson, 1990; Hotz-Boendermaker et al., 2008), altered cortical activations were found (Alkadhi et al., 2005; Cramer, Orr, Cohen, & Lacourse, 2007).

In particular, during motor imagery tasks, the primary and nonprimary motor cortices, and subcortical structures are more activated in individuals affected by SCI than control subjects (Alkadhi et al., 2005), even if the imagery does not involve paralyzed limbs (Alkadhi et al., 2005). Similarly, the minor activation of the primary motor cortex shown by healthy individuals during imagined movements compared to real movements, it is not present after SCI (Di Rienzo et al., 2014), suggesting that after SCI, the motor cortices are activated in the same way during motor imagery or real movements.

Experimental reports suggest that these effects lead to a different motor imagery strategy after SCI for subjective responses (Fiori et al., 2014). In fact, from behavioral investigations, motor imagery after SCI appears to be somatotopographically organized, with higher ability for the spared body parts, and it is impaired in presence of neuropathic pain (Scandola, Aglioti, Pozeg, et al., 2017).

## The space representation

The study of the space representation, since the seminal works by Rizzolatti and colleagues (Rizzolatti, Scandolara, Matelli, & Gentilucci, 1981) was divided in extrapersonal and peripersonal space, two spaces that can be broadly defined respectively as the space outside and within the reach of our limbs.

PPS representation is somatotopically organized: different extensions were found for the hands, trunk, head, and lower limbs (Schicke, Bauer, & Röder, 2009; Serino et al., 2015; Stone, Kandula, Keizer, & Dijkerman, 2017). It is modulated by several aspects of our everyday life, such as action (Brozzoli, Cardinali, Pavani, & Farnè, 2010), danger and fear (Sambo, Liang, Cruccu, & Iannetti, 2011), and stimuli valence (Spaccasassi, Romano, & Maravita, 2019).

PPS is somatotopically modulated by the presence of SCI, but this representation can be recovered with an action-based training. While the upper limbs PPS representation in paraplegics is preserved (Scandola, Aglioti, Bonente, Avesani, & Moro, 2016; Scandola et al., 2020; Sedda et al., 2018), the PPS around the lower limbs is shrunk (Scandola et al., 2016, 2020). However, applying passive mobilization on lower limbs can recover their PPS representation (Scandola et al., 2016).

The representation of the extrapersonal space is also modulated by action and metabolic energy. According to the Economy of Action principle (Proffitt, 2006), the perceptual representation of the space around us is modulated by our metabolic energy, in order to spare energetic resources, mostly when they are lacking. Therefore, when we are in a low physiological condition, or in a physically demanding situation, distances to be (implicitly) walked are perceived longer, and slopes are perceived stepper (Proffitt, 2006).

Indeed, experimental results showed that in estimating the slope of a hill or distance, people tend to overestimate when wearing a heavy backpack (Bhalla & Proffitt, 1999). These effects may be elicited by poor physical health, fatigue (Bhalla & Proffitt, 1999), while the opposite effect is appreciable after the assumption of energy beverages (Schnall, Zadra, & Proffitt, 2010). Thus, the characteristics of the perceptual representation of the extrapersonal space that are potentially linked to a bodily demanding action will be overestimated to prevent the action execution. Therefore, perceptual errors are natural, and they increase as the required effort increases.

In a virtual-reality task, participants affected by complete paraplegia saw a flag depicted on an ascending wheelchair ramp that could have different degrees of slant. Participants had to estimate the distance from the flag on their own wheelchair, or on a wheelchair never used before.

Distance estimation errors were modulated by the degrees of slant of the ramp only when they were seated on their own wheelchair, while on the other wheelchair, the estimation errors were constant (Scandola, Togni, et al., 2019), suggesting that only when they are on their wheelchair they are able to implicitly represent the action of reaching the flag, allowing a representation of space that follows the Economy of Action principle (Proffitt, 2006).

## The effects of rehabilitation on the body, action, and space representations

Physiotherapy interventions can significantly affect, not only from a rehabilitative point of view, but they can likewise be beneficial for the body, action, and space representations.

In particular, the effects of physiotherapy training on body schema representation were studied on a sample of 21 individuals affected by SCI. Before starting the training, and at the end of it, participants underwent a Mental Body Rotation paradigm (Parsons, 1987), with stimuli depicting a left or right hand, foot, or a body (Scandola, Dodoni, et al., 2019). Before the training, SCI participants and a group of healthy participants showed the same pattern for hands and bodies stimuli, while with feet stimuli, only the healthy group showed the bio-mechanical effect. Interestingly, the level of completeness of the lesion covaried with RTs with feet stimuli: slower RTs were associated with more complete lesions, suggesting a greater difficulty in processing stimuli concerning paralyzed body parts.

After the physiotherapy intervention, the bio-mechanical effect appeared also with feet stimuli in the SCI group, showing that a rehabilitative intervention has also effects on the body schema representation (Scandola, Dodoni, et al., 2019).

Another fascinating study, added to the normal physiotherapy rehabilitation a motor imagery training to tetraplegic patients and a control group. Results from magneto-encephalography showed that before the training in the SCI group, there was a compensatory recruitment, with a higher activation of more areas during motor imagery tasks than the control group. Immediately after the training, and after 2 additional months, the SCI group showed a recruitment similar to the control group (Di Rienzo, Guillot, Daligault, et al., 2014).

The impact of rehabilitative training, in particular when they involve the mobilization of paralyzed limbs, is evident also in the PPS representation. In 2016, a study from our group noticed that PPS in complete paraplegics individuals is shrunk around the feet, though a passive mobilization lasting 15 min can recover it (Scandola et al., 2016).

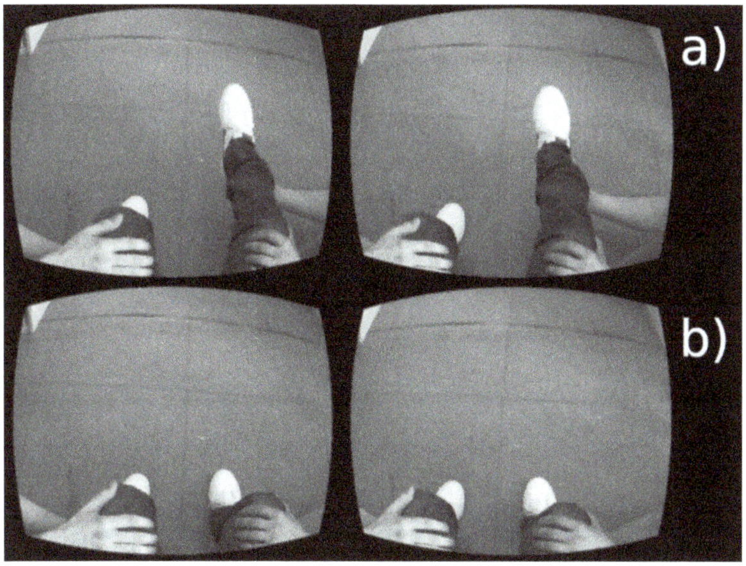


FIG. 5 Pictures from the videos used in Scandola et al. (2020). (A) Video in first-person perspective of passive mobilization. (B) Video in first-person perspective of no mobilization.

Importantly, this recovery is not present when the passive-mobilization is only visual (e.g., the lower limbs are not actually moved, but participants observe a virtual-reality video showing their legs being moved, see Fig. 5).

Results clearly showed that in complete paraplegic participants, the real motor component of the passive mobilization is needed for PPS representation recovery (Scandola et al., 2020). Connected with this latter finding, complete paraplegic individuals could detect when their legs were moved, even if they could not see them, notwithstanding their lack of proprioception (see Fig. 6), better than random guesses (Scandola et al., 2020). Moreover, PPS recovery was modulated by interoception sensitivity in participants affected by incomplete paraplegia: greater interoceptive sensitivity leads to a greater lower-limbs PPS recovery (Scandola et al., 2020).

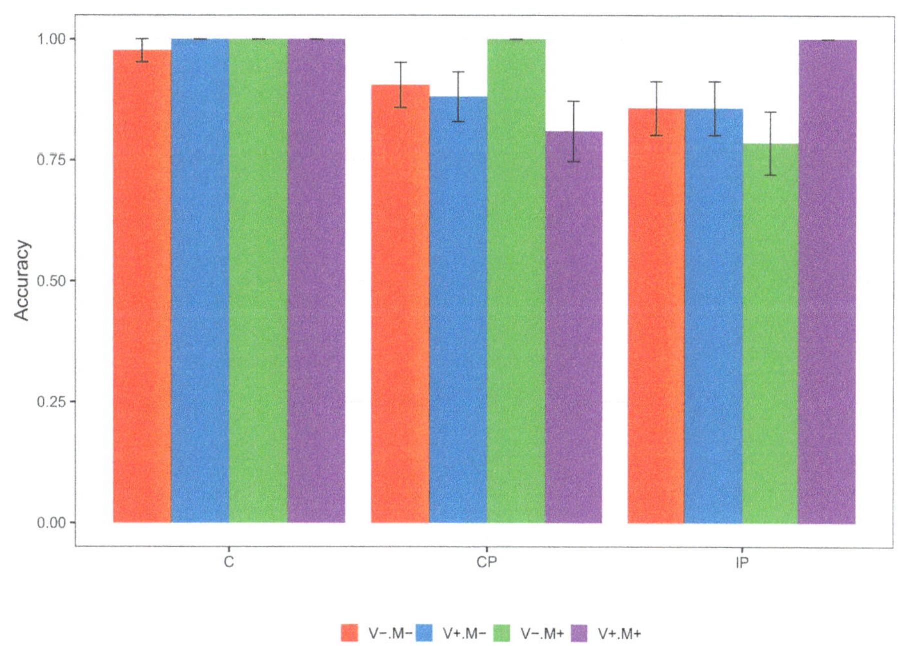


FIG. 6 Accuracies of the detection of motion of the lower legs during the 4 stimulations: V+ = video showing passive visual mobilization, *V-* = video showing no visual mobilization; *M+* = real mobilization; *M−* = no real mobilization. *C* = control group; *CP* = complete paraplegics group; *IP* = incomplete paraplegics group. *Adapted from Scandola, M., Aglioti, S. M., Lazzeri, G., Avesani, R., Ionta, S., & Moro, V. (2020). Visuo-motor and interoceptive influences on peripersonal space representation following spinal cord injury.* Scientific Reports, 10*(1), 1–16. https://doi.org/10.1038/s41598-020-62080-1.*

## Conclusions

The neuroplastic changes following SCI involve modifications of the body, space, and action representations (see Table 1 for a summary). Modifications in such representations are often considered as a cause of maladaptive plasticity that is related to phantom limb and neuropathic pain (Moore et al., 2000; Wrigley et al., 2009).

Modifications in the body, action, and space representations seem to follow the somato-topography rule: but only representations involving paralyzed limbs (Scandola, Aglioti, et al., 2019; Scandola et al., 2016, 2020, 2014).

Importantly, the possibility of new motor learnings and the action-dependent recovery of the representation of the space suggest top-down and bottom-up influences on above representations (Bloch et al., 2020; Pernigo et al., 2012; Scandola et al., 2016, 2020).

Also, BRs are impacted by bottom-up and top-down influences. For example, neuropathic and visceral pain sensations coming from the part of the body that is deafferented/deefferented seem to protect from body illusions, working as surrogate of normal BR (Scandola, Aglioti, Avesani, et al., 2017). Moreover, the wheelchair does not seem only the most iconic symbol representing the condition of individuals affected by SCI, but also a tool that is able to shape and define their BRs (Scandola, Togni, et al., 2019), and the extrapersonal space representation (Scandola, Togni, et al., 2019).

These results indicate that the body, space, and action representations rehabilitation is possible, and it might have beneficial neuroplastic effects on maladaptive plastic mechanisms (Di Rienzo, Guillot, Mateo, et al., 2014; Scandola et al., 2016, 2020). A further step might be a training focused on these representations, that side by side to a conventional physiotherapy training could improve rehabilitative outcomes.

## Applications to other areas of neuroscience

In this chapter, we have reviewed the neuroplastic modifications in the body, space, and action representations following complete or incomplete spinal cord injuries.

**TABLE 1 Summary of the main findings concerning body, action, and space representations in spinal cord injuries.**

| Domain | Subdomain | Finding | References |
|---|---|---|---|
| *Body representation* | *Explicit body ownership* | Visuo-tactile stimulations induce body ownership over a rubber body part, also when the tactile stimulus is administered to a body part that is different from the rubber one. This may be caused by the neuroplasticity involving the motor and somato-sensory cortexes, or by the mere vision of the rubber body part | Scandola et al. (2014), Pozeg et al. (2017), Tidoni et al. (2014) |
| | *Implicit body ownership* | Body Ownership is somatotopically modulated over the body, reduced over the paralyzed and anesthetized body parts, and it is extended to the wheelchair.<br>The use of a new wheelchair can impair Body Ownership over the whole body | Scandola, Togni, et al. (2019) |
| | *Body schema* | Complete SCI impairs body schema on the representation of the deafferented body parts (feet, but not hands) and, regardless the posture, whole-BR progressively deteriorates proportionally to SCI completeness, as a direct effect of the somato-sensation lack in the affected body-parts in SCI individuals<br>Body Schema can be recovered after physiotherapeutic interventions | Ionta et al. (2016), Scandola, Dodoni, et al. (2019) |
| | *Body illusions* | Various body representation illusions are reported for SCI individuals:<br>• Misproprioception of the paralyzed body: 77%<br>• Disownership-like sensation: 48%<br>• Somatoparaphrenia-like sensations: 46%<br>• Illusory motion of paralyzed body parts: 51%<br>These are reduced by the presence of visceral or neuropathic pain: pain sensations coming from the impaired body part are used by our cognitive system as a surrogate body representations in absence of other somatic sensations | Scandola, Aglioti, Avesani, et al. (2017) |

**TABLE 1 Summary of the main findings concerning body, action, and space representations in spinal cord injuries—cont'd**

| Domain | Subdomain | Finding | References |
|---|---|---|---|
| *Action representation* | *Action discrimination* | Complete paraplegics participants have slower reaction times and greater errors incidence, in body images that were different for the lower part of the body, than the control group of healthy participants<br>Agonistic wheelchair-basket players have a better performance with images changing only in the upper-part of the body than control healthy agonistic basketball players<br>A greater EEG observational contingent negative variation waveform (marker of action effect prediction) is observed during the evaluation of free throws videos, than control participants | Pernigo et al. (2012),<br>Özkan et al. (2019) |
| | *Action anticipation* | Complete paraplegics participants have worse performances in guessing the final outcome of rollerblades videos compared to the Skaters group, but also compared to the Physiotherapists group, inexperienced in skating, suggesting that abnormalities in the mirror system for the lower limbs affects the recognition of lower limbs' actions in a subtle, but still worsening way compared to a group that had no direct experience in the action | Scandola, Aglioti, et al. (2019) |
| | *Tendon vibration illusion* | Movement illusions are qualitatively different from the healthy control sample (e.g., as if the arm wanted to extend itself, or a sensation of pushing against something), reflecting different reorganization processes following SCI | Fusco et al. (2016) |
| | *Implicit learning* | In a study (Bloch et al., 2016) SCI participants showed a specific deficit in implicit learning of motor sequences<br>However, this result was not confirmed by the result of another study of the same group (Bloch et al., 2020) | Bloch et al. (2016),<br>Bloch et al. (2020) |
| | *Motor imagery* | During motor imagery tasks, the primary and non-primary motor cortices, and subcortical structures are more activated in individuals affected by SCI than control subjects.<br>SCI participants use a different strategy for motor imagery based on memory instead of on body representations, and it is impaired by neuropathic pain sensations.<br>Adding motor imagery tasks to the normal physiotherapy rehabilitation for 2 months leads to a normalization in the cortical activation of SCI participants | Alkadhi et al. (2005),<br>Di Rienzo, Guillot, Daligault, et al. (2014),<br>Fiori et al. (2014),<br>Scandola, Aglioti, Pozeg, et al. (2017) |
| *Space representation* | *Peripersonal space* | PPS in SCI is somatotopically impaired: the representation of upper-lesion body parts is spared, while the representation of lower-lesion body parts is shrunk.<br>The application of passive mobilization on lower limbs can induce a short-term recovery.<br>This passive mobilization must involve actual mobilization: the simple observation of movement is not enough | Scandola et al. (2016),<br>Scandola et al. (2020),<br>Sedda et al. (2018) |
| | *Extrapersonal space* | Distance estimation errors are modulated by the embodiment of the wheelchair in use. If the tool used to move in the surrounding space is embodied (i.e., they are using their own wheelchair), the distance estimation errors are modified by the difficulty in reaching that distance, as in healthy subjects that are able to internally represent the action. If the wheelchair is not embodied (i.e., they are using a never-used-before wheelchair), the action representation becomes impossible and purely visual strategies are then used in order to estimate space | Scandola, Togni, et al. (2019) |

The neuroscientists, or the neurologists interested in the modifications of these representations, and their interconnections with rehabilitative and theoretical or neurophysiological aspects, will be interested in these studies.

The study of the body, space, and action representations has often used studies on neurological patients (Buxbaum, Giovannetti, & Libon, 2000; Candia, Wienbruch, Elbert, Rockstroh, & Ray, 2003; Canzoneri, Marzolla, Amoresano, Verni, & Serino, 2013; Sirigu, Grafman, Bressler, & Sunderland, 1991; Soman, Steeves, & Lang, 2009; Tinazzi, Fiorio, Bertolasi, & Aglioti, 2004). However, SCI participants may represent an ideal model to explore to what extend sensorimotor deficits alter cognitive functions. In fact, in this population, all the postlesional changes in cognitive functions do not depend on brain damage or modifications in their body form but are exclusively related to the loss of sensory–motor afferences from as well as efferences to the disconnected body parts.

Furthermore, for the neuroscientists interested in rehabilitation of cognitive and motor skills, the above-reviewed studies here reviewed show trainings that are designed to counter maladaptive plasticity and neuropathic pain, and improve adaptive neuroplasticity (Mateo et al., 2015; Scandola et al., 2020; Scandola, Dodoni, et al., 2019).

## Mini-dictionary of terms

**Neuroplasticity**: the natural modifications of the brain areas, through growth and reorganization.

**Body ownership**: our sensation of owning a part of our body.

**Rubber hand illusion**: an experimental paradigm used to investigate body ownership, where a sensation of body ownership is elicited toward a rubber hand by means of synchronous visuo-tactile stimulation of the rubber hand and the participant's real hand.

**Body-view enhancement effect**: visual stimuli appearing on our body evoke faster reactions than visual stimuli appearing on the same space location, but on a neutral surface.

**Body schema**: the representation of the body parts position, in relation to the rest of the body.

**Mental Body Rotation Paradigm**: In this experimental paradigm, individuals are asked to identify the laterality of a hand or a foot presented on screen at different rotations. The underlying mechanisms of decision imply that participants rotate the body stimuli in a canonical position congruent with their body and therefore, response times (RTs) required to align series of gradually more rotated images to the vertical, congruent with the body position, are slower. This leads to a typical RTs pattern, characteristic of a normal body schema representation, called bio-mechanical effect.

**Motor imagery**: the act to internally represent an action, without its execution. It has to do with the activation of the same brain areas involved in the execution of the action, even though at a minor level.

**Peripersonal space**: the space within the reach of our limbs.

**Extrapersonal space**: the space outside the reach of our limbs.

**Temporal Occlusion Paradigm**: in this experimental paradigm, participants are asked to predict the outcome of an action seen in a video that has been occluded at a specific time-point, before the end of the action.

## Key facts of neuroplasticity following spinal cord injury

1. Brain modifications after spinal cord lesion starts within minutes from the lesion, as seen in animal models
2. The neuroplasticity leads to reduction of activity in the areas corresponding to paralyzed limbs
3. In the brain areas connected to the upper-lesion body parts the activity is increased
4. This neuroplasticity can cause phantom limb sensations
5. Maladaptive neuroplasticity can cause neuropathic pain
6. It is also the cause of modifications in body, action, and space representations

## Summary points

1. The neuroplastic modifications following a spinal cord injury change the body, action, and space representations with maladaptive consequences
2. The modifications in the body representations do not only involve the parts of the body below the lesion, but also the whole body
3. In individuals with chronic spinal cord injuries, that use the wheelchair for a long time, the wheelchair is part of their body representation and can define it
4. The action representation is impaired for the parts of the body below the lesion
5. The peripersonal space representation is somato-topographically modified: it only involves the parts of the body that are below the lesion
6. The body, action, and space representations can be rehabilitated using motor imagery and physiotherapeutic training.

## References


Aglioti, S. M., Cesari, P., Romani, M., & Urgesi, C. (2008). Action anticipation and motor resonance in elite basketball players. *Nature Neuroscience*, *11* (9), 1109–1116. https://doi.org/10.1038/nn.2182.

Aguilar, J., Humanes-Valera, D., Alonso-Calvino, E., Yague, J. G., Moxon, K. A., Oliviero, A., et al. (2010). Spinal cord injury immediately changes the state of the brain. *Journal of Neuroscience*, *30*(22), 7528–7537. https://doi.org/10.1523/JNEUROSCI.0379-10.2010.

Alkadhi, H., Brugger, P., Boendermaker, S. H., Crelier, G. R., Curt, A., Hepp-Reymond, M.-C., et al. (2005). What disconnection tells about motor imagery: Evidence from paraplegic patients. *Cerebral Cortex*, *15*(2), 131–140. https://doi.org/10.1093/cercor.

Avenanti, A., Candidi, M., & Urgesi, C. (2013). Vicarious motor activation during action perception: Beyond correlational evidence. *Frontiers in Human Neuroscience*, *7*(May), 185. https://doi.org/10.3389/fnhum.2013.00185.

Berlucchi, G., & Aglioti, S. M. (2010). The body in the brain revisited. *Experimental Brain Research. Experimentelle Hirnforschung. Expérimentation Cérébrale*, *200*(1), 25–35. https://doi.org/10.1007/s00221-009-1970-7.

Bhalla, M., & Proffitt, D. R. (1999). Visual–motor recalibration in geographical slant perception. *Journal of Experimental Psychology: Human Perception and Performance*, *25*(4), 1076–1096. Retrieved from http://psycnet.apa.org/journals/xhp/25/4/1076/.

Bloch, A., Shaham, M., Vakil, E., Ashkenazi, S. S., & Zeilig, G. (2020). Examining implicit procedural learning in tetraplegia using an oculomotor serial reaction time task. *PLoS One*, *15*(4), 1–13. https://doi.org/10.1371/journal.pone.0232124.

Bloch, A., Tamir, D., Vakil, E., & Zeilig, G. (2016). Specific deficit in implicit motor sequence learning following spinal cord injury. *PLoS One*, *11*(6), 1–13. https://doi.org/10.1371/journal.pone.0158396.

Bonda, E., Petrides, M., Frey, S., & Evans, A. (1995). Neural correlates of mental transformations of the body-in-space. *Proceedings of the National Academy of Sciences of the United States of America*, *92*(24), 11180–11184. https://doi.org/10.1073/pnas.92.24.11180.

Botvinick, M., & Cohen, J. (1998). Rubber hands "feel" touch that eyes see. *Nature*, *391*(6669), 756. https://doi.org/10.1038/35784.

Brozzoli, C., Cardinali, L., Pavani, F., & Farnè, A. (2010). Action-specific remapping of peripersonal space. *Neuropsychologia*, *48*(3), 796–802. https://doi.org/10.1016/j.neuropsychologia.2009.10.009.

Buxbaum, L. J., Giovannetti, T., & Libon, D. (2000). The role of the dynamic body schema in praxis: Evidence from primary progressive apraxia. *Brain and Cognition*, *44*(2), 166–191. https://doi.org/10.1006/brcg.2000.1227.

Calvo-Merino, B., Glaser, D. E., Grèzes, J., Passingham, R. E., & Haggard, P. (2004). Action observation and acquired motor skills: An fMRI study with expert dancers. *Cerebral Cortex*, *15*(8), 1243–1249. https://doi.org/10.1093/cercor/bhi007.

Candia, V., Wienbruch, C., Elbert, T., Rockstroh, B., & Ray, W. (2003). Effective behavioral treatment of focal hand dystonia in musicians alters somatosensory cortical organization. *Proceedings of the National Academy of Sciences of the United States of America*, *100*(13), 7942–7946. https://doi.org/10.1073/pnas.1231193100.

Canzoneri, E., Marzolla, M., Amoresano, A., Verni, G., & Serino, A. (2013). Amputation and prosthesis implantation shape body and peripersonal space representations. *Scientific Reports*, *3*, 2844. https://doi.org/10.1038/srep02844.

Cole, J. (2004). *Still lives: Narratives of spinal cord injury*. MIT Press.

Conomy, J. P. (1973). Disorders of body image after spinal cord injury. *Neurology*, *23*(8), 842. Retrieved from http://www.ncbi.nlm.nih.gov/pubmed/4737080.

Cramer, S. C., Orr, E. L. R., Cohen, M. J., & Lacourse, M. G. (2007). Effects of motor imagery training after chronic, complete spinal cord injury. *Experimental Brain Research*, *177*(2), 233–242. https://doi.org/10.1007/s00221-006-0662-9.

Curt, A., Bruehlmeier, M., Leenders, K. L., Roelcke, U., & Dietz, V. (2002). Differential effect of spinal cord injury and functional impairment on human brain activation. *Journal of Neurotrauma*, *19*(1), 43–51. https://doi.org/10.1089/089771502753460222.

Curt, A., Yengue, C. N., Hilti, L. M., & Brugger, P. (2011). Supernumerary phantom limbs in spinal cord injury. *Spinal Cord*, *49*(5), 588–595. https://doi.org/10.1038/sc.2010.143.

Decety, J. (1996). Neural representations for action. *Reviews in the Neurosciences*, *7*(4), 285–297. Retrieved from http://www.ncbi.nlm.nih.gov/pubmed/9044503.

Decety, J., & Boisson, D. (1990). Effect of brain and spinal cord injuries on motor imagery. *European Archives of Psychiatry and Clinical Neuroscience*, *240*(1), 39–43. https://doi.org/10.1007/BF02190091.

Di Rienzo, F., Guillot, A., Daligault, S., Delpuech, C., Rode, G., & Collet, C. (2014). Motor inhibition during motor imagery: A MEG study with a quadriplegic patient. *Neurocase*, *20*(5), 524–539. https://doi.org/10.1080/13554794.2013.826685.

Di Rienzo, F., Guillot, A., Mateo, S., Daligault, S., Delpuech, C., Rode, G., et al. (2014). Neuroplasticity of prehensile neural networks after quadriplegia. *Neuroscience*, *274*, 82–92. https://doi.org/10.1016/j.neuroscience.2014.05.021.

Fiori, F., Sedda, A., Ferrè, E. R., Toraldo, A., Querzola, M., Pasotti, F., et al. (2014). Motor imagery in spinal cord injury patients: Moving makes the difference. *Journal of Neuropsychology*, *8*(2), 199–215. https://doi.org/10.1111/jnp.12020.

Freund, P., Weiskopf, N., Ashburner, J., Wolf, K., Sutter, R., Altmann, D. R., et al. (2013). MRI investigation of the sensorimotor cortex and the corticospinal tract after acute spinal cord injury: A prospective longitudinal study. *Lancet Neurology*, *12*(9), 873–881. https://doi.org/10.1016/S1474-4422(13)70146-7.

Fusco, G., Tidoni, E., Barone, N., Pilati, C., & Aglioti, S. M. (2016). Illusion of arm movement evoked by tendon vibration in patients with spinal cord injury. *Restorative Neurology and Neuroscience*, *34*(5), 815–826. https://doi.org/10.3233/RNN-160660.

Ghosh, A., Peduzzi, S., Snyder, M., Schneider, R., Starkey, M., & Schwab, M. E. (2012). Heterogeneous spine loss in layer 5 cortical neurons after spinal cord injury. *Cerebral Cortex (New York, N.Y. : 1991)*, *22*(6), 1309–1317. https://doi.org/10.1093/cercor/bhr191.

Head, H., & Holmes, G. (1911). Sensory disturbances from cerebral lesions. *Brain*, *34*(2–3), 102–254. https://doi.org/10.1093/brain/34.2-3.102.

Henderson, L. A., Gustin, S. M., Macey, P. M., Wrigley, P. J., & Siddall, P. J. (2011). Functional reorganization of the brain in humans following spinal cord injury: Evidence for underlying changes in cortical anatomy. *The Journal of Neuroscience : The Official Journal of the Society for Neuroscience*, *31*(7), 2630–2637. https://doi.org/10.1523/JNEUROSCI.2717-10.2011.

Hotz-Boendermaker, S., Funk, M., Summers, P., Brugger, P., Hepp-Reymond, M.-C., Curt, A., et al. (2008). Preservation of motor programs in paraplegics as demonstrated by attempted and imagined foot movements. *Neuro Image*, *39*(1), 383–394. https://doi.org/10.1016/j.neuroimage.2007.07.065.

Ionta, S., Villiger, M., Jutzeler, C. R., Freund, P., Curt, A., & Gassert, R. (2016). Spinal cord injury affects the interplay between visual and sensorimotor representations of the body. *Scientific Reports*, *6*(1), 20144. https://doi.org/10.1038/srep20144.

Jackson, P. L., Lafleur, M. F., Malouin, F., Richards, C., & Doyon, J. (2001). Potential role of mental practice using motor imagery in neurologic rehabilitation. *Archives of Physical Medicine and Rehabilitation*, *82*(8), 1133–1141. https://doi.org/10.1053/apmr.2001.24286.

Jenkinson, P. M., Moro, V., & Fotopoulou, A. (2018). Definition: Asomatognosia. *Cortex*, *101*, 300–301. https://doi.org/10.1016/j.cortex.2018.02.001.

Jurkiewicz, M. T., Mikulis, D. J., Fehlings, M. G., & Verrier, M. C. (2010). Sensorimotor cortical activation in patients with cervical spinal cord injury with persisting paralysis. *Neurorehabilitation and Neural Repair*, *24*(2), 136–140. https://doi.org/10.1177/1545968309347680.

Kao, K.-L. C., & Goodale, M. A. (2009). Enhanced detection of visual targets on the hand and familiar tools. *Neuropsychologia*, *47*(12), 2454–2463. https://doi.org/10.1016/j.neuropsychologia.2009.04.016.

Kravitz, D. J., Saleem, K. S., Baker, C. I., & Mishkin, M. (2011). A new neural framework for visuospatial processing. *Nature Reviews Neuroscience*, *12* (4), 217–230. https://doi.org/10.1038/nrn3008.

Lundell, H., Christensen, M. S., Barthélemy, D., Willerslev-Olsen, M., Biering-Sørensen, F., & Nielsen, J. B. (2011). Cerebral activation is correlated to regional atrophy of the spinal cord and functional motor disability in spinal cord injured individuals. *NeuroImage*, *54*(2), 1254–1261. https://doi.org/10.1016/j.neuroimage.2010.09.009.

Makin, T. R., Holmes, N. P., & Zohary, E. (2007). Is that near my hand? Multisensory representation of peripersonal space in human intraparietal sulcus. *The Journal of Neuroscience : The Official Journal of the Society for Neuroscience*, *27*(4), 731–740. https://doi.org/10.1523/JNEUROSCI.3653-06.2007.

Mateo, S., Di Rienzo, F., Bergeron, V., Guillot, A., Collet, C., & Rode, G. (2015). Motor imagery reinforces brain compensation of reach-to-grasp movement after cervical spinal cord injury. *Frontiers in Behavioral Neuroscience*, *9*(September), 1–12. https://doi.org/10.3389/fnbeh.2015.00234.

Moore, C. I., Stern, C. E., Dunbar, C., Kostyk, S. K., Gehi, A., & Corkin, S. (2000). Referred phantom sensations and cortical reorganization after spinal cord injury in humans. *Proceedings of the National Academy of Sciences of the United States of America*, *97*(26), 14703–14708. https://doi.org/10.1073/pnas.250348997.

Nardone, R., Höller, Y., Brigo, F., Seidl, M., Christova, M., Bergmann, J., et al. (2013). Functional brain reorganization after spinal cord injury: Systematic review of animal and human studies. *Brain Research*, *1504*, 58–73. https://doi.org/10.1016/j.brainres.2012.12.034.

Nikulin, V. V., Hohlefeld, F. U., Jacobs, A. M., & Curio, G. (2008). Quasi-movements: A novel motor-cognitive phenomenon. *Neuropsychologia*, *46*(2), 727–742. https://doi.org/10.1016/j.neuropsychologia.2007.10.008.

Özkan, D. G., Pezzetta, R., Moreau, Q., Abreu, A. M., & Aglioti, S. M. (2019). Predicting the fate of basketball throws: An EEG study on expert action prediction in wheelchair basketball players. *Experimental Brain Research*, *237*(0123456789), 3363–3373. https://doi.org/10.1007/s00221-019-05677-x.

Panasiti, M. S., Pavone, E. F., & Aglioti, S. M. (2016). Electrocortical signatures of detecting errors in the actions of others: An EEG study in pianists, non-pianist musicians and musically naïve people. *Neuroscience*, *318*, 104–113. https://doi.org/10.1016/j.neuroscience.2016.01.023.

Parsons, L. M. (1987). Imagined spatial transformations of one's hands and feet. *Cognitive Psychology*, *19*(2), 178–241. https://doi.org/10.1016/0010-0285(87)90011-9.

Pernigo, S., Moro, V., Avesani, R., Miatello, C., Urgesi, C., & Aglioti, S. M. (2012). Massive somatic deafferentation and motor deefferentation of the lower part of the body impair its visual recognition: A psychophysical study of patients with spinal cord injury. *The European Journal of Neuroscience*, *36*(11), 3509–3518. https://doi.org/10.1111/j.1460-9568.2012.08266.x.

Pozeg, P., Palluel, E., Ronchi, R., Solca, M., Al-Khodairy, A. W., Jordan, X., et al. (2017). Virtual reality improves embodiment and neuropathic pain caused by spinal cord injury. *Neurology*, *89*(18), 1894–1903. https://doi.org/10.1212/WNL.0000000000004585.

Proffitt, D. R. (2006). Embodied perception and the economy of action. *Perspectives on Psychological Science*, *1*(2), 110–122. https://doi.org/10.1111/j.1745-6916.2006.00008.x.

Ramachandran, V. S., Stewart, M., & Rogers-Ramachandran, D. C. (1992). Perceptual correlates of massive cortical reorganization. *Neuroreport*, *3*(7), 583–586. Retrieved from http://www.ncbi.nlm.nih.gov/pubmed/1421112.

Rizzolatti, G., Cattaneo, L., Fabbri-Destro, M., & Rozzi, S. (2014). Cortical mechanisms underlying the organization of goal-directed actions and mirror neuron-based action understanding. *Physiological Reviews*, *94*(2), 655–706. https://doi.org/10.1152/physrev.00009.2013.

Rizzolatti, G., Scandolara, C., Matelli, M., & Gentilucci, M. (1981). Afferent properties of periarcuate neurons in macaque monkeys. II. Visual responses. *Behavioural Brain Research*, *2*(2), 147–163. https://doi.org/10.1016/0166-4328(81)90053-X.

Sambo, C. F., Liang, M., Cruccu, G., & Iannetti, G. D. (2011). Defensive peripersonal space: The blink reflex evoked by hand stimulation is increased when the hand is near the face. *Journal of Neurophysiology*, *44*. https://doi.org/10.1152/jn.00731.2011.

Scandola, M., Aglioti, S. M., Avesani, R., Bertagnoni, G., Marangoni, A., & Moro, V. (2017). Corporeal illusions in chronic spinal cord injuries. *Consciousness and Cognition*, *49*, 278–290. https://doi.org/10.1016/j.concog.2017.01.010.

Scandola, M., Aglioti, S. M., Avesani, R., Bertagnoni, G., Marangoni, A., & Moro, V. (2019). Anticipation of wheelchair and rollerblade actions in spinal cord injured people, rollerbladers, and physiotherapists. *PLoS One*, *14*(3). https://doi.org/10.1371/journal.pone.0213838, e0213838.

Scandola, M., Aglioti, S. M., Bonente, C., Avesani, R., & Moro, V. (2016). Spinal cord lesions shrink peripersonal space around the feet, passive mobilization of paraplegic limbs restores it. *Scientific Reports*, *6*(April), 24126. https://doi.org/10.1038/srep24126.

Scandola, M., Aglioti, S. M., Lazzeri, G., Avesani, R., Ionta, S., & Moro, V. (2020). Visuo-motor and interoceptive influences on peripersonal space representation following spinal cord injury. *Scientific Reports*, *10*(1), 1–16. https://doi.org/10.1038/s41598-020-62080-1.

Scandola, M., Aglioti, S. M., Pozeg, P., Avesani, R., & Moro, V. (2017). Motor imagery in spinal cord injured people is modulated by somatotopic coding, perspective taking, and post-lesional chronic pain. *Journal of Neuropsychology*, *11*(3), 305–326. https://doi.org/10.1111/jnp.12098.

Scandola, M., Dodoni, L., Lazzeri, G., Arcangeli, C. A., Avesani, R., Moro, V., et al. (2019). Neurocognitive benefits of physiotherapy for spinal cord injury. *Journal of Neurotrauma*. https://doi.org/10.1089/neu.2018.6123. neu.2018.6123.

Scandola, M., Tidoni, E., Avesani, R., Brunelli, G., Aglioti, S. M., & Moro, V. (2014). Rubber hand illusion induced by touching the face ipsilaterally to a deprived hand: Evidence for plastic "somatotopic" remapping in tetraplegics. *Frontiers in Human Neuroscience*, *8*, 404. https://doi.org/10.3389/fnhum.2014.00404.

Scandola, M., Togni, R., Tieri, G., Avesani, R., Brambilla, M., Aglioti, S. M., et al. (2019). Embodying their own wheelchair modifies extrapersonal space perception in people with spinal cord injury. *Experimental Brain Research*, *237*(10), 2621–2632. https://doi.org/10.1007/s00221-019-05618-8.

Schicke, T., Bauer, F., & Röder, B. (2009). Interactions of different body parts in peripersonal space: How vision of the foot influences tactile perception at the hand. *Experimental Brain Research*, *192*(4), 703–715. https://doi.org/10.1007/s00221-008-1587-2.

Schnall, S., Zadra, J. R., & Proffitt, D. R. (2010). Direct evidence for the economy of action: Glucose and the perception of geographical slant. *Perception*, *39*(4), 464–482. https://doi.org/10.1068/p6445.

Sedda, A., Ambrosini, E., Dirupo, G., Tonin, D., Valsecchi, L., Redaelli, T., et al. (2018). Affordances after spinal cord injury. *Journal of Neuropsychology*, 1–16. https://doi.org/10.1111/jnp.12151.

Sereno, M. I., & Huang, R.-S. (2006). A human parietal face area contains aligned head-centered visual and tactile maps. *Nature Neuroscience*, *9*(10), 1337–1343.

Serino, A., Noel, J.-P., Galli, G., Canzoneri, E., Marmaroli, P., Lissek, H., et al. (2015). Body part-centered and full body-centered peripersonal space representations. *Scientific Reports*, *5*(November), 18603. https://doi.org/10.1038/srep18603.

Sirigu, A., Grafman, J., Bressler, K., & Sunderland, T. (1991). Multiple representations contribute to body knowledge processing. Evidence from a case of autotopagnosia. *Brain: A Journal of Neurology*, *114*(Pt 1), 629–642. Retrieved from http://www.ncbi.nlm.nih.gov/pubmed/2004260.

Soman, T., Steeves, T. D. L., & Lang, A. E. (2009). Alien hand syndrome and dystonia in a pediatric patient. *Movement Disorders: Official Journal of the Movement Disorder Society*, *24*(10), 1557–1559. https://doi.org/10.1002/mds.22620.

Spaccasassi, C., Romano, D., & Maravita, A. (2019). Everything is worth when it is close to my body: How spatial proximity and stimulus valence affect visuo-tactile integration. *Acta Psychologica*, *192*(October 2018), 42–51. https://doi.org/10.1016/j.actpsy.2018.10.013.

Stone, K. D., Kandula, M., Keizer, A., & Dijkerman, H. C. (2017). Peripersonal space boundaries around the lower limbs. *Experimental Brain Research*. https://doi.org/10.1007/s00221-017-5115-0 (0123456789).

Tandon, S., Kambi, N., Lazar, L., Mohammed, H., & Jain, N. (2009). Large-scale expansion of the face representation in somatosensory areas of the lateral sulcus after spinal cord injuries in monkeys. *The Journal of Neuroscience*, *29*(38), 12009–12019. https://doi.org/10.1523/JNEUROSCI.2118-09.2009.

Tidoni, E., Fusco, G., Leonardis, D., Frisoli, A., Bergamasco, M., & Aglioti, S. M. (2015). Illusory movements induced by tendon vibration in right- and left-handed people. *Experimental Brain Research*, *233*(2), 375–383. https://doi.org/10.1007/s00221-014-4121-8.

Tidoni, E., Grisoni, L., Liuzza, M. T., & Aglioti, S. M. (2014). Rubber hand illusion highlights massive visual capture and sensorimotor face-hand remapping in a tetraplegic man. *Restorative Neurology and Neuroscience*, *32*(5), 611–622. https://doi.org/10.3233/RNN-130385.

Tinazzi, M., Fiorio, M., Bertolasi, L., & Aglioti, S. M. (2004). Timing of tactile and visuo-tactile events is impaired in patients with cervical dystonia. *Journal of Neurology*, *251*(1), 85–90. https://doi.org/10.1007/s00415-004-0282-x.

Wrigley, P. J., Press, S. R., Gustin, S. M., Macefield, V. G., Gandevia, S. C., Cousins, M. J., et al. (2009). Neuropathic pain and primary somatosensory cortex reorganization following spinal cord injury. *Pain*, *141*(1–2), 52–59. https://doi.org/10.1016/j.pain.2008.10.007.

Chapter 4

# Methods for treating pain and painful syndromes in spinal cord injury: Medications, therapies, interventions, and neuromodulation

James J. Bresnahan, James C. Pendleton, Benjamin R. Scoblionko, Patricia Orme, and Philip J. Koehler, III
*Department of Rehabilitation Medicine, Thomas Jefferson University Hospital, Philadelphia, PA, United States*

## List of abbreviations

**AB** able-bodied
**CNS** central nervous system
**DRGS** dorsal root ganglion stimulation
**IL** interleukin
**PAG** periaqueductal gray
**PNS** peripheral nerve stimulation
**PT** physical therapy
**RVM** rostral ventromedial medulla
**SNRI** serotonin-norepinephrine reuptake inhibitor
**SCI** spinal cord injury
**SCS** spinal cord stimulation
**tDCS** transcranial direct current stimulation
**TENS** transcutaneous electrical nerve stimulation

## Introduction

Pain frequently limits activities, decreases quality of life, and leads to significant psychological impairment independent of other functional deficits after SCI (Craig et al., 2015; Elliott & Frank, 1996; Finnerup et al., 2016). In general, SCI pain is not well understood from the cellular, molecular, and even spinal cord pathway levels (Siddall, Taylor, & Cousins, 1997).

Sixty-five percent to eighty percent of individuals with SCI live with chronic pain, and nearly one third rate this pain as severe (Cardenas, Bryce, Shem, Richards, & Elhefni, 2004; Finnerup et al., 2016; Siddall, Mcclelland, Rutkowski, & Cousins, 2003). About half of all individuals with SCI suffer from pain so severe that it interferes with function and rehabilitation independent of deficits related to their SCI ( Siddall & Loeser, 2001). It is generally believed that level of injury does not have an influence on the degree of post-SCI chronic pain. Penetrating injuries may be more likely to result in chronic pain (Richards, Stover, & Jaworski, 1990), and there is some evidence that incomplete lesions lead to more chronic pain than complete lesions (Beric, Dimitrijevic, & Lindblom, 1988; Davidoff, Roth, Guarracini, Sliwa, & Yarkony, 1987).

Pain after SCI can generally be divided into nociceptive, visceral, at-level neuropathic, and below-level neuropathic (Siddall et al., 1997). It is theorized that SCI pain may be fairly unique and related to relative serotonin, norepinephrine, and dopamine deficits due to loss of supraspinal inhibition (Boadas-Vaello et al., 2016; Bourne, Machado, & Nagel, 2014; Heinricher, Tavares, Leith, & Lumb, 2009). Additionally, there is a well-described inflammatory milieu that occurs after SCI and may contribute to an additional inflammatory basis of pain (Chambel, Tavares, & Cruz, 2020). Furthermore, post-SCI changes are noted in synaptic dendrites in individuals who experience significant pain, suggesting a dynamic neuroplastic component to pain (Benson, Reimer, & Tan, 2020). It is likely the unique combination central nervous system (CNS)

Diagnosis and Treatment of Spinal Cord Injury. https://doi.org/10.1016/B978-0-12-822498-4.00004-X

**TABLE 1** Classification of spinal cord injury-related pain.

| Classification | Definition |
|---|---|
| Nociceptive somatic pain | Pain from skin, subcutaneous tissues, bones, joints, connective tissue, muscle, and tendons. Often occurs from overuse, age-related breakdown, or acute injury |
| Nociceptive visceral pain | Pain from organs and supporting structures. Often occurs with internal injury or infection |
| At-level neuropathic pain | Pain within 2 levels of the segment of the spinal cord that is injured. Often central or peripheral (radicular) pain that has typical neuropathic qualities |
| Below-level neuropathic pain | Pain below the level of the injured spinal cord segment related to disruption of the typical neural connections |

*Original table*: These are the four generally agreed-upon classifications of pain as it related to the spinal cord-injured population.

**TABLE 2** Characteristics of different pain types.

| | Nociceptive pain | | |
|---|---|---|---|
| | **Somatic pain** | **Visceral pain** | **Neuropathic pain** |
| Pain origin | Skin, subcutaneous tissues, mucous membrane, joints, connective tissue, muscle, and bone | Organs, organ capsule or covering, connecting and supporting structures | Central or peripheral nerves |
| Location | Localized | Generalized or diffuse | Radiating or specific |
| Quality | Pinprick, stabbing, sharp, sore, aching | Cramping, throbbing, squeezing, pressure, or sharp | Burning, "pins and needles," tingling, electrical, or lancinating |
| Mechanism of pain | A-delta fiber activity<br>Located in the periphery | C Fiber activity<br>Involving deeper innervation | Nondermatomal (central), or dermatomal (periphery) |
| Clinical examples | • Sickle cell crisis<br>• Superficial burns and lacerations<br>• Stomatitis<br>• Intramuscular injections<br>• Spasticity-mediated pain<br>• Bone metastases | • Colic spasm<br>• Appendicitis<br>• Kidney stone<br>• Chronic pancreatitis<br>• IBS<br>• Angina<br>• Menstrual cramps | • Trigeminal neuralgia<br>• Avulsion neuralgia<br>• Posttraumatic neuralgia<br>• Radiculopathy<br>• Peripheral neuropathy<br>• Phantom limb<br>• Herpetic neuralgia |

*Original table*: Differentiating pain generators is important in determining a treatment algorithm. These are the general characteristics of different types of pain.

changes after SCI creates a "pro-pain" state, and not surprisingly, SCI pain is often refractory to medical treatments (Siddall & Loeser, 2001; Widerström-Noga, 2017b).

The treatment of SCI-related pain often comes down to using a multimodal approach with medications, physical interventions, treatment of underlying conditions like spasticity, treatment of acquired overuse syndromes, and potentially the use of neuromodulation interventions. The typical categories of SCI-related pain can be reviewed in Tables 1 and 2.

## Application to other areas of neuroscience

There is a growing emphasis on the connection between physical experiences and psychologic interpretation that goes beyond pain management. One particular area of interest has been the neural pathway between the amygdala, periaqueductal gray matter, and rostral ventromedial medulla (RVM). The amygdala stores emotional experiences. It is connected to the periaqueductal gray matter, which contains a high concentration of opioid, cannabinoid, and endorphin receptors (Basbaum & Fields, 1978). The RVM is not involved in generating an initial pain response but does help regulate the maintenance of neuropathic pain (Vera-Portocarrero et al., 2006). The ventral posteromedial nucleus plays a major role in central sensitization, a condition of the nervous system that is associated with development and maintenance of chronic pain that is caused by increased excitability of cell membranes, synaptic efficiency, and reduced inhibition of nociceptive pathways (Latremoliere & Woolf, 2009). The plasticity of neurons that undergo central sensitization has been the target of newer therapies, though significant mechanistic gaps in our understanding remain (Latremoliere & Woolf, 2009).

These pathways significantly influence pain processing, decision-making, avoidant behavior, and personality expression (Yin et al., 2020) and are thought to modulate opioid-induced analgesia (Samineni et al., 2017). Additionally, recent studies showed modulating this pathway impacts the extent of depression-associated pain in mice under chronic stress (Yin et al., 2020). Physical activity, such as Tai Chi and cycling, has been shown to modulate this opioid pathway, thereby reducing overall pain levels and reducing levels of circulating inflammatory markers (Liu et al., 2019). This concept is fundamental to the use of physical activity to combat pain in patients with depression.

## Main narrative text

### Physical therapy and exercise

In the AB population, the clinical treatment of pain generally progresses from noninterventional PT-based treatment to more aggressive interventional treatments such as injections and ultimately surgery if there continues to be minimal to no improvement and depending on the pain generator (Patrick, Emanski, & Knaub, 2014; Will, Bury, & Miller, 2018). There are important similarities and differences in treating pain after SCI compared to the AB population. The most common subset of pain in the SCI population is neuropathic pain, either at or below the level of injury (E. Widerström-Noga, 2017a) and generally is not as responsive to PT (Akyuz & Kenis, 2014). However, many individuals have concomitant nociceptive pain (Widerström-Noga, 2017a) that may be more receptive to traditional PT interventions.

Eighty percent of those with SCI pain found PT and exercise helpful to a large extent, second only to cannabis and alcohol in terms of treatment effect (Heutink, Post, Wollaars, & Van Asbeck, 2011). Other studies in the United States (Widerström-Noga & Turk, 2003) noted half of respondents who participated in PT reported "considerable" pain reduction and a study from Sweden (Norrbrink Budh & Lundeberg, 2004) noted very high satisfaction rates despite overall low levels of PT utilization. Telehealth PT using high-dose scapular stabilizing and rotator cuff strengthening program in SCI individuals with signs of impingement syndrome was effective in reducing shoulder pain by more than 50% after 24 weeks (Van Straaten, Cloud, Morrow, Ludewig, & Zhao, 2014).

Exercise has been used as an augmentative strategy for many chronic conditions including osteoarthritis, peripheral vascular disease, low back pain, and fibromyalgia (Buckelew et al., 1998; Ettinger, 1997; Hayden et al., 2019; Lane, Ellis, Watson, & Leng, 2014) and there is a linear relationship between frequency, duration, and intensity of exercise and chronic pain (Landmark, Romundstad, Borchgrevink, Kaasa, & Dale, 2011). There are few high-quality studies assessing exercise and pain in the SCI population. Norrbrink, Lindberg, Wahman, and Bjerkefors (2012) noted significant improvement in pain scales and number of days per week with pain in nociceptive and neuropathic pain levels after training on a double-poling ergometer three times per week for 10 weeks at 70%–100% peak heart rate during intervals.

Recently, due to increased consensus on the benefits of exercise after SCI, an SCI exercise guideline was released (Ginis et al., 2018). The PAG, which has high concentrations of opioid and cannabinoid receptors sends projections to the RVM and both areas are activated by aerobic exercise (Chen et al., 2018; Sluka, O'Donnell, Danielson, & Rasmussen, 2013; Stagg et al., 2011). Additionally, there is an antiinflammatory cascade involving downregulation of pro-inflammatory Interleukin (IL)-1B and Tumor necrosis factor-alpha and upregulation of antiinflammatory IL-6 and IL-10 (Chen, Li, Chen, Li, & Hung, 2012; Jankord & Jemiolo, 2004). Some of the potential improvement in pain as it relates to exercise may be a partial function of improved psychological well-being (Latimer, Martin Ginis, Hicks, & McCartney, 2004).

### Medications

Medications targeting neuropathic pain after SCI are the best studied. Pregabalin (Lyrica) and gabapentin (Neurontin) are antiepileptics and considered first-line medications in the treatment of neuropathic pain in patients with SCI injury (Loh et al., 2016). In 2012, pregabalin (Lyrica) became the first and remains the only Federal Drug Administration approved medication for neuropathic pain in patients with SCI in the United States. Its approval was based on two double blind placebo controlled randomized trials (Cardenas et al., 2013; Siddall et al., 2006). In the study by Siddall et al. (2006), 70 patients showed improvement in pain and sleep when compared to the control arm of 67 patients. Cardenas et al. (2013) studied 220 patients and exhibited improvement in pain 1 week into treatment. Mehta, McIntyre, Janzen, Loh, and Teasell (2016) published an updated systematic review from their 2010 publication (Teasell et al., 2010) and cited 13, level-1 randomized controlled trials supporting both the anticonvulsants pregabalin and gabapentin.

Antidepressants are also utilized in treating neuropathic pain in SCI. Like pregabalin (Lyrica) and gabapentin (Neurontin), the tricyclic antidepressants amitriptyline (Elavil), nortriptyline (Pamelor), imipramine (Tofranil), and desipramine (Norpramin) are first-line medications. These medications have only demonstrated efficacy in patients with concomitant depression (Mehta et al., 2016). More recently venlafaxine revealed similar effects to amitriptyline in subjects with pain and depression (Richards et al., 2015). Other antidepressants, such as duloxetine, have not proven to be effective in patients with pain related to SCI (Vranken et al., 2011). Taken together, the serotonin-norepinephrine reuptake inhibitors (SNRIs) duloxetine (Cymbalta) and venlafaxine (Effexor) are considered second-line medications.

Despite common use in practice, opioids have inferior efficacy than previously mentioned medications. Tramadol, a synthetic opioid with SNRI properties, has some efficacy and is a third-line medication (Loh et al., 2016; Norrbrink & Lundeberg, 2009). As a sole agent, tramadol should be employed before other opioids such as oxycodone. Oxycodone is considered a fourth-line medication and has only level 4 evidence supporting its use (Loh et al., 2016; Mehta et al., 2016).

The effects of cannabinoids are mediated via the $CB_1$ and $CB_2$ G-protein receptors, where $CB_1$ is widely expressed in the CNS (Zajicek & Apostu, 2011). Interestingly, $CB_1$ is upregulated following SCI (Knerlich-Lukoschus et al., 2011). The analgesic properties of cannabinoids are mediated through the PAG (Heinricher et al., 2009; Raichlen, Foster, Gerdeman, Seillier, & Giuffrida, 2012). Indirect effects on opiate, serotonin, NMDA, and GABA receptors link endocannabinoids to other pain-related pathways (Hill, 2015; Raichlen et al., 2012; Woolridge et al., 2005). A metaanalysis of five randomized control trials suggests that inhaled cannabis may provide short-term relief for about 20% of patients with neuropathic pain (Andreae et al., 2015). Future studies are needed to assess long-term effects of cannabinoids on neuropathic pain, specifically as it relates to SCI, as well as the optimal route of administration (Inglet et al., 2020). The mechanism of action for medications typically used for the treatment of SCI-related pain can be reviewed in Table 3.

**TABLE 3** Medication management of pain in spinal cord injury.

| Class of medication | Targeted category of pain | Mechanism of action | Monitoring |
|---|---|---|---|
| Anticonvulsant<br>• Gabapentin<br>• Pregabalin | Neuropathic | GABA analog | - Peripheral edema<br>- Weight gain<br>- Dose adjustment for renal impairment |
| Antidepressant, tricyclic<br>• Amitriptyline<br>• Nortriptyline | Neuropathic | Inhibit reuptake of serotonin and/or norepinephrine at the presynaptic neuronal membrane pump | - Behavior changes, suicidality<br>- Anticholinergic effects (e.g., constipation, urinary retention) |
| Antidepressant, serotonin/ norepinephrine reuptake inhibitor<br>• Duloxetine<br>• Venlafaxine | Neuropathic | Inhibits neuronal serotonin and norepinephrine reuptake; weakly inhibits dopamine reuptake | - Behavior changes, suicidality<br>- Avoid use in those with hepatic impairment<br>- Dose adjustment for renal impairment |
| Nonsteroidal antiinflammatory drug<br>• Ibuprofen<br>• Naproxen<br>• Celecoxib*<br>• Diclofenac<br>• Indomethacin<br>• Ketorolac<br>• Meloxicam<br>• Nabumetone | Nociceptive | Inhibits cyclooxygenase-1 and 2 (COX-1 and 2) enzymes, resulting in decreased formation of prostaglandin precursors; decrease proinflammatory cytokine levels | - Cardiovascular thrombotic events<br>- Gastrointestinal bleeding, ulceration, and perforation<br>- Avoid use in those with renal disease |

**TABLE 3** Medication management of pain in spinal cord injury—cont'd

| Class of medication | Targeted category of pain | Mechanism of action | Monitoring |
|---|---|---|---|
| Nonopioid analgesic<br>• Acetaminophen<br>• Paracetamol | Nociceptive | Not fully known, analgesic effects likely due to activation of CNS descending serotonergic inhibitory pathways | - Hepatotoxicity |
| Opioid<br>• Buprenorphine<br>• Codeine<br>• Fentanyl<br>• Hydrocodone<br>• Hydromorphone<br>• Methadone<br>• Nalbuphine<br>• Oxycodone<br>• Tramadol | Nociceptive | Binds to opiate receptors in the CNS, inhibiting ascending pain pathways | - CNS depression<br>- Respiratory depression<br>- Hypotension<br>- Constipation |
| Cannabis | Nociceptive | Activation of the endogenous endocannabinoid system through CB1 and CB2 receptors | - Psychoactive properties<br>- Dizziness<br>- - Nausea, vomiting<br>- Fatigue, drowsiness |
| Baclofen | Spasticity | GABA agonist at the GABA B receptor | - Lowers seizure threshold<br>- Consider dose adjustment for renal impairment<br>- CNS depression<br>- Urinary retention |
| Dantrolene | Spasticity | Inhibits the release of calcium from sarcoplasmic reticulum | - Peripherally acting on skeletal muscle<br>- Hepatotoxicity; monitor liver function closely |
| Alpha 2 adrenergic agonist<br>• Clonidine<br>• Tizanidine | Spasticity | Increasing presynaptic inhibition; reduces facilitation of spinal motor neurons | - Orthostatic hypotension<br>- Drowsiness, dizziness<br>- Bradycardia<br>- Monitor liver function |
| Benzodiazepine<br>• Diazepam | Spasticity | Facilitates GABA effects on the GABA A receptor | - Caution with concomitant use of opioids<br>- CNS depression<br>- Use with caution in those with renal and hepatic impairment and respiratory disease |

*Original table*: Outline of the mechanism of action of various types of pain medications commonly used for SCI pain.

## "Typical" nociceptive pain interventions

SCI individuals, especially those who use a wheelchair to traverse the community are predisposed to upper-body overuse injuries (Requejo et al., 2008) and almost three fourth of individuals with paraplegia have early osteoarthritis of the shoulders (Lal, 1998). Interestingly, shoulder pain and decreased passive range of motion is also present in the acute phase of SCI (Finley et al., 2020). Peripheral nerve stimulation (PNS) is a well-established treatment for poststroke shoulder pain (Wilson et al., 2018) but has not frequently been used for treating SCI-related pain. There is one case report showing improvement in shoulder pain after SCI with PNS use (Mehech, Mejia, Nemunaitis, Chae, & Wilson, 2018).

While literature is lacking with regard to specific interventions for non-SCI-related nociceptive pain, it should be treated just as it would be in the able-bodied population. For instance, tendinopathy is a common condition in the able-bodied population and also has a high prevalence in the SCI population due to overuse, as noted above. Traditionally tendinopathy has been treated with physical therapy, nonsteroidal antiinflammatory medications, and corticosteroid injections which can address the inflammatory component but not the degenerative component of the disorder. Orthobiologics such as platelet-rich plasma (PRP), autologous conditioned serum and stem cells have shown for various degrees of promise (Paoloni, DeVos, Hamilton, Murrell, & Orchard, 2011; Roberts & Rosenbaum, 2012; Wehling et al., 2007) should be considered in the SCI population as they would be in the AB population, if appropriate.

At least 65% of patients with chronic SCI have symptoms of spasticity (Adams & Hicks, 2005) and pain frequently coincides with spasticity (Finnerup, 2017; Holtz, Lipson, Noonan, Kwon, & Mills, 2017). In general, the pain associated with spasticity is felt to be nociceptive muscle pain, similar to that of tetanic contractions. The incidence of spasticity is correlated to incomplete injuries and higher level of injury (Holtz et al., 2017; Skold, Levi, & Seiger, 1999).

Oral antispasticity medications act systemically and can be divided into 3 categories based upon mechanism of action: (1) GABA receptor agonists (e.g., baclofen and diazepam), (2) alpha-2 receptor agonists (e.g., tizanidine), and (3) peripheral acting decouplers of excitation-contraction at the neuromuscular level (e.g., dantrolene) (Burchiel & K Hsu, 2001; Gracies, Nance, Elovic, McGuire, & Simpson, 1997; Kita & Goodkin, 2000; Rode, Maupas, Luaute, Courtois-Jacquin, & Boisson, 2003).

Injection of local chemodenervation agents results in impairment of the nerve (Lui, Sarai, & Mills, 2015). This technique minimizes systemic side effects associated with oral antispasticity medications and is the preferred for treating focal spasticity (Kirshblum, 1999; Lui et al., 2015). The chemodenervation agents most commonly used are phenol, ethanol, local anesthetics, and botulinum toxin (Lui et al., 2015). Botulinum toxin is the most widely used chemodenervation agent and similarly has a temporary effect. Phenol and ethanol exert a semi-permanent effect (Adams & Hicks, 2005; Kirshblum, 1999; Lui et al., 2015).

If oral drug treatment is ineffective in controlling spasticity or is otherwise not tolerated, consideration should be given to intrathecal delivery of baclofen (Burchiel & K Hsu, 2001; Gracies et al., 1997; Kita & Goodkin, 2000; Korenkov, Niendorf, Darwish, Glaeser, & Gaab, 2002). Bypassing the blood-spinal cord barrier allows as much as four times the concentration and lower dose of baclofen to be delivered to the spinal cord relative to an oral dose (Burchiel & K Hsu, 2001; Kita & Goodkin, 2000; Korenkov et al., 2002). Catheter and pump maintenance, appropriate dosage of baclofen, timely refill of the pump, and avoidance of abrupt interruption of baclofen administration are paramount to intrathecal pump management (Burchiel & K Hsu, 2001; Kita & Goodkin, 2000).

Surgery, typically in the form of rhizotomy or tendon lengthening, may be performed to improving function or correcting a deformity related to spasticity in refractory cases (Adams & Hicks, 2005; Ghai, Garg, Hooda, & Gupta, 2013).

## Neuromodulation

The presence of pain induces top-down modulation where the various cerebral regions act in unison to upregulate or downregulate pain signals (Martins & Tavares, 2017), mainly through the PAG and RVM (Heinricher et al., 2009). Serotonin, norepinephrine, and dopamine are released at the supraspinal and spinal levels to modulate pain levels and there is a relative deficiency in these peptides after SCI (Bourne et al., 2014).

There is also evidence for dendritic spine remodeling at the level of the spinal cord. After induction of neuropathic pain there is an adaptive and time-dependent response at the level of dorsal horns where neurons develop increased dendritic spine density, redistribute spines closer to the cell body, and increase the spine head surface area, which in turn alters excitatory potential propagation (Benson, Reimer, & Tan, 2020). Animal studies looking at dendritic spines pre- and postneuropathic injury note significant structural anomalies in the steady state fluctuations of spines (Benson et al., 2020). Taken in total, there is accumulating evidence that "pain memory" is at least in part a function carried out by the spinal cord itself and could partly explain the treatment resistant nature of neuropathic pain after SCI (Benson, Fenrich, et al., 2020; Benson, Reimer, et al., 2020; Ji, Kohno, Moore, & Woolf, 2003; Tan & Waxman, 2012; Tan, 2015). Disruption of these intrinsic neuromodulation pathways is postulated to be a unique factor in pain as a side effect of SCI and a differentiator of other peripheral sources of neuropathic pain (Benson, Reimer, et al., 2020; Bourne et al., 2014; Chambel et al., 2020; Heinricher et al., 2009).

Many have hypothesized on medications and interventions that may be of clinical utility given the aforementioned intrinsic neuromodulatory deficits present after SCI. The International Neuromodulation Society defines therapeutic neuromodulation as "the alteration of nerve activity through targeted delivery of a stimulus, such as electrical stimulation or chemical agents, to specific neurological sites in the body." Examples of neuromodulation that have been applied or could potentially be applied in the setting of SCI include SCS, dorsal root ganglion stimulation (DRGS), PNS, and transcranial direct current stimulation (tDCS).

### Spinal cord stimulation in SCI

Spinal cord stimulation (SCS) is the most common neuromodulatory intervention in the treatment of pain in patients with SCI. It was first trialed in patients with SCI almost 50 years ago. Previous studies have hypothesized multiple mechanism of action such as gait control theory, dorsal column retrograde stimulation, modulation of centralized pain processing, and even action at the spinal cord level with the alteration of neurotransmitter release (Arle, Carlson, Mei, Iftimia, & Shils, 2014). To date there have been no randomized trials examining efficacy of SCS in patients with SCI. A review of 27 studies cited an average of 30%–40% of pain relief (Lagauche, Facione, Albert, & Fattal, 2009) using various forms of SCS. Traditional SCS utilizes tonic 40–60 Hz stimulation placed on the epidural surface of the dorsal column, often resulting in paresthesia in the distribution that overlaps with the distribution of pain (Chakravarthy, Richter, Christo, Williams, & Guan, 2018). More recently, high-frequency SCS has been applied at low amplitudes resulting in paresthesia free subthreshold sensory activation (Chakravarthy et al., 2018) and has shown promising results for the treatment of refractory neuropathic pain in non-SCI individuals (Kapural et al., 2016). Interestingly, multiple studies looking at SCS have noted concomitant decreased spasticity (Aduddell, Romman, Burlison, & Doulatram, 2015; Dombovy-Johnson, Hunt, Morrow, Lamer, & Pittelkow, 2020; Harandi & Kapural, 2018; Li, Quidgley-Nevares, & Walker, 2011) and pain medication use (Aduddell et al., 2015; Delille, Ahmad, Baeza-Dager, & Mena, 2012; Dombovy-Johnson et al., 2020; Harandi & Kapural, 2018; Li et al., 2011). The potential for multimodal treatment is particularly interesting in this population. Both reviews suggest individuals with incomplete tetraplegia or paraplegia may benefit more than complete injuries (Dombovy-Johnson et al., 2020; Lagauche et al., 2009).

### Other forms of neuromodulation

Other forms of neuromodulation for pain are less common in SCI. tDCS over the motor cortex has shown modest decreases in pain (Ngernyam et al., 2015; Soler et al., 2010). However, both studies had small sample sizes and a study by Wrigley et al. (2013) found no improvement in longstanding neuropathic pain after SCI with the use of tDCS suggesting the effect is either less robust or potentially more beneficial in the acute to subacute phase. PNS, as noted previously, has been described as a treatment for post-SCI shoulder pain from chronic overuse injuries.

DRGS is a similar in nature to SCS except the leads are placed on the dorsal root ganglion instead of the epidural space. Stimulation of the DRG is thought to still work in a downstream effect on the spinal cord but has been particularly helpful with the treatment of regional neuropathic pain, such as complex regional pain syndrome (Harrison, Epton, Bojanic, Green, & FitzGerald, 2018). However, there is a case report of bilateral L2 DRGS improving spasticity and low back pain after SCI (Soloukey et al., 2020).

## Mini-dictionary of terms

Nociceptive pain: Pain from a noxious or potentially harmful stimulus. Nociceptive pain is well localized and corresponds to damage to skin, joints, muscle, bone, or connective tissue. It is typically described as sharp, achy, throbbing, etc.

Visceral pain: Pain from internal organs. Visceral pain is generally poorly localized and described as deep, squeezing, pressure, or aching.

At-level neuropathic pain: Pain in a dermatomal distribution that reflects damage to the spinal cord at the level of the individual's spinal cord injury (SCI). Occasionally, this can feel "band-like" or "squeezing," but often feels like "numbness," "pins-and-needles," or "tingling."

Below-level neuropathic pain: Pain that corresponds to damaged nerves in the spinal cord that typically provide sensation to an area below the level of the individual's SCI. Below-level neuropathic pain is typically described as "numbness," "pins-and-needles," or "tingling."

Neuromodulation: the alteration of nerve activity through targeted delivery of a stimulus, such as electrical stimulation or chemical agents, to specific neurological sites in the body.

Spinal cord stimulation (SCS): Implantable device where leads are placed in the epidural space overlying the dorsal columns. SCS devices send mild pulses of electricity at various frequencies depending on the device and settings that help dampen neuropathic pain.

Dorsal root ganglion stimulator (DRGS): Similar to an SCS but is instead implanted with the leads on the dorsal root ganglion. In general, DRGS is favored for regional pain.

Peripheral nerve stimulator (PNS): A permanent or temporary implantable lead that is placed in close proximity to a peripheral nerve. PNS devices stimulate mixed motor and sensory nerves at submaximal intensity with the goal of decreasing pain that is related to irritation, damage, or improper function of a single nerve.

Transcranial direct current stimulation (tDCS): constant, low direct current electrical stimulus that is applied to specific cortical areas through surface scalp electrodes.

Supraspinal pathways: Ascending and descending pathways that relay information to, from, and through the spinal cord. Disruption of different supraspinal pathways plays a critical role in many aspects of SCI, including pain.

## Key facts of pain after SCI

- Sixty-five percent to eighty percent of individuals with SCI live with chronic pain.
- One third of individuals with chronic pain after SCI rate this pain as severe and more than half note that pain interferes with their daily function independent of their other SCI deficits such as paralysis.
- Level of injury does not seem to influence the degree of associated pain, although there is some evidence that incomplete spinal cord injuries are more painful.
- Up to 80% of those with pain and SCI find PT to be helpful.
- Pregabalin is the only FDA-approved treatment for neuropathic SCI pain, although antiepileptic and antidepressant medications are commonly used and supported by literature.
- Neuromodulation is currently lacking evidence, but small studies are promising and a future direction of additional research.

## Summary points

- The PAG is a central pain relay center that modulates pain signals to the RVM and spinal cord effectively upregulating and downregulating pain signals.
- SCI disrupts this relay and leads to a relative deficiency in supraspinal inhibition and deficiency in spinal levels of serotonin, norepinephrine, and dopamine.
- Neuropathic pain after SCI is commonly treated with membrane-stabilizing agents.
- Nociceptive pain after SCI should be treated like nociceptive pain in the AB population.
- Neuromodulation is an interesting treatment possibility, although further research is needed.

## References


Adams, M. M., & Hicks, A. L. (2005). Spasticity after spinal cord injury. *Spinal Cord*, *43*(10), 577–586. https://doi.org/10.1038/sj.sc.3101757.

Aduddell, B., Romman, A., Burlison, S., & Doulatram, G. (2015). Use of spinal cord stimulator (SCS) in a case of refractory neuropathic pain due to syringomelia. In *Regional anesthesia and pain medicine conference: 14th annual pain medicine meeting of the American Society of Regional Anesth Pain Medicine, ASRA, 41(2)*.

Akyuz, G., & Kenis, O. (2014, March). Physical therapy modalities and rehabilitation techniques in the management of neuropathic pain. *American Journal of Physical Medicine and Rehabilitation*. https://doi.org/10.1097/PHM.0000000000000037.

Andreae, M. H., Carter, G. M., Shaparin, N., Suslov, K., Ellis, R. J., Ware, M. A., … Sacks, H. S. (2015). Inhaled Cannabis for chronic neuropathic pain: A Meta-analysis of individual patient data. *The Journal of Pain*, *16*(12), 1221–1232. https://doi.org/10.1016/j.jpain.2015.07.009.

Arle, J. E., Carlson, K. W., Mei, L., Iftimia, N., & Shils, J. L. (2014). Mechanism of dorsal column stimulation to treat neuropathic but not nociceptive pain: Analysis with a computational model. *Neuromodulation: Technology at the Neural Interface*, *17*(7), 642–655. https://doi.org/10.1111/ner.12178.

Basbaum, A. I., & Fields, H. L. (1978). Endogenous pain control mechanisms: Review and hypothesis. *Annals of Neurology*, *4*(5). https://doi.org/10.1002/ana.410040511.

Benson, C. A., Fenrich, K. K., Olson, K. L., Patwa, S., Bangalore, L., Waxman, S. G., & Tan, A. M. (2020). Dendritic spine dynamics after peripheral nerve injury: An intravital structural study. *Journal of Neuroscience*, *40*(22). https://doi.org/10.1523/JNEUROSCI.2858-19.2020.

Benson, C. A., Reimer, M. L., & Tan, A. M. (2020). Dendritic spines in the spinal cord: Live action pain. *Neuroscience Insights*. https://doi.org/10.1177/2633105520951164. SAGE Publications Ltd.

Beric, A., Dimitrijevic, M. R., & Lindblom, U. (1988). Clinical section central dysesthesia syndrome in spinal cord injury patients. *Pain*, *34*, 109–116.

Boadas-Vaello, P., Castany, S., Homs, J., Álvarez-Pérez, B., Deulofeu, M., & Verdú, E. (2016, May 1). Neuroplasticity of ascending and descending pathways after somatosensory system injury: Reviewing knowledge to identify neuropathic pain therapeutic targets. *Spinal Cord*. https://doi.org/10.1038/sc.2015.225. Nature Publishing Group.

Bourne, S., Machado, A. G., & Nagel, S. J. (2014, October 1). Basic anatomy and physiology of pain pathways. *Neurosurgery Clinics of North America*. https://doi.org/10.1016/j.nec.2014.06.001. W.B. Saunders.

Buckelew, S. P., Conway, R., Parker, J., Deuser, W. E., Read, J., Witty, T. E., … Kay, D. R. (1998). Biofeedback/relaxation training and exercise interventions for fibromyalgia: A prospective trial. *Arthritis and Rheumatism*, *11*(3), 196–209. https://doi.org/10.1002/art.1790110307.

Burchiel, K. J., & K Hsu, F. P. (2001, December). Pain and spasticity after spinal cord injury: Mechanisms and treatment. *Spine*. https://doi.org/10.1097/00007632-200112151-00024.

Cardenas, D. D., Bryce, T. N., Shem, K., Richards, J. S., & Elhefni, H. (2004). Gender and minority differences in the pain experience of people with spinal cord injury. *Archives of Physical Medicine and Rehabilitation*, *85*(11), 1774–1781. https://doi.org/10.1016/j.apmr.2004.04.027.

Cardenas, D. D., Nieshoff, E. C., Suda, K., Goto, S. I., Sanin, L., Kaneko, T., … Knapp, L. E. (2013). A randomized trial of pregabalin in patients with neuropathic pain due to spinal cord injury. *Neurology*, *80*(6), 533–539. https://doi.org/10.1212/WNL.0b013e318281546b.

Chakravarthy, K., Richter, H., Christo, P. J., Williams, K., & Guan, Y. (2018). Spinal cord stimulation for treating chronic pain: Reviewing preclinical and clinical data on paresthesia-free high-frequency therapy. *Neuromodulation*, *21*(1), 10–18. https://doi.org/10.1111/ner.12721.

Chambel, S. S., Tavares, I., & Cruz, C. D. (2020, July). Chronic pain after spinal cord injury: Is there a role for neuron-immune dysregulation? *Frontiers in Physiology*. https://doi.org/10.3389/fphys.2020.00748. Frontiers Media S.A.

Chen, Y. H., Lee, H. J., Lee, M. T., Wu, Y. T., Lee, Y. H., Hwang, L. L., … Chiou, L. C. (2018). Median nerve stimulation induces analgesia via orexininitiated endocannabinoid disinhibition in the periaqueductal gray. *Proceedings of the National Academy of Sciences of the United States of America*, *115*(45), E10720–E10729. https://doi.org/10.1073/pnas.1807991115.

Chen, Y. W., Li, Y. T., Chen, Y. C., Li, Z. Y., & Hung, C. H. (2012). Exercise training attenuates neuropathic pain and cytokine expression after chronic constriction injury of rat sciatic nerve. *Anesthesia and Analgesia*, *114*(6), 1330–1337. https://doi.org/10.1213/ANE.0b013e31824c4ed4.

Craig, A., Nicholson Perry, K., Guest, R., Tran, Y., Dezarnaulds, A., Hales, A., … Middleton, J. (2015). Prospective study of the occurrence of psychological disorders and comorbidities after spinal cord injury archives of physical medicine and rehabilitation. *Archives of Physical Medicine and Rehabilitation*, *96*, 1426–1460. https://doi.org/10.1016/j.apmr.2015.02.027.

Davidoff, G., Roth, E., Guarracini, M., Sliwa, J., & Yarkony, G. (1987). Function-limiting dysesthetic pain syndrome among traumatic spinal cord injury patients: A cross-sectional study. *Pain*, *29*, 39–48.

Delille, M., Ahmad, U. F., Baeza-Dager, J. M., & Mena, J. (2012). Neuropathic pain improvement in Incomplete Spinal cord injury after spinal cord stimulator trial: A case series. *PM & R: The Journal of Injury, Function, and Rehabilitation*, *1*, S325–S326.

Dombovy-Johnson, M. L., Hunt, C. L., Morrow, M. M., Lamer, T. J., & Pittelkow, T. P. (2020). Current evidence lacking to guide clinical practice for spinal cord stimulation in the treatment of neuropathic pain in spinal cord injury: A review of the literature and a proposal for future study. *Pain Practice*, *20*(3), 325–335. https://doi.org/10.1111/papr.12855.

Elliott, T. R., & Frank, R. G. (1996). Depression following spinal cord injury. *Archives of Physical Medicine and Rehabilitation*. https://doi.org/10.1016/S0003-9993(96)90263-4. W.B. Saunders.

Ettinger, W. H. (1997). A randomized trial comparing aerobic exercise and resistance exercise with a health education program in older adults with knee osteoarthritis. *JAMA*, *277*(1), 25. https://doi.org/10.1001/jama.1997.03540250033028.

Finley, M., Euiler, E., Trojian, T., Gracely, E., Schmidt-Read, M., Frye, S. K., … Geigle, P. R. (2020). Shoulder impairment and pain of individuals with newly acquired spinal cord injury compared to uninjured peers. *Spinal Cord Series and Cases*, *6*(1). https://doi.org/10.1038/s41394-020-0318-1.

Finnerup, N. B. (2017, December 1). Neuropathic pain and spasticity: Intricate consequences of spinal cord injury. *Spinal Cord*. https://doi.org/10.1038/sc.2017.70. Nature Publishing Group.

Finnerup, N. B., Jensen, M. P., Norrbrink, C., Trok, K., Johannesen, I. L., Jensen, T. S., & Werhagen, L. (2016). A prospective study of pain and psychological functioning following traumatic spinal cord injury. *Spinal Cord*, *54*(10), 816–821. https://doi.org/10.1038/sc.2015.236.

Ghai, A., Garg, N., Hooda, S., & Gupta, T. (2013). Spasticity—Pathogenesis, prevention and treatment strategies. *Saudi Journal of Anaesthesia*. https://doi.org/10.4103/1658-354X.121087.

Ginis, K. A. M., Van Der Scheer, J. W., Latimer-Cheung, A. E., Barrow, A., Bourne, C., Carruthers, P., … Goosey-Tolfrey, V. L. (2018). Evidence-based scientific exercise guidelines for adults with spinal cord injury: An update and a new guideline. *Spinal Cord*, *56*, 308–321. https://doi.org/10.1038/s41393-017-0017-3.

Gracies, J. M., Nance, P., Elovic, E., McGuire, J., & Simpson, D. M. (1997). Traditional pharmacological treatments for spasticity. Part II: General and regional treatments. *Muscle & Nerve Supplement*, *6*, S92–120.

Harandi, S., & Kapural, L. (2018). Four-extremity neurostimulation using two cervical octapolar leads and high frequency of 10 kHz. *Pain Practice*, *18*(2), 269–272. https://doi.org/10.1111/papr.12593.

Harrison, C., Epton, S., Bojanic, S., Green, A. L., & FitzGerald, J. J. (2018, April 1). The efficacy and safety of dorsal root ganglion stimulation as a treatment for neuropathic Pain: A literature review. *Neuromodulation*. https://doi.org/10.1111/ner.12685. Blackwell Publishing Inc.

Hayden, J. A., Wilson, M. N., Stewart, S., Cartwright, J. L., Smith, A. O., Riley, R. D., … Yeung, E. W. (2019). Exercise treatment effect modifiers in persistent low back pain: An individual participant data meta-analysis of 3514 participants from 27 randomised controlled trials. *British Journal of Sports Medicine*. https://doi.org/10.1136/bjsports-2019-101205. BMJ Publishing Group.

Heinricher, M. M., Tavares, I., Leith, J. L., & Lumb, B. M. (2009, April). Descending control of nociception: Specificity, recruitment and plasticity. *Brain Research Reviews*. https://doi.org/10.1016/j.brainresrev.2008.12.009. NIH Public Access.

Heutink, M., Post, M. W. M., Wollaars, M. M., & Van Asbeck, F. W. A. (2011). Chronic spinal cord injury pain: Pharmacological and non-pharmacological treatments and treatment effectiveness. *Disability and Rehabilitation*, *33*(5), 433–440. https://doi.org/10.3109/09638288.2010.498557.

Hill, K. P. (2015). Medical marijuana for treatment of chronic pain and other medical and psychiatric problems: A clinical review. *JAMA: The Journal of the American Medical Association*. https://doi.org/10.1001/jama.2015.6199. Article, United States: American Medical Association.

Holtz, K. A., Lipson, R., Noonan, V. K., Kwon, B. K., & Mills, P. B. (2017). Prevalence and effect of problematic spasticity after traumatic spinal cord injury. In *Vol. 98. Archives of physical medicine and rehabilitation* (pp. 1132–1138). United States: Elsevier BV. Article https://doi.org/10.1016/j.apmr.2016.09.124.

Inglet, S., Winter, B., Yost, S. E., Entringer, S., Lian, A., Biksacky, M., … Mortensen, W. (2020, November 2). Clinical data for the use of Cannabis-based treatments: A comprehensive review of the literature. *Annals of Pharmacotherapy*. https://doi.org/10.1177/1060028020930189.

Jankord, R., & Jemiolo, B. (2004). Influence of physical activity on serum IL-6 and IL-10 levels in healthy older men. *Medicine and Science in Sports and Exercise*, *36*(6), 960–964. https://doi.org/10.1249/01.MSS.0000128186.09416.18.

Ji, R. R., Kohno, T., Moore, K. A., & Woolf, C. J. (2003). Central sensitization and LTP: Do pain and memory share similar mechanisms? *Trends in Neurosciences*. https://doi.org/10.1016/j.tins.2003.09.017. Elsevier Ltd.

Kapural, L., Yu, C., Doust, M. W., Gliner, B. E., Vallejo, R., Sitzman, B. T., … Burgher, A. H. (2016). Comparison of 10-kHz high-frequency and traditional low-frequency spinal cord stimulation for the treatment of chronic Back and leg pain: 24-month results from a multicenter, randomized, controlled pivotal trial. *Neurosurgery*, *79*(5), 667–676. https://doi.org/10.1227/NEU.0000000000001418.

Kirshblum, S. (1999). Treatment alternatives for spinal cord injury related spasticity. *Journal of Spinal Cord Medicine*, *22*(3), 199–217. https://doi.org/10.1080/10790268.1999.11719570.

Kita, M., & Goodkin, D. E. (2000, March). Drugs used to treat spasticity. *Drugs*. https://doi.org/10.2165/00003495-200059030-00006.

Knerlich-Lukoschus, F., Noack, M., von der Ropp-Brenner, B., Lucius, R., Mehdorn, H. M., & Held-Feindt, J. (2011). Spinal cord injuries induce changes in $CB_1$ cannabinoid receptor and C-C chemokine expression in brain areas underlying circuitry of chronic pain conditions. *Journal of Neurotrauma*, *28*(4), 619–634. https://doi.org/10.1089/neu.2010.1652.

Korenkov, A. I., Niendorf, W. R., Darwish, N., Glaeser, E., & Gaab, M. R. (2002). Continuous intrathecal infusion of baclofen in patients with spasticity caused by spinal cord injuries. *Neurosurgical Review*, *25*(4), 228–230. https://doi.org/10.1007/s10143-002-0221-1.

Lagauche, D., Facione, J., Albert, T., & Fattal, C. (2009, March). The chronic neuropathic pain of spinal cord injury: Which efficiency of neuropathics stimulations? *Annals of Physical and Rehabilitation Medicine*. https://doi.org/10.1016/j.rehab.2008.12.010.

Lal, S. (1998). Premature degenerative shoulder changes in spinal cord injury patients. *Spinal Cord*, *36*(3), 186–189. https://doi.org/10.1038/sj.sc.3100608.

Landmark, T., Romundstad, P., Borchgrevink, P. C., Kaasa, S., & Dale, O. (2011). Associations between recreational exercise and chronic pain in the general population: Evidence from the HUNT 3 study. *Pain*, *152*(10), 2241–2247. https://doi.org/10.1016/j.pain.2011.04.029.

Lane, R., Ellis, B., Watson, L., & Leng, G. C. (2014, July). Exercise for intermittent claudication. *Cochrane Database of Systematic Reviews*. https://doi.org/10.1002/14651858.CD000990.pub3. John Wiley and Sons Ltd.

Latimer, A. E., Martin Ginis, K. A., Hicks, A. L., & McCartney, N. (2004). An examination of the mechanisms of exercise-induced change in psychological well-being among people with spinal cord injury. *Journal of Rehabilitation Research and Development*, *41*(5), 643–651. https://doi.org/10.1682/JRRD.2003.04.0043.

Latremoliere, A., & Woolf, C. J. (2009). Central sensitization: A generator of pain hypersensitivity by central neural plasticity. *The Journal of Pain*, *10*(9). https://doi.org/10.1016/j.jpain.2009.06.012.

Li, Q., Quidgley-Nevares, A., & Walker, R. (2011). Utility of spinal cord stimulator to modulate lower Limb spasticity and pain in a patient with spinal cord injury: A case report. *PM & R: The Journal of Injury, Function, and Rehabilitation*, *1*, S288–S289.

Liu, J., Chen, L., Chen, X., Hu, K., Tu, Y., Lin, M., … Kong, J. (2019). Modulatory effects of different exercise modalities on the functional connectivity of the periaqueductal grey and ventral tegmental area in patients with knee osteoarthritis: A randomised multimodal magnetic resonance imaging study. *British Journal of Anaesthesia*, *123*(4). https://doi.org/10.1016/j.bja.2019.06.017.

Loh, E., Guy, S. D., Mehta, S., Moulin, D. E., Bryce, T. N., Middleton, J. W., … Wolfe, D. (2016). The Can Pain SCI clinical practice guidelines for rehabilitation management of neuropathic Pain after spinal cord: Introduction, methodology and recommendation overview. *Spinal Cord*, *54*(S1), S1–S6. https://doi.org/10.1038/sc.2016.88.

Lui, J., Sarai, M., & Mills, P. B. (2015). Chemodenervation for treatment of limb spasticity following spinal cord injury: A systematic review. *Spinal Cord*, *53*(4), 252–264. https://doi.org/10.1038/sc.2014.241.

Martins, I., & Tavares, I. (2017). Reticular formation and pain: The past and the future. *Frontiers in Neuroanatomy*, *11*, 1–14. https://doi.org/10.3389/fnana.2017.00051.

Mehech, D., Mejia, M., Nemunaitis, G. A., Chae, J., & Wilson, R. D. (2018). Percutaneous peripheral nerve stimulation for treatment of shoulder pain after spinal cord injury: A case report. *Journal of Spinal Cord Medicine*, *41*(1), 119–124. https://doi.org/10.1080/10790268.2017.1293329.

Mehta, S., McIntyre, A., Janzen, S., Loh, E., & Teasell, R. (2016). Systematic review of pharmacologic treatments of pain after spinal cord injury: An update. *Archives of Physical Medicine and Rehabilitation*, *97*(8), 1381–1391. e1 https://doi.org/10.1016/j.apmr.2015.12.023.

Ngernyam, N., Jensen, M. P., Arayawichanon, P., Auvichayapat, N., Tiamkao, S., Janjarasjitt, S., … Auvichayapat, P. (2015). The effects of transcranial direct current stimulation in patients with neuropathic pain from spinal cord injury. *Clinical Neurophysiology*, *126*(2), 382–390. https://doi.org/10.1016/j.clinph.2014.05.034.

Norrbrink Budh, C., & Lundeberg, T. (2004). Non-pharmacological pain-relieving therapies in individuals with spinal cord injury: A patient perspective. *Complementary Therapies in Medicine*, *12*(4), 189–197. https://doi.org/10.1016/j.ctim.2004.10.003.

Norrbrink, C., Lindberg, T., Wahman, K., & Bjerkefors, A. (2012). Effects of an exercise programme on musculoskeletal and neuropathic pain after spinal cord injury—Results from a seated double-poling ergometer study. *Spinal Cord*, *50*(6), 457–461. https://doi.org/10.1038/sc.2011.160.

Norrbrink, C., & Lundeberg, T. (2009). Tramadol in neuropathic pain after spinal cord injury a randomized, double-blind, placebo-controlled trial. *Clinical Journal of Pain*, *25*(3), 177–184. https://doi.org/10.1097/AJP.0b013e31818a744d.

Paoloni, J., DeVos, R. J., Hamilton, B., Murrell, G. A. C., & Orchard, J. (2011). Platelet-rich plasma treatment for ligament and tendon injuries. *Clinical Journal of Sport Medicine*, *211*. 37–35.

Patrick, N., Emanski, E., & Knaub, M. A. (2014). Acute and chronic low back pain. *Medical Clinics of North America*. https://doi.org/10.1016/j.mcna.2014.03.005. W.B. Saunders.

Raichlen, D. A., Foster, A. D., Gerdeman, G. L., Seillier, A., & Giuffrida, A. (2012). Wired to run: Exercise-induced endocannabinoid signaling in humans and cursorial mammals with implications for the "runner's high". *Journal of Experimental Biology*, *215*(8), 1331–1336. https://doi.org/10.1242/jeb.063677.

Requejo, P. S., Mulroy, S. J., Haubert, L. L., Newsam, C. J., Gronley, J. A. K., & Perry, J. (2008, April). Evidence-based strategies to preserve shoulder function in manual wheelchair users with spinal cord injury. *Topics in Spinal Cord Injury Rehabilitation*. https://doi.org/10.1310/sci1304-86.

Richards, J. S., Bombardier, C. H., Wilson, C. S., Chiodo, A. E., Brooks, L., Tate, D. G., … Fann, J. R. (2015). Efficacy of venlafaxine XR for the treatment of pain in patients with spinal cord injury and major depression: A randomized, controlled trial. *Archives of Physical Medicine and Rehabilitation*, *96*(4), 680–689. https://doi.org/10.1016/j.apmr.2014.11.024.

Richards, J. S., Stover, S. L., & Jaworski, T. (1990). Effect of bullet removal on subsequent pain in persons with spinal cord injury secondary to gunshot wound. *Journal of Neurosurgery*, *73*(3), 401–404. https://doi.org/10.3171/jns.1990.73.3.0401.

Roberts, T. T., & Rosenbaum, A. J. (2012). Bone grafts, bone substitutes and orthobiologics: The bridge between basic science and clinical advancements in fracture healing. *Oranogenesis*, *8*(4), 114–124.

Rode, G., Maupas, E., Luaute, J., Courtois-Jacquin, S., & Boisson, D. (2003). Medical treatment of spasticity. *Neuro-Chirurgie*, *49*(2–3 Pt 2), 247–255.

Samineni, V. K., Grajales-Reyes, J. G., Copits, B. A., O'Brien, D. E., Trigg, S. L., Gomez, A. M., … Gereau, R. W. (2017). Divergent modulation of nociception by glutamatergic and GABAergic neuronal subpopulations in the periaqueductal gray. *eNeuro*, *4*(2). https://doi.org/10.1523/ENEURO.0129-16.2017.

Siddall, P. J., Cousins, M. J., Otte, A., Griesing, T., Chambers, R., & Murphy, T. K. (2006). Pregabalin in central neuropathic pain associated with spinal cord injury: A placebo-controlled trial. *Neurology*, *67*(10), 1792–1800. https://doi.org/10.1212/01.wnl.0000244422.45278.ff.

Siddall, P. J., & Loeser, J. D. (2001). Pain following spinal cord injury. *Spinal Cord*, *39*(2), 63–73. https://doi.org/10.1038/sj.sc.3101116.

Siddall, P. J., Taylor, D. A., & Cousins, M. J. (1997). Classification of pain following spinal cord injury. *Spinal Cord*, *35*(2), 69–75. https://doi.org/10.1038/sj.sc.3100365.

Siddall, P. J., Mcclelland, J. M., Rutkowski, S. B., & Cousins, M. J. (2003). A longitudinal study of the prevalence and characteristics of pain in the first 5 years following spinal cord injury. *Pain*, *103*, 249–257. https://doi.org/10.1016/S0.

Skold, C., Levi, R., & Seiger, A. (1999). Spasticity after traumatic spinal cord injury: nature, severity, and location. *Archives of Physical Medicine and Rehabilitation*, *80*(12), 1548–1557.

Sluka, K. A., O'Donnell, J. M., Danielson, J., & Rasmussen, L. A. (2013). Regular physical activity prevents development of chronic pain and activation of central neurons. *Journal of Applied Physiology*, *114*(6), 725–733. https://doi.org/10.1152/japplphysiol.01317.2012.

Soler, M. D., Kumru, H., Pelayo, R., Vidal, J., Tormos, J. M., Fregni, F., … Pascual-Leone, A. (2010). Effectiveness of transcranial direct current stimulation and visual illusion on neuropathic pain in spinal cord injury. *Brain*, *133*(9), 2565–2577. https://doi.org/10.1093/brain/awq184.

Soloukey, S., Drenthen, J., Osterthun, R., de Rooij, J. D., De Zeeuw, C. I., Huygen, F. J. P. M., & Harhangi, B. S. (2020). Bilateral L2 dorsal root ganglion stimulation suppresses lower limb spasticity following chronic motor complete spinal cord injury: A case report. *Brain Stimulation*, *13*(3), 637–639. https://doi.org/10.1016/j.brs.2020.02.005.

Stagg, N. J., Mata, H. P., Ibrahim, M. M., Henriksen, E. J., Porreca, F., Vanderah, T. W., & Malan, T. P. (2011). Regular exercise reverses sensory hypersensitivity in a rat neuropathic pain model: Role of endogenous opioids. *Anesthesiology*, *114*(4), 940–948. https://doi.org/10.1097/ALN.0b013e318210f880.

Tan, A. M., & Waxman, S. G. (2012). Spinal cord injury, dendritic spine remodeling, and spinal memory mechanisms. *Experimental Neurology*, *235*(1), 142–151. https://doi.org/10.1016/j.expneurol.2011.08.026.

Tan, A. M. (2015). Dendritic spine dysgenesis in neuropathic pain. In *Vol. 131*. *Progress in molecular biology and translational science* (pp. 385–408). Elsevier B.V. https://doi.org/10.1016/bs.pmbts.2014.12.001.

Teasell, R. W., Mehta, S., Aubut, J. A. L., Foulon, B., Wolfe, D. L., Hsieh, J. T. C., … Short, C. (2010). A systematic review of pharmacologic treatments of pain after spinal cord injury. *Archives of Physical Medicine and Rehabilitation*, *91*(5), 816–831. https://doi.org/10.1016/j.apmr.2010.01.022.

Van Straaten, M. G., Cloud, B. A., Morrow, M. M., Ludewig, P. M., & Zhao, K. D. (2014). Effectiveness of home exercise on pain, function, and strength of manual wheelchair users with spinal cord injury: A high-dose shoulder program with telerehabilitation. *Archives of Physical Medicine and Rehabilitation*, *95*(10), 1810–1817.e2. https://doi.org/10.1016/j.apmr.2014.05.004.

Vera-Portocarrero, L. P., Zhang, E. T., Ossipov, M. H., Xie, J. Y., King, T., Lai, J., & Porreca, F. (2006). Descending facilitation from the rostral ventromedial medulla maintains nerve injury-induced central sensitization. *Neuroscience*, *140*(4), 1311–1320. https://doi.org/10.1016/j.neuroscience.2006.03.016.

Vranken, J. H., Hollmann, M. W., Van Der Vegt, M. H., Kruis, M. R., Heesen, M., Vos, K., … Dijkgraaf, M. G. W. (2011). Duloxetine in patients with central neuropathic pain caused by spinal cord injury or stroke: A randomized, double-blind, placebo-controlled trial. *Pain*, *152*(2), 267–273. https://doi.org/10.1016/j.pain.2010.09.005.

Widerström-Noga, E. (2017a). Neuropathic pain and spinal cord injury: Phenotypes and pharmacological management. *Drugs*, *77*(9), 967–984. https://doi.org/10.1007/s40265-017-0747-8.

Wehling, P., Moser, C., Frisbie, D., McIlwraith, C. W., Kawcak, C. E., Krauspe, R., & Reinecke, J. A. (2007). Autologous conditioned serum in the treatment of orthopedic diseases: The orthokine therapy. *BioDrugs*, *21*(5), 323–332.

Widerström-Noga, E. (2017, June 1). Neuropathic pain and spinal cord injury: Phenotypes and pharmacological management. *Drugs*. https://doi.org/10.1007/s40265-017-0747-8. Springer International Publishing.

Widerström-Noga, E. G., & Turk, D. C. (2003, November). Types and effectiveness of treatments used by people with chronic pain associated with spinal cord injuries: Influence of pain and psychosocial characteristics. *Spinal Cord*. https://doi.org/10.1038/sj.sc.3101511.

Will, J. S., Bury, D. C., & Miller, J. A. (2018). Mechanical low back pain. *American Family Physician*, *98*(7), 421–428. Retrieved from https://pubmed-ncbi-nlm-nih-gov.proxy1.lib.tju.edu/30252425/.

Wilson, R. D., Bennett, M. E., Nguyen, V. Q. C., Bock, W. C., O'Dell, M. W., Watanabe, T. K., … Chae, J. (2018). Fully implantable peripheral nerve stimulation for hemiplegic shoulder pain: A multi-site case series with two-year follow-up. *Neuromodulation*, *21*(3), 290–295. https://doi.org/10.1111/ner.12726.

Woolridge, E., Barton, S., Samuel, J., Osorio, J., Dougherty, A., & Holdcroft, A. (2005). Cannabis use in HIV for pain and other medical symptoms. *Journal of Pain and Symptom Management*, *29*(4), 358–367. https://doi.org/10.1016/j.jpainsymman.2004.07.011.

Wrigley, P. J., Gustin, S. M., McIndoe, L. N., Chakiath, R. J., Henderson, L. A., & Siddall, P. J. (2013). Longstanding neuropathic pain after spinal cord injury is refractory to transcranial direct current stimulation: A randomized controlled trial. *Pain*, *154*(10), 2178–2184. https://doi.org/10.1016/j.pain.2013.06.045.

Yin, W., Mei, L., Sun, T., Wang, Y., Li, J., Chen, C., … Zhang, Z. (2020). A central amygdala–ventrolateral periaqueductal gray matter pathway for pain in a mouse model of depression-like behavior. *Anesthesiology*, *132*(5). https://doi.org/10.1097/ALN.0000000000003133.

Zajicek, J. P., & Apostu, V. I. (2011). Role of cannabinoids in multiple sclerosis. *CNS Drugs*, *25*(3), 187–201. https://doi.org/10.2165/11539000-000000000-00000.

Section B

# Clinical features of spinal injury

Chapter 5

# Factors contributing to pressure injuries in traumatic spinal cord injury

Gabrielle Gour-Provençal[a] and Jehane H. Dagher[a,b]

[a]*Montreal Rehabilitation Institute, Gingras Lindsay Pavillon, Montreal, QC, Canada;* [b]*Physical Medicine and Rehabilitation, McGill University Health Centre, Montreal General Hospital Site, Montreal, QC, Canada*

## Abbreviations

**ISNCSCI** International Standards for Neurological Classification of Spinal Cord Injury
**NPIAP** National Pressure Injury Advisory Panel
**PI** pressure injury
**SCI** spinal cord injury
**TSCI** traumatic spinal cord injury

## Introduction

A pressure injury (PI), also known as pressure ulcer, pressure sore, bedsore, or decubitus ulcer, is defined as a circumscribed injury to the skin and/or underlying tissue due to pressure, or shear combined with pressure, causing ischemia, cell death, and tissue necrosis (European Pressure Ulcer Advisory Panel, National Pressure Injury Advisory Panel and Pan Pacific Pressure Injury Alliance, 2019; Ham, Schoonhoven, Schuurmans, & Leenen, 2017). It may present as an open wound or intact skin. The most commonly used pressure injury classification system, based on the extent of tissue involvement, is the one developed by the National Pressure Injury Advisory Panel (NPIAP) (Table 1) (Fig. 1) (European Pressure Ulcer Advisory Panel, National Pressure Injury Advisory Panel and Pan Pacific Pressure Injury Alliance, 2019). In order to accurately describe PI, staging can only be done after proper cleansing of the wound and exposure debridement of slough or eschar, thus sometimes postponing final staging of certain PI where necrotic tissue is present in the wound bed (European Pressure Ulcer Advisory Panel, National Pressure Injury Advisory Panel and Pan Pacific Pressure Injury Alliance, 2019). PI assessment should also include dimensions, tissue quality, quantity and quality of exudate, wounds edges, presence of tunneling or undermining, peri-wound condition, odor and signs of infections, or other complications (European Pressure Ulcer Advisory Panel, National Pressure Injury Advisory Panel and Pan Pacific Pressure Injury Alliance, 2019).

**TABLE 1** Pressure injury stages according to the National Pressure Injury Advisory Panel (NPIAP).

| Stage | Description |
|---|---|
| Stage 1 | Intact skin with a circumscribed zone of non-blanchable erythema |
| Stage 2 | Partial-thickness skin damage where the dermis is exposed |
| Stage 3 | Full-thickness skin damage where the fat is exposed, granulation tissue is present and wounds edges are rolled |
| Stage 4 | Full-thickness skin damage where fascia, muscle, tendon, ligament, cartilage, or bone is exposed |
| Unstageable | Full-thickness skin damage where a Stage 3 or Stage 4 is suspected but cannot be determined due to presence of necrotic tissues |
| Deep tissue pressure injury | Intact or damaged skin with a circumscribed non-blanchable purple or dark red zone, or presence of epidermal separation exposing a dark wound bed or blood-filled blister |

Pressure injury classification system, based on the extent of tissue involvement, developed by the National Pressure Injury Advisory Panel (NPIAP) (European Pressure Ulcer Advisory Panel, National Pressure Injury Advisory Panel and Pan Pacific Pressure Injury Alliance, 2019).

Diagnosis and Treatment of Spinal Cord Injury. https://doi.org/10.1016/B978-0-12-822498-4.00005-1

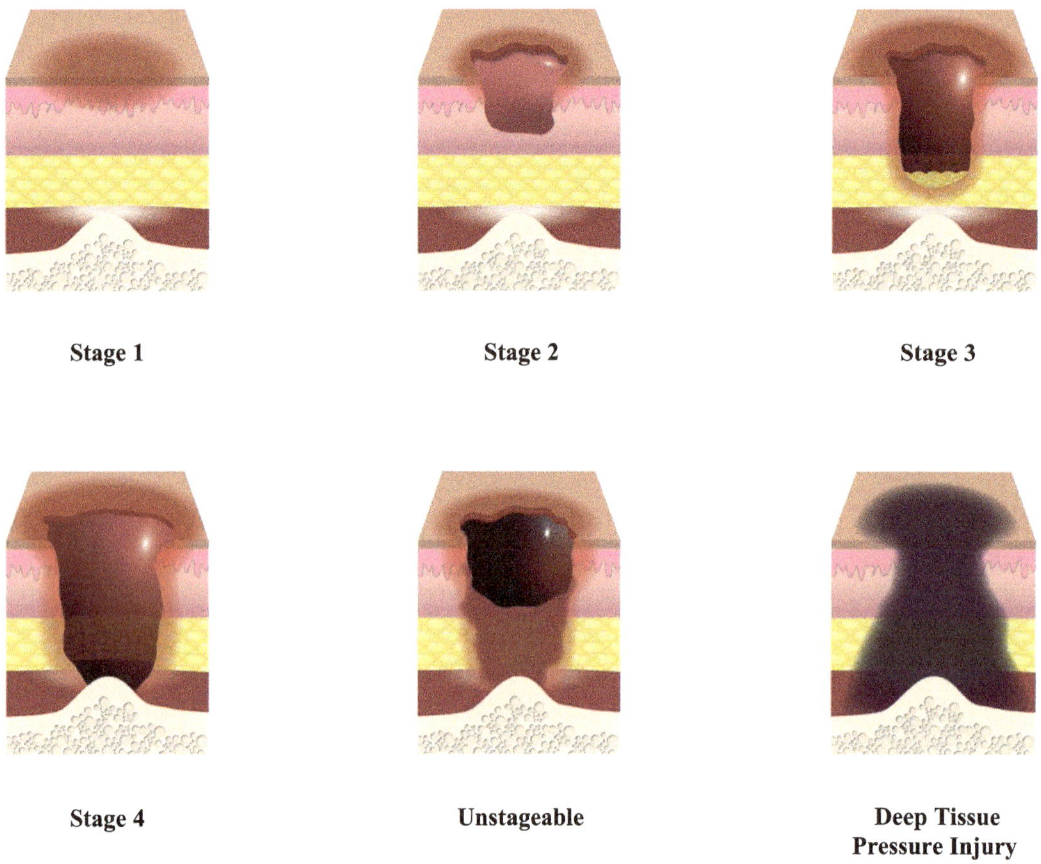


FIG. 1 Pressure injury stages. Schematic drawings of the different pressure injury stages (illustrated by Bréval Le Mestique).

Due to risk factors inherent to their condition, such as decreased sensation and mobility limitations, spinal cord injury (SCI) patients are at higher risk of developing a PI when compared to the general population (European Pressure Ulcer Advisory Panel, National Pressure Injury Advisory Panel and Pan Pacific Pressure Injury Alliance, 2019; Houghton & Campbell, 2013). Their occurrence can have significant physical, psychological, and functional consequences and may hinder independence, self-esteem, social well-being, and overall quality of life (Kisala, Tulsky, Choi, & Kirshblum, 2015; Street, Noonan, Cheung, Fisher, & Dvorak, 2015). Their prevention is therefore essential and should be an integral part of the medical and rehabilitation goals.

## Pathophysiology of pressure injuries

### Pressure and shear

Pressure and shear forces, especially over bony prominences or due to medical devices or other objects, are the principal factors associated with the development of PI. As a result of these forces, the loaded soft tissues will deform, causing strains and stresses which, if sustained, may lead to tissue damage (European Pressure Ulcer Advisory Panel, National Pressure Injury Advisory Panel and Pan Pacific Pressure Injury Alliance, 2019). Sustained soft tissue deformation leads to hypoxia, reduced lymphatic flow as well as reduced nutrient supply and removal of metabolic waste products which in turn causes cell death and tissue damage (Gawlitta, Li, et al., 2007; Gawlitta, Oomens, Bader, Baaijens, & Bouten, 2007; Gray, Worsley, Voegeli, & Bader, 2016). Pressure intensity and duration as well as tissue tolerance and vascular perfusion also play a significant role (European Pressure Ulcer Advisory Panel, National Pressure Injury Advisory Panel and Pan Pacific Pressure Injury Alliance, 2019).

Both low mechanical load for a long period and high load for a short period of time may result in tissue damage (Breuls, Bouten, Oomens, Bader, & Baaijens, 2003; Gawlitta, Li, et al., 2007; Gawlitta, Oomens, et al., 2007). To date, no universal safe tissue interface pressure threshold has been established, thus highlighting the importance of evaluating and addressing all possible PI risk factors as tissue interface pressure measurement alone is insufficient (Coleman et al., 2014; Groah, Schladen, Pineda, & Hsieh, 2015; Oomens, Loerakker, & Bader, 2010).

### Tissue tolerance

Different tissue types have varying degrees of tolerance to deformation with muscle tissues being more susceptible to damage than fat and skin (Salcido et al., 1994). The orientation of shear forces is also important as skin is more resistant to forces occurring in alignment with collagen fiber bundles than when they are applied perpendicular to fiber bundles (Destrade, Gilchrist, Prikazchikov, & Saccomandi, 2008).

In addition, soft tissue tolerance may be affected by many factors such as perfusion, micro-climate as well as patients and tissue characteristics (European Pressure Ulcer Advisory Panel, National Pressure Injury Advisory Panel and Pan Pacific Pressure Injury Alliance, 2019). For example, trauma victims have an increased risk of developing PI, with 45.8% of them developing within 48 h of admission (Ham et al., 2017). Because of the risk factors inherent to their condition, such as immobility, decreased autonomic control, hyper catabolic responses, and lack of protective response due to sensorimotor deficits (Houghton & Campbell, 2013; Michel et al., 2012), patients with traumatic spinal cord injury (TSCI) are almost 14 times more likely to develop a PI than other trauma patients (Grigorian et al., 2017).

## Prevalence, impact, and cost of pressure injuries in patients with SCI

Pressure injuries (PI) are one of the most common complications following SCI with up to 80% of patients developing a PI at some point during their lifetime (Garber, Rintala, Hart, & Fuhrer, 2000; Kirshblum & Lin, 2018). Compared to other rehabilitation phases, the acute hospitalization represents the period with the highest PI risk (Brienza, Krishnan, Karg, Sowa, & Allegretti, 2018) with a prevalence ranging from 2.7% to 57% when compared to 15% to 54% in the chronic SCI stage (Brienza et al., 2018; Gelis et al., 2009b; van der Wielen, Post, Lay, Glasche, & Scheel-Sailer, 2016). Factors such as severe neurological deficits, altered level of consciousness, and multiple concomitant traumatic injuries lead to prolonged periods of immobility and decreased general health status, putting patients at higher risk of PI during acute care (Ham et al., 2017).

### Most common locations and severity

During acute care and rehabilitation, when patients spend more time in a supine position, the most commonly affected areas include the ischium (28%), the sacrum (17%–27%), trochanters (12%–19%), and heels (9%–18%) (Kruger, Pires, Ngann, Sterling, & Rubayi, 2013). During the chronic stage, when patients spend more time in a sitting position, the ischium or perineum (48.3%), sacrum (37.2%), and trochanters (14.5%) are more susceptible to PI (Guihan et al., 2016).

Molano et al. have shown that in their cohort, the majority of PIs occurring during acute care were stage 2 and 3 (Molano Alvarez et al., 2004). For their part, Powers et al. have shown a predominance of stages 1 and 2 (Powers, Daniels, McGuire, & Hilbish, 2006). In a recent study, Ham et al. suggested that the number of stage 1 PI may be underestimated due to the absence of skin loss, making their identification sometimes more difficult and by the fact that they rapidly progress to more severe stages if not addressed (Ham et al., 2017).

### Impact

Patients with TSCI who develop a PI during acute hospitalization have higher complication rates, significantly longer length of stay, and higher risk of recurrence (Klotz, Joseph, Ravaud, Wiart, & Barat, 2002; Street et al., 2015; Verschueren et al., 2011). Complications of PI such as wound infection, cellulitis, septic arthritis, osteomyelitis, sepsis, and malignant transformation (Marjolin's ulcers) may also occur (Houghton & Campbell, 2013). In some cases, PI can lead to surgical procedures, amputation, and sometimes even death due to an infection (Houghton & Campbell, 2013). Indeed, 7%–8% of patients with TSCI who develop a PI die from associated complications (Richards, Waites, Chen, Kogos, & Schmitt, 2004). PI also limit long-term functional outcome by interfering with rehabilitation (Hastings, Ntsiea, & Olorunju, 2015; Post, Dallmeijer, Angenot, van Asbeck, & van der Woude, 2005). Following the acute hospitalization, patients are at higher risks of developing a PI during the first year post-SCI or more than 25 years post-injury (National Spinal Cord Injury Statistical Center, 2019). After urinary tract infections, PI represent the second most common cause of re-hospitalization during the first year post-TSCI as well as in the following years, accounting for 11.3% of re-hospitalizations at 1 year, 14.6% at 5 years, 17.5% at 10 years, and 21.3% at 20 years post injury (National Spinal Cord Injury Statistical Center, 2019).

### Cost

It has been estimated that one-quarter of the cost of care for patients with SCI is associated with PI and thus represent a significant preventable financial burden. In 2013, in Canada, the occurrence of PI resulted in an additional $18,758 to the cost of acute hospitalization following TSCI (White et al., 2017). In the United Kingdom, PI treatment ranges from $2000 to $18,000 (Bennett, Dealey, & Posnett, 2004) while in the United States, annual health care costs per patient was estimated to be $73,021 higher for patients with PI when compared to patients without PI (Stroupe et al., 2011). Finally, a recent systematic review showed that interventions to prevent PI occurrence were considerably less expensive than the interventions for their treatment (Demarré et al., 2015).

## Structural and physiological changes following SCI

Following SCI, multiple structural and physiological changes occur, decreasing the body's ability to maintain skin integrity, increasing the risk of tissue breakdown, and thus putting patients at higher risk of developing PI (Table 2).

### Skin

Thinning of skin at load-bearing sites, decrease in skin distensibility in the remaining skin tissues, decrease in fibroblast activity, and increase in collagen catabolism are seen in patients with SCI with greater changes in skin distensibility,

**TABLE 2 Structural and physiological changes following SCI.**

| | |
|---|---|
| Skin changes | – Skin thinning at load-bearing sites<br>– Decrease in skin distensibility in the remaining skin tissues<br>– Decreased fibroblast activity<br>– Greater collagen catabolism<br>– Decrease in type I to type III collagen ratio in skin below injury level |
| Disuse-induced muscle atrophy | – Fibers thinning<br>– Decrease in numbers of slow-twitch fibers and increase in fast-twitch fibers below the level of injury<br>– Increase in intra-muscular fat causing greater intra-muscular shear stresses at interfaces between muscle and intra-muscular fat tissues |
| Bone changes | – Flattening of ischial tuberosities due to chronic exposure to sitting mechanical loading coupled with cortical bone mass loss, shifting load transfer from weight-bearing ischial tuberosities to overlying soft tissues |
| Changes in macro- and micro-vasculature | – Vasodilation and impairment in tissue perfusion below injury level<br>– Decrease in the density of both alpha- and beta-adrenergic receptors in the skin below injury level resulting in abnormal vascular response<br>– Decrease in number and size of capillaries feeding skeletal muscle fibers<br>– Decrease in lower limbs cutaneous blood when sitting |
| Chronic inflammation and immune function | – Decrease healing response by depression of immune system function<br>– Chronic tissue inflammation decreases tissue tolerance and repair capacity further increasing tissue breakdown risk |
| Temperature dysregulation | – Increase in resting skin temperature and decrease in skin temperature reactivity in lower limbs |
| Sensory impairments | – Sensory deficits resulting in lack of protective response allowing the progression from ischemia to tissue breakdown and necrosis |

Summary of the multiple structural and physiological changes occurring following SCI predisposing to pressure injuries (Carda, Cisari, & Invernizzi, 2013; Cotie, Geurts, Adams, & MacDonald, 2011; Cruse et al., 2000; Deitrick, Charalel, Bauman, & Tuckman, 2007; European Pressure Ulcer Advisory Panel, National Pressure Injury Advisory Panel and Pan Pacific Pressure Injury Alliance, 2019; Gefen, 2018; Gorgey & Dudley, 2007; Kirshblum & Lin, 2018; Kruger et al., 2013; Lachenbruch, Tzen, Brienza, Karg, & Lachenbruch, 2013; Linder-Ganz et al., 2008; Michel et al., 2012; Park, Seo, Han, & Lee, 2011; Patterson, Cranmer, Fisher, & Engel, 1993; Rappl, 2008; Rodriguez, Claus-Walker, Kent, & Stal, 1986; Rodriguez & Markowski, 1995; Ruschkewitz & Gefen, 2011; Scelsi, 2001; Schwab, Zhang, Kopp, Brommer, & Popovich, 2014; Sopher, Nixon, Gorecki, & Gefen, 2011; Thorfinn, Sjöberg, & Lidman, 2002; Wilczweski et al., 2012; Wu & Bogie, 2013; Yalcin, Akyuz, Onder, Unalan, & Degirmenci, 2013; Zeevi, Levy, Brauner, & Gefen, 2018).

elasticity, and viscoelasticity observed in patients with longer time elapsed since SCI (Park et al., 2011; Yalcin et al., 2013). A decrease in type I to type III collagen ratio is also seen in the skin below the level of injury which may further contribute to skin fragility following SCI by decreasing tensile strength (Rodriguez & Markowski, 1995).

## Muscle and bone

In addition to normally occurring age-related muscle atrophy (Roubenoff, 1999), patients with SCI undergo significant disuse-induced muscle atrophy below the level of injury (Gorgey & Dudley, 2007) with greater atrophy seen at the level of ischial tuberosities (Wu & Bogie, 2013). Muscle atrophy decreases natural protective cushioning over bony prominences, increasing PI risk (Linder-Ganz & Gefen, 2009). Thinning of the fibers, decrease in the numbers of slow-twitch fibers, and increase in fast-twitch fibers may be seen as early as 4–6 weeks following TSCI (Carda et al., 2013). An increase in intra-muscular fat is also observed (Gorgey & Dudley, 2007), causing greater intra-muscular shear stresses at interfaces between muscle and intra-muscular fat tissues (Sopher et al., 2011; Wu & Bogie, 2013), thus further increasing PI risk. In addition, due to chronic exposure to sitting mechanical loading coupled with cortical bone mass loss, ischial tuberosities tend to flatten, shifting load transfer from weight-bearing ischial tuberosities to overlying soft tissues (Linder-Ganz et al., 2008).

## Macro- and micro-vasculature

Changes in macro- and micro-vasculature also occur partly due to alteration in autonomic control. First, a loss of vascular tone is seen below the lesion resulting in vasodilation and thus decreasing vascular resistance and impairing tissue perfusion (Rappl, 2008). Also, a decrease in the density of both alpha- and beta-adrenergic receptors is observed in the skin below the level of injury in patients with longer-standing injuries resulting in abnormal vascular response (Rodriguez et al., 1986). When compared to patients without SCI, lower limbs cutaneous blood flow has been shown to decrease by at least 50% in patients with SCI when sitting (Deitrick et al., 2007). A decrease in number and size of capillaries feeding skeletal muscle fibers is also seen, making these muscles more susceptible to ischemia, especially during episodes of low blood pressure (Ruschkewitz & Gefen, 2011; Scelsi, 2001). Moreover, when compared to healthy controls, patients with SCI tend to have significantly higher sitting pressure and significantly lower transcutaneous oxygen tension levels, followed by slower reactive hyperemia when unloaded, translating into a lower rate of reperfusion (Patterson et al., 1993; Thorfinn et al., 2002). This therefore suggests a longer recovery time is required for patients with SCI following loading periods to allow appropriate tissue re-oxygenation before reloading the area.

## Inflammation and immune function

A chronic systemic inflammatory response occurs following trauma and may predispose patients with SCI to PI and decrease healing response by depressing immune system function (Cruse et al., 2000; Schwab et al., 2014). Compromise in tissue perfusion paired with chronic tissue inflammation decrease tissue tolerance and repair capacity further increasing tissue breakdown risk (Gefen, 2018).

## Temperature

Due to loss of autonomic control, temperature dysregulation is also common in patients with SCI. Cotie et al. have shown that an increase in resting skin temperature and a reduced skin temperature reactivity are seen in the lower limbs of patients with SCI (Cotie et al., 2011). In addition, urinary and/or fecal incontinence is common in patients with SCI further increasing the risk of moisture and maceration (Kirshblum & Lin, 2018; Wilczweski et al., 2012). An increase in body temperature, especially when coupled with excessive humidity and maceration, has a profound effect on tissue tolerance to damage and thus must not be overlooked especially when addressing support surfaces and medical devices (Lachenbruch et al., 2013; Wilczweski et al., 2012; Zeevi et al., 2018).

## Sensory impairments

Finally, sensory deficits associated with SCI results in a lack of protective response due to decreased awareness of tissue injury thus allowing the progression from ischemia to tissue breakdown and necrosis (European Pressure Ulcer Advisory

Panel, National Pressure Injury Advisory Panel and Pan Pacific Pressure Injury Alliance, 2019; Kruger et al., 2013; Michel et al., 2012).

## Risk assessment tools

The purpose of a risk assessment tool is to identify and stratify individuals who are more susceptible to PI occurrence in order to better evaluate and address risk factors and the adequacy of preventive measures. Although the SCIPUS and SCIPUS-A scales, which were specifically developed to assess patients with SCI during acute hospitalization and rehabilitation, show promise, they haven't demonstrated acceptable accuracy and, thus at this point, cannot be recommended (Mortenson & Miller, 2008). The two most common scales used to assess PI risk are the Braden Scale and the Norton Scale. The Braden Scale measures six domains: sensory perception, moisture, activity, mobility, nutrition, and friction/shear. Each domain is scored on a scale of 1–4 except for friction/shear which is score from 1 to 3. The total score thus ranges from 6 to 23, with lower score meaning greater risk for skin breakdown. Preventative measures should be put in place in patients with a score of 18 or less. The Norton Scale measures five domains: physical condition, mental condition, activity, mobility, and continence. Each domain is scored on a scale of 1–4 which are then summed to give a total score ranging from 5 (worst prognosis) to 20 (best prognosis). Similarly, preventive measures should be put in place in patients with a score of 16 or less (Mortenson & Miller, 2008).

## Risk factors associated with pressure injuries in patients with traumatic spinal cord injury

As discussed, susceptibility, severity, and extent of PI depend on type, magnitude, and duration of mechanical load, but also on the mechanical properties of the tissue, tissue and bone morphology, tissue repair capacity as well as transport and thermal properties of tissues (Coleman et al., 2014). In addition to the structural and physiological changes occurring following an SCI described in the previous section, many other risk factors have been shown to influence key components of PI development thus making patients with TSCI even more vulnerable to PI. These may be classified as non-modifiable factors if they cannot be addressed and optimized, and modifiable factors if they can be reduced or changed (Table 3).

### Non-modifiable factors

#### *Patient characteristics*

Whether older age (Brienza et al., 2018; Di Prinzio et al., 2019; Gelis et al., 2009a, 2009b; Ham et al., 2017; Marin et al., 2013; Wilczweski et al., 2012) and sex (Ash, 2002; Di Prinzio et al., 2019; Gelis et al., 2009a, 2009b; Sheerin et al., 2005; Wilczweski et al., 2012) are significant risk factors to PI development is still debated in the literature. However, it has been shown that 10–20 years post injury, SCI accelerates the physical and functional decline associated with aging which may further increase PI risk (Menter & Hudson, 1995). Although some studies have shown that race may be associated with PI occurrence (Chen et al., 2005; Guihan et al., 2008; Saunders et al., 2010), a recent study found that when controlled for socioeconomic status and healthcare access, race becomes a non-significant risk factor (Saunders et al., 2012). Indeed, patients with SCI with lower household income ( Jorge et al., 2018; Saunders et al., 2010; Saunders et al., 2012), who are unemployed (Klotz et al., 2002; Krause et al., 2001; Krause & Broderick, 2004), who have lower educational levels (Chen et al., 2005; Klotz et al., 2002; Krause et al., 2001), and who are single (Chen et al., 2005; Gelis et al., 2009b; Krause et al., 2001) tend to be at higher risk of developing a PI.

#### *Injury characteristics and associated conditions*

The completeness of the TSCI (according to the International Standards for Neurological Classification of Spinal Cord Injury (ISNCSCI) (Kirshblum et al., 2011)) is significantly associated with greater PI risk (Brienza et al., 2018; Di Prinzio et al., 2019; Gelis et al., 2009a, 2009b; Gour-Provencal et al., 2020; Grigorian et al., 2017; Joseph & Nilsson Wikmar, 2016; Li et al., 2017; Marin et al., 2013; Marion et al., 2017; Verschueren et al., 2011). Higher levels of neurological injury may also be a higher risk of PI however this is still debated in the literature (Ash, 2002; Chen et al., 2005; Di Prinzio et al., 2019; Gelis et al., 2009b; Gour-Provencal et al., 2020; Grigorian et al., 2017; Krause & Broderick, 2004; Marin et al., 2013; Marion et al., 2017; Sheerin et al., 2005; Verschueren et al., 2011). Higher levels and complete injuries are associated with greater sensorimotor deficits, decreased autonomic control impairing micro-vascular response, greater risk of developing nutritional deficiencies, and more severe mobility limitations, all of which predispose to PI development

**TABLE 3 Risk factors associated with pressure injuries in patients' traumatic spinal cord injury.**

| Non-modifiable risk factors | Modifiable risk factors |
|---|---|
| Older age (still debated) | Obesity or being underweight |
| Lower household income<br>Unemployment<br>Lower educational levels<br>Single | Smoking and alcohol abuse (still debated)<br>Substance abuse |
| Greater severity of injury (motor complete vs incomplete) | Dysthymic symptoms<br>Schizophrenia<br>Personality disorders<br>Neurocognitive changes<br>Delirium |
| Higher neurological level of injury (tetraplegia vs paraplegia) (still debated) | Medication use for pain, insomnia, spasticity, or stress |
| Poorer functional status | Longer admission delay between trauma site and the emergency department<br>Longer time spent with immobilization device (>6 h)<br>Longer surgical duration (>6 h) |
| Greater trauma severity and presence of concomitant traumatic injuries | Longer acute care length of stay |
| Hypotension, hypovolemic, or hemorrhagic shock, with prolonged periods of mean arterial pressure of <70 mmHg | Care management in a non-SCI-specialized center<br>Surgical delay of >24 h |
| Mechanical ventilation | Medical complications |
| Patients with longer-standing TSCI, >25 years post SCI | Hypoalbuminemia<br>Malnutrition<br>Anemia (still debated) |
| Previous history of PI (stage 3 or 4 or requiring surgery) | Immobility in the frail<br>Poor bed and wheelchair positioning |
| Presence of comorbidities (diabetes, metabolic syndrome, cardiovascular disease, peripheral vascular disease, pulmonary diseases, and renal disease) | No access to appropriate support surfaces<br>Inappropriate turning in bed schedule<br>Poor compliance<br>Little or non-knowledge of PI prevention strategies |

Summary of the non-modifiable and modifiable risk factors associated with pressure injuries in patients traumatic spinal cord injury (Ash, 2002; Bourassa-Moreau, Mac-Thiong, Feldman, Thompson, & Parent, 2013; Brienza et al., 2018; Chen, Devivo, & Jackson, 2005; DeJong et al., 2014; Di Prinzio et al., 2019; European Pressure Ulcer Advisory Panel, National Pressure Injury Advisory Panel and Pan Pacific Pressure Injury Alliance, 2019; Garber et al., 2000; Gater, 2007; Gelis et al., 2009a, 2009b; Gour-Provencal, Mac-Thiong, Feldman, Bégin, & Richard-Denis, 2020; Grigorian et al., 2017; Groah et al., 2015; Guihan et al., 2008, 2016; Ham et al., 2017; Ham, Schoonhoven, Schuurmans, & Leenen, 2014; Hatchett, Mulroy, Eberly, Haubert, & Requejo, 2016; Houghton & Campbell, 2013; Jorge, White, & Agarwal, 2018; Joseph & Nilsson Wikmar, 2016; Klotz et al., 2002; Krause & Broderick, 2004; Krause, Vines, Farley, Sniezek, & Coker, 2001; Krishnan et al., 2017; Le Fort, Espagnacq, Perrouin-Verbe, & Ravaud, 2017; Li, DiPiro, Cao, Szlachcic, & Krause, 2016; Li, DiPiro, & Krause, 2017; Marin, Nixon, & Gorecki, 2013; Marion et al., 2017; Mathew, Samuelkamaleshkumar, Radhika, & Elango, 2013; Michel et al., 2012; Morita, Yamada, Watanabe, & Nagahori, 2015; Parent, Barchi, LeBreton, Casha, & Fehlings, 2011; Ploumis et al., 2011; Rabadi & Vincent, 2011; Richard-Denis, Thompson, Bourassa-Moreau, Parent, & Mac-Thiong, 2016; Saunders & Krause, 2010; Saunders, Krause, & Acuna, 2012; Saunders, Krause, Peters, & Reed, 2010; Scelsi, 2001; Sheerin, Gillick, & Doyle, 2005; Smith, Guihan, LaVela, & Garber, 2008; Tate, Forchheimer, Krause, Meade, & Bombardier, 2004; van der Wielen et al., 2016; Verschueren et al., 2011; Wilczweski et al., 2012).

(Brienza et al., 2018; Ham et al., 2017; Kruger et al., 2013; Mathew et al., 2013; Michel et al., 2012). Similarly, patients with SCI with poorer functional status tend to be more susceptible to PI (Garber et al., 2000; Mathew et al., 2013; Saunders et al., 2010; Verschueren et al., 2011). According to the National Spinal Cord Injury Statistical Center patients with complete tetraplegia have the highest risk of developing a PI during acute care and rehabilitation, followed by patients with complete paraplegia, incomplete tetraplegia and incomplete paraplegia (National Spinal Cord Injury Statistical Center, 2006).

Greater trauma severity (Ham et al., 2014, 2017) as well as the presence of concomitant traumatic injuries such as fractures, organ trauma, and traumatic brain injury increase the risk of developing a PI, as they are generally associated with lower tissue perfusion pressure, longer surgical times, and a more precarious health status (Ash, 2002; European Pressure Ulcer Advisory Panel, National Pressure Injury Advisory Panel and Pan Pacific Pressure Injury Alliance, 2019; Ham et al., 2014, 2017). Hypotension, hypovolemic or hemorrhagic shock, with prolonged periods of mean arterial pressure of less than 70 mmHg during hospitalization is also associated with PI onset (Gelis et al., 2009a; Wilczweski et al., 2012). PI incidence significantly increases in patients requiring mechanical ventilation (Brienza et al., 2018; Ham et al., 2014).

Patients with longer-standing TSCI, especially those more than 25 years post injury, have a higher risk of PI occurrence (Gelis et al., 2009b; van der Wielen et al., 2016) and a greater susceptibility to PI management requiring recurrent hospital admissions (Goodman, Schindler, Washington, Bogie, & Ho, 2014). Similarly, patients with SCI with a previous history of PI, especially those with stage 3 or 4 or those requiring surgery, are at higher risk of PI recurrence (DeJong et al., 2014; Le Fort et al., 2017; Verschueren et al., 2011).

### *Comorbidities*

The presence of comorbidities such as diabetes, metabolic syndrome, cardiovascular disease, peripheral vascular disease, pulmonary diseases, and renal disease has also been associated with PI onset (Ash, 2002; Di Prinzio et al., 2019; Gelis et al., 2009b; Guihan et al., 2008, 2016; Houghton & Campbell, 2013; Li et al., 2016; Marin et al., 2013; Michel et al., 2012).

## Modifiable factors

### *Lifestyle habits, medication, and psychological factors*

Some patients with SCI are at higher risk of obesity (Gater, 2007; Hatchett et al., 2016; Scelsi, 2001), resulting in greater compressive forces on weight-bearing sites, increasing PI susceptibility (Elsner & Gefen, 2008; Sopher, Nixon, Gorecki, & Gefen, 2010). Patients who are underweight are also at higher risk of PI (Krause et al., 2001).

Although smoking has been shown to have adverse effects on PI healing (Lane, Selleck, Chen, & Tang, 2016) by decreasing arterial oxygen tension, it is still debated in the literature whether it is a significant risk factor for PI occurrence (Di Prinzio et al., 2019; Gelis et al., 2009a, 2009b; Krause et al., 2001; Li et al., 2016, 2017; Marin et al., 2013) similar to alcohol abuse (Krause et al., 2001; Li et al., 2016, 2017; Tate et al., 2004). Substance abuse may increase the risk of self-negligent behaviors and thus has been observed to be a significant PI risk factor (Tate et al., 2004).

Patients with SCI and dysthymic symptoms, schizophrenia, personality disorders, neurocognitive changes, or delirium are all at higher risk of PI development (Di Prinzio et al., 2019; Gelis et al., 2009b; Krause & Broderick, 2004; Smith et al., 2008). The use of prescription medication for pain, insomnia, spasticity or stress may also increase PI susceptibility (Di Prinzio et al., 2019; Krause & Broderick, 2004; Li et al., 2017; Saunders & Krause, 2010).

### *Patient management and care*

Admission delay between trauma site and the emergency department, longer time spent with an immobilization device (backboard, cervical collar) as well as longer duration of surgery have all been shown to drastically increase PI occurrence, especially after 6 h (European Pressure Ulcer Advisory Panel, National Pressure Injury Advisory Panel and Pan Pacific Pressure Injury Alliance, 2019; Ham et al., 2014; Parent et al., 2011). These factors lead to longer periods of immobility, decreasing tissue oxygenation over bony prominences and increasing temperature and humidity under the devices, thereby increasing PI susceptibility (Ham et al., 2014). It has also been suggested that patients who wear immobilization devices for longer periods of time tend to be patients with more severe injuries further increasing PI risk (Ham et al., 2014).

Receiving care in a SCI-specialized center and a surgical delay of less than 24 h are shown to be protective factors (Bourassa-Moreau et al., 2013; Parent et al., 2011; Ploumis et al., 2011; Richard-Denis et al., 2016). This can be explained by the fact that the SCI-specialized centers comprise qualified nurses and therapists, who better understand the specifics of the overall management of patients with SCI (Parent et al., 2011). Shorter acute care length of stay has also been associated with fewer PI (Ploumis et al., 2011; van der Wielen et al., 2016).

The occurrence of medical complications such as urinary tract infections and pneumonia also increases PI risk (Brienza et al., 2018; Gour-Provencal et al., 2020; Joseph & Nilsson Wikmar, 2016; Krishnan, Vodovotz, et al., 2017) as they cause decreased tissue oxygenation, decreased mobility, and an impaired inflammatory response, thereby compromising the patient's immunity and further predisposing them to PI (Brienza et al., 2018; Krishnan, Karg, Boninger, & Brienza, 2017; Krishnan, Vodovotz, et al., 2017). Hypoalbuminemia and malnutrition have also been associated with higher PI risk while low hemoglobin is still debated (Di Prinzio et al., 2019; Marin et al., 2013; Rabadi & Vincent, 2011).

### *Prevention strategies and patient compliance*

Immobility in the frail, poor bed and wheelchair positioning and repositioning have been shown to increase PI risk (Groah et al., 2015). Access to appropriate support surface (Morita et al., 2015), turning in bed schedule, knowledge, and compliance to PI prevention strategies may be protective factors (Garber et al., 2000; Mathew et al., 2013; Morita et al., 2015).

## Conclusion

Despite medical advances and improvement in knowledge through the years, pressure injuries remain a serious challenge throughout the acute hospitalization, rehabilitation, and community reintegration stages for both patients with TSCI and medical professionals. The repercussions are significant on the physical, psychological, functional, and financial states, and thus their prevention is crucial. Better identifying and addressing factors contributing to their occurrence is the first step toward improving patient outcome and quality of life. Although various factors have been identified as potential PI risk factors, no single component can explain PI occurrence on its own but rather a complex interaction of different factors, therefore highlighting the importance of a comprehensive management and care plan.

## Applications to other areas of neuroscience

As reviewed in this chapter, due to risk factors inherent to their condition, patients with traumatic spinal cord injury (TSCI) are at higher risk of developing a pressure injury (PI). A recent retrospective analysis using data from the National Trauma Data Bank showed that, when adjusted for covariates, spinal cord injury was the strongest predictor of PI occurrence in adult trauma patients (OR 13.77, 95% CI 13.25–14.31, $P < 0.001$) (Grigorian et al., 2017). On univariable analysis, the odds (OR (95% CI)) of developing a PI in patients admitted between 2007 and 2015 was 1.77 (1.73–1.81) in patients with traumatic brain injury, 2.48 (2.34–2.62) in patients with cerebrovascular accident and 13.73 (13.22–14.25) in patients with SCI (Grigorian et al., 2017). Thus, similar to patients with SCI, individuals with traumatic brain injury, stroke, neurodegenerative diseases (such as dementia, Alzheimer's disease, and Parkinson's disease) or any other neurological conditions resulting in altered level of sensation, mobility, or mental status also are at increased risk of developing a PI ( Jaul, Barron, Rosenzweig, & Menczel, 2018). Furthermore, the combination of neurological impairments with associated complicating conditions such as polypharmacy, malnutrition, and anemia further increases PI susceptibility ( Jaul et al., 2018). Fife et al. showed that patients who are admitted to the neurologic intensive care unit are at higher risk of developing a PI especially if they have Braden scores of $\leq 16$ and are underweight (Fife et al., 2001). These results may suggest that the factors contributing to PI in patients with TSCI may also be applicable to other neurological conditions and thus shouldn't be overlooked.

## Mini-dictionary of terms

**Atrophy:** wasting or loss of muscle tissue.
**Catabolism:** metabolic pathways resulting in the breakdown of complex molecules into smaller particles.
**Debridement:** removal of dead or unhealthy tissues in a wound.
**Ischemia:** decrease blood flow to tissues causing reduction in oxygen and nutrient supplies.
**Maceration:** prolonged exposure to moisture resulting in softening and breaking down of skin.
**Necrosis:** premature death of cells and tissues.
**Pressure:** force vector that is perpendicular to the skin.
**Shear:** force vector that is tangential to the tissue contact surface.
**Tunneling:** channels extending from the initial injury site to subcutaneous tissues.
**Undermining:** extension of the injury in subcutaneous tissues under the wound borders.

## Key facts of pressure injuries

- A pressure injury (PI) is defined as a circumscribed injury to the skin and/or underlying tissue due to pressure, or shear combined with pressure.
- Loaded soft tissues deform, causing strains and stresses which, if sustained, may lead to tissue damage.

- Muscles are more susceptible to damage than fat and skin as they have lower tolerance to deformation.
- Patients with sensory deficits are more prone to PI due to their lack of protective responses caused by decreased awareness of tissue injury.
- Complications of PI include wound infection, cellulitis, septic arthritis, osteomyelitis, sepsis, malignant transformation (Marjolin's ulcers), surgical procedures, amputation, and death from infection.

## Summary points

- Susceptibility and extent of pressure injuries (PI) depend on type, magnitude, and duration of mechanical load, the mechanical properties of the tissue, tissue and bone morphology, tissue repair capacity as well as circulation and thermal properties of tissues.
- To date, no universal safe tissue interface pressure threshold has been established.
- Following spinal cord injury (SCI), multiple structural and physiological changes occur, decreasing the body's ability to maintain skin integrity, increasing the risk of tissue breakdown and thus increasing the risk of developing PI.
- Patients are at higher risks of developing a PI during the first year post-SCI or more than 25 years post injury.
- Patients with traumatic spinal cord injury (TSCI) who develop a PI during acute hospitalization have higher complication rates, significantly longer length of stay, higher risk of recurrence, and poorer long-term functional outcome.
- Non-modifiable risk factors include more severe TSCI, poorer functional outcome, greater trauma severity and the presence of concomitant traumatic injuries, hypotension, mechanical ventilation, longer time elapsed since TSCI, previous history of PI, lower household income, unemployment, being single, lower educational levels, and presence of comorbidities.
- Modifiable risk factors include obesity, substance abuse, dysthymic symptoms, schizophrenia, personality disorders, neurocognitive changes, delirium, medication use for pain, insomnia, spasticity or stress, longer admission delay from trauma site to emergency department, longer time spent with immobilization device, longer surgical duration, longer acute care length of stay, care management in a non-SCI-specialized center, surgical delay of >24 h, medical complications, hypoalbuminemia, malnutrition, immobility in the frail, poor bed and wheelchair positioning, no access to appropriate support surfaces, poor compliance, and little or non-knowledge of PI prevention strategies.
- Although various factors have been identified as potential PI risk factors, no single component can explain PI occurrence on its own but rather a complex interaction of different factors, therefore highlighting the importance of a comprehensive management and care plan.

## References


Ash, D. (2002). An exploration of the occurrence of pressure ulcers in a British spinal injuries unit. *Journal of Clinical Nursing*, *11*(4), 470–478.

Bennett, G., Dealey, C., & Posnett, J. (2004). The cost of pressure ulcers in the UK. *Age and Ageing*, *33*(3), 230–235. https://doi.org/10.1093/ageing/afh086.

Bourassa-Moreau, E., Mac-Thiong, J. M., Feldman, D. E., Thompson, C., & Parent, S. (2013). Non-neurological outcomes after complete traumatic spinal cord injury: The impact of surgical timing. *Journal of Neurotrauma*, *30*(18), 1596–1601. https://doi.org/10.1089/neu.2013.2957.

Breuls, R. G., Bouten, C. V., Oomens, C. W., Bader, D. L., & Baaijens, F. P. (2003). Compression induced cell damage in engineered muscle tissue: An in vitro model to study pressure ulcer aetiology. *Annals of Biomedical Engineering*, *31*(11), 1357–1364. https://doi.org/10.1114/1.1624602.

Brienza, D., Krishnan, S., Karg, P., Sowa, G., & Allegretti, A. L. (2018). Predictors of pressure ulcer incidence following traumatic spinal cord injury: A secondary analysis of a prospective longitudinal study. *Spinal Cord*, *56*(1), 28–34. https://doi.org/10.1038/sc.2017.96.

Carda, S., Cisari, C., & Invernizzi, M. (2013). Sarcopenia or muscle modifications in neurologic diseases: A lexical or patophysiological difference? *European Journal of Physical and Rehabilitation Medicine*, *49*(1), 119–130.

Chen, Y., Devivo, M. J., & Jackson, A. B. (2005). Pressure ulcer prevalence in people with spinal cord injury: Age-period-duration effects. *Archives of Physical Medicine and Rehabilitation*, *86*(6), 1208–1213. https://doi.org/10.1016/j.apmr.2004.12.023.

Coleman, S., Nixon, J., Keen, J., Wilson, L., McGinnis, E., Dealey, C., … Nelson, E. A. (2014). A new pressure ulcer conceptual framework. *Journal of Advanced Nursing*, *70*(10), 2222–2234. https://doi.org/10.1111/jan.12405.

Cotie, L. M., Geurts, C. L., Adams, M. M., & MacDonald, M. J. (2011). Leg skin temperature with body-weight-supported treadmill and tilt-table standing training after spinal cord injury. *Spinal Cord*, *49*(1), 149–153. https://doi.org/10.1038/sc.2010.52.

Cruse, J. M., Lewis, R. E., Dilioglou, S., Roe, D. L., Wallace, W. F., & Chen, R. S. (2000). Review of immune function, healing of pressure ulcers, and nutritional status in patients with spinal cord injury. *Journal of Spinal Cord Medicine*, *23*(2), 129–135.

Deitrick, G., Charalel, J., Bauman, W., & Tuckman, J. (2007). Reduced arterial circulation to the legs in spinal cord injury as a cause of skin breakdown lesions. *Angiology*, *58*(2), 175–184. https://doi.org/10.1177/0003319707300353.

DeJong, G., Hsieh, C. H., Brown, P., Smout, R. J., Horn, S. D., Ballard, P., & Bouchard, T. (2014). Factors associated with pressure ulcer risk in spinal cord injury rehabilitation. *American Journal of Physical Medicine & Rehabilitation*, *93*(11), 971–986. https://doi.org/10.1097/phm.0000000000000117.

Demarré, L., Van Lancker, A., Van Hecke, A., Verhaeghe, S., Grypdonck, M., Lemey, J., … Beeckman, D. (2015). The cost of prevention and treatment of pressure ulcers: A systematic review. *International Journal of Nursing Studies*, *52*(11), 1754–1774. https://doi.org/10.1016/j.ijnurstu.2015.06.006.

Destrade, M., Gilchrist, M. D., Prikazchikov, D. A., & Saccomandi, G. (2008). Surface instability of sheared soft tissues. *Journal of Biomechanical Engineering*, *130*(6). https://doi.org/10.1115/1.2979869, 061007.

Di Prinzio, M. F., Argento, F. J., Barbalaco, L., Cazurro, M. E., Pereyra, A., Sanchez Correa, C., … Annaratone, A. (2019). Risk factors for the development and recurrence of pressure ulcers in patients with spinal cord injury: A systematic review. *Revista de la Facultad de Ciencias Médicas de la Universidad Nac. de Córdoba*, *76*(4), 242–256. https://doi.org/10.31053/1853.0605.v76.n4.24906.

Elsner, J. J., & Gefen, A. (2008). Is obesity a risk factor for deep tissue injury in patients with spinal cord injury? *Journal of Biomechanics*, *41*(16), 3322–3331. https://doi.org/10.1016/j.jbiomech.2008.09.036.

European Pressure Ulcer Advisory Panel, National Pressure Injury Advisory Panel and Pan Pacific Pressure Injury Alliance. (2019). Prevention and treatment of pressure ulcers/injuries: Clinical practice guideline. In E. Haesler (Ed.), *The international guideline* EPUAP/NPIAP/PPPIA.

Fife, C., Otto, G., Capsuto, E. G., Brandt, K., Lyssy, K., Murphy, K., & Short, C. (2001). Incidence of pressure ulcers in a neurologic intensive care unit. *Critical Care Medicine*, *29*(2), 283–290.

Garber, S. L., Rintala, D. H., Hart, K. A., & Fuhrer, M. J. (2000). Pressure ulcer risk in spinal cord injury: Predictors of ulcer status over 3 years. *Archives of Physical Medicine and Rehabilitation*, *81*(4), 465–471. https://doi.org/10.1053/mr.2000.3889.

Gater, D. R., Jr. (2007). Obesity after spinal cord injury. *Physical Medicine and Rehabilitation Clinics of North America*, *18*(2), 333–351. vii https://doi.org/10.1016/j.pmr.2007.03.004.

Gawlitta, D., Li, W., Oomens, C. W., Baaijens, F. P., Bader, D. L., & Bouten, C. V. (2007). The relative contributions of compression and hypoxia to development of muscle tissue damage: An in vitro study. *Annals of Biomedical Engineering*, *35*(2), 273–284. https://doi.org/10.1007/s10439-006-9222-5.

Gawlitta, D., Oomens, C. W., Bader, D. L., Baaijens, F. P., & Bouten, C. V. (2007). Temporal differences in the influence of ischemic factors and deformation on the metabolism of engineered skeletal muscle. *Journal of Applied Physiology (Bethesda, MD: 1985)*, *103*(2), 464–473. https://doi.org/10.1152/japplphysiol.01374.2006.

Gefen, A. (2018). The future of pressure ulcer prevention is here: Detecting and targeting inflammation early. *EWMA Journal*, *19*(2), 7–13.

Gelis, A., Dupeyron, A., Legros, P., Benaim, C., Pelissier, J., & Fattal, C. (2009a). Pressure ulcer risk factors in persons with SCI: Part I: Acute and rehabilitation stages. *Spinal Cord*, *47*(2), 99–107. https://doi.org/10.1038/sc.2008.107.

Gelis, A., Dupeyron, A., Legros, P., Benaim, C., Pelissier, J., & Fattal, C. (2009b). Pressure ulcer risk factors in persons with spinal cord injury part 2: The chronic stage. *Spinal Cord*, *47*(9), 651–661. https://doi.org/10.1038/sc.2009.32.

Goodman, B. L., Schindler, A., Washington, M., Bogie, K. M., & Ho, C. H. (2014). Factors in rehospitalisation for severe pressure ulcer care in spinal cord injury/disorders. *Journal of Wound Care*, *23*(4), 165–166. 168, 170–162 passim 10.12968/jowc.2014.23.4.165.

Gorgey, A. S., & Dudley, G. A. (2007). Skeletal muscle atrophy and increased intramuscular fat after incomplete spinal cord injury. *Spinal Cord*, *45*(4), 304–309. https://doi.org/10.1038/sj.sc.3101968.

Gour-Provencal, G., Mac-Thiong, J. M., Feldman, D. E., Bégin, J., & Richard-Denis, A. (2020). Decreasing pressure injuries and acute care length of stay in patients with acute traumatic spinal cord injury. *Journal of Spinal Cord Medicine*, 1–9. https://doi.org/10.1080/10790268.2020.1718265.

Gray, R. J., Worsley, P. R., Voegeli, D., & Bader, D. L. (2016). Monitoring contractile dermal lymphatic activity following uniaxial mechanical loading. *Medical Engineering & Physics*, *38*(9), 895–903. https://doi.org/10.1016/j.medengphy.2016.04.020.

Grigorian, A., Sugimoto, M., Joe, V., Schubl, S., Lekawa, M., Dolich, M., … Nahmias, J. (2017). Pressure ulcer in trauma patients: a higher spinal cord injury level leads to higher risk. *Journal of the American College of Clinical Wound Specialists*, *9*(1–3), 24–31. e21 https://doi.org/10.1016/j.jccw.2018.06.001.

Groah, S. L., Schladen, M., Pineda, C. G., & Hsieh, C. H. (2015). Prevention of pressure ulcers among people with spinal cord injury: A systematic review. *PM & R: Journal of Injury, Function, and Rehabilitation*, *7*(6), 613–636. https://doi.org/10.1016/j.pmrj.2014.11.014.

Guihan, M., Garber, S. L., Bombardier, C. H., Goldstein, B., Holmes, S. A., & Cao, L. (2008). Predictors of pressure ulcer recurrence in veterans with spinal cord injury. *Journal of Spinal Cord Medicine*, *31*(5), 551–559. https://doi.org/10.1080/10790268.2008.11754570.

Guihan, M., Sohn, M. W., Bauman, W. A., Spungen, A. M., Powell-Cope, G. M., Thomason, S. S., … Bates-Jensen, B. M. (2016). Difficulty in identifying factors responsible for pressure ulcer healing in veterans with spinal cord injury. *Archives of Physical Medicine and Rehabilitation*, *97*(12), 2085–2094. e2081 https://doi.org/10.1016/j.apmr.2016.05.025.

Ham, W., Schoonhoven, L., Schuurmans, M. J., & Leenen, L. P. (2014). Pressure ulcers from spinal immobilization in trauma patients: A systematic review. *Journal of Trauma and Acute Care Surgery*, *76*(4), 1131–1141. https://doi.org/10.1097/ta.0000000000000153.

Ham, H. W., Schoonhoven, L. L., Schuurmans, M. M., & Leenen, L. L. (2017). Pressure ulcer development in trauma patients with suspected spinal injury; the influence of risk factors present in the Emergency Department. *International Emergency Nursing*, *30*, 13–19. https://doi.org/10.1016/j.ienj.2016.05.005.

Hastings, B. M., Ntsiea, M. V., & Olorunju, S. (2015). Factors that influence functional ability in individuals with spinal cord injury: A cross-sectional, observational study. *South African Journal of Physiotherapy*, *71*(1), 235. https://doi.org/10.4102/sajp.v71i1.235.

Hatchett, P. E., Mulroy, S. J., Eberly, V. J., Haubert, L. L., & Requejo, P. S. (2016). Body mass index changes over 3 years and effect of obesity on community mobility for persons with chronic spinal cord injury. *Journal of Spinal Cord Medicine*, *39*(4), 421–432. https://doi.org/10.1080/10790268.2015.1133482.

Houghton, P. E., & Campbell, K. (2013). *Canadian best practice guidelines for the prevention and management of pressure ulcers in people with spinal cord injury: A resource handbook for clinicians*. Ontario Neurotrauma Foundation.

Jaul, E., Barron, J., Rosenzweig, J. P., & Menczel, J. (2018). An overview of co-morbidities and the development of pressure ulcers among older adults. *BMC Geriatrics*, *18*(1), 305. https://doi.org/10.1186/s12877-018-0997-7.

Jorge, A., White, M. D., & Agarwal, N. (2018). Outcomes in socioeconomically disadvantaged patients with spinal cord injury: A systematic review. *Journal of Neurosurgery. Spine*, *29*(6), 680–686. https://doi.org/10.3171/2018.5.Spine171242.

Joseph, C., & Nilsson Wikmar, L. (2016). Prevalence of secondary medical complications and risk factors for pressure ulcers after traumatic spinal cord injury during acute care in South Africa. *Spinal Cord*, *54*(7), 535–539. https://doi.org/10.1038/sc.2015.189.

Kirshblum, S. C., Burns, S. P., Biering-Sorensen, F., Donovan, W., Graves, D. E., Jha, A., … Waring, W. (2011). International standards for neurological classification of spinal cord injury (revised 2011). *Journal of Spinal Cord Medicine*, *34*(6), 535–546. https://doi.org/10.1179/204577211x13207446293695.

Kirshblum, S., & Lin, V. W. (2018). *Spinal cord medicine*. Springer Publishing Company.

Kisala, P. A., Tulsky, D. S., Choi, S. W., & Kirshblum, S. C. (2015). Development and psychometric characteristics of the SCI-QOL Pressure Ulcers scale and short form. *Journal of Spinal Cord Medicine*, *38*(3), 303–314. https://doi.org/10.1179/2045772315y.0000000017.

Klotz, R., Joseph, P. A., Ravaud, J. F., Wiart, L., & Barat, M. (2002). The Tetrafigap Survey on the long-term outcome of tetraplegic spinal cord injured persons: Part III. Medical complications and associated factors. *Spinal Cord*, *40*(9), 457–467. https://doi.org/10.1038/sj.sc.3101317.

Krause, J. S., & Broderick, L. (2004). Patterns of recurrent pressure ulcers after spinal cord injury: Identification of risk and protective factors 5 or more years after onset. *Archives of Physical Medicine and Rehabilitation*, *85*(8), 1257–1264. https://doi.org/10.1016/j.apmr.2003.08.108.

Krause, J. S., Vines, C. L., Farley, T. L., Sniezek, J., & Coker, J. (2001). An exploratory study of pressure ulcers after spinal cord injury: Relationship to protective behaviors and risk factors. *Archives of Physical Medicine and Rehabilitation*, *82*(1), 107–113. https://doi.org/10.1053/apmr.2001.18050.

Krishnan, S., Karg, P. E., Boninger, M. L., & Brienza, D. M. (2017). Association between presence of pneumonia and pressure ulcer formation following traumatic spinal cord injury. *Journal of Spinal Cord Medicine*, *40*(4), 415–422. https://doi.org/10.1080/10790268.2016.1180099.

Krishnan, S., Vodovotz, Y., Karg, P. E., Constantine, G., Sowa, G. A., Constantine, F. J., & Brienza, D. M. (2017). Inflammatory mediators associated with pressure ulcer development in individuals with pneumonia after traumatic spinal cord injury: A pilot study. *Archives of Physical Medicine and Rehabilitation*, *98*(9), 1792–1799. https://doi.org/10.1016/j.apmr.2016.12.018.

Kruger, E. A., Pires, M., Ngann, Y., Sterling, M., & Rubayi, S. (2013). Comprehensive management of pressure ulcers in spinal cord injury: Current concepts and future trends. *Journal of Spinal Cord Medicine*, *36*(6), 572–585. https://doi.org/10.1179/2045772313y.0000000093.

Lachenbruch, C., Tzen, Y. T., Brienza, D. M., Karg, P. E., & Lachenbruch, P. A. (2013). The relative contributions of interface pressure, shear stress, and temperature on tissue ischemia: A cross-sectional pilot study. *Ostomy/Wound Management*, *59*(3), 25–34.

Lane, C. A., Selleck, C., Chen, Y., & Tang, Y. (2016). The impact of smoking and smoking cessation on wound healing in spinal cord-injured patients with pressure injuries: A retrospective comparison cohort study. *Journal of Wound, Ostomy, and Continence Nursing*, *43*(5), 483–487. https://doi.org/10.1097/won.0000000000000260.

Le Fort, M., Espagnacq, M., Perrouin-Verbe, B., & Ravaud, J. F. (2017). Risk analyses of pressure ulcer in tetraplegic spinal cord-injured persons: A French long-term survey. *Archives of Physical Medicine and Rehabilitation*, *98*(9), 1782–1791. https://doi.org/10.1016/j.apmr.2016.12.017.

Li, C., DiPiro, N. D., Cao, Y., Szlachcic, Y., & Krause, J. (2016). The association between metabolic syndrome and pressure ulcers among individuals living with spinal cord injury. *Spinal Cord*, *54*(11), 967–972. https://doi.org/10.1038/sc.2016.53.

Li, C., DiPiro, N. D., & Krause, J. (2017). A latent structural equation model of risk behaviors and pressure ulcer outcomes among people with spinal cord injury. *Spinal Cord*, *55*(6), 553–558. https://doi.org/10.1038/sc.2017.9.

Linder-Ganz, E., & Gefen, A. (2009). Stress analyses coupled with damage laws to determine biomechanical risk factors for deep tissue injury during sitting. *Journal of Biomechanical Engineering*, *131*(1). https://doi.org/10.1115/1.3005195, 011003.

Linder-Ganz, E., Shabshin, N., Itzchak, Y., Yizhar, Z., Siev-Ner, I., & Gefen, A. (2008). Strains and stresses in sub-dermal tissues of the buttocks are greater in paraplegics than in healthy during sitting. *Journal of Biomechanics*, *41*(3), 567–580. https://doi.org/10.1016/j.jbiomech.2007.10.011.

Marin, J., Nixon, J., & Gorecki, C. (2013). A systematic review of risk factors for the development and recurrence of pressure ulcers in people with spinal cord injuries. *Spinal Cord*, *51*(7), 522–527. https://doi.org/10.1038/sc.2013.29.

Marion, T. E., Rivers, C. S., Kurban, D., Cheng, C. L., Fallah, N., Batke, J., … Street, J. T. (2017). Previously identified common post-injury adverse events in traumatic spinal cord injury-validation of existing literature and relation to selected potentially modifiable comorbidities: A prospective Canadian cohort study. *Journal of Neurotrauma*, *34*(20), 2883–2891. https://doi.org/10.1089/neu.2016.4933.

Mathew, A., Samuelkamaleshkumar, S., Radhika, S., & Elango, A. (2013). Engagement in occupational activities and pressure ulcer development in rehabilitated South Indian persons with spinal cord injury. *Spinal Cord*, *51*(2), 150–155. https://doi.org/10.1038/sc.2012.112.

Menter, R. R., & Hudson, L. M. (1995). Spinal cord injury clinical outcomes. In S. L. Stover, J. A. De Lisa, & G. G. Whiteneck (Eds.), *Spinal cord injury: Clinical outcomes from the model systems* Aspen Publishers.

Michel, J. M., Willebois, S., Ribinik, P., Barrois, B., Colin, D., & Passadori, Y. (2012). As of 2012, what are the key predictive risk factors for pressure ulcers? Developing French guidelines for clinical practice. *Annals of Physical and Rehabilitation Medicine*, *55*(7), 454–465. https://doi.org/10.1016/j.rehab.2012.08.003.

Molano Alvarez, E., Murillo Perez Mdel, A., Salobral Villegas, M. T., Dominguez Caballero, M., Cuenca Solanas, M., & Garcia Fuentes, C. (2004). Pressure sores secondary to immobilization with cervical collar: A complication of acute cervical injury. *Enfermería Intensiva*, *15*(3), 112–122.

Morita, T., Yamada, T., Watanabe, T., & Nagahori, E. (2015). Lifestyle risk factors for pressure ulcers in community-based patients with spinal cord injuries in Japan. *Spinal Cord*, *53*(6), 476–481. https://doi.org/10.1038/sc.2015.18.

Mortenson, W. B., & Miller, W. C. (2008). A review of scales for assessing the risk of developing a pressure ulcer in individuals with SCI. *Spinal Cord*, *46*(3), 168–175. https://doi.org/10.1038/sj.sc.3102129.

National Spinal Cord Injury Statistical Center. (2006). *2006 annual statistical report*. University of Alabama at Birmingham.

National Spinal Cord Injury Statistical Center. (2019). *The 2019 annual statistical report for the spinal cord injury model systems*. Birmingham, Alabama: National Spinal Cord Injury Statistical Center.

Oomens, C. W., Loerakker, S., & Bader, D. L. (2010). The importance of internal strain as opposed to interface pressure in the prevention of pressure related deep tissue injury. *Journal of Tissue Viability*, *19*(2), 35–42. https://doi.org/10.1016/j.jtv.2009.11.002.

Parent, S., Barchi, S., LeBreton, M., Casha, S., & Fehlings, M. G. (2011). The impact of specialized centers of care for spinal cord injury on length of stay, complications, and mortality: A systematic review of the literature. *Journal of Neurotrauma*, *28*(8), 1363–1370. https://doi.org/10.1089/neu.2009.1151.

Park, J. W., Seo, C. H., Han, S. H., & Lee, Y. G. (2011). Sympathetic influence on biomechanical skin properties after spinal cord injury. *Spinal Cord*, *49*(2), 236–243. https://doi.org/10.1038/sc.2010.95.

Patterson, R. P., Cranmer, H. H., Fisher, S. V., & Engel, R. R. (1993). The impaired response of spinal cord injured individuals to repeated surface pressure loads. *Archives of Physical Medicine and Rehabilitation*, *74*(9), 947–953.

Ploumis, A., Kolli, S., Patrick, M., Owens, M., Beris, A., & Marino, R. J. (2011). Length of stay and medical stability for spinal cord-injured patients on admission to an inpatient rehabilitation hospital: A comparison between a model SCI trauma center and non-SCI trauma center. *Spinal Cord*, *49*(3), 411–415. https://doi.org/10.1038/sc.2010.132.

Post, M. W., Dallmeijer, A. J., Angenot, E. L., van Asbeck, F. W., & van der Woude, L. H. (2005). Duration and functional outcome of spinal cord injury rehabilitation in the Netherlands. *Journal of Rehabilitation Research and Development*, *42*(3 Suppl. 1), 75–85.

Powers, J., Daniels, D., McGuire, C., & Hilbish, C. (2006). The incidence of skin breakdown associated with use of cervical collars. *Journal of Trauma Nursing*, *13*(4), 198–200.

Rabadi, M. H., & Vincent, A. S. (2011). Do vascular risk factors contribute to the prevalence of pressure ulcer in veterans with spinal cord injury? *Journal of Spinal Cord Medicine*, *34*(1), 46–51. https://doi.org/10.1179/107902610x12923394765652.

Rappl, L. M. (2008). Physiological changes in tissues denervated by spinal cord injury tissues and possible effects on wound healing. *International Wound Journal*, *5*(3), 435–444. https://doi.org/10.1111/j.1742-481X.2007.00360.x.

Richard-Denis, A., Thompson, C., Bourassa-Moreau, E., Parent, S., & Mac-Thiong, J. M. (2016). Does the acute care spinal cord injury setting predict the occurrence of pressure ulcers at arrival to intensive rehabilitation centers? *American Journal of Physical Medicine & Rehabilitation*, *95*(4), 300–308. https://doi.org/10.1097/phm.0000000000000381.

Richards, S., Waites, K., Chen, Y., Kogos, S., & Schmitt, M. M. (2004). The epidemiology of secondary conditions following spinal cord injury. *Topics in Spinal Cord Injury Rehabilitation*, *10*(1), 15–29. https://doi.org/10.1310/yxm3-85yw-8vug-nnur.

Rodriguez, G. P., Claus-Walker, J., Kent, M. C., & Stal, S. (1986). Adrenergic receptors in insensitive skin of spinal cord injured patients. *Archives of Physical Medicine and Rehabilitation*, *67*(3), 177–180. https://doi.org/10.1016/0003-9993(86)90073-0.

Rodriguez, G., & Markowski, J. (1995). Changes in skin morphology and its relationship to pressure ulcer incidence in spinal cord injury [ACRM abstract]. *Archives of Physical Medicine and Rehabilitation*, *76*(6), 593.

Roubenoff, R. (1999). The pathophysiology of wasting in the elderly. *Journal of Nutrition*, *129*(1S Suppl), 256s–259s. https://doi.org/10.1093/jn/129.1.256S.

Ruschkewitz, Y., & Gefen, A. (2011). Cellular-scale transport in deformed skeletal muscle following spinal cord injury. *Computer Methods in Biomechanics and Biomedical Engineering*, *14*(5), 411–424. https://doi.org/10.1080/10255842.2010.529804.

Salcido, R., Donofrio, J. C., Fisher, S. B., LeGrand, E. K., Dickey, K., Carney, J. M., … Liang, R. (1994). Histopathology of pressure ulcers as a result of sequential computer-controlled pressure sessions in a fuzzy rat model. *Advances in Wound Care*, *7*(5), 23–24. 26, 28 passim.

Saunders, L. L., & Krause, J. S. (2010). Personality and behavioral predictors of pressure ulcer history. *Topics in Spinal Cord Injury Rehabilitation*, *16*(2), 61–71. https://doi.org/10.1310/sci1602-61.

Saunders, L. L., Krause, J. S., & Acuna, J. (2012). Association of race, socioeconomic status, and health care access with pressure ulcers after spinal cord injury. *Archives of Physical Medicine and Rehabilitation*, *93*(6), 972–977. https://doi.org/10.1016/j.apmr.2012.02.004.

Saunders, L. L., Krause, J. S., Peters, B. A., & Reed, K. S. (2010). The relationship of pressure ulcers, race, and socioeconomic conditions after spinal cord injury. *Journal of Spinal Cord Medicine*, *33*(4), 387–395.

Scelsi, R. (2001). Skeletal muscle pathology after spinal cord injury: Our 20 year experience and results on skeletal muscle changes in paraplegics, related to functional rehabilitation. *BAM-PADOVA*, *11*(2), 75–86.

Schwab, J. M., Zhang, Y., Kopp, M. A., Brommer, B., & Popovich, P. G. (2014). The paradox of chronic neuroinflammation, systemic immune suppression, autoimmunity after traumatic chronic spinal cord injury. *Experimental Neurology*, *258*, 121–129. https://doi.org/10.1016/j.expneurol.2014.04.023.

Sheerin, F., Gillick, A., & Doyle, B. (2005). Pressure ulcers and spinal-cord injury: Incidence among admissions to the Irish national specialist unit. *Journal of Wound Care*, *14*(3), 112–115. https://doi.org/10.12968/jowc.2005.14.3.26750.

Smith, B. M., Guihan, M., LaVela, S. L., & Garber, S. L. (2008). Factors predicting pressure ulcers in veterans with spinal cord injuries. *American Journal of Physical Medicine & Rehabilitation*, *87*(9), 750–757. https://doi.org/10.1097/PHM.0b013e3181837a50.

Sopher, R., Nixon, J., Gorecki, C., & Gefen, A. (2010). Exposure to internal muscle tissue loads under the ischial tuberosities during sitting is elevated at abnormally high or low body mass indices. *Journal of Biomechanics*, *43*(2), 280–286. https://doi.org/10.1016/j.jbiomech.2009.08.021.

Sopher, R., Nixon, J., Gorecki, C., & Gefen, A. (2011). Effects of intramuscular fat infiltration, scarring, and spasticity on the risk for sitting-acquired deep tissue injury in spinal cord injury patients. *Journal of Biomechanical Engineering*, *133*(2). https://doi.org/10.1115/1.4003325, 021011.

Street, J. T., Noonan, V. K., Cheung, A., Fisher, C. G., & Dvorak, M. F. (2015). Incidence of acute care adverse events and long-term health-related quality of life in patients with TSCI. *Spine Journal*, *15*(5), 923–932. https://doi.org/10.1016/j.spinee.2013.06.051.

Stroupe, K., Manheim, L., Evans, C., Guihan, M., Ho, C., Li, K., … Smith, B. (2011). Cost of treating pressure ulcers for veterans with spinal cord injury. *Topics in Spinal Cord Injury Rehabilitation*, *16*(4), 62–73. https://doi.org/10.1310/sci1604-62.

Tate, D. G., Forchheimer, M. B., Krause, J. S., Meade, M. A., & Bombardier, C. H. (2004). Patterns of alcohol and substance use and abuse in persons with spinal cord injury: Risk factors and correlates. *Archives of Physical Medicine and Rehabilitation*, *85*(11), 1837–1847.

Thorfinn, J., Sjöberg, F., & Lidman, D. (2002). Sitting pressure and perfusion of buttock skin in paraplegic and tetraplegic patients, and in healthy subjects: A comparative study. *Scandinavian Journal of Plastic and Reconstructive Surgery and Hand Surgery*, *36*(5), 279–283. https://doi.org/10.1080/028443102320791824.

van der Wielen, H., Post, M. W., Lay, V., Glasche, K., & Scheel-Sailer, A. (2016). Hospital-acquired pressure ulcers in spinal cord injured patients: Time to occur, time until closure and risk factors. *Spinal Cord*, *54*(9), 726–731. https://doi.org/10.1038/sc.2015.239.

Verschueren, J. H., Post, M. W., de Groot, S., van der Woude, L. H., van Asbeck, F. W., & Rol, M. (2011). Occurrence and predictors of pressure ulcers during primary in-patient spinal cord injury rehabilitation. *Spinal Cord*, *49*(1), 106–112. https://doi.org/10.1038/sc.2010.66.

White, B. A. B., Dea, N., Street, J. T., Cheng, C. L., Rivers, C. S., Attabib, N., … Dvorak, M. F. (2017). The economic burden of urinary tract infection and pressure ulceration in acute traumatic spinal cord injury admissions: Evidence for comparative economics and decision analytics from a matched case-control study. *Journal of Neurotrauma*, *34*(20), 2892–2900. https://doi.org/10.1089/neu.2016.4934.

Wilczweski, P., Grimm, D., Gianakis, A., Gill, B., Sarver, W., & McNett, M. (2012). Risk factors associated with pressure ulcer development in critically ill traumatic spinal cord injury patients. *Journal of Trauma Nursing*, *19*(1), 5–10. https://doi.org/10.1097/JTN.0b013e31823a4528.

Wu, G. A., & Bogie, K. M. (2013). Not just quantity: Gluteus maximus muscle characteristics in able-bodied and SCI individuals—Implications for tissue viability. *Journal of Tissue Viability*, *22*(3), 74–82. https://doi.org/10.1016/j.jtv.2013.03.003.

Yalcin, E., Akyuz, M., Onder, B., Unalan, H., & Degirmenci, I. (2013). Skin thickness on bony prominences measured by ultrasonography in patients with spinal cord injury. *Journal of Spinal Cord Medicine*, *36*(3), 225–230. https://doi.org/10.1179/2045772312y.0000000088.

Zeevi, T., Levy, A., Brauner, N., & Gefen, A. (2018). Effects of ambient conditions on the risk of pressure injuries in bedridden patients-multi-physics modelling of microclimate. *International Wound Journal*, *15*(3), 402–416. https://doi.org/10.1111/iwj.12877.

Chapter 6

# Venous thromboembolism in spinal cord injury—Prophylaxis, diagnosis and treatment

Rodrigo Lanna de Almeida
*Department of Spinal Cord Injury, Heidelberg University, Heidelberg, Germany*

## Abbreviations

| | |
|---|---|
| **aPTT** | activated partial thromboplastin time |
| **CSCM** | Consortium for Spinal Cord Medicine |
| **CT** | computed tomography |
| **CTPA** | CT pulmonary angiography |
| **CUS** | compression ultrasonography |
| **DVT** | deep venous thrombosis |
| **GCS** | graduated compression stockings |
| **IPCD** | intermittent pneumatic compression device |
| **IVCF** | inferior vena cava filter |
| **LMWH** | low-molecular-weight heparin |
| **MRI** | magnetic resonance imaging |
| **PE** | pulmonary embolism |
| **PESI** | pulmonary embolism severity index |
| **SCI** | spinal cord injury |
| **UH** | unfractionated heparin |
| **VKA** | vitamin K antagonists |
| **VTE** | venous thromboembolism |

## Introduction

The classical Virchow's postulate links the occurrence of venous thromboembolism (VTE) to the presence of a disruption of the circulatory flux, a damage of the endothelial tissue, or a hypercoagulability state. Immobility and palsy are prominent risk factors for VTE and are frequently seen in clinical rules used for its diagnosis (Duffett, Castellucci, & Forgie, 2020). Individuals with a spinal cord injury (SCI) present frequently not only with impaired mobility due to the associated neurological motor deficit, but endothelial injuries and a hypercoagulability condition may also arise as a consequence of the lesion mechanism, such as trauma or neoplasms, as well as secondary complications like infections or pressure ulcers (Agarwal & Mathur, 2009; Chung et al., 2014) (Fig. 1). Thus, SCI is associated with an increased risk of venous thromboembolism in its acute phase that may extend to its chronic phase (Casas, Sánchez, Arias, & Masip, 1977; Godat, Kobayashi, Chang, & Coimbra, 2015; Todd et al., 1976).

Many advances have been made regarding risk stratification for prophylaxis and diagnostic algorithms for VTE in the general population. The proposed strategies and clinical rules however have not been validated for individuals with SCI, posing a challenge for physicians involved in the care of such patients.

## Epidemiology

The reported prevalence of VTE in spinal cord injury is variable. A study by Todd et al. (1976) based on screening in asymptomatic patients with serial scintigraphy with labeled fibrinogen identified clots in 100% of the 20 studied patients.

**Diagnosis and Treatment of Spinal Cord Injury. https://doi.org/10.1016/B978-0-12-822498-4.00006-3**

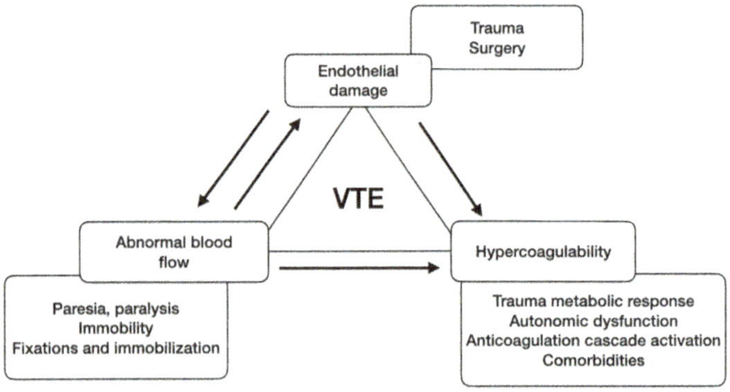


FIG. 1 Virchow's triad in spinal cord injury. This schematic representation illustrates how conditions frequently present in SCI exert a direct influence on VTE risk. *VTE*, venous thromboembolism; *SCI*, spinal cord injury. *(Based on Piran, S., & Schulman, S. (2019). Thromboprophylaxis in patients with acute spinal cord injury: A narrative review.* Semin Thromb Hemost, *45(2), 150–156. © Georg Thieme Verlag KG.)*

Other studies that are based on clinical symptoms or less sensitive screening methods report a prevalence of 10% to 15% for deep venous thrombosis (DVT) and 5% to 10% for pulmonary embolism (PE).

The risk appears to be the highest in the first 12 weeks after the injury (Agarwal & Mathur, 2009; Casas et al., 1977), with a retrospective cohort study by Chung et al. (2014) reporting a 17 times risk increase for DVT and 3.5 times for EP within 3 months after an SCI in comparison to age- and sex-matched individuals from the general population. A study by Green (1992), however, didn't find an increased risk in the first 3 days after the lesion. It is postulated that factors like the flaccid paralysis occurring during the spinal shock phase and the need for immobilization secondary to multiple trauma may be of significance in the augmented risk during the acute stage of SCI (Fig. 2). Moreover, the dysregulation of the autonomic nervous system may lead to an imbalance of the hemostatic and fibrinolytic systems, and there is evidence of increased platelet reactivity to collagen and an imbalance in factor VIII to factor VIII-C ration (Kelly, Yoder, Tang, & Wakefield, 2010).

Published data on the incidence of VTE during the sub-acute and chronic phases of SCI are conflicting. There appears to be a reduction in this risk, approaching that of the general population, although somewhat higher. A study by Giorgi Pierfranceschi et al. (2013) reported an incidence of 34.4 VTE events per 100,000 patient-year during the first 90 days which was reduced to 0.3 events per 100,000 patient-year thereafter. Data from a systematic review indicate a PE incidence ranging from 0.5% to 6.0% and a DVT incidence between 2.0% and 8.0% in the sub-acute phase in patients under different prophylaxis strategies (Alabed, Belci, Van Middendorp, Al Halabi, & Meagher, 2016). The above-mentioned study by Chung et al. (2014) reported a cumulative risk of 1.19 times after the first year of the SCI, and studies evaluating individuals in the chronic stage demonstrated DVT incidences as low as 0.55 per 10,000 patients-day (de Almeida, Rodrigues, et al., 2019) and as high as 8% (Mackiewicz-Milewska et al., 2016) in patients admitted for rehabilitation treatment.

Without adequate treatment, the estimated mortality rates may reach 8% for DVT and 25% for PE. In studies performed with individuals with SCI, VTE may be responsible for up to 3% of all the deaths in those admitted for rehabilitation

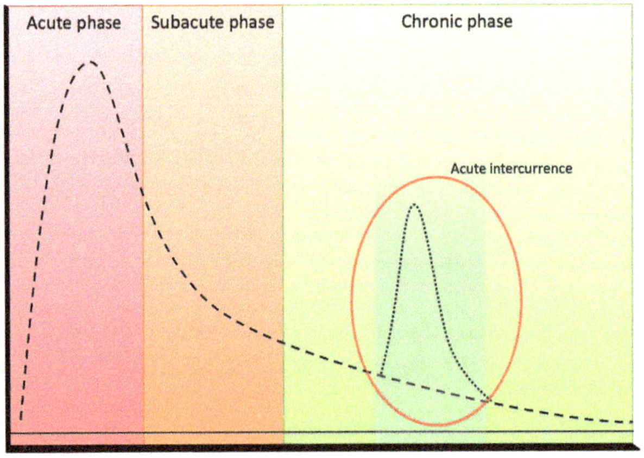


FIG. 2 Venous thromboembolism risk in spinal cord injury. VTE risk in SCI *(dashed lined)*: The risk is higher during the acute phase and decreases thereafter. Intercurrent clinical or surgical conditions may transiently elevate the risk during the chronic phase *(dotted line)*. The *solid line* represents the average VTE risk in the general population. *VTE*, venous thromboembolism; *SCI*, spinal cord injury.

(Green et al., 1994). In addition, individuals affected by VTE may present further clinical complications such as post-thrombotic syndrome, pulmonary hypertension, and recurrent VTE events (Kearon, 2003).

## Screening

The rationale for screening for asymptomatic VTE is based on its high incidence in SCI, therefore being a probable and preventable cause of death during the acute phase. The identification of an asymptomatic thrombus could then lead to immediate treatment, which could hinder further complications. Proposed approaches involve performing serial lower limb duplex ultrasound (Kadyan, Clinchot, & Colachis, 2004), D-Dimer testing, or both used in combination (Kumagai et al., 2020; Masuda, Ueta, Shiba, & Iwamoto, 2015).

Even though the majority of thrombi have an origin in the lower limb circulation, up to half resolve spontaneously within 72 h, and only around a sixth of them lead to disturbance in the deep vein circulation (Kearon, 2003). Screening studies have been successful in demonstrating high sensitivity and in identifying a high number of asymptomatic VTE. As the majority of these were designed as diagnostic and not therapeutic studies, there is uncertainty about the clinical benefits of this strategy which may well lead to overdiagnosis and consequently overtreatment. Considering the bleeding risk associated with therapeutic anti-coagulation, patients might be exposed to unnecessary use of a dangerous medication.

## Prophylaxis

### Acute phase

Since the 1970s, uncontrolled studies proposed the use of pharmacological prophylaxis in acute SCI (Casas et al., 1977). Different approaches involving the use of unfractionated heparin (UH), vitamin K antagonists (VKA), low-molecular-weight heparin (LMWH), intermittent pneumatic compression (IPC) and electrostimulation alone or in different combinations have been tested. In comparison to mechanical prophylaxis alone, the use of pharmacological prophylaxis may lead to an absolute risk reduction of up to 15% in the incidence of VTE during the acute phase of SCI (Halim, Chhabra, Arora, & Kumar, 2014).

The major advantage of LMWH in comparison to UH is related to an average of 4% absolute reduction in the risk of bleeding (Investigators, 2003; Thumbikat, Poonnoose, Balasubrahmaniam, Ravichandran, & McClelland, 2002). Enoxaparin was the most frequently used LMWH in clinical trials involving individuals with SCI, but dosing strategies and duration of treatment were greatly varied, from 20 to 40 mg once a day, to 30 mg twice a day, from as little as 14 days up to 8 weeks or longer times based on patient mobility (Frisbie & Sasahara, 1981; Green et al., 1990; Merli et al., 1988; Merli, Crabbe, Doyle, Ditunno, & Herbision, 1992; Spivack & Aisen, 1997; Thumbikat et al., 2002), preventing these studies from being combined in a meta-analysis. Until now, we could not find any study comparing the same dosing strategy with different durations, allowing a more precise evaluation of the risk-benefit profiles of longer or shorter prophylaxis strategies.

Current guidelines agree that the pharmacological prophylaxis should be initiated as soon as possible during the acute phase, preferably within the first 72 h of the injury, taking into consideration the clinical status and, most importantly, the assessment of the risk of bleeding of the individual patient (CSCM, 2016; Fehlings et al., 2017). Recommendations regarding the duration of prophylaxis continue to be at least 8 and up to 12 weeks, which may be prolonged based on the assessment of the individual risk of VTE.

### Sub-acute and chronic phase

Studies comparing prophylaxis strategies during the chronic phase of SCI are scarce. It is important to emphasize that in poor resource areas with limited access to specialized care, many individuals with SCI may only be admitted to inpatient rehabilitation already in the chronic phase. The delayed access to rehabilitation services may lead to early complications, such as structured articular mobility restrictions, pressure ulcers, recurrent infections, and obesity, which may further influence the risk of developing a VTE episode ( Jorge, White, & Agarwal, 2018).

As discussed above, the risk of VTE decreases after the acute phase. The use of pharmacological prophylaxis is usually safe but encompasses nevertheless an elevation of the patient's basal risk of bleeding. Minor bleedings (hematomas at the injection site) may occur in up to 64% of patients during rehabilitation (Eichinger et al., 2018), and major bleeding events have been reported in individuals with SCI receiving prophylactic doses of LMWH (de Almeida, Gonzaga, Beraldo, & Amado, 2019). We suggest taking into consideration the individual patient risk factors, such as comorbidity, mobility, independence, concomitant use of other medications, presence of infections and near surgical procedures to decide whether or not to initiate the pharmacological prophylaxis. Moreover, as suggested by the Consortium for Spinal Cord Medicine (CSCM) Clinical Practice Guidelines (2016), services involved in the care of individuals with SCI should have established policies regarding thromboprophylaxis which should be periodically revised and audited.

### Non-pharmacological methods

Maintenance of adequate hydration, early mobilization, physical rehabilitation and promoting independence are classically described as general methods that should be applied to every patient. Intermittent pneumatic compression devices (IPCD) may be used as an alternative prophylaxis method if the risk of complications with the use of pharmacological methods is deemed to be prohibitive (CSCM, 2016). Only a small case series evaluated IPCD without a pharmacological method in individuals with SCI, reporting a 43% incidence of DVT mostly detected by screening, and only 18% of these being a proximal DVT (Chung, Lee, Kim, & Eoh, 2011). These devices are used continuously and, therefore, may interfere with the acute rehabilitation program.

A meta-analysis demonstrated that graduated compression stocking (GCS) may be effective in preventing DVT in hospitalized patients, especially those submitted to surgical and orthopedic interventions. This benefit was also seen in those receiving background thromboprophylaxis. Only two studies have used LMWH as background prophylaxis and no difference was found with the addition of GCS. Furthermore, none of the included studies compared GCS alone versus a pharmacological method (Sachdeva, Dalton, & Lees, 2018). Moreover, these devices may cause skin lesions and should not be used in conditions such as peripheral arterial disease and decompensated heart failure.

Current evidence doesn't support the prophylactic use of inferior vena cava filters (IVCF), and there may be an increased risk of VTE in individuals with SCI who underwent an IVCF implantation, even when receiving pharmacological prophylaxis (Gorman, Qadri, & Rao-Patel, 2009).

## Diagnosis

The usual approach in the individual with suspected DVT or PE involves a clinical evaluation to estimate his pre-test probability based on signs, symptoms and risk factors, followed by a diagnostic strategy to rule in or rule out the diagnosis. The utility of diagnostic algorithms associated with clinical rules such as the Wells' criteria for DVT or PE, the Geneva score for PE or the PERC have been demonstrated in diverse clinical scenarios (Tritschler, Kraaijpoel, Le Gal, & Wells, 2018), but as of yet, are not validated for individuals with SCI.

Some of the clinical factors evaluated by these rules may have different significance and presentation in SCI (Table 1). Paresis or paralysis may be a major risk factor during the acute phase of an SCI, but this may not extend to the chronic phase. Benign postural lower limb edema is a frequent clinical sign in these individuals and may impair the clinical evaluation of a suspected DVT. Furthermore, the autonomic dysfunction in lesions above T6 may impair tachycardic responses (Krassioukov et al., 2012), as evaluated in some scores and pain complaints may be absent due to neurological impairment. In this scenario, it is reasonable to maintain a high level of clinical suspicion.

The test of choice for confirming or excluding the diagnosis will be based on the previously estimated pre-test probability. Moreover, the clinician should also consider how invasive the test is, the availability of the method, and trained staff for performing and interpreting the results and how high the clinical suspicion of DVT or PE is.

### D-dimer

The D-dimer is a fibrin degradation product that is positive not only in VTE. Its high sensitivity—especially in quantitative and semi-quantitative tests—in general patients makes it a helpful test for ruling out the diagnosis of both DVT and PE in those with a low pre-test probability (Tritschler et al., 2018). This high sensitivity has been demonstrated in low-quality evidence studies in patients with SCI (Roussi et al., 1999) and other neurological conditions under rehabilitation treatment (Akman, Cetin, Bayramoglu, Isiklar, & Kilinc, 2004), but not as part of a diagnostic algorithm in these populations. The low specificity associated with the test is not enough to confirm the diagnosis; therefore, a positive result will lead to further testing. Despite its high sensitivity in patients presenting a high pre-test probability, a negative result won't be enough to safely rule out the diagnosis.

**TABLE 1** Variables in clinical rules that may be influenced by spinal cord injury.

| *Wells' criteria for DVT* |
|---|
| Asymmetric leg swelling |
| Whole limb edema |
| Tenderness along the deep venous system |
| *Wells' criteria for PE* |
| Heart rate above 100 bpm |
| *Geneva score* |
| Heart rate above 100 bpm (original score) |
| Heart rate above 94 bpm (revised and simplified scores) |
| Limb pain (revised and simplified scores) |
| *PERC rule* |
| Heart rate above 100 bpm |
| Unilateral edema |

The table displays some of the risk factors present on commonly used clinical rules that may be difficult to evaluate due to neurological and autonomic dysfunction in SCI.

## Lower limb ultrasonography

Compression ultrasound examination (CUS) is a highly sensitive and specific non-invasive test that may be performed as a point-of-care diagnostic strategy by trained physicians (Tritschler et al., 2018). Based on its widespread availability, non-invasiveness and ease of performance, CUS is the method of choice for the evaluation of patients with moderate or high pre-test DVT probability. This can be performed in strategies that evaluate the proximal venous circulation up to the level of the popliteal vein, requiring a follow-up examination after 1 week if negative, or a complete scan from the common femoral vein down to the deep calf veins, without a need for retesting (Wells, Ihaddadene, Reilly, & Forgie, 2018). Even though there is no specific study on the accuracy of this method in patients in SCI patients, ultrasonography seems like a reasonable method due to its availability and good performance of point-of-care examinations in comparison with expert radiologist evaluation, which may avoid delays in the diagnostic process (Fischer et al., 2019). For this same reason, lower limb sonography may be an acceptable strategy as a first examination to rule in PE in patients with a higher likelihood of this diagnosis and with clinical signs of DVT (Wells et al., 2018), but an isolated negative test is not enough to exclude it.

## Lower limb venography

Venography, computerized tomography (CT), and magnetic resonance (MRI) venography are invasive methods requiring the use of contrast. Venography is considered the "gold standard" for the evaluation of the limb venous circulation. Even though a normal venography allows safe withholding of anti-coagulation, it is a limited method due to its restricted availability, costs, discomfort, and difficulties in obtaining standardized and good-quality images and proper evaluation (Bates et al., 2012). CT and MRI venography are promising methods that have demonstrated sensitivity and specificity around 95% in meta-analysis, but with high heterogeneity between studies (Abdalla et al., 2015; Thomas, Goodacre, Sampson, & van Beek, 2008). Therefore, they are not considered first-line diagnostic methods but are useful in situations where ultrasonography can't safely exclude the diagnosis, such as thrombosis in proximal or pelvic veins that are not compressible or in obese patients (Tritschler et al., 2018; Wells et al., 2018).

## CT pulmonary angiography

CT pulmonary angiography (CTPA) has replaced ventilation/perfusion scintigraphy as the method of choice for the diagnosis of PE (Tritschler et al., 2018). With a high sensitivity and specificity, CTPA allows a fast and comprehensive evaluation of the pulmonary circulation and is largely available (Konstantinides et al., 2019). It also allows the evaluation of pulmonary images which are useful for differential diagnosis. The constantly evolving technology and quality of images have led to a continual increase in sensitivity and, consequently, in the reported incidence of PE in the general population. This elevation occurred despite the current higher awareness of VTE and adherence to prophylaxis strategies and, according to studies performed in the United States, no concurrent elevation in age-adjusted mortality was observed, which suggests the occurrence of overdiagnosis (Dobler, 2019).

Small thrombus formation is not an infrequent event. One of the physiological functions of the normal lung may be to serve as a clot filter avoiding venous thrombi from reaching the systemic arterial circulation (Swan, Hitchen, Klok, & Thachil, 2019). The physician should, therefore, be aware of the potential risk involved in over-testing patients for PE, for instance, screening patients with a diagnosis of DVT for an asymptomatic PE.

## Treatment

Anti-coagulation is the cornerstone of the treatment of VTE (Di Nisio, van Es, & Büller, 2016). The primary goal of the treatment is to avoid acute complications such as further embolization of deep vein originated thrombi, further impairment of pulmonary function or pulmonary circulation with consequent cardiovascular instability (Burgazli et al., 2013). Adequate quality therapeutic anti-coagulation may also improve late outcomes, such as post-thrombotic syndrome (Ten Cate-Hoek, 2018). Affected individuals should receive treatment to achieve rapid anti-coagulation with posterior long-term maintenance with adequate time in therapeutic range (TTR). A summary of commonly used medications is presented in Table 2.

### Initial management

Initial treatment aims to quickly achieve effective anti-coagulation, which can be done both with parenteral and oral agents. Before initiating any anti-coagulant, the risk of bleeding as well as hemodynamic conditions of the individual should be assessed.

**TABLE 2** Commonly used anti-coagulant agents.

| | Class | Dosing | Effect reversal |
|---|---|---|---|
| *Parenteral agents* | | | |
| Heparin | UH | Bolus: 80 μg/kg IV<br>Maintenance: 18 μg/kg h IV<br>Adjust dose according to aPTT | Protamine |
| Enoxaparin | LMWH | 1 mg/kg 12/12 h SC | May benefit from protamine use |
| Fondaparinux | Pentasaccharide | <50 kg: 5 mg SC daily<br>50–100 kg: 7.5 mg SC daily<br>>100 kg: 10 mg SC daily | May benefit from andexanet alfa (off label) or activated prothrombin complex concentrate |
| *Oral agents* | | | |
| Apixaban | Factor Xa inhibitor | 10 mg twice daily for 7 days followed by 5 mg twice daily | – Andexanet alfa (off label for edoxaban reversal) |
| Rivaroxaban | | 15 mg twice daily for 21 days followed by 20 mg daily | |
| Edoxaban | | 60 mg once daily | |
| Dabigatran | Direct thrombin inhibitor | 150 mg twice daily | – Idarucizumab |
| Warfarin | Vitamin K antagonist | 2–10 mg once daily for 2 days.<br>Adjust according to INR | – Vitamin K (intravenous preparations)<br>– Four-factor prothrombin complex concentrate<br>– Fresh-frozen plasma |

Characteristics of some of the anti-coagulants commonly used in clinical practice. Detailed information about anti-coagulation reversal is beyond the scope of this chapter. We recommend referring to the manufacturer's information for detailed dosing and administration orientation. aPTT, activated partial thromboplastin time; *INR*, international normatized ratio; *IV*, intravenous; *LMWH*, low-molecular-weight heparin; *SC*, subcutaneous; *UH*, unfractionated heparin.

There's no validated score for bleeding risk estimation in individuals with SCI receiving anti-coagulant treatment; therefore, a thorough assessment of comorbidities, previous traumas, and surgeries is mandatory (Table 3). For hemodynamically unstable individuals and those who may need urgent or emergent surgery or that may benefit from thrombolytic therapy, the use of short-acting and easily reversible anti-coagulants should be preferred.

*Unfractionated heparin (UH)* is a polysaccharide solution with an average molecular weight of 15,000 Da, most of them with a low binding affinity with the anti-thrombin molecule (Onishi, St Ange, Dordick, & Linhardt, 2016). It is administered intravenously initially as a bolus followed by a continuous infusion with monitoring of the activated partial thromboplastin time (aPTT) and infusion adjustments every 4 to 6 h (Hull et al., 1992; Raschke, Reilly, Guidry, Fontana, & Srinivas, 1993). It is the therapy of choice for individuals with chronic renal disease as well as for those in whom the need for rapid interruption and reversal of the anti-coagulant effect is a concern, which can be achieved with the use of the antidote protamine (Di Nisio et al., 2016).

*Low-molecular-weight heparin (LMWH)* is obtained through the enzymatic breakdown of heparin polysaccharides, potentiating its affinity to the anti-thrombin molecule and, consequently, its ability to inhibit factor Xa (Onishi et al., 2016). In comparison to UH, LMWH treatment is associated with a lower occurrence of thrombotic complications and major hemorrhages (Robertson & Jones, 2017). Other advantages are facilitated fixed weight-based dosing, a longer half-life facilitating dose scheduling, and a lower risk of heparin-induced thrombocytopenia (Weitz, 1997).

*Pentasaccharides*, such as Fondaparinux, are ultralow-molecular-weight heparins with a high binding affinity to anti-thrombin that strongly and indirectly inhibits factor Xa. Its major advantage is the facilitated dosing scheme based on a weight range. According to a meta-analysis, pentasaccharides may be as effective and safe when compared to standard treatments (Brandao, Junqueira, Rollo, & Sobreira, 2017). Studies appraised in this meta-analysis used different types of comparison treatments including UH or LMWH followed by oral vitamin K antagonists (VKA) which may have influenced the final results.

*Novel oral anti-coagulants (NOACs)* were developed aiming to eliminate the difficulties encountered when managing individuals using VKAs. They have a rapid onset, predictable therapeutic effect, a fixed-dose schedule, do not require laboratory monitoring nor dose adjustments, and appear to be safer regarding bleeding risk compared to VKAs (Oliphant, Jacobs, Kabra, & Das, 2013). Their action can be directed to thrombin (Dabigatran) or factor Xa inhibition (Apixaban, Edoxaban, and Rivaroxaban).

Only Rivaroxaban and Apixaban were evaluated as initial and long-term therapy without previous administration of LMWH. Rivaroxaban was evaluated both in a DVT (Bauersachs et al., 2010) and on a PE trial (Büller et al., 2012), each one with more than 3000 individuals, whereas only a third of the 5400 studied individuals on the Apixaban trial had PE (Agnelli et al., 2013). Overall, they were shown to be non-inferior to standard treatment, with a trend to less bleeding events (Gómez-Outes, Terleira-Fernández, Lecumberri, Suárez-Gea, & Vargas-Castrillón, 2014), however, to this moment, they are still more expensive in comparison to VKAs.

## Anti-coagulation after initial management

After initial management, anti-coagulant treatment is maintained for 3 months or longer depending on the clinical presentation (whether DVT or PE), the individual ability to comply with the treatment, risk of recurrence, bleeding risk, costs, and clinical intercurrences. Oral agents are the preferred strategy, but LMWH and pentasaccharides are reasonable alternatives.

*Vitamin K antagonists* are the most frequently used oral anti-coagulant agents in the outpatient setting. Despite the necessity of constant laboratory monitoring and their narrow therapeutic range (Oliphant et al., 2013), they present lowest monthly cost considering the medication alone, are widely available and clinicians have been managing individuals using these agents for more than 60 years (Thean & Alberghini, 2016). The quality of the anti-coagulation is measured by the time in therapeutic range (TTR). A TTR below 65% is associated with a higher risk of both thromboembolic and bleeding events. Strategies to improve the TTR include management in anti-coagulation clinics, use of decision support systems, and point-of-care or self-testing (Schein et al., 2016). Adequate support through anti-coagulation clinics was shown to be effective even in poor resource and low level of formal education settings (Costa, Ferreira, Valacio, & Vieira Moreira, 2011).

*NOACs* are also used in long-term coagulation. Besides the above mentioned Apixaban and Rivaroxaban, the factor Xa inhibitor Edoxaban and the direct thrombin inhibitor Dabigatran have also been tested for the chronic anti-coagulation in individuals with VTE with results lying within the non-inferiority margin for VTE recurrence. There's no study with a direct comparison between these agents. A network meta-analysis pooling studies of the four agents failed to identify any specific benefit regarding VTE recurrence or VTE-related death, but there was a trend toward a reduction of clinically relevant bleeding with apixaban and dabigatran (Cohen et al., 2015). Patients using NOACs probably have a better TTR; nevertheless, the clinician should be aware that these medications are not indicated for patients with reduced renal function

**TABLE 3** Risk factors for bleeding with anti-coagulant therapy.

| Risk factors[a] |
|---|
| Age > 65 years |
| Age > 75 years |
| Previous bleeding |
| Cancer |
| Metastatic cancer |
| Renal failure |
| Liver failure |
| Thrombocytopenia |
| Previous stroke |
| Diabetes |
| Anemia |
| Antiplatelet therapy |
| Poor anti-coagulant control |
| Comorbidity and reduced functional capacity |
| Recent surgery[b] |
| Frequent falls |
| Alcohol abuse |

| | Estimated absolute risk of major bleeding (%) | | |
|---|---|---|---|
| **Categorization of risk of bleeding[c]** | **Low risk (0 risk factors)[d]** | **Moderate risk (1 risk factor)[d]** | **High risk (≥2 risk factors)[d]** |
| *Anti-coagulation 0 to 3 months* | | | |
| Baseline risk (%) | 0.6 | 1.2 | 4.8 |
| Increased risk (%) | 1 | 2 | 8 |
| Total risk (%) | 1.6[e] | 3.2 | 12.8[f] |
| *Anti-coagulation after first 3 months[g]* | | | |
| Baseline risk (%/year) | 0.3[h] | 0.6 | ≥ 2.5 |
| Increased risk (%/year) | 0.5 | 1 | ≥ 4 |
| Total risk (%/year) | 0.8[i] | 1.6[i] | ≥ 6.5 |

[a]*The increase in bleeding associated with a risk factor will vary with (1) severity of the risk factor, (2) temporal relationships, and (3) how effectively a previous cause of bleeding was corrected.*
[b]*Important for parenteral anti-coagulation (e.g., first 10 days), but less important for long-term or extended anti-coagulation.*
[c]*Although there is evidence that risk of bleeding increases with the prevalence of risk factors, this categorization scheme has not been validated. Furthermore, a single risk factor, when severe, will result in a high risk of bleeding (e.g., major surgery within the past 2 days, severe thrombocytopenia).*
[d]*Compared with low-risk patients, moderate-risk patients are assumed to have a twofold risk and high-risk patients an eightfold risk of major bleeding.*
[e]*The 1.6% corresponds to the average of major bleeding with initial UFH or LMWH therapy followed by VKA therapy. We estimated baseline risk by assuming a 2.6 relative risk of major bleeding with anti-coagulation (refer to footnote g).*
[f]*Consistent with frequency of major bleeding observed in previous studies.*
[g]*It's estimated that anti-coagulation is associated with a 2.6-fold increase in major bleeding based on comparison of extended anti-coagulation with no extended anti-coagulation. The relative risk of major bleeding during the first 3 months of therapy may be greater than during extended VKA therapy because (1) the intensity of anti-coagulation with initial parenteral therapy may be greater than with VKA therapy; (2) anti-coagulant control will be less stable during the first 3 months; and (3) predispositions to anti-coagulant-induced bleeding may be uncovered during the first 3 months of therapy. However, studies of patients with acute coronary syndromes do not suggest a ≥ 2.6 relative risk of major bleeding with parenteral anti-coagulation (e.g., UFH or LMWH) compared with control.*
[h]*Estimated baseline risk of major bleeding for low-risk patients (and adjusted up for moderate- and high-risk groups as per footnote d).*
[i]*Consistent with frequency of major bleeding during prospective studies of extended anti-coagulation for VTE.*

From Kearon, C., et al. (2012). Antithrombotic therapy for VTE disease: Antithrombotic Therapy and Prevention of Thrombosis, 9th ed: American College of Chest Physicians evidence-based clinical practice guidelines. *Chest, 141*, e419S. 

and are more expensive than VKAs, and specific antidotes for anti-coagulation reversal are not widely available and are high-cost agents.

## Special situations

Due to its ease of administration, LMWH may also be used for long term anti-coagulation in special situations, being for a long time the agent of choice in individuals with concomitant active neoplastic disease, as this strategy has demonstrated a lower VTE recurrence rate in clinical trials (Kahale et al., 2018). Recent studies have compared the agents Apixaban, Edoxaban and Rivaroxaban to LMWH for long term management of cancer associated VTE in a non-inferiority design. These agents were non-inferior to LMWH for the prevention of recurrent VTE, but there are concerns about a possible increased risk of bleeding especially in upper gastrointestinal cancer, acute leukemia, and primary or metastatic central nervous system tumors. Furthermore, NOACs may interact with different chemotherapeutic agents, therefore demanding a thorough evaluation of patients' bleeding risk, therapeutic plan as well as patients' preferences and prognosis (Duffett et al., 2020).

LMWH is also the preferred agent for pregnant women or women exposed to the possibility of pregnancy, due to a possible risk of fetal complications with other agents (Kearon et al., 2016).

ICVF may be an alternative for patients with VTE, whose clinical situation doesn't permit anti-coagulation, but there is no current evidence of benefit in the association of IVCF with anti-coagulation (Mismetti et al., 2015).

## Outpatient management

The management of VTE was facilitated after the advent of the NOACs due to their ease of administration and lack of necessity for laboratory monitoring. Moreover, Apixaban and Rivaroxaban are also approved for the acute management of DVT and PE, which could permit the prescription of oral medication in the emergency department followed by discharge and outpatient management. This strategy may be associated with reduced costs, less exposure to nosocomial resistant microorganisms, and reduced use of available beds. On the other hand, there is the risk of delayed treatment of complications associated with progression of the thrombotic event or bleeding.

Outpatient treatment is an acceptable strategy both for DVT and PE in those patients whose embolic and bleeding risks are deemed to be low and have access to a medical center for surveillance or treatment of complications. The pulmonary embolism severity index (PESI) and the Hestia criteria (Table 4) have been used to assess disease severity in PE (Wolf et al., 2018), but these were not validated in individuals with SCI. Moreover, both the SCI associated autonomic dysfunction, risk of autonomic dysreflexia, reduced respiratory reserve in patients with higher lesions, and sensory impairment may influence the patient's risk and should be taken into consideration. A shared decision-making strategy where individuals are informed in detail about possible risks and benefits should be used.

## Duration of anti-coagulation

To this moment no clinical trial was conducted specifically with individuals with SCI to determine the optimal duration of anti-coagulation. Guidelines directed at the general population agree on a treatment duration of at least 3 months for DVT, distal isolated DVT, and PE (Kearon et al., 2016; Konstantinides et al., 2019; Mazzolai et al., 2018). The decision of extending this period should be made on an individual basis, considering the risk of recurrence, especially the presence or resolution of provoking factors, and the risk of bleeding.

VTE occurring during the acute phase of SCI (i.e., during the period of higher thromboembolic risk) may be considered a provoked episode. During the chronic phase of SCI, individuals may be exposed in different periods to risk factors for VTE, such as bed rest due to fractures, pressure ulcers, or surgery. However, defining the most adequate treatment in otherwise healthy and active individuals with chronic SCI remains challenging.

Proposed strategies for this decision involving D-Dimer testing, evaluation of residual thrombosis (Carrier et al., 2011), clinical rules such as the HERDOO2, the DASH score, and the Vienna prediction model have demonstrated diverging results, lack external validation (Di Nisio et al., 2016; Tritschler et al., 2018) and individuals with SCI are underrepresented in the derivation and internal validations cohorts. Testing for inherited or acquired thrombophilia is not recommended after the first episode of provoked or unprovoked VTE and even a known thrombophilia may not necessarily grant an indication of long-term anti-coagulation after a VTE related to a strong provoking factor (Duffett et al., 2020).

**TABLE 4 Pulmonary embolism severity index and Hestia rule.**

| Criteria | Points assigned |
|---|---|
| *Pulmonary embolism severity index (PESI)* | |
| Age in years (per year) | 1 |
| Male sex | 10 |
| History of cancer | 30 |
| History of heart failure | 10 |
| History of chronic lung disease | 10 |
| Systolic blood pressure < 100 mmHg | 30 |
| Pulse rate ≥ 110 per minute | 20 |
| Arterial oxygen saturation < 90% | 20 |
| Respiratory rate ≥ 30 per minute | 20 |
| Temperature < 36°C | 20 |
| Altered mental status | 60 |
| *Score interpretation* | |
| Class I, very low risk | <65 |
| Class II, low risk | 66–85 |
| Class III, intermediate risk | 86–105 |
| Class IV, high risk | 106–125 |
| Class V, very high risk | >125 |
| *Hestia rule* | |
| Is the patient hemodynamically unstable?[a] | |
| Is thrombolysis or embolectomy necessary? | |
| Active bleeding or high risk of bleeding?[b] | |
| >24 h of oxygen supply to maintain oxygen saturation > 90%? | |
| Is pulmonary embolism diagnosed during anti-coagulant treatment? | |
| Severe pain needing intravenous pain medication for >24 h? | |
| Medical or social reason for treatment in the hospital for >24 h (infection, malignancy, no support system)? | |
| Does the patient have a creatinine clearance of <30 mL/min?[c] | |
| Does the patient have severe liver impairment?[d] | |
| Is the patient pregnant? | |
| Does the patient have a documented history of heparin induced thrombocytopenia? | |

HESTIA interpretation: If the answer to one of the questions is yes, in-hospital treatment is recommended. If the answer to all the questions is no, home treatment is recommended.

[a]*Including but not restricted to the following criteria: systolic blood pressure < 100 mmHg with heart rate > 100 beats/min; condition requiring admission to an intensive care unit.*

[b]*Gastrointestinal bleeding in the preceding 14 days, recent stroke (<4 weeks ago), recent operation (<2 weeks ago), bleeding disorder or thrombocytopenia (platelet count < 75.109/L), uncontrolled hypertension (systolic blood pressure > 180 mmHg or diastolic blood pressure > 110 mmHg).*

[c]*According to the Cockroft-Gault formula.*

[d]*Left to the discretion of the physician.*

Based on Roy, P. M., Moumneh, T., Penaloza, A., & Sanchez, O. (2017). Outpatient management of pulmonary embolism. *Thrombosis Research, 155*, 92–100. 

## Conclusion

VTE is a frequent complication of SCI and adequate prevention and management is a challenge for the multi-disciplinary team involved in the care of these individuals. There is a current lack of SCI-directed evidence, which is an exciting starting point for future research. There are remaining questions that need to be addressed regarding the optimal duration of pharmacological prophylaxis and anti-coagulation treatment, especially in individuals with chronic SCI.

## Applications to other areas of neuroscience

In this chapter, we review the clinical management of venous thromboembolism including its epidemiology, prophylaxis, diagnosis, and management. Venous thromboembolism is a possible complication in individuals presenting with paresis or plegia, especially those with an acute presentation, such as acute stroke, brachial or lumbar plexus lesion.

The occurrences of deep venous thrombosis or pulmonary embolism are factors that may delay and interfere with the rehabilitation process and expose patients to a treatment that is not risk-free (Akman et al., 2004). Institution of adequate prophylaxis, defining precisely the ideal moment to initiate or interrupt pharmacological prophylaxis, is important in order to prevent not only venous thromboembolism but also iatrogenic bleeding. These are challenges that are also present when dealing with individuals with Guillain-Barré syndrome, acute ischemic or hemorrhagic stroke, and traumatic brain injury.

As medical knowledge is constantly evolving with the addition of new evidence from clinical research, the concepts presented here aim to help the healthcare provider in the process of clinical reasoning and decision making.

## Mini-dictionary of terms

**Acute phase**: Comprises the first 12 weeks after a spinal cord injury.
**Chronic phase**: Spinal cord injury of 6 months or more.
**Novel oral anti-coagulants**: Also known as non-vitamin K antagonist oral anti-coagulants or NOACs are oral anti-coagulants that directly inhibit factor Xa or thrombin. They do not require laboratory monitoring.
**Clinical rules or clinical prediction rules**: Scores derived from risk factors, clinical characteristics and laboratory values that focus on probability estimation, in this chapter the probability of a venous thromboembolism diagnosis or recurrence.
**Pre-test probability**: Refers to the estimated probability of having a diagnosis before a test result is available.
**Post-thrombotic syndrome**: A consequence of valvular dysfunction on the venous circulation after a thrombotic event. Symptoms include limb edema, pain, skin pigmentation, and in severe grades venous ulcers.

## Key facts of venous thromboembolism in spinal cord injury

- VTE is a frequent complication of SCI, especially during the acute phase.
- It was one of the major causes of death during acute treatment.
- VTE Prophylaxis has been researched since the 1970s, and most VTE studies in individuals with SCI are directed at prophylaxis strategies.
- Neurological deficits and autonomic dysfunction may interfere with the diagnostic process.
- Both VTE-associated and anti-coagulant-associated bleeding may cause delays in rehabilitation.

## Summary points

- Pharmacological prophylaxis with LMWH should be used at least in the first 3 months after the SCI.
- The decision about pharmacological prophylaxis in patients with chronic SCI should be made on an individual basis.
- A high level of suspicion for the diagnosis should be maintained, but screening methods for VTE are not recommended.
- Classical clinical rules for the diagnosis of DVT and PE are not validated for individuals with SCI.
- Anticoagulation with parenteral or oral agents is the cornerstone of the treatment and should be used for at least 3 months.
- Prolongation of treatment should be made on an individual basis.
- Bleeding may occur even on prophylactic doses of anti-coagulants and sensorial deficit may delay the diagnosis.

## References


Abdalla, G., Fawzi Matuk, R., Venugopal, V., Verde, F., Magnuson, T. H., Schweitzer, M. A., & Steele, K. E. (2015). The diagnostic accuracy of magnetic resonance venography in the detection of deep venous thrombosis: A systematic review and meta-analysis. *Clinical Radiology*, *70*(8), 858–871.

Agarwal, N. K., & Mathur, N. (2009). Deep vein thrombosis in acute spinal cord injury. *Spinal Cord*, *47*(10), 769–772.

Agnelli, G., Buller, H. R., Cohen, A., Curto, M., Gallus, A. S., Johnson, M., ... AMPLIFY Investigators. (2013). Oral apixaban for the treatment of acute venous thromboembolism. *The New England Journal of Medicine*, *369*(9), 799–808.

Akman, M. N., Cetin, N., Bayramoglu, M., Isiklar, I., & Kilinc, S. (2004). Value of the D-dimer test in diagnosing deep vein thrombosis in rehabilitation inpatients. *Archives of Physical Medicine and Rehabilitation*, *85*(7), 1091–1094.

Alabed, S., Belci, M., Van Middendorp, J. J., Al Halabi, A., & Meagher, T. M. (2016). Thromboembolism in the sub-acute phase of spinal cord injury: A systematic review of the literature. *Asian Spine Journal*, *10*(5), 972–981.

Bates, S. M., Jaeschke, R., Stevens, S. M., Goodacre, S., Wells, P. S., Stevenson, M. D., ... Guyatt, G. H. (2012). Diagnosis of DVT: Antithrombotic therapy and prevention of thrombosis, 9th ed: American College of Chest Physicians Evidence-Based Clinical Practice Guidelines. *Chest*, *141* (2 Suppl), e351S–e418S.

Bauersachs, R., Berkowitz, S. D., Brenner, B., Buller, H. R., Decousus, H., Gallus, A. S., ... EINSTEIN Investigators. (2010). Oral rivaroxaban for symptomatic venous thromboembolism. *The New England Journal of Medicine*, *363*(26), 2499–2510.

Brandao, G. M., Junqueira, D. R., Rollo, H. A., & Sobreira, M. L. (2017). Pentasaccharides for the treatment of deep vein thrombosis. *Cochrane Database of Systematic Reviews*, *12*, CD011782.

Büller, H. R., Prins, M. H., Lensin, A. W., Decousus, H., Jacobson, B. F., Minar, E., ... EINSTEIN–PE Investigators. (2012). Oral rivaroxaban for the treatment of symptomatic pulmonary embolism. *The New England Journal of Medicine*, *366*(14), 1287–1297.

Burgazli, K. M., Atmaca, N., Mericliler, M., Parahuleva, M., Erdogan, A., & Daebritz, S. H. (2013). Deep vein thrombosis and novel oral anticoagulants: A clinical review. *European Review for Medical and Pharmacological Sciences*, *17*(23), 3123–3131.

Carrier, M., Rodger, M. A., Wells, P. S., Righini, M., Gal, L. E., & G. (2011). Residual vein obstruction to predict the risk of recurrent venous thromboembolism in patients with deep vein thrombosis: A systematic review and meta-analysis. *Journal of Thrombosis and Haemostasis*, *9*(6), 1119–1125.

Casas, E. R., Sánchez, M. P., Arias, C. R., & Masip, J. P. (1977). Prophylaxis of venous thrombosis and pulmonary embolism in patients with acute traumatic spinal cord lesions. *Paraplegia*, *15*(3), 209–214.

Chung, S. B., Lee, S. H., Kim, E. S., & Eoh, W. (2011). Incidence of deep vein thrombosis after spinal cord injury: A prospective study in 37 consecutive patients with traumatic or nontraumatic spinal cord injury treated by mechanical prophylaxis. *The Journal of Trauma*, *71*(4), 867–870 (discussion 870–861).

Chung, W. S., Lin, C. L., Chang, S. N., Chung, H. A., Sung, F. C., & Kao, C. H. (2014). Increased risk of deep vein thrombosis and pulmonary thromboembolism in patients with spinal cord injury: A nationwide cohort prospective study. *Thrombosis Research*, *133*(4), 579–584.

Cohen, A. T., Hamilton, M., Mitchell, S. A., Phatak, H., Liu, X., Bird, A., ... Batson, S. (2015). Comparison of the novel oral anticoagulants apixaban, dabigatran, edoxaban, and rivaroxaban in the initial and long-term treatment and prevention of venous thromboembolism: Systematic review and network Meta-analysis. *PLoS One*, *10*(12), e0144856.

Consortium for Spinal Cord Medicine. (2016). Prevention of venous thromboembolism in individuals with spinal cord injury: Clinical practice guidelines for health care providers, 3rd ed. *Topics in Spinal Cord Injury Rehabilitation*, *22*(3), 209–240.

Costa, G. L., Ferreira, D. C., Valacio, R. A., & Vieira Moreira, M. A. C. (2011). Quality of management of oral anticoagulation as assessed by time in therapeutic INR range in elderly and younger patients with low mean years of formal education: A prospective cohort study. *Age and Ageing*, *40*(3), 375–381.

de Almeida, R. L., Gonzaga, B. P. M., Beraldo, P. S. S., & Amado, V. M. (2019). Case report—Gluteal hematoma in two spinal cord patients on enoxaparin for venous thromboembolism prophylaxis: Evidence needed for a wiser choice. *Spinal Cord Series and Cases*, *5*, 36.

de Almeida, R. L., Rodrigues, C. C., Melo, E., Silva, C. A., Beraldo, P. S. S., & Amado, V. M. (2019). Comparison of two pharmacological prophylaxis strategies for venous thromboembolism in spinal cord injury patients: A retrospective study. *Spinal Cord*, *57*, 890–896. https://doi.org/10.1038/s41393-019-0293-1.

Di Nisio, M., van Es, N., & Büller, H. R. (2016). Deep vein thrombosis and pulmonary embolism. *Lancet*, *388*(10063), 3060–3073.

Dobler, C. C. (2019). Overdiagnosis of pulmonary embolism: Definition, causes and implications. *Breathe (Sheffield, England)*, *15*(1), 46–53.

Duffett, L., Castellucci, L. A., & Forgie, M. A. (2020). Pulmonary embolism: Update on management and controversies. *BMJ*, *370*, m2177.

Eichinger, S., Eischer, L., Sinkovec, H., Wittgruber, G., Traby, L., Kammer, M., ... Wildburger, R. (2018). Risk of venous thromboembolism during rehabilitation of patients with spinal cord injury. *PLoS One*, *13*(3), e0193735.

Fehlings, M. G., Tetreault, L. A., Aarabi, B., Anderson, P., Arnold, P. M., Brodke, D. S., ... Harrop, J. S. (2017). A clinical practice guideline for the management of patients with acute spinal cord injury: Recommendations on the type and timing of anticoagulant thromboprophylaxis. *Global Spine Journal*, *7*(3 Suppl), 212S–220S.

Fischer, E. A., Kinnear, B., Sall, D., Kelleher, M., Sanchez, O., Mathews, B., ... Olson, A. P. J. (2019). Hospitalist-operated compression ultrasonography: A point-of-care ultrasound study (HOCUS-POCUS). *Journal of General Internal Medicine*, *34*(10), 2062–2067.

Frisbie, J. H., & Sasahara, A. A. (1981). Low dose heparin prophylaxis for deep venous thrombosis in acute spinal cord injury patients: A controlled study. *Paraplegia*, *19*(6), 343–346.

Giorgi Pierfranceschi, M., Donadini, M. P., Dentali, F., Ageno, W., Marazzi, M., Bocchi, R., & Imberti, D. (2013). The short- and long-term risk of venous thromboembolism in patients with acute spinal cord injury: A prospective cohort study. *Thrombosis and Haemostasis*, *109*(1), 34–38.

Godat, L. N., Kobayashi, L., Chang, D. C., & Coimbra, R. (2015). Can we ever stop worrying about venous thromboembolism after trauma? *Journal of Trauma and Acute Care Surgery*, *78*(3), 475–480 (discussion 480-471).

Gómez-Outes, A., Terleira-Fernández, A. I., Lecumberri, R., Suárez-Gea, M. L., & Vargas-Castrillón, E. (2014). Direct oral anticoagulants in the treatment of acute venous thromboembolism: A systematic review and meta-analysis. *Thrombosis Research*, *134*(4), 774–782.

Gorman, P. H., Qadri, S. F., & Rao-Patel, A. (2009). Prophylactic inferior vena cava (IVC) filter placement may increase the relative risk of deep venous thrombosis after acute spinal cord injury. *The Journal of Trauma*, *66*(3), 707–712.

Green, D. (1992). Prophylaxis of thromboembolism in spinal cord-injured patients. *Chest*, *102*(6 Suppl), 649S–651S.

Green, D., Chen, D., Chmiel, J. S., Olsen, N. K., Berkowitz, M., Novick, A., … Tolotta, M. (1994). Prevention of thromboembolism in spinal cord injury: Role of low molecular weight heparin. *Archives of Physical Medicine and Rehabilitation*, *75*(3), 290–292.

Green, D., Lee, M. Y., Lim, A. C., Chmiel, J. S., Vetter, M., Pang, T., … Meyer, P. R. (1990). Prevention of thromboembolism after spinal cord injury using low-molecular-weight heparin. *Annals of Internal Medicine*, *113*(8), 571–574.

Halim, T. A., Chhabra, H. S., Arora, M., & Kumar, S. (2014). Pharmacological prophylaxis for deep vein thrombosis in acute spinal cord injury: An Indian perspective. *Spinal Cord*, *52*(7), 547–550.

Hull, R. D., Raskob, G. E., Rosenbloom, D., Lemaire, J., Pineo, G. F., Baylis, B., … Brant, R. (1992). Optimal therapeutic level of heparin therapy in patients with venous thrombosis. *Archives of Internal Medicine*, *152*(8), 1589–1595.

Investigators, Spinal Cord Injury Thromboprophylaxis Investigators. (2003). Prevention of venous thromboembolism in the acute treatment phase after spinal cord injury: A randomized, multicenter trial comparing low-dose heparin plus intermittent pneumatic compression with enoxaparin. *The Journal of Trauma*, *54*(6), 1116–1124 (discussion 1125-1116).

Jorge, A., White, M. D., & Agarwal, N. (2018). Outcomes in socioeconomically disadvantaged patients with spinal cord injury: A systematic review. *Journal of Neurosurgery. Spine*, *29*(6), 680–686.

Kadyan, V., Clinchot, D. M., & Colachis, S. C. (2004). Cost-effectiveness of duplex ultrasound surveillance in spinal cord injury. *American Journal of Physical Medicine & Rehabilitation*, *83*(3), 191–197.

Kahale, L. A., Hakoum, M. B., Tsolakian, I. G., Matar, C. F., Terrenato, I., Sperati, F., … Akl, E. A. (2018). Anticoagulation for the long-term treatment of venous thromboembolism in people with cancer. *Cochrane Database of Systematic Reviews*, *6*, CD006650.

Kearon, C. (2003). Natural history of venous thromboembolism. *Circulation*, *107*(23 Suppl. 1), I22–I30.

Kearon, C., Akl, E. A., Ornelas, J., Blaivas, A., Jimenez, D., Bounameaux, H., … Moores, L. (2016). Antithrombotic therapy for VTE disease: CHEST guideline and expert panel report. *Chest*, *149*(2), 315–352.

Kelly, B. M., Yoder, B. M., Tang, C. T., & Wakefield, T. W. (2010). Venous thromboembolic events in the rehabilitation setting. *PM & R : The Journal of Injury, Function, and Rehabilitation*, *2*(7), 647–663.

Konstantinides, S. V., Meyer, G., Becattini, C., Bueno, H., Geersing, G. J., Harjola, V. P., … Zamorano, J. L. (2019). ESC guidelines for the diagnosis and management of acute pulmonary embolism developed in collaboration with the European Respiratory Society (ERS): The task force for the diagnosis and management of acute pulmonary embolism of the European Society of Cardiology (ESC). *The European Respiratory Journal*, *54*(3), 1–68. https://doi.org/10.1183/13993003.01647-2019. PMID: 31473594.

Krassioukov, A., Biering-Sørensen, F., Donovan, W., Kennelly, M., Kirshblum, S., Krogh, K., … Autonomic Standards Committee of the American Spinal Injury Association/International Spinal Cord Society. (2012). International standards to document remaining autonomic function after spinal cord injury. *The Journal of Spinal Cord Medicine*, *35*(4), 201–210.

Kumagai, G., Wada, K., Kudo, H., Asari, T., Ichikawa, N., & Ishibashi, Y. (2020). D-dimer monitoring combined with ultrasonography improves screening for asymptomatic venous thromboembolism in acute spinal cord injury. *The Journal of Spinal Cord Medicine*, *43*(3), 353–357.

Mackiewicz-Milewska, M., Jung, S., Kroszczyński, A. C., Mackiewicz-Nartowicz, H., Serafin, Z., Cisowska-Adamiak, M., … Rość, D. (2016). Deep venous thrombosis in patients with chronic spinal cord injury. *The Journal of Spinal Cord Medicine*, *39*(4), 400–404.

Masuda, M., Ueta, T., Shiba, K., & Iwamoto, Y. (2015). D-dimer screening for deep venous thrombosis in traumatic cervical spinal injuries. *The Spine Journal*, *15*(11), 2338–2344.

Mazzolai, L., Aboyans, V., Ageno, W., Agnelli, G., Alatri, A., Bauersachs, R., … Brodmann, M. (2018). Diagnosis and management of acute deep vein thrombosis: A joint consensus document from the European Society of Cardiology working groups of aorta and peripheral vascular diseases and pulmonary circulation and right ventricular function. *European Heart Journal*, *39*(47), 4208–4218.

Merli, G. J., Crabbe, S., Doyle, L., Ditunno, J. F., & Herbision, G. J. (1992). Mechanical plus pharmacological prophylaxis for deep vein thrombosis in acute spinal cord injury. *Paraplegia*, *30*(8), 558–562.

Merli, G. J., Herbison, G. J., Ditunno, J. F., Weitz, H. H., Henzes, J. H., Park, C. H., & Jaweed, M. M. (1988). Deep vein thrombosis: Prophylaxis in acute spinal cord injured patients. *Archives of Physical Medicine and Rehabilitation*, *69*(9), 661–664.

Mismetti, P., Laporte, S., Pellerin, O., Ennezat, P. V., Couturaud, F., Elias, A., … PREPIC2 Study Group. (2015). Effect of a retrievable inferior vena cava filter plus anticoagulation vs anticoagulation alone on risk of recurrent pulmonary embolism: A randomized clinical trial. *JAMA*, *313*(16), 1627–1635.

Oliphant, C. S., Jacobs, A., Kabra, R., & Das, P. (2013). Novel oral anticoagulants for the prevention and treatment of thromboembolism. *Future Cardiology*, *9*(6), 849–861.

Onishi, A., St Ange, K., Dordick, J. S., & Linhardt, R. J. (2016). Heparin and anticoagulation. *Frontiers in Bioscience*, *21*, 1372–1392.

Raschke, R. A., Reilly, B. M., Guidry, J. R., Fontana, J. R., & Srinivas, S. (1993). The weight-based heparin dosing nomogram compared with a "standard care" nomogram. A randomized controlled trial. *Annals of Internal Medicine*, *119*(9), 874–881.

Robertson, L., & Jones, L. E. (2017). Fixed dose subcutaneous low molecular weight heparins versus adjusted dose unfractionated heparin for the initial treatment of venous thromboembolism. *Cochrane Database of Systematic Reviews*, *2*, CD001100.

Roussi, J., Bentolila, S., Boudaoud, L., Casadevall, N., Vallée, C., Carlier, R., … Bussel, B. (1999). Contribution of D-dimer determination in the exclusion of deep venous thrombosis in spinal cord injury patients. *Spinal Cord*, *37*(8), 548–552.

Sachdeva, A., Dalton, M., & Lees, T. (2018). Graduated compression stockings for prevention of deep vein thrombosis. *Cochrane Database of Systematic Reviews*, *11*, CD001484.

Schein, J. R., White, C. M., Nelson, W. W., Kluger, J., Mearns, E. S., & Coleman, C. I. (2016). Vitamin K antagonist use: Evidence of the difficulty of achieving and maintaining target INR range and subsequent consequences. *Thrombosis Journal*, *14*, 14.

Spivack, S. B., & Aisen, M. L. (1997). A comparison of low molecular weight heparin and low dose unfractionated heparin prophylaxis in subacute myelopathy. *The Journal of Spinal Cord Medicine*, *20*(4), 402–405.

Swan, D., Hitchen, S., Klok, F. A., & Thachil, J. (2019). The problem of under-diagnosis and over-diagnosis of pulmonary embolism. *Thrombosis Research*, *177*, 122–129.

Ten Cate-Hoek, A. J. (2018). Prevention and treatment of the post-thrombotic syndrome. *Research and Practice in Thrombosis and Haemostasis*, *2*(2), 209–219.

Thean, D., & Alberghini, M. (2016). Anticoagulant therapy and its impact on dental patients: A review. *Australian Dental Journal*, *61*(2), 149–156.

Thomas, S. M., Goodacre, S. W., Sampson, F. C., & van Beek, E. J. (2008). Diagnostic value of CT for deep vein thrombosis: Results of a systematic review and meta-analysis. *Clinical Radiology*, *63*(3), 299–304.

Thumbikat, P., Poonnoose, P. M., Balasubrahmaniam, P., Ravichandran, G., & McClelland, M. R. (2002). A comparison of heparin/warfarin and enoxaparin thromboprophylaxis in spinal cord injury: The Sheffield experience. *Spinal Cord*, *40*(8), 416–420.

Todd, J. W., Frisbie, J. H., Rossier, A. B., Adams, D. F., Als, A. V., Armenia, R. J., … Tow, D. E. (1976). Deep venous thrombosis in acute spinal cord injury: A comparison of 125I fibrinogen leg scanning, impedance plethysmography and venography. *Paraplegia*, *14*(1), 50–57.

Tritschler, T., Kraaijpoel, N., Le Gal, G., & Wells, P. S. (2018). Venous thromboembolism: Advances in diagnosis and treatment. *JAMA*, *320*(15), 1583–1594.

Weitz, J. I. (1997). Low-molecular-weight heparins. *The New England Journal of Medicine*, *337*(10), 688–698.

Wells, P. S., Ihaddadene, R., Reilly, A., & Forgie, M. A. (2018). Diagnosis of venous thromboembolism: 20 years of Progress. *Annals of Internal Medicine*, *168*(2), 131–140.

Wolf, S. J., Hahn, S. A., Nentwich, L. M., Raja, A. S., Silvers, S. M., Brown, M. D., & American College of Emergency Physicians Clinical Policies Subcommittee (Writing Committee) on Thromboembolic Disease. (2018). Clinical policy: Critical issues in the evaluation and management of adult patients presenting to the emergency department with suspected acute venous thromboembolic disease. *Annals of Emergency Medicine*, *71*(5), e59–e109.

Chapter 7

# Osteoporosis-related fractures: What they are and how they occur following spinal cord injury

Ana Paula Silva Champs[a], Gustavo Correa Netto de Melo[a], Gabriela Afonso Galante Maia[a], Fabiana Gonçalves Oliveira[a], Antonio Lopes da Cunha Júnior[a], and Maria Marta Sarquis Soares[b]

[a]*SARAH Network Rehabilitation Hospitals, Belo Horizonte, Minas Gerais, Brazil;* [b]*Internal Medicine Department, School of Medicine, Universidade Federal de Minas Gerais, Belo Horizonte, Minas Gerais, Brazil*

## Abbreviations

| | |
|---|---|
| **25(OH)D** | 25 hydroxyvitamin D |
| **aBMD** | areal bone mineral density |
| **AIS** | ASIA Impairment Scale |
| **BMD** | bone mineral density |
| **BMI** | Bone Mass Index |
| **BWSTT** | body-weight exercises supported treadmill training |
| **CTX-I** | type I collagen C-telopeptide |
| **DEXA** | dual-energy X-ray absorptiometry |
| **EMS** | electromyostimulation |
| **ES** | electrical stimulation |
| **FAO** | bone-specific alkaline phosphatase |
| **FES** | functional electrical stimulation |
| **FRAX** | fracture risk assessment tool |
| **pQCT** | quantitative computed tomography |
| **PTH** | parathyroid hormone |
| **SCI** | spinal cord injury |
| **WHO** | World Health Organization |

## Introduction

In 2017, approximately 22 million people around the world were living with a spinal cord injury (SCI), a rate increase of 14.21% in the last 10 years (GBD 2016 Traumatic Brain Injury and Spinal Cord Injury Collaborators, 2019). The better access to medical care resources such as emergency medical services and rehabilitation departments led to longer survival for this population (Frontera & Mollett, 2017). Henceforth, many consequences were related to SCI such as neurogenic bladder, neurogenic bowel, neuropathic pain, spasticity, and osteoporosis (Craven, Robertson, Delparte, Ashe, & Janice, 2008; Frontera and Mollett, 2017).

Osteoporosis affects the health of individuals with SCI because it increases the risk of fragility fractures with consequences related to fractures such as ulcers, immobility, depression and even death (Carbone et al., 2013; Jiang, Dai, & Jiang, 2006; Morse et al., 2009). Osteoporosis after SCI is defined as an excessive bone resorption after SCI and bone fragility fractures are defined as fractures caused by a trauma that would be insufficient for a normal bone to fracture (Craven et al., 2008; Zleik et al., 2019) or by minimally loaded situations (McPherson et al., 2014; Shields et al., 2005).

The impact of fractures on the lives of individuals with SCI suggests that most patients will be hospitalized after the fracture and undergone to surgery, and complications as worse performance in activities of daily living and ambulation after the fracture are mostly described. These issues mean increased disability and costs for the health system (Carbone et al., 2013).

Diagnosis and Treatment of Spinal Cord Injury. https://doi.org/10.1016/B978-0-12-822498-4.00007-5

Special attention should be paid to the prevention of fractures, and appropriate treatments should be instigated to prevent excessive bone loss and futher investigations should be performed. Most of the physicians are not educated to measure risk factors for bone fragility fractures. Dual-energy X-ray absorptiometry (DEXA) should be performed in patients with SCI, likely before the occurrence of fractures. Bone remodeling markers could give us a clue of the resorption rate and treatments to prevent further loss should be consider.

## Physiopathology

Bone demineralization after SCI occurs in a rate loss of approximately 4% per month in the first year (Battaglino, Lazzari, Garshick, & Morse, 2012; Bauman, Spungen, Morrison, Zhang, & Schwartz, 2005; Bauman, Wecht, et al., 2005), primarily in the knees. Demineralization begins in the early days, peaks around the 10th to 16th weeks and stabilizes around the 16th to 24th months (Battaglino et al., 2012; Bauman, Spungen, et al., 2005; Bauman, Wecht, et al., 2005). Recent studies suggest that the bone loss extends 3 to 8 years after SCI (Haider, Lobos, Simonian, Schnitzer, & Edwards, 2018). Previous studies demonstrated bone demineralization greater than 20% in the hip, 37% to 52% in the distal femur and 36% to 70% in proximal tibia after 1 to 3 years after SCI (Garland, Adkins, & Stewart, 2005; Haider et al., 2018; Troy & Morse, 2015).

Bone loss associated with SCI is related to immobility and metabolic changes (Bauman & Spungen, 2000; McPherson et al., 2014; Uebelhart, Demiaux-Domenech, Roth, & Chantraine, 1995). Individuals with SCI exhibit primarily hypercalcemia, hypercalciuria, hyperparathyroidism, decreased levels of osteocalcin, and higher levels of sclerotin. The loss of bone mineral density is 2 to 4 times greater than the loss that occurs in an immobilized individual without SCI (Uebelhart et al., 1995).

Other related factors may contribute to this bone loss, such as vitamin D deficiency and the use of methylprednisolone, anti-convulsant drugs, and psychotropic substances. Previous studies discovered that 14% to 32% of individuals with SCI were deficient in vitamin D (Bauman, Zhong, & Schwartz, 1995; Koutrakis et al., 2019). Even in tropical countries, vitamin D deficiency is a common health problem (Champs, Maia, Oliveira, de Melo, & Soares, 2020).

## Fractures

The consequence of osteoporosis in individuals with SCI is fracture due to bone fragility. It usually occurs in transfers and activities with minimal or no trauma, such as low-speed falls, torsional stress (Akhigbe et al., 2015; Bethel et al., 2016). Studies (Akhigbe et al., 2015; Bethel et al., 2016; Champs et al., 2020) detected a specific time between SCI and fracture was 9 to 10 years, average age of first fracture for women 50+ and for men 40 + years-old (Champs et al., 2020). Fractures may occur in 25% to 46% of these individuals over their lifetimes (Bethel et al., 2016; Craven, Robertson, McGillivray, & Adachi, 2009; Morse et al., 2009). Fractures are primarily related to torsional forces during transfer, passive mobilization, compressive forces, and falls (Akhigbe et al., 2015; Bethel et al., 2016; Champs et al., 2020). It seems the most frequent fracture cause is due to falls from a wheelchair or even their own height while walking or standing (Champs et al., 2020). This result is not surprising because falls are a common health problem in individuals with SCI, and the prevention of falls should be a priority. Another study found transfer as the main cause of fracture (Akhigbe et al., 2015).

The risk of fracture after SCI is different from the risk of osteoporosis (Abderhalden et al., 2017; Cirnigliaro et al., 2017), individuals with tetraplegia have more osteoporosis than those with paraplegia, however, individuals with paraplegia have a higher frequency of fractures due to exposure to falls (Champs et al., 2020; Craven et al., 2009; Morse et al., 2009). Most places of fractures are in the tibia and/or distal fibula, distal femur, or hip/proximal femur and proximal tibia (Akhigbe et al., 2015; Bethel et al., 2016; Morse et al., 2009). It seems hip/proximal femur and distal tibia/fibula are as common as the distal femur (Champs et al., 2020). However, individuals who are ambulatory and wheelchair dependent may have different areas of fractures. Ambulatory individuals fracture more often the distal tibia/fibula, and wheelchair-bound patients fractured the distal and proximal femur (Champs et al., 2020). A second (15%) and third (2%) fracture are experienced by patients (Champs et al., 2020).

The impact of fractures on the lives of people with SCI suggests that most of the patients are hospitalized after the fracture, and some had complications. Almost one-third of the participants reported worse performance in activities of daily living and ambulation after the fracture had healed. Individuals with osteoporosis-related fractures after SCI had several complications. Most individuals were hospitalized, and half of them underwent surgery (Champs et al., 2020). These issues mean increased disability and costs for the health system.

## Risk factors

Previous studies showed risk factors (Bethel et al., 2016; Garland et al., 2005; Morse et al., 2009) and the following ones were described associated with increased fracture criteria: AIS (ASIA Impairment Scale) A or B; age 40+ years old; SCI longer than 3 years; age at SCI of 16 years or less; three servings of coffee per day; smoking; women; family history of osteoporotic fractures; low bone mineral density; low weight Bone Mass Index (BMI) $<19$ kg/cm$^2$; alcohol intake greater than 30 g/day; paraplegia, and the use of corticosteroids (Craven et al., 2008, 2009). Studies by Craven developed protocols conducive to treatment based on these risk factors. Other aspects related to fractures are 25 hydroxyvitamin D [25(OH)D] levels less than 20 ng/mL (Koutrakis et al., 2019).

## Diagnosis

### Laboratory

Bone formation markers are useful in the management of SCI patients. Procollagen type I N-propeptide reflects osteoblastic activity, bone-specific alkaline phosphatase, which is a bone formation marker, is elevated on the first year after injury. Osteocalcin, another bone remodeling marker is reduced in the first year others (Bauman and Spungen, 2000; Uebelhart et al., 1995).

Bone resorption markers such as deoxypyridinoline, pyridinoline, sclerostin, and type I collagen C-telopeptide (CTX-I) can be used. CTX-I is increased in the second week after SCI and peaks between 2 and 4 months, and its levels can remain increased for up to 5 years (Bauman and Spungen, 2000; Uebelhart et al., 1995).

The immobilization resulting from acute SCI stimulates osteoclastic bone resorption and calcium homeostasis markers as parathyroid hormone (PTH) is reduced in the first year after injury, urinary calcium is elevated, serum calcium is increased, while ionized calcium is generally normal. These alterations peak from the 3rd to the 10rd month after SCI (Bauman and Spungen, 2000; Uebelhart et al., 1995).

### Dual-energy X-ray absorptiometry

Dual-energy X-ray absorptiometry (DEXA) for patients with SCI is a confirmatory test to predict the future risk of fracture and also serves for ongoing monitoring of bone health and to access bone loss (McPherson et al., 2014) by the bone mineral density (BMD) measurements. It is precise and safe (Shields et al., 2005) since there were strong associations among different observers and it emits a very low level of radiation (0.1 mSV), that is, 1/10–1/30 less radiation than a chest X-ray. DEXA calculates the measurement of bone mineral density in g/cm$^2$ and protocols for calculating bone mineral density for patients with SCI already exist. Some studies suggested that low BMD in the hip predicted fracture in individuals with SCI (Abderhalden et al., 2017) and other authors suggested that DEXA measurement of the distal femur and proximal tibia, with specific protocols for the evaluation of these sites (Cirnigliaro et al., 2017; Garland et al., 2005; Morse et al., 2009) and for knee fracture thresholds of 0.78 g/cm$^2$ which a fracture may start to occur and fracture breakpoint of 0.49 g/cm$^2$ (Garland et al., 2005; Fig. 1).

The measurement of BMD of the knees is extremely important for monitoring bone loss, as the hip and spine do not show the highest losses of BMD after SCI and because osteoporosis is most severe below the level of the injury (Morse, Biering-Soerensen, et al., 2019; Morse, Troy, et al., 2019). It can be even more useful in those patients where assessing the spine and hip is not possible, such as cases with spine surgeries or hip heterotopic ossifications. The limitations for measuring BMD in the knees are fractures and orthopedic surgeries. In cases of flexion deformities, the assessment of bone mineral density can be performed in the lateral scans (Fig. 2).

### Quantitative computed tomography

Quantitative computed tomography (pQCT) is also a safe method and evaluates density, volume and geometry, differentiates cortical bone and trabecular bone, but it is difficult to establish prediction, and have plus radiation, and aBMD $<114$ mg/cm$^3$ in the femur and aBMD $<72$ mg/cm$^3$ in the tibia are described as fracture thresholds (Garland et al., 2005; Liu et al., 2000).

## Treatment

It is still under debate the best way to prevent fragility fractures after SCI (Morse, Biering-Soerensen, et al., 2019; Morse, Troy, et al., 2019). A diet rich in fruits and vegetables, daily doses of calcium of 1000 to 1200 mg vitamin D,

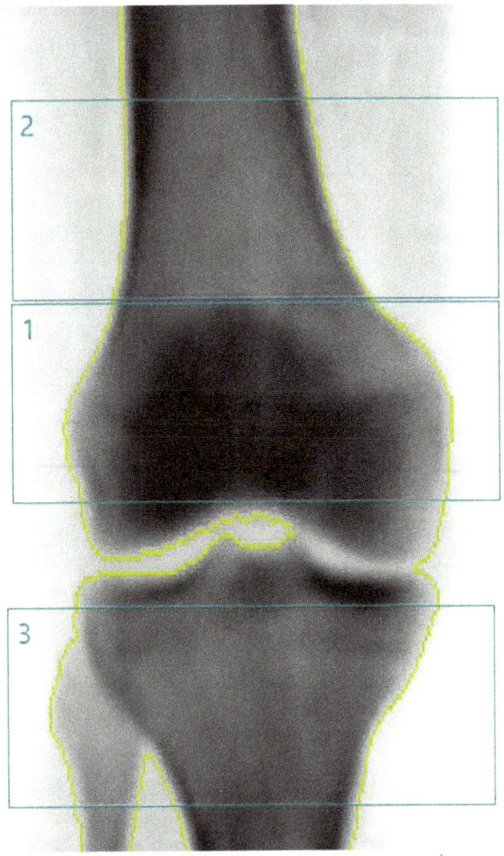


**FIG. 1** Dual-energy X-ray absorptiometry (DEXA) scan image of the knee. The boxes delineate the regions of the affected area scanned at the distal femur (*blue box* 1: distal femoral epiphysis and *blue box* 2: distal femoral metaphysis) and the proximal tibial (*blue box* 3: proximal tibial epiphysis).

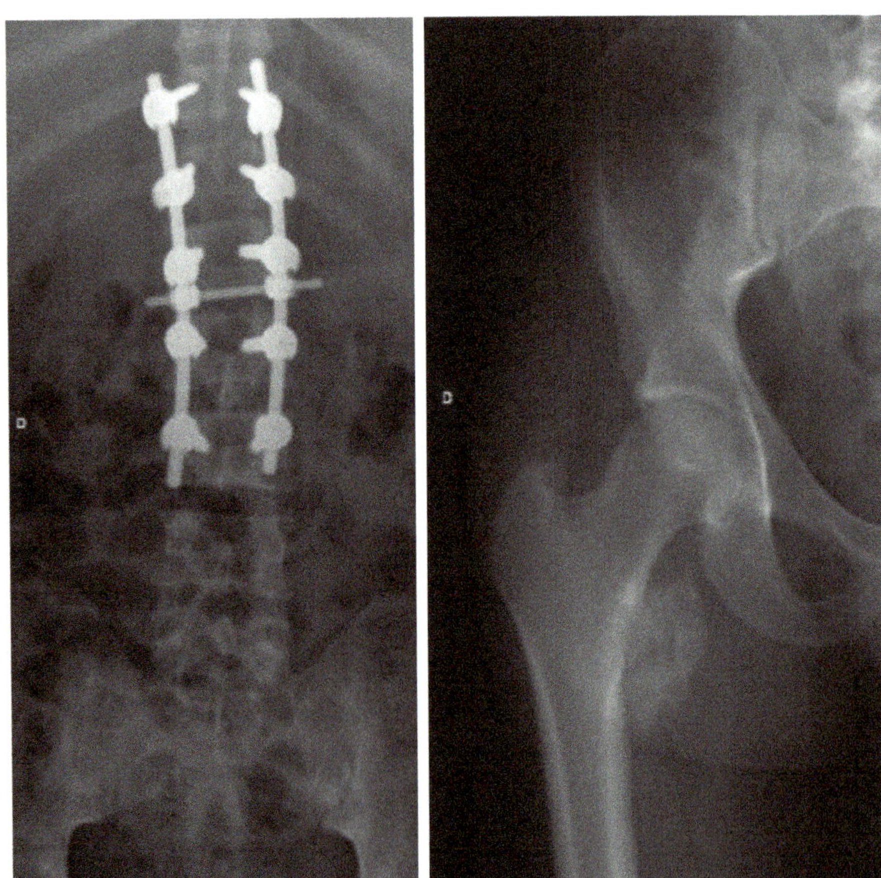


**FIG. 2** Digital X rays scans of the spine and the hip. Spine surgeries and heterotopic ossifications in the hip artificially increase bone mineral density.

200 to 1000 iU/day aiming a serum level of 30 ng/dL, physical exercises, stop smoking and drinking alcoholic beverages and avoid over doses of coffee are contributive recommendations (Bauman et al., 2015; Koutrakis et al., 2019; Morse, Biering-Soerensen, et al., 2019; Morse, Troy, et al., 2019).

## Pharmacologic therapy

Studies showed the efficacy of bisphosphonates in daily doses of alendronate 10 mg (orally) and calcium 500 mg to decrease bone turnover in people with SCI (Zehnder et al., 2004). Bisphosphonates were more effective if 25-OH > 33 ng/mL and vitamin D deficiency should be investigated because of osteomalacia risk (Morse, Biering-Soerensen, et al., 2019; Morse, Troy, et al., 2019). However, it is a medication with around 50% adherence despite being generally well-tolerated (Morse, Biering-Soerensen, et al., 2019; Morse, Troy, et al., 2019). The association of calcium 500 mg, alendronate, 70 mg weekly and vitamin D 400 U starting 10 days after SCI and use for twelve months to 24 months prevented the decrease of BMD in the distal tibial epiphysis in patients with SCI (Gilchrist et al., 2007; Zehnder et al., 2004). In contrast, the isolated use of 500 mg of calcium daily for the same period didn't prevent it (Gilchrist et al., 2007).

Pamidronate 60 mg usage, intravenously 4 times a year associated with dietary calcium (700 mg/day) and vitamin D, didn't prevent bone loss in the long term, in patients with spinal cord injury with complete motor injury (Bauman, Spungen, et al., 2005; Bauman, Wecht, et al., 2005). Zoledronic acid administered once a year, intravenously, showed a significant reduction in bone loss of lumbar spine, hip and trochanter with some side effects as myalgia, fever and nasal congestion (Shapiro et al., 2007). Another study described zoledronic acid administration failed to prevent bone loss at the knee in persons with acute spinal cord injury (Bauman et al., 2015).

Many complications are associated with bisphosphonates as hypocalcemia, gastrointestinal effects, osteomalacia, osteonecrosis of the jaw (1/10,000 patients) after more than 4 years of use, atrial fibrillation, hepatotoxicity, teratogenicity for women of childbearing age and it cannot be used in chronic renal failure.

Vitamin D is also effective (Bauman, Spungen, et al., 2005; Bauman, Wecht, et al., 2005). Vitamin D treatment determined significant reduction of urinary bone resorption markers despite neutral effects on bone formation markers (Bauman, Spungen, et al., 2005; Bauman, Wecht, et al., 2005).

Teriparatide reduced the risk of vertebral fractures at a dose of 20 μg daily, subcutaneously, when associated with robotic treatment. Teriparatide alone, did not improve BMD either on the spine or on the hip for patients with SCI (Chain, Koury, & Bezerra, 2012).

Denosumab inhibits osteoclast formation, prevents bone resorption, and reduces risk of spine fractures. A study by Gifre et al. (2016) treated 14 patients with SCI with denosumab, calcium, and vitamin D with improvement of bone mineral density in 12 months. Despite that, there are still poor evidence for people with SCI.

There is no consensus on the follow-up after staring pharmacologic therapy. It is recommended to repeat DEXA after 1 to 2 years after start of treatment (Morse, 2019).

Treatment with bisphosphonate has been done for 4 to 5 years and should be consider drug holiday after 3 to 5 years of use, because of its prolonged effect (Morse, Biering-Soerensen, et al., 2019; Morse, Troy, et al., 2019).

## Rehabilitation/non-pharmacological therapy

One of the possible causes of osteoporosis after SCI is the inability to perform weight bearing activities and loss of the strengthening effects of bone stress related to muscular contraction and gravity. In this way, rehabilitation is an option for osteoporosis prevention and treatment after SCI focus on stimulation muscle contraction and weight bearing in order to stimulate the mechanical stress, to which bone is exposed to daily activities among healthy people (Craven et al., 2008; Dolbow et al., 2011).

The mechanostat theory, based on Wolff's law, states that mechanical loading influences bone structure by changing its mass (amount of bone) and architecture (its arrangement) to provide a structure that resists to habitual loads. There are strains within bone that are kept within certain limits by adding and removing bone tissue that could promote bone strengthening, depending on the applied forces. The paralysis secondary SCI causes a decrease in the mechanical load and results in bone tissue loss (Cirnigliaro et al., 2017; Dolbow et al., 2011).

### *Standing and walking*

Supported standing with standing frames, tilt tables, long leg braces, wheelchair standing have long been used as a treatment for reducing and or delaying osteoporosis after SCI. It is traditionally incorporated in rehabilitation programs, even without proof of its effectiveness (Goktepe, Tugcu, Yilmaz, Alaca, & Gunduz, 2008).

There are three good quality studies that evaluate the influence of weight bearing activities in the acute phase of SCI, with controversial results (Alekna, Tamulaitiene, Sinevicius, & Juocevicius, 2008). It was observed that an early mobilized subject showed no or insignificant loss of trabecular bone when compared with non-interventional groups (De Bruin et al., 1999). In the same other way, two studies has found that patients who performed daily standing or walking exercises for more than 1 h, 5 times per week had significantly higher BMD in the lower extremities after 2 years, in comparison to those who did not perform standing movements (Alekna et al., 2008).

For the chronic phase, the results seemed more uniform, with very little evidence of any gain in BMD when the first year after injury had passed (Biering-Sørensen, Hansen, & Lee, 2009). Goktepe et al. found no significant effect on BMD in chronic patients after daily standing for less or more than 1 h for a period of 4 years (Goktepe et al., 2008). In the same way, a study by Kunkel et al. (1993) did not detect changes in fracture risk at femoral neck after standing in a frame program for 45 min twice a day for 5 months. Dudley-Javoroski et al. compared the effect of bone compressive loads using 0% body weight (no standing), 40% body weight (passive standing—"low dose"), and 150% body weight (quadriceps stimulation in supported stance—"high dose"). After a 3 year training protocol, 3 times a week, the high dos group had an attenuated BMD decline when compared to passive standing or no standing individuals. There were no differences between low-dose and no standing group, suggesting that BMD for participants doing passive stance is the same as those who performed no standing (Dudley-Javoroski & Shields, 2013).

Gait training with walking frame, orthosis, body-weight exercises supported treadmill training (BWSTT), and exoskeleton is among the activities of a rehabilitation program with the main purpose to facilitate ambulation SCI persons. Besides that, there are some studies investigating its benefits on bone loss after SCI, but up to now, there is no good quality evidence in favor of this kind of intervention. Carvalho et al. (2006) showed that a BWSTT (30%–50% weight relief) with neuromuscular electrical stimulation, twice a week, for 20 min, for 6 months, in tetraplegic individuals led to elevate bone formation and reduce bone resorption markers, although the BMD did not enough. Another study with 12 months of BWSTT training did not increase bone density in individuals with chronic incomplete SCI, but BMD did not decrease at fracture-prone sites (Giangregorio et al., 2006). Other studies involving ambulatory orthosis also observed that walking did not improve BMD (Ogilvie, Bowker, & Rowley, 1993; Thoumie et al., 1995).

Karelis et al., in a case report, noticed little benefits on locomotor training using a robotic exoskeleton 3 times/week, for 6 weeks. A previous case report study did not find BMD benefits after robotic-assisted BWSTT (Lokomat) for 1 h activity, 3 times a week, in a period of 3 months, in an acute SCI. It is important to be careful about fractures occurring during this activity as cited by Zleik et al. Some studies suggested that low BMD may be a relative contraindication for the use of these devices (Zleik et al., 2019).

The bone loss prevention/treatment needs intensive body loading and for a long period in order to maintain body mass. Although there is no BMD cut-off value to preclude patients from participating in rehab interventions, an assessment of fracture risk is indicated before activities such as standing and walking to ensure its safety (Cirnigliaro et al., 2017; Craven et al., 2008). It is recommended to avoid exercises that result in torsion of the distal lower extremity or that have a risk of falling (Craven et al., 2008).

Although practices for increased mobility and weight-bearing are clearly important to skeletal health in SCI, there is not sufficient evidence in their capability to maintain BMD of the lower extremity (Zleik et al., 2019). In summary, there is low evidence that more than 5 h per week standing exercises could preserve tibial BMD in acute (less than 1 year) SCI cases (Soleyman-Jahi et al., 2018).

### *Electrical stimulation*

Electrical stimulation (ES) and Functional Electrical Stimulation (FES) involve the use of surface or implanted electrodes to stimulate muscular activity. Whereas both methods typically employ cyclical patterns of electrical stimulation that simulate natural muscular activity, FES is directed toward the attainment of purposeful movement such as cycling or walking.

FES is the most extensively investigated approach since it combines the benefits from electrical stimulation and mechanical loading of the lower extremity long bones. Although FES intervention has been demonstrated to improve muscle atrophy, solid evidence from a large-scale study is still lacking regarding its influence on sub-lesional BMD attenuation.

FES studies enrolled participants with both acute and chronic injuries and are therefore difficult to classify as pure prevention or treatment interventions. In one review study, Bauman and Cardozo (2015) described that FES training administered at the time of acute injury appears to markedly reduced bone loss at the site applied or, if administered to those with chronic SCI, was also efficacious, albeit with a reduced magnitude of effect, at the place to which the load was applied. Soleyman-Jahi et al. (2018) investigated in a literature review the prevention- and evidence-based treatments for osteoporosis related to SCI. They discovered low-quality evidence indicating that ES provided no significant effects. Very low-quality evidence did not show any benefit for low-intensity (3 days per week) cycling with FES in chronic SCI.

Chang et al. (2013) carried out a meta-analysis study to investigate whether bisphosphonate administration or functional electrical stimulation (FES) training could effectively decrease bone mineral. They noticed that FES cycling did not significantly decrease BMD loss in acute SCI individuals. Whereas, for people with a chronic SCI, BMD demonstrated a significant increase near the site of maximal mechanical loading. Furthermore, the studies employing FES $\geq$5 days per week were likely to have better effectiveness than studies using FES $\leq$3 days per week (Chang et al., 2013).

In the acute phase, there is a non-randomized trial reporting that FES cycling provided 1 to 3 months after SCI can partially reduce BMD loss in distal femur (Lai et al., 2010). In addition, the effect on the attenuation of bone loss in the distal femur diminished once FES cycling was discontinued (Lai et al., 2010). Eser et al. (2003) measured the effect of an FES cycling intervention on bone mineral density (BMD) of the tibia in recently injured SCI people in a non-randomized clinical trial. The intervention consisted of 30-min FES cycling three times a week for 6 months. They concluded that FES cycling was not effective for improving bone mass at the tibial mid shaft during the first year after injury.

For the chronic phase, the studies are conflicting, but the investigations that show improvement seem to be those with a longer period of training, which is 12 months or more, 5 times/week (Belanger, Stein, Wheeler, Gordon, & Leduc, 2000; Chen et al., 2005). It is also evident that all these studies measured their improvement corresponding to the trabecular bone, in particular in the distal femur or the proximal tibia.

FES cycling studies that reported a positive effect on bone parameters used protocols of at least 3 sessions per week for 6 months in duration. The increase in bone parameters was in areas directly affected by the stimulated muscles. Although a study showed that the FES cycling intervention needed to be maintained or bone gains were lost (Chen et al., 2005). Frotzler et al. (2009) found that BMD was preserved in the distal locations of some participants in 12 months. They concluded that the high-volume FES-induced cycle training had clinical relevance as it could partially reverse bone loss and thus may reduce fracture risk at this fracture-prone site.

The use of ES to clinical care has been fraught with difficulties due to the labor intensive nature of the present approaches and the appreciation that any effect on bone is rapidly lost, when the ES training is either reduced in frequency or terminated (Belanger et al., 2000; Eser et al., 2003; Lai et al., 2010). Considerations for FES efficacy in osteoporosis in SCI should consider attention to the duration, frequency, and power of the output (Chang et al., 2013; Dudley-Javoroski & Shields, 2013). Guidelines include intensities with loads of 1–1.5 times body weight, sessions of 3 or more per week for several months to 1 or more years, and safety considerations to prevent fractures (Dolbow et al., 2011). Craven et al. (2008) considered that ES and FES should be used with caution in patients with combined hip and knee flexion contractures of >30, a prior lower extremity fracture, severe lower extremity spasticity, and/or significant ankle plantar flexion contractures.

Finally, although the use of FES cycling looks promising, the limited availability of cycle ergometry for home or longitudinal use may limit its generalization, if therapy cannot be maintained outside a clinical trial environment. The quality of evidence available in this field is poor. Some results suggest physical activity could attenuate bone loss (Chain et al., 2012; Goktepe et al., 2004). Studies reported benefits on arm bone mass maintenance but did not prevent demineralization in the lower body (Goktepe et al., 2004). The type of activity and training intensity were not well-described, and many studies used a self-reported physical activity as a reference.

## *Physical activity*

The quality of evidence available in this field is poor. Some results suggest physical activity could attenuate bone loss (Chain et al., 2012; Goktepe et al., 2004). Two of them reported benefits on arm bone mass maintenance but did not prevent demineralization in the lower body (Goktepe et al., 2004). The type of activity and training intensity were not well-described, and many studies used a self-reported physical activity as a reference. Miyahara et al. found that the earlier the athlete started sports after injury, the higher the BMD of the legs. Further, a longer period of athletic career after restarting was significantly related to higher leg BMD.

## *Vibration*

The mechanical vibration used to prevent and/or treat the loss of BMD mixed results (Alizadeh-Meghrazi, Masani, Popovic, & Craven, 2012; Asselin, Spungen, Muir, Rubin, & Bauman, 2011). Similar to FES, the specifications of the chosen full-body vibration platform, including intensity and frequency of vibration, signal transmission, and joint position and angle, could play a role in effectiveness (Alizadeh-Meghrazi et al., 2012; Asselin et al., 2011).

### *Ultrasound*

Ultrasound does not seem to play a role in preventing or treating osteoporosis. The only randomized study by Warden et al. (2001) did not indicate any benefit of low-intensity pulsed ultrasound to prevent decreased BMD. Therapeutic pulsed ultrasound was applied to the heel of individuals with spinal cord injury for 20 min a day, 5 times a week for a period of 6 consecutive weeks. The contralateral heel was simultaneously treated with inactive United States. No improvement was observed in any of the parameters evaluated ($P > .05$). According to the authors, this finding may be related mainly to the inability of the US to effectively penetrate the external cortex of the bone due to its acoustic properties.

### Combined treatments

There are few evaluated studies that combined interventions for the treatment of bone loss in chronic SCI. These studies assess the concomitant administration of pharmacological therapy with non-pharmacological rehabilitation interventions (Gordon, Wald, & Schnitzer, 2013; Morse, Biering-Soerensen, et al., 2019; Morse, Troy, et al., 2019). Morse, Biering-Soerensen, et al. (2019) and Morse, Troy, et al. (2019) investigated the combination of zoledronate (ZA) with FES-rowing. The results demonstrated that the osteogenic response to FES-rowing was dose-dependent and combination therapy with ZA and FES-row training had therapeutic potential to improve bone quality, and perhaps reduced fracture risk at the most common fracture site following SCI. Gordon et al. (2013) evaluated the response of bone to recombinant parathyroid hormone in combination with weight bearing. The results did not show a statistically significant increase in BMD of the lumbar spine (Gordon et al., 2013).

## Conclusion

Special attention should be paid to prevent fractures, and appropriate treatment for preventing excessive bone loss and investigations should be performed in rehabilitation hospitals for patients with SCI. Most of the physicians are not educated to measure these risk factors. DEXA should be performed, mainly after an osteoporotic fracture and in fact DEXA should be performed before the occurrence of fracture to prevent such event. Bone remodeling markers (like CTX-I) could give us a clue of the resorption rate and treatment could prevent further loss. Consider identifying risk factors and establishing prevention programs and appropriate treatment. Campaigns to prevent falls and stimulate the appropriate identification and treatment of osteoporosis in individuals with SCI must be promoted.

The analysis of literature for treatments still show mixed results, although there is no guideline, treatment with pharmacological therapy should be considered for patients with SCI, risk factors for fractures and DEXA criteria for osteoporosis. Use of oral bisphosphonates as alendronate or intravenous zoledronic acid, have showen efficacy but further studies should be performed.

The studies suggest standardization of the osteometabolic evaluation of patients with SCI, it is recommended undergo to laboratory evaluation before treatment, 6 months after beginning pharmacological therapy and repeated on annual basis. DEXA should be performed before treatment, 1 year after and every 2 years. Discussions with patients about side effects of these medications are recommended. It is important to emphasize that the protection of typical fractures is greater than the side effects when used for the appropriate length of time.

The effects of non-pharmacological measures are insufficient as a sole modality for osteoporosis prevention and treatment. Several literature reviews were conducted along the time and realized that studies on the use of physical therapies to prevent bone demineralization were inconclusive. The detection of an effective clinical intervention in this study area will require rigorous minimization of bias through a randomized controlled clinical trial design and likely require the involvement of multiple centers. It will also be dependent on appropriate measurement, with adequate intensity and duration of the intervention to detect a treatment effect. However, given the difficulty in conducting large-scale randomized controlled trials to address these issues, many of these gaps can best be addressed by using a multi-faceted approach including a combination of literature synthesis, large longitudinal observational studies and an expert opinion. Any prospective intervention is likely to benefit from early timing after acute SCI.

Before a treatment prescription, it is important to take into consideration timing since SCI, type, frequency, intensity and duration of intervention and the patient's possibilities to follow-up with a long-term treatment. Exercise and pharmacological prescriptions have to be safe. The patient has to be informed about the treatment evidence, risks, benefits and burdens of each alternative (including no intervention) and, together, make a shared decision, depending on their goals, preferences and concerns.

## Applications to other areas of neuroscience

In this chapter, we have reviewed the effects of osteoporosis after spinal cord injury and related fractures.

Osteoporosis after SCI is defined as an excessive bone resorption after SCI and bone fragility fractures caused by a trauma that would be insufficient for a normal bone to fracture (Craven et al., 2008; Zleik et al., 2019) or by minimally loaded situations (McPherson et al., 2014; Shields et al., 2005).

Osteoporosis is the most common silent disease in humans whose complication is fragility fractures. Osteoporosis is a treatable disease before fracture occurrence, and even if the first fracture occurred, there is benefit to prevent a new event.

Osteoporosis investigations should be performed for spinal cord injury, traumatic brain injury, neuromuscular disorders, and others because osteoporosis-related fractures in patients with chronic disabilities are a complication with consequences in activities of daily living and rehabilitation (Champs et al., 2020).

Dual-energy X-ray absorptiometry is an important method for diagnosis and should be performed in all patients with chronic disabilities (Garland et al., 2005).

The effects of non-pharmacological treatments are under debate, and studies showed the efficacy of bisphosphonates and Vitamin D to prevent fragility fractures (Bauman, Spungen, et al., 2005; Bauman, Wecht, et al., 2005; Gilchrist et al., 2007; Zehnder et al., 2004).

Campaigns to prevent falls are needed because falls are the first event related to fragility fractures (Champs et al., 2020).

## Mini-dictionary of terms

Osteoporosis: According to the WHO, osteoporosis is defined as a low bone mass and micro-architectural deterioration of bone tissue with a consequent increase in bone fragility and susceptibility to fracture.

Fragility fractures: These are fractures caused by a force that usually doesn't cause a fracture as torsional forces during transfer, compressive forces, unknown, fall from wheelchair, self-passive mobilization, exercises, findings on X-ray, orthostatism or walking training, and others. Fractures because of fall from greater than standing height, sports injuries, and motor vehicle/motor cycle accidents are excluded.

Spinal cord injury: Damage of the spinal cord can be traumatic because of motor vehicle/motor cycle accidents, falls, and others or can be non-traumatic because of virus, tumors or demyelinating disorders, and others.

Dual-energy X-ray absorptiometry (DEXA): It calculates the measurement of bone mineral density in g/cm$^2$, is a confirmatory test to predict the future risk of fracture, and also serves for ongoing monitoring of bone health and to access bone loss by the bone mineral density (BMD) measurements.

Bisphosphonates: These are drugs capable to inhibit bone resorption.

## Key facts of osteoporosis after spinal cord injury

- Bone demineralization after SCI occurs in a rate loss of approximately 4% per month in the first year, and bone loss can be extended 3 to 8 years after SCI.
- Bone loss associated with SCI is related to immobility and metabolic changes.
- Other related factors may contribute to this bone loss, such as vitamin D deficiency and the use of methylprednisolone, anti-convulsant drugs, and psychotropic substances.
- Studies demonstrated bone demineralization greater than 20% to 70% in lower limbs after SCI.

## Key facts of fragility fractures after spinal cord injury

- The consequence of osteoporosis in individuals with SCI is fracture due to bone fragility.
- It usually occurs in transfers, falls and activities with minimal or no trauma.
- Fractures may occur in 25% to 46% of these people with SCI over their lifetimes.
- Most places of fractures are in the tibia and/or distal fibula, distal femur, or hip/proximal femur and proximal tibia.

## Key facts of diagnosis of osteoporosis after spinal cord injury

- Osteometabolic evaluation of patients with SCI is recommended for evaluation before treatment, 6 months after beginning pharmacological therapy and repeated on annual basis.

- DEXA should be performed before treatment, 1 year after and every 2 years.
- DEXA measurement of the distal femur and proximal tibia, with specific protocols for the evaluation of these sites, is recommended.

## Key facts of treatment for osteoporosis-related fractures after spinal cord injury

- Treatment with pharmacological therapy should be considered for patients with SCI and risk factors plus DEXA criteria for osteoporosis.
- The effects of non-pharmacological measures are insufficient as a sole modality for osteoporosis prevention and treatment.
- Discussions with patients about side effects of these medications are recommended.
- The patient has to be informed by a healthcare professional about the treatment evidence, risks, benefits, and burdens of each alternative (including no intervention), and together make a shared decision, depending on their goals, preferences, and concerns.

## Summary points

- Osteoporosis affects the health of individuals with SCI because it increases the risk of fragility fractures with consequences.
- Most sites of fractures are in the tibia and/or distal fibula, distal femur, or hip/proximal femur and proximal tibia.
- Fractures are primarily related to torsional forces during transfer, passive mobilization, compressive forces, and falls.
- Special attention should be paid to the prevention of fractures, appropriate treatments should be instigated to prevent excessive bone loss, and investigations should be performed.
- Campaigns to prevent falls and stimulate the appropriate identification and treatment of osteoporosis in individuals with SCI must be promoted.
- Dual-energy X-ray absorptiometry (DEXA) is an important method for diagnosis.
- Treatment with pharmacological therapy for people with SCI with risk factors for SCI population should be considered.
- The patient has to be informed about the diagnosis and treatment options and make a shared decision for treatment, depending on their goals, preferences, and concerns.

## References


Abderhalden, L., Weaver, F. M., Bethel, M., Demirtas, H., Burns, S., Svircev, J., et al. (2017). Dual-energy X-ray absorptiometry and fracture prediction in patients with spinal cord injuries and disorders. *Osteoporosis International*, *28*(3), 925–934.

Akhigbe, T., Chin, A. S., Svircev, J. N., Hoenig, H., Burns, S. P., Weaver, F. M., et al. (2015). A retrospective review of lower extremity fracture care in patients with spinal cord injury. *The Journal of Spinal Cord Medicine*, *38*(1), 2–9.

Alekna, V., Tamulaitiene, M., Sinevicius, T., & Juocevicius, A. (2008). Effect of weight-bearing activities on bone mineral density in spinal cord injured patients during the period of the first two years. *Spinal Cord*, *46*(11), 727–732.

Alizadeh-Meghrazi, M., Masani, K., Popovic, M. R., & Craven, B. C. (2012). Whole-body vibration during passive standing in individuals with spinal cord injury: Effects of plate choice, frequency, amplitude, and subject's posture on vibration propagation. *PM & R : The Journal of Injury, Function, and Rehabilitation*, *4*(12), 963–975.

Asselin, P., Spungen, A. M., Muir, J. W., Rubin, C. T., & Bauman, W. A. (2011). Transmission of low-intensity vibration through the axial skeleton of persons with spinal cord injury as a potential intervention for preservation of bone quantity and quality. *The Journal of Spinal Cord Medicine*, *34*(1), 52–59.

Battaglino, R. A., Lazzari, A. A., Garshick, E., & Morse, L. R. (2012). Spinal cord injury-induced osteoporosis: Pathogenesis and emerging therapies. *Current Osteoporosis Reports*, *10*(4), 278–285.

Bauman, W. A., & Cardozo, C. P. (2015). Osteoporosis in individuals with spinal cord injury. *PM & R : The Journal of Injury, Function, and Rehabilitation*, *7*(2), 188–201.

Bauman, W. A., Cirnigliaro, C. M., La Fountaine, M. F., Martinez, L., Kirshblum, S. C., & Spungen, A. M. (2015). Zoledronic acid administration failed to prevent bone loss at the knee in persons with acute spinal cord injury: An observational cohort study. *Journal of Bone and Mineral Metabolism*, *33*(4), 410–421.

Bauman, W. A., & Spungen, A. M. (2000). Metabolic changes in persons after spinal cord injury. *Physical Medicine and Rehabilitation Clinics of North America*, *11*(1), 109–140.

Bauman, W. A., Spungen, A. M., Morrison, N., Zhang, R. L., & Schwartz, E. (2005). Effect of vitamin D analog on leg bone mineral density in patients with chronic spinal cord injury. *Journal of Rehabilitation Research and Development*, *42*(5), 625–634.

Bauman, W. A., Wecht, J. M., Kirshblum, S., Spungen, A. M., Morrison, N., Cirnigliaro, C., et al. (2005). Effect of pamidronate administration on bone in patients with acute spinal cord injury. *Journal of Rehabilitation Research and Development*, *42*(3), 305–313.

Bauman, W. A., Zhong, Y. G., & Schwartz, E. (1995). Vitamin D deficiency in veterans with chronic spinal cord injury. *Metabolism, Clinical and Experimental*, *44*(12), 1612–1616.

Belanger, M., Stein, R. B., Wheeler, G. D., Gordon, T., & Leduc, B. (2000). Electrical stimulation: Can it increase muscle strength and reverse osteopenia in spinal cord injured individuals? *Archives of Physical Medicine and Rehabilitation*, *81*(8), 1090–1098.

Bethel, M., Bailey, L., Weaver, F., Harmon, R. L., Priebe, M. M., Le, B., et al. (2016). A historical study of appendicular fractures in veterans with traumatic chronic spinal cord injury: 2002-2007. *The Journal of Spinal Cord Medicine*, *39*(6), 686–692.

Biering-Sørensen, F., Hansen, B., & Lee, B. S. (2009). Non-pharmacological treatment and prevention of bone loss after spinal cord injury: A systematic review. *Spinal Cord*, *47*(7), 508–518.

Carbone, L. D., Chin, A. S., Burns, S. P., Svircev, J. N., Hoenig, H., Heggeness, M., et al. (2013). Morbidity following lower extremity fractures in men with spinal cord injury. *Osteoporosis International*, *24*(8), 2261–2267.

Carvalho, D. C., Garlipp, C. R., Bottini, P. V., Afaz, S. H., Moda, M. A., & Cliquet, A., Jr. (2006). Effect of treadmill gait on bone markers and bone mineral density of quadriplegic subjects. *Brazilian Journal of Medical and Biological Research*, *39*(10), 1357–1363.

Chain, A., Koury, J. C., & Bezerra, F. F. (2012). Physical activity benefits bone density and bone-related hormones in adult men with cervical spinal cord injury. *European Journal of Applied Physiology*, *112*(9), 3179–3186.

Champs, A. P. S., Maia, G. A. G., Oliveira, F. G., de Melo, G. C. N., & Soares, M. M. S. (2020). Osteoporosis-related fractures after spinal cord injury: A retrospective study from Brazil. *Spinal Cord*, *58*(4), 484–489.

Chang, K. V., Hung, C. Y., Chen, W. S., Lai, M. S., Chien, K. L., & Han, D. S. (2013). Effectiveness of bisphosphonate analogues and functional electrical stimulation on attenuating post-injury osteoporosis in spinal cord injury patients—A systematic review and meta-analysis. *PLoS One*, *8*(11), e81124.

Chen, S. C., Lai, C. H., Chan, W. P., Huang, M. H., Tsai, H. W., & Chen, J. J. (2005). Increases in bone mineral density after functional electrical stimulation cycling exercises in spinal cord injured patients. *Disability and Rehabilitation*, *27*(22), 1337–1341.

Cirnigliaro, C. M., Myslinski, M. J., La Fountaine, M. F., Kirshblum, S. C., Forrest, G. F., & Bauman, W. A. (2017). Bone loss at the distal femur and proximal tibia in persons with spinal cord injury: Imaging approaches, risk of fracture, and potential treatment options. *Osteoporosis International*, *28*(3), 747–765.

Craven, C., Robertson, L., Delparte, J., Ashe, M., & Janice, J. E. (2008). Sublesional osteoporosis prevention, detection, and treatment: A decision guide for rehabilitation clinicians treating patients with spinal cord injury. *Critical Reviews in Physical and Rehabilitation Medicine*, *20*(4), 277–321.

Craven, B. C., Robertson, L. A., McGillivray, C. F., & Adachi, J. D. (2009). Detection and treatment of sublesional osteoporosis among patients with chronic spinal cord injury. *Topics in Spinal Cord Injury Rehabilitation*, *14*(4), 1–22.

De Bruin, E. D., Frey-Rindova, P., Herzog, R. E., Dietz, V., Dambacher, M. A., & Stüssi, E. (1999). Changes of tibia bone properties after spinal cord injury: Effects of early intervention. *Archives of Physical Medicine and Rehabilitation*, *80*(2), 214–220.

Dolbow, D. R., Gorgey, A. S., Daniels, J. A., Adler, R. A., Moore, J. R., & Gater, D. R., Jr. (2011). The effects of spinal cord injury and exercise on bone mass: A literature review. *NeuroRehabilitation*, *29*(3), 261–269.

Dudley-Javoroski, S., & Shields, R. K. (2013). Active-resisted stance modulates regional bone mineral density in humans with spinal cord injury. *The Journal of Spinal Cord Medicine*, *36*(3), 191–199. https://doi.org/10.1179/2045772313Y.0000000092.

Eser, P., de Bruin, E. D., Telley, I., Lechner, H. E., Knecht, H., & Stussi, E. (2003). Effect of electrical stimulation-induced cycling on bone mineral density in spinal cord-injured patients. *European Journal of Clinical Investigation*, *33*(5), 412–419.

Frontera, J. E., & Mollett, P. (2017). Aging with spinal cord injury: An update. *Physical Medicine and Rehabilitation Clinics of North America*, *28*(4), 821–828.

Frotzler, A., Coupaud, S., Perret, C., Kakebeeke, T. H., Hunt, K. J., & Eser, P. (2009). Effect of detraining on bone and muscle tissue in subjects with chronic spinal cord injury after a period of electrically-stimulated cycling: A small cohort study. *Journal of Rehabilitation Medicine*, *41*(4), 282–285.

Garland, D., Adkins, R., & Stewart, C. (2005). Fracture threshold and risk for osteoporosis and pathologic fractures in individuals with spinal cord injury. *Topics in Spinal Cord Injury Rehabilitation*, *11*(1), 61–69.

GBD 2016 Traumatic Brain Injury & Spinal Cord Injury Collaborators. (2019). Global, regional, and national burden of traumatic brain injury and spinal cord injury, 1990-2016: A systematic analysis for the global burden of disease study 2016. *The Lancet Neurology*, *18*(1), 56–87.

Giangregorio, L. M., Webber, C. E., Phillips, S. M., Hicks, A. L., Craven, B. C., Bugaresti, J. M., et al. (2006). Can body weight supported treadmill training increase bone mass and reverse muscle atrophy in individuals with chronic incomplete spinal cord injury? *Applied Physiology, Nutrition, and Metabolism*, *31*(3), 283–291.

Gifre, L., Vidal, J., Carrasco, J. L., Muxi, A., Portell, E., Monegal, A., et al. (2016). Denosumab increases sublesional bone mass in osteoporotic individuals with recent spinal cord injury. *Osteoporosis International*, *27*(1), 405–410.

Gilchrist, N. L., Frampton, C. M., Acland, R. H., Nicholls, M. G., March, R. L., Maguire, P., et al. (2007). Alendronate prevents bone loss in patients with acute spinal cord injury: A randomized, double-blind, placebo-controlled study. *The Journal of Clinical Endocrinology and Metabolism*, *92*(4), 1385–1390.

Goktepe, A. S., Tugcu, I., Yilmaz, B., Alaca, R., & Gunduz, S. (2008). Does standing protect bone density in patients with chronic spinal cord injury? *The Journal of Spinal Cord Medicine*, *31*(2), 197–201.

Goktepe, A. S., Yilmaz, B., Alaca, R., Yazicioglu, K., Mohur, H., & Gunduz, S. (2004). Bone density loss after spinal cord injury: Elite paraplegic basketball players vs. paraplegic sedentary persons. *American Journal of Physical Medicine & Rehabilitation*, *83*(4), 279–283.

Gordon, K. E., Wald, M. J., & Schnitzer, T. J. (2013). Effect of parathyroid hormone combined with gait training on bone density and bone architecture in people with chronic spinal cord injury. *PM & R : The Journal of Injury, Function, and Rehabilitation*, *5*(8), 663–671.

Haider, I. T., Lobos, S. M., Simonian, N., Schnitzer, T. J., & Edwards, W. B. (2018). Bone fragility after spinal cord injury: Reductions in stiffness and bone mineral at the distal femur and proximal tibia as a function of time. *Osteoporosis International*, *29*(12), 2703–2715.

Jiang, S. D., Dai, L. Y., & Jiang, L. S. (2006). Osteoporosis after spinal cord injury. *Osteoporosis International*, *17*(2), 180–192.

Koutrakis, N. E., Goldstein, R. L., Walia, P., Polak, M. M., Lazzari, A. A., Tun, C. G., et al. (2019). Vitamin D, diet, and lifestyle in a chronic SCI population. *Spinal Cord*, *57*(2), 117–127.

Kunkel, C. F., Scremin, A. M., Eisenberg, B., Garcia, J. F., Roberts, S., & Martinez, S. (1993). Effect of "standing" on spasticity, contracture, and osteoporosis in paralyzed males. *Archives of Physical Medicine and Rehabilitation*, *74*(1), 73–78.

Lai, C. H., Chang, W. H., Chan, W. P., Peng, C. W., Shen, L. K., Chen, J. J., et al. (2010). Effects of functional electrical stimulation cycling exercise on bone mineral density loss in the early stages of spinal cord injury. *Journal of Rehabilitation Medicine*, *42*(2), 150–154.

Liu, C. C., Theodorou, D. J., Theodorou, S. J., Andre, M. P., Sartoris, D. J., Szollar, S. M., et al. (2000). Quantitative computed tomography in the evaluation of spinal osteoporosis following spinal cord injury. *Osteoporosis International*, *11*(10), 889–896.

McPherson, J. G., Edwards, W. B., Prasad, A., Troy, K. L., Griffith, J. W., & Schnitzer, T. J. (2014). Dual energy X-ray absorptiometry of the knee in spinal cord injury: Methodology and correlation with quantitative computed tomography. *Spinal Cord*, *52*(11), 821–8255.

Morse, L. (2019). Osteoporosis prophylaxis in acute SCI. *Spinal Cord Series and Cases*, *5*(1), 27.

Morse, L. R., Battaglino, R. A., Stolzmann, K. L., Hallett, L. D., Waddimba, A., Gagnon, D., et al. (2009). Osteoporotic fractures and hospitalization risk in chronic spinal cord injury. *Osteoporosis International*, *20*(3), 385–392.

Morse, L. R., Biering-Soerensen, F., Carbone, L. D., Cervinka, T., Cirnigliaro, C. M., Johnston, T. E., et al. (2019). Bone mineral density testing in spinal cord injury. ISCD official position. *Journal of Clinical Densitometry*, *22*(4), 554–566.

Morse, L. R., Troy, K. L., Fang, Y., Nguyen, N., Battaglino, R., Goldstein, R. F., et al. (2019). Combination therapy with zoledronic acid and FES-row training mitigates bone loss in paralyzed legs: Results of a randomized comparative clinical trial. *JBMR Plus*, *3*(5), e10167.

Ogilvie, C., Bowker, P., & Rowley, D. I. (1993). The physiological benefits of paraplegic orthotically aided walking. *Paraplegia*, *31*(2), 111–115.

Shapiro, J., Smith, B., Beck, T., Ballard, P., Dapthary, M., BrintzenhofeSzoc, K., et al. (2007). Treatment with zoledronic acid ameliorates negative geometric changes in the proximal femur following acute spinal cord injury. *Calcified Tissue International*, *80*(5), 316–322.

Shields, R. K., Schlechte, J., Dudley-Javoroski, S., Zwart, B. D., Clark, S. D., Grant, S. A., et al. (2005). Bone mineral density after spinal cord injury: A reliable method for knee measurement. *Archives of Physical Medicine and Rehabilitation*, *86*(10), 1969–1973.

Soleyman-Jahi, S., Yousefian, A., Maheronnaghsh, R., Shokraneh, F., Zadegan, S. A., Soltani, A., et al. (2018). Evidence-based prevention and treatment of osteoporosis after spinal cord injury: A systematic review. *European Spine Journal*, *27*(8), 1798–1814.

Thoumie, P., Le Claire, G., Beillot, J., Dassonville, J., Chevalier, T., Perrouin-Verbe, B., et al. (1995). Restoration of functional gait in paraplegic patients with the RGO-II hybrid orthosis. A multicenter controlled study. II: Physiological evaluation. *Paraplegia*, *33*(11), 654–659.

Troy, K. L., & Morse, L. R. (2015). Measurement of bone: Diagnosis of SCI-induced osteoporosis and fracture risk prediction. *Topics in Spinal Cord Injury Rehabilitation*, *21*(4), 267–274.

Uebelhart, D., Demiaux-Domenech, B., Roth, M., & Chantraine, A. (1995). Bone metabolism in spinal cord injured individuals and in others who have prolonged immobilisation. A review. *Paraplegia*, *33*(11), 669–673.

Warden, S. J., Bennell, K. L., Matthews, B., Brown, D. J., McMeeken, J. M., & Wark, J. D. (2001). Efficacy of low-intensity pulsed ultrasound in the prevention of osteoporosis following spinal cord injury. *Bone*, *29*(5), 431–436.

Zehnder, Y., Risi, S., Michel, D., Knecht, H., Perrelet, R., Kraenzlin, M., et al. (2004). Prevention of bone loss in paraplegics over 2 years with alendronate. *Journal of Bone and Mineral Research*, *19*(7), 1067–1074.

Zleik, N., Weaver, F., Harmon, R. L., Le, B., Radhakrishnan, R., Jirau-Rosaly, W. D., et al. (2019). Prevention and management of osteoporosis and osteoporotic fractures in persons with a spinal cord injury or disorder: A systematic scoping review. *The Journal of Spinal Cord Medicine*, *42*(6), 735–759.

Chapter 8

# Understanding the effects of prolonged cervical spinal cord compression on the brain. Current knowledge and future challenges

Ángela Bernabéu-Sanz[a,b] and Eduardo Fernández-Jover[b,c]

[a]*Magnetic Resonance Department, Inscanner SL, Alicante, Spain;* [b]*Anatomy and Histology Department, Institute of Bioengineering, University Miguel Hernández, Elche, Spain;* [c]*Centro de Investigación Biomédica en Red (CIBER BBN), Madrid, Spain*

## Abbreviations

| | |
|---|---|
| **BOLD** | blood oxygen level-dependent |
| **CS** | cervical spondylosis |
| **CST** | corticospinal tract |
| **CSM** | cervical spondylosis myelopathy |
| **DTI** | diffusion tensor imaging |
| **FA** | fractional anisotropy |
| **FC** | functional connectivity |
| **GM** | gray matter |
| **fMRI** | functional magnetic resonance imaging |
| **MD** | mean diffusivity |
| **MR** | magnetic resonance |
| **MRI** | magnetic resonance imaging |
| **M1** | primary motor cortex |
| **RSNs** | resting-state networks |
| **RS-fMRI** | resting-state functional MR imaging |
| **SCI** | spinal cord injury |
| **SMC** | sensorimotor cortex |
| **SMA** | supplementary motor area |
| **S1** | primary sensory cortex |
| **SMG** | supramarginal gyri |
| **WM** | white matter |

## Introduction

Compression of the cervical spinal cord can be caused by a wide range of causes, but the most common is cervical spondylosis (CS), a chronic degenerative process that affects the vertebral bodies and inter-vertebral discs of the cervical spine (Table 1). CS is frequently found in asymptomatic adults after the fifth decade of life, but may progress to disk herniation, bone spur formation, compression of the spinal cord, and cervical spondylotic myelopathy (CSM) (Kelly, Groarke, Butler, Poynton, & Byrne, 2012).

The compression of the spinal cord by CS is a slow and progressive process caused by the reduction of the spinal canal diameter by dynamic and static factors (de Oliveira Vilaça et al., 2016). In CS, any flexion or extension of the cervical spine causes the intermittent compression of the spinal cord. This progressive compression leads to several changes such as apoptosis of the anterior horn neurons, ischemia, gradual spongy necrosis, demyelination of the white matter (WM), and Wallerian degeneration (Fig. 1; Grabher, Mohammadi, David, & Freund, 2017). Despite it may affect one single segment of the

Diagnosis and Treatment of Spinal Cord Injury. https://doi.org/10.1016/B978-0-12-822498-4.00008-7

**TABLE 1** Summary of the main causes of spinal cord compression.

| Traumatic causes | Atraumatic causes |
|---|---|
| Vehicular accidents | Degenerative spondylosis |
| Falls | Intervertebral disk herniation |
| Violence | Metastatic disease |
| Sports | Primary spinal cord tumor |
| Recreation activities | Spinal epidural abscess |
| | Spinal epidural hematoma |
| | Scoliosis |
| | Rheumatoid arthritis |
| | Bone diseases |

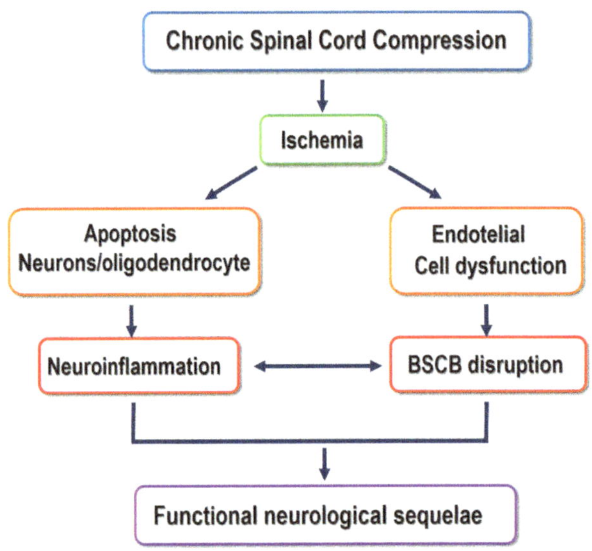


FIG. 1 Representative scheme of the pathophysiology of cervical spondylotic myelopathy. Chronic compression of the spinal cord causes chronic ischemia that damages oligodendrocytes and neurons eliciting an inflammatory response, endothelial cell loss, and disruption of the blood-spinal cord barrier (BSCB). Neuronal loss and axonal damage lead to neurological impairment. *Unpublished figure.*

spine, multi-level involvement is very frequent, generally at the lower segments. Contrary to the lumbar spine, which encloses only a small part of the spinal cord, stenosis of the cervical spinal canal can induce a direct compromise of the whole spinal cord and severe neurological deficits.

Characteristically, in CS, the severity of compression and deficits varies greatly across individuals. Some patients experience a benign form of the disease with stable periods lasting from months to years, while others develop CSM with neurological deficits and a considerable deterioration over time (Baron & Young, 2007; Edwards, Riew, Anderson, Hilibrand, & Vaccaro, 2003). If CSM fully develops, the symptoms are progressive and in a stepwise manner (de Oliveira Vilaça et al., 2016), with a wide range of clinical manifestations including weakening or sensory loss in one or more limbs, gait unsteadiness, fine motor deficits, and non-specific neck and shoulder pain with or without radiculopathy.

Diagnosis of CSM is based on physical examination and clinical imaging, usually with Magnetic Resonance Imaging (MRI). MRI allows to evaluate the vertebral discs, ligaments, subarachnoid space, spinal cord, and quantify the spinal stenosis. Still, it lacks sensitivity to provide information about the integrity of the spinal cord, presents a poor correlation with neurological and functional impairment, and does not provide reliable prognostic information to predict surgical outcomes (Martin et al., 2016). Recent advances in MR techniques have started to overcome these limitations by providing invaluable information about CSM pathology, with several studies reporting that prolonged cervical spine compression affects not only local structures but also distally from the site of injury. For instance, numerous brain changes, such as cortical and subcortical atrophy, functional reorganization of the sensorimotor cortex (SMC), and connectivity impairment, have been reported (Bernabéu-Sanz, Mollá-Torró, López-Celada, Moreno-López, & Fernández-Jover, 2020; Chen et al., 2017). These observations have had a great impact on the understanding of CSM, framing the disease as multi-factorial and complex, with its pathophysiology still being poorly understood. In this sense, a clear limitation to advancing in the field of CSM pathology has been the lack of reliable animal models.

Understanding the sequence of structural and functional changes of CSM at the brain level, and defining their effects on the clinical outcome is of crucial importance to better understand the disease, to prevent its fatal consequences, and to develop the best evidence-based rehabilitation therapies. In the next sections, we will summarize the current knowledge on the effects of prolonged spinal cord compression on the brain (Figs. 2–5).

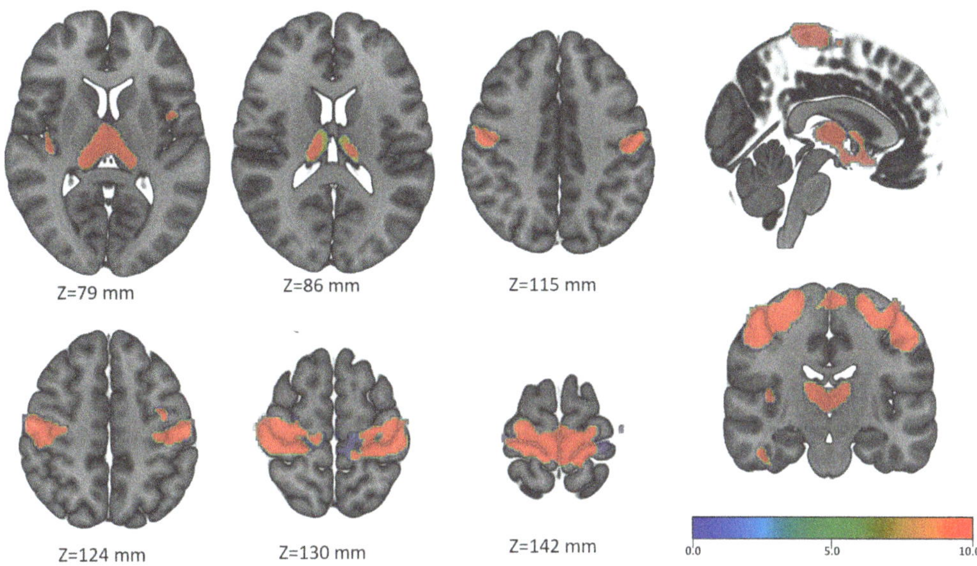


FIG. 2 Representation of the regional areas with significant cortical volume reduction in a group of patients with prolonged cervical spine compression. The results were obtained from a voxel-based morphometry analysis in a group of 64 subjects (32 cervical spondylosis patients vs. 32 healthy controls, $P < .05$ corrected), suggesting that prolonged cervical spine compression led to thalami, sensorimotor cortex, and supplementary motor area atrophy. *Unpublished figure*.

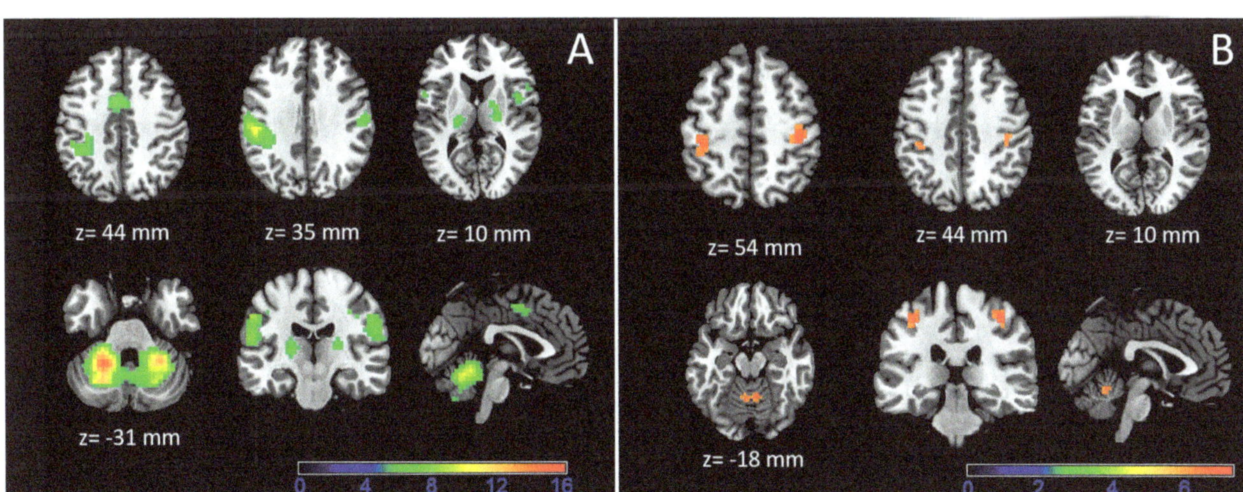


FIG. 3 Functional MRI results showing the differences in the brain activation pattern in a group of cervical spondylosis patients ($n = 32$) (A) and age-matched healthy controls ($n = 32$) (B) during a bilateral finger-tapping task. The activation maps correspond to a second-level analysis in axial, coronal, and sagittal planes in the MNI space ($P < .05$ corrected, $k = 10$). The patients presented a higher number of activated cortical and subcortical areas than healthy controls. *Unpublished figure*.

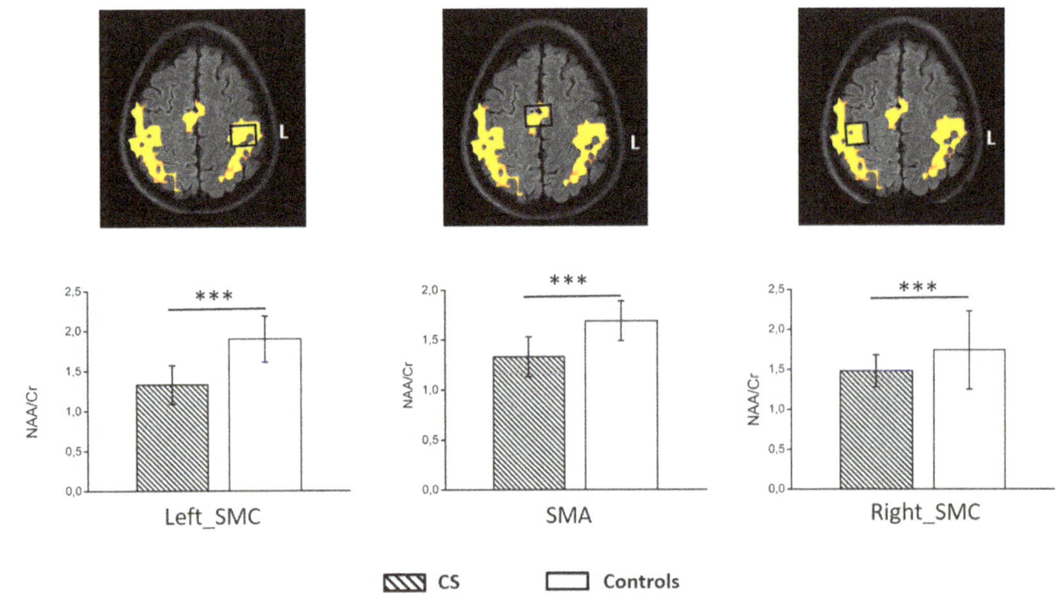


FIG. 4 MR-spectroscopy results of the motor cortex of cervical spondylosis patients (CS, $n$=32) and healthy controls (HC, $n$=32). The volume of interest was located on the activated area in both the primary sensorimotor cortex (SMC) and supplementary motor area (SMA) after a bilateral finger-tapping fMRI task paradigm (see *upper image* as example of voxel location for measurements). Graphs represent the NAA/ Cr (mean ± SD) values obtained after quantification in each group. A significant reduction was observed in all areas compared to controls suggesting significant neuronal/axonal damage. *** represents $P < .001$ after unpaired $t$-test. *NAA*, N-acetylaspartate; *Cr*, creatine. *Unpublished figure.*

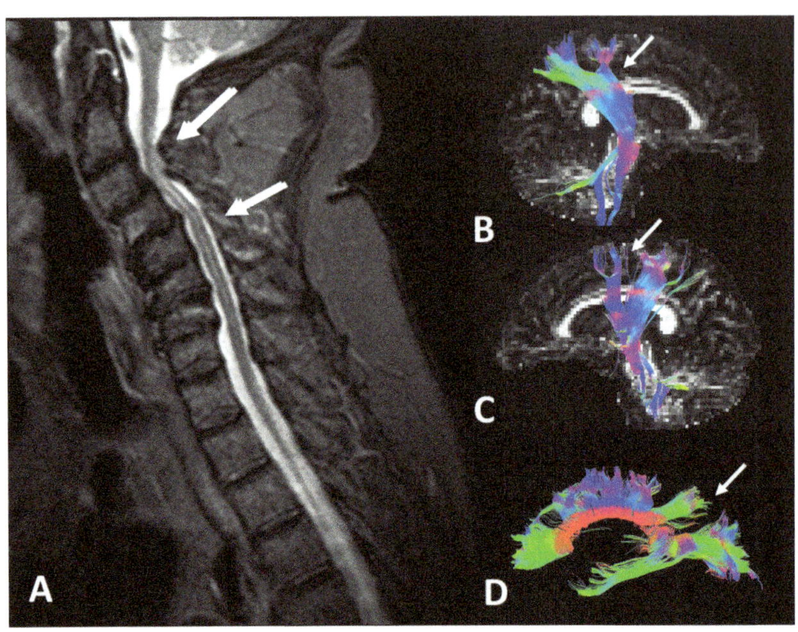


FIG. 5 Tractography results obtained in a patient with cervical spondylosis myelopathy. (A) Sagittal STIR of a 69-year-old male with cervical spondylosis and myelopathy at C2-C5 levels, significant cervical lordosis loss, and several diffuse herniated discs from C2 to C7. Clinically the patient only presented bilateral manual clumsiness. No cervicalgia or cervicobrachialgia was present. (B, C, D) Tractography results suggested significant bilateral and symmetrical axonal damage affecting both corticospinal tracts at the distal-anterior regions (B, C) and the *corpus callosum* (A) at the superior and posterior parietal projection fibers of both hemispheres *(see arrows)*. Commissural fibers were preserved. *Unpublished figure.*

## Cervical spine compression causes brain atrophy

There is increasing evidence supporting that spinal cord injury (SCI) may cause destructive cortical changes ranging from atrophy of projecting neurons to cell death (Freund et al., 2013; Hains, Black, & Waxman, 2003). In animal models, after SCI, the disruption of the motor efferents and sensory afferents results in atrophic changes of the neural sensorimotor systems (Hains et al., 2003). Indeed, several studies have demonstrated that SCI can induce cortical atrophy on the supraspinal network, whose degree correlates with the spinal cord atrophy (Freund et al., 2011, 2013) and the functional status (Freund et al., 2011).

There are still a few works exploring the effect of prolonged spinal cord compression by CS in the brain. However, the results are consistent reporting cortical atrophy in the primary SMC, with additional volume changes at the thalami, specifically at the pulvinar nuclei (Bernabéu-Sanz et al., 2020; Woodworth, Holly, Mayer, Salamon, & Ellingson, 2019). Furthermore, supplementary brain areas related to the execution of complex movements (such as gait, grasp, and fine hand-motor coordination) are also atrophied (Bernabéu-Sanz et al., 2020; Woodworth et al., 2019). Metabolic studies using MR-spectroscopy on CSM patients have also reported neuronal damage and impaired metabolism on the SMC that is not recovered or may worsen after surgery in spite of clinical improvement (Aleksanderek et al., 2017; Goncalves, Stevens, Doyle-Pettypiece, Bartha, & Duggal, 2016).

In a recent study regarding chronic CS, a consistent pattern of brain changes associated with neurological function and neck pain was observed (Woodworth et al., 2019). Specifically, cortical thinning in the superior frontal regions, anterior cingulate, insula, and precuneus were present. Furthermore, atrophy in the putamen, BA3a, and primary motor regions when worsening neurological symptoms was reported. Taking into account the role of BA3a as an afferent receiver of proprioception (Panchuelo et al., 2020), the observed atrophy in this study may explain the deficits in proprioception (gait and hand function) reported in CSM patients (Theodore, 2020).

CS patients may also present brain changes to chronic pain in the same fashion as other chronic pain conditions. Specifically, a cortical thinning within the insula and anterior cingulate cortex has been reported (Freund et al., 2011). Finally, a faster rate of cortical thinning and atrophy compared to age-matched healthy subjects in CS patients has been observed (Woodworth et al., 2019).

The significance of these findings is that despite CS has been traditionally considered as a benign disease typically present in the old-age population, it may, if not properly monitored, harm the brain inducing progressive atrophy. In this sense, the CS-related brain atrophy may become a potential obstacle to clinical recovery after surgery, even if significant repair is achieved at the site of compression.

## Cervical spine compression causes neural plasticity to preserve function

One of the most fascinating characteristics of the central nervous system is its ability to self-repair, reorganize after injury, and compensate for functional loss. This mechanism, known as neural plasticity (or functional reorganization), is present in numerous brain conditions (Cohen, Quarta, Bravi, Granato, & Minciacchi, 2017) and has also been observed in SCI. The exact mechanism is not fully understood but may occur through synaptic modifications of pre-existing connections and/or by the development of new circuitries (Holly, Dong, Albistegui-DuBois, Marehbian, & Dobkin, 2007).

Several studies have shown that SCI, although it does not directly involve cortical neurons, may alter the functional activation pattern of the SMC. These observations have been achieved with different physiological and imaging techniques such as functional MRI (fMRI). fMRI is a non-invasive imaging modality that allows to detect changes in brain activation over a period of time while a specific task is being performed. It can be used to localize areas of brain activation secondary to several cognitive and motor tasks, allowing to map the cortical representation of neural activation (Buchbinder, 2016).

In CSM patients, fMRI studies have shown the existence of significant changes in the cortical activation pattern during sensorimotor tasks. These changes consist of the expansion of the cortical representation for the affected limb by the inclusion of adjacent motor territories (Dong et al., 2008). Recruitment of supplementary areas outside the SMC including *globus pallidus*, left thalamus, caudate nucleus, and both cerebellum hemispheres have been reported. Additionally, decreased activation in SMA and both SMC, as well as increased activation of both cerebellum hemispheres, has been observed (Bernabéu-Sanz et al., 2020). Interestingly, the displacement and activation of the sensorimotor areas in CSM appeared in the vicinity of the areas with gray matter (GM) atrophy, suggesting that the displacement and recruitment of new cortical areas may be mediated by adaptive changes to GM atrophy (Henderson, Gustin, Macey, Wrigley, & Siddall, 2011). This functional reorganization is closely associated with the severity of the SCI (Zhou et al., 2015), helping to maintain a residual motor function, and contributing to its restoration if the disease progression is halted (Dong et al.,

2008). Indeed, these adaptive changes may also explain why some CSM patients with significant SCI are able to perform motor activities with apparently minor neurological deficits.

Comparison of pre-operative and post-operative studies has shown that surgical decompression of the cervical spine leads to cortical reorganization, probably by the recovery of conduction in the preserved axons at the spinal cord (Green et al., 2015). After surgical decompression, the altered SMC recruitment pattern gradually disappears (Holly et al., 2007), and everything progresses toward an activation pattern similar to the observed in healthy controls (Dong et al., 2008). There is an increased activation in the contralateral primary motor cortex (M1) concomitant with the improvement of the motor function and the extent of spinal cord expansion (Bhagavatula et al., 2016; Tam, Barry, Bartha, & Duggal, 2010). Also, a gradual disappearance of the ipsilateral primary SMC activation in addition to a progressive increase in the contralateral SMC and SMA is observed (Dong et al., 2008). This normalization continues over time and can be enhanced by activity-dependent mechanisms related to task practice and learning.

In summary, according to recent studies, CSM patients present compensatory mechanisms that consist of an increase and expansion of the cortical representation of the motor areas, especially those devoted to the hands. After the surgery, there is a tendency to normalization of the sensorimotor representation. However, the reported results are not completely consistent. This is probably because CSM patients form a heterogeneous group with different levels of impairment, different degrees of SCI, and different natural histories. In addition, there is no uniformity in the timing of imaging after the surgery. Still, the reported observations provide direct evidence of neuronal plasticity in the SMC of CSM patients that may play a main role in the post-operative motor functional recovery.

## Cervical spine compression changes the functional connectivity of the brain

Few studies have analyzed the alterations of the whole-brain network in CSM. These studies have been performed with resting-state functional MR imaging (RS-fMRI), a new technique that allows evaluating the functional relationship between brain areas, and enables the visualization of brain functional connectivity (FC) between spatially separated brain regions. RS-fMRI is a robust method that uses the endogenous brain activity detectable with blood oxygen level-dependent (BOLD) to identify areas that are interacting at rest. With RS-fMRI spontaneous low-frequency BOLD-fluctuations in neural activity (known as amplitude of low-frequency fluctuation) are detected, allowing to establish the different anatomically correlated functional networks (Smitha et al., 2017). As a result, the resting-state networks (RSNs) maps are obtained which represent topographies of functionally connected regions across the brain, helping to detect functionally relevant changes after injury to the central nervous system, if present.

In CSM, RS-fMRI studies have reported that compression of the cervical spinal cord alters the connectivity pattern of the sensorimotor and visual cortices of the brain. Specifically, an overall increase in the FC in the sensorimotor cortex and supplementary areas (precuneus, superior frontal gyrus, and anterior cingulate), and between the cerebellum and the pre- and post-central gyri has been reported. This increased FC correlates with the worsening of the neurological symptoms, suggesting that the altered FC between the SMC and supplementary areas might represent compensatory mechanisms for accomplishing motor tasks (Woodworth, Holly, Salamon, & Ellingson, 2018).

In a recent study, increased brain activity in the primary SMC and decreased brain activity in visual cortices and both supramarginal gyri (SMG) were observed. Characteristically after surgical decompression, the brain activity of the sensorimotor and visual cortices normalized (Takenaka et al., 2019). In another study by Chen et al., the authors only reported decreased neural activity in the primary visual cortex of CSM patients and an increased FC between the visual cortex and the posterior cingulate lobe (Chen et al., 2018). These observations suggested that the visual cortices were impaired, which accounted for the blurred vision observed in the patients of the study. Interestingly, the visual impairment also improved after surgical decompression. Considering that the cingulate gyri are related to the dorsal attention network, that in turn is involved in the control of visual attention (Fox, Corbetta, Snyder, Vincent, & Raichle, 2006), the authors concluded that the enhanced FC in the visual attention network was an adaptive mechanism to compensate for the diminished activity observed on the primary visual cortex.

In another study, increased FC between the left thalamus, the visual system (bilateral lingual gyrus, cuneus), and the right cerebellum posterior lobe, that correlated with the sensory scores (Peng, Tan, He, & Ou, 2020), were found. Moreover, decreased FC between the sensory and motor cortices was observed. The authors concluded that the increased FC in the visual-related network acted as a compensatory mechanism to the decreased somatosensory transmission information detected in the patients. Likewise, increased FC between the brainstem and visual cortex in both posterior parietal cortices and its association with neurological impairment and neck disability has been described in CSM (Wang, Laiwalla, Salamon, Ellingson, & Holly, 2020).

While the altered FC in the SMC is in line with previous fMRI literature during sensory and motor tasks, the observation of increased FC in the visual cortex is surprising as CS is not a primary visual disorder. In this sense, Wang et al. propose that the observed increased FC in the visual cortex may be secondary to the need for more visual input to optimize body movements, mechanics, and balance (Wang et al., 2020). This observation is in line with the work of Chen et al., where the authors suggest a compensatory compensation for the insufficient input from the proprioceptive system (Chen et al., 2018).

In summary, brain plasticity is present in CSM patients to overcome functional deficits. The main changes are long-term FC reorganizations in the sensorimotor regions and areas assisting in the regulation of movement and sensory perception. This altered connectivity between various sub-networks may have prognostic potential and/or therapeutic implications in CS patients.

## Cervical spine compression also damages the brain WM

Progressive chronic compression of the spinal cord also results in many pathological changes in the WM that are detectable using diffusion tensor imaging (DTI). DTI is an MRI technique that allows the evaluation of micro-structural changes in the WM by measuring the motility and orientation (anisotropy) of water molecules in tissues. DTI evaluates several parameters, the most important are: (1) mean diffusivity (MD), and (2) fractional anisotropy (FA). FA is a scalar value between 0 and 1 that evaluates the preferred direction of which diffusion occurs in the central nervous system and is the most sensitive parameter to detect SCI (Guan et al., 2017). FA is largely attributed to the transverse axonal barriers in WM (i.e., myelin sheath, axon membrane) restricting the water diffusion to a preferred longitudinal orientation. Reduced FA values indicate a disturbance in the directional orientation of nerve fibers and are associated with reduced myelin integrity (Chen et al., 2017).

In CSM, most of the DTI studies have been centered on the local changes at the spinal cord, reporting differences in diffusion characteristics of the water molecules at the site of compression, that are associated with increased neurological dysfunction (Ellingson, Salamon, Grinstead, & Holly, 2014; Guan et al., 2017; Martin et al., 2016). Specifically, decreased FA values have been found in stenotic segments (Guan et al., 2017) in the lateral and posterior column of WM suggesting significant myelin damage (Cui et al., 2014). Likewise, decreased FA values and increased MD values have been found in CS patients in the early stages of the disease even in cases without visible myelopathy in the MRI examination (Banaszek, Bladowska, Szewczyk, Podgórski, & Sąsiadek, 2014); suggesting that DTI presents higher sensitivity in myelopathy detection compared to conventional MRI.

In the brain, current studies in SCI have mostly analyzed the effect of traumatic etiology. In these reports the results show WM damage in supraspinal areas along the corticospinal tracts in the medulla, pons, midbrain, posterior limbs of the internal capsule (Freund et al., 2011, 2013; Guleria et al., 2008), and both centrum semiovale (Koskinen et al., 2014). Furthermore, the CST damage correlates with the brain functional reorganization, the lower cross-sectional spinal cord area, and the upper limb ability (Freund et al., 2012). Characteristically, this WM damage starts early after the traumatic SCI, with a pronounced and progressive decrease of the spinal cord area concomitant with a reduction in the cranial CST and GM volume at the SMC (Freund et al., 2013). These changes are assumed to be induced by retrograde degeneration along the CST axons that over time results in shrinkage of the soma of corticospinal projecting neurons. Overall, the studies in traumatic SCI suggest widespread changes in WM micro-structure in the CST and in areas adjacent to the SMC that in turn associate with the clinical state of the patients.

Although there are few reports on CS, similar results have been found suggesting that CS causes anterograde and retrograde degeneration of the spinal WM tracts, resulting in axonal loss and demyelination that may extend cranially to cerebral regions. For instance, micro-structural changes in the ventral horns above the level of injury that correlated with diffusivity changes in the CST have been observed (Grabher et al., 2017). Similarly, Wen et al. observed WM damage at the region of the CST that correlated with the myelopathy severity in CSM patients (Wen, Cui, Mak, Luk, & Hu, 2014). Thus, both studies suggest a direct association between neurodegeneration affecting the motor system at the spinal cord, the CST integrity, and clinical impairment.

Characteristically, in CS patients, significant WM damage at both CST independently of the affected limb and *corpus callosum* has been reported. At the corpus callosum, significant damage to the superior and posterior parietal projection fibers, areas related to motor and touch-stimulation (Fabri, Pierpaoli, Barbaresi, & Polonara, 2014) were observed. Tractography results on CS patients also suggested a common pattern of WM damage, being affected the anterior-distal segments, regions directly linked to the motor cortex. The preserved regions at both CST were those arising from S1, which are more posterior and medial than those of M1 and thus less vulnerable to compression. Besides, a direct association between the damage at the CST, the clinical scores, and the evolution after surgery was observed (Bernabéu-Sanz et al., 2020). These observations suggest that CST metrics with DTI may be used as a potential biomarker in the clinical setting to determine the impact of the disease and predict outcomes (Bernabéu-Sanz et al., 2020).

The CST is the largest descending pathway that connects the motor cortex to the brain stem and spinal cord and is mainly responsible for the movement of the distal extremities, particularly fine motor activities of the hand ( Jang, Medicine, & Medicine, 2014). It is particularly vulnerable to the negative effects of cervical spinal compression suffering early demyelination. Recent reports on SCI both of traumatic and compressive origin suggest that the predominant pathological mechanism on both conditions is the demyelination of the ascending and descending CST fibers, leading to a progressive loss of axonal conductivity (Dong et al., 2008), with additional connectivity impairment between the SMC and the brainstem (Wang et al., 2020). The impact of the CST damage in the brain cortex has been previously described in other pathologies such as stroke. In these patients, whereas the degree of CST injury strongly predicts the resulting motor impairment (Lotze et al., 2012; Stinear et al., 2007), the neurological function tends to be preserved by the reorganization and recruitment of new pathways as well as other motor network nodes to optimize the residual neurological function (Pantano, Petsas, Tona, & Sbardella, 2015; Ward, 2011). In this sense, it is believed that most of the axotomized CST neurons are able to maintain the sensory-motor circuits through axonal sprouting (Zdunczyk et al., 2018). However, if CST neurons are damaged, whittling down the population by spinal compression over time may lead to some critical threshold to impairment. This could be the reason why some CSM patients do not present a significant clinical improvement of their symptoms after surgery.

In summary, spinal cord compression leads to significant WM damage mainly at the corticospinal tracts leading to a progressive loss of axonal conductivity that extends cranially to the brain. This reduced connectivity associates with the neurological impairment and may constitute a potential barrier for motor recovery after surgery.

## Conclusion and future perspectives

As the population ages, the incidence of CSM will increase, but the knowledge of the effects of CSM in the brain is still sparse. On the other hand, there are no high-evidence studies providing well-founded recommendations for the best treatment modality. Considering that CSM has a potentially devastating outcome more studies are needed to better clarify the evolution of the disease and its consequences. Furthermore, to improve clinical outcomes the strategies should be equally targeted on the recruited adjacent cortex and the primary SMC. In this framework, we propose that increased collaborations among clinicians, experts in neuroimaging techniques, and basic researchers will enhance our ability for a better management of the disease.

## Applications to other areas of neuroscience

Cervical spondylosis myelopathy (CSM) is the major cause of disability in the elderly and, despite significant ongoing research, current therapeutic approaches have limited effectiveness and are far from satisfactory (Shiban & Meyer, 2014). One potential barrier is that current treatments generally assume intact neural motor systems for driving limb movements, but we should consider that this condition can lead to neuronal damage distal to the spinal lesion and cerebral reorganization. Therefore, the brain atrophy and the damage of cortical pathways in CSM could become a potential obstacle for achieving full and effective function recovery, even after successful surgical repair at the site of compression.

Advanced neuroimaging techniques can be applied to detect specific secondary brain damage in CSM. However, brain reorganization is not a static process, but rather very dynamic. Cortical reorganization starts early after spinal cord injury, and its impact over time will depend on a complex balance between spinal recovery and supraspinal reorganization (Freund et al., 2013). Thus, it is crucial to better understand how the brain reorganizes after the injury, as well as develop cost-effective therapies to maximize functional recovery. In this framework, therapies designed to enhance brain plasticity could provide functional benefits for these patients both before and following surgery (Sharma, Classen, & Cohen, 2013).

The findings of a progressive degenerative pattern in spinal cord compression suggest the need to properly monitor the disease. Advanced Magnetic Resonance techniques may present a main role in non-invasive assessment to better predict the patient's outcome as well as to identify the patients that will most likely benefit from different therapies and prioritize which patients will precise surgery more urgently.

## Mini-dictionary of terms

**Cervical spondylosis**: Term to define the progressive degenerative changes that affect all the components of the cervical spine (i.e., inter-vertebral discs, facet joints, joints of Luschka, ligamenta flava, and laminae).

**Cervical spondylosis myelopathy**: Term to define injury to the spinal cord due to severe compression caused by cervical spondylosis.

**Corticospinal tract**: White matter motor pathway that connects the cerebral cortex with the lower motor neurons and interneurons in the spinal cord.

**Diffusion tensor imaging**: MR technique based on the diffusion of water molecules of the tissues that allows to obtain information about the white matter architecture and integrity.

**MR-spectroscopy**: MR technique that allows to obtain the chemical composition of tissues, it is used to explore in vivo tissue metabolism and to detect and monitor several pathologies.

**Functional magnetic resonance imaging**: A technique that allows to obtain the brain activation pattern of a subject while performing a specific task. It is commonly used in the clinical setting to map cortical-activated areas before surgery to preserve their function. It is also widely used in research to study differences in brain activation patterns.

**Functional reorganization**: Also known as neural plasticity, neuroplasticity, or brain plasticity. The ability of the central nervous system to adapt in response to lesions or changes in the environment. This process could be accomplished through the recruitment of new/different neural networks, changes in the connectivity pattern, and the formation of new synapses.

**Resting-state functional magnetic resonance imaging**: A technique that allows to obtain the connectivity pattern of functional brain areas at rest that are spatially separated.

**Sensorimotor cortex:** Brain area located at the pre-central and post-central gyri which contain the primary motor and somatosensory areas of the brain.

**Supplementary motor area**: Brain area located in the midline cortex, anterior to the primary motor cortex that is responsible for movement control and coordination.

## Key facts of spinal cord compression

- The most common cause is cervical spondylosis which is frequently found in asymptomatic adults after the fifth decade of life.
- Cervical spondylosis may progress to cervical spondylosis myelopathy, a progressive condition that causes incomplete spinal cord injury and associated neurological impairment in a stepwise manner.
- At present most of the studies have been focused on the local changes at the cervical spine, showing apoptosis of the anterior horn neurons, gradual spongy necrosis, demyelination of the white matter, and Wallerian degeneration.
- Cervical spondylosis may present some similarities with traumatic spinal cord injury.
- Cervical spondylosis leads to adaptive changes in the brain to compensate for the functional loss, causing a wide range of clinical manifestations.

## Key facts of advanced MRI techniques

- Present high sensitivity to detect neural damage secondary to cervical spondylosis.
- Allow a better understanding of the consequences of cervical spine compression in the brain.
- May be useful for identifying those patients at risk of severe impairment.
- May help the non-invasive assessment of cervical spondylosis, to better predict outcomes and to identify the patients that most likely will benefit from different therapies.

## Summary points of cervical spine compression

- Leads to brain atrophy of the sensorimotor cortex, thalami, supplementary motor areas, and a faster rate of cortical thinning and atrophy related to age.
- Causes functional reorganization of the sensorimotor cortex that is more significant for the hand.
- Alters the FC of the brain at the sensorimotor cortex and visual cortex to compensate for the functional loss.
- Leads to white matter demyelination of corticospinal tracts, which extend cranially to the brain cortex mainly at the primary motor area.
- Causes myelin damage at the corpus callosum, mainly at the superior and posterior parietal projection fibers that are related to motor and touch-stimulation.

## References


Aleksanderek, I., McGregor, S., Stevens, T., Goncalves, S., Bartha, R., & Duggal, N. (2017). Cervical spondylotic myelopathy: Metabolite changes in the primary motor cortex after surgery. *Radiology*, *282*(3), 817–825. https://doi.org/10.1148/radiol.2016152083.

Banaszek, A., Bladowska, J., Szewczyk, P., Podgórski, P., & Sąsiadek, M. (2014). Usefulness of diffusion tensor MR imaging in the assessment of intramedullary changes of the cervical spinal cord in different stages of degenerative spine disease. *European Spine Journal*, *23*(7), 1523–1530. https://doi.org/10.1007/s00586-014-3323-x.

Baron, E. M., & Young, W. (2007). Cervical spondylotic myelopathy: A brief review of its pathophysiology, clinical course, and diagnosis. *Neurosurgery*, *60*(1 Suppl), 35–41.

Bernabéu-Sanz, A., Mollá-Torró, J. V., López-Celada, S., Moreno-López, P., & Fernández-Jover, E. (2020). MRI evidence of brain atrophy, white matter damage, and functional adaptive changes in patients with cervical spondylosis and prolonged spinal cord compression. *European Radiology*, *30*(1), 357–369. https://doi.org/10.1007/s00330-019-06352-z.

Bhagavatula, I. D., Shukla, D., Sadashiva, N., Saligoudar, P., Prasad, C., & Bhat, D. I. (2016). Functional cortical reorganization in cases of cervical spondylotic myelopathy and changes associated with surgery. *Neurosurgical Focus*, *40*(6), 1–7. https://doi.org/10.3171/2016.3.FOCUS1635.

Buchbinder, B. R. (2016). Functional magnetic resonance imaging. In *Vol. 135. Handbook of clinical neurology* (pp. 61–92). Elsevier B.V. https://doi.org/10.1016/B978-0-444-53485-9.00004-0.

Chen, Z., Wang, Q., Liang, M., Zhao, R., Zhu, J., Xiong, W., et al. (2018). Visual cortex neural activity alteration in cervical spondylotic myelopathy patients: A resting-state fMRI study. *Neuroradiology*, *60*(9), 921–932. https://doi.org/10.1007/s00234-018-2061-x.

Chen, Y. Y., Zhang, X., Lin, X. F., Zhang, F., Duan, X. H., Zheng, C. S., et al. (2017). DTI metrics can be used as biomarkers to determine the therapeutic effect of stem cells in acute peripheral nerve injury. *Journal of Magnetic Resonance Imaging*, *45*(3), 855–862. https://doi.org/10.1002/jmri.25395.

Cohen, E. J., Quarta, E., Bravi, R., Granato, A., & Minciacchi, D. (2017). Neural plasticity and network remodeling: From concepts to pathology. In *Vol. 344. Neuroscience* (pp. 326–345). Elsevier. https://doi.org/10.1016/j.neuroscience.2016.12.048.

Cui, J. L., Li, X., Chan, T. Y., Mak, K. C., Luk, K. D. K., & Hu, Y. (2014). Quantitative assessment of column-specific degeneration in cervical spondylotic myelopathy based on diffusion tensor tractography. *European Spine Journal*, *24*(1), 41–47. https://doi.org/10.1007/s00586-014-3522-5.

de Oliveira Vilaça, C., Orsini, M., Leite, M. A. A., de Freitas, M. R. G., Davidovich, E., Fiorelli, R., et al. (2016). Cervical spondylotic myelopathy: What the neurologist should know. *Neurology International*, *8*(4), 69–73. Press Publications https://doi.org/10.4081/ni.2016.6330.

Dong, Y., Holly, L. T., Albistegui-Dubois, R., Yan, X., Marehbian, J., Newton, J. M., et al. (2008). Compensatory cerebral adaptations before and evolving changes after surgical decompression in cervical spondylotic myelopathy: Laboratory investigation. *Journal of Neurosurgery. Spine*, *9*(6), 538–551. https://doi.org/10.3171/SPI.2008.10.0831.Compensatory.

Edwards, C. C., Riew, K. D., Anderson, P. A., Hilibrand, A. S., & Vaccaro, A. F. (2003). Cervical myelopathy: Current diagnostic and treatment strategies. *The Spine Journal*, *3*(1), 68–81. https://doi.org/10.1016/S1529-9430(02)00566-1.

Ellingson, B. M., Salamon, N., Grinstead, J. W., & Holly, L. T. (2014). Diffusion tensor imaging predicts functional impairment in mild-to-moderate cervical spondylotic myelopathy. *The Spine Journal : Official Journal of the North American Spine Society*, *14*(11), 2589–2597. https://doi.org/10.1016/j.spinee.2014.02.027.

Fabri, M., Pierpaoli, C., Barbaresi, P., & Polonara, G. (2014). Functional topography of the corpus callosum investigated by DTI and fMRI. *World Journal of Radiology*, *6*(12), 895–906. https://doi.org/10.4329/wjr.v6.i12.895.

Fox, M., Corbetta, M., Snyder, A., Vincent, J., & Raichle, M. (2006). Spontaneous neuronal activity distinguishes human dorsal and ventral attention system. *Proceedings of the National Academy of Sciences of the United States of America*, *103*(26), 10046–10051.

Freund, P., Weiskopf, N., Ashburner, J., Wolf, K., Sutter, R., Altmann, D. R., et al. (2013). MRI investigation of the sensorimotor cortex and the corticospinal tract after acute spinal cord injury: A prospective longitudinal study. *The Lancet Neurology*, *12*(9), 873–881. https://doi.org/10.1016/S1474-4422(13)70146-7.

Freund, P., Weiskopf, N., Ward, N. S., Hutton, C., Gall, A., Ciccarelli, O., et al. (2011). Disability, atrophy and cortical reorganization following spinal cord injury. *Brain*, *134*(6), 1610–1622. https://doi.org/10.1093/brain/awr093.

Freund, P., Wheeler-Kingshott, C. A., Nagy, Z., Gorgoraptis, N., Weiskopf, N., Friston, K., et al. (2012). Axonal integrity predicts cortical reorganisation following cervical injury. *Journal of Neurology, Neurosurgery and Psychiatry*, *83*(6), 629–637. https://doi.org/10.1136/jnnp-2011-301875.

Goncalves, S., Stevens, T. K., Doyle-Pettypiece, P., Bartha, R., & Duggal, N. (2016). N-acetylaspartate in the motor and sensory cortices following functional recovery after surgery for cervical spondylotic myelopathy. *Journal of Neurosurgery: Spine*, *25*(4), 436–443. https://doi.org/10.3171/2016.2.SPINE15944.

Grabher, P., Mohammadi, S., David, G., & Freund, P. (2017). Neurodegeneration in the spinal ventral horn prior to motor impairment in cervical spondylotic myelopathy. *Journal of Neurotrauma*, *34*(15), 2329–2334. https://doi.org/10.1089/neu.2017.4980.

Green, A., Cheong, P. W. T., Fook-Chong, S., Tiruchelvarayan, R., Guo, C. M., Yue, W. M., et al. (2015). Cortical reorganization is associated with surgical decompression of cervical spondylotic myelopathy. *Neural Plasticity*, *2015*. https://doi.org/10.1155/2015/389531.

Guan, L., Chen, X., Hai, Y., Ma, X., He, L., Wang, G., et al. (2017). High-resolution diffusion tensor imaging in cervical spondylotic myelopathy: A preliminary follow-up study. *NMR in Biomedicine*, *30*(10), 1–11. https://doi.org/10.1002/nbm.3769.

Guleria, S., Gupta, R. K., Saksena, S., Chandra, A., Srivastava, R. N., Husain, M., et al. (2008). Retrograde Wallerian degeneration of cranial corticospinal tracts in cervical spinal cord injury patients using diffusion tensor imaging. *Journal of Neuroscience Research*, *86*(10), 2271–2280. https://doi.org/10.1002/jnr.21664.

Hains, B. C., Black, J. A., & Waxman, S. G. (2003). Primary cortical motor neurons undergo apoptosis after axotomizing spinal cord injury. *Journal of Comparative Neurology*, *462*(3), 328–341. https://doi.org/10.1002/cne.10733.

Henderson, L. A., Gustin, S. M., Macey, P. M., Wrigley, P. J., & Siddall, P. J. (2011). Functional reorganization of the brain in humans following spinal cord injury: Evidence for underlying changes in cortical anatomy. *Journal of Neuroscience*, *31*(7), 2630–2637. https://doi.org/10.1523/JNEUROSCI.2717-10.2011.

Holly, L., Dong, Y., Albistegui-DuBois, R., Marehbian, J., & Dobkin, B. (2007). Cortical reorganization in patients with cervical spondylotic myelopathy. *Journal of Neurosurgery. Spine*, *6*(6), 544–551. https://doi.org/10.3171/spi.2007.6.6.5.Cortical.

Jang, S. H., Medicine, P., & Medicine, C. (2014). The corticospinal tract from the viewpoint of BRain rehabilitation. *Journal of Rehabilitation Medicine*, *46*(31), 193–199. https://doi.org/10.2340/16501977-1782.

Kelly, J. C., Groarke, P. J., Butler, J. S., Poynton, A. R., & Byrne, J. M. O. (2012). The natural history and clinical syndromes of degenerative cervical spondylosis. *Advances in Orthopedics*, 2012. https://doi.org/10.1155/2012/393642.

Koskinen, E. A., Hakulinen, U., Brander, A. E., Luoto, T. M., Ylinen, A., & Öhman, J. E. (2014). Clinical correlates of cerebral diffusion tensor imaging findings in chronic traumatic spinal cord injury. *Spinal Cord*, *52*(3), 202–208. https://doi.org/10.1038/sc.2013.163.

Lotze, M., Beutling, W., Loibl, M., Domin, M., Platz, T., Schminke, U., et al. (2012). Contralesional motor cortex activation depends on ipsilesional corticospinal tract integrity in well-recovered subcortical stroke patients. *Neurorehabilitation and Neural Repair*, *26*(6), 594–603. https://doi.org/10.1177/1545968311427706.

Martin, A. R., Aleksanderek, I., Cohen-Adad, J., Tarmohamed, Z., Tetreault, L., Smith, N., et al. (2016). NeuroImage: Clinical translating state-of-the-art spinal cord MRI techniques to clinical use: A systematic review of clinical studies utilizing DTI, MT, MWF, MRS, and fMRI. *NeuroImage. Clinical*, *10*, 192–238. https://doi.org/10.1016/j.nicl.2015.11.019. YNICL.

Panchuelo, R. M. S., Eldeghaidy, S., Marshall, A., McGlone, F., Francis, S. T., & Favorov, O. (2020). A nociresponsive specific area of human somatosensory cortex within BA3a: BA3c? *NeuroImage*, *221*. https://doi.org/10.1016/j.neuroimage.2020.117187.

Pantano, P., Petsas, N., Tona, F., & Sbardella, E. (2015). The role of fMRI to assess plasticity of the motor system in MS. *Frontiers in Neurology*, *6*, 55. https://doi.org/10.3389/fneur.2015.00055.

Peng, X., Tan, Y., He, L., & Ou, Y. (2020). Alterations of functional connectivity between thalamus and cortex before and after decompression in cervical spondylotic myelopathy patients: A resting-state functional MRI study. *Neuroreport*, 365–371. https://doi.org/10.1097/WNR.0000000000001346.

Sharma, N., Classen, J., & Cohen, L. G. (2013). Neural plasticity and its contribution to functional recovery. In *Vol. 110. Handbook of clinical neurology* (pp. 3–12). Elsevier B.V. https://doi.org/10.1016/B978-0-444-52901-5.00001-0.

Shiban, E., & Meyer, B. (2014). Treatment considerations of cervical spondylotic myelopathy. *Neurology: Clinical Practice*, *4*(4), 296–303. https://doi.org/10.1212/CPJ.0000000000000050.

Smitha, K., Raja, K., Arun, K., Rajesh, P., Thomas, B., Kapilamoorthy, T., et al. (2017). Resting state fMRI: A review on methods in resting state connectivity analysis and resting state networks. *The Neuroradiology Journal*, *30*(4), 305–317.

Stinear, C. M., Barber, P. A., Smale, P. R., Coxon, J. P., Fleming, M. K., & Byblow, W. D. (2007). Functional potential in chronic stroke patients depends on corticospinal tract integrity. *Brain*, *130*(1), 170–180. https://doi.org/10.1093/brain/awl333.

Takenaka, S., Kan, S., Seymour, B., Makino, T., Sakai, Y., Kushioka, J., et al. (2019). Towards prognostic functional brain biomarkers for cervical myelopathy: A resting-state fMRI study. *Scientific Reports*, *9*(1), 1–8. https://doi.org/10.1038/s41598-019-46859-5.

Tam, S., Barry, R. L., Bartha, R., & Duggal, N. (2010). Changes in functional magnetic resonance imaging cortical activation after decompression of cervical spondylosis: Case report. *Neurosurgery*, *67*(3), 2–7. https://doi.org/10.1227/01.NEU.0000374848.86299.17.

Theodore, N. (2020). Degenerative cervical spondylosis. *New England Journal of Medicine*, *383*(2), 159–168. Massachussetts Medical Society https://doi.org/10.1056/NEJMra2003558.

Wang, C., Laiwalla, A., Salamon, N., Ellingson, B. M., & Holly, L. T. (2020). Compensatory brainstem functional and structural connectivity in patients with degenerative cervical myelopathy by probabilistic tractography and functional MRI. *Brain Research*, *1749*(May), 147129. https://doi.org/10.1016/j.brainres.2020.147129.

Ward, N. (2011). Assessment of cortical reorganisation for hand function after stroke. *The Journal of Physiology*, *589*(Pt 23), 5625–5632. https://doi.org/10.1113/jphysiol.2011.220939.

Wen, C. Y., Cui, J. L., Mak, K. C., Luk, K. D. K., & Hu, Y. (2014). Diffusion tensor imaging of somatosensory tract in cervical spondylotic myelopathy and its link with electrophysiological evaluation. *The Spine Journal*, *14*(8), 1493–1500. https://doi.org/10.1016/j.spinee.2013.08.052.

Woodworth, D. C., Holly, L. T., Mayer, E. A., Salamon, N., & Ellingson, B. M. (2019). Alterations in cortical thickness and subcortical volume are associated with neurological symptoms and neck pain in patients with cervical spondylosis. *Clinical Neurosurgery*, *84*(3), 588–597. https://doi.org/10.1093/neuros/nyy066.

Woodworth, D. C., Holly, L. T., Salamon, N., & Ellingson, B. M. (2018). Resting-state functional magnetic resonance imaging connectivity of the brain is associated with altered sensorimotor function in patients with cervical spondylosis. *World Neurosurgery*, *119*, 1–10. https://doi.org/10.1016/j.wneu.2018.07.257.

Zdunczyk, A., Schwarzer, V., Mikhailov, M., Bagley, B., Rosenstock, T., Picht, T., et al. (2018). The corticospinal reserve capacity: Reorganization of motor area and excitability as a novel pathophysiological concept in cervical myelopathy. *Clinical Neurosurgery*, *83*(4), 810–818. https://doi.org/10.1093/neuros/nyx437.

Zhou, F., Wu, L., Liu, X., Gong, H., Luk, K. D. K., & Hu, Y. (2015). Characterizing thalamocortical disturbances in cervical spondylotic myelopathy: Revealed by functional connectivity under two slow frequency bands. *PLoS One*, *10*(6), 1–13. https://doi.org/10.1371/journal.pone.0125913.

Chapter 9

# Spasticity in spinal cord injury

Ramiro Palazón-García
*Physical Medicine and Rehabilitation Department, Hospital Nacional de Parapléjicos, Toledo, Spain*

## List of abbreviations

**ADL** activities of daily living
**AIS** ASIA impairment scale
**FES** functional electrical stimulation
**SC** spinal cord
**SCI** spinal cord injury

## Introduction

Spasticity is a common complication after spinal cord injury (SCI) that occurs in the context of an upper motor neuron injury syndrome, and may develop even years after the acute injury leading to further loss of function (Harrington & Bockenek, 2011; Rekand, Hagen, & Grønning, 2012).

Spasticity was defined by Lance (1980) as "a motor disorder characterized by a velocity dependent increase in tonic stretch reflexes (muscle tone) with exaggerated tendon jerks, resulting from hyper excitability of the stretch reflex, as one component of the upper motoneuron syndrome." Dynamic phenomena were not included, so the Ability Network (an international panel of clinical experts to develop management algorithms to guide and standardize assessment, treatment and outcomes in SCI people with disabling spasticity) recommends Pandyan definition (Burns et al., 2016): "Disordered sensorimotor control, resulting from an upper motor neuron injury, presenting as intermittent or sustained involuntary activation of muscles."

Considering every SCI people regardless of the time of evolution, neurological level or extent, the prevalence of spasticity is approximately 65% according to the studies of Harrington and Bockenek (2011). Some studies have been conducted on SCI complications; Joseph, Ismail, Naicker, Ying, and Mohammad (2009) demonstrated an incidence of 25%–37% in acute traumatic SCI, New, Rawicki, and Bailey (2002) showed an incidence of 15%–36% in acute nontraumatic SCI, and in papers by Adams and Hicks (2005) or Rekand et al. (2012), incidence of chronic SCI is estimated about 65%–93%. There is no significant correlation among neurological level, ASIA Impairment Scale (AIS) grade, and spasticity, but Adams and Hicks (2005) related severe spasticity with cervical levels and grade A, and a highest prevalence of spasticity among people with grade C.

## Pathophysiology of spasticity

The stretch reflex is a monosynaptic reflex that originates in the muscle spindles with any stimulus (not just stretch) and travels through an afferent pathway to the spinal cord (SC) where it synapses with the corresponding alpha motor neuron and contracts the muscle fibers where it is has produced the stimulus. If the stimulus is prolonged, the Golgi tendon organs are activated, and via interneurons Ib, the previous contraction can be relaxed. Studies by Whitlock (1990) show that spasticity is generally due to a lowered threshold of phasic or tonic stretch reflexes; when inhibitory signals are lost due to spinal cord damage, the segmental reflexes are released and become hyperactive. In addition, other spinal mechanisms are postulated, such as fusimotor hyperexcitability, axonal sprouting, reduction of presynaptic inhibition, and reduction of reciprocal inhibition.

Roy and Edgerton (2012) described that clinical features of spasticity depend on the greater or lesser loss of the ability to voluntarily modulate the level of activity of a given motor pool and the capacity of the interneurons that project to these motor pools to recruit the appropriate combination of them.

Diagnosis and Treatment of Spinal Cord Injury. https://doi.org/10.1016/B978-0-12-822498-4.00009-9

Besides, there are plastic alterations in affected muscles. Basically, spasticity causes fibrosis, atrophy of muscle fibers, decrease in the elastic properties, decrease in the number of sarcomeres, accumulation of connective tissue, and alteration of contractile properties toward tonic muscle characteristics.

## Manifestations of spasticity in spinal injuries

According to Decq (2003), two classification criteria can be used: the place where the stimulus for spasticity is located, which differentiates between intrinsic if the stimulus emerges within the central nervous system, and extrinsic if the afferences come from other structures such as the skin, muscles, and joints; also if there is no resultant movement, a tonic or static component is described, and if there is any movement, a phasic or dynamic component is described. So, we can distinguish three main types of spasticity:

(a) Intrinsic tonic spasticity: hypertonia.
(b) Intrinsic phasic spasticity: clonus and osteotendinous hyperreflexia.
(c) Extrinsic spasticity: spasms.

The presence of triggers was described by Phadke, Balasubramanian, Ismail, and Boulias (2013); these exacerbating factors can cause a patient with a SCI and a normal or decreased tone to present intense or frequent spasms. The most important triggers are neurogenic bowel (constipation) and bladder problems (hyperreflexia). The supine position is associated with more intense hypertonia and increased likelihood to provoke spasms, and according with Sadeghi, Mclvor, Finlayson, and Sawatzky (2016), the most frequent clinical sign is usually spasms when making any transfer. Other factors such as pregnancy, cold, circadian rhythm, pressure ulcers, menstrual cycle, stress, and even tight clothing also increase spasticity, while acute and severe infections (sepsis) and syringomyelia can cause both an increase of spasticity as a sudden absence of spasticity.

The ability to achieve functional goals may be impaired due to spasticity and can even make the length of hospital stay last longer, according to Richard-Denis, Nguyen, and Mac-Thiong (2020).

## Assessment of spasticity and its consequences

### Tone measurement

The most widely used method is the Modified Ashworth Scale (Table 1) because, although it is a Likert-type scale and therefore has been criticized for its possible lack of reliability, multiple studies have been carried out and showed that can be considered the most appropriate due to its reliability and test-retest agreement (Baunsgaard, Nissen, Christensen, & Biering-Sørensen, 2016; Meseguer-Henarejos, Sánchez-Meca, López-Pina, & Hernández, 2018). The Tardieu scale is based on the pendulum test and consists of performing the stretching of a muscle between two points and two speeds, but in SCI, it is still considered only as complementary.

### Assessment of dynamic phenomena

The frequency and intensity of spasms can be measured using the Penn spasm intensity or frequency scales, but there is a more specific scale for SCI developed by Benz, Hornby, Bode, Scheidt, and Schmit (2005), called the Spinal Cord Assessment Tool Spastic reflexes (SCATS, Table 2).

**TABLE 1** Modified Ashworth Scale (Bohannon & Smith, 1987).

| |
|---|
| 0. No increase in tone |
| 1. Slight increase in tone with a catch, or minimal resistance at the end of the range of movement (ROM) |
| 1+. Slight increase in tone with a catch, followed with minimal resistance throughout the remainder (less than half) of the ROM |
| 2. Marked increase in tone through most of the ROM, but limb is easily moved |
| 3. Considerable increase in tone; passive movement difficult |
| 4. Limb rigid or contracted |

**TABLE 2** Spinal Cord Assessment Tool for Spastic reflexes (SCATS).

| Clonus of ankle plantar flexors with rapid passive dorsiflexion of foot |
|---|
| 0. No reaction |
| 1. Mild: clonus <3 s |
| 2. Moderate: clonus lasts between 3 and 10 s |
| 3. Severe: clonus >10 s |
| Flexor spasms in response to pinprick on foot plantar surface with leg and hip in full extension |
| 0. No reaction |
| 1. Mild: extension of great toe or <10 degrees of hip/knee flexion |
| 2. Moderate: 10–30 degrees of hip/knee flexion |
| 3. Severe: >30 degrees of hip/knee flexion |
| Extensor spasms of quadriceps muscle after extension of leg from a position of hip/knee flexion |
| 0. No reaction |
| 1. Mild: spasms last <3 s |
| 2. Moderate: spasms last between 3 and 10 s |
| 3. Severe: spasms last >10 s |

## Functional assessment

From the Ability Network (Nene et al., 2018), it is recommended to measure as functional results:

- Joint alterations can be rated by goniometry.
- Interferences of spasticity with activities of daily living are assessed using the Measure of Independence in SCI version III (SCIM III).
- The ability to walk is estimated through Walking Index in SCI scale (WISCI II).
- The impact on the quality of life can also be analyzed, especially with specific scales such as Spinal Cord Injury Spasticity Evaluation Tool (SCI-SET, Table 3) or Patient-Reported Impact of Spasticity Measure (PRISM, Table 4). Generic scales are often not very suitable for SCI, since walking well or performing standing activities are considered signs of a

**TABLE 3** Spinal Cord Injury Spasticity Evaluation Tool (Adams, Ginis, & Hicks, 2007).

| −3. Extremely problematic |
|---|
| −2. Moderately problematic |
| −1. Somewhat problematic |
| 0. No effect |
| 1. Somewhat helpful |
| 2. Moderately helpful |
| 3. Extremely helpful |
| Evaluation of spasticity effects on components of daily life over a 7 days-period: showering, dressing/undressing, transfers, sitting positioning, preparation of meals, eating, drinking, small hand movements (writing, use of computer), household chores, recreational activities, enjoyment of social outings, standing/weight-bear, walking, stability/balance, muscle fatigue, flexibility of joints, therapy/exercise routine, manual or power wheelchair use, lying positioning, change positions in bed, getting to sleep, quality of sleep, sex life, annoyment feeling, embarrassing feeling, social comfort, physical comfort, pain, concern with feeling, concern with getting injured, concern with accidentally injuring someone else, concentration, control over one's own body, need for help. |

**TABLE 4** Patient-Reported Impact of Spasticity Measure (Cook et al., 2007).

| Evaluation of spasticity effects on life experiences over a 7-day period |
|---|
| 0. Never true for me |
| 1. Rarely true for me |
| 2. Sometimes true for me |
| 3. Often true for me |
| 4. Very often true for me |

good quality of life; however, scales such as the SF-36 or the Euro-QOL are still recommended by the Ability Network (Ertzgaard, Nene, Kiekens, & Burns, 2019).

### Subjective evaluation

Generally, patients can evaluate their own spasticity with a numerical rating scale, but they can describe their spasticity through specific signs.

### Other types of assessment

Ultrasound appearance of spastic muscles can be used, and there are studies (e.g., Hara et al., 2018) that compare the liability of improvement of poststroke spastic muscles treated with botulinum toxin according to its echo intensity (Heckmatt scale), confirming that increased intensity has worse prognostic, since it corresponds to a more chronic and severe spasticity. There is also no useful correlation with neurophysiological parameters.

## Treatment of spasticity

### Treatment indications

Spasticity may be useful to improve sitting or standing balance, facilitate activities of daily living (ADL) and transfers, maintain muscle mass and venous return. According with Stevenson (2010), treatment is necessary when we find functional disadvantages, and is aimed at preventing joint limitations, reducing pain, facilitating positioning, treating pressure ulcers, hygiene, and facilitating ADLs and transfers.

## Modalities of treatment in spinal injury

Since the intensity of spasticity can vary in intensity throughout the same day and eradication treatment is not possible, the management of spasticity should focus on EDUCATION; the patient must learn what spasticity is, its signs and symptoms and its consequences, and how they can be prevented, starting with avoiding exacerbating factors.

(a) **Control of triggers**. The first point in treatment is to eliminate the triggering factors, since although they do not cause spasticity, they are usually present when it appears or worsens.

(b) **Positioning**. The most appropriate posture is sitting; in the wheelchair, the patient must have hips, ankles, and knees flexed at 90 degrees and a slight inclination of the seat, in bed it is recommended lateral decubitus with more forced flexion of the hip and knee of the lower limb located just above the bed. You may improve spasticity using standing devices.

(c) **Physical modalities**. Studies by Gebler (1990) have shown that most of them have a short-lasting effect to be significant, and is based on the maintenance of the myotatic reflex by prolonging a stimulus (stretching, cold) and therefore the stimulation of the Golgi reflex.
- Stretching. It should be held for at least 1 min to achieve muscle relaxation and increase ranges of motion. It can be done through self-stretching, casting, splinting, or within a physical therapy protocol.
- Cold or heat application.

- Orthotics and splints. They are used to allow long-term stretching, but they should only be used in moderate spasticity or to maintain outcomes from another previous treatment.
- Muscle-strengthening. In this way, we mitigate the functional repercussions that traction of the antagonist spastic muscle can have on any muscle.
- Electrical stimulation. The repeated application of tetanic contractions by high frequencies >2500 Hz can fatigue the spastic muscle, but this effect is short-lasting, and even long-term may lead to increase spasticity, especially in incomplete injuries. Functional electrical stimulation (FES) is used to strengthen antagonist muscles.
- Robotic systems can work both by causing prolonged stretching and by strengthening antagonists, showed by Alcobendas-Maestro et al. (2012).
- Passive cycling is postulated as a treatment for strengthening weak muscles. In a study by Krause, Szecsi, and Straube (2008), it is shown that there may be improvement in spasticity but the significance is demonstrated with FES compared to passive cycling. The rest of the studies collected in the metaanalysis by Phadke et al. (2019) show that passive cycling can only achieve subjective improvement but objectively there are no effects or there is even a worsening in spasticity.
- Other modalities. Huang et al. (2019) performed a randomized controlled trial on kinesiotaping for spasticity in stroke patients resulting in short-lasting improvement, but there are no trials within SCI people. Shockwaves have been used in poststroke spasticity, and a systematic review by Dymarek et al. (2020) confirms moderate improvement, but it has not been studied in SCI either.

(d) **Oral drugs**. Its characteristics have been studied by Kita and Goodkin (2000), and its specific use for SCI patients has been described by Taricco, Pagliacci, Telaro, and Adone (2006).
- Baclofen. It is considered the gold standard drug because it acts by binding to receptors just in the spinal cord, since it is a structural analog of gamma-amino-butyric acid (GABA, an inhibitory neurotransmitter) and an agonist of GABA-B receptors. In adults, you can start with 15 mg daily in 3 doses and increase every 4–7 days 15 mg daily. Start dosage can be 15 mg/day divided in 3 doses and titration may be 15 mg each 4–7 days; in older patients or with comorbidities, the titration should be slower, and in children, dosage starts with 2.5 mg/day to a maximum dose of 30 mg/day in children 2–7 years and 60 mg/day in children older than 7. Side effects are dose-dependent; the most frequent are fatigue and drowsiness, it can also cause muscle weakness, confusion, dizziness, hypotension, or constipation.
- Diazepam. It binds to GABA-A receptors and causes presynaptic inhibition in the spinal cord. Side effects include sedation and cognitive impairment. Its main indication would be night spasms, prescribing 5 mg daily.
- Clonazepam. It is another benzodiazepine that is typically used when spasms predominate, the starting dose is 0.5 mg daily and requires a slower titration because of its side effects (withdrawal syndrome may be more dangerous).
- Tizanidine. It is an α2 agonist that acts by inhibiting the release of excitatory amino acids in SC interneurons. The most frequent side effects are drowsiness, dizziness, hypotension, or xerostomia, the starting dose is 4 mg daily. The effectiveness of tizanidine, usually considered as a second-line drug, is backed by the largest study.
- Clonidine. It is also an α2 agonist, but it causes hypotension more frequently, so it is not recommended in tetraplegic patients.

(e) **Intrathecal medication**. Studies by McIntyre et al. (2014) and Khurana and Garg (2014) show that baclofen pumps may be considered when high dosages of baclofen are required or there is no response with any other treatment. A programmable pump is implanted into the abdomen, from where a catheter conveys the baclofen into the intrathecal space, usually at the lumbar level since there is greater concentration of GABA receptors in the lumbar spinal cord. Prior to implantation of the pump, all candidates undergo a trial via lumbar puncture, with a dose of 50 μg of baclofen; tone and spasms are measured each 2 h, and if there is positive outcome (spasticity improves), we proceed to gradually withdraw oral drugs and then increase the dose of baclofen in the pump up to the optimum dose (average is 400–500 μg/day), which is kept refilling the reservoir pump at regular intervals (3–6 months). Most adverse effects occur during the titration phase and include drowsiness, headache, nausea, weakness, and hypotension; other complications may be due to mechanical problems (dislodgment, disconnection, kinking, blockage), pump failure, or infection. Any interruption of intrathecal baclofen delivery may result in a severe baclofen withdrawal syndrome that is usually characterized by a sudden increase of spasticity, pruritus, hyperthermia, autonomic dysregulation, seizures, coma, rhabdomyolysis, disseminated intravascular coagulation, and multisystem organ failure.

(f) **Local injections**. They are indicated in focal spasticity. Local phenol injections are discussed in the corresponding chapter. Studies by Palazón-García, Alcobendas-Maestro, Esclarin-Ruz, and Benavente-Valdepeñas (2019) have shown that focal spasticity is described in some cases of AIS C or D injuries. Botulinum toxin serotype A is produced by the *Clostridium botulinum* bacterium, and is a metalloprotease which, in nerve endings, proteolitically cleaves synaptosomal associated protein (SNAP-25) to inhibit the fusion of the synaptic vesicle with the presynaptic membrane of the axon terminal, and thus ultimately relax the muscle Botulinum toxin can also be used if there are muscles that cause worse functional repercussions (e.g., hip adductors to allow bladder indwelling).

(g) **Surgery**. In SCI, it is not used as an etiological treatment, but it can be applied as a correction of complications (tenotomies and tendon transfers).

(h) **Other treatments**. Studies by Krause and Straube (2003) have reported that repetitive magnetic stimulation at the lumbar nerve roots can decrease lower limbs spasticity in SCI patients, and Korzhova et al. (2018) found that repetitive transcranial magnetic stimulation may be effective against spasticity in SCI patients, even more than against cerebral spasticity.

## Outcome measures after treatment

It is performed by means of comparison with previous measurements, the global tolerance to the treatment, and a subjective assessment by the patient. It is also possible to assess the achievement of the objectives set before the treatment, or Patient Global Impression of Change scale, as stated in protocol by Alcobendas-Maestro, Palazón-García, Vargas-Baquero, and Esclarin-Ruz (2015).

## Management protocol for spasticity in spinal cord injury

There are studies on the application of one or other treatments such as that of Holtz, Szefer, Noonan, and Mills (2018) but there are no specific protocols or algorithms of management nowadays for spasticity in SCI. Most of the systematic reviews and metaanalysis have been performed on spasticity due to stroke or several mixed pathologies, so no evidence can be shown for any technique for SCI as stated by Taricco et al. (2006) and Khan, Amatya, Bensmail, and Yelnik (2019). A protocol for this treatment is then proposed:

(a) **Positioning and exercise**: The first step is to promote an adequate positioning and educate the patient in the need to exercise daily; during acute in-patient phase, this exercise will be performed in the context of physical therapy, and then in chronic phase perform only mobilizations or stretching as maintenance.

(b) **Control of triggering factors**: It can be considered as the first part of the treatment because it is the first factor to be taken into account when spasticity appears or is decompensated.

(c) **Baclofen**: If spasticity must be treated, the first-line drug is baclofen. We start with a dosage of 15 mg/day divided into 3 doses (lower dosage for children, elderly people or concomitant drugs), and in 3–4 days, it can be increased to 30 mg/day; we must wait 2 weeks to significantly evaluate the effect.
   - If the response to treatment is partial, we increase to 45 mg/day and if now the response is complete, the same dosage is maintained.
   - If there is no response for 4 weeks, we can go directly to next point by withdrawing baclofen.
   - If there is a partial response after 3 weeks with baclofen at 45–60 mg/day and spasticity remains with much functional interference, a second-line drug is added, and this depends on the spasticity features.

(d) **Second-line drugs**:
   - Only phasic spasticity. If the only symptom not controlled by baclofen is night spasms diazepam is added in a single nightly dose. If there are spasms or other dynamic phenomena all day long then clonazepam is added.
   - Predominance of hypertonia. In general, the drugs preferred as adjuvants to baclofen are tizanidine or clonidine, considering that if the patient has a cervical SCI, we only choose tizanidine.
   - Focal spasticity. If there are less than four muscle groups involved in spasticity, botulinum toxin is used. If spasticity is initially generalized but after administration of baclofen or others only focal spasticity persists, botulinum toxin is also indicated.

(e) **Third-line treatment**. If we are using two of the drugs mentioned earlier with a dose close to the maximum allowed or effective, we can add one of the others as third drug (e.g., if we use baclofen and tizanidine, we can associate diazepam). If one of the drugs mentioned earlier cannot be added as a third drug (elderly patients, prevention of adverse events), we can test other drugs with less recognized efficacy or focus on physical modalities (e.g., hydrotherapy).

(f) **Failure of all the treatments**. If we are using three or four oral drugs, we have every exacerbating factor controlled, the patient performs physical activity, and despite everything spasticity is still intense or disabling, we will consider the intrathecal baclofen pump. We must perform a previous baclofen test dose (positive if instillation of a 50 μg bolus produces improvement of spasticity), there must be permeable intrathecal flow, the patient must be motivated and psychologically stable, and there must have been a minimum time of evolution of the lesion (usually 9–12 months).

## Applications to other areas of neuroscience

**Differences between spasticity of cerebral and spinal origin** (Alcobendas-Maestro et al., 2015; Harrington & Bockenek, 2011).

- The most frequent presentation of spasticity when the origin is spinal is generalized and diffuse, while focal spasticity is more frequent when the origin is cerebral.
- Regarding hypertonia, patients with SCI develop a more intense spasticity, and the clasp knife phenomenon is more frequent.
- The most affected muscles in SCI are extensors (antigravity muscles), especially in the lower extremities, while people with stroke or cerebral palsy show specific synergies (hemiparesis, diparesis) with greater involvement of flexor muscles.
- Regarding intrinsic phasic spasticity, clonus of the plantar flexors is more frequent when the origin is spinal, and patellar clonus is rare if the origin of the spasticity is not cerebral.
- Extrinsic spasticity is more common in SCI; lower limb extensor spasms are the most prevalent spastic sign in SCI. The most important stimulus in SCI is the extension of the hip (especially the last 20 degrees). The highest concentration of mechanoreceptors in people with cerebrovascular diseases is in the knees.
- Spinal spasticity is further exacerbated by visceral diseases.
- SCI patients show shortening of muscles due to prolonged postures, and the most frequent affected muscles are hamstrings that are shortened by keeping the sitting in the wheelchair. However, stroke patients can maintain a posture due to spasticity when they develop spastic dystonia, whose principal example is elbow flexion posture.

## Mini-dictionary of terms

Antagonist muscle produces an opposite movement. Since antagonist muscles produce an opposing joint torque to the agonist muscles, this torque can be used to control the motion.

Antigravity muscle acts to counterbalance the pull of gravity and to maintain an upright posture. Most important antigravity muscles are quadriceps femoris and triceps surae.

Catch is a sudden appearance of increased muscle activity as a response to a fast passive stretch, which provokes an abrupt stop.

Clasp knife phenomenon occurs when there is an initial resistance to passive movement of a joint and this tone reduces suddenly and then the limb can move quite freely through its range of motion.

Clonus is an involuntary rhythmic muscle contraction that can cause distal joint oscillation.

Hypertonia is an involuntary increase in muscle resistance to passive stretching.

Osteotendinous hyperreflexia is an exaggerated muscular response to percussion of the tendons.

Spasms are abrupt, involuntary muscle contractions.

Spinal shock is the temporary loss of spinal reflex activity that appears as an initial phase when SCI starts abruptly (as occurs in traumatic or vascular injuries).

Tonus (muscle tone) is a continuous and passive partial contraction of the muscles that works as preparatory for voluntary movement.

## Key facts of "diagnosis of spasticity in spinal cord injury"

When a traumatic SCI is starting, the appearance of spasticity means the ending of the spinal shock.

Spasticity cannot be found in lower motoneuron syndromes. The clinical course of cauda equina or conus medullaris syndromes is like a second motoneuron syndrome, and therefore they never develop spasticity.

When the cause of the SCI is vascular, the duration of spinal shock is usually longer.

The presence of neurogenic bladder or bowel is associated with normal or even flaccid tone but frequent and spontaneous spasms.

We must make a differential diagnosis between spasticity and other tone disorders; resistance to passive movement found in extrapyramidal stiffness, paratonia, or simulation is velocity-independent.

## Key facts of "treatment of spasticity in spinal cord injury"

Physical modalities are often considered the mainstay of treatment for spasticity, but the duration of the effect is too short, so these treatments are useless. It may be beneficial to strengthen the muscles antagonistic to the spastic ones.
Although there is only one trial with only six SCI patients that compared oral baclofen with placebo, baclofen has been reported to be effective for reducing spasticity (particularly spasms) and is considered the first-line treatment for spasticity.
Since the proportion of elderly SCI patients is increasing, the dosage of baclofen is becoming more difficult due to its effects on cognitive abilities.
Tizanidine is the drug with the most important and reliable studies but it is usually associated with the treatment of stroke spasticity, where it is the only drug allowed.
Clonazepam has not been supported by any published study (not even observational), but it is found in all spasticity treatment protocols when the clinical features are spasms.

## Summary points

Spasticity is defined as a disordered sensorimotor control, resulting from an upper motor neuron injury, presenting as intermittent or sustained involuntary activation of muscles. It occurs in 65% of SCI patients.
Spinal cord damage breaks the connection between the brain and the lower areas, so inhibitory signals are not received, and this causes a release and hyperactivity of segmental reflexes.
Spasticity in spinal cord injuries is characterized by being generalized and very intense.
The diagnosis of spasticity is eminently clinical and is made by confirming the clinical symptoms and signs.
The assessment is carried out by means of the measurements of the tone (Modified Ashworth Scale), dynamic phenomena (Spinal Cord Assessment Tool for Spastic reflexes), and of the functional repercussions.
Treatment is required only if the spasticity causes functional disadvantage and the main point is the education of the patient as to the recognition of the factors that can exacerbate the spasticity and the interventions to reduce its intensity.
Baclofen is the first-choice drug for the treatment of spasticity in spinal cord injuries.

## References

Adams, M. M., Ginis, K. A., & Hicks, A. L. (2007). The spinal cord injury spasticity evaluation tool: Development and evaluation. *Archives of Physical Medicine and Rehabilitation*, *88*, 1185–1192.

Adams, H. H., & Hicks, A. L. (2005). Spasticity after spinal cord injury. *Spinal Cord*, *43*, 577–586.

Alcobendas-Maestro, M., Esclarín-Ruz, A., Casado-López, R. M., Muñoz-González, A., Pérez-Mateos, G., González-Valdizán, E., et al. (2012). Lokomat robotic- assisted versus overground training within 3 to 6 months of incomplete spinal cord lesion: Randomized controlled trial. *Neurorehabilitation and Neural Repair*, *26*, 1058–1063.

Alcobendas-Maestro, M., Palazón-García, R., Vargas-Baquero, E., & Esclarin-Ruz, A. (2015). Guía de práctica clínica para el tratamiento de la espasticidad espinal con toxina botulínica. *Rehabilitación (Madr)*, *49*, 38–44.

Baunsgaard, C. B., Nissen, U. V., Christensen, K. B., & Biering-Sørensen, F. (2016). Modified Ashworth scale and spasm frequency scale in spinal cord injury: Reliability and correlation. *Spinal Cord*, *54*, 702–708.

Benz, E. N., Hornby, T. G., Bode, R. K., Scheidt, R. A., & Schmit, B. D. (2005). A physiologically based clinical measure for spastic reflexes in spinal cord injury. *Archives of Physical Medicine and Rehabilitation*, *86*, 52–59.

Bohannon, R. W., & Smith, M. B. (1987). Interrater reliability of a modified Ashworth scale of muscle spasticity. *Physical Therapy*, *67*, 206–207.

Burns, A. S., Lanig, I., Grabljevec, K., New, P. W., Bensmail, D., Ertzgaard, P., et al. (2016). Optimizing the management of disabling spasticity following spinal cord damage: The ability network- an international initiative. *Archives of Physical Medicine and Rehabilitation*, *97*, 2222–2228.

Cook, K. F., Teal, C. R., Engebretson, J. C., Hart, K. A., Mahoney, J. S., Robinson-Whelen, S., et al. (2007). Development and validation of patient reported impact of spasticity measure (PRISM). *Journal of Rehabilitation Research and Development*, *44*, 363–371.

Decq, P. (2003). Physiopathologie de la spasticité. *Neurochirurgie*, *49*, 163–184.

Dymarek, R., Ptaszkowski, K., Ptaszkowska, L., Kowal, M., Sopel, M., Taradaj, J., et al. (2020). Shock waves as a treatment modality for spasticity reduction and recovery improvement in post-stroke adults—Current evidence and qualitative systematic review. *Clinical Interventions in Aging*, *15*, 9–28. https://doi.org/10.2147/CIA.S221032.

Ertzgaard, P., Nene, P., Kiekens, C., & Burns, A. S. (2019). A review and evaluation of patient-reported outcome measures for spasticity in persons with spinal cord damage: Recommendations from the ability network—An international initiative. *The Journal of Spinal Cord Medicine*, *13*, 1–11.

Gebler, K. (1990). Physical modalities. In M. B. Glenn, & J. Whyte (Eds.), *The practical management of spasticity in children and adults* (pp. 118–146). Philadelphia: Lea & Febiger.

Hara, T., Abo, M., Hara, H., Kobayashi, K., Shimamoto, Y., Shibata, Y., et al. (2018). Effects of botulinum toxin A therapy and multidisciplinary rehabilitation on lower limb spasticity classified by spastic muscle echo intensity in post-stroke patients. *The International Journal of Neuroscience*, *128*, 412–420.

Harrington, A. L., & Bockenek, W. L. (2011). Spasticity. In D. I. Campagnolo, & S. Kirshblum (Eds.), *Spinal cord medicine* (pp. 265–281). Philadelphia: Lippincott & Williiams.

Holtz, K. A., Szefer, E., Noonan, V. K., & Mills, P. B. (2018). Treatment patterns of in-patient spasticity medication use after traumatic spinal cord injury: A prospective cohort study. *Spinal Cord*, *56*, 1176–1183.

Huang, Y. C., Chen, P. C., Tso, H. H., Yang, Y. C., Ho, T. L., & Leong, C. P. (2019). Effects of kinesio taping on hemiplegic hand in patients with upper limb post-stroke spasticity: A randomized controlled pilot study. *European Journal of Physical and Rehabilitation Medicine*, *55*, 551–557.

Joseph, L. H., Ismail, O. H., Naicker, A. S., Ying, P. P., & Mohammad, A. R. (2009). Three-year study of spinal cord injury outcomes and related secondary complications in a tertiary Centre: A retrospective analysis. *Archives of Medical Science*, *5*, 190–194.

Khan, F., Amatya, B., Bensmail, D., & Yelnik, A. (2019). Non-pharmacological interventions for spasticity in adults: An overview of systematic reviews. *Annals of Physical and Rehabilitation Medicine*, *62*, 265–273.

Khurana, S. R., & Garg, D. S. (2014). Spasticity and the use of intrathecal baclofen in patients with spinal cord injury. *Physical Medicine and Rehabilitation Clinics of North America*, *25*, 655–669.

Kita, M., & Goodkin, D. E. (2000). Drugs used to treat spasticity. *Drugs*, *59*, 487–495.

Korzhova, J., Sinitsyn, D., Chervyakov, A., Poydasheva, A., Zakharova, M., Suponeva, N., et al. (2018). Transcranial and spinal cord magnetic stimulation in treatment of spasticity. A literature review and meta- -analysis. *European Journal of Physical and Rehabilitation Medicine*, *54*, 75–84.

Krause, P., & Straube, A. (2003). Repetitive magnetic and functional electrical stimulation reduce spastic tone increase in patients with spinal cord injury. *Supplements to Clinical Neurophysiology*, *56*, 220–225.

Krause, P., Szecsi, J., & Straube, A. (2008). Changes in spastic muscle tone increase in patients with spinal cord injury using functional electrical stimulation and passive leg movements. *Clinical Rehabilitation*, *22*, 627–634.

Lance, J. W. (1980). Symposium. In R. G. Feldman, R. R. Young, & W. P. Koella (Eds.), *Spasticity: Disordered motor control* (pp. 485–495). Chicago: Year Book Medical Pubs.

McIntyre, A., Mays, R., Mehta, S., Janzen, S., Townson, A., Hsieh, J., et al. (2014). Examining the effectiveness of intrathecal baclofen on spasticity in individuals with chronic spinal cord injury: A systematic review. *The Journal of Spinal Cord Medicine*, *37*, 11–18.

Meseguer-Henarejos, A. B., Sánchez-Meca, J., López-Pina, J. A., & Hernández, R. C. (2018). Inter- and intra-rater reliability of the modified Ashworth scale: A systematic review and meta- analysis. *European Journal of Physical and Rehabilitation Medicine*, *54*, 576–590.

Nene, A. V., Campos, A. R., Grabljevec, K., Lopes, A., Skoog, B., & Burns, A. (2018). Clinical assessment of spasticity in people with spinal cord damage: Recommendations from the ability network, an international initiative. *Archives of Physical Medicine and Rehabilitation*, *99*, 1917–1926.

New, P. W., Rawicki, B., & Bailey, M. J. (2002). Nontraumatic spinal cord injury: Demographic characteristics and complications. *Archives of Physical Medicine and Rehabilitation*, *83*, 996–1001.

Palazón-García, R., Alcobendas-Maestro, M., Esclarin-Ruz, A., & Benavente-Valdepeñas, A. M. (2019). Treatment of spasticity in spinal cord injury with botulinum toxin. *The Journal of Spinal Cord Medicine*, *42*, 281–287.

Phadke, C. P., Balasubramanian, C. K., Ismail, F., & Boulias, C. (2013). Revisiting physiologic and psychologic triggers that increase spasticity. *American Journal of Physical Medicine & Rehabilitation*, *92*, 357–369.

Phadke, C. P., Vierira, L., Mathur, S., Cipriano, G., Ismail, F., & Boulias, C. (2019). Impact of passive leg cycling in persons with spinal cord injury: A systematic review. *Top Spinal Cord Review*, *25*, 83–96.

Rekand, T., Hagen, E. M., & Grønning, M. (2012). Spasticity following spinal cord injury. *Tidsskr Nor Legeforen*, *132*, 970–973.

Richard-Denis, A., Nguyen, B. H., & Mac-Thiong, J. M. (2020). The impact of early spasticity on the intensive functional rehabilitation phase and community reintegration following traumatic spinal cord injury. *The Journal of Spinal Cord Medicine*, *43*, 435–443. https://doi.org/10.1080/10790268.

Roy, R. R., & Edgerton, V. R. (2012). Neurobiological perspective of spasticity as occurs after spinal cord injury. *Experimental Neurology*, *235*, 116–122.

Sadeghi, M., McIvor, J., Finlayson, H., & Sawatzky, B. (2016). Static standing, dynamic standing and spasticity in individuals with spinal cord injury. *Spinal Cord*, *54*, 376–382.

Stevenson, V. L. (2010). Rehabilitation in practice: Spasticity management. *Clinical Rehabilitation*, *24*, 293–304.

Taricco, M., Pagliacci, M. C., Telaro, E., & Adone, R. (2006). Pharmacological interventions for spasticity following spinal cord injury: Results of a Cochrane systematic review. *Europa Medicophysica*, *42*, 5–15.

Whitlock, J. A. (1990). Neurophysiology of spasticity. In M. B. Glenn, & J. Whyte (Eds.), *The practical management of spasticity in children and adults* (pp. 8–33). Philadelphia: Lea & Febiger.

Chapter 10

# Fall circumstances, consequences, assessments, and interventions to manage fall risk among individuals living with spinal cord injury

L.A Rice, L. Abou, R. Yarnot, E. Wong, K. Chhatwani, and L. Hawari

*Department of Kinesiology and Community Health, University of Illinois at Urbana-Champaign, Urbana, IL, United States*

## List of abbreviations

| | |
|---|---|
| **10-MWT** | 10-Meter walking test |
| **6-MWT** | 6-Minute Walking Test |
| **ADL** | activities of daily living |
| **BBS** | Berg balance scale |
| **CHIEF** | Craig Hospital Inventory of Environmental Factors |
| **FIST** | function in sitting test |
| **FOF** | fear of falling |
| **FES—I** | The Falls Efficacy Scale International |
| **mFRT** | modified functional reach test |
| **PEI** | photo-elicitation interviews |
| **SCI** | spinal cord injury |
| **SCI-FCS** | Spinal Cord Injury-Fall Concerns Scale |
| **SCIM III** | Spinal Cord Independence Measure III |
| **TAI** | transfer assessment instrument |
| **TUG** | timed up and go |
| **WISCI II** | Walking Index for SCI |
| **WST** | Wheelchair Skills Test |

## Introduction

Falls are a common scenario, among other impairments and conditions, experienced by individuals living with a spinal cord injury (SCI) (see Table 1). Between 31% and 82% of individuals living with SCI will experience at least one fall in a year (Khan et al., 2019). This chapter examines the frequency of falls, circumstances associated with falls, consequences of falls, methods to evaluate fall risk, and intervention programs to manage fall risk.

## Fall frequency

Falls are common among people living with SCI who are nonambulatory (use a wheelchair or scooter full time) and those who ambulate. A fall is defined as an event in which an individual unexpectedly makes contact with the ground, floor, or some other lower level (Lamb et al., 2005). Due to differences in functional abilities, it is important to consider differences in fall frequency between those who ambulate and those who do not.

A systematic review and meta-analysis performed by Khan et al. in 2019 found that 34% to 82% of individuals living with SCI who are able to ambulate fell at least once over a 6- to 12-month period, and 31% to 73% of individuals living with SCI who are nonambulatory fell at least once in the same time span (Khan et al., 2019) (see Table 2).

Diagnosis and Treatment of Spinal Cord Injury. https://doi.org/10.1016/B978-0-12-822498-4.00010-5

**TABLE 1 Common impairments, conditions, and scenarios associated with spinal cord injury.**

| |
|---|
| Spinal shock |
| Paralytic ileus |
| Deep venous thrombosis and pulmonary embolus |
| Autonomic dysreflexia |
| Falls |
| Pressure ulcers |
| Spasticity |
| Postural hypotension |
| Bladder, bowel, and sexual function |
| Osteoporosis |

A summary is provided of impairments, conditions, and scenarios commonly associated with spinal cord injury.

**TABLE 2 Frequency of falls among ambulatory and nonambulatory individuals living with spinal cord injury (Khan et al., 2019).**

| Frequency | % of Ambulatory individuals with SCI | % of Nonambulatory individuals with SCI |
|---|---|---|
| At least 1 fall | 34%–82% | 31%–73% |
| Recurrent falls (more than 1 fall) | 28%–68% | 30%–41% |

The frequency of falls among individuals living with Spinal Cord Injury (SCI) who ambulate and nonambulatory individuals are presented along with data on the frequency of the population, both ambulatory and nonambulatory, that sustain recurrent falls.
(Data source: Khan, A., Pujol, C., Laylor, M., Unic, N., Pakosh, M., Dawe, J., et al. (2019). Falls after spinal cord injury: A systematic review and meta-analysis of incidence proportion and contributing factors. *Spinal Cord, 57*(7), 526–539. https://doi.org/10.1038/s41393-019-0274-4).

Ambulatory individuals are more likely to fall compared to individuals who do not ambulate ( Jorgensen et al., 2016). The odds of falling for individuals who ambulate are more than 29 times higher compared to individuals who do not ambulate. Jorgensen et al. proposed that falls are less frequent among nonambulatory individuals due to the larger base of support provided by the wheelchair.

## Recurrent falls

Among nonambulatory individuals with SCI, Jorgensen et al. reported that men experienced recurrent falls (more than one fall) 3.1 times more often than women within a year. This finding could be explained by the difference in how men and women manage risk. As age increases, the odds of falling more than once among nonambulatory individuals with SCI decreased. The increased odds of falling among younger adults may be due to activity level and involvement in activities in the workplace that increase exposure to more potential fall situations than older adults ( Jorgensen et al., 2016).

Between 30% and 41% of nonambulatory individuals had recurrent falls within a 6- to 12-month period (Khan et al., 2019). Recurrent falls were also reported 14% more often among individuals with greater functional mobility independence, likely because more mobile individuals can engage in more intense physical activity and are therefore more likely to fall ( Jorgensen et al., 2016).

Most ambulatory individuals with SCI report falling more than once (Wirz & van Hedel, 2018). Within a 6- to 12-month period, 28%–68% of ambulatory individuals with SCI experienced multiple falls (Khan et al., 2019). Jorgensen et al. reported that individuals who could recover by themselves following a fall experienced 4.7 times more recurrent falls than those who could not get up by themselves. The increased frequency could be related to increased confidence in the ability to recover from a fall among those who are able to recover independently. Increased confidence may result in taking greater risks and exposure to more situations that lead to a fall ( Jorgensen et al., 2016).

## Fall circumstances

The circumstances in which individuals living with SCI are likely to encounter a fall vary widely. Compared with other populations, including older adults, much less is known about fall risk factors among individuals living with SCI (Khan et al., 2019). Further research is needed to comprehensively understand risk factors associated with falls.

To systematically examine the known risk factors associated with falls, the Biological, Behavioral, Social & Economic, and Environmental (BBSE) Model is used to classify fall risk factors and highlight the multi-factorial nature of falls (Institute of Medicine Committee., 2001). Fig. 1 displays a visual representation of the BBSE.

### Biological risk factors

Biological contributors of falls among both ambulatory and nonambulatory individuals living with SCI include many factors. The most common contributors include muscle weakness, loss of balance, spasticity, and muscle spasms (Brotherton, Krause, & Nietert, 2007b; Jorgensen et al., 2016; Musselman, Arnold, Pujol, Lynd, & Oosman, 2018; Nelson et al., 2010). Lower levels of functional mobility have also been linked to an increased risk of falling (Wannapakhe, Arayawichanon, Saengsuwan, & Amatachaya, 2014). However, Nelson et al. found that persons with more function are also at greater risk because they have a greater exposure to situations where falls may occur, making them more likely to engage in risk-taking behaviors (Nelson et al., 2010). Other biological factors that may increase the risk of falls summarized by Khan et al. include: male sex (wheelchair users only), an increased number of comorbidities, trunk weakness, fatigue, walking asymmetry, shorter duration since sustaining SCI, narcolepsy, and reduced sensation (Khan et al., 2019). Age can also be a risk factor. Among 94 participants, individuals who reported a high frequency of falls were significantly younger. This may be because these individuals are more active and have greater levels of enthusiasm and energy (Amatachaya, Pramodhyakul, Wattanapan, & Eungpinichpong, 2015). In addition, reports of pain in the 2 months prior to the fall can affect the performance of activities of daily living and increase fall risk (Nelson et al., 2010). Studies have also revealed that individuals living with incomplete spinal cord injury, injuries in which partial sensation and/or motor function persists below the level of injury, experience significantly more falls than individuals with complete injuries (Amatachaya, Wannapakhe, Arrayawichanon, Siritarathiwat, & Wattanapun, 2011).

### Behavioral risk factors

In a systematic review and meta-analysis, Khan et al. found that the most commonly reported behavioral attributions associated with falls included inattention or distraction, ambulation, fear of falling, a history of previous falls, and not using the safety features of a wheelchair or walking aid (Khan et al., 2019). Risk-taking behaviors are also associated with falls (Jannings, 2017), which is theorized to be the reason why men are three times more likely to fall than women (Jorgensen et al., 2016). Nelson et al. also found alcohol abuse increased the risk of falls as well as the use of a shorter

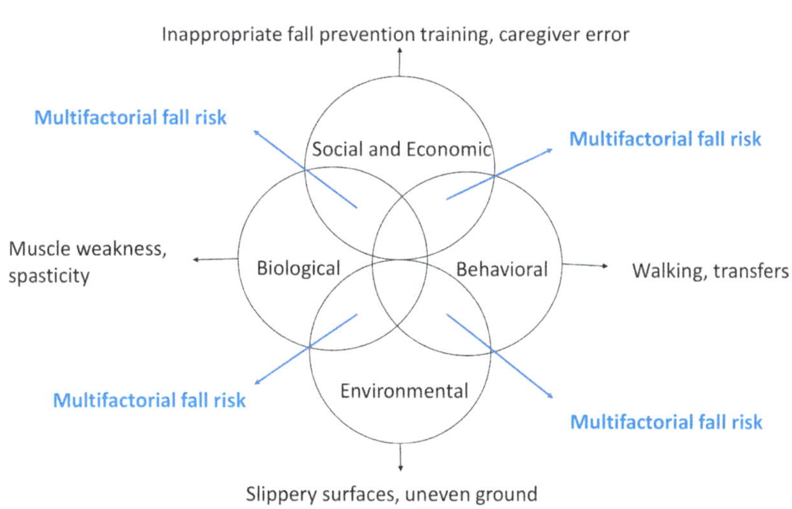


FIG. 1 Biological, Behavioral, Social and Economic, and Environmental Fall Model. The figure provides a visual representation of the biological, behavioral, social and economic, and environmental fall model. The model provides a framework for the classification of circumstances associated with falls. Please note how the various components intersect. This represents the multi-factorial nature of falls and the need to consider various factors when designing programs to manage fall risk. *(Data source: Institute of Medicine (US) Committee on Health and Behavior: Research. (2001). The National Academies Collection: Reports funded by National Institutes of Health. In Health and behavior: The interplay of biological, behavioral, and societal influences. National Academies Press (US) Copyright © 2001, National Academy of Sciences. https://doi.org/10.17226/9838).*

wheelchair. A shorter wheelchair allows for greater maneuverability; however, the design makes the chair less stable overall (Nelson et al., 2010). Finally, individuals with a lower quality of life were found to fall more frequently ( Jorgensen et al., 2016).

### Environmental risk factors

Environmental contributors to falls exist both within and outside the home. Hazards such as railings, stairs, heavy doors, and slippery floors are identified as increasing the risk of falls ( Jorgensen et al., 2016). Among nonambulatory individuals, Sung et al. reported that the bathroom was the most common location for falls to occur, and 43% of falls were associated with pushing or driving the wheelchair on rough surfaces (Sung, Trace, Peterson, Sosnoff, & Rice, 2019). Forslund et al. found that 65% of falls occurred indoors, and 47% of falls occurred between the hours of 9 AM and 6 PM (Forslund et al., 2017).

### Social and economic

Social and economic factors were also identified as fall contributors. Nelson et al. found that a higher level of education was associated with falls (Nelson et al., 2003). In addition, lack of fall prevention training (Musselman et al., 2018) and caregiver error ( Jannings, 2017) were also found to contribute to falls. Table 3 summarizes fall risk factors according to the BBSE.

## Consequences of falls

Falls experienced by individuals living with SCI can result in a variety of physical and psychosocial consequences for the faller, as well as high costs to society at large (see Table 4).

### Physical consequences of falls

A systematic review and meta-analysis conducted by Khan et al. (2019) revealed that the frequency of reported falls which resulted in injuries ranged from 6% to 62% in ambulatory individuals with SCI and 13%–38% in nonambulatory individuals with SCI, though the frequency of total falls experienced (both injurious and noninjurious) appears to be similar between ambulatory and nonambulatory individuals with SCI (see Table 5).

While the majority of physical injuries sustained as a result of a fall were considered minor, such as bruises, cuts, or abrasions, approximately 15% of falls result in more severe injury, such as bone fractures (Amatachaya et al., 2011; Brotherton, Krause, & Nietert, 2007a), loss of consciousness (Wannapakhe et al., 2014; Wannapakhe, Arrayawichanon, Saengsuwan, & Amatachaya, 2015), or concussion (Forslund et al., 2017). Other moderate injuries reported as a result

**TABLE 3 The top risk factors for falls classified by the biological, behavior, social and economic, and environmental model found in a systematic review and meta-analysis by Khan et al. (2019).**

| Biological | Behavioral | Environmental | Social and economic factors |
|---|---|---|---|
| • Muscle weakness<br>• Impaired balance<br>• Greater function/ mobility<br>• Muscle spasms/ spasticity<br>• Lower function/ mobility | • Inattention/distraction during movement<br>• Ambulating<br>• Fear of falling<br>• Previous History of Falls<br>• Not using protective straps/safety belt/ wheel locks | • Hazards in environment<br>• Obstacles<br>• Stairs<br>• Uneven ground<br>• Doors | • Higher education levels<br>• Lack of fall prevention training<br>• Caregiver error |

The top risk factors for falls, organized using the biological, behavioral, social & economic, and environmental model, as reported by Khan et al. in a 2019 systematic review.
(Data source: Khan, A., Pujol, C., Laylor, M., Unic, N., Pakosh, M., Dawe, J., et al. (2019). Falls after spinal cord injury: A systematic review and meta-analysis of incidence proportion and contributing factors. *Spinal Cord, 57*(7), 526–539. https://doi.org/10.1038/s41393-019-0274-4).

**TABLE 4 Consequences of falls experienced by ambulatory and nonambulatory individuals living with spinal cord injury.**

| | |
|---|---|
| Physical injuries consequences | Bruises, cuts, abrasions, bone fractures, concussions, death |
| Psychosocial consequences | Fear of falling, feelings of embarrassment or frustration; deconditioning, decreased community participation |
| Societal consequences | Cost to health-care systems, increased dependence on caregivers |

A summary is provided of the physical, psychosocial, and societal consequences of falls experienced by both ambulatory and nonambulatory individuals living with spinal cord injury.
(Data source: Amatachaya, S., Wannapakhe, J., Arrayawichanon, P., Siritarathiwat, W., & Wattanapun, P. (2011). Functional abilities, incidences of complications and falls of patients with spinal cord injury 6 months after discharge. *Spinal Cord, 49*(4), 520–524. https://doi.org/10.1038/sc.2010.163; Amatachaya, S., Pramodhyakul, W., Wattanapan, P., & Eungpinichpong, W. (2015). Ability of obstacle crossing is not associated with falls in independent ambulatory patients with spinal cord injury. *Spinal Cord, 53*(8), 598–603. https://doi.org/10.1038/sc.2015.22; Brotherton, S. S., Krause, J. S., & Nietert, P. J. (2007). Falls in individuals with incomplete spinal cord injury. *Spinal Cord, 45*(1), 37–40. https://doi.org/10.1038/sj.sc.3101909; Forslund, E. B., Jorgensen, V., Franzen, E., Opheim, A., Seiger, A., Stahle, A., et al. (2017). High incidence of falls and fall-related injuries in wheelchair users with spinal cord injury: A prospective study of risk indicators. *Journal of Rehabilitation Medicine, 49*(2), 144–151. https://doi.org/10.2340/16501977-2177; Jannings, W. (2017). A quality improvement project to investigate the circumstances of lower limb fractures in non-ambulant persons with spinal cord injury. *Journal of the Australasian Rehabilitation Nurses Association, 20*(August), 14–18.; Musselman, K. E., Arnold, C., Pujol, C., Lynd, K., & Oosman, S. (2018). Falls, mobility, and physical activity after spinal cord injury: An exploratory study using photo-elicitation interviewing. *Spinal Cord Series and Cases, 4*, 39. https://doi.org/10.1038/s41394-018-0072-9; Verma, S. K., Willetts, J. L., Corns, H. L., Marucci-Wellman, H. R., Lombardi, D. A., & Courtney, T. K. (2016). Falls and fall-related injuries among community-dwelling adults in the United States. *PLoS One, 11*(3), e0150939. https://doi.org/10.1371/journal.pone.0150939; Wannapakhe, J., Arayawichanon, P., Saengsuwan, J., & Amatachaya, S. (2014). Changes of functional ability in patients with spinal cord injury with and without falls during 6 months after discharge. *Physical Therapy, 94*(5), 675–681. https://doi.org/10.2522/ptj.20130260).

**TABLE 5 Injuries reported among fallers living with SCI (Khan et al., 2019).**

| Population | Frequency of injuries |
|---|---|
| Ambulatory individuals | 6%–62% |
| Nonambulatory individuals | 13%–38% |

A summary is reported of the injuries experienced by individuals living with spinal cord injury, as reported by Khan et al. in a 2019 systematic literature review and meta-analysis. Note the increased reports of injuries among ambulatory individuals.

of a fall sometimes include muscle tears (Amatachaya et al., 2015), sprains, bloody nose, and numbness (Nelson et al., 2010). In severe cases, injuries sustained from a fall can even result in death (Nelson et al., 2010).

Ambulatory individuals with SCI experience a greater number of minor injuries as a result of a fall compared to nonambulatory individuals (Wannapakhe et al., 2015). They also have a greater incidence of fractures as a result of falls compared to healthy older adults (Brotherton et al., 2007a). Ambulatory women with SCI report more injurious falls than men of the same population, similarly to women without SCI compared to their male peers. Ambulatory individuals with SCI who have a history of recurrent falls are also at an increased risk of experiencing an injurious fall compared to those with no such history ( Jorgensen et al., 2017).

## Psychosocial consequences of falls

### *Fear of falling*

While physical consequences of falling are more easily identified, there also exist psychosocial consequences of falls that heavily influence individuals with SCI.

Many individuals living with SCI who experience a fall, report an increase in fear of falling (Amatachaya et al., 2015; Musselman et al., 2018). Fear of falling (FOF) is defined as an ongoing concern about experiencing a fall which may ultimately limit the performance of daily activities (Legters, 2002). Developing a FOF is a common occurrence after experiencing a fall; however, a FOF can develop even when there is no history of falls. While a moderate amount of FOF may prompt individuals to be more mindful of their environment when performing tasks, FOF often leads individuals to unnecessarily avoid performing activities which they are physically capable of doing (Tinetti, Speechley, & Ginter, 1988). This can result in the deconditioning of the body which may put individuals with SCI at a further increased risk of falling and experiencing more severe consequences of falls when activity by necessity occurs, such as reaching for an item while getting dressed or transferring to a toilet (Brotherton et al., 2007b). Fear of falling may be caused by post-fall–related experiences, such as injury, or underlying factors that enhance fall risk, such as having poor balance, muscle weakness, and impaired sensation, which may decrease fall-related self-efficacy (Jorgensen et al., 2017).

Fear of falling is also a risk factor for recurrent falls in ambulatory individuals with SCI (Phonthee, Saengsuwan, Siritaratiwat, & Amatachaya, 2013) and has been found to be associated with decreased community participation (Peterson, Cho, & Finlayson, 2007; Tinetti, Speechley, & Ginter, 1988) and lower quality of life (Nelson et al., 2010). Sung et al. found that 65% of nonambulatory individuals with SCI reported a fear of falling. Of those who reported a FOF, 26% also indicated that they limited their activities because of this fear (Sung, Peterson, Shen, Sosnoff & Rice, 2021). Friedman, Munoz, West, Rubin, and Fried (2002) found that there was a "spiraling effect" between increasing fall occurrences, FOF, and functional decline. Ambulatory individuals with SCI who have better balance appear to show less fear of falling (Wirz, Müller, & Bastiaenen, 2010), suggesting that interventions targeted at improving balance may help to decrease FOF.

#### *Negative feelings and decreased community participation*

In combination with increased fear of falling, other feelings expressed by individuals with SCI after falling include feelings of embarrassment (Jannings, 2017) and frustration (Musselman et al., 2018), even if they were not physically injured. While these emotions, combined with FOF, may motivate an individual to be more cautious in the future, of great concern is the frequently reported decrease in community participation and social interactions by as many as 45% of fallers with SCI (Musselman et al., 2018).

### Societal consequences of falls

In addition to the physical and psychosocial burden of falls, there also exists a significant societal burden. In 2010, among U.S. adults, it was estimated that fall-related injuries cost health-care systems 111 billion dollars (Verma et al., 2016). Saunders, DiPiro, Krause, Brotherton, and Kraft (2013) found that approximately 20% of falls experienced by ambulatory individuals with SCI resulted in the need for medical attention. Fall-related injuries can also lead to an increased dependence on others for assistance with daily activities. This can lead to increased caregiver burden and costs over a lifetime (Forslund et al., 2017).

## Assessment of fall risk

The assessment of fall risk among people living with SCI is of high importance to foster management of fall risk. Methods used to assess fall risk include clinical measures (Srisim, Saengsuwan, & Amatachaya, 2015), self-report strategy (Sung et al., 2019), and photo-elicitation techniques (Singh et al., 2020).

It is important to note that fall risk factors in people with SCI have shown to be specific, individualized, and highly dependent on the specific characteristics of the injury including the type of injury (complete or incomplete), chronicity of injury, and the level of injury. Other factors, such as the functional ability of the person with SCI, including personal characteristics and the surrounding environment, also determine the level of risk of falls. Therefore, fall risk assessments differ between ambulatory and nonambulatory people with SCI (see Fig. 2).

### Assessment among ambulatory people with SCI

Most studies on fall risk assessment have focused on individuals living with SCI who ambulate (Abou, de Freitas, Palandi, & Ilha, 2018). Previous research indicates that biological factors are the primary risk factors associated with falls among ambulatory people and include level of functional independence, physical fitness, gait velocity, reaction time, limits of stability, and lower extremity muscle strength (Abou et al., 2018). These biological factors are measurable in clinical settings through the performance of clinical measures including the Berg Balance Scale (BBS), the Spinal cord Independence Measure (SCIM) III, and the Timed Up and Go (TUG). Abou, Ilha, Romanini, and Rice (2019) reported that the BBS is the most appropriate clinical balance measure and has the ability to effectively predict future falls among ambulatory people with SCI. However, since falls are multi-factorial, there is no consensus on the relationship between level of mobility and fall risk in this population. Some studies reported a significant relationship between a higher level of mobility and increased risk of falling (Amatachaya et al., 2018; Phonthee, Saengsuwan, & Amatachaya, 2013), while other studies related lower level of mobility to an increased risk of falling (Srisim et al., 2015). Undoubtedly, other biological factors such as sex, type

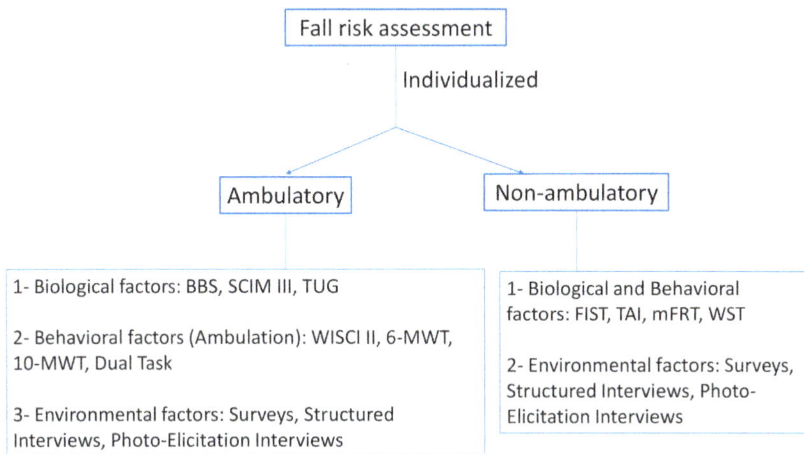


FIG. 2 Summary of fall risk assessments among people living with spinal cord injury. This table depicts the various assessments to examine fall risk among individuals living with spinal cord injury categorized by the primary target area using the biological, behavioral, social and economic, and environmental model. Abbreviations: *6-MWT*: 6-minute walking test, *10-MWT*: 10-meter walking test, *BBS*: Berg balance scale, *FIST*: function in sitting test, *mFRT*: modified functional reach test, *SCIM III*: spinal cord independence measure III, *TUG*: timed up and go, *WISCI II*: Walking Index for SCI*TAI*: transfer assessment instrument, *WST*: Wheelchair Skills Test.

of injury, fatigue, pain, sensorial impairment, time since SCI, trunk weakness, and the presence of comorbidities also contribute to an increased risk of falling.

Ambulating itself has been reported as one of the most common behavioral risks of falling along with inattention or distraction (Khan et al., 2019). This is because ambulating may easily lead to a loss of balance when performing an additional task (i.e., ambulation and a cognitive task) (Amatachaya et al., 2018). In clinical settings, ambulation can be assessed by use of the Walking Index for SCI (WISCI-II), the 6-Minute Walking Test (6-MWT), dual-task obstacle crossing (Amatachaya et al., 2018), and the 10-Meter Walking Test (10-MWT) (Moore et al., 2018). Amatachaya et al. showed that failing to cross an obstacle while performing a color-word Stroop task during the 10-MWT and the TUG is significantly associated with future risk of falls (Amatachaya, Promkeaw, Arayawichanon, Thaweewannakij, & Amatachaya, 2020). However, the relationship between ambulation outcome measures and falls still need to be further explored. Other behavioral factors that increase the risk of falling including a fear of falling, a history of previous falls, and not using a walking aid should be assessed with ambulation (Khan et al., 2019).

Environmental factors that increase the risk of falling both within and outside of the home can be assessed using different techniques. Hazards in the environment such as stairs, ramps or hills, and uneven or slippery surfaces can be assessed using surveys, semistructured interviews, and photo-elicitation interviews (PEI) (Brotherton et al., 2007a; Musselman et al., 2018). The Craig Hospital Inventory of Environmental Factors (CHIEF) is a comprehensive measure to evaluate environmental barriers (Whiteneck et al., 2004). However, compared with surveys and semistructured interviews, PEI provides more details about the environmental factors related to falls. PEI allows participants to take photographs of situations, places, people, hazards in the environment, and obstacles that may lead them to falls. Also, through PEI, participants are actively involved in identifying factors that lead them to falls (Clark-IbáÑez, 2004).

Finally, social and economic fall risk factors should also be considered and need further exploration among ambulatory people with SCI through surveys and interviews. Among other populations, including older adults, the level of education, fall prevention training, caregiver and family training have been associated with falls (Deandrea et al., 2010). These factors should be taken into consideration when evaluating an individual's risk for falls.

## Assessment among nonambulatory people with SCI

Few studies have focused on the exploration of fall risk factors among nonambulatory people with SCI (Sung et al., 2019). Performance of activities of daily living constitutes the main behavioral fall risk in this population. The influence of muscle spasticity and poor balance can negatively affect the safe performance of activities of daily living while using a wheelchair (Abou et al., 2018). Activities of daily living including transfers from/to different surfaces, wheelchair propulsion, and reaching for an object are commonly associated with falls (Sung et al., 2019) can be assessed with clinical measures. The Function in Sitting Test (FIST) (Abou, Sung, et al., 2019), Wheelchair Skills Test (Keeler, Kirby, Parker, McLean, & Hayden, 2019), Transfer Assessment Instrument (TAI) (Worobey, Rigot, Hogaboom, Venus, & Boninger, 2018; Worobey, Zigler, et al., 2018), and modified Functional Reach Test (Lynch, Leahy, & Barker, 1998) can be used to assess sitting balance, wheelchair skills, transfers, and functional reach, respectively, among nonambulatory people with SCI. Additionally, a smartphone-based assessment has been recently developed to assess sitting balance among nonambulatory people with SCI (Frechette, Abou, Rice, & Sosnoff, 2020). However, the relationship between the aforementioned outcome measures and falls still needs to be established.

Similar to ambulatory people with SCI, environmental factors that constitute important fall risk factors among nonambulatory people with SCI can be assessed using surveys, semistructured interviews, and PEI (Brotherton et al., 2007a; Musselman et al., 2018). PEI may foster the identification of environmental factors associated with falls in the bathroom, street, bedroom, living room, and kitchen. In addition, PEI may provide clinicians with more details about environmental conditions including wet and slippery surfaces, low lighting, and uneven ground which were also reported as conditions that increase the risk of falling among nonambulatory people with SCI (Sung et al., 2019).

### Assessing fear of falling

Several approaches have been used by researchers to assess FOF in individuals with SCI. While some studies have utilized single-item questions, such as directly asking participants if they are worried or concerned about falling ( Jorgensen et al., 2017; Sung et al., 2019), others have used more comprehensive assessment tools to identify levels of concern about falling in relation to specific activities. The Falls Efficacy Scale-International (FES-I) (Yardley, Beyer, Hauer, Kempen, & Piot-Ziegler, 2005) was designed for ambulatory elderly individuals but has been used in studies of other ambulatory individuals including those with SCI (Wirz et al., 2010). Boswell-Ruys, Harvey, Delbaere, and Lord (2010) modified the FES-I to be more appropriate to assess FOF among nonambulatory individuals with SCI, creating the Spinal Cord Injury-Fall Concerns Scale (SCI-FCS). Assessment tools such as the FES-I and SCI-FCS allow for a more comprehensive understanding of FOF and can help in the creation of tailored interventions to address warranted and unwarranted FOF.

## Intervention program to manage fall risk

As a result of the high frequency and significant consequences associated with falls among individuals living with SCI, effective management of fall risk is critical for the health and well-being of individuals living with SCI.

Rice, Sung, Keane, Peterson, and Sosnoff (2019) examined the efficacy of a one-time, 1:1, 45-min intervention among 18 full-time manual wheelchair users living with SCI. The intervention focused on the refinement of transfer skills and implementation of an exercise program to improve upper extremity strength and core stability. After the initial education, participants were given an opportunity to practice the skills and discuss individual fall concerns. Participants were then asked to integrate the education at home for a period of 12 weeks. After exposure to the intervention, fall incidence significantly decreased and seated balance significantly improved. Improvements were also seen with regards to quality of life.

Amatachaya et al. (2020) examined the effects of walking training on a walking track with different surfaces, including grass, soft areas, and pebbles, compared to standard overground walking training among individuals living with SCI who ambulate independently, with or without an assistive device. Participants engaged in walking training on the walking track (intervention) or received standard overground walking training (control) 5 days per week, 30 min per day for 4 weeks. In both groups, participants walked at a self-selected speed. Six months after exposure to the intervention fewer participants in the intervention group ($n = 5$, 18%) experienced falls compared to the control group ($n = 12$, 46%).

While the literature is sparse related to specific fall management programs for individuals living with SCI, other peer-reviewed intervention programs have been developed to manage common circumstances associated with falls. The studies presented in Table 6 contain details on intervention programs that target risk factors associated with falls including poor transfer skills (Worobey, Rigot, et al., 2018; Worobey, Zigler, et al., 2018), wheelchair and scooter skills (Keeler et al., 2019), and dynamic balance (Chisholm et al., 2017; Khurana et al., 2017; Qi et al., 2018; Williams et al., 2020).

## Application to other areas of neuroscience

In this chapter, we have examined falls among individuals living with SCI. This is an expanding topic of concern that frequently impacts other individuals living with a variety of neurological impairments. For example, falls are also very common among individuals living with Multiple Sclerosis and can result in similarly significant consequences.

When assessing which individuals are at high risk for falls and circumstances associated with falls, it is important to consider that falls are multi-factorial. While some of the information presented in this book chapter can be carried over to other populations of individuals living with neurological impairments, it is important to note the various factors that

**TABLE 6 Intervention programs to manage fall risk among ambulatory and nonambulatory individuals living with spinal cord injury.**

| Author | Area of focus | Population | Key intervention components | Outcome |
|---|---|---|---|---|
| Rice (Rice et al., 2019) | Comprehensive fall management | Full-time manual wheelchair users | Transfer education and exercise to increase upper extremity strength and core stability | After exposure to the intervention: Decreased fall frequency and improved seated balance |
| Amatachaya (Amatachaya et al., 2020) | Comprehensive fall management | Independent ambulators or individuals who used an assistive device to ambulate | Walking on a track with different surfaces (grass, soft areas, pebbles) vs standard overground walking training | 18% of intervention participants experienced a fall vs control group |
| Worobey (Worobey, Rigot, et al., 2018; Worobey, Zigler, et al., 2018) | Transfer skills | Individuals who perform a sitting pivot transfer | In-person management of individual transfer skill deficits vs online skill management | Improvement seen in transfer quality through both in-person and online education |
| Keeler (Keeler et al., 2019) | Wheelchair skills | Full-time manual and power wheelchair users | Various instruction methods (1:1, group, peer-led, etc.) to improve wheelchair skill capacity and quality | Various formats studied result in improvements in wheelchair skill capacity and quality among manual and power wheelchair users |
| Williams (Williams et al., 2020) | Dynamic balance | C6-T12, AIS A-C | Arm crank ergometer training | After exposure to intervention, static sitting balance significantly improved |
| Qi (Qi et al., 2018) | Balance control and quality of life | C6-L1, AIS B-D | Tai Chi | Individuals who practiced Tai Chi had greater static sitting balance and increased quality of life compared to a control group |
| Khurana (Khurana, Walia, & Noohu, 2017) | Dynamic balance | T6-T12 | Virtual reality game training | Virtual reality game-based training resulted in greater improvements in balance and functional performance compared to real-world, task-specific training |
| Chisholm (Chisholm, Alamro, Williams, & Lam, 2017) | Dynamic balance | C7-T4, AIS A-B | Robotic training | Overground robotic training resulted in improved postural stability vs treadmill-based robotic gait training |

A summary is provided of peer-reviewed intervention programs designed to comprehensively manage fall risk and target risk factors associated with falls. References of individual studies are noted within the table.

influence falls. For example, wheelchair propulsion or driving over rough surfaces is reported as a common circumstance associated with falls among individuals living with SCI. This information will apply to other populations who use wheelchairs and scooters on a full-time basis. However, other factors more specific to SCI, such as the influence of spasticity or lack of sensation, may further increase the individuals' risk of falling. As a result, it is important for a clinician to consider the various factors that influence falls when examining fall risk and developing intervention programs to manage fall risk.

Among both individuals living with SCI and other neurological impairments, research related to this topic is limited. In recent years, research has begun to uncover important points including the frequency of falls and circumstances associated with falls. However, compared to research focused on older adults, the literature is very limited. Most significantly there is limited information regarding appropriate ways to assess fall risk and interventions to manage falls among individuals living with SCI and other populations affected by neurological impairments. It is important to consider that this is an evolving field of study and further research is needed.

## Mini-dictionary of terms

Ambulatory: A term referring to individuals who are able to walk by themselves with or without an assistive device, such as a cane or walker.

Balance: A generic term describing the dynamics of body posture to prevent falling. Balance refers to the ability of the body to maintain the center of gravity within the limits of stability as determined by the base of support.

Deconditioning: A decline in the function of body parts and systems as a result of inactivity or lack of use.

Dual-task: A paradigm in experimental neuropsychology that requires an individual to perform two tasks simultaneously. A common dual-task example used in clinical experiments consists of walking and carrying on a conversation.

Fall: An event in which an individual unintentionally makes contact with some lower surface, such as the ground or floor.

Fear of falling: A fear or concern that a fall will occur. Fear of falling may develop as a result of a fall or independent of a fall event. An individual may limit their performance of activities they enjoy doing in their home or community that they remain capable to doing.

Gait training: Strategy to improve ambulation skills focused on restoring the pattern of walking to develop biomechanically efficient gait. Assistive devices may be used.

Nonambulatory: A term referring to individuals who are unable to walk without assistance from another person or a mobile device, such as a wheelchair.

Photo-Elicitation Interviews: An interview strategy in which participants are asked to take photos of areas associated with falls in order to foster a more in-depth discussion.

Quality of life: The degree to which an individual is healthy, comfortable, and able to participate in and enjoy life. This can be challenging to assess because nearly everyone places different values on these measures.

## Key facts of fall frequency and fall circumstances

- Falls are a common health concern for individuals living with SCI.
- Between 31% and 82% of individuals living with SCI will experience at least one fall over a 6- to 12-month period of time.
- Individuals who ambulate fall more often compared to individuals who use a wheelchair or scooter.
- A wide variety of factors, such as muscle weakness, being distracted, and pushing or driving a wheelchair on a rough surface contribute to the occurrence of a fall.
- Falls often occur due to a combination of many different factors. For example, a fall may occur as a result of pushing a wheelchair on a rough surface while talking with a friend (being distracted).

## Key facts of consequences of falls

- Falls may result in physical injuries and have psychosocial consequences, as well as societal costs.
- Up to 62% of individuals living with SCI may experience an injury as a result of a fall.
- The majority of physical injuries are minor; however, significant injuries, such as fractures or concussions, may occur.
- Fear of falling can lead to a spiraling effect, in which someone who is afraid of falling stops performing activities, leading to deconditioning, which in turn puts them at greater risk of falling in the future.

## Summary points

- This chapter examines fall frequency, consequences of falls, fall circumstances, methods to assess fall risk, and interventions to manage fall risk.

- In a 6- to 12-month period, 34%–82% of ambulatory individuals and 31%–73% of nonambulatory individuals living with SCI experience at least one fall.
- Between 28% and 68% of ambulatory and 30% and 41% of nonambulatory individuals will experience recurrent falls.
- The circumstances associated with falls among ambulatory and nonambulatory individuals are diverse; however, some of the most common circumstances include falls due to the influence of muscle weakness or spasticity and inattention or distraction.
- Falls may occur both inside and outside of the home and were more common among individuals who were working or studying.
- Falls can result in significant consequences, including physical injuries, such as bruising, fractures, or head injuries.
- Falls can also result in the development of a fear of falling that can limit engagement in desired activities in the home and community.
- Falls can be evaluated using a combination of clinical assessments, self-report surveys, or photo-elicitation interviews.
- Interventions to manage fall risk show promise; however, additional studies are needed to develop comprehensive intervention programs.

## References


Abou, L., de Freitas, G. R., Palandi, J., & Ilha, J. (2018). Clinical instruments for measuring unsupported sitting balance in subjects with spinal cord injury: A systematic review. *Topics in Spinal Cord Injury Rehabilitation*, *24*(2), 177–193. https://doi.org/10.1310/sci17-00027. Spring.

Abou, L., Ilha, J., Romanini, F., & Rice, L. A. (2019). Do clinical balance measures have the ability to predict falls among ambulatory individuals with spinal cord injury? A systematic review and meta-analysis. *Spinal Cord*. https://doi.org/10.1038/s41393-019-0346-5.

Abou, L., Sung, J., Sosnoff, J. J., & Rice, L. A. (2019). Reliability and validity of the function in sitting test among nonambulatory individuals with spinal cord injury. *The Journal of Spinal Cord Medicine*, 1–8. https://doi.org/10.1080/10790268.2019.1605749.

Amatachaya, S., Pramodhyakul, W., Wattanapan, P., & Eungpinichpong, W. (2015). Ability of obstacle crossing is not associated with falls in independent ambulatory patients with spinal cord injury. *Spinal Cord*, *53*(8), 598–603. https://doi.org/10.1038/sc.2015.22.

Amatachaya, S., Promkeaw, D., Arayawichanon, P., Thaweewannakij, T., & Amatachaya, P. (2020). Various surfaces benefited functional outcomes and fall incidence in individuals with spinal cord injury: a randomized controlled trial with prospective data follow-up. *Archives of Physical Medicine and Rehabilitation*. https://doi.org/10.1016/j.apmr.2020.08.009.

Amatachaya, S., Srisim, K., Thaweewannakij, T., Arrayawichanon, P., Amatachaya, P., & Mato, L. (2018). Failures in dual-task obstacle crossing could predict risk of future fall in independent ambulatory individuals with spinal cord injury. *Clinical Rehabilitation*. https://doi.org/10.1177/0269215518788913, 269215518788913.

Amatachaya, S., Wannapakhe, J., Arrayawichanon, P., Siritarathiwat, W., & Wattanapun, P. (2011). Functional abilities, incidences of complications and falls of patients with spinal cord injury 6 months after discharge. *Spinal Cord*, *49*(4), 520–524. https://doi.org/10.1038/sc.2010.163.

Boswell-Ruys, C. L., Harvey, L. A., Delbaere, K., & Lord, S. R. (2010). A falls concern scale for people with spinal cord injury (SCI-FCS). *Spinal Cord*, *48* (9), 704–709.

Brotherton, S. S., Krause, J. S., & Nietert, P. J. (2007). Falls in individuals with incomplete spinal cord injury. *Spinal Cord*, *45*(1), 37–40. https://doi.org/10.1038/sj.sc.3101909.

Brotherton, S. S., Krause, J. S., & Nietert, P. J. (2007b). A pilot study of factors associated with falls in individuals with incomplete spinal cord injury. *The Journal of Spinal Cord Medicine*, *30*(3), 243–250. https://doi.org/10.1080/10790268.2007.11753932.

Chisholm, A. E., Alamro, R. A., Williams, A. M., & Lam, T. (2017). Overground vs treadmill-based robotic gait training to improve seated balance in people with motor-complete spinal cord injury: A case report. *Journal of Neuroengineering and Rehabilitation*, *14*(1), 27. https://doi.org/10.1186/s12984-017-0236-z.

Clark-IbáÑez, M. (2004). Framing the social world with photo-elicitation interviews. *American Behavioral Scientist*, *47*(12), 1507–1527. https://doi.org/10.1177/0002764204266236.

Deandrea, S., Lucenteforte, E., Bravi, F., Foschi, R., La Vecchia, C., & Negri, E. (2010). Risk factors for falls in community-dwelling older people: A systematic review and meta-analysis. *Epidemiology*, *21*(5), 658–668. https://doi.org/10.1097/EDE.0b013e3181e89905.

Forslund, E. B., Jorgensen, V., Franzen, E., Opheim, A., Seiger, A., Stahle, A., et al. (2017). High incidence of falls and fall-related injuries in wheelchair users with spinal cord injury: A prospective study of risk indicators. *Journal of Rehabilitation Medicine*, *49*(2), 144–151. https://doi.org/10.2340/16501977-2177.

Frechette, M. L., Abou, L., Rice, L. A., & Sosnoff, J. J. (2020). The validity, reliability, and sensitivity of a smartphone-based seated postural control assessment in wheelchair users: A pilot study. *Frontiers in Sports and Active Living*. https://doi.org/10.3389/fspor.2020.540930.

Friedman, S. M., Munoz, B., West, S. K., Rubin, G. S., & Fried, L. P. (2002). Falls and fear of falling: Which comes first? A longitudinal prediction model suggests strategies for primary and secondary prevention. *Journal of the American Geriatrics Society*, *50*(8), 1329–1335.

Institute of Medicine (US) Committee on Health and Behavior: Research. (2001). The National Academies Collection: Reports funded by National Institutes of Health. In *Health and behavior: The interplay of biological, behavioral, and societal influences* National Academies Press (US). Copyright © 2001, National Academy of Sciences https://doi.org/10.17226/9838.

Jannings, W. (2017). A quality improvement project to investigate the circumstances of lower limb fractures in non-ambulant persons with spinal cord injury. *Journal of the Australasian Rehabilitation Nurses Association*, *20*(August), 14–18.

Jorgensen, V., Butler Forslund, E., Franzen, E., Opheim, A., Seiger, A., Stahle, A., et al. (2016). Factors associated with recurrent falls in individuals with traumatic spinal cord injury: A multicenter study. *Archives of Physical Medicine and Rehabilitation*, *97*(11), 1908–1916. https://doi.org/10.1016/j.apmr.2016.04.024.

Jorgensen, V., Butler Forslund, E., Opheim, A., Franzen, E., Wahman, K., Hultling, C., et al. (2017). Falls and fear of falling predict future falls and related injuries in ambulatory individuals with spinal cord injury: A longitudinal observational study. *Journal of Physiotherapy*, *63*(2), 108–113. https://doi.org/10.1016/j.jphys.2016.11.010.

Keeler, L., Kirby, R. L., Parker, K., McLean, K. D., & Hayden, J. A. (2019). Effectiveness of the wheelchair skills training program: A systematic review and meta-analysis. *Disability and Rehabilitation. Assistive Technology*, *14*(4), 391–409. https://doi.org/10.1080/17483107.2018.1456566.

Khan, A., Pujol, C., Laylor, M., Unic, N., Pakosh, M., Dawe, J., et al. (2019). Falls after spinal cord injury: A systematic review and meta-analysis of incidence proportion and contributing factors. *Spinal Cord*, *57*(7), 526–539. https://doi.org/10.1038/s41393-019-0274-4.

Khurana, M., Walia, S., & Noohu, M. M. (2017, Summer). Study on the effectiveness of virtual reality game-based training on balance and functional performance in individuals with paraplegia. *Topics in Spinal Cord Injury Rehabilitation*, *23*(3), 263–270. https://doi.org/10.1310/sci16-00003.

Lamb, S. E., Jørstad-Stein, E. C., Hauer, K., Becker, C., & Prevention of Falls Network Europe and Outcomes Consensus Group. (2005). Development of a common outcome data set for fall injury prevention trials: The prevention of falls network Europe consensus. *Journal of the American Geriatrics Society*, *53*(9), 1618–1622. https://doi.org/10.1111/j.1532-5415.2005.53455.x.

Legters, K. (2002). Fear of falling. *Physical Therapy*, *82*(3), 264–272. https://www.ncbi.nlm.nih.gov/pubmed/11869155.

Lynch, S. M., Leahy, P., & Barker, S. P. (1998). Reliability of measurements obtained with a modified functional reach test in subjects with spinal cord injury. *Physical Therapy*, *78*(2), 128–133. http://www.ncbi.nlm.nih.gov/pubmed/9474105.

Moore, J. L., Potter, K., Blankshain, K., Kaplan, S. L., O'Dwyer, L. C., & Sullivan, J. E. (2018). A core set of outcome measures for adults with neurologic conditions undergoing rehabilitation: A CLINICAL PRACTICE GUIDELINE. *Journal of Neurologic Physical Therapy*, *42*(3), 174–220. https://doi.org/10.1097/NPT.0000000000000229.

Musselman, K. E., Arnold, C., Pujol, C., Lynd, K., & Oosman, S. (2018). Falls, mobility, and physical activity after spinal cord injury: An exploratory study using photo-elicitation interviewing. *Spinal Cord Series And Cases*, *4*, 39. https://doi.org/10.1038/s41394-018-0072-9.

Nelson, A., Ahmed, S., Harrow, J., Fitzgerald, S., Sanchez-Anguiano, A., & Gavin-Dreschnack, D. (2003). Fall-related fractures in persons with spinal cord impairment: A descriptive analysis [Research Support, non-U.S. Gov't]. *SCI Nursing*, *20*(1), 30–37. Spring http://www.ncbi.nlm.nih.gov/pubmed/14626016.

Nelson, A. L., Groer, S., Palacios, P., Mitchell, D., Sabharwal, S., Kirby, R. L., et al. (2010, Aug). Wheelchair-related falls in veterans with spinal cord injury residing in the community: A prospective cohort study [multicenter study research support, U.S. Gov't, Non-P.H.S.]. *Archives of Physical Medicine and Rehabilitation*, *91*(8), 1166–1173. https://doi.org/10.1016/j.apmr.2010.05.008.

Peterson, E. W., Cho, C. C., & Finlayson, M. L. (2007). Fear of falling and associated activity curtailment among middle aged and older adults with multiple sclerosis. *Multiple Sclerosis Journal*, *13*(9), 1168–1175.

Phonthee, S., Saengsuwan, J., & Amatachaya, S. (2013). Falls in independent ambulatory patients with spinal cord injury: Incidence, associated factors and levels of ability. *Spinal Cord*, *51*(5), 365–368. https://doi.org/10.1038/sc.2012.147.

Phonthee, S., Saengsuwan, J., Siritaratiwat, W., & Amatachaya, S. (2013). Incidence and factors associated with falls in independent ambulatory individuals with spinal cord injury: A 6-month prospective study. *Physical Therapy*, *93*(8), 1061–1072. https://doi.org/10.2522/ptj.20120467.

Qi, Y., Zhang, X., Zhao, Y., Xie, H., Shen, X., Niu, W., et al. (2018). The effect of wheelchair Tai Chi on balance control and quality of life among survivors of spinal cord injuries: A randomized controlled trial. *Complementary Therapies in Clinical Practice*, *33*, 7–11. https://doi.org/10.1016/j.ctcp.2018.07.004.

Rice, L. A., Sung, J. H., Keane, K., Peterson, E., & Sosnoff, J. J. (2019). A brief fall prevention intervention for manual wheelchair users with spinal cord injuries: A pilot study. *The Journal of Spinal Cord Medicine*, 1–9. https://doi.org/10.1080/10790268.2019.1643070.

Saunders, L., DiPiro, N., Krause, J., Brotherton, S., & Kraft, S. (2013). Risk of fall-related injuries among ambulatory participants with spinal cord injury. *Topics in Spinal Cord Injury Rehabilitation*, *19*, 259–266.

Singh, H., Scovil, C. Y., Yoshida, K., Oosman, S., Kaiser, A., Craven, C., et al. (2020). Factors that influence the risk of falling after spinal cord injury: A qualitative photo-elicitation study with individuals that use a wheelchair as their primary means of mobility. *BMJ Open*, *10*(2). https://doi.org/10.1136/bmjopen-2019-034279, e034279.

Srisim, K., Saengsuwan, J., & Amatachaya, S. (2015). Functional assessments for predicting a risk of multiple falls in independent ambulatory patients with spinal cord injury. *The Journal of Spinal Cord Medicine*, *38*(4), 439–445. https://doi.org/10.1179/2045772313Y.0000000186.

Sung, J., Peterson, E., Shen, S., Sosnoff, J., & Rice, L. A. (2021). Fear of falling, community participation and quality of life among community-dwelling people who use wheelchairs full time. *Archives of Physical Medicine and Rehabilitation*, *102*, 1140–1146.

Sung, J., Trace, Y., Peterson, E. W., Sosnoff, J. J., & Rice, L. A. (2019). Falls among full-time wheelchair users with spinal cord injury and multiple sclerosis: A comparison of characteristics of fallers and circumstances of falls. *Disability and Rehabilitation*, *41*(4), 389–395. https://doi.org/10.1080/09638288.2017.1393111.

Tinetti, M. E., Speechley, M., & Ginter, S. F. (1988). Risk factors for falls among elderly persons living in the community. *New England Journal of Medicine*, *319*(26), 1701–1707.

Verma, S. K., Willetts, J. L., Corns, H. L., Marucci-Wellman, H. R., Lombardi, D. A., & Courtney, T. K. (2016). Falls and fall-related injuries among community-dwelling adults in the United States. *PLoS One*, *11*(3). https://doi.org/10.1371/journal.pone.0150939, e0150939.

Wannapakhe, J., Arayawichanon, P., Saengsuwan, J., & Amatachaya, S. (2014). Changes of functional ability in patients with spinal cord injury with and without falls during 6 months after discharge. *Physical Therapy*, *94*(5), 675–681. https://doi.org/10.2522/ptj.20130260.

Wannapakhe, J., Arrayawichanon, P., Saengsuwan, J., & Amatachaya, S. (2015). Medical complications and falls in patients with spinal cord injury during the immediate phase after completing a rehabilitation program. *The Journal of Spinal Cord Medicine*, *38*(1), 84–90. https://doi.org/10.1179/2045772313y.0000000173.

Whiteneck, G. G., Harrison-Felix, C. L., Mellick, D. C., Brooks, C. A., Charlifue, S. B., & Gerhart, K. A. (2004). Quantifying environmental factors: A measure of physical, attitudinal, service, productivity, and policy barriers. *Archives of Physical Medicine and Rehabilitation*, *85*(8), 1324–1335. https://doi.org/10.1016/j.apmr.2003.09.027.

Williams, A. M. M., Chisholm, A. E., Lynn, A., Malik, R. N., Eginyan, G., & Lam, T. (2020). Arm crank ergometer "spin" training improves seated balance and aerobic capacity in people with spinal cord injury. *Scandinavian Journal of Medicine & Science in Sports*, *30*(2), 361–369. https://doi.org/10.1111/sms.13580.

Wirz, M., Müller, R., & Bastiaenen, C. (2010). Falls in persons with spinal cord injury: Validity and reliability of the berg balance scale. *Neurorehabilitation and Neural Repair*, *24*(1), 70–77.

Wirz, M., & van Hedel, H. J. A. (2018). Balance, gait, and falls in spinal cord injury. *Handbook of Clinical Neurology*, *159*, 367–384. https://doi.org/10.1016/B978-0-444-63916-5.00024-0.

Worobey, L. A., Rigot, S. K., Hogaboom, N. S., Venus, C., & Boninger, M. L. (2018). Investigating the efficacy of web-based transfer training on independent wheelchair transfers through randomized controlled trials. *Archives of Physical Medicine and Rehabilitation*, *99*(1), 9–16. e10 https://doi.org/10.1016/j.apmr.2017.06.025.

Worobey, L. A., Zigler, C. K., Huzinec, R., Rigot, S. K., Sung, J., & Rice, L. A. (2018, Summer). Reliability and validity of the revised transfer assessment instrument. *Topics in Spinal Cord Injury Rehabilitation*, *24*(3), 217–226. https://doi.org/10.1310/sci2403-217.

Yardley, L., Beyer, N., Hauer, K., Kempen, G., & Piot-Ziegler, C. T. (2005). Development and initial validation of the falls efficacy scale international (FES-I). *Age and Ageing*, *34*, 614–619.

Chapter 11

# Infections and spinal cord injury: Covid-19 and beyond

Raquel Madroñero-Mariscal[a,b], Ángel Arévalo-Martín[c], Francisco Gutiérrez-Henares[d], Miguel Rodríguez-Cola[e], Melchor Alvarez de Mon[f,g], and Elisa López-Dolado[h,i]

[a]*Rehabilitation Department, Fundación del Lesionado Medular and Hospital Universitario Infanta Leonor, Madrid, Spain;* [b]*Laboratory of Interfaces for Neural Repair, Hospital Nacional de Parapléjicos, Toledo, Spain;* [c]*Laboratory of Neuroinflammation, Hospital Nacional de Parapléjicos, Toledo, Spain;* [d]*Rehabilitation Department, Hospital Nacional de Parapléjicos, Toledo, Spain;* [e]*Internal Medicine Department, Hospital Nacional de Parapléjicos, Toledo, Spain;* [f]*Hospital Universitario Príncipe de Asturias, Madrid, Spain;* [g]*School of Department of Medicine and Medical Specialties, School of Medicine, Universidad de Alcalá (UAH), Alcalá de Henares, Spain;* [h]*Rehabilitation Department, Laboratory of Interfaces for Neural Repair (LINER), Hospital Nacional de Parapléjicos, Toledo, Spain;* [i]*Department of Medicine and Medical Specialties, School of Medicine, Universidad de Alcalá (UAH), Alcalá de Henares, Spain*

## List of abbreviations

| | |
|---|---|
| **ARDS** | acute respiratory distress syndrome |
| **bDMARDs** | biological disease-modifying anti-rheumatic drugs |
| **CT** | computerized tomography |
| **Covid-19** | coronavirus disease 2019 |
| **GCs** | glucocorticoids |
| **HPA** | hypothalamic-pituitary-adrenal |
| **IUC** | indwelling urethral catheter |
| **ICU** | intensive care unit |
| **ISNCSCI** | International Standards for Neurological Classification of the Spinal Cord Injury |
| **JAKi** | Janus kinase inhibitors |
| **MRI** | magnetic resonance imaging |
| **NICE** | National Institute for Clinical Excellence |
| **NE** | norepinephrine |
| **PI** | pulmonary infections |
| **PEP** | positive expiratory pressure |
| **PPE** | personal protective equipment |
| **ISC** | self-catheterization |
| **SARS-CoV-2** | severe acute respiratory syndrome coronavirus 2 |
| **SPC** | suprapubic catheter |
| **SCI** | spinal cord injury |
| **SCIP** | spinal cord-injured patients |
| **SNS** | sympathetic nervous system |
| **SPNs** | sympathetic preganglionic neurons |
| **UTI** | urinary tract infections |
| **WHO** | World Health Organization |

## Introduction

An epidemic of severe acute respiratory syndrome coronavirus 2 (SARS-CoV-2) started in Wuhan in December 2019 and quickly spread across the world. The 2019 coronavirus disease (Covid-19) was declared a pandemic on March 11, 2020, by the World Health Organization (WHO). On February 7th, 2021, 105,962,538 cases were confirmed worldwide with 2,313,136 deaths ( Johns Hopkins Coronavirus Resource Center, 2020). This new SARS-CoV-2 has brought higher levels

**Diagnosis and Treatment of Spinal Cord Injury. https://doi.org/10.1016/B978-0-12-822498-4.00011-7**

of illness, deaths, and fear to our planet than any other virus in current history. In this context, it is mandatory to determine the effect in people with spinal cord injury (SCI).

SCI induces numerous chronic disorders that put these individuals at a high risk of severe Covid-19 prognosis. Specifically, the SCI population presents higher rates of hypertension, SCI-induced immunosuppression, and, if the metameric level is T8 or above, respiratory failure with continuous or episodic hypoxemia due to respiratory muscle weakness. Sympathetic denervation following SCI compromises body temperature regulation, as a result of dysautonomia, which not only complicates the early diagnosis of Covid-19 which places them at risk of a poor prognosis, but also makes it difficult to control infection transmission to other patients and/or healthcare staff.

In general population, clinical symptoms appear after an incubation period of around 5 days (Zhu et al., 2020), presented in order of frequency as: fever (87.9%), dry cough (67.7%), asthenia (38.1%), dyspnea (18.6%), pharyngeal pain and odynophagia (13.9%), headache (13.6%), arthromyalgia (14.8%), chills (11.4%), nausea or vomiting (5%), nasal congestion (4.8%), and anosmia and diarrhea (3.7%). Approximately 80% of patients present as asymptomatic or with few symptoms including dry cough and fever or low-grade fever (Guan et al., 2020), but the remaining 20% of cases develop around the 7th day after the onset, severe hypoxemic respiratory failure that progresses to Acute Respiratory Distress Syndrome (ARDS) (15% of the total) and even to multi-organ failure (5% of the total), requiring mechanical ventilation and admission to an Intensive Care Unit (ICU) (Huang et al., 2020). Up to 18% of patients who request consultation for Covid-19 compatible symptoms do not present radiological alterations. When these appear, the most frequent clinical sign is unilateral or bilateral ground glass opacification with an interstitial pneumonia pattern or bilateral patched consolidation, evident on plain radiography and on chest computerized tomography (CT) examination (56.4%) (Hosseiny et al., 2020; Zu et al., 2020). At diagnosis, more than 80% of patients present lymphopenia (Guan et al., 2020), which is more severe and accompanied by neutrophilia in patients requiring ICU admission (Huang et al., 2020). The main cause of death after Covid-19 infection is respiratory failure and fulminant myocarditis, (Ruan et al., 2020). In these patients, a highly exaggerated inflammatory response has been described known as hyperinflammatory syndrome, with a massive release of cytokines into the bloodstream (Mehta et al., 2020; Ritchie & Singanayagam, 2020) similar to that observed in hemophagocytic syndromes caused by other virosis (Ramos-Casals et al., 2014). A series of clinical factors have been described, associated with symptoms of greater morbidity and mortality: male gender, age above 60 years—especially above 80 years old, where the fatality rate is close to 16%—and the presence of previous comorbidities such as: hypertension, ischemic heart disease, diabetes mellitus, lung disease, and various disease entities that lead to immunosuppression, such as autoimmune diseases or cancer (Zhou et al., 2020). Analytical markers associated with a higher risk of mortality are sustained lymphopenia and elevation of D-dimer, IL-6, ferritin, LDH, and troponin (Zhou et al., 2020).

The aim of this chapter is to analyze the clinical presentation of Covid-19 in spinal cord-injured patients (SCIP). The authors pretend to make pause to understand if this emergent disease, which is deadly hitting our general population, behaves in the same or different way as in these especial patients, in order to understand if the SCI condition is acting as a risk factor for morbidity or not, and why. For this purpose, we believe the immune system plays a significant role in infection, especially from our knowledge of other infections.

## Immunosuppression due to spinal cord injury

Among non-specialist the most evident consequences of SCI is loss of mobility and sensibility. However, in spite of their dramatic effect on patient quality of life, this is only the tip of an iceberg compared to the other lifelong neurological and non-neurological sequelae. These patients have an immunodepressed state characterized by changes in the number and function of immune cells that render them highly susceptible to infections (Bracchi-Ricard et al., 2016; Brommer et al., 2016; Kopp et al., 2017). Indeed, these are the leading cause of morbidity and mortality after SCI (Thietje et al., 2011). In this chapter, we review why immunodepression occurs after SCI and how this affects the ability of patients to cope with the current Covid-19 pandemic and other infectious diseases.

### Pathophysiology of spinal cord injury-induced immunodepression

The physiological responses to physical or psychological stress are mediated by the coordinated response of the sympathetic nervous system (SNS) and the hypothalamic-pituitary-adrenal (HPA) axis. Both systems mediate the so-called fly-or-fight responses by setting the whole organism to cope with stressful stimuli at the expenses of arresting functions that are not indispensable for immediate survival. This physiological and evolutionary adapted response may, however, turn into a maladaptive chronic situation. Spinal cord injuries are a source of tremendous physical and psychological stress that unfortunately exemplifies the effect on the immune system of maladaptive SNS and HPA responses.

## Dysregulation of the sympathetic nervous system

The main mechanism driving immunodepression after SCI is the dysregulation of the autonomic sympathetic nervous system. The cell bodies of the sympathetic preganglionic neurons (SPNs) are localized in the lateral horn of the spinal cord, in the intermedio-lateral nucleus. Classically, it has been considered as sympathetic the neurons located from T1 to the first lumbar segments (L2 – 3), although recently, the parasympathetic neurons located in the lateral horn of the sacral segments have been shown to have phenotypical and ontogenetic characteristics of SPNs (Espinosa-Medina et al., 2016). In any case, SPNs synapse in sympathetic ganglia majorly with post-ganglionic noradrenergic neurons, which in turn release norepinephrine (NE) in the target organs, though a minority of post-ganglionic sympathetic neurons are cholinergic, as those innervating sweat glands (Dale & Feldberg, 1934). The timing and intensity of SPNs activation is controlled by nuclei located in the brainstem, pons and hypothalamus that project inhibitory descendants into the spinal cord (Dénes et al., 2005). After a severe SCI, these supraspinal projections are interrupted and, thus, SPNs loss their inhibitory inputs. Consequently, when these neurons are activated, it may give raise to an exaggerated or more durable sympathetic response and, thus, to an excessive release of NA in the target organs. In addition, the sublesional spinal circuitry involved in the control of preganglionic neurons undergoes a remodeling process that favors the connectivity between activating interneurons and SPNs (Ueno et al., 2016). Overall, the resultant sympathetic overactivation below the lesion level underlies a plethora of autonomic dysfunctions after SCI, including the suppression of the immune system.

The sympathetic nervous system innervates both the primary and the secondary lymphoid organs, and modulates all the immune processes, from hematopoiesis to immune responses ( Jung et al., 2017; Noble, Brennan, & Popovich, 2018). Immune cells sense NE majorly through beta-2-adrenergic receptors, although monocytes/macrophages also express lower levels of alpha-1- and alpha-2-adrenergic receptors (Kavelaars, 2002). The overall effect of adrenergic receptor activation on physiological conditions over innate immune cells is anti-inflammatory (Meltzer et al., 2004; Nance & MacNeil, 2001). On B cells, NE participates in T-cell mediated IgG production (Alaniz et al., 1999; Kohm & Sanders, 2001), while the effects of NE over T lymphocytes is more complex and varies according to the cellular stage of maturation and timing of exposure. Depending on the stage of maturation of T CD8$^{+}$ cells, exposure to NE may either promote or decrease their cytolytic activity (Nance & Sanders, 2007). Similarly, exposing Th1 cells to NE before their activation decreases the synthesis of IL-2 and IFN-gamma, while exposing them to NE after activation increases the synthesis of IFN-gamma (Nance & Sanders, 2007). In any case, the overall effect of NE over adaptive immune responses is also considered to be anti-inflammatory, inhibiting Th1 cell differentiation and promoting Treg-suppressive activity (Elenkov, Wilder, Chrousos, & Vizi, 2000; Guereschi et al., 2013).

After SCI, immune cells may be subjected to persistent or more intense exposure to NE, which results in immunosuppression (Prass et al., 2003). A factor that may dramatically affect the sympathetic dysregulation-induced immunodepression is the lesion level, being lesions at T5 or above associated to a greater impairment of the immune system both in patients and in experimental animal models (Brommer et al., 2016; Lucin et al., 2007; Lucin et al., 2009). This observation coincides with the fact that most of the sympathetic innervation of immune organs arise from preganglionic neurons located below T5. The innervation of lymph nodes is not resolved, but it is considered to originate from spinal segments close to their location in the body (Nance & Sanders, 2007). Thymus receives most of its innervation from above T5, specifically between T1 and T7 (Trotter et al., 2007). In any case, the sympathetic dysregulation induced by SCI suppresses immune cells in the decentralized lymphoid organs. This statement is supported by direct experimental evidence showing that immune suppression effects depend on beta-2-adrenergic stimulation and may be reverted by receptor antagonists (Table 1). Indirect evidence is further provided by the observation of a relationship between lesion level and extent of immunodepression (Table 2).

**TABLE 1** Effects of SCI over immune cells with experimental evidence of dependence on beta-2-adrenergic receptor stimulation.

| Cell type | Effect | References |
|---|---|---|
| B cells | (1) Impaired primary humoral responses<br>(2) Apoptosis | (1) Lucin et al. (2007), Oropallo et al. (2012)<br>(2) Lucin et al. (2009) |
| T cells | Apoptosis | Lucin et al. (2009) |
| CD8$^{+}$ T cells | T cell exhaustion | Zha et al. (2014) |

Experimental evidence shows that SCI-induced immune suppression effects depend on beta-2-adrenergic stimulation and may be reverted by receptor antagonist. *SCI: spinal cord injury.*

**TABLE 2** Effects of SCI over immune cells dependent on lesion level.

| Cell type | Effect | References |
|---|---|---|
| B cells | Decreased number | Lucin et al. (2007), Held, Steward, Blanc, and Lane (2010), Zhang et al. (2013) |
| T cells | Decreased number | Lucin et al. (2009) |
| $CD4^+$ T cells | Decreased number | Zhang et al. (2013) |
| NK cells | Decreased number | Herman et al. (2018) |
| Dendritic cells | Decreased number | Iversen et al. (2000) |

The observation of a relationship between the injury level and severity and the extent of immunodepression highlights the effects of SCI over immune cells. *SCI: spinal cord injury.*

There is evidence of the contribution of sympathetic overactivation to immune depression after SCI results from autonomic dysreflexia crises, potentially triggered by any sustained sensory stimuli entering the spinal cord below the lesion level. This occurs mainly in patients with severe lesions above T5 because these injuries render without inhibitory supraspinal inputs the SPNs that control abdominal circulation (Weaver et al., 2002). As a consequence, SPNs overactivation occurs and results in vasoconstriction and hypertension. As a counteractive measure it is triggered a vagal response that induces vasodilatation (only successful above the lesion level) and bradycardia, which may be life-threatening if derives into a cardiac arrest. It has been experimentally proved that in these crises, the overactive sympathetic response further aggravates immunodepression in animals with high thoracic (T3) SCI (Zhang et al., 2013), which may help to explain why infections result in morbidity and mortality among patients with tetraplegia compared to patients with paraplegia.

Another potential symptom of immunodepression driven by the autonomous nervous system is the "inflammatory reflex" (Pavlov & Tracey, 2017). When the vague nerve is stimulated, spleen macrophages are polarized into an anti-inflammatory profile. Evidence shows that the spleen does not receive parasympathetic innervation. However, the vagus nerve synapses with sympathetic post-ganglionic neurons in the celiac ganglion, eliciting the release of NE in the spleen. In response, T lymphocytes produce acetylcholine, which acting on alpha-7-nicotinic receptors expressed by macrophages inhibits the production of TNF-alpha, IL-1 beta and IL-18 by these cells. The contribution of the inflammatory reflex to SCI-induced immunodepression is not elucidated, but it seems reasonable to believe that it could be triggered by the vague nerve reflex that takes place during autonomic dysreflexia crises.

### *Dysregulation of the hypothalamic-pituitary-adrenal axis*

The activation of the HPA axis after SCI is well documented by the increase of blood serum glucocorticoids (GCs) in patients and in experimental animals. Besides the direct immunosuppressive actions of GCs on immune cells, glucocorticoids (GCs) and NE synergize to modulate immune cell function. GCs increase the affinity and persistence in the cell membrane of beta-2-adrenergic receptors (Davies & Lefkowitz, 1981; Mak et al., 1995) while NE potentiates GCs signaling (Rangarajan, Umesono, & Evans, 1992). As such, when cortisol and the NE analog terbutaline are added together, they synergize to induce apoptosis of B and T cells in vitro (Lucin et al., 2009). Notably, the spleen atrophy and the decrease in the number of splenocytes induced by an experimental SCI in T3 is partially reverted by the join administration of antagonists of beta-2-adrenergic and GC-receptors (Lucin et al., 2009). In addition, it should also be considered that adrenal glands are innervated by the SNS, which induces the release of GCs.

## Leading infection diseases in spinal cord-injured patients

Infection is one of the leading complications of inpatients, especially severe for SCIP due to the risk of death when it occurs a progression to a bloodstream infection. These patients have a greater predisposition to suffer infectious diseases because of their required use of medical devices such as intravenous and urinary catheters or ventilators. Special conditions suffered by these patients such as lack of mobility or bladder and respiratory dysfunctions contribute to a higher level of infection (Esclarín de Ruz et al., 2009).

There are two main features that make diagnosis of infection particularly challenging in SCIP. First, their loss of feeling can mask standard symptoms which make its diagnosis more difficult to diagnosis. Second, there are also some other described conditions that can simulate an infection, but they are really the manifestation of another disease with no infection as it occurs in hyperthermia, abdominal retention, and pulmonary embolism or collapsed lung (Esclarín de Ruz et al., 2009).

In general terms for treatment, it is strongly recommended to use narrow-spectrum and less toxic antibiotics to avoid eliminating patient's native bacterial flora. It is also important to choose an appropriate directed antibiotic therapy to avoid the appearance of resistant bacteria.

The major infections that suffer SCIP are explained below (Table 3).

## Urinary tract infections (UTI)

UTI are the most frequent cause of infections (95%) in SCIP. They increase the risk of morbidity and mortality because of interruptions of rehabilitation treatment, required visits to the emergency room, and progression to a sepsis. The latter can make SCIP more likely to develop an immunodeficiency syndrome that facilitates the appearance of recurrent UTI (Craven et al., 2019; Esclarín de Ruz et al., 2009). *Escherichia coli*, *Klebsiella pneumoniae*, and *Proteus mirabilis* are the most frequently involved pathogens causing an UTI (Skelton-Dudley et al., 2019).

### *Physiopathology and etiology*

Neurogenic bladder presents overdistension and high pressure due to the interruption of its voiding process. This produces an ischemia of the bladder's wall, which causes bacteria to invade its submucosal layer, and starts the UTI. If this process repeatedly occurs, collagen forms a scared tissue that replaces muscle fibers from the bladder's wall, which decreases bladder's wall compliance. This leads to increase ischemia of the bladder's wall and bladder's overdistension (Fig. 1) (Linsenmeyer, 2018).

To avoid high intravesical pressure and overdistension of the bladder's wall, more than a 60% of SCIP need a medical device to facilitate their bladder voiding such as an intermittent self-catheterization (ISC), an indwelling urethral catheter (IUC) or a suprapubic catheter (SPC). ISC is the gold standard method because it allows a cyclic emptying and filling of the bladder, comparable to the physiological bladder's functioning. Some studies as Kinnear et al., 2020, suggest that ISC is associated with lower rates of UTI when comparing with other medical devices. Despite that, SCIP who use ISC are likely to suffer UTI when they drink plenty of fluids or they do not practice the voiding method regularly, because of an overdistension of the bladder's wall. It becomes worse when patients have an overactive bladder and/or a decreased bladder function (Kinnear et al., 2020).

**TABLE 3 Principle infection features in SCIP.**

| | UTI | PI | Covid-19 |
|---|---|---|---|
| Frequency | It is the most frequent cause of infection (95%) in SCIP | • The 50% of SCIP who suffer acute tetraplegia develop pneumonia during acute hospitalization<br>• PI and respiratory failure are the major causes of death in this population | Unknown but lower than expected |
| Pathogens | • *Escherichia coli*<br>• *Klebsiella pneumoniae*<br>• *Proteus mirabilis* | • Pneumococcus, Pseudomonas (outpatients)<br>• No data (inpatients) | SARS-CoV-2 |
| Etiology | • Changes due to neurogenic bladder produce an overdistension of the bladder, high intravesical pressure and bladder compliance decrease that can lead to an UTI<br>• Medical devices that are used for facilitating bladder voiding can lead to an UTI as well | • The dysfunction of inspiratory and expiratory muscles, the increased risk of dysphagia disorders and the changes inside lung that produces a neurogenic pulmonary edema can lead to a PI<br>• Medical devices to assist with breathing can lead to a PI as well | • Immunosuppression induced by SCI<br>• Neuromuscular respiratory failure due to respiratory muscles weakness and clearance secretions' decrease in SCI levels of injury above T6–T8 |

*Continued*

**TABLE 3 Principle infection features in SCIP—cont'd**

| | UTI | PI | Covid-19 |
|---|---|---|---|
| Diagnosis | • It is necessary to differentiate between:<br>- Asymptomatic bacteriuria/ bladder colonization:<br>  - $\geq 10^2$ cfu/mL bacteriuria at sediment<br>  - or any bacteria detected at sediment when using IUC<br>  - no symptoms<br>- [a]UTI:<br>  - bacteriuria (with one primary organism detected at culture)<br>  - one bladder symptom at least:<br>    - hematuria<br>    - new/increased incontinence<br>    - change in urine odor/clarity<br>    - pyuria<br>  - one general symptom at least:<br>    - fever<br>    - acute autonomic dysreflexia<br>    - acutely increased spasticity<br>    - sepsis<br>    - persistent fatigue<br>    - vomiting<br>• When symptoms appear, urine and blood cultures are required<br>• When symptoms persist, it may be necessary to carry out additional test as ultrasounds, cystoscopy or kidney scan, to rule out complications (hydronephrosis, vesicoureteral reflux, urinary lithiasis or kidney abscesses)<br>• Blood biomarkers (procalcitonin and interleukin-6) has shown poor evidence in ITU detection | • Clinical symptoms: fever, dyspnea, secretions<br>• Chest X-ray that shows infiltrates within lungs<br>• Alterations in blood gas analysis<br>• Leukocytosis<br>• Culture and sputum Gram stain:<br>– when NOT wearing an artificial airway: Sputum culture (>25 polymorphonuclear leukocytes and < 10 epithelial cells per field)<br>– when wearing an artificial airway:<br>  - bronchoalveolar lavage sample: $>10^4$ cfu/mL<br>  - tracheal aspiration sample: $>10^6$ cfu/mL | • Mild symptoms of onset: hypoxia, secretion clearance impairment could be the only symptoms of onset; but SCIP can also present fever, dyspnea, fatigue, anosmia, ageusia, arthromyalgias, and headache<br>• Close observation to neurofunctional outcomes, especially with the help of the International Standards for Neurological Classification of the Spinal Cord Injury (ISNCSCI) Worksheet, is needed to know if this infection produces sensory and motor deficits in these patients<br>• Whenever those symptoms are suspected, it is needed to make a RT-PCR SARS-CoV-2 from nasopharyngeal or bronchial aspiration test to confirm the infected condition<br>• Other paraclinical tests: Chest X-ray (to explore infiltrates at lungs), gas blood analysis (to explore ventilation perfusion state), and a blood analysis (to rule out lymphopenia, increase of D-dimer, ferritin and IL-6<br>• After 7–14 days from the onset of the symptoms, if patient presents a favorable clinical evolution, a serological SARS-CoV-2 should be done to look for IgM and IgG levels |
| Treatment | • The choice of the most appropriate antibiotic treatment depends on culture results<br>• Long-term antibiotic treatment (from 7 to 14 days) is recommended | • Prompt empirical antibiotic treatment is needed<br>- [b]Community acquired pneumonia<br>  - empiric anti-pseudomonal coverage is recommended<br>  - optimize secretion mobilization:<br>    • Multi-modal treatment<br>    • Quad coughing<br>    • Mechanical insufflator-exsufflator<br>- [b]Early hospital-acquired pneumonia<br>  - empiric antibiotic treatment depends on most common pathogens for each hospital, ICU and anti-microbial susceptibility patterns of each community<br>• It is recommended to use agents from a different antibiotic class than the patient has recently received | • The same as general population<br>• In these patients, respiratory rehabilitation is especially important due to their condition of respiratory muscle weakness (secretions' clearance therapies, lower and upper active-assisted or passive kinesiotherapy, and relaxation exercises, and a positive expiratory pressure exercises with a Threshold PEP device to optimize the alveolar recruitment)<br>• It is mandatory that Physiotherapists use PPE for self-protection |

**TABLE 3 Principle infection features in SCIP—cont'd**

| | UTI | PI | Covid-19 |
|---|---|---|---|
| Prevention | • Minimize urinary retention: anti-cholinergic medications or onabotulinum toxin A (Botox) injections for overactive bladders<br>• Change bacterial flora of the bladder:<br>- urine acidification (methenamine hippurate)<br>- bacterial interference<br>• Strengthen the host:<br>- probiotics<br>- proanthocyanidins (cranberry extracts)<br>- immune biotherapy<br>• Reduce biofilm formation:<br>- D-mannose<br>- nanoparticles<br>- to cover the catheter with hydrophilic material<br>- bacteriophages | • [b]Outpatients:<br>- annual Influenza and Pneumococcal polysaccharide vaccines<br>- aggressive mobilization of secretions<br>• [b]Inpatients:<br>- standard precautions<br>- hand washing<br>- gloving and gowning<br>- to perform tracheal and nasopharyngeal respiratory techniques and cares under aseptic conditions<br>- to consider non-invasive positive-pressure ventilation instead of invasive ventilation<br>- to perform orotracheal rather than nasotracheal intubation whenever possible<br>- to clear secretions<br>- to avoid saliva or respiratory secretions microaspiration by raising the head of the bed from 30 to 45 degrees<br>- to perform pre-operative breathing exercises and post-operative incentive spirometry<br>- to remove patients out of bed as soon as possible<br>- to prevent hospital-acquired, ventilator associated, and healthcare-associated pneumonia | • Isolation and hygienic measures, the same as in general population.<br>• Caregivers and health workers need to use PPE for self-protection. |

Pathogens, frequency, etiology, diagnosis, treatment, and prevention strategies of the main infections that affect SCIP. *Covid-19*: coronavirus disease 2019; *ICU*: intensive care unit; *PEP*: positive expiratory pressure.; *PI*: pulmonary infections; *PPE*: personal protective equipment; *SARS-CoV-2*: severe acute respiratory syndrome coronavirus 2; *SCIP*: spinal cord-injured patients; *UTI*: urinary tract infections.

[a]*Modified from: Craven, B. C., et al. (2019). Conception and development of urinary tract infection Indicators to advance the quality of spinal cord injury rehabilitation: SCI-high project.* Journal of Spinal Cord Medicine, *42(sup1), 205–214.*

[b]*Modified from: Burns, S. P. (2007). Acute respiratory infections in persons with spinal cord injury.* Physical Medicine and Rehabilitation Clinics of North America, *18(2), 203–16.*

## *Diagnosis*

First, it is important to provide an early UTI detection. For this purpose, it is necessary to differentiate between asymptomatic bacteriuria and symptomatic UTI. The first one appears when there is a count of $10^2$ cfu/mL or higher at the urine sediment, and when any bacteria are detected with the use of IUC. In this situation, the patient shows no symptoms, and it is considered a colonization of the bladder. This situation is different from an UTI because colonization does not need to be treated as it may play a protective role to the bladder flora (Kirshblum et al., 2011; Linsenmeyer, 2018). To make the diagnosis of UTI, the Experts in Urohealth and/or UTI recognition and management and the SCI-High Project Team at Craven et al. (2019), propose the following criteria: bacteriuria should be confirmed with at least one germ in the urine culture, at least one bladder symptom (hematuria, change in urine odor and clarity, new or increased incontinence, pyuria) and at least one general symptom (fever, sepsis, acute autonomic dysreflexia, persistent fatigue, acutely increased spasticity, vomiting, etc.).

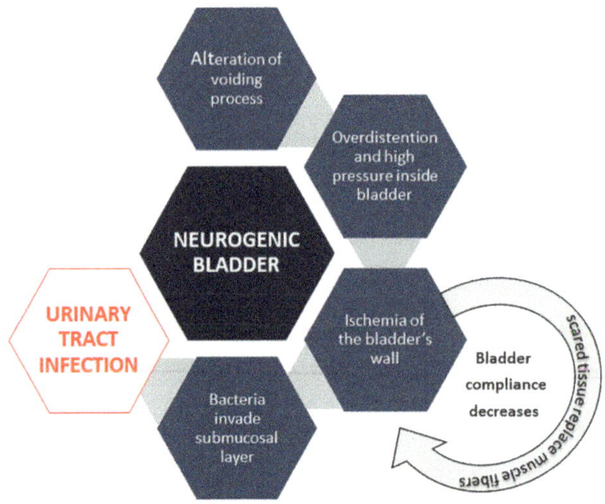


FIG. 1 Physiopathology of urinary tract infections. Step-by-step summary of the sequence of events leading to urinary tract infections (UTI) in SCI neurogenic bladder.

When those symptoms appear, it is necessary to take a urine and a blood culture in order to prescribe an appropriate antibiotic treatment to reduce antibiotic resistance and other associated complications in these patients.

It may be useful to practice a urine sediment because the majority of patients wearing a IUC present leukocyturia (50 leukocytes or more per field). The absence of leukocyturia excludes the diagnosis of UTI. This test has a high negative predictive value. Pyuria is only significant when there is also leukocyturia (Esclarín de Ruz et al., 2009).

If, in spite of an appropriate antibiotic treatment, symptoms persist, it may be necessary to carry out some additional tests such as ultrasound examination, cystoscopy, or kidney scan, to look for a structural cause or a complication being the leading cause of the persistent UTI (hydronephrosis, vesicoureteral reflux, urinary lithiasis, or kidney abscesses).

Blood biomarkers have shown poor evidence in the detection of UTI. Procalcitonin has been described as a biomarker in the diagnosis of UTI and pyelonephritis in pediatric population, but not in adults. IL-6 could help to differentiate between UTI and pyelonephritis, but further validation is needed yet (Skelton-Dudley et al., 2019).

## *Treatment and prevention*

SCIP develop complicated UTI because of the urinary tract dysfunction. Because of this reason, it is necessary to follow a long-term antibiotic treatment from 7 to 14 days. To avoid the increased likelihood of anti-microbial resistance and to reduce the administration of wide spectrum antibiotics, it is necessary to perform a urine culture to select the most appropriate antibiotic treatment (Esclarín de Ruz et al., 2020).

Though some studies advise using prophylactic antibiotics as nitrofurantoin to prevent catheter associated UTI (Chew et al., 2019); the truth is that more consistent evidence is required to make this recommendation in a safe way to compensate the risk of resistances appearance derived from the use of these antibiotics (Skelton-Dudley et al., 2019), because there are some other studies like Morton et al. (2002), which show that using prophylactic antibiotics is worthless as it does not decrease the number of UTI, and even may be harmful, as it reduces asymptomatic bacteriuria while increasing antibiotic resistance.

To prevent recurrent UTI in SCIP it is desirable to improve patients' conditions minimizing urinary retention and optimizing urinary voiding, using anti-cholinergic medications or onabotulinum toxin A (Botox) when they have an overactive bladder. Also, it is important to correct anatomical problems like urethral stricture or bladder calculi (Linsenmeyer, 2018).

There are other prevention strategies for SCIP with recurrent UTI. The most commonly used are: (1) changing the bacterial flora of the bladder with urine acidification; or (2) strengthening the host by administering probiotics, proanthocyanidins such as cranberry extracts or D-mannose (to reduce biofilm formation on the bladder and catheter wall). In the foreseeable future, clinicians will also have nanoparticles capable of inhibiting biofilm formation by *E. coli* and *Staphylococcus aureus*, but this promising strategy is currently in the pre-clinical phase. To cover the catheter with hydrophilic material before its introduction to the bladder delays and decreases the number of UTI in acute SCI (Cardenas et al., 2011) by blocking the biofilm catheter formation. The use of bacteriophages that are responsible for producing a depolymerase enzyme that breaks the extracellular polysaccharide matrix of the biofilms on urinary catheter, are also useful when the patient has developed antibiotic resistance because, when antibiotic-resistant bacteria become resistant to the phage, in some cases, the bacterial sensitivity to antibiotics is restored. Also, immune biotherapy uses the administration of antigenic components to be recognized as "danger signals" which contribute to activate the innate immune system as a prophylactic method (Linsenmeyer, 2018).

## Pulmonary infections (PI)

The 50% of SCIP who suffer acute tetraplegia develop pneumonia during acute hospitalization and rehabilitation (Schilero et al., 2005). PI and respiratory failure are major causes of death in this population (Yong et al., 2012). Review study by Burns (2007) has identified Pneumococcus as the first leading pathogen causing pneumonia in outpatients with SCI, while Pseudomonas is the second one. Conversely, there are no published data on etiologic pathogens that cause pneumonia during acute hospitalization and rehabilitation for these patients. It is assumed that the etiology of a nosocomial pneumonia depends on the type of hospital, ICU, and anti-microbial susceptibility patterns of each community (Díaz, Martín-Loeches, & Vallés, 2013).

Review study by Burns (2007) also shows that in ventilator-dependent patients, more than 90% of patients suffering from pneumonias and requiring hospitalizations have previously developed upper respiratory tract infections.

### *Physiopathology and etiology*

Depending on the level and severity of the SCI, there are different mechanisms that are thought to be involved in causing pneumonia. In complete lesions above C3, the respiratory muscles are paralyzed. In C3–C5 lesions, despite varying degrees of phrenic nerve injury, respiratory neuromuscular activation is preserved but dysfunctional. Lesions below C5 retain some degree of ventilatory function due to intercostal and abdominal muscles, but above all, both diaphragms are preserved. These three situations involve inspiratory and expiratory hypofunction, which leads to dyspnea, decreased vital capacity, ineffective cough and problems with mucociliary clearance (Yong et al., 2012), and also they are more likely to suffer dysphagia disorders, especially due to tracheostomy, rigid fixation and high complete lesions due to the delay of mechanical ventilation weaning (Abel, Ruf, & Spahn, 2004). Apart from that, patients with low thoracic SCI, who retain good muscular respiratory function, may develop neurogenic pulmonary edema due to pulmonary vein constriction, increased pulmonary capillary hydrostatic pressure, alveolar damage and increased capillary permeability due to transient sympathetic discharge, which is underdiagnosed and provides the right environment for developing PI (Yong et al., 2012) (Fig. 2).

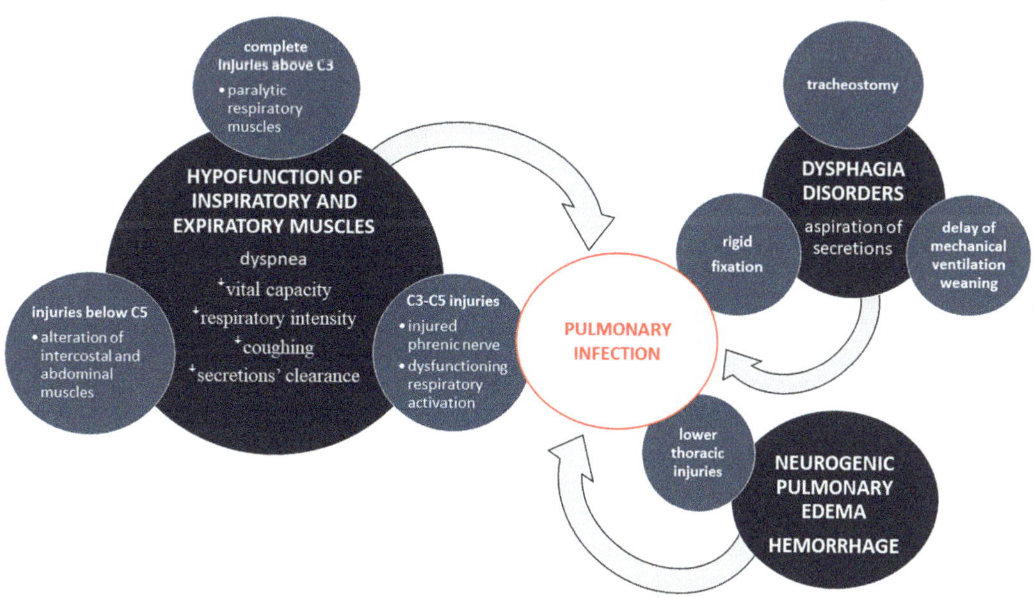


FIG. 2 Physiopathology of pulmonary infections. SCI-induced inspiratory and expiratory muscles hypofunction, dysphagia disorders, and neurogenic pulmonary edema can lead to develop a pulmonary infection in spinal cord-injured patients.

### Diagnosis

Fever, respiratory symptoms as dyspnea or secretions, chest X-rays with infiltrates within lungs and alterations in blood gas analysis, and leukocytosis determine a diagnosis of pneumonia. To administrate an appropriate antibiotic treatment, it is necessary to make a culture and a sputum Gram stain.

In cases where a patient is not wearing an artificial airway, it is sufficient to make a sputum culture. A sample of >25 polymorphonuclear leukocytes and <10 epithelial cells per field is considered representative of the lower respiratory tract. In cases where patients are wearing an artificial airway, it is necessary to take a sample by bronchoalveolar lavage or a tracheal aspiration. An infection is diagnosed when there are $>10^4$ cfu/mL in bronchoalveolar lavage sample or $>10^6$ cfu/mL in tracheal aspiration sample (Esclarín de Ruz et al., 2020).

### Treatment and prevention

While obtaining a culture, and before administrating a specific antibiotic treatment it is necessary to use a prompt empirical antibiotic treatment for avoiding respiratory infection progression. The choice of empiric antibiotic coverage depends on the most common pathogens for each hospital, ICU and anti-microbial susceptibility patterns of each community.

Burns (2007) recommends the management of community acquired pneumonia to consider empiric anti-pseudomonal coverage. It is also recommended to optimize secretion mobilization by multi-modal treatment, quad coughing or mechanical insufflator-exsufflator. To choose the empiric antibiotic coverage for SCIP with an early hospital-acquired pneumonia, it is recommended to use the same medications used for hospital-acquired pneumonia in patients without SCI, as the etiologic organisms for pneumonia that develop during acute care in SCIP are still undescribed. It is recommended to include agents from a different antibiotic class than the patient has received recently.

To prevent respiratory infections in outpatients with spinal cord injury, Burns (2007) recommends annual Influenza and Pneumococcal polysaccharide vaccines, and aggressive mobilization of secretions. In the case of SCI inpatients, the same author also recommends using the standard precautions (hand washing, use of gloves and gowns), performing the various tracheal and nasopharyngeal respiratory techniques and cares under aseptic conditions, as well as considering non-invasive positive pressure ventilation instead of invasive ventilation, opting for orotracheal instead of nasotracheal intubation whenever possible, clear secretions, avoid saliva or respiratory secretions microaspiration by raising the head of the bed from 30 to 45 degrees, perform pre-operative breathing exercises and post-operative incentive spirometry, and remove patients out of bed as soon as possible, to prevent hospital-acquired, ventilator-associated and healthcare-associated pneumonia.

There are other less frequent but also important infections that can affect SCIP such as pressure ulcer infections, surgery wound infections, intra-abdominal abscesses causing acute abdomen, sepsis… that are not explained here but expert SCI physicians need to know deeply to provide the acutest diagnosis and the best care and treatment to these patients. The latter review of the most frequent infections of SCIP (UTI) and the more severe ones (PI), pretend to help to achieve the greater understanding of the new Covid-19 infection in these patients.

## COVID-19 disease: A new infectious normality in spinal cord-injured patients?

As it is well established, SCIP can be quite fragile. When Covid-19 pandemic started, we feared that they would have to face poor evolution and harsh prognosis, because of the presence of immunosuppression and induced autoimmunity, but above all, because of the respiratory muscles' weakness in all those cases with injury levels above T6–T8 (López-Dolado & Gil-Agudo, 2020). In addition, the mortality rate from influenza pneumonia was known to be higher than that of the general population (Soden et al., 2000). However, this grim prognosis does not seem to be confirmed. In fact, it has been reported that SCIP with concomitant Covid-19 infection show mild initial symptoms and better outcomes, even being aged over 60 and/or presenting clinically severe phenotypes (D'Andrea et al., 2020; Rodríguez-Cola et al., 2020), (Table 3). This apparently "indifferent" behavior toward Covid-19 infection has also been described in some autoimmune osteoarthropathies such as rheumatoid arthritis, also in those patients treated with biological disease-modifying anti-rheumatic drugs (bDMARDs) or Janus kinase inhibitors (JAKi) (Stradner et al., 2020). Data regarding Covid-19 and SCI are sparse, and further studies are needed to investigate the course of Covid-19 in SCIP, whether or not they have comorbidities and whether they are acute or chronic, but two simple although important points must be taken into account. First of all, the typical immune response involved in viral infections is sensibly different than in bacterial ones. Viruses are intracellular pathogens while bacteria are extracellular, so the role of CD8s and NK is quite different in both cases (see above). Second, as important as the pathogen in clearing up any infection is the immune system that fights against it, so in this regard, the SCIP cannot be considered as a homogeneous population. Factors such as advanced age, the presence of comorbidities or long-term SCI chronic stages would negatively influence the immune response and could conditionate different risk levels (Monroe et al., 2019). Finally, with the available data to date, it is not possible to support or reject that Covid-19 infection is less symptomatic or severe in SCIP, since the use of drugs that dampen the local pulmonary inflammatory response is common in this population and the available data mostly come from case reports or at most from case series, valuable work for sure, but with restricted levels of evidence.

Another interesting aspect is the thrombogenic nature of Covid-19 shared by SCI. Using proteomic analysis to look for diagnostic biomarkers of Covid-19 infection in a cohort of SCIP, a specific role of heparin response to Covid-19 infection was found, supported by a significant correlation with proteins implied in coagulation/platelet activation. In this study, SCIP with Covid-19 receive from the moment of diagnosis prophylactic heparin, a common practice with these persons when bedridden is needed to avoid typical thrombotic and embolic complications (Calvo et al., 2020). Interestingly, heparin has demonstrated therapeutic potential against SARS-CoV-2 infection as a competitive inhibitor in general population Covid-19 patients.

It is well established that close to half of the hospitalized Covid-19 patients exhibit neurologic manifestations at the onset of the infection. To the author's knowledge, no study has been found that specifically analyzes this aspect in SCIP, although myalgias, headache, anosmia, and ageusia have frequently been found (Liotta et al., 2020). Sensory and motor deficits, barely been reported in the general population, have not been detected at all in SCIP, which does not mean that they have not appeared, but that we have not been able to detect them. Close observation to the neurofunctional outcome of the Covid-19 SCIP in the near and further future will permit to address this aspect.

The prevalence of Covid-19 infection among acute or chronic SCIP is unknown, but judging by the number of published cases, it seems lower than that of the general population (D'Andrea et al., 2020; López-Dolado & Gil-Agudo, 2020; Rodríguez-Cola e al., 2020). Part of the explanation could lie in the lower number of social contacts experienced by individuals who have reduced mobility and disability, in addition to the hygienic measures usually and constantly taken by their caregivers, which are similar to the measures that protect general population from Covid-19. The risk of transmission for the caregivers is nevertheless, another matter. Because being paucisymptomatic, Covid-19 infection could go unnoticed, turning the SCIP into a reservoir that would substantially increase the risk of contagion their caregivers. Therefore, caregivers and health providers must have access to adequate personal protective equipment (PPE) (Cossarizza et al., 2020). On top of that, both patients and their caregivers have experienced this pandemic, anxiety, loneliness and sadness, emotions that, if not addressed, will have negative impact on the health of all those groups (Taylor et al., 2020). The current pandemic has severely restricted SCIPs' access to health services, either for clinical control and follow-up, or to receive rehabilitative interventions. It is our vocation and obligation to use all the tools at our disposal, like the telemedicine platforms to maintain a safe, economical method of patient care.

Despite the burden of this worldwide disease there is no specific drug available to combat Covid-19. Vaccine efficacy depends largely upon the vaccine target and platform. Generally, the clinical development of a new vaccine requires many years of efforts in this area. However, having the results of previous efforts related to SARS-CoV and MERS vaccine research available, this time it has been possible to develop several vaccines, yet almost ready for their massive use in general population. There are no data to support the preferential vaccination of SCIP if the expressiveness and severity of the Covid-19 infection is taken into account. However, the fact that they can become virus reservoirs, and therefore, a source of transmission, is a compelling reason to recommend its extensive vaccination. Former experiences during previous pandemics, such as H1V1 or influenza, highlighted the need to make every effort not just to get the vaccine but to inform the target population to achieve sufficient vaccination rates. All possible efforts to ensure that accurate and clear information reaches the target SCIP must be done, since it is known that the better the information available on the disease and the vaccine, the higher the vaccination rate (LaVela et al., 2012). The medical community serving SCIP constitutes an exceptional group for quality information on Covid-19 vaccination to permeate (Locatelli et al., 2013). Thus, the current pandemic has shown us that from now on we should include in our health education programs all necessary information to protect SCIP against emerging pathogens and, among them, a better knowledge of available vaccines.

## Applications to other areas of neuroscience

In this chapter we have reviewed the implication of the alteration of immune system in spinal cord-injured patients, and how this process leads the different infections that can suffer these patients. We also interlink the role of this immunosuppressed induced condition, with the current SARS-Cov-2 pandemic.

A better understanding about the interaction between the nervous system and the immune system could be reached by comprehending how SARS-Cov-2 virus affects patients with central nervous system diseases, such as in the spinal cord-injured patients, as Covid-19 influences the nervous system functioning and the immunity status as well. Due to its neurotrophic feature, the deep knowledge of its structure, function, infection mechanism and characteristics, may help to improve the design of the current viral vectors applied at Neuroscience.

It should be an increasingly common requirement at the educational and clinical practice to include the study of the effects on our immune system in order to achieve the acutest and better understanding of every oldest known and new diseases. These diseases develop in a shared territory between neuroscience and immunology, so the deep knowledge of this space "in between" is a challenge, not only for physicians and immunologists, but also for neuroscientists, the spearhead of a successful knowledge in the research of central nervous system diseases and disabilities.

## Mini-dictionary of terms

**Respiratory syndrome coronavirus 2**: Infection disease caused by Coronavirus that originates a severe damage to the lungs that involves a respiratory insufficiency, even in some cases, a breathing failure, requiring artificial ventilation supply.

**Immunodepression**: State of weakness of the immune system that delays and makes difficult to develop a prompt and active defense response against pathogens that invade human body. It has been demonstrated that spinal cord injury is a systemic disease which precipitates systemic changes that affect the immune system in this way.

**Sympathetic nervous system**: Part of the autonomic nervous system made up of neurons located from the first thoracic to the first lumbar segments which are presumed to carry out within its multiple functions, to innervate the primary and secondary lymphoid organs and modulate all the immune processes. Any process affecting this sympathetic nervous system, as spinal cord injury, could result in an alteration of the immune response of a patient.

**Hypothalamic-pituitary-adrenal axis**: Complex hormonal communicating system between the brain and different body glands that participate in different body responses, as the stress responses, energy metabolism, or the modulation of immune responses, as sympathetic nervous system does as well. In addition, adrenal glands, which are part of this axis, are innervated by the sympathetic nervous system. If a patient suffers a spinal cord injury, it will result in an alteration of all those responses by damaging the sympathetic nervous system and by altering this axis as well.

**Bloodstream infection**: Presence of pathogen organisms in the blood, having come directly from outside or being the spreading of a previously local infection. This is a severe infectious disease that can threat patient's life.

**Native bacterial flora**: Specific species of bacterial organisms that normally dwell different parts of human body without causing a disease. In fact, the loss of these bacterial organisms could imply the invasion from other pathogens that can lead an infection, so this native bacterial flora usually plays a protective role.

**Neurogenic bladder**: Dysfunction of the bladder due to a neurological damage that implies an overdistension of the bladder's wall and an increased high pressure inside due to the alteration of its voiding process. Therefore, patient suffers incontinence and urine retention that may need medications and/or medical devices to facilitate and normalize their bladder's voiding.

**Recurrent urinary tract infection**: Repeated infection of the urinary tract that does not disappear despite adequate anti biotic treatment. In spinal cord-injured patients, it should make consider the presence of a urinary complication as hydronephrosis, vesicoureteral reflux, urinary lithiasis, or kidney abscesses. First, it will be necessary to rule out these disorders to treat the infection successfully, later on.

**Biofilm**: Adherence of pathogens to the wall of artificial products that are in direct contact to body tissues, such as medical devices, that are able to produce a protective layer that make them resistant to standard antibiotic treatments. When it appears in a catheter and it remains unnoticed, the catheter biofilm becomes a leading cause of recurrent urinary tract infection and appearance of resistant bacterial pathogens.

**Dysphagia disorders**: Alteration of the deglutition process that implies the risk of passing food through airway by the larynx. It is a leading cause of pneumonia, which is considered to be the severest infection in spinal cord-injured patients because it can threat their life.

**Neurogenic pulmonary edema**: Fluid invasion of the alveolar spaces from the lungs due to an inflammatory cascade which is thought to be the result of the immune system's damage suffered in severe neurological injured patients. The resulting pulmonary edema could behave as an adequate environment for developing a pulmonary infection in these patients.

**Telemedicine and telerehabilitation**: Way of supplying a medical attention, or medical, and/or rehabilitation treatment that replace the classic patient's physical attendance at the place of the doctor visit, by a telephone or video consultation, which avoids patients' movements from home or familiar environment. It provides an effective source of access to medical healthcare to chronically affected patients as spinal cord-injured patients, who may find impairments to achieve transportations, especially during pandemic restriction period.

## Key facts of "infections and spinal cord injury: Covid-19 and beyond"

### Key facts of "immunosuppression due to spinal cord injury"

- Spinal cord injury is a systemic disease.
- Spinal cord injury induces an alteration of the immune system.
- This alteration of the immune system leads patients to suffer from infections.
- The better understanding of the modifications produced by the alteration of the immune system enables physicians and scientists to be able to appreciate the clinical presentations of these diseases, and patient's responses to treatments, as well.

### Key facts of "Leading infection diseases in spinal cord-injured patients"

- Infection is one of the leading complications of spinal cord-injured patients because this disease can threat their life.
- Urinary tract infections are the most frequent cause of infection in these patients because they present a neurogenic bladder and they usually need to use medical devices to facilitate their bladder's voiding.
- Its acute diagnosis will be of great importance to establish a long-term and appropriate antibiotic treatment to avoid antimicrobial resistances and further complications.
- Pulmonary infections are the most severe cause of infection in these patients because they can lead to a respiratory insufficiency and breathing failure that can strongly threat patients' life.
- While waiting for etiologic organisms' results and before starting an adequate antibiotic treatment guided by those results, with the only clinical suspicion, physicians need to start an empirical antibiotic treatment, to avoid the progression of the infection.

### Key facts of "Covid-19 disease"

- Most of the published studies show a tendency of milder initial symptoms and a less severe evolution of the Covid-19 disease in spinal cord-injured patients, but currently further validation is needed to support or reject that Covid-19 infection is less symptomatic or severe in this population.
- Covid-19 disease has shown a thrombogenic nature in spinal cord-injured patients that must be treated with heparins.
- In spite of coronavirus' neurotropism, it has not been detected a sensory or motor deterioration in spinal cord-injured patients due to Covid-19 disease yet; but close observation to the neurofunctional outcomes should be done to ensure this affirmation.
- As Covid-19 disease could go unnoticed, turning the spinal cord-injured patients into a reservoir that could increase the contagion of their caregivers, medical community serving these patients should do every possible effort to provide appropriate healthcare educational programs with all the information needed to protect them from emerging pathogens that include the better knowledge of the available vaccines.
- Telemedicine has demonstrated to be a useful and effective tool to provide access to medical healthcare to chronically affected patients as spinal cord-injured patients, especially during this pandemic restriction period.

## Summary points

- Spinal cord injury induces not only a loss of mobility and sensibility, but also numerous chronic disorders.
- Spinal cord-injured patients suffer a dysregulation of the sympathetic nervous system and the hypothalamic-pituitary-adrenal axis, which causes an alteration of all their immune processes.
- The combination of this situation with other locally impaired conditions provides a suitable environment for developing an infection.
- Urinary tract infections are the most frequent infections in these patients, because of the presence of a neurogenic bladder and the use of catheters to facilitate its voiding.
- Pulmonary infections are the severest ones, because of the respiratory muscle weakness, dysphagia disorders, pulmonary edema, and the use of ventilators to assist with breathing.
- The deep understanding of the physiopathology of these infections should help us to understand its appropriate diagnosis, treatment, and methods of prevention.

- The pandemic Covid-19, which is deadly hitting our general population, seems to show a tendency of milder initial symptoms and a less severe evolution in spinal cord-injured patients, but currently further validation is needed to support or reject it.
- The altered immune response could play a critical role in the clinical presentation of these patients.
- Telemedicine has demonstrated to be a useful and effective tool to provide access to medical healthcare to these chronically affected patients, especially during this pandemic restriction period.

## References


Abel, R., Ruf, S., & Spahn, B. (2004). Cervical spinal cord injury and deglutition disorders. *Dysphagia*, *19*(2), 87–94.

Alaniz, R. C., et al. (1999). Dopamine β-hydroxylase deficiency impairs cellular immunity. *Proceedings of the National Academy of Sciences of the United States of America*, *96*(5), 2274–2278.

Bracchi-Ricard, V., et al. (2016). Chronic spinal cord injury attenuates influenza virus-specific antiviral immunity. *Journal of Neuroinflammation*, *13*(1), 1–13.

Brommer, B., et al. (2016). Spinal cord injury-induced immune deficiency syndrome enhances infection susceptibility dependent on lesion level. *Brain*, *139*(3), 692–707.

Burns, S. P. (2007). Acute respiratory infections in persons with spinal cord injury. *Physical Medicine and Rehabilitation Clinics of North America*, *18*(2), 203–216.

Calvo, E., et al. (2020). Why does COVID-19 affect patients with spinal cord injury milder? A case-control study: Results from two observational cohorts. *Journal of Perinatal Medicine*, *10*(4), E182.

Cardenas, D. D., et al. (2011). Intermittent catheterization with a hydrophilic-coated catheter delays urinary tract infections in acute spinal cord injury: A prospective, randomized, multicenter trial. *PM & R : The Journal of Injury, Function, and Rehabilitation*, *3*(5), 408–417.

Chew, A. B., et al. (2019). Long-term prescribing of nitrofurantoin for urinary tract infections (UTI) in veterans with spinal cord injury (SCI). *The Journal of Spinal Cord Medicine*, *42*(4), 485–493.

Cossarizza, A., et al. (2020). Handling and processing of blood specimens from patients with COVID-19 for safe studies on cell phenotype and cytokine storm. *Cytometry. Part A: The Journal of the International Society for Analytical Cytology*, *97*(7), 668–673.

Craven, B. C., et al. (2019). Conception and development of urinary tract infection indicators to advance the quality of spinal cord injury rehabilitation: SCI-high project. *The Journal of Spinal Cord Medicine*, *42*(sup1), 205–214.

D'Andrea, S., et al. (2020). Clinical features and prognosis of COVID-19 in people with spinal cord injury: A case–control study. *Spinal Cord Series And Cases*, *6*(1), 69.

Dale, H. H., & Feldberg, W. (1934). The chemical transmission of secretory impulses to the sweat glands of the cat. *The Journal of Physiology*, *82*(1), 121–128.

Davies, A. O., & Lefkowitz, R. J. (1981). Agonist-promoted high affinity state of the β-adrenergic receptor in human neutrophils: Modulation by corticosteroids. *Journal of Clinical Endocrinology and Metabolism*, *53*(4), 703–708.

Dénes, Á., et al. (2005). Central autonomic control of the bone marrow: Multisynaptic tract tracing by recombinant pseudorabies virus. *Neuroscience*, *134*(3), 947–963.

Díaz, E., Martín-Loeches, I., & Vallés, J. (2013). Neumonía nosocomial [Nosocomial pneumonia]. *Enfermedades Infecciosas y Microbiología Clínica*, *31*(10), 692–698.

Elenkov, I. J., Wilder, R. L., Chrousos, G. P., & Vizi, E. S. (2000). The sympathetic nerve—An integrative interface between two supersystems: The brain and the immune system. *Pharmacological Reviews*, *52*(4), 595–638.

Esclarín de Ruz, A., et al. (2009). *Lesión medular. Enfoque multidisciplinario* (1st ed.). Spain: Panamericana (Chapter 18).

Esclarín de Ruz, A., et al. (2020). *Lesión medular. Enfoque multidisciplinario* (2nd ed.). Spain: Panamericana (Chapter 20).

Espinosa-Medina, I., et al. (2016). The sacral autonomic outflow is sympathetic. *Science*, *354*(6314), 893–897.

Guan, W. J., et al. (2020). China medical treatment expert Group for Covid-19. Clinical characteristics of coronavirus disease 2019 in China. *The New England Journal of Medicine*, *382*(18), 1708–1720.

Guereschi, M. G., et al. (2013). Beta2-adrenergic receptor signaling in CD4+ Foxp3+ regulatory T cells enhances their suppressive function in a PKA-dependent manner. *European Journal of Immunology*, *43*(4), 1001–1012.

Held, K. S., Steward, O., Blanc, C., & Lane, T. E. (2010). Impaired immune responses following spinal cord injury lead to reduced ability to control viral infection. *Experimental Neurology*, *226*(1), 242–253.

Herman, P., et al. (2018). Persons with chronic spinal cord injury have decreased natural killer cell and increased toll-like receptor/inflammatory gene expression. *Journal of Neurotrauma*, *35*(15), 1819–1829.

Hosseiny, M., et al. (2020). Radiology perspective of coronavirus disease 2019 (COVID-19): Lessons from severe acute respiratory syndrome and Middle East respiratory syndrome. *AJR. American Journal of Roentgenology*, *214*(5), 1078–1082.

Huang, C., et al. (2020). Clinical features of patients infected with 2019 novel coronavirus in Wuhan, China. *Lancet (London, England)*, *395*(10223), 497–506.

Iversen, P. O., et al. (2000). Depressed immunity and impaired proliferation of hematopoietic progenitor cells in patients with complete spinal cord injury. *Blood*, *96*(6), 2081–2083.

Johns Hopkins Coronavirus Resource Center. (2020). *Coronavirus COVID-19 Global Cases by the Center for Systems*. Science and Engineering (CSSE) at Johns Hopkins University (JHU). https://www.arcgis.com/apps/opsdashboard/index. html#/bda7594740fd40299423467b48e9ecf6. Accessed the 7th of February, 2021).

Jung, W. C., et al. (2017). It takes nerve to fight back: The significance of neural innervation of the bone marrow and spleen for immune function. *Seminars in Cell and Developmental Biology*, *61*, 60–70.

Kavelaars, A. (2002). Regulated expression of α-1 adrenergic receptors in the immune system. *Brain, Behavior, and Immunity*, *16*(6), 799–807.

Kinnear, N., et al. (2020). The impact of catheter-based bladder drainage method on urinary tract infection risk in spinal cord injury and neurogenic bladder: A systematic review. *Neurourology and Urodynamics*, *39*(2), 854–862.

Kirshblum, S., et al. (2011). *Spinal cord medicine* (2nd ed.). Philadelphia, PA: Lippicott Williams & Wilikins (Chapter 13).

Kohm, A. P., & Sanders, V. M. (2001). Norepinephrine and β2-adrenergic receptor stimulation regulate CD4+ T and B lymphocyte function in vitro and in vivo. *Pharmacological Reviews*, *53*(4), 487–525.

Kopp, M., et al. (2017). Long-term functional outcome in patients with acquired infections after acute spinal cord injury. *Neurology*, *88*(9), 892–900.

LaVela, S. L., et al. (2012). Factors associated with H1N1 influenza vaccine receipt in a high-risk population during the 2009-2010 H1N1 influenza pandemic. *Topics in spinal cord injury rehabilitation*, *18*(4), 306–314.

Linsenmeyer, T. A. (2018). Catheter-associated urinary tract infections in persons with neurogenic bladders. *The Journal of Spinal Cord Medicine*, *41*(2), 132–141.

Liotta, E. M., et al. (2020). Frequent neurologic manifestations and encephalopathy-associated morbidity in Covid-19 patients. *Annals of Clinical Translational Neurology*, *7*(11), 2221–2230.

Locatelli, S. M., et al. (2013). Influenza infection control guidance provided to staff at veterans affairs facilities for veterans with spinal cord injury during a pandemic. *The Journal of Spinal Cord Medicine*, *36*(6), 666–671.

López-Dolado, E., & Gil-Agudo, A. (2020). Lessons learned from the coronavirus disease 2019 (Covid-19) outbreak in a monographic center for spinal cord injury. *Spinal Cord*, *58*(5), 517–519.

Lucin, K., et al. (2007). Impaired antibody synthesis after spinal cord injury is level dependent and is due to sympathetic nervous system dysregulation. *Experimental Neurology*, *207*(1), 75–84.

Lucin, K. M., et al. (2009). Stress hormones collaborate to induce lymphocyte apoptosis after high level spinal cord injury. *Journal of Neurochemistry*, *110* (5), 1409–1421.

Mak, J. C. W., et al. (1995). Protective effects of a glucocorticoid on downregulation of pulmonary β2-adrenergic receptors in vivo. *Journal of Clinical Investigation*, *96*, 99–106.

Mehta, P., et al. (2020). COVID-19: Consider cytokine storm syndromes and immunosuppression. *Lancet*, *395*(10229), 1033–1034.

Meltzer, J. C., et al. (2004). Stress-induced suppression of in vivo splenic cytokine production in the rat by neural and hormonal mechanisms. *Brain, Behavior, and Immunity*, *18*(3), 262–273.

Monroe, D., et al. (2019). Clinical associations with telomere length in chronic spinal cord injury. *Spinal Cord*, *57*(12), 1084–1093.

Morton, S. C., et al. (2002). Antimicrobial prophylaxis for urinary tract infections in persons with spinal cord dysfunction. *Archives of Physical Medicine and Rehabilitation*, *83*(1), 129–138.

Nance, D. M., & MacNeil, B. (2001). *Immunoregulation by the sympathetic nervous system*. Elsevier.

Nance, D. M., & Sanders, V. M. (2007). Autonomic innervation and regulation of the immune system (1987-2007). *Brain, Behavior, and Immunity*, *21*(6), 736–745.

Noble, B. T., Brennan, F. H., & Popovich, P. G. (2018). The spleen as a neuroimmune interface after spinal cord injury. *Journal of Neuroimmunology*, *321*, 1–11.

Oropallo, M. A., et al. (2012). Chronic spinal cord injury impairs primary antibody responses but spares existing humoral immunity in mice. *Journal of Immunology (Baltimore, Md.: 1950)*, *188*(11), 5257–5266.

Pavlov, V. A., & Tracey, K. J. (2017). Neural regulation of immunity: Molecular mechanisms and clinical translation. *Nature Neuroscience*, *20*(2), 156–166.

Prass, K., et al. (2003). Stroke-induced immunodeficiency promotes spontaneous bacterial infections and is mediated by sympathetic activation reversal by Poststroke T helper cell type 1–like Immunostimulation. *The Journal of Experimental Medicine*, *198*(5), 725–736.

Ramos-Casals, M., et al. (2014). Adult haemophagocytic syndrome. *Lancet*, *383*(9927), 1503–1516.

Rangarajan, P. N., Umesono, K., & Evans, R. M. (1992). Modulation of glucocorticoid receptor function by protein kinase a. *Molecular Endocrinology*, *6* (9), 1451–1457.

Ritchie, A. I., & Singanayagam, A. (2020). Immunosuppression for hyperinflammation in COVID-19: A double-edged sword? *Lancet*, *395*(10230), 1111.

Rodríguez-Cola, M., et al. (2020). Clinical features of coronavirus disease 2019 (COVID-19) in a cohort of patients with disability due to spinal cord injury. *Spinal Cord Series and Cases*, *6*(1), 39. https://doi.org/10.1038/s41394-020-0288-3.

Ruan, Q., et al. (2020). Clinical predictors of mortality due to COVID-19 based on an analysis of data of 150 patients from Wuhan, China. *Intensive Care Medicine*, *46*(5), 846–848.

Schilero, G., et al. (2005). Assessment of airway caliber and bronchodilator responsiveness in subjects with spinal cord injury. *Chest*, *127*(1), 149–155.

Skelton-Dudley, F., et al. (2019). Spinal cord injury creates unique challenges in diagnosis and Management of Catheter-Associated Urinary Tract Infection. *Topics in spinal cord injury rehabilitation*, *25*(4), 331–339. fall.

Soden, R. J., et al. (2000). Causes of death after spinal cord injury. *Spinal Cord*, *38*(10), 604–610.

Stradner, M. H., et al. (2020). Rheumatic musculoskeletal diseases and COVID-19 a review of the first 6 months of the pandemic. *Frontiers in Medicine*, *7*, 562142.

Taylor, S., et al. (2020). Fear and avoidance of healthcare workers: An important, under-recognized form of stigmatization during the COVID-19 pandemic. *Journal of Anxiety Disorders*, *75*, 102289.

Thietje, R., et al. (2011). Mortality in patients with traumatic spinal cord injury: Descriptive analysis of 62 deceased subjects. *The Journal of Spinal Cord Medicine*, *34*(5), 482–487.

Trotter, R. N., et al. (2007). Transneuronal mapping of the CNS network controlling sympathetic outflow to the rat thymus. *Autonomic Neuroscience : Basic & Clinical*, *131*(1–2), 9–20.

Ueno, M., et al. (2016). Silencing spinal interneurons inhibits immune suppressive autonomic reflexes caused by spinal cord injury. *Nature Neuroscience*, *19*(6), 784–787.

Weaver, L. C., et al. (2002). Central mechanisms for autonomic dysreflexia after spinal cord injury. *Progress in Brain Research*, *137*, 83–95.

Yong, T., et al. (2012). Pulmonary edema and hemorrhage, possible causes of pulmonary infection and respiratory failure in the early stage of lower spinal cord injury. *Medical Hypotheses*, *79*(3), 299–301.

Zha, J., et al. (2014). Chronic thoracic spinal cord injury impairs CD8+ T-cell function by up-regulating programmed cell death-1 expression. *Journal of Neuroinflammation*, *11*, 65.

Zhang, Y., et al. (2013). Autonomic Dysreflexia causes chronic immune suppression after spinal cord injury. *Journal of Neuroscience*, *33*(32), 12970–12981.

Zhou, F., et al. (2020). Clinical course and risk factors for mortality of adult inpatients with COVID-19 in Wuhan, China: A retrospective cohort study. *Lancet*, *395*(10229), 1054–1062.

Zhu, N., et al. (2020). China novel coronavirus investigating and research team. A novel coronavirus from patients with pneumonia in China, 2019. *The New England Journal of Medicine*, *382*(8), 727–733.

Zu, Z. Y., et al. (2020). Coronavirus disease 2019 (COVID-19): A perspective from China. *Radiology*, *296*(2), E15–E25.

Chapter 12

# Sleep problems in spinal cord injury

David J. Berlowitz[a,b] and Marnie Graco[a,b]

[a]*Department of Physiotherapy, The University of Melbourne, Austin Health, Heidelberg, Melbourne, VIC, Australia;* [b]*Institute for Breathing and Sleep, Austin Health, Heidelberg, Melbourne, VIC, Australia*

## List of abbreviations

| | |
|---|---|
| **AASM** | American association of sleep medicine |
| **AHI** | apnea hypopnea index |
| **AQOL** | assessment of quality of life |
| **BMI** | body mass index |
| **COSAQ** | randomized controlled trial of CPAP for OSA in Quadriplegia |
| **CPAP** | continuous positive airway pressure |
| **CSA** | central sleep apnea |
| **EEG** | electroencephalogram |
| **EMG** | electromyogram |
| **EOG** | electrooculogram |
| **OSA** | obstructive sleep apnea |
| **PASAT** | paced auditory serial addition task |
| **$P_{CRIT}$** | critical closing pressure of the upper airway during sleep |
| **PLM** | periodic leg movement |
| **PLMD** | periodic leg movement disorder |
| **RHT** | retinohypothalamic tract |
| **RLS** | restless leg syndrome |
| **SCI** | spinal cord injury |
| **SCN** | suprachiasmatic nuclei |
| **SDB** | sleep-disordered breathing |
| **3T MRI** | 3-Tesla magnetic resonance imaging |

## Introduction

Spinal cord injury (SCI) is a significant global cause of mortality and morbidity. In 2007, the global incidence was estimated to be 23 cases per million per annum, ranging from 15 cases per million in Australia to 29 cases per million in sub-Saharan Africa (Lee, Cripps, Fitzharris, & Wing, 2014).

Most people with SCI sustain their injuries in their second or third decade of life. Because of improvements in both the acute management and long-term supportive care, their life expectancy after injury has increased significantly. As such, secondary diseases and impairments that are a direct consequence of the SCI have significant impacts for many years. Sleep disorders, especially sleep-disordered breathing (SDB), periodic and/or restless leg syndromes (PLM/RLS), and circadian rhythm disorders, are all prevalent disorders after SCI. This chapter focuses on these particular sleep disorders because while insomnia, generalized fatigue, and other sleep-associated disorders affect people living with SCI just as the general population, the increased prevalence and altered pathogenicity due to the spinal lesion per se of SDB, PLMs, and circadian rhythm disorders after SCI deserves special consideration (Sankari, Badr, Martin, Ayas, & Berlowitz, 2019). Population prevalence, known physiological differences in causation, response to therapies, and longer-term consequences of sleep disorders in these sleep disorders after SCI are discussed.

**Diagnosis and Treatment of Spinal Cord Injury. https://doi.org/10.1016/B978-0-12-822498-4.00012-9**

## Sleep-disordered breathing

Sleep-disordered breathing is an umbrella term that includes obstructive sleep apnea (OSA), central sleep apnea (CSA), and sleep-related hypoventilation. OSA is repeated partial (narrowing/hypopnea) or complete (occlusion/apnea) closure and interruption to ventilation at the level of the upper airway during sleep. The drive to breath is generally maintained during OSA, whereas CSA is characterized by periods of diminished or absent ventilatory drive with associated reductions in ventilation. Both OSA and CSA result in periodic oxygen desaturation and fragmentation of sleep typically resulting excessive daytime sleepiness. Hypoventilation refers to periods during which arterial carbon dioxide homeostasis is not maintained because minute ventilation inadequately matches metabolism. The consequence of hypoventilation is a raised arterial carbon dioxide. All three disorders may exist separately or together in the same person and to various degrees over, and indeed within, sleep periods (Guilleminault, Tilkian, & Dement, 1976). The most common risk factors for OSA in the general population are being male and having a raised body mass index (BMI).

There is very little published research into respiratory sleep disorders in paraplegia. While little data reduces our ability to be certain, it is believed that the pathophysiology, risk factors and prevalence of respiratory sleep disorders in those living with paraplegia is not substantially different to the general population. Conversely, tetraplegia has been studied extensively. Despite substantial overall respiratory muscle weakness, (Berlowitz & Tamplin, 2013) tetraplegia is not routinely associated with hypoventilation, but rather with OSA (Berlowitz et al., 2019). The upper airway muscles derive motor innervation from the cranial nerves, predominantly the hypoglossal, rather than from the spinal segmental nerves. Hence, an injury that disrupts, for example, C6 should not have any effect on the genioglossus. Yet while OSA is highly prevalent (Fig. 1), hypoventilation appears rare at a population level. In a recent large, multi-center randomized controlled trial of continuous positive airway pressure (CPAP) for SDB in acute tetraplegia (Berlowitz et al., 2013), only eight of the 212 otherwise eligible participants (3.8%) were excluded due to hypoventilation ($PaCO_2 > 45$ mmHg) (Berlowitz et al., 2019). Hypoventilation has been reported, particularly at sleep onset in physiological studies (Bascom, Sankari, Goshgarian, & Badr, 2015) and in association with a reduced ventilatory reserve for carbon dioxide

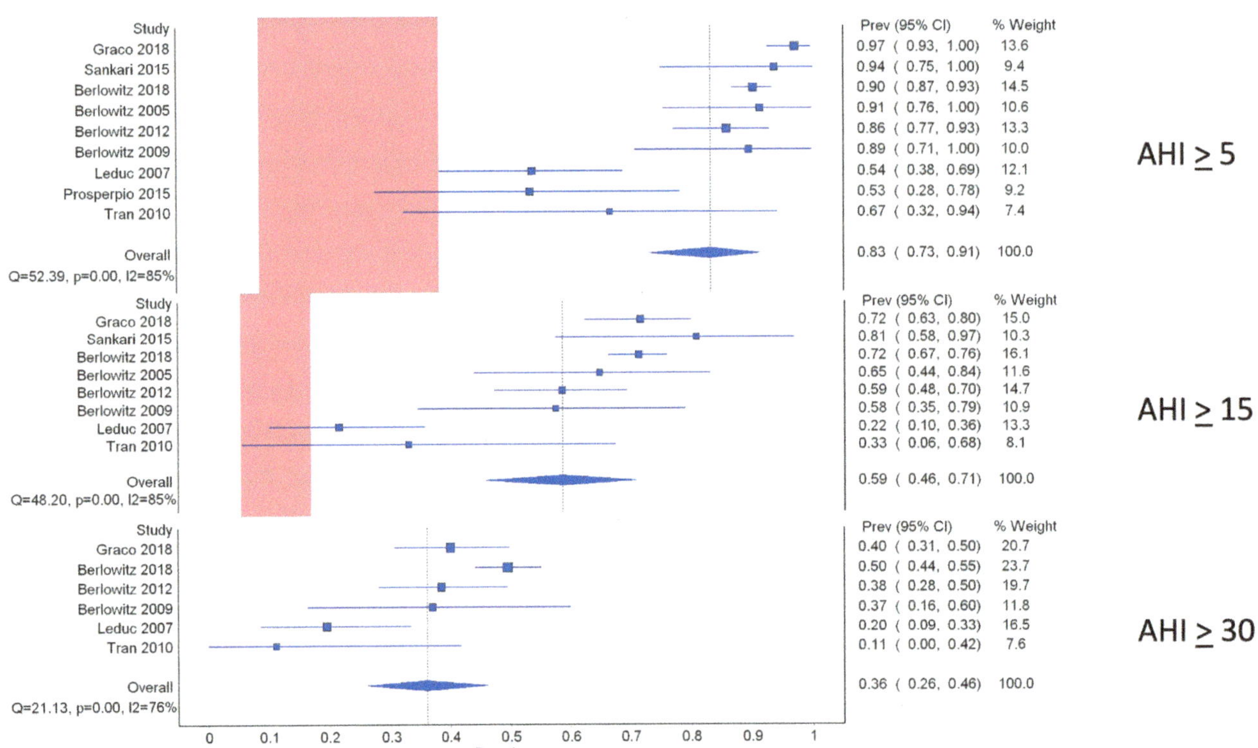


FIG. 1 Prevalence estimates and their relationships to general population data for mild, moderate, and severe sleep-disordered breathing in tetraplegia. Forest plots of individual study data, pooled estimates of prevalence in studies of people with tetraplegia, and ranges of general population estimates at mild (AHI > 5), moderate (AHI > 15), and severe (AHI > 30) levels of SDB severity (Graco, McDonald, Green, Jackson, & Berlowitz, 2021). These graphs illustrate the substantially higher prevalence of sleep-disordered breathing in tetraplegia across all ranges of severity. Solid bands across the mild and moderate data ranges are general population estimates ranges (Senaratna et al., 2017). *AHI* = apnea hypopnea index (measured in events per hour).

(Sankari, Bascom, Chowdhuri, & Badr, 2014), but there is little evidence for increased hypoventilation prevalence across all those living with tetraplegia. Clinically however, if a raised carbon dioxide is detected in a particular patient, this relatively rare sign demands immediate clinical attention.

## Prevalence of sleep-disordered breathing after tetraplegia

The prevalence of disordered breathing during sleep is estimated to be between 9% and 38%, and increases with age (Senaratna et al., 2017). In previous cross-sectional population studies of sleep disorders in patients with tetraplegia, the prevalence of SDB has been estimated to be between 27% and 97% (Giannoccaro et al., 2013; Graco et al., 2018). However, a recent meta-analysis of the published population estimates in tetraplegia presented the pooled estimates at various cut-off severities (Graco et al., 2021). Twelve articles across 20 years were included, nine of which presented data amenable to meta-analysis (combined $n = 630$). The reported SDB prevalence rates from the 12 studies ranged from 46% to 97%. Following meta-analysis the mean prevalence of at least mild (apnea hypopnea index AHI $\geq$ 5 events per hour), moderate (AHI $\geq$ 15), and severe (AHI $\geq$ 30) SDB were 83% (95% CI = 73–91), 59% (46–71), and 36% (26–46), respectively. As illustrated in Fig. 1, the prevalence of SDB after cervical SCI is clearly elevated at all levels of severity compared with general population estimates (Senaratna et al., 2017).

## Sleep-disordered breathing is an acute and direct consequence of cervical SCI

As noted above, the upper airway is innervated directly from the brain via the cranial nerves, not by the spinal segmental nerves. Notwithstanding this, an acute cervical spinal cord injury appears to directly compromise upper airway patency during sleep within weeks of the injurious event. Berlowitz et al. studied all patients presenting to the specialized SCI unit in Melbourne, Australia, over an 18-month period and undertook full, bedside polysomnography to determine when SDB appeared after injury (Berlowitz, Brown, Campbell, & Pierce, 2005). As illustrated in Fig. 2, approximately 10% of the cohorts were at increased risk of the SDB prior to injury as estimated by a prediction algorithm that incorporates signs and symptoms. Within 2 weeks of injury, the measured prevalence of clinically important disease had increased to 60% and peaked at over 80% 3 months after initial injury. Most people with SDB have an insidious onset of the disorder over a number of years as they age and gain weight. In contrast, tetraplegia appears to be the only model of acute SDB in humans. As such, the pathophysiology in acute tetraplegia is likely different from the general population and is not yet fully understood (Sankari et al., 2019).

## Sleep-disordered breathing after tetraplegia has clinically important impacts

Alongside the cardinal SDB signs of nightly intermittent hypoxia, repeated arousals from sleep, and excessive daytime sleepiness, SDB is also associated with increased cardiovascular risk and clinically important degrees of neurocognitive dysfunction. In the general population, numerous reviews have described the deleterious impact of SDB on neuropsychological functioning and demonstrated impairment in the domains of attention, information processing, executive function, memory, and learning (Bucks, Olaithe, & Eastwood, 2013). While SDB in the general population is associated with

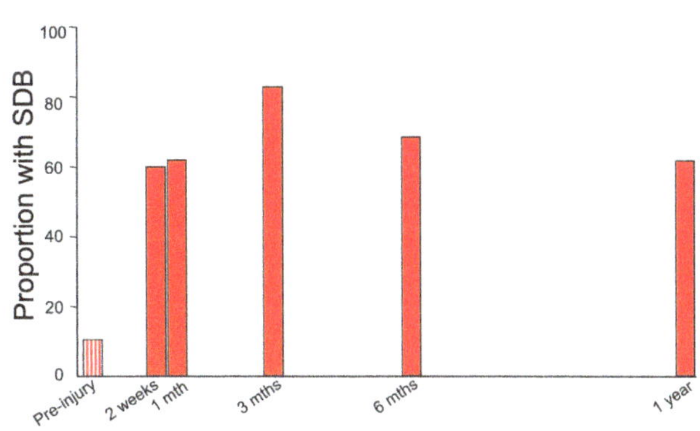


FIG. 2 Proportion of those with sleep-disordered breathing in a prospective, acute cohort study. Solid bar graphs refer to the proportion of the cohort with SDB at each assessment. The pre-injury proportion with an AHI score greater than 10 (vertical striped bar) was estimated using a prediction algorithm that incorporates, age, sex, body mass index and (participant recalled) symptoms to give a likelihood of an AHI > 10 (Berlowitz, Brown, Campbell and Pierce, 2005). *SDB* = sleep-disordered breathing as defined by a respiratory disturbance index (RDI) > 10 events per hour.

hypertension, altered diurnal blood pressure patterns and heart rate variability (HRV), these relationships are not as uniformly observed in those with tetraplegia and SDB. Goh and colleagues demonstrated that tetraplegia is associated with relative nocturnal hypertension and "reverse dipping" or a loss of the usual reduction in blood pressure overnight (Goh et al., 2018). In contrast, Sankari et al. have demonstrated that those with tetraplegia and SDB are also more likely to have diagnosed hypertension and cardiovascular disease (Sankari, Martin, & Badr, 2015). However, Fang et al. were unable to show any difference in the 24-h, night-time, and daytime blood pressure patterns in those with tetraplegia and mild, moderate, severe or no SDB (Fang, Goh, O'callaghan, & Berlowitz, 2018).

SDB has been associated with worse quality of life in tetraplegia. A study investigating the relationships between quality of life and sleep disorders in chronic tetraplegia measured both health-related quality of life (Assessment of Quality of Life) and SDB severity with a full, portable home-based sleep study (Berlowitz, Spong, Gordon, Howard, & Brown, 2012; Spong, Graco, Brown, Schembri, & Berlowitz, 2015). As illustrated in Fig. 3, quality of life and health utility scores were worse in the group with complete lesions compared with incomplete lesions. This single co-morbidity reduces health-utility by an approximately five times the minimally clinically important difference. As such, any effective treatment is likely to have an important impact on those living with both tetraplegia and SDB.

In patients living with chronic SCI, Sajkov et al. (1998) published neuropsychological evaluation data that were performed concurrently with a diagnostic sleep study in 40 people who had been living with tetraplegia for many years (McEvoy et al., 1995). The neuropsychological functions most affected were verbal attention and concentration, immediate and short-term memory, cognitive flexibility, internal scanning and working memory. These impairments are similar to those observed in the able bodied with OSA, which suggested that even if the pathogenesis of SDB is different immediately after SCI, the cognitive consequences of living with SDB for decades appear similar.

As noted earlier, our group recently completed a large, multi-center randomized controlled trial of CPAP for SDB in acute tetraplegia (Berlowitz et al., 2013; Berlowitz et al., 2019). Baseline data from that study was utilized to investigate the relationship between apnea severity and neuropsychological function in patients with acute-onset tetraplegia and SDB dysfunction (Schembri, Spong, Graco, Berlowitz, & Team, 2017). Study enrolment analyses showed more severe sleep apnea was significantly associated with poorer attention, information processing, and immediate recall. Higher pre-injury intelligence and being younger reduced the associations with sleep-disordered breathing; however, these protective factors were insufficient to counter the damage to attention, immediate recall, and information processing associated with the sleep-disordered breathing. As illustrated in Table 1, these deficits in attention and information processing, as measured by the PASAT, were clinically important, being well within the range of values reported in moderate to severe isolated head

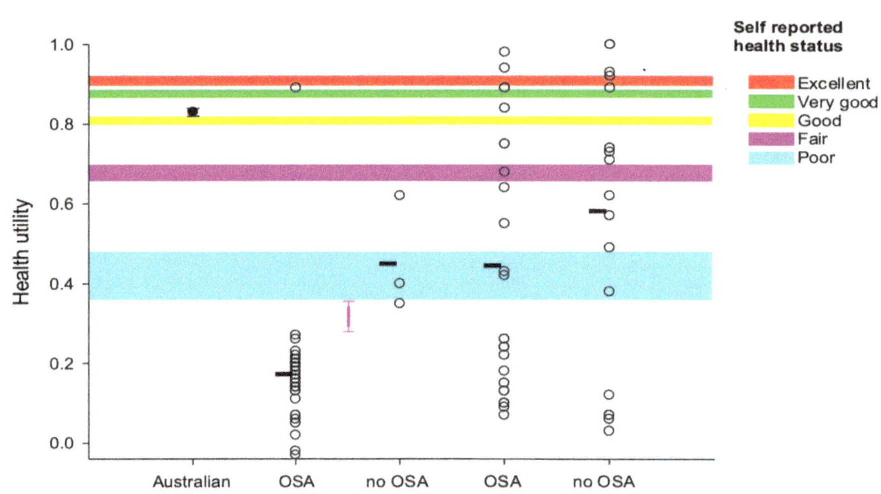


FIG. 3 Tetraplegia severity and obstructive sleep apnea independently modify health status. Colored bands represent the Australian population averages for categories of self-reported health status. Health utility ranges from one (self-rated "perfect health"), through zero (self-rated health state equivalent to death), and down to a health status of −0.04 (self-reported health status that is perceived to be "worse than death"). The Australian population average (95% confidence interval) score is located between "good" and "very good" (Hawthorne & Osbourne, 2005). The pink vertical bar represents the minimal clinically important difference in health status (0.06) and the associated confidence interval of this difference. People with tetraplegia are categorized by the presence or not of sleep apnea and of motor and sensory complete spinal injury (ASIA impairment scale A vs B, C, or D). Open circles represent individual participant scores, and the horizontal bars represent group mean values (Berlowitz et al., 2012). *OSA* = obstructive sleep apnea as defined by an apnea hypopnea index (AHI) > 10 events per hour.

**TABLE 1 Attention and Information processing performance as measured by the PASAT performance in acute tetraplegia compared with other patient groups.**

| Group | PASAT score mean or range of reported means |
|---|---|
| Healthy controls | 56–74 |
| Chronic fatigue syndrome | 62 |
| Multiple Sclerosis | 45–54 |
| Mild traumatic brain injury | 39–57 |
| Moderate to severe traumatic brain injury | 19–51 |
| *Mild-moderate SDB in acute tetraplegia* | *48* |
| *Severe SDB in acute tetraplegia* | *33* |
| *Overall baseline COSAQ* | *41* |
| *Overall end study COSAQ* | *49* |

Scores are represented as the percentage of correct responses, and thus higher percentages indicate better performance. Data in italics for SDB in acute tetraplegia are taken from the CPAP for OSA in Quadriplegia (COSAQ) trial (Berlowitz et al., 2013; Berlowitz et al., 2019; Schembri et al., 2017). Comparison values are from a review of the range of values for the PASAT across a number of research study samples (Tombaugh, 2006).

injury (Tombaugh, 2006). Attention and information processing as measured on the PASAT naturally decline as we age, but the reduction seen in those with severe SDB compared with milder SDB was equivalent to an additional 31 years of ageing.

## Why do people with tetraplegia have sleep-disordered breathing, predominantly obstructive sleep apnea?

It is apparent that SDB in SCI is not simply due to peri-accident neuroanatomical compromise; however, the exact etiology remains unknown. Obesity, as measured by increased weight, body mass index (BMI), waist, abdominal, and neck girth, is associated with OSA in both SCI and the general population (Berlowitz et al., 2012; Graco et al., 2021). Obesity results in upper airway narrowing in both lateral pharyngeal tissues and the tongue in the general population, as well as those with SCI (Kim et al., 2014; O'donoghue et al., 2018). Truncal obesity reduces lung volume and distal tethering of the trachea which in turn leads to decreased upper airway caliber during sleep (Owens, Malhotra, Eckert, White, & Jordan, 2009). In chronic SCI, several authors have observed associations between OSA prevalence and increasing age, BMI and neck circumference, (Berlowitz et al., 2012; Burns, Kapur, Yin, & Buhrer, 2001; McEvoy et al., 1995; Stockhammer et al., 2002) but these relationships appear weaker in the immediate period after injury (Berlowitz et al., 2005).

Many people living with SCI are prescribed medications (sedatives, muscle relaxants, and narcotics) that can affect breathing, especially during sleep. Loud snoring (a typical OSA sign) has been associated with the use of anti-spasticity medications and obesity in SCI in one cross-sectional study (Ayas et al., 2001). There have been no controlled studies in SCI of the effect of medication provision or withdrawal on SDB severity; however, a smaller proportion of those in the intervention (CPAP treatment for SDB) arm of the COSAQ study were prescribed baclofen at end-study (Berlowitz et al., 2019), and a non-significant reduction in baclofen prescription rates was observed in those whose SDB improved over time in another cohort trial (Berlowitz et al., 2005).

People with tetraplegia and those with lesions above T6, also develop increased nasal and total upper airway resistance due to increased parasympathetic tone. This increased nasal resistance may contribute to upper airway narrowing by increasing negative (collapsing) pressure in the upper airway during inspiration (Gainche et al., 2016; Wijesuriya et al., 2019). While this increased upper airway resistance can be acutely reduced with topical sympathomimetic application, no difference in SDB was observed in a pilot, cross-over randomized controlled trial of topical sympathomimetic application (phenylephrine) in people with tetraplegia and SDB (Wijesuriya et al., 2019).

Our research group in Melbourne, Australia and colleagues in Detroit, USA have undertaken the majority of these physiological sleep studies and both centers have used comparable physiological measures ensuring that our data are complementary. By comparing people with tetraplegia and SDB with control participants (no SCI) that had SDB, a number of putative physiological factors have been examined. The reflex response of genioglossus to negative upper airway pressure was reduced and delayed in people with both OSA and SCI compared to those with OSA but without SCI (Wijesuriya et al.,

2018). This reflex is an important protective response of the upper airway and a reduction predisposes the upper airway to collapse during sleep. The critical closing pressure of the upper airway ($P_{CRIT}$) is determined by first stabilizing the upper airway during sleep with small amounts of continuous positive airway pressure and then reducing that external pressure until the upper airway collapses. Sankari et al. demonstrated that the $P_{CRIT}$ in tetraplegia and SDB is not different to non-disabled people with SDB (Sankari, Bascom, & Badr, 2014). Data examining arousal responses and muscle recruitment remain equivocal.

The upper airway is a collapsible tube where patency is largely dependent on the surrounding soft tissue, phasic and tonic muscle tone, fat pad volume, and extraluminal tissue pressure from blood, lymph, and extracellular water. A series of magnetic resonance imaging studies in people with SDB and tetraplegia have compared the structure and function of the upper airway with non-disabled controls and observed that while SCI OSA patients have heterogeneous pharyngeal dilator muscle responses to the negative pressures occurring during inspiration, as a group they are more similar to able bodied OSA patients than healthy controls of similar age and BMI (Hatt et al., 2020; O'donoghue et al., 2018).

## Clinical management of sleep-disordered breathing in people with SCI

Despite the high prevalence of SDB and its detrimental effects on health and quality of life, surveys indicate that less than 25% of people with SCI are diagnosed and treated for the disorder (Burns et al., 2001; Sankari et al., 2015). The "gold-standard" method for diagnosing SDB is a Level I, overnight, attended polysomnography (PSG). Clinical practice guidelines recommend PSG for people with SCI and signs and symptoms of SDB (Consortium for Spinal Cord Medicine, 2005). However, access to this test, which requires an overnight stay in a sleep laboratory, is often poor. At least two studies have investigated the use of limited channel studies to detect SDB in SCI with results demonstrating moderate accuracy of this less intensive method (Bauman et al., 2016; Graco et al., 2018). These limited channel sleep studies may provide alternatives to PSG, particularly when access to sleep specialist services is poor.

Treatments for SDB in SCI are essentially the same as for the general population. Continuous positive airway pressure (CPAP) is the first-line treatment for SDB. CPAP provides a continuous positive airway pressure of typically between 4 and 20 mmHg, via a mask, to prevent upper airway collapse during sleep. Our COSAQ study, described earlier, is the first and only controlled trial to investigate the effects of CPAP in people with SCI. It demonstrated that CPAP effectively improved daytime sleepiness in people with acute tetraplegia, although there was no effect of CPAP on neurocognition (Berlowitz et al., 2019). CPAP effectiveness is limited by poor adherence and acceptance of the therapy in both SCI and the general population. When defined as an average of at least 4 h per night, CPAP adherence in people with tetraplegia and SDB has been estimated at approximately 25% (Graco et al., 2019; Graco et al., 2019). A qualitative study seeking to understand the barriers and enablers to CPAP use among people with tetraplegia found that both the burdens and the benefits of using CPAP were substantial, and patients actively weighed these burdens and benefits to decide whether to continue with the therapy (Graco, Green, et al., 2019).

Other potential treatments for SDB in SCI include bi-level positive airway pressure (PAP), mandibular advancement splints (MAS) and positional therapy. Bi-level PAP is a more expensive PAP therapy than CPAP, which delivers a higher inspiratory than expiratory pressure and is commonly used for hypoventilation disorders. Some clinicians prefer to use it to treat SDB in patients with more severe SCI (i.e., high tetraplegia) and greater respiratory compromise, and in those with a predominance of central than obstructive events. However, there is no research evidence to support the use of bi-level PAP over CPAP for treating SDB in SCI, and vice-versa. MAS are commonly prescribed in the general population to those who are unable to tolerate CPAP or as a first-line therapy to treat mild OSA. The mouthguard-like device is worn on the upper and lower jaw and is designed to pull the lower jaw forwards to open the airway while simultaneously increasing tension in the soft tissues and preventing collapse. Research has demonstrated that MAS devices can effectively improve the symptoms of OSA in the general population (Iftikhar et al., 2017). As yet, there have been no published studies investigating MAS in SCI.

Recent qualitative research aiming to describe and understand the management of SDB in SCI rehabilitation centers has identified that the management of SDB is highly varied in SCI (Graco, Berlowitz, & Green, 2019). The most common management pathway for people with SCI suspected of having SDB involves referral from the primary care or rehabilitation doctor to a specialist sleep/respiratory physician for investigation and management (Graco, Berlowitz, & Green, 2019). However, this care model often presents significant access barriers to people with tetraplegia and may contribute to the low rates of detection and treatment seen in this population (Graco, Green, et al., 2019). The same study identified three spinal cord injury (SCI) rehabilitation centers that had developed an "in-house" SDB management model in response to poor access to specialist sleep services. In each center, a multi-disciplinary team provides screening, diagnosis and treatment of uncomplicated SDB using portable equipment and without direct specialist sleep service involvement. This demonstrates that it is feasible for multi-disciplinary SCI rehabilitation teams to independently manage un-complicated SDB Graco, Berlowitz, & Green, 2019) and further research investigating the safety and effectiveness of SDB management provided by the rehabilitation team is urgently needed.

## How does SCI alter the circadian rhythm?

Following a complete cervical SCI, afferent somatic and autonomic fiber pathways below the SCI are disrupted and the efferent sympathetic innervation of the pineal gland via the superior cervical ganglion is interrupted. This anatomical disruption alters the circadian rhythmicity of melatonin regulation and production in tetraplegia and those with paraplegia lesions associated with autonomic dysfunction, typically T6 and higher. Under normal conditions, light striking the retina stimulates afferent pathways through the suprachiasmatic nuclei (SCN) that in turn pass to the superior cervical ganglion and onward to the pineal gland to inhibit melatonin production. Darkness removes this inhibition, and melatonin is produced. The fibers from the SCN travel alongside those from the sympathetic system to exit the spine between the first and sixth thoracic levels. The SCN to pineal pathway then ascends *outside* the spinal canal to the superior cervical ganglion, back into the cortex, and on to innervate the pineal gland. This circuitous pathway exposes melatonin regulation to disruption after SCI (Fig. 4).

In 2000, Zeitzer et al. demonstrated that complete tetraplegia is associated with near abolition of circadian melatonin rhythmicity and a markedly reduction in circulating levels of the hormone (Zeitzer, Ayas, Shea, Brown, & Czeisler, 2000). The same authors also demonstrated that this dysrhythmia was specific to melatonin and did not extend to other circadian modified rhythms such as cortisol or thyroid stimulating hormone. Despite these clear abnormalities of circadian control in SCI and the ready availability of melatonin as a therapeutic, surprisingly little research has explored the clinical utility of exogenous melatonin after SCI. A small pilot study (Spong, Kennedy, Brown, Armstrong, & Berlowitz, 2013) and two small randomized controlled, cross-over trials have demonstrated that exogenous melatonin administration in tetraplegia is safe and appears to be associated with some subjective sleep improvement (Spong et al., 2013; Zeitzer, Ku, Ota, & Kiratli, 2014). Exogenous melatonin administration in tetraplegia has also been shown to normalize clock gene expression in

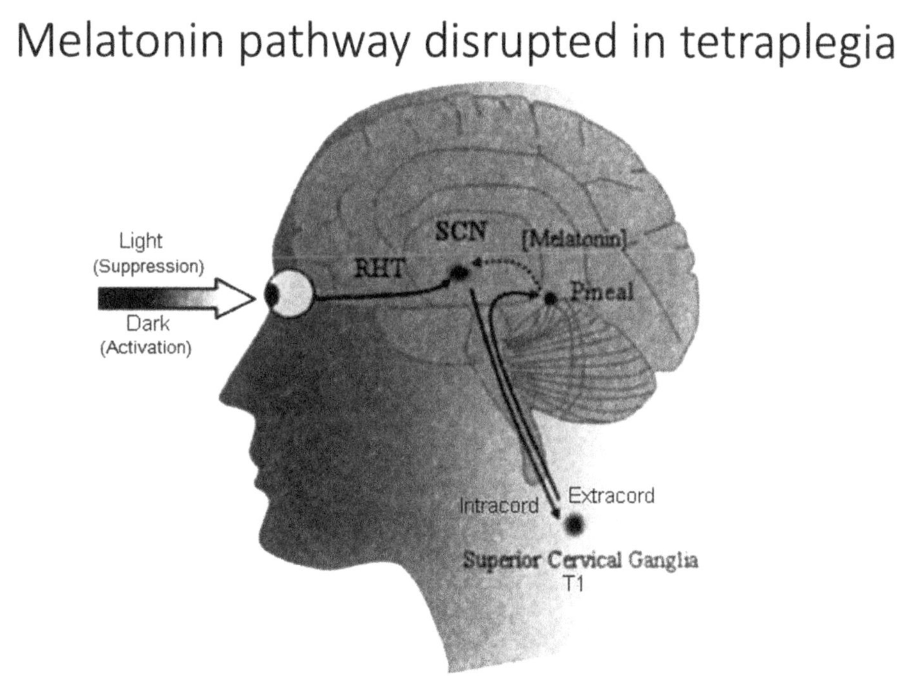


FIG. 4 Tetraplegia-associated interruption of Melatonin secretion. Light striking the retina stimulates the intrinsically photosensitive retinal ganglion cells which project monosynaptically, to the SCN via the optic nerve. The pathway continues to the superior cervical ganglion and onward to the pineal gland to inhibit melatonin production. Darkness removes this inhibition and thus melatonin is produced. The fibers from the SCN exit the spine between the first and sixth thoracic levels. The SCN to Pineal pathway then ascends outside the spinal canal to the superior cervical ganglion, back into the cortex, and on to innervate the pineal gland. A spinal cord injury above the first thoracic level will likely interrupt this pathway. *RHT* = retinohypothalamic tract. *SCN* = suprachiasmatic nuclei.

peripheral blood (Kostovski et al., 2015; Kostovski et al., 2018); however, the clinical significance of these findings is unclear.

## Periodic leg movements of sleep in SCI

Periodic leg movements (PLM) are characterized by periodic episodes of repetitive and highly stereotyped limb movements during sleep. The associated restless leg syndrome (RLS) is characterized by uncomfortable leg sensations usually prior to sleep onset that causes an almost irresistible urge to move. Both conditions can result in considerable disruption to sleep quality and are typically associated with excessive daytime sleepiness. Up to 80% of RLS patients also experience PLMS (periodic leg movement syndrome), as defined by five or more PLM events per hour of sleep (Ferri et al., 2015). Periodic leg movement disorder (PLMD) appears to be another distinct sleep disorder that is more common in people living with SCI. The high prevalence of SDB in SCI, particularly in tetraplegia, confounds studies of PLMs because the AASM PLM scoring rules exclude events that occur from 0.5 s prior to or following a scored respiratory event. As such, co-existent SDB would theoretically tend to underestimate the severity of PLM, yet studies examining both SDB and PLM have consistently found an elevated rate of both. For example, Peters et al. reviewed 262 clinical and research sleep studies in people with tetraplegia and reported a SDB prevalence of 88% and a PLMD prevalence of 59% (PLMs per hour >15) (Peters et al., 2018).

While the precise etiology of PLMs in the general population is unclear, evidence of substantia nigra involvement and the clinical effectiveness of dopaminergic agents, has made them the mainstay of therapy and the belief that PLMs are centrally generated is the mainstream view. PLMs are predominantly observed in non-rapid eye movement (NREM) sleep in the general population whereas after SCI PLMs are apparent throughout sleep, including during rapid eye movement (REM) sleep. Indeed, PLMs have even been observed during wake after SCI (Peters et al., 2018). Interventions which alter cortical arousal in SCI do not necessarily result in a reduction in PLMs and similarly, inhibition of PLMs does not reduce sleep disruption (Manconi et al., 2012; Salminen et al., 2013). Taken together, these findings suggest that PLMs after SCI arise peripherally rather than centrally. Regardless of whether PLMs are a central and/or peripheral phenomenon in SCI, dopaminergic agents are still effective. A clinical case series of people with SCI referred for refractory "spasticity" despite maximal anti-spasticity therapy recently demonstrated that most in fact had untreated PLMs (Levy et al., 2018). Sleep studies demonstrated a substantial reduction in PLMs after pramipexole therapy.

## Applications to other areas of neuroscience

In this chapter, we reviewed the prevalent sleep disorders in SCI, sleep-disordered breathing, circadian rhythm dysfunction, and periodic leg movements and examined the barriers to effective diagnosis and treatment. While all three disorders are well described in the non-injured population, the pathogenesis of these after SCI provides unique insights into the underlying neuroscience. It appears that tetraplegia is the only model of acute OSA in humans. As such, the comparative upper airway physiology experiments we detailed extend the phenotyping approach to understanding the individual personal determinants of the "end-organ dysfunction" we describe as OSA. In particular, differences in ventilatory control and afferent reflex activation in tetraplegia provide novel opportunities for exploration of neuropathophysiology. Similar to OSA, tetraplegia provides a unique model of disruption to the hypothalamic/melatonin/pineal axis of circadian control. Experiments such as those involving exogenous melatonin administration and light therapy with subsequent assaying of melatonin metabolism and peripheral clock gene expression will undoubtedly provide novel insights into neurohumoral circadian control, including in the non-injured. The periodic leg movement disorder had been characterized as a central dysfunction, with much of therapy focused on cortical dopaminergic augmentation. The high prevalence of PLMD after SCI refutes the "central-only" hypothesis and perhaps similarly to the OSA data, point to PLMD as another form of "end-organ dysfunction" with multiple phenotypic pathways to expression. SCI perhaps reveals pathogenic pathways not readily observable in the non-injured. Lastly, the examination of barriers to symptom recognition, diagnosis, and treatment of sleep disorders in SCI highlights issues that are likely common to other neurological and neuromuscular disorders such as stroke, multiple sclerosis, etc.

## Mini-dictionary of terms

Periodic leg movements (PLM). These are periodic episodes of repetitive and highly stereotyped limb movements during sleep.

Restless leg syndrome (RLS). The RLS is associated with PLM and is characterized by uncomfortable leg sensations usually prior to sleep onset that causes an almost irresistible urge to move. Both PLM and RLS can co-exist in the same person and result in considerable disruption to sleep quality and subsequent excessive daytime sleepiness.

Circadian rhythm. The circadian rhythm is the natural, internal process that regulates the sleep–wake cycle, is regulated by the circadian clock and repeats roughly every 24 h.

Sleep-disordered breathing (SDB). SDB is an umbrella term representing a group of sleep disorders involving the respiratory system and characterized by repetitive periods of total or partial cessation of airflow during sleep.

Obstructive sleep apnea (OSA). OSA is a sub-category of SDB, characterized by repetitive periods of complete or partial obstruction of the upper airway during sleep resulting in complete or reduced airflow.

Central sleep apnea (CSA). CSA is another sub-category of SDB, characterized by the absence of respiratory effort at the time of complete/partial reduction in airflow.

## Key facts of sleep

- Sleep is composed of four distinct stages. During normal sleep, a person cycles through these stages four to six times.
- We do most of our dreaming in REM (Rapid Eye Movement) sleep, which is Stage 4 sleep. REM sleep makes up between 20% and 25% of total sleep in healthy adults.
- During REM sleep all skeletal muscles, apart from the diaphragm and the ocular muscles, are essentially paralyzed.
- Teenagers move their circadian phase later in the day and need almost as much sleep as toddlers.
- Body temperature during sleep drops by half to one degree during sleep before rising again as we wake.
- The most common sleep disorders in humans are insomnia, sleep apnea, restless less syndrome and narcolepsy. It is estimated that 20%–40% of adults have a sleep disorder.

## Summary points

- Acute cervical spinal cord injury results in sleep-disordered breathing.
- Sleep-disordered breathing in tetraplegia is predominately obstructive sleep apnea rather than central sleep apnea or hypoventilation.
- Acutely, obstructive sleep apnea impairs attention and information processing which improves over time; CPAP does not change recovery in neurocognitive function, but it improves sleepiness.
- Sleep-disordered breathing in tetraplegia is highly prevalent (83% mild, 59% moderate, and 36% severe).
- Sleep-disordered breathing in tetraplegia reduces health-related quality of life by approximately 30%.
- Obstructive sleep apnea in tetraplegia is not due to an upper airway that is smaller, more collapsible, or moves abnormally compared.
- The upper airway has a higher resistance in people with tetraplegia and OSA and that can be reduced, but larger studies are needed and phenylepherine is likely not to be the agent of choice.
- Melatonin supplementation may have subjective benefits.
- Translatable, novel treatment models of diagnosis and care for sleep disorders after SCI must be developed and tested for non-inferiority of person-centered outcomes.

## References

Ayas, N. T., Epstein, L. J., Lieberman, S. L., Tun, C. G., Larkin, E. K., Brown, R., et al. (2001). Predictors of loud snoring in persons with spinal cord injury. *The Journal of Spinal Cord Medicine*, *24*, 30–34.

Bascom, A. T., Sankari, A., Goshgarian, H. G., & Badr, M. S. (2015). Sleep onset hypoventilation in chronic spinal cord injury. *Physiological Reports*, *3*.

Bauman, K. A., Kurili, A., Schotland, H. M., Rodriguez, G. M., Chiodo, A. E., & Sitrin, R. G. (2016). Simplified approach to diagnosing sleep-disordered breathing and nocturnal hypercapnia in individuals with spinal cord Injury. *Archives of Physical Medicine and Rehabilitation*, *97*, 363–371.

Berlowitz, D. J., Ayas, N., Barnes, M., Brown, D. J., Cistulli, P. A., Geraghty, T., et al. (2013). Auto-titrating continuous positive airway pressure treatment for obstructive sleep apnoea after acute quadriplegia (COSAQ): Study protocol for a randomized controlled trial. *Trials*, *14*, 181.

Berlowitz, D. J., Brown, D. J., Campbell, D. A., & Pierce, R. J. (2005). A longitudinal evaluation of sleep and breathing in the first year after cervical spinal cord injury. *Archives of Physical Medicine and Rehabilitation*, *86*, 1193–1199.

Berlowitz, D. J., Schembri, R., Graco, M., Ross, J. M., Ayas, N., Gordon, I., et al. (2019). Positive airway pressure for sleep-disordered breathing in acute quadriplegia: A randomised controlled trial. *Thorax*, *74*, 282–290.

Berlowitz, D. J., Spong, J., Gordon, I., Howard, M. E., & Brown, D. J. (2012). Relationships between objective sleep indices and symptoms in a community sample of people with tetraplegia. *Archives of Physical Medicine and Rehabilitation*, *93*, 1246–1252.

Berlowitz, D. J., & Tamplin, J. (2013). Respiratory muscle training for cervical spinal cord injury. *Cochrane Database of Systematic Reviews, 7*, CD008507.

Bucks, R. S., Olaithe, M., & Eastwood, P. (2013). Neurocognitive function in obstructive sleep apnoea: A meta-review. *Respirology, 18*, 61–70.

Burns, S. P., Kapur, V., Yin, K. S., & Buhrer, R. (2001). Factors associated with sleep apnea in men with spinal cord injury: A population-based case-control study. *Spinal Cord, 39*, 15–22.

Consortium for Spinal Cord Medicine. (2005). Respiratory management following spinal cord injury: A clinical practice guideline for health-care professionals. *The Journal of Spinal Cord Medicine, 28*, 259–293.

Fang, X., Goh, M. Y., O'callaghan, C., & Berlowitz, D. (2018). Relationship between autonomic cardiovascular control and obstructive sleep apnoea in persons with spinal cord injury: A retrospective study. *Spinal Cord Series And Cases, 4*, 29.

Ferri, R., Proserpio, P., Rundo, F., Lanza, A., Sambusida, K., Redaelli, T., et al. (2015). Neurophysiological correlates of sleep leg movements in acute spinal cord injury. *Clinical Neurophysiology, 126*, 333–338.

Gainche, L., Berlowitz, D. J., Leguen, M., Ruehland, W. R., O'donoghue, F. J., Trinder, J., et al. (2016). Nasal resistance is elevated in people with tetraplegia and is reduced by topical sympathomimetic administration. *Journal of Clinical Sleep Medicine, 12*, 1487–1492.

Giannoccaro, M. P., Moghadam, K. K., Pizza, F., Boriani, S., Maraldi, N. M., Avoni, P., et al. (2013). Sleep disorders in patients with spinal cord injury. *Sleep Medicine Reviews, 17*, 399–409.

Goh, M. Y., Millard, M. S., Wong, E. C. K., Berlowitz, D. J., Graco, M., Schembri, R. M., et al. (2018). Comparison of diurnal blood pressure and urine production between people with and without chronic spinal cord injury. *Spinal Cord, 56*, 847–855.

Graco, M., Berlowitz, D. J., & Green, S. E. (2019). Understanding the clinical management of obstructive sleep apnoea in tetraplegia: A qualitative study using the theoretical domains framework. *BMC Health Services Research, 19*(1), 405.

Graco, M., Green, S. E., Tolson, J., Stevens, B., Barnes, M., Rigoni, A., et al. (2019). Worth the effort? Weighing up the benefit and burden of continuous positive airway pressure therapy for the treatment of obstructive sleep apnoea in chronic tetraplegia. *Spinal Cord, 57*, 247–254.

Graco, M., McDonald, L., Green, S. E., Jackson, M. L., & Berlowitz, D. J. (2021). Prevalence of sleep-disordered breathing in people with tetraplegia-a systematic review and meta-analysis. *Spinal Cord, 59*, 474–484.

Graco, M., Schembri, R., Cross, S., Thiyagarajan, C., Shafazand, S., Ayas, N. T., et al. (2018). Diagnostic accuracy of a two-stage model for detecting obstructive sleep apnoea in chronic tetraplegia. *Thorax, 73*, 864–871.

Graco, M., Schembri, R., Ross, J., Green, S. E., Booker, L., Cistulli, P. A., et al. (2019). Continuous positive airway pressure use for obstructive sleep apnea in acute, traumatic tetraplegia. *Archives of Physical Medicine and Rehabilitation, 100*, 2276–2282.

Guilleminault, C., Tilkian, A., & Dement, W. (1976). The sleep apnoea syndromes. *Annual Reviews in Medicine, 27*, 465–484.

Hatt, A., Brown, E., Berlowitz, D. J., O'donoghue, F., Meaklim, H., Connelly, A., et al. (2020). Tetraplegic obstructive sleep apnoea patients dilate the airway similarly to able-bodied obstructive sleep apnoea patients. *The Journal of Spinal Cord Medicine*, 1–11.

Hawthorne, G., & Osbourne, R. (2005). Population norms and meaningful differences for the Assessment of Quality of Life (AQoL) measure. *Australian and New Zealand Journal of Public Health, 29*, 136–142.

Iftikhar, I. H., Bittencourt, L., Youngstedt, S. D., Ayas, N., Cistulli, P., Schwab, R., et al. (2017). Comparative efficacy of Cpap, MADs, exercise-training, and dietary weight loss for sleep apnea: A network meta-analysis. *Sleep Medicine, 30*, 7–14.

Kim, A. M., Keenan, B. T., Jackson, N., Chan, E. L., Staley, B., Poptani, H., et al. (2014). Tongue fat and its relationship to obstructive sleep apnea. *Sleep, 37*, 1639–1648.

Kostovski, E., Dahm, A. E., Mowinckel, M. C., Stranda, A., Skretting, G., Osterud, B., et al. (2015). Circadian rhythms of hemostatic factors in tetraplegia: A double-blind, randomized, placebo-controlled cross-over study of melatonin. *Spinal Cord, 53*, 285–290.

Kostovski, E., Frigato, E., Savikj, M., Dahm, A., Sandset, P. M., Mowinckel, M. C., et al. (2018). Normalization of disrupted clock gene expression in males with tetraplegia: A crossover randomized placebo-controlled trial of melatonin supplementation. *Spinal Cord, 56*, 1076–1083.

Lee, B. B., Cripps, R. A., Fitzharris, M., & Wing, P. C. (2014). The global map for traumatic spinal cord injury epidemiology: Update 2011, global incidence rate. *Spinal Cord, 52*, 110–116.

Levy, J., Hartley, S., Mauruc-Soubirac, E., Leotard, A., Lofaso, F., Quera-Salva, M. A., et al. (2018). Spasticity or periodic limb movements? *European Journal of Physical and Rehabilitation Medicine, 54*, 698–704.

Manconi, M., Ferri, R., Zucconi, M., Bassetti, C. L., Fulda, S., Aricò, D., et al. (2012). Dissociation of periodic leg movements from arousals in restless legs syndrome. *Annals of Neurology, 71*, 834–844.

McEvoy, R. D., Mykytyn, I., Sajkov, D., Flavell, H., Marshall, R., Antic, R., et al. (1995). Sleep apnoea in patients with quadriplegia. *Thorax, 50*, 613–619.

O'donoghue, F. J., Meaklim, H., Bilston, L., Hatt, A., Connelly, A., Jackson, G., et al. (2018). Magnetic resonance imaging of the upper airway in patients with quadriplegia and obstructive sleep apnea. *Journal of Sleep Research, 27*, e12616.

Owens, R. L., Malhotra, A., Eckert, D. J., White, D. P., & Jordan, A. S. (2009). The influence of end-expiratory lung volume on measurements of pharyngeal collapsibility. *Journal of Applied Physiology, 00755*, 2009.

Peters, A. E. J., Van Silfhout, L., Graco, M., Schembri, R., Thijssen, D., & Berlowitz, D. J. (2018). Periodic limb movements in tetraplegia. *The Journal of Spinal Cord Medicine, 41*, 318–325.

Sajkov, D., Marshall, R., Walker, P., Mykytyn, I., Mcevoy, R. D., Wale, J., et al. (1998). Sleep apnoea related hypoxia is associated with cognitive disturbances in patients with tetraplegia. *Spinal Cord, 36*, 231–239.

Salminen, A. V., Manconi, M., Rimpilä, V., Luoto, T. M., Koskinen, E., Ferri, R., et al. (2013). Disconnection between periodic leg movements and cortical arousals in spinal cord Injury. *Journal of Clinical Sleep Medicine : Jcsm : Official Publication of the American Academy of Sleep Medicine, 9*, 1207–1209.

Sankari, A., Badr, M. S., Martin, J. L., Ayas, N. T., & Berlowitz, D. J. (2019). Impact of spinal cord Injury on sleep: Current perspectives. *Nature and Science of Sleep*, *11*, 219–229.

Sankari, A., Bascom, A. T., & Badr, M. S. (2014). Upper airway mechanics in chronic spinal cord injury during sleep. *Journal of Applied Physiology (1985)*, *116*, 1390–1395.

Sankari, A., Bascom, A. T., Chowdhuri, S., & Badr, M. S. (2014). Tetraplegia is a risk factor for central sleep apnea. *Journal of Applied Physiology (1985)*, *116*, 345–353.

Sankari, A., Martin, J. L., & Badr, M. (2015). A retrospective review of sleep-disordered breathing, hypertenstion and cardiovascular diseases in spinal cord injury patients. *Spinal Cord*, *53*, 496–497.

Schembri, R., Spong, J., Graco, M., Berlowitz, D. J., & Team, C. S. (2017). Neuropsychological function in patients with acute tetraplegia and sleep disordered breathing. *Sleep*, *40*.

Senaratna, C. V., Perret, J. L., Lodge, C. J., Lowe, A. J., Campbell, B. E., Matheson, M. C., et al. (2017). Prevalence of obstructive sleep apnea in the general population: A systematic review. *Sleep Medicine Reviews*, *34*, 70–81.

Spong, J., Graco, M., Brown, D. J., Schembri, R., & Berlowitz, D. J. (2015). Subjective sleep disturbances and quality of life in chronic tetraplegia. *Spinal Cord*, *53*, 636–640.

Spong, J., Kennedy, G. A., Brown, D. J., Armstrong, S. M., & Berlowitz, D. J. (2013). Melatonin supplementation in patients with complete tetraplegia and poor sleep. *Sleep Disorders*, *2013*, 8.

Stockhammer, E., Tobon, A., Michel, F., Eser, P., Scheuler, W., Bauer, W., et al. (2002). Characteristics of sleep apnea syndrome in tetraplegic patients. *Spinal Cord*, *40*, 286–294.

Tombaugh, T. N. (2006). A comprehensive review of the paced auditory serial addition test (PASAT). *Archives of Clinical Neuropsychology*, *21*, 53–76.

Wijesuriya, N. S., Eckert, D. J., Jordan, A. S., Schembri, R., Lewis, C., Meaklim, H., et al. (2019). A randomised controlled trial of nasal decongestant to treat obstructive sleep apnoea in people with cervical spinal cord injury. *Spinal Cord*, *57*, 579–585.

Wijesuriya, N. S., Gainche, L., Jordan, A. S., Berlowitz, D. J., Leguen, M., Rochford, P. D., et al. (2018). Genioglossus reflex responses to negative upper airway pressure are altered in people with tetraplegia and obstructive sleep apnoea. *The Journal of Physiology*, *596*, 2853–2864.

Zeitzer, J. M., Ayas, N. T., Shea, S. A., Brown, R., & Czeisler, C. A. (2000). Absence of detectable melatonin and preservation of cortisol and thyrotropin rhythms in tetraplegia. *The Journal of Clinical Endocrinology and Metabolism*, *85*, 2189–2196.

Zeitzer, J. M., Ku, B., Ota, D., & Kiratli, B. J. (2014). Randomized controlled trial of pharmacological replacement of melatonin for sleep disruption in individuals with tetraplegia. *The Journal of Spinal Cord Medicine*, *37*, 46–53.

## Further reading

Bucks, R. S., Olaithe, M., Rosenzweig, I., & Morrell, M. J. (2017). Reviewing the relationship between OSA and cognition: Where do we go from here? *Respirology*, *22*, 1253–1261.

Section C

# Diagnosis and evaluation

Chapter 13

# Biomarkers in spinal cord injury: A highlight on prognostic insights

Leonardo Fonseca Rodrigues[a] and Tania Cristina Leite de Sampaio e Spohr[b,c]

[a]*Hospital São Vicente de Paulo, Rio de Janeiro, RJ, Brazil;* [b]*Instituto Estadual do Cérebro Paulo Niemeyer, Rio de Janeiro, RJ, Brazil;* [c]*Centogene GmbH, Rostock, Germany*

## Abbreviations

| | |
|---|---|
| **APRIL** | a proliferation-inducing ligand |
| **ASIA** | American Spine Injury Association |
| **AZGP** | zinc alpha 2 glycoprotein |
| **B2M** | β2 microglobulin |
| **BAFF** | B-cell activating factor |
| **BCMA** | B-cell maturation antigen |
| **BSCB** | blood spinal cord barrier |
| **CathD** | cathepsin D |
| **CCL2** | chemokine ligand 2 |
| **CD95L** | CD95 (Fas/APO-1)-ligand |
| **CNS** | central nervous system |
| **CSF** | cerebral spinal fluid |
| **DEmiRNAs** | differently expressed miRNAs |
| **EAE** | experimental autoimmune encephalomyelitis |
| **ENO2** | enolase 2 |
| **EV** | extracellular vesicle |
| **FGF** | fibroblast growth factor |
| **GBM** | glioblastoma |
| **GDNF** | glial cell-derived neurotrophic factor |
| **GFAP** | glial fibrillary-acidic protein |
| **HMGB1** | high mobility group 1 box |
| **HP** | haptoglobin |
| **IGF** | insulin-like growth factor |
| **IL** | interleukin |
| **INF** | interferon |
| **IP** | C-X-C motif chemokine ligand 10 |
| **MBP** | myelin basic protein |
| **MCP** | monocyte chemotactic protein |
| **MIF** | macrophage migration inhibitor |
| **miRNAs** | microRNAs |
| **MMPs** | metalloproteinases |
| **MS** | multiple sclerosis |
| **NFL** | neurofilaments |
| **NSE** | neuron-specific enolase |
| **PEA15** | phosphoprotein enriched in astrocytes 15 |
| **pNF-H** | phosphorylated form of neurofilament heavy chain |
| **S100β** | glial-specific calcium-binding β protein |
| **SBDP** | spectrin breakdown products |
| **SCI** | spinal cord injury |
| **TF** | transferrin |

Diagnosis and Treatment of Spinal Cord Injury. https://doi.org/10.1016/B978-0-12-822498-4.00013-0

| | |
|---|---|
| **TGF** | growth factor transformer |
| **TPI** | triosephosphate isomerase |
| **UCH-L1** | ubiquitin carboxy-terminal hydrolase L1 |
| **VEGF** | vascular endothelial growth factor |

## Introduction

Spinal cord injury (SCI) is a broad term that includes changeable grades of neurological deficits and is still considered an incurable condition. In spite of substantial advances in surgical and medical treatment, just around 1% of SCI patients experience complete healing. Unfortunately, almost 45% of the them suffer from serious neurological loss, which includes incomplete or complete tetraplegia, with or without respiratory compromise (Spinal Cord Injury (SCI), 2016). Therefore, SCI usually has devastating vocational, social, and physical implications for patients and caregivers (Ahuja et al., 2017).

The patient's outcome after initial SCI is still very difficult to predict, because the tools we have today for both evaluating the gravity of spinal cord tissue destruction and better assessing the recovery of SCI patients are still restricted (Rodrigues, Moura-Neto, & Spohr, 2018). Imaging examinations of the spinal cord cannot accurately predict the prognosis of the disease. Computed tomography shows only fractures or dislocations, while magnetic resonance imaging is not highly sensitive to edema of the spinal cord, be it related or not to SCI. Furthermore, laboratory tests of blood biochemistry such as lactate dehydrogenase and protein kinase C are not specific for SCI, as other non-neurological damage can also cause changes in blood parameters, having, therefore, no clinical application (Ding et al., 2019). There's no doubt that the sooner we evaluate the extent of the lesion and start treating the patient, the better the outcome is. Nowadays, the best predictive value of neurological assessment of the SCI patient is defined by typical clinical neurological examinations during the patient's first evaluation (Rodrigues et al., 2018). So the identification of biomarkers that can predict lesions is crucial to helping assess damage severity, prognosis, and therapeutic outcomes (Rodrigues et al., 2018).

In this chapter, we will highlight the most recent research on the expression of biomarkers in the pathophysiology of the SCI and try to correlate to diagnosis and prognosis.

## Pathophysiology of SCI

It is well known that SCI occurs in two mechanisms that progress through three stages. Both mechanisms that lead to acute SCI are now very well characterized and recognized as primary (mechanical) and secondary injuries. The primary injury is caused by the initial mechanical trauma of surrounding tissues, followed by contusion and compression of the spinal cord, which induces damage to nerve cells, blood vessels, and myelin (Rodrigues et al., 2018). The secondary lesion is triggered by the primary lesion and is classified in three phases, namely the acute phase, the sub-acute (or intermediate) phase, and the chronic phase, according to the pathomechanism and post-injury time (Bareyre & Schwab, 2003; Rodrigues et al., 2018).

The acute stage begins with the mechanical injury and extends to the first 48 h after the initial traumatic injury. In this stage, the patient develops some neurological deficits directly associated with the mechanical injury. This is when innumerable pathophysiological processes begin, leading to the initial inflammatory response to the injury and causing patients to develop "spinal shock" (Albayar et al., 2019; Rodrigues et al., 2018). It is well established that, at this stage, crucial events are induced by vascular rupture, ischemia, and hemorrhage by the action of inflammatory growth factors which affects neurons and glial cells, including excessive production of free radicals, ionic dysregulation, inflammatory response, and excitotoxicity, inducing pathological changes (Kwon, Tetzlaff, Grauer, Beiner, & Vaccaro, 2004; Rodrigues et al., 2018). At this stage, primary care should be provided for the patients, including protection of the airways, maintenance of the respiratory function, and use of hemodynamic support (Albayar et al., 2019).

It is, however, often difficult to define the baseline severity of the injury, as the patient is often in an unstable clinical condition due to spinal shock (Hachisuka et al., 2014). In order to reduce inflammation and neurological deficits, some centers use high-dose corticosteroids in the first 8 h after the injury, following the NASCIS II protocol (Witiw & Fehlings, 2015). Nevertheless, damage to glial cells, neurons and oligodendrocytes caused by the presence of excessive amounts of excitatory neurotransmitters (glutamate, aspartate) cannot be prevented. Furthermore, neurons die during all phases of the injury due to necrosis and astrocytes. Neurons and oligodendrocytes also die due to apoptotic mechanisms (Beattie, Hermann, Rogers, & Bresnahan, 2002).

The secondary or sub-acute stage lasts up to 2 weeks after the damage. At this stage, the patient usually recovers from the initial spine shock. However, neurological loss as well as other potential complications associated with the original trauma continue, as nerve cell dysfunction and blood supply problems derived from the acute stage can get even worse

(Albayar et al., 2019). In this phase, reactive gliosis increases both the content of glial fibrillary-acidic protein (GFAP) and the size of astrocytes to form the glial scar, In this phase, an intense inflammatory response begins, whereby phagocytic cells are recruited (Bareyre & Schwab, 2003; Rodrigues et al., 2018). The glial scar, with its firmly fused processes, forms an inhibitory agglomerate in the form of a mesh that plays a fundamental role in interrupting axonal regeneration (Ahuja et al., 2017). However, despite limiting regeneration after an injury to the central nervous system (CNS), the uncontrolled proliferation of reactive astrocytes suppresses the formation of aberrant synapses at the injury site and also contributes to both the restoration of the integrity of the blood spinal cord barrier (BSCB) and the reconstruction of microenvironment homeostasis (Faulkner et al., 2004). This mechanism is important for the removal of edemas, in addition to reducing the infiltration of cells of the immune system, helping, thus to limit the spread of the lesion (Wanner et al., 2013).

The chronic SCI phase, whose characterization requires further investigation, follows the sub-acute SCI phase. At this stage, patients may experience partial neurological recovery, which will depend on the extent of the lesion and the general condition of each patient. In addition, patients may also have maturation of adaptive mechanisms or the onset of more delayed neurological symptoms such as urinary bladder dysfunction, neuropathic pain, lipodystrophy, musculoskeletal atrophy, dysautonomia, and abnormal skeletal postures (Eckert & Martin, 2017). This phase may prolong from days to years after the shock, and is characterized by Wallerian degeneration, apoptosis and scarring that cause functional impairment (Bareyre & Schwab, 2003; Rodrigues et al., 2018). At this stage, it is also normal for patients to develop **syringomyelia*** due to the formation of the glial scar. Therapeutic strategies generally focus on increasing the regeneration of damaged axons and remyelination through different mechanisms, whether through pharmacological methods or cell therapy (Kim, Ha, & Kim, 2017).

## Biomarkers of SCI

First, it is important to define the word **biomarker***. Biomarkers, also known as "biological markers," are related to a wide sub-category of medical signs which can be measured with precision and reproducibility. Thus, biomarkers are, for purposes of interpretation, quantifiable results of biological mechanisms (Strimbu & Tavel, 2010). A biomarker of tissue injury should ideally be abundant, preferably (or exclusively) produced in the tissue of interest, and to be naturally present in small concentrations in blood and other body fluids (Laterza et al., 2009; Rodrigues et al., 2018). It is also important to have good biomarkers that can contribute as a measure of valuable therapeutic responses. They are highly valuable in the decision-making process of seeking additional clinical evaluation of a drug, and in helping to define critical parameters such as the intervention time window, drug dose and monitoring schedule for a clinical trial. It is also important to have good biomarkers that are acutely related to biochemical changes made or sustained during all phases of SCI. It is finally very important to validate all the potential biomarkers candidates.

Currently, there are neither drugs for treatment nor specific laboratory tests for the diagnosis of SCI. Therefore, in order to determine the extent of SCI as soon as possible, it is urgent to further the discovery and use of specific biomarkers for SCI. This could help guide doctors and researchers toward the discovery of a new object of intervention that can help prevent and decrease disability resulting from SCI.

### SCI in the acute stage

In the acute stage, there is a change in the permeability of the BSCB, due to the formation of glial scarring and the disturbance of endothelial cells. These cells secrete various inflammatory cytokines and some are currently being investigated as potential SCI biomarkers. High concentrations of interleukin-6 (IL-6), interleukin-8 (IL-8), and monocyte chemotactic protein (MCP)-1 (also known as chemokine ligand 2 (CC motif) (CCL2)) have been expressed in CSF. They have been shown to be dependent on the severity of the injury, when patients presented with American Spinal Injury Association grade A (ASIA A) injury had higher levels of these cytokines compared to ASIA B and C patients (Kwon et al., 2010).

Recently, a study with 15 ASIA A and B patients demonstrated that IL-10, IL-1rα, IL-9, IL-16, and IL-18 were positively regulated 24 h after SCI, and that interferon-gamma (INF-γ) and IL-13 have been downregulated (Fernández et al., 2020). Upregulation of the high mobility group 1 box (HMGB1) and the inflammatory cytokine macrophage migration inhibitor (MIF) factor were also identified, both occurring regardless of the degree or severity of injury (Kwon et al., 2019). Several studies also analyzed the transient expression of some proteins during the acute phase. Some structural proteins in glial cells and neurons are usually modulated in the CSF and in the serum of patients with acute SCI. Proteins such as GFAP, Phosphorylated Neurofilament Heavy (pNF-H), light chains of neurofilaments (NFL, 68 kDa), neuron-specific

**TABLE 1** Cytokines and proteins highly expressed in the acute stage.

| Expression in CSF | Cytokines | Proteins |
|---|---|---|
| Upregulated | IL-6, IL-8, IL-10, IL-1rα, IL-9, IL-16, IL-18MCP-1 (CCL2), HMGB1, MIF | GFAP, S100β, NSE, p-NF-H, NFL, UCH-L1, SBDP, MBP |
| Downregulated | INF-γ, IL-13 | |

Table based on de Mello Rieder et al. (2019), Du et al. (2018), Fernández et al. (2020), Ghosh and Pal (2019), Kwon et al. (2019, 2010, 2017), Leister et al. (2020), and Moghieb et al. (2016).

enolase (NSE), glial-specific calcium-binding β protein (S100β), tau, ubiquitin carboxy-terminal hydrolase L1 (UCH-L1), αII spectrin breakdown products (SBDP), myelin basic protein (MBP), transferrin, cathepsin D, triosephosphate isomerase-1, and astrocytic phosphoprotein PEA-15 were analyzed in patients with acute SCI (de Mello Rieder et al., 2019; Du et al., 2018; Kwon et al., 2010, 2017; Leister et al., 2020).

GFAP is an astrocyte-specific intermediate filament protein associated with glial scar formation, and astrocyte protein S100β in adults is generally elevated due to damage to the nervous system. It has been shown that both GFAP and S100β are present at higher levels in CSF and serum 24 h after injury, depending on severity, and that both can serve as a prognostic biomarker (Kwon et al., 2017; Leister et al., 2020). Inverse results were observed with the tau protein, which is an intracellular neuron protein involved in the stabilization of microtubules. A negative correlation was observed between CSF and serum tau concentrations and the severity of spinal cord injury 24 h after the injury, suggesting that tau may score as a prognostic biomarker for neurological improvement after acute spinal cord injury (Kwon et al., 2010; Leister et al., 2020).

NSE, p-NF-H, NFL have been found to be elevated in the serum and CSF of patients with acute SCI. However, when p-NF-H decreases to normal values, it allows the prediction of a favorable outcome (Ghosh & Pal, 2019). UCH-L1, SBDP, MBP have also been found to be transiently elevated in the CSF of patients with SCI and show a correlation between severity and recovery (Moghieb et al., 2016) (see Table 1).

## SCI in the sub-acute stage

In the sub-acute stage, during the formation of the **glial scar***, astrocytes increase the expression of not only GFAP, but also of nestin and vimentin. It has also been reported that, at this stage, astrocytes also secrete various **cytokines*** and growth factors such as vascular endothelial growth factor (VEGF), fibroblast growth factor (FGF)-β, glial cell-derived neurotrophic factor (GDNF), and growth factor transformer (TGF)-β, which are speculated to promote in later stages the differentiation, proliferation and migration of oligodendrocyte precursor cells. TGF-β1 is suggested to be implicated in scar formation after spinal cord injury (Kim et al., 2017; Rodrigues et al., 2018).

Recently, a study showed that patients with SCI had a decrease in both serum TGF-β1, an insulin-like growth factor 1 (IGF-1), and in sCD95L, but there was an upregulation 12 weeks after the injury correlated with the absence of neurological recovery (Ferbert et al., 2017).

CD95(Fas/APO-1)-ligand (CD95L) protein during the sub-acute phase is cleaved and a portion of it is released into the peripheral blood. This protein induces apoptosis in CD95-sensitive cells, especially at the sub-acute stages when apoptotic activity increases. This protein has the potential of acting as a biomarker at the sub-acute stage, and mainly of indicating the apoptotic effect destructive to the spinal cord tissue and the consequent neurological loss (Biglari et al., 2013). At this stage, the phagocytic inflammatory cells normally release cytokines that could be used as biomarkers. One of those cytokines is IL-1β, which has shown a significant fall at 1 and 4 weeks post injury in patients who experienced less progress (Biglari et al., 2015). Several metalloproteinases (MMPs) are upregulated during the sub-acute phase, and specially MMP-8 and -9, which have been pointed as a useful indicator for recovery potential (Albayar et al., 2019; Moghaddam et al., 2017). The C-X-C motif chemokine ligand 10 (IP-10) was found increased in serum at 3–6 h after SCI. However, there was a peak after 7 days of trauma at the sub-acute phase (Hassanshahi et al., 2013). Other proteins were pointed as potential biomarkers in the sub-acute phase, among them transferrin (TF), cathepsin D (CathD), triosephosphate isomerase-1 (TPI-1), phosphoprotein enriched in astrocytes 15 (PEA15), zinc alpha 2 glycoprotein (AZGP) and Haptoglobin (HP). They were found in human CSF from SCI patients, whose AZGP and HP protein levels decline in the CSF as the secondary injury process progresses (Moghieb et al., 2016; Sengupta et al., 2014) (see Table 2).

**TABLE 2** MMPs, cytokines, proteins and growth factors highly expressed in the sub-acute stage.

| Expression in CSF | Metalloproteinase | Proteins | Growth factors | Cytokines |
|---|---|---|---|---|
| Upregulated | MMP-8 and 9 | Nestin, Vimentin, TF, CathD, TPI-1, PEA15 | GFAP, VEGF, FGF-β, GDNF, TGF-β 1[a], IGF-1[a], CD95L[a] | IP-10 |
| Downregulated | | AZGP, HP | | |

[a]*Upregulation 12 weeks after the injury is correlated with the absence of neurological recovery.*
*Table based on Albayar et al. (2019), Biglari et al. (2013, 2015), Ferbert et al. (2017), Hassanshahi et al. (2013), Kim et al. (2017), Moghaddam et al. (2017), Moghieb et al. (2016), Rodrigues et al. (2018), and Sengupta et al. (2014).*

## SCI in the chronic stage

During the chronic phase, glial cells and neurons continue to die from apoptosis, compromising the channel and the functions of the receptor, and scarring and demyelination accompany Wallerian degeneration. All of these processes lead to deficits and are signs that the injury has matured (Norenberg, Smith, & Marcillo, 2004; Rodrigues et al., 2018).

It has been shown that two potent pro-inflammatory cytokines, namely macrophage migration inhibiting factor (MIF) and high mobility group box 1 (HMGB1), which are present in large quantities in both types of immune and neuronal cells, were found increased in patients with chronic spinal cord injury, regardless of the severity or level of the injury (Kwon et al., 2019; Papatheodorou et al., 2017; Stein et al., 2013). Studies have also shown that, at this stage, patients have elevated levels of IL-2, tumor necrosis factor alpha (TNFα), C-reactive protein, IL-6 and interleukin-1 receptor antagonist (IL-1RA), the two latter also being related to urinary tract infections, neuropathic pain or comorbidities such as pressure ulcers (Davies, Hayes, & Dekaban, 2007; Frost, Roach, Kushner, & Schreiber, 2005; Kwon et al., 2019).

It has also been shown hat in the chronic phase there is an elevated expression of some autoimmunity-promoting cytokines such as B-cell maturation antigen (BCMA), B-cell activating factor (BAFF), and A proliferation-inducing ligand (APRIL), and that some proteins that were found elevated in acute SCI were also increasingly present in people with chronic SCI such as β2 microglobulin (B2M), enolase 2 (ENO2), and S100A9 (Saltzman et al., 2013; Streijger et al., 2018). These studies demonstrated that some biomarkers could be used as good candidates to evaluate biochemical changes made or sustained during all stages of the SCI in order to help validate the potential impact on the clinical care of patients throughout the treatment (see Table 3).

# microRNAs as biomarkers

Recently, microRNAs (**miRNAs***) have been identified as good candidates as biomarker in blood in various pathologies, including in pancreatic adenocarcinomas, Alzheimer's disease, low-grade gliomas, ectopic pregnancy, breast cancer, and ovarian cancer among others (Rodrigues et al., 2018). The advantages of using miRNAs as biomarkers are that they are tissue specific, very stable in fluids, and have a phylogenetic relationship. In addition, they are present in all systems and have specific expression patterns at the developmental- and tissue-specific levels (Farh et al., 2005). MiRNAs downregulate the translation of RNA proteins by binding to the 3′-UTR of their target mRNAs. They are endogenously derived, short (usually 18–22 nucleotides in length), and generally induce mRNA degradation or translational repression (Bartel, 2004). The changes in the expression of miRNAs after SCI can be classified into three groups: (1) decreased miRNAs; (2) increased miRNAs; and (3) bidirectional (decreased or increased) miRNAs. There is little research related to miRNA and SCI in humans. Most of the studies have been conducted in rats, mice, or pigs. Efforts are currently being made to

**TABLE 3** Cytokines and proteins highly expressed in the chronic stage.

| Expression in CSF | Proteins | Cytokines |
|---|---|---|
| Upregulated | B2M, ENO2, S100A9 | MIF, HMGB1, IL-2, IL-6,[a] Il-1RA,[a] BCMA, BAFF, APRIL |

[a]*Also related to urinary tract infections, neuropathic pain or comorbidities such as pressure ulcers.*
*Table based on Davies et al. (2007b), Frost et al. (2005), Kwon et al. (2019), Norenberg et al. (2004), Papatheodorou et al. (2017), Rodrigues et al. (2018), Saltzman et al. (2013), Stein et al. (2013), and Streijger et al. (2018).*

validate in humans data that have been observed in animal models. There is no doubt that possible targets for modified miRNAs after SCI include genes that encode components involved in oxidative stress, inflammation, and apoptosis. The problem is that SCI patients represent very heterogeneous groups according to injury mechanism, severity and location, and the age and genetic background of the patients.

Recently, 190 detected miRNAs were found to be differentially expressed in the CSF of patients with all three SCI severities (ASIA A, B, and C). It was also demonstrated that these miRNAs showed changes in CSF and serum depending on the severity of the injury (Tigchelaar et al., 2019) (see Table 4).

Moreover, it was also shown that CSF miRNA expression at 24 h post injury within patients with ASIA A differs between those patients who will, and those who will not convert to ASIA B, and also differs from those that will not convert to C or D at 6 months post injury (Tigchelaar et al., 2019).

**TABLE 4** **Most differentially expressed miRNAs in cerebrospinal fluid after 24 h of the injury.**[a]

| miRNA | Regulation | AISA |
|---|---|---|
| miR-219a-2-3p | ↑ | A |
| miR-9-5p | ↑ | A |
| miR-9-3p | ↑ | A |
| miR-129-5p | ↑ | A |
| miR-219a-5p | ↑ | B |
| miR-1246 | ↑ | A |
| miR-760 | ↑ | A |
| miR-410-3p | ↑ | A |
| miR-485-5p | ↑ | B |
| miR-323b-3p | ↑ | A |
| miR-124-3p | ↑ | A |
| miR-488-3p | ↑ | A |
| miR-320b | ↑ | B |
| miR-433-3p | ↑ | B |
| miR-1298-3p | ↑ | A |
| miR-211-5p | ↑ | B |
| miR-92b-3p | ↑ | B |
| miR-128-3p | ↑ | A |
| miR-1910-5p | ↑ | C |
| miR-10b-3p | ↑ | B |
| miR-3605-3p | ↑ | B |
| miR-125b-1-3p | ↑ | B |
| miR-584-5p | ↑ | B |
| miR-338-5p | ↑ | B |
| miR-21-3p | ↑ | C |
| miR-1307-5p | ↑ | C |
| miR-23b-3p | ↓ | C |
| miR-195-5p | ↓ | C |

[a] *Table based on Tigchelaar et al. (2019).*

It was shown that a total of 19 miRNAs were differently expressed in human serum, and that 16 of them were specific only to the ASIA A group. The miRNA miR-133a-3p demonstrated the most significant change in expression, and it is already associated with injury severity. The miRNAs miR-208 and miR-499 associated with skeletal muscle mass regulation were demonstrated not only to be regulated in human serum after the SCI but also to have similar severity-dependent expression patterns (Boon et al., 2015; Tigchelaar et al., 2019).

However, the take-home message of the work from the group of Tigchelaar is that, although 190 miRNAs were detected in the CSF and 19 in serum after evaluating 44 individuals that met one of the three ASIA A, B and C criteria, only one of them, namely miR-423, was shown to be one of the strongest prognosticators of neurologic progress at 6 months post injury in humans (Tigchelaar et al., 2019). It is already known that miR-423 is related to neuronal apoptosis, as it regulates caspase-3 activity (Li et al., 2014).

## Exosomes

Although "free circulating" miRNAs can be used as biomarkers, they have low concentration and little stability in other body fluids (Ding et al., 2019). Therefore, the best way to search for miRNAs is in the small extracellular vesicles (EVs) secreted by most of the eukaryotic cells, called exosomes.

Exosomes are EVs derived from the 30 to 120 nm endocytic membrane and are present in many or perhaps in all body fluids. They participate in cell-to-cell communication and distribution of proteins and RNA, as they transport a diversity of proteins, lipids and nucleic acids. Exosomal miRNAs in body fluids have many advantages over free miRNAs. It has already been shown that the molecular components of exosomes, especially exosomal proteins and miRNAs, are promising as new biomarkers for clinical diagnosis (Lin et al., 2015).

Recently, it was shown in a rat model of acute SCI that there were changes in the expression of miRNA in circulating exosomes associated with the pathological mechanism of acute SCI by their target-regulated pathways. Identified serum exosomal miR-130a-3p, miR-125b-5p, and miR-152-3p are distinguishing and easily detectable diagnostic markers in acute SCI. In the same study, other differently expressed miRNAs (DEmiRNAs) and their expression were found to be either up- or downregulated, depending on the activation or inhibition of signaling pathways with response to the injury (Ding et al., 2020) (see Tables 5 and 6).

**TABLE 5 miRNAs identified in exosomes in acute SCI.**

| miRNA | Expression in SCI | Actions already known |
|---|---|---|
| miR-152-3p | Increased in the injured area after SCI | • Diagnosis, treatment, and prognosis of tumors<br>• Dengue infection<br>• Type 2 diabetic nephropathy<br>• Alzheimer's disease |
| miR-130a-3p | Significantly increased in the SCI | • Regulate the insulin sensitivity and hepatic steatosis<br>• Regulate cell migration and invasion in gemcitabine-resistant hepatoma cells<br>• Suppress cell migration and invasion in human gastric carcinoma<br>• Suppress cell viability, proliferation, and invasion in nasopharyngeal carcinoma<br>• Increase cisplatin resistance in non-small-cell lung cancer<br>• Regulate gemcitabine resistance in cholangiocarcinoma<br>• Attenuate activation, and induces apoptosis of hepatic stellate cells |
| microRNA-125b | SIGNIFICANTLY decreased in the acute phase of SCI | • Significantly upregulated in primary glioblastomas<br>• Highly expressed in mouse models of cardiac hypertrophy and idiopathic end-stage failing human hearts<br>• Decreased in primary neuroblastoma tumors, psoriasis, and atopic eczema<br>• Important negative regulator of p53 and p53-induced apoptosis |

Table based on Ding et al. (2019, 2020) and Lin et al. (2015).

**TABLE 6** DemiRNA targets that were identified in acute SCI.

| Expression | Most enriched pathways of the targets of DEmiRNAs |
|---|---|
| Upregulated | • Axon guidance<br>• Melanogenesis<br>• Wnt signaling pathway<br>• Long-term potentiation |
| Downregulated | • ECM receptor interaction<br>• Focal adhesion |

Table based on Ding et al. (2019, 2020) and Lin et al. (2015).

# Conclusions

SCI can cause permanent disability, and yet, despite advances in clinical management, many patients are left with substantial neurological impairment. Presently, prevention is the only procedure we can take to avoid the primary mechanical damage to the spinal cord (Rodrigues et al., 2018). There is no doubt that these patients need to be correctly diagnosed, but an effective treatment for SCI hasn't been found yet.

The fact is that the sooner patients are treated and the magnitude of their lesion is defined, the better their chance of a good outcome. However, the problem of using proteins as biomarkers is that they should be frequently collected from the CSF in order to characterize the severity of the injury, but collecting CSF through a lumbar puncture is unquestionably invasive (Kwon et al., 2017; Rodrigues et al., 2018). Further bioinformatics analysis and functional research will be of great help in clarifying the role of exosomes, microRNAs and proteins in the pathological process of spinal cord injury and in judging whether they can be used as diagnostic markers (Ding et al., 2019). Unfortunately, the identification of biomarkers to predict SCI prognosis and treatment still needs much more research.

## Applications to other areas of neuroscience

The research concerning the use of biomarkers as predictor for the treatment of SCI and several other devastating diseases such as cancer, cardiovascular diseases, Alzheimer's and Parkinson has been increasing since the last decade (Califf, 2018).

Recently, the search for good biomarkers has been focused on exosomal biomarkers, as they support comparable or superior sensitivity and specificity associated to their excellent stability (Lin et al., 2015). Using an animal model, it was demonstrated that urine exosomes may be a specific and sensitive source of miRNA biomarkers for experimental autoimmune encephalomyelitis (EAE) disease and Multiple Sclerosis (MS). It was possible to identify miRNAs in urine, plasma and spinal cord exosomes in a period before the onset of the disease, suggesting their early involvement in the pathology of EAE and MS (Singh, Deshpande, Suhail, Rattan, & Giri, 2016).

In addition to exosomal proteins and RNAs, exosomal lipids have also been shown to have a diagnostic potential. In patients with glioblastoma (GBM), several exosomes and miRNAs have been identified as being useful for diagnostic and/or prognostic biomarkers (Saadatpour et al., 2016). But unfortunately, exosomal biomarkers are still at an early stage of discovery/development and their potential value in clinical diagnosis has not been fully analyzed yet (Lin et al., 2015).

# Mini-dictionary of terms

***Biomarkers***—Biomarkers, also known as "biological markers," are related to a wide sub-category of medical signs, which can be measured with precision and reproducibility. Therefore, biomarkers are often determined and evaluated to examine pathogenic processes, normal biological processes, or pharmacologic responses to a therapeutic intervention. In medicine, they can be used in clinical settings for guiding treatment based on their application: predictive, prognostic, or diagnostic.

***Syringomyelia*** is an illness in which a cyst or cavity forms within the spinal cord. The cyst, also called a syrinx, can grow and become quite large over time, destroying the spinal cord. The damage may result in paralysis, loss of feeling, weakness, and rigidity in the shoulders, back, and extremities. The illness may also induce a loss of the ability to feel extremes of cold or hot, especially in the hands. It may also lead to a cape-like bilateral loss of temperature sensation and pain along the arms and upper chest.

***miRNAs*** are small non-coding RNAs molecules containing about 22 nucleotide which are found in all species including plants, animals and some viruses. MiRNAs function in silencing and post-transcriptional regulation of gene expression, via base-pairing with complementary sequences within mRNA molecules. As a result, these mRNA molecules are silenced by one or more of the subsequent ways (1) cleavage of the mRNA strand into two pieces; (2) destabilization of the mRNA through shortening of its poly(A) tail, and (3) less efficient translation of the mRNA into proteins by ribosomes.

***glial scar*** is formed by reactive astrocytes that proliferate and have processes that are firmly fused, making an inhibitory meshlike bundle that has the valuable task of acting as an obstacle for axonal regeneration, and therefore being considered the main origin for the restricted regeneration after a CNS damage.

***Cytokines*** are a broad category of small proteins (~5–20 kDa) that are very important in cell signaling. They are peptides that cannot cross the lipid bilayer of cells to enter the cytoplasm, so they act through cell surface receptors and are especially important in the immune system. Cytokines can act in an autocrine, paracrine and endocrine fashion way. Cytokines include interleukins, chemokines, lymphokines, interferons, and tumor necrosis factors, but generally not hormones or growth factors (despite some overlap in the terminology). They are produced by a broad range of cells, including immune cells like macrophages, B lymphocytes, T lymphocytes and mast cells, as well as endothelial cells, fibroblasts, and various stromal cells. Cytokines are important in health and disease, specifically in host immune responses to inflammation, infection, cancer, trauma, sepsis, and reproduction.

## Key facts of biomarkers spinal cord injury

- Spinal cord injury (SCI) is still considered an incurable condition, resulting in a heterogeneous recovery and an uncertain prognosis.
- The sooner patients are treated and the magnitude of their lesion is defined, the better their chance of a good outcome.
- Biomarkers, also known as "biological markers," are part of the biological response to injury, and are related to a wide sub-category of medical signs which can be measured with precision and reproducibility.
- Identification of biomarkers that can predict lesions is crucial to helping assess damage severity, prognosis, and therapeutic outcomes.
- Good biomarkers that can contribute as a measure of therapeutic responses are highly valuable in the decision-making process of seeking additional clinical evaluation of a drug, and in helping to define critical parameters such as the intervention time window, dose and monitoring schedule for a clinical trial.

## Summary points

- Biomarkers are part of the biological response to injury.
- It is important to understand the degree or stage of the injury to focus on treatment.
- The change of proteins in some fluids is consistent, but their distribution in peripheral fluids is not the same.
- Non-invasive examinations with reproducible results are better to assess the injury.
- MiRNAs are tissue specific, very stable in fluids and have a phylogenetic relationship.
- Exosomes participate in cell-to-cell communication and the distribution of proteins and RNA, as they transport a diversity of proteins, nucleic acids and lipids.
- As exosomes are distributed in body fluids, exosomal miRNAs have many advantages over free miRNAs, being thus promising as new biomarkers for clinical diagnosis.

## Acknowledgments

This work was supported by the National Institute for Translational Neuroscience (INNT) of the Ministry of Science and Technology; the Brazilian Federal Agency for the Support and Evaluation of Graduate Education (CAPES) of the Ministry of Education; the National Council for Scientific and Technological Development (CNPq); the Carlos Chagas Filho Research Support Foundation (FAPERJ); and the Ary Frauzino Foundation for Cancer Research.

## References

Ahuja, C. S., Nori, S., Tetreault, L., Wilson, J., Kwon, B., Harrop, J., et al. (2017). Traumatic spinal cord injury—Repair and regeneration. *Clinical Neurosurgery*, *80*(3), S22–S90.

Ahuja, C. S., Wilson, J. R., Nori, S., Kotter, M. R. N., Druschel, C., Curt, A., et al. (2017). Traumatic spinal cord injury. *Nature Reviews. Disease Primers*, *3*(1), 17018.

Albayar, A. A., Roche, A., Swiatkowski, P., Antar, S., Ouda, N., Emara, E., et al. (2019). Biomarkers in spinal cord injury: Prognostic insights and future potentials. *Frontiers in Neurology*, *10*, 27. https://doi.org/10.3389/fneur.2019.00027. eCollection 2019.

Bareyre, F. M., & Schwab, M. E. (2003). Inflammation, degeneration and regeneration in the injured spinal cord: Insights from DNA microarrays. *Trends in Neurosciences*, *26*(10), 555–563.

Bartel, D. P. (2004). MicroRNAs: Genomics, biogenesis, mechanism, and function. *Cell*, *116*(2), 281–297.

Beattie, M. S., Hermann, G. E., Rogers, R. C., & Bresnahan, J. C. (2002). Cell death in models of spinal cord injury. *Progress in Brain Research*, *137*, 37–47.

Biglari, B., Büchler, A., Swing, T., Biehl, E., Roth, H. J., Bruckner, T., et al. (2013). Increase in soluble CD95L during subacute phases after human spinal cord injury: A potential therapeutic target. *Spinal Cord*. https://doi.org/10.1038/sc.2012.139.

Biglari, B., Swing, T., Child, C., Büchler, A., Westhauser, F., Bruckner, T., et al. (2015). A pilot study on temporal changes in IL-1β and TNF-α serum levels after spinal cord injury the serum level of TNF-α in acute SCI patients as a possible marker for neurological remission. *Spinal Cord*, *53*, 510–514.

Boon, H., Sjögren, R. J. O., Massart, J., Egan, B., Kostovski, E., Iversen, P. O., et al. (2015). MicroRNA-208b progressively declines after spinal cord injury in humans and is inversely related to myostatin expression. *Physiological Reports*, *3*(11), e12622. https://doi.org/10.14814/phy2.12622. Epub 2015 Nov 24.

Califf, R. M. (2018). Biomarker definitions and their applications. *Experimental Biology and Medicine*, *243*(3), 213–221. https://doi.org/10.1177/1535370217750088.

Davies, A. L., Hayes, K. C., & Dekaban, G. A. (2007). Clinical correlates of elevated serum concentrations of cytokines and autoantibodies in patients with spinal cord injury. *Archives of Physical Medicine and Rehabilitation*, *88*(11), 1384–1393. https://doi.org/10.1016/j.apmr.2007.08.004. https://doi.org/S0003-9993(07)01342-1 [pii].

de Mello Rieder, M., Oses, J. P., Kutchak, F. M., Sartor, M., Cecchini, A., Rodolphi, M. S., et al. (2019). Serum biomarkers and clinical outcomes in traumatic spinal cord injury: Prospective cohort study. *World Neurosurgery*, *122*, e1028–e1036. https://doi.org/10.1016/j.wneu.2018.10.206. Epub 2018 Nov 7.

Ding, S. Q., Chen, Y. Q., Chen, J., Wang, S. N., Duan, F. X., Shi, Y. J., et al. (2020). Serum exosomal microRNA transcriptome profiling in subacute spinal cord injured rats. *Genomics*, *12*(6), 5086–5100. https://doi.org/10.1016/j.ygeno.2019.09.021. Epub 2020 Sep 9.

Ding, S. Q., Chen, J., Wang, S. N., Duan, F. X., Chen, Y. Q., Shi, Y. J., et al. (2019). Highlight article: Identification of serum exosomal microRNAs in acute spinal cord injured rats. *Experimental Biology and Medicine*, *244*(14), 1149–1161. https://doi.org/10.1177/1535370219872759. Epub 2019 Aug 26.

Du, W., Li, H., Sun, J., Xia, Y., Zhu, R., Zhang, X., et al. (2018). The prognostic value of serum neuron specific enolase (NSE) and s100b level in patients of acute spinal cord injury. *Medical Science Monitor*, *24*, 4510–4515. https://doi.org/10.12659/MSM.907406.

Eckert, M. J., & Martin, M. J. (2017). Trauma: Spinal cord injury. *Surgical Clinics of North America*, *97*(5), 1031–1045. https://doi.org/10.1016/j.suc.2017.06.008.

Farh, K. K., Grimson, A., Jan, C., Lewis, B. P., Johnston, W. K., Lim, L. P., et al. (2005). The widespread impact of mammalian microRNAs on mRNA repression and evolution. *Science*, *310*(5755), 1817–1821.

Faulkner, J. R., Herrmann, J. E., Woo, M. J., Tansey, K. E., Doan, N. B., & Sofroniew, M. V. (2004). Reactive astrocytes protect tissue and preserve function after spinal cord injury. *The Journal of Neuroscience*, *24*(9), 2143–2155.

Ferbert, T., Child, C., Graeser, V., Swing, T., Akbar, M., Heller, R., et al. (2017). Tracking spinal cord injury: Differences in cytokine expression of IGF-1, TGF-B1, and sCD95l can be measured in blood samples and correspond to neurological remission in a 12-week follow-up. *Journal of Neurotrauma*, *34*(3), 607–614.

Fernández, M., Baldassarro, V. A., Capirossi, R., Montevecchi, R., Bonavita, J., Cescatti, M., et al. (2020). Possible strategies to optimize a biomarker discovery approach to correlate with neurological outcome in patients with spinal cord injury: A pilot study. *Journal of Neurotrauma*, *37*(3), 431–440. https://doi.org/10.1089/neu.2018.6362. Epub 2019 Aug 1.

Frost, F., Roach, M. J., Kushner, I., & Schreiber, P. (2005). Inflammatory C-reactive protein and cytokine levels in asymptomatic people with chronic spinal cord injury. *Archives of Physical Medicine and Rehabilitation*, *86*(2), 312–317. https://doi.org/10.1016/j.apmr.2004.02.009.

Ghosh, A., & Pal, R. (2019). Spinal cord injury: Point-of-care biomarkers for better prognosis. *Journal of Neurosciences in Rural Practice*, *10*(3), 390–391. https://doi.org/10.1055/s-0039-1698032. Epub 2019 Oct 7.

Hachisuka, S., Kamei, N., Ujigo, S., Miyaki, S., Yasunaga, Y., & Ochi, M. (2014). Circulating microRNAs as biomarkers for evaluating the severity of acute spinal cord injury. *Spinal Cord*, *52*(8), 596–600.

Hassanshahi, G., Amin, M., Shunmugavel, A., Vazirinejad, R., Vakilian, A., Sanji, M., et al. (2013). Temporal expression profile of CXC chemokines in serum of patients with spinal cord injury. *Neurochemistry International*, *63*(5), 363–367. https://doi.org/10.1016/j.neuint.2013.07.012. Epub 2013 Aug 6.

Kim, Y. H., Ha, K. Y., & Kim, S. I. (2017). Spinal cord injury and related clinical trials. *Clinics in Orthopedic Surgery*, *9*(1), 1–9.

Kwon, B. K., Bloom, O., Wanner, I. B., Curt, A., Schwab, J. M., Fawcett, J., et al. (2019). Neurochemical biomarkers in spinal cord injury. *Spinal Cord*, *57*(10), 819–831. https://doi.org/10.1038/s41393-019-0319-8. Epub 2019 Jul 4.

Kwon, B. K., Stammers, A. M., Belanger, L. M., Bernardo, A., Chan, D., Bishop, C. M., et al. (2010). Cerebrospinal fluid inflammatory cytokines and biomarkers of injury severity in acute human spinal cord injury. *Journal of Neurotrauma*, *27*(4), 669–682.

Kwon, B. K., Streijger, F., Fallah, N., Noonan, V. K., Bélanger, L. M., Ritchie, L., et al. (2017). Cerebrospinal fluid biomarkers to stratify injury severity and predict outcome in human traumatic spinal cord injury. *Journal of Neurotrauma*, *34*(3), 567–580.

Kwon, B. K., Tetzlaff, W., Grauer, J. N., Beiner, J., & Vaccaro, A. R. (2004). Pathophysiology and pharmacologic treatment of acute spinal cord injury. *The Spine Journal*, *4*(4), 451–464.

Laterza, O. F., Lim, L., Garrett-Engele, P. W., Vlasakova, K., Muniappa, N., Tanaka, W. K., et al. (2009). Plasma microRNAs as sensitive and specific biomarkers of tissue injury. *Clinical Chemistry*, *55*(11), 1977–1983.

Leister, I., Haider, T., Mattiassich, G., Kramer, J. L. K., Linde, L. D., Pajalic, A., et al. (2020). Biomarkers in traumatic spinal cord injury—Technical and clinical considerations: A systematic review. *Neurorehabilitation and Neural Repair*.

Li, M. M., Jiang, T., Sun, Z., Zhang, Q., Tan, C. C., Yu, J. T., et al. (2014). Genome-wide microRNA expression profiles in hippocampus of rats with chronic temporal lobe epilepsy. *Scientific Reports*, *4*, 4734. https://doi.org/10.1038/srep04734.

Lin, J., Li, J., Huang, B., Liu, J., Chen, X., Chen, X. M., et al. (2015). Exosomes: Novel biomarkers for clinical diagnosis. *Science World Journal*, *2015*, 657086. https://doi.org/10.1155/2015/657086. Epub 2015 Jan 27.

Moghaddam, A., Heller, R., Daniel, V., Swing, T., Akbar, M., Gerner, H. J., et al. (2017). Exploratory study to suggest the possibility of MMP-8 and MMP-9 serum levels as early markers for remission after traumatic spinal cord injury. *Spinal Cord*, *55*(1), 8–15. https://doi.org/10.1038/sc.2016.104. Epub 2016 Jul 5.

Moghieb, A., Bramlett, H. M., Das, J. H., Yang, Z., Selig, T., Yost, R. A., et al. (2016). Differential neuroproteomic and systems biology analysis of spinal cord injury. *Molecular & Cellular Proteomics*, *15*(7), 2379–2395. https://doi.org/10.1074/mcp.M116.058115. Epub 2016 May 5.

Norenberg, M. D., Smith, J., & Marcillo, A. (2004). The pathology of human spinal cord injury: Defining the problems. *Journal of Neurotrauma*, *21*(4), 429–440.

Papatheodorou, A., Stein, A., Bank, M., Sison, C. P., Gibbs, K., Davies, P., et al. (2017). High-mobility group box 1 (HMGB1) is elevated systemically in persons with acute or chronic traumatic spinal cord injury. *Journal of Neurotrauma*, *34*(3), 746–754. https://doi.org/10.1089/neu.2016.4596. Epub 2016 Oct 26.

Rodrigues, L. F., Moura-Neto, V., & Spohr, T. (2018). Biomarkers in spinal cord injury: From prognosis to treatment. *Molecular Neurobiology*, *55*(8), 6436–6448.

Saadatpour, L., Fadaee, E., Fadaei, S., Nassiri Mansour, R., Mohammadi, M., Mousavi, S. M., et al. (2016). Glioblastoma: Exosome and microRNA as novel diagnosis biomarkers. *Cancer Gene Therapy*, *23*(12), 415–418. https://doi.org/10.1038/cgt.2016.48. Epub 2016 Nov 11.

Saltzman, J. W., Battaglino, R. A., Salles, L., Jha, P., Sudhakar, S., Garshick, E., et al. (2013). B-cell maturation antigen, a proliferation-inducing ligand, and B-cell activating factor are candidate mediators of spinal cord injury-induced autoimmunity. *Journal of Neurotrauma*, *30*(6), 434–440. https://doi.org/10.1089/neu.2012.2501. Epub 2013 Mar 28.

Sengupta, M. B., Basu, M., Iswarari, S., Mukhopadhyay, K. K., Sardar, K. P., Acharyya, B., et al. (2014). CSF proteomics of secondary phase spinal cord injury in human subjects: Perturbed molecular pathways post injury. *PLoS One*, *9*(10), e110885. https://doi.org/10.1371/journal.pone.0110885. eCollection 2014.

Singh, J., Deshpande, M., Suhail, H., Rattan, R., & Giri, S. (2016). Targeted stage-specific inflammatory microRNA profiling in urine during disease progression in experimental autoimmune encephalomyelitis: Markers of disease progression and drug response. *Journal of Neuroimmune Pharmacology*, *11*(1), 84–97. https://doi.org/10.1007/s11481-015-9630-0. Epub 2015 Aug 16.

Spinal Cord Injury (SCI). (2016). 2016 facts and figures at a glance. *The Journal of Spinal Cord Medicine*, *39*(4), 493–494. https://doi.org/10.1080/10790268.2016.1210925.

Stein, A., Panjwani, A., Sison, C., Rosen, L., Chugh, R., Metz, C., et al. (2013). Pilot study: Elevated circulating levels of the proinflammatory cytokine macrophage migration inhibitory factor in patients with chronic spinal cord injury. *Archives of Physical Medicine and Rehabilitation*, *94*(8), 1498–1507. https://doi.org/10.1016/j.apmr.2013.04.004. Epub 2013 Apr 22.

Streijger, F., So, K., Manouchehri, N., Gheorghe, A., Okon, E. B., Chan, R. M., et al. (2018). A direct comparison between norepinephrine and phenylephrine for augmenting spinal cord perfusion in a porcine model of spinal cord injury. *Journal of Neurotrauma*, *35*(12), 1345–1357. https://doi.org/10.1089/neu.2017.5285. Epub 2018 Mar 28.

Strimbu, K., & Tavel, J. A. (2010). What are biomarkers? *Current Opinion in HIV and AIDS*, *5*(6), 463–466. https://doi.org/10.1097/COH.0b013e32833ed177.

Tigchelaar, S., Gupta, R., Shannon, C. P., Streijger, F., Sinha, S., Flibotte, S., et al. (2019). MicroRNA biomarkers in cerebrospinal fluid and serum reflect injury severity in human acute traumatic spinal cord injury. *Journal of Neurotrauma*, *36*(15), 2358–2371. https://doi.org/10.1089/neu.2018.6256. Epub 2019 May 14.

Wanner, I. B., Anderson, M. A., Song, B., Levine, J., Fernandez, A., Gray-Thompson, Z., et al. (2013). Glial scar borders are formed by newly proliferated, elongated astrocytes that interact to corral inflammatory and fibrotic cells via STAT3-dependent mechanisms after spinal cord injury. *The Journal of Neuroscience*, *33*(31), 12870–12886. https://doi.org/10.1523/JNEUROSCI.2121-13.2013.

Witiw, C. D., & Fehlings, M. G. (2015). Acute spinal cord injury. *Journal of Spinal Disorders and Techniques*, *28*(6), 202–210. https://doi.org/10.1097/BSD.0000000000000287.

Chapter 14

# Quality of life tools for spinal cord–injured people

Sebastian Salvador-De La Barrera, Ibone Gimenez-Jiménez, Manuel Astray-Lopaz, Antonio Rodríguez-Sotillo, and M. Elena Ferreiro-Velasco

*Unidad de Lesionados Medulares. Hospital de A Coruña, A Coruña, Spain*

## Abbreviations

| | |
|---|---|
| **ADL** | activities of daily living |
| **ASIA** | American Spinal Injury Association |
| **BPI** | brief pain inventory |
| **CHART** | Craig Handicap Assessment and Reporting Technique |
| **FSFI** | female sexual function index |
| **HRQOL** | health-related quality of life |
| **ICF** | International Classification of Functioning disability and health |
| **IIEF** | International Index of Erectile Function |
| **ISCOS** | International Spinal Cord Society |
| **LISAT** | life satisfaction questionnaire |
| **NIH** | National Institutes of Health |
| **ParQOL** | Participation and Quality of life toolkit |
| **PRISM** | Patient Reported Interference of Spasticity Measure |
| **PROMIS** | Patient Reported Outcomes Measurement Information System |
| **QLI** | quality of life index |
| **QOL** | quality of life |
| **QWB** | quality of well-being scale |
| **SCI** | spinal cord injury/spinal cord injured |
| **SF-36** | Medical Outcomes Study 36-item short form health survey |
| **SWLS** | satisfaction with life scale |
| **WHO** | World Health Organization |
| **WHOQOL** | WHO quality of life instrument |

## Introduction

From the second half of the last century, advances in the study and medical care aimed to help people with spinal cord injury (SCI) have resulted in greater life expectancy in this population (Hill, Noonan, Sakakibara, & Miller, 2010). SCI sequelae are well known, as well as other specific associated medical problems that these patients have to overcome, which frequently interfere with everyday life. Likewise, people with SCI require specific equipment to enable the integration into their social and home environments and adapt to daily life, i.e., from mobility devices to incontinence products, as well as help from other people for basic activities of daily living (ADL). All these circumstances cause great psychological pressure with which they have to live and result in comorbidities, e.g., anxiety, depression, distortion of body image, self-esteem, sense of failure, etc.

Moreover, there are different barriers that interfere particularly with the quality of life (QOL) of people with disabilities (Bökel et al., 2020). In 2001, after various periods of study and reflection on models for evaluating people with distinct functional capabilities, the World Health Organization (WHO) developed the International Classification of Functioning (ICF) (WHO: https://apps.who.int/iris/handle/10665/42407) that addresses the medical condition from a biopsychosocial perspective and seeks to integrate a person's functional capacity with the execution of activities and social participation (Dijkers, 2005).

Diagnosis and Treatment of Spinal Cord Injury. https://doi.org/10.1016/B978-0-12-822498-4.00014-2

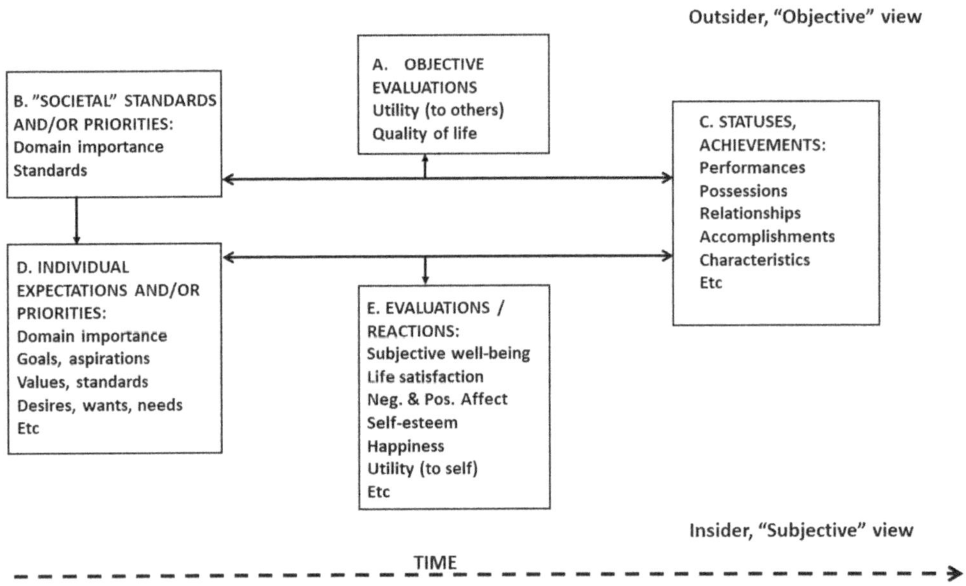


FIG. 1 Dijkers' Model of approach to quality of life. Approach to quality of life according to a "objective" or "Subjective" view, which shows various factors that influence the researcher's perception in their social environment, or the individual's and their subjective standards. *(With permission from: Dijkers MP. (2003) Individualization in quality of life measurement: instruments and approaches. Arch Phys Med Rehabil,84 Suppl 2: Fig. 1, page S4.).*

However, despite the work done, there is no agreement in the scientific literature on the concept of QOL to be studied in handicapped people and/or with chronic diseases, as in the case of this population (Hill et al., 2010; Post, 2014). In their assessment, researchers additionally need to define the concept of health, which affects the daily life of the people subject of study.

Since the establishment of the concept of *health* by the WHO in 1947, "*a state of complete physical, mental and social well-being and not merely the absence of disease or infirmity*", QOL was correlated to states of subjective well-being, although later, some objective aspects were included (Post, 2014).

Thus, the definition of *quality of life* in medicine and in psychology has been the subject of controversy on many occasions. On the one hand, a subjective approach is adopted by the person subject of study, in which there should be a consistency between the person's expectations and achievements to perceive satisfaction in their lives. On the other hand, there is an objective approach in which case the observer establishes certain parameters that characterize a satisfactory QOL from different points of view (Dijkers, 2003) (Fig. 1) one of them being the medical condition considered by the researcher (Hill et al., 2010; Miller et al., 2010).

If we consider the previously exposed ideas, different angles can be used when assessing QOL in people with SCL: subjective assessment of QOL by the person (Moons, Budts, & De Geest, 2006), its interrelation with the social environment in which the person lives, physical components—mainly impairments due to physical sequelae—mood, overall health status, or by specifying the difficulties in the management or assumption of system dysfunction that affect having a satisfactory social life—family-related, work-related, and education.

## Definitions: Quality of life and health-related quality of life

QOL is a multi-dimensional concept that includes subjective and objective factors, for which reason it is defined differently in each academic field, social group, or individuals. In fact, the WHO defines QOL as "*individual's perception of their position in life in the context of the culture and value system in which they live and in relation to their goals, expectations, standards, and concerns*" (The WHOQOL Group, 1998). This conveys a subjective view linked to the social context.

For an integrative assessment of QOL, an approach from a variable number of dimensions or *domains* is accepted, each of which refers to an area of vital aspects that differs from the others (Dijkers, 2005). To complete the definition of *health* proposed by the WHO, at least three dimensions are included: physical function, mental state, and ability to have social interactions (Post, 2014), which are usually found in the instruments that measure QOL from a subjective view by the affected person and the objective perspective of the observer.

Thus, approaching QOL in medicine is easily associated with the concept of health-related quality of life (HRQOL), which according to Dijkers is part of an objective view and makes reference to *those components in the person's QOL directly or indirectly affected by the disease, lesions, and treatment side effects* (Dijkers, 1997), for which reason the "medical condition" of the person is included, as pointed out by Post (2014).

On a higher level, the interaction of the health domains with the ability to interact with the surroundings and function of the person is contemplated in the structure of the ICF (Post & Noreau, 2005).

## Implementation in the study of quality of life in people with spinal cord injury

Defining the concept of QOL and its assessment from an objective view is subject of discussion and lack of agreement between researches (Tate & Forchheimer, 2014), due to the complexity of SCI physical sequelae, along with the outcome of rehabilitation and evidence that people with SCI may enjoy a full life integrated in their social environment. The instruments are usually developed based on values imposed on the participants, and sometimes there is little relation between the values communicated by the concerned individuals and those believed to be relevant by the researches and even by the caregivers (Fuhrer, Rintala, Hart, Clearman, & Young, 1992). As stated by Tate (Tate, Kalpakjian, & Forchheimer, 2002), subjects affected by SCI distinguish between having spinal cord injury–related sequelae and their health status ("I am healthy").

For this reason, the impact certain sequelae have on various QOL domains is examined in instruments assessing HRQOL in people with SCI-related sequelae, along with the assessment of global QOL, the latter form a subjective view as it can only be determined by the affected person (Moons et al., 2006).

### Databases

There are numerous academic institutions and knowledge bases trying to cluster the different instruments for measuring abilities, sequelae, or dysfunctions, and their interaction with the subject's life regarding physical aspects, function, mental, social, and of QOL in general. Concerning SCI, the SCIRE project (Miller et al., 2010; www.scireproject.com) is of great relevance. Revisions of the different fields of action are carried out in the project, and recommendations of different assessment instruments given, with regular updates, reviews on their validity, and diffusion in the medical literature. Likewise, numerous instruments for assessing health conditions secondary to SCI are found in the resources grouped in the Participation and Quality of Life Toolkit (ParQOL; www.parQOL.com), but in this case, they are related with Dijkers's QOL model and provide use and access instructions.

From a perspective focused on the person, the project developed under the auspice of the National Institutes of Health in the United States of America as of 2004, the Patient-Reported Outcomes Measurement Information System (PROMIS: www.healthmeasures.net), groups numerous useful instruments for assessing different elements linked with HRQOL that have shown a correlation with aspects collected by the ICF (Wong et al., 2018). The mission of PROMIS is to create a health model to transmit the health status communicated by the subject (Post, 2014), so that a global assessment on the person that is useful for the observer can be made compatible with the standards of the ICF. Its use in subjects with neurological disorders concentrates in Neuro-QOL (Tulsky et al., 2015). A specific HRQOL for SCI is the SCI-QOL, inside the PROMIS project, that is undergoing validation with a large sample of persons with SCI in United States of America (Tate & Forchheimer, 2014).

However, the use of many instruments is complex in clinical practice, and consequently clinical researches focus on a limited number of them. When choosing a tool, we must consider which is the aspect of QOL we want to assess: the subjective perspective of the person, the relationship with the social and cultural environment, specific health domains, effect of a specific sequelae or dysfunction, comparison or not with other populations—either by cultural setting or different type of disability or function—specific validation for SCI or general, etc. (Geyh, Fellinghauer, Kirchberger, & Post, 2010; Tate & Forchheimer, 2014).

Several reviews of the instruments and methods used for assessing QOL in people with SCI have been done (Hill et al., 2010) (Tables 1 and 2).

Below we discuss the most commonly used instruments in the medical literature, clustering them as global QOL and specific HRQOL for certain SCI-related sequelae tools (Tables 3 and 4). Despite the numerous instruments used in this population, there are no agreed standard measures and some show deficiencies in validity and sensitivity to change in comparative studies.

**TABLE 1** Objective QOL Instruments.

| Objective tools | | | | | | | | | |
|---|---|---|---|---|---|---|---|---|---|
| Name | Description | Items | Domains (No. of questions) | Scoring and response | Measurement model | Time taken (min) | Alternative languages | Item bias | Interviewer (I) or self-administration |
| Patient reported impact of spasticity measure (PRISM) | Addresses negative and positive impact of spasticity on QOL in SCI population | 41 | Societal avoidance/ anxiety (11); psychological agitation (5); daily activities (6); need for assistance/ positioning (5); impact on activities +/– (4); need for intervention (5); social embarrassment (5) | 5-point Likert scale<br>Sub-scale scores are obtained by averaging item scores and dividing by number of items<br>Higher is unhealthier | "Relative to the range of the scale, more persons scored in lower ranges (indicating less impact)" | Not available | None | Developed in SCI population | SA |
| Quality of well-being questionnaire-SA (QWB-SA) | Point-in-time preference measure designed to measure HRQOL | 71 | Symptoms and problems (58); mobility, physical activity, social activity (13) | 0.0—death to 1.0—optimum health<br>Higher is healthier | No floor and ceiling effects | <20 | Various | Face validity supports low bias | SA |
| Qualiveen | Disease-specific perspective on QOL in SCI for urinary disorders | 30 | Limitations (9); constraints (8); fears (8); feelings (5) | 5-point Likert scales (0–4); total range 0–100<br>Tabulate total<br>Higher is unhealthier | Minimal | Not available | Various | Developed in SCI population | I or SA |
| Sickness Impact Profile (SIP68) | Generic health status measure, measures health-related changes in behavior associated with the carrying out of daily activities | 68 | Somatic autonomy (17); mobility control (12); mobility range (10); social behavior (12); emotional stability (6); psychological autonomy and communication (11) | No = 0, yes = 1; total range 0–68<br>Yes responses are tabulated; can divide into SIPSOM, SIPPSY, SIPSOC sub-scales<br>Higher is unhealthier | Ceiling effects MR (31.3%), ES (54%), & PAC (53.8%) domains | ˥15–20 | Dutch | Scoring on one question adjusted for SCI | I or SA |

| | | | | | | | | | |
|---|---|---|---|---|---|---|---|---|---|
| Short Form 36 (SF-36) | Addresses basic human values relevant to QOL and well-being through individual domains and two global components | 36 | Physical functioning (10); role limitations because of physical health problems (4); bodily pain (2); general health (5); vitality (4); social functioning (2); role limitations because of emotional problems (3); mental health (5) | Total range 0–100 Norm-based score is 50, with standard deviation 10. Can be divided into two summary scores: physical and mental, calculated by a system of + and − weights on domain scores Higher is healthier | Floor and ceiling effect PF (12.2%–24.2%; free—29.7%), RLP (28.1%–36.3%; 22.5%–54.4%); ceiling effect SF (free—33.5%), RLE (63.8%–75.3%) | <15 | Available in most major languages | Face validity supports low bias | I or SA |
| Short form 36 veterans/SCI (SF-36 V) | Version of SF-36 designed for use in the disabled population. Physical functioning section has been modified to accommodate SCI population | 36 | Physical functioning (8) | Each domain converted to range of 0–100 Divided into two summary scores: physical and mental components, are norm based; general population score 50 and standard deviation of 10 Higher is healthier | "Descriptive responses to the items showed a floor effect on many of the items" | Not Available | None | PF domain altered in SCI population | I or SA |
| Short form 12 (SF-12) | A shortened version of the SF-36, used in population studies | 12 | Physical functioning (2); role limitations because of physical health problems (2); bodily pain (1); general health (1); vitality (1); social functioning (1); role limitations because of emotional problems (2); | Total range 0–100 Divided into two summary scores: physical and mental components, are norm based; general population score 50 and standard | Summary scores free of floor and ceiling effects | 2–4 | Available in most major languages | Face validity supports low bias | I or SA |

*Continued*

**TABLE 1** Objective QOL Instruments—cont'd

| | | | | Objective tools | | | | | |
|---|---|---|---|---|---|---|---|---|---|
| Name | Description | Items | Domains (No. of questions) | Scoring and response | Measurement model | Time taken (min) | Alternative languages | Item bias | Interviewer (I) or self-administration |
| | | | general mental health (2) | deviation of 10<br>Higher is healthier | | | | | |
| Short form 6-disability (SF-6D) | A six-dimensional health state classification based on the SF-36 | 11 | Physical functioning (3); role limitation physical (1); role limitations emotional (1); social functioning (1); pain (2); mental health (2); vitality (1) | 0.29 (the worst health state)—1.00 (perfect or full health)<br>Preference-based weights were assigned to each response; negative responses detract from perfect health score<br>Higher is healthier | Floor effect PF (37%); ceiling effect RLP and E (55%), SF (50%), P (36%), MH (35%) | Not available | Available in most major languages | Face validity supports low bias | I or SA |

Reproduction of table showing the objective instruments of QOL according to MR Hill. Adapted with permission from: Hill, M., Noonan, V., Sakakibara, B. and the SCIRE research team. (2010). Quality of life instruments and definitions in individuals with spinal cord injury: a systematic review. Spinal Cord, 48, 438–450.

**TABLE 2** Subjective QOL Instruments.

| Subjective tools | | | | | | | | | |
|---|---|---|---|---|---|---|---|---|---|
| **Name** | **Description** | **Items** | **Domains (No. of questions)** | **Scoring and response** | **Measurement model** | **Time taken (min)** | **Alternative languages** | **Item bias** | **Interviewer (I) or self-administration** |
| Quality of life index (QLI) | Measures subjective QOL in terms of satisfaction | 32–37 | Health and functioning (8); social and economic (8); psychological and spiritual (8); family (8) | 6-point Likert scales for both importance and satisfaction sub-sections<br>Weigh satisfaction (S) scores with corresponding importance (I) scores; 3.5 subtracted from S scores, raw I score used. Total and sub-scale scores tabulated<br>Higher is healthier | None | ~10 | None for SCI version | SCI version evaluated by SCI patients | I or SA |
| Quality of life profile for adults with physical disabilities (QOLP-PD) | A holistic approach to QOL that empowers individuals in which elements of QOL are the same for people with and without disabilities, but each may address issues differently; questions address aspects of daily life | 102 | Being (32): physical, psychological, spiritual; belonging (37): physical, social, community; becoming (33): practical, leisure, growth | 5-point Likert item scales 1—not at all satisfied to 5—extremely satisfied and 1—not important to 5—very important<br>Weigh satisfaction and importance scores for each item; three points are subtracted from S scores, raw I scores used<br>Higher is healthier | None reported | Not available | None | Developed in disabled and SCI population | I or SA |

*Continued*

**TABLE 2** Subjective QOL Instruments—cont'd

| Subjective tools | | | | | | | | | |
|---|---|---|---|---|---|---|---|---|---|
| Name | Description | Items | Domains (No. of questions) | Scoring and response | Measurement model | Time taken (min) | Alternative languages | Item bias | Interviewer (I) or self-administration |
| Satisfaction with life survey (SWLS) | Addresses life satisfaction as a whole, reflecting a global perspective of individual's values | 5 | In most ways my life is close to ideal; the conditions of my life are excellent; I am satisfied with my life; so far I am getting the important things I want in life; if I could live my life over, I would change almost nothing | 7-point Likert scale<br>Global score is computed<br>Higher is healthier | Floor effects seen on some items | <5 | Various | Face validity supports low bias; some questions | I or SA |
| Sense of well-being index (SWBI) | Measures subjective QOL for people with disabilities in work rehabilitation | 26 | Physical well-being and associated feelings about self (6); psychological well-being (7); family and social well-being (6); financial well-being (8) | 4-point Likert scale<br>Tabulate total and domain scores<br>Higher is healthier | None reported | "Brief" | None | Altered for and evaluated in SCI population | I or SA |
| World Health Organization quality of life (WHOQOL-BREF) | Instrument that conceptually fits with the WHO definition of QOL | 26 | Physical health/capacity (7); psychological health/well-being (6); social relationships (3); environment (8); overall QOL (1); general health (1) | 5-point Likert scale<br>Domain scores calculated by multiplying mean of facet scores by four, and transformed onto a scale of 0–100<br>Higher is healthier | None; floor effect in one item: mobility (29.7%) | Not reported | Available in most major languages | Face validity supports low bias | SA |

Reproduction of the continuation of the table showing the subjective instruments of QOL according to MR Hill. Adapted with permission from: Hill, M., Noonan, V., Sakakibara, B. and the SCIRE research team. (2010). Quality of life instruments and definitions in individuals with spinal cord injury: a systematic review. Spinal Cord, 48, 438–450.

**TABLE 3** Global instruments.

| Instrument | Authorship | Available | Validated SCI | Items global QOL or HRQOL Yes Y (n)/No N | Mainly Subjective S/Objective O |
|---|---|---|---|---|---|
| CHART | Whiteneck 1992 | http://tbims.org/combi/chart/ | Yes | N | O |
| SF-36 | Ware 1992 | https://www.rand.org/health-care/surveys_tools/mos/36-item-short-form.html | Yes | N | O |
| QLI | Ferrans & Powers 1985 | http://qli.org.uic.edu/questionaires/questionnairehome.htm | Yes | Y (10%) | S |
| SWLS | Diener 1985 | https://eddiener.com/scales/7 | Yes | Y (100%) | S |
| LISAT-9 | Fugl-Meyer 1991 | *http://scireproject.com/* | Yes | Y (1) | S |
| QWB-SA | Andresen 1999 | https://hoap.ucsd.edu/qwb-info/ | No | N | O |
| WHOQOL-BREF | WHOQOL Group 1998 | https://www.who.int/healthinfo/survey/whoqol-qualityoflife/en/ | Yes | Y (2) | S |
| EQ-5-5 L; EQ-5-3 L | EuroQol Group 1990 | https://euroqol.org/support/how-to-obtain-eq-5d/ | No | Y (1) | O |

Characteristics of global instruments reported in the text. The author that originally reported each one, and on-line resource are shown. The fact that items about global QOL are asked in the questionnaire correlates with subjective view of the instrument.

**TABLE 4** Specific instruments for associated conditions in SCI.

| Instrument | Measure | Authorship | Available at | Validated for SCI |
|---|---|---|---|---|
| CHART | Disability | Whiteneck et al., 1992 | http://tbims.org/combi/chart/ | Yes |
| PRISM | Spasticity | Cook KF et al., 2007 | https://pubmed.ncbi.nlm.nih.gov/18247233/ | Yes |
| SCI-SET | Spasticity | Adams et al., 2007 | https://www.archives-pmr.org/action/showPdf?pii=S0003-9993%2807%2900426-1 | Yes |
| BPI | Pain | Cleeland CS, 1994 | *Long form:* https://www.mdanderson.org/content/dam/mdanderson/documents/Departments-and-Divisions/Symptom-Research/BPI-Long_English_SAMPLE.pdf *Short form:* https://www.mdanderson.org/content/dam/mdanderson/documents/Departments-and-Divisions/Symptom-Research/BPI-SF_English-24h_Original_SAMPLE.pdf | Yes |
| MPI | Pain | Widerström-Noga et al., 2002 | https://www.archives-pmr.org/article/S0003-9993(02)96227-1/fulltext#secd11784441e1566 | Yes |

*Continued*

**TABLE 4** Specific instruments for associated conditions in SCI—cont'd

| Instrument | Measure | Authorship | Available at | Validated for SCI |
|---|---|---|---|---|
| MPRCQ2 | Pain | Nielson WR et al., 2003 | https://www.ncbi.nlm.nih.gov/pmc/articles/PMC2758642/pdf/nihms125144.pdf | Yes |
| WUSPI | Pain | Curtis KA et al., 1995 | | Yes |
| NBDS | Neurogenic bowel dysfunction | Krogh K et al. 2005 | Long form: https://www.coloplast.ca/Global/3_Bladder%20and%20Bowel/Pictures/Peristeen%20HCP/Peristeen_NBD_Score_Questionaire_A4.pdf | Yes |
| FIQOL | Fecal incontinence | Rockwood et at, 2000 | https://pubmed.ncbi.nlm.nih.gov/10813117/ | No |
| FICQOL | Fecal incontinence and constipation | Nanigian et al., 2008 | https://pubmed.ncbi.nlm.nih.gov/18721959/ | Yes (Spina Bifida) |
| Qualiveen-30 | Neurogenic bladder dysfunction | Costa, P, 2001 | https://pubmed.ncbi.nlm.nih.gov/11173948/ | Yes |
| SF-Qualiveen | Neurogenic bladder dysfunction | Bonniaud,V 2008 | https://pubmed.ncbi.nlm.nih.gov/18950816/ | Yes |
| SCI-QOL Bladder Complications | Urinary tract infections | Tulsky,D·S, 2015 | https://www.ncbi.nlm.nih.gov/pmc/articles/PMC4445020/ | Yes |
| SIAS | Sexuality | Kreuter,M 1994 | https://pubmed.ncbi.nlm.nih.gov/7885719/ | Yes |
| FSFI | Sexual function in women | Rosen, R.C. 2000 | https://pubmed.ncbi.nlm.nih.gov/10782451/ | No |
| IIEF | Sexual function in males | Rosen,R.C. 1997 | https://pubmed.ncbi.nlm.nih.gov/9187685/ | No |

Specific instruments that are used for measuring the impact of some conditions in the QOL of the person. The author that spread it is displayed. Abbreviations: PRISM, Patient Reported Impact of Spasticity Measure; SCI-SET, Spinal Cord Injury Spasticity Evaluation Tool; BPI, Pain Brief Inventory; MPI, Multi-dimensional Pain Inventory; MPRCQ2, Multi-dimensional Pain Readiness to Change Questionnaire; WUSPI, Wheelchair Users Shoulder Pain Index; NBDS, Neurogenic Bowel Dysfunction Score; FIQOL, Fecal Incontinence Quality of Life Scale; FICQOL, Fecal Incontinence and Constipation Quality of Life; SIAS, Sexual Interest, Activity and Satisfaction; FSFI, Female Sexual Function Index; IIEF, International Index of Erectile Function.

There are also many other tools for evaluating different aspects of people with SCI in relation with their QOL, as measures of health status, physical function, mental function, and participation. We display some examples of tools referenced in the literature that are useful in SCI population studies (Table 5).

## Integrative quality of life assessment instruments

The **Medical Outcomes Study 36-Item Short Form Health Survey (SF-36)** was developed in the early 90s (Ware Jr. & Sherbourne, 1992). It is a generic scale that provides a profile of the health status (physical and mental components) and has been useful for assessing HRQOL in the general population and also in people with SCI. This scale has been validated and used in several languages and a short version of the SF-36 has been constructed (SF12) with similar validity (Iorio-Morin et al., 2018). However, all the above-mentioned is not an impediment to point out there are certain difficulties for its use in SCI persons, as some of the items are not easy to apply in this population due to mobility issues (Andresen, Gravitt, Aydelotte, & Podgorski, 1999; Tate et al., 2002) and modifications were done adapted to this objective, either by reformulating the items (Froehlich-Grobe, Andresen, Caburnay, & White, 2008) or adapting new items to the disability (SF-36 V) (Kazis et al., 2004; Luther et al., 2006). Nonetheless, it is one of the most widely used reference instruments,

**TABLE 5** General instruments referenced for SCI assessments.

| Instrument | Measure | Assessing | Refer to |
|---|---|---|---|
| Sickness Impact Profile | Health status | Person's perception of their health status in six sub-scales with respect to their disease impact. | Bergner M et al. 1981 https://physio-pedia.com/File:SIP.png |
| Nottingham Health Profile | Health status | Person's perception of emotional, social and physical health problems. | Hunt SM et al. 1981 https://doi.org/10.1016/0271-7123(81)90005-5 |
| CHART | Participation | Ability in six domains with respect to maintain typical community roles | Whiteneck GG et al., 1992 http://tbims.org/combi/chart/ |
| FIM | Disability | Records the severity of disability of rehabilitation patients in six areas | Linacre JM et al. 1994 https://www.physio-pedia.com/File:Functional_Independence_Measure.jpg |
| SCIM-III | Disability | Independence in ADL, mobility and bladder & bowel function specific for SCI people | Itzkovich M et al. 2007 https://scireproject.com/wp-content/uploads/SCIM_Toolkit_Printable-1.pdf |
| Neuro-QoL Lower Extremity Function | Function | Ability or difficulty for mobility in different environment | healthmeasures.net https://www.healthmeasures.net/index.php?option=com_instruments&view=search&Itemid=977 |
| Patient Health Questionnaire | Mental function | Diagnostic criteria for major depressive disorder | Kroenke K et al. 2001 https://www.icsi.org/guideline/depression/appendix-a-patient-health-questionnaire-phq-9/ |
| Hospital Depression and Anxiety Scale | Mental function | States of depression and anxiety in the setting of a hospital medical outpatient clinic | Bjelland I et al. 2002 https://www.researchgate.net/figure/The-standard-Hospital-anxiety-and-depression-scale-questionnaire-in-English_fig1_321034906 |
| Brief Symptom Inventory | Mental function | Global psychological distress, covering nine symptom dimensions of depression | Derogatis LR, Melisaratos N. 1983 http://psychcorp.pearsonassessments.com/HAIWEB/Cultures/en-us/Productdetail.htm?Pid=PAbsi |
| DASS-21 | Mental function | Screening for depression, anxiety and stress | Mitchell MC et al. 2008 https://maic.qld.gov.au/wp-content/uploads/2016/07/DASS-21.pdf |
| SCL CSQ | Mental function & general | Three domains about coping with SCI: acceptance of injury, social reliance and fighting spirit | Elfström ML et al. 2007 http://scireproject.com/wp-content/uploads/Clinician-Summary-v.5.0_SCL_CSQ.pdf |

There are several measurement instruments used in combination to QOL scales. Some representative tools are displayed, with the author that spread it. Inside the PROMIS-NEURO-QoL system, there are several measures that are useful in many function and health domains. Healthmeasures.net is referred for obtaining them. Abbreviations: CHART, Craig Handicap Assessment and Reporting Technique; FIM, Functional Independence Measurement; SCIM-III, Spinal Cord Independence Measurement v.III; DASS-21, Depression Anxiety Stress Scale; SCLCSQ, Spinal Cord Lesion-related Coping Strategies Questionnaire.

with high consistency and reproducibility. It measures the following domains: physical, mental, emotional, pain, vitality, social functioning, and health in general.

The **Quality of Life Index** (QLI) has been widely used since its proposal by Ferrans and Powers and has been adapted to SCI (May & Warren, 2002). It allows assessing general QOL with social, spiritual, and economic aspects, and has been applied to SCI populations (Mortenson, Noreau, & Miller, 2010), associated or not to other objective instruments, e.g., assessment of bladder dysfunction (Best, Ethans, Craven, Noreau, & Hitzig, 2017). It is available in numerous languages and thus very useful to evaluate this population, although its psychometric properties have not been completely established in SCI (Miller et al., 2010).

Another known instrument for global assessment is the **Satisfaction with Life Scale** (SWLS) (Pavot & Diener, 1993), validated for people with SCI (Dijkers, 1999) and widely used in different studies in subjects with SCI (Post, van Leeuwen, van Koppenhagen, & de Groot, 2012), including assessment of rehabilitation outcomes (Hicks et al., 2005). This tool is recommended for its shortness (five items) and easy-to-use in daily clinical practice; it has an acceptable consistency and reliability for comparative studies (Miller et al., 2010). Despite its wide use, larger psychometric studies should be performed in subjects with SCI (Geyh et al., 2010).

The **Life Satisfaction Questionnaire** (LISAT) developed by Fugl-Meyer in 1991, has been applied in several studies, mainly European, in its nine-item version, showing correct validity in relation with other QOL instruments (Post & Noreau, 2005; M. W. Post et al., 2012) and correct internal consistency (Geyh et al., 2010). However, its use in assessment studies with the SCI population is scarce.

The **Quality of Well-Being** (QWB) scale was used in numerous studies since its development by Andresen in 1998. Later, the self-administered version (QWB-SA) was created. However, this is an instrument with a high number of items (71) and there are few references on its validity in people with SCI (Andresen et al., 1999; Whitehurst, Engel, & Bryan, 2014).

During the decade of the 90s the WHO Quality of Life group developed the research of instruments for assessing QOL (The WHOQOL Group, 1998) in people affected by diverse conditions (AIDS, mental disorders, cancer, transplant recipients, etc.). Thus, in 1995 the **WHOQOL-100** questionnaire was created, and later its reduced version—the **WHOQOL-BREF**—was created (https://www.who.int/mental_health/publications/whoqol/en/). It has been applied on people with SCI since 2004 ( Jang, Hsieh, Wang, & Wu, 2004) and validated in different languages, together with the 26-item short version. This version (WHOQOL-BREF) has demonstrated good applicability in clinical diagnosis with sufficient psychometric characteristics, although some of its domains (social) have less consistency due to the low number of items (Skevington, Lotfy, & O'Connell, 2004). Many validation studies against other reference instruments such as the SF-36 have been carried out (Lin, Hwang, Chen, & Chiu, 2007), showing lower ceiling and floor percentages. Similarly, in general, its items are better accepted by people with SCI than the SF-36. Its validity for studying global QOL and its relationship with physical, psychological, and relation with the surroundings, makes it interesting for studying the QOL in people with SCI-related sequelae (Salvador-De La Barrera et al., 2018).

In the early 90s, the European EuroQol-Group (1990) developed an instrument for measuring HRQOL. Five assessment domains were established: mobility, daily activities and self-care, psychological functioning, social and role performance, and pain or other health problems. Furthermore, a grading of global QOL was added in the form of a visual analogue scale (VAS) (Devlin & Brooks, 2017). This instrument began to be widely used and two versions were made: the **EQ-5D** and the **EQ-3D** that differentiate in the number of levels that characterize each domain, based on the levels of difficulty or affectation (five or three degrees, respectively). The EuroQol instrument has been widely used, translated into several languages, and applied to people with SCI; however, it has not been sufficiently validated for this population, with the disadvantage that mobility items are drafted incorrectly for wheelchair users (Whitehurst, Noonan, Dvorak, & Bryan, 2012).

## Health-related quality of life assessment instruments for conditions resulting from spinal cord injury

Numerous instruments have been proposed to assess the effects the different sequelae and dysfunctions caused by devices have on QOL in people with SCI. Give the diversity of scales and instruments it is not possible to discuss all of them (Table 4). We have selected a sample of the most referenced ones in the medical literature, e.g., assessment of the impact of conditions such as bladder dysfunction, pain, spasticity, or sexual function.

- Assessment of the impact of neurogenic bladder on the quality of life of patients with spinal cord injury

The Qualiveen-30 questionnaire was developed by Costa (Costa et al., 2001) to specifically measure the impact of bladder dysfunction on QOL in spinal cord-injured patients. It is a 30-item subjective questionnaire comprising four domains, each considering different aspects derived from alterations in bladder emptying (limitations, constraints, fears, and feelings).

The validity of the questionnaire in patients with SCI has been tested in many studies (Bonniaud, Bryant, Parratte, & Guyatt, 2008a; Nikfallah et al., 2015) and stands out for its high reproducibility and internal consistency (Nikfallah et al., 2015). Furthermore, the questionnaire has been validated by comparing it to other QOL questionnaires, such as the Kings Health Questionnaire (Karapolat, Akkoç, Eyigör, & Tanıgör, 2018). The use of the Qualiveen-30 questionnaire has been extended to other medical specialties, e.g., multiple sclerosis (Milinis, Tennant, A Young, & Group, 2017).

In 2008, the short version of this questionnaire was developed, the SF-Qualiveen (Bonniaud et al., 2008a), with eight items and four domains.

Recently published reviews recommend Qualiveen-30 and SF-Qualiveen and identify them as the best options to assess the QOL in people suffering from neurogenic bladder after SCI (Best et al., 2017).

- Assessment of the impact of spasticity on the quality of life of people with spinal cord injury

Spasticity is a common impairment associated to SCI and considered a limiting factor by many patients (Levi, Hultling, Nash, & Seiger, 1995). To date, few scales are available for patients with SCI. However, in a systematic review carried out by Balioussis (Balioussis, Hitzig, Flett, Noreau, & Craven, 2014), the authors concluded that the Patient-Reported Impact of Spasticity Measure (PRISM) is superior to the Spasticity Evaluation Tool (SCI-SET), because besides including more psychometric data, PRISM is more sensitive for measuring the impact of spasticity and represents more affective values.

**PRISM** is a subjective scale developed by Cook (Cook et al., 2007), a 41-item tool divided into seven sub-scales. Currently it is only available in English and Serbian, has good reliability and reproducibility (Hill et al., 2010), and has been validated for patients with multiple sclerosis (Knezevic et al., 2015). Several studies support the usefulness of this measuring tool (Hill et al., 2010; Sweatman, Heinemann, Furbish, & Field-Fote, 2020; Whitehurst et al., 2012).

- Assessment of the impact of pain on the quality of life of patients with spinal cord injury

Although exact prevalence of chronic pain in patients with SCI is variable, this condition is considered a major limiting factor, causing huge impact on QOL (van Gorp, Kessels, Joosten, van Kleef, & Patijn, 2015).

The **Brief Pain Inventory (BPI)** is an instrument designed to assess the severity of pain and its impact on ADL. It was shaped based on McGill's pain questionnaire, translated to several languages, and may be completed as a self-report or interview (Cleeland & Ryan, 1994). This instrument is available in two formats: the BPI short (used for clinical trials and translations to other languages) and long (with additional descriptive items) forms (de Andrés Ares et al., 2015).

There are three modified versions of this inventory for the SCI population (7, 10, and 12 items). Five additional items make up the 12-item scale (self-care, recreational activities, social activities, communication, learning new information) and only the first three ones are used to complete the 10-item scale. Its clinical use has been confirmed in different studies (Dawu et al., 2019; Hand, Velozo, & Krause, 2018).

Regarding its reliability, internal consistency of the BPI is high for the three modified versions and as for the validity, it has high correlation with pain intensity numerical rating scales and with the Short Form-36 (SF-36) scale (Raichle, Osborne, Jensen, & Cardenas, 2006).

- Assessment of the impact on sexual dysfunction

Preserving sexual function is one of the high priority issues of subjects after SCI. Subjective scales were developed at the end of the 80s and early 90s to assess QOL associated to sexual function (The SCIRE Project; https://scireproject.com/outcome-measures/). However, in the systematic review carried out by Abrahmson (Abramson, McBride, Konnyu, Elliott, & Team, 2008) the authors pointed out that none of the scales have the ability to assess the most relevant aspects that define sexual health of this population group. On the same line are the recently published ISCOS guidelines, which establish the recommendations on data collection when analysis of sexual function is done segregated by gender: International SCI Male Sexual Function Basic Data Set and the International SCI Female Sexual and Reproduction Function Basic Data Set (Alexander et al., 2017).

At present, the most widely used sexual function scales in people with SCI are the International Index of Erectile Function (IIEF) and the Female Sexual Function Index (FSFI), although neither one is sufficiently validated in patients with spinal cord injury (S. Elliott & Querée, 2018) and cannot be considered QOL assessment instruments. Recently (2019), a questionnaire within the SCI-High Project has been developed—the SCI-High Sexual Health Patient Questionnaire—and is pending validation (S. Elliott et al., 2019).

Finally, it is important to point out the need for widely used, validated, and accepted instruments, and we support the initiative of the International SCI Data Sets Project of the international societies ASIA and ISCOS (Biering-Sørensen et al., 2006; Tate & Forchheimer, 2014). Although the SCI-QOL-Basic Data Set (QOL-BDS) questionnaire (Charlifue et al., 2012) is not widely distributed and there is scarce experience, it has shown sufficient consistency for its use in studies and data collection databases (Post et al., 2019).

## Conclusions

When assessing the QOL of people with SCI-related sequelae, the subjective view of the person must always be taken into consideration, choosing the instrument as per the global or specific aspect to be studied.

Global QOL assessment instruments are the most widely used covering many domains linked to the medical condition of the person. When studying HRQOL, an objective approach is done defining the domains possibly affected by SCI, without putting aside the subjective perspective of the person.

The use of instruments focused on a specific SCI-related condition or sequelae should be based on wide studies. Furthermore, there must be agreements for their validity in many social and cultural environments.

## Applications to other areas of neuroscience

Assessment of quality of life (QOL) in handicapped people is part of the rehabilitation and measurement of outcomes. In the case of spinal cord injuries (SCI), sequelae resulting from the lesion or disease affect the person's live and measuring instruments are essential. Simultaneously, the consequences on the person's psychological sphere have an effect on the subjective component. Thus, many of the studies performed to assess these instruments belong to the field of psychology.

On the other hand, major advances have been made on the knowledge and treatment of SCI-related sequelae, although their impact on QOL and acceptance by the affected subject, have not always been sufficiently assessed, e.g., the different alternatives for the management of bladder dysfunction (Best et al., 2017). Thus, the use of instruments that measure health-related QOL, when studying outcomes of new therapeutic methods for chronic sequelae, is of great importance.

Similarly, measuring QOL is recommended after rehabilitation and when using new therapies aimed at motor recovery (Hicks et al., 2005), such as robotic systems for walking and technical help for mobility, whose integration into daily life may affect his/her perspective of QOL. This assessment will also be of great help to show the advances of neurobiological therapies of spinal cord injury.

Usually, global QOL instruments are not specific for SCI and have been adapted to diverse neurological pathologies, which allow their comparison in many populations. Instruments that are specific for SCI are usually validated for subjects with multiple sclerosis (Bonniaud, Bryant, Parratte, & Guyatt, 2008b), among other conditions.

Finally, we reiterate the need of international validation and agreement regarding the instruments (D. Tate & Forchheimer, 2014), as this will allow making comparisons between diverse cultures and conditions.

## Dictionary of terms

**Quality of life (QOL):** Individual's perception of their position in life in the context of the culture and value systems in which they live and in relation to their goals, expectations, standards and concerns.

**Health:** a state of complete physical, mental and social well-being and not merely the absence of disease or infirmity.

**Health-related quality of life (HRQOL):** Those aspects of QOL that are affected by the state of physical or mental health.

**Disability:** an umbrella term, covering impairments, activity limitations and participation restrictions, reflecting an interaction between features of a person's body and features of the society in which he or she lives.

**Functioning**: is a term that denotes the positive aspects of the interaction between a person's health—body structures and their limitations—and his contextual (environmental and personal) factors.

**Domain**: is an area of knowledge, emotions, or abilities, which are distinguished from others, being measured or examined by a test.

**Spasticity:** disordered sensory-motor control, resulting from an upper motor neuron lesion, presenting as intermittent or sustained involuntary activation of muscle.

**Neuropathic pain:** Pain caused by a lesion or disease of the somatosensory nervous system.

**Neurogenic bladder:** abnormal or difficult function of the bladder, urethra, sphincter system in mature individuals, in the context of clinically confirmed relevant neurologic disorder.

**Validity:** ability of the instrument to measure the construct that it intends to measure and for which it was designed.

## Key facts of measuring quality of life in sci people

- Quality of life (QOL) is a subjective multi-dimensional concept that may be assessed with instruments not specific for people with spinal cord injuries (SCI).
- There is no universal agreement on the assessment of QOL in this population.
- For assessing the impact of SCI-related sequelae on the person's QOL, instruments aiming objective measures are used (Health-related QOL [HRQOL] instruments).
- Instruments for objective measurements of HRQOL must also include items for a subjective assessment of QOL by the person.
- Multi-center agreements should be reached regarding the numerous HRQOL measuring instruments to achieve greater diffusion.

## Summary


- Spinal cord injuries–related sequelae lead to a series of daily difficulties for the affected person that may further hinder his/her quality of life.
- Quality of life (QOL) is a *subjective* concept defined as "*individual's perception of their position in life in the context of the culture and value system in which they live and in relation to their goals, expectations, standards, and concerns.*"
- QOL may be viewed from a subjective perspective by the affected person or objective by the observer, then including QOL aspects affected by the disease or lesion (health-related quality of live—HRQOL).
- Global QOL questionnaires, such as WHOQOL, SWLS, QLI, SF-36, and EuroQol, are the most commonly used assessment tools.
- Specific instruments to assess the impact that certain conditions associated to spinal cord injuries (SCI) have on QOL have been developed, e.g., bladder dysfunction, spasticity, as well as others adapted from other populations, such as neuropathic pain and sexual dysfunction.
- It is important to establish international agreements regarding the use of validated instruments that can be compared in diverse populations for measuring different aspects of QOL in people with SCI.


## References


Abramson, C. E., McBride, K. E., Konnyu, K. J., Elliott, S. L., & Team, S. R. (2008). Sexual health outcome measures for individuals with a spinal cord injury: A systematic review. *Spinal Cord*, *46*(5), 320–324.

Alexander, M. S., New, P. W., Biering-Sørensen, F., Courtois, F., Popolo, G. D., Elliott, S., et al. (2017). International spinal cord injury male sexual function and female sexual and reproductive function basic data sets-version 2.0. *Spinal Cord Series And Cases*, *3*, 17050.

Andresen, E. M., Gravitt, G. W., Aydelotte, M. E., & Podgorski, C. A. (1999). Limitations of the SF-36 in a sample of nursing home residents. *Age and Ageing*, *28*(6), 562–566.

Balioussis, C., Hitzig, S. L., Flett, H., Noreau, L., & Craven, B. C. (2014). Identifying and classifying quality of life tools for assessing spasticity after spinal cord injury. *Topics in Spinal Cord Injury Rehabilitation*, *20*(3), 208–224.

Best, K. L., Ethans, K., Craven, B. C., Noreau, L., & Hitzig, S. L. (2017). Identifying and classifying quality of life tools for neurogenic bladder function after spinal cord injury: A systematic review. *The Journal of Spinal Cord Medicine*, *40*(5), 505–529.

Biering-Sørensen, F., Charlifue, S., DeVivo, M., Noonan, V., Post, M., Stripling, T., et al. (2006). International spinal cord injury data sets. *Spinal Cord*, *44*(9), 530–534.

Bökel, A., Dierks, M. L., Gutenbrunner, C., Weidner, N., Geng, V., Kalke, Y. B., et al. (2020). Perceived environmental barriers for people with spinal cord injury in Germany and their influence on quality of life. *Journal of Rehabilitation Medicine*, *52*(8), jrm00090.

Bonniaud, V., Bryant, D., Parratte, B., & Guyatt, G. (2008a). Development and validation of the short form of a urinary quality of life questionnaire: SF-Qualiveen. *The Journal of Urology*, *180*(6), 2592–2598.

Bonniaud, V., Bryant, D., Parratte, B., & Guyatt, G. (2008b). Qualiveen, a urinary-disorder specific instrument: 0.5 corresponds to the minimal important difference. *Journal of Clinical Epidemiology*, *61*(5), 505–510.

Charlifue, S., Post, M. W., Biering-Sorensen, F., Catz, A., Dijkers, M., Geyh, S., et al. (2012). International spinal cord injury quality of life basic data set. *Spinal Cord*, *50*(9), 672–675.

Cleeland, C. S., & Ryan, K. M. (1994). Pain assessment: Global use of the brief pain inventory. *Annals of the Academy of Medicine, Singapore*, *23*(2), 129–138.

Cook, K. F., Teal, C. R., Engebretson, J. C., Hart, K. A., Mahoney, J. S., Robinson-Whelen, S., et al. (2007). Development and validation of patient reported impact of spasticity measure (PRISM). *Journal of Rehabilitation Research and Development*, *44*(3), 363–371.

Costa, P., Perrouin-Verbe, B., Colvez, A., Didier, J., Marquis, P., Marrel, A., et al. (2001). Quality of life in spinal cord injury patients with urinary difficulties. Development and validation of qualiveen. *European Urology*, *39*(1), 107–113.

Dawu, W., Chao, H., Kaiting, L., Lan, S., Xiaohua, P., & Dingqun, B. (2019). Pain correlates with social integration in individuals with traumatic spinal cord injury: A cross-sectional survey. *Journal of Rehabilitation Medicine*, *51*(7), 506–512.

de Andrés Ares, J., Cruces Prado, L. M., Canos Verdecho, M. A., Penide Villanueva, L., Del Valle Hoyos, M., Herdman, M., et al. (2015). Validation of the short form of the brief pain inventory (BPI-SF) in Spanish patients with non-cancer-related pain. *Pain Practice*, *15*(7), 643–653.

Devlin, N. J., & Brooks, R. (2017). EQ-5D and the EuroQol group: Past, present and future. *Applied Health Economics and Health Policy*, *15*(2), 127–137.

Dijkers, M. (1997). Quality of life after spinal cord injury: A meta analysis of the effects of disablement components. *Spinal Cord*, *35*(12), 829–840.

Dijkers, M. P. (1999). Correlates of life satisfaction among persons with spinal cord injury. *Archives of Physical Medicine and Rehabilitation*, *80*(8), 867–876.

Dijkers, M. P. (2003). Individualization in quality of life measurement: Instruments and approaches. *Archives of Physical Medicine and Rehabilitation*, *84*(4 Suppl 2), S3–14.

Dijkers, M. P. (2005). Quality of life of individuals with spinal cord injury: A review of conceptualization, measurement, and research findings. *Journal of Rehabilitation Research and Development*, *42*(3 Suppl 1), 87–110.

Elliott, S., Jeyathevan, G., Hocaloski, S., O'Connell, C., Gulasingam, S., Mills, S., et al. (2019). Conception and development of sexual health indicators to advance the quality of spinal cord injury rehabilitation: SCI-high project. *The Journal of Spinal Cord Medicine*, *42*(sup1), 68–84.

Elliott, S., & Querée, M. (2018). Sexual and reproductive health following spinal cord injury. In J. J. Eng, R. W. Teasell, W. C. Miller, D. L. Wolfe, A. F. Townson, J. T. C. Hsieh, & M. Querée (Eds.), *Spinal cord injury rehabilitation evidence. Version 6.0* (pp. 1–133). Vancouver.

EuroQol-Group. (1990). EuroQol—a new facility fpr the measurement of health-related quality of life. *Health Policy*, *16*(3), 199–208.

Froehlich-Grobe, K., Andresen, E. M., Caburnay, C., & White, G. W. (2008). Measuring health-related quality of life for persons with mobility impairments: An enabled version of the short-form 36 (SF-36E). *Quality of Life Research*, *17*(5), 751–770.

Fuhrer, M. J., Rintala, D. H., Hart, K. A., Clearman, R., & Young, M. E. (1992). Relationship of life satisfaction to impairment, disability, and handicap among persons with spinal cord injury living in the community. *Archives of Physical Medicine and Rehabilitation*, *73*(6), 552–557.

Geyh, S., Fellinghauer, B. A., Kirchberger, I., & Post, M. W. (2010). Cross-cultural validity of four quality of life scales in persons with spinal cord injury. *Health and Quality of Life Outcomes*, *8*, 94.

Hand, B. N., Velozo, C. A., & Krause, J. S. (2018). Measuring the interference of pain on daily life in persons with spinal cord injury: A Rasch-validated subset of items from the brief pain inventory interference scale. *Australian Occupational Therapy Journal*, *65*(5), 405–411.

Hicks, A. L., Adams, M. M., Martin Ginis, K., Giangregorio, L., Latimer, A., Phillips, S. M., et al. (2005). Long-term body-weight-supported treadmill training and subsequent follow-up in persons with chronic SCI: Effects on functional walking ability and measures of subjective well-being. *Spinal Cord*, *43*(5), 291–298.

Hill, M. R., Noonan, V. K., Sakakibara, B. M., & Miller, W. C. (2010). Quality of life instruments and definitions in individuals with spinal cord injury: A systematic review. *Spinal Cord*, *48*(6), 438–450.

Iorio-Morin, C., Noonan, V. K., White, B., Noreau, L., Leblond, J., Dumont, F. S., et al. (2018). Quality of life and health utility scores among Canadians living with traumatic spinal cord injury—a National Cross-Sectional Study. *Spine (Phila Pa 1976)*, *43*(14), 999–1006.

Jang, Y., Hsieh, C. L., Wang, Y. H., & Wu, Y. H. (2004). A validity study of the WHOQOL-BREF assessment in persons with traumatic spinal cord injury. *Archives of Physical Medicine and Rehabilitation*, *85*(11), 1890–1895.

Karapolat, H., Akkoç, Y., Eyigör, S., & Tanıgör, G. (2018). Bladder-related quality of life in people with neurological disorders: Reliability and validity of the Turkish version of the King's health questionnaire in people with spinal cord injury. *Turkish Journal of Urology*, *44*(5), 411–417.

Kazis, L. E., Miller, D. R., Clark, J. A., Skinner, K. M., Lee, A., Ren, X. S., et al. (2004). Improving the response choices on the veterans SF-36 health survey role functioning scales: Results from the veterans health study. *The Journal of Ambulatory Care Management*, *27*(3), 263–280.

Knezevic, T., Konstantinovic, L., Rodic, S., Foti, C., Drulovic, J., Dackovic, J., et al. (2015). Validity and reliability of the Serbian version of patient-reported impact of spasticity measure in multiple sclerosis. *International Journal of Rehabilitation Research*, *38*(3), 199–205.

Levi, R., Hultling, C., Nash, M. S., & Seiger, A. (1995). The Stockholm spinal cord injury study: 1. Medical problems in a regional SCI population. *Paraplegia*, *33*(6), 308–315.

Lin, M. R., Hwang, H. F., Chen, C. Y., & Chiu, W. T. (2007). Comparisons of the brief form of the World Health Organization quality of life and short Form-36 for persons with spinal cord injuries. *American Journal of Physical Medicine & Rehabilitation*, *86*(2), 104–113.

Luther, S. L., Kromrey, J., Powell-Cope, G., Rosenberg, D., Nelson, A., Ahmed, S., et al. (2006). A pilot study to modify the SF-36V physical functioning scale for use with veterans with spinal cord injury. *Archives of Physical Medicine and Rehabilitation*, *87*(8), 1059–1066.

May, L. A., & Warren, S. (2002). Measuring quality of life of persons with spinal cord injury: External and structural validity. *Spinal Cord*, *40*(7), 341–350.

Milinis, K., Tennant, A., Young, C. A., & TONiC Study Group. (2017). Rasch analysis of SF-Qualiveen in multiple sclerosis. *Neurourology and Urodynamics*, *36*(4), 1161–1166.

Miller, W., Sakakibara, B., Noonan, V., Tawashy, A., Aubut, J., Connolly, S., et al. (2010). Outcome measures. In J. J. Eng, R. W. Teasell, W. C. Miller, D. L. Wolfe, A. F. Townson, J. T. C. Hsieh, & B. M. Sakakibara (Eds.), *Spinal cord injury rehabilitation evidence* (pp. 1–147). Vancouver.

Moons, P., Budts, W., & De Geest, S. (2006). Critique on the conceptualisation of quality of life: A review and evaluation of different conceptual approaches. *International Journal of Nursing Studies*, *43*(7), 891–901.

Mortenson, W. B., Noreau, L., & Miller, W. C. (2010). The relationship between and predictors of quality of life after spinal cord injury at 3 and 15 months after discharge. *Spinal Cord*, *48*(1), 73–79.

Nikfallah, A., Rezaali, S., Mohammadi, N., Abrishamkar, M., Rezaei, E., Sahraian, M. A., et al. (2015). Translation, cultural adaptation and validation of the Qualiveen-30 questionnaire in Persian for patients with spinal cord injury and multiple sclerosis. *Lower Urinary Tract Symptoms*, *7*(1), 42–49.

Pavot, W., & Diener, E. (1993). Review of the satisfaction with life scale. *Psychological Assessment*, *5*, 164–172.

Post, M. W. (2014). Definitions of quality of life: What has happened and how to move on. *Topics in Spinal Cord Injury Rehabilitation*, *20*(3), 167–180.

Post, M. W. M., Forchheimer, M. B., Charlifue, S., D'Andréa Greve, J. M., New, P. W., & Tate, D. G. (2019). Reproducibility of the international spinal cord injury quality of life basic data set: An international psychometric study. *Spinal Cord*, *57*(11), 992–998.

Post, M., & Noreau, L. (2005). Quality of life after spinal cord injury. *Journal of Neurologic Physical Therapy*, *29*(3), 139–146.

Post, M. W., van Leeuwen, C. M., van Koppenhagen, C. F., & de Groot, S. (2012). Validity of the life satisfaction questions, the life satisfaction questionnaire, and the satisfaction with life scale in persons with spinal cord injury. *Archives of Physical Medicine and Rehabilitation*, *93*(10), 1832–1837.

Raichle, K. A., Osborne, T. L., Jensen, M. P., & Cardenas, D. (2006). The reliability and validity of pain interference measures in persons with spinal cord injury. *The Journal of Pain*, *7*(3), 179–186.

Salvador-De La Barrera, S., Mora-Boga, R., Ferreiro-Velasco, M. E., Seoane-Pillado, T., Montoto-Marqués, A., Rodríguez-Sotillo, A., et al. (2018). A validity study of the Spanish-World Health Organization quality of life short version instrument in persons with traumatic spinal cord injury. *Spinal Cord*, *56*(10), 971–979.

Skevington, S. M., Lotfy, M., & O'Connell, K. A. (2004). The World Health Organization's WHOQOL-BREF quality of life assessment: Psychometric properties and results of the international field trial. A report from the WHOQOL group. *Quality of Life Research*, *13*(2), 299–310.

Sweatman, W. M., Heinemann, A. W., Furbish, C. L., & Field-Fote, E. C. (2020). Modified PRISM and SCI-SET spasticity measures for persons with traumatic spinal cord injury: Results of a Rasch analyses. *Archives of Physical Medicine and Rehabilitation*, *101*(9), 1570–1579.

Tate, D., & Forchheimer, M. (2014). Review of cross-cultural issues related to quality of life after spinal cord injury. *Topics in Spinal Cord Injury Rehabilitation*, *20*(3), 181–190.

Tate, D. G., Kalpakjian, C. Z., & Forchheimer, M. B. (2002). Quality of life issues in individuals with spinal cord injury. *Archives of Physical Medicine and Rehabilitation*, *83*(12 Suppl 2), S18–S25.

The WHOQOL Group. (1998). Development of the World Health Organization WHOQOL-BREF quality of life assessment. *Psychological Medicine*, *28*, 551–558.

Tulsky, D. S., Kisala, P. A., Victorson, D., Choi, S. W., Gershon, R., Heinemann, A. W., et al. (2015). Methodology for the development and calibration of the SCI-QOL item banks. *The Journal of Spinal Cord Medicine*, *38*(3), 270–287.

van Gorp, S., Kessels, A. G., Joosten, E. A., van Kleef, M., & Patijn, J. (2015). Pain prevalence and its determinants after spinal cord injury: A systematic review. *European Journal of Pain*, *19*(1), 5–14.

Ware, J. E., Jr., & Sherbourne, C. D. (1992). The MOS 36-item short-form health survey (SF-36). I. Conceptual framework and item selection. *Medical Care*, *30*(6), 473–483.

Whitehurst, D. G., Engel, L., & Bryan, S. (2014). Short form health surveys and related variants in spinal cord injury research: A systematic review. *The Journal of Spinal Cord Medicine*, *37*(2), 128–138.

Whitehurst, D. G., Noonan, V. K., Dvorak, M. F., & Bryan, S. (2012). A review of preference-based health-related quality of life questionnaires in spinal cord injury research. *Spinal Cord*, *50*(9), 646–654.

Wong, A. W. K., Lau, S. C. L., Fong, M. W. M., Cella, D., Lai, J. S., & Heinemann, A. W. (2018). Conceptual underpinnings of the quality of life in neurological disorders (Neuro-QoL): Comparisons of core sets for stroke, multiple sclerosis, spinal cord injury, and traumatic brain injury. *Archives of Physical Medicine and Rehabilitation*, *99*(9), 1763–1775.

Chapter 15

# S100b in spinal cord injury

**Andreas Zigouris, Georgios Kafritsas, George A. Alexiou, and Spyridon Voulgaris**
*Department of Neurosurgery, University Hospital of Ioannina, Ioannina, Greece*

## List of abbreviations

**CSF** cerebrospinal fluid
**SCI** spinal cord injury
**MRI** magnetic resonance imaging
**CNS** central nervous system
**MS** multiple sclerosis

## Introduction

Spinal cord injury (SCI) results in severe social–economic problems, because it is observed mostly among younger adults. The incidence of SCI has been reported at 10.5 cases per 100,000 people (Kumar, Lim, Mekary, et al., 2018). This type of injury occurs in young to middle-aged populations due to road traffic accidents, violence, and contact sports. Unfortunately, only 1% of SCI patients experience injuries that fully resolve. Almost half of cases suffer of severe neurological loss (para/tetraparesis, para/tetraplegia) with or without respiratory disturbances (Albayar, Roche, Swiatkowski, et al., 2019).

Several prognostic factors have been proposed according to the previously published studies. But there is still no consensus on which clinical diagnostic tests could be applied to reliably indicate the sub-clinical neurophysiological changes occurring in the spinal cord during the acute phase, and this affects the prediction of cell death and tissue destruction in SCI. Neurological examination provide a general indication of spinal cord neurological function, but are unreliable, especially during the acute phase (first 24 h) after SCI as a result of spinal shock in many patients (Cao, Yang, Liu, et al., 2008). MRI provides an accurate anatomical diagnosis, but it is impossible to distinguish neuronal tissue necrosis from post-traumatic edema. Most of the medical decisions are targeting the stabilization of the patients and preventing further injury, but no definitive treatment for the present state of the CNS trauma exists (Albayar et al., 2019). In these cases, where the real-time injury is the question in clinical diagnosis and prognosis, serum and CSF biomarker evaluation provides another option. A biomarker must be accurate, sensitive, specific, and provide high predictive value. The most widely used biomarker in such cases is S100b, a protein found in cells of neural crest derivation, which can be measured in the serum after disruption of the blood brain barrier (Wolf, Krall, Pajenda, et al., 2014, b). S100b is implicated in a very broad number of CNS conditions, and as such, it is perhaps one of the strongest candidate markers (Sen and Belli, 2007). In case of a CNS injury, accompanied by a nervous tissue and cellular damage, this structural protein is released from nervous cells and its concentration increase extra-cellularly—including CSF and blood. In these cases, S100b could be a potential biomarker of nervous tissue damage (Hajduková, Sobek, Prchalová, et al., 2015).

## S100b characteristics

The S-100 protein family constitutes a sub-group of $Ca^{+2}$-binding proteins. The term "S-100" was used because this protein was soluble in 100% ammonium sulfate and refers to a mixture of dimeric proteins consisting of two sub-units termed α and β (Sen and Belli, 2007). Three isoforms are known. S100a (αβ) is found in glial cells and melanocytes, S-100b (ββ) is present in high concentrations in glial cells and Schwann cells of the central and peripheral nervous systems as well as in Langerhans cells and cells of the anterior pituitary gland and S-100a0 (αα), which represents 5% of the S-100 protein in the brain (hippocampus, olfactory cells). The last is found outside the nervous system in slow-twitch muscle, heart, and kidney (Donato, 1991). The genes encoding the majority of human S100 proteins are organized in a cluster within the

Diagnosis and Treatment of Spinal Cord Injury. https://doi.org/10.1016/B978-0-12-822498-4.00015-4

chromosomal region 1q21, while some genes coding individual S100 proteins are located in other chromosomal regions, including 21q22 where, in particular, the gene for the S100b protein is located (Michetti, Corvino, Geloso, et al., 2012).

S100b is primarily an astrocytic protein and it was first detected in CSF at the acute phase of exacerbation of MS (Michetti et al., 2012). The majority of astrocytic S100b is located within the cytoplasm, 5%–7% is membrane bound. At least 80%–90% of the total S100b pool is found within the brain, the remainder being located in other non-neuronal tissues. S100b is thought to be involved in a number of $Ca^{+2}$-dependent cellular functions, including protein phosphorylation, enzyme activation, cell proliferation and differentiation, cytoskeletal dynamics, intra-cellular $Ca^{+2}$ homeostasis and protection against oxidative injury (Donato, 1991). In a low concentration it seems to have a neurotrophic effect, it stimulates the growth of neurons, and it increases their survival during development and also during an injury. Higher concentrations might be toxic and evoke cell death. S100b is metabolized and excreted by the kidneys and has a half-life of about 30 min. S100b is found at low levels in healthy patients and remains chemically stable for several hours after sampling (Donato, 1991, 2003).

## S100b and CNS pathology

S100b was regarded for almost two decades as specific to the nervous system (Michetti et al., 2012). Generally, S100b acts like a damage-associated molecular pattern (DAMP), which is released from damaged or activated cells under conditions of cell stress. Furthermore, a more complex role of S100b in the inflammatory processes has been described (Hajduková, Sobek, Prchalová, et al., 2015).

The concept of brain-specific proteins was first proposed in 1965 when S100 was characterized by Moore as a neuronal protein (Moore, 1965). Many studies followed this conclusion and found a clear role of S100b protein in maintenance and development of CNS. This role of S100b protein could be summarized in the following points (Donato, 1991, 2003):

- Regulation of enzymatic activities
- Interactions with cytoskeleton proteins
- Stimulation of neuronal differentiation, neurite extension activity, and pro-survival effects on neurons
- Protection of p53 from thermal denaturation and aggregation; stimulation of p53-dependent cell growth arrest and apoptosis (tumor-suppressor activity); inhibition of p53-dependent transcription activation (tumor-promoting activity) via disruption of the p53 tetramer
- Stimulation of astrocyte proliferation and apoptosis
- Stimulation of IL-6 secretion by neurons, NO secretion by astrocytes and microglia
- Modulation of synaptic plasticity

Over the years, different studies observed a varied grade of correlation between S100b overexpression in neuropathologic states in vivo and trophic/toxic effects in vitro. For example, overexpression of S100b protein has also been shown in cerebral palsy, Gilles de la Tourette syndrome, Down syndrome, Alzheimer's disease, schizophrenia, cerebral vasospasm after subarachnoid hemorrhage, astrocytomas, after head injury, and Creutzfeldt–Jacob disease (Sen and Belli, 2007). S100b protein could also represent a useful tool for early detection of intra-ventricular hemorrhage in the post-asphyxia period when clinical examination and cranial ultrasound might still be silent. It is reasonable to hypothesize that S100b accumulates in the extracellular space after astrocyte death or due to increased release by activated astrocytes, or after cellular disintegration of the damaged parenchyma. The finding that increased S100b is not accompanied by alterations in other structural markers of neural cell types, such as glial fibrillary acidic protein (GFAP), myelin basic protein (MBP) and neuron-specific enolase (NSE) has been regarded as an indication that S100b is actively released by astrocytes rather than leaked from damaged cells (Michetti et al., 2012). Under these conditions, the increased accumulation of S100b may become toxic due to its stimulatory effects on nitric oxide production by astrocytes and microglia (Sen and Belli, 2007).

S100b is normally present in cells derived from the neural crest but also from chondrocytes, adipocytes, myoepithelial cells, macrophages, Langerhans cells, dendritic cells, and keratinocytes (Benneker, Leitner, Martinolli, et al., 2008). Additionally, many studies have suggested extra-cranial sources of S100b, such as following cardiopulmonary bypass surgery, cardiac arrest, carbon monoxide and benzodiazepine poisoning, small and large bone fractures, abdominal injuries, and in fat tissue (Wolf, Krall, Pajenda, et al., 2014, b). Although in trauma patients larger extra-cranial soft tissue and bone injuries led also to an increase of S100b, confounding the interpretation of results, studies focused only on TBI cases revealed a high negative predictive power of S100b. They have found S100b elevation in CSF and peripheral blood after TBI, which is positively correlated with the severity of injury but negatively correlated with outcome. It was concluded that monitoring of S100b levels could contribute to early detection of patients at risk of secondary rises in intra-cranial pressure and subsequent fatal outcome (Sen and Belli, 2007).

## Biomarkers in SCI

The acute phase of SCI is evaluated by neurological examination and MRI study. These patients are classified based on image characteristics, severity and need for surgical intervention. In comparison with MRI, neurological findings are most predictive of outcome. As the degree and localization of injured and spared white matter primarily determine functioning after SCI, MRI has limited success as a prognostic tool (Leister, Haider, Mattiassich, et al., 2020). On the other hand, injury severity alone does not adequately predict the degree of spontaneous neurological recovery.

Spinal cord trauma causes an acute disruption of the spinal cord parenchyma, which is followed by a secondary axonal degeneration and further degeneration or death of nerve cells by either apoptosis or necrosis, processes that may last from days to weeks. As the spinal cord is surrounded by vessels and CSF, damage to the spinal cord releases proteins and metabolites from the nervous tissue into CSF and blood. The implementation of biomarkers in SCI provides a useful tool supporting clinical decision making in acute SCI, especially in severe injured unresponsive patients (Guez, Hildingsson, Rosengren, et al., 2003; Leister et al., 2020).

An ideal prognostic CNS biomarker should have all of the following properties: CNS specificity, rapid and profound release into blood or/and CSF after injury, easily assessed, predictability of serious injury from an early sample, correlation of biomarker value with injury severity, of low cost minimally influenced by other factors and reproducible (Vaage and Anderson, 2001). The most studied biomarkers for SCI patients and their characteristics are summarized in Table 1.

**TABLE 1 Characteristics of most studied biomarkers.**

| Biomarkers | Characteristics |
|---|---|
| Neurofilament proteins<br>✓ NF-L (light)<br>✓ NF-M (medium)<br>✓ NF-H (heavy) | • Components of neuronal cytoskeleton<br>• Control axonal signaling and transport<br>• Increase extra-cellular concentration upon neurological injuries<br>• NF-L correlated with severity of SCI and neurological outcome |
| Glia fibrillary protein (GFAP) | • Components of astrocytic cytoskeleton<br>• Development of astrocytic cytoskeleton<br>• Variable CSF levels according to type of spinal injury<br>• Post-traumatic CSF levels correlated with motor improvement |
| Cleaved Tau protein (C-Tau) | • Axonal microtubules<br>• Stabilization of cytoskeleton<br>• Axonal transport<br>• Doubtful role |
| S100b protein | • Calcium-binding protein in astrocytes<br>• Homeostatic balance<br>• Regulating calcium levels, cell proliferation and differentiation<br>• Enzymatic and metabolic<br>• Early increase correlated with severity of SCI and outcome |
| Neuron-specific enolase (NSE) | • Neuronal cytoplasm enzyme<br>• Homeostatic balance<br>• Intra-lesional levels increased in experimental studies<br>• Early increase correlated with severity of SCI and outcome |
| MAP-2 (microtubule-associated protein 2) | • Cytoskeletal proteins enriched in dendritic cells<br>• Determining and stabilizing neuronal morphology during neuron development<br>• Rapid loss post injury in experimental studies |
| Myelin basic protein | • Myelin sheath<br>• In rat models post-traumatic detection in regenerated oligodendrocytes and marked increase in CSF, serum and intra-lesionally |
| Matrix metalloproteinases (MMPs) | • Extra-cellular matrix<br>• Cell migration<br>• CNS repair<br>• Post-traumatic increase, in sub-acute phase correlated with outcome |

**TABLE 2 Characteristics of most studied inflammatory biomarkers.**

| Inflammatory biomarkers | Characteristics |
|---|---|
| Transforming growth factor beta (TGF-β) | • Regulation axonal extension, astrogliosis, and proteoglycans' accumulation<br>• After 12 weeks there was a correlation between the increased levels and the poor neurological improvement |
| Tumor necrosis factor-alpha (TNF-α) | • Released within a few hours after damage<br>• In 12 weeks post-injury CSF and serum levels correlated with neurological outcome |
| Insulin-like growth factor (IGF) | • In 12 weeks post-injury higher serum value is correlated with worse neurological improvement |
| Interleukins (ILs) | • Inflammatory process and glial scar tissue formation after SCI<br>• At first and fourth post-traumatic week decreased serum levels of IL-1B in patients with poorer neurological status |
| Soluble CD95 ligand (sCD95L) | • Upregulation responsible for apoptosis in oligodendrocytes and spinal cord neurons<br>• Increased serum levels during the sub-acute phase |

SCI is a well-known neuroinflammatory process starting with local ischemia and edema. In cellular level, ionic disturbances and mitochondrial dysfunction have been observed. These events lead to neuron death, disruption of blood-brain barrier (BBB), neutrophils and macrophages infiltration and finally scar formation into the spinal cord (Albayar et al., 2019). Based on this concept several inflammatory biomarkers were also studied from several authors as specific predictive tool in cases of SCI (Biglari, Swing, Child, et al., 2015; Ferbert, Child, Graeser, et al., 2017) (Table 2).

## S100b and SCI

The last 20 years there are several experimental and in vitro studies about the predictive role of S100b in tissue, serum or CSF after SCI, concluded that this biomarker could be a real-time indicator of early stages of spinal cord involvement (Table 3).

**TABLE 3 Studies of the last 20 years about the role of S100b in SCI.**

| Study | Experimental/ Human | Biological fluid | Results |
|---|---|---|---|
| Ma, Novikov, Karlsson, et al. (2001) | Rats | Serum | At 3, 12, and 72 h after SCI increased levels were identified |
| Cornefjord, Nyberg, Rosengren, et al. (2004) | Pigs | CSF | Non-significant differences were identified after the induced nerve root injury |
| Loy, Sroufe, Pelt, et al. (2005) | Rats | Serum | At 6 h after SCI increased levels |
| Cao et al. (2008) | Rats | Serum, CSF | At 2 h after SCI serum and CSF concentrations were elevated |
| Schultke, Griebel, and Juurlink (2010) | Rats | Serum | At 6, 12, and 24 h after the trauma S-100b levels were significantly higher in quercetin-treated animals |
| Kwon, Stammers, Belanger, et al. (2010) | Human | CSF, serum | The highest levels during the first 24 h were included in the final prediction model |
| Ersahin, Toklu, Erzik, et al. (2011) | Rats | Serum | Significant increase 1 week after SCI induction |
| Zhang, Huang, Su, et al. (2011) | Pigs | CSF, serum | S-100B was significantly increased in the serum and cerebrospinal fluid 3 h after trauma |

**TABLE 3 Studies of the last 20 years about the role of S100b in SCI—cont'd**

| Study | Experimental/ Human | Biological fluid | Results |
|---|---|---|---|
| Pouw, Kwon, Verbeek, et al. (2014) | Human | CSF | The mean S-100β concentration in motor complete patients was significantly higher compared with motor incomplete patients |
| Wolf, Krall, Pajenda, et al. (2014, 2014b) | Human | Serum | Patients with neurologic deficits had significantly higher S100b levels compared with the patients with vertebral fractures but without neurologic deficits within 24 h after injury |
| Dalkilic, Fallah, Noonan, et al. (2018) | Human | CSF | Significantly lower in patients, who experienced an improvement within 6 months after injury |
| Yang, Bramlett, Moghieb, et al. (2017) | Rats | CSF, serum | Significantly elevated levels within 24 h (acute phase) |
| Kwon, Streijger, Fallah, et al. (2017 | Human | CSF | Motor score improvement also was strongly correlated with the 24-h post-injury CSF levels |
| Du, Li, Sun, et al. (2018) | Human | Serum | Levels rose after the injury and reached a peak on the fourth day of injury, which was followed by a gradual decline 2 weeks after the injury |

The initial experimental approaches found high level of S100b in the first 3–6 h in CSF and serum, whereas in some other cases was recorded an increase even at 3–7 days after SCI (Cornefjord et al., 2004; Ersahin et al., 2011; Loy et al., 2005; Ma, Novikov, Karlsson, et al., 2001). An experimental study from Cao, Yang, Liu, et al., 2008 showed significantly elevated serum and CSF levels of S100b and NSE which were then considered to be the most specific biomarkers for the acute phase of SCI (Cao et al., 2008). Their data indicated that NSE and S100b protein level changes are time-dependent and correlated with higher neurological defects and injury severity. A significant increase in NSE and S100b protein levels in serum and CSF was observed at 2 h after injury, and in a stepwise manner they reached maximum levels at 6 h. Serum S100b levels were elevated as well as 24 h after SCI as it was observed from Schultke et al. (2010). In general, S-100b protein levels in animals' serum and CSF elevate following SCI and have diagnostic value. During the initial 6 h of SCI, the level of this protein is very high in CSF and serum, but with time passing, the serum level of this protein decreases and after 24 h, its rate does not differ from animals without SCI (Faridaalee and Khajeh, 2019). Marquardt et al. found that the serum levels of S100b in rats rose rapidly within a short time (24 h) and then declined gradually and reached the normal level around the 10th day of injury, an observation similar to study from Marquardt, Setzer, and Seifert (2006) and Zhang, Underwood, Landfield, et al. (2000).

S100b was studied as a predictor of initial spinal cord damage that reflects directly to the severity of SCI. Kwon et al. studied in vitro the correlation between elevation in IL-6, IL-8, MCP-1, Tau-Protein, S100b, and GFAP and severity of injury (Kwon et al., 2010). Using protein concentrations at 24 h post injury, the model accurately predicted the observed ASIA grade in 89% of patients and furthermore provided a better prediction of segmental motor recovery at 6 months post injury than the patients' baseline ASIA grade. In a CSF study of 16 SCI patients Pouw et al. found that mean levels of S100b were significantly higher in more severely injured patients according to ASIA scale, reinforcing the aforementioned findings of Pouw et al. (2014). Protein levels were comparable between males and females, and not associated with age which gives rise for a more objective assessment despite the heterogeneity in patients with SCI. Marquardt et al. observed also that S100b could have a predictive value in other cases of spinal cord pathology. Patients with epidural empyema and persistently increased S100b levels for a minimum of 3 days after operative decompression had unfavorable functional outcome, as measured by motor function. Patients with spinal metastasis and persistently increased S100b levels for a minimum of 10 days after operative decompression had unfavorable functional outcome, as measured by motor function (Marquardt, Setzer, & Seifert et al., 2004a, 2004b). Marquardt concluded that the only significant biomarker correlation was between S100b serum levels and fractures of the vertebral spine, as well as between S100b serum levels and spinal cord injuries with resulting neurological deficits. Animals with favorable outcome had either always normal levels or levels that were initially increased but normalized within 2 days, when the values of animals with unfavorable outcome were elevated throughout (Marquardt, Setzer, Theisen, et al., 2011). Du et al. found that S100b has a sensitivity of 71.79% and a specificity of 66.67% for the evaluation of neurologic functional recovery (Du et al., 2018). The cut-off value of 29.07 μg/L indicated that patients with more than this value had increased risk of poor prognosis. Serum levels were low in the well-recovered group but very high in the modest-recovered group. Patients' recovery was mainly associated with the extent of SCI damage at the time of injury. The serum levels of S100b increased after injury and reached a peak on the 4th day of injury. Then, the levels declined gradually, 2 weeks after the injury. Elevated S100b in human CSF after SCI was observed in patients with worse neurological outcome and lower concentration in cases of significant improvement after 6 months in a study from Dalkilic et al., when Kwon et al. reported a strong correlation between motor score improvement and CSF level of S100b in the first 24 h (Dalkilic, Fallah, Noonan, et al., 2018; Kwon et al., 2017). In the same year, an experimental study confirmed that highest levels of S100b in serum and CSF appeared in acute post-traumatic period (Yang et al., 2017).

Significant S100b serum level changes were seen in patients with vertebral spine fractures and spinal cord injury. When spinal fracture occurs, the fractured vertebrae affect the adjacent spinal cord which is a bundle of neurons, and S100b is then released from the injured spinal cord. A percentage of 5%–15% of cases with significant cervical injuries is missed on plain radiographs though were high-risk patients (Pouw, Hosman, van Middendorp, et al., 2009). CT or MRI is recommended in high-energy injuries and in patients with normal plain radiographs but continued symptoms. On the other hand, as it was reported in previous studies, there is a "false positive" range of biomarker elevations because it was observed that patients with a period of aortic clamping and patients with a whiplash trauma may have elevated biomarker levels in CSF without neurologic deficits. S100b can play a significant role in determining whether further evaluation is needed for patients with normal plain radiographs of spine even in absence of neurologic deficits (Pouw et al., 2009). Markedly increased levels of S100b in patients with spinal fracture might be evidence of spinal cord injury (Lee, Kim, Lee, et al., 2010). In a study of 19 patients with isolated vertebral fractures, Benneker detected elevated S100b serum levels, with a mean S100b value of 0.38 μg/L. Benneker concluded that the measurement of S100b serum levels has a high negative predictive power to rule out vertebral fractures (Benneker et al., 2008). In another study from Wolf et al. (S100b) serum levels was significantly higher in patients with vertebral spine fractures and additional presence of neurological deficits was correlated significantly with higher S100b serum levels (Wolf, Krall, Pajenda, et al., 2014, b).

## Limitations and perspectives

S100b protein could be a useful tool after SCI for early detection of the degree of spinal cord involvement and the grade of neurological outcome within a 6-month period. The presence of extra-cranial injuries where fat and bone tissue are involved contributes to elevated S100b in serum and may be over interpreted as a reliable positive predictor for injuries to the spine (Albayar et al., 2019; Benneker et al., 2008). Additionally, before the 72 h post-injury period, factors such as spinal shock and medical instability affect the reliability of the neurological examination (Pouw et al., 2014; Zhang et al., 2011). Noteworthy that there is a severe difficulty to obtain CSF sample from SCI patients as it is painful and is associated with a risk of epidural hematoma that may lead to further spinal cord compression.

The most recent study from Leister et al. included series of clinical trials, cohort studies, pilot studies based on the PRISMA (Preferred Reporting Items for Systematic Reviews and Meta-Analyses) statement and consists a detailed report for biomarkers in SCI (Leister et al., 2020). The correlation between S100b and severity of spinal cord injury according to neurological outcome could raise considerations for future respective studies such as:

- Correlation with later time imaging findings may reveal new pathways between structural changes and fluctuations of biomarkers (Zhang et al., 2011).
- Biomarkers that could be related with short and long-term complications after SCI, when there are at least 10 times more chronic SCI patients than new SCI cases (Albayar et al., 2019; Yousefifard, Sarveazad, Babahajian, et al., 2019).
- Use the protein not only as a biomarker but also as a potential therapeutic target for treatment of the pathological conditions in which it is involved (Sen and Belli, 2007).
- Increased knowledge about mechanisms of regulation of post-injury S100b release (Donato, 1991, 2003; Michetti et al., 2012).

## Applications to other areas of neuroscience

- When the patient suffers from TBI, the blood-brain barrier is disrupted, causing a leakage of proteins from the CSF with subsequent cerebral edema formation.
- The most studied protein biomarker of cerebral damage in TBI is S100b.
- All traumatic cerebral injuries have been shown to have increased S100b in serum, but focal injuries, such as cerebral contusions and subdural hematomas, present higher levels as compared to diffuse injuries.

- Patients without a steady decline of S100b in CSF and serum have been shown to present with worse outcome.
- S100b appears to be an important and useful predictor of functional outcome in moderate-to-severe TBI.

## Mini-dictionary of terms

- ***S100 protein***: Family of low-molecular-weight proteins with two calcium-binding sites and are involved in a variety of intra-cellular and extra-cellular functions.
- ***Spinal cord injury***: Traumatic damage to the spinal cord with or without bone and ligament involvement.
- ***Traumatic brain injury***: An alteration in normal brain function caused by an external force.
- ***Biomarker***: A measurable indicator of a medical condition which is evaluated using sample of blood, urine, CSF or soft tissue.
- ***Blood-brain barrier***: A barrier between capillaries (small blood vessels) and other components of brain and spinal cord (parenchyma, meninges, bone) that protects them from pathogens and toxins.

## Key facts of calcium-binding proteins

- Calcium in nervous system is involved in functions membrane excitability, signal transduction, neurotransmitters release, synaptic plasticity, cell-cycle regulation, cell migration, and axon growing.
- Calcium-binding proteins react to $Ca^{2+}$ transients and modulate the activity of proteins involved in both maintaining homeostatic conditions.
- Calmodulin, parvalbumin, calbindin, S100 proteins, calpain, and more than 200 proteins belong to this family.
- S100b protein has been measured in CSF and blood of patients with cerebral infarction, transient ischemic attack, hemorrhage, and head injury.
- S100b might be a selective biomarker to estimate the extent of nervous tissue damage and the prognosis of patients after SCI.

## Summary points

- Spinal cord injury (SCI) results in severe social-economic problems, because it is observed mostly among younger adults.
- Most of the medical decisions are targeting the stabilization of the patients and preventing further injury, but no definitive treatment for the present state of the CNS trauma exists.
- In these cases, where the real-time injury is the requested in clinical diagnosis and prognosis, serum and CSF biomarker examination provides a quantitative method.
- As the spinal cord is surrounded by vessels and CSF, damage to the spinal cord releases proteins and metabolites from the nervous tissue into CSF and blood. The implementation of biomarkers in SCI provides a useful tool supporting clinical decision making in acute SCI.
- S100b constitutes was regarded for almost two decades as specific to the nervous system.
- The last 20 years there are several experimental and in vitro studies about the predictive role of S100b in tissue, serum or CSF after SCI, concluded that this biomarker could be a real-time indicator of early stages of spinal cord involvement.

## References

Albayar, A. A., Roche, A., Swiatkowski, P., et al. (2019). Biomarkers in spinal cord injury: Prognostic insights and future potentials. *Frontiers in Neurology*, *29*, 10. Article 27.

Benneker, L. M., Leitner, C., Martinolli, L., et al. (2008). Isolated vertebral fractures give elevated serum protein S-100B levels. *Scandinavian Journal of Trauma, Resuscitation and Emergency Medicine*, *16*, 13.

Biglari, B., Swing, T., Child, C., et al. (2015). A pilot study on temporal changes in IL-1β and TNF-α serum levels after spinal cord injury: The serum level of TNF-α in acute SCI patients as a possible marker for neurological remission. *Spinal Cord*, *53*, 510–514.

Cao, F., Yang, X. F., Liu, W. G., et al. (2008). Elevation of neuron-specific enolase and S-100beta protein level in experimental acute spinal cord injury. *Journal of Clinical Neuroscience*, *15*(5), 541–544.

Cornefjord, M., Nyberg, F., Rosengren, L., et al. (2004). Cerebrospinal fluid biomarkers in experimental spinal nerve root injury. *Spine*, *29*, 1862–1868.

Dalkilic, T., Fallah, N., Noonan, V. K., et al. (2018). Predicting injury severity and neurological recovery after acute cervical spinal cord injury: A comparison of cerebrospinal fluid and magnetic resonance imaging biomarkers. *Journal of Neurotrauma*, *235*, 435–445.

Donato, R. (1991). Perspectives in S-100 protein biology. Review article. *Cell Calcium*, *12*(10), 713–726.

Donato, R. (2003). Intracellular and extracellular roles of S100 proteins. *Microscopy Research and Technique*, *60*(6), 540–551.

Du, W., Li, H., Sun, J., et al. (2018). The prognostic value of serum neuron specific enolase (NSE) and S100B level in patients of acute spinal cord injury. *Medical Science Monitor*, *24*, 4510–4515.

Ersahin, M., Toklu, H. Z., Erzik, C., et al. (2011). Ghrelin alleviates spinal cord injury in rats via its anti-inflammatory effects. *Turkish Neurosurgery*, *21*(4), 599–605.

Faridaalee, G., & Khajeh, F. K. (2019). Serum and cerebrospinal fluid levels of S-100b. Is a biomarker for spinal cord injury; a systematic review and Meta-analysis. *Archives of Academic Emergency Medicine*, *7*(1), e19.

Ferbert, T., Child, C., Graeser, V., et al. (2017). Tracking spinal cord injury: Differences in cytokine expression of IGF-1, TGF- B1, and sCD95l can be measured in blood samples and correspond to neurological remission in a 12-week follow-up. *Journal of Neurotrauma*, *34*(3), 607–614.

Guez, M., Hildingsson, C., Rosengren, L., et al. (2003). Nervous tissue damage markers in cerebrospinal fluid after cervical spine injuries and whiplash trauma. *Journal of Neurotrauma*, *20*(9), 853–858.

Hajduková, L., Sobek, O., Prchalová, D., et al. (2015). Biomarkers of brain damage: S100B and NSE concentrations in cerebrospinal fluid—A normative study. *BioMed Research International*, *2015*, 379071.

Kumar, R., Lim, J., Mekary, R. A., et al. (2018). Traumatic spinal injury: Global epidemiology and worldwide volume. *World Neurosurgery*, *113*, e345–e363.

Kwon, B. K., Stammers, A. M., Belanger, L. M., et al. (2010). Cerebrospinal fluid inflammatory cytokines and biomarkers of injury severity in acute human spinal cord injury. *Journal of Neurotrauma*, *27*(4), 669–682.

Kwon, B. K., Streijger, F., Fallah, N., et al. (2017). Cerebrospinal fluid biomarkers to stratify injury severity and predict outcome in human traumatic spinal cord injury. *Journal of Neurotrauma*, *34*, 567–580.

Lee, S. J., Kim, C. W., Lee, K. J., et al. (2010). Elevated serum S100b levels in acute spinal fracture without head injury. *Emergency Medicine Journal*, *27* (3), 209–212.

Leister, I., Haider, T., Mattiassich, G., et al. (2020). Biomarkers in traumatic spinal cord injury-technical and clinical considerations: A systematic review. *Neurorehabilitation and Neural Repair*, *34*(2), 95–110.

Loy, D. N., Sroufe, A. E., Pelt, J. L., et al. (2005). Serum biomarkers for experimental acute spinal cord injury: Rapid elevation of NSE and S-100b. *Neurosurgery*, *56*(2), 391–397 (discussion 391–397).

Ma, J., Novikov, L. N., Karlsson, K., et al. (2001). (2001). Plexus avulsion and spinal cord injury increase the serum concentration of S-100 protein: An experimental study in rats. *Scandinavian Journal of Plastic and Reconstructive Surgery and Hand Surgery*, *35*, 355–359.

Marquardt, G., Setzer, M., & Seifert, V. (2004a). Protein S-100b as serum marker for prediction of functional outcome in metastatic spinal cord compression. *Acta Neurochirurgica*, *146*, 449–452.

Marquardt, G., Setzer, M., & Seifert, V. (2004b). Protein S-100b for individual prediction of functional outcome in spinal epidural empyema. *Spine*, *29*, 59–62.

Marquardt, G., Setzer, M., & Seifert, V. (2006). Serum biomarkers for experimental acute spinal cord injury: Rapid elevation of neuron-specific enolase and S-100b. *Neurosurgery*, *58*, E590.

Marquardt, G., Setzer, M., Theisen, A., et al. (2011). Experimental subacute spinal cord compression: Correlation of serial S100B and NSE serum measurements, histopathological changes, and outcome. *Neurological Research*, *33*(4), 421–426.

Michetti, F., Corvino, V., Geloso, M. C., et al. (2012). The S100b protein in biological fluids: More than a lifelong biomarker of brain distress. *Journal of Neurochemistry*, *120*, 644–659.

Moore, B. W. (1965). A soluble protein characteristic of the nervous system. *Biochemical and Biophysical Research Communications*, *19*, 739–744.

Pouw, M. H., Hosman, A. J., van Middendorp, J. J., et al. (2009). Biomarkers in spinal cord injury. *Spinal Cord*, *47*(7), 519–525.

Pouw, M. H., Kwon, B. K., Verbeek, M. M., et al. (2014). Structural biomarkers in the cerebrospinal fluid within 24 h after a traumatic spinal cord injury: A descriptive analysis of 16 subjects. *Spinal Cord*, *52*, 428–433.

Schultke, E., Griebel, R. W., & Juurlink, B. H. (2010). Quercetin administration after spinal cord trauma changes S-100 levels. *The Canadian Journal of Neurological Sciences*, *37*(2), 223–228.

Sen, J., & Belli, A. (2007). S100b in neuropathologic states: The CRP of the brain? *Journal of Neuroscience Research*, *85*(7), 1373–1380.

Vaage, J., & Anderson, R. (2001). Biochemical markers of neurologic injury in cardiac surgery: The rise and fall of S100β. *The Journal of Thoracic and Cardiovascular Surgery*, *122*, 853–855.

Wolf, H., Krall, C., Pajenda, G., et al. (2014). Alterations of the biomarker S-100B and NSE in patients with acute vertebral spine fractures. *The Spine Journal*, *14*, 2918–2922.

Yang, Z., Bramlett, H. M., Moghieb, A., et al. (2017). Temporal profile and severity correlation of a panel of rat spinal cord injury protein biomarkers. *Molecular Neurobiology*, *55*(3), 2174–2184.

Yousefifard, M., Sarveazad, A., Babahajian, A., et al. (2019). Potential diagnosting and prognostic value of serum and CSF biomarkers in traumatic SCI: A systematic review. *Journal of Neurochemistry*, *149*, 317–330.

Zhang, B., Huang, Y., Su, Z., et al. (2011). Neurological, functional, and biomechanical characteristics after high-velocity behind armor blunt trauma of the spine. *Journal of Trauma*, *71*(6), 1680–1688.

Zhang, S. X., Underwood, M., Landfield, A., et al. (2000). Cytoskeletal disruption following contusion injury to the rat spinal cord. *Journal of Neuropathology & Experimental Neurology*, *59*(4), 287–296.

Chapter 16

# Assessment of postural control after spinal cord injury or disease: A narrative review

Kristin E. Musselman[a], Janelle Unger[b], and Jean-François Lemay[c]

[a]Department of Physical Therapy, Temerty Faculty of Medicine, University of Toronto, SCI Mobility Lab, Lyndhurst Centre, Toronto Rehabilitation Institute-University Health Network, Toronto, ON, Canada; [b]School of Physical Therapy, Western University, London, ON, Canada; [c]School of Rehabilitation, Faculty of Medicine, University of Montreal, Montreal, QC, Canada

## List of abbreviations

| | |
|---|---|
| **ABC Scale** | Activities-specific Balance Confidence Scale |
| **AIS** | American Spinal Injury Association Impairment Scale |
| **BBS** | Berg Balance Scale |
| **CB&M Scale** | Community Balance and Mobility Scale |
| **COM** | center of mass |
| **COP** | center of pressure |
| **FRT** | Functional Reach Test |
| **IMU** | inertial measurement units |
| **KAFO** | knee ankle foot orthoses |
| **L/E** | lower extremity |
| **mini-BESTest** | mini-Balance Evaluation Systems Test |
| **mFRT** | modified Functional Reach Test |
| **MMT** | manual muscle testing |
| **SCI/D** | spinal cord injury or disease |
| **SWAT** | Standing and Walking Assessment Tool |
| **WheelCon** | Wheelchair Use Confidence Scale v.3 |

## Introduction

Following spinal cord injury or disease (SCI/D), physical rehabilitation is focused on regaining independence for community living and maximizing quality of life. The multi-disciplinary rehabilitation team facilitates the retraining of daily activities (e.g., bathing, feeding) and mobility (e.g., walking, wheeling, transferring), and assists with the reintegration of vocational and recreational activities. The priorities of rehabilitation are also informed by people living with SCI/D, who have identified improving upper limb and trunk strength, sexual function, bowel and bladder function, and walking ability as being most important for life satisfaction (Anderson, 2004). Common to the safe and successful completion of all of these tasks is the requirement of postural control. Without postural control, movement is limited and independence compromised.

The term postural control is often used interchangeably with the terms balance or balance control. Indeed, these are similar and inter-related constructs. Postural control refers to one's ability to maintain, achieve, or restore a state of balance at any given time (Pollock, Durward, Rowe, & Paul, 2000). From a biomechanical perspective, balance refers to the relationship between one's center of mass and one's base of support (Maki & McIlroy, 1997). When the vertical projection of one's center of mass moves outside of one's base of support, muscle activity is required to change the base of support or reposition the center of mass within the base of support. Without this corrective muscle activity, a fall would result. Achieving a state of balance, or postural control, is required when maintaining a posture (e.g., sitting or standing), moving voluntarily, or reacting to an external perturbation (e.g., trip, push) (Pollock et al., 2000).

For many individuals living with SCI/D, postural control is an ongoing challenge due to the sensorimotor deficits associated with their injury. Falling is common after SCI/D; 78% (95% confidence interval: 73%–83%) of ambulators fall each year, while 69% (95% confidence interval: 60%–76%) of wheelchair users fall (Khan et al., 2019). Falls have a significant

Diagnosis and Treatment of Spinal Cord Injury. https://doi.org/10.1016/B978-0-12-822498-4.00016-6

detrimental impact on those living with SCI/D (Musselman, Arnold, Pujol, Lynd, & Oosman, 2018; Singh et al., 2020). Up to 62% of ambulators and 38% of wheelchair users have reported a physical injury from a fall, with most injuries being minor in nature (i.e., cuts, bruises) (Khan et al., 2019). The most common serious injuries include fractures and head trauma with a loss of consciousness (Khan et al., 2019). More than 50% of chronic wheelchair users with SCI/D experience osteoporosis ( Jensen et al., 2013); hence, these individuals are at high risk of sustaining a fall-related fracture ( Jannings, 2017).

In addition to physical injury, those who fall or who have a high fall risk may develop a fear of falling, defined as an enduring concern about falling that causes one to avoid activities despite being physically able to perform them (Brouwer, Musselman, & Culham, 2004). About 47% to 63% of people living with SCI/D report a fear of falling (Butler Forslund, Roaldsen, Hultling, Wahman, & Franzén, 2016; Chan et al., 2020). This fear often leads to a "post-fall syndrome" characterized by dependence, loss of autonomy, depression, anxiety, reduced mobility and restricted participation in daily activities (World Health Organization, 2007). Individuals living with SCI/D describe the harmful psychosocial impact of falls, including feelings of vulnerability and embarrassment, lost work productivity, interference with parenting, and a restriction of meaningful recreational activities (Singh et al., 2020). The constant threat of falling, and fear of the potential consequences of a fall, can have a significant impact on quality of life (Singh et al., 2020).

## Chapter objectives

Postural control is inherent to our ability to move and function safely, but is commonly impaired following SCI/D. Hence, improving postural control is a target of rehabilitation assessment and intervention across the continuum of care (i.e., from acute care to community rehabilitation). Regardless of neurological level of injury, severity of the SCI/D (i.e., as indicated by the American Spinal Injury Association Impairment Scale (AIS)), time since injury or mobility status, an evaluation of postural control is warranted for all individuals living with SCI/D. Yet, clinicians and researchers lack guidance on how to approach the assessment of postural control after SCI/D, especially given the heterogeneity in clinical presentation of this population. The primary aim of this narrative review is to outline an individualized and comprehensive approach to the assessment of postural control after SCI/D. The secondary aims are to: (1) describe common impairments in postural control following SCI/D, and (2) describe the current state of postural control assessment for individuals with SCI/D.

## Comprehensive assessment of postural control

More than 30 measures of postural control have been used in research studies with the SCI/D population (Arora, Oates, Lynd, & Musselman, 2020). Hence, the challenge of postural control assessment post-SCI/D is not the availability of measures, but rather deciding which measure(s) to use. In their systematic review, Arora, Oates, et al. (2020) considered the psychometric properties, clinical utility and comprehensiveness of each measure. A measure's comprehensiveness is a relatively new concept that reflects the complexity of postural control. Postural control requires the complex interaction of numerous musculoskeletal and neurophysiological mechanisms, which are effectively outlined in the Systems Framework for Postural Control (Horak, 2006; Sibley, Beauchamp, Van Ooteghem, Straus, & Jaglal, 2015). In the original framework, Horak (2006) described six systems that contribute to postural control and suggested that appropriate evaluation of each system would enable identification of an individual's specific impairments and compensations, promoting individualized treatments. In 2015, Sibley and colleagues led a revision of the Framework, resulting in the identification of nine contributing systems (see Table 1); functional stability limits, underlying motor systems, static stability, verticality, reactive postural control, anticipatory postural control, dynamic stability, sensory integration, and cognitive influences. This revised Systems Framework for Postural Control provides a basis for a comprehensive assessment of postural control for individuals with SCI/D.

Following SCI/D, impairments in postural control are expected. The spinal cord contains numerous descending motor spinal pathways that are involved in the regulation of postural control through activation of postural synergies (Beloozerova, Sirota, Orlovsky, & Deliagina, 2005; Zelenin, Beloozerova, Sirota, Orlovsky, & Deliagina, 2010). There are also a number of ascending pathways, which carry somatosensory information to the brain and allow the adaptation of postural control to the environment (Chiba, Takakusaki, Ota, Yozu, & Haga, 2016). The extent of the impairments in postural control due to the SCI/D depends on the neurological level and severity of injury. The resulting impairments are more pronounced in a sub-set of the postural control systems (see Table 1 and Fig. 1). All individuals with SCI/D are likely to experience impairments in functional stability limits, underlying motor systems (e.g., muscle strength and coordination), static stability and sensory integration. Research involving individuals with motor incomplete SCI/D (i.e., AIS C and D) demonstrated deficits in reactive postural control (Arora, Musselman, et al., 2020; Chan et al., 2020) and cognitive influences (Lemay et al., 2020; Tse et al., 2017). For dynamic stability, the overall findings are variable. Individuals with

**TABLE 1 Ten components of a comprehensive assessment of postural control.**

| | |
|---|---|
| (1) **Functional Stability Limits** (Horak, 2006; Sibley et al., 2015) | |
| Definition: | • The ability to move the COM within the base of support (base of support includes mobility aid, wheelchair). |
| Example Activities: | • Reaching in sitting or standing.<br>• Reaching for the armrest when transferring.<br>• Leaning for pressure relief in sitting. |
| Impact of SCI/D: | • Impaired<br>Decreased control over COM resulting in longer path to stability limit and smaller functional boundaries in sitting (AIS A, C, D) (Shin & Sosnoff, 2013) and standing (AIS C, D) (Houston, Unger, Lee, Masani, & Musselman, 2020; Lemay et al., 2014). |
| (2) **Underlying Motor Systems** (Horak, 2006; Sibley et al., 2015) | |
| Definition: | • Muscular and neuromuscular properties contributing to strength and coordination. |
| Example Activities: | • Leg and trunk strength and/or sequence of muscle activation to transfer from sitting to standing, climb stairs, take a step, etc.<br>• Trunk strength and/or sequence of muscle activation to lean in sitting. |
| Impact of SCI/D: | • Impaired<br>Varying degrees of decreased strength below level of SCI/D.<br>Reduced trunk-leg movement coordination in standing (AIS C, D) (Noamani, Lemay, Musselman, & Rouhani, 2020).<br>Role of trunk strength on postural control in sitting after SCI/D is unclear (AIS A) (Chen et al., 2003). |
| (3) **Static Stability** (Horak, 2006; Sibley et al., 2015) | |
| Definition: | • The ability to control the COM within a static base of support during sitting and standing. |
| Example Activities: | • Independent sitting without trunk or upper limb support.<br>• Independent standing without upper limb support. |
| Impact of SCI/D: | • Impaired<br>Increased postural sway in sitting (AIS A, B, C) (Milosevic, Gagnon, Gourdou, & Nakazawa, 2017) and standing with greater deficits without vision (AIS D) (Lemay et al., 2013). |
| (4) **Verticality** (Horak, 2006; Sibley et al., 2015) | |
| Definition: | • The ability to align the body with respect to gravity. |
| Example Activities: | • Standing, sitting or walking on an incline or decline. |
| Impact of SCI/D: | • Minimal<br>Increased variability in perception of vertical; however, reliance on visual and vestibular inputs compensates for decreased somatosensation (AIS A) ( Joassin, Bonniaud, Barra, Marquer, & Perennou, 2010). |
| (5) **Reactive Postural Control** (Horak, 2006; Sibley et al., 2015) | |
| Definition: | • The ability to return the COM within the base of support after an external perturbation using corrective motor strategies (e.g., ankle and hip strategies, reactive stepping). |
| Example Activities: | • Taking step(s) to recover from a slip on ice.<br>• Flexing/extending at the hips when balance has been lost posteriorly/anteriorly on a fixed base of support (e.g., balance beam, crowded subway platform). |
| Impact of SCI/D: | • Impaired<br>Decreased ability to execute single step responses during a forward fall in standing (AIS C, D) (Chan et al., 2020).<br>Able to maintain state of balance in standing during unexpected surface perturbations (AIS D) (Thigpen et al., 2009).<br>Decreased ability to increase margin of stability in response to a slip while walking (AIS D) (Arora et al., 2020). |
| (6) **Anticipatory Postural Control** (Horak, 2006; Sibley et al., 2015) | |
| Definition: | • The ability to shift the COM appropriately prior to discrete movements. |
| Example Activities: | • When turning the head to look behind the body, shifting body weight toward the direction of the turn.<br>• Walking with decreased speed, shorter step length, wider step width and a flatter foot-floor angle when on a slippery surface. |

*Continued*

**TABLE 1 Ten components of a comprehensive assessment of postural control—cont'd**

| | |
|---|---|
| Impact of SCI/D: | • Likely impaired with more severe SCI/D<br>• Minimal for less severe SCI/D<br><br>Able to modulate timing and amplitude of muscle activity in lower limb muscles in response to expected surface perturbations (AIS D) (Thigpen et al., 2009).<br>Able to use anticipatory strategies in a similar manner to individuals without SCI/D when walking on known slippery surface (AIS D) (Bone et al., 2021). |
| (7) **Dynamic Stability** (Horak, 2006; Sibley et al., 2015) | |
| Definition: | • The ability to control the COM while the base of support is changing (e.g., during walking). |
| Example Activities: | • Walking, transferring. |
| Impact of SCI/D: | • Likely impaired with more severe SCI/D<br>• Variable findings for less severe SCI/D<br><br>Increased inter-stride variability of gait parameters (e.g., cadence, gait cycle time) (AIS D) (Lemay et al., 2020).<br>Increased foot placement variability during treadmill walking (AIS D) (Day, Kautz, Wu, Suter, & Behrman, 2011).<br>Stabilizing and destabilizing forces during single support phase of gait suggest increased stability (AIS D) (Lemay, Duclos, Nadeau, Gagnon, & Desrosiers, 2014).<br>Increased stability during walking (i.e., decreased speed, shorter steps, greater percentage of time in double support) (Arora et al., 2019). |
| (8) **Sensory Integration** (Horak, 2006; Sibley et al., 2015) | |
| Definition: | • The ability to combine sensory input from multiple sources and re-weight when input is altered. |
| Example Activities: | • Transitioning between different surfaces when walking (e.g., pavement to sand).<br>• Transitioning between environments with varying degrees of light (e.g., bright theatre lobby to dark theatre). |
| Impact of SCI/D: | • Impaired<br><br>Increased reliance on visual input (i.e., decreased ability to re-weight sensory input) during standing (AIS D) (Lemay et al., 2013; Noamani, Lemay, Musselman, & Rouhani, 2021) and walking (AIS D) (Lemay et al., 2020) under different sensory conditions. |
| (9) **Cognitive Influences** (Horak, 2006; Sibley et al., 2015) | |
| Definition: | • The ability to maintain postural control while responding to commands during a task or attending to additional tasks (i.e., dual tasking). |
| Example Activities: | • Responding to an instruction to increase speed while walking.<br>• Attending to a cognitive load (e.g., conversing, mental math) while sitting, standing or walking. |
| Impact of SCI/D: | • Impaired<br><br>Cognitive dual-task decreased standing time (AIS C, D) (Tse, Carpenter, Liu-Ambrose, Chisholm, & Lam, 2017) and increased inter-stride variability of cadence when walking (AIS D) (Lemay et al., 2020). |
| (10) **Balance Confidence** (Hatch, Gill-Body, & Portney, 2003) | |
| Definition: | • The level of perceived confidence in the ability to maintain a state of balance while performing daily activities. |
| Example Activities: | • Having confidence to complete daily activities (e.g., get onto an escalator, reach for something on tiptoes, and walk on an uneven surface) without losing balance. |
| Impact of SCI/D: | • Impaired<br><br>Decreased balance confidence during standing and walking activities (AIS C, D) (Shah, Oates, Arora, Lanovaz, & Musselman, 2017).<br>Greater balance confidence associated with decreased velocity of COP in standing (AIS C, D) (Shah et al., 2017).<br>Decreased falls self-efficacy associated with increased postural sway in standing with bilateral KAFO (AIS A) (John, Cherian, & Babu, 2010). |

An original table providing the definition of each component of postural control along with example activities. A description of the observed impact of spinal cord injury or disease (SCI/D) on each component is summarized. *COM*, center of mass; *AIS*, American Spinal Injury Association Impairment Scale; *COP*, center of pressure; *KAFO*, knee ankle foot orthoses.

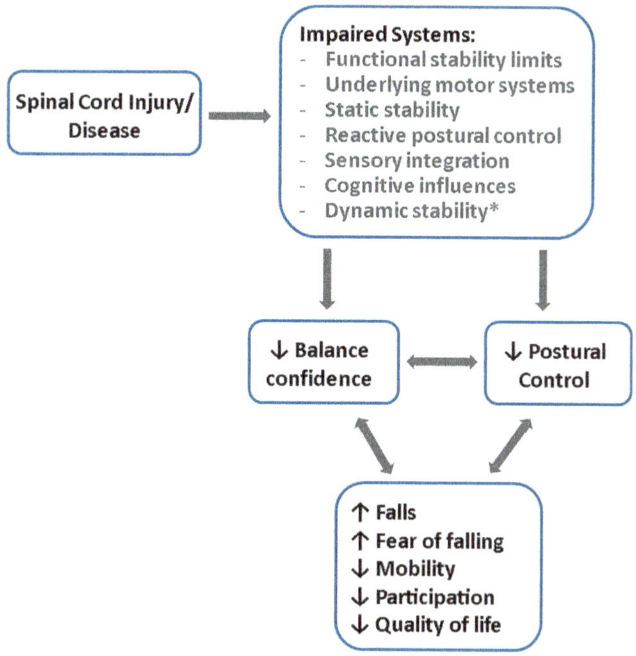


FIG. 1 Effect of spinal cord injury/disease on the systems of postural control. An original figure summarizing the systems of postural control that are negatively affected by spinal cord injury or disease (*conflicting research findings concerning dynamic stability). These impaired systems result in reduced postural control and balance confidence, with these two constructs influencing one another. Reduced postural control and balance confidence can negatively impact the occurrence and fear of falls, mobility, participation and quality of life, resulting in a cycle of decline whereby falls and decreased mobility/participation result in further decreases in postural control and confidence.

AIS D SCI/D show variability in gait parameters, suggesting compromised dynamic stability (Day et al., 2011; Lemay et al., 2020); however, they also show markers of greater stability during walking than age- and sex-matched able-bodied adults (Arora et al., 2019; Lemay, Duclos, et al., 2014). Arora et al. (2019) proposed that increased stability when walking may be a compensatory strategy to avoid losing balance and having to rely on an impaired reactive postural control system to prevent a fall. There do not appear to be impairments in anticipatory postural control for those individuals with the least severe SCI/D (i.e., AIS D) (Bone et al., 2021; Thigpen et al., 2009), but deficits may be evident in those with greater sensorimotor deficits. Lastly, SCI/D does not affect verticality ( Joassin et al., 2010), although deficits may occur in individuals with a relevant co-morbidity, such as a vestibular disorder or brain injury.

Not only do the above-mentioned physiological factors influence postural control, but so do psychological factors, such as balance confidence (Carpenter, Adkin, Brawley, & Frank, 2006). Balance confidence or self-efficacy refers to an individual's perceived self-confidence in maintaining a state of balance during relatively non-hazardous activities (Hatch et al., 2003). The Systems Framework for Postural Control does not account for the effect of this psychological construct on postural control. However, due to the associations between balance confidence and postural control after SCI/D ( John et al., 2010; Shah et al., 2017) (see Table 1 and Fig. 1), a measure of balance confidence should be included as the tenth component of a comprehensive assessment of postural control (see Fig. 2). This inclusion allows the clinician or researcher to evaluate the alignment between an individual's perceived balance confidence and physical abilities. For example,

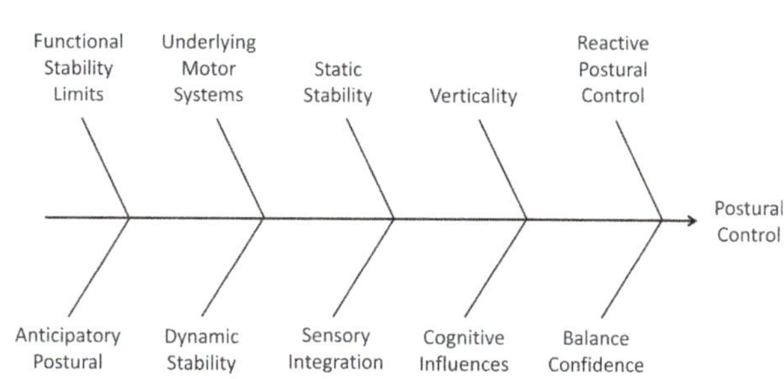


FIG. 2 Fishbone diagram of postural control after spinal cord injury/disease. An original figure highlighting ten components that impact postural control.

perceived balance confidence may be high while the physical ability to maintain a state of balance is low. These findings may suggest an individual lacks insight into his/her postural control deficits, resulting in a high fall risk due to an increased likelihood of risk-taking behaviors. Conversely, someone with low balance confidence, but capable of maintaining a state of balance during daily activities, may needlessly restrict their mobility and participation due to an exaggerated concern about falling. Hence, balance confidence adds a unique dimension to the comprehensive assessment of postural control.

## Current state of postural control assessment in clinical settings

Survey studies involving more than 400 Canadian physical therapists have shed light on the postural control assessment practices for adult populations (Oates et al., 2017; Sibley, Straus, Inness, Salbach, & Jaglal, 2011). While the researchers reported variation in how postural control was assessed clinically, they noted that several postural control systems were regularly evaluated (i.e., by >80% of respondents): static stability, dynamic stability and underlying motor systems (Sibley et al., 2011). In contrast, reactive postural control was evaluated by only 41% of respondents (Sibley et al., 2011). The most commonly used measures of postural control included movement observation, the Berg Balance Scale and the single-leg stance test (Oates et al., 2017; Sibley et al., 2011). In another survey by Sibley, Straus, Inness, Salbach, and Jaglal (2013), 94% of physical therapists agreed that quantifying impairments in postural control was critical for patient care; however, only 43% of respondents agreed that existing measures of postural control were adequate for clinical practice. Moreover, even fewer (21%) indicated that these measures evaluated postural control in a comprehensive way (Sibley et al., 2013), highlighting a key gap in current approaches to the assessment of postural control.

Several organizations have produced clinical recommendations regarding the assessment of postural control. The American Physical Therapy Association and the Academy of Neurological Physical Therapy supported the development of a core set of outcome measures for adults with neurological conditions (Moore et al., 2018). The recommended measures of postural control include the Berg Balance Scale for static and dynamic sitting and standing, the Functional Gait Assessment for dynamic balance, and the Activities-specific Balance Confidence (ABC) Scale for balance confidence. More specific to the SCI/D population, the Canadian Standing and Walking Module Group produced the Standing and Walking Assessment Tool (SWAT) (Musselman et al., 2019). The SWAT is a SCI/D-specific, staged approach to the standardized assessment of standing and walking ability post-SCI/D. Within the SWAT, the Berg Balance Scale is recommended for the majority of individuals with SCI/D (i.e., those who have at minimum the ability to stand with assistance) while the ABC Scale is recommended for those who ambulate, whether therapeutically with assistance or independently in the community (Musselman, Lemay, et al., 2019). In sum, these two practice guidelines are aligned and highlight appropriate measures of postural control for the SCI/D population. However, as outlined below, the Berg Balance Scale and ABC Scale on their own do not fulfill the requirement of a comprehensive and individualized approach to the assessment of postural control after SCI/D.

Arora, Oates, et al. (2020) concluded their systematic review with recommendations for the clinical assessment of postural control after SCI/D. They concluded that of the >30 measures reviewed "no single measure had high clinical utility, strong psychometric properties and comprehensiveness" (Arora, Oates, et al., 2020). The Berg Balance Scale and Functional Reach Test were highlighted for their well-established psychometric properties in the SCI/D population, while the mini-Balance Evaluation Systems Test (mini-BESTest) was identified as the most comprehensive measure that has been used with individuals with SCI/D. The mini-BESTest was developed from the original Systems Framework of Postural Control, and is one of the only clinical measures that evaluates reactive postural control (Arora, Oates, et al., 2020; Horak, Wrisley, & Frank, 2009). Since Arora, Oates, et al. (2020) completed the database searching for their review, the mini-BESTest has been shown to be a valid and reliable measure of postural control for individuals with motor incomplete SCI/D (Chan et al., 2019; Jørgensen, Opheim, Halvarsson, Franzén, & Roaldsen, 2017). Given the comprehensiveness, strong psychometric properties and high clinical utility of the mini-BESTest, its use among the SCI/D population is expected to increase.

## Instrumented assessments of postural control

In addition to the above-mentioned clinical measures, there are instrumented assessments of postural control involving force plates and inertial measurement units (IMU) that have primarily been used in research settings. These instruments can be adapted to facilitate assessment of the components of postural control, with the exception of balance confidence. In their review, Arora, Oates, et al. (2020) highlighted the use of instrumented assessments of postural control among the SCI/D population. Center of pressure-based measures of postural sway, quantified with force plates, were most commonly used (Arora, Oates, et al., 2020). Force plates consist of a rigid surface instrumented with multiple force transducers allowing the measurement of forces in three dimensions. Force plate-based measures of postural sway are considered the gold standard measure of postural control in standing (Ghislieri, Gastaldi, Pastorelli, Tadano, & Agostini, 2019). They are a valid and reliable tool for the evaluation of standing postural control among individuals with incomplete SCI/D (Tamburella, Scivoletto, Iosa, & Molinari, 2014). More specifically, force plates have been used to quantify the performance of functional stability limits (Houston et al., 2020; Lemay, Gagnon, et al., 2014), underlying motor systems (Lee et al., 2021), static stability (Lemay et al., 2013), anticipatory postural control (Arora et al., 2019), dynamic stability (Arora et al., 2019), sensory integration (Lemay et al., 2013) and cognitive influences (Tse et al., 2017). More recently, force plates have been used to assist with the identification of spatiotemporal deficits in reactive postural control (Chan et al., 2020), with some metrics showing adequate validity and reliability (Unger et al., 2020).

IMU are being increasingly used to quantify postural control during sitting (Perez-SanPablo, Quinzanos-Fresnedo, Lopez-Romero, & Quinones-Uriostegui, 2018), standing (Noamani et al., 2020, 2021) and walking (Lemay et al., 2020). IMU typically include accelerometers to measure linear acceleration and gyroscopes to measure angular velocity, and may include magnetometers to measure the amplitude and direction of magnetic fields (Ghislieri et al., 2019). Information characterizing postural control in static standing can be collected with a single IMU (Noamani et al., 2021); however, multiple sensors are needed to extract information about more complex and dynamic postural tasks (Lemay et al., 2020; Noamani et al., 2020). Their reliability and/or validity for use during sitting, standing and walking with individuals with SCI/D have been established (Lemay et al., 2020; Noamani et al., 2020; Perez-SanPablo et al., 2018). IMU can be used in a variety of contexts and for a variety of tasks, and therefore address a number of the components of postural control. For example, among the SCI/D population, IMU have been used to quantify underlying motor systems (Noamani et al., 2020), static stability (Noamani et al., 2020), dynamic stability (Lemay et al., 2020), sensory integration (Lemay et al., 2020; Noamani et al., 2021) and cognitive influences (Lemay et al., 2020).

Although IMU and force plates can facilitate a comprehensive assessment of postural control, these tools tend to have low clinical utility due to their high cost, high level of skill required to administer and interpret the measure, and in the case of force plates, their low portability (Arora, Oates, et al., 2020). As low-cost alternatives to these instrumented tools are developed, their use in clinical settings is likely to increase.

## Individualized and comprehensive assessment of postural control after SCI/D

When an individual with SCI/D begins a rehabilitation program, whether during the acute, sub-acute or chronic phase of recovery, a comprehensive and individualized approach to the assessment of postural control is warranted. Evaluation of the ten components of postural control (see Table 1 and Fig. 2) will ensure the initial assessment is comprehensive. A comprehensive approach will allow the identification of an individual's unique impairments in postural control, thereby identifying the components that should be addressed in therapy and monitored throughout the rehabilitation process at subsequent assessments.

In addition to being comprehensive, the assessment of postural control should be individualized to a person's level of mobility and function. The majority of clinical measures are not applicable to all levels of SCI/D severity. For example, the Berg Balance Scale has a ceiling effect for individuals with AIS D SCI/D (Chan et al., 2017; Jørgensen et al., 2017; Lemay & Nadeau, 2010). To select the measures of postural control most appropriate for an individual, mobility status should be considered. The above-mentioned SWAT is a standardized approach to the staging of mobility status after SCI/D that has been used at Canadian rehabilitation hospitals since 2015 (Musselman, Lemay, et al., 2019). The SWAT stages span all possible levels of mobility and postural control, from Non-independent Sitting Capacity (Stage 0) to Full Walking Capacity (Stage 4) (see Fig. 2). Stage 1 (Standing Capacity), Stage 2 (Therapeutic Walking Capacity (indoors)) and Stage 3 (Functional Walking (outdoors)) are each further divided into three stages, resulting in 12 stages that span the continuum of standing and walking ability. Definitions for each stage are provided to assist physical therapists with the staging process (see Table 2). The SWAT links these stages of mobility with well-established, clinical measures of gait (i.e., 10-m walk test, 6-min walk test) and balance (i.e., Berg Balance Scale, ABC Scale), aiding physical therapists in their selection of appropriate measures to use with an individual (Musselman, Lemay, et al., 2019). The validity and responsiveness of the SWAT stages have been established (Musselman et al., 2019; Musselman, Lemay, et al., 2019).

The SWAT can be used to determine which clinical measures to include in an individual's assessment of postural control. The recommended measures change as the SWAT stages progress from minimal sitting capacity to full community ambulation (see Fig. 3). The selection of clinical measures was guided by the above-mentioned, pre-existing recommendations of the Canadian SCI Standing and Walking Module Group, the American Physical Therapy Association and the Academy of Neurological Physical Therapy, as well as by the review completed by Arora, Oates, et al. (2020). The selected measures have high clinical utility, strong psychometric properties in the SCI/D population and are freely available. The following sections include descriptions of each measure along with an explanation of how the measure fits into a comprehensive assessment of postural control. The sections are organized according to three groupings of SWAT stages (Bayley

**TABLE 2 Standing and walking assessment tool stages of walking recovery.**

| SWAT stage | Classification | | Definition |
|---|---|---|---|
| 0.0 | Non-independent sitting capacity | | Individual is unable to sit independently of a seating system on a firm surface with hips and knees at 90 degrees and feet on the floor for 60 s without using arms to stabilize. |
| 0.5 | Independent sitting capacity | | Individual is able to sit independently of a seating system on a solid surface with hips and knees at 90 degrees and feet on the floor for 60 s without using arms to stabilize. |
| 1 | Standing capacity | | Individual cannot ambulate but may be able to stand with assistance |
| | 1A | Trace L/E movement | Unable to stand without total assistance of gait aid and/or orthoses and/or therapist(s). No voluntary L/E functional movement. (L/E MMT ≤ grade 1 in tibialis anterior, soleus, quads, and gluteus.) |
| | 1B | Voluntary Non-Functional L/E movement | Unable to stand independently/needs partial assistance of gait aid and/or orthoses (except bilateral KAFO) and/or therapist(s) to stand. The use of bilateral KAFO is not allowed. Voluntary L/E Movement (L/E MMT of grade 1+/2- to grade 3- in anti-gravity muscles). |
| | 1C | Voluntary Functional L/E movement | Able to stand independently with minimal assistance of gait aid for limited amount of time (<30 s). Orthoses are allowed except for bilateral KAFO. Voluntary L/E Movement. (L/E MMT of grade 3 or higher in most anti-gravity muscles except grade 1 tibialis anterior with an orthosis would fit this group.) |
| 2 | Therapeutic walking capacity (indoors) | | Individual is starting to ambulate with therapist assistance and gait aids/orthoses and progresses toward minimal assistance. |
| | 2A | Maximal assistance | Ability to stand and initiate reciprocal steps through voluntary L/E movement but requires maximal physical assistance (>50% of total effort) of at least one person and may include use of assistive devices and/or orthoses with the exception of bilateral KAFO. |
| | 2B | Moderate assistance | Ability to stand and initiate reciprocal steps through voluntary L/E movement but requires moderate physical assistance (25%–50% of total effort) of one person and may include use of assistive devices and/or orthoses with the exception of the bilateral KAFO. |
| | 2C | Minimal assistance | Ability to stand and initiate reciprocal steps through voluntary L/E movement but requires minimal physical assistance (<25% of total effort) of one person and may include use of assistive devices and/or orthoses with the exception of the bilateral KAFO. |
| 3 | Functional walking (outdoors) | | Individual is starting to ambulate without therapist assistance, but still requires gait aids/orthoses. Individual progresses to ambulating in the community. |
| | 3A | Supervised household ambulator | Ability to ambulate daily using reciprocal steps over ground for short distances (10–100 m) with supervision. Individual may use assistive devices and/or orthoses with the exception of bilateral KAFO. |
| | 3B | Independent household ambulator | Ability to ambulate daily using reciprocal steps over ground for short distances (10–100 m) independently. Individual may use assistive devices and/or orthoses with the exception of the bilateral KAFO. |
| | 3C | Community ambulator | Ability to ambulate daily using reciprocal steps over ground for long distances (>100 m) independently. Individual may use assistive devices and/or orthoses with the exception of the bilateral KAFO. |
| 4 | Full walking capacity | | Individual ambulates independently without therapist assistance or gait aids/orthoses. Independent ambulator—ability to ambulate full time daily at home and in the community without assistive devices, orthoses, or physical assistance. |

Classification and definition for each stage of the Standing and Walking Assessment Tool (SWAT). *SCI*, spinal cord injury; *L/E*, lower extremity; *MMT*, manual muscle testing; *KAFO*, knee ankle foot orthoses.
Reproduced with permission from the Canadian SCI Standing and Walking Module Group.

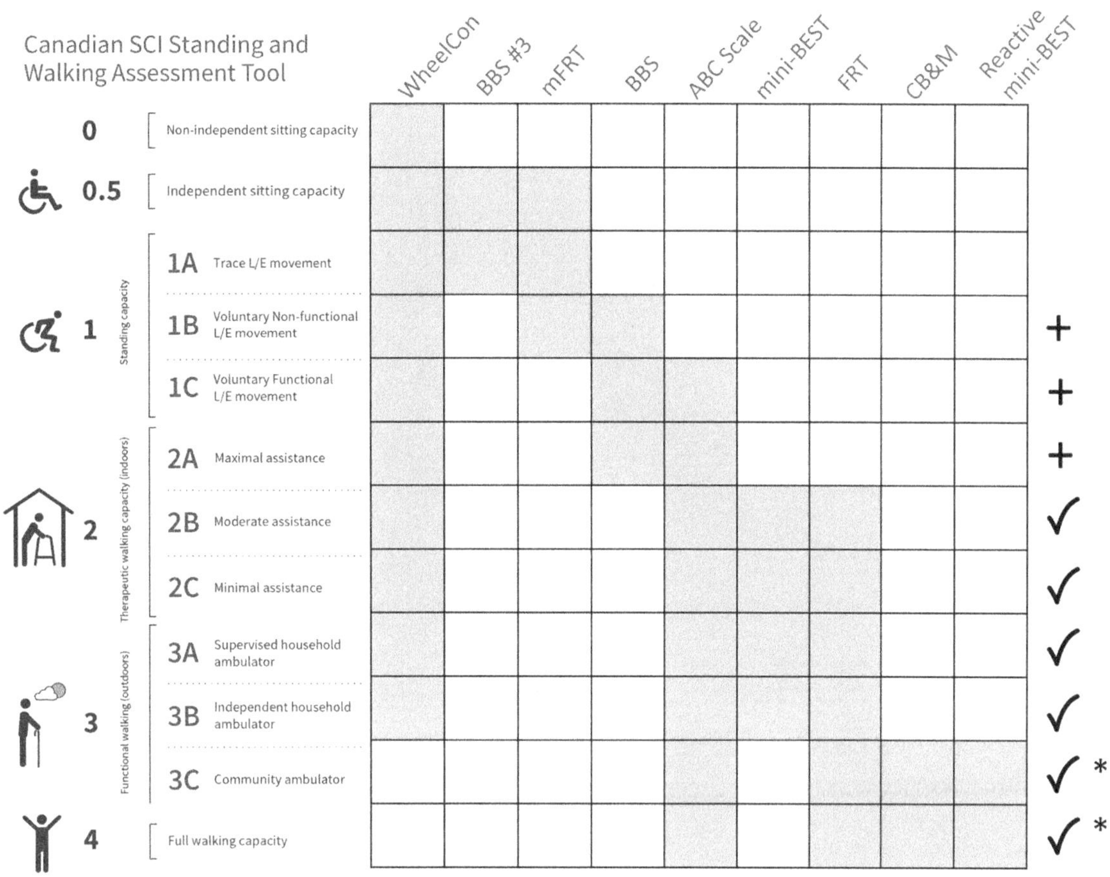


FIG. 3 Individualized and comprehensive approach to postural control assessment. The stages of the Standing and Walking Assessment Tool (left, reproduced with permission from the Canadian SCI Standing and Walking Module Group) are used to identify which clinical measures of postural control are appropriate for each stage. +Additional non-standardized measures of reactive postural control, cognitive influences and verticality in sitting could be added to increase comprehensiveness. √All components of postural control assessed by recommended measures. √*All components of postural control, with the exception of verticality, assessed by recommended measures. *BBS #3*, third item on the Berg Balance Scale (BBS); *Reactive mini-BEST*, Reactive sub-scale of the mini-Balance Evaluation Systems Test (mini-BESTest); *SCI*, spinal cord injury; *L/E*, lower extremity; *WheelCon*, Wheelchair Use Confidence Scale v.3; *mFRT*, modified Functional Reach Test; *ABC Scale*, Activities-specific Balance Confidence Scale; *FRT*, Functional Reach Test; *CB&M*, Community Balance and Mobility Scale.

et al., 2019): (1) wheelchair users with varying degrees of independence in sitting and standing (i.e., Stages 0 through 2A); (2) part-time ambulators who also use a wheelchair for mobility (i.e., Stages 2B through 3B); and (3) full-time community ambulators (i.e., Stages 3C and 4).

## Wheelchair users: Assessment of postural control

For individuals with SCI/D who are at a Stage 0 through 2A on the SWAT, the assessment of postural control is focused on sitting and standing activities, as appropriate for the individual's level of function (see Table 2). The clinical measures recommended for these SWAT stages address some, but not all, of the components of postural control (see Fig. 3 and Table 3).

### Stage 0

A meaningful postural control assessment for individuals lacking independent sitting capacity is limited to an evaluation of balance confidence. To gauge balance confidence in a wheelchair user, the Wheelchair Use Confidence Scale v.3 (WheelCon) can be used, with separate versions available for manual and power wheelchair users (Rushton, Miller, Kirby, Eng, & Yip, 2011). The questionnaire measures one's confidence in using a wheelchair according to six wheelchair-related topics: wheelchair activities, negotiating the physical environment, knowledge and problem solving,

**TABLE 3** Comprehensiveness of recommended clinical measures.

| | Components of postural control | | | | | | | | | |
|---|---|---|---|---|---|---|---|---|---|---|
| Clinical measure | Functional stability limits | Underlying motor systems | Static stability | Verticality | Reactive postural control | Anticipatory postural control | Dynamic stability | Sensory integration | Cognitive influences | Balance confidence |
| WheelCon | | | | | | | | | | ✓ |
| BBS #3 | | | ✓ | | | | | | | |
| mFRT | ✓ | ✓ | | | | ✓ | | | | |
| BBS | ✓ | ✓ | ✓ | | | ✓ | ✓ | ✓ | | |
| ABC Scale | | | | | | | | | | ✓ |
| Mini-BESTest | | ✓ | ✓ | ✓ | ✓ | ✓ | ✓ | ✓ | ✓ | |
| FRT | ✓ | ✓ | | | | ✓ | | | | |
| CB&M Scale | | ✓ | ✓ | | | ✓ | ✓ | ✓ | ✓ | |
| Reactive mini-BESTest | | | | | ✓ | | | | | |

Clinical measures recommended for a comprehensive and individualized approach to postural control assessment after spinal cord injury or disease are listed (left). √ The component of postural control is assessed by the clinical measure (Arora, Oates, et al., 2020; Sibley et al., 2015). *BBS #3*, third item on the Berg Balance Scale (BBS); *Reactive mini-BEST*, Reactive sub-scale of the mini-Balance Evaluation Systems Test (mini-BESTest); *WheelCon*, Wheelchair Use Confidence Scale v.3; *mFRT*, modified Functional Reach Test; *ABC Scale*, Activities-specific Balance Confidence Scale; *FRT*, Functional Reach Test; *CB&M*, Community Balance and Mobility Scale.

advocacy, managing emotions and managing social situations (Rushton et al., 2011). The WheelCon was shown to have excellent reliability, validity and responsiveness in a group of wheelchair users, 60% of whom had SCI/D (Rushton, Miller, Kirby, & Eng, 2013).

### Stages 0.5 and 1A

For those individuals with no standing capacity, but independent sitting capacity, the assessment of postural control is focused on a few appropriate components: (1) balance confidence related to wheelchair use, as measured with the WheelCon; (2) static stability in sitting, as measured with item three on the Berg Balance Scale, "Sitting with Back Unsupported but Feet Supported on Floor or on Stool" (Berg, Wood-Dauphinee, Williams, & Gayton, 1989); and, (3) functional stability limits, underlying motor systems and anticipatory postural control in sitting, as measured by the modified Functional Reach Test. The modified Functional Reach Test measures the maximum distance one can reach in a seated position. It has validity and reliability in both the motor complete (Lynch, Leahy, & Barker, 1998) and motor incomplete SCI/D (Field-Fote & Ray, 2010) populations.

### Stages 1B, 1C, and 2A

For individuals with minimal ability to step, a greater number of the components of postural control can be evaluated through inclusion of the entire Berg Balance Scale. The Berg Balance Scale evaluates balance control during 14 sitting and standing tasks, such as chair-to-chair transfers and reaching forward in standing (Berg et al., 1989). It is a valid and reliable measure of balance for the SCI/D population (Wirz, Muller, & Bastiaenen, 2010). Although the Berg Balance Scale increases the comprehensiveness of the assessment for wheelchair users, several components of postural control are missing from this clinical measure: reactive postural control, cognitive influences and verticality (Arora, Oates, et al., 2020). To address these gaps in assessment, non-standardized evaluations of reactive postural control and dual tasking in sitting could be added. If an individual presents with a co-morbid condition known to affect verticality, then a non-standardized evaluation of this postural control system in sitting could be included.

To evaluate balance confidence, the WheelCon is recommended for individuals staged at 1B through 2A. However, for individuals who are beginning to stand (i.e., Stages 1C and 2A), the ABC Scale should also be included to gauge balance confidence during upright activities. The ABC Scale asks individuals to rate their confidence, from 0% to 100%, in their ability to maintain balance while performing 16 standing and walking tasks (Powell & Myers, 1995). The ABC Scale has validity and reliability among individuals with chronic incomplete SCI/D (Shah et al., 2017).

## Part-time ambulators: Assessment of postural control

At Stages 2B through 3B, individuals with SCI/D can stand, but have varying degrees of ambulatory ability and are likely using a wheelchair to meet some of their mobility needs (see Table 2). At these SWAT stages, an assessment including all 10 components of postural control is achieved by combining two clinical scales (i.e., the mini-BESTest and the Functional Reach Test) and two questionnaires that query balance confidence (i.e., the WheelCon and the ABC Scale) (see Fig. 3 and Table 3).

For part-time ambulators, the mini-BESTest is the clinical measure of choice. The mini-BEST is the most comprehensive measure of postural control that has been used with the SCI/D population (Arora, Oates, et al., 2020). It is also known to have less of a ceiling effect than the Berg Balance Scale (Jørgensen et al., 2017). The mini-BESTest assesses four balance control systems: anticipatory, reactive postural control, sensory orientation and dynamic gait (Franchignoni, Horak, Godi, Nardone, & Giordano, 2010). The mini-BESTest and its sub-scales have validity and reliability among the chronic incomplete SCI/D population (Chan et al., 2019; Jørgensen et al., 2017). As the mini-BESTest evaluates all systems outlined in the modified Systems Framework for Postural Control with the exception of functional stability limits (Arora, Oates, et al., 2020), it is recommended that the Functional Reach Test is used as well for part-time ambulators. The Functional Reach Test is similar to the modified Functional Reach Test, except it is performed in standing.

## Full-time ambulators: Assessment of postural control

A comprehensive assessment of postural control that is appropriate for the high level of function of Stages 3C and 4 (see Table 2) is possible by combining a questionnaire of balance confidence (i.e., the ABC Scale) with three clinical measures (see Fig. 3 and Table 3). First, to evaluate functional stability limits, the Functional Reach Test is recommended. Second, to evaluate reactive postural control, the Reactive sub-scale of the mini-BEST may be used. Third, to evaluate the remaining components of postural control, with the exception of verticality, the Community Balance and Mobility Scale is recommended. The Community Balance and Mobility Scale is a 19-item scale that evaluates performance on challenging balance tasks requiring speed and accuracy; for example, tandem walking, hopping forward, and running with a controlled stop (Inness et al., 2011). The tasks are completed without any assistance or gait aid. It is valid and reliable measure among inpatients with primarily AIS D SCI/D (Chan et al., 2017). Lastly, if an evaluation of verticality is indicated based on an individual's medical history, item 9 on the mini-BESTest (i.e., "Incline—Eyes Closed") (Franchignoni et al., 2010) could be included in the assessment.

## Summary

Postural control assessment is a valued part of the rehabilitation of individuals living with SCI/D. Yet current approaches in clinical practice are unstandardized and rarely target all ten components of postural control. New measures are not needed to address these gaps; currently available clinical measures and questionnaires can be combined to create a comprehensive assessment of postural control. Instrumented tools, such as force plates and IMU, can be applied in a variety of contexts and tasks, contributing to a comprehensive assessment. The stages of the SWAT can be used to individualize the assessment to the individual's mobility status, providing clinicians and researchers with a standardized approach to the selection of clinical measures.

## Applications to other areas of neuroscience

An assessment of postural control is a critical part of the rehabilitation for individuals with spinal cord injury or disease, as well as individuals living with other neurological conditions, such as stroke, Multiple Sclerosis, Parkinson's disease, and cerebral palsy. Despite differences in the nature and location of neurological damage across conditions, the principles of comprehensiveness and individualization apply to the postural control assessment for all. Each neurological condition will have a unique clinical presentation that can be characterized through standardized assessment of all components of postural control.

Similarly, the measures recommended for the assessment of postural control (see Fig. 3) are not specific to spinal cord injury or disease; many of these measures were developed and validated for other clinical populations. Only the stages of the Standing and Walking Assessment Tool are specific to spinal cord injury; however, these stages were developed from classification systems used with other neurological populations, such as the Functional Ambulation Categories. Moreover, the broad categories of mobility status presented here (i.e., wheelchair users, part-time ambulators, full-time ambulators) have applicability to other neurological conditions.

## Mini-dictionary of terms

- **Anticipatory postural control**: The ability to shift one's center of mass appropriately prior to discrete movements.
- **Balance**: The term that describes the relationship between one's center of mass and one's base of support.
- **Balance confidence**: The level of perceived confidence in one's ability to maintain a state of balance while performing daily activities.
- **Cognitive influences**: The ability to maintain postural control while responding to commands during a task or attending to additional tasks (i.e., dual tasking).
- **Dynamic stability**: The ability to control one's center of mass while the base of support is changing (e.g., during walking).
- **Functional stability limits**: The ability to move one's center of mass within one's base of support.
- **Postural control**: One's ability to maintain, achieve or restore a state of balance when maintaining a posture, moving voluntarily, or reacting to an external perturbation.
- **Reactive postural control**: The ability to return one's center of mass within one's base of support after an external perturbation using corrective motor strategies (e.g., ankle and hip strategies, reactive stepping).
- **Sensory integration**: The ability to combine sensory input from multiple sources and re-weight when input is altered.
- **Static stability**: The ability to control one's center of mass within a static base of support during sitting and standing.
- **Underlying motor systems**: The muscular and neuromuscular properties that contribute to strength and coordination.
- **Verticality**: The ability to align one's body with respect to gravity.

## Key facts of postural control

- Postural control is the ability to maintain, achieve or restore a state of balance when maintaining a posture (e.g., sitting, standing), moving (e.g., walking) or reacting to a perturbation (e.g., slip, trip), thereby preventing a fall.
- Physiological factors, such as muscle strength and coordination, and psychological factors, such as balance confidence, influence postural control.
- Impairments in postural control are experienced by most individuals living with spinal cord injury or disease, resulting in frequent falls.
- Clinicians view assessment of postural control as important for patient care.
- Movement observation, standardized clinical scales and instrumented measures, such as force plates and inertial measurement units, can be used to evaluate postural control.

## Summary points

- Individuals with spinal cord injury or disease have impairments in postural control regardless of the neurological level or severity of injury.
- Following damage to the spinal cord, impairments in functional stability limits, underlying motor systems, static stability, sensory integration, reactive postural control, cognitive influences, dynamic stability and balance confidence are common.
- Clinicians and researchers lack a standardized and comprehensive approach to the assessment of postural control in individuals with spinal cord injury or disease.
- A comprehensive and individualized approach to the assessment of postural control is needed to tailor rehabilitation for each unique individual.
- A comprehensive assessment of postural control includes evaluation of ten components based on the Systems Framework of Postural Control combined with balance confidence.
- Instrumented measures, such as force plates and inertial measurement units, can contribute to comprehensive assessments of postural control.
- An individualized assessment is achieved by considering an individual's mobility status as outlined in the Stages of the Standing and Walking Assessment Tool.
- Based on an individual's Standing and Walking Assessment Tool Stage, clinical measures are combined to create a comprehensive assessment of postural control.

## References


Anderson, K. D. (2004). Targeting recovery: Priorities of the spinal cord-injured population. *Journal of Neurotrauma*, *21*, 1371–1383.

Arora, T., Musselman, K. E., Lanovaz, J., Linassi, G., Arnold, C., Milosavljevic, S., et al. (2020). Reactive balance responses to an unexpected slip perturbation in individuals with incomplete spinal cord injury. *Clinical biomechanics*, *78*, 105099.

Arora, T., Musselman, K. E., Lanovaz, J., Linassi, G., Arnold, C., Milosavljevic, S., et al. (2019). Walking stability during normal walking and its association with slip intensity among individuals with incomplete spinal cord injury. *Physical Medicine & Rehabilitation*, *11*, 270–277.

Arora, T., Oates, A., Lynd, K., & Musselman, K. E. (2020). Current state of balance assessment during transferring, sitting, standing and walking activities for the spinal cord injured population: A systematic review. *Journal of Spinal Cord Medicine*, *43*, 10–23.

Bayley, M. T., Kirby, R. L., Farahani, F., Titus, L., Smith, C., Routhier, F., et al. (2019). Development of wheeled mobility indicators to advance the quality of spinal cord injury rehabilitation: SCI-high project. *Journal of Spinal Cord Medicine*, *42*(Suppl 1), S130–S140.

Beloozerova, I. N., Sirota, M. G., Orlovsky, G. N., & Deliagina, T. G. (2005). Activity of pyramidal tract neurons in the cat during postural corrections. *Journal of Neurophysiology*, *93*, 1831–1844.

Berg, K., Wood-Dauphinee, S., Williams, J., & Gayton, D. (1989). Measuring balance in the elderly: Preliminary development of an instrument. *Physiotherapy Canada*, *41*, 304–311.

Bone, M. D., Arora, T., Musselman, K. E., Lanovaz, J. L., Linassi, G. A., & Oates, A. R. (2021). Investigating proactive balance control in individuals with incomplete spinal cord injury while walking on a known slippery surface. *Neuroscience Letters*, *749*, 135744.

Brouwer, B., Musselman, K., & Culham, E. (2004). Physical function and health status among seniors with and without a fear of falling. *Gerontology*, *6*, 135–141.

Butler Forslund, E., Roaldsen, K. S., Hultling, C., Wahman, K., & Franzén, E. (2016). Concerns about falling in wheelchair users with spinal cord injury – Validation of the Swedish version of the spinal cord injury falls concern scale. *Spinal Cord*, *54*, 115–119.

Carpenter, M. G., Adkin, A. L., Brawley, L. R., & Frank, J. S. (2006). Postural, physiological and psychological reactions to challenging balance: Does age make a difference? *Age and Ageing*, *35*, 298–303.

Chan, K., Guy, K., Shah, G., Golla, J., Flett, H., Williams, J., et al. (2017). Measuring balance in high-functioning individuals with incomplete spinal cord injury: The community balance and mobility scale. *Spinal Cord, 55*, 294–299.

Chan, K., Lee, J. W., Unger, J., Yoo, J., Masani, K., & Musselman, K. E. (2020). Reactive stepping after a forward fall in people living with incomplete spinal cord injury or disease. *Spinal Cord, 58*, 185–193.

Chan, K., Unger, J., Lee, J. W., Johnston, G., Constand, M., Masani, K., et al. (2019). Quantifying balance control after spinal cord injury: Reliability and validity of the mini-BESTest. *Journal of Spinal Cord Medicine, 42*(Sup 1), 141–148.

Chen, C. L., Yeung, K. T., Bih, L. I., Wang, C. H., Chen, M. I., & Chien, J. C. (2003). The relationship between siting stability and functional performance in patients with paraplegia. *Archives of Physical Medicine and Rehabilitation, 84*, 1276–1281.

Chiba, R., Takakusaki, K., Ota, J., Yozu, A., & Haga, N. (2016). Human upright posture control models based on multisensory inputs; in fast and slow dynamics. *Neuroscience Research, 2014*, 96–104.

Day, K. V., Kautz, S. A., Wu, S. S., Suter, S. P., & Behrman, A. L. (2011). Foot placement variability as a walking balance mechanism post-spinal cord injury. *Clinical biomechanics, 13*, 13.

Field-Fote, E. C., & Ray, S. S. (2010). Seated reach distance and trunk excursion accurately reflect dynamic postural control in individuals with motor-incomplete spinal cord injury. *Spinal Cord, 48*, 745–749.

Franchignoni, F., Horak, F., Godi, M., Nardone, A., & Giordano, A. (2010). Using psychometric techniques to improve the balance evaluation system's test: The mini-BESTest. *Journal of Rehabilitation Medicine, 42*, 323–331.

Ghislieri, M., Gastaldi, L., Pastorelli, S., Tadano, S., & Agostini, V. (2019). Wearable inertial sensors to assess standing balance: A systematic review. *Sensors (Basel), 19*, 4075.

Hatch, J., Gill-Body, K. M., & Portney, L. G. (2003). Determinants of balance confidence in community-dwelling elderly people. *Physical Therapy, 83*, 1072–1079.

Horak, F. B. (2006). Postural orientation and equilibrium: What do we need to know about neural control of balance to prevent falls? *Age and Ageing, 35* (Suppl_2), ii7–ii11.

Horak, F. B., Wrisley, D. M., & Frank, J. (2009). The balance evaluation systems test (BESTest) to differentiate balance deficits. *Physical Therapy, 89*, 484–498.

Houston, D., Unger, J., Lee, J. W., Masani, K., & Musselman, K. E. (2020). Functional electrical stimulation plus visual feedback balance training for standing balance performance among individuals with incomplete spinal cord injury: A case series. *Frontiers in Neurology—Neurorehabilitation, 11*, 680.

Inness, E. L., Howe, J., Niechwiej-Szwedo, E., Jaglal, S. B., McIlroy, W. E., & Verrier, M. C. (2011). Measuring balance and mobility after traumatic brain injury: Validation of the community balance and mobility scale (CB&M). *Physiotherapy Canada, 63*, 199–208.

Jannings, W. (2017). A quality improvement project to investigate the circumstances of lower limb fractures in non-ambulant persons with spinal cord injury. *Journal of the Australasian Rehabilitation Nurses Association, 20*, 14–18.

Jensen, M. P., Truitt, A. R., Schomer, K. G., Yorkston, K. M., Baylor, C., & Molton, I. R. (2013). Frequency and age effects of secondary health conditions in individuals with spinal cord injury: A scoping review. *Spinal Cord, 51*, 882–892.

Joassin, R., Bonniaud, V., Barra, J., Marquer, A., & Perennou, D. (2010). Somaesthetic perception of the vertical in spinal cord injured patients: A clinical study. *Annals of Physical and Rehabilitation Medicine, 53*, 568–574.

John, L. T., Cherian, B., & Babu, A. (2010). Postural control and fear of falling in persons with low-level paraplegia. *Journal of Rehabilitation Research & Development, 47*, 497–502.

Jørgensen, V., Opheim, A., Halvarsson, A., Franzén, E., & Roaldsen, K. S. (2017). Comparison of the berg balance scale and the mini-BESTest for assessing balance in ambulatory people with spinal cord injury: Validation study. *Physical Therapy, 97*, 677–687.

Khan, A., Pujol, C., Laylor, M., Unic, N., Pakosh, M., Dawe, J., et al. (2019). Falls after spinal cord injury: A systematic review and meta-analysis of incidence proportion and contributing factors. *Spinal Cord, 57*, 526–539.

Lee, J. W., Chan, K., Unger, J., Yoo, J., Musselman, K. E., & Masani, K. (2021). Inter-joint coordination between the ankle and hip joints during quiet standing in individuals with incomplete spinal cord injury. *Journal of Neurophysiology, 125*, 1681–1689.

Lemay, J. F., Duclos, C., Nadeau, S., Gagnon, D., & Desrosiers, É. (2014). Postural and dynamic balance while walking in adults with incomplete spinal cord injury. *Journal of Electromyography and Kinesiology, 24*, 739–746.

Lemay, J. F., Gagnon, D., Duclos, C., Grangeon, M., Gauthier, C., & Nadeau, S. (2013). Influence of visual inputs on quasi-static standing postural steadiness in individuals with spinal cord injury. *Gait & Posture, 38*, 357–360.

Lemay, J. F., Gagnon, D. H., Nadeau, S., Grangeon, M., Gauthier, C., & Duclos, C. (2014). Center-of-pressure total trajectory length is a complementary measure to maximum excursion to better differentiate multidirectional standing limits of stability between individuals with incomplete spinal cord injury and able-bodied individuals. *Journal of Neuroengineering and Rehabilitation, 11*, 8.

Lemay, J. F., & Nadeau, S. (2010). Standing balance assessment in AISA D paraplegic and tetraplegic participants: Concurrent validity of the berg balance scale. *Spinal Cord, 48*, 245–250.

Lemay, J. F., Noamani, A., Unger, J., Houston, D. J., Rouhani, H., & Musselman, K. E. (2020). Using wearable sensors to characterize gait after spinal cord injury: Evaluation of test–retest reliability and construct validity. *Spinal Cord, 59*, 1–9.

Lynch, S. M., Leahy, P., & Barker, S. P. (1998). Reliability of measurements obtained with a modified functional reach test in subjects with spinal cord injury. *Physical Therapy, 78*, 128–133.

Maki, B. E., & McIlroy, W. E. (1997). The role of limb movements in maintaining upright stance: The 'change-in support' strategy. *Physical Therapy, 77*, 488–507.

Milosevic, M., Gagnon, D. H., Gourdou, P., & Nakazawa, K. (2017). Postural regulatory strategies during quiet sitting are affected in individuals with thoracic spinal cord injury. *Gait & Posture*, *58*, 446–452.

Moore, J. L., Potter, K., Blankshain, K., Kaplan, S. L., O'Dwyer, L. C., & Sullivan, J. E. (2018). A core set of outcome measures for adults with neurologic conditions undergoing rehabilitation. *Journal of Neurologic Physical Therapy*, *42*, 174–220.

Musselman, K. E., Arnold, C., Pujol, C., Lynd, K., & Oosman, S. (2018). Falls, mobility and physical activity after spinal cord injury: An exploratory study using photo-elicitation interviewing. *Spinal Cord Series And Cases*, *4*, 39.

Musselman, K. E., Chan, K., Lemay, J. F., Walden, K., Verrier, M., & Gagnon, D. (2019). Responsiveness of the standing and walking assessment tool in inpatients with spinal cord injury. *Archives of Physical Medicine and Rehabilitation*, *100*, E198–E199.

Musselman, K. E., Lemay, J. F., Walden, K., Harris, A., Gagnon, D., & Verrier, M. (2019). The standing and walking assessment tool for individuals with spinal cord injury: A qualitative study of validity and clinical use. *Journal of Spinal Cord Medicine*, *42*(Sup1), 108–118.

Noamani, A., Lemay, J. F., Musselman, K. E., & Rouhani, H. (2020). Postural control strategy after incomplete spinal cord injury: Effect of sensory inputs on trunk-leg movement coordination. *Journal of Neuroengineering and Rehabilitation*, *17*, 141.

Noamani, A., Lemay, J. F., Musselman, K. E., & Rouhani, H. (2021). Characterization of standing balance after incomplete spinal cord injury: Alteration in integration of sensory information in ambulatory individuals. *Gait & Posture*, *83*, 152–159.

Oates, A., Arnold, C., Walker-Johnston, J., Van Ooteghem, K., Oliver, A., Yausie, J., et al. (2017). Balance assessment practices of Saskatchewan physiotherapists: A brief report of survey findings. *Physiotherapy Canada*, *69*, 217–225.

Perez-SanPablo, A. S., Quinzanos-Fresnedo, J., Lopez-Romero, J. C., & Quinones-Uriostegui, I. (2018). Reliability of posturographic measures obtained during instrumental assessment of trunk control using inertial sensors in individuals with spinal cord injury. *Annals of Physical and Rehabilitation Medicine*, *61*(Suppl), e84.

Pollock, A. S., Durward, B. R., Rowe, P. J., & Paul, J. P. (2000). What is balance? *Clinical Rehabilitation*, *14*, 402–406.

Powell, L. E., & Myers, A. M. (1995). The activities-specific balance confidence (ABC) scale. *Journal of Gerontology Series A Biological Sciences & Medical Sciences*, *50*, M28–M34.

Rushton, P. W., Miller, W. C., Kirby, R. L., & Eng, J. J. (2013). Measure for the assessment of confidence with manual wheelchair use (WheelCon-M) version 2.1: Reliability and validity. *Journal of Rehabilitation Medicine*, *45*, 61–67.

Rushton, P. W., Miller, W. C., Kirby, R. L., Eng, J. J., & Yip, J. (2011). Development and content validation of the wheelchair use confidence scale: A mixed-methods study. *Disability and Rehabilitation: Assistive Technology*, *6*, 57–66.

Shah, G., Oates, A. R., Arora, T., Lanovaz, J. L., & Musselman, K. E. (2017). Measuring balance confidence after spinal cord injury: The validity and reliability of the activities-specific balance confidence scale. *Journal of Spinal Cord Medicine*, *40*, 768–776.

Shin, S., & Sosnoff, J. J. (2013). Spinal cord injury and time to instability in seated posture. *Archives of Physical Medicine and Rehabilitation*, *94*, 1615–1620.

Sibley, K. M., Beauchamp, M. K., Van Ooteghem, K., Straus, S. E., & Jaglal, S. B. (2015). Using the systems framework for postural control to analyze the components of balance evaluated in standardized balance measures: A scoping review. *Archives of Physical Medicine and Rehabilitation*, *96*, 122–132.

Sibley, K. M., Straus, S. E., Inness, E. L., Salbach, N. M., & Jaglal, S. B. (2011). Balance assessment practices and use of standardized balance measures among Ontario physical therapists. *Physical Therapy*, *91*, 1583–1591.

Sibley, K. M., Straus, S. E., Inness, E. L., Salbach, N. M., & Jaglal, S. B. (2013). Clinical balance assessment: Perceptions of commonly-used standardized measures and current practices among physiotherapists in Ontario, Canada. *Implementation Science*, *8*, 33.

Singh, H., Scovil, C., Yoshida, K., Oosman, S., Kaiser, A., Jaglal, S. B., et al. (2020). Capturing the psychosocial impacts of falls from the perspectives of wheelchair users with spinal cord injury through photo-elicitation. *Disability and Rehabilitation*, *43*(19), 2680–2689.

Tamburella, F., Scivoletto, G., Iosa, M., & Molinari, M. (2014). Reliability, validity, and effectiveness of center of pressure parameters in assessing stabilometric platform in subjects with incomplete spinal cord injury: A serial cross-sectional study. *Journal of Neuroengineering and Rehabilitation*, *11*, 86.

Thigpen, M. T., Cauraugh, J., Creel, G., Day, K., Flynn, S., Fritz, S., et al. (2009). Adaptation of postural responses during different standing perturbation conditions in individuals with incomplete spinal cord injury. *Gait & Posture*, *29*, 113–118.

Tse, C. M., Carpenter, M. G., Liu-Ambrose, T., Chisholm, A. E., & Lam, T. (2017). Attentional requirements of postural control in people with spinal cord injury: The effect of dual task. *Spinal Cord*, *55*, 915–920.

Unger, J., Oates, A. R., Lanovaz, J., Chan, K., Lee, J. W., Theventhiran, P., et al. (2020). The measurement properties of the Lean-and-Release test in people with incomplete spinal cord injury or disease. *Journal of Spinal Cord Medicine*, 1–10. https://doi.org/10.1080/10790268.2020.1847562.

Wirz, M., Muller, R., & Bastiaenen, C. (2010). Falls in persons with spinal cord injury: Validity and reliability of the Berg Balance Scale. *Neurorehabilitation and Neural Repair*, *24*, 70–77.

World Health Organization. (2007). *WHO Global report on falls prevention in older age*. https://www.who.int/ageing/publications/Falls_prevention7March.pdf.

Zelenin, P. V., Beloozerova, I. N., Sirota, M. G., Orlovsky, G. N., & Deliagina, T. G. (2010). Activity of red nucleus neurons in the cat during postural corrections. *Journal of Neuroscience*, *30*, 14533–14542.

Section D

# Treatments: Experimental and clinical

Chapter 17

# Surgical management of acute spinal cord injury in emergency setting

Federico Fusini[a,b], Angela Coniglio[c], Alessandro Rava[b], Salvatore Risitano[b,d], Alessandro Massè[b], and Massimo Girardo[c]

[a]*Department of Orthopaedics and Traumatology, ASL CN1, Regina Montis Regalis Hospital, Mondovì, Cuneo, Italy;* [b]*Department of Orthopaedics and Traumatology, CTO Hospital, University of Turin, Turin, Italy;* [c]*Spine Surgery Unit, Department of Orthopaedics and Traumatology, CTO Hospital, Turin, Italy;* [d]*Department of Orthopaedics and Traumatology, Maggiore Hospital of Chieri, Chieri, Italy*

## List of abbreviations

| | |
|---|---|
| **AIS** | ASIA Impairment Scale |
| **ASIA score** | American Spinal Injury Association score |
| **CCS** | Central Cord Syndrome |
| **CES** | Cauda Equina Syndrome |
| **CMS** | Conus Medullary Syndrome |
| **CRHSCIR** | Canadian Rick Hansen Spinal Cord Injury Registry |
| **CSM** | cervical spondylotic myelopathy |
| **ESD** | early surgical decompression |
| **NACTN** | North American Clinical Trials Network for SCI |
| **SCI** | spinal cord injury |
| **SCIM** | spinal cord independence measure |
| **SD** | surgical decompression |
| **SDSF** | surgical decompression and segment fixation |
| **STACSIS study** | Surgical Timing in Acute Spinal Cord Injury study |

## Introduction

Historically, closed reduction with external immobilization was the milestone of treatment for acute SCI and only few cases were reserved for surgical management. However, in the last 2 decades, spine surgeons emphasized early surgical management of SCI (Hawryluk, Rowland, Kwon, & Fehlings, 2008; Rowland, Hawryluk, Kwon, & Fehlings, 2008; Wilson & Fehlings, 2011).

Preclinical studies suggested that persistent compression of spinal cord after primary injury is responsible for an additional secondary damage which might be avoided reducing tissue damage as soon as possible and improving outcomes (Dimar, Glassman, Raque, Zhang, & Shields, 1999; Fehlings & Perrin, 2006).

SDSF allows stabilization of vertebral column and decompression of spinal cord. Moreover, stabilization prevents any further trauma on spinal cord and allows early rehabilitation (Wilson et al., 2017).

Since vascular mechanisms of secondary injury after SCI were already discovered in the past, the practice of ESD has now biological support (Ahuja et al., 2017; Tator & Fehlings, 1991).

Removing spinal cord compression and restoring blood supply should reduce the biochemical process induced by ischemia due to local compression (Dolan, Tator, & Endrenyi, 1980; Guha, Tator, Endrenyi, & Piper, 1987; Rivlin & Tator, 1978).

Despite the increasing literature, there are still some unanswered questions about the optimal methods for ESD. We still don't know if all patients are going to benefit from ESD or if there are subgroups which are more likely to benefit than others

Diagnosis and Treatment of Spinal Cord Injury. https://doi.org/10.1016/B978-0-12-822498-4.00017-8

**TABLE 1** SCI severity.

| Classification of spinal cord injury (SCI) severity | | | |
|---|---|---|---|
| **Normal neurological condition** | **Incomplete SCI** | **Complete SCI** | **Clinical syndromes** |
| ASIA scale E | ASIA scale B, C or D | ASIA scale A | – Central Cord<br>– Anterior Cord<br>– Conus Medullaris<br>– Cauda Equina<br>– Brown-Sequard |

ASIA scale scores are paired with normal neurological status or incomplete or complete SCI. In the fourth column all special clinical syndromes related to SCI are reported.

(Table 1). We still don't know the appropriate time for surgical intervention balancing key priorities of efficacy and clinical feasibility.

In this narrative review, we will highlight the situation of last 20 years literature about ESD in patients with acute SCI focusing to which subgroups of SCI will benefit from ESD and the time of surgical intervention.

## Natural history of patients with acute cervical SCI

Approximately, one third of patients with cervical SCI has a complete injury. However, the incidence is double in patients with associated polytrauma (Stephan et al., 2015).

At 1 year from injury, 70% of patients with an initial ASIA (Table 2) score A injury will remain A, while about 30% of them will show some signs of improvement. Around 14% will improve to B, 9.5% to C, 6.5% to D, while none will improve to ASIA E (Wiberg & Hauge, 1988; Zariffa, Curt, & Steeves, 2012). On the other hand, patients with incomplete SCI have a much higher spontaneous recovery (Kirshblum, Millis, McKinley, & Tulsky, 2004).

Unfortunately, at 1 year after time of injury, patients with complete cervical SCI had a mean recovery ranging from 9.0 to 14 motor score points(Marino et al., 2011), calculating as motor score (ranging from 0 to 5 per muscle group), the strength for each muscle group below the lesion level (Maynard et al., 1997).

Neurological improvement will typically show within the first 3 months and then it will enter a steady state (Fawcett et al., 2007); even if in a little percentage of patients (about 9%) some little improvement can be showed even 5 years after injury (Ter Wengel, De Haan, Feller, Oner, & Vandertop, 2020).

## Natural history of patients with thoracic or thoracolumbar SCI

A complete neurological injury is reported ranging from 16.2% to 73.0% of patients with acute thoracic SCI (Nijendijk, Post, & Van Asbeck, 2014) and it is more frequent in patients with polytrauma (Stephan et al., 2015) (Fig. 1).

**TABLE 2** ASIA scale classification of neurological impairment.

| ASIA scale | Description |
|---|---|
| A | COMPLETE. No motor or sensory function is preserved in the sacral segments S4-S5 |
| B | INCOMPLETE. Sensory but not motor functions preserved under the level of lesion. It includes S4-S5 |
| C | INCOMPLETE. Motor function is preserved below the neurological level and more than 50% of key muscles below the neurological level have a muscle grade less than 3 |
| D | INCOMPLETE. Motor function is preserved below the neurological level and at least 50% of key muscles below the neurological level have a muscle grade greater than or equal to 3 |
| E | NORMAL. Motor and sensory function are normal |

ASIA scale scores are reported on the left and paired with the accurate description of signs and symptoms.

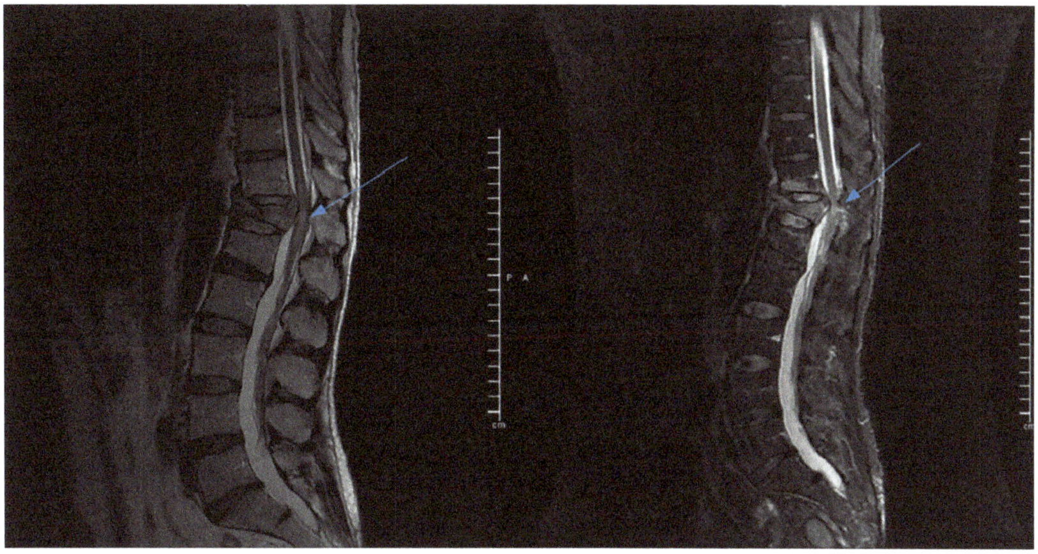

FIG. 1 Magnetic resonance imaging (MRI) about T12 fracture causing complete SCI. In this picture T2 *(left)* and STIR *(right)* MRI sequences of a SCI clinical case. The patient (46 years old, man) arrived after trauma occurred at home. On the left and on the right, marked by the *blue arrows*, hyperintensity can be seen (SCI lesion causing paraplegia).

At 1 year from injury, about 85% of patients with initial ASIA A will not improve, while about 15% of them will show 1 or more ASIA grade improvement. In patients with initial ASIA B, more than 80% of them will show some improvement (Fawcett et al., 2007; Zariffa et al., 2012).

Substantially, acute thoracic SCI will improve less than cervical or thoracolumbar SCI (Zariffa et al., 2011). In only 3.6% of patients with ASIA A grade at 1 year after injury, some minor neurological recovery can be recorded up till 5 years (Kirshblum et al., 2004).

Thoracolumbar fractures can be associated with conus medullary injury (fractures from T8-L2) and cauda equina injury (fractures below L1-L2) (Brouwers et al., 2017).

Both syndromes are difficult to be distinguished one by the other. CMS is often characterized by a symmetric sensory deficits, bladder and bowel disfunction, and mild lower extremity weakness. CES is often defined by asymmetric sensory and motor deficits. In 20% of patients a complete lack of sensory and motor function will be present below the level of injury (Pickett et al., 2006).

Thoracolumbar SCI will show the same rate of neurological recovery of complete cervical SCI (Zariffa et al., 2011) After 1 year from trauma, patients will experience a mean recovery of 4.5 motor score points below injury level (Dvorak et al., 2014).

## Surgical procedures

SCI are a wide spectrum of injuries, ranging from less to more severe spinal cord damage. For this reason, a singular approach to the spine is not correct and surgical choice should be individualized to each patient (Gargiulo et al., 2019; Girardo et al., 2018). The main goals of treatment are fixation of instability and decompression of the spinal cord (Fig. 2) and the choice of the correct surgical procedure is based on severity, level and mechanism of injury; location and extension of compression and surgeon preference.

Surgical approaches to the spine are classified as anterior or posterior. With both approaches stabilization and decompression can be achieved, but in more severe cases where a circumferential arthrodesis is needed, both approaches are combined together (staged or in single setting) (Bose, Osterholm, Northrup, & Cotler, 1985).

In cervical SCI, anterior approach is the most widely used allowing easily a ventral decompression and corpectomy, while a posterior approach is more appropriate for the thoracolumbar spine, where chest and abdomen represent an anatomical barrier.

In most cases, due to complex injury with disruption of anterior and posterior elements, anterior and posterior approach where combined together to perform a 360° arthrodesis.

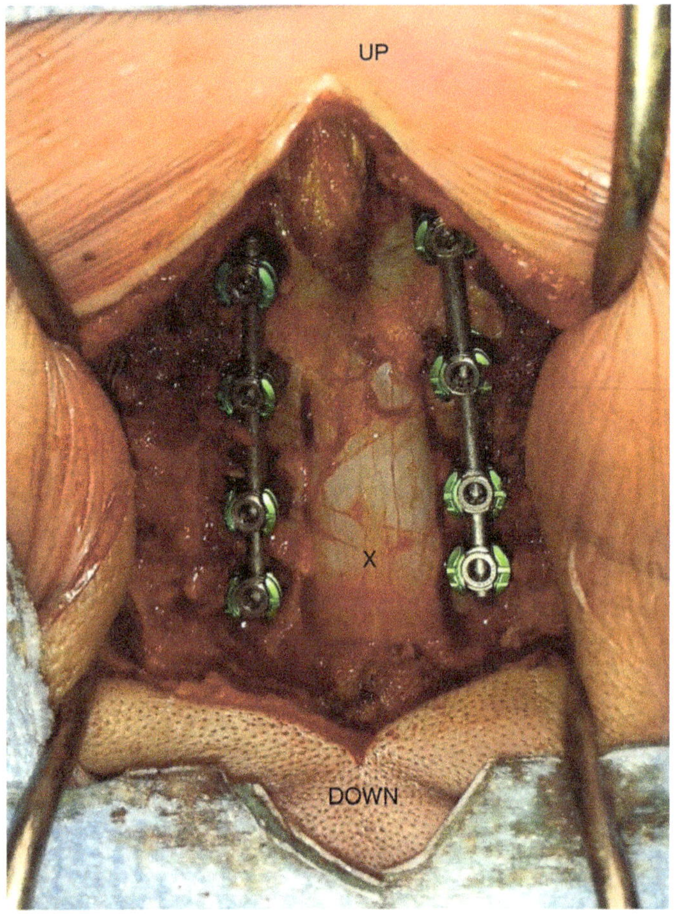


FIG. 2 Intraoperative picture about surgical decompression and fixation with screws and bars. In this picture the medulla (marked with "X") was decompressed and the cervical column was stabilized with screws and bars. "UP" and "DOWN" are referred to the patient head position.

Most common instrumentation devices used are pedicle screws with or without cement augmentation for thoracolumbar spine and pedicle or lateral masses screws for cervical spine. Usually, in acute traumatic SCI only one or two vertebral segments are involved and short segment stabilization is always preferred if possible. In flexion type injuries, posterior approach is usually preferred in the most of thoracolumbar injuries. In most cases, laminectomy, assure quick spinal decompression and also reduction with ligamentotaxis for thoracolumbar fractures (Feuchtbaum, Buchowski, & Zebala, 2016; Girardo et al., 2018) (Fig. 3).

## How to define early versus late surgery post SCI?

Several articles have examined a different post injury range of time thresholds to define early versus late surgery. However, how much time can be considered "early" in not already clearly established. To recommend one of these thresholds and to choose the most favorable moment, two main principles must be considered (Fehlings & Sekhon, 2001; Wilson et al., 2017):

(1) efficacy (the extent of time in which surgery results in better outcomes).
(2) feasibility (the extent of time in which it is possible to perform surgery before the said limit taking into account the practical realities of the prehospital and hospital environment).

## Which is the most efficacious threshold?

The theoretical objective of surgery is to begin a neuroprotective therapy performing a post injury treatment as soon as possible, preventing a secondary injury and favoring better recovery.

According to literature (Wilson et al., 2020), three common time thresholds can be found:

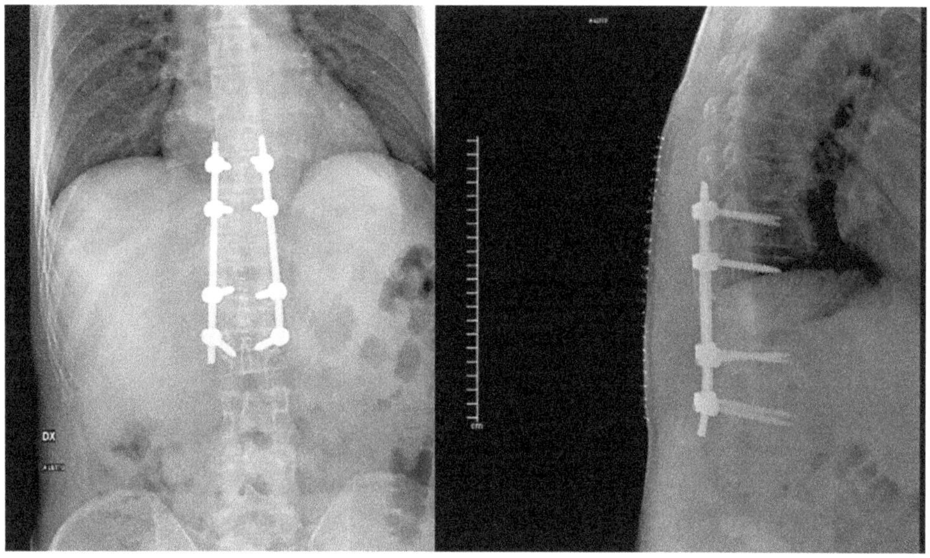

FIG. 3 X-rays about surgical treatment of SCI. About the same case of Fig. 1. The patient underwent ultra ESD with T10-L2 fixation with screws and bars.

(1) ultra-early threshold (8–12 h post SCI).
(2) early threshold (24 h post SCI).
(3) late threshold (48–72 h post SCI).

## Ultra-early threshold (8–12 h post SCI)

Jug et al. evaluated the neurological outcomes of 48 patients who underwent surgery before 8 h and between 8 and 24 h post-injury. After 6 months from injury, patients treated <8-h showed a greater median improvement in ASIA motor score and a higher likelihood of improve of at least 2 AIS grade ( Jug et al., 2015).

Grassner et al. evaluated the impact of surgery on functional outcomes in 70 patients who underwent SD prior to 8 h at 1 year follow up. The Authors analysis showed that patients receiving surgery prior to 8 h had superior scores and better motor outcomes at follow-up (Grassner et al., 2016).

In another study, 48 patients were retrospectively categorized into three groups depending on the timing of surgery: ultra-early (<12 h), early (12–24 h), and late (>24 h). Authors found a mean significant improvement of 1.3 AIS grade in ultra-early surgery group as compared with the mean improvement of 0.5 of patients who underwent early surgery (Burke et al., 2019).

## Early threshold (24 h post SCI)

STACSIS highlight the role of early surgery in cervical SCI, showing greater neurological recovery, with 2 AIS grade improvement at 6 months of follow up (Van Middendorp, Hosman, & Doi, 2013). On another study focused on thoracolumbar SCI, Rahimi-Movaghar et al. didn't reach the same conclusion. In fact, despite the study lack some power, the difference between early and late surgery were not statistically significant (Rahimi-Movaghar et al., 2009).

In the comprehensive study of Dvorak et al., early surgery for all anatomical level reached significantly greater motor recovery. However, the same effect was not demonstrated in complete SCI (Dvorak et al., 2015).

In the Ontario-based cohort study, early surgery showed a higher likelihood of a 2 AIS grade improvement with a trend of additional improvement in motor recovery (Wilson et al., 2012).

It must be acknowledged that the 24 h cut-off is biologically arbitrary and it has been proposed for practical purpose; nothing changes between 23 and 25 h. So, when interpreting literature, it is important to remind that surgery is more effective when performed earlier but 24 h is only a threshold.

### Late threshold (48–72 h post SCI)

Vaccaro et al. evaluated 64 SCI patients who received SD before 72 h after SCI or later than 5 days. At 1 year of follow up, the two groups didn't show any statistically difference in AIS grade conversion and motor score recovery (Vaccaro et al., 1997).

McKinley et al., reported the result of a study with a large sample of 779 patients from Model SCI Systems database and divided patients on the basis of SD prior or post 72 h. Authors found shorter hospital stay and reduced respiratory complications in patients who underwent SD before 72 h; however, no difference was found for neurological recovery (McKinley, Meade, Kirshblum, & Barnard, 2004).

Clohisy et al. evaluated 20 patients with thoracolumbar junction SCI. They found that patients surgically treated before 48 h showed grater improvements in neurological recovery than patients treated later. However, in this study patients who receive later surgery were treated about 2 months after injury (Clohisy, Akbarnia, Bucholz, Kenneth Burkus, & Backer, 1992).

## What threshold is feasible?

Despite the common objective is to promote early treatment, the reality of daily life trauma care, from prehospital support to emergency department access, carry on several barriers.

For many reasons, including hospital transport, patient stability and diagnostic delays, only 50% of patients would be eligible to undergo SD before 24 h while other 50% of patients, in most cases underwent SD between 24 and 48 h (Tator, Fehlings, Thorpe, & Taylor, 1999).

The analysis of the CRHSCIR reported that in only 39% of cervical SCI and 45% of thoracic SCI, spine surgeons were able to perform SD before 24 h from trauma. Some Authors suggested that to improve early surgery rate, it is necessary to reduce barriers to early surgery in order to optimize patient outcomes (Dvorak et al., 2015).

Despite about 90% of patients arrived at the site of definitive treatment within 6 h of trauma, only 50% of them are able to undergo surgical treatment within 24 h. Older age and intermediate stop in peripheral health centers, prior to arrival in specialized center, were important factors for delay of surgery (Wilson et al., 2016) (Battistuzzo et al., 2016).

To support this findings, Furlan et al. reported that those patients who receive an early treatment (<24 h) spent less time in intermediate center before the arrival to definitive center (Furlan, Tung, & Fehlings, 2013).

In common practice, when patients access emergency department, they should be screened for any life threatened pathology, which must be treated first, since SCI is commonly associated with high energy trauma. For that reason, it is difficult to believe that most surgeries could be performed before 24 h from injury.

## Does ESD offers the same benefit for all acute SCI?

SCI offers a wide spectrum of neurological impairment, from minimally to significant disabling, with a wide variability due to injury severity and neurological level of injury.

We will evaluate the current evidences about the efficacy of ESD in acute SCI based on level and severity of injury and time of surgical intervention.

## Complete versus incomplete SCI

In literature, only few studies have examined the role of ESD in complete and incomplete SCI. Dvorak et al. evaluated 888 patients recruited in the CRHSCIR. In this study, attention was focused on the significant role of early surgery (<24 h post SCI) on motor recovery and the Authors concluded as ESD was not associated with significant improvements in motor function. However, considering only patients with incomplete SCI graded as ASIA B-D, ESD was associated with an improvement of 6.3 points in motor score recovery at 6 months follow-up. As expected, no significant effect for ESD was observed in grade A patients (Dvorak et al., 2015).

Bourassa-Moreau et al. examined a cohort of SCI. He found a 34% of AIS improvement of ESD (<24 h) and 13% of improvement in late surgery (>24 h), with even a higher difference for cervical SCI (64% versus 36%). Moreover, they didn't observe any conversion in AIS grade for complete SCI between early and late surgery (Bourassa-Moreau et al., 2016).

More recently, Ma et al. analysed in their meta-analysis the effects of ultra ESD (<8 h) on patients with acute SCI. He found more significant improvement in AIS score than in patients treated with early (8–24 h) or late (>24 h) surgery.

Patients had more benefits in neurological recovery from ultra ESD with no difference in hospital stay and rate of complications (Ma et al., 2020).

On the other hand, ESD (<24 h after injury) reduced hospital stay, especially for patients with complete SCI (Dvorak et al., 2015).

All studies were in accordance regarding the benefits of ESD also for what concern the length of hospital stay for patients with complete SCI, while no other study support the benefit of ESD for incomplete SCI (Bourassa-Moreau, Mac-Thiong, Feldman, Thompson, & Parent, 2013; Mac-Thiong, Feldman, Thompson, Bourassa-Moreau, & Parent, 2012).

Unfortunately, there are no large prospective studies that confirmed the superior effects of ESD on neurological outcomes for incomplete versus complete SCI. These findings are supported by the hypothesis that for incomplete SCI, the primary energy of injury is not so powerful to cause a complete SCI, and there may be greater potential for neuroprotection (Dvorak et al., 2015) (Bourassa-Moreau et al., 2016).

## Effects of ESD on different neurological level of injury

One of the most important factors in patients with SCI is the neurological level of injury, which can influence patients' potential recovery (Wilson et al., 2012).

However, there are no many studies focused on specific neurological level and the effect of early or late surgery on neurological recovery. Commonly, cervical and thoracolumbar SCI are reported together and data are difficult to be extracted for each neurological level.

### Effect of SD in patients with cervical SCI

STASCIS is the largest study involving patients with only cervical SCI (Fehlings et al., 2012). Authors of this study analysed several cervical SCI undergoing either early (<24 h) or late (>24 h) SD and compared AIS grade changes at 6 months follow up. Both groups underwent at least 1 grade; however, patients treated with early surgery are more likely to improve AIS of 2 grade or more (odds ratio 2.57; $P = .01$), than patients who underwent late surgery. No differences between groups were found in term of complications, despite the trend is in favor of ESD (24% versus 30% respectively) (Van Middendorp, 2012).

More recently, Aarabi et al. evaluated 72 patients with traumatic cervical SCI with AIS grade A, B or C. Patients were stratified on the basis of ultra-early (<12 h), early (<24 h) and late (>24 h) surgery. Improvement of one grade or more was present in 55.6% of grade A, 60.9% of grade B, and 86.4% of grade C patients, but not in a significant manner. Authors also found other crucial predictive factors that could be more likely associated with injury recovery such as admission motor score and intramedullary lesion length (Aarabi et al., 2020).

### Effect of SD in patients with thoracic and thoracolumbar SCI

Rahimi-Movaghar et al. evaluated 35 patients with thoracic and thoracolumbar (T1-L1) SCI. Some underwent early (<24 h) while other late (>24 h) surgery. There was no difference between motor score recovery at 12 months of follow up, considering both complete and incomplete SCI (Rahimi-Movaghar et al., 2009).

In the NACTN for SCI, researchers evaluated 86 patients with SCI and neurological injury between T1 and L1. Early surgery (<24 h) was associated with additional seven points of ASIA motor recovery and a reduction of 60% in pulmonary complications (Wilson et al., 2018).

### Effect of SD on miscellaneous SCI

Some studies reported cervical, thoracic and thoracolumbar SCI as a unique entity; we report them in this section since it is not possible to revise the results and divide them in anatomical region.

In the study of Bourassa-Moreau et al., patients treated with early surgical decompression (<24 h) showed improved rates of AIS conversion among cervical AIS grade A, while the same was not reported for thoracolumbar trauma (Bourassa-Moreau et al., 2016).

The Ontario-based cohort study involved 84 patients with all levels of SCI. Patients treated surgically prior to 24 h didn't show a significant increasing trend in motor recovery at rehabilitation discharge. Authors found the level of injury to be a significant predicting factor for neurological recovery, but they didn't investigate if the effects of ESD vary with level of injury (Wilson et al., 2012).

In the study of Dvorak et al., despite the impact of surgery was not analysed, it appears that patients with incomplete SCI from all anatomical level benefit from ESD (Dvorak et al., 2015).

## The special case of Central Cord Syndrome

Central Cord Syndrome (CCS), the most frequent type of incomplete SCI, was first described by Schneider in 1954; it is characterized by involvement of the four limbs but with greater motor and sensitive deficits in upper extremities. Commonly, it is often secondary to low energy trauma especially in elderly patients where a pre-existing spinal stenosis (Schneider & Knighton, 1959; Schneider, Thompson, & Bebin, 1958) (Fig. 4).

The natural history of neurological recovery for CCS has been generally favorable. Most of patients well able to walk independently, without supports, with normal sphincteric function at long term follow-up (Dahdaleh et al., 2013). For that reason, conservative treatment had been historically the gold standard in CCS, while early operative intervention might derail this otherwise favorable natural history. This concerns find their foundation in Schneider's original description of CSS, where one patient had a worsening of neurologic injury after acute surgical treatment (Schneider, Cherry, & Pantek, 1954).

Despite early surgery is gaining popularity as the preferred approach for treatment of SCI, practices surrounding the time of surgery remains undefined. Currently, many surgeons continue to adhere to historical practice (Fehlings, Rabin, Sears, Cadotte, & Aarabi, 2010). In fact, in a recent survey, only 44% of attendants preferred to operate within the first 24 h from injury (Ter Wengel et al., 2018).

Currently, despite only few high-quality studies are about this topic, the best evidence found in international literature support the practice of early surgery.

Lenehan et al. evaluated the impact of surgical timing on clinical outcomes in patients with CCS. Patients were divided in early and late surgical group according to surgical intervention time, prior or post 24 h from injury. Patients treated early, experienced 6.3 points more of motor recovery and had about three times more of AIS grade conversion than late surgery. The same was observed for patients who underwent ESD with no significant perioperative morbidity (Lenehan et al., 2010).

Despite the huge historical significance, the surgical approach of Schneider, who performed a posterior direct intradural approach to cervical disc, would not be acceptable today. With better and improved surgical techniques and technologies, surgery can be performed effectively and safely.

Although natural history of CCS is favorable for ambulation, motor recovery and sphincter function, it must be acknowledge that some patients could suffer from neuropathic pain, ongoing spasticity and impairments in hand dexterity (Dvorak et al., 2005).

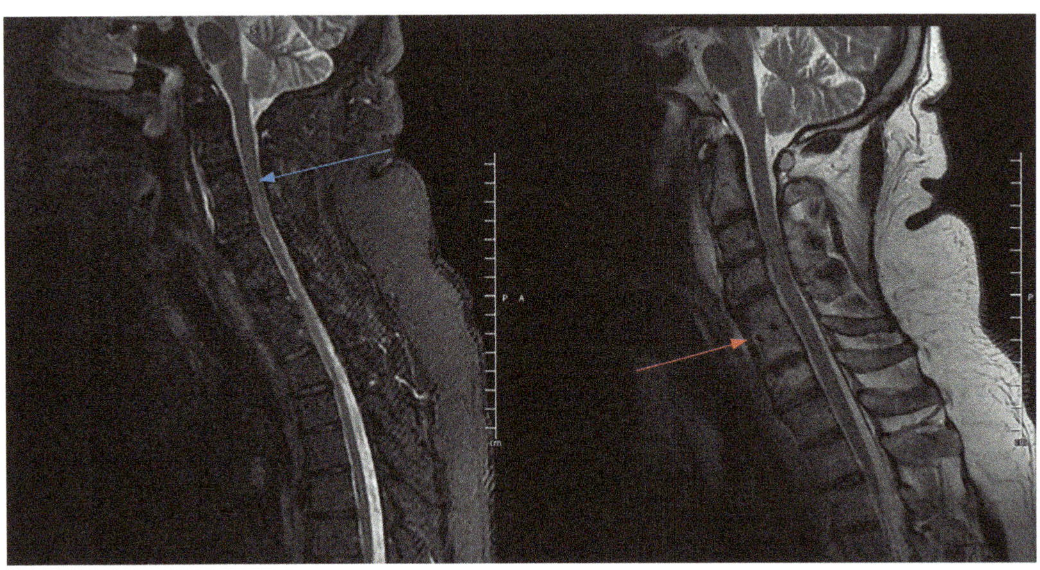

FIG. 4 Clinical case of Central Cord Syndrome (CCS). In this picture STIR *(left)* and T2 *(right)* MRI sequences of a CCS clinical case. The patient (50 years old; man) arrived after a motorcycle accident. On the left, marked by the *blue arrow*, hyperintensity can be seen (Central Cord Syndrome lesion). In this specific case the lesion occurred over a rigid cervical spine segment marked by the *red arrow* (previous arthrodesis).

## Conclusions

Surgical treatment of SCI is important to avoid the secondary mechanism of injury related with vascular impairment and ischemic damage. Natural history of complete cervical SCI is characterized by an absence of improvement for the majority of cases, while in patients with incomplete SCI, the chance of natural neurological improvement is significant higher. Complete thoracic SCI shares the same rate of improvement of complete cervical SCI; however, thoracolumbar SCI has lower motor score recovery than cervical SCI.

Surgical treatment is performed through SDSF; anterior, posterior or combined approach depends of the level of injury, extension of compression and surgeon preference.

According to literature, surgical treatment can be performed ultra-early, early or late, depending of time passed from injury. At the moment early treatments offers superior results.

Despite earlier is better, it's extremely difficult to undergo surgical intervention before 8 h from trauma. This difficulty is connected to transport from place of injury to reference center for spinal injury, injury assessment, patient stability and other life-threatening pathologies. For this reason, the most suitable and realistic time of intervention balancing surgical risks and better outcomes should be set before 24 h.

## Applications to other areas of neuroscience

In this chapter, we have reported a review about efficacy and effectiveness of SD for the treatment of SCI. This kind of surgery also has an important role for the treatment of CSM caused by spondylosis or ossification of the posterior longitudinal ligament (Boogaarts & Bartels, 2015). This pathology usually affects people aged over 50 years old causing progressive dysfunction and disability. The principal symptoms that may be found are weakness in upper and lower body, numbness, clumsiness, and difficulty handling. Posterior decompression is the most accepted treatment for CSM and it plays a crucial role like decompression for SCI (Singrakhia, Malewar, Singrakhia, & Deshmukh, 2017). The treatment aims to reduce stress and forces that attack spinal cord and medulla driving to ischemic troubles and lesions. Unlike decompression for SCI, the CSM one is usually performed posteriorly and may be performed with or without segment fixation. In fact, despite the indication of posterior decompression surgery without segment fixation is rarely performed; some Authors reported this technique as a valuable, easier and faster surgical approach in patients strictly selected on the base of preoperative cervical alignment or segmental stability (Gargiulo et al., 2019). However, the correct indication about fixing or not fixing is actually debated in literature.

## Mini-dictionary of terms

- Spinal Cord Injury: A damage to the spinal cord.
- Complete spinal cord injury: Damage to the spinal cord in which the nerves below the level of injury cannot communicate with the brain anymore
- Incomplete spinal cord injury: Damage to the spinal cord in which the nerves below the level of injury can still sent some signals along the spinal cord. The clinical pattern of movement and sensation may vary a lot among patients
- Tetraplegia: Also known as quadriplegia, it is the paralysis that results in the partial or total loss of use of all four limbs.
- Paraplegia: Paralysis that results in the partial or total loss of use of the lower limbs.
- Central Cord Syndrome: Most common form of SCI; characterized by loss of motion and sensation in arms while the lower limbs result to be normal
- Anterior Cord Syndrome: It is caused by ischemia of the anterior spinal artery that lead to the complete loss of function of the anterior spinal cord with motor function loss associated to loss of pain and temperature sensation.
- Cauda Equina Syndrome: Serious neurologic condition characterized by low back pain that radiates down the leg associated with numbness around anus and loss of bladder and anus control. This condition affects only lower motor neurons.
- Conus Medullaris Syndrome: It is a subset of SCI caused by injuries that involve the tract between T12 an L2. Very similar to Cauda Equina Syndrome (CES), but unlike CES a combined involvement of upper and lower motor neuron may be found.
- Brown-Sequard Syndrome: Rare neurological condition characterized by hemiplegia and hemianesthesia affecting two opposite side of the body

- Surgical Decompression: Surgical procedure that is usually performed to alleviate pain caused by pinched nerves. In case of SCI, surgeons may perform this procedure to alleviate the compression forces that damage the spinal cord.
- Posterior stabilization: Surgical procedure that consists in stabilize the spine through a posterior surgical access. The hardware most used are screws and bars.
- ASIA scale: Universal classification tool for spinal cord injuries based on standardized sensory and motor assessment.

## Key facts of surgical treatment of spinal cord injury

- Wait and see approach is not recommended
- SD granted oedema reduction, compression removal, blood supplies improvement and neuronal apoptosis inhibition
- ESD (<24 h) of SCI improved neurological symptoms
- Ultra ESD (<6 h) ensure better chance of spinal cord recovery; however, no later than 24 h decompression is recommended, since life treating condition could delay SD
- SD of Central Cord Syndrome is still controversial; the more recent literature support early surgical decompression

## Summary points

- In SCI the persistent compression derived by the first injury may inexorably drive to a second injury.
- Low ASIA scale scores at injury are related with worse prognosis. Approximately 85% of ASIA scale grade A patients at injury will not improve their neurological status at 1 year of follow up.
- Thoracic SCI improve less than cervical and lumbar ones.
- Surgery has two main objectives: to fix and to decompress.
- Three common time thresholds are described in literature: ultra-early threshold (8–12 h post SCI), early threshold (24 h post SCI) and late threshold (48–72 h post SCI).
- Anterior and posterior approaches to the spine may be adopted. Posterior ones are preferred in thoracolumbar spine.
- Earlier decompression surgery may guarantee better rate of neurological improvement.
- In CCS patients, decompression efficacy is debated.

## References


Aarabi, B., et al. (2020). Efficacy of ultra-early (<12 h), early (12-24 h), and late (>24-138.5 h) surgery with magnetic resonance imaging-confirmed decompression in American spinal injury association impairment scale grades A, B, and C cervical spinal cord injury. *Journal of Neurotrauma*, *37*(3), 448–457.

Ahuja, C. S., et al. (2017). Traumatic spinal cord injury. *Nature Reviews. Disease Primers*, *3*, 17018 (Nature Publishing Group).

Battistuzzo, C. R., et al. (2016). Early decompression following cervical spinal cord injury: Examining the process of care from accident scene to surgery. *Journal of Neurotrauma*, *33*(12), 1161–1169.

Boogaarts, H. D., & Bartels, R. H. M. A. (2015). Prevalence of cervical spondylotic myelopathy. *European Spine Journal*, *24*(Suppl 2), 139–141.

Bose, B., Osterholm, J. L., Northrup, B. E., & Cotler, J. M. (1985). Management of lumbar translocation injuries: Case reports. *Neurosurgery*, *17*(6), 958–961.

Bourassa-Moreau, É., Mac-Thiong, J. M., Feldman, D. E., Thompson, C., & Parent, S. (2013). Non-neurological outcomes after complete traumatic spinal cord injury: The impact of surgical timing. *Journal of Neurotrauma*, *30*(18), 1596–1601.

Bourassa-Moreau, É., et al. (2016). Do patients with complete spinal cord injury benefit from early surgical decompression? Analysis of neurological improvement in a prospective cohort study. *Journal of Neurotrauma*, *33*(3), 301–306.

Brouwers, E., Van De Meent, H., Curt, A., Starremans, B., Hosman, A., & Bartels, R. (2017). Definitions of traumatic conus medullaris and cauda equina syndrome: A systematic literature review. *Spinal Cord*, *55*(10), 886–890.

Burke, J. F., et al. (2019). Ultra-early (<12 hours) surgery correlates with higher rate of American spinal injury association impairment scale conversion after cervical spinal cord injury. *Neurosurgery*, *85*(2), 199–203.

Clohisy, J. C., Akbarnia, B. A., Bucholz, R. D., Kenneth Burkus, J., & Backer, R. J. (1992). Neurologic recovery associated with: Anterior decompression of spine fractures at the thoracolumbar junction (t12-l1). *Spine*, *17*(8 Suppl), 325–330.

Dahdaleh, N. S., et al. (2013). Evidence-based management of central cord syndrome. *Neurosurgical Focus*, *35*(1), E6.

Dimar, J. R., Glassman, S. D., Raque, G. H., Zhang, Y. P., & Shields, C. B. (1999). The influence of spinal canal narrowing and timing of decompression on neurologic recovery after spinal cord contusion in a rat model. *Spine*, *24*(16), 1623–1633.

Dolan, E. J., Tator, C. H., & Endrenyi, L. (1980). The value of decompression for acute experimental spinal cord compression injury. *Journal of Neurosurgery*, *53*(6), 749–755.

Dvorak, M. F., et al. (2005, October 15). Factors predicting motor recovery and functional outcome after traumatic central cord syndrome: A long-term follow-up. *Spine (Phila Pa 1976)*, *30*(20), 2303–2311.

Dvorak, M. F., et al. (2014). Minimizing errors in acute traumatic spinal cord injury trials by acknowledging the heterogeneity of spinal cord anatomy and injury severity: An observational Canadian cohort analysis. *Journal of Neurotrauma*, *31*(18), 1540–1547.

Dvorak, M. F., et al. (2015). The influence of time from injury to surgery on motor recovery and length of hospital stay in acute traumatic spinal cord injury: An observational Canadian cohort study. *Journal of Neurotrauma*, *32*(9), 645–654.

Fawcett, J. W., et al. (2007). Guidelines for the conduct of clinical trials for spinal cord injury as developed by the ICCP panel: Spontaneous recovery after spinal cord injury and statistical power needed for therapeutic clinical trials. *Spinal Cord*, *45*, 190–205.

Fehlings, M. G., Rabin, D., Sears, W., Cadotte, D. W., & Aarabi, B. (2010). Current practice in the timing of surgical intervention in spinal cord injury. *Spine*, *35*(21 Suppl), S166–S173.

Fehlings, M. G., & Sekhon, L. H. (2001). Acute interventions in spinal cord injury: What do we know, what should we do? *Clinical Neurosurgery*, *48*, 226–242.

Fehlings, M. G., et al. (2012). Early versus delayed decompression for traumatic cervical spinal cord injury: Results of the surgical timing in acute spinal cord injury study (STASCIS). *PLoS One*, *7*(2), e32037.

Fehlings, M. G., & Perrin, R. G. (2006). The timing of surgical intervention in the treatment of spinal cord injury: A systematic review of recent clinical evidence. *Spine (Phila Pa 1976)*, *31*, S28–S35.

Feuchtbaum, E., Buchowski, J., & Zebala, L. (2016). Subaxial cervical spine trauma. *Current Reviews in Musculoskeletal Medicine*, *9*(4), 496–504 (Humana Press Inc).

Furlan, J. C., Tung, K., & Fehlings, M. G. (2013). Process benchmarking appraisal of surgical decompression of spinal cord following traumatic cervical spinal cord injury: Opportunities to reduce delays in surgical management. *Journal of Neurotrauma*, *30*(6), 487–491.

Gargiulo, G., et al. (2019). Clinical comparison between simple laminectomy and laminectomy plus posterior instrumentation in surgical treatment of cervical myelopathy. *European Journal of Orthopaedic Surgery and Traumatology*, *29*(5), 975–982.

Girardo, M., Rava, A., Fusini, F., Gargiulo, G., Coniglio, A., & Cinnella, P. (2018). Different pedicle osteosynthesis for thoracolumbar vertebral fractures in elderly patients. *European Spine Journal*, *27*(Suppl 2), 198–205.

Grassner, L., et al. (2016). Early decompression (< 8 h) after traumatic cervical spinal cord injury improves functional outcome as assessed by spinal cord independence measure after one year. *Journal of Neurotrauma*, *33*(18), 1658–1666.

Guha, A., Tator, C. H., Endrenyi, L., & Piper, I. (1987). Decompression of the spinal cord improves recovery after acute experimental spinal cord compression injury. *Paraplegia*, *25*(4), 324–339.

Hawryluk, G. W. J., Rowland, J., Kwon, B. K., & Fehlings, M. G. (2008). Protection and repair of the injured spinal cord: A review of completed, ongoing, and planned clinical trials for acute spinal cord injury. *Neurosurgical Focus*, *25*(5), E14.

Jug, M., et al. (2015). Neurological recovery after traumatic cervical spinal cord injury is superior if surgical decompression and instrumented fusion are performed within 8 hours versus 8 to 24 hours after injury: A single center experience. *Journal of Neurotrauma*, *32*(18), 1385–1392.

Kirshblum, S., Millis, S., McKinley, W., & Tulsky, D. (2004). Late neurologic recovery after traumatic spinal cord injury. *Archives of Physical Medicine and Rehabilitation*, *85*(11), 1811–1817.

Lenehan, B., Fisher, C. G., Vaccaro, A., Fehlings, M., Aarabi, B., & Dvorak, M. F. (2010). The urgency of surgical decompression in acute central cord injuries with spondylosis and without instability. *Spine*, *35*(21 Suppl), S180–S186.

Ma, Y., Zhu, Y., Zhang, B., Wu, Y., Liu, X., & Zhu, Q. (2020). The impact of urgent (<8 hours) decompression on neurologic recovery in traumatic spinal cord injury: A meta-analysis. *World Neurosurgery*, *140*, e185–e194.

Mac-Thiong, J. M., Feldman, D. E., Thompson, C., Bourassa-Moreau, E., & Parent, S. (2012). Does timing of surgery affect hospitalization costs and length of stay for acute care following a traumatic spinal cord injury? *Journal of Neurotrauma*, *29*(18), 2816–2822.

Marino, R. J., Burns, S., Graves, D. E., Leiby, B. E., Kirshblum, S., & Lammertse, D. P. (2011). Upper- and lower-extremity motor recovery after traumatic cervical spinal cord injury: An update from the national spinal cord injury database. *Archives of Physical Medicine and Rehabilitation*, *92*(3), 369–375.

Maynard, F. M., et al. (1997). International standards for neurological and functional classification of spinal cord injury. *Spinal Cord*, *35*(5), 266–274.

McKinley, W., Meade, M. A., Kirshblum, S., & Barnard, B. (2004). Outcomes of early surgical management versus late or no surgical intervention after acute spinal cord injury. *Archives of Physical Medicine and Rehabilitation*, *85*(11), 1818–1825.

Nijendijk, J. H. B., Post, M. W. M., & Van Asbeck, F. W. A. (2014). Epidemiology of traumatic spinal cord injuries in the Netherlands in 2010. *Spinal Cord*, *52*(4), 258–263.

Pickett, G. E., et al. (2006). Epidemiology of traumatic spinal cord injury in Canada. *Spine (Phila Pa 1976)*, *31*, 799–805.

Rahimi-Movaghar, V., et al. (2009). The efficacy of surgical decompression before 24 hours versus 24 to 72 hours in patients with spinal cord injury from T1 to L1 – With specific consideration on ethics: A randomized controlled trial. *Trials*, *10*, 77.

Rivlin, A. S., & Tator, C. H. (1978). Effect of duration of acute spinal cord compression in a new acute cord injury model in the rat. *Surgical Neurology*, *10* (1), 39–43.

Rowland, J. W., Hawryluk, G. W. J., Kwon, B., & Fehlings, M. G. (2008). Current status of acute spinal cord injury pathophysiology and emerging therapies: Promise on the horizon. *Neurosurgical Focus*, *25*(5), E2.

Schneider, R. C., Cherry, G., & Pantek, H. (1954). The syndrome of acute central cervical spinal cord injury; with special reference to the mechanisms involved in hyperextension injuries of cervical spine. *Journal of Neurosurgery*, *11*(6), 546–577.

Schneider, R. C., & Knighton, R. (1959). Chronic neurological sequelae of acute trauma to the spine and spinal cord. III. The syndrome of chronic injury to the cervical spinal cord in the region of the central canal. *The Journal of Bone and Joint Surgery. American Volume*, *41-A*(5), 905–919.

Schneider, R. C., Thompson, J. M., & Bebin, J. (1958). The syndrome of acute central cervical spinal cord injury. *Journal of Neurology, Neurosurgery, and Psychiatry*, *21*(3), 216–227.

Singrakhia, M. D., Malewar, N. R., Singrakhia, S. M., & Deshmukh, S. S. (2017). Cervical laminectomy with lateral mass screw fixation in cervical spondylotic myelopathy: Neurological and sagittal alignment outcome: Do we need lateral mass screws at each segment? *Indian Journal of Orthopaedics*, *51*, 658–665.

Stephan, K., et al. (2015). Spinal cord injury - incidence, prognosis, and outcome: An analysis of the TraumaRegister DGU. *The Spine Journal*, *15*(9), 1994–2001.

Tator, C. H., & Fehlings, M. G. (1991). Review of the secondary injury theory of acute spinal cord trauma with emphasis on vascular mechanisms. *Journal of Neurosurgery*, *75*(1), 15–26.

Tator, C. H., Fehlings, M. G., Thorpe, K., & Taylor, W. (1999). Current use and timing of spinal surgery for management of acute spinal cord injury in North America: Results of a retrospective multicenter study. *Journal of Neurosurgery*, *91*(1 Suppl), 12–18.

Ter Wengel, P. V., De Haan, Y., Feller, R. E., Oner, F. C., & Vandertop, W. P. (2020). Complete traumatic spinal cord injury: Current insights regarding timing of surgery and level of injury. *Global Spine Journal*, *10*(3), 324–331 (SAGE Publications Ltd).

Ter Wengel, P. V., et al. (2018). Timing of surgery in traumatic spinal cord injury: A national, multidisciplinary survey. *European Spine Journal*, *27*(8), 1831–1838.

Vaccaro, A. R., et al. (1997). Neurologic outcome of early versus late surgery for cervical spinal cord injury. *Spine (Phila Pa 1976)*, *22*, 2609–2613.

Van Middendorp, J. J. (2012). Letter to the editor regarding: Early versus delayed decompression for traumatic cervical spinal cord injury: Results of the Surgical Timing in Acute Spinal Cord Injury Study (STASCIS). *The Spine Journal*, *12*(6), 540.

Van Middendorp, J. J., Hosman, A. J. F., & Doi, S. A. R. (2013). The effects of the timing of spinal surgery after traumatic spinal cord injury: A systematic review and meta-analysis. *Journal of Neurotrauma*, *30*(21), 1781–1794.

Wiberg, J., & Hauge, H. N. (1988). Neurological outcome after surgery for thoracic and lumbar spine injuries. *Acta Neurochirurgica*, *91*(3–4), 106–112.

Wilson, J. R., & Fehlings, M. G. (2011). Emerging approaches to the surgical management of acute traumatic spinal cord injury. *Neurotherapeutics*, *8*(2), 187–194.

Wilson, J. R., Voth, J., Singh, A., Middleton, J., Jaglal, S. B., Singh, J. M., … Fehlings, M. G. (2016). Defining the pathway to definitive care and surgical decompression after traumatic spinal cord injury: Results of a Canadian population-based cohort study. *Journal of Neurotrauma*, *33*(10), 963–971.

Wilson, J. R., Witiw, C. D., Badhiwala, J., Kwon, B. K., Fehlings, M. G., & Harrop, J. S. (2020). Early surgery for traumatic spinal cord injury: Where are we now? *Global Spine Journal*, *10*(1 Suppl), 84S–91S.

Wilson, J. R., et al. (2012). Early versus late surgery for traumatic spinal cord injury: The results of a prospective Canadian cohort study. *Spinal Cord*, *50*(11), 840–843.

Wilson, J. R., et al. (2017). Timing of decompression in patients with acute spinal cord injury: A systematic review. *Global Spine Journal*, *7*(3 Suppl), 95S–115S (SAGE Publications Ltd).

Wilson, J. R., et al. (2018). Natural history, predictors of outcome, and effects of treatment in thoracic spinal cord injury: A multi-center cohort study from the north American clinical trials network. *Journal of Neurotrauma*, *35*(21), 2554–2560.

Zariffa, J., Curt, A., & Steeves, J. D. (2012). Functional motor preservation below the level of injury in subjects with American spinal injury association impairment scale grade a spinal cord injuries. *Archives of Physical Medicine and Rehabilitation*, *93*(5), 905–907.

Zariffa, J., Kramer, J. L. K., Fawcett, J. W., Lammertse, D. P., Blight, A. R., Guest, J., … Steeves, J. D. (2011). Characterization of neurological recovery following traumatic sensorimotor complete thoracic spinal cord injury. *Spinal Cord*, *49*(3), 463–471.

Chapter 18

# Spinal cord epidural stimulation for autonomic nervous system control: A focus on improving bladder, bowel, and cardiovascular function

April N. Herrity[a,b], Sevda C. Aslan[a], Susan J. Harkema[a], and Charles H. Hubscher[c]

[a]*Department of Neurological Surgery, Kentucky Spinal Cord Injury Research Center, University of Louisville, Louisville, KY, United States;* [b]*Department of Physiology, University of Louisville, Louisville, KY, United States;* [c]*Department of Anatomical Sciences and Neurobiology, Kentucky Spinal Cord Injury Research Center, University of Louisville, Louisville, KY, United States*

## List of abbreviations

| | |
|---|---|
| **AD** | autonomic dysreflexia |
| **AIS** | American Spinal Injury Association Impairment Scale |
| **FDA** | Food and Drug Administration |
| **IDE** | Investigational Device Exemption |
| **L** | lumbar (referring to spinal cord level) |
| **scES** | spinal cord epidural stimulation |
| **SCI** | spinal cord injury |
| **T** | thoracic (referring to spinal cord level) |
| **S** | sacral (referring to spinal cord level) |

## Introduction

Spinal cord injury (SCI) affects nearly 2 million people in the United States (Reeve Foundation Paralysis Resource, 2021) and results in profound neurologic impairment with widespread deficits to sensorimotor and autonomic systems. While the lack of ambulation is most widely recognized, there are many internal systems that are disrupted after SCI due to the critical role of the spinal cord in coordinating bodily functions. As a result, individuals with SCI endure a myriad of complications that result in increased mortality, morbidity, hospitalization, high burden of care and health care costs, and a drastically lowered quality of life (Merritt, Taylor, Yelton, & Ray, 2019).

The autonomic nervous system serves to provide essential regulatory communication between the brain and internal organs, vascular supply, and numerous glands in order to maintain dynamic equilibrium and homeostasis of all systems. Sympathetic and parasympathetic systems (including the enteric system for the gastrointestinal tract) function to integrate complex reflex pathways between the target organ and integration centers within the nervous system (Gibbins, 2013). Thus, autonomic disruption extends across multiple systems and drastically impacts overall health and quality of life. Alterations of urogenital and bowel dysfunctions are consistently rated as top consumer priorities for recovery post-SCI (Anderson, 2004; Ditunno, Patrick, Stineman, & Ditunno, 2008; Simpson et al., 2012). Acutely post-SCI, bladder dysfunction often manifests as "flaccid paralysis," resulting in urinary retention necessitating bladder catheterization. Overtime, bladder dysfunction progresses into detrusor hyperreflexia and detrusor-sphincter dyssynergia, resulting in incontinence and incomplete emptying. Gastrointestinal disruptions after SCI are highly prevalent and include a wide-range of symptoms, including delayed gastric emptying, diminished propulsive transit, abdominal distention, pain, constipation, anal fissures, hemorrhoids, rectal prolapse, and sphincter/defecation disturbances (Krogh et al., 1997). Cardiovascular dysfunction associated with autonomic dysregulation contributes to increased morbidity and is the leading cause of mortality in acute and chronic SCI (Devivo, Krause, & Lammertse, 1999). The disruption of supraspinal control over sympathetic pathways

Diagnosis and Treatment of Spinal Cord Injury. https://doi.org/10.1016/B978-0-12-822498-4.00018-X

results in decreased sympathetic tone below the level of injury and impaired autonomic function and blood pressure regulation (Koh, Brown, Beightol, Ha, & Eckberg, 1994). One critical complication that occurs in individuals with lesions above T6 is autonomic dysreflexia (AD), an exaggerated sympathetic nervous system response to a noxious stimulus below the level of injury, including bladder/bowel distention, urinary tract infections, fecal impaction, and orgasms (Karlsson, 1999). As a result, increased cardiovascular dysregulation has a profound effect on the body as a whole and significantly interferes with motor and autonomic recovery post-injury.

## Bladder dysfunction after SCI

Long-term deficits in bladder function after SCI manifest as detrusor hyperreflexia (bladder contractions at low volumes, causing incontinence and smooth muscle hypertrophy), detrusor-sphincter dyssynergia (uncoordinated bladder and external urethral sphincter contractions, causing inefficient emptying and smooth muscle hypertrophy), decreased compliance (unable to store urine under appropriately low pressures) and loss of continence (Fig. 1) requiring lifelong management, maintenance, and health care visits (Stover, Devivo, & Go, 1999). Major urological concerns contributing to increased morbidity and mortality include repeated lower urinary tract infections that can lead to sepsis, chronic vesico-ureteral reflux and hydronephrosis with progression to renal failure as a result of high-intravesical pressures, and inter-related cardiovascular complications such as AD (Hagen, Faerestrand, Hoff, Rekand, & Gronning, 2011) that

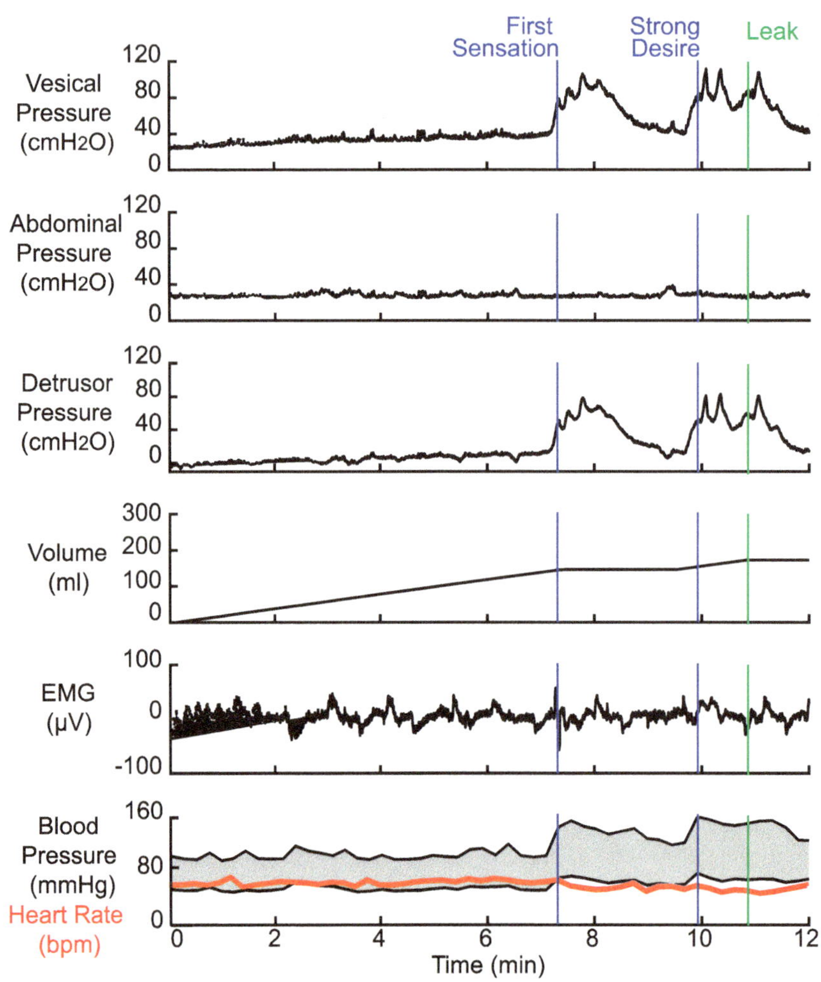


FIG. 1 Neurogenic bladder dysfunction. Representative cystometry recording, including vesical pressure ($cmH_2O$), abdominal pressure ($cmH_2O$), detrusor pressure ($cmH_2O$), volume of infused saline (Ml), surface electromyography (EMG, μV) of the external anal sphincter, continuous blood pressure (mmHg), and heart rate (red, bpm) responses in a 26-year-old male (C4, AIS B), who performs intermittent catheterization for bladder emptying. Multiple detrusor contractions are present, resulting in incontinence at 171 ml with detrusor leak point pressure of 58 $cmH_2O$. The rise in blood pressure was timed with the rise in detrusor pressure.

limits bladder storage (Hubscher et al., 2018). Current therapies to improve the efficiency of bladder voiding, management and continence after SCI include catheterization, pharmacologic and surgical interventions, functional electrical stimulation, as well as urethral stents. In particular, neurogenic detrusor overactivity and detrusor sphincter dyssynergia are commonly managed with anti-muscarinic drugs and Botox injections, and while these pharmacotherapy agents do not prevent or reverse the changes, they primarily focus on improving storage outcomes and do not address the emptying phase. Furthermore, anti-muscarinic agents can have unpleasant side effects such as dry mouth and constipation, exacerbating existing bowel dysfunction. The primary goal in managing neurogenic bladder is to protect the upper urinary tract and reduce the incidence of infections, followed by achieving continence. As such, when conservative management fails, surgeries to reduce high bladder pressures and chronic urinary incontinence include urinary diversion and lower urinary tract reconstruction approaches (Biers, Venn, & Greenwell, 2012). As urinary complications continue to impact long-term morbidity in this population, additional therapeutic and rehabilitative approaches are needed that aim to improve function by targeting the recovery of underlying impairments.

## Bowel dysfunction after SCI

Complications with bowel storage and evacuation are highly prevalent after SCI, with 95% of patients reporting issues with constipation (Glickman & Kamm, 1996). Also, 75% report episodes of fecal incontinence with 30% considering bowel disorders more burdensome than bladder (Krogh et al., 1997) and having a substantial negative impact on quality of life, social integration, and overall independence (Emmanuel, 2010). Neurogenic bowel dysfunction can be characterized by the presence of increased colonic wall and anal tone along with weak abdominal musculature, resulting in fecal retention and constipation. A significant risk of incontinence can occur from an inability to adequately engage the external anal sphincter and pelvic floor musculature (Stiens, Bergman, & Goetz, 1997). While gastrointestinal transit is primarily influenced by local enteric reflexes, motility patterns can be modulated by extrinsic innervation derived from sympathetic, parasympathetic and vagal inputs. Due to difficulties with elimination, prolonged colonic transit time, and impaired motor dexterity (Longo, Ballantyne, & Modlin, 1989), large amounts of time are devoted to bowel care programs (for some, up to 2 h) (Glickman & Kamm, 1996; Hubscher et al., 2018). Many individuals with SCI are also dependent on caregiver assistance. As a result, many choose to conduct their programs every other day (Kirshblum, Gulati, O'connor, & Voorman, 1998), which may increase the risk of constipation, fecal impaction, colorectal distension, and episodes of AD. Conservative bowel regimen approaches include the use of laxatives and stool softeners to obtain a desired consistency, followed by rectal stimulation with either suppositories, enemas, or digital stimulation. Digital stimulation involves rectal pressure with a gloved finger to activate the recto-colic reflex, generating coordinated colonic peristaltic activity and increased motility (Shafik, 1996). Consequently, there is tremendous interest to use neuromodulatory strategies to achieve same or better effects.

## Cardiovascular dysfunction and interactions with other autonomic systems after SCI

Cardiovascular diseases are the leading causes of morbidity and mortality among the SCI population (Wecht et al., 2020). Hemodynamic and metabolic cardiovascular effects occur in both the acute and chronic phases of SCI. Disruptions of sympathetic pathways to the heart and vasculature cause autonomic dysfunction and baroreflex impairment resulting in a reduced ability of cardiovascular system to maintain stable blood pressure after SCI. Dysregulation leads to persistent hypotension, bradycardia, orthostatic hypotension, and episodes of AD which drastically impacts quality of life by affecting health and disrupting engagement in daily activities (Krassioukov, Eng, Warburton, & Teasell, 2009). Combinations of sympathetic agonists and venous compression devices are the primary measures currently used for symptomatic management. Manifestations of the reduced quality of life experienced is evidenced by reductions in peripheral blood flow, resulting in an increased risk of pressure ulcers, parasympathetic imbalance increasing vasovagal syncope, fatigue and cognitive dysfunction, and symptomatic orthostasis (Wecht & Bauman, 2018). With time, post-injury blood pressure instability becomes more prominent and persists in many individuals with SCI for the rest of their lives (Katzelnick et al., 2019). AD, characterized by episodes of acute hypertension, is generated by unmodulated sympathetic reflexes below the injury level, and is typically accompanied by compensatory baroreceptor-mediated bradycardia (Karlsson, 1999). Distention or contraction of hollow organs, such as bladder and bowel, are the most common triggers of AD (Lindan, Joiner, Freehafer, & Hazel, 1980). Such below-level peripheral afferent stimuli reach the isolated spinal cord and activate a massive unmodulated sympathetic reflex causing widespread vasoconstriction, presenting as severe hypertension. With high-level SCI, the additional interruption of descending modulatory autonomic pathways that normally inhibit sympathetic preganglionic neurons during hypertension allows AD to persist until the stimulus is removed (Rabchevsky, 2006) (Fig. 2). Management

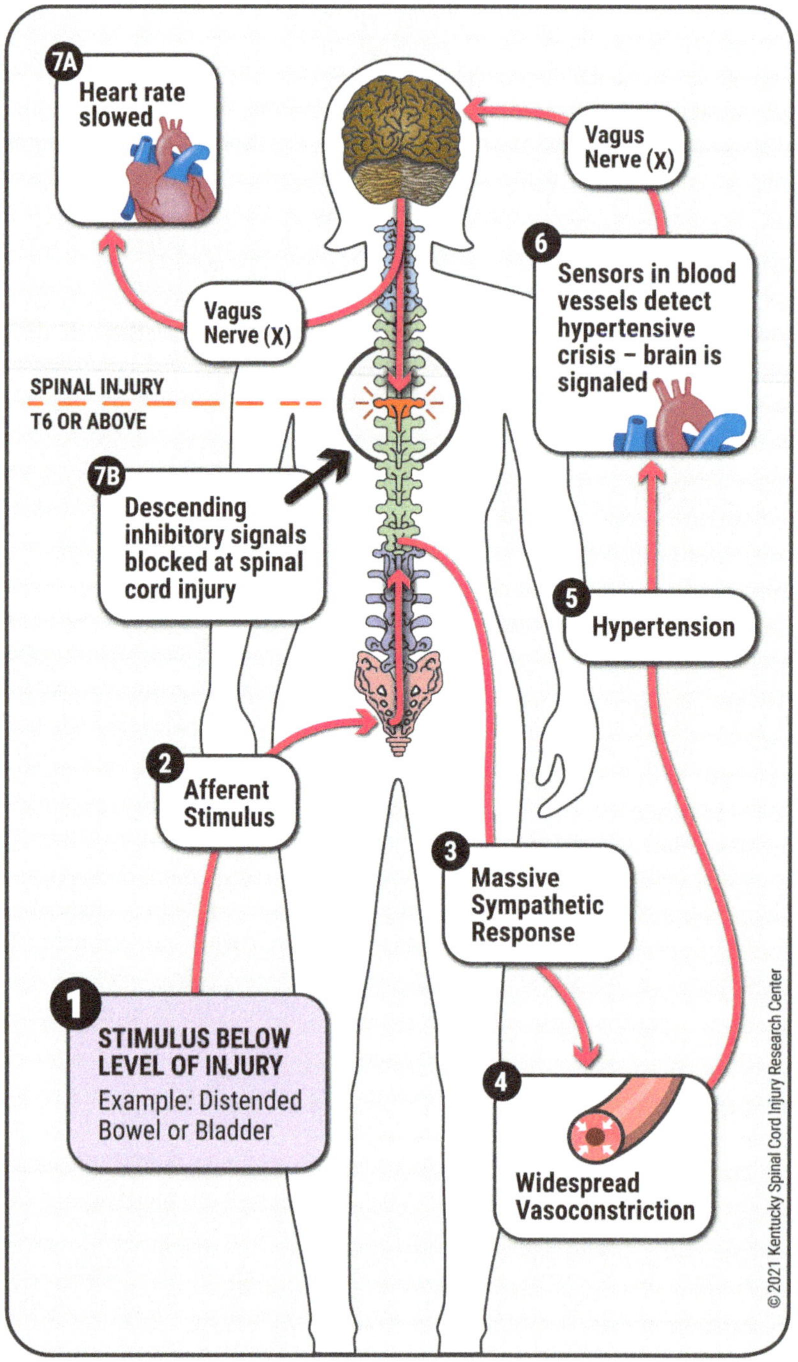


FIG. 2 Autonomic Dysreflexia after SCI. Cartoon diagram outlining the cascade of events leading up autonomic dysreflexia. Primary precipitating stimuli for autonomic dysreflexia include bladder and/or bowel distention.

and mitigation of unstable blood pressure is therefore a high consumer priority in SCI care; however, therapies available that have demonstrated the ability to resolve chronic cardiovascular dysfunction are lacking (Wecht et al., 2020; Wecht, Cirnigliaro, Azarelo, Bauman, & Kirshblum, 2015). With time, post-injury episodes of AD become more prominent and persist long-term. The ability to recover bladder function such as, increasing bladder capacity, minimizing detrusor instability and improving bladder pressure and emptying, is limited by such severe fluctuations in blood pressure. Furthermore, improvements in bowel function are challenged by regular occurrences of AD symptoms during bowel programs, evoked by distention (Inskip, Lucci, McGrath, Willms, & Claydon, 2018) (Fig. 3). The extreme dysregulation of the cardiovascular system underscores the importance of addressing these multi-faceted autonomic complications.

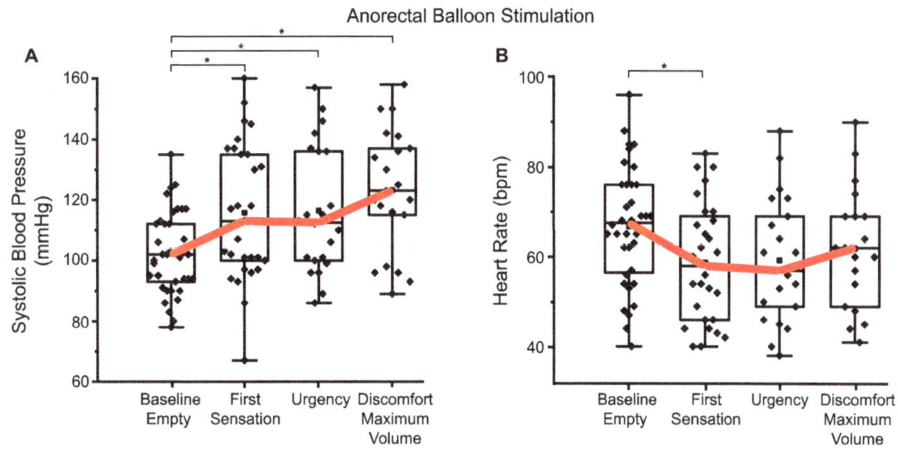


FIG. 3 Blood pressure responses to anorectal distention. Participants ($n = 37$) were asked to respond to sensations of rectal fullness associated with gradual inflation of a balloon catheter during anorectal manometry. Relative to baseline (prior to inflation), mean systolic blood pressure *(red line)* was significantly greater at first sensation ($n = 30$, $P < .01$), urgency ($n = 22$, $P < .05$), and maximum volume ($n = 21$, $P < .01$) (A), while mean heart rate *(red line)* was significantly lower at first sensation ($n = 30$, $P < .01$) relative to baseline (B). Note, while data are normally distributed, not every participant reported each sensation. The normative ranges for volume (mL) at which first, urgency, and maximum sensations of fullness occur are 10–60 mL, 10–100 mL, and 100–200 mL, respectively.

## Introduction of spinal cord stimulation

Melzack and Wall's publication of the Gate Theory (Melzack & Wall, 1965) transformed our understanding of pain mechanisms and the associated clinical management, leading to the realization that pain was the result of complex dynamic processes in the nervous system and not simply the result of activity in a hard-wired system. The concept that damage to the nervous system can itself result in chronic pain led to a gradual shift from ablative and destructive surgical approaches, such as cutting nerves, toward reversible neuromodulatory strategies such as therapeutic electrical stimulation of the dorsal spinal cord. The theory that nociceptive impulses, carried by Aδ and C fibers, could be inhibited by electrical stimulation of thick myelinated Aβ fibers provided the first scientific basis for the use of stimulation for pain. At that time, implanted electrode-induced electrical stimulation of the dorsal columns was initially introduced in 1967 in an effort to treat chronic intractable pain (Shealy, Mortimer, & Reswick, 1967). Its use was expanded to treat a wide variety of pain disorders following improvements in sensorimotor function in the multiple sclerosis population (Cook & Weinstein, 1973) and was first used to treat neuropathic pain in SCI in the early 1970s (Nashold Jr. & Friedman, 1972; Shealy, Mortimer, & Hagfors, 1970). Subsequently, the US Food and Drug Administration (FDA) approved the use of spinal cord stimulation in 1989 for the treatment of chronic intractable pain of the trunk or limbs and now accounts for the majority of all neuromodulation interventions (Barolat & Sharan, 2000).

Initial applications of spinal cord epidural stimulation (scES), which served to clinically control severe spasticity in SCI (Dimitrijevic, Illis, Nakajima, Sharkey, & Sherwood, 1986), led to the examination of whether scES had the capacity to drive locomotor capabilities through excitation of the lumbosacral motor circuitry (Waltz, Andreesen, & Hunt, 1987). By activating sophisticated spinal networks below the level of injury, scES functions by reactivating spared neural pathways and bridging supraspinal connectivity across the lesion with the goal of improving recovery after SCI. In persons with motor complete SCI, rhythmic, locomotor-like EMG activity and tonic motor patterns of the lower limbs from the supine position (in the absence of afferent input) were induced with scES (Dimitrijevic, Gerasimenko, & Pinter, 1998). Harkema and colleagues then demonstrated that a critical level of excitability, provided by scES and coupled with intense training that provides appropriate task-specific sensory cues, was necessary to facilitate standing and stepping, as well as intentional movement in clinically motor complete SCI (Angeli, Edgerton, Gerasimenko, & Harkema, 2014; Harkema et al., 2011). The combination of locomotor training (stepping and standing) plus scES not only enhanced coordinated and controlled voluntary motor behavior, but was also reported by individuals to improve physiologic outcomes such as bladder, bowel, and sexual function (Darrow et al., 2019; Harkema et al., 2011; Herrity et al., 2020), as depicted in Fig. 4. A summary of spinal cord stimulation approaches is provided in Table 1.

## Neuromodulation to improve bladder function

Many current and past neuromodulatory procedures have had variable success at sustained improvements of bladder function after SCI (McGee, Amundsen, & Grill, 2015; Wheeler et al., 2018). The Finetech-Brindley device (marketed

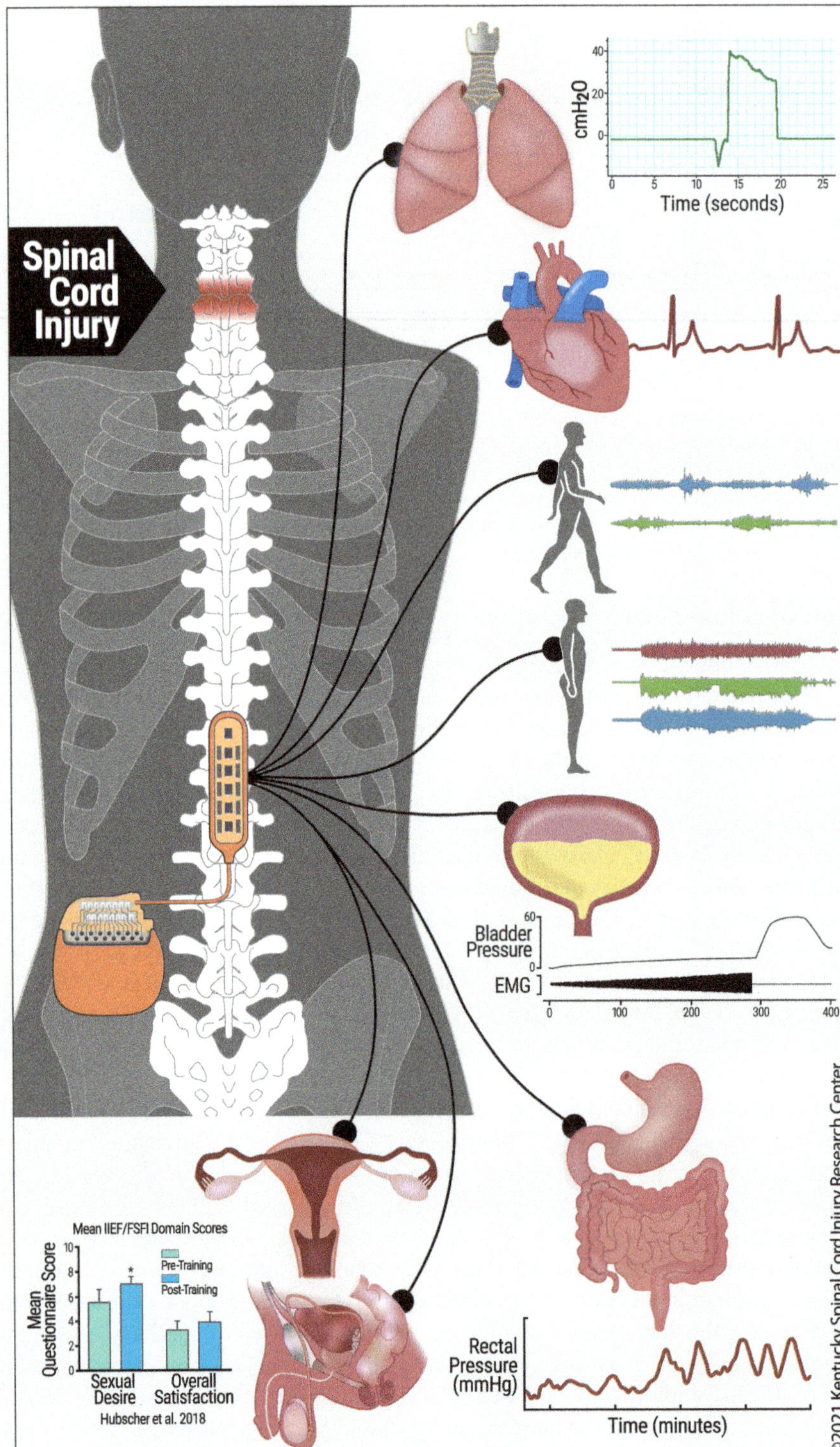


FIG. 4 Restoration of motor and autonomic function using scES. Cartoon illustration depicting some of the body systems targeted by scES. Device implantation in individuals with SCI currently involves placement of an electrode array at the lumbosacral segment of the spinal cord between L1 to S1. The location of the stimulation paddle electrode is guided by intra-operative fluoroscopy and electrophysiology mapping to identify segments of the spinal cord that respond to electrical stimulation. Following surgery, additional neurophysiological spatiotemporal mapping of various electrode configurations is performed.

**TABLE 1 Spinal cord stimulation approaches in pre-clinical and clinical studies targeting autonomic function after spinal cord injury.**

| STIM type | Visceral target | Study model | Author/date | Journal |
|---|---|---|---|---|
| Epidural | Bladder | Rats | Gad et al. (2014) | PLOS One |
| | | Rats | Gad et al. (2016) | Exp Neurol |
| | | Rats[a] | Chang et al. (2018) | Exp Neurol |
| | | Rats | Sysoev et al. (2020) | Front Sys Neurosci |
| | | Rats | Hoey et al. (2021) | Sci Reports |
| | | Rats[a] | Yousefpour and Erfanian (2021) | Sci Reports |
| | | Pigs[a] | Guiho et al. (2018) | IEEE Trans Neural Syst Rehabil Eng |
| | | Human | Meglio, Cioni, Amico, Ronzoni, and Rossi (1980) | Acta Neurochir (Wien) |
| | | Human | Loubser (1997) | J Pain Symptom manage |
| | | Human | Herrity et al. (2018) | Sci Reports |
| | | Human | Walter et al. (2018) | Front Physiol |
| | | Human | Schieferdecker, Neudorfer, El Majdoub, and Maarouf (2019) | Oper Neurosurg |
| | | Human | Darrow et al. (2019) | J Neurotrauma |
| | | Human | Herrity et al. (2020) | Front Sys Neurosci |
| | Bowel | Rats | Hoey et al. (2021) | Sci Reports |
| | | Pigs[a] | Guiho et al. (2018) | IEEE Trans Neural Syst Rehabil Eng |
| | | Human | Walter et al. (2018) | Front Physiol |
| | | Human | Darrow et al. (2019) | J Neurotrauma |
| | Cardiovascular | Rats, Rhesus Macaques | Squair et al. (2021) | Nature |
| | | Human | Richardson et al. (1979) | Neurosurgery |
| | | Human | Harkema, Wang, et al. (2018) | Front Human Neurosci |
| | | Human | Harkema, Legg Ditterline, et al. (2018) | JAMA Neurol |
| | | Human | Aslan et al. (2018) | Front Physiol |
| | | Human | West et al. (2018) | JAMA Neurol |
| | | Human | Legg Ditterline et al. (2020) | Clinical Auton Res |
| | | Human | Legg Ditterline et al. (2020) | Exp Physiol |
| | | Human | Legg Ditterline et al. (2020) | Front Neurosci |
| | | Human | Bloom et al. (2020) | Front Sys Neurosci |
| | | Human | Darrow et al. (2019) | J Neurotrauma |
| | | Human | Squair et al. (2021) | Nature |

*Continued*

**TABLE 1 Spinal cord stimulation approaches in pre-clinical and clinical studies targeting autonomic function after spinal cord injury—cont'd**

| STIM type | Visceral target | Study model | Author/date | Journal |
|---|---|---|---|---|
| Intra-spinal | Bladder | Mice | Abud et al. (2015) | AJP, Renal Physiology |
| | | Rats[a] | Blok and Holstege (1997) | Neurosci Lett |
| | | Cats[a] | Grill, Wang, Hadziefendic, and Haxhiu (1998) | Brain Res |
| | | Cats[a] | Fedirchuk and Shefchyk (1991) | Exp Br Res |
| | | Cats | Tai, Booth, de Groat, and Roppolo (2004) | Exp Neurol |
| | | Cats | Pikov et al. (2007) | J Neural Eng |
| | | Cats | Pikov et al. (2007) | J Neural Eng |
| | | Dogs | Walter, Robinson, Khan, Wheeler, and Wurster (1989) | Stereotact Funct Neurosurg |
| | | Human | Nashold Jr. and Friedman (1972) | Arch Surg |
| | | Human | Carlsson and Fall (1984) | Paraplegia |
| | Cardiovascular | Rats[a] | (Qin, Farber, Linderoth, Shahid, & Foreman, 2008) | Journal of Pain |
| | Bowel | Cats[a] | Tai, Booth, de Groat, and Roppolo (2001) | Brain Res |
| Transcutaneous | Bladder | Mice | Ahmed (2017) (trans-spinal) | J Neural Eng |
| | | Rhesus Macaques[a] | Havton, Christe, Edgerton, and Gad (2019) | Exp Neurol |
| | | Human | Gad, Kreydin, Zhong, Latack, and Edgerton (2018) | J Neurotrauma |
| | | Human | Gad et al. (2018) | Front Neurosci |
| | | Human | Niu, Bennett, Keller, Leiter, and Lu (2018) | Sci Reports |
| | | Human | Kreydin et al. (2020) | Front Sys Neurosci |
| | Bowel | Human | Tsai et al. (2009) (magnetic) | J Rehab Medicine |
| | | Human | Lin et al. (2001) (magnetic) | Arch Phys Med Rehab |
| | Cardiovascular | Rats | Sachdeva et al. (2021) | Neurotherapeutics |
| | | Human | Phillips et al. (2018) | J Neurotrauma |

[a]*Denotes non-SCI model.*

in the United States as VOCARE Bladder System), designed to stimulate the anterior sacral roots (S2–S4), was used to aid voiding via intermittent stimulation bursts that maintained constant bladder pressure while periodically relaxing the urethral sphincter, allowing urine to flow in between stimulus intervals (post-stimulus voiding) (Brindley, 1994). Despite the clinical improvements in bladder emptying, one of the major accompanying limitations of the device stems from the irreversibility of ablating intact neural roots. Thus, a shift toward continuous or conditional neuromodulation was proposed in order to preserve the potential for future treatment options in patients with SCI (Hansen et al., 2005). Continuous sacral neuromodulation (applied most commonly over the S3 nerve root) is now a widely accepted stimulation modality for patients with refractory non-neurogenic lower urinary tract dysfunction, including, non-obstructive chronic urinary retention, urgency-frequency syndrome, and urge incontinence (Kessler et al., 2007). Initially, its use was not intended for those with neurogenic bladder dysfunction, but has shown some benefit for reducing detrusor hyperreflexia in those with incomplete SCI (Kessler et al., 2010). Note that peripheral nerves have also served as promising, minimally invasive targets to modulate the bladder circuitry after SCI, especially for those who may not be appropriate candidates for sacral neuromodulation. However, the stimulation site is distant from the central nervous system and involves repeated applications that may limit long-term durability in the SCI population (Gaziev et al., 2013).

Lumbosacral scES may be an effective approach which mechanistically may involve indirect activation of the same neural networks for bladder function. Several pre-clinical studies have demonstrated that electrical stimulation at select spinal locations can modulate both reflex detrusor and external urethral sphincter activity to induce micturition in different conditions: non-injured rats (Sysoev et al., 2020), nerve crush injuries (Chang, Yeh, Ichiyama, Rodriguez, & Havton, 2018), and following SCI in rats (Abud, Ichiyama, Havton, & Chang, 2015; Hoey et al., 2021) and cats (Pikov, Bullara, & Mccreery, 2007). In the context of human studies using scES, unanticipated gains in bladder function were achieved following locomotor training, where stimulation was optimized for improving standing and stepping (Harkema et al., 2011). The effects of activity-dependent scES on bladder function was further tested in 10 participants enrolled in scES training studies relative to participants in usual care ($n = 10$) who continued their typical daily lives without any study-related change in routine (no intervention), which resulted in improved bladder storage outcomes (increased bladder capacity and compliance and reduced detrusor pressure) at the post-training time point. Even though scES parameters were not targeted to the urinary system and scES was not "on" during bladder testing, long-term gains in bladder capacity persisted a year after training (Herrity et al., 2020), indicating the existence of integrated locomotor and autonomic networks.

When stimulation was targeted for bladder function, key parameters were identified through scES configuration mapping at the caudal end of the array that improved reflexive voiding efficiency in a participant characterized with motor complete SCI (Herrity, Williams, Angeli, Harkema, & Hubscher, 2018). This configuration was also re-tested in four additional individuals (3 AIS A, 1 AIS B), improving bladder emptying in each one, using the same frequency found effective in the initial participant. In these examples, it is possible that sufficient excitation of the lumbosacral circuitry through scES can enable activation of both autonomic and motor spinal cord circuits that control the lower urinary tract, permitting timed detrusor contraction with adequate relaxation of the sphincters and pelvic floor to facilitate more efficient emptying. One important advantage of scES is its capability to simultaneously target multiple systems. Given the heterogeneous nature of neurogenic lower urinary tract dysfunction, it is likely that parameters may vary from participant to participant and configurations found effective for bladder storage may differ from those that improve bladder emptying. Use of multi-electrode scES arrays has the advantage of targeting cardiovascular-specific parameters to control blood pressure in response to bladder distention in susceptible individuals (those with injuries above T6) (Fig. 5), in addition to other patient-specific urological needs (i.e., those using indwelling suprapubic catheters vs. those performing intermittent catheterization).

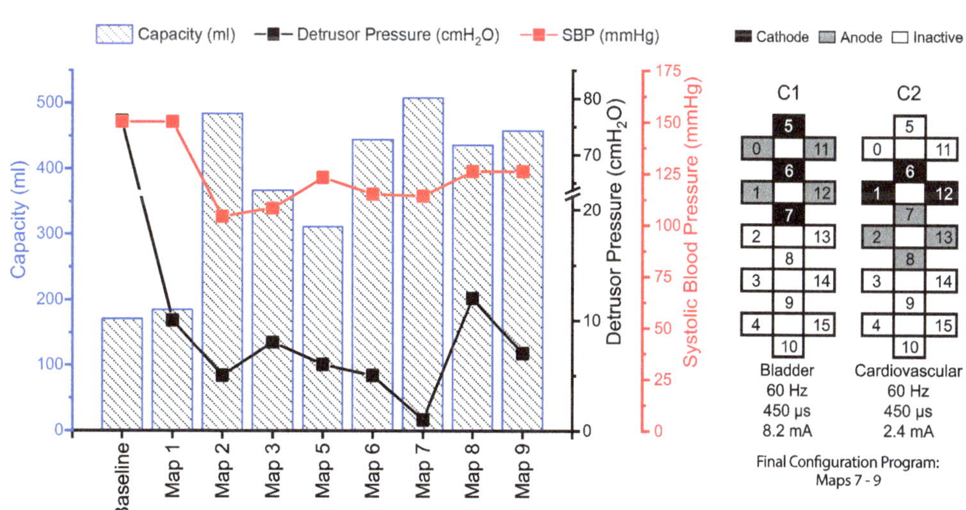


FIG. 5 Inter-system scES improves bladder function. Bladder mapping was conducted during urodynamics to identify specific stimulation parameters effective for bladder storage, which included maintaining bladder capacity *(blue)* within normative ranges established by the International Continence Society (300–600 mL) and maximum detrusor pressure < 40 $cmH_2O$ *(black line)* (C1, cohort 1). Note the integration of a second cohort (C2), used to stabilize blood pressure *(red line)* in response to bladder filling.

### Neuromodulation to improve bowel function

It has been demonstrated that electrical neuromodulation of gastrointestinal and autonomic networks can increase peristaltic activity to improve bowel motility and reduce constipation using a wide range of approaches including sacral nerve stimulation and wireless implanted stimulators within the GI system (Deng et al., 2018; Schiano Di Visconte et al., 2019). Vagal nerve stimulation of autonomic networks has also been shown to increase peristaltic activity and reduce inflammation in patients with inflammatory bowel disorders (Frøkjaer et al., 2016). Experimental pre-clinical studies evaluating the use of scES have demonstrated a role in reducing hypersensitivity associated with irritable bowel syndrome (Greenwood-Van Meerveld, Johnson, Foreman, & Linderoth, 2005) and modulating peristaltic activity in a rodent model of SCI (Hoey et al., 2021). In clinical studies with the Finetech-Brindley device, while stimulation of sacral anterior roots increased peristaltic activity in the distal colon and rectum, it also caused concomitant external anal sphincter contraction which prevented emptying during stimulation (Varma, Binnie, Smith, Creasey, & Edmond, 1986). Sacral nerve stimulation has primarily shown benefits in decreasing symptoms of constipation associated with neurogenic bowel dysfunction and decreasing episodes of incontinence in incomplete SCI (Lombardi & Del Popolo, 2009). In patients with gastroparesis, who have symptoms and pathophysiologic similarities to those with SCI (constipation, delayed emptying, prolonged transit time, autonomic dysfunction, abnormal inflammation, dysbiosis, metabolic abnormalities), direct gastric electrical stimulation improved symptoms of nausea and vomiting, constipation, decreased inflammation, and improved gastric motility (Abell et al., 2019). Compared to a conventional bowel routine, a 55%-fold reduction in bowel management time and a reduction in the neurogenic bowel dysfunction score from 15 (severe) to 8 (minor) was demonstrated using scES (pre-programmed for lower extremity and trunk activation) in a case report of one individual with motor complete SCI (Walter, Lee, Kavanagh, Phillips, & Krassioukov, 2018). A reduction in the time required to complete one's bowel program in one participant following the use of scES targeted for voluntary movement was also reported (Darrow et al., 2019). Despite these improvements, much remains unknown about the potential of and how best to target scES to improve bowel function after SCI. Further development of the parameters of stimulation and programming strategies for specifically targeting motility, maintaining continence and promoting evacuation are needed. Importantly, bowel distention is another main trigger of AD after SCI, and therefore, it is likely that stimulation will need to consider parameters to control blood pressure (similar to bladder) when scES is used to promote emptying during one's bowel program.

### Neuromodulation to improve cardiovascular function

Cardiovascular autonomic regulation is also a complex interplay of sympathetic and parasympathetic stimulation which is mediated by the spinal cord as the key integrator of complex signals including baroreceptors from the periphery and from supraspinal centers in the brainstem. In spinally intact individuals, lumbosacral scES can promote increased peripheral blood flow in those with peripheral vascular disease (Huber, Vaglienti, & Huber, 2000) and increase blood pressure in a model of central baroreflex failure (Yamasaki et al., 2006). Early evidence of the benefit of spinal stimulation for cardiovascular function in SCI is demonstrated in a study examining the use of percutaneous epidural stimulation to reduce spasticity in 5 subjects, which subsequently resolved AD in four of five individuals (Richardson, Cerullo, & Meyer, 1979). Our recent work has demonstrated that scES in the absence of descending input can modify the excitability of relevant spinal interneuron pools allowing them to respond to peripheral autonomic input and approximate normal cardiovascular control (Harkema et al., 2018). In the context of human locomotor studies in SCI, scES used to facilitate motor activity in individuals undergoing stand training also increased arterial blood pressure, mitigating orthostatic intolerance and stabilizing blood pressure for the duration of standing (Aslan et al., 2018). By stimulating the lumbosacral spinal cord with individual-specific cardiovascular scES configurations, blood pressure can be effectively and safely elevated and maintained within normative ranges from a chronic hypotensive state, without activation of lower limb skeletal muscles, in individuals with severe SCI (Harkema et al., 2018; Legg Ditterline et al., 2020; West et al., 2018). Daily training with scES parameters for cardiovascular function has also been shown to acutely alleviate orthostatic hypotension during orthostatic stress testing, demonstrating an immediate cardiovascular responsiveness to scES (Harkema, Legg Ditterline, et al., 2018).

## Future directions with scES and applications to other areas of neuroscience

Restoration of standing and stepping function as well as the regulation of autonomic function in chronic motor complete SCI has been achieved using the existing scES technology. However, the technology remains limited for full independent use in the home and community. The scES devices that are already FDA-approved (for use in chronic pain in non-SCI individuals), and IDE-approved for SCI, are not specifically designed to meet the complex needs of the SCI population. Upgrading the stimulator's programming and wireless communication platforms to generate a more seamless system capable of delivering multiple training paradigms across multiple physiological systems is needed. The current gap that places the burden on the user to adjust and monitor stimulation and physiological parameters remains one of the most important limiting factors in the effective utilization of this technology outside the laboratory. Thus, to improve long-term adaptations and recovery of autonomic function, integrating use of the technology in the natural environment is key.

From a clinical perspective, identifying those individuals who will potentially benefit from scES, including those with other neurological disorders such as multiple sclerosis (Abbate, Cook, & Atallah, 1977), as well as establishing clear guidelines and protocols regarding the optimization and long-term adjustments of stimulation paradigms, have also been identified as critical areas to advancing the technology (Solinsky, Specker-Sullivan, & Wexler, 2020). Also, given the invasive nature of the scES approach, use of transcutaneous stimulation may be effective in identifying those individuals who respond to scES should device implantation be a future option, as well as testing and determining the optimal spinal locations for different autonomic functions. For example, pilot studies with transcutaneous stimulation applied over interspinous segments of T11 and L1 during urodynamics indicate improvements in bladder storage and emptying in individuals with motor/sensory complete through motor incomplete SCI (Gad et al., 2018; Kreydin et al., 2020).

## Mini-dictionary of terms

- **Autonomic Dysreflexia (AD)**: A potentially dangerous syndrome that develops in individuals with cervical or upper thoracic spinal cord injury, resulting in acute, uncontrolled hypertension. AD is typically triggered by noxious or non-noxious peripheral or visceral stimulation below the level of SCI. The most common triggers of AD are from bladder and/or bowel distention.
- **Baroreflex:** a homeostatic response mechanism in which an elevation in blood pressure reflexively causes the heart rate to decrease and blood pressure to decrease.
- **Bowel program**: a regularly scheduled time dedicated to bowel emptying. The process usually begins with stimulating the anorectal region, typically using a suppository, enema, or by digital stimulation.
- **Detrusor-sphincter dyssynergia:** term used to describe a detrusor (bladder muscle) contraction with concomitant involuntary urethral sphincter activation, visualized during urodynamics, and typically present in individuals with neurologically induced bladder dysfunction.
- **Detrusor hyperreflexia:** characterized by the presence of involuntary uninhibited involuntary detrusor contractions, resulting in symptoms of urgency, frequency, and oftentimes, incontinence.
- **Gastroparesis:** impaired breakdown of food in the stomach and slowed gastric emptying resulting in primary symptoms of nausea, vomiting, and abdominal pain.
- **Hydronephrosis:** a condition characterized by excess fluid in a kidney due to the backup of urine, which can cause the kidney to swell.
- **Spinal cord epidural stimulation:** a neurostimulation device that delivers electrical signals through a lead implanted in the epidural space.
- **Urodynamics:** is an assessment of lower urinary tract function to objectively measure, explain and categorize how well the bladder, sphincters, and urethra are storing and releasing urine.
- **Vesico-ureteral reflux**: abnormal retrograde flow (reflux) of urine from the bladder up through one or more of the ureters and sometimes reaching the kidney.

## Key facts of autonomic dysfunction after SCI

- Regulation of autonomic function requires an intact neural axis between supraspinal centers and spinal cord autonomic pre-ganglionic neurons.
- Disruption in autonomic nervous system control after SCI results in multi-organ system dysfunction and a loss of overall bodily homeostasis.
- The recovery of bladder and bowel dysfunction may be challenged by interactions from the cardiovascular system, especially in individuals with higher level injuries (above T6) experiencing multiple episodes of AD.
- Autonomic dysfunction has a major impact on the health, well-being, and quality of life in individuals with SCI.

## Key facts of scES after SCI

- While initially focused on the motor system, the application of scES in motor complete SCI has more recently been shown to result in meaningful clinical improvements in autonomic function.
- The process of examining scES effects on bladder, bowel, and cardiovascular function begins with identifying specific stimulation parameters (location, rate, intensity, pulse width) during functional mapping that produce the desired outcome.
- Optimal improvements in autonomic function may necessitate the use of inter-system stimulation approaches in which an integration of configurations for multiple systems will enhance coordinated control (i.e., bladder storage and voiding with cardiovascular regulation).

## Summary points

- This chapter discusses the widespread impact of SCI on autonomic function and how scES can be used to enhance recovery of bladder, bowel, and cardiovascular function.
- Autonomic dysfunction negatively impacts heath, well-being, and overall quality of life in individuals with SCI.
- Initial applications of scES in the SCI population focused on treating severe spasticity, which subsequently led to the examination of whether scES had the capacity to drive locomotor capabilities through excitation of the lumbosacral motor circuitry.
- Numerous pre-clinical and human studies have demonstrated that scES can also be used to drive meaningful gains in bladder, bowel, and cardiovascular function after SCI.
- Activity-dependent recovery training in combination with scES can facilitate functional reorganization of lumbosacral networks, strengthening residual connections between supraspinal axons and the spinal circuitry involved in autonomic regulation.

## References

Abbate, A. D., Cook, A. W., & Atallah, M. (1977). Effect of electrical stimulation of the thoracic spinal cord on the function of the bladder in multiple sclerosis. *The Journal of Urology*, *117*, 285–288.

Abell, T. L., Kedar, A., Stocker, A., Beatty, K., Mcelmurray, L., Hughes, M., et al. (2019). Gastroparesis syndromes: Response to electrical stimulation. *Neurogastroenterology and Motility*, *31*, e13534.

Abud, E. M., Ichiyama, R. M., Havton, L. A., & Chang, H. H. (2015). Spinal stimulation of the upper lumbar spinal cord modulates urethral sphincter activity in rats after spinal cord injury. *American Journal of Physiology. Renal Physiology*, *308*, F1032–F1040.

Ahmed, Z. (2017). Effects of cathodal trans-spinal direct current stimulation on lower urinary tract function in normal and spinal cord injury mice with overactive bladder. *Journal of Neural Engineering*, *14*, 056002.

Anderson, K. D. (2004). Targeting recovery: Priorities of the spinal cord-injured population. *Journal of Neurotrauma*, *21*, 1371–1383.

Angeli, C. A., Edgerton, V. R., Gerasimenko, Y. P., & Harkema, S. J. (2014). Altering spinal cord excitability enables voluntary movements after chronic complete paralysis in humans. *Brain*, *137*, 1394–1409.

Aslan, S. C., Legg Ditterline, B. E., Park, M. C., Angeli, C. A., Rejc, E., Chen, Y., et al. (2018). Epidural spinal cord stimulation of lumbosacral networks modulates arterial blood pressure in individuals with spinal cord injury-induced cardiovascular deficits. *Frontiers in Physiology*, *9*, 565.

Barolat, G., & Sharan, A. D. (2000). Future trends in spinal cord stimulation. *Neurological Research*, *22*, 279–284.

Biers, S. M., Venn, S. N., & Greenwell, T. J. (2012). The past, present and future of augmentation cystoplasty. *BJU International*, *109*, 1280–1293.

Blok, B. F. M., & Holstege, G. (1997). Ultrastructural evidence for a direct pathway from the pontine micturition center to the parasympathetic preganglionic motoneurons of the bladder of the cat. *Neuroscience Letters*, *222*, 195–198.

Bloom, O., Wecht, J. M., Legg Ditterline, B. E., Wang, S., Ovechkin, A. V., Angeli, C. A., … Harkema, S. J. (2020). Prolonged targeted cardiovascular epidural stimulation improves immunological molecular profile: A case report in chronic severe spinal cord injury. *Frontiers in Systems Neuroscience*, *14*. 571011–571011.

Brindley, G. S. (1994). The first 500 patients with sacral anterior root stimulator implants: General description. *Paraplegia*, *32*, 795–805.

Carlsson, C. A., & Fall, M. (1984). Electrical stimulation of the conus medullaris for bladder emptying in a paraplegic. *Paraplegia*, *22*, 87–91.

Chang, H. H., Yeh, J. C., Ichiyama, R. M., Rodriguez, L. V., & Havton, L. A. (2018). Mapping and neuromodulation of lower urinary tract function using spinal cord stimulation in female rats. *Experimental Neurology*, *305*, 26–32.

Cook, A. W., & Weinstein, S. P. (1973). Chronic dorsal column stimulation in multiple sclerosis. Preliminary report. *New York State Journal of Medicine*, *73*, 2868–2872.

Darrow, D., Balser, D., Netoff, T. I., Krassioukov, A., Phillips, A., Parr, A., et al. (2019). Epidural spinal cord stimulation facilitates immediate restoration of dormant motor and autonomic supraspinal pathways after chronic neurologically complete spinal cord injury. *Journal of Neurotrauma*, *36*, 2325–2336.

Deng, Y., Dong, Y., Liu, Y., Zhang, Q., Guan, X., Chen, X., et al. (2018). A systematic review of clinical studies on electrical stimulation therapy for patients with neurogenic bowel dysfunction after spinal cord injury. *Medicine (Baltimore)*, *97*, e12778.

Devivo, M. J., Krause, J. S., & Lammertse, D. P. (1999). Recent trends in mortality and causes of death among persons with spinal cord injury. *Archives of Physical Medicine and Rehabilitation*, *80*, 1411–1419.

Dimitrijevic, M. R., Gerasimenko, Y., & Pinter, M. M. (1998). Evidence for a spinal central pattern generator in humans. *Annals of the New York Academy of Sciences*, *860*, 360–376.

Dimitrijevic, M. R., Illis, L. S., Nakajima, K., Sharkey, P. C., & Sherwood, A. M. (1986). Spinal cord stimulation for the control of spasticity in patients with chronic spinal cord injury: II. Neurophysiologic observations. *Central Nervous System Trauma*, *3*, 145–152.

Ditunno, P. L., Patrick, M., Stineman, M., & Ditunno, J. F. (2008). Who wants to walk? Preferences for recovery after SCI: A longitudinal and cross-sectional study. *Spinal Cord*, *46*, 500–506.

Emmanuel, A. (2010). Managing neurogenic bowel dysfunction. *Clinical Rehabilitation*, *24*, 483–488.

Fedirchuk, B., & Shefchyk, S. J. (1991). Effects of electrical stimulation of the thoracic spinal cord on bladder and external urethral sphincter activity in the decerebrate cat. *Experimental Brain Research*, *84*, 635–642.

Frøkjaer, J. B., Bergmann, S., Brock, C., Madzak, A., Farmer, A. D., Ellrich, J., et al. (2016). Modulation of vagal tone enhances gastroduodenal motility and reduces somatic pain sensitivity. *Neurogastroenterology and Motility*, *28*, 592–598.

Gad, P. N., Kreydin, E., Zhong, H., Latack, K., & Edgerton, V. R. (2018). Non-invasive neuromodulation of spinal cord restores lower urinary tract function after paralysis. *Frontiers in Neuroscience*, *12*, 432.

Gad, P. N., Roy, R. R., Zhong, H., Gerasimenko, Y. P., Taccola, G., & Edgerton, V. R. (2016). Neuromodulation of the neural circuits controlling the lower urinary tract. *Experimental neurology*, *285*, 182–189.

Gad, P. N., Roy, R. R., Zhong, H., Lu, D. C., Gerasimenko, Y. P., & Edgerton, V. R. (2014). Initiation of bladder voiding with epidural stimulation in paralyzed, step trained rats. *PLOS One*, *9*, e108184.

Gaziev, G., Topazio, L., Iacovelli, V., Asimakopoulos, A., Di Santo, A., De Nunzio, C., et al. (2013). Percutaneous Tibial nerve stimulation (PTNS) efficacy in the treatment of lower urinary tract dysfunctions: A systematic review. *BMC Urology*, *13*, 61.

Gibbins, I. (2013). Functional organization of autonomic neural pathways. *Organogenesis*, *9*, 169–175.

Glickman, S., & Kamm, M. A. (1996). Bowel dysfunction in spinal-cord-injury patients. *Lancet*, *347*, 1651–1653.

Greenwood-Van Meerveld, B., Johnson, A. C., Foreman, R. D., & Linderoth, B. (2005). Spinal cord stimulation attenuates visceromotor reflexes in a rat model of post-inflammatory colonic hypersensitivity. *Autonomic Neuroscience*, *122*, 69–76.

Grill, W. M., Wang, B., Hadziefendic, S., & Haxhiu, M. A. (1998). Identification of the spinal neural network involved in coordination of micturition in the male cat. *Brain Research*, *796*, 150–160.

Guiho, T., Delleci, C., Azevedo-Coste, C., Fattal, C., Guiraud, D., Vignes, J. R., & Bauchet, L. (2018). Impact of direct epispinal stimulation on bladder and bowel functions in pigs: A feasibility study. *Neurourology and Urodynamics*, *37*, 138–147.

Hagen, E. M., Faerestrand, S., Hoff, J. M., Rekand, T., & Gronning, M. (2011). Cardiovascular and urological dysfunction in spinal cord injury. *Acta Neurologica Scandinavica. Supplementum*, *124*, 71–78.

Hansen, J., Media, S., NØhr, M., Biering-SØrensen, F., Sinkjær, T., & Rijkhoff, N. J. M. (2005). Treatment of neurogenic detrusor overactivity in spinal cord injured patients by conditional electrical stimulation. *Journal of Urology*, *173*, 2035–2039.

Harkema, S., Gerasimenko, Y., Hodes, J., Burdick, J., Angeli, C., Chen, Y., et al. (2011). Effect of epidural stimulation of the lumbosacral spinal cord on voluntary movement, standing, and assisted stepping after motor complete paraplegia: A case study. *Lancet*, *377*, 1938–1947.

Harkema, S. J., Legg Ditterline, B., Wang, S., Aslan, S., Angeli, C. A., Ovechkin, A., et al. (2018). Epidural spinal cord stimulation training and sustained recovery of cardiovascular function in individuals with chronic cervical spinal cord injury. *JAMA Neurology*, *75*, 1569–1571.

Harkema, S. J., Wang, S., Angeli, C. A., Chen, Y., Boakye, M., Ugiliweneza, B., et al. (2018). Normalization of blood pressure with spinal cord epidural stimulation after severe spinal cord injury. *Frontiers in Human Neuroscience*, *12*, 83.

Havton, L. A., Christe, K. L., Edgerton, V. R., & Gad, P. N. (2019). Noninvasive spinal neuromodulation to map and augment lower urinary tract function in rhesus macaques. *Experimental Neurology*, *322*, 113033.

Herrity, A. N., Aslan, S. C., Ugiliweneza, B., Mohamed, A. Z., Hubscher, C. H., & Harkema, S. J. (2020). Improvements in bladder function following activity-based recovery training with epidural stimulation after chronic spinal cord injury. *Frontiers in Systems Neuroscience*, *14*, 614691.

Herrity, A. N., Williams, C. S., Angeli, C. A., Harkema, S. J., & Hubscher, C. H. (2018). Lumbosacral spinal cord epidural stimulation improves voiding function after human spinal cord injury. *Scientific Reports*, *8*, 8688.

Hoey, R. F., Medina-Aguiñaga, D., Khalifa, F., Ugiliweneza, B., Zdunowski, S., Fell, J., et al. (2021). Bladder and bowel responses to lumbosacral epidural stimulation in uninjured and transected anesthetized rats. *Scientific Reports*, *11*, 3268.

Huber, S. J., Vaglienti, R. M., & Huber, J. S. (2000). Spinal cord stimulation in severe, inoperable peripheral vascular disease. *Neuromodulation*, *3*, 131–143.

Hubscher, C. H., Herrity, A. N., Williams, C. S., Montgomery, L. R., Willhite, A. M., Angeli, C. A., et al. (2018). Improvements in bladder, bowel and sexual outcomes following task-specific locomotor training in human spinal cord injury. *PLoS One*, *13*, e0190998.

Inskip, J. A., Lucci, V. M., McGrath, M. S., Willms, R., & Claydon, V. E. (2018). A community perspective on bowel management and quality of life after spinal cord injury: The influence of autonomic dysreflexia. *Journal of Neurotrauma*, *35*, 1091–1105.

Karlsson, A. K. (1999). Autonomic dysreflexia. *Spinal Cord*, *37*, 383–391.

Katzelnick, C. G., Weir, J. P., Jones, A., Galea, M., Dyson-Hudson, T. A., Kirshblum, S. C., et al. (2019). Blood pressure instability in persons with SCI: Evidence from a 30-day home monitoring observation. *American Journal of Hypertension*, *32*, 938–944.

Kessler, T. M., Buchser, E., Meyer, S., Engeler, D. S., Al-Khodairy, A.-W., Bersch, U., et al. (2007). Sacral neuromodulation for refractory lower urinary tract dysfunction: Results of a Nationwide Registry in Switzerland. *European Urology*, *51*, 1357–1363.

Kessler, T. M., La Framboise, D., Trelle, S., Fowler, C. J., Kiss, G., Pannek, J., et al. (2010). Sacral neuromodulation for neurogenic lower urinary tract dysfunction: Systematic review and Meta-analysis. *European Urology*, *58*, 865–874.

Kirshblum, S. C., Gulati, M., O'connor, K. C., & Voorman, S. J. (1998). Bowel care practices in chronic spinal cord injury patients. *Archives of Physical Medicine and Rehabilitation*, *79*, 20–23.

Koh, J., Brown, T. E., Beightol, L. A., Ha, C. Y., & Eckberg, D. L. (1994). Human autonomic rhythms: Vagal cardiac mechanisms in tetraplegic subjects. *The Journal of Physiology*, *474*, 483–495.

Krassioukov, A., Eng, J. J., Warburton, D. E., & Teasell, R. (2009). A systematic review of the management of orthostatic hypotension after spinal cord injury. *Archives of Physical Medicine and Rehabilitation*, *90*, 876–885.

Kreydin, E., Zhong, H., Latack, K., Ye, S., Edgerton, V. R., & Gad, P. (2020). Transcutaneous electrical spinal cord neuromodulator (TESCoN) improves symptoms of overactive bladder. *Frontiers in Systems Neuroscience*, *14*, 1.

Krogh, K., Nielsen, J., Djurhuus, J. C., Mosdal, C., Sabroe, S., & Laurberg, S. (1997). Colorectal function in patients with spinal cord lesions. *Diseases of the Colon and Rectum*, *40*, 1233–1239.

Legg Ditterline, B. E., Aslan, S. C., Wang, S., Ugiliweneza, B., Hirsch, G. A., Wecht, J. M., et al. (2020). Restoration of autonomic cardiovascular regulation in spinal cord injury with epidural stimulation: A case series. *Clinical Autonomic Research*, *31*, 317–320.

Lin, V. W., Nino-Murcia, M., Frost, F., Wolfe, V., Hsiao, I., & Perkash, I. (2001). Functional magnetic stimulation of the colon in persons with spinal cord injury. *Archives of Physical Medicine and Rehabilitation*, *82*, 167–173.

Lindan, R., Joiner, E., Freehafer, A. A., & Hazel, C. (1980). Incidence and clinical features of autonomic dysreflexia in patients with spinal cord injury. *Paraplegia*, *18*, 285–292.

Lombardi, G., & Del Popolo, G. (2009). Clinical outcome of sacral neuromodulation in incomplete spinal cord injured patients suffering from neurogenic lower urinary tract symptoms. *Spinal Cord*, *47*, 486–491.

Longo, W. E., Ballantyne, G. H., & Modlin, I. M. (1989). The colon, anorectum, and spinal cord patient. A review of the functional alterations of the denervated hindgut. *Diseases of the Colon and Rectum*, *32*, 261–267.

Loubser, P. G. (1997). Adverse effects of epidural spinal cord stimulation on bladder function in a patient with chronic spinal cord injury pain. *Journal of Pain and Symptom Management*, *13*, 251–252.

McGee, M. J., Amundsen, C. L., & Grill, W. M. (2015). Electrical stimulation for the treatment of lower urinary tract dysfunction after spinal cord injury. *The Journal of Spinal Cord Medicine*, *38*, 135–146.

Meglio, M., Cioni, B., Amico, E. D., Ronzoni, G., & Rossi, G. F. (1980). Epidural spinal cord stimulation for the treatment of neurogenic bladder. *Acta Neurochirurgica*, *54*, 191–199.

Melzack, R., & Wall, P. D. (1965). Pain mechanisms: A new theory. *Science*, *150*, 971–979.

Merritt, C. H., Taylor, M. A., Yelton, C. J., & Ray, S. K. (2019). Economic impact of traumatic spinal cord injuries in the United States. *Neuroimmunology and Neuroinflammation*, *6*, 9.

Nashold, B. S., Jr., & Friedman, H. (1972). Dorsal column stimulation for control of pain. Preliminary report on 30 patients. *Journal of Neurosurgery*, *36*, 590–597.

Niu, T., Bennett, C. J., Keller, T. L., Leiter, J. C., & Lu, D. C. (2018). A proof-of-concept study of transcutaneous magnetic spinal cord stimulation for neurogenic bladder. *Scientific Reports*, *8*, 12549.

Phillips, A. A., Squair, J. W., Sayenko, D. G., Edgerton, V. R., Gerasimenko, Y., & Krassioukov, A. V. (2018). An autonomic neuroprosthesis: noninvasive electrical spinal cord stimulation restores autonomic cardiovascular function in individuals with spinal cord injury. *Journal of Neurotrauma*, *35*, 446–451.

Pikov, V., Bullara, L., & Mccreery, D. B. (2007). Intraspinal stimulation for bladder voiding in cats before and after chronic spinal cord injury. *Journal of Neural Engineering*, *4*, 356–368.

Qin, C., Farber, J. P., Linderoth, B., Shahid, A., & Foreman, R. D. (2008). Neuromodulation of thoracic intraspinal visceroreceptive transmission by electrical stimulation of spinal dorsal column and somatic afferents in rats. *The Journal of Pain*, *9*, 71–78.

Rabchevsky, A. G. (2006). Segmental organization of spinal reflexes mediating autonomic dysreflexia after spinal cord injury. In L. C. Weaver, & C. Polosa (Eds.), *Progress in brain research* Elsevier.

Reeve Foundation Paralysis Resource, C. (2021). *One degree of separation: Paralysis and spinal cord injury in the United States*. Christopher & Dana Reeve Foundation. http://www.christopherreeve.org/site/c.mtKZKgMWKwG/b.5184189/k.5587/Paralysis_Facts__Figures.htm.

Richardson, R. R., Cerullo, L. J., & Meyer, P. R. (1979). Autonomic hyper-reflexia modulated by percutaneous epidural neurostimulation: A preliminary report. *Neurosurgery*, *4*, 517–520.

Sachdeva, R., Nightingale, T. E., Pawar, K., Kalimullina, T., Mesa, A., Marwaha, A., et al..... (2021). Noninvasive neuroprosthesis promotes cardiovascular recovery after spinal cord injury. *Neurotherapeutics*, *18*, 1244–1256.

Schiano Di Visconte, M., Pasquali, A., Cipolat Mis, T., Brusciano, L., Docimo, L., & Bellio, G. (2019). Sacral nerve stimulation in slow-transit constipation: Effectiveness at 5-year follow-up. *International Journal of Colorectal Disease*, *34*, 1529–1540.

Schieferdecker, S., Neudorfer, C., El Majdoub, F., & Maarouf, M. (2019). A retrospective case series of high-frequency spinal cord stimulation (HF10-SCS) in neurogenic bladder incontinence. *Operative Neurosurgery (Hagerstown)*, *17*, 14–20.

Shafik, A. (1996). Recto-colic reflex: Role in the defecation mechanism. *International Surgery*, *81*, 292–294.

Shealy, C. N., Mortimer, J. T., & Hagfors, N. R. (1970). Dorsal column electroanalgesia. *Journal of Neurosurgery*, *32*, 560–564.

Shealy, C. N., Mortimer, J. T., & Reswick, J. B. (1967). Electrical inhibition of pain by stimulation of the dorsal columns: Preliminary clinical report. *Anesthesia and Analgesia*, *46*, 489–491.

Simpson, L. A., Eng, J. J., Hsieh, J. T. C., Wolfe, D. L., & Spinal Cord Injury Rehabilitation Evidence Scire Research, T. (2012). The health and life priorities of individuals with spinal cord injury: A systematic review. *Journal of Neurotrauma*, *29*, 1548–1555.

Solinsky, R., Specker-Sullivan, L., & Wexler, A. (2020). Current barriers and ethical considerations for clinical implementation of epidural stimulation for functional improvement after spinal cord injury. *The Journal of Spinal Cord Medicine*, *43*, 653–656.

Squair, J. W., Gautier, M., Mahe, L., Soriano, J. E., Rowald, A., Bichat, A., et al.…. (2021). Neuroprosthetic baroreflex controls haemodynamics after spinal cord injury. *Nature*, *590*, 308–314.

Stiens, S. A., Bergman, S. B., & Goetz, L. L. (1997). Neurogenic bowel dysfunction after spinal cord injury: Clinical evaluation and rehabilitative management. *Archives of Physical Medicine and Rehabilitation*, *78*, S86–102.

Stover, S. L., Devivo, M. J., & Go, B. K. (1999). History, implementation, and current status of the national spinal cord injury database. *Archives of Physical Medicine and Rehabilitation*, *80*, 1365–1371.

Sysoev, Y., Bazhenova, E., Lyakhovetskii, V., Kovalev, G., Shkorbatova, P., Islamova, R., et al. (2020). Site-specific neuromodulation of detrusor and external urethral sphincter by epidural spinal cord stimulation. *Frontiers in Systems Neuroscience*, *14*, 47.

Tai, C., Booth, A. M., de Groat, W. C., & Roppolo, J. R. (2001). Colon and anal sphincter contractions evoked by microstimulation of the sacral spinal cord in cats. *Brain Research*, *889*, 38–48.

Tai, C., Booth, A. M., de Groat, W. C., & Roppolo, J. R. (2004). Bladder and urethral sphincter responses evoked by microstimulation of S2 sacral spinal cord in spinal cord intact and chronic spinal cord injured cats. *Experimental Neurology*, *190*, 171–183.

Tsai, P. Y., Wang, C. P., Chiu, F. Y., Tsai, Y. A., Chang, Y. C., & Chuang, T. Y. (2009). Efficacy of functional magnetic stimulation in neurogenic bowel dysfunction after spinal cord injury. *Journal of Rehabilitation Medicine*, *41*, 41–47.

Varma, J. S., Binnie, N., Smith, A. N., Creasey, G. H., & Edmond, P. (1986). Differential effects of sacral anterior root stimulation on anal sphincter and colorectal motility in spinally injured man. *BJS (British Journal of Surgery)*, *73*, 478–482.

Walter, J. S., Robinson, C. J., Khan, T., Wheeler, J. S., & Wurster, R. D. (1989). Sacral epidural electrodes for voiding in the chronic spinal dog. *Stereotactic and Functional Neurosurgery*, *53*, 113–124.

Walter, M., Lee, A. H. X., Kavanagh, A., Phillips, A. A., & Krassioukov, A. V. (2018). Epidural spinal cord stimulation acutely modulates lower urinary tract and bowel function following spinal cord injury: A case report. *Frontiers in Physiology*, *9*, 1816.

Waltz, J. M., Andreesen, W. H., & Hunt, D. P. (1987). Spinal cord stimulation and motor disorders. *Pacing and Clinical Electrophysiology*, *10*, 180–204.

Wecht, J. M., & Bauman, W. A. (2018). Implication of altered autonomic control for orthostatic tolerance in SCI. *Autonomic Neuroscience*, *209*, 51–58.

Wecht, J. M., Cirnigliaro, C. M., Azarelo, F., Bauman, W. A., & Kirshblum, S. C. (2015). Orthostatic responses to anticholinesterase inhibition in spinal cord injury. *Clinical Autonomic Research*, *25*, 179–187.

Wecht, J. M., Harel, N. Y., Guest, J., Kirshblum, S. C., Forrest, G. F., Bloom, O., et al. (2020). Cardiovascular autonomic dysfunction in spinal cord injury: Epidemiology, diagnosis, and management. *Seminars in Neurology*, *40*, 550–559.

Wecht, J. M., Weir, J. P., Katzelnick, C. G., Chiaravalloti, N. D., Kirshblum, S. C., Dyson-Hudson, T. A., et al. (2020). Double-blinded, placebo-controlled crossover trial to determine the effects of midodrine on blood pressure during cognitive testing in persons with SCI. *Spinal Cord*, *58*, 959–969.

West, C. R., Phillips, A. A., Squair, J. W., Williams, A. M., Walter, M., Lam, T., et al. (2018). Association of epidural stimulation with cardiovascular function in an individual with spinal cord injury. *JAMA Neurology*, *75*, 630–632.

Wheeler, T. L., De Groat, W., Eisner, K., Emmanuel, A., French, J., Grill, W., et al. (2018). Translating promising strategies for bowel and bladder management in spinal cord injury. *Experimental Neurology*, *306*, 169–176.

Yamasaki, F., Ushida, T., Yokoyama, T., Ando, M., Yamashita, K., & Sato, T. (2006). Artificial baroreflex: Clinical application of a bionic baroreflex system. *Circulation*, *113*, 634–639.

Yousefpour, A., & Erfanian, A. (2021). A general framework for automatic closed-loop control of bladder voiding induced by intraspinal microstimulation in rats. *Scientific Reports*, *11*, 3424.

Chapter 19

# Treating spinal cord injury with implanted spinal cord stimulators

Hengze You, Yong-Ping Zheng, and Monzurul Alam
*Department of Biomedical Engineering, The Hong Kong Polytechnic University, Hong Kong*

## Abbreviations

**AIS** American Spinal Injury Association Impairment Scale
**CRPS** complex regional pain syndrome
**EES** epidural electrical stimulation
**EMG** electromyography
**FBSS** failed back surgery syndrome
**IPG** implantable pulse generator
**MRI** magnetic resonance imaging
**SCI** spinal cord injury
**SCS** spinal cord stimulation
**SCSr** spinal cord stimulator

## Introduction

### Spinal cord injury

Spinal cord injury (SCI), which literally means damage to the spinal cord, is a devastating and debilitating neurological condition, affecting millions of people worldwide. Traumatic SCI caused by external impacts accounts for the majority of all SCI cases and always accompanies a secondary injury that causes the further death of neurons and glial cells, leading to long-term neurological defects and paralysis. It can significantly harm the spinal cord's functions and burden patients, their families, and friends with substantial personal and socio-economic pressures. According to the 2020 SCI datasheet reported by the National Spinal Cord Injury Statistical Center, approximately 294,000 persons are living with SCI in the United States, with about 17,810 new cases every year, while the World Health Organization (WHO) had estimated that 250,000–500,000 people suffer an SCI each year around the world (Alam et al., 2020; Courtine & Sofroniew, 2019). Clearly, SCI has caused a significant burden on the healthcare system all over the globe.

Etiologically, SCI can be categorized into two types: traumatic SCI and non-traumatic SCI. Traumatic SCI is caused by external physical impacts such as vehicle accidents, sports injuries, violence, or falls. In a traumatic SCI, the external impact firstly causes bone, blood vessel, and ligament damage of the spine, then the debris and dislocation subsequently hurt and rip the spinal cord tissue, initiating the primary injury. After that, it leads to a secondary injury cascade (Fig. 1), resulting in the death of glial cells and neurons, and a long-term neurological defect (Ahuja et al., 2017). Differently, non-traumatic SCI is usually caused by other chronic or degenerative disorders such as tumors or infection, which however can induce a similar level of neurological deficit to a traumatic one. Despite its existence, non-traumatic SCI rarely happens compared with the traumatic SCI that accounts for over 90% of all SCI cases (Alizadeh, Dyck, & Karimi-Abdolrezaee, 2019).

With the improvement in emergency healthcare and medical services, many patients can survive from traumatic SCI and usually live up to a very optimistic life expectancy, however, frequently accompanied by terribly devastating disabilities. The injury cascade leads to complications such as inflammation, ischemia, and subsequently ruin in organization and structure of the spinal cord. Indeed, the clinical outcome is also heavily subject to the location and severity of the injury, since the patients may be partially or wholly paralyzed—depending on the lesion intensity—lose the sensory and motor functions below the level of the lesion, resulting in different extents of disability such as paraplegia and quadriplegia. If the

Diagnosis and Treatment of Spinal Cord Injury. https://doi.org/10.1016/B978-0-12-822498-4.00019-1

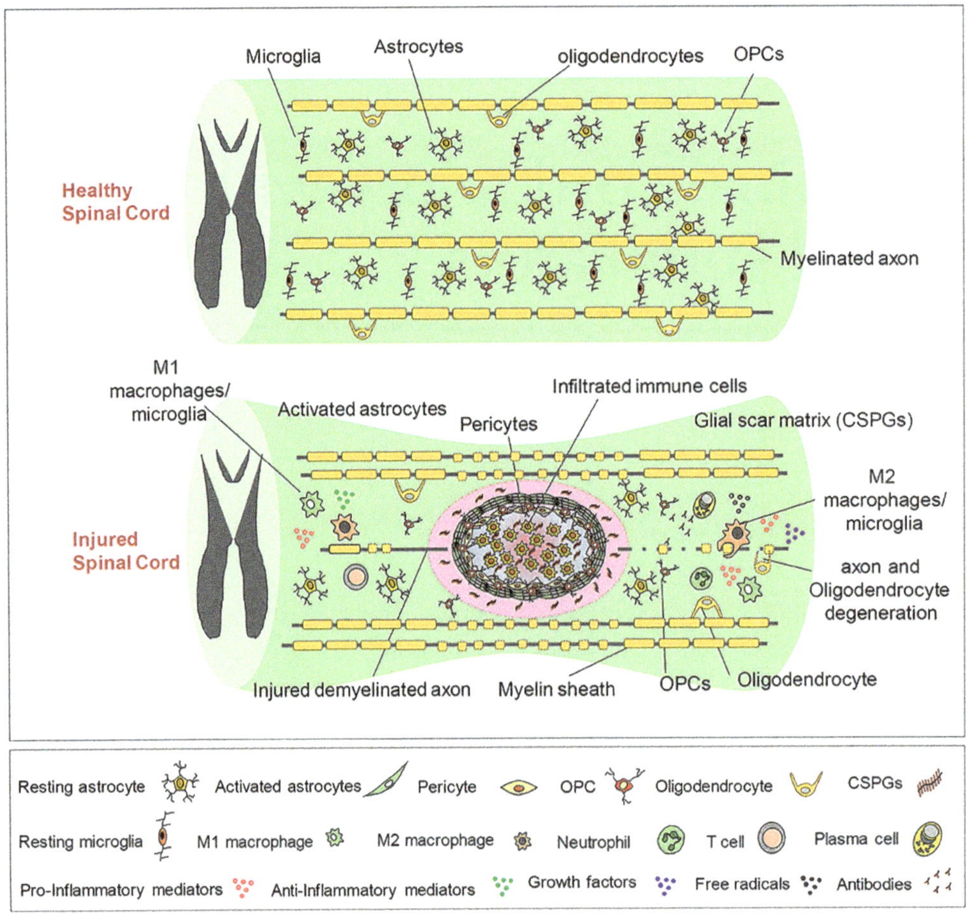


FIG. 1 Pathophysiology of traumatic spinal cord injury. In the healthy spinal cord, axons remain intact, are supported and protected by glia, astrocytes, or other components. After traumatic SCI, a secondary injury following the primary injury leads to the death of glial cells and neurons, initiating a series of devastating responses leading to motor and sensory function loss. *(Reprinted with permission from Alizadeh, A., Dyck, S. M., & Karimi-Abdolrezaee, S. (2019). Traumatic spinal cord injury: An overview of pathophysiology, models and acute injury mechanisms [review]. Frontiers in Neurology, 10(282). https://doi.org/10.3389/fneur.2019.00282.)*

sensory and motor function is totally lost below the injury level, it is clinically defined as a complete SCI; but if there is still any spare function left, it is called an incomplete SCI (Gaber & Brown, 2020). Regardless of different levels of disability, the loss of independence and ability returning to a job and daily life causes immense difficulties, extra works, and costs in caring for SCI patient's daily living. The lifetime cost for each SCI patient is estimated to be US$2.35 million (Alizadeh et al., 2019), which is definitely a considerable amount for many families, especially under the circumstance that the patient themselves cannot often work in their regular jobs anymore. Besides the loss of sensorimotor functions, chronic pain is another common problem, which frequently happens after SCI, with a rate of about 30%–80% and strikingly one third of which suffer from severe pain (Dijkers, Bryce, & Zanca, 2009; Widerström-Noga et al., 2008). Consequently, understanding SCI's pathophysiological mechanism and developing effective treatment to this condition is critical and highly significant.

## Treatments for SCI

Over the years, although substantial research and efforts have been contributed to study the SCI, there is still no known method to reverse or cure the damage of SCI at present. Fortunately, different strategies have been found to improve the lost sensorimotor functions, self-independence, and life quality of SCI patients; and other promising treatments are also in progress. Currently, available treatments for SCI are mainly in three aspects: intervention shortly after the injury happens (acute stage), compensatory strategies, and functional recovery in sub-acute and chronic stages (Roy, Harkema, & Edgerton, 2012). The primary purpose of interventions after acute SCI is to decompress the spinal cord and minimize secondary injury. The typical early intervention approach includes surgical and pharmacological interventions. Though applications on animal models had suggested beneficial outcomes of surgical interventions, its effect on human trials remains unclear (Silva, Sousa, Reis, & Salgado, 2014). Also, despite some successful cases, the ability of pharmacological interventions to improve neurological function remains controversial at present and requires more demonstrations in patients (del Popolo, Mencarini, Nelli, & Lazzeri, 2012). The compensatory strategies usually help improve functions by utilizing assistive devices or training modalities to shift the tasks to unaffected body parts. Compensatory strategies are commonly employed to complete SCI, which has relatively limited restoring capability of the spinal cord functions (Alam & He, 2014; Alam & Zheng, 2017). On the other hand, functional recovery strategies attempt to restore the functions after SCI by means such as stimulations or neurorehabilitation training. These strategies normally show better outcomes for less severe SCI since incomplete SCI has a higher chance of functional recovery; however, stimulation-induced motor function recovery after complete SCI is also heavily investigated (Dietz & Fouad, 2014; Gaber & Brown, 2020). A promising rehabilitative strategy to SCI in recent years is spinal cord stimulation (SCS), which has shown encouraging outcomes in not only functional recovery but also SCI pain relief even in patients who did not expect to improve any function with classical interventions (James, McMahon, Field-Fote, & Bradbury, 2018). In this chapter, we have discussed the principles behind the SCS-mediated rehabilitation of sensorimotor functions after SCI, introduced several notable spinal cord stimulators, and illustrated the procedures of their clinical application.

## Spinal cord stimulation

### Principles of SCS

SCS has been used to treat neuropathic pains for over 50 years. It is often applied to alleviate chronic pain for which conventional treatments have failed, such as congenital or acquired chronic low back and leg pains, complex regional pain syndrome (CRPS), failed back surgery syndrome (FBSS), and post-SCI pains, etc. (Giugno et al., 2017). SCS blocks pain but not cures it. SCS relieves pain by generating electric pulses at the spinal cord. Its working principle to block pain is the gate control theory of pain (Fig. 2). Gate control theory indicates that other inputs will shut the "nerve gate" for the painful input in the spinal cord, preventing pain signals from reaching the brain so that one does not feel the pain (Ireland, 2010). Based on this theory, the stimulator produces electrical pulses on the spinal cord's dorsal surface, which closes the "nerve gate" to pain signals, prevents them from being received by the brain, thus masks the neuropathic pains (Mammis, 2016).

Besides, SCS starts to serve as a promising tool for movement restoration after SCI. Both invasive and non-invasive SCS could be a powerful tool. The exact principle underlying the motor function improvement by SCS is largely unknown at present. Some findings suggest that even isolated from the brain the spinal circuitry is still able to generate locomotor activity driven by the externally applied stimulation (Courtine et al., 2009; Dimitrijevic, Gerasimenko, & Pinter, 1998; Shah et al., 2016). According to this principle, the SCS replaces the brain to generate stimuli for triggering locomotor-like activity for the paralyzed. So SCS with different patterns, amplitudes, frequencies, pulse widths at different locations may result in different locomotor activities. Theoretically, these motion activities could happen only if SCS is applied to the patient sustainedly. However, another explanation suggests that instead of replacing the brain, the SCS enables the brain

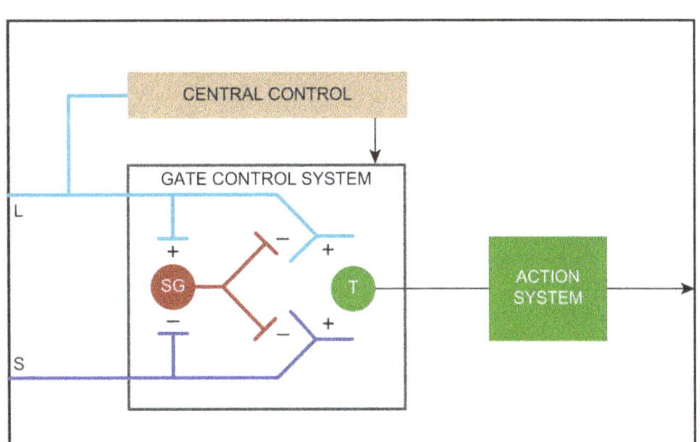


FIG. 2 The gate control theory of pain. The inhibitory neurons in substantia gelatinosa (SG, lamina II) are excited by the non-nociceptive input (L) and thus close the pain gate. *(Reprinted with permission from Braz, J., Solorzano, C., Wang, X., & Basbaum, A. I. (2014). Transmitting pain and itch messages: A contemporary view of the spinal cord circuits that generate gate control. Neuron, 82(3), 522–536. https://doi.org/10.1016/j.neuron.2014.01.018.)*

to recruit the spared and functionally silent nerves residue in SCI, so the reorganization of the residual neural pathway by the SCS promotes and improves locomotor functions (van den Brand et al., 2012). In this theory, the brain is still the command center of the limb movement, and the role of SCS is just to activate the residual neural pathway important for locomotion. So, it is possible that after a period of SCS, the locomotor improvement became long-term and sustained even after the SCS is removed.

### Functional rehabilitation with SCS

Though the detailed mechanisms remain largely unclear, many studies found that SCS has benefits on motor function recovery after incomplete or complete SCI, notably in patients' voluntary movement, standing, and stepping recovery as well as cardiovascular and bowel functions management (Table 1). Harkema et al. (2011) propose that tonic SCS could modulate spinal cord circuitry and enable patients with paralysis after SCI to undertake tasks such as standing and stepping. They reported a case in 2011 that a paraplegic patient was able to undergo a full-weight bearing standing and some leg movements during task-specific SCS (Harkema et al., 2011). The subject in the study had a clinically complete loss of voluntary motor function but incomplete sensory SCI, so the regain of some voluntary functions in this study is interpreted by that the residual but silent neuropathway was reactivated or the neuroplasticity was promoted by the SCS (Edgerton & Harkema, 2011). However, in a later study reported in 2014, they found that even for the patients diagnosed with a complete motor and sensory SCI, SCS can still help to proceed detectable leg movements after the paralysis for years (Angeli et al., 2014). This finding uncovers that SCS could be a potential rehabilitative tool for diagnostically complete SCI even a long time after the injury happens. These previous studies apply a relatively low stimulation frequency—usually below 30 Hz—for SCS, and patients still need some external assistance to stand. In 2015, they also reported that two motor-complete but sensory-incomplete SCI patients were able to process full-body weight standing without using any external assistance (Rejc et al., 2015). In this study, a higher stimulation frequency (25–60 Hz) relative to that in previous studies (<30 Hz) lends to an EMG pattern more effective for patients to stand. Also, a body weight-bearing sensory input could remarkably increase the EMG activity to sufficient for standing, while very little EMG activity was generated by lower limb muscles during epidural SCS. The research by Sayenko et al. also demonstrated that the stimulation parameters and sensory conditions significantly influence the SCS outcomes, showing different stimulation responses between standing and supination (Sayenko, Angeli, Harkema, Edgerton, & Gerasimenko, 2014). Besides lower limb movement recovery, SCS at the cervical level was also successfully applied to improve upper limb motor functions and promote voluntary hand movements for patients with tetraplegia (Lu et al., 2016; Qian, Ling, Zhong, Zheng, & Alam, 2020).

In the studies mentioned earlier, the efficacy of SCS on SCI motor function recovery has been witnessed; nonetheless, volitional movements were generated only if real-time task-specific SCS is being provided. Rejc et al. (2017) had progressively found the reemergence of muscle activation pattern sufficient for the standing of complete SCI patients without any real-time SCS after they underwent long-term task-specific motor training with SCS (Rejc et al., 2017). After then, a number of recent studies also showed that a period of SCS motor training enabled SCI patients to conduct independent standing and stepping over the ground (Angeli et al., 2018; Gill et al., 2018; Wagner et al., 2018). These achievements reveal that SCS could stimulate real-time motor activation and be a promising training tool to recover the independent motor functions for patients with complete chronic SCI (Fig. 3) (Cho, Squair, Bloch, & Courtine, 2019).

## Spinal cord stimulators

### Current spinal cord stimulators

There are currently various kinds of spinal cord stimulators on the market. This section briefly discussed several typical products and compared their properties and features (Table 2). Precision Spectra is an SCSr designed by Boston Scientific Corporation. Its function mainly focuses on pain relief. This product's working principle is similar to what was mentioned previously: generating impulses to mask the pain signals before they could reach the brain. Precision Spectra is featured with more coverage, high flexibility, and advanced control. It is a safe, drug-free, approved by the Food and Drug Administration (FDA) of the United States, and paresthesia-free pain management therapy that is clinically proven to help reduce chronic pains. The Spectra can be personalized for every individual's unique pain and can deliver more than one type of stimulation simultaneously, even at multiple areas, to find the optimum treatment for a particular pain. Firstly, Spectra introduces the world's first implantable pulse generator (IPG) with 32 contacts and a dedicated power source, providing more excellent coverage and more precise targeting to the spinal cord but not increasing the product size. Secondly, Spectra presents an unprecedented level of flexibility. The combination of the four-port design and the broad lead portfolio enables

**TABLE 1** The application of implanted SCSrs for SCI.

| Paper | *n* | Motor complete/ incomplete (AIS) | Spinal cord stimulator | Trainings | Therapy duration (months) | Therapy intensity (sessions/ week) | Total number of training sessions | Training outcomes or the improvement in ISNCSCI Motor Score: mean (range) |
|---|---|---|---|---|---|---|---|---|
| DiMarco, Geertman, Tabbaa, Nemunaitis, & Kowalski, (2020)) | 5 | Complete (A) | Finetech Medical's cough stimulation system + Ardiem Medical's wire lead electrodes | Bowel management (cough restore) training | 5.25 | 14–21 | 294–441 | Restore cough and improve bowel management |
| Darrow et al. (2019) | 2 | Complete (A) | Abbott's Tripole and Proclaim Elite + 16-electrode paddle array | Simultaneous stimulation and assessment | No training period | No training period | No training period | Restore some supraspinal control over motor function below the level of injury: volitional motor control, cardiovascular function, bowel-bladder synergy |
| Harkema et al. (2018) | 4 | Complete (A/B) | Medtronic's Restore ADVANCED + 16-electrode array | Spinal cord epidural stimulation for cardiovascular function | 0.5 | 2.5 | 5 | Activate mechanisms to elevate blood pressures to normal ranges from a chronic hypotensive state in SCI |
| Wagner et al. (2018) | 3 | Incomplete (C/D) | Medtronic's Activa RC + 16-electrode paddle array | Cycling + locomotor training | 5 | 5 | 108 | 10.3 (4–16) |
| Angeli et al. (2018) | 4 | Complete (A/B) | Medtronic's epidural arrays + neurostimulators + programming devices | Locomotor training | Range 6–20 | 7 | 168–560 | 0.25 (0–1) |
| Grahn et al. (2017) | 1 | Complete (A) | Medtronic's RestoreSensor SureScan MRI + 5–6-5 Specify 16-electrode array | Locomotor training | 0.5 | 4 | 8 | Enable volitional control of task-specific muscle activity, rhythmic muscle activity to produce step-like movements while side-lying, independent standing |
| Rejc, Angeli, Atkinson, and Harkema (2017) | 1 | Complete (B) | Medtronic's Restore ADVANCED + 5–6-5 Specify 16-electrode array | Activity-based training/task-specific training | 16 | 5 | 80 | From no volitional muscle activation to standing, stepping, volitional leg movement |

*Continued*

**TABLE 1** The application of implanted SCSrs for SCI—cont'd

| Paper | *n* | Motor complete/incomplete (AIS) | Spinal cord stimulator | Trainings | Therapy duration (months) | Therapy intensity (sessions/week) | Total number of training sessions | Training outcomes or the improvement in ISNCSCI Motor Score: mean (range) |
|---|---|---|---|---|---|---|---|---|
| Lu et al. (2016) | 3 | Complete (B) | Boston Scientific's Precision Plus + Artisan 16-electrode epidural array | Volitional hand function | 24 | 1–7 | 96–672 | 19.5 (16–23) |
| Rejc, Angeli, and Harkema (2015) | 4 | Complete (A/B) | Medtronic's Restore ADVANCED +5–6-5 Specify 16-electrode array | Stand training | 16 | 5 | 80 | Stand over-ground bearing full body weight without any external assistance |
| Angeli, Edgerton, Gerasimenko, and Harkema (2014) | 4 | Complete (A/B) | Medtronic's Restore ADVANCED + 16-electrode array | Locomotor training | 3–7 | 7 | 21–49 | Able to generate EMG activity and movement during ankle dorsiflexion in the presence of epidural stimulation |
| Possover (2014) | 4 | Complete/incomplete (B/C) | Medtronics Restore ADVANCED + four 4-array fine wire electrodes | Locomotor training | 12 | 3–5 | 156–260 | 12.8 (1 – 21) |
| Harkema et al. (2011) | 1 | Complete (B) | Medtronic's Restore ADVANCED + 16-electrode array | Standing + locomotor training | 7 | 1.7 | 170 | Full weight-bearing standing for 4.25 min, supraspinal control of some leg movements |
| Carhart, Jiping, Herman, Luzansky, and Willis (2004) | 1 | Incomplete (C) | X-trel 3470 implanted receiver + implanted quadripolar electrode leads (PISCES-Quad Plus, Model 3888) + external transmitter (X-trel, Model 3425) | Locomotor training | 7 | 5 | 140 | Reduction in sense of effort for over ground walking from 8/10 to 3/10 (Borg scale), and was able to double his walking speed reached maximum walking speeds of 0.35 m/s, and was able to ambulate over 325 m |

*AIS*, American Spinal Injury Association Impairment Scale.
The treatment regimens used experimentally in laboratory models or clinically in patients with spinal cord injury of which implanted spinal cord stimulators were applied: subject number (*n*), diagnosis, stimulators, training parameters, and outcomes.

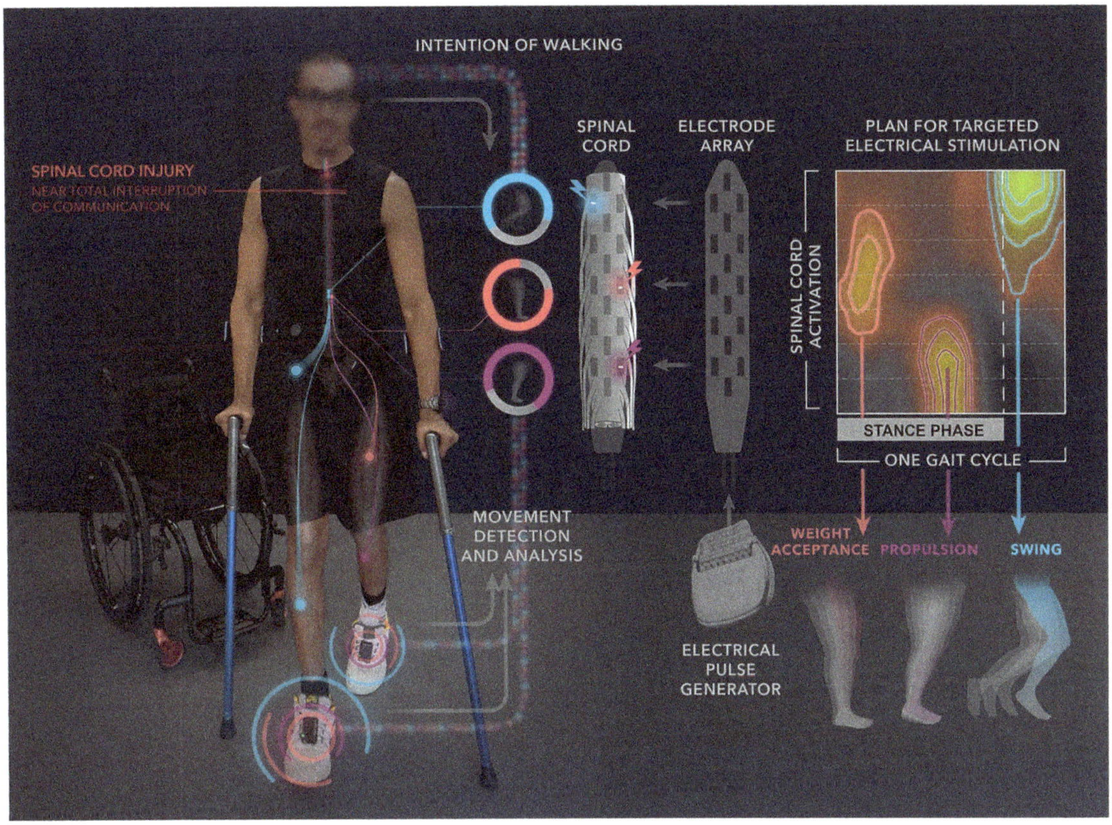


FIG. 3 Enabling over-ground walking with SCS in SCI patients. Stimulation bursts matching the spatial and temporal dynamics of the neural network reproduces natural activation of the spinal cord and enables locomotion in SCI. The spinal cord activation map was reconstructed based on the projection of electromyographic signals on an estimated location of the neural circuit in the spinal cord. *(Reprinted with permission from Cho, N., Squair, J. W., Bloch, J., & Courtine, G. (2019). Neurorestorative interventions involving bioelectronic implants after spinal cord injury. Bioelectronic Medicine, 5(1), 10. https://doi.org/10.1186/s42234-019-0027-x.)*

unmatched flexibility of lead configuration to adapt the possible change of the pain patterns in the future. Besides, Spectra also provides a series of advanced control technologies.

Proclaim XR is a recharge-free SCS system developed by Abbott Laboratories. A long-lasting battery of up to 10 years is a highlight in this SCSr. The battery longevity without any recharging is achieved by delivering low-dose stimulation. Proclaim XR applies BurstDR stimulation therapy, also known as burst stimulation, which imitates natural firing patterns of the human central nervous system to have better pain relief than tonic stimulation and potential benefits to various mechanisms of action (Kirketeig, Schultheis, Zuidema, Hunter, & Deer, 2019). BoldXR dosing protocol is used for patient-specific stimulation to ensure the optimal dose to have the best outcome and prevent the side effects of over-stimulation.

Restore by Medtronic is a family of FDA-approved SCSr that allows the patients to have an MRI scan. There are three IPGs with little differences under this series. Restoresensor Surescan stimulator features a customized pain management pattern and an automatic adjustment between low and high doses to achieve the optimal stimulation setting at the proper position. Restoreadvanced Surescan MRI stimulator specifically features the longest battery endurance among all Medtronic stimulators through a low dose programming. RestoreUltra SureScan allows safe access to the MRI scan, which is FDA-approved for 1.5-Tesla anywhere on the patient's body.

Omnia is a next-generation SCSr launched by Nevro Corporation. The selling point of Omnia is maximum versatility, able to deliver and combine all approved frequencies from 2 to 10,000 Hz, and claimed to have the broadest range of stimulation solution. It provides both direct neural pain inhabitation with a 10 kHz—high frequency (HF10) therapy and dorsal column stimulation at traditionally low frequency; or it can even pair high frequency with low frequency, HF10 with a burst to generate various blending waveforms for different individuals. Also, Omnia is claimed to have the fastest charging rate in SCSrs and is labeled with a full-body conditional MRI scan.

**TABLE 2** A comparison of commercial SCSrs.

| Device | Description | Specifications | Features |
|---|---|---|---|
|  **Boston Scientific**<br>Precision Spectra | Wider pain coverage with 32 contacts and 32 power sources. Flexible configurations realized by 4 ports and various leads | *Medical use*: Pain relief<br>*Max. freq.*: 1200 Hz<br>*Max. pulse wid.*: 1000 μs | Customized stim.<br>4 ports \| 32 electrodes<br>1.5 T MRI head only |
| 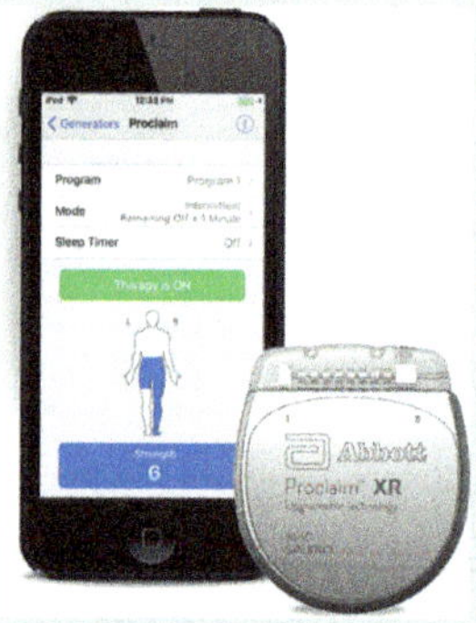 **Abbott Laboratories**<br>Proclaim XR | Low energy burst stimulation to prevent overstimulation. Long-lasting battery which can run up to 10 years without any recharging | *Medical use*: Pain relief<br>*Max. freq.*: 1200 Hz<br>*Max. pulse wid.*: 1000 μs | BurstDR + tonic stim.<br>2 ports \| 16 electrodes<br>1.5 T head and extremity |
| 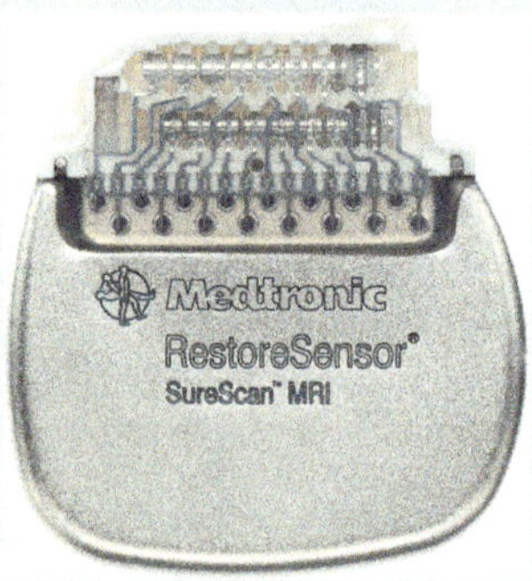 **Medtronic**<br>Restoresensor Surescan MRI | FDA approved full-body MRI scan eligibility. AdaptiveStim technology providing personalized pain management | *Medical use*: Pain relief<br>*Max. freq.*: 1.2 kHz<br>*Max. pulse wid.*: 1 ms | Low + high dose stim.<br>2 ports \| 16 electrodes<br>1.5 T MRI whole body |
|  **Nevro**<br>Omnia | All approved frequencies are available. Various waveforms, dual mechanisms available for more flexible therapy to relieve pain | *Medial use*: Pain relief<br>*Max. freq.*: 10 kHz<br>*Max. pulse wid.*: no info. | HF10 + Burst + Tonic stim.<br>No info. about electrodes<br>1.5-T MRI full body |

Four common commercial SCSrs and their basic information: Precision Spectra by Boston Scientific; Proclaim XR by Abbott Laboratories; Restoresensor Surescan MRI by Medtronic; and Omnia by Nevro.

## Surgical implantation of neurostimulators

For chronic use, an SCSr is needed to be implanted into the body, which is an invasive procedure that commonly requires surgery. Hence, SCS usually requires candidate evaluation before implantation. Usually, a patient receiving more than 50% pain relief in the trial with all other criteria reached is regarded as a suitable candidate; otherwise, they are not good candidates for the stimulator implantation (Mammis, 2016). In this section, different components of SCSr implantation procedures are introduced.

As mentioned in the previous section, an SCSr commonly have three main components: the leads, an implantable pulse generator (IPG), and an external wireless programmer (Fig. 4). There are also three main types of pulse generators: the conventional one requires another surgery to replace the battery once it runs out, the rechargeable one with a battery that can be recharged without any surgery, and the radiofrequency one with a battery outside the body.

In the trial stimulation, the injection site is first given a local anesthetic. Then insulated leads are inserted through a needle into the epidural space of the spinal cord, guided by fluoroscopy. The electrodes on the lead terminal are then

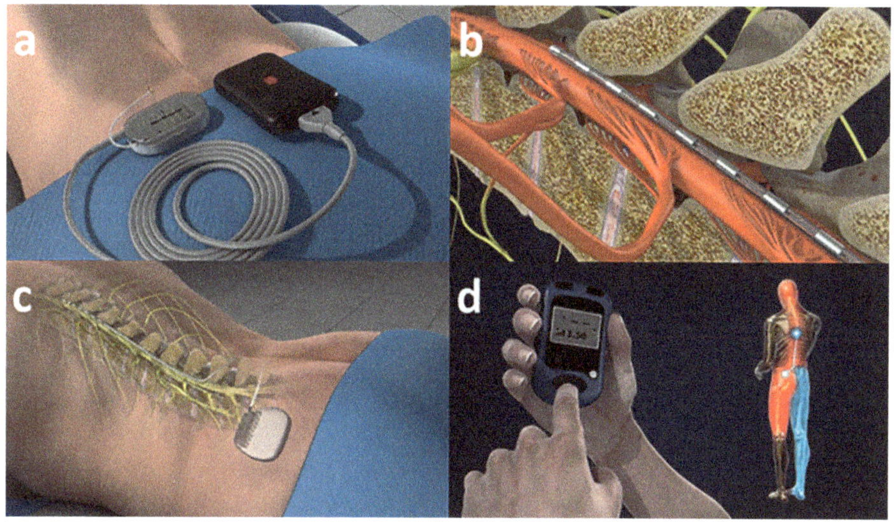


FIG. 4 Surgical steps of a spinal cord stimulation system. (A) Trial stimulation with a temporary SCSr. (B) Insertion of the lead electrodes into the epidural space. (C) Implantation of the IPG at the buttock. (D) Using a wireless programmer for operating the stimulator. *(Printed with permission from ViewMedica.)*

intermittently activated to generate electrical impulses to block the pain signals. During this procedure, the patient gives feedback to help the physician determine the best placement of the electrodes. The lead is then linked to an external trial stimulator for a trial lasting for about 1 week to evaluate SCS efficacy. If the SCS shows an acceptable result during the trial period, permanent implantation may be conducted. The trial leads are removed, and similarly, the patient is anesthetized, and the permanent leads are inserted into the epidural space. A small incision is then created, and the implantable pulses generator (IPG) is implanted in the buttocks. The leads are connected to the IPG. Finally, the external programmer can be used to turn on or off the generator and adjust the stimulation power level and other settings.

## Significance of SCS

Motor function deficit and chronic back and leg pain remain critical problems in spinal disorders, affecting millions of patients worldwide. They severely affect patients' self-independences and life qualities, yet traditional medical treatments such as pharmaceutical and surgical interventions often fail to achieve satisfactory results. In contrast, SCS shows significant performance in treating neuropathic pains for which conservative treatments have failed and promising efficacy on motor function improvement after clinically complete or incomplete SCI. The application of SCS on pain relief has been relatively mature, with many studies proving its effectiveness and various kinds of products available on the market. Furthermore, as the functional mechanism is well understood, emerging stimulations with different frequencies, waveforms, and patterns are used to achieve user-specific and more effective therapies.

SCS also demonstrates its significance in motor function restoration. In previous studies, SCS has been used to improve animal models and humans' motor function after an SCI (Wenger et al., 2016). The paraplegic patients were able to conduct voluntary leg movements such as standing, stepping over the ground, or bladder control via SCS. Closed-loop neurostimulation has also shown promises for optimal micturition for SCI (Peh et al., 2018). Sufficient muscle activities related to these motions could be measured during the SCS. Furthermore, after a period of task-specific training with SCS, the patients presented an improvement in standing and stepping independently, even when the SCS was removed. This indicates that the motor function restoration can persist beyond the SCS period. Neuroplasticity plays a significant role in these permanent recoveries (Alam, Rodrigues, Pham, & Thakor, 2016; Ling, Alam, & Zheng, 2019). Based on breakthroughs in the existed studies, SCS is promising to be developed into effective tools for functional restoration following SCI, as the functional mechanism behind them is currently being investigated more deeply.

## Current limitations of the neurostimulators

There has already been a wide variety of SCSrs available on the market at present. As mentioned in the previous section, companies such as Boston Scientific, Abbott Laboratories, Medtronic, and Nevro have offered their own SCSr with unique features. These stimulators present various characteristics, such as large pain coverage, long-lasting battery, MRI compatibility, flexible therapy, etc. However, the current stimulators still have some limitations (Alam et al., 2019; Li et al., 2020). One of them is the invasiveness of the stimulator implantation via surgery, implicating the risks of bleeding, allergic response, infection, etc. The trial stimulations indicate extra implantation; meanwhile, the change of patient's pain condition needs stimulation adjustment and might require additional surgeries for new stimulator implantations, especially for those stimulators unable to provide designable stimulation patterns. Another problem is the device-related problem such as lead migration, connection failure, or damage during use.

A more critical problem of all existed stimulators on the market is their application scope, designing purposely for pain relief. It means that the stimulation pattern, frequency, or waveform of the stimulators are not designed for functional restoration. Hence, SCSrs' application in SCI patients is mostly limited since the pain relief is more likely the emphasis in treating SCI compared with the restoration of function.

## Advanced neurostimulators

Besides the current SCSr mentioned above, a recently developed injectable stimulator developed by Stimwave Technologies, Freedom-8A, is a futuristic SCS system, which may overcome some limitations of the existed SCSrs (Billet, Wynendaele, & Vanquathem, 2018). Unlike traditional stimulators requiring bulky implantation, Freedom-8A is a wireless system with only one micro-size component to inject into the epidural space. Without a battery and wires accompanied, Freedom-8A is much smaller and does not require an IPG, which makes it less invasive and reduces the implants' paresthesia. Meanwhile, Freedom-8A can provide up to 64 electrode contacts, twofold of Precision Spectra mentioned previously, and 3.0 T full-body MRI compatibility. It also provides programmable stimulation with a frequency of up to 10 kHz. By integrating features of existed stimulators with its own novelties, Freedom-8A is offering a minimally invasive all-in-one SCS solution for SCI patients. A large clinical outcome of this stimulator is, however, yet to be reported.

GTX Medical, a newly founded spin-off from the Swiss federal institute of technology Lausanne (EPFL), is developing a new generation SCSr for targeted epidural SCS therapy to promote recovery of motor functions and neurologic control in adults with SCI and paralysis based on the recent research findings (Wagner et al., 2018). Their implantable Go-2 system, which is designed to deliver targeted epidural SCS therapy, has recently (June 2020) granted United States Food and Drug Administration (FDA) breakthrough device designation. The Company anticipates the first clinical trial for the complete Go-2 system in patients to take place in 2021.

## Prospects of the stimulation for spinal cord injury rehabilitation

A non-invasive SCS could be the solution to the problems that arise from the invasiveness of the SCSrs. Some studies found that besides the epidural SCS, transcutaneous SCS, which is non-invasive without requiring any surgeries, could also enable mobility after SCI and may confer similar outcomes to epidural SCS if suitable parameters are adopted (Gerasimenko, Kozlovskaya, & Edgerton, 2016; Ievins & Moritz, 2017). If the non-invasive transcutaneous SCS could achieve comparable efficacy on motor function improvement to the epidural SCS and even replace it in the future, a great number of complications caused by surgery will be readily solved.

Until today, most commercially available SCSrs are working on pain relief when the application on motor function restoration after SCI is still at the experimental stage. Although many previous results have demonstrated the improvement of motor function for paralysis following SCI during the stimulation period, the preferable breakthrough in the future may be that the SCS training could achieve sustained and unrelenting motion improvement, which means after a period of training the patients could persist the mobility even beyond the period of stimulation (Ievins & Moritz, 2017). More investigations on the mechanism behind SCS induced functional improvement are needed to achieve this goal. Also, the efficacy of SCS on the recovery of more complex movements, such as dexterity control, continued walking, balance, requires further research study and clinical trial (García, Serrano-Muñoz, Taylor, Avendaño-Coy, & Gómez-Soriano, 2019). Lastly, combining SCS with pharmaceutic and cellular interventions may also lead to effective treatment to improve voluntary movement after SCI in the future.

## Applications to other areas of neuroscience

In this chapter, we have reviewed the application of SCS by the SCSrs on improving motor functions and relieving chronic pain caused by SCI. We also discussed the advancements and limitations of the existing stimulators and then prospected on the futuristic stimulators in these processes.

The application of SCS also demonstrates high efficacy for pain relief in other neurological or spinal disorders, especially those that fail the conventional treatment approaches. These include failed back surgery syndrome (FBSS), complex regional pain syndrome (CRPS), others such as arachnoiditis, heart pain, nerve-related pain, peripheral vascular diseases, post-amputation pain, and visceral abdominal pain and perineal pain. Indeed, the sources of pain in the above disorders could be very different from the pain of the SCI. However, due to the nature of SCS pain relief, blocking the pain signals from reaching the brain by generating electric pulse, it is reasonable that SCS can be used for other chronic pain relief.

A recent study assessed a group of type I CRPS patients on pain relief and their life qualities after SCS implantation. The result shows that pain relief occurs in most subjects after SCS (Mammis, 2016). While another review on 80 cases of SCS treatments shows that at a 12 months follow-up, 40% of patients did not require medication anymore, while the rest claimed that the pain became manageable accompanied by using a first-line analgesic (Gopal, Fitzgerald, & McCrory, 2016). Similarly, various studies have shown evidence that SCS has clinical effectiveness on FBSS pain relief. One trial applies traditional SCS surgery, which is low-frequency SCS, for a group of subjects who have suffered from unrelenting post-spinal-surgery back pain for more than 3 months. The results collected over more than 2.5 years show about 50% pain relief for 50% of patients (Kumar et al., 2005; Kumar et al., 2007; North, Kidd, Farrokhi, & Piantadosi, 2005).

## Mini-dictionary of terms

**Spinal cord injury:** The damage to the spinal cord. It is often caused by trauma, and the severe one can result in paralysis.

**Neuromodulation:** The alteration of nerve activity via a certain stimulus such as electrical stimulation, electromagnetic stimulation, and chemical agents.

**Spinal cord stimulation:** The use of electric impulses to stimulate the region of the spinal cord for a certain purpose. Spinal cord stimulation can be either invasive or non-invasive.

**Spinal cord stimulator:** The device that delivers spinal cord stimulation.

**Complete SCI:** The spinal cord injury causing the total loss of communication between the brain and nerves below the point of the injury. It implicates paralysis.

**Incomplete SCI:** The spinal cord injury in which there is still residual motor or sensory communication between the brain and nerves below the point of the injury. It implicates a paresis.

**Neuropathway:** The connection formed by axons to enable signals to be sent from one region to another in the nervous system.

**Paresthesia:** An abnormal sensation. The cause of paresthesia could be various.

**Epidural space:** The space between the dura mater and the vertebral wall, just above the dural sac.

**Chronic pain:** The pain lasting for over 3 months, usually coming and going without clearly know reasons.

## Key facts of spinal cord injury

- Millions of patients are currently suffering from spinal cord injury (SCI) worldwide, and an estimated 768,473 new cases occur every year.
- There are mainly two types of SCI: traumatic and non-traumatic SCI, when over 90% are traumatic SCI.
- Primary and secondary injury cascades in traumatic SCI cause the death of neurons and glial cells, resulting in long-term neurological defects.
- SCI can cause disabilities: paraplegia and quadriplegia are commonly seen depending on the injury level, sometimes accompanied by pains.
- Complete SCI features the complete loss of sensory or motor functions below the injury site when incomplete SCI has limited function through the residual connection between the brain and nerves below the injury.

## Summary points

- Spinal cord injury causes paralysis and chronic pain, affecting millions of patients' self-independence and life quality worldwide.
- Therapeutic and surgical interventions show limited efficacy in restoring patients' functions after SCI.
- Spinal cord stimulation presents a stunning performance in treating chronic back or leg pain and promising effectiveness in improving motor functions after SCI.
- In some studies, patients with motor complete SCI were able to conduct locomotor functions such as weight-bearing standing and stepping over the ground during SCS.

- There are various kinds of spinal cord stimulators on the market targeting chronic pain relief. These stimulators possess large pain coverage, flexible and customized therapy, high-frequency stimulation, and MRI compatibility to achieve a better outcome and user experience.
- There are few spinal cord stimulators specially designed for motor function improvement on the market yet. Further understanding of the mechanism behind SCS will be needed to develop safe and effective stimulators for improving motor functions for SCI.

## References


Ahuja, C. S., Wilson, J. R., Nori, S., Kotter, M. R. N., Druschel, C., Curt, A., et al. (2017). Traumatic spinal cord injury. *Nature Reviews Disease Primers*, *3*(1), 17018. https://doi.org/10.1038/nrdp.2017.18.

Alam, M., & He, J. (2014). Lower-limb neuroprostheses: Restoring walking after spinal cord injury. In G. R. Naik, & Y. Guo (Eds.), *Emerging theory and practice in neuroprosthetics* (pp. 153–180). IGI Global. https://doi.org/10.4018/978-1-4666-6094-6.ch009.

Alam, M., Li, S., Ahmed, R. U., Yam, Y. M., Thakur, S., Wang, X.-Y., et al. (2019). Development of a battery-free ultrasonically powered functional electrical stimulator for movement restoration after paralyzing spinal cord injury. *Journal of Neuro Engineering and Rehabilitation*, *16*(1), 36. https://doi.org/10.1186/s12984-019-0501-4.

Alam, M., Ling, Y. T., Wong, A. Y. L., Zhong, H., Edgerton, V. R., & Zheng, Y.-P. (2020). Reversing 21 years of chronic paralysis via non-invasive spinal cord neuromodulation: A case study. *Annals of Clinical and Translational Neurology*, *7*(5), 829–838. https://doi.org/10.1002/acn3.51051.

Alam, M., Rodrigues, W., Pham, B. N., & Thakor, N. V. (2016). Brain-machine interface facilitated neurorehabilitation via spinal stimulation after spinal cord injury: Recent progress and future perspectives. *Brain Research*, *1646*, 25–33. https://doi.org/10.1016/j.brainres.2016.05.039.

Alam, M., & Zheng, Y.-P. (2017). Motor neuroprosthesis for injured spinal cord: Who is an ideal candidate? *Neural Regeneration Research*, *12*(11), 1809–1810. https://doi.org/10.4103/1673-5374.219041.

Alizadeh, A., Dyck, S. M., & Karimi-Abdolrezaee, S. (2019). Traumatic spinal cord injury: An overview of pathophysiology, models and acute injury mechanisms [review]. *Frontiers in Neurology*, *10*(282). https://doi.org/10.3389/fneur.2019.00282.

Angeli, C. A., Boakye, M., Morton, R. A., Vogt, J., Benton, K., Chen, Y., et al. (2018). Recovery of over-ground walking after chronic motor complete spinal cord injury. *New England Journal of Medicine*, *379*(13), 1244–1250. https://doi.org/10.1056/NEJMoa1803588.

Angeli, C. A., Edgerton, V. R., Gerasimenko, Y. P., & Harkema, S. J. (2014). Altering spinal cord excitability enables voluntary movements after chronic complete paralysis in humans. *Brain*, *137*(5), 1394–1409. https://doi.org/10.1093/brain/awu038.

Billet, B., Wynendaele, R., & Vanquathem, N. E. (2018). A novel minimally invasive wireless technology for neuromodulation via percutaneous intercostal nerve stimulation for post-herpetic neuralgia: A case report with short-term follow-up. *Pain Practice*, *18*(3), 374–379. https://doi.org/10.1111/papr.12607.

Carhart, M. R., Jiping, H., Herman, R., Luzansky, S. D., & Willis, W. T. (2004). Epidural spinal-cord stimulation facilitates recovery of functional walking following incomplete spinal-cord injury. *IEEE Transactions on Neural Systems and Rehabilitation Engineering*, *12*(1), 32–42. https://doi.org/10.1109/TNSRE.2003.822763.

Cho, N., Squair, J. W., Bloch, J., & Courtine, G. (2019). Neurorestorative interventions involving bioelectronic implants after spinal cord injury. *Bioelectronic Medicine*, *5*(1), 10. https://doi.org/10.1186/s42234-019-0027-x.

Courtine, G., Gerasimenko, Y., van den Brand, R., Yew, A., Musienko, P., Zhong, H., et al. (2009). Transformation of nonfunctional spinal circuits into functional states after the loss of brain input. *Nature Neuroscience*, *12*(10), 1333–1342. https://doi.org/10.1038/nn.2401.

Courtine, G., & Sofroniew, M. V. (2019). Spinal cord repair: Advances in biology and technology. *Nature Medicine*, *25*(6), 898–908. https://doi.org/10.1038/s41591-019-0475-6.

Darrow, D., Balser, D., Netoff, T. I., Krassioukov, A., Phillips, A., Parr, A., et al. (2019). Epidural spinal cord stimulation facilitates immediate restoration of dormant motor and autonomic supraspinal pathways after chronic neurologically complete spinal cord injury. *Journal of Neurotrauma*, *36*(15), 2325–2336. https://doi.org/10.1089/neu.2018.6006.

del Popolo, G., Mencarini, M., Nelli, F., & Lazzeri, M. (2012). Controversy over the pharmacological treatments of storage symptoms in spinal cord injury patients: A literature overview. *Spinal Cord*, *50*(1), 8–13. https://doi.org/10.1038/sc.2011.110.

Dietz, V., & Fouad, K. (2014). Restoration of sensorimotor functions after spinal cord injury. *Brain*, *137*(Pt 3), 654–667. https://doi.org/10.1093/brain/awt262.

Dijkers, M., Bryce, T., & Zanca, J. (2009). Prevalence of chronic pain after traumatic spinal cord injury: A systematic review. *Journal of Rehabilitation Research and Development*, *46*(1), 13–29. https://doi.org/10.1682/JRRD.2008.04.0053.

DiMarco, A. F., Geertman, R. T., Tabbaa, K., Nemunaitis, G. A., & Kowalski, K. E. (2020). Effects of lower thoracic spinal cord stimulation on bowel management in individuals with spinal cord injury. *Archives of Physical Medicine and Rehabilitation*. https://doi.org/10.1016/j.apmr.2020.09.394.

Dimitrijevic, M. R., Gerasimenko, Y., & Pinter, M. M. (1998). Evidence for a spinal central pattern generator in humans. *Annals of the New York Academy of Sciences*, *860*, 360–376. https://doi.org/10.1111/j.1749-6632.1998.tb09062.x.

Edgerton, V. R., & Harkema, S. (2011). Epidural stimulation of the spinal cord in spinal cord injury: Current status and future challenges. *Expert Review of Neurotherapeutics*, *11*(10), 1351–1353. https://doi.org/10.1586/ern.11.129.

Gaber, T., & Brown, M. (2020). Recent advances in neuromodulation for spinal cord injuries. *Progress in Neurology and Psychiatry*, *24*(1), 4–8. https://doi.org/10.1002/pnp.554.

García, A., Serrano-Muñoz, D., Taylor, J., Avendaño-Coy, J., & Gómez-Soriano, J. (2019). Transcutaneous spinal cord stimulation and motor rehabilitation in spinal cord injury: A systematic review. *Neurorehabilitation and Neural Repair*, *34*. https://doi.org/10.1177/1545968319893298, 154596831989329.

Gerasimenko, Y., Kozlovskaya, I., & Edgerton, V. R. (2016). Sensorimotor regulation of movements: Novel strategies for the recovery of mobility. *Human Physiology*, *42*(1), 90–102. https://doi.org/10.1134/S0362119716010096.

Gill, M. L., Grahn, P. J., Calvert, J. S., Linde, M. B., Lavrov, I. A., Strommen, J. A., et al. (2018). Neuromodulation of lumbosacral spinal networks enables independent stepping after complete paraplegia. *Nature Medicine*, *24*(11), 1677–1682. https://doi.org/10.1038/s41591-018-0175-7.

Giugno, A., Gulì, C., Basile, L., Graziano, F., Maugeri, R., Visocchi, M., et al. (2017). Spinal cord stimulation: An alternative concept of rehabilitation? *Acta Neurochirurgica. Supplement*, *124*, 15–18. https://doi.org/10.1007/978-3-319-39546-3_3.

Gopal, H., Fitzgerald, J., & McCrory, C. (2016). Spinal cord stimulation for FBSS and CRPS: A review of 80 cases with on-table trial of stimulation. *Journal of Back and Musculoskeletal Rehabilitation*, *29*, 7–13. https://doi.org/10.3233/BMR-150608.

Grahn, P. J., Lavrov, I. A., Sayenko, D. G., Van Straaten, M. G., Gill, M. L., Strommen, J. A., et al. (2017). Enabling task-specific volitional motor functions via spinal cord neuromodulation in a human with paraplegia. *Mayo Clinic Proceedings*, *92*(4), 544–554. https://doi.org/10.1016/j.mayocp.2017.02.014.

Harkema, S., Gerasimenko, Y., Hodes, J., Burdick, J., Angeli, C., Chen, Y., et al. (2011). Effect of epidural stimulation of the lumbosacral spinal cord on voluntary movement, standing, and assisted stepping after motor complete paraplegia: A case study. *Lancet (London, England)*, *377*(9781), 1938–1947. https://doi.org/10.1016/S0140-6736(11)60547-3.

Harkema, S. J., Wang, S., Angeli, C. A., Chen, Y., Boakye, M., Ugiliweneza, B., et al. (2018). Normalization of blood pressure with spinal cord epidural stimulation after severe spinal cord injury [original research]. *Frontiers in Human Neuroscience*, *12*(83). https://doi.org/10.3389/fnhum.2018.00083.

Ievins, A., & Moritz, C. T. (2017). Therapeutic stimulation for restoration of function after spinal cord injury. *Physiology*, *32*(5), 391–398. https://doi.org/10.1152/physiol.00010.2017.

Ireland, R. (2010). *A dictionary of dentistry*. Oxford University Press. https://doi.org/10.1093/acref/9780199533015.001.0001.

James, N., McMahon, S., Field-Fote, E., & Bradbury, E. (2018). Neuromodulation in the restoration of function after spinal cord injury. *Lancet Neurology*, *17*, 905–917. https://doi.org/10.1016/S1474-4422(18)30287-4.

Kirketeig, T., Schultheis, C., Zuidema, X., Hunter, C. W., & Deer, T. (2019). Burst spinal cord stimulation: A clinical review. *Pain Medicine*, *20*(Suppl. 1), S31–s40. https://doi.org/10.1093/pm/pnz003.

Kumar, K., North, R., Taylor, R., Sculpher, M., Van den Abeele, C., Gehring, M., et al. (2005). Spinal cord stimulation vs. conventional medical management: A prospective, randomized, controlled, multicenter study of patients with failed back surgery syndrome (PROCESS study). *Neuromodulation: Technology at the Neural Interface*, *8*(4), 213–218. https://doi.org/10.1111/j.1525-1403.2005.00027.x.

Kumar, K., Taylor, R. S., Jacques, L., Eldabe, S., Meglio, M., Molet, J., et al. (2007, Nov). Spinal cord stimulation versus conventional medical management for neuropathic pain: A multicentre randomised controlled trial in patients with failed back surgery syndrome. *Pain*, *132*(1–2), 179–188. https://doi.org/10.1016/j.pain.2007.07.028.

Li, S., Alam, M., Ahmed, R. U., Zhong, H., Wang, X.-Y., Ng, S., et al. (2020). Ultrasound-driven piezoelectric current activates spinal cord neurocircuits and restores locomotion in rats with spinal cord injury. *Bioelectronic Medicine*, *6*(1), 13. https://doi.org/10.1186/s42234-020-00048-2.

Ling, Y. T., Alam, M., & Zheng, Y. P. (2019). Spinal cord injury: Lessons about neuroplasticity from paired associative stimulation. *Neuroscientist*, *26*(3), 266–277. https://doi.org/10.1177/1073858419895461.

Lu, D. C., Edgerton, V. R., Modaber, M., AuYong, N., Morikawa, E., Zdunowski, S., et al. (2016). Engaging cervical spinal cord networks to reenable volitional control of hand function in tetraplegic patients. *Neurorehabilitation and Neural Repair*, *30*(10), 951–962. https://doi.org/10.1177/1545968316644344.

Mammis, A. (2016). *Spinal cord stimulation: Principles and practice*. Nova Science Publishers, Incorporated. https://books.google.com.hk/books?id=QSCFjgEACAAJ.

North, R. B., Kidd, D. H., Farrokhi, F., & Piantadosi, S. A. (2005). Spinal cord stimulation versus repeated lumbosacral spine surgery for chronic pain: A randomized, controlled trial. *Neurosurgery*, *56*(1), 98–106. discussion 106–107 https://doi.org/10.1227/01.neu.0000144839.65524.e0.

Peh, W. Y. X., Raczkowska, M. N., Teh, Y., Alam, M., Thakor, N. V., & Yen, S.-C. (2018). Closed-loop stimulation of the pelvic nerve for optimal micturition. *Journal of Neural Engineering*, *15*(6). https://doi.org/10.1088/1741-2552/aadee9, 066009.

Possover, M. (2014). Recovery of sensory and supraspinal control of leg movement in people with chronic paraplegia: A case series. *Archives of Physical Medicine and Rehabilitation*, *95*(4), 610–614. https://doi.org/10.1016/j.apmr.2013.10.030.

Qian, Q., Ling, Y. T., Zhong, H., Zheng, Y.-P., & Alam, M. (2020). Restoration of arm and hand functions via noninvasive cervical cord neuromodulation after traumatic brain injury: A case study. *Brain Injury*, *34*(13–14), 1771–1780. https://doi.org/10.1080/02699052.2020.1850864.

Rejc, E., Angeli, C. A., Atkinson, D., & Harkema, S. J. (2017). Motor recovery after activity-based training with spinal cord epidural stimulation in a chronic motor complete paraplegic. *Scientific Reports*, *7*(1), 13476. https://doi.org/10.1038/s41598-017-14003-w.

Rejc, E., Angeli, C., & Harkema, S. (2015). Effects of lumbosacral spinal cord epidural stimulation for standing after chronic complete paralysis in humans. *PLoS One*, *10*(7). https://doi.org/10.1371/journal.pone.0133998, e0133998.

Roy, R. R., Harkema, S. J., & Edgerton, V. R. (2012). Basic concepts of activity-based interventions for improved recovery of motor function after spinal cord injury. *Archives of Physical Medicine and Rehabilitation*, *93*(9), 1487–1497. https://doi.org/10.1016/j.apmr.2012.04.034.

Sayenko, D. G., Angeli, C., Harkema, S. J., Edgerton, V. R., & Gerasimenko, Y. P. (2014). Neuromodulation of evoked muscle potentials induced by epidural spinal-cord stimulation in paralyzed individuals. *Journal of Neurophysiology*, *111*(5), 1088–1099. https://doi.org/10.1152/jn.00489.2013.

Shah, P. K., Sureddi, S., Alam, M., Zhong, H., Roy, R. R., Edgerton, V. R., et al. (2016). Unique spatiotemporal neuromodulation of the lumbosacral circuitry shapes locomotor success after spinal cord injury. *Journal of Neurotrauma*, *33*(18), 1709–1723. https://doi.org/10.1089/neu.2015.4256.

Silva, N. A., Sousa, N., Reis, R. L., & Salgado, A. J. (2014). From basics to clinical: A comprehensive review on spinal cord injury. *Progress in Neurobiology*, *114*, 25–57. https://doi.org/10.1016/j.pneurobio.2013.11.002.

van den Brand, R., Heutschi, J., Barraud, Q., DiGiovanna, J., Bartholdi, K., Huerlimann, M., et al. (2012). Restoring voluntary control of locomotion after paralyzing spinal cord injury. *Science*, *336*(6085), 1182–1185. https://doi.org/10.1126/science.1217416.

Wagner, F. B., Mignardot, J. B., Le Goff-Mignardot, C. G., Demesmaeker, R., Komi, S., Capogrosso, M., et al. (2018). Targeted neurotechnology restores walking in humans with spinal cord injury. *Nature*, *563*(7729), 65–71. https://doi.org/10.1038/s41586-018-0649-2.

Wenger, N., Moraud, E. M., Gandar, J., Musienko, P., Capogrosso, M., Baud, L., et al. (2016). Spatiotemporal neuromodulation therapies engaging muscle synergies improve motor control after spinal cord injury. *Nature Medicine*, *22*(2), 138–145. https://doi.org/10.1038/nm.4025.

Widerström-Noga, E., Biering-Sørensen, F., Bryce, T., Cardenas, D. D., Finnerup, N. B., Jensen, M. P., et al. (2008). The international spinal cord injury pain basic data set. *Spinal Cord*, *46*(12), 818–823. https://doi.org/10.1038/sc.2008.64.

Chapter 20

# Bowel dysfunction in spinal cord injury

María Elena Ferreiro-Velasco[a], Antonio Montoto-Marqués[a,b], Mª Eugenia Díaz-Recarey[a], Renato Vilas-Boas[a], Sebastian Salvador-De La Barrera[a], and Antonio Rodríguez-Sotillo[a,b]

[a]*Unidad de Lesionados Medulares, Complejo Hospitalario Universitario A Coruña, A Coruña, Spain;* [b]*Facultad de Ciencias de la Salud, Universidad de A Coruña, A Coruña, España*

## Abbreviations

| | |
|---|---|
| **AD** | autonomic dysreflexia |
| **BC** | bowel care |
| **CTT** | colonic transit time |
| **NBD** | neurogenic bowel dysfunction |
| **DRF** | digital removal of feces |
| **DRS** | digital rectal stimulation |
| **EAS** | external anal sphincter |
| **ENS** | enteric nervous system |
| **LMN** | lower motor neuron |
| **NB** | neurogenic bowel |
| **SARS** | sacral anterior root stimulation |
| **SC** | spinal cord |
| **SCI** | spinal cord injury |
| **UMN** | upper motor neuron |

## Introduction

Spinal cord injury (SCI) results in loss of motor, sensory and autonomic functions, causing dysfunction of practically all body systems leading to long-term complications (Table 1). Neurogenic bowel (NB) is one of these main impairments after spinal cord injury. It occurs because of changes in innervation of the colon and rectum, as well as the loss of voluntary control of the anal sphincter, affecting colonic motility and defecation maneuvers. Between 80% and 97% of people with SCI suffer at least one NB-related symptom (Correa & Rotter, 2000; Ozisler, Koklu, Ozel, & Unsal-Delialioglu, 2015), mostly constipation and/or fecal incontinence, but also a prolonged and difficult defecation, sense of incomplete emptying, abdominal pain and distension, autonomic dysreflexia (AD), hemorrhoids, and rectal prolapse. Twenty-seven percent of the subjects considers that bowel dysfunction is a greater problem than bladder or sexual dysfunction, and 38% perceive it limits their quality of life (QoL) or social activities (Faaborg, Christensen, Finnerup, Laurberg, & Krogh, 2008).

The level of injury, severity, and presence or absence of sacral reflexes influences NB clinical manifestations and its management. Constipation and incontinence tend to become chronic and are stubborn toward currently available treatments, being a relevant source of complications. This, and the amount of time people with spinal injury spend on their bowel care, causes neurogenic bowel dysfunction (NBD) to have a very negative effect on QoL.

## Characteristics of neurogenic bowel in spinal cord injury

Nervous control of the colon is done through the enteric, autonomic, and central nervous systems (Chung & Emmanuel, 2006; Lynch, Antony, Dobbs, & Frizelle, 2001). The enteric nervous system (ENS) is key for generating colonic propulsive activity. It functions independently, but is modulated by the spinal cord (SC) through the sympathetic fibers arising in the intermediolateral column of T5 to L3 and parasympathetic fibers that originate in segments S2 to S4. Sympathetic fibers inhibit colon motility, while parasympathetic fibers enable the contraction of the left colon, sigmoid, and rectum. The motor neurons in S2–S4 innervate the external anal sphincter (EAS), the puborectalis muscle, and the muscles of the pelvic floor;

Diagnosis and Treatment of Spinal Cord Injury. https://doi.org/10.1016/B978-0-12-822498-4.00020-8

**TABLE 1 Principal medical issues after spinal cord injury.**

| |
|---|
| Respiratory complications[a]<br>Atelectasis, pneumonia and respiratory failure |
| Cardiovascular complications[b]<br>Orthostatic hypotension, autonomic dysreflexia, bradycardia |
| Neurogenic bladder |
| Neurogenic bowel |
| Sexual and fertility dysfunction |
| Spasticity[c] |
| Chronic pain<br>Nociceptive pain<br>Neuropathic pain: Above the level, at the level or below the level of injury |
| Secondary immunodeficiency[d]<br>Increased susceptibility to infections |
| Osteoporosis<br>Increased risk of low impact fractures |
| Neurogenic heterotopic ossification[e] |
| Pressure sores |

[a]*Due to paralysis of the phrenic nerve, intercostal muscles and/or abdominal muscles leaving to reduced lung capacity and ineffective cough.*
[b]*Due to loss of supraspinal sympathetic control in patients with injury at T6 and above.*
[c]*Velocity-dependent increase in muscle tone with exaggerated deep tendon reflexes that results from injury to upper motor neurons.*
[d]*Systemic dysfunction of macrophages, T cells, B cells, and natural killer cells due to disruption of central nervous system input to immune organs.*
[e]*Ectopic bone formation in the connective tissue around joints.*

they receive voluntary control from motor neurons of the cerebral cortex and sensitive information from mechanoreceptors of the pelvic floor. SCI causes disturbances in autonomic and central innervation, and consequently loss of voluntary control of defecation, and changes in intestinal motility and sphincter tone. Various studies report a slowing of total and/or segmental colonic transit (Correa & Rotter, 2000; Vallès, Vidal, Clavé, & Mearin, 2006), increased colonic wall tone, changes in rectal compliance (Trivedi, Kumar, & Emmanuel, 2016) and in sphincter tone and sensitivity, as well as lack of relaxation of the anal sphincter in response to rectal distension (Lynch, Anthony, Dobbs, & Frizelle, 2000). Although the function of the ENS remains, it has been shown there is loss of neurons and decrease of nerve fiber density in the myenteric plexus (den Braber-Ymker, Lammens, van Putten, & Nagtegaal, 2017) after SCI. The inflammatory process and changes of extrinsic innervation may alter neuromuscular structures of the colon and contribute to decrease bowel function (White & Holmes, 2019). Not much research has been carried out on the changes of gut microbiota; however, it may well be a contributing factor and thus a target of study for the development of new treatments (Zhang et al., 2018).

Based on the level of SCI, two bowel dysfunction patterns may be distinguished. Lesions over the medullary cone cause upper motor neuron NB syndrome or *hyperreflexic bowel*, in which case there is loss of supraspinal control, but reflex coordination and stool propulsion are preserved. This syndrome is characterized by the increase in the tone of colonic and anal wall, the sphincter remains its tone or it becomes hypertonic, and spasticity of pelvic floor muscles may occur. It associates with constipation and fecal retention; incontinence due to overflow or reflex relaxation of the anal sphincter may also develop. Lesions in the medullary cone and cauda equina cause lower motor neuron NB syndrome or *areflexic bowel*, in which case there is loss of supraspinal control, reflex activity mediated by the spinal cord, and the sphincter is hypotonic. Colonic peristalsis occurs only because of the activity of the ENS, which results in an increase of colonic transit time (CTT). The areflexic bowel is associated to constipation, less difficulty during defecation, and high risk of incontinence due to EAS atony.

## Clinical manifestations

Constipation, fecal incontinence, pain, and abdominal distension are the main symptoms of NB in people with SCI. *Constipation* is associated with slowing colonic transit, failure to increase intra-abdominal pressure during defecation, absence of anal relaxation during the defecation, and increase of rectal and sigmoid compliance (Trivedi et al., 2016; Vallès & Mearin, 2009). Immobility and use of drugs such as opioids or anti-cholinergics, also favor constipation (Paily, Preziosi, Trivedi, & Emmanuel, 2019). Mean time for stool evacuation is longer, with 35% of SCI patients requiring more than an hour (Engkasan & Sudin, 2013; Inskip, Lucci, McGrath, Willms, & Claydon, 2018). *Fecal incontinence* is associated with the loss of voluntary control of the anal sphincter, atonic or hypotonic sphincter, low rectal compliance, low anal pressure at rest, and uninhibited rectal contractions (Lynch et al., 2001; Trivedi et al., 2016; Vallès & Mearin, 2009). It has been more frequently reported in subjects with lower motor neuron NB than in upper motor neuron NB (Inskip et al., 2018). Inadequate bowel management and use of laxatives also favor incontinence.

*Abdominal pain* is a frequent symptom, emerging later than neuropathic pain (in average 4.2 years after the injury) (Siddall, McClelland, Rutkowski, & Cousins, 2003). Its prevalence is over 30% in chronic SCIs (Nielsen, Faaborg, Christensen, Krogh, & Finnerup, 2017). It characterizes for being diffuse; some patients describe it as a prick, itch, warmth, or burning. It is closely associated with constipation, it worsens or improves with bowel movement.

*Abdominal distension and flatulence* are symptoms linked with constipation and fecal impaction. *Pathologies of the rectum and anus*, such as hemorrhoids, prolapse, anal fissures, rectal ulcers, and rectal bleeding, are favored by digital rectal stimulation (DRS), manual rectal evacuation (MRE), and the use of suppositories and enemas. Fecal impaction, anal fissures, and defecation routine per se are stimuli that trigger *autonomic dysreflexia* (Faaborg et al., 2014). Arterial hypertension may occur without accompanying symptoms that alert of AD (silent AD) (Kirshblum, House, & O'connor, 2002). The relevance of this lies on the fact that recurrent AD episodes associate with immunosuppression, involvement of cardiac mechanics, cerebrovascular changes, and cognitive impairment after SCI (Eldahan & Rabchevsky, 2018).

It is easy to understand that NBD due to its clinical manifestations and complications, negatively affect physical and psychological well-being, with a significant association between NBD severity and impact on QoL (Pires et al., 2018).

## Diagnosis

The *medical history* is key to know the symptoms and their severity, patients' pathological background, and the medication being received. It must include the presence of sensation of defecation, diet, physical activity, frequency of bowel movements, stool consistency, and tested strategies for bowel management. The Bristol scale is a useful instrument to assess stool consistency in a fast and simple way (Lewis & Heaton, 1997). The time spent in bowel toileting is particularly relevant; reducing this time is one of the aims of treatment. Recording a diary of defecation by de patient will allow knowing the response to therapeutic measures. Uniformity in data collection is important when comparing results. The International SCI Bowel Function Basic Data Set (Krogh et al., 2017) gathers the basic data on bowel function following SCI that may be used in clinical practice.

*Physical examination* includes general, neurological, and abdominal inspection, as well as of the anorectal area. Other aspects to assess are the functional ability, balance in sitting position, and presence of spasticity. *Examination of the anorectal area* allows assessing the sensitivity, anal sphincter tone, and voluntary contraction. The presence of anal and bulbocavernosus sacral reflexes are indicative of anorectal reflex activity.

General *blood tests* are very useful to assess patients' general condition, detect warning signs, or endocrine and metabolic disorders that may contribute to bowel dysfunction.

*Plain abdominal X-ray* is useful to assess the amount and distribution of stools in the colon and detect megacolon, fecal impaction, and volvulus.

*Colonoscopy* has the same indication as for the general population, although it requires a more intensive gut preparation than in subjects who are neurologically intact (Song, Svircev, Teng, Dominitz, & Burns, 2018).

Other tests not used routinely are the *measurement of CTT* (total and segmental) and *anorectal manometry*, which may help identify the primary physiopathological mechanism of the symptoms.

### Assessment scales

Multiple scales allow measuring the symptoms in bowel dysfunction, primarily constipation and incontinence, as well as the impact it has on QoL, although few have been validated for the SCI population. The Cleveland Clinic Constipation Scoring System (Agachan, Chen, Pfeifer, Reissman, & Wexner, 1996) and the Wexner Continence Grading Scale (Jorge & Wexner, 1993), are frequently used in studies with SCI patients. The 36-Item Short Form Survey (Ware & Sherbourne, 1992) and the Fecal Incontinence Quality of Life questionnaire (Rockwood et al., 2000) are widely used instruments in QoL studies.

The Neurogenic Bowel Dysfunction Score (Table 2) is a scale developed and validated specifically for adult SCI patients (Krogh, Christensen, Sabroe, & Laurberg, 2006). It allows scoring the clinical symptoms of colorectal and anal

**TABLE 2** Neurogenic bowel dysfunction score.

| Items | Points |
|---|---|
| (1) Frequency of defecation | |
| Daily (0)<br>Two to six times every week (1)<br>Less than once a week (6) | |
| (2) Time used for each defecation | |
| 0–30 min (0)<br>31–60 min (3)<br>More than 1 h (7) | |
| (3) Uneasiness, headache, or perspiration during defecation | |
| No (0)<br>Yes (2) | |
| (4) Regular use of tablets against constipation | |
| No (0)<br>Yes (2) | |
| (5) Regular use of drops against constipation | |
| No (0)<br>Yes (2) | |
| (6) Digital stimulation or evacuation of the anorectum | |
| Less than once every week (0)<br>Once or more every week (6) | |
| (7) Frequency of fecal incontinence | |
| Less than once every month (0)<br>One to four times every month (6)<br>One to six times every week (7)<br>Daily (13) | |
| (8) Medication against fecal incontinence | |
| No (0)<br>Yes (4) | |
| (9) Flatus incontinence | |
| No (0)<br>Yes (2) | |
| (10) Perianal skin problems | |
| No (0)<br>Yes (3) | |
| Total NBD score (range 0–47) | |
| **NBD score** | **Bowel dysfunction** |
| 0–6 | Very minor |
| 7–9 | Minor |
| 10–13 | Moderate |
| 14 or more | Severe |

10-Item score scale. A score of 0–47, the higher score indicates more severe bowel symptoms.
Modified from Krogh, K., Christensen, P., Sabroe, S., & Laurberg, S. (2006). Neurogenic bowel dysfunction score. *Spinal Cord, 44*(10), 625–631.

dysfunction; the importance of each item was established based on the strength of its association with self-perceived QoL. The greater the scoring, the greater the impact on QoL (Adriaansen, van Asbeck, van Kuppevelt, Snoek, & Post, 2015).

## Conservative treatment of the neurogenic bowel

### General recommendations

First, it is convenient to check the medications the patient is taking and make adjustments on those that might be contributing to the symptoms. Analgesics, anti-cholinergics, anti-depressants, and anti-psychotics, as well as iron supplements and aluminum compounds, which favor constipation. Excessive use of laxatives favors fecal incontinence, abdominal pain, and flatulence. Assessing the eating habits of the patient is also important.

### Diet

Stool volume and consistency can be regulated through diet. Low *fluid* intake increases the probability of suffering constipation. On the other hand, high intake (>2 L/day), may be associated with prolonged bowel care (Engkasan & Sudin, 2013). Recommendation on fluid intake should be tailored and balanced with the type of bladder management.

*Fiber* helps increase fecal mass and its capacity to retain water. Consequently, colonic transit is accelerated and feces become softer. Although there are no accepted recommendations for SCI, an initial daily intake of 15 g of dietary fiber is suggested, monitoring the effect, and if necessary, increase the amount of fiber slowly and gradually, taking into account that high-fiber diets (30 g/day) may increase CTT in these patients (Cameron, Nyulasi, Collier, & Brown, 1996).

### Physical exercise

Regular *physical exercise* may improve constipation symptoms, as it affects bowel motility, speeding up the colonic and gut transit time, and stimulates abdominal muscles helping move fecal matter toward the rectum (Gao et al., 2019). In people with SCI, standing seems to improve bowel function, although this has not been demonstrated in clinical trials (Kwok et al., 2015).

### Abdominal massage

The mechanisms suggested to explain the effect of abdominal massage on the bowel are the increase of abdominal pressure and improvement of peristalsis. It may be performed during bowel routine to help pass stool. Fifteen minutes of *abdominal massage* following digital stimulation, decreases CTT, abdominal distension and fecal incontinence, and increases the frequency of defecation (Ayaş, Leblebici, Sözay, Bayramoğlu, & Niron, 2006). Similarly, *electrical stimulation of the abdominal muscles* during the bowel care program shortens the time until the feces come out and total time of bowel care (Korsten et al., 2004).

### Mechanical techniques to ease bowel movements

***Digital rectal stimulation*** (DRS) dilates the anal canal relaxing the puborectalis muscle, which reduces the resistance of stool to come out, triggers rectal contraction, and increases contractile activity of the descending and sigmoid colon (Korsten et al., 2007). ***Manual rectal evacuation*** (MRE) of feces is the most frequent form of bowel evacuation in some series (Coggrave, Norton, & Wilson-Barnett, 2009; Savic, Frankel, Jamous, Soni, & Charlifue, 2018). It is often the only effective way to empty the bowel in patients with areflexic bowel. In subjects with reflex function, MRE may be necessary to complete the emptying process (Coggrave et al., 2009) (Table 3).

### Pharmacological treatment

Drugs that act on the colon are primarily used, aiming to increase its motility or change the texture of the stools (Table 4). Rectal administration of medication is used to ease defecation.

**TABLE 3** Interventions for defecation.

| Mechanical interventions | | | |
|---|---|---|---|
| Digital rectal stimulation | | | |
| Manual evacuation (digital removal of feces) | | | |
| **Pharmacological interventions** | | | |
| | **Presentation** | **Mechanism of action** | **Characteristics** |
| Glycerine | Suppositories, microenemas (5–10 mL) | Stimulant and osmotic laxative | Effective in 15–30 min<br>Side effects: Irritation, stinging, anal itching |
| Bisacodyl (with a polyethylene glycol base or a hydrogenated vegetable oil base) | Suppositories | Stimulant laxative | Effective in 10–30 min<br>Side effects: Abdominal pain and diarrhea |
| Docusate sodium or calcium | Microenemas (5–10 mL) | Stool softener | Effective in 2–15 min |
| Sodium citrate, sodium lauryl sulfoacetate, sorbitol | Microenemas (5–10 mL) | Osmotic laxative | Effective in 5–15 min<br>Side effects: Abdominal pain and diarrhea |
| Sodium phosphate | Enemas (80, 140, 250 mL) | Saline laxative | Effective in 1–5 min<br>Side effects: Abdominal bloating and pain, vomiting, renal impairment |

Mechanical and pharmacological techniques to trigger defecation. Unpublished.

**TABLE 4** Pharmacological treatment of neurogenic bowel.

| | Mechanism of action | Active principle | Onset of action | Side effects |
|---|---|---|---|---|
| *Oral laxatives* | | | | |
| Bulk-forming laxatives | Retain fluid in the stool increasing its weight and consistency, stimulate peristalsis and soften fecal matter | Psyllium<br>Methylcellulose | 24–72 h | Abdominal bloating, and flatulence |
| Stool softeners laxatives | Tensioactive agents which soften and lubricate fecal matter | Docusate sodium | 24–72 h | Inhibition of gallbladder secretion, gastric mucosal lesions, anorexia, nausea and vomiting |
| | | Liquid paraffin | 6–8 h | Irritation of rectal mucosa, aspiration pneumonitis |
| Osmotic laxatives | Create an osmotic gradient drawing fluid into the intestinal lumen and promoting peristalsis and stool propulsion | Lactitol<br>Lactulose | 1–3 days | Cramping, abdominal bloating, and flatulence |
| | | Polyethylene glycol (macrogol) | 1–4 days | Diarrhea, nausea, abdominal bloating and pain, flatulence |
| | | Magnesium and sodium salts | 0.5–3 h | Electrolyte imbalance |
| Stimulant laxatives | Stimulate intestinal mucosal, decrease fluid and electrolyte absorption | Bisacodyl | 6–12 h | Cramping, electrolyte imbalance with chronic use |
| | | Sodium picosulfate | 6–9 h | |
| | | Sennosides A and B | 8–12 h | Cramping, nausea, abdominal bloating and pain |

**TABLE 4 Pharmacological treatment of neurogenic bowel—cont'd**

| | Mechanism of action | Active principle | Onset of action | Side effects |
|---|---|---|---|---|
| *Prokinetic agents* | | | | |
| | Highly-selective serotonin 5-HT4 receptor agonist | Prucalopride | | Peak plasma concentration achieved in 2–3 h<br>Requires dose adjustment in renal failure<br>Side effects: headache, nausea, diarrhea, abdominal pain |
| | Inhibits the hydrolysis of acetylcholine by competing with acetylcholine for attaching to acetylcholinesterase | Neostigmine | | Bradycardia, bronchoconstriction |

Main oral laxative and prokinetic drugs for the treatment of constipation, including its active principle, mechanism and starting action, and the most frequent side effects. Unpublished.

### *Rectal laxatives*

Suppositories or micro-enemas, alone or in combination with DRS and MRE, are used to trigger defecation. Patients whose rectal-anal reflexes are preserved respond to pharmacological stimulation, although the effectiveness and response time may vary greatly. The most commonly used chemical stimulants are *bisacodyl* and *glycerin*. Large-volume *phosphate enemas* are not routinely used for the chronic management of constipation in SCI patients, although they may be useful in cases of fecal impaction (Table 3).

### *Oral laxatives*

Although they have been used for a long time, there is limited available evidence on their effectiveness and safety (Coggrave, Norton, & Cody, 2014). Bulk-forming and non-saline osmotic are the most commonly used oral laxatives. The choice depends on several factors such as its mechanism of action, speed of onset and duration of effect.

### *Prokinetics*

Prokinetics promote bowel transit by reducing the passage time of stools through the bowel. *Metoclopramide* and *erythromycin* have no effect on colonic motility. They are frequently used for treating gastric dilatation and paralytic ileus, characteristic of the spinal shock phase.

Reduction in CTT has been shown with *prucalopride*, as well as improvement in the frequency of bowel evacuations in SCI patients. No significant improvement was shown for the duration of bowel care, episodes of fecal incontinence, or need for digital evacuation (Krogh et al., 2002).

In people with SCI, *neostigmine* given with glycopyrrolate improves the evacuation and reduces the time of bowel care (Korsten et al., 2005; Rosman et al., 2008). Their cardiovascular and pulmonary adverse effects and parenteral administration limit a chronic use of these types of medication. Recently, good results were reported with transdermal delivery by iontophoresis (Korsten et al., 2018).

## Other drugs

*Fampridine* increases the days of bowel movements in patients with incomplete SCI (Cardenas et al., 2007). *Oxymetazoline* topical gel may help treating fecal incontinence (Barak, Gecse, & Takács, 2019). Other drugs such as *botulinum toxin, colchicine, lubiprostone, and linaclotide* have shown to be effective in chronic constipation and NB of different etiology, although it has not been studied in SCI patients.

## Bowel program

The establishment of an early routine bowel program is considered the best practice for SCI patients (Qi, Middleton, & Malcolm, 2018; Stoffel, Van der Aa, Wittmann, Yande, & Elliott, 2018). The aim is to achieve regular, predictable, effective evacuation, as well as to eliminate or minimize incontinence episodes and other secondary complications of NB. Between 56% and 84.8% of the patients will achieve a satisfactory bowel management with the program (Adriaansen et al., 2015; Correa & Rotter, 2000; Engkasan & Sudin, 2013).

The bowel program is the establishment of an evacuation routine following a scheme that sets the moment during the day and frequency at which the routine is going to be carried out. Furthermore, it includes dietary recommendations, oral laxatives, and use of rectal stimulation and/or manual bowel evacuation. The program is different for each patient based on the main symptom, the characteristics of the SCI and presence or absence of sacral reflexes. Most patients will require more than one intervention to reach an effective bowel management. The key for a successful program lies largely on education and training.

The first recommendation is to schedule a regular *frequency and time*. In general, it is advisable to empty the bowel every 48 h to achieve a minimum of three bowel movement every week (Coggrave, Burrows, & Durand, 2006; Correa & Rotter, 2000). A lower frequency or irregular care contributes to constipation and impaction episodes. Patients with areflexic NB, with high risk of incontinence, may need to empty their bowel on a daily basis or even twice a day.

The patient has to establish the time based on his/her needs. Routine is a key element in the program and must be repeated at the same time. It is recommended to perform it 20–30 min after the main meal to take advantage of bowel's increased physiological motility after a meal (gastrocolic reflex) (Suttor et al., 2009).

The patient should preferably be in a sitting position in the toilet and an adapted environment is essential. The need of technical aids will be assessed, i.e., a bathtub stool, side handles, and a toilet seat riser. When the patient has a fragile equilibrium, the program will be carried out in the bed, preferably in the left lateral decubitus position (Ozisler et al., 2015).

For optimum stool consistency and to promote colonic transit, dietary adjustments or an oral laxative may be necessary.

For the act of defecation, a triggering stimulus is usually needed (DRS, suppositories, or micro-enema) (Table 3). MRE may also be necessary as the only emptying method, before the administration of a suppository or DRS, or subsequently, to complete the emptying. Some physical activity can be recommended before the programmed bowel movement, as well as abdominal massage during the process.

There are some differences in the program between reflex and areflexic NB patients (Table 5). Thus, in subjects with reflex NB bowel movement may be achieved with chemical rectal stimulants, DRS, or a combination of both. In some cases, taking a stimulant laxative before the program may be of help, although it is not advisable to do this as a routine. In patients in which sacral reflexes are not preserved, digital rectal and pharmacological stimulation are ineffective and MRE should be used. Stools should be firm in these patients to facilitate manual evacuation and decrease the risk of incontinence in cases of sphincter hypotony or atony. Anal plugs may be effective when there is fluid or gas loss in small volumes.

A well-established program must achieve the shortest possible time spend in bowel care. If it takes more than an hour, it must be assessed to identify which elements need to be corrected. Patients must know how SCI affects the function of their bowel. NBD must be part of the regular follow-up during SCI chronic phase.

**TABLE 5** Bowel care in neurogenic bowel.

| | Hyperreflexic neurogenic bowel | Areflexic neurogenic bowel |
|---|---|---|
| | Presence of sacral reflexes<br>Injury above the medullary cone | Absence of sacral reflexes<br>Injury in medullary cone and cauda equina |
| Frequency | Daily or alternate days | Daily or twice a day |
| Consistency of feces | Soft, Bristol 4 | Hard, Bristol 3 |
| Defecation | DRS, suppository, microenema | Manual rectal evacuation |
| Supporting measures | After meals, exercise, abdominal massage | After meals, exercise, abdominal massage |

Outline bowel care program according to hyperreflexic vs areflexic neurogenic bowel. *DRS*, digital rectal stimulation. Unpublished.

### Anorectal biofeedback

Biofeedback is a reeducation technique through feedback of bowel movement and continence. This treatment modality is scarcely use in NBD, but may be beneficial in people with incomplete motor SCI (Mazor, Jones, Andrews, Kellow, & Malcolm, 2016).

### Transanal irrigation

Transanal irrigation (TAI) allows performing a programmed defecation and is a technique for treating bowel dysfunction in SCI patients. Some studies show that it improves constipation, incontinence, shortens the time spent in bowel care, reduces the intake of medications, and improves QoL (Christensen et al., 2006; Faaborg et al., 2009). Good results have been reported in 68% and 63% of patients with incontinence and constipation, respectively (Del Popolo et al., 2008).

The most used systems are cones or rectal balloon catheters, a recipient for water and an infusion set-up. The *infusion* may be carried out through manual pumping (Fig. 1A) or using an electronic control unit (Fig. 1B) in which the volume of water to be infused, the speed of the infusion, and the size of the balloon are programed.

The *irrigation fluid* is usually warm tap water (36–38°C). The instilled volume may vary, generally starting with 500 mL and increased up to 1000 mL.

As with the classical bowel care program, stool consistency should be optimized if necessary. The recommendations to establish a routine at a specific time of the day and perform the irrigation 20–20 min after the meal are maintained. Initially, it is advisable to carry out this on a daily basis and reduce the frequency considering the preferences and needs of the patient.

Absolute *contraindications* for TAI include anal or rectal stenosis, colorectal cancer, inflammatory bowel disease, and recent colonic surgery (Emmanuel et al., 2013). Up to 48% of subjects my suffer *adverse effects* such as abdominal pain or discomfort, rectal bleeding, fatigue, general malaise, sweating, perianal discomfort, nauseas, tremors, headache, and facial redness (Faaborg et al., 2009). Rectum perforation is rare, although it is a potential risk (Christensen et al., 2016; Faaborg et al., 2009). TAI is also a trigger of AD.

## Surgical treatment of the neurogenic bowel

### Electrical stimulation

Both, neurostimulation and neuromodulation, have been investigated for the treatment of NBD. The goal of ***sacral neuromodulation*** is to adjust and reestablish the balance between the inhibitory and facilitatory reflexes that control the functional activity of the pelvic floor. Some studies suggest this therapy improves bowel function in patients with incomplete SCI (Chen & Liao, 2015; Lombardi, Del Popolo, Cecconi, Surrenti, & Macchiarella, 2010). To date, a widespread use cannot be recommended. The effect on bowel function of ***stimulating the pudendal nerve*** has also been assessed in patients with complete cauda equina lesions (George et al., 2014). Short-term results seem encouraging, showing improvement of constipation and incontinence. ***Sacral anterior root stimulation*** requires a sacral dorsal root rhizotomy with subsequent implantation of a stimulator in the anterior sacral roots S2–S4. Although this system was designed for treating neurogenic bladder in SCI, some patients use it as a method for bowel emptying with reported good results (Rasmussen et al., 2015; Vallès, Rodríguez, Borau, & Mearin, 2009).

### Antegrade continence enema

A continence tube is created, generally an appendicostomy, which allows intermittent catheterization for colonic irrigation. This procedure has been used in SCI patients, showing improvement of QoL associated to fecal incontinence (Smith & Decter, 2015).

### Colostomy and ileostomy

An invasive procedure, not exempt from complications, which should be indicated to selected patients when other more conservative treatment methods fail.

The most common indications are constipation and unmanageable fecal incontinence, extended bowel care time, AD or pain during defecation; it is also performed to facilitate the healing of pressure sores or infections in the perineum or abdomen. Left-side colostomies maintain better the colonic surface and prevent excessive loss of water, facilitate stoma care, and there is much lower risk of leakage (Safadi, Rosito, Nino-Murcia, Wolfe, & Perkash, 2003). With right-side

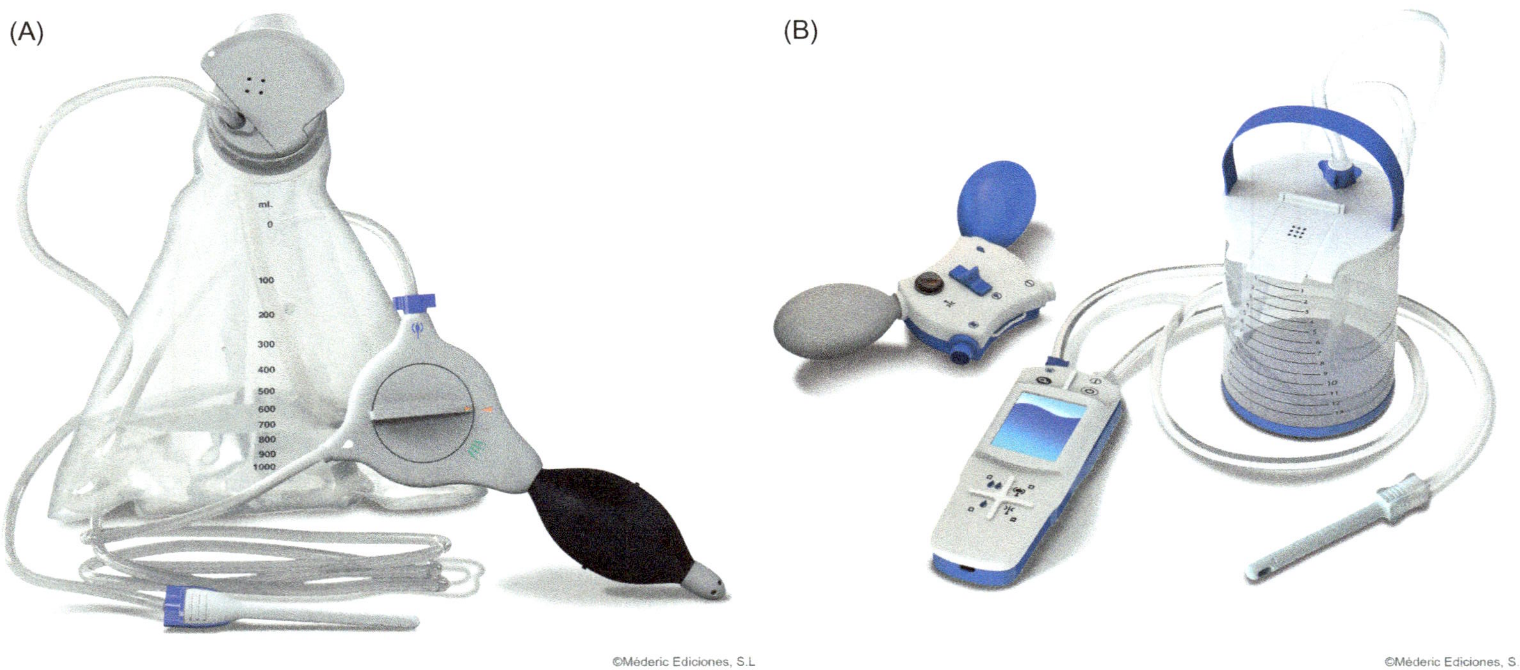


FIG. 1 Transanal irrigation devices. (A) Transanal irrigation device with manual pumping; (B) transanal irrigation device using an electronic control unit. *(With permission of Mederic Ediciones S. L.)*

ileostomies or colostomies, there is less emptying problems. Results from different studies report that they simplify and reduce bowel care time, decrease the number of bowel dysfunction-related hospitalizations, and provide equal or better QoL (Rosito, Nino-Murcia, Wolfe, Kiratli, & Perkash, 2002). Patients are usually satisfied with the surgery and a high percentage communicates that he/she would have undergone surgery earlier (Safadi et al., 2003). Despite the good results, this method is scarcely used; between 2.4% and 8% have undergone a colostomy aimed at bowel care (Adriaansen et al., 2015; Coggrave et al., 2009; Pardee, Bricker, Rundquist, MacRae, & Tebben, 2012).

## Conclusions

With the current knowledge on NBD, its therapeutic management remains empirical, based on a step-up approach going from conservative to more invasive options (Fig. 2). It can be predicted that advances in pharmacological research and control of nervous system function through electrical stimulation will provide in-between treatment options.

## Applications to other areas of neuroscience

Bowel disturbance is one of the sequelae of spinal cord injury that most affects the life of these patients; however, it is possibly the less studied condition. Advancement in developing successful treatments for neurogenic bowel dysfunction has been slow in part due to the absence of reliable and validated tools to assess neurogenic bowel dysfunction-related symptoms and complications, both from the perspective of the patient and the physician. Therefore, the standardization of validated scales aimed at this group of patients and the further development of quality of life measurement tools specifically associated with intestinal dysfunction will help advance the evaluation, research and better management of neurogenic bowel dysfunction.

On the other hand, it is necessary to develop and improve tools such as colonic transit time, manometry and others, that provide more objective assessment, needed in therapeutic decision-making and in research studies on new treatment approaches.

Although spinal cord injury is the condition most frequently associated with neurogenic bowel dysfunction, any patient suffering from central nervous system disease may experience symptoms of intestinal dysfunction. Neurological conditions such as multiple sclerosis, cerebral palsy, stroke, or Parkinson's disease, placing the individual at risk for constipation and incontinence. Although the presentation of intestinal dysfunction in these conditions can vary, the underlying cause is also the damage to the nervous control of the bowel. Advances in the knowledge of physiopathology and management of bowel dysfunction in SCI may help improve bowel care in other neurological pathologies causing neurogenic bowel dysfunction.

Finally, accurate spinal cord injury bowel dysfunction understanding, and improvement of diagnostic methods may contribute, without doubt, to the assessment of new therapeutic approaches for the treatment of spinal cord injury currently being developed, such as cell therapies or medullary stimulation.

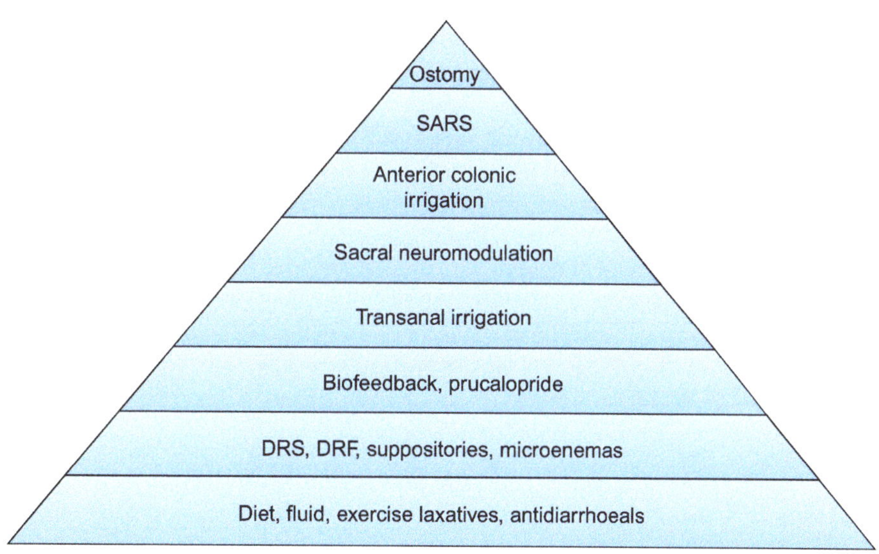


FIG. 2 Treatment options in neurogenic bowel dysfunction management. The pyramid represents a stepwise treatment from conservative measures to more invasive measures. *SARS*, sacral anterior root stimulation; *DRS*, digital rectal stimulation; *DRF*, digital removal of feces. Unpublished.

## Dictionary of terms

**Abdominal massage:** Massage applied to the abdomen in a clockwise direction using the hand.
**Anal reflex:** Visible contraction of anus in response to pinprick of surrounding skin.
**Autonomic dysreflexia:** Abnormal sympathetic nervous system response to a noxious stimulus below the level of injury in individuals with spinal cord injury above the seven thoracic levels. The guiding symptom is arterial hypertension.
**Bowel care:** Number of different interventions combined in a bowel routine to promote effective and timely fecal evacuation.
**Bulbocavernosus reflex:** Contraction of anus in response to pressure on glans penis or clitoris.
**Colonic transit time:** Time it takes for stool to pass through the colon.
**Dietary fiber:** Carbohydrate polymers with 10 or more monomeric units, which are not hydrolyzed by the endogenous enzymes in the small intestine of humans.
**Digital rectal stimulation:** Technique performed by inserting lubricated finger through the anal canal into the rectum and slowly rotating the finger in a circular movement maintaining contact with rectal mucosa.
**Digital removal of feces:** Technique performed by insertion of a single gloved, lubricated finger into the rectum to break up or remove stool.
**Neurogenic bowel or neurogenic bowel dysfunction**: Colonic dysfunction as result of central neurological disease or damage.
**Transanal irrigation:** Procedure that involves introducing water into de colon through the anus to evacuate the distal colon and rectum.

## Key facts of neurogenic bowel dysfunction

- Neurogenic bowel affects to a greater or lower degree most spinal cord injured patients; around 50% perceive restrictions in their quality of live or social activities.
- Spinal cord injury causes colon and rectum innervation changes that lead to loss of voluntary control of defecation, bowel motility impairments, and changes in sphincter tone.
- Fecal incontinence, constipation, and a combination of both are the primary manifestations.
- Based on the level of spinal cord injury, two neurogenic bowel dysfunction patterns are distinguished: hyperreflexic and areflexic.
- The diagnostic assessment is mainly clinical. Assessment scales allow estimating the seriousness of the symptoms and response to treatment.
- Neurogenic bowel dysfunction treatment is largely empirical.

## Summary points

- Slower colonic transit, increased colonic tone, changes in rectal compliance, of the tone and sensitivity of the sphincter, and lack of relaxation of the sphincter, are the mechanisms responsible of neurogenic bowel symptoms in spinal cord injury.
- Secondary disturbances in enteric nervous system, colonic neuromuscular structures, and bowel microbiota may be contributing factors.
- Establishing a bowel emptying routine together with a technique that triggers defecation, remains the treatment of choice.
- Optimizing stool consistency through diet and/or a bulk-forming or osmotic laxative may be necessary.
- Transanal irrigation is a treatment modality for effective bowel emptying in cases in which control is not achieved with the bowel program.
- Invasive treatments are an option in selected cases. Colostomy may improve the quality of life in these patients.

## References

Adriaansen, J. J., van Asbeck, F. W., van Kuppevelt, D., Snoek, G. J., & Post, M. W. (2015). Outcomes of neurogenic bowel management in individuals living with a spinal cord injury for at least 10 years. *Archives of Physical Medicine and Rehabilitation*, *96*(5), 905–912.

Agachan, F., Chen, T., Pfeifer, J., Reissman, P., & Wexner, S. D. (1996). A constipation scoring system to simplify evaluation and management of constipated patients. *Diseases of the Colon and Rectum*, *39*(6), 681–685.

Ayaş, S., Leblebici, B., Sözay, S., Bayramoğlu, M., & Niron, E. A. (2006). The effect of abdominal massage on bowel function in patients with spinal cord injury. *American Journal of Physical Medicine & Rehabilitation*, *85*(12), 951–955.

Barak, N., Gecse, K. B., & Takács, I. (2019). Topical oxymetazoline for fecal incontinence in patients with spinal cord injury: A double-blind randomized controlled crossover study. *Diseases of the Colon and Rectum*, *62*(2), 234–240.

Cameron, K. J., Nyulasi, I. B., Collier, G. R., & Brown, D. J. (1996). Assessment of the effect of increased dietary fibre intake on bowel function in patients with spinal cord injury. *Spinal Cord*, *34*(5), 277–283.

Cardenas, D. D., Ditunno, J., Graziani, V., Jackson, A. B., Lammertse, D., Potter, P., et al. (2007). Phase 2 trial of sustained-release fampridine in chronic spinal cord injury. *Spinal Cord*, *45*(2), 158–168.

Chen, G., & Liao, L. (2015). Sacral neuromodulation for neurogenic bladder and bowel dysfunction with multiple symptoms secondary to spinal cord disease. *Spinal Cord*, *53*(3), 204–208.

Christensen, P., Bazzocchi, G., Coggrave, M., Abel, R., Hultling, C., Krogh, K., et al. (2006). A randomized, controlled trial of transanal irrigation versus conservative bowel management in spinal cord-injured patients. *Gastroenterology*, *131*(3), 738–747.

Christensen, P., Krogh, K., Perrouin-Verbe, B., Leder, D., Bazzocchi, G., Petersen Jakobsen, B., et al. (2016). Global audit on bowel perforations related to transanal irrigation. *Techniques in Coloproctology*, *20*(2), 109–115.

Chung, E. A., & Emmanuel, A. V. (2006). Gastrointestinal symptoms related to autonomic dysfunction following spinal cord injury. *Progress in Brain Research*, *152*, 317–333.

Coggrave, M., Burrows, D., & Durand, M. A. (2006). Progressive protocol in the bowel management of spinal cord injuries. *British Journal of Nursing*, *15* (20), 1108–1113.

Coggrave, M., Norton, C., & Cody, J. D. (2014). Management of faecal incontinence and constipation in adults with central neurological diseases. *Cochrane Database of Systematic Reviews*, (1), CD002115.

Coggrave, M., Norton, C., & Wilson-Barnett, J. (2009). Management of neurogenic bowel dysfunction in the community after spinal cord injury: A postal survey in the United Kingdom. *Spinal Cord*, *47*(4), 323–330. quiz 331–323.

Correa, G. I., & Rotter, K. P. (2000). Clinical evaluation and management of neurogenic bowel after spinal cord injury. *Spinal Cord*, *38*(5), 301–308.

Del Popolo, G., Mosiello, G., Pilati, C., Lamartina, M., Battaglino, F., Buffa, P., et al. (2008). Treatment of neurogenic bowel dysfunction using transanal irrigation: A multicenter Italian study. *Spinal Cord*, *46*(7), 517–522.

den Braber-Ymker, M., Lammens, M., van Putten, M. J., & Nagtegaal, I. D. (2017). The enteric nervous system and the musculature of the colon are altered in patients with spina bifida and spinal cord injury. *Virchows Archiv*, *470*(2), 175–184.

Eldahan, K. C., & Rabchevsky, A. G. (2018). Autonomic dysreflexia after spinal cord injury: Systemic pathophysiology and methods of management. *Autonomic Neuroscience*, *209*, 59–70.

Emmanuel, A. V., Krogh, K., Bazzocchi, G., Leroi, A. M., Bremers, A., Leder, D., et al. (2013). Consensus review of best practice of transanal irrigation in adults. *Spinal Cord*, *51*(10), 732–738.

Engkasan, J. P., & Sudin, S. S. (2013). Neurogenic bowel management after spinal cord injury: Malaysian experience. *Journal of Rehabilitation Medicine*, *45*(2), 141–144.

Faaborg, P. M., Christensen, P., Finnerup, N., Laurberg, S., & Krogh, K. (2008). The pattern of colorectal dysfunction changes with time since spinal cord injury. *Spinal Cord*, *46*(3), 234–238.

Faaborg, P. M., Christensen, P., Krassioukov, A., Laurberg, S., Frandsen, E., & Krogh, K. (2014). Autonomic dysreflexia during bowel evacuation procedures and bladder filling in subjects with spinal cord injury. *Spinal Cord*, *52*(6), 494–498.

Faaborg, P. M., Christensen, P., Kvitsau, B., Buntzen, S., Laurberg, S., & Krogh, K. (2009). Long-term outcome and safety of transanal colonic irrigation for neurogenic bowel dysfunction. *Spinal Cord*, *47*(7), 545–549.

Gao, R., Tao, Y., Zhou, C., Li, J., Wang, X., Chen, L., et al. (2019). Exercise therapy in patients with constipation: A systematic review and meta-analysis of randomized controlled trials. *Scandinavian Journal of Gastroenterology*, *54*(2), 169–177.

George, A. T., Dudding, T. C., Gurmany, S., Kamm, M. A., Nicholls, R. J., & Vaizey, C. J. (2014). Pudendal nerve stimulation for bowel dysfunction in complete cauda equina syndrome. *Annals of Surgery*, *259*(3), 502–507.

Inskip, J. A., Lucci, V. M., McGrath, M. S., Willms, R., & Claydon, V. E. (2018). A community perspective on bowel management and quality of life after spinal cord injury: The influence of autonomic dysreflexia. *Journal of Neurotrauma*, *35*(9), 1091–1105.

Jorge, J. M., & Wexner, S. D. (1993). Etiology and management of fecal incontinence. *Diseases of the Colon and Rectum*, *36*(1), 77–97.

Kirshblum, S. C., House, J. G., & O'connor, K. C. (2002). Silent autonomic dysreflexia during a routine bowel program in persons with traumatic spinal cord injury: A preliminary study. *Archives of Physical Medicine and Rehabilitation*, *83*(12), 1774–1776.

Korsten, M. A., Fajardo, N. R., Rosman, A. S., Creasey, G. H., Spungen, A. M., & Bauman, W. A. (2004). Difficulty with evacuation after spinal cord injury: Colonic motility during sleep and effects of abdominal wall stimulation. *Journal of Rehabilitation Research and Development*, *41*(1), 95–100.

Korsten, M. A., Lyons, B. L., Radulovic, M., Cummings, T. M., Sikka, G., Singh, K., et al. (2018). Delivery of neostigmine and glycopyrrolate by iontophoresis: A nonrandomized study in individuals with spinal cord injury. *Spinal Cord*, *56*(3), 212–217.

Korsten, M. A., Rosman, A. S., Ng, A., Cavusoglu, E., Spungen, A. M., Radulovic, M., et al. (2005). Infusion of neostigmine-glycopyrrolate for bowel evacuation in persons with spinal cord injury. *American Journal of Gastroenterology*, *100*(7), 1560–1565.

Korsten, M. A., Singal, A. K., Monga, A., Chaparala, G., Khan, A. M., Palmon, R., et al. (2007). Anorectal stimulation causes increased colonic motor activity in subjects with spinal cord injury. *Journal of Spinal Cord Medicine*, *30*(1), 31–35.

Krogh, K., Christensen, P., Sabroe, S., & Laurberg, S. (2006). Neurogenic bowel dysfunction score. *Spinal Cord*, *44*(10), 625–631.

Krogh, K., Emmanuel, A., Perrouin-Verbe, B., Korsten, M. A., Mulcahey, M. J., & Biering-Sørensen, F. (2017). International spinal cord injury bowel function basic data set (version 2.0). *Spinal Cord*, *55*(7), 692–698.

Krogh, K., Jensen, M. B., Gandrup, P., Laurberg, S., Nilsson, J., Kerstens, R., et al. (2002). Efficacy and tolerability of prucalopride in patients with constipation due to spinal cord injury. *Scandinavian Journal of Gastroenterology*, *37*(4), 431–436.

Kwok, S., Harvey, L., Glinsky, J., Bowden, J. L., Coggrave, M., & Tussler, D. (2015). Does regular standing improve bowel function in people with spinal cord injury? A randomised crossover trial. *Spinal Cord*, *53*(1), 36–41.

Lewis, S. J., & Heaton, K. W. (1997). Stool form scale as a useful guide to intestinal transit time. *Scandinavian Journal of Gastroenterology*, *32*(9), 920–924.

Lombardi, G., Del Popolo, G., Cecconi, F., Surrenti, E., & Macchiarella, A. (2010). Clinical outcome of sacral neuromodulation in incomplete spinal cord-injured patients suffering from neurogenic bowel dysfunctions. *Spinal Cord*, *48*(2), 154–159.

Lynch, A. C., Anthony, A., Dobbs, B. R., & Frizelle, F. A. (2000). Anorectal physiology following spinal cord injury. *Spinal Cord*, *38*(10), 573–580.

Lynch, A. C., Antony, A., Dobbs, B. R., & Frizelle, F. A. (2001). Bowel dysfunction following spinal cord injury. *Spinal Cord*, *39*(4), 193–203.

Mazor, Y., Jones, M., Andrews, A., Kellow, J. E., & Malcolm, A. (2016). Anorectal biofeedback for neurogenic bowel dysfunction in incomplete spinal cord injury. *Spinal Cord*, *54*(12), 1132–1138.

Nielsen, S. D., Faaborg, P. M., Christensen, P., Krogh, K., & Finnerup, N. B. (2017). Chronic abdominal pain in long-term spinal cord injury: A follow-up study. *Spinal Cord*, *55*(3), 290–293.

Ozisler, Z., Koklu, K., Ozel, S., & Unsal-Delialioglu, S. (2015). Outcomes of bowel program in spinal cord injury patients with neurogenic bowel dysfunction. *Neural Regeneration Research*, *10*(7), 1153–1158.

Paily, A., Preziosi, G., Trivedi, P., & Emmanuel, A. (2019). Anti-muscarinic drugs increase rectal compliance and exacerbate constipation in chronic spinal cord injury: Anti-muscarinic drug effect on neurogenic bowel. *Spinal Cord*, *57*, 662–668.

Pardee, C., Bricker, D., Rundquist, J., MacRae, C., & Tebben, C. (2012). Characteristics of neurogenic bowel in spinal cord injury and perceived quality of life. *Rehabilitation Nursing*, *37*(3), 128–135.

Pires, J. M., Ferreira, A. M., Rocha, F., Andrade, L. G., Campos, I., Margalho, P., et al. (2018). Assessment of neurogenic bowel dysfunction impact after spinal cord injury using the International Classification of Functioning, Disability and Health. *European Journal of Physical and Rehabilitation Medicine*, *54*(6), 873–879.

Qi, Z., Middleton, J. W., & Malcolm, A. (2018). Bowel dysfunction in spinal cord injury. *Current Gastroenterology Reports*, *20*(10), 47.

Rasmussen, M. M., Kutzenberger, J., Krogh, K., Zepke, F., Bodin, C., Domurath, B., et al. (2015). Sacral anterior root stimulation improves bowel function in subjects with spinal cord injury. *Spinal Cord*, *53*(4), 297–301.

Rockwood, T. H., Church, J. M., Fleshman, J. W., Kane, R. L., Mavrantonis, C., Thorson, A. G., et al. (2000). Fecal Incontinence Quality of Life Scale: Quality of life instrument for patients with fecal incontinence. *Diseases of the Colon and Rectum*, *43*(1), 9–16. discussion 16–17.

Rosito, O., Nino-Murcia, M., Wolfe, V. A., Kiratli, B. J., & Perkash, I. (2002). The effects of colostomy on the quality of life in patients with spinal cord injury: A retrospective analysis. *Journal of Spinal Cord Medicine*, *25*(3), 174–183.

Rosman, A. S., Chaparala, G., Monga, A., Spungen, A. M., Bauman, W. A., & Korsten, M. A. (2008). Intramuscular neostigmine and glycopyrrolate safely accelerated bowel evacuation in patients with spinal cord injury and defecatory disorders. *Digestive Diseases and Sciences*, *53*(10), 2710–2713.

Safadi, B. Y., Rosito, O., Nino-Murcia, M., Wolfe, V. A., & Perkash, I. (2003). Which stoma works better for colonic dysmotility in the spinal cord injured patient? *American Journal of Surgery*, *186*(5), 437–442.

Savic, G., Frankel, H. L., Jamous, M. A., Soni, B. M., & Charlifue, S. (2018). Long-term bladder and bowel management after spinal cord injury: A 20-year longitudinal study. *Spinal Cord*, *56*(6), 575–581.

Siddall, P. J., McClelland, J. M., Rutkowski, S. B., & Cousins, M. J. (2003). A longitudinal study of the prevalence and characteristics of pain in the first 5 years following spinal cord injury. *Pain*, *103*(3), 249–257.

Smith, P. H., & Decter, R. M. (2015). Antegrade continence enema procedure: Impact on quality of life in patients with spinal cord injury. *Spinal Cord*, *53*(3), 213–215.

Song, S. H., Svircev, J. N., Teng, B. J., Dominitz, J. A., & Burns, S. P. (2018). A safe and effective multi-day colonoscopy bowel preparation for individuals with spinal cord injuries. *Journal of Spinal Cord Medicine*, *41*(2), 149–156.

Stoffel, J. T., Van der Aa, F., Wittmann, D., Yande, S., & Elliott, S. (2018). Neurogenic bowel management for the adult spinal cord injury patient. *World Journal of Urology*, *36*(10), 1587–1592.

Suttor, V. P., Ng, C., Rutkowski, S., Hansen, R. D., Kellow, J. E., & Malcolm, A. (2009). Colorectal responses to distension and feeding in patients with spinal cord injury. *American Journal of Physiology. Gastrointestinal and Liver Physiology*, *296*(6), G1344–G1349.

Trivedi, P. M., Kumar, L., & Emmanuel, A. V. (2016). Altered colorectal compliance and anorectal physiology in upper and lower motor neurone spinal injury may explain bowel symptom pattern. *American Journal of Gastroenterology*, *111*(4), 552–560.

Vallès, M., & Mearin, F. (2009). Pathophysiology of bowel dysfunction in patients with motor incomplete spinal cord injury: Comparison with patients with motor complete spinal cord injury. *Diseases of the Colon and Rectum*, *52*(9), 1589–1597.

Vallès, M., Rodríguez, A., Borau, A., & Mearin, F. (2009). Effect of sacral anterior root stimulator on bowel dysfunction in patients with spinal cord injury. *Diseases of the Colon and Rectum*, *52*(5), 986–992.

Vallès, M., Vidal, J., Clavé, P., & Mearin, F. (2006). Bowel dysfunction in patients with motor complete spinal cord injury: Clinical, neurological, and pathophysiological associations. *American Journal of Gastroenterology*, *101*(10), 2290–2299.

Ware, J. E., & Sherbourne, C. D. (1992). The MOS 36-item short-form health survey (SF-36). I. Conceptual framework and item selection. *Medical Care*, *30*(6), 473–483.

White, A. R., & Holmes, G. M. (2019). Investigating neurogenic bowel in experimental spinal cord injury: Where to begin? *Neural Regeneration Research*, *14*(2), 222–226.

Zhang, C., Zhang, W., Zhang, J., Jing, Y., Yang, M., Du, L., et al. (2018). Gut microbiota dysbiosis in male patients with chronic traumatic complete spinal cord injury. *Journal of Translational Medicine*, *16*(1), 353.

Chapter 21

# Management of neurogenic lower urinary tract dysfunction due to spinal cord injury

Noritoshi Sekido
*Department of Urology, Toho University Ohashi Medical Center, Meguro-City, Tokyo, Japan*

## Abbreviations

| | |
|---|---|
| **AC** | augmentation cystoplasty |
| **AD** | autonomic dysreflexia |
| **AUS** | artificial urinary sphincter |
| **BNR** | bladder neck reconstruction |
| **BNS** | bladder neck sling |
| **BOO** | bladder outlet obstruction |
| **BoNT-A** | type A botulinum toxin |
| **CIC** | clean intermittent catheterization |
| **DO** | detrusor overactivity |
| **DSD** | detrusor sphincter dyssynergia |
| **DU** | detrusor underactivity |
| **FVC** | frequency volume chart |
| **HCC** | hydrophilic-coated catheter |
| **LUT** | lower urinary tract |
| **NLUTD** | neurogenic lower urinary tract dysfunction |
| **PAG** | periaqueductal gray |
| **PMC** | pontine micturition center |
| **PVR** | postvoid residual |
| **QOL** | quality of life |
| **PVC** | polyvinyl catheter |
| **SARS** | sacral anterior root stimulation |
| **SCI** | spinal cord injury |
| **SNM** | sacral neuromodulation |
| **UI** | urinary incontinence |
| **UTI** | urinary tract infection |
| **UUT** | upper urinary tract |
| **VUDS** | video-urodynamic study |

## Introduction

Since Sir Ludwig Guttmann introduced sterile intermittent catheterization in the 1940s at Stoke Mandeville Hospital (Guttmann & Frankel, 1966), management of lower urinary tract (LUT) dysfunction has been an important part of the clinical practice of patients with spinal cord injury (SCI). This chapter begins with a brief overview of normal LUT function and pathophysiological aspects of neurogenic LUT dysfunction (NLUTD) due to SCI. Then, it outlines the key points of assessment as well as treatment of LUT dysfunction according to the pathway of management of patients with NLUTD due to SCI, as shown in Fig. 1.

**Diagnosis and Treatment of Spinal Cord Injury. https://doi.org/10.1016/B978-0-12-822498-4.00021-X**

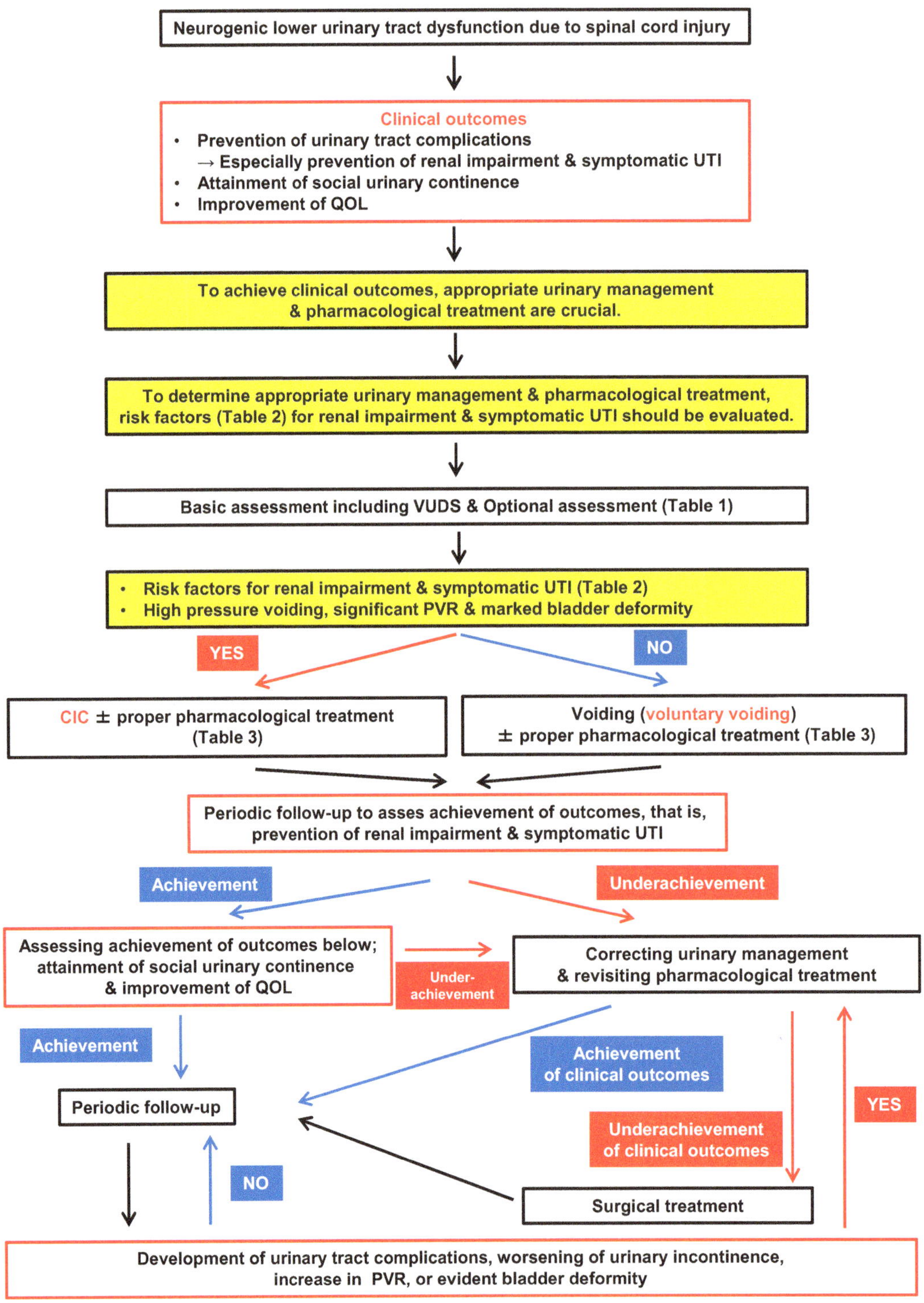


FIG. 1 Clinical pathway of management of patients with neurogenic lower urinary tract dysfunction due to spinal cord injury. Please see "Pathway of NLUTD management" section for details. To achieve the prevention of renal impairment and symptomatic UTI, proper urinary management and pharmacological treatment are determined, based on the basic and optional assessment. Then, further treatment will be considered if these outcomes are not achieved, or obtaining social urinary continence is warranted. *CIC*, clean intermittent catheterization; *PVR*, postvoid residual; *QOL*, quality of life; *UTI*, urinary tract infection; *VUDS*, video-urodynamic study.

## Physiology of storage and emptying

The urinary tract consists of the upper urinary tract (UUT) and LUT. While UUT includes the kidneys and ureters, LUT includes the urinary bladder and urethra with the external urethral sphincter. During the storage phase, the bladder gradually distends, which activates myelinated A-delta afferent fibers mainly entering S2 to S4 (de Groat, Griffiths, & Yoshimura, 2015). Then, information about bladder filling ascends the spinal cord and stimulate sympathetic nerves innervating the LUT as well as the pudendal nerve innervating the external urethral sphincter, while eventually reaching the periaqueductal gray (PAG), from which the information is relayed to the relevant areas of the brain (Griffiths, 2015). Ultimately, the medial prefrontal cortex probably determines urine storage or bladder emptying (Griffiths). Once a decision is made to empty the bladder, the medial prefrontal cortex activates the pontine micturition center (PMC) via PAG (Griffiths).

In terms of the peripheral nervous system during the storage phase, the sympathetic nerves from T10 to L1 relax the detrusor via β3-adrenoceptors and contract smooth muscle from bladder neck to proximal urethra via α1-adrenoceptors (de Groat et al., 2015). In addition, the pudendal nerve from S2 to S4 (Onuf's nucleus) controls external urethral sphincter activity (de Groat et al., 2015). These nerve activities result in urine storage at low pressure and of sufficient volume without urinary incontinence (UI). During the emptying phase, once PMC is activated, the parasympathetic nerves from S2 to S4 contract the detrusor via muscarinic receptors (M3 receptors), while sympathetic and pudendal nerves are inhibited, resulting in residual-free voiding with good urinary flow without high pressure (de Groat et al., 2015). The coordination between the detrusor and external urethral sphincter during the emptying phase requires intact neural pathways between PMC and the sacral cord.

## Pathophysiology of NLUTD due to SCI

### Classification of NLUTD due to SCI

NLUTD due to SCI is divided into two types: supra-sacral NLUTD due to damage above the sacral cord, and sacral NLUTD due to damage to the sacral cord (Sekido et al., 2020). During the acute phase of SCI (spinal shock phase), LUT in the supra-sacral NLUTD reveals an acontractile detrusor with a closed bladder outlet, which lasts about for a few weeks to 3 months. However, a recent report warned that unfavorable urodynamic findings leading to high intravesical pressure were found in two-thirds of patients even during the spinal shock phase (Bywater, Tornic, Mehnert, & Kessler, 2018). In the chronic phase, the supra-sacral NLUTD is characterized by detrusor overactivity (DO) with a discordant external urethral sphincter (detrusor sphincter dyssynergia, DSD), resulting in high-pressure storage as well as emptying. DO can causes reflex UI. In addition, besides DSD during the emptying phase, detrusor underactivity (DU) due to disruption of signal transmission from supra-spinal centers to the sacral cord may contribute to emptying dysfunction (Osman et al., 2014). The sacral NLUTD is characterized by impaired detrusor contractility (DU or acontractile detrusor) with an incompetent urethral closure mechanism (intrinsic sphincter deficiency, ISD) during storage and inadequate urethral relaxation (nonrelaxing sphincter) during emptying, resulting in refractory stress UI as well as incomplete bladder emptying. The bladder sensation is reduced or absent regardless of the level of injury.

### Association between neurological level of SCI and NLUTD

The neurological level of SCI is represented by the lowest segment of the cord with residual function, which corresponds to the upper limit of the extent of SCI. On the other hand, as described above, NLUTD is classified according to whether the lower limit of SCI involves the sacral cord. Therefore, it is difficult to accurately predict the NLUTD condition based on the neurological level of SCI regardless of its completeness of SCI, and a video-urodynamic study (VUDS) is required for the assessment of the status of NLUTD after the end of the spinal shock phase (see "Basic and Optional assessments" section).

### Elderly SCI

Elderly SCI is characterized by incomplete cervical SCI without bone injury, and central cord syndrome (Chan et al., 2018). Central cervical SCI damages the function of the upper extremities much more than the lower extremities, and the patients often recover well except for impaired hand dexterity. Theresultant impairment in hand dexterity poses serious problems for implementing clean intermittent self-catheterization. While the incidence of LUT dysfunction was reported to be low (Merriam, Taylor, Ruff, & McPhail, 1986), unfavorable urodynamic findings such as DSD were higher than previously reported, and some patients manifested urinary tract complications (Smith, Kraus, Nickell, & Boone, 2000). Therefore, in patients with NLUTD due to central cord injury, VUDS should be performed to assess NLUTD, and regular urological surveillance is mandatory. Moreover, it should be noted that aging in SCI patients worsens their daily life activities, and eventually converts clean intermittent self-catheterization to indwelling catheterization (Chan et al., 2018). Besides this, renal impairment due to NLUTD or diseases other than NLUTD, urolithiasis, and cognitive impairment due to an excessive anti-cholinergic load should be carefully monitored in these patients (Chan et al., 2018; Welk & McArthur, 2020).

## Overview of management of patients with NLUTD due to SCI

### Clinical outcomes of management of NLUTD

The critical clinical outcomes or goals of the clinical practice in NLUTD due to SCI are prevention of urinary tract complications [e.g., renal impairment (renal dysfunction/UUT deterioration), symptomatic urinary tract infection (UTI), urolithiasis, gross hematuria, urethral stricture, urethrocutaneous fistula, bladder cancer, and catheter obstruction], attainment of social urinary continence, and improvement of quality of life (QOL) (Sekido et al., 2020).

### Pathway of NLUTD management

Recently, patient-reported outcomes have become more important in the field of NLUTD than before, and QOL has been increasingly included in clinical outcomes of treatment for NLUTD (Adriaansen et al., 2017). Even though improvement of QOL is crucial, prevention of urinary tract complications, some of which are closely related to the patient's survival, remains the most important clinical outcome to be achieved in patients with NLUTD due to SCI. Therefore, as shown in Fig. 1, in order to prevent urinary tract complications, the first step is to determine the appropriate urinary management and pharmacological treatment aimed at preventing urinary tract complications, especially renal impairment and symptomatic UTI. Once the appropriate urinary management and pharmacological treatment has been determined, periodic follow-up assesses whether this outcome has been achieved. If the clinical outcomes of prevention of renal impairment and symptomatic UTI are not achieved, correcting urinary management as well as revisiting pharmacological treatment should be considered. If these outcomes are achieved, obtaining social urinary continence and enhancing QOL are the subsequent important clinical outcomes. To this end, urinary management and pharmacological treatment should be optimized without compromising the outcomes of prevention of renal impairment and symptomatic UTI. Because UI is associated with poor QOL, strategies to attain social continence should be actively addressed with careful consideration of the individual patient's preference, characteristics, and social environment. When maximum conservative treatment fails to achieve the above outcomes, surgical intervention becomes an option in carefully selected patients.

## Basic and optional assessments

All the patients with NLUTD due to SCI should be assessed by basic assessment (Table 1) (Sekido et al., 2020). The basic assessment includes a history taking (Zlatev, Shem, & Elliott, 2018), assessment of QOL (Costa et al., 2001; Pinder et al., 2012; Welk et al., 2018), focused physical and neurological findings (Sekido et al., 2020), a frequency volume chart (FVC) or a bladder diary, urinalysis, blood tests to screen for renal dysfunction (Goto, Kawasaki, Takemoto, Abe, & Namima, 2018), a morphological evaluation of the urinary tract usually assessed by abdominal ultrasound, measurement of postvoid residual (PVR) and uroflow in patients with voiding (Sekido et al., 2020), and VUDS (Kreydin et al., 2018; Sekido et al., 2020). In addition to the basic assessment, optional assessment may be performed, which includes static and/or dynamic renal scintigrams in cases of renal dysfunction or vesico-ureteral reflux, and cystourethroscopy in cases of suspicious of urethral pathology, anatomical bladder outlet obstruction (BOO), bladder stones, or bladder cancer (Sekido et al., 2020). VUDS and cystourethroscopy should be performed with caution against developing autonomic dysreflexia (AD) during examinations (Liu, Zhou, Biering-Sørensen, & Krassioukov, 2017). Performing the basic as well as optional assessment identifies the presence or absence of risk factors for renal impairment, symptomatic UTI, UI, and QOL (Table 2) (Sekido et al., 2020). In addition, the presence or absence of high-pressure voiding due to functional BOO, significant PVR (>100 mL) due to BOO or DU/acontractile detrusor, and marked bladder deformity defined as grade 2 (bladder having mild pseudodiverticula) or higher by Ogawa's classification should be noted (Sekido et al., 2020).

**TABLE 1 Basic assessment.**

| Evaluations | Comments |
|---|---|
| History taking | • The level of injury: neurological findings due to SCI are essential in determining urinary management, as residual upper extremity function and the ability to hold a sitting position or open the legs are important factors for clean intermittent self-catheterization.<br>• Urinary management during and after the acute phase<br>• The presence or absence of symptomatic UTI, urinary incontinence, AD<br>• Concomitant medications<br>• The status of anorectal and sexual function<br>• Current co-morbid diseases<br>• Past medical and surgical histories |
| Assessment of QOL | • Qualiveen 30<br>• The Neurogenic Bladder Symptom Score<br>• Intermittent Self-Catheterization Questionnaire |
| Focused physical and neurological examinations of the lumbosacral region | • Sensation in the perianal area<br>• Ability to voluntarily contract the external anal sphincter<br>• The presence or absence of anal reflex (S2–S5) and bulbospongiosus reflex (S2–S4) |
| Frequency volume chart (or bladder diary) | Useful for<br>• Evaluating bladder storage function<br>• Detecting nocturnal polyuria<br>• Confirming proper frequency/interval of CIC<br>• Measuring treatment outcomes |
| Urinalysis | Also need urine bacteriology in case of suspected symptomatic UTI |
| Blood chemistry | • Serum creatinine or cystatin C<br>• For early detection of renal impairment, cystatin C is more useful than serum creatinine<br>• Prostate-specific antigen for middle-aged or older male patients, if indicated |
| Transabdominal ultrasonography | Screening test for upper and lower urinary tract abnormalities, such as<br>• Hydroureteronephrosis<br>• Bladder wall thickening/trabeculation<br>• Urolithiasis<br>• Prostatic enlargement |
| Postvoid residual and uroflowmetry[a] in patients with voiding | • Screening test for lower urinary tract dysfunction in patients managed with voiding |
| Video-urodynamic study | • Indispensable functional examination for NLUTD due to SCI, regardless of complete or incomplete injury, because neurological levels of injury do not always predict the nature of NLUTD<br>• Identification of risk factors for renal impairment, symptomatic UTI, and UI to provide convincing evidence for urinary management and pharmacological treatment<br>• Assessment of AD during the micturition cycle<br>• "As low as reasonably achievable (ALARA)" should be kept in mind |

The basic assessment provides clues for the optional assessment and evidence for urinary management and pharmacological treatment.
*AD*, autonomic dysreflexia; *CIC*, clean intermittent catheterization; *NLUTD*, neurogenic lower urinary tract dysfunction; *QOL*, quality of life; *SCI*, spinal cord injury; *UTI*, urinary tract infection.
[a]*Only patients enabled to urinate on a uroflowmeter.*

## Urinary management (Table 3)

### Urinary management during acute phase

During an acute phase, an indwelling catheterization should be changed to CIC as soon as the general condition of the patient becomes stable and urine output decreases to approximately 1.5 to 2.0 L per day, because long-term indwelling catheterization will cause catheter-associated complications (Welk et al., 2018).

**TABLE 2 Risk factors for clinical outcomes.**

| | Risk factors |
|---|---|
| Renal impairment (renal dysfunction[a]/upper urinary tract deterioration[b]) | • Video-urodynamic findings: decreased maximal bladder capacity, high pressure, and/or sustained detrusor overactivity, detrusor sphincter dyssynergia, decreased bladder compliance, high detrusor leak point pressure<br>• Urinary management: indwelling urethral catheterization, bladder reflex triggering<br>• Level and severity of the injury: tetraplegia, complete paralysis |
| Symptomatic urinary tract infection[c] | • Video-urodynamic findings: vesico-ureteral reflux secondary to lower urinary tract dysfunction, that is, decreased maximal bladder capacity, detrusor overactivity, detrusor sphincter dyssynergia, decreased bladder compliance, high detrusor leak point pressure<br>• Urinary management: indwelling urethral catheterization<br>• Level and severity of injury: cervical cord injury, A or B on American Spinal Injury Association impairment scale |
| Urinary incontinence | • Video-urodynamic findings: detrusor overactivity, decreased bladder compliance, intrinsic sphincter deficiency |
| Quality of life | • Negative impact: indwelling catheterization, clean intermittent catheterization by attendants<br>• Positive impact: fewer worries about and fewer restrictions arising from urinary incontinence, less concerns about urinary tract infection as well as deterioration of bladder function |

The presence or absence of risk factors is relevant to management of lower urinary tract dysfunction.
[a]*Renal dysfunction: increased serum creatinine level, decreased glomerular filtration rate, or renal scarring on renal scintigraphy.*
[b]*Upper urinary tract deterioration: hydroureteronephrosis, vesico-ureteral reflux, febrile upper urinary tract infection, or upper urinary tract stones.*
[c]*Symptomatic urinary tract infection: significant bacteriuria, pyuria, and symptoms suggestive of urinary tract infection, such as discomfort and/or pain in renal or supra-pubic area in incomplete injury, cloudiness of urine, worsening of urine odor, development or worsening of urinary incontinence or leakage around the catheter, pain on micturition, worsening of spasticity discomfort/lethargy/anorexia, autonomic dysreflexia, and fever or chills, without any obvious causes for those findings and symptoms.*

## Urinary management during chronic phase

The appropriate urinary management is guided by the presence or absence of risk factors for renal impairment and symptomatic UTI, as well as high-pressure voiding, significant PVR, and marked bladder deformity.

### *CIC*

Patients who have risk factors and/or high-pressure voiding, PVR, or bladder deformity should be managed with CIC (Romo et al., 2018). Based on the VUDS findings, the physician should determine the frequency, timing, and interval of CIC. Proper education and instruction with adequate follow-up are of paramount importance in the long-term compliance and persistence of CIC as well as avoiding complications such as urethral trauma and stricture in male patients, gross hematuria, and symptomatic UTI (Biardeau & Corcos, 2016). In addition, clean intermittent self-catheterization is preferred to CIC by caregivers in terms of maintaining QOL (Adriaansen et al., 2017). In patients managed with CIC, treatment for asymptomatic pyuria/bacteriuria (asymptomatic UTI) with antibiotics is not recommended because there is a dearth of clear evidence supporting the use of antibiotics for prevention of symptomatic UTI and the emergence of drug-resistant bacteria is a major concern (Sekido et al., 2020). There are several different types of catheters available for CIC ( Jeong & Oh, 2019; Li et al., 2013; McClurg et al., 2017; Sun, Comiter, & Elliott, 2018) and please see "Key Facts of Catheters for Clean Intermittent Catheterization."

### *Voiding*

Patients who do not have the above hostile findings and can void voluntarily can be managed with voluntary voiding, which means that patients are able to store an adequate volume of urine at low pressure and to void without bladder reflex triggering or bladder expression (Sekido et al., 2020). Voiding with bladder reflex triggering is indicated only in carefully selected male cervical cord injured patients with good reflex bladder contractility who cannot be managed with CIC due to the patient's disabilities and/or the lack of caregiver supports (Romo et al., 2018; Sekido et al., 2020). Transurethral

**TABLE 3 Urinary management.**

| Urinary management | Descriptions |
|---|---|
| Voiding | • Transurethral bladder emptying without any catheterization |
| Voluntary (spontaneous) voiding | • Coordinated physiologic voluntary voiding, that is, voluntary detrusor contraction simultaneously with voluntary urethral relaxation<br>• No need of triggering or Crede/Valsalva maneuvers |
| Bladder reflex triggering | • Need of triggers such as percussion on supra-pubic area to induce micturition reflex<br>• Often concomitant sphincterotomy to overcome DSD |
| Bladder expression | • Need of Crede and/or Valsalva maneuvers to empty bladder<br>• Preferable to be replaced by CIC |
| CIC | • Usually per urethra but, in carefully selected patients unable to perform transurethral catheterization, per abdominal continent catheterizable channel<br>• Low pressure and residual-free bladder emptying within low-pressure storage, if properly performed<br>• Need of adherence to appropriate frequency/interval as well as individualized fluid intake to avoid bladder over-distension or high storage pressure<br>• May be performed even by male and female patients with cervical cord injury at the level of C5B and C6B, respectively, with great difficulty, but in practice more residual functions are necessary. |
| Indwelling catheterization | • Long term bladder drainage method with indwelling catheters<br>• The last resort due to risk of a variety of catheter-associated complications |
| Indwelling urethral catheterization | • Transurethral catheter drainage<br>• 14–16 Fr caliber catheters with 5–10 mL balloon replaced every 2–6 weeks |
| Indwelling supra-pubic catheterization | • Percutaneous catheter drainage<br>• Frequent need of spinal or general anesthesia and supra-pubic cystotomy at catheter insertion<br>• Optional urinary management in patients with high cervical SCI<br>• Avoidance of urethral complications<br>• Use of larger (18–22 Fr) caliber catheters with short tips, resulting in less incidence of catheter blockage than indwelling urethral catheterization |

This table describes the key urinary management for patients with spinal cord injury.
*CIC*, clean intermittent catheterization; *DSD*, detrusor sphincter dyssynergia.

sphincterotomy is often needed to achieve low-pressure storage as well as voiding without significant PVR and to prevent UUT deterioration (Takahashi, Kimoto, & Eto, 2018). Bladder expression with manual compression of the supra-pubic area (Crede maneuver) or straining (Valsalva maneuver) is not recommended as voiding because bladder expression is potentially urodynamically unsafe for the urinary tract due to the presence of a nonrelaxing urethral sphincter and should be replaced by CIC (Sekido et al., 2020).

### *Indwelling catheterization*

Generally speaking, indwelling catheterization is not recommended for urinary management of patients with SCI due to a constellation of unavoidable catheter-associated complications, such as symptomatic UTI, urolithiasis, catheter blockage/obstruction, AD, and bladder cancer (Igawa, Wyndaele, & Nishizawa, 2008; Sekido et al., 2020). This type of urinary management is mainly indicated for patients with cervical SCI who cannot be managed with CIC (Sorokin & De, 2015). Suprapubic catheterization may be better than urethral catheterization in terms of freedom from urethral or genital complications such as urethral trauma, urethral diverticula, urethrocutaneous fistula, epididymitis, or intrascrotal abscess (Feifer & Corcos, 2008; Hunter, Bharmal, & Moore, 2013; Igawa, et al., 2008). On the other hand, the intermittent (temporary) use of an indwelling urethral catheter is an option for patients with nocturnal polyuria and/or difficulty in performing CIC away from home (Ozawa et al., 2005; Sekido et al., 2020; Sturm, Cantrell, Durbin-Johnson, & Kurzrock, 2020). For example, a specialized indwelling catheter for this purpose (Fig. 2) has been approved in Japan, and it is recommended that indwelling time should be kept to a minimum, up to half a day at the most (Sekido et al., 2020). For this reason, proper patient education, including the fact that it is not an alternative to CIC, is essential.

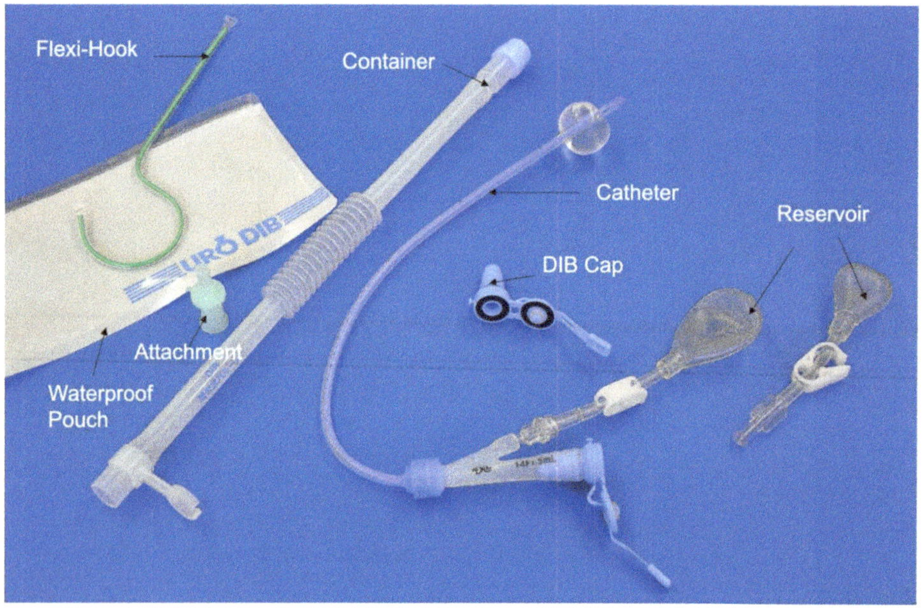


FIG. 2 Specialized indwelling catheter for intermittent (temporary) use approved in Japan. To inflate and indwell the balloon in bladder, sterile water from the reservoir to the balloon is sent. To contract the balloon and remove the catheter, the sterile water in the balloon is sent back to the reservoir. DIB Cap is used for intermittent drainage, especially away from home. Kindly provided by courtesy of DIB International Co., Ltd., Tokyo, JAPAN.

## Pharmacological treatment

### To facilitate urine storage

Anti-cholinergic drugs, β3-adrenoceptor agonist, or a combination of these drugs suppresses DO and increases bladder compliance and bladder capacity, which ameliorates bladder storage dysfunction (Han, Cho, Jung, Jang, & Lee, 2019; Sekido et al., 2020; Stothers, Tsang, Nigro, Lazare, & Macnab, 2016). Therefore, these drugs are recommended for patients who have the risk factors for renal impairment and symptomatic UTI or UI (Sekido et al., 2020).

### To facilitate bladder emptying

α1-Adrenernoceptor antagonists may reduce the resistance at the bladder outlet (Yasuda, Yamanishi, Kawabe, Ohshima, & Morita, 1996). Therefore, these drugs may be used in patients managed by (voluntary) voiding for their potential to improve voiding efficiency despite the lack of robust evidence of these drugs.

The use of parasympathomimetic drugs [muscarinic receptor agonists (e.g., bethanechol) and peripheral cholinesterase inhibitors (e.g., distigmine)] that may improve bladder contractility are not recommended because there is a dearth of evidence on these drugs for patients with NLUTD due to SCI and a possibility of worsening of functional BOO (Barendrecht, Oelke, Laguma, & Michel, 2007).

## Assessment of clinical outcomes (periodic follow-up)

### Proposal for follow-up protocol

Periodic follow-up is needed to ensure that clinical outcomes are achieved; it includes a medical interview and urinalysis every 3 to 6 months, combined with FVC or the bladder diary especially in patients with CIC, and transabdominal ultrasonography to evaluate the urinary tract annually (Kreydin et al., 2018; Sekido et al., 2020). Follow-up protocols of measurement of PVR and uroflow, and VUDS have not been established; however, these examinations should be performed within 3 to 6 months if urinary management and/or pharmacological treatments are changed. Moreover, these examinations should be promptly performed if urinary tract complications develop, UI is worsened, PVR increases, or bladder deformity becomes evident (Kreydin et al., 2018; Sekido et al., 2020). VUDS can tailor urodynamic study without fluoroscopy to the individual patient's condition.

### Proposal for repeated urodynamic study

Regarding urodynamic study, The International Continence Society Urodynamic Committee proposed that "The panel recommends performing the first exam as soon as possible after the end of spinal shock phase and to perform another urodynamic testing at 6 and 12 months after the injury (level of evidence 4, expert opinions). Then, a yearly exam is preferable in SCI patients with cervical and thoracic injury for the first 5 years, deferring it to one in 2 years after attaining a low-pressure reservoir with continence and complete emptying. In case of lumbar injury, urodynamics could be done yearly for the first 2 years and once in 2 years thereafter (level of evidence 4, expert opinions)" (Schurch et al., 2018). In addition, the Committee emphasized that "Maximal attention should be paid to patients with incomplete suppression of detrusor overactivity, in those who empty their bladders by reflex voiding, to poor bladder compliance and in the presence of detrusor sphincter dyssynergia" (Schurch et al., 2018).

### Proposal for surveillance for bladder cancer

Although the development of bladder cancer associated with long-term indwelling catheterization is a serious problem, recent reports suggest that the incidence of bladder cancer is not as high as previously thought (Elliott, 2015), and routine screening for bladder cancer using cystourethroscopy or urine cytology in asymptomatic patients is not recommended (Gui-Zhong & Li-Bo, 2017). On the other hand, it should be kept in mind that early detection of bladder cancer in patients with NLUTD is difficult and cancer-specific survival rates are low (Welk, McIntyre, Teasell, Potter, & Loh, 2013). Therefore, not only in patients under long-term indwelling catheterization, but also in all patients with NLUTD due to SCI, it is essential to try to investigate each case on an individual basis, without underestimating the suspicious signs of bladder cancer, such as gross hematuria, worsening of AD, refractory recurrent UTI, and abnormal thickening of the bladder wall on transabdominal ultrasound (Sekido et al., 2020).

## Surgical treatment

### Transurethral intra-detrusor injection of type A botulinum toxin (BoNT-A)

Several good quality randomized controlled trials have demonstrated the efficacy and safety of transurethral intra-detrusor injection of BoNT-A, thus establishing this minimally invasive treatment as an excellent option for refractory DO (Cruz et al., 2011; Ginsberg et al., 2012). BoNT-A hampers the fusion between a synaptic vesicle containing acetylcholine and a presynaptic membrane, leading to suppression of DO. BoNT-A decreases the number of UI episodes and detrusor pressure as well as increases bladder capacity and bladder compliance. The most frequent complications are symptomatic UTI and, in patients managed with voiding, increased PVR requiring CIC.

### Other surgical treatments

The surgical therapies dealt with below should be performed by experienced neurourologists in the specialized centers for NLUTD. In addition, robot-assisted laparoscopic techniques for augmentation cystoplasty (AC), bladder neck surgery, and urinary diversion in patients with NLUTD are still considered to be in their infancy, but they show promise in terms of reducing the invasiveness of the procedures (Cohen, Pariser, Anderson, Pearce, & Gundeti, 2015; Gargollo, 2015).

#### *AC*

When all the maximum conservative and minimally invasive treatments fail, AC remains an important option for the improvement of progressive UUT deterioration, recurrent symptomatic UTI, and intractable UI that are attributed to refractory DO and low bladder compliance (Hoen et al., 2017). Usually, the bladder is augmented by a detubularized intestinal segment (enterocystoplasty) with the most commonly used segment being ileum (ileocystoplasty). The major problems with AC are the various frequently occurring peri- and postoperative complications. Peri-operative complications include prolonged ileus, wound infection, thromboembolism, and hemorrhage (Hoen et al., 2017). Long-term postoperative complications include urolithiasis, bowel dysfunction, electrolyte-metabolic disturbances, bladder perforation, and malignant neoplasm (Hoen et al., 2017). AC without the use of an intestinal segment is called auto-augmentation (detrusor myectomy), in which significant area of the detrusor is resected, leaving the mucosal layer intact. Although less invasive than enterocystoplasty, in terms of creating a low-pressure and high-volume environment for urine storage, auto-augmentation generally has poorer outcomes than enterocystoplasty (Wyndaele et al., 2018).

### Anti-incontinence surgery

As surgical treatments for ISD in patients with NLUTD, bladder neck reconstruction (BNR), bladder neck sling (BNS), artificial urinary sphincter (AUS), and bladder neck closure with an abdominal continent catheterizable channel have been reported, although optimal surgical procedures remain controversial (Myers et al., 2016; Phé et al., 2017; Wyndaele et al., 2018). BNR is often selected for ISD associated with congenital anomalies, such as bladder exstrophy, and there have been few reports on the efficacy and safety for ISD associated with SCI (Myers et al., 2016). BNS can be conducted in both male and female patients, but the majority of reports are of a pubovaginal sling in female patients with some established efficacy and safety (Myers et al., 2016; Wyndaele et al., 2018). Not only in female patients but also in male patients with NLUTD, AUS is preferably placed at the bladder neck to probable decrease erosion rates (Myers et al., 2016; Wyndaele et al., 2018). Bladder neck closure is the last resort in cases of intractable UI resulting from severe ISD that would be difficult to treat with other anti-incontinence procedures, or in cases where previous anti-incontinence procedures have failed (Wyndaele et al., 2018). Moreover, mid-urethral synthetic sling procedures such as tension-free vaginal tape and transobturator tape in female patients as well as a perineal mesh sling in male patients with NLUTD have recently been reported (El-Azab & El-Nashar, 2015; Myers et al., 2016; Pannek & Wöllner, 2017). The postoperative complications include persistent UI, difficulty in catheterization, urolithiasis, device malfunction and device infection in AUS, vesico-cutaneous fistula and stomal stenosis in bladder neck closure with an abdominal continent catheterizable channel, and tape or mesh erosion in procedures using synthetic materials (Wyndaele et al., 2018).

### Urinary diversion

Incontinent urinary diversion as well as a continent urinary reservoir is regarded as the last resort for carefully selected patients with NLUTD (Sekido et al., 2020). Although preservation of renal function, decreased incidence of symptomatic UTI, and attainment of social continence are achieved, a variety of peri- and postoperative complications may be encountered, which needs to be appropriately addressed (Sekido et al., 2020).

### Sacral neuromodulation (SNM)

SNM may be a treatment of choice for carefully selected patients with incomplete SCI with refractory overactive bladder as well as nonobstructive urinary retention (Kessler et al., 2010). Electrode(s) for stimulation are usually inserted into the S3 foramen unilaterally or bilaterally, and stimulation to somatic afferents may bring reorganization of neural pathways to the central nervous system, resulting in suppression as well as enhancement of the micturition reflex. Moreover, recent studies indicated that implementing SNM during the acute phase of SCI above T12 may inhibit development of DO and DSD in the chronic phase (Sievert et al., 2010). Adverse events include new pain/undesirable change in stimulation, pain at the implantable pulse generator site, and adverse change in bowel function (Kessler et al., 2010).

### Sacral anterior root stimulation (SARS)

SARS with dorsal rhizotomy may be a treatment of choice to attain voiding and urinary continence in extremely carefully selected patients with supra-sacral NLUTD due to complete SCI (Wyndaele et al., 2018). Although stimulation of S2 to S4 anterior roots excites both detrusor and the external urethral sphincter, once the stimulation is turned off, the external sphincter relaxes, while detrusor contraction continues, which means that voiding occurs after stimulation (poststimulus voiding). SARS not only reestablishes voiding, but also ameliorates UI, decreases the frequency of symptomatic UTI and AD, and prevents UUT deterioration through inhibition of DO due to dorsal rhizotomy (Wyndaele et al., 2018). Adverse events include device malfunction and leakage of cerebrospinal fluid (Wyndaele et al., 2018).

### Transurethral sphincterotomy and others

As mentioned in "Urinary management" section, sphincterotomy for DSD is often performed to improve voiding efficiency and to prevent UUT deterioration in patients managed by bladder reflex triggering. Complications include insufficient relief of DSD leading to repeated procedures, hemorrhage, and urethrocutaneous fistula (Sekido et al., 2020; Wyndaele et al., 2018). Less invasive procedures such as urethral stents and intra-sphincteric injection of type A botulinum toxin, which is an off-label use, have been reported (Chancellor et al., 1999; Mehta et al., 2012), but have not become widespread yet.

## Applications to other areas of neuroscience

Management for NLUTD due to spinal cord injury is the basis for all NLUTDs or non-NLUTDs (Groen et al., 2016). In terms of urinary management, CIC become the standard urinary tract management for NLUTD in which voluntary voiding is not urodynamically safe. For pharmacological treatment, anti-cholinergics and β3 adrenoreceptor agonists are used to ameliorate the neurogenic overactive bladder associated with stroke and Parkinson's disease. BoNT-A has established efficacy and safety in the treatment of refractory DO due to multiple sclerosis (Cruz et al., 2011; Ginsberg et al., 2012).

Moreover, a plenty of basic research using the SCI model has played a major role in elucidating the nervous system involved in LUT function. In normal animals, myelinated A-delta afferent fibers activate the micturition reflex, whereas in chronic SCI animals, micturition is induced by unmyelinated C afferent fibers, resulting in development of DO (de Groat et al., 2015). This contrasts with the pathophysiology of supra-spinal DO seen in stroke and Parkinson's disease, which is caused by loss of forebrain inhibition and/or enhancement of forebrain activation to PAG and PMC (de Groat et al., 2015). Recently, it was reported that transurethral intra-detrusor injection of BoNT-A reduced transient receptor potential vanilloid receptor 1 receptors and purinergic subtype 2X receptors on C fibers, which is considered to be a contributing factor for ameliorating DO (de Groat et al., 2015).

## Mini-dictionary of terms

Bladder compliance: An index of the distensibility of the bladder during the storage phase. The normal bladder has very good distensibility, ensuring urine storage at low pressure and of sufficient capacity.

Detrusor: Smooth muscle of the bladder, which relaxes during the storage phase but contracts during the emptying phase.

Detrusor leak point pressure: The lowest detrusor pressure at leakage without detrusor contraction or increased abdominal pressure.

Detrusor overactivity: Involuntary detrusor contraction during the storage phase, which can be phasic or sustained.

Detrusor sphincter dyssynergia: Involuntary external urethral sphincter contraction concomitant with detrusor overactivity during the storage phase or voiding detrusor contraction during the emptying phase in patients with relevant neurological disorders.

Detrusor underactivity: Inadequate detrusor contractile strength and/or duration.

Frequency volume chart: A chart describing the time and volume of each micturition by voiding or clean intermittent catheterization. If relevant information such as the fluid intake, frequency and degree of, and circumstances at urinary incontinence, is also included, the chart is called a "Bladder diary."

Intrinsic sphincter deficiency: Incompetent urethral closure mechanism, which usually caused by anatomical or functional abnormality of the urethral smooth muscle and/or external urethral sphincter.

Neurogenic lower urinary tract dysfunction: Lower urinary tract dysfunction caused by relevant neurological disorders.

Nonrelaxing urethral sphincter: During the emptying phase, the external urethral sphincter is not relaxed but obstructed, resulting in reduced urinary flow.

Postvoid residual: Urine volume left in the bladder after voiding.

Urodynamic study: Examination for precise assessment of lower urinary tract function during storage and emptying, which simultaneously measures intravesical and abdominal pressure, infused volume of fluid, and often electromyography of external urethral sphincter, concomitant with uroflowmetry during the emptying phase. Detrusor pressure is calculated by subtracting abdominal pressure from intravesical pressure.

Uroflowmetry: Measurement of urine volume (mL) excreted per unit of time (s) using a specific device called a uroflowmeter.

Video-urodynamic study: Performing a urodynamic study under fluoroscopy, which enables functional as well as morphological evaluation.

## Key facts of catheters for clean intermittent catheterization

Catheter for clean intermittent catheterization (CIC) is divided into two types: disposable and reusable.

The disposable catheter is popular worldwide and is divided into two types: polyvinyl (PVC) and hydrophilic-coated (HCC), although disposable catheters are frequently reused.

At CIC, PVC requires a lubricant for smooth catheterization, while HCC does not, which increases patient convenience.

HCC decreases the incidence of urinary tract infection and gross hematuria (surrogate for urethral trauma), improves the quality of life, and is cost-effective, although long-term outcomes remain controversial.

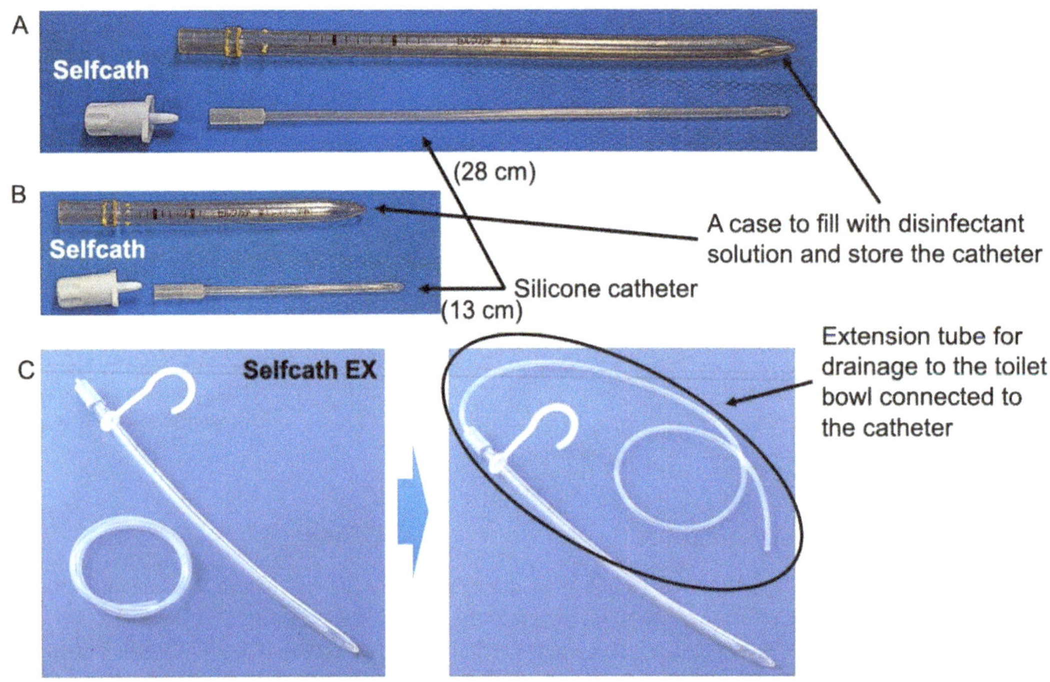


FIG. 3 Examples of reusable silicone catheters approved in Japan. The patients carry the reusable silicone catheter in a case filled with disinfectant solution. (A) for males; (B) for females; (C) for wheelchair patients. Kindly provided by courtesy of Fuji Systems Corporation, Tokyo, Japan.

The cost of HCC is an obstacle to its widespread use, especially in developing countries.

Disposable catheter waste may become an issue in the future in terms of environmental impact.

A reusable catheter is available in some countries, for example in Japan, and some patients use a reusable catheter at home, while using a disposable catheter away from home.

In Japan, several kinds of reusable silicon catheters (Fig. 3) have been approved and widely used for many years, but they should be changed monthly according to the product description, and is stored in a glycerine solution containing 0.025% benzalkonium chloride that is changed daily.

Some reusable catheters are designed to be more convenient for wheelchair patients (Fig. 3).

## Summary points

- Management of lower urinary tract dysfunction is an important part of the clinical practice of patients with spinal cord injury.
- The critical clinical outcomes to be achieved are prevention of urinary tract complications, especially renal impairment and symptomatic urinary tract infection, attainment of social urinary continence, and improvement of quality of life.
- Video-urodynamic study plays an important role in identifying risk factors for renal impairment, symptomatic urinary tract infection, and urinary incontinence.
- Appropriate urinary management and pharmacological treatment aimed at preventing renal impairment and symptomatic urinary tract infection are determined depending on the presence or absence of risk factors for these urinary tract complications.
- If these outcomes are achieved, urinary management and pharmacological treatment to obtain social urinary continence and enhance the quality of life should be optimized without compromising the outcomes of prevention of renal impairment and symptomatic urinary tract infection.
- If maximum conservative treatment fails to achieve the above outcomes, surgical treatment becomes an option.

## References

Adriaansen, J. J., van Asbeck, F. W., Tepper, M., Faber, W. X., Visser-Meily, J. M., de Kort, L. M., et al. (2017). Bladder-emptying methods, neurogenic lower urinary tract dysfunction and impact on quality of life in people with long-term spinal cord injury. *The Journal of Spinal Cord Medicine*, *40*, 43–53.

Barendrecht, M. M., Oelke, M., Laguma, M. P., & Michel, M. C. (2007). Is the use of parasympathomimetics for treating an underactive urinary bladder evidence-based? *BJU International*, *99*, 749–752.

Biardeau, X., & Corcos, J. (2016). Intermittent catheterization in neurologic patients: Update on genitourinary tract infection and urethral trauma. *Annals of Physical and Rehabilitation Medicine*, *59*, 125–129.

Bywater, M., Tornic, J., Mehnert, U., & Kessler, T. M. (2018). Detrusor acontractility after acute spinal cord injury-myth or reality? *The Journal of Urology*, *199*, 1565–1570.

Chan, L. W., Griebling, T. L., Arnold, E. P., Chu, P. S., New, P. W., & Wagg, A. (2018). Special considerations in the urological management of the older spinal cord injury patient. *World Journal of Urology*, *36*, 1603–1611.

Chancellor, M. B., Bennett, C., Simoneau, A. R., Finocchiaro, M. V., Kline, C., Bennett, J. K., et al. (1999). Sphincteric stent versus external sphincterotomy in spinal cord injured men: Prospective randomized multicenter trial. *The Journal of Urology*, *161*, 1893–1898.

Cohen, A. J., Pariser, J. J., Anderson, B. B., Pearce, S. M., & Gundeti, M. S. (2015). The robotic appendicovesicostomy and bladder augmentation: The next frontier in robotics, are we there? *The Urologic Clinics of North America*, *42*, 121–130.

Costa, P., Perrouin-Verbe, B., Colvez, A., Didier, J., Marquis, P., Marrel, A., et al. (2001). Quality of life in spinal cord injury patients with urinary difficulties. Development and validation of qualiveen. *European Urology*, *39*, 107–113.

Cruz, F., Herschorn, S., Aliotta, P., Brin, M., Thompson, C., Lam, W., et al. (2011). Efficacy and safety of onabotulinumtoxinA in patients with urinary incontinence due to neurogenic detrusor overactivity: A randomized, double-blind, placebo-controlled trial. *European Urology*, *60*, 742–750.

de Groat, W. C., Griffiths, D., & Yoshimura, N. (2015). Neural control of the lower urinary tract. *Comprehensive Physiology*, *5*, 327–396.

El-Azab, A. S., & El-Nashar, S. A. (2015). Midurethral slings versus the standard pubovaginal slings for women with neurogenic stress urinary incontinence. *International Urogynecology Journal*, *26*, 427–432.

Elliott, S. P. (2015). Screening for bladder cancer in individuals with spinal cord injury. *The Journal of Urology*, *193*, 1880–1881.

Feifer, A., & Corcos, J. (2008). Contemporary role of suprapubic cystostomy in treatment of neuropathic bladder dysfunction in spinal cord injured patients. *Neurourology and Urodynamics*, *27*, 475–479.

Gargollo, P. C. (2015). Robotic-assisted bladder neck repair: Feasibility and outcomes. *The Urologic Clinics of North America*, *42*, 111–120.

Ginsberg, D., Gousse, A., Keppenne, V., Sievert, K. D., Thompson, C., Lam, W., et al. (2012). Phase 3 efficacy and tolerability study of onabotulinumtoxinA for urinary incontinence from neurogenic detrusor overactivity. *The Journal of Urology*, *187*, 2131–2139.

Goto, T., Kawasaki, Y., Takemoto, J., Abe, Y., & Namima, T. (2018). Evaluating estimated glomerular filtration rates of creatinine and cystatin C for male patients with chronic spinal cord injury. *Spinal Cord*, *56*, 447–452.

Griffiths, D. (2015). Neural control of micturition in humans: A working model. *Nature Reviews. Urology*, *12*, 695–705.

Groen, J., Pannek, J., Castro Diaz, D., Del Popolo, G., Gross, T., Hamid, R., et al. (2016). Summary of European Association of Urology (EAU) guidelines on neuro-urology. *European Urology*, *69*, 324–333.

Gui-Zhong, L., & Li-Bo, M. (2017). Bladder cancer in individuals with spinal cord injuries: A meta-analysis. *Spinal Cord*, *55*, 340–345.

Guttmann, L., & Frankel, H. (1966). The value of intermittent catheterisation in the early management of traumatic paraplegia and tetraplegia. *Paraplegia*, *4*, 63–84.

Han, S. H., Cho, I. K., Jung, J. H., Jang, S. H., & Lee, B. S. (2019). Long term efficacy of mirabegron add on therapy to antimuscarinic agents in patients with spinal cord injury. *Annals of Rehabilitation Medicine*, *43*, 54–61.

Hoen, L., Ecclestone, H., Blok, B. F. M., Karsenty, G., Phé, V., Bossier, R., et al. (2017). Long-term effectiveness and complication rates of bladder augmentation in patients with neurogenic bladder dysfunction: A systematic review. *Neurourology and Urodynamics*, *36*, 1685–1702.

Hunter, K. F., Bharmal, A., & Moore, K. N. (2013). Long-term bladder drainage: Suprapubic catheter versus other methods: A scoping review. *Neurourology and Urodynamics*, *32*, 944–951.

Igawa, Y., Wyndaele, J. J., & Nishizawa, O. (2008). Catheterization: Possible complications and their prevention and treatment. *International Journal of Urology*, *15*, 481–485.

Jeong, S. J., & Oh, S. J. (2019). Recent updates in urinary catheter products for the neurogenic bladder patients with spinal cord injury. *Korean Journal of Neurotrauma*, *15*, 77–87.

Kessler, T. M., Framboise, D. L., Trelle, S., Fowler, C. J., Kiss, G., Pannek, J., et al. (2010). Sacral neuromodulation for neurogenic lower urinary tract dysfunction: Systematic review and meta-analysis. *European Urology*, *58*, 865–874.

Kreydin, E., Welk, B., Chung, D., Clemens, Q., Yang, C., Danforth, T., et al. (2018). Surveillance and management of urologic complications after spinal cord injury. *World Journal of Urology*, *36*, 1545–1553.

Li, L., Ye, W., Ruan, H., Yang, B., Zhang, S., & Li, L. (2013). Impact of hydrophilic catheters on urinary tract infections in people with spinal cord injury: Systematic review and meta-analysis of randomized controlled trials. *Archives of Physical Medicine and Rehabilitation*, *94*, 782–787.

Liu, N., Zhou, M. W., Biering-Sørensen, F., & Krassioukov, A. V. (2017). Cardiovascular response during urodynamics in individuals with spinal cord injury. *Spinal Cord*, *55*, 279–284.

McClurg, D., Coyle, J., Long, A., Moore, K., Cottenden, A., May, C., et al. (2017). A two phased study on health care professionals' perceptions of single or multi-use of intermittent catheters. *International Journal of Nursing Studies*, *72*, 83–90.

Mehta, S., Hill, D., Foley, N., Hsieh, J., Ethans, K., Potter, P., et al. (2012). A meta-analysis of botulinum toxin sphincteric injections in the treatment of incomplete voiding after spinal cord injury. *Archives of Physical Medicine and Rehabilitation*, *93*, 597–603.

Merriam, W. F., Taylor, T. K., Ruff, S. J., & McPhail, M. J. (1986). A reappraisal of acute traumatic central cord syndrome. *Journal of Bone and Joint Surgery. British Volume (London)*, *68*, 708–713.

Myers, J. B., Mayer, E. N., Lenherr, S., & Neurogenic Bladder Research Group (NBRG.org). (2016). Management options for sphincteric deficiency in adults with neurogenic bladder. *Translational Andrology and Urology*, *5*, 145–157.

Osman, N. I., Chapple, C. R., Abrams, P., Dmochowski, R., Haab, F., Nitti, V., et al. (2014). Detrusor underactivity and the underactive bladder: A new clinical entity? A review of current terminology, definitions, epidemiology, aetiology, and diagnosis. *European Urology*, *65*, 389–398.

Ozawa, H., Uematsu, K., Ohmori, H., Kondo, A., Iwatsubo, E., & Takasaka, S. (2005). Long-term usefulness and safety of the contemporary balloon catheter. *Nihon Hinyokika Gakkai Zasshi*, *96*, 541–547.

Pannek, J., & Wöllner, J. (2017). Treatment of stress urinary incontinence in men with spinal cord injury: Minimally invasive = minimally effective? *Spinal Cord*, *55*, 739–742.

Phé, V., Boissier, R., Blok, B. F. M., Del Popolo, G., Musco, S., Castro-Diaz, D., et al. (2017). Continent catheterizable tubes/stomas in adult neuro-urological patients: A systematic review. *Neurourology and Urodynamics*, *36*, 1711–1722.

Pinder, B., Lloyd, A. J., Elwick, H., Denys, P., Marley, J., & Bonniaud, V. (2012). Development and psychometric validation of the intermittent self-catheterization questionnaire. *Clinical Therapeutics*, *34*, 2302–2313.

Romo, P. G. B., Smith, C. P., Cox, A., Averbeck, M. A., Dowling, C., Beckford, C., et al. (2018). Non-surgical urologic management of neurogenic bladder after spinal cord injury. *World Journal of Urology*, *36*, 1555–1556.

Schurch, B., Iacovelli, V., Averbeck, M. A., Stefano, C., Altaweel, W., & Finazzi, A. (2018). Urodynamics in patients with spinal cord injury: A clinical review and best practice paper by a working group of The International Continence Society Urodynamics Committee. *Neurourology and Urodynamics*, *37*, 581–591.

Sekido, N., Igawa, Y., Kakizaki, H., Kitta, T., Sengoku, A., Takahashi, S., et al. (2020). Clinical guidelines for the diagnosis and treatment of lower urinary tract dysfunction in patients with spinal cord injury. *International Journal of Urology*, *7*, 276–288.

Sievert, K. D., Amend, B., Gakis, G., Toomey, P., Badke, A., Kaps, H. P., et al. (2010). Early sacral neuromodulation prevents urinary incontinence after complete spinal cord injury. *Annals of Neurology*, *67*, 74–84.

Smith, C. P., Kraus, S. R., Nickell, K. G., & Boone, T. B. (2000). Video urodynamic findings in men with the central cord syndrome. *The Journal of Urology*, *164*, 2014–2017.

Sorokin, I., & De, E. (2015). Options for independent bladder management in patients with spinal cord injury and hand function prohibiting intermittent catheterization. *Neurourology and Urodynamics*, *34*, 167–176.

Stothers, L., Tsang, B., Nigro, M., Lazare, D., & Macnab, A. (2016). An integrative review of standardized clinical evaluation tool utilization in anticholinergic drug trials for neurogenic lower urinary tract dysfunction. *Spinal Cord*, *54*, 1114–1120.

Sturm, R. M., Cantrell, A. B., Durbin-Johnson, B., & Kurzrock, E. A. (2020). Use of a closed diurnal indwelling catheter for patients with spinal cord disorders: A pilot study. *Urology*, *142*, 237–242. https://doi.org/10.1016/j.urology.2020.04.033. Epub 2020 Apr 19 32320790.

Sun, A. J., Comiter, C. V., & Elliott, C. S. (2018). The cost of a catheter: An environmental perspective on single use clean intermittent catheterization. *Neurourology and Urodynamics*, *37*, 2204–2208.

Takahashi, R., Kimoto, Y., & Eto, M. (2018). Long-term urodynamic follow-up after external sphincterotomy in patients with spinal cord injury. *Neurourology and Urodynamics*, *37*, 2625–2631.

Welk, B., Lenherr, S., Elliott, S., Stoffel, J., Presson, A. P., Zhang, C., et al. (2018). The neurogenic bladder symptom score (NBSS): A secondary assessment of its validity, reliability among people with a spinal cord injury. *Spinal Cord*, *56*, 259–264.

Welk, B., & McArthur, E. (2020). Increased risk of dementia among patients with overactive bladder treated with an anticholinergic medication compared to a beta-3 agonist: A population-based cohort study. *BJU International*, *126*, 183–190.

Welk, B., McIntyre, A., Teasell, R., Potter, P., & Loh, E. (2013). Bladder cancer in individuals with spinal cord injuries. *Spinal Cord*, *51*, 516–523.

Welk, B., Schneider, M. P., Thavaseelan, J., Traini, L. R., Curt, A., & Kessler, T. M. (2018). Early urological care of patients with spinal cord injury. *World Journal of Urology*, *36*, 1537–1544.

Wyndaele, J. J., Birch, B., Borau, A., Burks, F., Castro-Diaz, D., Chartier-Kastler, E., et al. (2018). Surgical management of the neurogenic bladder after spinal cord injury. *World Journal of Urology*, *36*, 1569–1576. 2018.

Yasuda, K., Yamanishi, T., Kawabe, K., Ohshima, H., & Morita, T. (1996). The effect of urapidil on neurogenic bladder: A placebo controlled double-blind study. *The Journal of Urology*, *156*, 1125–1130.

Zlatev, D. V., Shem, K., & Elliott, C. S. (2018). Predictors of long-term bladder management in spinal cord injury patients-upper extremity function may matter most. *Neurourology and Urodynamics*, *37*, 1106–1112.

Chapter 22

# Bed support surfaces for preventing pressure injuries after spinal cord injury

Annie Levasseur[a], Jean-Marc Mac-Thiong[b], and Andréane Richard-Denis[c]

[a]*Department of Biomedical Sciences, University of Montreal, Research Center CIUSSS Nord-de-l'Île-de-Montréal, Montreal, QC, Canada;* [b]*Department of Surgery, Faculty of Medicine, University of Montreal, Hôpital du Sacré-Coeur de Montréal, CIUSSS Nord-de-l'Île-de -Montréal, Montreal, QC, Canada;* [c]*Department of Medicine, Faculty of Medicine, University of Montreal, Hôpital du Sacré-Coeur de Montréal, CIUSSS Nord-de-l'Île-de-Montréal, Montreal, QC, Canada*

## List of abbreviations

| | |
|---|---|
| **NPIAP** | National pressure injury advisory panel |
| **PI** | Pressure injury |
| **SCI** | Spinal cord injury |

## Introduction

After a spinal cord injury (SCI), individuals often remain with limitations that will impede mobility and functional independence. Presence and severity of the neurological deficits associated with SCI represent important risks for developing pressure injuries (PI). During hospitalization, pain, associated injuries, medical complications, spasticity, joint contractures, and surgical interventions may further lead to prolonged bed rest and contribute to increased occurrence of PI. Achieving an optimal plan for preventing PI can be difficult in practice, particularly for individuals requiring total assistance for mobilization and repositioning in bed. The selection of a specialized support surface is thus a critical component of a comprehensive plan for PI prevention and treatment. Support surface refers to "a specialized device for pressure redistribution designed for management of tissue loads, micro-climate, and/or other therapeutic functions" as defined by the National Pressure Injury Advisory Panel (NPIAP S31, 2007). In this document, support surface pertains to bed, mattress, overlay, and integrated bed system. At the end of this chapter, the reader will be able to:

(1) Describe the different forces that are involved in the pathogenesis of PI in the SCI population.
(2) Provide an overview of the main types and features of support surfaces to prevent and/or treat PI.
(3) Propose a simple decision table for guiding healthcare providers to select a support surface for the prevention and treatment of PI in hospitalized SCI patients.

## Pressure injuries (PI): Definition, epidemiology and impacts

Defined by the National Pressure Injury Advisory Panel (NPIAP) as a "localized damage to the skin and/or underlying tissue over a bony prominence," pressure injuries (PI) (also called bedsores, decubitus ulcers, pressure sores) represent one of the most common medical complications experienced by hospitalized patients (NPIAP, 2021). No other preventable event occurs as frequently as PI, occurring at a rate of 2%–40% of all acute care hospitalization in the United States and Canada (NPIAP, 2021; Cullum, McInnes, Bell-Syer, & Legood, 2004). PI can occur throughout the continuum of care, with a prevalence of 9.7%, 11.8%, and 12.0% for all patients hospitalized in acute care, in patient rehabilitation, and long-term care (nursing home), respectively (NPIAP, 2021).

In the lying position, PIs usually develop at the sacrum (17%–27%), heel (9%–18%), malleoli (4.6%), trochanters (1.4%), and scapulae or occiput (<5%) areas (Bhattacharya & Mishra, 2015; Kruger, Pires, Ngann, Sterling, & Rubayi, 2013; Talley Group, 2021). It is however proposed that prevention and management of heel PIs may be best managed independently from the bed support surface (Norton, Coutts, & Sibbald, 2011). The occurrence of PI is

**Diagnosis and Treatment of Spinal Cord Injury. https://doi.org/10.1016/B978-0-12-822498-4.00022-1**

particularly concerning for individuals with SCI as it interferes with rehabilitative care, community reintegration, and quality of life (Consortium for Spinal Cord Medicine, 2014; Cullum et al., 2004; McInnes et al., 2015). PIs are associated with higher morbidity and mortality and represent an enormous growing financial burden for healthcare systems, with a patient care cost per PI of 20,900 US$ to 151,700 US$, totalizing 26.8 billion US$ in 2019 (Chan et al., 2013; NPIAP, 2021). Individuals with SCI are among the most vulnerable populations for PI, occurring in more than 95% of them during their lifetime (Cowan et al., 2019). Recent literature estimates that hospital-acquired PI among SCI patients ranges from 29.7% to 49.2% (Tran, McLaughlin, Li, & Philips, 2016). According to Fuhrer, Garber, Rintala, Clearman, and Hart (1993), the risk to develop PI is higher for individuals with severe spinal cord injury.

## Pathogenesis of pressure injuries following spinal cord injury

The pathophysiology of PI involves different extrinsic factors, paralleling the severity of the neurological impairments related to the SCI. Pressure is acknowledged to be the main factor and is defined as the amount of force applied *perpendicular* to a surface per unit area of application (International Review, 2010). In addition, shearing and friction represent stresses that are exerted *parallel* to the area of application and significantly contribute to the capillary damage (Agrawal & Chauhan, 2012; International Review, 2010) (Fig. 1). The resultant reduction in blood flow reduces the oxygen and nutrients being delivered to the tissues, while simultaneously limiting the removal of metabolic waste products (Agrawal & Chauhan, 2012; Bhattacharya & Mishra, 2015). Neurological impairments associated with severe SCI also lead to difficulties in identifying painful stimulus associated with prolonged and/or excessive compression, to which must be added the difficulty to make independent postural adjustments necessary to restore tissue perfusion (Bhattacharya & Mishra, 2015; Grey & Harding, 2006). Without the relief of compression on soft tissues, ischemia persisting for more than 2 hours may lead to necrosis of subcutaneous tissues (Kuffler, 2015). Inadequate maneuvers while transferring, repositioning, and positioning in bed may also lead to excessive shearing and friction stresses to the skin, further contributing to the development of PI (Agrawal & Chauhan, 2012). Immobility is also associated with the build-up of temperature at the interface, increasing local inflammation biomarkers, associated with PI occurrence (Yoshimura et al., 2015).

Dysfunction of the autonomic nervous system may also contribute the pathophysiology of PI following SCI, by altering the micro-vascular response below the level of the injury. Blood flow regulation relies on the rhythmic alteration of blood vessels constriction controlled by central neurogenic, local myogenic and metabolic mechanisms, which can be altered following SCI (Brienza, Geyer, & Jan, 2005). Impaired myogenic and neurogenic responses of the vascular smooth muscle due to chronic denervation also lead to an inappropriate local blood flow and metabolic responses to pressure load and unload. This phenomenon may be exacerbated in individuals with endothelial dysfunction (for instance, caused by smoking) and in the elderly, for which the loss of elastin and degradation of the collagen matrix put the skin at further risk of deformation of the blood vessels under loading pressure (Phillips, Ainslie, Krassioukov, & Warburton, 2013).

Moreover, 70%–84% of individuals with SCI experience sphincter dysfunction (neurogenic bladder & bowel), which may lead to higher moisture and irritation of the skin, fostering the development of PI (Agrawal & Chauhan, 2012). Nutrition status, comorbidities and body weight should also be assessed following SCI. Obesity and diabetes may reduce the skin ability to dissipate local heat and cutaneous temperature regulation under loading. On the other hand, muscle atrophy associated with SCI may make bony prominences to emerge, leading to increased local pressure and friction/shear in bed. Spasms and joint contractures associated with SCI should also be assessed, as a potential cause of excessive friction and shear and hindrance for proper positioning and repositioning in bed (increased pressure).

## Support surfaces: An approach for prevention and treatment of pressure injuries

### Biomechanical properties of support surfaces during loading

Support surfaces use three main mechanisms to prevent PI during loading: (1) pressure redistribution; (2) micro-climate, and (3) horizontal stiffness management. Their performance levels on these important features allow differentiating between the variety of available products and help in selecting a proper support surface that matches patient's needs. During loading, the pressure at a specific location will decrease as the contact area increases. Based on this concept, support surfaces aim to redistribute the pressure by using two basic principles (Consortium for Spinal Cord Medicine, 2014; International Review, 2010; McInnes et al., 2015; NPIAP, 2021) (Fig. 2):

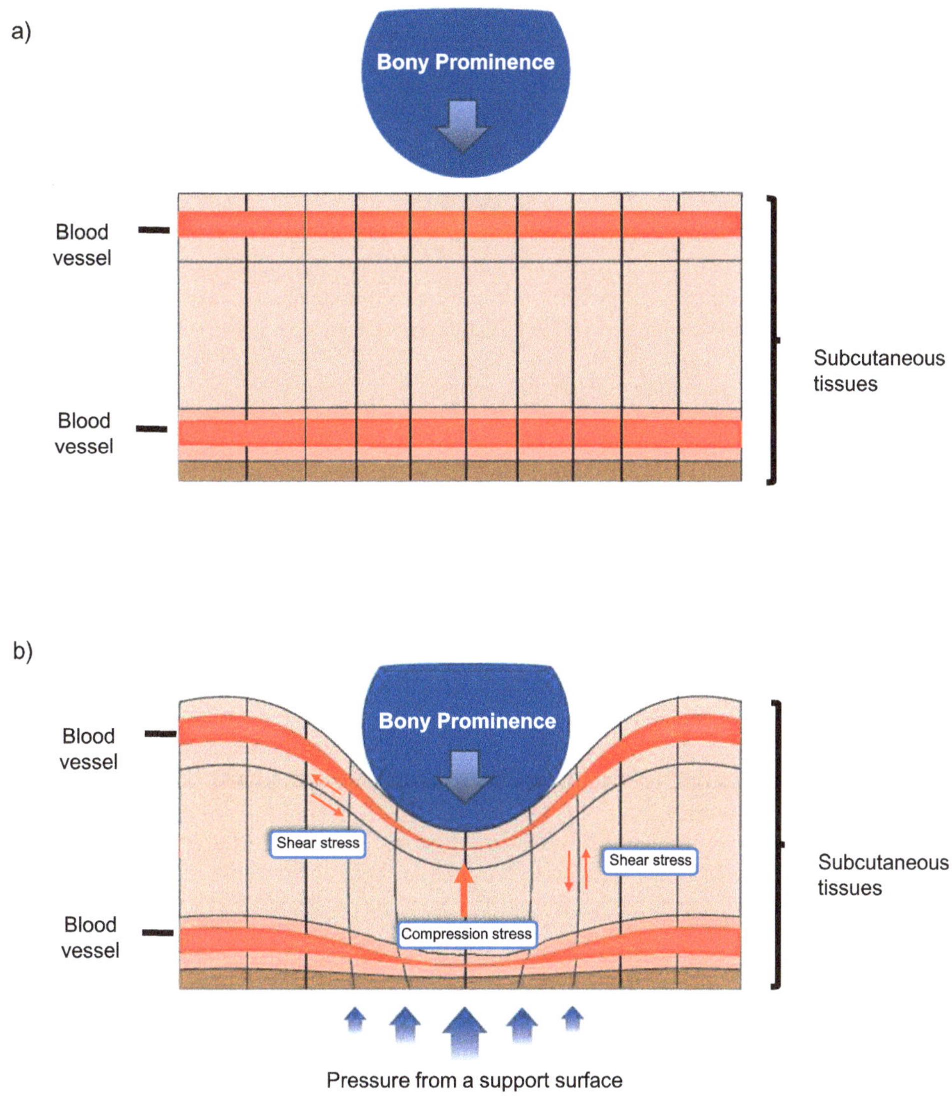


FIG. 1 Compression and shear stresses resulting from an even pressure distribution. (A) Forces applied by the bony prominences, (B) exerted compression and shear stress to subcutaneous tissues that contribute to their damage. *(Adapted from Wounds International (Pressure ulcer prevention: pressure, shear, friction, and micro-climate in context. A consensus document.))*

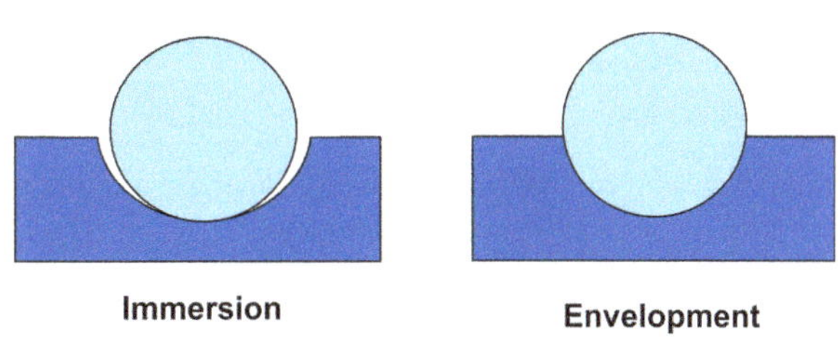


FIG. 2 Illustration of pressure mode distribution. This figure illustrates the ability of a support surface to let the body sink (immersion) or mold its contour (envelopment) to redistribute pressure.

(1) *Envelopment*, referring to the ability of a support surface to conform, so to fit and mold around irregularities in the body. It is generally measured in laboratories by indenter tests providing average pressures on each depth level (mm Hg).
(2) *Immersion*, referring to the penetration (sinking) of the body into a support surface, measured by the depth of penetration (mm).

A higher envelopment and immersion performance of a support surface is associated with greater pressure redistribution, but at the expense of higher instability of the surface making it more difficult for a patient to reposition and/or get out of bed (NPIAP, 2021).

The *micro-climate* is defined as the temperature and humidity at the body interface with the support surface and plays an important role in the development of PI, independently of the average peak pressure (Yoshimura et al., 2015). High skin temperature leads to an increased cutaneous stiffness under loading, higher inflammation, and metabolic activity, further leading to tissue damage (Reger, Adams, Maklebust, & Sahgal, 2001). Moisture can accumulate on the skin increasing the coefficient of friction at the interface while diluting the skin acidity, which was shown to reduce its antibacterial properties (Reger et al., 2001; Tomova-Simitchieva, Lichterfeld-Kottner, Blume-Peytavi, & Kottner, 2018). Micro-climate is measured in laboratories using the local temperature, relative humidity, evaporative capacity and/or dissipating properties of the support surface. However, support surfaces that provide great micro-climate management may also cause patients to feel cold or dehydrate.

The *horizontal stiffness* refers to shear and friction stresses that are generated when the gravity pulls the body down on a support surface, which reacts by pushing back. Horizontal stiffness is measured and reported in Newtons of force. A stiff support surface exerts high resisting forces on the tissues in response to a given amount of displacement, while this force is lower for softer support surfaces. However, the latter can contribute to sliding down in the bed and even to falls (NPIAP, 2021).

## Main configurations and classification of support surfaces

Support surfaces are commercially available in three main configurations: (1) mattress, (2) mattress overlay (on top of the patient's mattress), and (3) integrated bed system (combining bed frame and a support surface into a single unit). Individuals with SCI with limited mobility may particularly benefit from specialized bedframes (manual, semi-electric, and electric), which allow adjustments of height and head/foot elevation. These adjustments may promote functional independence, enhance sleep quality, and manage different medical conditions associated with SCI (e.g., autonomic dysreflexia, orthostatic hypotension, and/or spasticity).

Support surfaces can be classified into two main groups (reactive or active) based on the technology used (Hougton et al., 2013; International Review, 2010). Costs related to the different types of support surface technologies are important to consider, as the most expensive does not necessarily signify the "best" mattress for a given patient. Specific properties and performances should be analyzed based on a holistic assessment of the patient's characteristics.

## Reactive support surfaces

Reactive support surfaces are powered or non-powered surfaces with the ability to adjust its load distribution properties solely in response to an applied load (Hougton et al., 2013; NPIAP, 2021). Thus, reactive support surfaces do not independently change its load distribution. Consequently, reactive support surfaces provide a constant pressure at the interface, unless the patient actually moves. Different types of reactive support surfaces are available, differing by their structure and material used, providing different properties.

### *Non-powered reactive support surface*

#### Foam

In its simplest form, a non-powered reactive support surface consists of a single block of foam covered by a plastic or nylon cover. Foam has a flexible, cellular material structure, which may be available in different densities. Foam mattress work reactively to provide pressure redistribution by spreading the load across the whole interface surface area (NPIAP, 2021). However, their capacity to immerse and envelop the body is limited by their stiffness and density (Hougton et al., 2013). Due to its low cost, foam mattresses are widely used in hospitals (Berthe, Bustillo, Mélot, & de Fontaine, 2007). Within this type of support surface, high specification foam may incorporate multiple constructs and/or be compartmentalized into different sections with distinct properties (Hougton et al., 2013; International Review, 2010; NPIAP, 2021). Softer foam mattresses will provide higher level of immersion as compared to stiffer foams, but necessarily need to be thick enough to avoid "bottoming out," referring to the deformation beyond critical immersion whereby the effective pressure redistribution is lost (International Review, 2010; NPIAP, 2021). Foam mattresses can progressively degrade and lose their capability of returning to its original forms or thickness, typically wearing out after 3 years (Brienza et al., 2005; Consortium for Spinal Cord Medicine, 2014). Cullum et al. (2004) suggested that high specification foam mattresses were more effective than "standard" hospital foam mattresses in moderate to high-risk patients in terms of pressure relief. Bueno de Camargo et al. (2018) suggested that the use of a viscoelastic support surface reduced the incidence of PI (stage 2) in critically ill patients when compared to standard mattress. Moreover, when non-air reactive mattresses are used, the use of contoured postural management components may enhance its pressure redistribution properties (Hosking, 2017).

#### Gel

Gel mattresses consist of a semi-solid system of solid aggregates, colloidal dispersion or polymers exhibiting elastic properties (NPIAP, 2021). Gel mattresses can range from hard to soft, providing distinctive immersion, envelopment, and horizontal properties. Gel mattresses are generally easy to clean up; however, they may be heavy. Temperature and moisture control may be challenging with the use of non-air non-powered reactive support surfaces, due to their lack of aeration and poor heat conduction properties (Brienza et al., 2005). Tomova-Simitcheva et al. showed in 2017 that a reactive gel support surface (featuring a breathable cover and an open gel column) may exhibit similar effectiveness with an alternating pressure mattress (with low-air-loss function) on the transepidermal water loss, temperature, and erythema at the sacral and heel skin, both being superior as compared to a basic foam mattress.

#### Air-filled

The use of reactive air-filled mattress overlay is an alternative in patients at some risk of PI development. Indeed, Conner and Clack (1993) suggested that air mattress overlay plus three-inch foam increases its area of contact and reduced tissue shear when compared with foam overlays, using in vivo measurements derived from pelvic CT scans. Matsuo et al. (2011) demonstrated that in air support surfaces, the internal pressure adjustment (inflation level) significantly influenced pressure redistribution properties, with a lower internal pressure resulting in greater immersion and contact area, resulting in a decreased maximum interface pressure. Similarly, Higer and James (2016) demonstrated that the reactive air support surface was the most effective pressure-redistributing material for pediatric occipital pressure, displaying the lowest interface pressure and the most homogeneous pressure distribution compared to foam, fluidized and gel surfaces.

### *Powered reactive support surface*

In a more complex form, reactive support surfaces use a powered flow of air to assist in the management of temperature and humidity of the skin (NPIAP, 2021). Low-air-loss support surfaces are composed of multiple air chambers throughout. However, in contrast to alternating air pressure support (discussed below), low-air-loss surface displays tiny holes at its top surface continually blowing air out, while maintaining a specific inflation level in the chambers (Fig. 3) (Brienza et al., 2005; Hougton et al., 2013; International Review, 2010; McInnes et al., 2015). The pressure at which the support surface must be inflated, and the rate for blowing out the air, is based on the individual's height and weight (Brienza et al.,

| High / Active Technologies | Mechanism of action | Effect on properties during loading |
|---|---|---|
| Alternating air pressure | 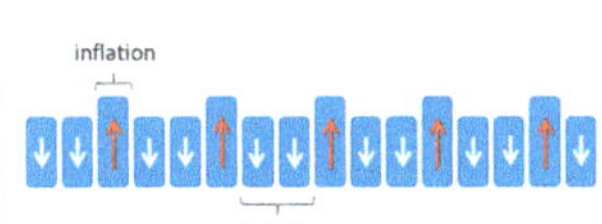 | - Pressure redistribution via powered cyclic changes in loading and unloading characterized by specific parameters<br>- Immersion and envelopment |
| Low-air-loss | 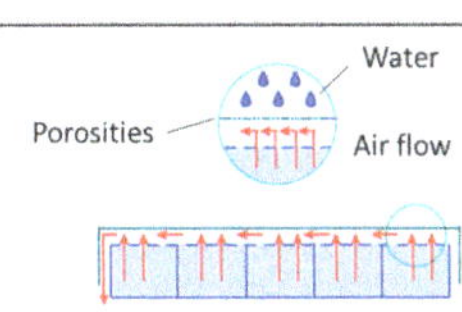 | - Immersion and envelopment<br>- Microclimate management |
| Air fluidized | 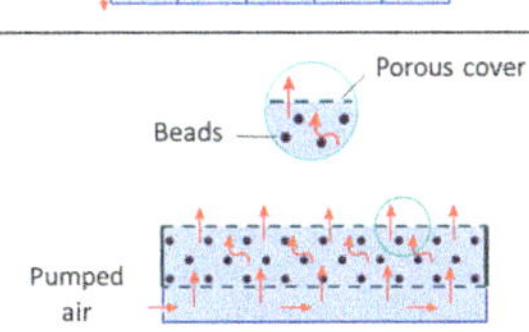 | - Immersion and envelopment<br>- Microclimate management |

FIG. 3 Description of active support surface technologies. This figure aims to summarize the active technologies and illustrates their mechanisms of action. *PR*, pressure redistribution; *SR*, shear reduction; *MC*, microclimate management; *T*, impact on self-transferring ability. *(Adapted from Wounds International (Pressure ulcer prevention: pressure, shear, friction and microclimate in context. A consensus document.))*

2005), and is adjusted to provide the desired level of immersion/envelopment. Temperature and moisture control at the skin surface is also improved, as compared to alternating air pressure system and non-powered support surfaces, most likely due to the evaporation effect caused by air circulating from the permeable cover (International Review, 2010; Reger et al., 2001). However, the optimal skin temperature and moisture level for PI prevention is still unknown, and this technology represents higher costs compared to non-powered systems (Reger et al., 2001).

A comparison cohort study by Black et al. in 2012 has suggested that critically ill patients placed on low-air-loss mattresses with micro-climate management in surgical intensive care unit had lower PI (stage 2) incidence than those placed on an integrated power air pressure redistribution beds (air alternating pressure) (Black, Berke, & Urzendowski, 2012). Lippodolt et al. in 2014 showed that low-air-loss technology significantly reduced the interface pressure at 0, 10, 30, and 45 degrees of backrest elevation in comparison with foam and air suspension support surfaces (air fluidized). It was thus suggested that low-air-loss technology could be an additional useful tool to help prevent skin breakdown at the sacrum as a compromise between seemingly incompatible demands of skin integrity and prevention of ventilator-associated pneumonia (Lippoldt, Pernicka, & Staudiner, 2014). However, this study did not assess horizontal stiffness.

Reger et al. also suggested in 2001 that measurements of support interface climate change might allow for selective grouping of low-air-loss surfaces according to their rate of moisture evaporation and resulting temperature reduction. Combined, these characteristics can effectively describe the performance of any low-air-loss support system and may be used to define standards of performance (Reger et al., 2001). A meta-analysis completed by Shi, Dumville, and Cullum (2018) suggested that hybrid low-air-loss air surfaces may have the highest probability of being the most effective intervention for PI prevention. However, the authors remain uncertain as to the true ranking of the different support surfaces because the certainty of evidence was very low (Shi et al., 2018). It was also suggested that the performance of low-air-loss support surfaces may also be influenced by their age and should thus be assessed in the quality assessment of these technologies (Black et al., 2012).

## Active support surfaces

In opposition to reactive support surfaces, active support surfaces are powered support surface that have the ability to periodically change the load distribution regardless of the applied load (individual's position or movement) (McInnes et al., 2015). This technology mostly rely on its ability to provide pressure redistribution via cyclic changes in loading and unloading as characterized by frequency, duration, amplitude and rate of change parameters (Cullum et al., 2004; McInnes et al., 2015; NPIAP, 2021). Because of this feature, powered support surfaces are generally beneficial to individuals with low mobility. Furthermore, the inflation level can be adjusted to provide the desired level of immersion/envelopment and horizontal stiffness with the body. As illustrated in Fig. 3, active support surfaces can be stratified into two groups according to their mechanism of action (alternating air pressure and air fluidized).

### *Alternating air pressure*

Alternating air pressure support surfaces are composed of multiple air chambers throughout, in which the air is pumped alternately to inflate and deflate the chambers (Fig. 3) (Brienza et al., 2005; Consortium for Spinal Cord Medicine, 2014; Hougton et al., 2013; McInnes et al., 2015). As a result, the pressure is periodically relieved and redistributed by changing the location of the contact areas between the individual's body and the support surface (Brienza et al., 2005; Hougton et al., 2013; McInnes et al., 2015). Cyclic redistribution of loading helps to restore blood flow and re-establish blood supply to the soft tissues (Vanderwee, Grypdonck, & Defloor, 2007). However, it was recently hypothesized that unloading may be associated with reperfusion-induced inflammatory process (Xiao, Wu, & Mak, 2014). Accordingly, the main limitation of air alternating support surfaces remain the lack of evidence pertaining to the ideal frequency, duration and amplitude of cell's inflation/deflation for preventing and treating (Cullum et al., 2004; McInnes et al., 2015). However, a study by Rithalia and Gonsalkorale (1998) showed that the maximum contact pressure on the sacrum were significantly lower on devices for which inflation pressure was adjusted according to the body mass of the subject.

Hickerson, Slugocki, Thaker, Dunkan, and Bishop (2004) investigated the pressure-relieving ability of three specialized support surfaces (low-air-loss, air fluidized, pegasus airwave (air alternating pressure)) compared to a standard foam hospital mattress, by measuring the total body tissue interface pressure. The air alternating pressure surface outperformed the other tested support surfaces by decreasing the overall pressure (Hickerson et al., 2004). A meta-analysis completed by Shi et al. in 2018 suggested a moderate certainty evidence that powered active (alternating) air surfaces and powered hybrid air surfaces (offering both reactive and active pressure redistribution modes) probably reduce PI incidence compared with standard hospital surfaces by 58% and 78% on average, respectively.

An interesting study was completed by Jan, Brienza, Boninger, and Brenes (2011), aiming to guide the selection of parameters of commercial alternating pressure support surfaces for the prevention of PI in the SCI population. In this study, sacral skin response (using laser Doppler flowmetry) was compared for different pressure interface in alternating air pressure and continuous (reactive) air systems. Results showed that alternating pressure (at a low-interface pressure at 0 mmHg and high-interface pressure at 60 mmHg with a cycle time of 5 min) enhanced skin perfusion of weight bearing tissues as compared with constant pressure (with an interface pressure of 30 mmHg) in people with SCI.

## Air fluidized

Air fluidized is a feature of a support surface that provides pressure redistribution by forcing air through a granular medium (e.g., beads) producing a fluid state (NPIAP, 2021; Reddy, Gill, & Rochon, 2006). This high-technology (also called high-air-loss) may provide the greatest level of immersion and envelopment of any support surfaces (Brienza et al., 2005; International Review, 2010; McInnes et al., 2015). According to Holzapfel (1993), almost two-third of the body is immersed using this type of support surface, which represents a large contact surface distribution (Holzapfel, 1993). In addition to its high-pressure redistribution quality, air fluidized support surfaces also provide an effective micro-climate management, by letting warm air escape from the top surface (Brienza et al., 2005; International Review, 2010). Rothenberg et al. examined in 2014 the heel blood flow between standard hospital foam mattress bed, viscoelastic foam mattress and an air-fluidized bed. Authors reported that only the air-fluidized bed resulted in a maintained blood circulation (Rothenberg et al., 2014). Previous randomized controlled studies have shown the benefit of this system for individuals with stage III and IV PI in comparison with other non-fluidized support surfaces (International Review, 2010). However, it also showed its risk of dehydration and dry skin (also indirectly associated with PI). One systematic review reported significant reductions in PI size (of any stage and any setting) with air-fluidized devices (McInnes, Jammali-Blasi, Bell-Syer, & Leung, 2018). Authors however concluded that there was a limited evidence for the effectiveness of air-fluidized as opposed to "low-tech" devices in the treatment of existing PI. While the body of evidence was qualified as "good" by Cullum, Deeks, Sheldon, Song, and Fletcher (2000), it was deemed as low quality by McInnes (McInnes et al., 2018). Air fluidized technology is also one of the most expensive support surface system.

## Emerging technologies

Adjustable rotational beds consist in integrated bed systems featuring computerized lateral rotation (tilting) of the support surface for assisting patients for periodical repositioning in bed. In addition to promoting functional independence, these systems may also facilitate caregiver interventions (e.g., turning and transferring individuals, examining the skin, assisting in bedding equipment changes, and providing physical/respiratory therapies). Different tilting options can be provided, such as rotational beds and lateral rotating beds. On the other hand, these systems are expensive. Moreover, while pressure redistribution in this system mainly relies on the repositioning of the patient, these systems may potentially generate higher shear and friction stresses than conventional technologies. However, this remains to be defined in the clinical and laboratory settings.

## Positioning, surface material and bed making

Positioning of the bed and surface materials are important factors to consider as they influence support surfaces performance during loading. A recent study by Boyle et al. (2020) identified, using a computational model of the weight-bearing pelvis that applying a sufficient supporting lateral pressure may counteract the under-body pressure. Although this study was completed in seated individuals, this strategy is yet to be determined in the lying position (support surfaces) (Boyle et al., 2020). Tran et al. in 2016 investigated the impact of the repositioning frequency on shear stress and overall healthcare cost by decreased utilization of medical personnel. The authors concluded that the traditional repositioning frequency of every 2 hours may be equal in effectiveness to a frequency of 4 hours, and it was suggested that the frequency of repositioning be individualized based on the patient's risk factors, needs and support surface utilized (Tran et al., 2016). Current literature doesn't show that any method of positioning is preferred, and therefore, 30-degree tilted side-lying position, the prone position, and the 30-degree semi-Fowler position, are all considered acceptable (Tran et al., 2016).

Current evidence suggests that surface material impacts on micro-climate, temperature, and humidity at the interface. Posada-Moreno et al. (2011) observed that temperatures were lower in all risk areas that had no support surface protector and were greater when the surfaces were in contact with protector material, with increases up to 2.13°C. Recent studies suggest that silk-like fabrics may be effective in reducing shear and subsequent PI when compared with cotton or cotton-blend fabrics (Tran et al., 2016). Iuchi et al. (2014) showed that friction and bed making method do not significantly reduce maximum interface pressure, although the hammock effect (forces generated by tension along the loaded interface) can be reduced through the use of stretchable bed sheets on the support surface (Iuchi et al., 2014). Another study proposed by Williamson et al. (2013) concluded that putting additional linens or underpads on low-air-loss surfaces may adversely affect skin temperature and moisture, thereby reducing the PI prevention potential of these surfaces.

## Global evidence on the effectiveness of support surfaces

Globally, studies assessing support surfaces are heterogeneous in terms of settings, participants' baseline skin status, and follow-up durations. Moreover, over half of the studies have serious or very serious study limitations, reflecting the global uncertainty of evidence pertaining to the efficacy of support surfaces technologies to prevent and treat PI. It was also reported that most studies assessing support surfaces in PI prevention mostly involved participants aged over 55 years old, suggesting that the current evidence may generally apply for the older population (Shi et al., 2018). Overall, with the current state of evidence, it is thus impossible to determine the most effective support surface for either prevention or treatment. Thorough assessment of the patient's characteristics is required for selecting a support surface that will best serve the patient's needs.

## Decision-making for selecting proper support surfaces

Proper risk assessment and implementation of prevention strategies for PI are crucial to providing comprehensive care in the SCI population while reducing healthcare costs (Tran et al., 2016). The following section aims to propose a simple algorithm to identify an appropriate support surface for patients with SCI in hospital settings. This section considers recommendations of the S31 NPIAP committee for "using standards to choose wisely" (NPIAP, 2021), suggesting these important steps:

- *Carefully consider the patient population and needs*

  SCI individuals may be highly vulnerable to PI occurrence, according to their level and severity of the injury, which is particularly true in hospitals setting. Accordingly, particular attention should be given to the following characteristics at admission: presence and severity of motor and sensory function, risk of sphincter incontinence, current or past PI, presence of obesity or bony prominence(s), comorbidities, age, smoking status, and nutritional status. The current mobility (transfer and repositioning) and functional status are also crucial to assess. Secondary conditions related to the SCI and medical complications should also be identified, as they may limit proper positioning and/or repositioning of the patient in bed (orthostatic hypotension, dysphagia, mechanical ventilation, spasticity, joint contractures, etc.).

  Individuals with severe (motor-complete) tetraplegia (or high paraplegia) may be at risk of autonomic dysreflexia and poikilothermia (defined as the ability to regulate core body temperature due to the SCI), leading to variability of the body temperature and higher risk of elevated temperature/moisture. The authors thus recommend that the temperature of low-air-loss and air fluidized support surfaces be carefully monitored in individuals with SCI at risk of poikilothermia. Finally, as suggested by Norton et al. (2011), heels will be excluded of the proposed decision table, as the prevention and management of PI at this location may be best managed independently from the support surface.
- *Determine the type of support surface that will be evaluated and request relevant testing data for the chosen surface from manufacturers*

  After determining the patient's profile and needs, selection of the support surface's type can be completed by matching their specific properties (ability of immersion, envelopment, horizontal stiffness, and to manage micro-climate) during loading. Table 1 (adapted from Norton et al., 2011) provides a simple decision table to guide the multi-disciplinary team to select a support surface's type based on relevant characteristics of the SCI population in a hospital context (for prevention or treatment of PI). One can also refer to the manufacturer's data and current literature to compile support surface's performance characteristics on pressure redistribution and micro-climate properties to provide an overall ranking for the most appropriate support surface type, considering costs and hospital resources. Finally, collaborate with the hospital's administration regarding procurement of recommended support surface, keeping in mind that PIs are a common and resource-intensive challenge for hospitals worldwide and prevention strategies are significantly less costly than treatment regimen.

**TABLE 1** Decision table for selecting an appropriate support surface type in the SCI population based on the patient's mobility level in bed (vertically) and important risk factors of PI development during hospitalization (horizontally).

| | | PI risk assessment (sphincter incontinence, level and severity of sensory and motor dysfunction, etc.) | | | |
|---|---|---|---|---|---|
| | | Prevention | | Treatment | |
| | | No other risk factors<br>No PI | Other risk factors +<br>No PI | PI + | Severe/refractory/ multiple PI or cannot be positioned off |
| Ability to change position in bed | Total assist | Non-powered reactive or active (APA) | Active (APA) | Powered reactive or active (APA) | Powered reactive or active (AF to consider) |
| | Moderate assist | Non-powered reactive | Reactive or active (APA) | Powered reactive or active (APA) | Powered reactive or active (AF to consider) |
| | Independent with or without device | Regular hospital mattress[a] | Non-powered reactive | Reactive | Powered reactive or active (AF to consider) |

[a]*Only category where standard hospital mattress may be acceptable.*
*This table serves as a guidance to the multi-disciplinary team* with experience in SCI care *to select a support surface's type based on relevant characteristics of the SCI population in a hospital context for prevention or treatment of PI.* PI, *pressure injury;* APA, *air pressure alternating;* AF, *air fluidized; Powered reactive (low-air-loss feature); Non-powered reactive (gel, air-filled).*
Adapted from Norton, L., Coutts, P., & Sibbald, R.G. (2011). Beds: Practical pressure management for surfaces and mattresses. *Advances in Skin & Wound Care, 24* (7), 324–32.

## Conclusion

Despite advances in the prevention and management of pressure injury (PI) in the last years, PIs remain a severe and frequent medical complication associated to SCI. Support surface is a critical part of the comprehensive plan for the prevention and treatment of PI. Various types of support surfaces have emerged over the years, this chapter aimed at providing information to guide healthcare professionals in selecting an appropriate support surface based on the patient's characteristics and needs following a spinal cord injury (SCI) in hospital settings.

## Applications to other areas of neuroscience

Pressure injuries (PI) are preventable medical complications involving all populations with limited mobility and/or independence. Individuals with spinal cord injury (SCI) represent one of the most vulnerable populations for PI throughout the continuum of care and particularly during hospitalization. Similarly, individuals affected by any neurological condition (e.g., traumatic brain injury, stroke, neurocognitive disorder, etc.) affecting their mobility will have an increased risk of developing PI. The level of mobility, neurological status, comorbidities, history of PI and multiple other factors (such as the presence of humidity due to bladder/bowel dysfunction and nutrition level) are important to consider when assessing patient's risk factors of PI. Selection of an appropriate support surface is one of the cornerstones of the comprehensive plan for the prevention and treatment of PI. In this chapter, we have reviewed the epidemiology and pathogenesis of pressure injury as well as its global impact for the spinal cord injury population, particularly in the hospital context. The effects of support surfaces on skin properties (envelopment, immersion, horizontal stiffness and micro-climate management) during loading were also discussed. The main configuration and classification of support surfaces (types, technologies and features) were described and discussed in a clinical context, based on the current clinical and biomechanical studies. Specific characteristics, limitations, and effectiveness of each support surface are also presented. Strategies in the positioning, repositioning in bed, surface material and bed making were also discussed as they may influence the support surface's properties. Along with providing an overview of the main types and features of support surfaces available for the prevention and treatment of PI, this chapter proposes a simple and comprehensive algorithm for the selection an appropriate support surface for the prevention and treatment of PI specific to the SCI population in a clinical context.

## Mini-dictionary of terms

*The following terms and definitions are provided based on the National Pressure Injury Advisory Panel Support Surface Initiation (S31) "Terms and definitions related to support surfaces" Ver. 01/29/2007, revised 12/27/2018; 11/19/2019. Retrieved at: https://npiap.com/page/S3I.*

**Active support surface**: A powered support surface, with the capability to change its load distribution properties, with or without applied load.
**Mechanical load**: Force distribution acting on a surface.
**Pressure redistribution**: The ability of a support surface to distribute load over the contact areas of the human body.
**Reactive support surface**: A powered or non-powered support surface with the capability to change its load distribution properties only in response to applied load.
**Support surface**: A specialized device for pressure redistribution designed for management of tissue loads, microclimate, and/or other therapeutic functions. Support surfaces include but are not limited to mattresses, integrated bed systems, mattress replacements or overlays, or seat cushions and seat cushion overlays.

## Key facts of pressure injuries after spinal cord injury

- Individuals with spinal cord injuries represent a population highly vulnerable for pressure injuries, which represent one of the leading causes for unplanned re-hospitalizations (Chen, Devivo, & Jackson, 2005).
- Although pressure injuries are evitable, they represent a major public health problem, as their occurrence is associated with severe comorbidities and enormous costs of care (as much as 25% of the total direct costs of care) for the spinal cord injury population.
- There are several choices of surfaces using various technologies, as a crucial part of the strategy for prevention and treatment of pressure injuries following a spinal cord injury.
- The selection of a bed and mattress in individuals at risk of pressure injury development should be based on a multidisciplinary approach, including the patient, considering specific risk factors of pressure injuries.
- Further research is essential to improve pressure redistribution effectiveness and aid in the decision-making process. These achievements may represent important steps for the treatment and prevention of pressure injuries.

## Summary points

- Pressure injuries (PI) are of a main concern for individuals with spinal cord injury (SCI). No other preventable event occurs as frequently, occurring at a rate of 2%–40% of all acute care hospitalization in United States and Canada. The recent literature estimates that hospital-acquired PI among SCI patients ranges from 29.7% to 49.2%, which represent enormous costs for healthcare systems.
- The selection of a support surface is a critical component of a comprehensive plan for PI prevention and treatment. There are two main groups of pressure-redistribution support surfaces for the prevention and treatment of PI: (1) reactive and (2) active, which are then stratified based on their mechanism of action and features.
- Overall, with the current state of evidence, it is impossible to determine the most effective support surface for either prevention or treatment. It is thus suggested to assess patient's characteristics in a holistic way before selecting a support surface based on its various properties during loading.
- Selection of support surface may be managed by a multi-disciplinary team, ideally including the patient (if possible), and must take into account many factors, such as the level of mobility, medical status previous or current pressure injuries, and other specific risk factors of pressure injuries.
- Despites the efforts devoted to the development of emerging technologies, further research is warranted in both clinical and fundamental context, to better determine and improve the efficiency of pressure redistributing technologies for prevention and treatment of pressure injuries.

## References

Agrawal, K., & Chauhan, N. (2012). Pressure ulcers: Back to the basics. *Indian Journal of Plastic Surgery*, *45*(2), 244–254.

Berthe, J. V., Bustillo, A., Mélot, C., & de Fontaine, S. (2007). *Acta Chirurgica Belgica*, *107*(2), 155–161.

Bhattacharya, S., & Mishra, R. K. (2015). Pressure ulcers: Current understanding and newer modalities of treatment. *Indian Journal of Plastic Surgery*, *48*(1), 4–16.

Black, J., Berke, C., & Urzendowski, G. (2012). Pressure ulcer incidence and progression in critically ill subjects: Influence of low air loss mattress versus powered air pressure redistribution mattress. *Journal of Wound, Ostomy, and Continence Nursing*, *39*(3), 267–273.

Boyle, C. J., Carpanen, D., Pandelani, T., Higgins, C. A., Masen, M. A., & Masouros, S. D. (2020). Lateral pressure equalisation as a principle for designing support surfaces to prevent deep tissue pressure ulcers. *PLoS One*, *15*(1), e0227064.

Brienza, D. M., Geyer, M. J., & Jan, Y.-K. (2005). A comparison of changes in rhythms of sacral skin blood flow in response to heating and indentation. *Archives of Physical Medicine and Rehabilitation*, *86*(6), 1245–1251.

Bueno de Camargo, W. H., Pereira, R. D., Tanita, M. T., Heko, L., Grion, I. C., Festi, J., et al. (2018). The effect of support surfaces on the incidence of pressure injuries in critically ill patients: A randomized controlled trial. *Critical Care Research and Practice*, *2018*, 3712067.

Chan, B. C., Nanwa, N., Mittmann, N., Bryant, D., Coyte, P. C., & Hougton, P. E. (2013). The average cost of pressure ulcer management in a community dwelling spinal cord injury population. *International Wound Journal*, *10*(4), 431–440.

Chen, Y., Devivo, M., & Jackson, A. B. (2005). Pressure ulcer prevalence in people with spinal cord injury: Age-period-duration effects. *Archives of Physical Medicine Rehabilitation*, *86*(6), 1208–1213.

Conner, L. M., & Clack, J. W. (1993). In vivo (CT scan) comparison of vertical shear in human tissue caused by various support surfaces. *Decubitus*, *6*(2), 20–23. 26-8.

Consortium for Spinal Cord Medicine. (2014). *Presssure ulcer prevention and treatment following spinal cord injury: A clinical practice guideline for health-care professionals (second edition)*. Paralyzed Veterans of America. (March 1[st] 2021). Retrieved at: https://pva.org/research-resources/publications/clinical-practice-guidelines/.

Cowan, L. J., Ahn, H., Flores, M., Yarrow, J., Barks, L. S., Garvan, C., et al. (2019). Pressure ulcer prevalence by level of paralysis in patients with spinal cord injury in lon-term care. *Advances in Skin & Wound Care*, *32*(3), 122–130.

Cullum, N., Deeks, J., Sheldon, T. A., Song, F., & Fletcher, A. W. (2000). Beds, mattresses and cushions for pressure sore prevention and treatment. *Cochrane Database of Systematic Reviews*, (2), CD001735.

Cullum, N., McInnes, E., Bell-Syer, S. E. M., & Legood, R. (2004). Support surfaces for pressure ulcer prevention. *Cochrane Database of Systematic Reviews*, *3*, CD001735.

Fuhrer, M. J., Garber, S. L., Rintala, D. H., Clearman, R., & Hart, K. A. (1993). Pressure ulcer in community-resident persons with spinal cord injury: Prevalence and risk factors. *Archives of Physical Medicine and Rehabilitation*, *74*(11), 1172–1177.

Grey, J. E., & Harding, K. G. (2006). Pressure ulcers. *BMJ*, *332*(7539), 472–475.

Hickerson, W. L., Slugocki, M., Thaker, R. L., Dunkan, R., & Bishop, J. F. (2004). Comparison of total body tissue interface pressure of specialized pressure relieving mattresses. *Journal of Long-Term Effects of Medical Implants*, *14*(2), 81–94.

Higer, S., & James, T. (2016). Interface pressure mapping pilot study to select surfaces that effectively redistribute pediatric occipital pressure. *Journal of Tissue Viability*, *25*(1), 41–49.

Holzapfel, S. K. (1993). Support surfaces and their use in the prevention and treatment of pressure ulcers. *Journal of ET Nursing*, *20*(6), 251–260.

Hosking, J. (2017). A pilot study comparing custom contoured and planal support surfaces for pressure ulcer risk over the heels for night time postural management using interface pressure mapping and disconfort scores. *Journal of Tissue Viability*, *26*(3), 189–195.

Hougton, P., Campbell, K., & CPG Panel. (2013). *Canadian best practice guidelines for the prevention and management of pressure ulcers in spinal cord injury. A resource handbook for Clinicians*. (March 1[st] 2021), Retrieved at: http://www.onf.org.

International Review. (2010). *Pressure ulcer prevention: Pressure, shear, friction and microclimate in control*. England: Wounds International Ltd. (march 1[st] 2021), Retrieved at: https://www.woundsinternational.com/resources/details/international-review-pressure-ulcer-prevention-pressure-shear-friction-and-microclimate-context.

Iuchi, T., Nakajima, Y., Fukuda, M., Matsuo, J., Okamoto, H., Sanada, H., et al. (2014). Using an extreme bony prominence anatomical model to examine the influence of bed sheet materials and bed making methods on the distribution of pressure on the support surface. *Journal of Tissue Viability*, *23*, 60–68.

Jan, Y.-K., Brienza, D. M., Boninger, M. L., & Brenes, G. (2011). Comparison of skin perfusion response with alternating and constant pressures in people with spinal cord injury. *Spinal Cord*, *49*(1), 136–141.

Kruger, E. A., Pires, M., Ngann, Y., Sterling, M., & Rubayi, S. (2013). Comprehensive management of pressure ulcers in spinal cord injury: Current concepts and future trends. *The Journal of Spinal Cord Medicine*, *36*(6), 572–585.

Kuffler, D. P. (2015). Improving the ability to eliminate wounds and pressure ulcers. *Wound Repair and Regeneration*, *23*(3), 312–317.

Lippoldt, J., Pernicka, E., & Staudiner, T. (2014). Interface pressure at different degrees of backrest elevation with various types of pressure-redistribution surfaces. *American Journal of Critical Care*, *23*(2), 119–126.

Matsuo, J., Sugama, J., Sanada, H., Okuwa, M., Nakatani, T., Konya, C., et al. (2011). Development and validity of a new model for assessing pressure redistribution properties of support surfaces. *Journal of Tissue Viability*, *20*(2), 55–66.

McInnes, E., Jammali-Blasi, A., Bell-Syer, S. E. M., Dumville, J. C., Middleton, V., & Cullum, N. (2015). Support surfaces for pressure ulcer prevention. *Cochrane Database of Systematic Reviews*, *9*, CD001735.

McInnes, E., Jammali-Blasi, A., Bell-Syer, S. E. M., & Leung, V. (2018). Support surfaces for treating pressure ulcers. *Cochrane Database of Systematic Reviews*, *10*(10), CD009490.

National Pressure Injury Advisory Panel (NPIAP). (2021). (March 1[st] 2021). Retrieved at: https://npiap.com/.

National Pressure Injury Advisory Panel (NPIAP) Support Surface Standards Initiative (S31). (2007). *Terms and definitions related to support surfaces*. NPIAP. (March 1st 2020). Retrieved at: https://npiap.com/page/S3I.

Norton, L., Coutts, P., & Sibbald, R. G. (2011). Beds: Practical pressure management for surfaces and mattresses. *Advances in Skin & Wound Care*, *24*(7), 324–332.

Phillips, A. A., Ainslie, P. N., Krassioukov, A. V., & Warburton, D. E. R. (2013). Regulation of cerebral blood flow after spinal cord injury. *Journal of Neurotrauma*, *30*(18), 1551–1563.

Posada-Moreno, P., Losa Igesias, M. E., Becerro de Bengoa Vallejo, R., Soriano, I. O., Zaragoza-Garcia, I., & Martinez-Rincon, C. (2011). Influence of different bed support surface covers on skin temperature. *Contemporary Nurse*, *39*(2), 206–220.

Reddy, M., Gill, S. S., & Rochon, P. A. (2006). Preventing pressure ulcers: A systematic review. *JAMA*, *293*(8), 974–984.

Reger, S. I., Adams, T. C., Maklebust, J. A., & Sahgal, V. (2001). Validation test for climate control on air-loss supports. *Archives of Physical Medicine and Rehabilitation*, *82*(5), 597–603.

Rithalia, S. V., & Gonsalkorale, M. (1998). Assessment of alternating air mattress using a time-based interface pressure threshold technique. *Journal of Rehabilitation Research and Development*, *35*(2), 225–230.

Rothenberg, J., Krauss, S., Held, M., Bender, D., Schaller, H.-E., Rahmanian-Schwarz, A., et al. (2014). A quantitative analysis of microcirculation in sore-prone pressure areas on conventional and pressure relief hospital mattresses using laser Doppler flowmetry and tissus spectropotometry. *Journal of Tissue Viability*, *23*(4), 129–136.

Shi, C., Dumville, J. C., & Cullum, N. (2018). Support surfaces for pressure ulcer prevention: A network meta-analysis. *PLoS One*, *13*(2), e0192707.

Talley Group. (2021). *The role of support surfaces in pressure ulcer prevention and treatment*. Retrieved at: https://www.talleygroup.com.

Tomova-Simitchieva, T., Lichterfeld-Kottner, A., Blume-Peytavi, U., & Kottner, J. (2018). Comparing the effects of 3 different pressure ulcer prevention support surfaces on the structure and function of heel and sacral skin: An exploratory cross-over trial. *International Wound Journal*, *15*(3), 429–437.

Tran, J. P., McLaughlin, J. M., Li, R. T., & Philips, L. G. (2016). Prevention of pressure ulcers in the acute care setting: New innovations and technologies. *Plastic and Reconstructive Surgery*, *138*, 232S. Journal of Biomechanics, 47(7), 1658–64.

Vanderwee, K., Grypdonck, M., & Defloor, T. (2007). Alternating pressure air mattresses as prevention for pressure ulcers: A literature review. *International Journal of Nursing Studies*, *45*(5), 784–801.

Williamson, R., Lachenbruch, C., & VanGilder. (2013). A laboratory study examining the impact of linen use on low-air-loss support surface heat and water vaporisation transmission rates. *Ostomy/Wound Management*, *59*(8), 32–41.

Xiao, D. Z. T., Wu, S. Y. Q., & Mak, A. F. T. (2014). *Accumulation of loading damage and unloading reperfusion injury—Modeling of the propagation of deep tissue ulcers*.

Yoshimura, M., Nakagami, G., Lizaka, S., Yoshida, M., Uehata, Y., Kohna, M., et al. (2015). Microclimate is an independent risk factor for the development of intraoperatively acquired pressure ulcers in the park-bench position: A prospective observational study. *Wound Repair and Regeneration*, *23*(6), 939–947.

Chapter 23

# Nerve and tendon transfers in tetraplegia: A new narrative

Federico Fusini[a,b], Paolo Titolo[c], Stefano Artiaco[b], Bruno Battiston[c], Salvatore Risitano[b,d], and Alessandro Massè[b]

[a]Department of Orthopaedics and Traumatology, ASL CN1, Regina Montis Regalis Hospital, Mondovì, Cuneo, Italy; [b]Department of Orthopaedics and Traumatology, CTO Hospital, University of Turin, Turin, Italy; [c]Department of Orthopaedics and Traumatology, Hand and Upper limb Surgery Unit, CTO Hospital, Turin, Italy; [d]Department of Orthopaedics and Traumatology, Maggiore Hospital of Chieri, Chieri, Italy

## List of abbreviations

| | |
|---|---|
| **SCI** | spinal cord injury |
| **ICSHT** | international classification of surgery for the hand in tetraplegia |
| **PIN** | posterior interosseous nerve |
| **AIN** | anterior interosseous nerve |
| **ECRL** | extensor carpi radialis longus |
| **ECRB** | extensor carpi radialis brevis |
| **BR** | brachioradialis |
| **FDS** | flexor digitorum superficialis |
| **FDP** | flexor digitorum profundus |
| **FLP** | flexor pollicis longus |
| **EPL** | extensor pollicis longus |
| **ECD** | extensor communis digitorum |
| **PT** | pronator teres |
| **CMC** | carpometacarpal |
| **IP** | interphalangeal |
| **EPI** | extensor proprius indicis |
| **EDM** | extensor digitorum minimi |
| **FCU** | flexor carpi ulnaris |
| **FP** | flexor pollicis |
| **MP** | metacarpophalangeal |

## Introduction

The most common causes of spinal cord injury (SCI) are road accidents and falls from a height, with a prevalence for male patients between 16 and 30 years old (Titolo et al., 2019) and an incidence between 10 and 80 new cases per millions of people annually worldwide (Kumar, Osman, & Chowdhury, 2017).

More than 50% of all spinal cord injuries involve cervical spine leading tetraplegia as main clinical feature, with loss of effective upper-limb function.

The lack of hand function is the sequela which mostly affected the quality of life in patients with tetraplegia and always requires support in their daily-life activities and mobility from other people (Anderson, 2004; Bednar & Woodside, 2018; Waters, Sie, Gellman, & Tognella, 1996). Furthermore, it is reported as more desired than bowel, bladder and sexual function, standing and pain control (Anderson, 2004; Brown, 2011). In this scenario, tendon transposition is a reliable surgical technique able to restore active movement and strength of a damaged anatomical segment; however, the shortage of functional muscles above the segment lesion available for transfer is the main problem treating cervical SCI (Titolo et al.,

Diagnosis and Treatment of Spinal Cord Injury. https://doi.org/10.1016/B978-0-12-822498-4.00023-3

2019). Similarly, nerve transfer required the adjustment of a healthy donor nerve to a denervated anatomical district to restore function nerve-target, but this means to sacrifice a potential useful nervous structure (Limthongthang, Bachoura, Songcharoen, & Osterman, 2013).

In our review, we are going to illustrate the current available techniques of tendon and nervous transfer in order to restore hand and forearm function combining nerve and tendon transfer in patients with tetraplegia.

## Classifications

The most useful and accepted classification for upper-limb lesion in tetraplegia is the:

The Classification for Surgery of the Hand in Tetraplegia (ICSHT), where, the most common patterns of injuries are classified following the number of functional muscles existing below the elbow.

A functional muscle is described as muscle graded 4 or more according to the Muscle Grading System ( James, 2007) (Table 1), and the ICSHT principles are reported in the literature as the "gold standard" to choose the most appropriate reconstructive technique mainly based on tendon transfer and tenodesis procedure (Table 2).

It's important to select the most appropriate donor muscle in order to preserve function of anatomical region without creating another functional deficit after tendon transfer which can be used alone or in combination with tenodesis and arthrodesis showing interesting and reproducible results (Bednar & Woodside, 2018).

**TABLE 1** International Classification for Surgery of the Hand in Tetraplegia (ICSHT).

| Motor GROUP | Characteristics | Function |
|---|---|---|
| 0 | No muscle below elbow suitable for transfer | |
| 1 | BR | Flexion and supination of the elbow |
| 2 | ECRL | Extension of the wrist |
| 3 | ECRB | Extension of the wrist |
| 4 | PT | Pronation of the wrist |
| 5 | FCR | Flexion of the wrist |
| 6 | Finger extensors | Extrinsic extension of the fingers |
| 7 | Thumb extensor | Extrinsic extension of the thumb |
| 8 | Partial digital flexors | Extrinsic flexion of the fingers |
| 9 | Lacks only intrinsics | |
| 10 | Exceptions | |

BR brachioradialis, ECRL extensor carpi radialis longus, ECRB extensor carpi radialis brevis, PT pronator teres, FCR flexor carpi radialis.

**TABLE 2** Muscle Grading System of the British Medical Research Council, muscle function ranged from 0 (no contraction) to 5 (normal power).

| GROUPS | Description |
|---|---|
| 0 | No contraction |
| 1 | Flicker or trace of contraction |
| 2 | Active movement with gravity eliminated |
| 3 | Active movement against gravity |
| 4 | Active movement against gravity and resistance |
| 5 | Normal power |

According to function importance, a priority order to choose the correct sequence of donor muscles (Limthongthang et al., 2013; Zlotolow, 2011) is well established:

(1) wrist extension recovery,
(2) pinch recovery,
(3) grasp recovery,
(4) finger and thumb extension recovery,
(5) intrinsic muscles function recovery.

When transfer options are ended, the remaining functions are reached using tenodesis and arthrodesis (Titolo et al., 2019).

In the setting of arm dysfunction, nerve transfer is also a successful surgical procedure approaching proximal brachial plexus injury with avulsion of nerve roots or more peripheral nerve injuries. Nerve transfers are commonly used where anatomic repair of original motor nerve is not possible as well as where it's possible to rapidly restore function by rerouting expendable donor nerves (de Mendonça Cardoso, Gepp, Lima, & Gushiken, 2020). Recently, some Authors introduced nerve transfers in tetraplegia, reporting interesting outcomes (Mooney, Hewitt, & Hahn, 2020; Titolo et al., 2019). According to our previous work (Titolo et al., 2019), the current authors propose a new surgical strategy based on classical tendon transfer surgery combined with nerve transfer techniques (Table 3). The use of both surgical approaches allows avoiding a frequent concern about the use of nerve transfer in reconstructive surgery: the risk of using sources of "predictable" results for "unpredictable" results.

In this proposed technique, the choice of the most appropriate nerve donor site is crucial in order to avoid the subtraction of useful muscles; in this way, it is always possible to perform a second stage surgery according to classical tenodesis or tendon transfer surgeries.

## Indications and timing

The standard patient candidate to combined nerves and tendons transfer is a patient with cervical spine injury, with incomplete or partial paralysis of the upper limb and a stabilized upper extremity motor function.

Their general condition should be considered stable, with good pain control, no spasticity and infection free. The passive range of motion could be almost complete and they should be very motivated for the long postoperative rehabilitation (Fox et al., 2019).

Patients should be correctly informed before surgery, especially about the difference in terms of recovery time: nerve transfer needs a long time before they can show their results, and often they need a second stage procedure to achieve the set objectives. Great motivation and realistic objectives are strongly requested to obtain good improvement in hand function (Gohritz et al., 2007).

Conversely, contraindications for nerve and tendon transfer surgery which could lead to poor results, are (Lee & Wolfe, 2012) spasticity contractures, chronic pain problems, psychological instability with unrealistic expectations and insufficient motivation.

However, there is no consensus about timing of nerve transfers in tetraplegia. Unfortunately, within the year from injury, all possible spontaneous recovery has already occurred; for this reason, it is recommended to perform tendon transfer after 1 year from SCI. Nevertheless, the neurological recovery usually occurs in the first 6 months from trauma (Lee & Wolfe, 2012).

It is usually accepted that terminal muscular atrophy is established after 1 year of denervation, with no possibility of function restoration (Fu & Gordon, 1995a, 1995b). According to these premises, Lamb and Chan observed that if a muscle is completely paralyzed during the initial assessment and it will remain paralyzed 1 month later, it will be rare to observe any significant improvement, even 1 year later (Lamb & Chan, 1983). On the contrary, Bertelli et al. reported functional muscle reinnervation after distal nerve transfers performed 18 months after SCI (Bertelli & Ghizoni, 2013).

Furthermore, Fox et al. described two type of lesions with different indications about the appropriate surgical time for intervention (Fox, 2016) (Fig. 1).

First, time-independent lesions: clear-cut SCI with a little zone of cell damage located at the anterior horn of the spinal cord. Peripheral nerve transfer can restore segment movement by reconnecting functional motor unit to brain. This transfer switches the motor control from an expendable donor to a muscle group distal to lesion level, bypassing the site of lesion. It can be done at any time after trauma.

Second, time-dependent lesions: injury site is more extensive; roots can be involved as a peripheral nerve lesion. To restore the voluntary control or to reinnervate muscles that would become atrophic nerve transfer should be used.

**TABLE 3** Combining Nerve and Tendon Strategy in tetraplegia.

Group 0

| Primary procedure | Positive outcome? | Secondary procedure |
|---|---|---|
| • Teres minor to triceps nerve transfer | NO | Posterior deltoid to triceps tendon transfer |
| • Brachialis to ECRL nerve transfer | YES | Flexor pollicis longus tenodesis + Moberg key pinch procedure |

**Group 1 (BR → M 4)**

| Primary procedure | Positive outcome? | Secondary procedure |
|---|---|---|
| • Teres minor to triceps nerve transfer | NO | Posterior deltoid to triceps tendon transfer |
| • Brachialis to AIN/FDS nerve transfer | YES | Extensor digitorum communis tenodesis + Extensor pollicis longus tenodesis |
| | NO | Flexor pollicis longus tenodesis + Moberg key pinch procedure |
| • BR to ECRB tendon transfer | | |

**Group 2 (ERCL → M 4)**

| Primary procedure | Positive outcome? | Secondary procedure |
|---|---|---|
| • Teres minor to triceps nerve transfer | NO | Posterior deltoid to triceps tendon transfer |
| • Supinator to PIN nerve transfer | NO | Extensor digitorum communis tenodesis + Extensor pollicis longus tenodesis |
| • Brachialis to AIN/FDS nerve transfer | YES | BR to opposition |
| | NO | BR to FPL tendon transfer |

**Group 3 (ECRB → M 4)**

| Primary procedure | Positive outcome? | Secondary procedure |
|---|---|---|
| • Teres minor to triceps nerve transfer | NO | Posterior deltoid to triceps tendon transfer |
| • Supinator to PIN nerve transfer | NO | Extensor digitorum communis tenodesis + Extensor pollicis longus tenodesis |
| • Brachialis to AIN/FDS nerve transfer | YES | BR to opposition |
| | NO | BR to FPL tendon transfer + Tenodesis FDP 2° to FDP 3°-4°-5° |
| • ECRL to FDP (3°-4°-5° finger) tendon transfer | | |

**Group 4 (PT → M 4)**

| Primary procedure | Positive outcome? | Secondary procedure |
|---|---|---|
| • Supinator to PIN nerve transfer | NO | Extensor digitorum communis tenodesis + Extensor pollicis longus tenodesis |
| • Brachialis to AIN/FDS nerve transfer | YES | BR to opposition |
| | NO | BR to FPL tendon transfer + Tenodesis FDP 2° to FDP 3°-4°-5° |
| • ECRL to FDP (3°-4°-5° finger) tendon transfer | | |

**TABLE 3** Combining Nerve and Tendon Strategy in tetraplegia—cont'd

| Group 0 | | |
|---|---|---|
| **Primary procedure** | **Positive outcome?** | **Secondary procedure** |
| **Group 5 (FRC → M 4)** | | |
| **Primary procedure** | **Positive outcome?** | **Secondary procedure** |
| • Supinator to PIN nerve transfer | NO | Extensor digitorum communis tenodesis + Extensor pollicis longus tenodesis |
| • Brachialis to AIN/FDS nerve transfer | YES | BR to opposition |
| | NO | BR to FPL tendon transfer + Tenodesis FDP 2° to FDP 3°-4°-5° |
| • ECRL to FDP (3°-4°-5° finger) tendon transfer | | |
| **Group 6 (EDC → M 4)** | | |
| **Primary procedure** | **Positive outcome?** | **Secondary procedure** |
| • Brachialis to AIN/FDS nerve transfer | YES | BR to opposition |
| | NO | BR to FPL tendon transfer + Tenodesis FDP 2° to FDP 3°-4°-5° |
| • ECRL to FDP (3°-4°-5° finger) tendon transfer | | |
| • EPL tenodesis | | |
| **Group 7 (EPL → M 4)** | | |
| **Primary procedure** | **Positive outcome?** | **Secondary procedure** |
| • Brachialis to AIN/FDS nerve transfer | YES | EDM to APB or EIP to APB |
| | NO | BR to FPL tendon transfer + Tenodesis FDP 2° to FDP 3°-4°-5° + EDM to APB or EIP to APB |
| • ECRL to FDP (3°-4°-5° finger) tendon transfer | | |
| **Group 8 (partial finger flexion)** | | |
| **Primary procedure** | **Positive outcome?** | **Secondary procedure** |
| • ECRB to AIN nerve transfer | YES | EPI/EDM to opposition |
| | NO | BR to FPL tendon transfer + Tenodesis FDP 2° to FDP 3°-4°-5° + EPI/EDM to opposition |
| **Group 9 (intrinsic deficit)** | | |
| **Primary procedure** | **Positive outcomes** | **Secondary procedure** |
| • Intrinsic reconstruction (Zancolli lasso/house intrinsic procedure) | YES | None |
| • Opponensplasty | NO | |

A new proposal of surgical treatment based on ICSHT group. In case of failure of the proposed treatment, a salvage procedure (actual treatment) is always possible.

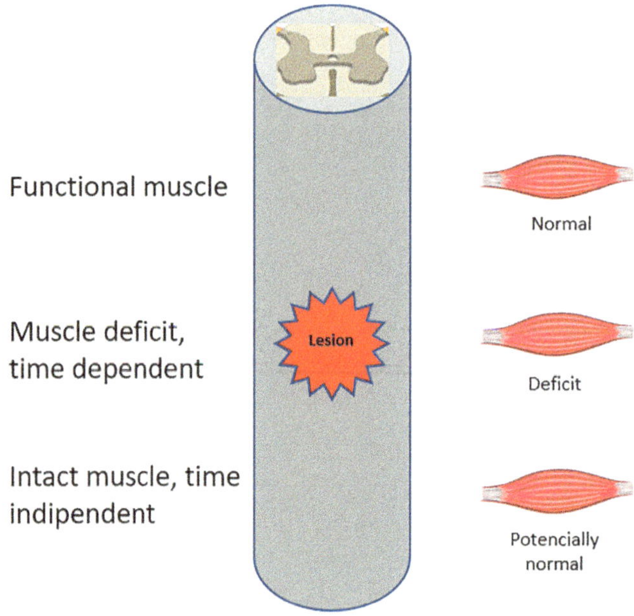


FIG. 1 Time-dependent and time-independent injury based on the level of spinal cord lesion.

Denervated muscles might benefit from early surgery before the onset of muscular atrophy, while paralyzed muscles leave more time for surgery (Cain, Gohritz, Fridén, & van Zyl, 2015; Fox et al., 2018).

Taking into account all these factors, an ideal time for surgery should be performed between 6 and 12 months from trauma, before the onset of irreversible muscle atrophy. Procedures could be anticipated only if muscles denervation is recorded by electromyography (EMG) (Bertelli, Ghizoni, & Tacca, 2011).

In the following paragraphs, we will describe primary and secondary procedures for each group of ICSHT classification (Table 3).

## Group 0

*Neurological level of injury:* C5.

*Residual motor function:* no muscle of grade 4 below the elbow.

*Current reconstructive strategy:* In this group, tendon transfer procedures can be performed to restore elbow extension:

- Moberg's procedure based on posterior deltoid to triceps transfer.
- Zancolli's procedure based on biceps to triceps transfer (Arnet, Muzykewicz, Fridén, & Lieber, 2013; Moberg, 1990; Muzykewicz, Arnet, Lieber, & Fridén, 2013). Conventional surgery considers useless any tendons transfers to restore wrist and hand function in this group. The restoration of wrist extension is really important, since tenodesis can allow thumb-to-index key pinch, due to wrist motion. It is also important because it increases patient autonomy, and it improves some daily life activities such as eating, personal care and wheelchair propulsion.

In one larger series, groups 0 and 1 are often together and represent almost 30% of the lesions (Lee & Wolfe, 2012).

*Our proposal:* It is possible to treat this condition with nerve transfer:

- teres minor to triceps nerve transfer.
- brachialis to extensor carpi radialis longus nerve transfer.

The transfer of teres minor motor branch to triceps long head motor branch allows patients to restore elbow extension without any deficit to the donor site area (Bertelli et al., 2011). Infraspinatus muscle is able to compensate external rotation deficit due to the teres minor nerve transposition. Moreover, teres minor transfer is not typically used in tetraplegia since this transfer doesn't preclude Moberg's procedure as a second surgery in case of failure or to improve a partial recovery.

Bertelli also proposed the use of the posterior division of the axillary nerve to the triceps long and medial head nerve (Bertelli & Ghizoni, 2013); however, if this procedure fails, it is not possible to use posterior deltoid tendon transfer to triceps (Moberg procedure) as salvage procedure.

It is extremely important to fully evaluate shoulder function and to explore any partial deficit of deltoid or teres minor muscle function, since if a deficit is present, it is not possible to use deltoid to restore elbow extension.

Fridén suggested the selective nerve transfer of brachialis to extensor carpi radialis longus (ECRL) to restore wrist extension (Fridén & Gohritz, 2012). Brachialis muscle is primarily innervated by a single branch of the musculocutaneous nerve deriving from the spinal roots C5/6; if biceps is working, this branch can be used as a donor. However, this will not allow a secondary Zancolli's procedure to restore elbow extension.

If brachialis transfer will have positive outcome, the lesion will upgrade from type 0 to type 1 and it is possible to perform flexor pollicis longus tenodesis and Moberg procedure for passive key pinch reconstruction (Wolfe et al., n.d.).

## Group 1

*Neurological level of injury:* C5.

*Residual motor function:* only brachioradialis (BR) is graded as 4 or more. Elbow flection is still possible since brachioradialis and biceps are functioning.

*Current reconstructive strategy:* As for group 0, Moberg or Zancolli's procedure can be used to restore elbow extension. The transfer of BR tendon to radial wrist extensors is able to provide enough strength to extend wrist against resistance.

*Our proposal*: It is possible to transfer:

- Teres minor to triceps nerve.
- Brachialis to anterior interosseous nerve (AIN) and flexor digitorum superficialis (FDS) nerve transfer.
- BR to extensor carpi radialis brevis (ECRB) tendon transfer.
- BR to ECRB tendon transfer is able to restore active wrist extension and passive key pinch reconstruction thanks to flexor pollicis longus (FLP) tenodesis with some tricks (Hentz, House, McDowell, & Moberg, 1992; House, Gwathmey, & Lundsgaard, 1976; Johanson, 2016).

About 10 years ago, Mackinnon described the brachialis to AIN transfer in patients with spinal cord injury (Mackinnon, Yee, & Ray, 2012). In this way, it is possible to improve intrinsic hand muscles function, which are very important to increase grip strength and ability, allowing patients to self-feeding (van Zyl et al., 2014).

If it is possible to partially restore flexor pollicis longus function, this will improve tenodesis effects grasp and provide extra strength and holding power as described by Fox et al. They used brachialis to AIN and FDS nerve transfer for this purpose (Fox et al., 2015a, 2015b, 2018).

If these procedures have good results, with restoration of pinch function, a second stage procedure can be planned to increase grip strength. ECD and EPL can improve hand opening and grasp function. In case of failure, FLP tenodesis can be performed anyway at any time.

## Group 2

Neurological level of injury: C6.

Residual motor function: both BR and ECRL graded 4 or more; however, it is difficult to discriminate the differential strength between ECRL (group 2) and ECRB (group 3). It is possible to discriminate between the two groups with different tests, but none of them seems to be really effective:

- Bean's sign (Mohammed, Rothwell, Sinclair, Willems, & Bean, 1992).
- Surgical exploration (Hentz et al., 1992; Moberg, 1990).
- Pronator Teres (PT) as marker of function (Allieu, 2002).

Current reconstructive strategy: surgery aims to obtain an effective key-grip thumb function, and it is usually obtained with House one-stage active key-pinch reconstruction. This technique involves:

- BR to FLP tendon transfer.
- FLP split and arthrodesis of IP joint of the thumb (Henz procedure).
- Carpometacarpal (CMC) fusion.
- Tenodesis of EPL to balance the flection tendency (House, Comadoll, & Dahl, 1992).

Other authors preferred to perform ECRL tendon transfer, only if ECRB has normal strength, while they suggest to avoid this procedure if strength is graded 4 or less (Hentz et al., 1992).

Our proposal: The purpose of nerve transfer in this group is to obtain a functional and effective pinch to allow and active fingers flexion and extension, along with elbow extension:

- Teres minor to triceps nerve transfer.
- Supinator to posterior interosseous nerve (PIN) transfer.
- Brachialis to AIN transfer.
- FDS nerve transfer.

Motor branches for supinator come out from PIN, which is the distal motor division of radial nerve for the forearm. When the lesion is at C6 level, some muscles innervated from radial nerve are still functioning. Supinator is innervated from neuronal cell above the site of lesion and it could be used to partially restore fingers and thumb extension.

Supinator is always functional when there are strong wrist extensors, and a sacrifice of supinator will not affects supination function since it will be allowed by biceps (Bertelli, Tacca, Ghizoni, Kechele, & Santos, 2010).

The objective of nerve transfers is to allow reinnervation of ECD, EPL, abductor pollicis longus and extensor pollicis brevis, in order to preserve trapeziometacarpal joint mobility and avoid tenodesis of abductor pollicis longus. Moreover, the reinnervation of extensor carpi ulnaris allows a secondary transfer to flexor carpi ulnaris to stabilize the wrist during finger extension (Bertelli et al., 2010).

In long time tetraplegia with no response of PIN from electric stimulation, Bertelli also proposed a free gracilis muscle transfer for thumb and finger extension with innervation supplied by supinator (Bertelli & Ghizoni, 2016).

- In case of failure of teres minor motor branch to triceps long head motor branch, it is possible to perform a secondary salvage procedure with posterior deltoid tendon to triceps transfer (Moberg procedure).
- In case of failure of supinator to PIN transfer, it is possible to undertake a salvage procedure with ECD and EPL tenodesis
- If nerve branch of brachialis to AIN transfer reached good functions, the procedure could be completed by BR tendon transfer with a tendinous graft in order to strength thumb opposition (House et al., 1976).

In case of failure, it is always possible to perform the House one-stage active key pinch reconstruction to restore key-pinch.

## Group 3

*Neurological level of injury:* C6.

*Residual motor function:* both ECRL and ECRB are graded 4 or more.

*Current reconstructive strategy:* it is based on Zancolli's procedure (Zancolli, 1979), which consists of two steps:

(1) restoration of extension function
(2) restoration of flexion function

For the first stage it is suggested to perform an extensor tenodesis (ECD and EPL), a "lasso procedure" and a thumb IP joint fusion. The Zancolli's "lasso procedure" aimed to prevent MP joint hyperextension and clawing of the fingers during extension; FDS is used to perform MP joint flexion (Wolfe et al., n.d.).

In the second stage, ECRL to FDP and BR to FLP tendon transfer could be performed.

Some author prefer to treat this group of lesions as group 2 in order to prevent the loss of wrist strength using House's one stage active key pinch reconstruction along with a "lasso procedure" if the index finger doesn't flex enough to allow the key pinch (Mohindra, Gogna, Sangwan, Gaba, & Kundu, 2017).

*Our proposal:* combination of nerve and tendon transfer:

- teres minor to triceps nerve transfer,
- supinator to PIN nerve transfer,
- brachialis to AIN and FDS nerve transfer,
- ECRL tendon to FDP tendon transfer for 3-4-5 finger.

In case of failure of teres minor nerve transfer, the salvage procedure consists in Moberg procedure, while if supinator to PIN nerve transfer fails, to restore thumb and finger extension it is possible to perform ECD and EPL tenodesis.

It also possible to transfer ECRB nerve to AIN, as described by Bertelli, and if it fails, a salvage procedure with brachioradialis and brachialis tendon transfer for finger flexion (Bertelli, 2015).

Recently, experiments on ECRB nerve to AIN transfer, showed recovery of fingers flexion. This could be related to intermuscular nerve connections between AIN and ulnar nerve (Bertelli & Ghizoni, 2017).

We prefer to spare ERCB for secondary tendon transfer, and we prefer to perform brachialis nerve to AIN transfer, since ERCB tendon transfer is a very reliable and suitable technique to obtain finger flexion.

While waiting for extension function recovery (due to supinator to PIN transfer), finger flection imbalance can be temporarily managed with a radial nerve palsy splint, or, in case of nerve transfer failure, with extensor tenodesis as secondary procedure.

According to our experience, if the brachialis nerve to AIN transfer has a good outcome, an independent index finger pinch function could be obtained without the need to primarily transfer ECRL to FDP. Thumb opposition can be improved with BR tendon transfer (House et al., 1976).

If brachialis nerve to AIN transfer fails, it is possible to perform tenodesis of index finger FDP to other fingers (3rd–4th–5th). Moreover, it is possible to perform additional procedures, depending on secondary recovery:

- FLP splint.
- arthrodesis of IP joint.
- CMC fusion.

## Group 4

*Neurological level of injury:* C7.

*Residual motor function:* PT strength with grade 4 or more.

PT could be considered for tendon transfer, but it is necessary to pay attention to its important role in wheelchair propulsion.

*Current reconstructive strategy:* House reconstructive strategy of a two-stage reconstruction of grasp and release function.

Extensor phase is performed first and consists in three surgical steps:

- CMC fusion and ELP tenodesis.
- ECD tenodesis to the distal radius or BR to ECD transfer.
- Intrinsic tenodesis using a free tendon graft routed to the lumbrical canals (House et al., 1976).

Flexion phase is performed later and it is divided into 2 different steps:

(1) BR or PT transfer to FLP
(2) ECRL to FDP

It is also possible to improve opposition with BR or PT with FDS graft.

If PT instead of BR is used for FLP tendon transfer, BR could be used to restore adduction—opponens function, using paralyzed FDS, which can be considered an "in situ tendon graft"(Wolfe et al., n.d.).

*Our proposal:* same as group 3, but in this case, it is not necessary to transfer teres minor to triceps since in this group triceps strength is normal.

In alternative, it is possible to perform PT to FDS nerve transfer (Titolo et al., 2019).

Two branches of the median nerve innervate the PT and the transfer of 1 branch to ECRL in high radial nerve palsy is able to restore almost complete wrist extension (grade 4), preserving at least a grade 4 pronation strength (García-López, Navarro, Martinez, & Rojas, 2014).

It is possible to perform:

- supinator nerve to PIN transfer.
- brachialis nerve to AIN and FDS nerve transfer.
- tendon transfer of ECRL to FDP of the 3rd–4th–5th finger.

Since the lesion is more caudal, BR, ECRL and PT are available for tendon transfer. Usually, ECRB should be preserve in order to stabilize wrist joint (Lee & Wolfe, 2012; Wolfe et al., n.d.).

In radial nerve palsy, pronation of the forearm can be achieved by pronator quadratus, FCR, FDP, and FDS contraction. For that reason, it is difficult to retrieve which is the real pronation deficit in group 4.

In normal hand, complete pronosupination movement can be achieved with BR, extensor carpi ulnaris and EDM (Bertelli & Ghizoni, 2017; Bertelli, Ghizoni, & Tacca, 2016); however, in this group only brachioradialis would still produce a partial pronation in case of deficit of PT. Since PT is fundamental for wheelchair propulsion, we suggest to avoid PT or FDS nerve transfer. It is possible to perform a secondary House intrinsic procedure.

## Group 5

*Neurological level of injury:* C7.

*Residual motor function:* FRC graded 4 or more.

It is possible to transfer supinator to PIN nerve, brachialis nerve to AIN and FDS, while it is possible to transfer the tendon of ECRL to FDP of the 3rd–4th–5th finger.

*Current reconstructive strategy:* same of group 4.

*Our proposal:* same of group 4. FRC can be transferred to PIN or AIN (García-López et al., 2014), but in our practice, we prefer to spare FRC in place to preserve the maximal wrist control possible (House et al., 1992).

## Group 6

*Neurological level of injury:* C7.

*Residual motor function:* ECD graded 4 or more. Extensor phase is not necessary in this type of patients and surgery is focused on flexor phase only.

Patients have a "flat hand" because digital extension is not opposed by antagonist muscles, with lack of active grasp. Since the lesion is at C7 level and not above, patients have a perfect control on shoulder, elbow and wrist to control the position of the hand in space, so they will benefit from grasp and reconstructive surgery.

*Current reconstructive strategy:* same of flexion phase of previous two groups.

*Our proposal:*

- Brachialis to AIN and FDS nerve transfer.
- EPL tenodesis and ECRL to FDP of the 3rd–4th–5th finger tendon transfer.

The EPL is very weak or totally absent, while finger active extension is strong. For this reason, it is useful to perform a tenodesis of EPL or a transfer to EDC to improve thumb extension.

Some authors proposed to transfer ECRB (extensor carpi radialis brevis) terminal motor branch to FLP (flexor pollicis longus) nerve to reconstruct thumb flexion and brachialis tendon transfer with graft to restore fingers flexion (Bertelli et al., 2012; Bourrel, 1974; Mangus, 1973; Zancolli, 1979).

We suggest to spare ECRL nerve and to use a tendon transfer to obtain fingers flexion. It is also possible to perform a House intrinsic procedure to complete the surgical procedure.

## Group 7

*Neurological level of injury*: C8.

*Residual motor function:* EPL graded as 4 or more. Fingers and thumb extensors are complete functioning and only flexor phase reconstruction is needed.

*Current reconstructive strategy:*

- ECRL to FDP tendon transfer.
- BR or PT to FLP tendon transfer to restore active pinch as for the previous group.

A possible alternative is based on Brachialis to AIN and FDS nerve transfer and ECRL to FDP of the 3rd–4th–5th finger tendon transfer.

Thumb control can be improved with CMC fusion and opponens plasty depending on thumb control and patient's preference. A Lasso procedure can be also performed to improve digital grasp (Gupta, Consul, & Swamy, 2015).

*Our proposal:* same steps of group 6 but avoiding EPL tenodesis. Moreover, we perform also Extensor Indicis Proprius (EPI) or EDM opponens plasty. A further intrinsic reconstruction procedures as suggested by Zancolli (Zancolli, 1979) and Bourrel (Bourrel, 1974) can be added to achieve the desired results.

## Group 8

*Neurological level of injury*: C8 level.

*Residual motor function*: intact FCU and partial FDP. The flexors of 4th and 5th fingers are usually stronger than those of first three fingers.

*Current reconstructive strategy:* aim of surgical intervention is to restore fingers flexion using Zancolli's two stage reconstruction: extrinsic transfers and intrinsic reconstruction.

- Extrinsic transfer:
  - side to side suture of the four profundus tendons and half of the index FDP is woven together with other tendons to ensure right tension;
  - a secondary additional part is the BR or PT to FLP transfer.
- Intrinsic reconstruction:
  - Zancolli lasso technique,
  - opponensplasty with EPI or EDM plus CMC fusion or intrinsic tenodesis as described by House et al., 1976.

To achieve an independent flexion of the index it is possible to perform ECRB to FP of the 2nd digit tendon transfer. Intrinsic function reconstruction may always be performed as a second step procedure.

*Our proposal:* ECRB nerve branch pro AIN nerve transfer (Allieu, 2002; House et al., 1992).

If the procedure fails, House's one-stage active pinch reconstruction can be performed anyway along with Henz procedures. The flexion tendency will be balanced by CMC fusion and EPL tenodesis, while complete hand function will be restored through tenodesis of FDP of second finger to 3rd–4th–5th finger FDP.

## Group 9

*Neurological level of injury*: C8-T1.

*Residual motor function:* FDS and other extrinsic muscle for fingers and thumb flexion, with lack of intrinsic muscle function.

*Current reconstructive strategy and our proposal:* surgical intervention aim to restore intrinsic nerve function. Opponensplasty and intrinsic reconstructions can be performed according to the intrinsic imbalance present (Titolo et al., 2019).

A resulting claw deformity of fingers with hyperextension of MP joints can be treated with Zancolli passive or active lasso procedure (Gupta et al., 2015; Zancolli, 1979), while House intrinsic tenodesis can be used to treat PIP flexion deformity (House et al., 1992).

## Conclusions

Thanks to new advances in the field of nerve transfer, it is possible to fuse the old concepts of tenodesis and tendon transfer for tetraplegic patients with the new approaches based on nerve surgery to improve upper-limb function.

The fusion between old and new approaches could widen therapeutic choices for this kind of patients, sparing some tendon transfers as salvage procedure which could always be useful in case of combined approach failure.

## Applications to other areas of neuroscience

Nerve and tendon transfers are a new surgical approach based on the combination of both techniques. Each technique has some indication and contraindication, which can preclude the chance to get the best result. The principle of tendon transfer is based on the transposition of an expendable tendon to achieve a primary function; however, this is not always possible since higher is the medullary lesion, less functional muscle-nerve unit you have. On the other hand, nerve transfer offers the chance to reinnervate muscle distal to the site of lesion; however, there is no guarantee of success.

Despite recovery of hand function being the most important goal for patients with tetraplegia, combined nerve and tendon transfer can also offer some choice of treatment in patients with lower-limb nerve injury (Park & Casale, 2020).

The recovery of leg motor function due to a combination of nerve and tendon transfer could represent a new perspective in the recovery of walking and standing function (Crowe, Mosca, Osorio, Lewis, & Tse, 2020).

## Mini-dictionary of terms

**Spinal Cord Injury.** Lesion of the central nervous system with loss of function below the level of injury.
**Cervical Spine.** Upper region of the spine deputy to control of motor function of the upper limb.

**Complete Spinal Cord Injury**. Lesion involving whole spinal cord with complete loss of function.
**Incomplete Spine Cord Injury.** Lesion involving a part of spinal cord with preservation of some function.
**Tetraplegia.** Loss of function of whole body below the level of injury with loss of function in both upper and lower limbs.
**Paraplegia.** Loss of function of whole body below the level of injury with conservation of function in upper limbs.
**Tendon Transfer.** Surgical procedure where the tendon insertion is moved, while the origin remains in the same location. It involves redistribution of muscle power, and generally, tendons are transferred from lesser to more important functions.
**Nerve transfer.** Surgical procedure used in nerve injury with complete loss of muscle function. It involves sacrificing nerve with less important function or redundant branches of a nerve and using it to innervate a target muscle.
**Tenodesis.** Surgical procedure of fixation of the distal end of a tendon to a bone.
**Arthrodesis.** Surgical procedure of fusion of a joint with loss of mobility but increase of stability.
**Salvage procedure.** Procedure to achieve an acceptable result in case of failure of other surgical procedure.

## Key facts of nerve and tendon transfers in tetraplegia: A new narrative

### Key facts of nerve and tendon transfer

Restore motor function of upper limb in patients suffering from tetraplegia is very important to gain some level of autonomy.
In time-independent lesion, surgery can be performed at any time.
In time-dependent lesion, surgery should be performed as earlier as possible.
Generally, surgery should be performed between 6 and 12 months from injury.
Tendon transfer is the current surgical strategy used to treat tetraplegic upper limb, with fast recovery of function, but with some limitation.
Nerve transfer is a new surgical approach to restore motor function; however, the results are less fast and certain.
Combination of nerve and tendon transfer can overcome the limits of each technique, assuring better recovery of function.
In case of failure, a conventional surgical approach is always possible as salvage procedure.

## Summary points

- Usually, SCI occurs after major trauma due to fall from height or motor vehicle accident.
- The restoration of upper-limb function offers patients with tetraplegia some kind of daily living autonomy.
- More distal is the level of injury, more muscles are spared and can be used for tendon and nerve transfer, and more functions are achieved.
- Usually, surgery should be performed between 6 and 12 months from injury for both time-dependent and -independent injury.
- Combination of nerve and tendon transfer is able to overcome the limits of each technique.
- Arthrodesis procedure should be considered to complete nerve and tendon transfer procedures to achieve the best results.
- In case of failure, the current approach of tendon transfer and tenodesis could always be performed as a salvage procedure.

## References

Allieu, Y. (2002). General indications for functional surgery of the hand in tetraplegic patients. *Hand Clinics*, *18*(3), 413–421.

Anderson, K. D. (2004). Targeting recovery: Priorities of the spinal cord-injured population. *Journal of Neurotrauma*, *21*(10), 1371–1383. https://doi.org/10.1089/neu.2004.21.1371.

Arnet, U., Muzykewicz, D. A., Fridén, J., & Lieber, R. L. (2013). Intrinsic hand muscle function, part 1: Creating a functional grasp. *The Journal of Hand Surgery*, *38*(11), 2093–2099. https://doi.org/10.1016/j.jhsa.2013.08.099.

Bednar, M. S., & Woodside, J. C. (2018). Management of Upper Extremities in tetraplegia. *Journal of the American Academy of Orthopaedic Surgeons*, *26*(16), e333–e341. https://doi.org/10.5435/JAAOS-D-15-00465.

Bertelli, J. A. (2015). Transfer of the radial nerve branch to the extensor carpi radialis brevis to the anterior interosseous nerve to reconstruct thumb and finger flexion. *The Journal of Hand Surgery*, *40*(2), 323–328.e2. https://doi.org/10.1016/j.jhsa.2014.10.060.

Bertelli, J. A., & Ghizoni, M. F. (2013). Single-stage surgery combining nerve and tendon transfers for bilateral upper limb reconstruction in a tetraplegic patient: Case report. *The Journal of Hand Surgery*, *38*(7), 1366–1369. https://doi.org/10.1016/j.jhsa.2013.03.056.

Bertelli, J. A., & Ghizoni, M. F. (2016). Nerve and free Gracilis muscle transfers for thumb and finger extension reconstruction in long-standing tetraplegia. *The Journal of Hand Surgery*, *41*(11), e411–e416. https://doi.org/10.1016/j.jhsa.2016.08.003.

Bertelli, J. A., & Ghizoni, M. F. (2017). Nerve transfers for restoration of finger flexion in patients with tetraplegia. *Journal of Neurosurgery: Spine*, *26*(1), 55–61. https://doi.org/10.3171/2016.5.SPINE151544.

Bertelli, J. A., Ghizoni, M. F., & Tacca, C. P. (2011). Transfer of the teres minor motor branch for triceps reinnervation in tetraplegia. *Journal of Neurosurgery*, *114*(5), 1457–1460. https://doi.org/10.3171/2010.12.JNS101519.

Bertelli, J. A., Ghizoni, M. F., & Tacca, C. P. (2016). Results of wrist extension reconstruction in C5–8 brachial plexus palsy by transferring the pronator quadratus motor branch to the extensor carpi radialis brevis muscle. *Journal of Neurosurgery*, *124*(5), 1442–1449. https://doi.org/10.3171/2015.3.JNS142428.

Bertelli, J. A., Mendes Lehm, V. L., Tacca, C. P., Winkelmann Duarte, E. C., Ghizoni, M. F., & Duarte, H. (2012). Transfer of the distal terminal motor branch of the extensor carpi radialis brevis to the nerve of the flexor pollicis longus: An anatomic study and clinical application in a tetraplegic patient. *Neurosurgery*, *70*(4), 1011–1016. discussion 1016 https://doi.org/10.1227/NEU.0b013e3182367642.

Bertelli, J. A., Tacca, C. P., Ghizoni, M. F., Kechele, P. R., & Santos, M. A. (2010). Transfer of supinator motor branches to the posterior interosseous nerve to reconstruct thumb and finger extension in tetraplegia: Case report. *The Journal of Hand Surgery*, *35*(10), 1647–1651. https://doi.org/10.1016/j.jhsa.2010.07.012.

Bourrel, P. (1974). Restoration of opposition of the thumb. Apropos of 52 cases. *Annales de Chirurgie*, *28*(4), 309–317.

Brown, J. (2011). Nerve transfers in tetraplegia I: Background and technique. *Surgical Neurology International*, *2*(1), 121. https://doi.org/10.4103/2152-7806.84392.

Cain, S., Gohritz, A., Fridén, J., & van Zyl, N. (2015). Review of upper extremity nerve transfer in cervical spinal cord injury. *Journal of Brachial Plexus and Peripheral Nerve Injury*, *10*(01), e34–e42. https://doi.org/10.1055/s-0035-1558427.

Crowe, C. S., Mosca, V. S., Osorio, M. B., Lewis, S. P., & Tse, R. W. (2020). Partial tibial nerve transfer for foot drop from deep peroneal palsy: Lessons from three pediatric cases. *Microsurgery*. https://doi.org/10.1002/micr.30650.

de Mendonça Cardoso, M., Gepp, R., Lima, F. L., & Gushiken, A. (2020). Intercostal to musculocutaneous nerve transfer in patients with complete traumatic brachial plexus injuries: Case series. *Acta Neurochirurgica*. https://doi.org/10.1007/s00701-020-04433-3.

Fox, I. K. (2016). Nerve transfers in tetraplegia. *Hand Clinics*, *32*(2), 227–242. https://doi.org/10.1016/j.hcl.2015.12.013.

Fox, I. K., Davidge, K. M., Novak, C. B., Hoben, G., Kahn, L. C., Juknis, N., … Mackinnon, S. E. (2015a). Nerve transfers to restore upper extremity function in cervical spinal cord injury: Update and preliminary outcomes. *Plastic and Reconstructive Surgery*, *136*(4), 780–792. https://doi.org/10.1097/PRS.0000000000001641.

Fox, I. K., Davidge, K. M., Novak, C. B., Hoben, G., Kahn, L. C., Juknis, N., … Mackinnon, S. E. (2015b). Use of peripheral nerve transfers in tetraplegia: Evaluation of feasibility and morbidity. *Hand (New York, N.Y.)*, *10*(1), 60–67. https://doi.org/10.1007/s11552-014-9677-z.

Fox, I., Hoben, G., Komaie, G., Novak, C., Hamm, R., Kahn, L., … James, A. (2019). Nerve transfer surgery in cervical spinal cord injury: A qualitative study exploring surgical and caregiver participant experiences. *Disability and Rehabilitation*. https://doi.org/10.1080/09638288.2019.1669225.

Fox, I. K., Novak, C. B., Krauss, E. M., Hoben, G. M., Zaidman, C., Ruvinskaya, R., … Mackinnon, S. E. (2018). The use of nerve transfers to restore upper extremity function in cervical spinal cord injury. *PM & R: The Journal of Injury, Function, and Rehabilitation*. https://doi.org/10.1016/j.pmrj.2018.03.013.

Fridén, J., & Gohritz, A. (2012). Brachialis-to-extensor carpi radialis longus selective nerve transfer to restore wrist extension in tetraplegia: Case report. *The Journal of Hand Surgery*, *37*(8), 1606–1608. https://doi.org/10.1016/j.jhsa.2012.05.005.

Fu, S. Y., & Gordon, T. (1995a). Contributing factors to poor functional recovery after delayed nerve repair: Prolonged axotomy. *The Journal of Neuroscience : The Official Journal of the Society for Neuroscience*, *15*(5 Pt 2), 3876–3885. Retrieved from http://www.ncbi.nlm.nih.gov/pubmed/7751952.

Fu, S. Y., & Gordon, T. (1995b). Contributing factors to poor functional recovery after delayed nerve repair: Prolonged denervation. *The Journal of Neuroscience : The Official Journal of the Society for Neuroscience*, *15*(5 Pt 2), 3886–3895. Retrieved from http://www.ncbi.nlm.nih.gov/pubmed/7751953.

García-López, A., Navarro, R., Martinez, F., & Rojas, A. (2014). Nerve transfers from branches to the flexor carpi radialis and pronator teres to reconstruct the radial nerve. *The Journal of Hand Surgery*, *39*(1), 50–56. https://doi.org/10.1016/j.jhsa.2013.10.011.

Gohritz, A., Fridén, J., Herold, C., Aust, M., Spies, M., & Vogt, P. M. (2007). Ersatzoperationen bei Ausfall motorischer Funktionen an der Hand. *Der Unfallchirurg*, *110*(9), 759–776. https://doi.org/10.1007/s00113-007-1322-7.

Gupta, V., Consul, A., & Swamy, M. (2015). Zancolli lasso procedure for correction of paralytic claw hands. *Journal of Orthopaedic Surgery*, *23*(1), 15–18. https://doi.org/10.1177/230949901502300104.

Hentz, V. R., House, J., McDowell, C., & Moberg, E. (1992). Rehabilitation and surgical reconstruction of the upper limb in tetraplegia: An update. *The Journal of Hand Surgery*, *17*(5), 964–967.

House, J. H., Comadoll, J., & Dahl, A. L. (1992). One-stage key pinch and release with thumb carpal-metacarpal fusion in tetraplegia. *The Journal of Hand Surgery*, *17*(3), 530–538.

House, J. H., Gwathmey, F. W., & Lundsgaard, D. K. (1976). Restoration of strong grasp and lateral pinch in tetraplegia due to cervical spinal cord injury. *The Journal of Hand Surgery*, *1*(2), 152–159.

James, M. A. (2007). Use of the Medical Research Council muscle strength grading system in the upper extremity. *The Journal of Hand Surgery*, *32*(2), 154–156. https://doi.org/10.1016/j.jhsa.2006.11.008.

Johanson, M. E. (2016). Rehabilitation after surgical reconstruction to restore function to the upper limb in tetraplegia: A changing landscape. *Archives of Physical Medicine and Rehabilitation*, *97*(6), S71–S74. https://doi.org/10.1016/j.apmr.2016.03.015.

Kumar, N., Osman, A., & Chowdhury, J. R. (2017). Traumatic spinal cord injuries. *Journal of Clinical Orthopaedics and Trauma*, *8*(2), 116–124. https://doi.org/10.1016/j.jcot.2017.06.022.

Lamb, D. W., & Chan, K. M. (1983). Surgical reconstruction of the upper limb in traumatic tetraplegia. A review of 41 patients. *The Journal of Bone and Joint Surgery*, *65*(3), 291–298.

Lee, S. K., & Wolfe, S. W. (2012). Nerve transfers for the upper extremity: New horizons in nerve reconstruction. *The Journal of the American Academy of Orthopaedic Surgeons*, *20*(8), 506–517. https://doi.org/10.5435/JAAOS-20-08-506.

Limthongthang, R., Bachoura, A., Songcharoen, P., & Osterman, A. L. (2013). Adult brachial plexus injury. *Orthopedic Clinics of North America*, *44*(4), 591–603. https://doi.org/10.1016/j.ocl.2013.06.011.

Mackinnon, S. E., Yee, A., & Ray, W. Z. (2012). Nerve transfers for the restoration of hand function after spinal cord injury. *Journal of Neurosurgery*, *117*(1), 176–185. https://doi.org/10.3171/2012.3.JNS12328.

Mangus, D. J. (1973). Flexor pollicis longus tendon transfer for restoration of opposition of the thumb. *Plastic and Reconstructive Surgery*, *52*(2), 155–159.

Moberg, E. (1990). Surgical rehabilitation of the upper limb in tetraplegia. *Spinal Cord*, *28*(5), 330–334. https://doi.org/10.1038/sc.1990.43.

Mohammed, K. D., Rothwell, A. G., Sinclair, S. W., Willems, S. M., & Bean, A. R. (1992). Upper-limb surgery for tetraplegia. *The Journal of Bone and Joint Surgery*, *74*(6), 873–879.

Mohindra, M., Gogna, P., Sangwan, S. S., Gaba, S., & Kundu, Z. S. (2017). The scope of upper limb surgery for tetraplegics: Role of tendon transfers and universal cuff. *Acta Orthopaedica et Traumatologica Turcica*, *51*(5), 367–371. https://doi.org/10.1016/j.aott.2017.06.002.

Mooney, A., Hewitt, A. E., & Hahn, J. (2020). Nothing to lose: A phenomenological study of upper limb nerve transfer surgery for individuals with tetraplegia. *Disability and Rehabilitation*. https://doi.org/10.1080/09638288.2020.1750716.

Muzykewicz, D. A., Arnet, U., Lieber, R. L., & Fridén, J. (2013). Intrinsic hand muscle function, part 2: Kinematic comparison of 2 reconstructive procedures. *The Journal of Hand Surgery*, *38*(11), 2100–2105.e1. https://doi.org/10.1016/j.jhsa.2013.08.098.

Park, J. S., & Casale, M. J. (2020). Posterior Tibial tendon transfer for common peroneal nerve injury. *Clinics in Sports Medicine*. https://doi.org/10.1016/j.csm.2020.07.003. W.B. Saunders.

Titolo, P., Fusini, F., Arrigoni, C., Isoardo, G., Conforti, L., Artiaco, S., … Battiston, B. (2019, April 9). Combining nerve and tendon transfers in tetraplegia: A proposal of a new surgical strategy based on literature review. *European Journal of Orthopaedic Surgery and Traumatology*. https://doi.org/10.1007/s00590-018-2352-2. Springer-Verlag France.

van Zyl, N., Hahn, J. B., Cooper, C. A., Weymouth, M. D., Flood, S. J., & Galea, M. P. (2014). Upper limb reinnervation in C6 tetraplegia using a triple nerve transfer: Case report. *The Journal of Hand Surgery*, *39*(9), 1779–1783. https://doi.org/10.1016/j.jhsa.2014.06.017.

Wolfe, S. W., Hotchkiss, R. N., Pederson, W. C., Kozin, S. H., Cohen, M. S., & Revision of: Green, D. P (n.d.). Green's operative hand surgery.

Waters, R. L., Sie, I. H., Gellman, H., & Tognella, M. (1996). Functional hand surgery following tetraplegia. *Archives of Physical Medicine and Rehabilitation*, *77*(1), 86–94.

Zancolli, E. (1979). *Structural and dynamic bases of hand surgery*. Lippincott.

Zlotolow, D. A. (2011). The role of the upper extremity surgeon in the Management of Tetraplegia. *The Journal of Hand Surgery*, *36*(5), 929–935. https://doi.org/10.1016/j.jhsa.2011.03.001.

Chapter 24

# GEMINI-supported spinal cord transplantation for the treatment of chronic spinal paralysis: Overview and initial clinical translation

X.P. Ren[a,b], P. Henderson[c], C.Y. Kim[a,b,d], and S. Canavero[a,b]

[a]*Global Initiative to Cure Paralysis (GICUP), Department of Orthopedics, Ruikang Hospital, Nanning, China;* [b]*Heaven/Gemini Global initiative, Turin, Italy;* [c]*Microsurgical Research Society, Icahn School of Medicine at Mount Sinai, New York, NY, United States;* [d]*Konkuk University School of Medicine, Seoul, South Korea*

## Introduction

There is at present no biological cure for chronic paralysis following spinal cord injury (SCI). A medley of experimental approaches have been tried over the years, and, despite some improvement in some patients, no full recovery has been observed (Ren, Kim, & Canavero, 2019). As underscored by Illis,

> *"it would be difficult to find any other branch of science with over a century of such sterile endeavour. In effect, there has been repetition of the same idea, albeit with different techniques, that is, looking at the lesion site. Are we sentenced to repeating the same experiments in the hope of expecting a different result?"*
>
> (Illis, 2012).

Work conducted mainly by US neurosurgeon L. Walter Freeman more than half a century ago suggests that a permanent, biological cure is possible, at least in several cases of chronic SCI, by removing *en bloc* the "*lesion site*," i.e., the most damaged portion of the spinal cord, and connecting the two free undamaged ends, after spinal (vertebral) shortening (Freeman, 1963; see also Canavero & Ren, 2016a). Transplantation of a healthy segment of cord is an alternative (see below).

In both scenarios, a technology that allows the functional reconnection of the severed ends is necessary. In 2013, the GEMINI spinal cord fusion (SCF) protocol was proposed with this goal in mind (Canavero, 2013, 2015).

Here, we will briefly review the GEMINI SCF protocol and lay out the various alternatives for its ongoing deployment in the clinical setting.

## The GEMINI spinal cord fusion protocol: Overview

The GEMINI spinal cord fusion protocol is predicated on four pillars (Canavero, 2015; Canavero & Ren, 2020; Canavero, Ren, Kim, & Rosati, 2016):

1. Sharp severance of the spinal cord with an ultra-thin blade, with its attendant minimal tissue damage, as compared to clinical SCI (<10 N vs 26,000 N)
2. Exploitation of the gray matter internuncial sensori-motor "highway" (so-called cortico-trunco-reticulo-propiospinal pathway; C-TRPS) re-bridged by sprouting connections between the two re-apposed cord stumps. This "highway" runs in parallel to corticospinal descending fibers and is co-responsible for sensori-motor transmission through the cord (see also Kostyuk & Vasilenko, 1979; Shik, 1983).
3. Application of "fusogens/sealants": these "seal" the thin layer of injured cells in the gray matter, both neuronal, glial and vascular; simultaneously they fuse a certain number of axons in the white matter. The pro-regenerative early scar is not inhibited, but necrosis/cavitation is.

Diagnosis and Treatment of Spinal Cord Injury. https://doi.org/10.1016/B978-0-12-822498-4.00024-5

4. Acceleration of axonal outgrowth by electrical spinal cord stimulation straddling the fusion point and concomitant motor cortex stimulation.

We will briefly review the technology involved. An in-depth discussion can be found elsewhere (Canavero et al., 2016; Canavero & Ren, 2020).

## GEMINI: Fusogens

Cell fusion is the process by which the membranes and intracellular contents of any two cells are permanently fused. There are multiple situations where this occurs naturally—fertilization is one very fundamental example, but there is also increasing evidence that cell fusion can be induced for clinical benefit. The most striking clinical application of this is the potential for reconstitution of axon integrity and electrical conductivity through the fusion of axon membranes; this can be performed in order to reconnect the two cut ends of a severed axon, or to create membranous continuity between the ends of two axons that have been newly placed in apposition (such as in a allogeneic transplant). While close spatial proximity is required, and usually achieved through some sort of physical (often micro-surgical) manipulation, the scale at which this occurs is small and relies on a chemical manipulation to complete the fusion at the level of the cellular membrane. The substances that are capable of performing this function are heterogeneous in composition and function, and often referred to as "fusogens" (Ryan & Henderson, 2020).

Axon membrane fusion is valuable in any attempt to manage nerve injury because it allows for the possibility of immediate reconstitution of electrical conductivity and therefore nerve function. An outcome on such a timeline is impossible given the current standard methods for managing nerve injury (approximation via epineurial repair), which accept distal Wallerian degeneration and the need to wait for axon regeneration from the point of injury to the end organ (which is traditionally thought to progress at the rate of approximately 1 mm per day). Therefore, fusogens represent a quantum leap in the way that nerves are managed, as they allow for the possibility of true axon *repair*, as opposed to the traditional methods, which at best harness axon *regeneration* (and accordingly the outcomes after this treatment approach are partial at best, and measured on the order of months to years).

There are two general mechanisms by which fusogens are thought to function: cell aggregation and membrane modification (Ryan & Henderson, 2020). Substances that function only via cell aggregation lead to increasingly closer apposition of the two axon ends (often by extraction of intervening water molecules). This group includes chitosan, dextran sulfate, small organic molecules, and lipids. Substances that function only via membrane modification alter membrane charges to create a more favorable electrochemical environment for apposition and ultimately fusion to occur. This group includes cations such as $Ca^{2+}$ and $Mg^{2+}$, sodium nitrate, and H-alpha-7. Polyethylene glycol (PEG) is a substance that is thought to function by a combination of both mechanisms, and for that reason is it is the fusogen that, to date, has been used most widely and most successfully in both research and clinical settings.

PEG is a hydrophilic polymer that is inexpensive and commonly encountered, as it has proven to play a role in many different medical, biological, and industrial scenarios. Its potential utility as a means of promoting cellular fusion, however, was first recognized in the 1970s after it was shown to be able to induce immortalized hybridoma cells and fused mammalian erythrocytes (Ren et al., 2019). The ability to induce axon membrane fusion (either in repair of cut ends of the same axon, or the fusion to ends of a different axons) was first realized and progressively tested beginning in the 1980s, starting with invertebrate models (Bittner, Sengelaub, Trevino, et al., 2016), and gradually progressing into rodents and then large mammals (fully reviewed in Canavero & Ren, 2020). It has successfully restored complete digital nerve function after traumatic injury in small human studies (Bamba, Waitayawinyu, Nookala, et al., 2016; see also De Medinaceli & Merle, 1991) and has been effective at restoring neurologic function at the level of the spinal cord in multiple animal models (Ren et al., 2019; fully reviewed in Canavero & Ren, 2020).

In addition to the cell fusion mechanisms of cell aggregation and membrane modification, PEG has been shown to have other effects that are beneficial for maximizing nerve function. First, it is neuroprotective due to its ability to inhibit the formation of the mitochondrial permeability transition pore; second, it may reduce the neuronal membrane tension and thereby improve membrane fluidity; third, it causes the release of neurotrophin-3, which attracts neural stem cells and promotes their differentiation into axons. However, PEG does not lead to the fusion or repair of connective tissue and therefore does not provide any structural strength and in very rare cases may be associated with allergic reactions (see references in Ren et al., 2019).

## GEMINI: Electrical stimulation

The GEMINI SCF protocol includes the positioning of an extradural stimulating 16-contact paddle straddling the point(s) of fusion followed by continuous stimulation (15–60 Hz, 5–9 V) during the entire rehabilitation period (Canavero, 2015;

Canavero et al., 2016; Canavero & Ren, 2020). Spinal cord stimulation (SCS) is paired to non-invasive transcranial magnetic stimulation (TMS) of the motor cortex (M1) (Canavero & Ren, 2016a, 2016b).

Electrical stimulation (ES) serves two primary purposes:

1. Promotion/Acceleration of axonal outgrowth and plasticity across the fusion interface. Shirres first emphasized the ability of electricity to stimulate spinal cord regeneration (see full details in Ren et al., 2019). ES has been shown to accelerate axonal outgrowth (neural sprouting) at sites of injury, including the cord, in animal studies (e.g., Thornton, Mehta, Morad, et al., 2018; see also Jack, Hurd, Martin, & Fouad, 2020, Ju, Park, Kim, Kim, et al., 2020). Of note, ES is co-additive to PEG in SCI (Wang et al., 2016).
2. Activation/Facilitation of propriospinal neurons caudal to the fusion interface.

Following complete spinal transection in man, the neuronal networks responsible for locomotion (i.e., central pattern generators, CPG) residing in the spinal cord are intact, but fail to produce limb movements: the excitability of the spinal cord is too depressed to enable the coordinated recruitment of motor neuron pools. Consequently, enabling robust levels of activity during rehabilitation is critical to steer activity-dependent plasticity in the trained circuitry. This is achieved by ES. ES replaces missing sources of excitation to reactivate spinal circuits, and thus enable motor control. In particular, following chemical fusion, sprouting across the fusional interface restarts the flow of electrical transmission between the fused cords. ES amplifies voluntary commands from the brain and thus supports their propagation to the CPGs at cervical and lumbar levels (Canavero & Ren, 2016a, 2016b; Canavero & Ren, 2020; Minassian & Hofstoetter, 2016). In turn, these integrate supraspinal and sensory information into the execution of purposeful movements.

Over the past 20 years, on the heels of Dimitrijevic's pioneering work in the last decades of the XX century (Dimitrijevic, 1983), multiple independent laboratories have shown that the delivery of continuous electrical stimulation (tonic) over the lumbar spinal cord immediately reestablishes intentional control over the activity of previously paralyzed leg muscles, even more than a decade after the occurrence of the SCI. Continuous ES also restores full weight-bearing standing and facilitates stepping (see reviews in James, McMahon, Field-Fote, & Bradbury, 2018; Calvert, Grahn, Zhao, & Lee, 2019; Cho, Squair, Bloch, & Courtine, 2019; Jack et al., 2020; see also chapters in this volume). ES of the spinal cord (SCS) is known to elicit activity locally and rostrally up to the cortex (as shown in chronic pain: see Canavero & Bonicalzi, 2018, pp. 465–474).

Spatiotemporal SCS protocols have been touted as an improvement over continuous stimulation (reviewed in Cho et al., 2019). However, no trials exist that compare continuous with spatiotemporal stimulation in homogeneous patients. Serotonergic agonists may further potentiate the effects of ES (Calvert et al., 2019; Cho et al., 2019).

ES can be employed in conjunction with specialized multidirectional robotic body weight support systems to steer activity-dependent plasticity in response to training (Cho et al., 2019).

Although clinical experience is limited to ES of the lumbar CPG, similar results are expected for the cervical cord (Canavero & Ren, 2016a, 2016b; Sunshine et al., 2013).

Importantly, since recovery of supraspinal control over limb movements is *directly correlated with the amount of spared tissues*, ES in the present context can only be deployed *after* chemical fusion.

As mentioned, M1 is concomitantly stimulated to promote corticospinal tract axonal outgrowth and plasticity; timing of paired stimuli is leveraged to produce plasticity during the rehabilitative phase (Canavero & Ren, 2016a, 2016b; Jack et al., 2020; Song, Amer, Ryan, & Martin, 2015; Zareen, Shinozaki, Ryan, et al., 2017). Stimulation of M1 can be carried out both invasively and noninvasively (Canavero, 2009, 2014). In the present context, noninvasive cortical stimulation (NICS) by means of TMS is the preferred modality. M1 ES promotes sprouting of corticofugal fibers and modulates post-injury neuroplasticity (Canavero, 2009; Canavero & Ren, 2016a, 2016b). M1 orchestrates recovery after SCI, developing new routes through the C-TRPS (Canavero & Ren, 2020; see also Cho et al., 2019).

## Clinical translation

Based on experimental evidence (reviewed in Ren et al., 2019), PEG turns out to be the fusogen of choice: it is most effective when applied locally and acutely after cord transection and stump apposition. Cord necrosis and cavitation are greatly reduced or curbed. Studies show that the molecular weight of PEG greatly affects the final outcome, with PEG 600 and PEG1500 as the ones evincing the best results in animal models of cord transection (see in Canavero & Ren, 2020).

### Gemini hydrogelation

In the simplest approach, the damaged cord segment (or at least half of it) is removed, up to its border with rostral and caudal healthy tissue and the gap is filled with a PEG hydrogel. PEG can be cross-linked to form porous hydrogels, which can serve as biocompatible matrices that can closely mimic the Extra-Cellular Matrix. PEG hydrogels possess both high water content/porosity and solidity. Injectable PEG, by in situ gelling, can conform geometrically to the defect (see Ren et al., 2019). Regenerating propriospinal fibers course through this hydrogel "superhighway," as demonstrated in rodent studies after many months (Estrada, Brazda, Schmitz, et al., 2014). Certainly, the use of PEG alone cannot completely mimic the three-dimensional porous structure of the spinal cord and would allow the upper and lower fiber bundles to grow in mismatched or even misplaced channels or pores, which is not the case with the next two approaches (see next section). Moreover, this approach would need many months for recovery. However, microspheres loaded with neurotrophins (e.g., BDNF and GDNF) could be embedded for slow release to accelerate this regrowth (see Ren et al., 2019).

## Resection-apposition (Freeman-GEMINI approach)

As mentioned, Walter Freeman suggested the severance-reapposition approach to chronic SCI: he surgically removed the damaged segment of the cord in dogs creating a gap, performed a complete *en bloc* vertebrectomy thus shortening the spine, brought the two fresh cord stumps in contact with fresh plasma and sutured the dura tightly: walking animals resulted after several months (Freeman, 1963; see also Heimburger, 2005, 2006). Notably, he observed direct electrophysiological conductance across the apposed stumps and provided histological evidence of axonal regeneration across the sectional interface (see also Beneš, Druga, Rokyta, & Stastný, 1991 and De Medinaceli & Wyatt, 1993; review in Ren et al., 2019). There is experimental evidence that the transected spinal cord in rabbits too can regenerate with concomitant return of motor, sensory and sphincter function when adequate apposition is obtained (Murray, Ugray, & Graves, 1965). Others carried out Freeman's surgery in dogs either immediately (Derlon, Roy-Camille, Saillant, Poirier, & Pichon, 1978) or 7 days after full section of the cord (Lumb & Nornes, 1983). However, Street (1967) carried out Freeman's procedure in paraplegic patients without success.

Spine-shortening vertebral osteotomy (a.k.a. vertebral column resection) is a surgical technique for correcting severe spinal deformities, treating congenital spinal anomalies, e.g., cord tethering, traumatic spine dislocations, and spine tumors at both cervical and thoraco-lumbar levels (Aoun, Elguindy, Barrie, et al., 2018; Hsieh, Stapleton, Moldavskiy, et al., 2010; Mody, Bravo Iñiguez, Armstrong, et al., 2016; Qiu, Yang, Ma, et al., 2015; Steinberg, Wali, Martin, et al., 2017). In the Freeman-GEMINI variant (Canavero & Ren, 2016a, 2016b; Ren et al., 2019), section of the damaged segment of the cord is performed at the moment of removing the vertebral body; the two ends are further trimmed so that no undue pressure is exerted on either stump by pressure vectors (too much as opposed to modest-pressure would lead to squeezing and local ischemia, jeopardizing the result). PEG is applied at this moment. The vertebra is removed and stabilization carried out simultaneously. The two spinal cord ends can be kept in apposition by some kind of micro-connector system: one incorporating a micro-channel system, through which PEG circulates, has been proposed (Brazda, Voss, Estrada, et al., 2013). Systems like this allow a tension-free, precise apposition of sharply transected nerve spinal cord stumps, as required by GEMINI. Spinal cord stimulators are epidurally positioned.

## Spinal cord transplantation (Shirres-GEMINI approach)

In 1905, Shirres reported his attempt to graft a segment of healthy canine cord in a human paraplegic patient: clear signs of neuroregeneration were observed on autoptic material, with initial sensory recuperation at 3 months; autopsy showed signs of neuroregeneration (see Ren et al., 2019). Forty years later, Woolsey, Minckler, Rezende, and Klemme (1944) operated on a 16-year-old male with complete loss of motor and sensory function after he was shot in his right shoulder with the bullet reaching the superior border of T4. Following laminectomy, the injured spinal cord was completely transected and replaced with a cadaveric spinal cord that had been fixed in 10% formalin for 12 days, and cleaned and sterilized with running and distilled water and 70% alcohol. No improvement in the patient's condition was noted, and the patient died almost 4 months after the surgery. Autopsy showed exceptional preservation of the transplanted graft, although with restricted regeneration and limited tissue reaction. The preservation was attributed to the preoperative use of formalin, and no explanations or related conclusion on the microscopic findings could be made.

Today, human cord grafts/transplants are once again a consideration given the availability of an effective spinal fusion protocol, i.e., GEMINI, which would be key to interfacing the donor segment with the recipient's cord stumps (Canavero, Ren, & Kim, 2021). In this variant, PEG would neuroprotect the graft/transplant until vascularization from the healthy ends of the patient. PEG hydrogels support the formation of vascularized tissue in vivo and PEG has been shown to promote angiogenesis in an SCI model (see Ren et al., 2019). The donor segment is harvested from a brain-dead organ donor at the same level of injury. All the roots would be reanastomosed (a process akin to reimplantation of avulsed nerve roots:

Carlstedt, James, & Risling, 2017, see (Canavero et al., 2021)). However, until tolerogenic protocols are developed for this contingency, immunosuppressants would be likely needed and this adds toxicity.

Depending on the length of the segment, surgical restoration of the vascular supply on one side may be required, but remains challenging with current techniques (Canavero et al., 2021).

Alternatively, a vascularized segment of the anatomically undamaged healthy cord caudal or rostral to the injury epicenter represents a suitable, interim alternative to fill the gap, as are bundles of peripheral nerves, as suggested by past animal experiments (Derlon, Roy-Camille, Lechevalier, Bisserie, & Coston, 1983; Sugar & Gerald, 1940) (modified Shirres-GEMINI variant).

Three points should be emphasized:

1. the hemicord used as a bridge in chronic SCI has undergone years of plastic readjustments (as the entire CNS: Isa, 2017), some maladaptive (e.g., in the case of cord central pain: Canavero & Bonicalzi, 2018). This might interfere with the process of recovery in some cases. However, it is well known that so-called entrenched plasticity is in fact reversible (discussed in Canavero & Bonicalzi, 2011).
2. Roots from the hemigraft have to be cut, which causes sensorimotor disruption in their respective territories of innervation. Studies of dorsal rhizotomies in human chronic pain patients verify that nearby roots take over and compensate for the resulting deficits (discussed in Canavero & Bonicalzi, 2007). Motor deficits would affect chest and abdominal wall muscles in a limited area.
3. The entire anatomical layout is disrupted. However, as mentioned (see above), it is the gray matter that supports locomotor recovery, even if white matter bundles are misaligned and not in continuity. This bridging segment is basically a gray matter bridge through which sensorimotor transmission is restored.

In all cases, PEG and electrical stimulation are deployed. PEG alone leads to remarkable recovery, but this is expected to be boosted by spinal cord and motor cortex stimulation (see above; reviewed in Canavero et al., 2016; Canavero & Ren, 2020).

## Current human trial

*CANINE TRIAL*: A preliminary canine study of the modified Shirres-GEMINI approach has been conducted on four female Beagles in China. In this proof-of-principle canine trial, paralysis following removal of an entire 3 cm long segment of the cord was at least partially reversed by a pedicled autograft of the animal's hemicord treated with PEG at the contact interfaces (Fig. 1). At 6 months the final average motor score (canine 0–19 Basso-Bresnahan-Beatty scale, cBBB) was roughly in line with our previous canine PEG studies of a complete spinal cord transection (medians: respectively 11.5 vs 11: Table 1) (reviewed in Ren et al., 2019). The main differences were a slower onset of symptom reversal (17 days vs 3 days) and lower final top scores (12 vs 18). This is likely due to the need for re-growing fibers to cross two, instead of one, sectional levels and more prolonged neuroplastic rearrangement than a simple section. Sphincter control was not regained at 6 months. DTI and histology data verified a vigorous regrowth of fibers across the lesional interfaces (Figs. 2 and 3).

*HUMAN TRIAL*: On the heels of the above study, a clinical trial was approved by the Medical Ethics Committee of Ruikang Hospital Affiliated to Guangxi University of Traditional Chinese Medicine. Inclusion criteria are: (1) Age < 50 -years old; (2) Thoracic or thoraco-lumbar traumatic-only SCI; (3) Interval from injury >12 months; (4) Absence of neuropsychiatric conditions; (5) Normal cardiopulmonary function. Surgical removal of any fixation devices was carried out in consenting patients. All patients were then submitted to Magnetic Resonance Imaging (MRI), Diffusion Tensor Imaging (DTI), Computerized Tomography (CT), Somatosensory Evoked Potential (SEP), and Motor Evoked Potential (MEP). Initial enrollment included 13 paraplegic patients (9 men, 3 women) classified as American Spinal Injury Association (ASIA) grade A. The goal of this study was to test the two arms (chemical and electrical) of the GEMINI SCF protocol singly (chemical only) or combined and compare long-term results and speed of recovery of both cord autografts and peripheral nerve bridges. Ten male (age range: 14–46 years) and three female (age range: 26–47) patients received the surgery in 2020. Eight underwent cord autotransplants (SCI levels: T2 up to T11) and five peripheral nerve grafting (SCI levels: T10–L2).

*PROCEDURE*: After positioning the patients in the prone position, general anesthesia was induced. The skin and muscles overlying the thoracic and lumbar spinal column were incised. A laminectomy was performed at spinal cord injury level with a cutting ultrasonic scalpel (BoneScalpel®, Misonix, USA) to expose the dura mater, which was then cut open to expose the spinal cord. The cord autotransplant was performed (see Fig. 1A and 4(1/2)). The SCI epicenter was removed leaving a gap between two fresh spinal cord stumps and sent out for histological assessment. After measuring the gap, a hemicord segment of equal length along with the spared posterior spinal artery (PSA) was gently split from the rest of the cord either rostrally or caudally (depending on the level of SCI) and turned over to bridge the gap. The PSA ensured

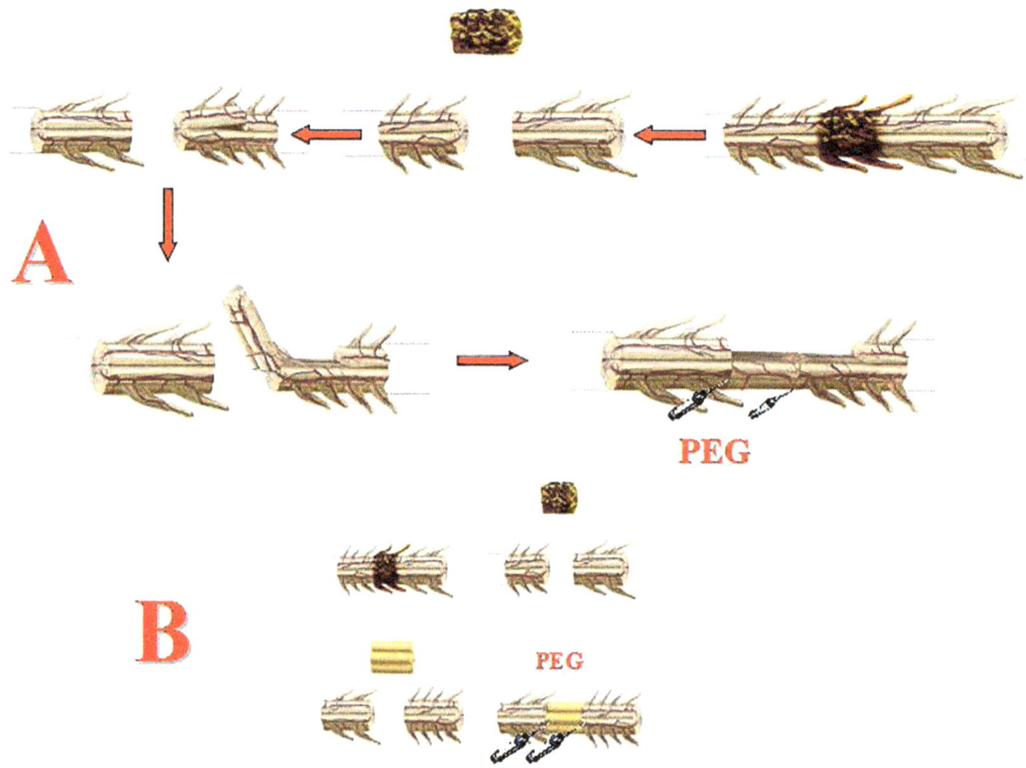


FIG. 1 (A) Modified Shirres-GEMINI protocol: surgical technique. The spinal injury epicenter, as assessed radiologically and then visually during surgery, is removed *en bloc*, leaving a gap bordered by undamaged cord ends. A pedicled (vascularized) hemicord of adequate length is split along the sagittal plane and subsequently turned over to fill the gap. PEG is applied at this moment at both ends. After dural closure with a graft, electrical stimulators are stitched onto the dura. (B) Whenever an autograft is not possible, a bundle of segments from the patient's own sural nerve is positioned inside the gap.

continual vascularization of the cord bridge. At the points of contact between the cord transplant and the stumps PEG600 (10 mL, Sigma-Aldrich/Merck, Germany) was topically injected. Finally, the dura mater was sutured with an artificial dural patch (Guanhao Biotech, China). A 3-hole silicone drainage tube was placed into the wound, and the wound was sutured by layers.

Whenever a cord autograft was not feasible (distal thoracic and lumbar), *free sural nerve transplantation (FSNT)* was performed. The skin and subcutaneous fat overlying the sural nerve of one leg were incised and the sural nerve excised and split into several segments of the same length, which were then assembled into one bundle. The spinal cord was exposed and treated similarly to the PCAT protocol (Fig. 1B). The sural nerve bundle was inserted into the gap and sutured under the operating microscope with 7–0 threads; PEG600 (10 mL) Sigma-Aldrich/Merck, Germany) was topically applied at the two bridge-stump interfaces.

Spinal cord stimulation employed four MEDTRONIC (USA) Model 3778 Pisces-Octad electrodes positioned to cover the two fusional interfaces (Fig. 4(3)). These were initially positioned in the last patients of this series and will be installed in all future patients.

The entire operation (both versions) took about 4 h to complete. Somatosensory evoked potentials (SEP) and motor evoked potentials (MEP) were recorded at preoperatively and 5 min, 1, 2, 3, 4 weeks postoperatively and analyzed with NIM-ECLIPSE® System (Medtronic, USA). All patients were scanned with MRI and DTI using a 1.5 T MRI system in the supine position preoperatively and at 4 weeks postoperatively. Post-operative spinal angiography is a consideration to assess graft vascularization, but possible toxicity from the contrast medium limits its use to selected cases.

Two weeks after surgery, rehabilitation began employing either an assistive robotic exoskeleton (AiWalker, Beijing Ai-Robotics Technology Co. Ltd., China) 30 min *bid*, or an intelligent auxiliary mobile robot (BangBang, Shanghai Bangbang Robot Co. Ltd., China) or a paraplegic standing frame, 40 min *bid*, or a rehabilitation bicycle 1 h *bid*.

*PRELIMINARY RESULTS*: The postoperative period was uneventful in all cases (no fever, incisional infections, or cerebrospinal fluid leaks). No adverse reaction of any kind to PEG600 was observed. Immunohistochemistry evinced complete

**TABLE 1** cBBB scores of four female Beagles submitted to the modified Shirres-GEMINI variant.

| Days | Transplant +PEG | | | | Mean |
|---|---|---|---|---|---|
| −1 | 19 | 19 | 19 | 19 | 19 |
| 0 | 0 | 0 | 0 | 0 | 0 |
| 3 | 0 | 0 | 0 | 0 | 0 |
| 10 | 0 | 0 | 0 | 0 | 0 |
| 17 | 0 | 0 | 1 | 1 | 0.5 |
| 21 | 0 | 0 | 3 | 4 | 1.75 |
| 25 | 1 | 1 | 4 | 6 | 3 |
| 29 | 4 | 4 | 5 | 6 | 4.75 |
| 32 | 7 | 6 | 6 | 6 | 6.25 |
| 37 | 9 | 8 | 7 | 9 | 8.25 |
| 45 | 10 | 8 | 9 | 10 | 9.25 |
| 53 | 12 | 8 | 10 | 11 | 10.25 |
| 59 | 12 | 9 | 10 | 12 | 10.75 |
| 180 | 12 | 11 | 10 | 12 | 11.25 |

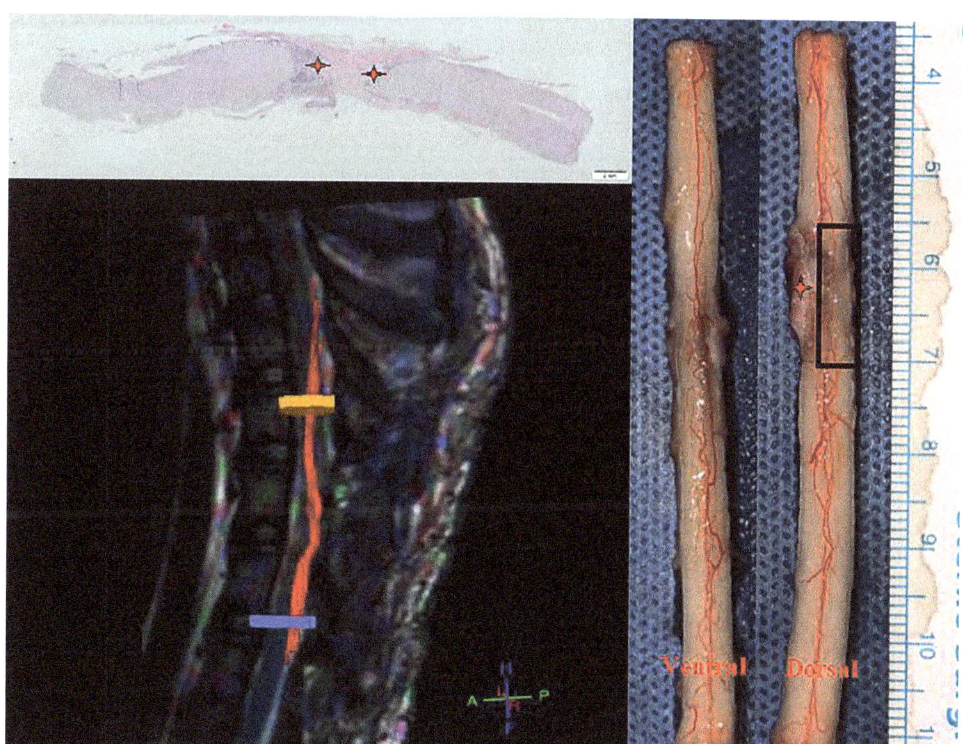


FIG. 2 Canine trial of the modified Shirres-GEMINI variant. Left upper half: sagittal view of the grafted cord (H&E staining): notice the absence of cysts at the points of fusion (stars). Right half: autoptic view of the cords: notice the graft perfectly enmeshed with the healthy cords (black rectangle); on the side, the hemigap is filled by scar. Left lower half: Diffusion Tensor Imaging (DTI) image at 6 months: new fibers extend beyond the fusional interfaces of the graft.

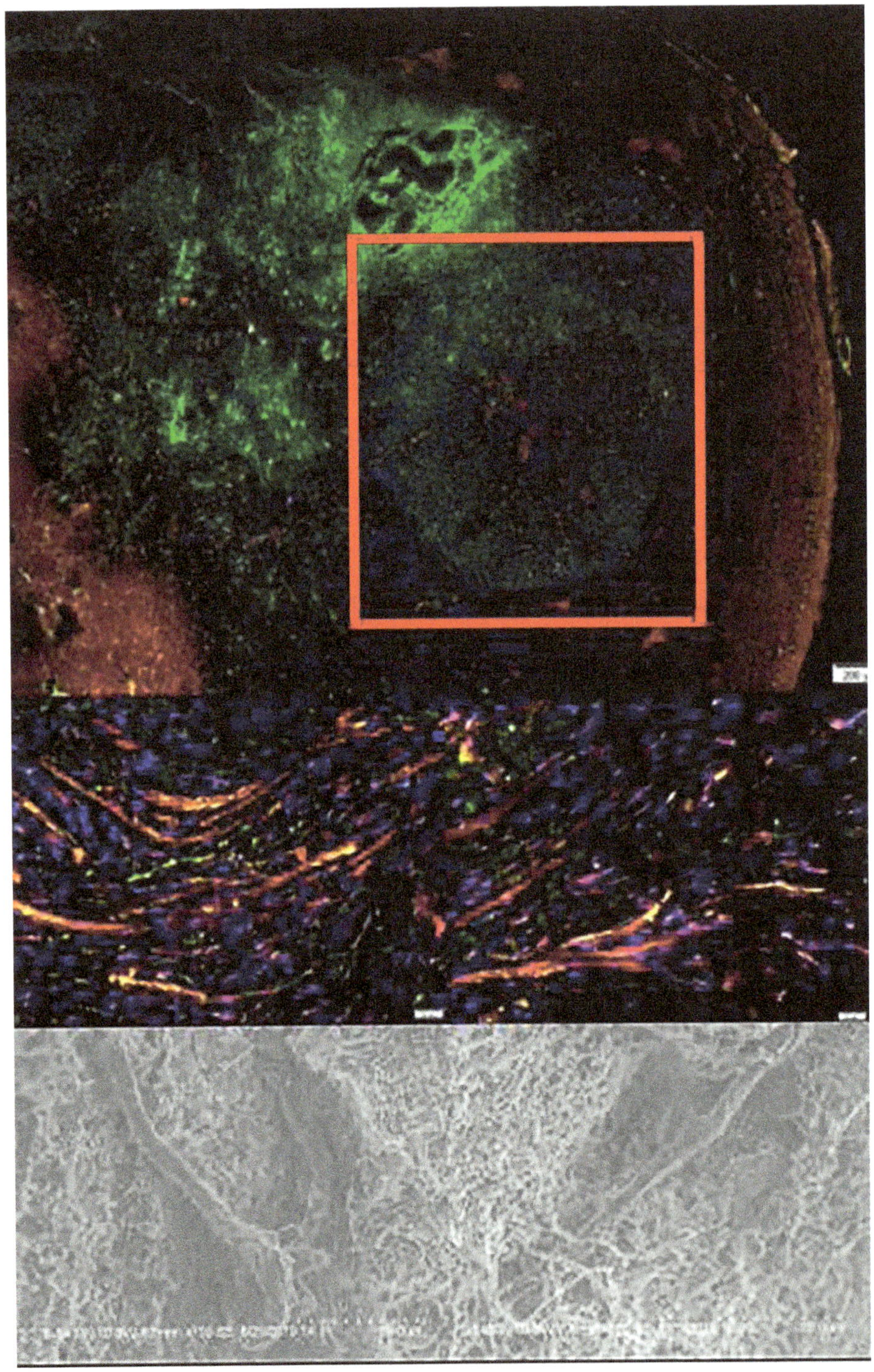

FIG. 3 Canine trial of the modified Shirres-GEMINI variant. Neurofilament immunofluorescence of the graft (red square) and fusional interfaces (middle): notice profuseness of fiber regrowth. Lower third: Scanning Electron Microscope images of sprouting fibers across the fusional interfaces.

absence of surviving fibers in the removed injured segment and correct removal of all damaged tissues. In line with the canine study, no sign of motor recovery was evinced for several days after the surgery. In patients treated with PEG only, the first signs of recovery were visible at 2 weeks, as one cord-transplant patient was able to move the knee, ankle and toe joints of both lower extremities voluntarily; a nerve-bridge patient did the same at 3 weeks. By month 4, several patients displayed voluntary control of the legs; initial supported ambulation was observed in some cases (Fig. 5). DTI and MEPs showed anatomical and physiological recovery (Fig. 5).

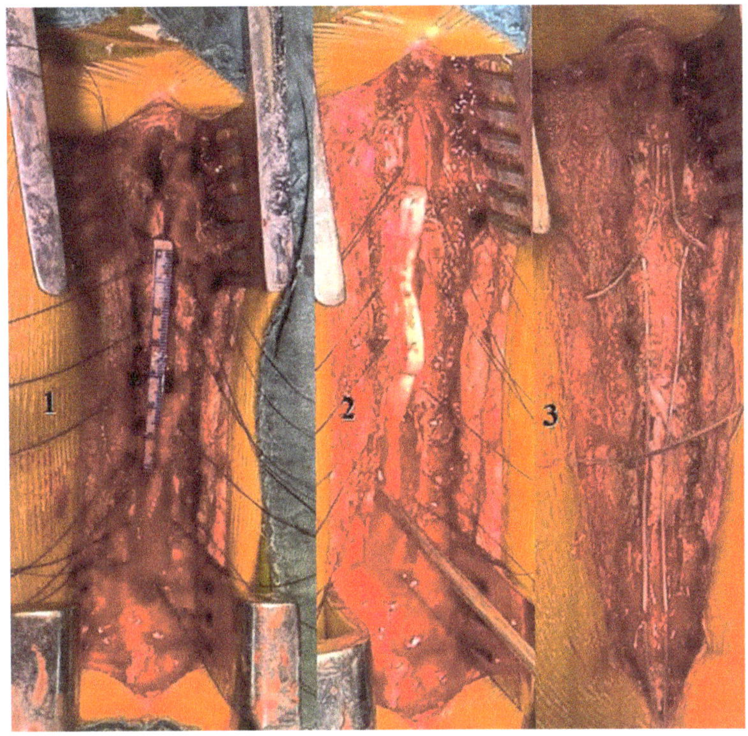


FIG. 4 Shirres-GEMINI operative view in a patient. Post-resection gap (T6–7), length: 5 cm (1); pedicled hemisegment in place (2); four MEDTRONIC octopolar leads are positioned on the dural graft (3).

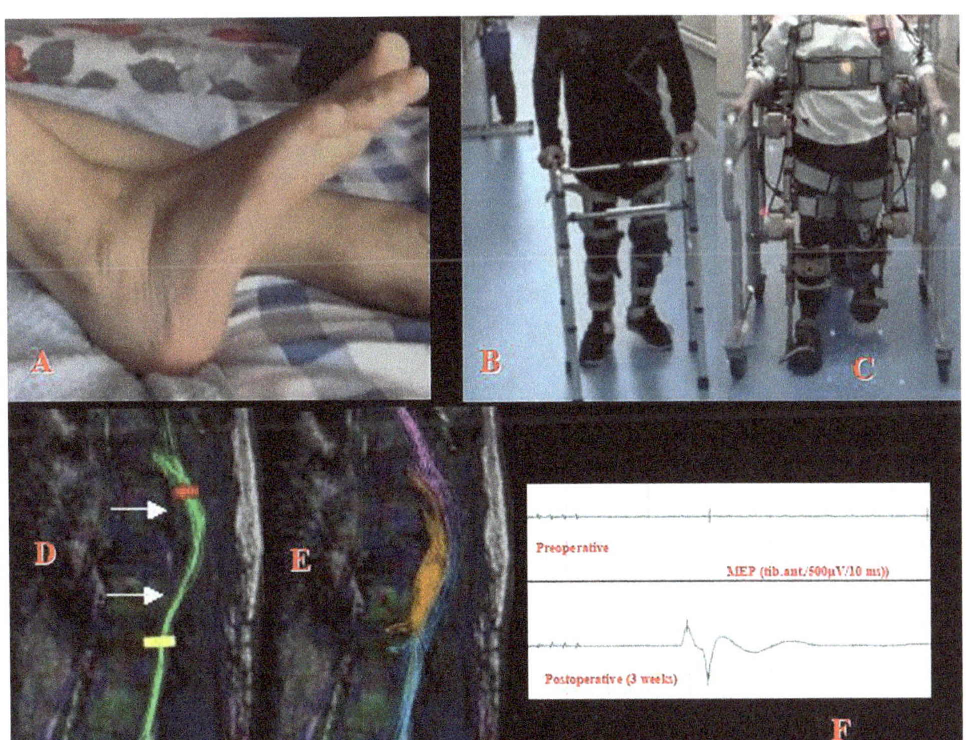


FIG. 5 Evidence of motor recovery in patients. (A) At 1 month, lower limbs can be flexed voluntarily. (B) At 2 months, patients can take the first steps with assistance. (C) Patient undergoing rehabilitation inside a robotic harness. (D and E) DTI showing fiber regrowth across the cord autograft. (F) Recovery of motor evoked potentials.

Spinal cord stimulation was conducted by stimulating with either the upper or the lower electrodes or combined. After a few weeks, stimulation with the upper electrodes could elicit motor activity in the legs, proving the viability of the anatomical bridge; patients experienced voluntary control over the legs at high voltage (5–9 V). As expected, stimulation with the lower electrodes activated leg movements immediately after the surgery, again at high voltages. Combination stimulation was carried out at low voltage (3 V) cranially and high voltage caudally (max 9 V). Frequencies assessed ranged from low (20 Hz) to moderately high (60 Hz); pulse widths were tested between 60 and 450 μs. Preliminary observations suggest that electrical stimulation does indeed accelerate recovery.

## Conclusion

The decades-old idea that SCI can be cured by removing the most heavily damaged segment of a cord and bringing the healthy tissue above and below the epicenter in direct—or indirect contact, through a viable graft/transplant (either cord or peripheral nerve segments)—is now being translated clinically on the heels of the introduction of the GEMINI spinal cord fusion protocol in 2013. The initial results of the first clinical trial of GEMINI-powered spinal cord autografting are promising.

## Applications to other areas of neuroscience

In this chapter, we have reviewed the GEMINI spinal cord fusion protocol in its translation to spinal cord injury. The GEMINI protocol exploits fusogens such as polyethylene glycol (PEG) and spinal electrical stimulation to re-start nerve conduction across transected spinal cords. Animal studies including in rodents, canines, and primates confirmed its efficacy. The GEMINI protocol has been developed to enable head and brain transplants. Head transplants are now been assessed in man, while brain transplants are in the experimental phase. Both have been developed primarily for life extension, and once tolerogenic protocols that avoid immunosuppression become available, they will revolutionize human society as we know it. Cosmetic body transplants will also become an option. The GEMINI protocol can theoretically be applied to several neuroconditions, e.g., multiple sclerosis, in which focal damage in the cord is primarily responsible for motor disability. Transplanting segments of healthy cords could reverse arm and gait impairment. The availability of a powerful nerve function restoration protocol can actually be envisioned as a treatment for several brain ailments. For instance, a possible scenario deals with transplantation of segments of healthy brain (from donor, fresh cadavers, or organoids) as replacement of diseased brain tissue, e.g., after stroke or major trauma, including prolonged unconsciousness states.

## Mini-dictionary of terms

1. PEG: Polyethylene Glycol is a widely used chemical with membrane fusogenic properties. Once applied, it leads to the reconstitution of damaged cell membranes.
2. Spinal Cord Electrical Stimulation: this is a widely employed technique to control chronic pain, with the additional benefit of kick-starting depressed neural activity in the cord movement generating centers after spinal injury.
3. GEMINI: this is the technology that re-establishes anatomophysiologic continuity of a transected spinal cord. It combines fusogens and electrical stimulation of the nervous system.
4. Shirres-GEMINI approach: this surgery combines *en bloc* removal of the most injured segment of the cord, replacement with a pedicled hemicord segment from above or below the extirpation area and application of the GEMINI spinal cord fusion protocol at both ends of the grafted segment.
5. Freeman-GEMINI approach: this surgery combines en bloc removal of the most injured segment of the cord, shortening of the vertebral column and direct fusion of the stumps under GEMINI conditions.

## Key facts of the GEMINI spinal cord fusion protocol for spinal cord injury

1. GEMINI was introduced with the goal of enabling head and brain transplants in 2013
2. GEMINI exploits special substances called fusogens, such as polyethylene glycol (PEG) combined with spinal electrical stimulation to force two ends of a transected cord to re-establish communication.
3. Fusogens have the ability to refuse damaged cell membranes, including neuronal and axonal membranes.

4. Electricity stimulates nerve fiber regrowth and "powers up" depressed neuronal circuits after spinal cord injury.
5. The GEMINI spinal cord fusion (SCF) protocol has been shown to reverse motor paralysis after spinal cord transection in rodents, canines and primates

## Summary points

1. There is at present no biological cure for chronic paralysis following spinal cord injury (SCI). Chronic disability after spinal cord injury remains challenge.
2. Many experimental treatments including stem cells have not led to a long-sought cure
3. In the 1940s to 1950s, surgeons believed that the first step toward a cure would require the extirpation of the most damaged segment of the cord. The result would be the creation of two undamaged stumps whose ends would have to be connected somehow.
4. The GEMINI spinal cord fusion protocol was introduced in 2013 to mend transected spinal cords in the context of head/brain transplants
5. By combining 3 and 4, a permanent, biological cure becomes: after removing the most damaged portion of the spinal cord *en bloc*, the two ends of the cord can be brought together after spinal (vertebral) shortening (Freeman-GEMINI approach) or alternatively, the post-extirpation gap is filled with a segment of normal cord (Shirres-GEMINI approach)
6. We describe these two approaches and report the preliminary results of the first clinical trial testing autologous cord grafting in chronic SCI patients.

## References


Aoun, S. G., Elguindy, M., Barrie, U., et al. (2018). Four-level vertebrectomy for en bloc resection of a cervical chordoma. *World Neurosurgery*, *118*, 316–323.

Bamba, R., Waitayawinyu, T., Nookala, R., et al. (2016). A novel therapy to promote axonal fusion in human digital nerves. *Journal of Trauma and Acute Care Surgery*, *81*, S2177.

Beneš, V., Jr., Druga, R., Rokyta, R., & Stastný, J. (1991). Vertebrectomy and its possible use in repair of spinal cord injuries by neural transplantation. *Restorative Neurology and Neuroscience*, *2*, 255–260.

Bittner, G. D., Sengelaub, D. R., Trevino, R. C., et al. (2016). The curious ability of polyethyne glycol fusion technologies to restore lost behaviors after nerve severance. *Journal of Neuroscience Research*, *94*, 207–230.

Brazda, N., Voss, C., Estrada, V., et al. (2013). A mechanical microconnector system for restoration of tissue continuity and long-term drug application into the injured spinal cord. *Biomaterials*, *34*, 10056–10064.

Calvert, J. S., Grahn, P. J., Zhao, K. D., & Lee, K. H. (2019). Emergence of epidural electrical stimulation to facilitate sensorimotor network functionality after spinal cord injury. *Neuromodulation*, *22*, 244–252.

Canavero, S. (Ed.). (2009). Textbook of therapeutic cortical stimulation. In New York: Nova Science Pub. Co.

Canavero, S. (2013). HEAVEN: The head anastomosis venture project outline for the first human head transplantation with spinal linkage (GEMINI). *Surgical Neurology International*, *4*, S335–S342.

Canavero, S. (Ed.). (2014). *Textbook of cortical brain stimulation*. Berlin: DeGruyter Open.

Canavero, S. (2015). The "Gemini" spinal cord fusion protocol: Reloaded. *Surgical Neurology International*, *6*, 18.

Canavero, S., & Bonicalzi, V. (2007). *Central pain syndrome*. New York: Cambridge University Press.

Canavero, S., & Bonicalzi, V. (2011). *Central pain syndrome* (2nd ed.). Cambridge: Cambridge University Press.

Canavero, S., & Bonicalzi, V. (2018). *Central pain syndrome* (3rd ed.). Springer-Nature.

Canavero, S., & Ren, X. (2016a). Houston, GEMINI has landed: Spinal cord fusion achieved. *Surgical Neurology International*, *7*, S626–S628.

Canavero, S., & Ren, X. P. (2016b). The spark of life: Engaging the cortico-truncoreticulo-propriospinal pathway by electrical stimulation. *CNS Neuroscience & Therapeutics*, *22*, 260–261.

Canavero, S., & Ren, X. P. (Eds.). (2020). *The technology of head transplantation*. New York: NOVA Science Pub. Co.

Canavero, S., Ren, X., & Kim, C.-Y. (2021). Heterologous spinal cord transplantation in man. *Surgical Neurology International*, *12*, 295.

Canavero, S., Ren, X., Kim, C. Y., & Rosati, E. (2016). Neurologic foundations of spinal cord fusion (GEMINI). *Surgery*, *160*, 11–19.

Carlstedt, T., James, N., & Risling, M. (2017). Surgical reconstruction of spinal cord circuit provides functional return in humans. *Neural Regeneration Research*, *12*, 1960–1963.

Cho, N., Squair, J. W., Bloch, J., & Courtine, G. (2019). Neurorestorative interventions involving bioelectronic implants after spinal cord injury. *Bioelectronic Medicine*, *5*, 10.

De Medinaceli, L., & Merle, M. (1991). Applying "cell surgery" to nerve repair: A preliminary report on the first ten human cases. *Journal of Hand Surgery (British)*, *16*, 499–504.

De Medinaceli, L., & Wyatt, R. J. (1993). A method for shortening of the rat spine and its neurologic consequences. *Journal of Neural Transplantation & Plasticity*, *4*, 39–52.

Derlon, J. M., Roy-Camille, R., Lechevalier, B., Bisserie, M., & Coston, A. (1983). Delayed spinal cord anastomosis. In C. C. Kao, R. P. Bunge, & P. J. Reier (Eds.), *Spinal cord reconstruction* (pp. 223–234). New York: Raven Press.

Derlon, J. M., Roy-Camille, R., Saillant, G., Poirier, J., & Pichon, F. (1978). Experimental transection of the spinal cord. *Neurochirurgie*, *24*, 103–111.

Dimitrijevic, M. R. (1983). Neurophysiological evaluation and epidural stimulation in chronic spinal cord injury patients. In C. C. Kao, R. P. Bunge, & P. J. Reier (Eds.), *Spinal cord reconstruction* (pp. 465–474). New York: Raven Press.

Estrada, V., Brazda, N., Schmitz, C., et al. (2014). Long-lasting significant functional improvement in chronic severe spinal cord injury following scar resection and polyethylene glycol implantation. *Neurobiology of Disease*, *67*, 165–179.

Freeman, L. W. (1963). Observation on the regeneration of spinal axons in mammals. In *Proceedings, X congreso latinoamericano de neurochirurgia* (pp. 135–144). Brazil: Editorial Don Bosco.

Heimburger, R. F. (2005). Return of function after spinal cord transection. *Spinal Cord*, *43*, 438–440.

Heimburger, R. F. (2006). Is there hope for return of function in lower extremities paralyzed by spinal cord injury? *Journal of the American College of Surgeons*, *202*, 1001–1004.

Hsieh, P. C., Stapleton, C. J., Moldavskiy, P., et al. (2010). Posterior vertebral column subtraction osteotomy for the treatment of tethered cord syndrome: Review of the literature and clinical outcomes of all cases reported to date. *Neurosurgical Focus*, *29*, E6.

Illis, L. S. (2012). Central nervous system regeneration does not occur. *Spinal Cord*, *50*, 259–263.

Isa, T. (2017). The brain is needed to cure spinal cord injury. *Trends in Neurosciences*, *40*, 625–636.

Jack, A., Hurd, C., Martin, J. H., & Fouad, K. (2020). Electrical stimulation as a tool to promote plasticity of the injured spinal cord. *Journal of Neurotrauma*, *37*, 1933–1953.

James, N. D., McMahon, S. B., Field-Fote, E. C., & Bradbury, E. J. (2018). Neuromodulation in the restoration of function after spinal cord injury. *Lancet Neurology*, *17*, 905–917.

Ju, C., Park, E., Kim, T., Kim, T., et al. (2020). Effectiveness of electrical stimulation on nerve regeneration after crush injury: Comparison between invasive and non-invasive stimulation. *PLoS One*, *15*, e0233531.

Kostyuk, P. G., & Vasilenko, D. A. (1979). Spinal interneurons. *Annual Review of Physiology*, *41*, 115–126.

Lumb, W. V., & Nornes, H. O. (1983). Vertebral resection and spinal cord reapposition. In C. C. Kao, R. P. Bunge, & P. J. Reier (Eds.), *Spinal cord reconstruction* (pp. 209–222). New York: Raven Press.

Minassian, K., & Hofstoetter, U. S. (2016). Spinal cord stimulation and augmentative control strategies for leg movement after spinal paralysis in humans. *CNS Neuroscience & Therapeutics*, *22*, 262–270.

Mody, G. N., Bravo Iñiguez, C., Armstrong, K., et al. (2016). Early surgical outcomes of en bloc resection requiring vertebrectomy for malignancy invading the thoracic spine. *The Annals of Thoracic Surgery*, *101*, 231–236.

Murray, G., Ugray, E, & Graves, A. (1965). Regeneration in injured spinal cord. *American Journal of Surgery*, *109*, 406–409.

Qiu, F., Yang, J. C., Ma, X. Y., et al. (2015). Relationship between spinal cord volume and spinal cord injury due to spinal shortening. *PLoS One*, *10*, e0127624.

Ren, X., Kim, C., & Canavero, S. (2019). Bridging the gap: Spinal cord fusion as a treatment of chronic spinal cord injury. *Surgical Neurology International*, *10*, 51.

Ryan, I., & Henderson, P. W. (2020). Gemini: Overview of fusogens. In S. Canavero, & X. P. Ren (Eds.), *The technology of head transplantation*. New York: Nova Science Publishers (Chapter 4).

Shik, M. L. (1983). Action of the brainstem locomotor region on spinal stepping generators via propriospinal pathways. In C. C. Kao, R. P. Bunge, & P. J. Reier (Eds.), *Spinal cord reconstruction* (pp. 421–434). New York: Raven Press.

Song, W., Amer, A., Ryan, D., & Martin, J. H. (2015). Combined motor cortex and spinal cord neuromodulation promotes corticospinal system functional and structural plasticity and motor function after injury. *Experimental Neurology*, *277*, 46–57.

Steinberg, J. A., Wali, A. R., Martin, J., et al. (2017). Spinal shortening for recurrent tethered cord syndrome via a lateral retropleural approach: A novel operative technique. *Cureus*, *9*, e1632.

Street, D. M. (1967). Traumatic paraplegia treated by vertebral resection, excision of spinal cord lesion, suture of the spinal cord and interbody fusion. In *Proceedings of the annual clinical spinal cord injury conference (Long Beach, CA, USA)*, *16* (pp. 92–103).

Sugar, O., & Gerald, R. W. (1940). Spinal cord regeneration in the rat. *Journal of Neurophysiology*, *3*, 1–19.

Sunshine, M. D., Cho, F. S., Lockwood, D. R., Fechko, A. S., Kasten, M. R., & Moritz, C. T. (2013). Cervical intraspinal microstimulation evokes robust forelimb movements before and after injury. *Journal of Neural Engineering*, *10*, 036001.

Thornton, M. A., Mehta, M. D., Morad, T. T., et al. (2018). Evidence of axon connectivity across a spinal cord transection in rats treated with epidural stimulation and motor training combined with olfactory ensheathing cell transplantation. *Experimental Neurology*, *309*, 119–133.

Wang, A., Zhang, G., Xiaochen, W., Zhang, C., Tao, S., & Huo, X. (2016). Combination of applied electric field and polyethylene glycol effectively enhance functional recovery in acute spinal cord injury of rats. In *Paper presented at: 2016 Asia-Pacific international symposium on electromagnetic compatibility (APEMC)* (pp. 17–21).

Woolsey, D., Minckler, J., Rezende, N., & Klemme, R. (1944). Human spinal cord transplantation. *Experimental Medicine and Surgery*, *2*, 93–102.

Zareen, N., Shinozaki, M., Ryan, D., et al. (2017). Motor cortex and spinal cord neuromodulation promote corticospinal tract axonal outgrowth and motor recovery after cervical contusion spinal cord injury. *Experimental Neurology*, *297*, 179–189.

Chapter 25

# Chondroitinase ABC I as a novel candidate for reducing damage in spinal cord injury

Akram Shirdel[a] and Khosrow Khalifeh[b,c]

[a]*Department of Biochemistry, Faculty of Biological Sciences, Tarbiat Modares University, Tehran, Iran;* [b]*Department of Biology, Faculty of Sciences, University of Zanjan, Zanjan, Iran;* [c]*Department of Biotechnology, Research Institute of Modern Biological Techniques, University of Zanjan, Zanjan, Iran*

## Abbreviations

| | |
|---|---|
| **cABC I** | chondroitinase ABC I |
| **CNS** | central nervous system |
| **CS** | chondroitin sulfate |
| **CSPG** | chondroitin sulfate proteoglycan |
| **CSPGs** | chondroitin sulfate proteoglycans |
| **DS** | dermatan sulfate |
| **DSPG** | dermatan sulfate proteoglycan |
| **ECM** | extracellular matrix |
| **GAGs** | glycosaminoglycans |
| **GBM** | glioblastoma |
| **HA** | hyaluronic acid |
| **Hep** | heparin |
| **HepS** | heparan sulfate |
| **HS** | heparin sulfate |
| **KS** | keratan sulfate |
| **PEG** | polyethylene glycol (PEG) |
| **SCI** | spinal cord injury |

## Molecular perspectives

### Glycosaminoglycans and proteoglycans

Extracellular matrices (ECMs) and cell surfaces of the mammalian tissues contain repeating disaccharide units known as glycosaminoglycans (GAGs). These linear polysaccharides can be sulfated in various positions to obtain a negatively charged nature that is essential for molecular recognition and binding. The corresponding reactions are catalyzed by the enzymes of the sulfotransferase family. Based on the type of disaccharide units, GAGs are categorized into several groups, including hyaluronic acid (HA), chondroitin sulfate/dermatan sulfate (CS/DS), heparin/heparin sulfate (Hep/HS), and keratan sulfate (KS) (Gandhi & Mancera, 2008).

A typical chondroitin sulfate disaccharide unit in human, chondroitin-6-sulfate (C6S) with the primary configuration containing a uronic acid (Glucuronic acid) and amino sugar (Galactosamine) is shown in Fig. 1A. As shown in Fig. 1B; GAG chains can covalently bind at their reducing ends to a serine or asparagine residue of a target protein, forming a proteoglycan ( Jackson, Busch, & Cardin, 1991). On a higher level, a biopolymer such as hyaluronic acid acts as a backbone structure to assemble similar or various types of proteoglycans (Fig. 1C).

These macromolecules are synthesized in all mammalian cells and based on the type and function; they may be secreted into the ECM, localized into the plasma membrane, or stored in secretory vesicles (Couchman, 2010; Dick, Akslen-Hoel, Grøndahl, Kjos, & Prydz, 2012; Iozzo & Schaefer, 2015; Yanagishita, 1993). Due to the wide variety of functions, any

Diagnosis and Treatment of Spinal Cord Injury. https://doi.org/10.1016/B978-0-12-822498-4.00025-7

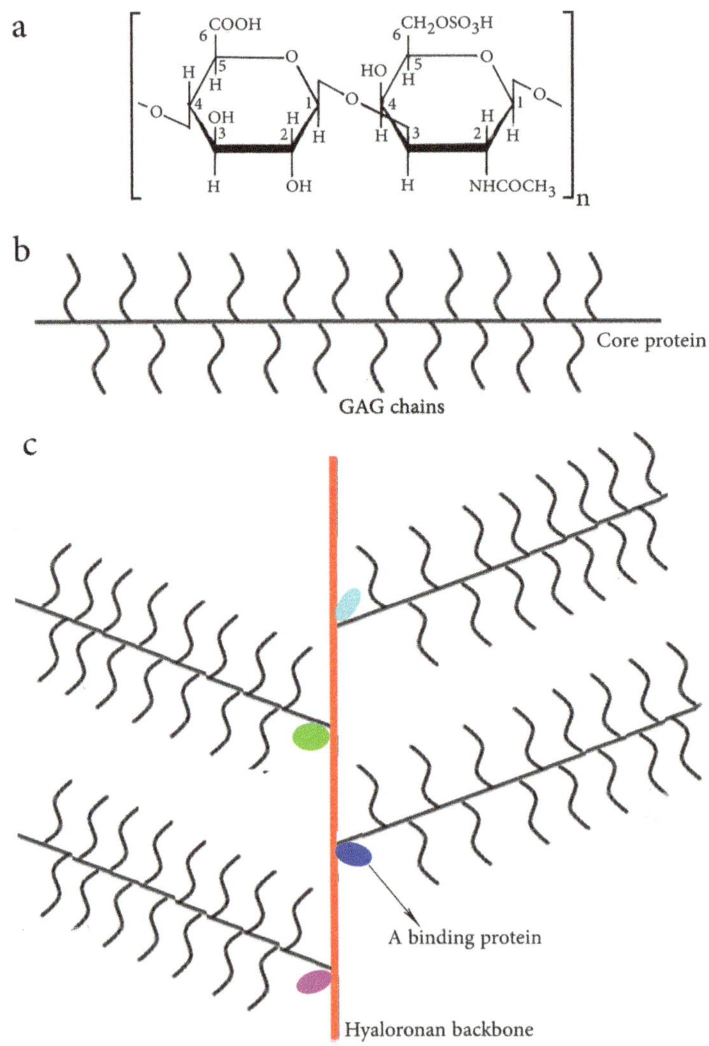


FIG. 1 Structural organization of Glycosaminoglycans and Proteoglycans. (A) The structure of a disaccharide unite in chondroitin-6-sulfate (C6S). D-glucuronic acid has a β1–3 linkage to *N*-acetyl-β-D-galactosamine. The connection between disaccharide units in the polymerized GAG is β1–4. Galactosamine can be variously sulfated at either 4 or 6 positions to give C4S or C6S, respectively. (B) The polymerized GAG chains are attached to the target protein forming a proteoglycan. The GAG chains may be the same or of other types. (C) The hyaluronic acid serves as a backbone to assemble similar or different types of proteoglycans. Specific proteins, such as growth factors and cytokines, can reversibly bind to this macromolecule.

defects in the synthesis and processing of proteoglycans have important clinical manifestations (Schwartz & Domowicz, 2018). Regarding the protein binding affinity and biocompatibility of GAGs, they are proposed as a tool for localized drug-delivery, using enzymes as drug (Hachim, Whittaker, Kim, & Stevens, 2019).

## Distribution and functions of GAGs and proteoglycans

The GAGs play essential roles in fundamental biological processes, and wide distribution of these biopolymers in animals demonstrates that they have conserved functions in organisms (DeAngelis, 2002). In the normal state, individual tissues of the human body contain a defined combination of GAGs, some of which are tissue-specific, and others may be temporally observe during growth, development, and pathogenic conditions.

The ability to absorb water molecules and their spatial structure enables GAGs and proteoglycans to provide a suitable scaffold that is essential to maintain the shape, cohesive properties, and mechanical features of tissues (Beldowski, Mazurkiewicz, Topoliński, & Małek, 2019). Accordingly, in tissue-based diseases, damage to the structure of proteoglycan is a central event (Hardingham & Bayliss, 1990). Some of the GAGs can bind into several specific proteins such as cytokines, chemokines, growth factors, morphogens, etc., to form biologically effective types of macromolecules (Imberty, Lortat-Jacob, & Pérez, 2007). These interactions are crucial in regulating of some critical biological processes, including cell migration (Merle, Durussel, Delmas, & Clézardin, 1999), regulation of blood coagulation and homeostasis (Bourin & Lindahl, 1993; Lasne, Jude, & Susen, 2006), growth factor control (Lander & Selleck, 2000), cell adhesion, and signaling (Werz & Schachner, 1988), inflammation (Gill, Wight, & Frevert, 2010), and pathogens attachment (García et al., 2016). They are involved in multicellular processes and have critical roles in development.

The GAGs are also involved in the inhibition of smooth muscle cell proliferation (Clowes & Karnowsky, 1977). It has been reported that proteoglycans may have essential roles in normal glomerular cell function and progressive renal disease (Templeton, 1992). They are considered as the first defensive line at the bladder's luminal surface against urinary tract infections, that created by pathogenic *Escherichia coli* among postmenopausal women (Anand et al., 2012). It has postulated that the accumulation of specific GAGs may form a physical barrier to the passage of immunocompetent cells from the mother to the fetus (Calatroni & di Ferrante, 1969). The changes in the composition and molecular structure of placental GAGs during pregnancy could alter the molecular transport rate through placental connective tissue and affect the rate of fetal growth (Lee, Jamieson, & Schafer, 1973). It has been reported that GAGs are involved in tumor development, angiogenesis and metastasis in certain types of cancers (Muir, De Winter, Verhaagen, & Fawcett, 2019).

## CSPGs in the nervous system

It has revealed that CSPGs and DSPGs are dominant proteoglycans in CNS-ECM (Iozzo & Schaefer, 2015). Chondroitin sulfate proteoglycans (CSPGs) are those CS-GAGs that are covalently linked to a core protein (Imberty et al., 2007). The GAG side chains in CSPGs are of different lengths, and the disaccharide units of CSPGs have either *N*-acetylglucosamine or *N*-acetylgalactosamine as well as uronic acid (see Fig. 1). Hence, depending on the length and the type of disaccharide unit and the position of sulfate moiety (positions 4 or 6), different types of CSPGs have identified, some of them have important tasks in CNS (Quraishe, Forbes, & Andrews, 2018). In CNS-ECM, various types of CSPGs bind to linear non-sulfated polysaccharides such as hyaluronan to produce a specific combination of proteoglycans, which is collectively known as lectican family proteoglycans. They include Neurocan, Brevican, Aggrecan, Perlecan, Testican, Versican, and Biglycan ( Jones, Margolis, & Tuszynski, 2003; Siebert, Conta Steencken, & Osterhout, 2014). Some of the GAGs are attached to the transmembrane proteins and form the transmembrane family of proteoglycans. A schematic representation

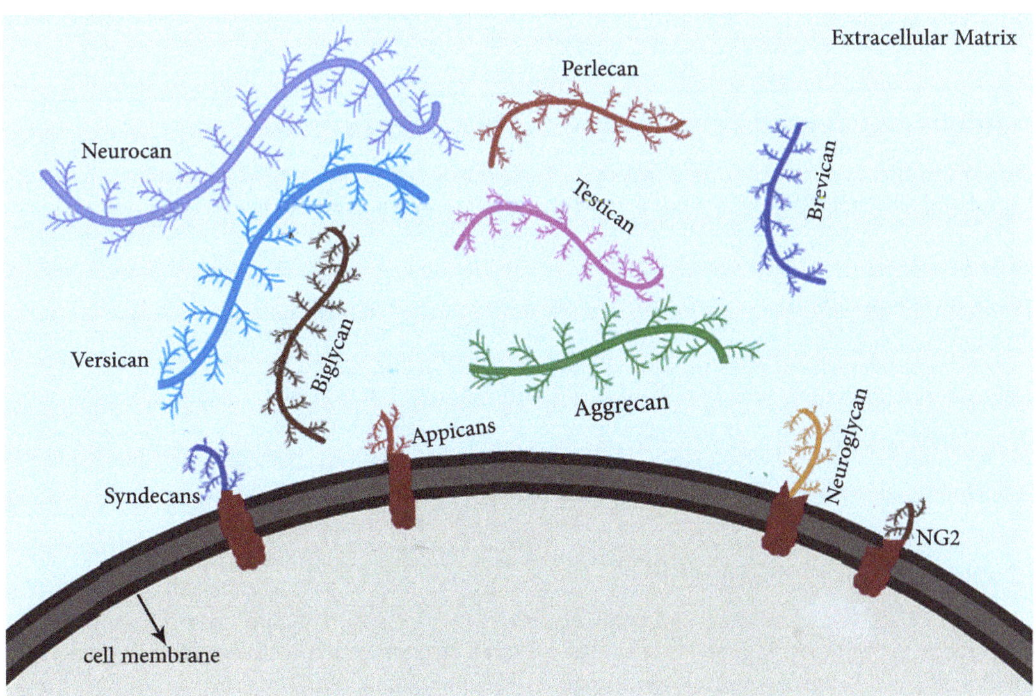


FIG. 2 A schematic representation of the proteoglycan families found in the extracellular environment and the membrane of the central nervous system. The proteoglycans in the extracellular matrix of central nervous system (CNS) including, Aggrecan, Versican, Neurocan, and Brevican, are members of the Lectican family. Members of syndecan family along with Apican, Neuroglycan, and NG2 are transmembrane proteoglycans.

of various proteoglycans in the cell surface and ECM of nervous tissues is provided in Fig. 2 (Schwartz & Domowicz, 2018). Structurally, the CNS-ECM proteoglycans provide the viscoelastic properties, maintain ions and water content and normal osmotic pressure, and dictate proper tissue organization. It has shown that the development of the nervous system in mammals involves coordinated cellular interactions, such as patterning and routing of migrating neural cells, axon pathfinding, and synapse formation, that is due to the proper organization of the nervous ECM (Silver & Silver, 2014; Tessier-Lavigne & Goodman, 1996). Hence, the time-dependent concentration of CSPGs relative to other adhesive matrix molecules at the various developmental stage, define the normal state of the CNS-ECM (Galtrey & Fawcett, 2007).

### Implication of CSPGs in pathogen conditions

It has been reported that CSPGs have implications in the processes of tumor metastasis by affecting endothelial migration and adhesion (Denholm, Lin, & Silver, 2001). In the case of brain cancer, changing the pattern of interaction of CS with ligands, growth factor receptors, extracellular matrix components, and structural proteins is accompanied by the progress of Glioblastoma (GBM) (Wade et al., 2013). It has also revealed that neurodegenerative disorders are related to defects in the synthesis/degradation of proteoglycans that affect the structural integrity of nervous tissues (DeWitt, Richey, Praprotnik, Silver, & Perry, 1994). Genome-wide association studies have shown that mental disorders such as bipolar character and depression are related to unbalance in CNS-ECM constituents (Cichon et al., 2011; Shi et al., 2011).

One of the important clinical manifestations after SCI includes the rapid increasing of various types of CSPGs in ECM. It is believed that this event is a defensive mechanism that seals off the injury to prevent bacterial penetration. The corresponding injury response in damaged CNS is called reactive gliosis or glial scarring (Fawcett & Asher, 1999). The glial scar also has other positive effects such as reducing the inflammation area, modulating immune activity, and helping to restore the blood–brain barrier (Lin, Kwok, Crespo, & Fawcett, 2008; Rolls, Shechter, & Schwartz, 2009; Tuinstra, Ducommun, Briley, & Shea, 2013). However, despite these positive impacts, the glial scar as a physical barrier reveals a negative role in axonal regrowth, sprouting, and synaptic reorganization following SCI (Cregg et al., 2014; Siebert et al., 2014). Production of CSPGs occurs after the cute phase in approximately 48 h following SCI (Katoh, Yokota, & Fehlings, 2019). It also prevents oligodendrocyte maturation and remyelination (Karus et al., 2016). Accordingly, medical intervention to control the concentration of CSPGs and limit the scar size in CNS-ECM has become central to the treatment of spinal cord injuries, and management of the CSPGs concentration during this period can result in the successful treatment of SCI in subsequent phases. There are numerous reports that carbohydrate and proteoglycan degrading enzymes can be used as a therapeutic drug to prevent CSPGs chains from accumulating in the site of SCI (Bradbury et al., 2002; García-Alías, Barkhuysen, Buckle, & Fawcett, 2009; Rosenzweig et al., 2019; Tauchi et al., 2012).

## Glycosaminoglycan degrading enzymes: A brief description

Regarding the various functions of GAG-proteoglycans, it is evident that creating a delicate balance between the biosynthesis and degradation of heteropolysaccharides is crucial in maintaining the normal health conditions of tissues. As mentioned in Implication of CSPGs in pathogen conditions section, the formation of proteoglycan-rich glial scar around traumatic injury of SCI inhibits long-term axonal recovery predominantly by CSPGs moieties, and degrading enzymes is considered as therapeutic drugs to overcome this medicinal challenge. Two types of enzymes can depolymerize GAG chains including lyases (EC 4.2.2.-) and hydrolases (EC 3-2-1.-). The lyases degrade GAGs by eliminative cleavage, while hydrolases-based degradation is carried out by hydrolytic cleavage. Based on the position of the cutting site, the pattern of enzymatic action is exolytic or endolytic mode (Linhardt, 2001) (Ernst, Langer, Cooney, & Sasisekharan, 1995). In vivo degradation of GAGs is carried out by corresponding hydrolases such as keratinase, hyaluronic acid hydrolase, and heparin hydrolase. The GAG degrading lyases have mainly bacterial sources, and are categorized into three main groups, including heparinases, chondroitinases, and hyaluronidases.

### Structural features of chondroitinase ABC I

According to CAZy (Carbohydrate-Active enZYmes Database), the tertiary structure of cABC I include $(\alpha/\alpha)_6$ toroid + anti-parallel β-sheet. As shown in Fig. 3, it is composed of three distinctive domains, including N- and C-terminals, that are symmetrically attached to the central (catalytic) domain (Huang et al., 2003). The catalytic residues are located in the central domain, and include His-501, Tyr-508, Arg-560, and Glu-653 (Prabhakar et al., 2005). The peripheral domains are mainly made up of well-defined organized β-sheets, and the central domain has mainly helix conformation. The movement of N- and C-terminal domains around the catalytic region leads to the creation of open and closed conformation that is

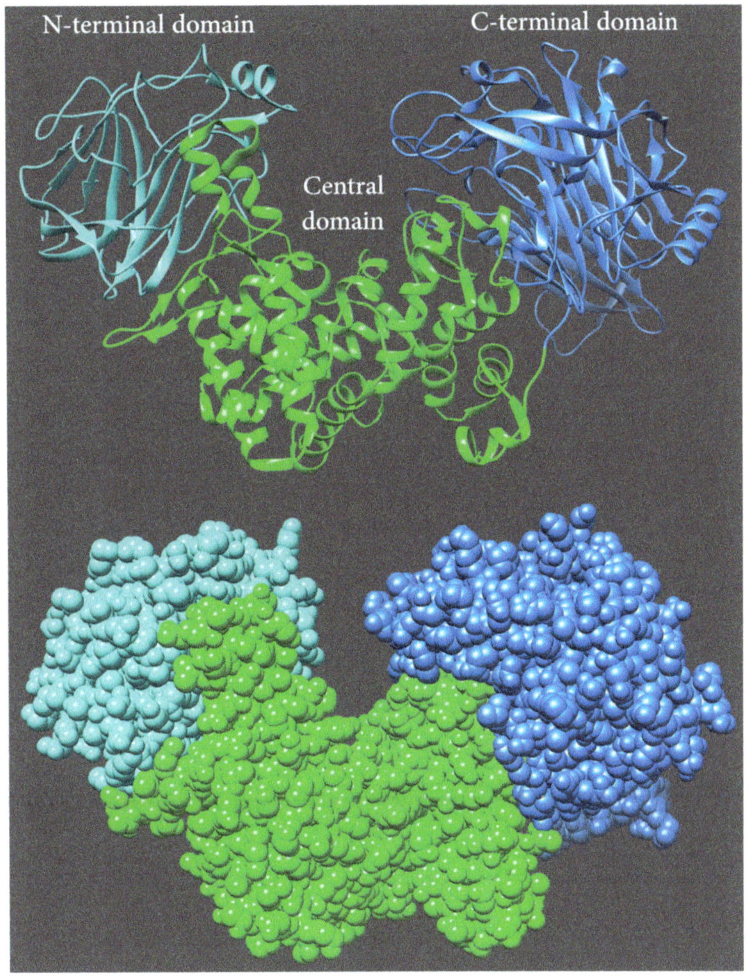


FIG. 3 The Tertiary structure of chondroitinase ABC I. Ribbon (Upper panel) and sphere (lower panel) representation of cABC I. The structure was depicted by the UCSF Chimera program using the crystal structure of cABC I (PDB ID:1HN0).

essential for the catalytic activity of the enzyme. The enzyme cleaves its substrates at the glycosidic bond via β-elimination of 1,4-hexosaminidic bonds to produce Δ4,5-unsaturated disaccharides, and a reducing end polysaccharide. Spectroscopically, product formation can be monitored as an increase in absorbance at 232 nm as a function of time.

## Chondroitinase ABC I and spinal cord injury

By convention, A, B, and C in chondroitinase ABC (cABC) nomenclature, refers to chondroitin 4-sulfate, dermatan sulfate, and chondroitin 6-sulfate, respectively. According to BRENDA—The Comprehensive Enzyme Information database (www.brenda-enzymes.org), cABC I from *Proteus vulgaris* (EC 4.2.2.20) is a broad specific endolytic enzyme that degrades a variety of glycosaminoglycans of the chondroitin-sulfate- and dermatan-sulfate type. It can also acts on hyaluronan at a much lower rate. Fortunately, these substrates are the main constituents of CNS GAGs that are formed by activated astrocytes following SCI. Hence, the broad specificity of cABC I toward a combination of GAGs can be considered as a positive feature to degrade the GAG chains and can restore the right condition for nerve regeneration following SCI.

In vivo delivery of cABC I to reduce the size of glial scar and removing the inhibitory environment for the treatment of SCI was first reported by Lemons et al. in 1999 (Lemons, Howland, & Anderson, 1999). Four years later, research groups from the University of Cambridge and the University of London delivered cABC to the lesioned dorsal columns of adult rats and observed a marked functional recovery of injured CNS (Bradbury et al., 2002). To analyze the effects of cABC on axon regeneration, Lin and co-workers made unilateral nigrostriatal lesions in adult rats and found that injection of cABC can digest hyaluronan and reduce neurocan in a limited area around the injection region. At the same time, the majority of GAGs are inaccessible to cABC digestion. They suggest that a single injection of cABC can produce an environment conducive to CNS repair for a long-term over 10 days (Lin et al., 2008). In 2011, Fawcett et al. showed that using cABC combined with rehabilitation is effective in functional recovery of forelimb function in rats with chronic SCI. They used cABC I 4 weeks after SCI (Wang, Ichiyama, Zhao, Andrews, & Fawcett, 2011). These studies, together with other investigations, have confirmed the ability of cABC I to help axon regeneration following SCI (Kasinathan, Volety, & Josyula, 2016; Muir et al., 2019).

## Chondroitinase ABC I for enzyme therapy: Challenges and perspectives

*Proteus Vulgaris* as the bacterial source of cABC I presents in soil and intestinal tract of the human body. It uses acidic polysaccharides along with epithelial cells in the intestinal lumen as its nutrient source. Accordingly, its optimum temperature for the activity is in accordance with the normal temperature of the human body. However, the time-dependent concentration of cABC I in *P. vulgaris* is low, which is essential for its solubility. In other words, protein aggregation is an entropy-dependent process, and increasing the concentration of the enzyme molecules for practical uses, particularly at 37 °C, leads to protein aggregation due to the higher probability of effective molecular collisions (Mohammadyari, Shirdel, Jafarian, & Khalifeh, 2019). Additionally, the enzyme adaptation has been occurred not only on the temperature but also on the other environmental factors of *P. vulgaris* such as pH and ionic strengths that are different from those found in the new therapeutic environment. Accordingly, changing the environmental conditions for the enzyme is accompanied by decreasing its half-life and lower catalytic efficiency due to its low conformational stability. So, more concentration of the enzyme is needed to have efficient catalysis. Nonetheless, as mentioned above, there is a critical concentration of the enzyme above that it undergoes molecular aggregation. Moreover, using a high concentration of the recombinant enzyme per injection or increasing the number of enzyme injections at the site of SCI for more efficient catalysis may have immunogenicity responses (Muir et al., 2019). Considering the instability of cABC I due to protein aggregation and decreased half-life as well as medical challenges following repeatedly administer of the recombinant enzyme, various enzyme stabilization and enzyme delivery methods, should be developed to efficient use of the enzyme for SCI treatment. Furthermore, the formulation of the enzyme appropriate for long-term storage permits the clinicians to access the enzyme in the shortest time and a usable form.

### Protein engineering strategies toward functional drug delivery

The enzyme can be synthesized, extracted, and purified with the enzyme technology methods. The classical procedures for enhancing the enzyme stability involve setting up right environmental conditions using suitable solvents containing various types of osmolytes and additives as stabilizers. It was shown that chemical modification of the recombinant enzymes with Polyethylene glycol (PEG) could increase the protein half-life and reduces the possibility of immunological side effects due to its bacterial origin (Greenwald, 2001). Additionally, site-directed mutagenesis has intensively used to logically replace critical amino acids on the structure of the protein with new residues to improve the functional and structural features of the enzyme. For example, it was revealed that increasing the repulsive electrostatic interactions between enzyme molecules via changing of surface-exposed residues by site-directed mutagenesis can reduce aggregation (Mohammadyari et al., 2019). Shoichet et al. used PEG-chemical modification procedure and site-directed mutagenesis strategy to produce improved engineered enzyme for SCI treatment. They demonstrated that representative PEGylated mutants of cABC could be locally delivered at the site of SCI in a more stabilized form (Hettiaratchi et al., 2019). Drug nano-carriers such as nano-structured lipid carriers (NLCs) are also potential candidates for the efficiently deliver of the enzyme toward its target. Enzyme immobilization using biocompatible nano-structures can protect the enzyme molecules from biodegradation and leads to the gradual release of the enzyme (Xia et al., 2015; Zuidema, Gilbert, & Osterhout, 2016). These are only examples of ongoing research studies on the humanization and methodological setting for using cABC I as a drug. Details of such studies have been reviewed elsewhere (Kasinathan et al., 2016; Muir et al., 2019).

## Combinatorial therapy with chondroitinase ABC I

It is worth mentioning that SCI is accompanied by several events including proliferation of astrocytes, accumulation of macrophages and microglia, production of growth-inhibitory matrix around SCI region, abnormality in signal transduction, etc. These phenomena result in multiple clinical consequences. Therefore, SCI should be simultaneously targeted with various medical interventions (Rolls et al., 2009). In other words, cABC I should be used besides other therapeutic agents to achieve a satisfactory outcome. Efficient delivery of cABC as a drug requires the determination of the therapeutic window of enzyme utilization during treatment time. As well as timing, the threshold of enzyme concentration influences the efficacy of enzyme therapy. It was shown that the expression of corresponding proteins increases following SCI within the first 24 h, and the maximum level of CSPGs is generally observed approximately 2 weeks after SCI. However, the gradual increase of CSPGs over time is occurred with different rates according to the type of proteoglycan ( Jones et al., 2003). Finally, locomotor recovery depends on optimized drug-based treatments combined with rehabilitation.

## Applications to other areas of neuroscience

Regarding the diverse roles of CSPGs in the physiological and pathological conditions of the CNS-ECM, cABC I can be used to control the right composition of the CSPGs according to the abnormal states of the nervous tissue. It had demonstrated that the efficiency of nerve recovery through peripheral nerve graft (PNG) increased in combined with cABC I and acidic fibroblast growth factor usage (DePaul, Lin, Silver, & Lee, 2015). The enzyme can also be used to eliminate the inhibitory function of CSPGs on the remyelination, which is essential in the treatment of multiple sclerosis (Lau et al., 2012). Investigating the experimental models has revealed that cABC I offers the opportunity to improve the outcome of Alzheimer's disease treatment via restoration of extracellular matrix plasticity in the synaptic region of hippocampal (Végh et al., 2014). It would be beneficial in the case of Parkinson's disease and for patients that are involved in phobias and anxiety (Moon, Asher, Rhodes, & Fawcett, 2001; Thompson et al., 2018). There are pieces of evidence that cABC I is helpful in the treatment of glioblastoma through sensitizing the glioblastoma cells to cancer drugs and enhancing the spread of oncolytic viruses as well as changing the pattern of the interaction of CSPGs with specific proteins (Dmitrieva et al., 2011; Jaime-Ramirez et al., 2017; Wade et al., 2013).

## Mini-dictionary of terms

**Glycosaminoglycan:** Linear polysaccharide made up of repeating disaccharide units. Based on the type of sugar building blocks, various glycosaminoglycans are found in extracellular matrices of tissues.

**Proteoglycan:** A highly glycosylated protein produced by attachment of a given glycosaminoglycan to a given target protein. A combination of various types of proteoglycans is found in extracellular matrices. The time-dependent concentration of proteoglycans is changed due to the developmental stage or pathogenic conditions of the cells and tissues.

**Chondroitin sulfate proteoglycans:** Specific types of proteoglycans that are sulfated in representative positions of their sugar moiety. They are one of the main constituents of the extracellular matrix of the nervous system.

**CAZy:** stands for "Carbohydrate-Active enZYmes Database." It is the specialized database that describes the families of structurally-related enzymes and carbohydrate-binding modules that are involved in the breakdown, biosynthesis, or modification of carbohydrates and glycoconjugates. It can be found at http://www.cazy.org/

**BRENDA:** It is a bioinformatics database containing the functional and structural information of the known enzymes. It can be found at www.brenda-enzymes.org

**Glioblastoma:** A type of cancer originated from the proliferation of astrocytes in the brain or spinal cord.

**Conformational stability:** In statistical thermodynamics, the conformational stability of a given enzyme refers to the ratio of the population of the native structure to denature states of the enzyme molecules at equilibrium condition. When a single molecule is considered, it is the structural integrity of an enzyme molecule that is the difference energy of the unfolded and folded states of an enzyme molecule. In both paradigms, increasing conformational stability leads to the activity improvement of the enzyme.

**Catalytic efficiency:** The minimum concentration of the substrate molecule required for an enzyme to reach its maximum catalysis capacity.

**Half-life:** The time required for an enzyme to lose 50% of its original activity. Increasing the conformational stability leads to increasing the half-life of an enzyme.

**Protein Aggregation:** is a process in which protein molecules accumulate and form molecular aggregates. Upon aggregation, the protein loses its activity. It occurred by attractive intermolecular interactions, mainly the hydrophobic one, between surface-exposed hydrophobic patches. Based on the identity of hydrophobic interaction, protein aggregation depends on the molecular collisions, which enhance by increasing the temperature and the concentration of the protein molecule. Proteins with surfaced-exposed hydrophobic patches are more prone to aggregation.

**Domain:** In protein science, domain is defined as distinct structural unite in the three dimensional structure of proteins. In the simplest form, the whole structure of a small and compact protein is considered as a domain. Some proteins have two or more domains. Domains are responsible for the structural integrity and stability of proteins. They are also involved in the functional features of the proteins.

**Recombinant DNA technology:** Recombinant DNA technology includes insertion of a given DNA sequence from an organism to the other one. In protein engineering studies, the host organism may be an engineered bacterium that is used for the expression of a recombinant protein which is of value to biotechnology.

## Key facts of chondroitinase ABC I

Chondroitinase ABC I is a non-hydrolytic degrading enzyme that cleaves the glycosidic bonds of polysaccharides.

- It has three distinctive domains, where the peripheral ones are symmetrically arranged around the central catalytic domain.
- Naturally, this enzyme is expressed in *Proteus Vulgaris*. This bacterium is found in soil and intestinal tract of the human body.
- The enzyme and its mutated variants can be produced by recombinant DNA technology, and using engineered bacteria as protein express systems.
- Chondroitinase ABC I as a polysaccharide lyase has broad substrate specificity.
- Its substrates constitute the main glycosaminoglycans of extracellular matrix of the central nervous system.

## Key facts of the therapeutic potential of chondroitinase ABC I

- The concentration of corresponding substrates of chondroitinase ABC I increases in the extracellular matrix upon spinal cord injury, which acts as inhibitors against axon regeneration.
- The enzyme can be used to degrade the glycosaminoglycans around the traumatic region of the spinal cord.
- Its conformational stability and half-life reduce in the environmental conditions around the region of SCI.
- To use as a drug, the conformational stability, and biocompatibility of the enzyme should be increased via protein engineering strategies.
- Appropriate drug delivery methods should be developed to efficient use of chondroitinase ABC I at the site of spinal cord damage.

## Summary points

- The extracellular matrix of the central nervous system contains various types of glycosaminoglycans in the form of proteoglycans.
- A fine balance between the concentrations of different proteoglycans is established in normal physiological conditions.
- Glial scar is proposed as the main inhibitor for axon regeneration following spinal cord injury.
- Glial scar is composed of various types of concentrated glycosaminoglycans in the region of spinal cord damage.
- Degradation of glycosaminoglycans at the site of spinal cord damage combined with other therapeutic strategies can help axons to recover.
- The broad substrate specificity of cABC I in accord with the composition of glial scar makes the enzyme a suitable therapeutic drug to restore tissue plasticity and axon regeneration.
- Using a high concentration of the enzyme per injection leads to molecular aggregation.
- Increasing the number of injection of the enzyme may results in immunological side effects.
- Protein engineering and drug delivery procedures are needed to produce, store, and therapeutically use of chondroitinase ABC I.
- Enzyme therapy should be used in predefined periods of time.
- Combinatorial approaches and other medical interventions along with intensive rehabilitation are being developed for locomotor recovery.

## References

Anand, M., Wang, C., French, J., Isaacson-Schmid, M., Wall, L. L., & Mysorekar, I. U. (2012). Estrogen affects the glycosaminoglycan layer of the murine bladder. *Female Pelvic Medicine & Reconstructive Surgery*, *18*(3), 148–152. https://doi.org/10.1097/SPV.0b013e31824b76bd.

Beldowski, P., Mazurkiewicz, A., Topoliński, T., & Małek, T. (2019). Hydrogen and water bonding between glycosaminoglycans and phospholipids in the synovial fluid: Molecular dynamics study. *Materials (Basel)*, *12*(2060), 1–11. https://doi.org/10.3390/ma12132060.

Bourin, M., & Lindahl, U. (1993). Glycosaminoglycans and the regulation of blood coagulation. *Biochemical Journal*, *289*(2), 313–330. https://doi.org/10.1042/bj2890313.

Bradbury, E. J., Moon, L. D. F., Popat, R. J., King, V. R., Bennett, G. S., Patel, P. N., … McMahon, S. B. (2002). Chondroitinase ABC promotes functional recovery after spinal cord injury. *Nature*, *416*, 636–640. https://doi.org/10.1038/416636a.

Calatroni, A., & di Ferrante, N. (1969). The glycosaminoglycans of human term placenta. *Carbohydrate Research*, *10*(4), 535–548. https://doi.org/10.1016/S0008-6215(00)80122-6.

Cichon, S., Mühleisen, T. W., Degenhardt, F. A., Mattheisen, M., Miró, X., Strohmaier, J., … Nöthen, M. M. (2011). Genome-wide association study identifies genetic variation in neurocan as a susceptibility factor for bipolar disorder. *American Journal of Human Genetics*, *88*, 372–381. https://doi.org/10.1016/j.ajhg.2011.01.017.

Clowes, A. W., & Karnowsky, M. J. (1977). Suppression by heparin of smooth muscle cell proliferation in injured arteries. *Nature*, *265*(5595), 625–626. https://doi.org/10.1038/265625a0.

Couchman, J. R. (2010). Transmembrane signaling proteoglycans. *Annual Review of Cell and Developmental Biology*, *26*, 89–114. https://doi.org/10.1146/annurev-cellbio-100109-104126.

Cregg, J. M., DePaul, M. A., Filous, A. R., Lang, B. T., Tran, A., & Silver, J. (2014). Functional regeneration beyond the glial scar. *Experimental Neurology*, 197–207. https://doi.org/10.1016/j.expneurol.2013.12.024.

DeAngelis, P. L. (2002). Evolution of glycosaminoglycans and their glycosyltransferases: Implications for the extracellular matrices of animals and the capsules of pathogenic bacteria. *Anatomical Record*, *268*(3), 317–326. https://doi.org/10.1002/ar.10163.

Denholm, E. M., Lin, Y. Q., & Silver, P. J. (2001). Anti-tumor activities of chondroitinase AC and chondroitinase B: Inhibition of angiogenesis, proliferation and invasion. *European Journal of Pharmacology*, *416*(3), 213–221. https://doi.org/10.1016/S0014-2999(01)00884-6.

DePaul, M. A., Lin, C. Y., Silver, J., & Lee, Y. S. (2015). Peripheral nerve transplantation combined with acidic fibroblast growth factor and chondroitinase induces regeneration and improves urinary function in complete spinal cord transected adult mice. *PLoS One*, *10*. https://doi.org/10.1371/journal.pone.0139335, e0139335.

DeWitt, D. A., Richey, P. L., Praprotnik, D., Silver, J., & Perry, G. (1994). Chondroitin sulfate proteoglycans are a common component of neuronal inclusions and astrocytic reaction in neurodegenerative diseases. *Brain Research*, *656*(1), 205–209. https://doi.org/10.1016/0006-8993(94)91386-2.

Dick, G., Akslen-Hoel, L. K., Grøndahl, F., Kjos, I., & Prydz, K. (2012). Proteoglycan synthesis and Golgi Organization in polarized epithelial cells. *Journal of Histochemistry and Cytochemistry*, *60*(12), 936–965. https://doi.org/10.1369/0022155412461256.

Dmitrieva, N., Yu, L., Viapiano, M., Cripe, T. P., Chiocca, E. A., Glorioso, J. C., & Kaur, B. (2011). Chondroitinase ABC I-mediated enhancement of oncolytic virus spread and antitumor efficacy. *Clinical Cancer Research*, *17*, 1362–1372. https://doi.org/10.1158/1078-0432.CCR-10-2213.

Ernst, S., Langer, R., Cooney, C. L., & Sasisekharan, R. (1995). Enzymatic degradation of glycosaminogliycans. *Critical Reviews in Biochemistry and Molecular Biology*, *30*(5), 387–444.

Fawcett, J. W., & Asher, R. A. (1999). The glial scar and central nervous system repair. *Brain Research Bulletin*, *49*(6), 377–391. https://doi.org/10.1016/S0361-9230(99)00072-6.

Galtrey, C. M., & Fawcett, J. W. (2007). The role of chondroitin sulfate proteoglycans in regeneration and plasticity in the central nervous system. *Brain Research Reviews*, *54*, 1–18. https://doi.org/10.1016/j.brainresrev.2006.09.006.

Gandhi, N. S., & Mancera, R. L. (2008). The structure of glycosaminoglycans and their interactions with proteins. *Chemical Biology and Drug Design*, *72*, 455–482. https://doi.org/10.1111/j.1747-0285.2008.00741.x.

García, B., Merayo-Lloves, J., Martin, C., Alcalde, I., Quirós, L. M., & Vazquez, F. (2016). Surface proteoglycans as mediators in bacterial pathogens infections. *Frontiers in Microbiology*, *7*(220), 1–11. https://doi.org/10.3389/fmicb.2016.00220.

García-Alías, G., Barkhuysen, S., Buckle, M., & Fawcett, J. W. (2009). Chondroitinase ABC treatment opens a window of opportunity for task-specific rehabilitation. *Nature Neuroscience*, *12*(9), 1145–1151. https://doi.org/10.1038/nn.2377.

Gill, S., Wight, T. N., & Frevert, C. W. (2010). Proteoglycans: Key regulators of pulmonary inflammation and the innate immune response to lung infection. *Anatomical Record*, *293*(6), 968–981. https://doi.org/10.1002/ar.21094.

Greenwald, R. B. (2001). PEG drugs: An overview. *Journal of Controlled Release*, *74*(1–3), 159–171. https://doi.org/10.1016/S0168-3659(01)00331-5.

Hachim, D., Whittaker, T. E., Kim, H., & Stevens, M. M. (2019). Glycosaminoglycan-based biomaterials for growth factor and cytokine delivery: Making the right choices. *Journal of Controlled Release*, *313*, 131–147. https://doi.org/10.1016/j.jconrel.2019.10.018.

Hardingham, T., & Bayliss, M. (1990). Proteoglycans of articular cartilage: Changes in aging and in joint disease. *Seminars in Arthritis and Rheumatism*, *20*(3 (1)), 12–33. https://doi.org/10.1016/0049-0172(90)90044-G.

Hettiaratchi, M. H., O'Meara, M. J., Teal, C. J., Payne, S. L., Pickering, A. J., & Shoichet, M. S. (2019). Local delivery of stabilized chondroitinase ABC degrades chondroitin sulfate proteoglycans in stroke-injured rat brains. *Journal of Controlled Release*. https://doi.org/10.1016/j.jconrel.2019.01.033.

Huang, W., Lunin, V. V., Li, Y., Suzuki, S., Sugiura, N., Miyazono, H., & Cygler, M. (2003). Crystal structure of Proteus vulgaris chondroitin sulfate ABC lyase I at 1.9 Å resolution. *Journal of Molecular Biology*, *328*(3), 623–634. https://doi.org/10.1016/S0022-2836(03)00345-0.

Imberty, A., Lortat-Jacob, H., & Pérez, S. (2007). Structural view of glycosaminoglycan-protein interactions. *Carbohydrate Research*, *342*(3–4), 430–439. https://doi.org/10.1016/j.carres.2006.12.019.

Iozzo, R. V., & Schaefer, L. (2015). Proteoglycan form and function: A comprehensive nomenclature of proteoglycans. *Matrix Biology*, *42*, 11–55. https://doi.org/10.1016/j.matbio.2015.02.003.

Jackson, R. L., Busch, S. J., & Cardin, A. D. (1991). Glycosaminoglycans: Molecular properties, protein interactions, and role in physiological processes. *Physiological Reviews*, *71*(2), 481–539. https://doi.org/10.1152/physrev.1991.71.2.481.

Jaime-Ramirez, A. C., Dmitrieva, N., Yoo, J. Y., Banasavadi-Siddegowda, Y., Zhang, J., Relation, T., … Kaur, B. (2017). Humanized chondroitinase ABC sensitizes glioblastoma cells to temozolomide. *The Journal of Gene Medicine*, *19*(3), 1–18. https://doi.org/10.1002/jgm.2942.

Jones, L. L., Margolis, R. U., & Tuszynski, M. H. (2003). The chondroitin sulfate proteoglycans neurocan, brevican, phosphacan, and versican are differentially regulated following spinal cord injury. *Experimental Neurology*, *182*(2), 399–411. https://doi.org/10.1016/S0014-4886(03)00087-6.

Karus, M., Ulc, A., Ehrlich, M., Czopka, T., Hennen, E., Fischer, J., … Faissner, A. (2016). Regulation of oligodendrocyte precursor maintenance by chondroitin sulphate glycosaminoglycans. *Glia*, *64*, 270–286. https://doi.org/10.1002/glia.22928.

Kasinathan, N., Volety, S. M., & Josyula, V. R. (2016). Chondroitinase: A promising therapeutic enzyme. *Critical Reviews in Microbiology*, *42*(3), 474–484. https://doi.org/10.3109/1040841X.2014.959893.

Katoh, H., Yokota, K., & Fehlings, M. G. (2019). Regeneration of spinal cord connectivity through stem cell transplantation and biomaterial scaffolds. *Frontiers in Cellular Neuroscience*, *13*(June), 1–22. https://doi.org/10.3389/fncel.2019.00248.

Lander, A. D., & Selleck, S. B. (2000). The elusive functions of proteoglycans: In vivo veritas. *Journal of Cell Biology*, *148*(2), 227–232. https://doi.org/10.1083/jcb.148.2.227.

Lasne, D., Jude, B., & Susen, S. (2006). From normal to pathological hemostasis. *Canadian Journal of Anesthesia*, *53*, 52–511. https://doi.org/10.1007/bf03022247.

Lau, L. W., Keough, M. B., Haylock-Jacobs, S., Cua, R., Döring, A., Sloka, S., … Yong, V. W. (2012). Chondroitin sulfate proteoglycans in demyelinated lesions impair remyelination. *Annals of Neurology*, *72*, 419–432. https://doi.org/10.1002/ana.23599.

Lee, T. Y., Jamieson, A. M., & Schafer, I. A. (1973). Changes in the composition and structure of glycosaminoglycans in the human placenta during development. *Pediatric Research*, *7*(12), 965–977. https://doi.org/10.1203/00006450-197312000-00005.

Lemons, M. L., Howland, D. R., & Anderson, D. K. (1999). Chondroitin sulfate proteoglycan immunoreactivity increases following spinal cord injury and transplantation. *Experimental Neurology*, *160*, 51–65. https://doi.org/10.1006/exnr.1999.7184.

Lin, R., Kwok, J. C. F., Crespo, D., & Fawcett, J. W. (2008). Chondroitinase ABC has a long-lasting effect on chondroitin sulphate glycosaminoglycan content in the injured rat brain. *Journal of Neurochemistry*, *104*(2), 400–408. https://doi.org/10.1111/j.1471-4159.2007.05066.x.

Linhardt, R. J. (2001). Analysis of glycosaminoglycans with polysaccharide lyases. *Current Protocols in Molecular Biology*. https://doi.org/10.1002/0471142727.mb1713bs48 (edited by Frederick M. Ausubel .. [et al.]).

Merle, B., Durussel, L., Delmas, P. D., & Clézardin, P. (1999). Decorin inhibits cell migration through a process requiring its glycosaminoglycan side chain. *Journal of Cellular Biochemistry*, *75*(3), 538–546.

Mohammadyari, H., Shirdel, S. A., Jafarian, V., & Khalifeh, K. (2019). Designing and construction of novel variants of Chondroitinase ABC I to reduce aggregation rate. *Archives of Biochemistry and Biophysics*, *668*, 46–53. https://doi.org/10.1016/j.abb.2019.05.013.

Moon, L. D. F., Asher, R. A., Rhodes, K. E., & Fawcett, J. W. (2001). Regeneration of CNS axons back to their target following treatment of adult rat brain with chondroitinase ABC. *Nature Neuroscience*, *4*, 465–466. https://doi.org/10.1038/87415.

Muir, E., De Winter, F., Verhaagen, J., & Fawcett, J. (2019). Recent advances in the therapeutic uses of chondroitinase ABC. *Experimental Neurology*, *321*, 1–16. https://doi.org/10.1016/j.expneurol.2019.113032.

Prabhakar, V., Raman, R., Capila, I., Bosques, C. J., Pojasek, K., & Sasisekharan, R. (2005). Biochemical characterization of the chondroitinase ABC I active site. *Biochemical Journal*, *390*, 395–405. https://doi.org/10.1042/BJ20050532.

Quraishe, S., Forbes, L. H., & Andrews, M. R. (2018). The extracellular environment of the CNS: Influence on plasticity, sprouting, and axonal regeneration after spinal cord injury. *Neural Plasticity*, 1–18. https://doi.org/10.1155/2018/2952386.

Rolls, A., Shechter, R., & Schwartz, M. (2009). The bright side of the glial scar in CNS repair. *Nature Reviews Neuroscience*, *10*, 235–241. https://doi.org/10.1038/nrn2591.

Rosenzweig, E. S., Salegio, E. A., Liang, J. J., Weber, J. L., Weinholtz, C. A., Brock, J. H., … Tuszynski, M. H. (2019). Chondroitinase improves anatomical and functional outcomes after primate spinal cord injury. *Nature Neuroscience*, *22*(8), 1269–1275. https://doi.org/10.1038/s41593-019-0424-1.

Schwartz, N. B., & Domowicz, M. S. (2018). Proteoglycans in brain development and pathogenesis. *FEBS Letters*, *592*, 3791–3805. https://doi.org/10.1002/1873-3468.13026.

Shi, J., Potash, J. B., Knowles, J. A., Weissman, M. M., Coryell, W., Scheftner, W. A., … Levinson, D. F. (2011). Genome-wide association study of recurrent early-onset major depressive disorder. *Molecular Psychiatry*, *16*, 193–201. https://doi.org/10.1038/mp.2009.124.

Siebert, J. R., Conta Steencken, A., & Osterhout, D. J. (2014). Chondroitin sulfate proteoglycans in the nervous system: Inhibitors to repair. *BioMed Research International*, 1–15. https://doi.org/10.1155/2014/845323.

Silver, D. J., & Silver, J. (2014). Contributions of chondroitin sulfate proteoglycans to neurodevelopment, injury, and cancer. *Current Opinion in Neurobiology*, 171–178. https://doi.org/10.1016/j.conb.2014.03.016.

Tauchi, R., Imagama, S., Natori, T., Ohgomori, T., Muramoto, A., Shinjo, R., … Kadomatsu, K. (2012). The endogenous proteoglycan-degrading enzyme ADAMTS-4 promotes functional recovery after spinal cord injury. *Journal of Neuroinflammation*, *9*(53), 1–10. https://doi.org/10.1186/1742-2094-9-53.

Templeton, D. M. (1992). Proteoglycans in cell regulation. *Critical Reviews in Clinical Laboratory Sciences*, *29*(2), 141–184. https://doi.org/10.3109/10408369209114599.

Tessier-Lavigne, M., & Goodman, C. S. (1996). The molecular biology of axon guidance. *Science*, *274*, 1123–1133. https://doi.org/10.1126/science.274.5290.1123.

Thompson, E. H., Lensjø, K. K., Wigestrand, M. B., Malthe-Sørenssen, A., Hafting, T., & Fyhn, M. (2018). Removal of perineuronal nets disrupts recall of a remote fear memory. *Proceedings of the National Academy of Sciences of the United States of America*, *115*, 607–612. https://doi.org/10.1073/pnas.1713530115.

Tuinstra, H. M., Ducommun, M. M., Briley, W. E., & Shea, L. D. (2013). Gene delivery to overcome astrocyte inhibition of axonal growth: An in vitro model of the glial scar. *Biotechnology and Bioengineering*, *110*(3), 947–957. https://doi.org/10.1002/bit.24750.

Végh, M. J., Heldring, C. M., Kamphuis, W., Hijazi, S., Timmerman, A. J., Li, K. W., … van Kesteren, R. E. (2014). Reducing hippocampal extracellular matrix reverses early memory deficits in a mouse model of Alzheimer's disease. *Acta Neuropathologica Communications*, *2*(76), 1–11. https://doi.org/10.1186/s40478-014-0076-z.

Wade, A., Robinson, A. E., Engler, J. R., Petritsch, C., James, C. D., & Phillips, J. J. (2013). Proteoglycans and their roles in brain cancer. *FEBS Journal*, *280*(10), 2399–2417. https://doi.org/10.1111/febs.12109.

Wang, D., Ichiyama, R. M., Zhao, R., Andrews, M. R., & Fawcett, J. W. (2011). Chondroitinase combined with rehabilitation promotes recovery of forelimb function in rats with chronic spinal cord injury. *Journal of Neuroscience*, *31*(25), 9332–9344. https://doi.org/10.1523/JNEUROSCI.0983-11.2011.

Werz, W., & Schachner, M. (1988). Adhesion of neural cells to extracellular matrix constituents. Involvement of glycosaminoglycans and cell adhesion molecules. *Developmental Brain Research*. https://doi.org/10.1016/0165-3806(88)90101-0.

Xia, T., Ni, S., Li, X., Yao, J., Qi, H., Fan, X., & Wang, J. (2015). Sustained delivery of chondroitinase ABC by poly(propylene carbonate)-chitosan micron fibers promotes axon regeneration and functional recovery after spinal cord hemisection. *Brain Research*, *1624*, 469–478. https://doi.org/10.1016/j.brainres.2015.08.018.

Yanagishita, M. (1993). Function of proteoglycans in the extracellular matrix. *Pathology International*, *43*(6), 283–293. https://doi.org/10.1111/j.1440-1827.1993.tb02569.x.

Zuidema, J. M., Gilbert, R. J., & Osterhout, D. J. (2016). Nanoparticle technologies in the spinal cord. *Cells, Tissues, Organs*, *202*, 102–115. https://doi.org/10.1159/000446647.

Chapter 26

# Phenol neurolysis for spasticity management in people with spinal cord injury

Radha Korupolu[a,b] and Lavina Jethani[a]

[a]*Physical Medicine and Rehabilitation, University of Texas Health Science Centre, Houston, TX, United States;* [b]*TIRR Memorial Hermann Hospital, Houston, TX, United States*

## Abbreviations

| | |
|---|---|
| **BCR** | brachioradialis |
| **BUE** | bilateral upper extremity |
| **EMG** | electromyography |
| **FCR** | flexor carpi radialis |
| **FCU** | flexor carpi ulnaris |
| **FDP** | flexor digitorum profundus |
| **FDS** | flexor digitorum superficialis |
| **ITB** | intrathecal baclofen |
| **OT** | occupational therapists |
| **PT** | physical therapists |
| **PWC** | power wheelchair |
| **ROM** | range of motion |
| **SCI** | spinal cord injury |

## Introduction

Spasticity is one of the most common secondary complications following a complete or incomplete spinal cord injury (SCI). After SCI, the prevalence of spasticity ranges from 65% to 78%, and 35% to 49% have problematic spasticity (Holtz, Lipson, Noonan, Kwon, & Mills, 2017; Maynard, Karunas, & Waring 3rd., 1990). Currently, problematic spasticity is managed through a combination of therapeutic modalities, oral medications, botulinum toxin injections, phenol neurolysis, and surgical procedures (Table 1). Pharmaceutical options include medications delivered orally, via local injections, or through intrathecal baclofen (ITB) pumps. Botulinum toxin injections and chemical neurolysis with phenol or alcohol are the first-line treatment options for problematic focal spasticity. However, after the introduction of botulinum toxins for the treatment of spasticity, the use of phenol/alcohol chemical neurolysis for spasticity management declined over the period. Potential causes of the decline in neurolysis use include laborious techniques that required more precise skill in the pre-ultrasound era. There is a concern for the risk of loss of sensation, dysesthesias, and weakness. However, a review of existing evidence suggests that chemical neurolysis with phenol and alcohol has demonstrated effectiveness in eliminating clonus, decreasing spasticity, improving joint range of motion, decreasing painful spasms, and improving function (Akkaya et al., 2010; Gunduz, Kalyon, Dursun, Mohur, & Bilgic, 1992; Jang et al., 2004; Karri, Mas, Francisco, & Li, 2017; Karri, Zhang, & Li, 2020). In our experience, ultrasound guidance and electrical stimulation improved these procedures' efficiency and quality.

Poorly controlled severe spasticity after SCI can result in multiple complications, including the development of joint contractures (Fig. 1), inability to participate in therapy, skin breakdown, decreased function, poor sleep, and poor quality of

**Diagnosis and Treatment of Spinal Cord Injury. https://doi.org/10.1016/B978-0-12-822498-4.00026-9**

**TABLE 1 Current treatments approach for management of spasticity in spinal cord injury[a].**

| | |
|---|---|
| Physical Modalities and Therapeutics | Passive or active stretching, Splinting, serial casting, Heat and cold application, Electrical stimulation, Vibration therapy, Orthoses, etc. |
| Commonly prescribed oral medications | Baclofen, Tizanidine, Diazepam, Dantrolene, Clonidine, etc. |
| Local interventions | Diagnostic nerve blocks: lidocaine, bupivacaine<br>Neurolysis: phenol, alcohol<br>Chemodenervation (botulinum toxins serotypes A and B) |
| Surgical interventions | Tendon lengthening procedures, Tendon transfers<br>Intrathecal baclofen pump implantation, Dorsal rhizotomy<br>Peripheral neurectomy |

[a]*These are commonly prescribed treatments for spasticity and not an exclusive list of treatment options.*

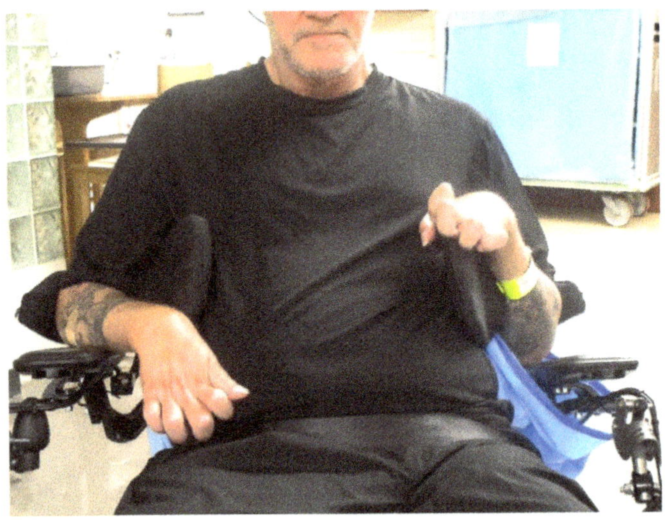

FIG. 1 Joint contracture and deformities due to delayed spasticity management.

life (Holtz et al., 2017; Rekand, Hagen, & Gronning, 2012; Sezer, Akkus, & Ugurlu, 2015). In people with cervical SCI, we are often faced with managing problematic spasticity in all four extremities. In this scenario, a combination of treatment approaches, including phenol or alcohol neurolysis, may help manage spasticity effectively and prevent complications due to poorly controlled spasticity (Escaldi, 2018; Viel, Pellas, Ripart, Pelissier, & Eledjam, 2005). Besides, there are many other case scenarios where neurolysis can be a very useful tool to manage spasticity effectively. Such scenarios include patient refusal of ITB therapy, contraindications to ITB therapy due to ongoing infectious process, hypersensitivity to baclofen, and insufficient resources to pursue botulinum toxin injections or ITB therapy.

Additionally, botulinum toxin and intrathecal baclofen therapies are not available in many parts of the world to date. Untreated spasticity can lead to significant limitations in function and limit recovery (Tibbett, Widerstrom-Noga, Thomas, & Field-Fote, 2019; van Cooten, Snoek, Nene, de Groot, & Post, 2015). Therefore, it is essential to incorporate cost-effective alternative therapies such as phenol neurolysis to treat spasticity. This chapter will describe the following with regard to phenol neurolysis: mechanism, techniques, dosing, risks, and benefits. It will demonstrate the procedure's application in people with SCI with a case study. These procedures are similar to those for alcohol neurolysis.

## Phenol or alcohol neurolysis history

The usefulness of chemical neurolysis or nerve block was first recognized in 1863 by Luton to treat pain due to phenol's and alcohol's ability to block nerve conduction (D'Souza & Hooten, 2021). There were many reports of loss of motor function after intrathecal injections of phenol for pain management in the 1950s (Kelly & Gautier-Smith, 1959; Liversedge & Maher, 1960). Between 1964 and 1967, Khalili et al. published their experience of performing peripheral nerve blocks in patients diagnosed with a stroke, cerebral palsy, and spinal cord injury (Khalili & Betts, 1967). They reported improved spasticity in the muscles following nerve block by a 2%–3% phenol solution to mixed or motor nerves (Khalili & Betts, 1967). They also reported several unanticipated side effects, including loss of sensation in <1% instances following a nerve block. Subsequently, peripheral motor point blocks were performed to minimize sensory side effects under electromyography (EMG) and electrical stimulation guidance (Copp, Harris, & Keenan, 1970; Halpern & Meelhuysen, 1966).

## Mechanism of action

Phenol, also known as carbolic acid ($C_6H_5OH$), is a benzene ring. In a solid state, it exists as a crystal. It is water-soluble at room temperature at ≤6.7% concentration. Phenol is typically used in 3%–6% concentration (Escaldi, 2018). When injected at a concentration of 5% and above, phenol denatures protein, which results in tissue necrosis and axonal degeneration (D'Souza & Hooten, 2021; Escaldi, 2018; Felsenthal, 1974). The effects of neurolysis last longer with 5%–6% concentration. Phenol at <3% concentration causes demyelination and some axonal destruction, resulting in shorter-lasting effects (Horn, Singh, & Dabrowski, 2015). The phenol application to the nerve trunk or motor point results in a short-term anesthetic effect that lasts for few hours. Short-term and immediate local anesthetic effects result in instant spasticity relief (D'Souza & Warner, 2021; Zhang, Darji, Francisco, & Li, 2021). However, long-term and full effects may take 7–9 days after injections due to the time required for axonal degeneration (Zhang et al., 2021). The duration of effects is dependent on the site of injection and concentration of phenol used (Halpern & Meelhuysen, 1967; Sung et al., 2001). Following injection at the proximal portion of a nerve using a higher concentration of phenol, the nerve takes 6–9 months to re-generate, whereas injections at the nerve's distal portion takes 3–4 months to recover. Repeated neurolysis can result in permanent nerve damage and muscle atrophy (Horn et al., 2015).

## Phenol neurolysis technique and dose

Extensive knowledge of the anatomy of mixed, motor, and sensory nerve pathways and muscles innervated by individual motor branches is essential to prevent serious side effects and improve outcomes. Phenol neurolysis can be performed with anatomic localization, electrical stimulation guidance, or ultrasound guidance. Often, these techniques are combined to maximize the efficiency of the procedure and minimize errors in localizing targets. In some cases, first performing a temporary nerve block with 1% lidocaine prior to phenol neurolysis may help assess the efficacy and potential loss of function that can be expected with the long-lasting procedure. The injection technique and materials required for performing a motor point block and peripheral neurolysis are similar. Teflon-coated electromyography needles of 25- to 27-gauge, surface electrodes, and a portable electrical stimulator are required for performing injections with electrical stimulation guidance. Anatomical localization along with electrical stimulation allows rapid localization of motor points or peripheral nerves. The use of a lower amount of current (≤1 mA) and stronger contractions suggests closer proximity to motor points or target nerve branches. Ultrasound guidance combined with electrical stimulation can decrease side effects and improve procedure efficiency and efficacy. In a study, injections performed under combined ultrasound and electrical stimulation guidance resulted in a lower phenol dose compared to electrical stimulation guidance alone (Karri et al., 2017; Kaymak, Kara, Gurcay, Aydin, & Ozcakar, 2019; Matsumoto, Berry, Yung, Matsumoto, & Munin, 2018). Motor points typically require multiple injection sites for a reduction in spasticity. Phenol neurolysis of proximal nerve trunk requires injection at one to two sites, and electrical stimulation results in stronger muscle contraction. However, ultrasound guidance requires additional equipment and skill. Typically, at each injection site, 0.2–0.5 mL of 5%–6% phenol is injected. However, the phenol amount per nerve or site may vary based on the practitioner's expertise and guidance used for injections (Karri et al., 2017).

Immediate spasticity relief experienced due to phenol's anesthetic effect during the procedure can be used to titrate the dose when injecting motor points. There are no clear studies about the target recommended dose, but the current literature recommends no more than 1.0–1.2 g in total (e.g., 20 mL of 5%–6% concentration) (D'Souza & Warner, 2021; Gaid, 2012; Horn et al., 2015). However, a study reported safe use of up to 30 mL of 6% phenol (Karri et al., 2017). Systemic doses of 8.5 g and above can result in serious side effects, including cardiovascular and central nervous system dysfunction (Escaldi, 2018; Horn et al., 2015).

The decision to inject the proximal portion of nerve vs distal motor points depends on various factors, including time from injury, complete vs incomplete injury, spasticity management goals, and current functional status. A careful evaluation and discussion of the risks and benefits of various options for spasticity management should be discussed. Before neurolysis, evaluate for any potential for nerve transfers to improve function. It is also essential to remember that while preserving motor points is vital for potential neurorecovery, poorly controlled spasticity can prevent participation in therapies, result in joint contractures, and significantly limit recovery. Current evidence suggests a significant improvement in spasticity in the upper and lower extremity muscles following chemical neurolysis (Awad, 1972;

Chang & Boudier-Reveret, 2020; Chua & Kong, 2001; Copp et al., 1970; Easton, Ozel, & Halpern, 1979; Elovic et al., 2009; Jang et al., 2004; Karri et al., 2017, 2020; Keenan, Tomas, Stone, & Gersten, 1990; Kong & Chua, 2002; Lee, Kim, Kim, Ro, & Lee, 2020).

## Upper extremity targets

A list of the phenol neurolysis of upper extremity targets in people with SCI is presented in Table 2. Our center's data suggest that medial and lateral pectoral nerve neurolysis and musculocutaneous motor point block are the most common upper extremity targets for phenol neurolysis in people with SCI. Spasticity of shoulder adductors and shoulder internal rotators muscles can be very severe and challenging in people with spinal cord injury. Stretching is often limited due to pain in the shoulder and inadequate response to botulinum toxin injections. Phenol neurolysis to medial and lateral pectoral and thoracodorsal nerves can successfully decrease spasticity in these muscle groups and improve shoulder range of motion. The pectoral nerves can be easily located with ultrasound guidance between the pectoralis major and pectoralis minor (Fig. 2). Due to the proximity of these nerves to the lungs and pleura, ultrasound guidance is recommended. The musculocutaneous nerve is a mixed nerve and supplies sensation to the lateral forearm by its terminal branch known as the lateral cutaneous nerve of the forearm. The musculocutaneous nerve can be localized near the pectoralis major tendon's distal insertion between the coracobrachialis and the short head of the biceps on the arm's medial side (Keenan et al., 1990). In people with incomplete injury, a motor point block to target muscle is recommended to prevent weakness of the elbow flexors and loss of sensation over the forearm (Sirico, Zappia, Di Meglio, Castaldo, & Nurzynska, 2020). The motor point block to distal muscle groups (finger flexors and wrist flexors) of the upper extremity or mixed nerves can be targeted with ultrasound or electrical stimulation guidance. However, we must isolate the target motor branches or perform motor point blocks to prevent unintended sensory loss and weakness in other muscle groups supplied by the target nerve (Karri et al., 2020; Sirico et al., 2020).

## Lower extremity targets

A list of the phenol neurolysis of lower extremity targets in people with SCI is presented in Table 3. The obturator nerve and tibial motor points are the lower extremities' most frequent targets of phenol (Ghai et al., 2013). Though the obturator nerve is a mixed nerve, its sensory branch supplies only a small area over the knee's medial aspect. The obturator nerve can be localized 1–2 cm below and lateral to the pubic tubercle and medial to femoral vessels in the thigh's upper part. Sciatic motor branches to the hamstrings can be localized at two locations; one is located at 20% and the other at 33% of the distance from the ischial tuberosity on a line drawn from the ischial tuberosity to lateral femoral condyle (Seidel, Seidel, Gans, & Dijkers, 1996). The tibial motor points can be located 1–2 cm medial and lateral of a point located 1–2 cm below the popliteal crease's midpoint (Chua & Kong, 2001). The quadriceps are richly innervated with multiple motor points located over the muscles' mid-belly. The practitioner should be cautious while injecting motor branches of the sciatic nerve, femoral nerve, and tibial nerve as these are mixed nerves. A diagnostic lidocaine motor point can help identify cases in which people rely on their spasticity for ambulation prior to phenol motor point blocks. Accessing the motor point to

**TABLE 2** Upper extremity targets for phenol neurolysis.

| Spasticity pattern | Main target muscles | Nerve block or motor point block |
|---|---|---|
| Shoulder adduction and internal rotation | Pectoralis major, Latissimus dorsi | Medial and lateral pectoral nerves, thoracodorsal nerves |
| Elbow flexors | Brachialis, Biceps brachii, BCR | Musculocutaneous motor points to brachialis and biceps<br>Radial motor branches to BCR |
| Elbow extensors | Triceps lateral head, long head, and medial head | Radial motor branches to triceps |
| Pronation | Pronator teres | Median motor points to pronator teres |
| Wrist flexors | FCR and FCU | Median nerve motor branches to FCR and ulnar motor branches to FCU |
| Finger flexors | FDS and FDP | Respective median and ulnar motor branches to FDS and FDP |

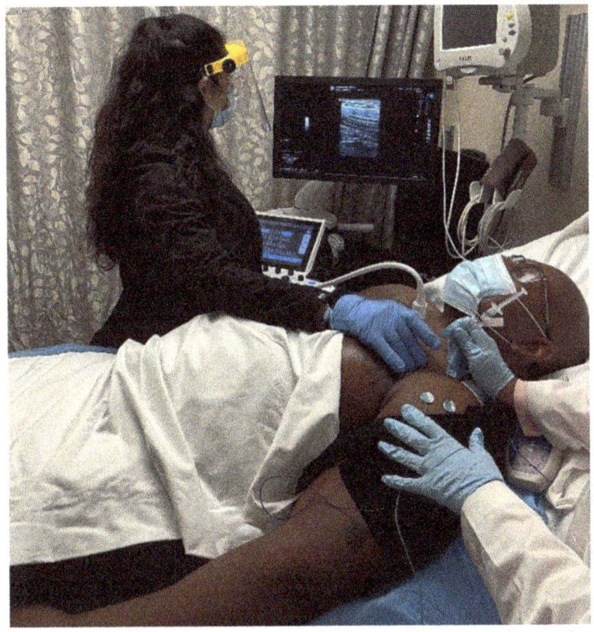

FIG. 2 Lateral pectoral neurolysis under ultrasound and electrical stimulation guidance.

**TABLE 3** Lower extremity targets for phenol neurolysis.

| Spasticity pattern | Main target muscles | Nerve block or *Motor* point block |
|---|---|---|
| Hip adductors | Adductor Magnus, Adductor Longus, Adductor Brevis | Obturator nerve |
| Hip Flexors | Iliacus, Sartorius, Rectus femoris | Femoral motor branches to iliacus, sartorius, rectus femoris |
| Knee flexors | Hamstrings | Sciatic motor branches to biceps femoris, semimembranosus, semitendinosus |
| Knee extensors | Quadriceps | Femoral motor branches to quadriceps |
| Ankle plantar flexors | Gastrocnemius, soleus | Tibial motor branches to gastrocnemius and soleus |

the psoas major can be challenging due to the abdominal viscera and blood vessels in the vicinity. However, there are case reports in people with stroke and SCI describing successful phenol lumbar nerve block and psoas motor point block to treat hip flexor spasticity (Awad, 1972; Koyama, Murakami, Suzuki, & Suzaki, 1992).

## Adverse effects

Risks associated with any injection include infection at the site of injection, pain, bleeding, and bruising. Incomplete nerve (mixed or sensory) destruction may cause dysesthesias. A complete nerve block is required to get relief from dysesthesias. A retrospective study evaluating 293 phenol neurolysis procedures reported the following incidence of adverse events; 4% pain. 2.7% inflammation, 0.7% dysesthesias, and 0.7% hypotension (Karri et al., 2017). Other complications include swelling and loss of sensation. Swelling, pain at the injection site, hypotension, and inflammation all resolve within a few days. Accidental intravascular injection of phenol can cause systemic side effects, including flushing, nausea, vomiting, cardiovascular and central nervous system depression, resulting in cardiac arrest and coma (Lee et al., 2020). Systematic side effects are uncommon in doses less than 100 mg (D'Souza & Warner, 2021).

## Case study

A 49-year-old African American male patient sustained a pedestrian motor vehicle accident, resulting in incomplete quadriplegia (C4 neurological level), pictured in Fig. 3. After his injury, he had a 2-month stay in an intensive care unit. He was discharged to a skilled nursing facility. He presented to our SCI center 8 months post-injury with several SCI-associated complications, including a stage 4 sacral pressure injury, multiple joint contractures, and severe generalized spasticity. He was bed-bound since the time of injury. After admission to our inpatient rehabilitation facility, oral medications for spasticity control were optimized with minimal benefits. Given significant spasticity in all four extremities, he needed multiple interventions to manage his spasticity to improve his function. Phenol motor point blocks to the bilateral shoulder abductors, elbow flexors, and hip adductors were performed. We also recommended intrathecal baclofen therapy. However, he needed immediate relief from his severe spasticity as it interfered with his positioning in bed and his wheelchair. It also negatively affected his hygiene surrounding the sacral pressure injury, impeding its healing. He also benefitted from botulinum toxin injections to the bilateral pronators, wrist flexors, and finger flexors. These interventions decreased spasticity and improved active and passive ROM.

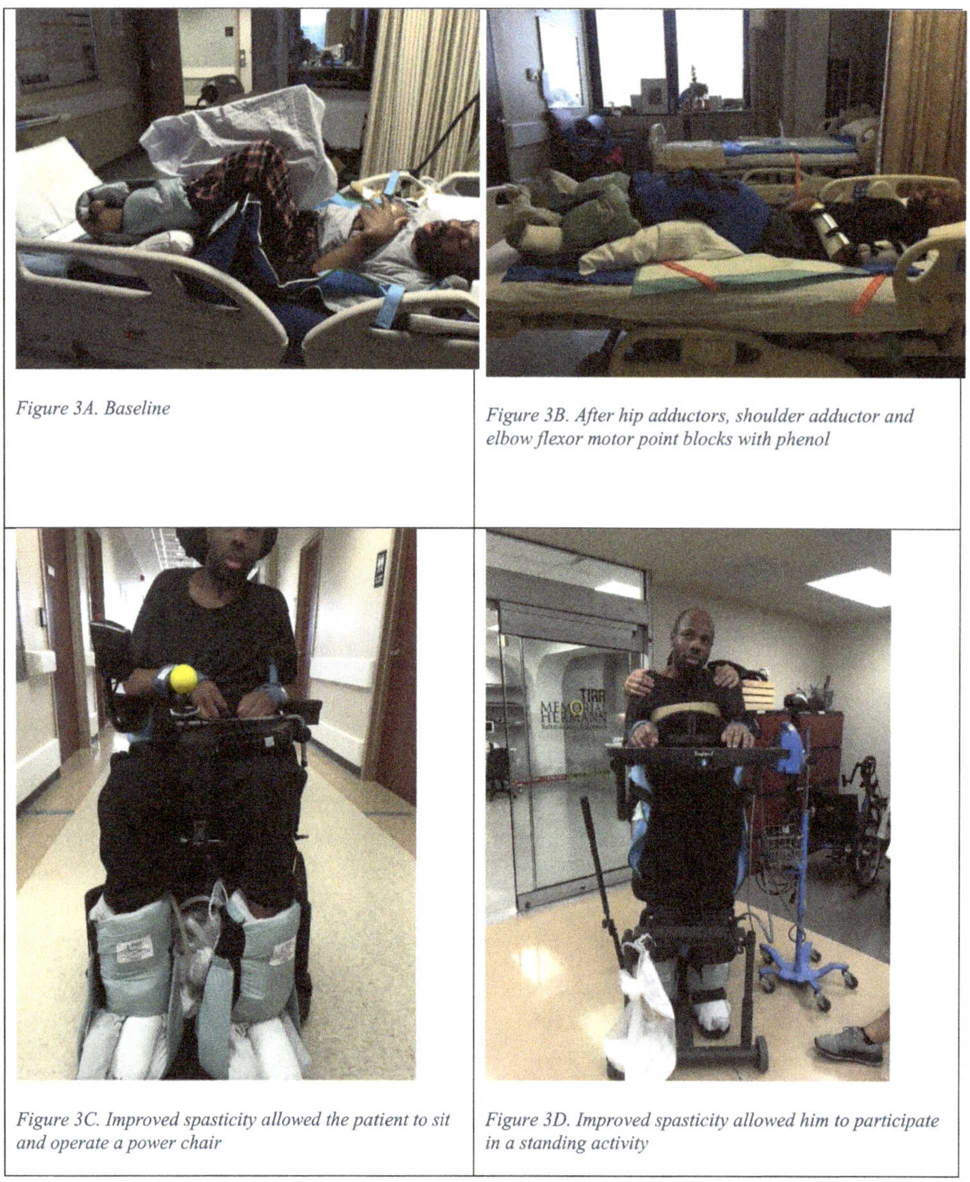

*Figure 3A. Baseline*

*Figure 3B. After hip adductors, shoulder adductor and elbow flexor motor point blocks with phenol*

*Figure 3C. Improved spasticity allowed the patient to sit and operate a power chair*

*Figure 3D. Improved spasticity allowed him to participate in a standing activity*

FIG. 3 Effects of phenol neurolysis and spasticity management in a person with SCI.

Physical therapists (PT) and occupational therapists (OT) worked with this patient to manage his spasticity to gain range of motion and prevent further contractures through stretching, joint mobilization, splinting, positioning, and weight-bearing. OT and PT collaborated to create a bed positioning program to facilitate prolonged stretch of spastic muscles and relieve pressure over the sacral wound. His therapy program was updated and modified as the patient gained range of motion in joints through spasticity management efforts. The patient's mobility improved, allowing us to order a power wheelchair (PWC) to improve his access to his environment. Switch access enabled the patient to perform his weight shifts independently and communicate with his friends and family through a Bluetooth joystick connection to a tablet. During physical therapy, the patient worked on prolonged stretching in various positions, including supine, prone, and standing in a standing frame. PT also worked with the patient on safety and endurance with PWC mobility and independence with instruction of all assistance needed for functional mobility and positioning. Improvements in posture allowed for better positioning in bed, and sitting in a PWC resulted in a healed sacral pressure injury without surgical intervention. Improved function achieved was translated into being able to operate a PWC, perform independent pressure reliefs via power tilt, access a tablet with assistive devices, and feeding himself with set up. Caregiver burden for transfers and positioning decreased from 2 people to 1 person.

## Conclusion/summary

Severe spasticity can be debilitating and often challenging to manage if appropriate interventions are delayed. A comprehensive team approach to spasticity management can prevent complications and improve quality of life after SCI. In complex situations, a combination of multiple treatment options is required to improve function and independence. Phenol neurolysis can be a very useful tool in these tricky situations. Currently, limited evidence and our experience suggest that chemical neurolysis is a safe, useful tool for spasticity management in people with SCI. Besides, phenol neurolysis utilization has been shown to be associated with no procedure-related severe complications and only a small incidence of observed adverse effects. Ultrasound guidance along with electrical stimulation can reduce the total required dose and minimize procedure-related side effects.

## Applications to other areas of neuroscience

Phenol or alcohol neurolysis has been used to manage spasticity in various upper motor neuron disorders. Other than spinal cord injury, the most common diagnoses for which this technique is used are stroke, brain injury, and multiple sclerosis (Karri et al., 2017; Kong & Chua, 1999, 2002; Mas, DiTommaso, & Li, 2019). It is often used in concert with botulinum toxin injections, especially when multiple areas require treatment to avoid using more than the recommended amount of either medication (Karri et al., 2020). Phenol neurolysis can also be used to treat spasticity in children. It is frequently performed to treat hip adductor and elbow flexor spasticity in children with cerebral palsy. Children may not tolerate this procedure, especially if multiple nerves are injected. For these reasons, phenol neurolysis is often performed under general anesthesia in children (Gormley, Krach, & Piccini, 2001). Few case reports of successful use of phenol neurolysis manage dystonia have been reported in the literature (Garcia Ruiz & Sanchez Bernardos, 2000; Kim, Lee, Ko, Ko, & Chung, 2003; Takeuchi, Chuma, & Mano, 2004). Phenol neurolysis has been used for decades in the management of persistent and intractable pain in various conditions. Phenol neurolysis or nerve blocks are frequently used to manage pain in cancer and other diagnoses resulting in chronic pain (Filippiadis et al., 2019; Koyyalagunta & Burton, 2010). It has been reportedly used to manage severe chronic nonmalignant pain to target pain generators, including the lumbar sympathetic chain, medial branches to the facet joints of the spine, the sacroiliac joint, the intercostal nerves, the occipital nerves, and the paracoccygeal area (Welksler et al., 2007).

## Mini-dictionary of terms

**Neurolysis**: Targeted mechanical or chemical injury to the nerve.

**Dysesthesia**: Abnormal, painful sensation due to damage to the peripheral or central nervous system.

**Motor point**: Location where motor branches of the nerve enter the muscle.

**Nerve block**: Injection of a chemical agent to cause temporarily or permanently block the nerve conduction to treat pain, spasticity, etc.

**Demyelination**: Damage to the myelin sheath, a protective outer layer of the nerve, which helps in nerve conduction.

**Spasticity**: Muscle tightness or spasms due to upper motor neuron injury such as spinal cord injury, stroke.

**Motor nerve**: Peripheral nerve fibers that transmit signals from the brain and spinal cord to the muscles.

**Sensory nerve**: Peripheral nerve fibers that carry sensory information to the spinal cord and brain.
**Mixed nerve**: A peripheral nerve that carries both sensory and motor fibers.
**Intrathecal baclofen pump**: A system that delivers baclofen into intrathecal space.

## Key facts of phenol

- Phenol is a chemical, also known as carbolic acid ($C_6H_5OH$), a benzene ring.
- At room temperature, phenol dissolves in water.
- Phenol is used at low concentrations in household cleaners as a disinfectant and antiseptic.
- Phenol is also used as a spray in a concentration of 1.4% for pain relief from sore throat.
- The usefulness of chemical neurolysis or nerve block was first recognized in 1863 for treating pain.
- Accidental exposure of large quantities of phenol can be lethal to humans.

## Summary points

- Poorly controlled severe spasticity after spinal cord injury can result in multiple complications, including the development of joint contractures, inability to participate in therapy, skin breakdown, decreased function, poor sleep, and poor quality of life.
- Phenol neurolysis is a cost-effective tool to manage spasticity, which requires skill to localize to the target nerve or motor points.
- Phenol neurolysis can be performed with anatomic localization, electrical stimulation guidance, or ultrasound guidance. Often these techniques are combined to maximize the efficiency of the procedure and minimize errors in localizing targets.
- 5%–6% phenol is typically used for spasticity management with effects lasting from 3 to 9 months.
- Medial pectoral nerves, lateral pectoral nerves, and musculocutaneous motor branches are common targets of the upper extremity for treating shoulder adductor and elbow flexor spasticity.
- Obturator nerve and tibial motor nerves are frequently injected targets of the lower extremity to address hip adductor and ankle plantar flexor spasticity.
- Local adverse effects include pain, dysesthesias, loss of sensation, and swelling.
- Systemic severe side effects can result from the intravascular injection of phenol, resulting in cardiac arrest and coma.

## References


Akkaya, T., Unlu, E., Alptekin, A., Gumus, H. I., Umay, E., & Cakci, A. (2010). Neurolytic phenol blockade of the obturator nerve for severe adductor spasticity. *Acta Anaesthesiologica Scandinavica*, *54*(1), 79–85. https://doi.org/10.1111/j.1399-6576.2009.02130.x.

Awad, E. A. (1972). Phenol block for control of hip flexor and adductor spasticity. *Archives of Physical Medicine and Rehabilitation*, *53*(12), 554–557.

Chang, M. C., & Boudier-Reveret, M. (2020). Management of elbow flexor spasticity with ultrasound-guided alcohol neurolysis of the musculocutaneous nerve. *Acta Neurologica Belgica*, *120*(4), 983–984. https://doi.org/10.1007/s13760-020-01300-x.

Chua, K. S., & Kong, K. H. (2001). Clinical and functional outcome after alcohol neurolysis of the tibial nerve for ankle-foot spasticity. *Brain Injury*, *15*(8), 733–739. https://doi.org/10.1080/02699050010009775.

Copp, E. P., Harris, R., & Keenan, J. (1970). Peripheral nerve block and motor point block with phenol in the management of spasticity. *Proceedings of the Royal Society of Medicine*, *63*(9), 937–938.

D'Souza, R. S., & Hooten, W. M. (2021). *Neurolytic blocks*. Treasure Island, FL: StatPearls.

D'Souza, R. S., & Warner, N. S. (2021). *Phenol nerve block*. Treasure Island, FL: StatPearls.

Easton, J. K., Ozel, T., & Halpern, D. (1979). Intramuscular neurolysis for spasticity in children. *Archives of Physical Medicine and Rehabilitation*, *60*(4), 155–158.

Elovic, E. P., Esquenazi, A., Alter, K. E., Lin, J. L., Alfaro, A., & Kaelin, D. L. (2009). Chemodenervation and nerve blocks in the diagnosis and management of spasticity and muscle overactivity. *PM & R: The Journal of Injury, Function, and Rehabilitation*, *1*(9), 842–851. https://doi.org/10.1016/j.pmrj.2009.08.001.

Escaldi, S. (2018). Neurolysis: A brief review for a fading art. *Physical Medicine and Rehabilitation Clinics of North America*, *29*(3), 519–527. https://doi.org/10.1016/j.pmr.2018.03.005.

Felsenthal, G. (1974). Pharmacology of phenol in peripheral nerve blocks: A review. *Archives of Physical Medicine and Rehabilitation*, *55*(1), 13–16.

Filippiadis, D. K., Tselikas, L., Tsitskari, M., Kelekis, A., de Baere, T., & Ryan, A. G. (2019). Percutaneous neurolysis for pain management in oncological patients. *Cardiovascular and Interventional Radiology*, *42*(6), 791–799. https://doi.org/10.1007/s00270-019-02185-x.

Gaid, M. (2012). Phenol nerve block for the management of lower limb spasticity. *Advances in Clinical Neuroscience & Rehabilitation*, *12*(3), 23–25.

Garcia Ruiz, P. J., & Sanchez Bernardos, V. (2000). Intramuscular phenol injection for severe cervical dystonia. *Journal of Neurology*, *247*(2), 146–147. https://doi.org/10.1007/pl00007799.

Ghai, A., Sangwan, S. S., Hooda, S., Garg, N., Kundu, Z. S., & Gupta, T. (2013). Evaluation of interadductor approach in neurolytic blockade of obturator nerve in spastic patients. *Saudi Journal of Anaesthesia*, *7*(4), 420–426. https://doi.org/10.4103/1658-354X.121074.

Gormley, M. E., Jr., Krach, L. E., & Piccini, L. (2001). Spasticity management in the child with spastic quadriplegia. *European Journal of Neurology*, *8* (Suppl. 5), 127–135. https://doi.org/10.1046/j.1468-1331.2001.00045.x.

Gunduz, S., Kalyon, T. A., Dursun, H., Mohur, H., & Bilgic, F. (1992). Peripheral nerve block with phenol to treat spasticity in spinal cord injured patients. *Paraplegia*, *30*(11), 808–811. https://doi.org/10.1038/sc.1992.156.

Halpern, D., & Meelhuysen, F. E. (1966). Phenol motor point block in the management of muscular hypertonia. *Archives of Physical Medicine and Rehabilitation*, *47*(10), 659–664.

Halpern, D., & Meelhuysen, F. E. (1967). Duration of relaxation after intramuscular neurolysis with phenol. *JAMA*, *200*(13), 1152–1154.

Holtz, K. A., Lipson, R., Noonan, V. K., Kwon, B. K., & Mills, P. B. (2017). Prevalence and effect of problematic spasticity after traumatic spinal cord injury. *Archives of Physical Medicine and Rehabilitation*, *98*(6), 1132–1138. https://doi.org/10.1016/j.apmr.2016.09.124.

Horn, L. J., Singh, G., & Dabrowski, E. R. (2015). In A. Brashear, & E. Elovic (Eds.), *Chemoneurolysis with phenol and alcohol: A "dying art" that merits revival* (pp. 109–128). New York: Springer Publishing Company.

Jang, S. H., Ahn, S. H., Park, S. M., Kim, S. H., Lee, K. H., & Lee, Z. I. (2004). Alcohol neurolysis of tibial nerve motor branches to the gastrocnemius muscle to treat ankle spasticity in patients with hemiplegic stroke. *Archives of Physical Medicine and Rehabilitation*, *85*(3), 506–508. https://doi.org/10.1016/s0003-9993(03)00468-4.

Karri, J., Mas, M. F., Francisco, G. E., & Li, S. (2017). Practice patterns for spasticity management with phenol neurolysis. *Journal of Rehabilitation Medicine*, *49*(6), 482–488. https://doi.org/10.2340/16501977-2239.

Karri, J., Zhang, B., & Li, S. (2020). Phenol neurolysis for management of focal spasticity in the distal upper extremity. *PM & R: The Journal of Injury, Function, and Rehabilitation*, *12*(3), 246–250. https://doi.org/10.1002/pmrj.12217.

Kaymak, B., Kara, M., Gurcay, E., Aydin, G., & Ozcakar, L. (2019). Selective peripheral neurolysis using high frequency ultrasound imaging: A novel approach in the treatment of spasticity. *European Journal of Physical and Rehabilitation Medicine*, *55*(4), 522–525. https://doi.org/10.23736/S1973-9087.18.05295-4.

Keenan, M. A., Tomas, E. S., Stone, L., & Gersten, L. M. (1990). Percutaneous phenol block of the musculocutaneous nerve to control elbow flexor spasticity. *The Journal of Hand Surgery*, *15*(2), 340–346. https://doi.org/10.1016/0363-5023(90)90120-g.

Kelly, R. E., & Gautier-Smith, P. C. (1959). Intrathecal phenol in the treatment of reflex spasms and spasticity. *Lancet*, *2*(7112), 1102–1105. https://doi.org/10.1016/s0140-6736(59)90095-9.

Khalili, A. A., & Betts, H. B. (1967). Peripheral nerve block with phenol in the management of spasticity. Indications and complications. *JAMA*, *200*(13), 1155–1157.

Kim, J. S., Lee, K. S., Ko, Y. J., Ko, S. B., & Chung, S. W. (2003). Idiopathic foot dystonia treated with intramuscular phenol injection. *Parkinsonism & Related Disorders*, *9*(6), 355–359. https://doi.org/10.1016/s1353-8020(02)00118-9.

Kong, K. H., & Chua, K. S. (1999). Neurolysis of the musculocutaneous nerve with alcohol to treat poststroke elbow flexor spasticity. *Archives of Physical Medicine and Rehabilitation*, *80*(10), 1234–1236. https://doi.org/10.1016/s0003-9993(99)90021-7.

Kong, K. H., & Chua, K. S. (2002). Intramuscular neurolysis with alcohol to treat post-stroke finger flexor spasticity. *Clinical Rehabilitation*, *16*(4), 378–381. https://doi.org/10.1191/0269215502cr508oa.

Koyama, H., Murakami, K., Suzuki, T., & Suzaki, K. (1992). Phenol block for hip flexor muscle spasticity under ultrasonic monitoring. *Archives of Physical Medicine and Rehabilitation*, *73*(11), 1040–1043.

Koyyalagunta, D., & Burton, A. W. (2010). The role of chemical neurolysis in cancer pain. *Current Pain and Headache Reports*, *14*(4), 261–267. https://doi.org/10.1007/s11916-010-0123-9.

Lee, K. H., Kim, D. H., Kim, Y. H., Ro, S. H., & Lee, J. (2020). Neurolytic abdominal wall blocks with alcohol for intractable gastrostomy site pain in a cancer patient—A case report. *Korean Journal of Anesthesiology*, *73*(3), 247–251. https://doi.org/10.4097/kja.19041.

Liversedge, L. A., & Maher, R. M. (1960). Use of phenol in relief of spasticity. *British Medical Journal*, *2*(5191), 31–33. https://doi.org/10.1136/bmj.2.5191.31.

Mas, M. F., DiTommaso, C., & Li, S. (2019). Phenol neurolysis for the management of shoulder spasticity in early recovery from traumatic brain injury: A case report. *PM & R: The Journal of Injury, Function, and Rehabilitation*, *11*(1), 90–93. https://doi.org/10.1016/j.pmrj.2018.06.002.

Matsumoto, M. E., Berry, J., Yung, H., Matsumoto, M., & Munin, M. C. (2018). Comparing electrical stimulation with and without ultrasound guidance for phenol neurolysis to the musculocutaneous nerve. *PM & R: The Journal of Injury, Function, and Rehabilitation*, *10*(4), 357–364. https://doi.org/10.1016/j.pmrj.2017.09.006.

Maynard, F. M., Karunas, R. S., & Waring, W. P., 3rd. (1990). Epidemiology of spasticity following traumatic spinal cord injury. *Archives of Physical Medicine and Rehabilitation*, *71*(8), 566–569.

Rekand, T., Hagen, E. M., & Gronning, M. (2012). Spasticity following spinal cord injury. *Tidsskrift for den Norske Lægeforening*, *132*(8), 970–973. https://doi.org/10.4045/tidsskr.10.0872.

Seidel, P. M., Seidel, G. K., Gans, B. M., & Dijkers, M. (1996). Precise localization of the motor nerve branches to the hamstring muscles: An aid to the conduct of neurolytic procedures. *Archives of Physical Medicine and Rehabilitation*, *77*(11), 1157–1160. https://doi.org/10.1016/s0003-9993(96)90140-9.

Sezer, N., Akkus, S., & Ugurlu, F. G. (2015). Chronic complications of spinal cord injury. *World Journal of Orthopedics*, *6*(1), 24–33. https://doi.org/10.5312/wjo.v6.i1.24.

Sirico, F., Zappia, M., Di Meglio, F., Castaldo, C., & Nurzynska, D. (2020). Can we reduce the risk of adverse effects in selective neurolysis of the motor branches of musculocutaneous nerve? Evidence-based anatomy approach to the rescue of patients with muscle spasticity. *European Journal of Physical and Rehabilitation Medicine*, *56*(1), 126–127. https://doi.org/10.23736/S1973-9087.19.05875-1.

Sung, D. H., Han, T. R., Park, W. H., Je Bang, H., Kim, J. M., Chung, S. H., et al. (2001). Phenol block of peripheral nerve conduction: Titrating for optimum effect. *Archives of Physical Medicine and Rehabilitation*, *82*(5), 671–676. https://doi.org/10.1053/apmr.2001.21975.

Takeuchi, N., Chuma, T., & Mano, Y. (2004). Phenol block for cervical dystonia: Effects and side effects. *Archives of Physical Medicine and Rehabilitation*, *85*(7), 1117–1120. https://doi.org/10.1016/j.apmr.2003.10.012.

Tibbett, J., Widerstrom-Noga, E. G., Thomas, C. K., & Field-Fote, E. C. (2019). Impact of spasticity on transfers and activities of daily living in individuals with spinal cord injury. *The Journal of Spinal Cord Medicine*, *42*(3), 318–327. https://doi.org/10.1080/10790268.2017.1400727.

van Cooten, I. P., Snoek, G. J., Nene, A. V., de Groot, S., & Post, M. W. (2015). Functional hindrance due to spasticity in individuals with spinal cord injury during inpatient rehabilitation and 1 year thereafter. *Spinal Cord*, *53*(9), 663–667. https://doi.org/10.1038/sc.2015.41.

Viel, E., Pellas, F., Ripart, J., Pelissier, J., & Eledjam, J. J. (2005). Peripheral neurolytic blocks and spasticity. *Annales Françaises d'Anesthèsie et de Rèanimation*, *24*(6), 667–672. https://doi.org/10.1016/j.annfar.2005.03.015.

Welksler, N., Klein, M., Gurevitch, B., Rozentsveig, V., Rudich, Z., Brill, S., et al. (2007). Phenol neurolysis for sever chronic nonmalignant pain: Is the old also obsolete? *Pain Medicine*, *8*(4), 332–337. https://doi.org/10.1111/j.1526-4637.2006.00228.x.

Zhang, B., Darji, N., Francisco, G. E., & Li, S. (2021). The time course of onset and peak effects of phenol neurolysis. *American Journal of Physical Medicine & Rehabilitation*, *100*(3), 266–270. https://doi.org/10.1097/PHM.0000000000001563.

Chapter 27

# Anti-repulsive guidance molecule: An antibody treatment in spinal cord injury

Takahide Itokazu[a,b] and Toshihide Yamashita[a,b]

[a]*Department of Molecular Neurosciences, Graduate School of Medicine, Osaka University, Suita, Japan;* [b]*Department of Neuro-Medical Science, Graduate School of Medicine, Osaka University, Suita, Japan*

## Abbreviations

| | |
|---|---|
| **BBB** | blood brain barrier |
| **BMP** | bone morphogenetic protein |
| **CNS** | central nervous system |
| **CST** | corticospinal tract |
| **DCC** | deleted in colorectal cancer |
| **GPI** | glycosylphosphatidylinositol |
| **MCAO** | middle cerebral artery occlusion |
| **MPTP** | 1-methyl-4-phenyl-1,2,3,6-tetrahydropyridine |
| **MS** | multiple sclerosis |
| **NMO** | neuromyelitis optica |
| **PD** | Parkinson's disease |
| **RGC** | retinal ganglion cell |
| **RGM** | repulsive guidance molecule |
| **Rho GEF** | Rho guanine nucleotide exchange factor |
| **SCI** | spinal cord injury |
| **SN** | substantia nigra |
| **TBI** | traumatic brain injury |
| **UEMS** | upper extremity motor score |

## Introduction

The repulsive guidance molecule (RGM) is a membrane-associated glycosylphosphatidylinositol (GPI)-anchored glycoprotein, which was identified as an axon guidance molecule in the retinotectal system (Monnier et al., 2002). By analyzing the development of the retinotectal projection system in chicks—a method frequently used to study the molecular basis of topographic projections, including axon guidance cues—Monnier et al. found that RGM is expressed in a spatial gradient manner along the anterior–posterior axis in the tectum (Fig. 1A). Importantly, in growth cone collapse assays, temporal, but not nasal, retinal growth cones collapsed when confronted with RGM-decorated membranes (Fig. 1B). Similarly, in the stripe assay (Walter, Kern-Veits, Huf, Stolze, & Bonhoeffer, 1987), RGM guides axonal elongation only for temporal retinal axons (Fig. 1C). These findings strongly suggest that RGM can act as an axon-specific repulsive guidance molecule (Monnier et al., 2002).

However, the receptor mechanism for RGM function was not defined at that time. In 2004, Rajagopalan et al. and Matsunaga et al. reported that neogenin mediates the biological effects of RGM as its cognate receptor (Matsunaga et al., 2004; Rajagopalan et al., 2004). Strikingly, neogenin was revealed to be expressed in a temporal–nasal gradient in the retina (Rajagopalan et al., 2004); thus, the repulsiveness of temporal, not nasal, retinal axons by RGM is well explained by RGM–neogenin interactions (Fig. 1A). RGM is also found to act as a co-receptor for bone morphogenetic proteins (BMPs) and modulates BMP signaling (Babitt et al., 2005).

Since then, RGM signaling in the central nervous system (CNS) has been extensively studied and a wide range of functions has been reported; RGM plays multiple roles in neuronal development [e.g., neural tube closure (Matsunaga

Diagnosis and Treatment of Spinal Cord Injury. https://doi.org/10.1016/B978-0-12-822498-4.00027-0

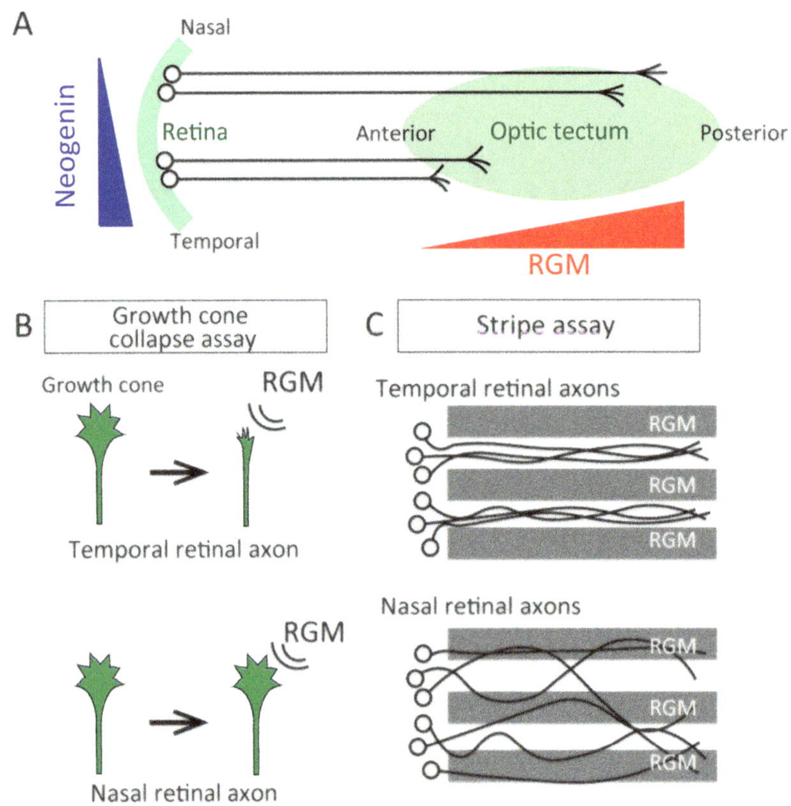


FIG. 1 RGM is an inhibitory axon guidance molecule. (A) Schematic illustration of retinotectal projections and expression patterns of RGMa and neogenin. Nasal retinal ganglion cell (RGC) axons project to the posterior part of the optic tectum, whereas temporal RGC axons project to the anterior part of the tectum. RGM is expressed in an increasing anterior–posterior gradient manner in the tectum, and neogenin is expressed along an elevating nasal–temporal gradient in the retina. (B) Representation of growth cone collapse assay. RGM exerts growth cone collapse effects only for temporal retinal axons. (C) Representation of stripe assay. The gray colored stripe indicates membrane fractions from RGM-transfected cells, whereas the white colored stripe represents membrane fractions from mock-transfected cells.

et al., 2004)], and is also involved in the pathophysiology of various CNS diseases, as discussed below. Moreover, external to the CNS, RGM has been shown to regulate endochondral bone formation (Zhou et al., 2010), iron homeostasis (Babitt et al., 2006), and inflammation (Körner et al., 2019). In mammals, at least three homologues of RGM, RGMa, RGMb (also known as DRAGON), and RGMc (also known as hemojuvelin, HFE2), have been identified. Among them, RGMa is most closely related to chicken RGM (80% homology), and most of the critical findings in CNS development/disease are obtained from RGMa research. Thus, in this review, we mainly focus on the functions of RGMa.

## Characteristics of RGMa

### Ligand–receptor interactions and downstream signaling of RGMa

As mentioned above, RGM binds to neogenin, a transmembrane protein originally isolated from chick cerebellum as a homolog of deleted in colorectal cancer (DCC: a receptor for the axon guidance molecule netrin-1). RGM also binds to BMP morphogens as a co-receptor and modulates BMP signaling. On the other hand, neogenin is also known as a netrin-1 receptor. How these proteins interact with each other and how downstream signaling is regulated are not fully understood. However, recent studies, including crystal structure analysis (Bell et al., 2013; Healey et al., 2015; Malinauskas, Peer, Bishop, Mueller, & Siebold, 2020), provide direct insight into their interactions.

The major binding site of RGM for neogenin is the carboxyl-terminal domain of RGM (C-RGM) (Bell et al., 2013; Itokazu, Fujita, Takahashi, & Yamashita, 2012). Interestingly, two C-RGMs act as a molecular staple bringing two neogenin receptors together, and this unique architecture is proposed to be important for the subsequent activation of downstream signaling (Bell et al., 2013; Siebold, Yamashita, Monnier, Mueller, & Pasterkamp, 2017). On the other hand, the amino-terminal domain of RGM (N-RGM) has been shown to interact with BMP, and this binding is suggested to link BMP and neogenin signaling by comprising a BMP–RGM–neogenin complex (Healey et al., 2015). RGM–BMP interactions have been thought to potentiate BMP signaling via the canonical SMAD pathway. However, a recent report revealed that RGM can compete with growth differentiation factor 5 (GDF5), a member of the BMP family, and thus act as an inhibitor of BMP signaling (Malinauskas et al., 2020). Therefore, how intracellular signaling is regulated through interactions involving N-RGM and BMPs is presumed to be context-dependent. Further studies are warranted to determine the physiological roles of RGM–BMP interactions in each developmental/pathological condition.

RGM is thought to transduce signals in both *cis* and *trans* manner. Its distinctive functions, such as inducing growth cone collapse or axon repulsion, are induced in *trans*, that is via cell-to-cell interactions. As the RGM extracellular domain is thought to be cleaved at, and secreted from, cell membranes, *trans*-signaling can be achieved by a gradient of soluble RGM (Fig. 2). However, it is unknown whether long-range signal transduction is possible.

Although a detailed picture of RGM signaling has not yet been obtained, several key aspects have been uncovered. One of the most important downstream signaling pathways involves RhoA activation by RGM, because it is widely accepted that Rho-GTPases play key roles in axon guidance and neurite growth. Indeed, the inhibition of Rho kinase, a downstream effector of RhoA, abolishes the inhibitory effects of RGMa on neurite outgrowth (Hata et al., 2006). These authors also revealed that Unc5B, a member of the netrin receptor family, interacts with neogenin as a co-receptor for RGMa and activates RhoA through LARG, a member of the Rho guanine nucleotide exchange factor (RhoGEF) subfamily (Hata, Kaibuchi, Inagaki, & Yamashita, 2009). In addition, the involvement of Ras activity was also evaluated because of its well-known functions as a mediator of growth cone collapse and neurite retraction (Elowe, Holland, Kulkarni, & Pawson, 2001). As expected, RGMa–neogenin binding has been revealed to inactivate Ras activity, leading to growth cone collapse (Endo & Yamashita, 2009).

More recently, detailed investigations unveiled more complex mechanisms surrounding RGM–neogenin signaling regulation, including the proteolytic processing of RGMa (Tassew, Charish, Seidah, & Monnier, 2012, well summarized in Siebold et al., 2017), ectodomain shedding of neogenin (van Erp et al., 2015), and ϒ-secretase cleavage of the neogenin intracellular domain (Banerjee et al., 2016). Further research is needed to elucidate how these multiple signaling pathways are regulated and integrated in vivo.

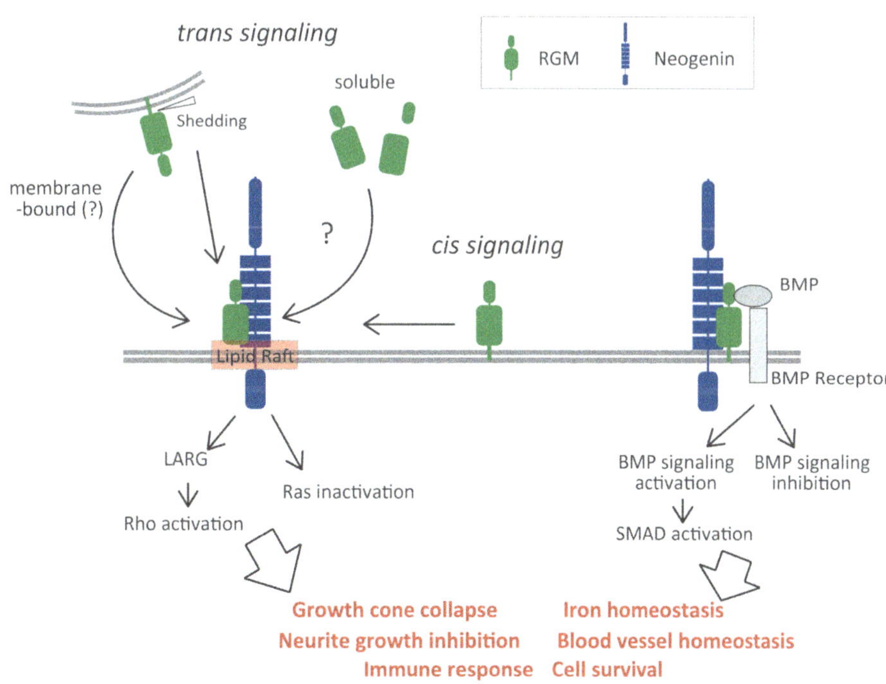


FIG. 2 RGM–neogenin interactions and downstream signaling. Schematic illustration of possible ligand-receptor interactions involving RGMa. The inhibitory effect for neurite outgrowth is brought about by *trans*-signaling; that is, neurogenin-expressing axons are repelled by cell-bound RGMa or via an RGMa gradient established by extracellular shedding. Neogenin localization to lipid rafts—membrane domains enriched in cholesterol, sphingolipids, and protein receptors, present a platform for cellular signaling events, including ligand:receptor interactions—has been proposed to be indispensable for signal transduction.

## RGMa in spinal cord injury (SCI)

### RGMa expression after SCI

Given that RGMa, as an inhibitory protein, is involved in neural circuit formation during development by modulating the pathfinding activity of growing axons, it is possible that RGMa exerts negative effects on axonal regeneration. To test this concept in vivo, RGMa expression in the spinal cord of adult animals under normal and SCI conditions has been addressed.

Immunohistochemical analysis of rat spinal cord using anti-C-RGMa antibody revealed its localization to neurons and oligodendrocytes under normal conditions (Hata et al., 2006; Schwab, Conrad, et al., 2005). After SCI, RGMa expression around the lesion site is upregulated; with expression by neurons, oligodendrocytes, certain reactive astrocytes, activated microglia, and infiltrating leukocytes (Hata et al., 2006; Schwab, Conrad, et al., 2005). These observations support the possibility that RGMa is an inhibitory protein against axonal regeneration after SCI; therefore, RGMa neutralization is considered to be a potential therapeutic strategy to attenuate neural deficits induced by SCI. An increase in RGMa around the SCI lesion site is also observed in primates (Nakagawa et al., 2019). In that study, RGMa was reported to be mainly expressed by Iba-1-positive microglia/macrophages.

Importantly, RGMa expression patterns in the human spinal cord have also been reported. In the uninjured spinal cord, RGMa is expressed at low levels by neurons and glial cells. However, immunoreactivity for RGMa in the spinal cord is significantly increased following SCI, and the localization of RGMa in neurons, axons, and glial cells has been confirmed (Mothe et al., 2017). On the other hand, the RGMa receptor neogenin is expressed by neurons in the spinal cord in both rats and humans (Mothe et al., 2017).

Taken together, RGMa upregulation induced by SCI is common across divergent species, suggesting the significance of RGMa in the pathophysiology of CNS injury.

### RGMa inhibition promotes recovery after SCI in rodents

The first direct evidence of the therapeutic effects of RGMa inhibition on SCI was reported in 2006 (Hata et al., 2006). To test the hypothesis that upregulated RGMa plays a negative role in axonal regeneration in vivo, Hata et al. generated an RGMa-neutralizing antibody (directed against RGM residues 309–322). After receiving Th9/10 dorsal transection, which resulted in both dorsal (main) and lateral corticospinal tract (CST) injury, rats were treated with the RGMa-neutralizing antibody via an osmotic mini-pump with catheters placed near the lesion site, and locomotor functions were monitored. As expected, the Basso–Beattie–Bresnahan (BBB) locomotion score (Basso, Beattie, & Bresnahan, 1995) was significantly improved by RGMa inhibition. The authors also performed anatomical tracing studies and observed a lesser degree of CST retraction, a greater number of collateral formations, and a significant amount of axon extension through the lesion site (Fig. 3A). This enhanced anatomical regeneration temporally correlated with functional recovery.

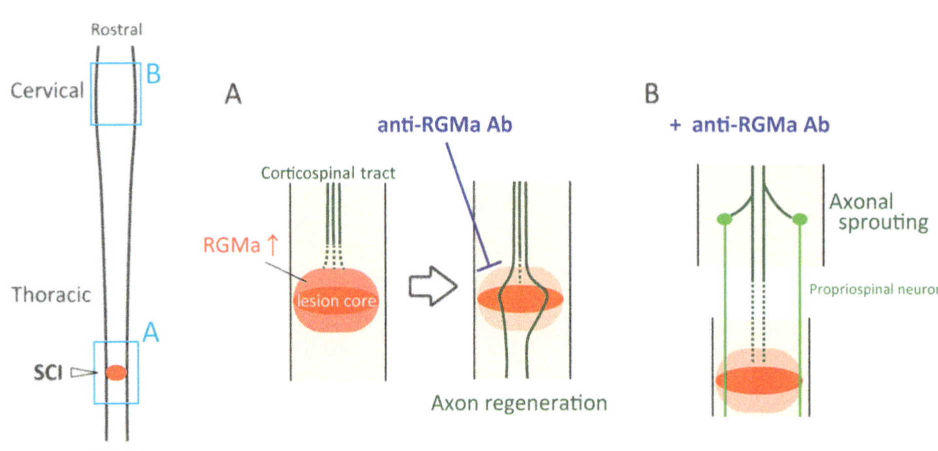


FIG. 3 RGMa neutralization enhances axonal regeneration and neural circuit reorganization after spinal cord injury in rodents. RGMa neutralization promotes several types of neuroplasticity after spinal cord injury. (A) RGMa is upregulated around the lesion site and limits axonal regeneration. Anti-RGMa treatment results in axonal growth of the corticospinal tract (CST). (B) Anti-RGMa treatment is suggested to enhance neural synaptic rearrangement in the cervical spinal cord, distant from the lesion site. Collaterals from hind-limb axons possibly make connections to descending long propriospinal neurons to compensate for the damaged circuit (Bareyre et al., 2004).

Using the same experimental setting, Kyoto et al. tested the possibility that neural circuit plasticity might be enhanced by anti-RGMa antibody treatment—even in lesions distant from the injury site (Kyoto et al., 2007). The results showed that antibody treatment facilitated synapse formation at the level of the cervical spinal cord (Fig. 3B), which may contribute to the rerouting of hind-limb CST connections through propriospinal neurons (Bareyre et al., 2004; Fink & Cafferty, 2016; Oudega & Perez, 2012). This observation may be explained by the notion that the adult CNS offers an inhibitory environment for adoptive reorganization (i.e., neural circuit remodeling) and RGMa is one of its major components; therefore, RGMa inhibition can contribute to the establishment of reorganization-permissive environments.

On the other hand, Tassew et al. explored strategies to inhibit neogenin (Tassew et al., 2014). During developmental stages, neogenin is known to act as a dependence receptor that induces cell death in the absence of RGM (Matsunaga et al., 2004). Therefore, targeting neogenin instead of RGMa may be beneficial, not only for neural circuit reorganization, but also for neuronal survival after SCI. Based on the observation that the receptor functions of neogenin require its localization in lipid rafts, the authors addressed whether blocking neogenin raft association improves SCI outcomes. They found that interactions involving the neogenin 4Ig domain and the N-terminal regions of RGMa (N-RGMa) are critical for the attraction of neogenin to lipid rafts, and recombinant 4Ig treatment effectively abolishes their interactions. Thus, they treated SCI rats with 4Ig (intravenous injection or local intrathecal infusion). As expected, axonal regeneration and functional recovery (as assessed by BBB scores and ladder walk tests) were significantly enhanced. Furthermore, this treatment attenuated neuronal loss in perilesional neurons.

Based on this evidence, humanized monoclonal antibodies specific for N-RGMa were generated (Demicheva et al., 2015), and their therapeutic potential in impact-compression SCI was tested (Mothe et al., 2017). To provide a clinically applicable strategy, the authors examined the efficacy of RGMa inhibition by systemic administration of humanized anti-RGMa antibody and revealed that this treatment effectively enhanced functional recovery (as assessed by BBB scores, ladder walk tests, and footprint analysis). Moreover, by histological and anatomical analysis, it was also confirmed that anti-RGMa antibody treatment enhances neuronal survival and axonal regeneration. Interestingly, anti-RGMa antibody treatment also attenuated neuropathic pain after SCI. Although effects on microglia/macrophage were suggested, the precise mechanisms involved remain unknown. In a recent paper, Mothe et al. also reported that delayed administration (~3 h after SCI) of anti-N-RGMa antibody could still effectively enhance recovery of motor functions, with increased neuronal sparing and axonal plasticity (Mothe et al., 2020). Interestingly, the recovery of bladder function was also enhanced, indicating that anti-RGMa treatment exerts broad beneficial effects in terms of SCI convalescence.

Taken together, the effectiveness of the strategy for RGMa/neogenin signaling inhibition has been repeatedly confirmed in rodent SCI models (Table 1). In the next chapter, evidence from primates is discussed.

**TABLE 1** Summary of pre-clinical studies.

| Author (Year) | Species | Method of SCI | Intervention | Route of Administration | Functional Outcome | Suggested Mechanism |
|---|---|---|---|---|---|---|
| Hata et al. (2006) | Rat | Dorsal hemisection | anti RGMa-C antibody | Intrathecal administration via osmotic minipump | Locomotor recovery ↑ (BBB score) | CST regeneration/sprouting ↑ |
| Kyoto et al. (2007) | Rat | Dorsal hemisection | anti RGMa-C antibody | Intrathecal administration via osmotic minipump | *not tested* | Synapse formation of CST in the cervical SC ↑ |
| Tassew et al. (2014) | Rat | Impact compression | 4Ig peptide (prevent RGMa-Neogenin interaction) | Intraspinal injection and continuous intrathecal derivery | Motor function ↑ (BBB score, Motor subscore, Ladder walk test) | Axonal regeneration ↑<br>Neuronal loss ↓ |
| Mothe et al. (2017) | Rat | Impact compression | anti-RGMa-N human antibody | Intravenous administration | Locomotor recovery ↑ (BBB score, Motor subscore, gait analysis)<br>Neuropathic pain ↓ | Perilesional neuronal survival ↑<br>CST axonal growth ↑<br>Neuronal survival ↑ |
| Chen et al. (2019) | Lamprey | Transection | Neogenin antisense morpholino oligo | Retrograde transport from lesion site | Restoration of locomotor activity | Survival of reticulospinal neurons ↑ |
| Mothe et al. (2020) | Rat | Impact compression | elezanumab (human monoclonal RGMa antibody) | Intravenous delayed administration | 3h post-SCI administration:<br>Locomotor recovery ↑ (BBB score, Motor subscore)<br>Bladder function ↑ | CST axonal plasticity ↑<br>Serotonergic axonal plasticity ↑<br>Perilesional neuronal sparing ↑ |
| Yang et al. (2020) | Rat | Compression | sh-RGMa (lentiviral vector) | Direct injection to the spinal cord | Locomotor recovery ↑ (BBB score) | Nerve regeneration ↑ |
| Nakagawa et al. (2019) | Macaque monkey | Unilateral lesion at C6/C7 | anti-human RGMa-C antibody | intrathecal administration via osmotic minipump | Manual dexterity ↑ | Axonal sprouting of CST fiber ↑<br>CST fiber - motoneuron connection ↑ |

Inhibition of RGMa/neogenin signaling promotes functional recovery after SCI (Chen & Shifman, 2019; Kyoto, Hata, & Yamashita, 2007; Mothe et al., 2017, 2020; Nakagawa, Ninomiya, Yamashita, & Takada, 2019; Tassew et al., 2014; Yang & Sun, 2020).

## RGMa inhibition promotes recovery after SCI in primates

Considering the clinical application of potential therapies to SCI patients, differences in possible mechanisms underlying neural circuit remodeling between higher primates and rodents should be taken into account. For example, the CST in rodents mainly descends to the dorsal funiculus, whereas in primates, the CST passes through the lateral and ventral portions of the spinal cord. More importantly, direct corticomotoneuronal connections, which are thought to be critically involved in manual dexterity in primates, are lacking in rodents (Lemon, 1993). Restoring hand dexterity remains a key clinical challenge and addressing whether potential treatments have beneficial effects in terms of skilled motor behavior in primates is of enormous importance.

Therefore, to test the effects of anti-RGMa antibody treatment in higher primates, Nakagawa et al. conducted SCI experiments in rhesus monkeys (Nakagawa et al., 2019). Unilateral spinal cord lesions were performed by transection at C6/C7 (Fig. 4A), which led to the complete loss of lateral CST fibers below the lesion. Then, a neutralizing antibody against C-RGMa was continuously delivered to the lesion site via an osmotic pump. Manual dexterity was assessed by a reaching/grasping task (Nakagawa, Ninomiya, Yamashita, & Takada, 2015) and the modified Brinkman Board Test (Freund et al., 2009). In both tests, monkeys were required to grasp food pellets placed in small horizontal or vertical slots (Fig. 4B). The results of these tests conformed a drastic improvement in the motor performance of the animals in the anti-RGMa antibody-treated group.

Further, the anatomical basis for this improvement was investigated. As summarized in Fig. 4C, anti-RGMa antibody treatment enhanced axonal sprouting of CST fibers arising from the motor cortex in the spinal cord at C7-Th1. Moreover, numbers of biotinylated dextran amine (BDA)-labeled direct contacts of CST fibers with wheat germ agglutinin (WGA)-labeled motor neurons (i.e., corticomotoneuronal direct connections) at the levels of C7, C8, and Th1 were significantly increased. These findings suggest that sprouting CST fibers can find appropriate target motoneurons to promote the efficient recovery of motor functions. Furthermore, inactivation of the contralesional motor cortex by muscimol injection in the anti-RGMa antibody-treated animals was performed. The otherwise recovered manual dexterity was then completely abolished.

Taken together, it is reasonable to assume that improved functional recovery of manual dexterity can be achieved by enhancement of axonal reorganization induced by anti-RGMa antibody treatment.

## Clinical studies

Based on preclinical findings, two clinical trials using anti-RGMa antibody are, as at October 2020, underway. One of these, MT-3921, involves a monoclonal antibody against the C fragment of RGMa and is under investigation for the

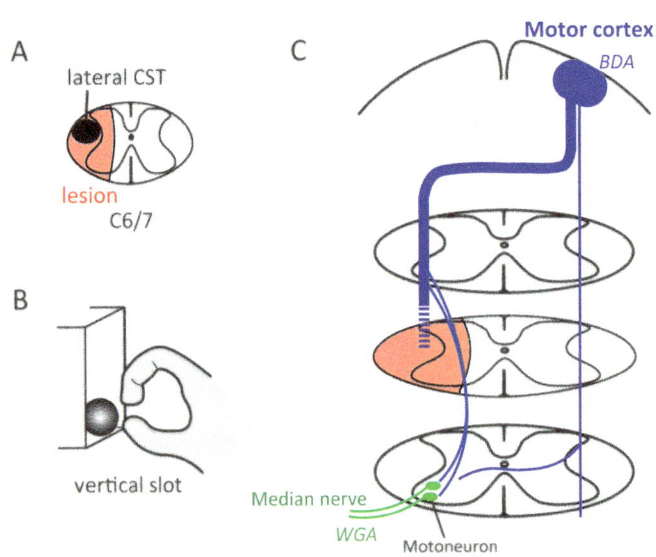


FIG. 4 RGMa neutralization enhances neural circuit reorganization after SCI in primates. (A) Schematic representation of spinal cord lesions in a study by Nakagawa et al. (2019). (B) Behavioral analysis of manual dexterity. Motor performance was greatly recovered by anti-RGMa antibody treatment. This treatment also advanced the commencement of recovery following SCI. (C) RGMa neutralization enhances axonal sprouting of corticospinal tract (CST) fibers. CST axons are labeled by anterograde injection of anterograde tracer (BDA; biotinylated dextran amine). For motoneuron labeling through median nerves, wheat germ agglutinin (WGA) was directly injected into the median nerve.

treatment of acute SCI (NCT04096950, www.clinicaltrials.gov, sponsored by Mitsubishi Tanabe Pharma Development America, Inc.). Another is elezanumab (ABT-555), a monoclonal antibody against the N fragment of RGMa. This antibody is also planned for patients with acute traumatic SCI with C4, C5, C6, or C7 (neurological level) injury (NCT04295538, www.clinicaltrials.gov, sponsored by AbbVie). These trials will determine whether the beneficial effects of RGMa neutralization are also observed in patients suffering from CNS injury.

## RGMa in other CNS disorders

Increasing evidence suggests that RGMa is also involved in various CNS disorders, including multiple sclerosis (MS), stroke, traumatic brain injury (TBI), neuromyelitis optica (NMO), and Parkinson's disease (PD). Especially in MS, RGMa inhibition is considered one of the most promising therapeutic strategies (Demicheva et al., 2015; Muramatsu et al., 2011; Tanabe, Fujita, Ikuma, & Yamashita, 2018; Tanabe & Yamashita, 2014). Using an experimental autoimmune encephalomyelitis (EAE) model, Muramatsu et al. showed that anti-RGMa antibody treatment attenuates the symptoms of EAE mice by modulating T cell responses. Particularly impressive is the effect of anti-RGMa treatment on peripheral blood mononuclear cells (PBMCs) obtained from MS patients; this treatment leads to a reduction in their proliferative responses and pro-inflammatory cytokine production (Muramatsu et al., 2011). In another study, RGMa was found to be highly expressed by Th17 cells and mediated neurodegeneration through direct interaction with its receptor, neogenin, which is expressed by neurons (Tanabe & Yamashita, 2014). Demicheva et al. performed histological analysis of human autopsied MS brain tissue and revealed that RGMa is expressed in MS lesions and by infiltrating lymphocytes. To test neural regeneration capacity, they employed a rat targeted-EAE model (established by Kerschensteiner et al., 2004) involving the spinal cord and optic nerve and demonstrated that anti-RGMa treatment promotes both functional recovery and regenerative axonal growth, accompanied by enhanced remyelination and decreased activation of microglia/macrophages (Demicheva et al., 2015). Taken together, anti-RGMa antibody is expected to attenuate MS symptoms by modulating immune responses and axonal regeneration.

The effects of anti-RGMa antibody treatment in NMO were also reported. Harada et al. established a localized NMO model by injecting NMO-IgG obtained from AQP4-Ab–positive NMO patients, into the thoracic spinal cord of rats. After NMO-IgG injection, animals developed motor dysfunction in the tail and hind-limbs, accompanied by the loss of astrocytes in lesion areas, and systemic injection of anti-RGMa antibody effectively attenuated motor deficit, microglial activation, and IL-17A+ T-cell infiltration (Harada et al., 2018).

Several lines of evidence also indicate that RGMa/neogenin pathway inhibition is also a promising strategy to treat patients with cerebral infarction. Based on the observation that RGMa is upregulated in the infarct and peri-infarct (penumbral) lesions of the brain in patients suffering from cerebral infarction (Schwab, Monnier, et al., 2005), Shabanzadeh et al. tested the effect of anti-RGMa antibody in a middle cerebral artery occlusion (MCAO) model in rats and confirmed that such treatment effectively attenuated the infarct area and improved the functional outcomes (Shabanzadeh et al., 2015). RGMa is also suggested to be involved in the blood–brain barrier (BBB) dysfunction (Li et al., 2018), angiogenesis (Wang et al., 2018), and glial scar formation (Zhang et al., 2018) after stroke (MCAO). Recently, Yao et al. revealed that astrocytic neogenin is involved in blood vessel homeostasis. Astrocyte-specific neogenin depletion leads to an increase in the number of blood vessels with leaky BBB in the mouse cortex. Moreover, they showed that neogenin knock-out impairs BMP2-induced netrin-1 expression in astrocytes, which causes a dysregulation of blood vessel homeostasis and function (Yao et al., 2020). Though RGMa involvement was not tested in this study, RGMa may play a direct role in BBB maintenance.

RGMa is also suggested to be involved in PD. Based on the observation that RGMa is upregulated in the substantia nigra (SN) of human PD brains (Bossers et al., 2009), Korecka et al. addressed whether RGMa affects the dopaminergic system, and revealed that AAV-mediated overexpression of RGMa in murine SN results in selective degeneration of dopaminergic neurons with strong microglial activation (Korecka et al., 2017). These results indicate that upregulated RGMa in the SN of PD patients negatively affects midbrain dopaminergic neurons, possibly by inducing microglial accumulation and activation. Indeed, by using 1-methyl-4-phenyl-1,2,3,6-tetrahydropyridine (MPTP) mouse model, Oda et al. showed that the loss of dopaminergic neurons and accumulation of microglia/macrophages in the SN were attenuated by anti-RGMa antibody treatment (Oda et al., 2021).

In summary, RGMa has been shown to regulate a wide range of pathological events, including immune responses, neuronal survival, and BBB maintenance, following CNS damage. Because SCI also shares these pathological conditions, anti-RGMa antibody treatment might exert its therapeutic effects by modulating various pathological processes involved in SCI (Table 2).

TABLE 2 Therapeutic potential of RGMa inhibition in other CNS disorders.

| | Author (Year) | Animal model | RGMa inhibition | Outcome / Suggested Mechanism |
|---|---|---|---|---|
| Multiple Sclelosis | Muramatsu et al. (2011) | EAE (mouse) | anti RGMa-C antibody (i.t.) | Clinical score ↓, Inflammatory-cell infiltration ↓ / T cell response ↓ |
| | Tanabe et al. (2014) | Th17-induced EAE (mouse) | anti RGMa-C antibody (i.p.) | Clinical score ↓ / Th17 cell-mediated neurodegeneration ↓ |
| | Demicheva et al. (2015) | targeted-EAE (rat) | anti human RGMa-N antibody (i.v.) | Clinical score ↓ / Axonal regeneration ↑ |
| | Tanabe et al. (2018) | NOD-EAE (mouse) as SPMS model | anti human RGMa-C antibody (i.v.) | Secondary progression (clinical score) ↓ / Demyelination ↓, Axonal degeneration ↓ |
| Neuromyelitis Optica | Harada et al. (2018) | NMO-IgG injection into the spinal cord (rat) | anti human RGMa-C antibody (i.v.) | Clinical score ↓, Astrocyte loss ↓ / Inflammatory-cell infiltration ↓ |
| Cerebral Infarction | Shabanzadeh et al. (2015) | MCAO (rat) | 41g peptide (prevent RGMa-Neogenin interaction), anti RGMa antibody (i.v.) | Motor function ↑ / Infarct volume ↓, Neuronal death ↓ |
| | Li et al. (2018) | MCAO (rat) | sh-RGMa (adenovirus) | Motor function ↑ / Infarct volume ↓, BBB integrity ↑ |
| | Zhang et al. (2018) | MCAO (rat) | 6FNIII (RGMa function-blocking peptide) sh-RGMa (adenovirus) (i.c.v.) | Motor function ↑ / TGFβ1 dependent reactive astrogliosis & glial scar formation ↓ |
| Parkinson's disease | Korecka et al. (2017) | AAV-mediated RGMa overexpression in SN (mouse) | - | dopaminergic neuron loss, motor impairment |
| | Oda et al. (2021) | MPTP model (mouse) | anti human RGMa-C antibody (i.v.) | dopaminergic neuron loss ↓, microglia/macrophages activation ↓ |

EAE, experimental autoimmune encephalomyelitis; NOD-EAE mouse, experimental autoimmune encephalomyelitis in non-obese diabetic mouse; SPMS, secondary progressive multiple sclerosis; MCAO, middle cerebral artery occlusion; SN, substantia nigra; MPTP, 1-methyl-4-phenyl-1,2,3,6-tetrahydropyridine; i.t., intrathecal administration; i.p., intraperitoneal administration; i.v., intravenous administration; i.c.v., intracerebroventricular administration.

## Conclusions

Over the past decade, experimental studies have greatly enhanced our understanding of neural circuit reorganization after SCI, and various neuro-restorative therapies have been proposed. In particular, a great deal of effort has been invested to identify a biological agent that can modulate the inhibitory function in axonal regeneration/reorganization, with anti-RGMa-neutralizing antibody proving to be one of the most promising candidates with great success in pre-clinical studies. Indeed, various forms of neuroplasticity, including axonal regeneration and neural circuit reorganization, have been shown to be promoted by RGMa inhibition. Moreover, anti-RGMa antibody appears to exert multiple mechanisms of action in SCI, in addition to axonal regeneration, improving neuronal viability, modulating inflammatory responses, or regulating BBB dysfunction. Hopefully, successful clinical trials of this unique therapeutic antibody will open a new window for SCI therapeutics.

## Applications to other areas of neuroscience

RGMa has been shown to regulate a wide range of pathological events, including axonal plasticity, immune responses, neuronal survival, and the blood–brain barrier (BBB) maintenance, following central nervous system (CNS) damage. Because most neurological disorders share these pathological conditions to some extent, anti-RGMa antibody treatment might be beneficial for many kinds of CNS diseases.

Indeed, in preclinical studies, RGMa is suggested to be critically involved in multiple sclerosis (MS), stroke, traumatic brain injury (TBI), neuromyelitis optica (NMO), and Parkinson's disease.

In particular, in MS, RGMa inhibition is considered one of the most promising therapeutic strategies. In experimental autoimmune encephalomyelitis (EAE) model mice, anti-RGMa antibody treatment attenuates neurological symptoms by modulating T cell responses. RGMa has also been suggested to be involved in Th17-cell–induced neurodegeneration in EAE.

Inhibition of the RGMa/neogenin pathway has also been proposed as a promising strategy to treat patients with cerebral infarction. RGMa is upregulated in the infarct and peri-infarct lesions, and anti-RGMa antibody treatment improves the functional outcomes in middle cerebral artery occlusion (MCAO) model rats. In the MCAO model, involvement of RGMa in BBB dysfunction, angiogenesis, and glial scar formation has been suggested.

In the main text, we discussed the applications of anti-RGMa treatment for other neurological disorders in greater detail.

## Mini-dictionary of terms

Retinotectal projection: This projection is organized in a retinotopic manner, and is frequently used to study the molecular mechanisms underlying neural circuit formation based on chemoaffinity theory.

RhoA: A member of the Rho family of GTPases. Rho proteins play a critical role in actin cytoskeletal reorganization and are thus important for cell motility. Rho activity is tightly regulated by local guanine nucleotide exchange factors (GEFs).

Dorsal transection: One of the major SCI models in rodents. The main CST of rodents descends the dorsal funiculus; thus, severe injury to this area leads to complete loss of the main CST on both sides.

Targeted EAE: As the conventional EAE model shows disseminated inflammatory lesions, testing the relationship between functional deficits and structural damage is difficult. To overcome this limitation, a localized EAE model was generated.

BMP: BMPs are a group of morphogens that belong to the transforming growth factor-β (TGF-β) superfamily, and have diverse functions in the processes of embryogenesis, development, and adult tissue homeostasis.

## Key facts of axonal sprouting

CST axons have the capacity to spontaneously sprout after SCI.

Axonal sprouting is one of the mechanisms involved in neural circuit reorganization.

Axonal sprouting occurs both rostrally and caudally to the lesion and occurs from both injured and intact CST axons.

Instead of long-range regeneration, CST axons sprout over short distances and establish synaptic connections with descending propriospinal neurons.

Such newly formed connections can serve as bypass circuits and contribute to functional recovery.

Rehabilitative training or neutralization of neurite outgrowth inhibitory proteins such as Nogo-A and RGMa are thought to enhance axonal sprouting.

## Summary points

- RGM was identified as an axon guidance molecule of the retinotectal system.
- RGM induces growth cone collapse and neurite growth inhibition.
- RGM exerts its neurite growth inhibitory effect by interacting with its receptor, neogenin.
- RGMa is upregulated in the injured spinal cord after SCI in rodents, monkeys, and humans.
- In rodents, anti-RGMa-neutralizing antibody treatment enhances axonal regeneration/reorganization and functional recovery after SCI.
- In monkeys, RGMa-neutralizing antibody treatment enhances neural circuit reorganization and functional recovery of hand dexterity.
- RGMa is also involved in the regulation of the immune system, gliosis, blood–brain barrier integrity, and angiogenesis.
- Two lines of clinical trials using antibodies against RGMa are currently underway.

## References

Babitt, J. L., Huang, F. W., Wrighting, D. M., Xia, Y., Sidis, Y., Samad, T. A., et al. (2006 May). Bone morphogenetic protein signaling by hemojuvelin regulates hepcidin expression. *Nature Genetics*, *38*(5), 531–539.

Babitt, J. L., Zhang, Y., Samad, T. A., Xia, Y., Tang, J., Campagna, J. A., et al. (2005 August 19). Repulsive guidance molecule (RGMa), a DRAGON homologue, is a bone morphogenetic protein co-receptor. *The Journal of Biological Chemistry*, *280*(33), 29820–29827.

Banerjee, P., Harada, H., Tassew, N. G., Charish, J., Goldschneider, D., Wallace, V. A., et al. (2016 March). Υ-secretase and LARG mediate distinct RGMa activities to control appropriate layer targeting within the optic tectum. *Cell Death and Differentiation*, *23*(3), 442–453.

Bareyre, F. M., Kerschensteiner, M., Raineteau, O., Mettenleiter, T. C., Weinmann, O., & Schwab, M. E. (2004 March). The injured spinal cord spontaneously forms a new intraspinal circuit in adult rats. *Nature Neuroscience*, *7*(3), 269–277.

Basso, D. M., Beattie, M. S., & Bresnahan, J. C. (1995 February). A sensitive and reliable locomotor rating scale for open field testing in rats. *Journal of Neurotrauma*, *12*(1), 1–21.

Bell, C. H., Healey, E., van Erp, S., Bishop, B., Tang, C., Gilbert, R. J. C., et al. (2013 July 5). Structure of the repulsive guidance molecule (RGM)-neogenin signaling hub. *Science*, *341*(6141), 77–80.

Bossers, K., Meerhoff, G., Balesar, R., van Dongen, J. W., Kruse, C. G., Swaab, D. F., et al. (2009 January). Analysis of gene expression in Parkinson's disease: Possible involvement of neurotrophic support and axon guidance in dopaminergic cell death. *Brain Pathology*, *19*(1), 91–107.

Chen, J., & Shifman, M. I. (2019 June 1). Inhibition of neogenin promotes neuronal survival and improved behavior recovery after spinal cord injury. *Neuroscience*, *408*, 430–447.

Demicheva, E., Cui, Y. F., Bardwell, P., Barghorn, S., Kron, M., Meyer, A. H., et al. (2015 March 24). Targeting repulsive guidance molecule A to promote regeneration and neuroprotection in multiple sclerosis. *Cell Reports*, *10*(11), 1887–1898.

Elowe, S., Holland, S. J., Kulkarni, S., & Pawson, T. (2001 November). Downregulation of the Ras-mitogen-activated protein kinase pathway by the EphB2 receptor tyrosine kinase is required for ephrin-induced neurite retraction. *Molecular and Cellular Biology*, *21*(21), 7429–7441.

Endo, M., & Yamashita, T. (2009 May 20). Inactivation of Ras by p120GAP via focal adhesion kinase dephosphorylation mediates RGMa-induced growth cone collapse. *The Journal of Neuroscience*, *29*(20), 6649–6662.

Fink, K. L., & Cafferty, W. B. (2016 April). Reorganization of intact descending motor circuits to replace lost connections after injury. *Neurotherapeutics*, *13*(2), 370–381.

Freund, P., Schmidlin, E., Wannier, T., Bloch, J., Mir, A., Schwab, M. E., et al. (2009 March). Anti-Nogo-A antibody treatment promotes recovery of manual dexterity after unilateral cervical lesion in adult primates—Re-examination and extension of behavioral data. *The European Journal of Neuroscience*, *29*(5), 983–996.

Harada, K., Fujita, Y., Okuno, T., Tanabe, S., Koyama, Y., Mochizuki, H., et al. (2018 January 8). Inhibition of RGMa alleviates symptoms in a rat model of neuromyelitis optica. *Scientific Reports*, *8*(1), 34.

Hata, K., Fujitani, M., Yasuda, Y., Doya, H., Saito, T., Yamagishi, S., et al. (2006 April 10). RGMa inhibition promotes axonal growth and recovery after spinal cord injury. *The Journal of Cell Biology*, *173*(1), 47–58.

Hata, K., Kaibuchi, K., Inagaki, S., & Yamashita, T. (2009 March 9). Unc5B associates with LARG to mediate the action of repulsive guidance molecule. *The Journal of Cell Biology*, *184*(5), 737–750.

Healey, E. G., Bishop, B., Elegheert, J., Bell, C. H., Padilla-Parra, S., & Siebold, C. (2015 June). Repulsive guidance molecule is a structural bridge between neogenin and bone morphogenetic protein. *Nature Structural & Molecular Biology*, *22*(6), 458–465.

Itokazu, T., Fujita, Y., Takahashi, R., & Yamashita, T. (2012). Identification of the neogenin-binding site on the repulsive guidance molecule A. *PLoS One*, *7*(3), e32791.

Kerschensteiner, M., Stadelmann, C., Buddeberg, B. S., Merkler, D., Bareyre, F. M., Anthony, D. C., et al. (2004 April). Targeting experimental autoimmune encephalomyelitis lesions to a predetermined axonal tract system allows for refined behavioral testing in an animal model of multiple sclerosis. *The American Journal of Pathology*, *164*(4), 1455–1469.

Korecka, J. A., Moloney, E. B., Eggers, R., Hobo, B., Scheffer, S., Ras-Verloop, N., et al. (2017 September 27). Repulsive guidance molecule a (RGMa) induces neuropathological and behavioral changes that closely resemble Parkinson's disease. *The Journal of Neuroscience*, *37*(39), 9361–9379.

Körner, A., Schlegel, M., Kaussen, T., Gudernatsch, V., Hansmann, G., Schumacher, T., et al. (2019 February 7). Sympathetic nervous system controls resolution of inflammation via regulation of repulsive guidance molecule A. *Nature Communications*, *10*(1), 633.

Kyoto, A., Hata, K., & Yamashita, T. (2007 December). Synapse formation of the cortico-spinal axons is enhanced by RGMa inhibition after spinal cord injury. *Brain Research*, *1186*, 74–86.

Lemon, R. N. (1993 May). The G. L. Brown prize lecture. Cortical control of the primate hand. *Experimental Physiology*, *78*(3), 263–301.

Li, M., Wen, Y., Zhang, R., Xie, F., Zhang, G., & Qin, X. (2018 September). Adenoviral vector-induced silencing of RGMa attenuates blood-brain barrier dysfunction in a rat model of MCAO/reperfusion. *Brain Research Bulletin*, *142*, 54–62.

Malinauskas, T., Peer, T. V., Bishop, B., Mueller, T. D., & Siebold, C. (2020 July 7). Repulsive guidance molecules lock growth differentiation factor 5 in an inhibitory complex. *Proceedings of the National Academy of Sciences of the United States of America*, *117*(27), 15620–15631.

Matsunaga, E., Tauszig-Delamasure, S., Monnier, P. P., Mueller, B. K., Strittmatter, S. M., Mehlen, P., et al. (2004 August). RGM and its receptor neogenin regulate neuronal survival. *Nature Cell Biology*, *6*(8), 749–755.

Monnier, P. P., Sierra, A., Macchi, P., Deitinghoff, L., Andersen, J. S., Mann, M., et al. (2002 September 26). RGM is a repulsive guidance molecule for retinal axons. *Nature*, *419*(6905), 392–395.

Mothe, A. J., Coelho, M., Huang, L., Monnier, P. P., Cui, Y. F., Mueller, B. K., et al. (2020 September). Delayed administration of the human anti-RGMa monoclonal antibody elezanumab promotes functional recovery including spontaneous voiding after spinal cord injury in rats. *Neurobiology of Disease*, *143*, 104995.

Mothe, A. J., Tassew, N. G., Shabanzadeh, A. P., Penheiro, R., Vigouroux, R. J., Huang, L., et al. (2017 September 5). RGMa inhibition with human monoclonal antibodies promotes regeneration, plasticity and repair, and attenuates neuropathic pain after spinal cord injury. *Scientific Reports*, *7*(1), 10529.

Muramatsu, R., Kubo, T., Mori, M., Nakamura, Y., Fujita, Y., Akutsu, T., et al. (2011 April). RGMa modulates T cell responses and is involved in autoimmune encephalomyelitis. *Nature Medicine*, *17*(4), 488–494.

Nakagawa, H., Ninomiya, T., Yamashita, T., & Takada, M. (2015 July 1). Reorganization of corticospinal tract fibers after spinal cord injury in adult macaques. *Scientific Reports*, *5*, 11986.

Nakagawa, H., Ninomiya, T., Yamashita, T., & Takada, M. (2019 February 1). Treatment with the neutralizing antibody against repulsive guidance molecule-a promotes recovery from impaired manual dexterity in a primate model of spinal cord injury. *Cerebral Cortex*, *29*(2), 561–572.

Oda, W., Fujita, Y., Baba, K., Mochizuki, H., Niwa, H., & Yamashita, T. (2021 February 15). Inhibition of repulsive guidance molecule-A protects dopaminergic neurons in a mouse model of Parkinson's disease. *Cell Death & Disease*, *12*(2), 181.

Oudega, M., & Perez, M. A. (2012 August 15). Corticospinal reorganization after spinal cord injury. *The Journal of Physiology*, *590*(16), 3647–3663.

Rajagopalan, S., Deitinghoff, L., Davis, D., Conrad, S., Skutella, T., Chedotal, A., et al. (2004 August). Neogenin mediates the action of repulsive guidance molecule. *Nature Cell Biology*, *6*(8), 756–762.

Schwab, J. M., Conrad, S., Monnier, P. P., Julien, S., Mueller, B. K., & Schluesener, H. J. (2005 March). Spinal cord injury-induced lesional expression of the repulsive guidance molecule (RGM). *The European Journal of Neuroscience*, *21*(6), 1569 1576.

Schwab, J. M., Monnier, P. P., Schluesener, H. J., Conrad, S., Beschorner, R., Chen, L., et al. (2005 October). Central nervous system injury-induced repulsive guidance molecule expression in the adult human brain. *Archives of Neurology*, *62*(10), 1561–1568.

Shabanzadeh, A. P., Tassew, N. G., Szydlowska, K., Tymianski, M., Banerjee, P., Vigouroux, R. J., et al. (2015 May 7). Uncoupling neogenin association with lipid rafts promotes neuronal survival and functional recovery after stroke. *Cell Death & Disease*, *6*(5), e1744.

Siebold, C., Yamashita, T., Monnier, P. P., Mueller, B. K., & Pasterkamp, R. J. (2017 May). RGMs: Structural insights, molecular regulation, and downstream signaling. *Trends in Cell Biology*, *27*(5), 365–378.

Tanabe, S., Fujita, Y., Ikuma, K., & Yamashita, T. (2018 October 17). Inhibiting repulsive guidance molecule-A suppresses secondary progression in mouse models of multiple sclerosis. *Cell Death & Disease*, *9*(11), 1061.

Tanabe, S., & Yamashita, T. (2014 November 20). Repulsive guidance molecule-A is involved in Th17-cell-induced neurodegeneration in autoimmune encephalomyelitis. *Cell Reports*, *9*(4), 1459–1470.

Tassew, N. G., Charish, J., Seidah, N. G., & Monnier, P. P. (2012 February 14). SKI-1 and Furin generate multiple RGMa fragments that regulate axonal growth. *Developmental Cell*, *22*(2), 391–402.

Tassew, N. G., Mothe, A. J., Shabanzadeh, A. P., Banerjee, P., Koeberle, P. D., Bremner, R., et al. (2014 August 21). Modifying lipid rafts promotes regeneration and functional recovery. *Cell Reports*, *8*(4), 1146–1159.

van Erp, S., van den Heuvel, D. M. A., Fujita, Y., Robinson, R. A., Hellemons, A. J. C. G. M., Adolfs, Y., et al. (2015 December 7). Lrig2 negatively regulates ectodomain shedding of axon guidance receptors by ADAM proteases. *Developmental Cell*, *35*(5), 537–552.

Walter, J., Kern-Veits, B., Huf, J., Stolze, B., & Bonhoeffer, F. (1987 December). Recognition of position-specific properties of tectal cell membranes by retinal axons in vitro. *Development*, *101*(4), 685–696.

Wang, Y., Zhang, R., Xing, X., Guo, J., Xie, F., Zhang, G., et al. (2018 January 1). Repulsive guidance molecule a suppresses angiogenesis after ischemia/reperfusion injury of middle cerebral artery occlusion in rats. *Neuroscience Letters*, *662*, 318–323.

Yang, W., & Sun, P. (2020 September 18). Promoting functions of microRNA-29a/199B in neurological recovery in rats with spinal cord injury through inhibition of the RGMA/STAT3 axis. *Journal of Orthopaedic Surgery and Research*, *15*(1), 427.

Yao, L. L., Hu, J. X., Li, Q., Lee, D., Ren, X., Zhang, J. S., et al. (2020 November). Astrocytic neogenin/netrin-1 pathway promotes blood vessel homeostasis and function in mouse cortex. *The Journal of Clinical Investigation*, *3*, 132372.

Zhang, R., Wu, Y., Xie, F., Zhong, Y., Wang, Y., Xu, M., et al. (2018 August). RGMa mediates reactive astrogliosis and glial scar formation through TGFβ1/Smad2/3 signaling after stroke. *Cell Death and Differentiation*, *25*(8), 1503–1516.

Zhou, Z., Xie, J., Lee, D., Liu, Y., Jung, J., Zhou, L., et al. (2010 July 20). Neogenin regulation of BMP-induced canonical Smad signaling and endochondral bone formation. *Developmental Cell*, *19*(1), 90–102.

Chapter 28

# Mitochondrial biogenesis for the treatment of spinal cord injury

Epiphani C. Simmons[a,b], Natalie E. Scholpa[a,c], Joshua D. Crossman[a,c], and Rick G. Schnellmann[a,b,c,d,e,f]

[a]*Department of Pharmacology and Toxicology, College of Pharmacy, University of Arizona, Tucson, AZ, United States;* [b]*Department of Neurosciences, College of Medicine, University of Arizona, Tucson, AZ, United States;* [c]*Southern Arizona VA Health Care System, Tucson, AZ, United States;* [d]*College of Pharmacy, University of Arizona, Tucson, AZ, United States;* [e]*Southwest Environmental Health Science Center, University of Arizona, Tucson, AZ, United States;* [f]*Center for Innovation in Brain Science, University of Arizona, Tucson, AZ, United States*

## Abbreviations

| | |
|---|---|
| **5-HTR** | 5-hydroxytryptamine receptor |
| **7,8-DHF** | dihydroxyflavone |
| **Ac** | acetyl group |
| **AICAR** | 5-aminoimidazole-4-carboxamide ribonucleotide |
| **Akt** | protein kinase B |
| **AMPK** | adenosine monophosphate-activated kinase |
| **AP-1** | activating protein 1 |
| **ATP** | adenosine triphosphate |
| **BSCB** | blood–spinal cord barrier |
| **cGMP** | cyclic guanosine monophosphate |
| **CNS** | central nervous system |
| **CO** | carbon monoxide |
| **cyt *c*** | cytochrome *c* |
| **DEX** | dexmedetomidine |
| **DRP1** | dynamin-related protein 1 |
| **EAAT** | excitatory amino acid transporter |
| **eNOS** | endothelial nitric oxide synthase |
| **ETC** | electron transport chain |
| **FDA** | Food and Drug Administration |
| **FIS1** | fission 1 |
| **Glu-R** | glutamate receptor |
| **GPCR** | G protein-coupled receptor |
| **HO-1** | heme oxygenase 1 |
| **IL-6** | interleukin 6 |
| **MAPK** | mitogen-activated protein kinase |
| **MB** | mitochondrial biogenesis |
| **Mdivi-1** | mitochondrial division inhibitor 1 |
| **Mfn** | mitofusin |
| **Mn** | manganese |
| **mPTP** | mitochondrial permeability transition pore |
| **mtDNA** | mitochondrial DNA |
| **Nrf** | nuclear respiratory factor |
| **NSC** | neuronal stem cell |
| **OPA1** | optic atrophy 1 |
| **OXPHOS** | oxidative phosphorylation |
| **P** | phosphate group |
| **PDE** | phosphodiesterase |

Diagnosis and Treatment of Spinal Cord Injury. https://doi.org/10.1016/B978-0-12-822498-4.00028-2

| | |
|---|---|
| **PGC-1α** | peroxisome proliferator-activated receptor-gamma coactivator 1 alpha |
| **PI3K** | phosphoinositide-3 kinase |
| **PINK1** | PTEN-induced kinase 1 |
| **PPAR** | peroxisome proliferator-activated receptor |
| **ROS** | reactive oxygen species |
| **SCI** | spinal cord injury |
| **SIRT1** | sirtuin-1 |
| **TFAM** | mitochondrial transcription factor A |
| **TMP** | tetramethylpyrazine |
| **TNFα** | tumor necrosis factor alpha |
| **TZDs** | thiazolidinediones |
| **U** | ubiquitin group |
| **β-AR** | β-adrenergic receptor |

## Introduction

Neuronal integrity is dependent on mitochondrial homeostasis and function, resulting in the central nervous system (CNS) being particularly sensitive to mitochondrial dysfunction (Golpich et al., 2017). Spinal cord injury (SCI) results in an intricate pathology involving heterogeneous cell types with unique roles in injury and recovery. Following SCI, mitochondria are dysfunctional, resulting in an array of consequences including decreased mitochondrial respiration and adenosine triphosphate (ATP) production, depolarization of the mitochondrial membrane, mitochondrial DNA (mtDNA) fragmentation, oxidative stress, compromised calcium homeostasis, altered mitochondrial dynamics, and cell death (Simmons, Scholpa, & Schnellmann, 2020). These cellular dysfunctions contribute to the secondary injury cascade of SCI, exacerbating injury, and hindering recovery.

Mitochondrial biogenesis (MB) is an intricate process involving the generation of new, functional mitochondria (Whitaker, Corum, Beeson, & Schnellmann, 2016). In recent years, there has been an increase in published data documenting that pharmacological induction of MB restores mitochondrial and organ function following various pathological events in vivo, including SCI (Simmons, Scholpa, & Schnellmann, 2020). These findings, in conjunction with the plethora of evidence indicating that restoring mitochondrial homeostasis promotes recovery following SCI, speak to the potential for this treatment strategy. Additionally, in vitro reports have elucidated cell type-specific consequences of mitochondrial dysfunction and MB in SCI-relevant cell types, which could aid in the future development of targeted therapies.

Importantly, there exists an increasing number of U.S. Food and Drug Administration (FDA)-approved pharmaceuticals capable of MB induction, demonstrating the clinical applicability of this approach (Simmons, Scholpa, & Schnellmann, 2020). This perspective will briefly review and explore CNS cell type-specific mitochondrial dysfunction and MB, as well as describe current therapeutic strategies employing inducers of MB post-SCI.

## Mitochondrial dysfunction

Mitochondrial dysfunction results from alterations to homeostatic mitochondrial processes, leading to deficient energy metabolism, primarily through decreased oxidative phosphorylation (OXPHOS) and ATP synthesis. Examples of such alterations include inadequate mitochondrial number and/or mass, altered membrane potential, mitochondrial DNA (mtDNA) mutation/fragmentation, defective electron transport chain (ETC) activity, increased production of reactive oxygen species (ROS), intracellular calcium dysregulation, and impaired mitochondrial dynamics and mitophagy (Scholpa & Schnellmann, 2017; Simmons, Scholpa, & Schnellmann, 2020).

Due to high energy demand within the CNS, mitochondrial dysfunction and ensuing loss of ATP can prevent the function of various ATPases ($H^+$, $Ca^{2+}$, $Na^+/K^+$-ATPase) required for effective neurotransmission, thereby deregulating cellular ion gradients. Mitochondrial dysfunction can also disrupt calcium buffering and signaling, which is crucial for neuronal synapses, leading to calcium overload and excitotoxicity (Vos, Lauwers, & Verstreken, 2010). In SCI pathology, activated astrocytes and glial cells release pro-inflammatory cytokines, resulting in mitochondrial dysfunction and ultimately apoptosis (Simmons, Scholpa, & Schnellmann, 2020). Additionally, SCI is characterized by vasculature disruption, leading to loss of blood flow and local ischemia, which contributes to oxidative stress, mitochondrial dysfunction and the propagation of cell death observed during secondary injury (Fig. 1) (Scholpa & Schnellmann, 2017). Therefore, evidence suggests that restoring mitochondrial function could be an effective strategy to mediate multiple facets of injury progression and aid in preventing further cell death.

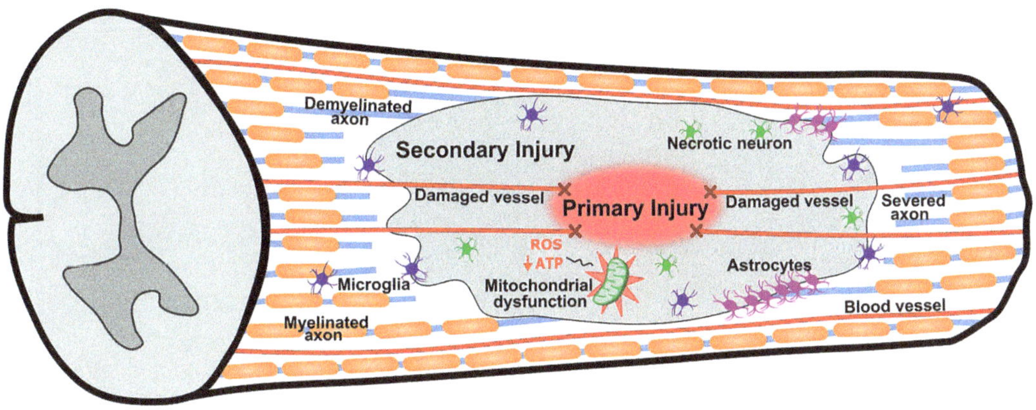


**FIG. 1** Spinal cord injury pathology. Damage after spinal cord injury (SCI) is a combination of initial trauma and secondary injury. The primary injury disrupts spinal cord vasculature, reducing local oxygen delivery, decreasing mitochondrial function and adenosine triphosphate (ATP) synthesis, and increasing reactive oxygen species (ROS) production. Additional hallmarks of secondary injury after SCI include neuronal cell death, axon severing and demyelination, microglia activation and glial scar formation. *Duplicated with permission from Scholpa, N. E. & Schnellmann, R. G. (2017). Mitochondrial-based therapeutics for the treatment of spinal cord injury: Mitochondrial biogenesis as a potential pharmacological target.* The Journal of Pharmacology and Experimental Therapeutics, 363, *303–313.*

## Mitochondrial biogenesis

Although many studies exist assessing mitochondria-targeted strategies for treatment of SCI, the majority address singular aspects of downstream mitochondrial dysfunction, such as increasing antioxidant defenses or inhibiting opening of the mitochondrial permeability transition pore (mPTP) (Scholpa & Schnellmann, 2017). In contrast, MB is the production of new, functional mitochondria via the growth and division of pre-existing mitochondria, which could therefore address multiple, if not all, facets of mitochondrial dysfunction (Fig. 2). This complex process involves the cooperation of multiple cellular pathways, requiring the synthesis of mtDNA, transcription and translation of mitochondrial- and nuclear-encoded proteins and ultimately assembly of ETC complexes (Whitaker et al., 2016).

### Regulation of MB

Coordination of the nuclear and mitochondrial genomes is modulated by transcriptional coactivators, with the most relevant to MB being the "master regulator" peroxisome proliferator-activated receptor-gamma coactivator-1 alpha (PGC-1$\alpha$) (Liang & Ward, 2006). Activation of PGC-1$\alpha$ can be initiated via multiple post-translational modifications, including deacetylation by sirtuin-1 (SIRT1) and phosphorylation by adenosine monophosphate-activated kinase (AMPK) and p38-mitogen-activated protein kinase (MAPK) (Ventura-Clapier, Garnier, & Veksler, 2008). Additionally, agonism of G-protein-coupled receptors (GPCRs), such as 5-hydroxytryptamine (5-HTRs) and $\beta$-adrenergic receptors ($\beta$-ARs), can activate the protein kinase B (Akt)/endothelial nitric oxide synthase (eNOS)/cyclic guanosine monophosphate (cGMP) pathway leading to activation of PGC-1$\alpha$, translocation to the nucleus and, subsequently, induction of MB (Fig. 3) (Garrett, Whitaker, Beeson, & Schnellmann, 2014; Gibbs, Garrett, Beeson, & Schnellmann, 2018; Wills et al., 2012). Importantly, studies have shown that PGC-1$\alpha$ is decreased in the spinal cord after SCI, indicative of disrupted MB (Hu, Lang, Cao, Zhang, & Lu, 2015; Scholpa, Williams, et al., 2019).

PGC-1$\alpha$ regulates MB through interactions with transcription factors such as peroxisome proliferator-activated receptors (PPARs) and nuclear respiratory factors (Nrf1/2), among others (Liang & Ward, 2006; Schreiber et al., 2004). PPARs are involved in the expression of fatty acid oxidation and Krebs cycle enzymes and OXPHOS components (Lin, Handschin, & Spiegelman, 2005; Schreiber et al., 2004). Activation of Nrf1/2 induces transcription of mitochondrial transcription factor A (TFAM) (Gureev, Shaforostova, & Popov, 2019), which translocates to the mitochondria, where it activates mitochondrial gene expression and mtDNA replication (Virbasius & Scarpulla, 1994). Coordination of these transcription factors via PGC-1$\alpha$ culminates in the induction of MB.

Intimately related to MB is mitochondrial dynamics, namely fusion and fission. Fusion is the joining of two mitochondria mediated by mitofusins (Mfn1/2) and optic atrophy 1 (OPA1), whereas fission initiates cleavage and division of mitochondria, and is mediated, in part, by dynamin-related protein 1 (DRP1) and outer membrane receptor fission 1 (FIS1). Dysfunctional mitochondria contain impaired proteins, damaged membranes and fragmented mtDNA, increasing

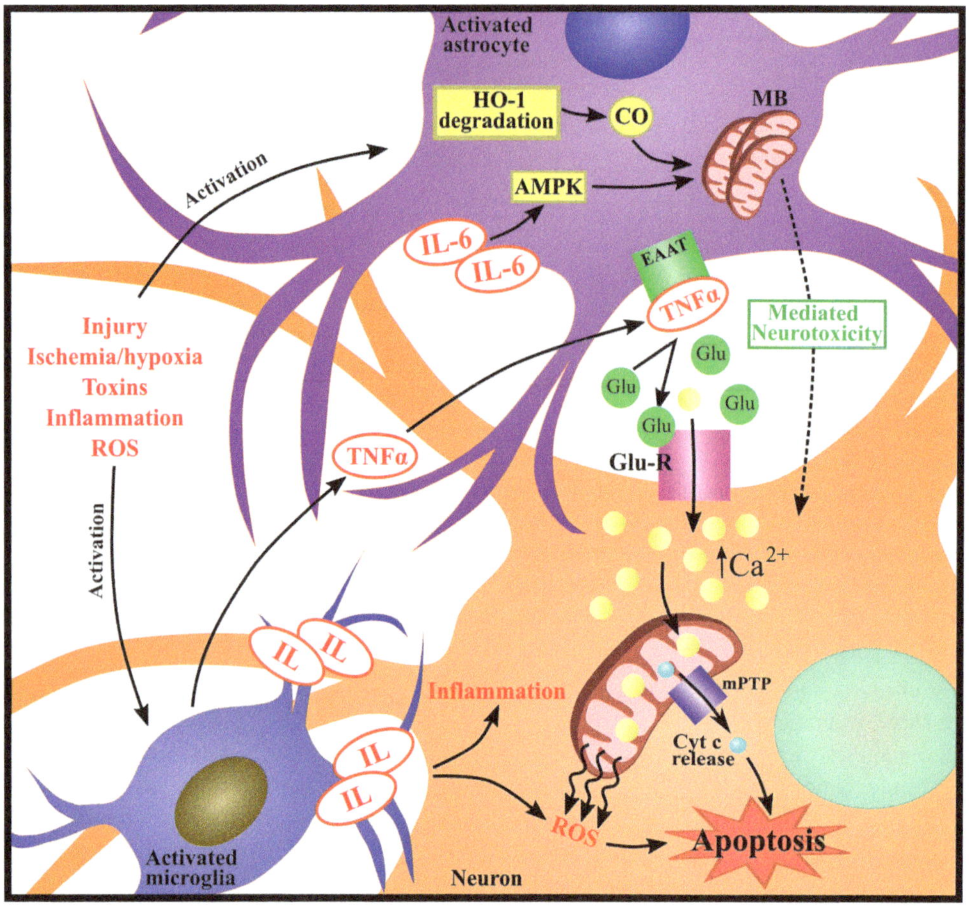


FIG. 2 Role of neuroinflammation on mitochondrial dysfunction. Various stressors can trigger activation of astrocytes and microglia, releasing cytokines and chemokines, promoting calcium overload, apoptosis and neurotoxicity. Conversely, ischemic injury increases astrocyte CO levels via HO-1 degradation. CO as well as IL-6 can promote MB and mediate neurotoxicity. AMPK, adenosine monophosphate-activated kinase; $Ca^{+}$, calcium; Cyt *c*, cytochrome *c*; EAAT, excitatory amino acid transporter; Glu, glutamate; Glu-R, glutamate receptor; HO-1, heme oxygenase 1; IL, interleukins; IL-6, interleukin-6; mPTP, mitochondrial permeability transition pore; ROS, reactive oxygen species; TNFα, tumor necrosis factor alpha. *Modified with permission from Simmons, E. C., Scholpa, N. E., & Schnellmann, R. G. (2020). Mitochondrial biogenesis as a therapeutic target for traumatic and neurodegenerative CNS diseases.* Experimental Neurology, 329, *113309.*

fission mechanisms and mitophagy, which is the selective degradation of damaged mitochondria by autophagosomes mediated, in part, by PTEN-induced kinase 1 (PINK1) and E3-ubiquitin ligase (Parkin) interaction (Fig. 4) (Simmons, Scholpa, & Schnellmann, 2020). Conversely, promotion of fusion mechanisms has been implicated in MB and recovery (Scholpa & Schnellmann, 2017). Proper balance of MB, dynamics and mitophagy is critical for mitochondrial function and response to various stressors, including SCI.

## Cell type-specific mitochondrial dysfunction and MB

SCI is a complex pathology involving heterogeneous cell types with distinct roles in the progression of injury and recovery. Many in vivo studies exist evaluating the global effects of mitochondrial-based therapies for SCI, while few reports exist detailing cell type-specific effects. Understanding the role of mitochondrial dysfunction and MB in relevant cell types can uncover not only underexplored mechanisms, but also potential therapeutic strategies following SCI.

### Neurons

In addition to their primary role in bioenergetics, neuronal mitochondria contribute to regulation of synaptic transmission, calcium homeostasis, neuronal excitability and response to stress (Fanibunda et al., 2019). Neurons are dependent on efficient ATP regulation and have limited capacity to buffer oxidative stress. As such, neurons are particularly vulnerable to

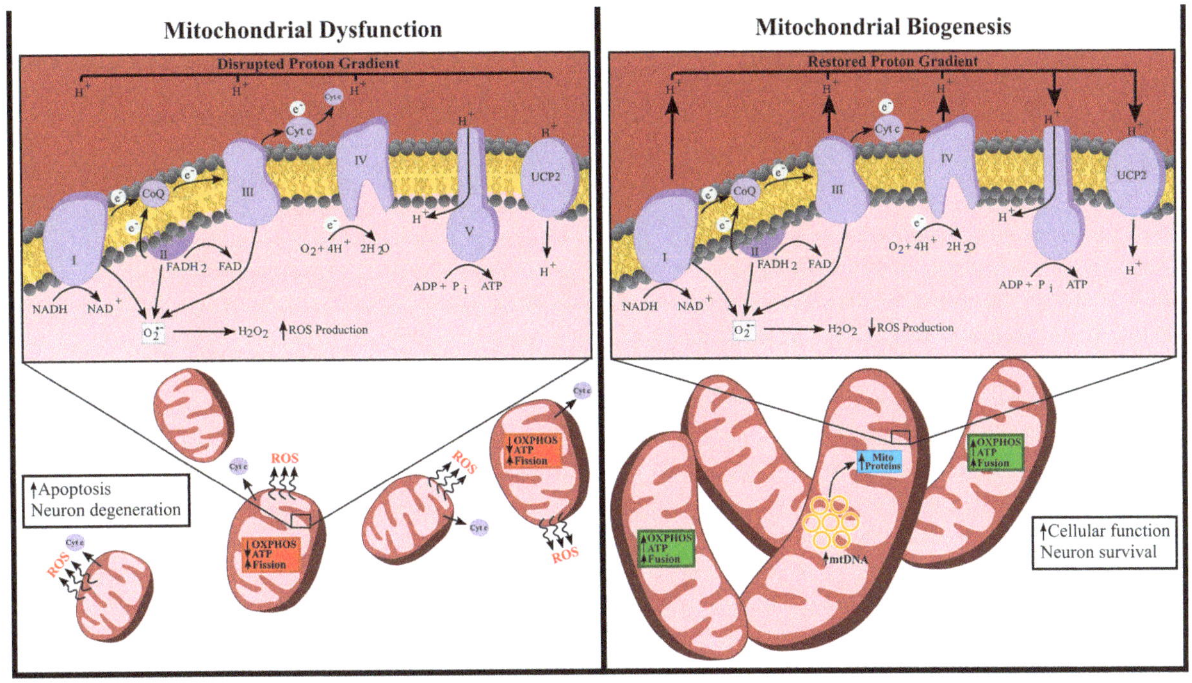


FIG. 3 Mitochondrial dysfunction and biogenesis. Mitochondrial dysfunction is characterized by decreased expression of OXPHOS proteins, impaired mitochondrial membrane potential, reduced ATP production, enhanced mitochondrial fission, ROS production and cyt *c* release. Mitochondrial dysfunction is often paired with increased apoptosis and neural degeneration. Conversely, mitochondrial biogenesis is characterized by mitochondrial fusion, reduced ROS, restoration of membrane potential and increased expression of OXPHOS proteins. CoQ, coenzyme Q; COX1, cytochrome *c* oxidase subunit 1; Cyt *c*, cytochrome *c*; $e^-$, electron; FAD, Flavin adenine dinucleotide; NAD$^+$, Nicotinamide adenine dinucleotide; $O_2^-$, superoxide; OXPHOS, oxidative phosphorylation; $P_i$, inorganic phosphate; ROS, reactive oxygen species; UCP2, uncoupling protein 2. *Modified with permission from Simmons, E. C., Scholpa, N. E., & Schnellmann, R. G. (2020). Mitochondrial biogenesis as a therapeutic target for traumatic and neurodegenerative CNS diseases.* Experimental Neurology, 329, *113309.*

mitochondrial dysfunction, resulting in excitotoxicity, calcium overload, axon demyelination and cell death (Scholpa & Schnellmann, 2017).

In the presence of mitochondrial dysfunction in vitro, MB compounds can improve mitochondrial homeostasis and neuronal survival. For example, mitochondrial division inhibitor-1 (Mdivi-1), a DRP1 inhibitor and inducer of MB, improved mitochondrial function and neural conductance in cultured spinal cord neurons following glutamate-induced ischemia-reperfusion. Treatment also reduced oxidative stress markers, neuronal injury and apoptosis in this model (Liu et al., 2015). Serotonin, a neurotransmitter implicated in mitochondrial and functional recovery from SCI, enhanced mtDNA content, mitochondrial concentration and mitochondrial function, while demonstrating neuroprotective effects against oxidative stress in cultured rodent cortical neurons (Fanibunda et al., 2019).

MB is also implicated in stem cell differentiation and neural fate commitment. Treatment with bezafibrate, a PPAR agonist, evoked up-regulation of MB-related genes and improved cell viability, mitochondrial membrane potential and cell number during late-stage differentiation in human-induced pluripotent stem cells (Augustyniak et al., 2019). Furthermore, inherent increases in mtDNA content and MB-related gene and protein expression exist during differentiation of human neuronal stem cells (NSCs) into motor neurons (O'brien, Keeney, & Bennett Jr., 2015). NSC transplantation is being developed to promote recovery of the neural network after SCI (Zhu, Uezono, Yasui, & Nakashima, 2018); the reliance of motor neuron differentiation on MB, however, suggests that differentiating NSCs may be vulnerable to mitochondrial dysfunction after injury (O'brien et al., 2015). Therefore, concurrent treatment employing NSCs and MB compounds may enhance successful motor neuron differentiation, thereby improving NSC-induced neurogenesis and recovery post-SCI.

## Astrocytes

Integral within the CNS, astrocytes maintain neuronal energy, metabolism and structural support, modulation of synaptic transmission, regulation of intercellular ion concentration, vasomodulation and promotion of the myelinating activity of

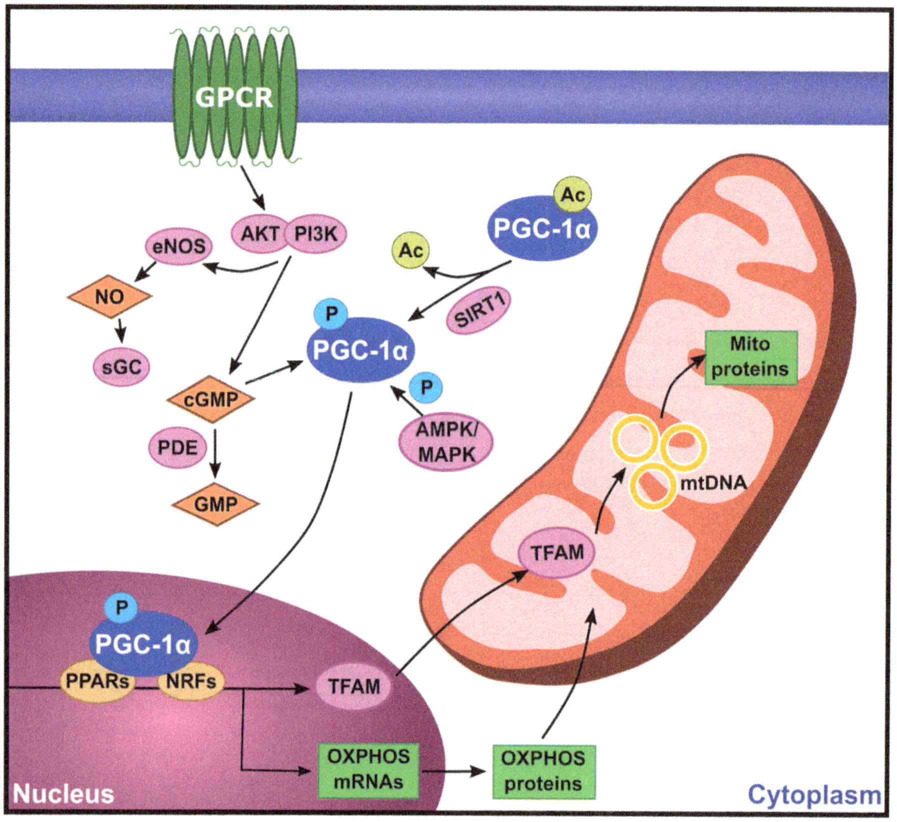


FIG. 4 Regulation of mitochondrial biogenesis. Mitochondrial biogenesis (MB) is a highly regulated process involving diverse pathways. Pharmacological agents can augment MB by targeting different aspects of these pathways, all culminating in increased expression of mitochondrial genes and MB. Ac, acetyl group; AKT, protein kinase B; AMPK, adenosine monophosphate-activated kinase; cGMP, cyclic guanosine monophosphate; eNOS, endothelial nitric oxide synthase; MAPK, mitogen activated protein kinase; mtDNA, mitochondrial DNA; NO, nitric oxide; NRF, nuclear respiratory factor; OXPHOS, oxidative phosphorylation; P, phosphate; PDE, phosphodiesterase; PGC-1α: peroxisome proliferator-activated receptor-gamma coactivator 1 alpha; PI3K, phosphoinositide-3 kinase; PPAR, peroxisome proliferator-activated receptor; sGC, soluble guanylate cyclase; SIRT1, sirtuin 1; TFAM, mitochondrial transcription factor A. *Modified with permission from Scholpa, N. E. & Schnellmann, R. G. (2017). Mitochondrial-based therapeutics for the treatment of spinal cord injury: Mitochondrial biogenesis as a potential pharmacological target.* The Journal of Pharmacology and Experimental Therapeutics, 363, *303–313*.

oligodendrocytes. Following SCI, astrocytes undergo complex morphological, gene expression and functional changes (Falnikar, Li, & Lepore, 2015). Given that defects in astrocyte activity and metabolism are associated with a variety of neuropathological disorders (Choi et al., 2016), maintaining the metabolic activity of astrocytes is likely crucial for neuronal function post-SCI.

Aβ1–42, a toxic peptide aggregate found in Alzheimer's disease pathology, increases MB in astrocytes, while inducing lipid peroxidation, apoptosis and cell death in neurons. Despite these deleterious neuronal effects, increased neuronal survival, restored mitochondrial homeostasis, improved resistance to oxidative damage and modulation of the neuroinflammatory response were documented when neurons and astrocytes were co-cultured (Aguirre-Rueda et al., 2015). The neuroprotective nature of astrocytic MB reported in this study is compelling, revealing a potential therapeutic target following SCI.

Manganese (Mn) targets astrocytes and promotes the expression of pro-inflammatory cytokine interleukin-6 (IL-6) and tumor necrosis factor alpha (TNFα), leading to mitochondrial dysfunction and neurotoxicity in vitro. In mouse cerebral astrocytes, Mn-induced inflammation led to decreased mitochondrial mass, excessive fission, as well as impaired basal and ATP-linked mitochondrial respiration (Sarkar et al., 2018). Interestingly, the antioxidant mito-apocynin attenuated this inflammatory response and restored mitochondrial mass (Sarkar et al., 2018). While Mn toxicity is often observed following occupational or environmental exposure (Harischandra et al., 2019), this impaired astrocytic mitochondrial profile presents similarly to that following SCI (McAvoy & Kawamata, 2019). As such, future studies should investigate the therapeutic potential of mito-apocynin in mediating astrocytic dysfunction following SCI.

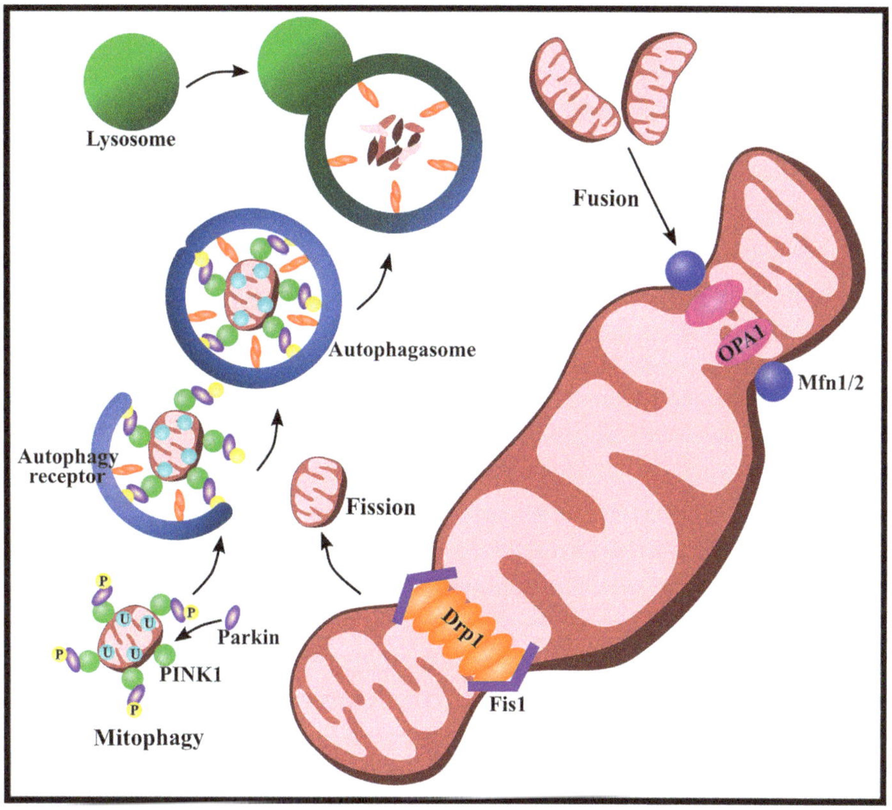


FIG. 5 Mitochondrial dynamics. Mitochondrial fusion is mediated by outer (Mfn1/2) and inner membrane (Opa1) proteins. Fission is initiated by recruitment of DRP1 to the mitochondrial membrane, where it interacts with receptors (Fis1), inducing membrane splitting. Excessive fission may promote mitophagy, which is activated by destabilization of PINK1, which recruits, phosphorylates (P) and activates Parkin, promoting the formation of autophagosomes, which will fuse with a lysosome leading to degradation. Drp1, dynamin-related protein 1; Fis1, fission 1; Mfn1/2, mitofusin 1/2; OPA1, optic atrophy 1; PINK1, PTEN-induced kinase 1. *Duplicated with permission from Simmons, E. C., Scholpa, N. E., & Schnellmann, R. G. (2020). Mitochondrial biogenesis as a therapeutic target for traumatic and neurodegenerative CNS diseases.* Experimental Neurology, 329, *113309*.

In addition to pharmacological induction, internally driven MB can occur as a programmed response to aid in the recovery of astrocytes following oxidative insults (Fig. 5) (Chen et al., 2018; Choi et al., 2016). Oxidant injuries stimulate the expression of heme oxygenase-1 (HO-1) in astrocytes. Carbon monoxide (CO), a by-product of HO-1-driven heme degradation, regulates energy metabolism and can mitigate tissue injury and inflammation in neurological diseases. In addition, HO-1-derived CO increases MB, as evidenced by increased PGC-1α and ATP synthesis in astrocytes (Choi et al., 2016). Therefore, astrocytic HO-1 induction during ischemic events may be endogenously protective via stimulation of MB, thereby hindering propagation of ischemic damage. Furthermore, reports demonstrate that endogenous and exogenous CO can promote MB in cerebral astrocytes in vitro (Piantadosi, Carraway, Babiker, & Suliman, 2008). As such, CO-mediated MB in astrocytes may be protective following SCI.

In cultures of rat cerebral astrocytes, septic conditions induced neuroinflammation and triggered protective mechanisms to increase MB and ATP at early time points. Additionally, treatment with IL-6, a cytokine mainly produced by astrocytes, enhanced MB through IL-6/AMPK signaling under septic conditions (Chen et al., 2018), suggesting that astrocyte-specific MB in the presence of neuroinflammation may intrinsically combat mitochondrial dysfunction and promote recovery. Although increased IL-6 levels following SCI are considered inherently negative, future studies should explore the potential neuroprotective effects of astrocytic IL-6-induced MB following injury.

## Endothelial cells

Endothelial cells are integral components of the blood–spinal cord barrier (BSCB) and are specifically organized to facilitate selective permeability, limiting blood-derived molecules from entering the spinal cord. SCI often results in catastrophic destruction of the BSCB, leading to enhance permeability and allowing neuroinflammation to propagate secondary injury. Conversely, angiogenesis and restoration of BSCB integrity has been associated with improved pathological and behavioral outcomes. Mitochondrial quality control mechanisms, including mitochondrial dynamics and mitophagy, are needed to maintain angiogenic and vasodilator functions of endothelial cells (Kluge, Fetterman, & Vita, 2013). Though mitochondrial content in endothelial cells is relatively modest, angiogenesis is an energy-demanding process essential for SCI recovery and has been closely linked to MB pathways (Kluge et al., 2013; Kumar, Ropper, Lee, & Han, 2017). A comprehensive review detailing key aspects of mitochondrial function in the endothelium can be found in Kluge et al. (2013) and Eelen et al. (2018).

Therapeutic strategies targeting increased activity of PGC-1α and AMPK consistently demonstrate improved endothelial function. The glycolysis inhibitor 2-deoxyglucose activates AMPK, inducing autophagy and protecting against cell death of endothelial cells in vitro (Kluge et al., 2013). In addition, thiazolidinediones reduce ROS while activating PGC-1α and MB in endothelial cells (Kluge et al., 2013). Metformin, an inhibitor of IL-1β and activator of AMPK, impedes mPTP opening and endothelial apoptosis, thereby preventing endothelial dysfunction (Kluge et al., 2013). Resveratrol also induces MB and reduces hydrogen peroxide levels of endothelial cells in culture (Kluge et al., 2013). These studies suggest that MB strategies may increase recovery following SCI, in part by improving mitochondrial function in the endothelium. In support of this, pharmacological activation of MB following SCI has been shown to improve BSCB integrity, correlating with enhanced recovery in vivo (Simmons, Scholpa, Cleveland, & Schnellmann, 2020).

## Mitochondrial biogenesis mediates consequences of SCI

### Mitochondrial homeostasis and functional recovery

PGC-1α is decreased in the injured spinal cord in rodent contusion SCI models (Hu et al., 2015; Hu, Lang, Zhang, Ni, & Lu, 2016; Scholpa, Williams, et al., 2019), indicating loss of mitochondrial function and decreased MB. Lentiviral overexpression of PGC-1α in the injured cord has been shown to decrease neuronal cell death and improve recovery (Hu et al., 2015, 2016). Furthermore, treatment with pharmacological agents that induce PGC-1α expression have similar restorative effects (Hu et al., 2015; Scholpa, Williams, et al., 2019; Simmons, Scholpa, Cleveland, et al., 2020).

Tetramethylpyrazine (TMP), a plant extract, is neuroprotective following SCI in rats via induction of PGC-1α and MB, leading to improved locomotor capability and decreased neural apoptosis (Hu et al., 2015). Treatment with the $\beta_2$-AR agonist formoterol beginning 8 h post-SCI increased PGC-1α and downstream protein expression in the injured cord, improved locomotor function and decreased lesion volume in a mouse contusion SCI model (Scholpa, Simmons, Tilley, & Schnellmann, 2019; Scholpa, Williams, et al., 2019). Daily administration of the 5-$HT_{1F}$ receptor agonist LY344864 after SCI similarly increased PGC-1α expression, as well as improved markers of functional recovery and decreased BSCB leakage (Simmons, Scholpa, Cleveland, et al., 2020). Of note, formoterol is FDA-approved for the treatment of asthma and the 5-$HT_{1F}$ receptor agonist lasmiditan was recently approved for the treatment of migraines (Lamb, 2019), indicating the potential applicability of this SCI treatment strategy.

### Body composition

In addition to local dysfunction in the spinal cord, SCI can decrease peripheral mitochondrial activity, with decreased mitochondrial size, protein expression and function reported in skeletal muscle after injury (Mccully, Mulcahy, Ryan, & Zhao, 2011; Scelsi, Marchetti, Poggi, Lotta, & Lommi, 1982). Paralysis following SCI can lead to disuse atrophy and rapid loss of skeletal muscle function and metabolic activity, which can increase the risk of developing metabolic diseases such as type II diabetes (Gorgey et al., 2019; O'brien & Gorgey, 2016). In fact, while pneumonia and septicemia are the leading causes of death post-SCI, mortality rates for metabolic and musculoskeletal disorders are on the rise (National Spinal Cord Injury Statistical Center, 2020).

Besides its canonical role as an inducer of MB, PGC-1α is also involved in activation of muscle hypertrophy and downregulation of atrophy (Qin, Pan, Wu, Bauman, & Cardozo, 2010; Scholpa, Simmons, et al., 2019). Therefore, induction of PGC-1α and increased MB can address multiple SCI-induced deficits. In support of this, exercise training induces PGC-1α in the skeletal muscle of rodents and humans (Baar et al., 2002; Pilegaard, Saltin, & Neufer, 2003), and increases mitochondrial mass and function in paralyzed muscle (O'brien & Gorgey, 2016). Transgenic overexpression of PGC-1α has also been found to protect against muscle atrophy and promote MB (Dinulovic et al., 2016; Sandri et al., 2006). In addition to improving functional recovery, daily formoterol treatment after SCI increased gastrocnemius muscle mass, corresponding to increased hypertrophy and decreased expression of atrophy markers in mice (Scholpa, Simmons, et al., 2019). Formoterol treatment also restored body weight to that of pre-injury (Scholpa, Simmons, et al., 2019), indicating reestablishment of body composition, which is fundamentally altered following SCI.

### Axonal growth

Limited recovery within the spinal cord post-SCI is largely due to lack of axonal growth (Meves & Zheng, 2014). In addition to axons' poor intrinsic growth capabilities, the presence of growth-inhibiting molecules and lack of growth-promoting factors make the CNS environment unconducive to axonal growth after injury. Unfortunately, while multiple attempts have been made to encourage axonal growth, success has been limited (Meves & Zheng, 2014). Neuronal growth requires a substantial amount of energy to not only synthesize the necessary substrates, but also transport these substrates to distal axonal locations. PGC-1α has been demonstrated to regulate mitochondrial density and energy production in both neuronal cell bodies and axons (Wareski et al., 2009). Overexpression of PGC-1α led to a twofold increase in total number of mitochondria in axons, particularly in the distal end furthest from the cell body, corresponding with axonal growth acceleration (Vaarmann et al., 2016). Additionally, suppression of PGC-1α also suppresses neuronal growth, further solidifying the relationship between axonal growth and MB (Vaarmann et al., 2016).

### Inflammation

Mitochondria are integral to regulating innate immunity and the inflammatory response (Mohanty, Tiwari-Pandey, & Pandey, 2019) and as such, inflammatory disorders are associated with mitochondrial dysfunction (Missiroli et al., 2020). Within 15 min, SCI initiates inflammation, which plays a central role in the regulation of secondary injury (Zhang, Yin, Xu, Wu, & Chen, 2012). Inflammation can lead to neuron and oligodendrocyte apoptosis, formation of the glial scar and ultimately loss of neuronal function (Zhang et al., 2012). Therefore, it is hypothesized that mitigating inflammation could limit secondary injury progression.

During early inflammation, mitochondrial dysfunction is induced by pro-inflammatory cytokines such as TNFα and interleukins (Cherry & Piantadosi, 2015), with the degree of mitochondrial dysfunction being roughly correlated with clinical outcomes and recovery in humans and rodents (Brealey et al., 2004; Cherry & Piantadosi, 2015). Restoration of mitochondrial function and homeostasis is dependent upon rapid induction of quality control mechanisms, such as MB (Cherry & Piantadosi, 2015). In fact, activation of the inflammatory pathway directly impacts mitochondrial function via up-regulation of MB (Fernandez-Marcos & Auwerx, 2011). Inflammatory receptors activate multiple pathways within a cell, including MAPK and Akt pathways, which in turn lead to the activation of activating protein-1 (AP-1) and p38, ultimately increasing PGC-1α (Cherry & Piantadosi, 2015; Fernandez-Marcos & Auwerx, 2011). Additionally, Akt phosphorylates Nrf1, which induces TFAM activation and the transcription of MB genes (Cherry & Piantadosi, 2015).

While long-term respirometric dysfunction is unsuitable for cell survival, early inflammation-induced mitochondrial dysfunction may be advantageous, attenuating ROS production and decreasing ATP availability, preventing the appropriation of host cell energetics and, ultimately, the spread of infection (Cherry & Piantadosi, 2015). Furthermore, macrophages and microglia have been suggested to be neuroprotective after injury, and complete depletion of neutrophils severely impairs functional recovery after SCI (Okada, 2016; Rust & Kaiser, 2017). Additionally, TNFα signaling appears to be necessary for remyelination (Arnett et al., 2001), and certain parts of the glial scar have been shown to support axon regeneration post-SCI (Anderson et al., 2016). Consequently, therapeutic strategies aimed at mitigating inflammation after SCI via MB induction must take into consideration the potential benefit of the inflammatory response.

Importantly, recent evidence indicates that the SCI-induced inflammatory response is not exclusive to the spinal cord, but is systemic, leading to widespread organ damage (Sun et al., 2016). Treatment with the mitochondrial-targeted peptide SS-31 improved mitochondrial function, controlled the inflammatory response and alleviated lung damage in a mouse model of SCI (Zhu, Li, He, Zhou, & Jiang, 2017). Therefore, systemic pharmacological induction of MB has the potential to mitigate widespread inflammation following SCI.

## Limitations of MB

While MB can aid in restoring cellular homeostasis and recovery from injury, limitations do exist. Many disease pathologies, including SCI, present a heterogeneous mix of mitochondrial health, particularly at early stages of disease progression. Although pre-clinical models have consistently demonstrated MB induction can attenuate mitochondrial dysfunction and improve recovery, there exists no way to selectively target healthy mitochondria for replication. As such, MB induction within dysfunctional mitochondria may propagate replication of mutated mtDNA, resulting in increased injury severity and disease progression (Simmons, Scholpa, & Schnellmann, 2020).

## Applications to other areas of neuroscience

Considering central nervous system (CNS) function is highly dependent on mitochondrial homeostasis and energetics, it is unsurprising that brain and spinal cord systems are sensitive to mitochondrial dysfunction (Golpich et al., 2017). Both acute and neurodegenerative diseases are characterized by mitochondrial dysfunction and impaired mitochondrial biogenesis (MB). This dysregulation often leads to defective electron transport chain function and reduced adenosine triphosphate production, thereby promoting neuronal death. Remarkably, pharmacological activation of MB has been shown to restore mitochondrial function and homeostasis, and promote recovery across a variety of CNS pathologies (Table 1). MB-related therapies have proven efficacious in restoring mitochondrial homeostasis and cognitive capabilities across pre-clinical models of injury, including stroke and traumatic brain injury (Gibbs, Scholpa, Beeson, & Schnellmann, 2018; Simmons, Scholpa, & Schnellmann, 2020). In addition, mitochondrial dysfunction and MB is well-studied in neurodegenerative diseases including Alzheimer's disease, Parkinson's disease, amyotrophic lateral sclerosis, Huntington's disease and multiple sclerosis, with several known MB-inducing pharmaceuticals currently in clinical trials (Gibbs, Scholpa, et al., 2018; Hayashi et al., 2017; Simmons, Scholpa, & Schnellmann, 2020). As such, attenuation of mitochondrial dysfunction through induction of MB is an encouraging therapeutic strategy in several areas of neuroscience. For comprehensive reviews detailing MB in traumatic and neurodegenerative diseases, see Golpich et al. (2017), Sebastian, Palacin, and Zorzano (2017), and Simmons, Scholpa, and Schnellmann (2020).

**TABLE 1** Partial list of pharmacological inducers of MB for the treatment of CNS-related diseases.

| Disease | Mitochondrial dysfunction | Tested MB treatments |
|---|---|---|
| ACUTE/TRAUMATIC | | |
| Spinal Cord Injury | ↓mitochondrial homeostasis<br>↑mtDNA mutations/fragmentation<br>↓OXPHOS proteins<br>↓ATP synthesis<br>↓ATP-dependent cellular processes<br>↓Ion homeostasis<br>↑mitophagy<br>↑Oxidative damage<br>↓Calcium buffering<br>↑mPTP opening<br>↑Cell death | **Formoterol**: ↑PGC-1α: ↓cell death, ↑function<br>**LY344864**: ↑PGC-1α: ↓cell death, ↑function |
| Traumatic Brain Injury | | **Quercetin**: ↑PGC-1α, ↓cell death<br>**DEX**: ↑PGC-1α, ↓cell death, ↑behavior<br>**7,8-DHF**: ↑PGC-1α, ↑function<br>**SS-31**: ↑PGC-1α, ↑function, ↑behavior |
| Stroke | | **Daidzein**: ↓cell death, antioxidant<br>**Metformin**: ↓cell death, ↑behavior<br>**Melatonin**: ↑mitophagy, antioxidant<br>**Resveratrol**: ↑neuron survival |
| CHRONIC/NEURODEGENERATIVE | | |
| Alzheimer's Disease | ↓mitochondrial homeostasis<br>↑mtDNA mutations/fragmentation<br>↓OXPHOS proteins<br>↓ATP synthesis<br>↓mitophagy<br>↑Oxidative damage<br>↑Cell death | **TZDs**: ↑behavior<br>**Resveratrol**: ↑PGC-1α, ↑behavior, ↑pAKT<br>**AICAR**: ↑behavior, ↑pAKT<br>**Melatonin**: ↑PGC-1α, ↑function, ↓Aβ |
| Parkinson's Disease | | **Bezafibrate**: ↑PGC-1α, ↑behavior<br>**Resveratrol**: ↑PGC-1α, ↑function, ↓cell death<br>**LY344864**: ↑PGC-1α, ↑function, ↓cell death<br>**Triterpenoids**: ↑Nrf2, antioxidant |
| Huntington's Disease | | **Resveratrol**: ↑function, ↓cell death |
| Amyotrophic lateral sclerosis | | **Resveratrol**: ↑function, ↓cell death |
| Huntington's Disease | | **Resveratrol**: ↑function, ↓cell death |
| Multiple Sclerosis | | **Dimethyl fumarate**: ↑Nrf2, ↑function<br>**Metformin**: ↑AMPK, ↑function |

Arrows indicated increase (↑) and decrease (↓). AMPK, adenosine monophosphate-activated kinase; mPTP: mitochondrial permeability transition pore; mtDNA, mitochondrial DNA; Nrf2, nuclear respiratory factor 2; OXPHOS, oxidative phosphorylation; pAKT, phosphorylated protein kinase B; PGC-1α: peroxisome proliferator-activated receptor-gamma coactivator 1 alpha.
Modified with permission from Simmons, E. C., Scholpa, N. E., & Schnellmann, R. G. (2020). Mitochondrial biogenesis as a therapeutic target for traumatic and neurodegenerative CNS diseases. *Experimental Neurology, 329*, 113309..

## Mini-dictionary

- Oxidative Phosphorylation: Metabolic pathway involving mechanisms on the inner mitochondrial membrane including electron transport chain activity to produce adenosine triphosphate.
- Mitochondrial Dysfunction: Disruption of mitochondrial homeostasis, including impairment of oxidative phosphorylation, reduced mitochondrial number and mass, altered membrane potential, intracellular calcium dysregulation and mutated/fragmented mitochondrial DNA.
- PGC-1α: The master regulator of mitochondrial biogenesis, co-activates the transcription of several genes necessary for mitochondrial function.
- Mitophagy: Selective degradation of mitochondria by autophagy.
- Mitochondrial Biogenesis: Complex process by which cells increase mitochondrial mass and function.
- Mitochondrial Dynamics: Morphological cycles of fission and fusion executed by mitochondria, crucial for maintenance of function.
- Neuroinflammation: Inflammation of the nervous system as a response to stressors, such as injury.
- Oxidative Stress: Imbalance between free radicals and antioxidants often present after a traumatic injury such as spinal cord injury.
- Neurotoxicity: Damage caused by endogenous or exogenous neurotoxins resulting in altered activity of the nervous system and neuronal death.
- Ischemia: Local impairment of perfusion often propagating biological dysfunctions including mitochondrial dysfunction and cell death.

## Key facts

Activation of mitochondrial biogenesis (MB) restores mitochondrial function in spinal cord injury-related cell types:

- Protects against oxidative stress and improves mitochondrial function in neurons
- Prevents neuronal injury and improves neuronal conductance and cell survival
- Mediates pro-inflammatory assaults and resulting neurotoxicity in astrocytes
- Protects against mitochondrial dysfunction while promoting endothelial cell survival
- Internally driven astrocytic mitochondrial biogenesis can be a programmed response to oxidative stress, hindering propagation of ischemic damage

Pharmacological activation of MB promotes recovery of mitochondrial homeostasis and function after spinal cord injury:

- Improves mitochondrial dynamics and mitophagy
- Restores mitochondrial composition and oxidative buffering capacity
- Recovers mitochondrial respiration and adenosine triphosphate production
- Mediates inflammatory response and neurotoxicity
- Improves blood–spinal cord barrier integrity
- Promotes axonal growth and myelination
- Improves body composition
- Improves locomotor capabilities

## Summary points

- Mitochondrial dysfunction is a hallmark of spinal cord injury (SCI), propagating the progression of secondary injury, neuronal cell death and loss of function.
- Mitochondrial biogenesis inducers restore mitochondrial function in cultures of CNS-derived neurons, astrocytes and endothelial cells.
- Pharmacological activation of mitochondrial biogenesis mitigates several consequences of SCI including axonal regeneration, blood–spinal cord barrier integrity, altered body composition and locomotor capabilities in pre-clinical models of SCI.
- Therapies targeting mitochondrial biogenesis induction promote functional recovery across multiple traumatic and neurodegenerative CNS-related diseases.
- Several FDA-approved pharmaceuticals can induce mitochondrial biogenesis; therefore, repurposing these compounds for SCI treatment could be a relatively expeditious process.

## References

Aguirre-Rueda, D., Guerra-Ojeda, S., Aldasoro, M., Iradi, A., Obrador, E., Ortega, A., et al. (2015). Astrocytes protect neurons from Aβ1-42 peptide-induced neurotoxicity increasing TFAM and PGC-1 and decreasing PPAR-γ and SIRT-1. *International Journal of Medical Sciences*, *12*, 48–56.

Anderson, M. A., Burda, J. E., Ren, Y., Ao, Y., O'shea, T. M., Kawaguchi, R., et al. (2016). Astrocyte scar formation aids central nervous system axon regeneration. *Nature*, *532*, 195–200.

Arnett, H. A., Mason, J., Marino, M., Suzuki, K., Matsushima, G. K., & Ting, J. P. (2001). TNF alpha promotes proliferation of oligodendrocyte progenitors and remyelination. *Nature Neuroscience*, *4*, 1116–1122.

Augustyniak, J., Lenart, J., Gaj, P., Kolanowska, M., Jazdzewski, K., Stepien, P. P., et al. (2019). Bezafibrate upregulates mitochondrial biogenesis and influence neural differentiation of human-induced pluripotent stem cells. *Molecular Neurobiology*, *56*, 4346–4363.

Baar, K., Wende, A. R., Jones, T. E., Marison, M., Nolte, L. A., Chen, M., et al. (2002). Adaptations of skeletal muscle to exercise: Rapid increase in the transcriptional coactivator PGC-1. *The FASEB Journal*, *16*, 1879–1886.

Brealey, D., Karyampudi, S., Jacques, T. S., Novelli, M., Stidwill, R., Taylor, V., et al. (2004). Mitochondrial dysfunction in a long-term rodent model of sepsis and organ failure. *American Journal of Physiology. Regulatory, Integrative and Comparative Physiology*, *286*, R491–R497.

Chen, X. L., Wang, Y., Peng, W. W., Zheng, Y. J., Zhang, T. N., Wang, P. J., et al. (2018). Effects of interleukin-6 and IL-6/AMPK signaling pathway on mitochondrial biogenesis and astrocytes viability under experimental septic condition. *International Immunopharmacology*, *59*, 287–294.

Cherry, A. D., & Piantadosi, C. A. (2015). Regulation of mitochondrial biogenesis and its intersection with inflammatory responses. *Antioxidants & Redox Signaling*, *22*, 965–976.

Choi, Y. K., Park, J. H., Baek, Y. Y., Won, M. H., Jeoung, D., Lee, H., et al. (2016). Carbon monoxide stimulates astrocytic mitochondrial biogenesis via L-type Ca(2+) channel-mediated PGC-1α/ERRα activation. *Biochemical and Biophysical Research Communications*, *479*, 297–304.

Dinulovic, I., Furrer, R., Di Fulvio, S., Ferry, A., Beer, M., & Handschin, C. (2016). PGC-1α modulates necrosis, inflammatory response, and fibrotic tissue formation in injured skeletal muscle. *Skeletal Muscle*, *6*, 38.

Eelen, G., De Zeeuw, P., Treps, L., Harjes, U., Wong, B. W., & Carmeliet, P. (2018). Endothelial cell metabolism. *Physiological Reviews*, *98*, 3–58.

Falnikar, A., Li, K., & Lepore, A. C. (2015). Therapeutically targeting astrocytes with stem and progenitor cell transplantation following traumatic spinal cord injury. *Brain Research*, *1619*, 91–103.

Fanibunda, S. E., Deb, S., Maniyadath, B., Tiwari, P., Ghai, U., Gupta, S., et al. (2019). Serotonin regulates mitochondrial biogenesis and function in rodent cortical neurons via the 5-HT(2A) receptor and SIRT1-PGC-1α axis. *Proceedings of the National Academy of Sciences of the United States of America*, *116*, 11028–11037.

Fernandez-Marcos, P. J., & Auwerx, J. (2011). Regulation of PGC-1α, a nodal regulator of mitochondrial biogenesis. *The American Journal of Clinical Nutrition*, *93*, 884S–890S.

Garrett, S. M., Whitaker, R. M., Beeson, C. C., & Schnellmann, R. G. (2014). Agonism of the 5-hydroxytryptamine 1F receptor promotes mitochondrial biogenesis and recovery from acute kidney injury. *Journal of Pharmacology and Experimental Therapeutics*, *350*, 257.

Gibbs, W. S., Garrett, S. M., Beeson, C. C., & Schnellmann, R. G. (2018). Identification of dual mechanisms mediating 5-hydroxytryptamine receptor 1F-induced mitochondrial biogenesis. *American Journal of Physiology. Renal Physiology*, *314*, F260–F268.

Gibbs, W. S., Scholpa, N. E., Beeson, C. C., & Schnellmann, R. G. (2018). Pharmacological activation of mitochondrial biogenesis for the treatment of various pathologies. In Y. Will, & J. A. Dykens (Eds.), *Mitochondrial dysfunction caused by drugs and environmental toxicants* (pp. 569–592). Wiley Online Books.

Golpich, M., Amini, E., Mohamed, Z., Azman Ali, R., Mohamed Ibrahim, N., & Ahmadiani, A. (2017). Mitochondrial dysfunction and biogenesis in neurodegenerative diseases: Pathogenesis and treatment. *CNS Neuroscience & Therapeutics*, *23*, 5–22.

Gorgey, A. S., Witt, O., O'brien, L., Cardozo, C., Chen, Q., Lesnefsky, E. J., et al. (2019). Mitochondrial health and muscle plasticity after spinal cord injury. *European Journal of Applied Physiology*, *119*, 315–331.

Gureev, A. P., Shaforostova, E. A., & Popov, V. N. (2019). Regulation of mitochondrial biogenesis as a way for active longevity: Interaction between the Nrf2 and PGC-1α signaling pathways. *Frontiers in Genetics*, *10*, 435.

Harischandra, D. S., Ghaisas, S., Zenitsky, G., Jin, H., Kanthasamy, A., Anantharam, V., et al. (2019). Manganese-induced neurotoxicity: New insights into the triad of protein misfolding, mitochondrial impairment, and neuroinflammation. *Frontiers in Neuroscience*, *13*, 654.

Hayashi, G., Jasoliya, M., Sahdeo, S., Saccà, F., Pane, C., Filla, A., et al. (2017). Dimethyl fumarate mediates Nrf2-dependent mitochondrial biogenesis in mice and humans. *Human Molecular Genetics*, *26*, 2864–2873.

Hu, J., Lang, Y., Cao, Y., Zhang, T., & Lu, H. (2015). The neuroprotective effect of tetramethylpyrazine against contusive spinal cord injury by activating PGC-1α in rats. *Neurochemical Research*, *40*, 1393–1401.

Hu, J., Lang, Y., Zhang, T., Ni, S., & Lu, H. (2016). Lentivirus-mediated PGC-1α overexpression protects against traumatic spinal cord injury in rats. *Neuroscience*, *328*, 40–49.

Kluge, M. A., Fetterman, J. L., & Vita, J. A. (2013). Mitochondria and endothelial function. *Circulation Research*, *112*, 1171–1188.

Kumar, H., Ropper, A. E., Lee, S. H., & Han, I. (2017). Propitious therapeutic modulators to prevent blood-spinal cord barrier disruption in spinal cord injury. *Molecular Neurobiology*, *54*, 3578–3590.

Lamb, Y. N. (2019). Lasmiditan: First approval. *Drugs*, *79*, 1989–1996.

Liang, H., & Ward, W. F. (2006). PGC-1α: A key regulator of energy metabolism. *Advances in Physiology Education*, *30*, 145–151.

Lin, J., Handschin, C., & Spiegelman, B. M. (2005). Metabolic control through the PGC-1 family of transcription coactivators. *Cell Metabolism*, *1*, 361–370.

Liu, J. M., Yi, Z., Liu, S. Z., Chang, J. H., Dang, X. B., Li, Q. Y., et al. (2015). The mitochondrial division inhibitor mdivi-1 attenuates spinal cord ischemia-reperfusion injury both in vitro and in vivo: Involvement of BK channels. *Brain Research*, *1619*, 155–165.

McAvoy, K., & Kawamata, H. (2019). Glial mitochondrial function and dysfunction in health and neurodegeneration. *Molecular and Cellular Neurosciences*, *101*, 103417.

Mccully, K. K., Mulcahy, T. K., Ryan, T. E., & Zhao, Q. (2011). Skeletal muscle metabolism in individuals with spinal cord injury. *Journal of Applied Physiology*, *111*, 143–148.

Meves, J. M., & Zheng, B. (2014). Extrinsic inhibitors in axon sprouting and functional recovery after spinal cord injury. *Neural Regeneration Research*, *9*, 460–461.

Missiroli, S., Genovese, I., Perrone, M., Vezzani, B., Vitto, V. A. M., & Giorgi, C. (2020). The role of mitochondria in inflammation: From cancer to neurodegenerative disorders. *Journal of Clinical Medicine*, *9*, 740.

Mohanty, A., Tiwari-Pandey, R., & Pandey, N. R. (2019). Mitochondria: The indispensable players in innate immunity and guardians of the inflammatory response. *Journal of Cell Communication and Signaling*, *13*, 303–318.

National Spinal Cord Injury Statistical Center. (2020). *Facts and figures at a glance*. Birmingham, Alabama: University of Alabama at Birmingham.

O'brien, L. C., & Gorgey, A. S. (2016). Skeletal muscle mitochondrial health and spinal cord injury. *World Journal of Orthopedics*, *7*, 628–637.

O'brien, L. C., Keeney, P. M., & Bennett, J. P., Jr. (2015). Differentiation of human neural stem cells into motor neurons stimulates mitochondrial biogenesis and decreases glycolytic flux. *Stem Cells and Development*, *24*, 1984–1994.

Okada, S. (2016). The pathophysiological role of acute inflammation after spinal cord injury. *Inflammation and Regeneration*, *36*, 20.

Piantadosi, C. A., Carraway, M. S., Babiker, A., & Suliman, H. B. (2008). Heme oxygenase-1 regulates cardiac mitochondrial biogenesis via Nrf2-mediated transcriptional control of nuclear respiratory factor-1. *Circulation Research*, *103*, 1232–1240.

Pilegaard, H., Saltin, B., & Neufer, P. D. (2003). Exercise induces transient transcriptional activation of the PGC-1alpha gene in human skeletal muscle. *The Journal of Physiology*, *546*, 851–858.

Qin, W., Pan, J., Wu, Y., Bauman, W. A., & Cardozo, C. (2010). Protection against dexamethasone-induced muscle atrophy is related to modulation by testosterone of FOXO1 and PGC-1alpha. *Biochemical and Biophysical Research Communications*, *403*, 473–478.

Rust, R., & Kaiser, J. (2017). Insights into the dual role of inflammation after spinal cord injury. *The Journal of Neuroscience*, *37*, 4658–4660.

Sandri, M., Lin, J., Handschin, C., Yang, W., Arany, Z. P., Lecker, S. H., et al. (2006). PGC-1alpha protects skeletal muscle from atrophy by suppressing FoxO3 action and atrophy-specific gene transcription. *Proceedings of the National Academy of Sciences of the United States of America*, *103*, 16260–16265.

Sarkar, S., Malovic, E., Harischandra, D. S., Ngwa, H. A., Ghosh, A., Hogan, C., et al. (2018). Manganese exposure induces neuroinflammation by impairing mitochondrial dynamics in astrocytes. *Neurotoxicology*, *64*, 204–218.

Scelsi, R., Marchetti, C., Poggi, P., Lotta, S., & Lommi, G. (1982). Muscle fiber type morphology and distribution in paraplegic patients with traumatic cord lesion. Histochemical and ultrastructural aspects of rectus femoris muscle. *Acta Neuropathologica*, *57*, 243–248.

Scholpa, N. E., & Schnellmann, R. G. (2017). Mitochondrial-based therapeutics for the treatment of spinal cord injury: Mitochondrial biogenesis as a potential pharmacological target. *The Journal of Pharmacology and Experimental Therapeutics*, *363*, 303–313.

Scholpa, N. E., Simmons, E. C., Tilley, D. G., & Schnellmann, R. G. (2019). Beta2-adrenergic receptor-mediated mitochondrial biogenesis improves skeletal muscle recovery following spinal cord injury. *Experimental Neurology*, *322*, 113064.

Scholpa, N. E., Williams, H., Wang, W., Corum, D., Narang, A., Tomlinson, S., et al. (2019). Pharmacological stimulation of mitochondrial biogenesis using the Food and Drug Administration-approved β(2)-Adrenoreceptor agonist formoterol for the treatment of spinal cord injury. *Journal of Neurotrauma*, *36*, 962–972.

Schreiber, S. N., Emter, R., Hock, M. B., Knutti, D., Cardenas, J., Podvinec, M., et al. (2004). The estrogen-related receptor alpha (ERRalpha) functions in PPARgamma coactivator 1alpha (PGC-1alpha)-induced mitochondrial biogenesis. *Proceedings of the National Academy of Sciences of the United States of America*, *101*, 6472–6477.

Sebastian, D., Palacin, M., & Zorzano, A. (2017). Mitochondrial dynamics: Coupling mitochondrial fitness with healthy aging. *Trends in Molecular Medicine*, *23*, 201–215.

Simmons, E. C., Scholpa, N. E., Cleveland, K. H., & Schnellmann, R. G. (2020). 5-Hydroxytryptamine 1F receptor agonist induces mitochondrial biogenesis and promotes recovery from spinal cord injury. *The Journal of Pharmacology and Experimental Therapeutics*, *372*, 216–223.

Simmons, E. C., Scholpa, N. E., & Schnellmann, R. G. (2020). Mitochondrial biogenesis as a therapeutic target for traumatic and neurodegenerative CNS diseases. *Experimental Neurology*, *329*, 113309.

Sun, X., Jones, Z. B., Chen, X. M., Zhou, L., So, K. F., & Ren, Y. (2016). Multiple organ dysfunction and systemic inflammation after spinal cord injury: A complex relationship. *Journal of Neuroinflammation*, *13*, 260.

Vaarmann, A., Mandel, M., Zeb, A., Wareski, P., Liiv, J., Kuum, M., et al. (2016). Mitochondrial biogenesis is required for axonal growth. *Development*, *143*, 1981–1992.

Ventura-Clapier, R., Garnier, A., & Veksler, V. (2008). Transcriptional control of mitochondrial biogenesis: The central role of PGC-1α. *Cardiovascular Research*, *79*, 208–217.

Virbasius, J. V., & Scarpulla, R. C. (1994). Activation of the human mitochondrial transcription factor A gene by nuclear respiratory factors: A potential regulatory link between nuclear and mitochondrial gene expression in organelle biogenesis. *Proceedings of the National Academy of Sciences of the United States of America*, *91*, 1309–1313.

Vos, M., Lauwers, E., & Verstreken, P. (2010). Synaptic mitochondria in synaptic transmission and organization of vesicle pools in health and disease. *Frontiers in Synaptic Neuroscience*, *2*, 139.

Wareski, P., Vaarmann, A., Choubey, V., Safiulina, D., Liiv, J., Kuum, M., et al. (2009). PGC-1{alpha} and PGC-1{beta} regulate mitochondrial density in neurons. *The Journal of Biological Chemistry*, *284*, 21379–21385.

Whitaker, R. M., Corum, D., Beeson, C. C., & Schnellmann, R. G. (2016). Mitochondrial biogenesis as a pharmacological target: A new approach to acute and chronic diseases. *Annual Review of Pharmacology and Toxicology*, *56*, 229–249.

Wills, L. P., Trager, R. E., Beeson, G. C., Lindsey, C. C., Peterson, Y. K., Beeson, C. C., et al. (2012). The β2-adrenoceptor agonist formoterol stimulates mitochondrial biogenesis. *Journal of Pharmacology and Experimental Therapeutics*, *342*, 106.

Zhang, N., Yin, Y., Xu, S.-J., Wu, Y.-P., & Chen, W.-S. (2012). Inflammation & apoptosis in spinal cord injury. *The Indian Journal of Medical Research*, *135*, 287–296.

Zhu, L. L., Li, M. Q., He, F., Zhou, S. B., & Jiang, W. (2017). Mitochondria targeted peptide attenuates mitochondrial dysfunction, controls inflammation and protects against spinal cord injury-induced lung injury. *Cellular Physiology and Biochemistry*, *44*, 388–400.

Zhu, Y., Uezono, N., Yasui, T., & Nakashima, K. (2018). Neural stem cell therapy aiming at better functional recovery after spinal cord injury. *Developmental Dynamics*, *247*, 75–84.

Chapter 29

# Exploring the exogenous and endogenous effects of melatonin on spinal cord injury

Dewan Md. Sumsuzzman[a,b], Kanghui Park[c], and Yonggeun Hong[a,b]

[a]Department of Rehabilitation Science, Graduate School of Inje University, Gimhae, Gyeongsangnam-do, Republic of Korea; [b]Department of Physical Therapy, College of Medical Healthcare & Engineering, Inje University, Gimhae, Gyeongsangnam-do, Republic of Korea; [c]Department of Physical Therapy, Dong-ju College, Busan, Republic of Korea

## List of abbreviations

| | |
|---|---|
| **COX** | cyclooxygenase |
| **GSH** | glutathione |
| **IL-1α** | interleukin-1α |
| **IL-1β** | interleukin-1β |
| **iNOS** | inducible nitric oxide synthase |
| **L/D** | light/dark |
| **L/L** | light/light |
| **LPO** | lipid peroxidation |
| **MDA** | malondialdehyde |
| **MT** | melatonin |
| **NF-κB** | nuclear factor-κB |
| **NG-2** | neuron/glial antigen 2 |
| **SCI** | spinal cord injury |
| **SOD** | superoxide dismutase |
| **TBI** | traumatic brain injury |
| **TNF-α** | tumor necrosis factor-α |

## Introduction

Spinal cord injury (SCI) is a catastrophic incident with a global mortality rate ranging from 4.4% to 16.7%, often leading to neurological disability (Yamazaki, Kawabori, Seki, & Houkin, 2020). Traumatic injury to the spinal cord induces death in a number of local neurons and glia at the lesion site that cannot be recovered or regenerated. The mechanisms of secondary injury begin immediately after the primary insult and continue for weeks or months via a diverse array of pathophysiological processes, including inflammation, excitotoxicity, and oxidative cell damage (Alizadeh, Dyck, & Karimi-Abdolrezaee, 2019). These secondary insults lead to further destruction of neuronal and glial cells and to massive extension of the damage whereby the paralysis can extend to higher segments.

Disruption of circadian rhythm is a common feature in SCI individuals. Circadian misalignment can increase neuronal death, resulting in deterioration of sensorimotor functions and cognitive deficits (Li et al., 2016). In experimental studies, SCI was shown to lead to wide-ranging circadian rhythm disruption, including dysregulated rhythms of body temperature, and inflammatory gene expression (Gaudet et al., 2018). Besides, circulating levels of serum melatonin (MT) has been altered in patients with cervical SCI (Fatima, Sharma, & Verma, 2016), suggesting alteration of diurnal rhythms negatively influencing the SCI recovery process.

As SCI results in permanent or long-term disability and poor quality of life, it imposes an enormous financial burden on society in terms of increased healthcare costs and decreased productivity (Chan, Cadarette, Wodchis, Krahn, & Mittmann, 2019). Although efforts are currently being pursued to develop novel therapeutics to combat the pathogenesis of SCI, the

**Diagnosis and Treatment of Spinal Cord Injury. https://doi.org/10.1016/B978-0-12-822498-4.00029-4**

effects of these interventions are generally limited. For example, exercise interventions, which are widely used to improve functional recovery following SCI, have an inadequate ability to improve motor function after SCI (Petrosyan, Alessi, Hunanyan, Sisto, & Arvanian, 2015).

MT is a neurohormone that is synthesized and released from the pineal gland in a 24-h diurnal pattern and exerts its actions in peripheral tissues. This pineal hormone is best known for maintaining the 24-h internal clock and has excellent antioxidant capacity (Sumsuzzman, Choi, Khan, & Hong, 2020). MT exerts beneficial effects by altering the levels of oxidative stress markers, including malondialdehyde (MDA), glutathione (GSH), superoxide dismutase (SOD), and myeloperoxidase, which are generally reported to show abnormalities with progression of SCI (Aghazadeh, Azarnia, Shirazi, Mahdavi, & Zangii, 2007; Erten et al., 2003). MT has other biological functions in SCI, including reduction of pro-inflammatory molecules (Zhang et al., 2019) and regulation of autophagy (Li et al., 2019). Recently, we demonstrated that a significant reduction in endogenous MT levels disrupted neural remodeling, and homeostasis of endogenous MT levels contributed to excitatory synaptic formation and axonal outgrowth in a rodent model of SCI (Hong et al., 2019), indicating MT replacement strategies improve the neuronal repair process following SCI.

In this chapter, we explore the promising molecular mechanisms of MT intervention that may be beneficial in the management of SCI. We discuss MT and exercise combination therapy in SCI. Finally, we also discuss the crucial role of endogenous MT, which may influence the recovery process after SCI.

## Effects of exogenous melatonin on secondary injury after spinal cord injury

Neuronal death caused by primary injury from SCI cannot be prevented. Therefore, current research focuses on preventing the secondary injury cascade to mitigate progressive tissue damage, representing a novel strategy for the management of SCI pathology. The timeline of major pathological events of secondary injury and timeline-specific therapeutic strategies are shown in Fig. 1. In this section, we will discuss recent findings on the effects of exogenous MT on SCI, particularly the neuroprotective mechanism of action of MT in secondary injury events.

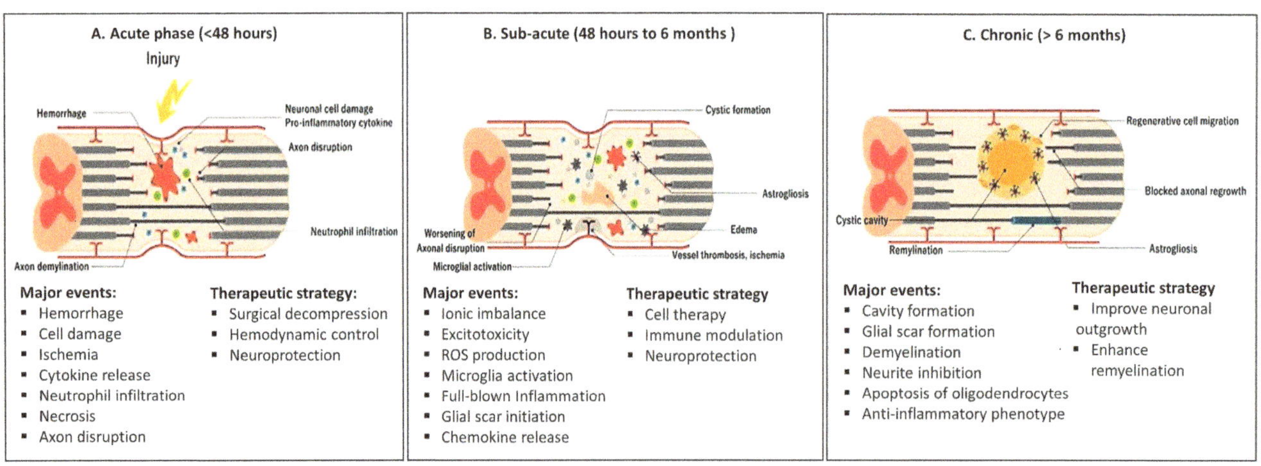


FIG. 1 Timeline of the major pathological events after spinal cord injury (SCI) and the best therapeutic strategies for SCI. The secondary injury mechanisms are initiated after the primary insult and sustained over a long period. The events that occur after primary insults are divided into the acute (<48 h), sub-acute (2 days–6 months), and chronic (>6 months) phases. (A) In the acute phase, hemorrhage, ischemia, and edema cause cell dysfunction and death. In addition, the inflammatory cells infiltrate into the injured spinal cord, resulting in the release of pro-inflammatory cytokines. These excessive inflammatory mediators can disrupt the blood-spinal cord barrier, further exacerbating the injury. (B) In the sub-acute phase, glutamate excitotoxicity, excessive ROS generation, and the full-blown inflammatory response cause further cell death and damage to the spinal cord architecture. Importantly, glial scar formation initiates this phase. (C) In the chronic phases, axons continuously degenerate, and the glial scar matures, inhibiting the neuronal regeneration process. The inflammatory phenotype shifts from pro-inflammatory to anti-inflammatory in this phase. Cavity formation also appears, which suppresses axonal regrowth. During the acute and sub-acute phases of SCI, multiple pathological cascades trigger the cell death process, leading to greater damage than that caused by the primary insult. Therefore, the pathophysiological events associated with the acute and sub-acute phases are the ultimate targets for therapeutic interventions. *ROS*, reactive oxygen species; *SCI*, spinal cord injury. *(Reprinted from Yamazaki, K., Kawabori, M., Seki, T., & Houkin, K. (2020). Clinical trials of stem cell treatment for spinal cord injury.* International Journal of Molecular Sciences, 21, *3994. According to the Creative Commons License (CC-BY-NC-ND).)*

## Effects of exogenous melatonin on oxidative stress

Pre-clinical research in animal models has reported that increased production of reactive oxygen species (ROS) and the consequent oxidative stress are crucial pathological events in SCI, which lead to neurological deficits. Oxidative stress is a hallmark of the secondary injury in SCI. Much of the current literature on secondary injury after SCI has focused particularly on oxidative stress because down-regulation of its detrimental effects is considered a key strategy for therapeutic interventions (Cristante, de Barros Filho, Marcon, Letaif, & da Rocha, 2012). After SCI, oxidative stress may increase the level of lipid peroxidation (LPO) (McDonald & Sadowsky, 2002), as determined by quantifying the levels of MDA, GSH, and SOD. Electron resonance spectrometry was used to measure MDA levels and provided the first lines of evidence for ROS production in SCI. Seligman et al. (1977) reported that the level of MDA was significantly increased within the first 5 h following SCI. A recent study showed that the MDA level increased markedly after 1 day and peaked at 7 days after SCI (Song et al., 2013). SOD activity was decreased at 1 day following SCI (Song et al., 2013). Previous studies showed that MT and its derivatives scavenge free radicals and induce the activities of various antioxidant enzymes. Erşahin et al. (2012) investigated the impact of MT in a rat model of standard weight drop-induced moderate SCI and showed that MT significantly restored the MDA and GSH levels. Immediately after laminectomy, treatment of SCI rats with MT once daily for 10 days also resulted in restored GSH levels (Erol et al., 2008). In addition, the pathophysiological mechanism was proposed to involve iron-catalyzed LPO contributing to autodestruction in the injured spinal cord regions (Emerit, Beaumont, & Trivin, 2001). Previous research established that ferrous iron can initiate brain LPO (Behrmann, Bresnahan, Beattie, & Shah, 1992). After SCI, the levels of free iron and MDA were significantly increased but markedly decreased by MT treatment (Liu, Tang, Yang, & Xiao, 2004). Topsakal et al. (2003) investigated the effects of MT, prostaglandin E1, and oxytetracycline on LPO and antioxidant activities in an experimental model of SCI. The results showed that experimentally induced SCI reduced erythrocyte SOD and plasma GSH activities and increased tissue and blood MDA levels (Topsakal et al., 2003). These physiological alterations were inhibited by MT, prostaglandin E1, and oxytetracycline treatments to varying degrees, with MT showing the greatest effect (Topsakal et al., 2003). Taken together, pre-clinical trials suggest that MT may exert neuroprotective effects on SCI by reducing oxidative stress and promoting functional recovery after SCI (Fig. 2).

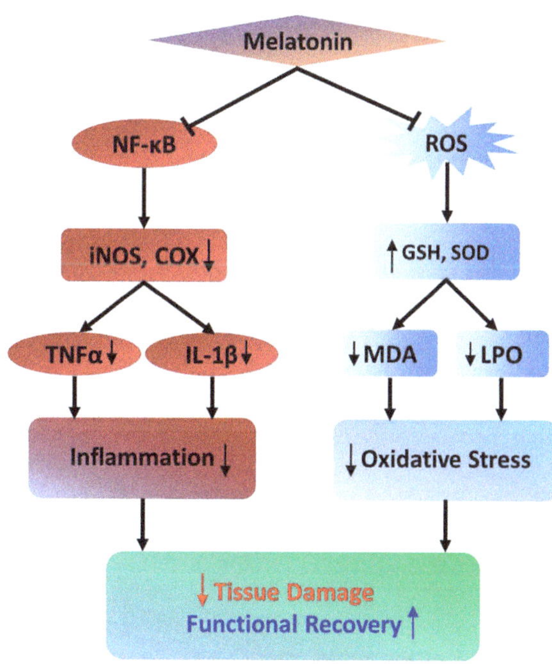


FIG. 2 Schematic representation of the potential mechanisms of action of melatonin in spinal cord injury. Increased production of ROS, MDA, LPO, and consequent oxidative stress is a crucial pathological event in SCI. Melatonin can scavenge free radicals, increase the levels of antioxidant enzymes such as SOD and GSH, and reduce the levels of MDA and LPO following SCI. Moreover, inflammation can directly promote tissue damage and weaken functional recovery after SCI by generating a wide array of inflammatory molecules. However, exogenous melatonin can reduce the levels of inflammatory molecules, such as TNF-α and IL-1β, possibly by inhibiting NF-κB. Melatonin also reduces the expression of iNOS and COX. *COX*, cyclooxygenase; *IL-1β*, interleukin-1β; *iNOS*, nitric oxide synthase; *LPO*, lipid peroxidation; *SOD*, superoxide dismutase; *GSH*, glutathione; *MDA*, malondialdehyde; *NF-κB*, nuclear factor-κ light chain enhancer of activated B cells; *ROS*, reactive oxygen species; *SCI*, spinal cord injury; *TNF-α*, tumor necrosis factor-α.

## Effects of exogenous melatonin on inflammation

Within a few hours following SCI, pro-inflammatory cytokines, such as interleukin (IL)-1α, tumor necrosis factor-α (TNF-α), and IL-1β, are activated (Ritz & Hausmann, 2008). After SCI, up-regulation of inducible nitric oxide synthase (iNOS) results in excessive production of nitric oxide (NO), which is abundant in astrocytes and microglial cells (Conti et al., 2007). These excessive levels of NO further stimulate the synthesis of several pro-inflammatory cytokines (Conti et al., 2007). These pro-inflammatory molecules are associated with activation of local microglia and astrocytes, which promote secondary injury after SCI. Several studies have postulated that MT may contribute to reduced inflammation after SCI. Previously, our group reported that iNOS mRNA expression was significantly lower in the MT treatment groups than SCI groups in the region of the injured spinal cord (Park et al., 2010). Moreover, a recent study showed that treatment with MT markedly inhibited the accumulation and proliferation of microglia and astrocytes in the injured spinal cord and suppressed TNF-α, IL-1β, and iNOS expression after SCI (Yang, Bao, Chen, & He, 2020). Those findings may explain the cellular and molecular mechanisms by which MT exerts its neuroprotective effects and contributes to functional recovery after SCI. In a mouse model of severe crush SCI, the inflammatory response appeared to be significantly attenuated at 14 days after MT treatment (10 mg/kg) (Krityakiarana et al., 2016). Immunohistochemical analysis also showed that MT treatment markedly reduced IL-1β and neuron/glial antigen 2 levels (Krityakiarana et al., 2016). The results of that study suggested that MT can decrease the expression and release of pro-inflammatory molecules, thereby inhibiting tissue damage from the secondary inflammatory response. Furthermore, TNF-α may play a pivotal role in the acute phase of SCI. Haddadi and Fardid (2015) conducted a study in radiation-induced SCI rats to assess the efficacy of MT on TNF-α expression and reported that TNF-α expression was markedly increased in the irradiated group compared with the normal group 3 weeks after injury. The most striking result was that the oral MT treatment group (100 mg/kg) showed significantly decreased TNF-α expression compared with the radiation group, suggesting that oral MT impedes the up-regulation of TNF-α expression after radiation-induced SCI (Haddadi & Fardid, 2015). Activation of the nuclear factor-κB (NF-κB) signaling has also been implicated in the induction of inflammation and is considered one of the pathophysiological causes of the spinal cord inflammatory response following SCI (Mao, Wang, Qiao, & Wang, 2010). NF-κB further generates ROS, cyclooxygenase, and iNOS, which synergistically induce inflammation (Mao et al., 2010). Several lines of evidence suggest that MT inhibits the expression of NF-κB and attenuates the production of pro-inflammatory cytokines (Szczepanik, 2007). Recent research on MT suggested that this pineal hormone reduces secondary injury severity and neuronal death after SCI (Schiaveto-de-Souza, da Silva, Defino, & Del Bel, 2013). Another study conclusively showed that MT treatment effectively decreased inflammation and tissue injury in experimental SCI (Genovese et al., 2005). Overall, MT may exert neuroprotective effects on SCI by reducing inflammation and tissue damage after SCI (Fig. 2).

## Effects of exogenous melatonin on autophagy

Autophagy plays a significant role in improving the neurological recovery process and enhances neurological function within a short time following acute SCI (Shen et al., 2017). Surprisingly, inhibition of autophagy markedly increased apoptosis in experimental SCI rats (Gao et al., 2016; Wang, Liu, Muharram, Wu, & Lin, 2014). In contrast, enhancing autophagy at the lesion site of the spinal cord significantly inhibited apoptosis (Wang et al., 2016), suggesting that increased levels of autophagy can reduce apoptosis, contribute to the neurological recovery process, and improve neurological function after SCI. A systematic review and network meta-analysis reported that MT improved locomotor recovery by inhibiting neuronal cell apoptosis in a rat model of SCI (Yang et al., 2016). Similarly, Shen et al. (2017) demonstrated that MT impeded neuronal cell apoptosis and encouraged locomotor recovery by activating the Wnt/β-catenin pathway following SCI. Previously, our group examined the endogenous and exogenous effects of MT on autophagic markers (beclin-1, LC3) in a contusion SCI rat model (Park et al., 2012). In that experiment, all groups were exposed to one of the following lighting conditions: 12/12 h light/dark (L/D) or 24/0 h constant light (LL). Western blot analysis showed that the transformation ratio of LC3-I to LC3-II protein was very low or undetectable in the normal control group (Park et al., 2012). Interestingly, the ratio was elevated after injury in a time-dependent manner, particularly in the SCI group under the L/D condition. At the site of the spinal cord lesion, the LC3-II protein level was increased at 3 days and peaked at 28 days post-SCI. In addition, no significant difference in beclin-1 expression was observed under the L/D condition, but its expression was significantly elevated under the LL condition. Surprisingly, exogenous MT treatment markedly decreased these effects under both L/D and LL conditions. Those findings suggest that autophagy-related cell death was increased in the spinal cord, and MT treatment accelerated motor recovery by down-regulation of LC3-II and beclin-1 (Park et al., 2012). These results corroborate with Kanno, Ozawa, Sekiguchi, and Itoi (2009), who also reported that the beclin-1 protein level was significantly increased at the lesion site after SCI. However, as mentioned above, up-regulation of autophagy inhibited apoptosis in the spinal cord, inducing neurological repair following SCI. A recent study revealed that treatment with MT markedly increased the levels of autophagy-activated proteins, such as beclin-1 and LC3, but decreased the levels of apoptosis-activated proteins, such as caspase-3, caspase-9, and bax, in neurons after SCI (Li et al., 2019). Those findings indicate that MT can improve locomotor function by increasing autophagy as well as decreasing apoptosis after SCI in rats, probably via the PI3K/AKT/mTOR signaling pathway (Li et al., 2019). In summary, MT can increase autophagy, improve neuronal survival, and enhance locomotor function, while decreasing apoptosis in the spinal cord following SCI via the PI3K/AKT/mTOR signaling pathway. Alternatively, MT may improve neuronal survival by down-regulating autophagy-mediated apoptosis after SCI (Fig. 3). Further studies are required to clarify the differential effects of MT on autophagy in several experimental models of SCI.

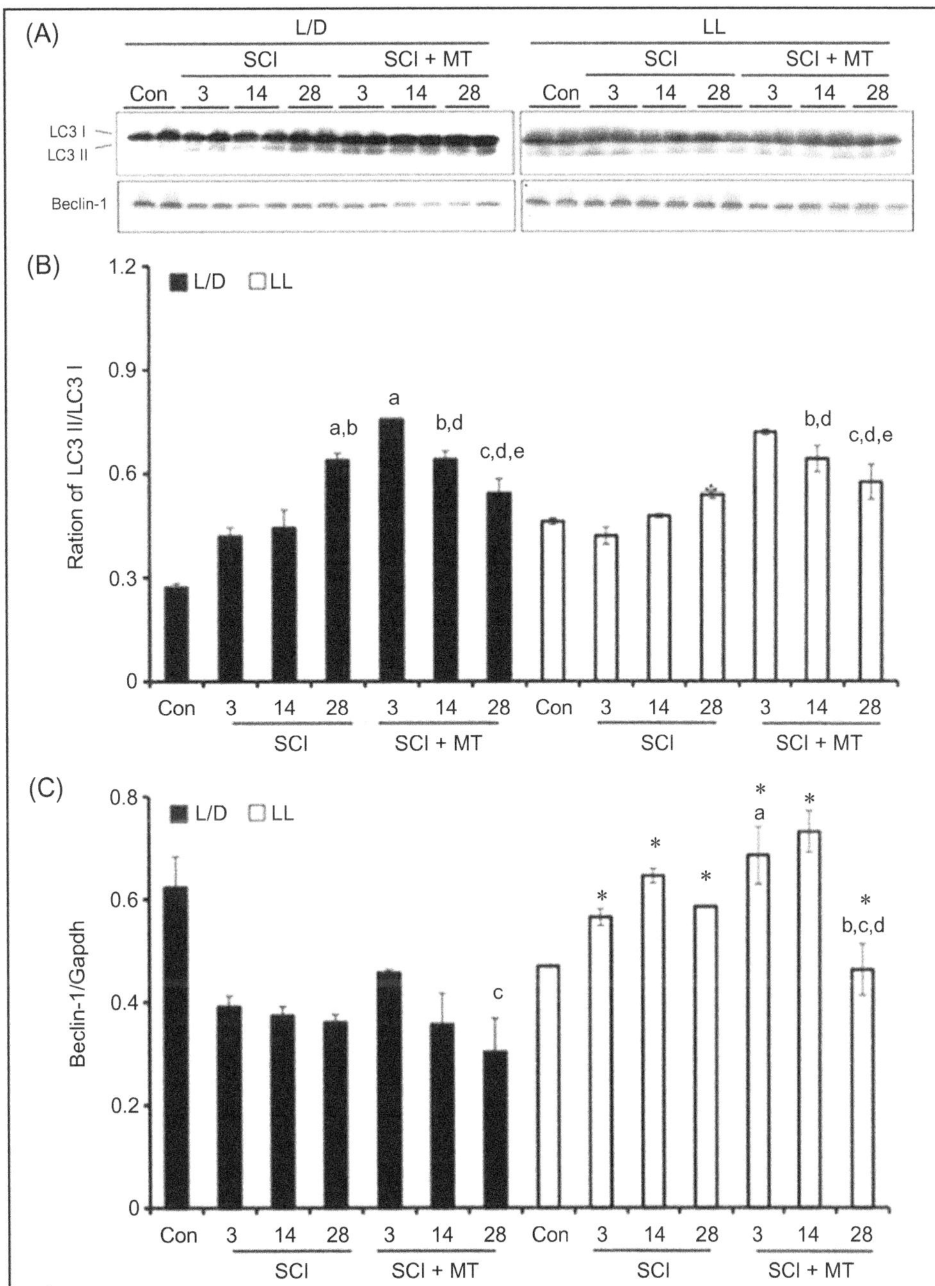


**FIG. 3** Effects of exogenous melatonin on autophagy in the injured spinal cord after spinal cord injury. (A–C) Western blot analysis indicated that the autophagy-activating proteins (LC3-II and beclin-1) were decreased by melatonin treatment. *(Republished with permission from Park, S., Lee, S.-K., Park, K., Lee, Y., Hong, Y., Lee, S., Jeon, J.-C., Kim, J.-H., Lee, S.-R., Chang, K.-T., & Hong, Y. (2012). Beneficial effects of endogenous and exogenous melatonin on neural reconstruction and functional recovery in an animal model of spinal cord injury.* Journal of Pineal Research, 52, *107–119. Wiley. Permission conveyed through Copyright Clearance Center, Inc.)*

## Combination treatment strategy for spinal cord injury

Although the single-intervention strategy has been effective in pre-clinical trials, the recovery process remains ineffective, and targeting a single barrier has been unsuccessful for improving functional recovery in cases of SCI in clinical trials. In contrast, combination treatment strategies that address diverse aspects of the pathology of SCI have been developed to achieve functional recovery after SCI (Hong et al., 2010; Lee et al., 2015). Most recent pre-clinical and clinical trials on functional recovery after SCI have focused on several combination treatment strategies, including (i) antioxidants plus exercise and (ii) glucocorticoids plus antioxidants (Hong et al., 2010). In this section, we discuss recent studies focusing on the beneficial effects of MT plus exercise on SCI.

### Melatonin combined with exercise

Based on recent studies of SCI, the combination of MT and exercise was hypothesized to show synergistic effects, accelerating motor function recovery. As indicated previously, MT reduces LPO and increases the levels of several antioxidant enzymes after SCI (Erol et al., 2008; Erşahin et al., 2012; Liu et al., 2004), indicating that MT itself is a super antioxidant. On the other hand, exercise interventions can influence structural and functional changes in the cerebral cortex, spinal cord, and skeletal muscle, thereby promoting neural and muscular function after SCI. Exercise interventions can accelerate nerve regeneration with functional recovery (Doyle & Roberts, 2006; Park et al., 2015) and sustain the functionality of spinal cord neurons (Fu, Wang, Deng, & Li, 2016). Exercise can enhance neurological function following SCI but inadequate to improve motor function (Petrosyan et al., 2015). These limitations may be due to insufficient neurotrophic factor production by exercise training. In addition to exercise, therefore, several combination strategies have been investigated. Many studies have shown that antioxidants, such as MT, combined with exercise can considerably improve functional recovery following SCI and alleviate spasticity in rats compared with single interventions alone (Fig. 4). The benefit of this combination strategy is that MT can diminish the side-effects related to exercise, such as fatigue and impairment (Park, Park, Lee, Chang, & Hong, 2013). Recently, we reported synergistic effects of MT and exercise on neuronal repair and functional recovery in a contusion-induced SCI animal model (Park et al., 2010), suggesting excessive production of iNOS accelerates secondary damage to spinal tissue, which may be reversed by MT combined with exercise. The results of that study showed that MT combined with exercise markedly elevated hind limb movement, decreased iNOS expression, and increased the proportion of motor neurons in the ventral horn compared with exercise intervention alone (Park et al., 2010). Previously, we also reported synergistic effects of MT and exercise in a rodent SCI model, with notable behavioral enhancement, histological recovery, and elevated numbers of BrdU/nestin double-positive endogenous neural stem/progenitor cells (Lee et al., 2014). This study further demonstrated that MT and exercise intervention had a synergistic effect on nestin expression in the rostral perilesion site, suggesting that nestin may be associated with functional recovery. Using Golgi-Cox analysis, we showed that MT combined with exercise elevated dendritic spine density (Lee et al., 2014). Previously, our group also comprehensively reviewed the therapeutic effects of MT combined with exercise on SCI. In this review, we suggest that MT combined with exercise would not only enhance functional recovery but also alleviate the secondary tissue damage after SCI (Hong et al., 2010). Finally, we propose that future studies should explore the effects of MT and other antioxidants in combination with exercise on SCI recovery, representing a novel strategy for faster recovery after SCI (Hong et al., 2010). Finally, the combination of MT and exercise decreases the secondary injury related to SCI, and thus MT can reduce the side-effects associated with exercise.

## Role of endogenous melatonin in spinal cord injury

Poor sleep quality is commonly observed in individuals with SCI, particularly those with tetraplegia, probably due to disruption of circadian rhythmicity (Fatima et al., 2016). SCI also causes circadian rhythm disruption such as dysregulated rhythms of body temperature, and inflammatory gene expression (Gaudet et al., 2018). Misalignment of circadian rhythmicity also alters endogenous MT secretion (Fatima et al., 2016), which further negatively affects the SCI recovery process. In contrast, the levels of endogenous MT were shown to increase under conditions of 24-h darkness, and the elevated endogenous MT levels subsequently showed anti-inflammatory and anti-oxidative effects in lesioned spinal cords, resulting in faster neural remodeling, such as axonal outgrowth, excitatory synapse formation, and oligodendrogenesis (Hong et al., 2019). To better understand, the roles of impaired circadian rhythmicity and its effects on sleep in SCI patients, (Verheggen et al., 2012) analyzed evening onset of MT patterns in patients with cervical (tetraplegia) or thoracic (paraplegia) SCI and revealed a higher prevalence of poor sleep quality in tetraplegia (83%) and paraplegia (75%) compared with able-bodied controls (20%). Additionally, they reported that even though the concentration of endogenous MT started to increase from the evening in both the control and paraplegia groups, this increase was completely absent in the tetraplegia

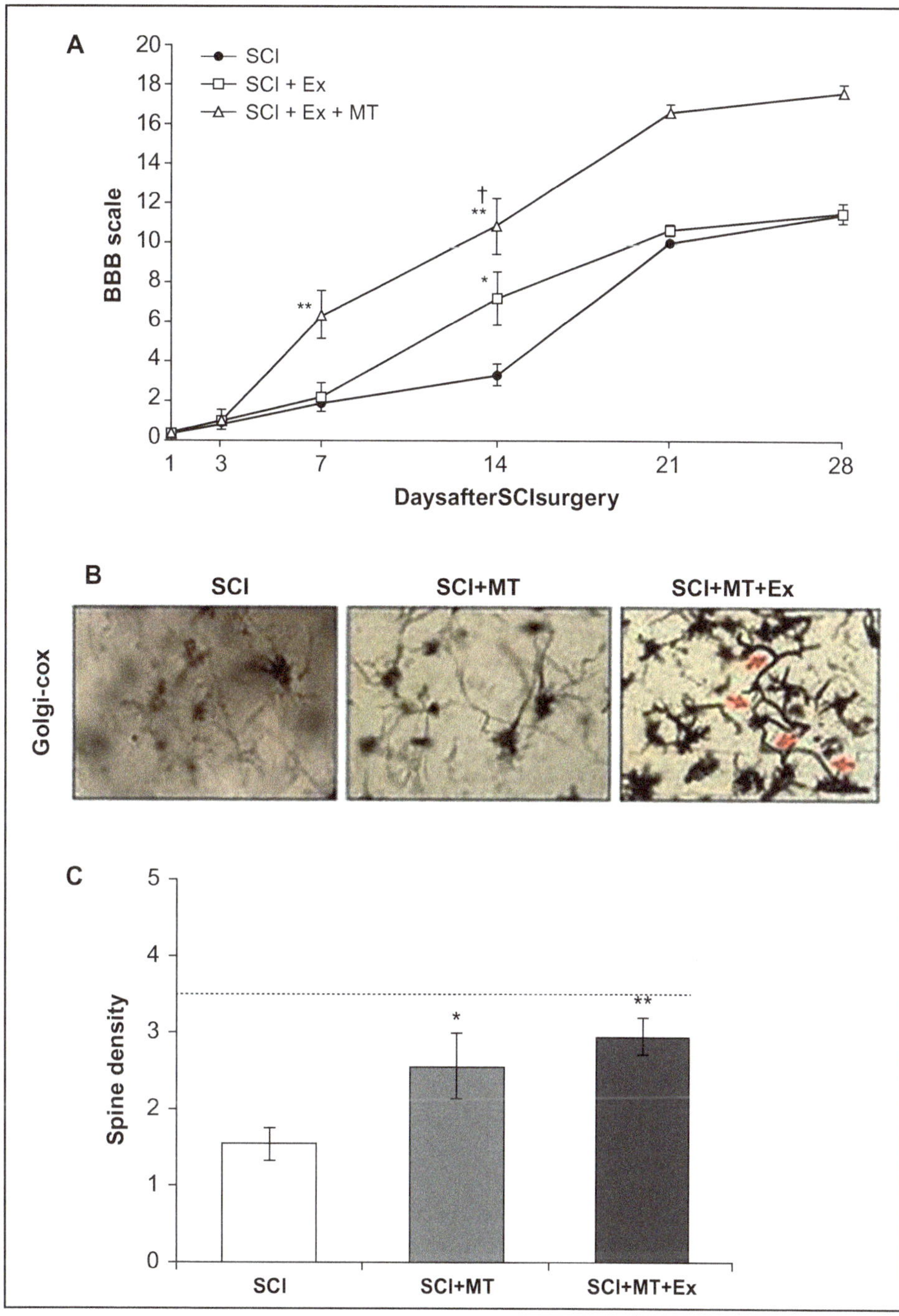


FIG. 4 Combined effects of melatonin and exercise on functional recovery after spinal cord injury. (A) The effects of melatonin combined with exercise on locomotor function in SCI. (B) Golgi-Cox staining indicated that melatonin and exercise promoted neurite outgrowth and production of new vessels *(red arrow)* after SCI. (C) After SCI, the combination of melatonin and exercise greatly increased dendritic spine density compared with melatonin alone. *(Republished with permission from Park, K., Lee, Y., Park, S., Lee, S., Hong, Y., Lee, S.K, & Hong, Y. (2010). Synergistic effect of melatonin on exercise-induced neuronal reconstruction and functional recovery in a spinal cord injury animal model.* Journal of Pineal Research, 48, *270–281. Wiley; and Reprinted from Lee, Y., Lee, S., Lee, S.-R., Park, K., Hong, Y., Lee, M., Park, S., Jin, Y., Chang, K.-T., & Hong, Y. (2014). Beneficial effects of melatonin combined with exercise on endogenous neural stem/progenitor cells proliferation after spinal cord injury.* International Journal of Molecular Sciences, 15, *2207–2222. According to the Creative Commons License (CC-BY-NC-ND).)*

group (Verheggen et al., 2012). Recently, Thøfner Hultén et al. (2018) also demonstrated that MT secretion was decreased to a greater extent in individuals with cervical SCI compared with thoracic SCI (Fig. 5). Patients with SCI were reported to show dysregulated circadian variation in thermoregulatory control ( Jones et al., 2014). Elevation of MT during the evening was reported to be inversely correlated with core and skin temperature in able-bodied controls, particularly in paraplegic

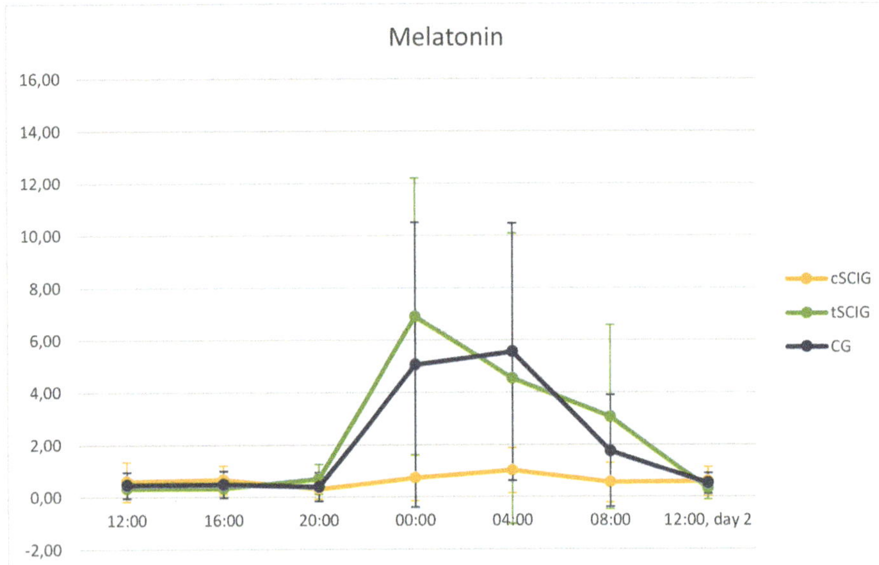


FIG. 5 Saliva melatonin levels in individuals with spinal cord injury. *cSCIG*, cervical spinal cord injury group; *tSCI*, thoracic spinal cord injury group; *CG*, control group. *(Republished with permission from Thøfner Hultén, V. D., Biering-Sørensen, F., Jørgensen, N. R., & Jennum, P. J. (2018). Melatonin and cortisol in individuals with spinal cord injury.* Sleep Medicine, 51, *92–98. Elsevier. Permission conveyed through Copyright Clearance Center, Inc.)*

individuals, while this correlation was absent in tetraplegic individuals during the same period ( Jones et al., 2014). In clinical trials, individuals with tetraplegia were more likely to lose their nocturnal production of MT than were those with mild paralysis (Zeitzer, Ayas, Wu, Czeisler, & Brown, 2005), indicating that neurologically complete cervical SCI induces complete loss of nighttime MT production. In a study involving rats with SCI, Gezici, Karakaş, Ergün, and Gündüz (2010) reported that MT secretion began immediately after SCI but slowed substantially 2 to 6 h later. In an animal study, complete lesion of the lower cervical spinal cord was related to insufficient MT secretion; on the other hand, paraplegic rats with complete lesion of the upper thoracic spinal cord showed normal MT secretion (Gezici et al., 2010). The absence of nighttime MT secretion after cervical SCI may partially explain the observed sleep disturbance. These findings raise the possibility that MT replacement therapy may improve sleep in patients with tetraplegia (Scheer et al., 2006). The endogenous MT levels in individuals with complete cervical SCI and healthy controls are illustrated in Table 1.

## Clinical application of melatonin in spinal cord injury

Several randomized controlled trials (RCTs) have been conducted to assess the effects of exogenous MT treatment on endogenous MT levels and sleep outcomes in individuals with cervical SCI. Most RCTs showed that endogenous MT levels generally increase with MT supplementation, although there were no impacts on subjective or objective parameters of sleep (Fatima et al., 2016; Kostovski et al., 2015; Spong et al., 2014; Spong, Kennedy, Brown, Armstrong, & Berlowitz, 2013; Zeitzer, Ku, Ota, & Kiratli, 2014). Generally, those trials involved MT supplementation long after the initial injury, with most of the studies conducted at least 16 years after the presenting injury. In this regard, it is hypothesized that MT administration shortly after the initial injury may improve sleep quality. However, Kostovski et al. (2017) reported that the fluctuation of MT concentration alters the expression of several clock genes in individuals with tetraplegia, while exogenous MT treatment may align the expression of these clock genes. Previous studies evaluated the clinical effects of exogenous MT only on sleep in patients with complete cervical SCI. There have been no reports regarding the effects of exogenous MT on neurological recovery or pain scores after complete cervical SCI, and therefore future studies are urgently required. Furthermore, characterization of endogenous MT levels during different phases of SCI is also important to optimize the timing of MT supplementation, as previous trials were performed several years after the initial injury, when the window of beneficial clinical effects may have been missed.

## Conclusion

Experimental studies have indicated that exogenous MT may restore neurological function and have neuroprotective effects on SCI. Several cellular and molecular mechanisms, such as reduction of oxidative stress, attenuation of iNOS

**TABLE 1** Endogenous melatonin levels in individuals with cervical spinal cord injury.

| Author (Year) | Country | Patients characteristic | Detection method (Sample medium) | Day-time MT levels (*P*-value), Time | Night-time MT levels (*P*-value), Time |
|---|---|---|---|---|---|
| Li, Jiang, Wang, Jiao, and Pang (1989) | China | 1. Sex: 7 M:1F<br>2. Age: 30.1<br>3. Location: C3–C7 | RIA (Blood) | 1. SCI: 11.6 ± 1.9 pg/mL<br>2. Healthy: 16.3 ± 1.9 pg/mL<br>($P = .0004$)<br>Time: 14:00 PM | 1. SCI: 11.6 ± 3.3 pg/mL<br>2. Healthy: 67.4 ± 15.6 pg/mL<br>($P < .0001$)<br>Time: 2:00 AM |
| Verheggen et al. (2012) | Netherlands | 1. Sex: All male<br>2. Age: 43<br>3. Location: C4–C7 | ELISA (Saliva) | NA | 1. SCI: 2.41 ± 1.25 pg/mL<br>2. Healthy: 10.62 ± 4.59 pg/mL<br>($P = .0018$)<br>Time: 11:00 PM |
| Fatima et al. (2016) | India | 1. Sex: All male<br>2. Age: 35.7<br>3. Location: cervical | ELISA (Blood) | 1. SCI: 16.9 pg/mL<br>2. Healthy: 14.4 pg/mL<br>($P > .05$)<br>Time: 12:00 PM | 1. SCI: 49.3 ± 19.5 pg/mL<br>2. C: 65.6 ± 10.6 pg/mL<br>($P < .01$)<br>Time: 12:00 AM |
| Kostovski et al. (2015) | Norway | 1. Sex: All male<br>2. Age: 46<br>3. Location: C5–C8 | ELISA (Blood) | 1. SCI: 5.85 pg/mL<br>2. Healthy: 36.49 pg/mL<br>(NA)<br>Time: 1:00 PM | 1. SCI: 4.62 pg/mL<br>2. Healthy: 49.78 pg/mL<br>(NA)<br>Time: 1:00 AM |
| Thøfner Hultén, Biering-Sørensen, Jørgensen, and Jennum (2018) | Denmark | 1. Sex: NA<br>2. Age: 45.9<br>3. Location: C4–C7 | ELISA (Saliva) | 1. SCI: 0.59 ± 0.76 pg/mL<br>2. Healthy: 0.44 ± 0.50 pg/mL<br>($P = .72$)<br>Time: 12:00 PM | 1. SCI: 0.73 ± 0.88 pg/mL<br>2. Healthy: 5.04 ± 5.45 pg/mL<br>($P = .01$)<br>Time: 12:00 AM |

*MT*, melatonin; *M*, male; *F*, female; *RIA*, radioimmunoassay; *SCI*, spinal cord injury; *C*, cervical; *NA*, not available.

and pro-inflammatory cytokines, and regulation of autophagy, may be involved in neuroprotection on SCI. The beneficial effects of MT have been shown to be greater when combined with exercise, particularly for neural regeneration and functional recovery after SCI. However, additional studies are needed to determine the proper timing and intensity of MT and exercise for use in clinical practice. Future studies should also explore the effects of MT and other antioxidants and/or traditional anti-inflammatory drugs in combination with exercise on recovery after SCI, which would represent a novel strategy for faster recovery after SCI. Both pre-clinical and clinical trials have shown that endogenous MT production and release are altered and play significant roles in the SCI recovery process. Therefore, further studies are also needed to understand the cellular and molecular mechanisms underlying the pathophysiological effects of altered endogenous MT levels and to identify novel targets for the development of clinical therapies that can be used in SCI patients.

## Applications to other areas of neuroscience

The pathophysiological mechanisms of SCI are similar to those of various neuronal disorders, such as ischemic stroke and traumatic brain injury. For example, excessive ROS production is a significant contributor to secondary injury in stroke. Similarly, most pre-clinical research in animal models has also demonstrated that increased production of ROS and consequential development of oxidative stress are crucial pathological events in SCI, which lead to neurological deficits. The considerable level of ROS further stimulates production of abundant pro-inflammatory molecules, including IL-1α, -1β, and TNF-α, which further provoke both SCI and stroke pathology. Moreover, in SCI, the activation of autophagy in the injured spinal cord may be a double-edged sword for neuronal survival similar to ischemic stroke. As similar pathophysiological mechanisms are involved in both cases, MT, which shows neuroprotective and neuroregenerative effects in SCI, may be helpful for neural regeneration and functional recovery after ischemic stroke. As for SCI, the pathophysiological process of stroke is multifactorial, and therefore combination treatment strategies that address diverse aspects of stroke pathology are required. In this chapter, we highlighted that the combination of MT and exercise exerts synergistic effects to reduce the secondary injury pathologies related to SCI. This combination strategy may also be more effective and promote faster recovery after stroke.

## Mini-dictionary of terms

- **Autophagic flux:** A process to quantify the rate of autophagic degradation.
- **Circadian misalignment:** An incorrect arrangement of sleep/wake pattern.
- **Paraplegia:** A state of sensory and motor function loss following injury to the thoracic region of the spinal cord.
- **Secondary injury:** Catastrophic molecular events in cells and tissues, beginning immediately after the primary insult and continuing for hours to weeks, resulting in further damage to cells and tissues.
- **Tetraplegia:** A state of sensory and motor function loss following injury to the cervical region of the spinal cord.

## Key facts

### Key facts of melatonin and exercise combination strategy for spinal cord injury

- Melatonin and exercise exert synergistic effects on neuronal repair and functional recovery after spinal cord injury.
- Melatonin can diminish the side effects of exercise.
- Melatonin combined with exercise can result in faster motor recovery compared with exercise intervention alone.
- Melatonin combined with exercise intervention triggers nestin expression.
- In spinal cord injury, nestin may be associated with the functional recovery process.

### Key facts of endogenous melatonin levels after spinal cord injury

- The secretion of melatonin appears to be decreased to a greater extent in individuals with cervical than thoracic spinal cord lesions.
- The concentration of endogenous melatonin started increasing from the evening in both the able-bodied and paraplegia groups, whereas this increase was completely absent in the tetraplegia group.
- The evening increase in melatonin concentration was inversely correlated with core body temperature in the able-bodied and paraplegia groups, but this correlation was completely absent in tetraplegic individuals.
- The absence of nighttime melatonin secretion after cervical spinal cord injury partially explains the sleep disturbance.
- In clinical trials, endogenous melatonin levels generally increase with exogenous melatonin supplementation.
- Elevated endogenous melatonin causes faster neural remodeling, such as axonal outgrowth and formation of excitatory synapses, after experimental spinal cord injury.

## Summary points

- Exogenous melatonin treatment notably decreased LPO and restored the levels of antioxidant enzymes, including GSH and SOD.
- Melatonin treatment effectively decreased the levels of pro-inflammatory molecules, such as TNF-α, IL-1β, and iNOS.
- Melatonin can decrease apoptosis by increase autophagy via the PI3K/AKT/mTOR signaling pathway.
- The combination of melatonin and exercise resulted in faster functional recovery.
- The endogenous melatonin secretion pattern was significantly altered following cervical spinal cord injury.

## References

Aghazadeh, S., Azarnia, M., Shirazi, A., Mahdavi, S. R., & Zangii, B. M. (2007). Melatonin as a protective agent in spinal cord damage after gamma irradiation. *Reports of Practical Oncology and Radiotherapy*, *12*, 95–99.

Alizadeh, A., Dyck, S. M., & Karimi-Abdolrezaee, S. (2019). Traumatic spinal cord injury: An overview of pathophysiology, models and acute injury mechanisms. *Frontiers in Neurology*, *10*, 282.

Behrmann, D. L., Bresnahan, J. C., Beattie, M. S., & Shah, B. R. (1992). Spinal cord injury produced by consistent mechanical displacement of the cord in rats: Behavioral and histologic analysis. *Journal of Neurotrauma*, *9*, 197–217.

Chan, B. C.-F., Cadarette, S. M., Wodchis, W. P., Krahn, M. D., & Mittmann, N. (2019). The lifetime cost of spinal cord injury in Ontario, Canada: A population-based study from the perspective of the public health care payer. *The Journal of Spinal Cord Medicine*, *42*, 184–193.

Conti, A., Miscusi, M., Cardali, S., Germanò, A., Suzuki, H., Cuzzocrea, S., & Tomasello, F. (2007). Nitric oxide in the injured spinal cord: Synthases cross-talk, oxidative stress and inflammation. *Brain Research Reviews*, *54*, 205–218.

Cristante, A. F., de Barros Filho, T. E. P., Marcon, R. M., Letaif, O. B., & da Rocha, I. D. (2012). Therapeutic approaches for spinal cord injury. *Clinics (São Paulo, Brazil)*, *67*, 1219–1224.

Doyle, L. M. F., & Roberts, B. L. (2006). Exercise enhances axonal growth and functional recovery in the regenerating spinal cord. *Neuroscience*, *141*, 321–327.

Emerit, J., Beaumont, C., & Trivin, F. (2001). Iron metabolism, free radicals, and oxidative injury. *Biomedicine & Pharmacotherapy*, *55*, 333–339.

Erol, F. S., Kaplan, M., Tiftikci, M., Yakar, H., Ozercan, I., Ilhan, N., & Topsakal, C. (2008). Comparison of the effects of octreotide and melatonin in preventing nerve injury in rats with experimental spinal cord injury. *Journal of Clinical Neuroscience*, *15*, 784–790.

Erşahin, M., Özdemir, Z., Özsavcı, D., Akakın, D., Yeğen, B.Ç., Reiter, R. J., & Sener, G. (2012). Melatonin treatment protects against spinal cord injury induced functional and biochemical changes in rat urinary bladder. *Journal of Pineal Research*, *52*, 340–348.

Erten, S. F., Kocak, A., Ozdemir, I., Aydemir, S., Colak, A., & Reeder, B. S. (2003). Protective effect of melatonin on experimental spinal cord ischemia. *Spinal Cord*, *41*, 533–538.

Fatima, G., Sharma, V. P., & Verma, N. S. (2016). Circadian variations in melatonin and cortisol in patients with cervical spinal cord injury. *Spinal Cord*, *54*, 364–367.

Fu, J., Wang, H., Deng, L., & Li, J. (2016). Exercise training promotes functional recovery after spinal cord injury. *Neural Plasticity*, *2016*, 4039580.

Gao, S., Zhang, Z.-M., Shen, Z.-L., Gao, K., Chang, L., Guo, Y., … Wang, A. M. (2016). Atorvastatin activates autophagy and promotes neurological function recovery after spinal cord injury. *Neural Regeneration Research*, *11*, 977–982.

Gaudet, A. D., Fonken, L. K., Ayala, M. T., Bateman, E. M., Schleicher, W. E., Smith, E. J., … Watkins, L. R. (2018). Spinal cord injury in rats disrupts the circadian system. *eNeuro*, *5*, ENEURO.0328-18.2018.

Genovese, T., Mazzon, E., Muià, C., Bramanti, P., De Sarro, A., & Cuzzocrea, S. (2005). Attenuation in the evolution of experimental spinal cord trauma by treatment with melatonin. *Journal of Pineal Research*, *38*, 198–208.

Gezici, A. R., Karakaş, A., Ergün, R., & Gündüz, B. (2010). Rhythms of serum melatonin in rats with acute spinal cord injury at the cervical and thoracic regions. *Spinal Cord*, *48*, 10–14.

Haddadi, G. H., & Fardid, R. (2015). Oral administration of melatonin modulates the expression of tumor necrosis factor-α (TNF-α) gene in irradiated rat cervical spinal cord. *Reports of Practical Oncology and Radiotherapy*, *20*, 123–127.

Hong, Y., Jin, Y., Park, K., Choi, J., Kang, H., Lee, S.-R., & Hong, Y. (2019). Elevated serum melatonin under constant darkness enhances neural repair in spinal cord injury through regulation of circadian clock proteins expression. *Journal of Clinical Medicine*, *8*, 135.

Hong, Y., Palaksha, K. J., Park, K., Park, S., Kim, H.-D., Reiter, R. J., & Chang, K. T. (2010). Melatonin plus exercise-based neurorehabilitative therapy for spinal cord injury. *Journal of Pineal Research*, *49*, 201–209.

Jones, H., Eijsvogels, T. M. H., Nyakayiru, J., Verheggen, R. J. H. M., Thompson, A., Groothuis, J. T., … Thijssen, D. H. (2014). Within-subject correlations between evening-related changes in body temperature and melatonin in the spinal cord injured. *Chronobiology International*, *31*, 157–165.

Kanno, H., Ozawa, H., Sekiguchi, A., & Itoi, E. (2009). Spinal cord injury induces upregulation of Beclin 1 and promotes autophagic cell death. *Neurobiology of Disease*, *33*, 143–148.

Kostovski, E., Dahm, A. E. A., Mowinckel, M. C., Stranda, A., Skretting, G., Østerud, B., … Iversen, P. O. (2015). Circadian rhythms of hemostatic factors in tetraplegia: A double-blind, randomized, placebo-controlled cross-over study of melatonin. *Spinal Cord*, *53*, 285–290.

Kostovski, E., Frigato, E., Dahm, A., Skretting, G., Mowinkel, M.-C., Sandset, P. M., & Bertolucci, C. (2017). Melatonin modifies peripheral blood cell oscillators in humans. *The FASEB Journal*, *31*(S1), 614.3.

Krityakiarana, W., Sompup, K., Jongkamonwiwat, N., Mukda, S., Pinilla, F. G., Govitrapong, P., & Phansuwan-Pujito, P. (2016). Effects of melatonin on severe crush spinal cord injury-induced reactive astrocyte and scar formation. *Journal of Neuroscience Research*, *94*, 1451–1459.

Lee, S., Park, S., Won, J., Lee, S.-R., Chang, K.-T., & Hong, Y. (2015). The incremental induction of neuroprotective properties by multiple therapeutic strategies for primary and secondary neural injury. *International Journal of Molecular Sciences*, *16*, 19657–19670.

Lee, Y., Lee, S., Lee, S.-R., Park, K., Hong, Y., Lee, M., … Hong, Y. (2014). Beneficial effects of melatonin combined with exercise on endogenous neural stem/progenitor cells proliferation after spinal cord injury. *International Journal of Molecular Sciences*, *15*, 2207–2222.

Li, D., Ma, S., Guo, D., Cheng, T., Li, H., Tian, Y., … Wang, J. (2016). Environmental circadian disruption worsens neurologic impairment and inhibits hippocampal neurogenesis in adult rats after traumatic brain injury. *Cellular and Molecular Neurobiology*, *36*, 1045–1055.

Li, Y., Guo, Y., Fan, Y., Tian, H., Li, K., & Mei, X. (2019). Melatonin enhances autophagy and reduces apoptosis to promote locomotor recovery in spinal cord injury via the PI3K/AKT/mTOR signaling pathway. *Neurochemical Research*, *44*, 2007–2019.

Li, Y., Jiang, D. H., Wang, M. L., Jiao, D. R., & Pang, S. F. (1989). Rhythms of serum melatonin in patients with spinal lesions at the cervical, thoracic or lumbar region. *Clinical Endocrinology*, *30*, 47–56.

Liu, J.-B., Tang, T.-S., Yang, H.-L., & Xiao, D.-S. (2004). Antioxidation of melatonin against spinal cord injury in rats. *Chinese Medical Journal*, *117*, 571–575.

Mao, L., Wang, H., Qiao, L., & Wang, X. (2010). Disruption of Nrf2 enhances the upregulation of nuclear factor-kappaB activity, tumor necrosis factor-α, and matrix metalloproteinase-9 after spinal cord injury in mice. *Mediators of Inflammation*, *2010*, 238321.

McDonald, J. W., & Sadowsky, C. (2002). Spinal-cord injury. *Lancet, 359*(9304), 417–425.

Park, K., Lee, S., Hong, Y., Park, S., Choi, J., Chang, K.-T., … Hong, Y. (2015). Therapeutic physical exercise in neural injury: Friend or foe? *Journal of Physical Therapy Science, 27*, 3933–3935.

Park, K., Lee, Y., Park, S., Lee, S., Hong, Y., Lee, S. K., & Hong, Y. (2010). Synergistic effect of melatonin on exercise-induced neuronal reconstruction and functional recovery in a spinal cord injury animal model. *Journal of Pineal Research, 48*, 270–281.

Park, S., Lee, S.-K., Park, K., Lee, Y., Hong, Y., Lee, S., … Hong, Y. (2012). Beneficial effects of endogenous and exogenous melatonin on neural reconstruction and functional recovery in an animal model of spinal cord injury. *Journal of Pineal Research, 52*, 107–119.

Park, S., Park, K., Lee, Y., Chang, K.-T., & Hong, Y. (2013). New prophylactic and therapeutic strategies for spinal cord injury. *Journal of Lifestyle Medicine, 3*, 34–40.

Petrosyan, H. A., Alessi, V., Hunanyan, A. S., Sisto, S. A., & Arvanian, V. L. (2015). Spinal electro-magnetic stimulation combined with transgene delivery of neurotrophin NT-3 and exercise: Novel combination therapy for spinal contusion injury. *Journal of Neurophysiology, 114*, 2923–2940.

Ritz, M.-F., & Hausmann, O. N. (2008). Effect of 17beta-estradiol on functional outcome, release of cytokines, astrocyte reactivity and inflammatory spreading after spinal cord injury in male rats. *Brain Research, 1203*, 177–188.

Scheer, F. A. J. L., Zeitzer, J. M., Ayas, N. T., Brown, R., Czeisler, C. A., & Shea, S. A. (2006). Reduced sleep efficiency in cervical spinal cord injury; association with abolished night time melatonin secretion. *Spinal Cord, 44*, 78–81.

Schiaveto-de-Souza, A., da Silva, C. A., Defino, H. L. A., & Del Bel, E. A. (2013). Effect of melatonin on the functional recovery from experimental traumatic compression of the spinal cord. *Brazilian Journal of Medical and Biological Research, 46*, 348–358.

Seligman, M. L., Flamm, E. S., Goldstein, B. D., Poser, R. G., Demopoulos, H. B., & Ransohoff, J. (1977). Spectrofluorescent detection of malonaldehyde as a measure of lipid free radical damage in response to ethanol potentiation of spinal cord trauma. *Lipids, 12*, 945–950.

Shen, Z., Zhou, Z., Gao, S., Guo, Y., Gao, K., Wang, H., & Dang, X. (2017). Melatonin inhibits neural cell apoptosis and promotes locomotor recovery via activation of the Wnt/β-catenin signaling pathway after spinal cord injury. *Neurochemical Research, 42*, 2336–2343.

Song, Y., Liu, J., Zhang, F., Zhang, J., Shi, T., & Zeng, Z. (2013). Antioxidant effect of quercetin against acute spinal cord injury in rats and its correlation with the p38MAPK/iNOS signaling pathway. *Life Sciences, 92*, 1215–1221.

Spong, J., Kennedy, G. A., Tseng, J., Brown, D. J., Armstrong, S., & Berlowitz, D. J. (2014). Sleep disruption in tetraplegia: A randomised, double-blind, placebo-controlled crossover trial of 3 mg melatonin. *Spinal Cord, 52*, 629–634.

Spong, J., Kennedy, G. A., Brown, D. J., Armstrong, S. M., & Berlowitz, D. J. (2013). Melatonin supplementation in patients with complete tetraplegia and poor sleep. *Sleep Disorders, 2013*, 1–8.

Sumsuzzman, D. M., Choi, J., Khan, Z. A., & Hong, Y. (2020). Protective effects of melatonin against severe burn-induced distant organ injury: A systematic review and meta-analysis of experimental studies. *Antioxidants, 9*, 1196.

Szczepanik, M. (2007). Melatonin and its influence on immune system. *Journal of Physiology and Pharmacology, 58*, 115–124.

Thøfner Hultén, V. D., Biering-Sørensen, F., Jørgensen, N. R., & Jennum, P. J. (2018). Melatonin and cortisol in individuals with spinal cord injury. *Sleep Medicine, 51*, 92–98.

Topsakal, C., Kilic, N., Ozveren, F., Akdemir, I., Kaplan, M., Tiftikci, M., et al. (2003). Effects of prostaglandin E1, melatonin, and oxytetracycline on lipid peroxidation, antioxidant defense system, paraoxonase (PON1) activities, and homocysteine levels in an animal model of spinal cord injury. *Spine, 28*, 1643–1652.

Verheggen, R. J. H. M., Jones, H., Nyakayiru, J., Thompson, A., Groothuis, J. T., Atkinson, G., et al. (2012). Complete absence of evening melatonin increase in tetraplegics. *The FASEB Journal, 26*, 3059–3064.

Wang, C., Liu, C., Gao, K., Zhao, H., Zhou, Z., Shen, Z., et al. (2016). Metformin preconditioning provide neuroprotection through enhancement of autophagy and suppression of inflammation and apoptosis after spinal cord injury. *Biochemical and Biophysical Research Communications, 477*, 534–540.

Wang, Z.-Y., Liu, W.-G., Muharram, A., Wu, Z.-Y., & Lin, J.-H. (2014). Neuroprotective effects of autophagy induced by rapamycin in rat acute spinal cord injury model. *Neuroimmunomodulation, 21*, 257–267.

Yamazaki, K., Kawabori, M., Seki, T., & Houkin, K. (2020). Clinical trials of stem cell treatment for spinal cord injury. *International Journal of Molecular Sciences, 21*, 3994.

Yang, L., Yao, M., Lan, Y., Mo, W., Sun, Y., Wang, J., et al. (2016). Melatonin for spinal cord injury in animal models: A systematic review and network meta-analysis. *Journal of Neurotrauma, 33*, 290–300.

Yang, Z., Bao, Y., Chen, W., & He, Y. (2020). Melatonin exerts neuroprotective effects by attenuating astro- and microgliosis and suppressing inflammatory response following spinal cord injury. *Neuropeptides, 79*, 102002.

Zeitzer, J. M., Ayas, N. T., Wu, A. D., Czeisler, C. A., & Brown, R. (2005). Bilateral oculosympathetic paresis associated with loss of nocturnal melatonin secretion in patients with spinal cord injury. *The Journal of Spinal Cord Medicine, 28*, 55–59.

Zeitzer, J. M., Ku, B., Ota, D., & Kiratli, B. J. (2014). Randomized controlled trial of pharmacological replacement of melatonin for sleep disruption in individuals with tetraplegia. *The Journal of Spinal Cord Medicine, 37*, 46–53.

Zhang, Y., Liu, Z., Zhang, W., Wu, Q., Zhang, Y., Liu, Y., et al. (2019). Melatonin improves functional recovery in female rats after acute spinal cord injury by modulating polarization of spinal microglial/macrophages. *Journal of Neuroscience Research, 97*, 733–743.

Chapter 30

# High-intensity interval training in individuals with spinal cord injury

Byron Lai[a], Jereme Wilroy[b], and Gordon Fisher[c]

[a]Department of Pediatrics, Division of Pediatric Rehabilitation Medicine, University of Alabama at Birmingham, Birmingham, AL, United States; [b]Department of Physical Medicine and Rehabilitation, University of Alabama at Birmingham, Rehabilitation Center, Birmingham, AL, United States; [c]Department of Human Studies, Nutrition Obesity Research Center, Center for Exercise Medicine, University of Alabama at Birmingham, Birmingham, AL, United States

## Abbreviations

| | |
|---|---|
| **AGREE** | Appraisal of Guidelines, Research and Evaluation |
| **HIIT** | high-intensity interval training |
| **MIT** | moderate-intensity continuous training |
| **NSCISC** | National Spinal Cord Injury Statistical Center |
| **SCI** | spinal cord injury |
| **SIT** | sprint interval training |
| **T2D** | type 2 diabetes |
| **$VO_2$ max** | maximal oxygen consumption |

## Introduction

This chapter summarizes the public health issue of physical inactivity (i.e., lack of participation in health-enhancing exercise behavior) among people with spinal cord injury (SCI). The purpose of this chapter is to provide the reader with an understanding of the importance of exercise and potentially advantageous types of exercise for people with SCI, with a particular focus on high-intensity interval training (HIIT). The chapter is broken into four sections. The first section will present current incidence and prevalence data as well as healthcare burden statistics on individuals with SCI. The second section highlights cardiometabolic comorbidities associated with SCI, low levels of exercise, and difficulties (i.e., barriers) associated with participation. The third section summarizes benefits, guidelines, and strategies to promote exercise participation. The fourth section highlights recent findings on HIIT interventions among people with SCI and developing research in this area.

## Section 1: Background and overview

### Incidence and prevalence

The National Spinal Cord Injury Statistical Center (NSCISC) estimates 291,000 people are living with an SCI in the United States with 17,730 new cases each year, or 54 per million (National Spinal Cord Injury Statistical Center, 2019). Other studies have found similar incidence rates, ranging from 52 to 56.4 per million ( Jain et al., 2015); however, due to varying methodologies, other reports include prevalence estimates of 1.5 million to 2.6 million (GBD 2016 Traumatic Brain Injury and Spinal Cord Injury Collaborators, 2016). For decades, the typical person acquiring SCI was 29; however, that has increased to 43, likely due to an increase in the etiology of falls among an aging population. People with newly acquired SCI are typically male (78%) and white (59.5%) with leading causes of injury vehicle crashes (39.3%) and falls (31.8%). The majority of injuries are incomplete, either tetraplegia (47.6%) or paraplegia (19.9%) followed by complete paraplegia (19.6%) and complete tetraplegia (12.3%).

Diagnosis and Treatment of Spinal Cord Injury. https://doi.org/10.1016/B978-0-12-822498-4.00030-0

### Healthcare and economic burden

Healthcare burden is largely dependent upon the degree of severity of the SCI with hospitalizations in the first year of injury, which includes the bulk of the direct cost estimated at 1.7 billion in the United States (Mahabaleshwarkar & Khanna, 2014). For example, an individual with tetraplegia and vent-dependent has an average healthcare expenditure over 1.2 million dollars in the first year after injury with almost $200,000 in subsequent years, whereas someone with paraplegia cost estimates include $550,381 for the first year and $72,909 in subsequent years (National Spinal Cord Injury Statistical Center, 2019). The first year after following injury is the costliest and the most critical to the life of the person with SCI, as life expectancy increases significantly after living past the first year of injury. For individuals injured at 20 years old, estimated life loss is 7 years among individuals with paraplegia and 48 years of life loss among individuals with higher injury levels and vent dependency. In addition to direct cost and life loss, indirect cost, such as loss of earnings and productivity, is estimated at $76,237 per year.

Depending on the level of injury and personal response, many individuals do not return to full-time work or even part-time work. Data from NSCISC show 66% of individuals working at time of injury with only 17% working 1 year out and 23% working 10 years after injury. Lastly, length of stay for rehabilitation after sustaining an SCI has been reduced from 98 to 34 days over the last 50 years (The 2018 Annual Statistical Report for The Spinal Cord Injury Model Systems, 2018). Post-acute rehabilitation offers people with SCI the benefit of individualized, intensive training before returning home. However, shorter length of stay and lack of intensive programming for the home has been linked to higher incidence of rehospitalization due to secondary health conditions, as well as increased discharge to institutional settings rather than the home, all of which drive healthcare cost (Eastwood, Hagglund, Ragnarsson, Gordon, & Marino, 1999; Rimmer, 2012; Tschoepe, Benfield, Mercer, & Posey, 2019).

## Section 2: Cardiometabolic comorbidities associated with SCI

This section will present cardiometabolic comorbidities associated with SCI. There is greater prevalence of these comorbidities due to deconditioning and the myriad of barriers to exercise for this population. Similar to other disability groups, people with SCI encounter three different categories of conditions. This includes the associated conditions, such as paralysis and its impacts motor functioning, sensation, and bodily processes (e.g., bladder functioning); secondary health conditions, such as neuropathic pain, urinary tract infections, pressure ulcers, and social isolation; and chronic conditions (i.e., heart disease, obesity).

### Associated risk factors

People with SCI have reduced physical functioning due to loss of motor functioning and loss of muscle mass, which result in deleterious cardiometabolic effects. People with SCI have high rates of secondary health conditions, which are defined by the Institute of Medicine as preventable conditions secondary to the associated condition. These secondary conditions (e.g., pain, pressure ulcer) lead to a downward spiral in health and function for people with SCI and in the aggregate, can impose substantial limitations in rates of participation including employment, social engagement, and performing activities of daily living (Ehde et al., 2003; Ginis et al., 2010; Latimer, Martin Ginis, Craven, & Hicks, 2006; Martin Ginis & Hicks, 2005; Rajan, McNeely, Warms, & Goldstein, 2008; Tian et al., 2013; Turner, Cardenas, Warms, & McClellan, 2001). The NCISC reported in 2018 that within the first year of sustaining an SCI, 36.5% of patients are re-hospitalized due to infections associated with the urinary, respiratory, or integumentary system, with an average length of hospitalization of 19 days (The 2018 Annual Statistical Report for The Spinal Cord Injury Model Systems, 2018). Additionally, Skelton, Hoffman, Reyes, and Burns (2015) reported that within the first year of sustaining an SCI, 57% used emergency room care. Following rehabilitation, health trajectories in people with SCI are impacted by the onset and course of these secondary conditions and their "additive" effect on changes in health and function (Adriaansen et al., 2016). Some of the highest rates of physiological secondary conditions in people with SCI include excess weight gain (i.e., obesity) (Gorgey & Gater, 2007; Laughton, Buchholz, Martin Ginis, & Goy, 2009; Rajan et al., 2008; Tian et al., 2013), extremely low fitness levels (Ginis et al., 2010; Latimer et al., 2006; Martin Ginis & Hicks, 2005), and pain (neuropathic and musculoskeletal) (Ehde et al., 2003; Turner et al., 2001; Fig. 1).

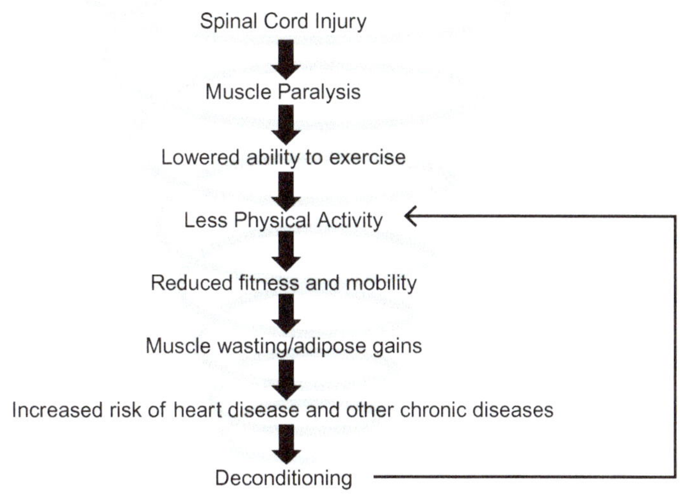


FIG. 1 Downward spiral of deconditioning among people with SCI.

### Health disparities, barriers, and SCI

People with SCI have higher rates of health disparities, which includes lower levels of physical activity. Data show that around half of people with SCI obtain little to no aerobic physical activity, which has resulted in people with SCI at three times more likely to have chronic conditions, such as heart disease, obesity, and stroke than the general population (Carroll et al., 2014; Ginis et al., 2010). This is largely due to the reduced physical functioning and sedentary lifestyles acquired by many individuals with SCI due to the issues of deconditioning and secondary conditions discussed in the above section (Galea, 2012; Rimmer, Schiller, & Chen, 2012). This is also related to the low rates of employment previously mentioned, as well as other lower quality of life indicators among people with SCI. There is an extreme challenge faced by people with SCI to lead an active and healthy lifestyle because of muscle loss, reduced motor functioning, and health, environmental, and logistical barriers (Galea, 2012; Martin Ginis, Ma, Latimer-Cheung, & Rimmer, 2016). Deconditioning is only one aspect of a plethora of exercise barriers that span each level of the social ecological model for people with SCI. Many systematic reviews have highlighted the issue of barriers to exercise among people with SCI (Martin Ginis et al., 2016). At the intrapersonal level, people with SCI have low levels of self-efficacy, fear of injuring themselves, and experience an increase in musculoskeletal pain, as well as the addition of neuropathic pain, due to the SCI. On the interpersonal level, there is a lack of social support for individuals with SCI resulting in cases of learned helplessness, which is detrimental to gaining independence. Next, there is a lack of awareness and knowledge about supporting exercise behavior among people with SCI at the organizational level and lack of accessible fitness centers, exercise programs, and equipment at the community level. Lastly, the largest barrier to exercise has been transportation with many dependent upon public transportation.

## Section 3: SCI exercise benefits, guidelines, and participation

This section will cover the current state-of-the-science regarding exercise benefits and guidelines for SCI. To date, there are three topic areas in SCI exercise research that are receiving substantial attention: (1) exercise benefits, (2) exercise guidelines, and (3) participation issues. The first subsection, exercise benefits, will summarize the anticipated benefits that people with SCI can obtain from exercise participation. The second subsection will briefly overview issues participation in exercise research. The third subsection will summarize the latest guidelines or recommendations to exercise for people with SCI.

### Exercise benefits

Exercise research for SCI is a relatively new area of research inquiry that emerged around the 1980s. Because of this, there is limited high-level evidence that confirms precise physiological benefits of exercise among this population.

Unfortunately, the relatively small prevalence of SCI has made it difficult for researchers to confirm the benefits of exercise in SCI within large high-quality trials, such as randomized controlled trials. This lack of trials has resulted in few systematic reviews for SCI compared with other disability groups (e.g., stroke and multiple sclerosis) and, ultimately, no meta-analyses. Therefore, while the benefits for SCI are likely similar to those observed in the general population, there is a lack of research evidence supporting interventions that can lead to actual benefits in SCI. This makes it difficult for clinicians to prescribe exercise and improve participation. Nevertheless, we hereby present the latest high-quality evidence of benefits in SCI as of the time of this publication.

In 2017, a high-quality systematic review of exercise studies for people SCI reported that exercise could improve fitness cardiometabolic health (Ginis et al., 2018; van der Scheer et al., 2017). The review included a total of 211 studies where exercise interventions were delivered. The review included non-randomized studies (non-randomized trials, case studies, etc.) in addition to randomized studies, which means that we should interpret not only their findings but also the confidence they had in their findings. There were insufficient high-quality studies that included people with acute SCI (≤12 months post-injury). Thus, the findings were presented only for people with chronic SC (189/211 studies). The key findings are presented below:

Benefits summary: (Ginis et al., 2018; van der Scheer et al., 2017)

- Low to moderate confidence evidence demonstrates that exercising 3 to 5 times per week at a moderate to vigorous intensity can improve cardiorespiratory fitness, muscular strength and metabolic health (body composition and cardiovascular risk factors).
- Different doses of exercise result in different health benefits.
- Doses of lower training durations and frequencies can improve fitness (i.e., cardiorespiratory fitness and muscular strength/power).
- Whereas doses with higher training durations and frequencies appear to be necessary to result in cardiometabolic health improvements.

These findings had important implications and informed the development of the latest SCI exercise guidelines.

Of note, the published studies primarily base their evidence off of short-term clinical studies. In other words, the studies are conducted in highly controlled laboratory or clinical settings, as opposed to real-world changes in the lifestyle of the person with SCI after the study. A scoping review examined whether exercise benefits could be sustained after completing the formal intervention of an intervention for adults with disabilities (Lai et al., 2019). The report identified that, from exercise trials published from 2006 to 2016, only one study for SCI found a sustained benefit toward a health outcome that lasted throughout a follow-up period (a period after completing a formal intervention component). Consequently, there is a need to identify programs that not only improve health and function in the short term for SCI, but also demonstrate long-term benefits, which will ultimately require strategies that promote long-term participation.

## Exercise guidelines

Given the infancy of the field, exercise guidelines or recommendations designed specifically for SCI are quite new and still evolving. The latest and most widely acknowledged exercise guidelines were created in 2017 by a large international team of highly respected disability-exercise scientists, led by Dr. Kathleen Martin Ginis the University of British Columbia, Canada and Victoria Goosey-Tolfrey at Loughborough University, United Kingdom (Ginis et al., 2018; van der Scheer et al., 2017). The guidelines were split into two different categories, fitness benefits and cardiometabolic health, as stated below:

Guidelines summary: (Ginis et al., 2018; van der Scheer et al., 2017)

- Aerobic exercise should be performed 2 to 3 sessions per week at a moderate to vigorous intensity for 20 to 40 min (at minimum, 20 min).
- Strength exercise should be performed 2 to 3 times per week and each session should include 3 sets of 10 repetitions for large muscle groups (weight should be set at 50% to 80% of the person's maximum weight; at minimum 3 sets should be performed in a week).

When prescribing these guidelines to people with SCI, there are a few important concepts to note. The duration component for aerobic exercise necessary to anticipate health benefits is, at minimum, 20 min. If improving health is the primary goal, people with SCI should aim to exceed the SCI guidelines, particularly since guidelines for SCI are lower than those for the general adult population. The US Department of Health and Human Services recommends the general adult population and adults with disabilities should aim to achieve at least 150 to 300 min of moderate-intensity aerobic exercise per week; or 75 to 150 min of vigorous intensity aerobic exercise (US Department of Health and Human Services Physical Activity Guidelines for Americans, 2018). The collective evidence suggests that the SCI exercise guidelines should be considered as a minimum recommendation or a starting point for exercise promotion. The long-term goal should be to progress toward higher volumes (frequency of exercise sessions per week and session durations) of exercise overtime. As a final note, there will certainly be many people with SCI who will have difficulty achieving or may never achieve the SCI guidelines (barriers discussed in the last subsection). Although achieving the guidelines is an important target for exercise prescription, know that some exercise is always better than none: gains can be made from even low levels of exercise participation, particularly in people who are physically deconditioned or new to exercise.

### Exercise participation

Although there is an encouraging amount of literature demonstrating benefits, ultimately, published studies have not demonstrated much success with identifying programs that can increase exercise participation. Specifically, there have been no evidence-based programs successful at maintaining exercise participation over a long term (Lai et al., 2018) and, most importantly, there is substantial difficulty with getting people with SCI enrolled into an exercise intervention. A review study (Lai et al., 2018) found the mean sample size of randomized controlled trials published between the years 2006 to 2016 was 25 people; 14 people per study group when they were randomized to a treatment group and control group. This means exercise interventions may have poor generalizability (i.e., findings may not be able to be projected to the population as a whole), and we must be careful when interpreting study findings. This limitation is currently the largest factor that impedes the growth of exercise research in SCI.

To address this issue, we would urge researchers to use theory-based strategies to enroll people with SCI into exercise interventions or programs. Behavioral change strategies that are framed upon theories (e.g., social cognitive theory) have been found to be successful for promoting physical activity throughout an intervention for people with SCI, albeit for short-to-mid terms (Ma & Martin Ginis, 2018). Conceivably, theory-based strategies should also frame how we enroll people with SCI into exercise interventions. Our team conducted a qualitative investigation that generated a theory for enhancing the likelihood of success with enrollment (Lai et al., 2020). The theory posits that wheelchair users with SCI ($n = 33$) conceive three core concerns with enrolling into an exercise intervention: (1) they first assess their capability to participate in the program due to scheduling, transportation, and secondary health conditions; (2) they mentally balance anticipated benefits versus the difficulty of starting the program; and (3) they assess the desirability of the program characteristics based upon their needs and preferences. Successful recruitment will depend on a recruiters ability to address these three concerns while presenting time-efficient, accessible, and desirable programs (Lai et al., 2020).

## Section 4: Health benefits of high-intensity interval training

The primary cited reason for not participating in regular exercise in non-disabled individuals is perceived "lack of time" (Lai et al., 2021; Trost, Owen, Bauman, Sallis, & Brown, 2002). With this in mind, individuals with SCI face even greater challenges as they have few, if any options to participate in physical activity. Three of the most frequently reported barriers to participation, are structural (accessibility to exercise facilities, exercise equipment), time-related (getting dressed/undressed, unreliable transportation services), and lack of knowledgeable exercise personnel (Scelza, Kalpakjian, Zemper, & Tate, 2005). Thus, it is imperative that novel exercise strategies are identified to address these barriers in individuals with SCI.

### High-intensity interval training in able-bodied individuals

One mode of training that may overcome some of these barriers and provide similar cardiometabolic health benefits is low-volume high-intensity interval training (HIIT). HIIT can be performed many different ways, but usually consists of brief intervals of higher intensity exercise performed between (~80%–200% of $VO_2$ max) separated by rest intervals of lower intensity exercise or complete rest (Gillen et al., 2016; MacInnis & Gibala, 2016). Furthermore, interval training can be broken down to HIIT, which consists of repeated bouts of exercise performed at workloads between the ventilatory threshold and $VO_2$ max, or sprint interval training (SIT), which is performed at workloads greater than $VO_2$ max (Buchheit & Laursen, 2013; Gibala, Gillen, & Percival, 2014). At this point, there are insufficient data to support the idea that individuals with SCI are able to perform SIT, which requires maximal or supramaximal efforts for the duration of the interval. Thus, for simplicity, we will refer to both of these terms as HIIT when referring to the effects on individuals with SCI throughout this chapter. A growing body of evidence in non-disabled individuals, from our group (Fisher et al., 2015) and others (Burgomaster et al., 2008; Martins et al., 2015; Richards et al., 2010; Whyte, Gill, & Cathcart, 2010) has demonstrated the potential for low-volume high-intensity interval training (HIIT) to provide comparable or superior improvements in cardiometabolic health outcomes compared to continuous moderate-intensity training (MIT) requiring 60% to 80% greater time commitment. For example, we recently found similar improvements in % fat, blood lipids, and insulin sensitivity in obese males following 6 weeks of training despite HIIT only requiring 20% of the total time commitment as MIT. Importantly, this mode of exercise has been shown to be well tolerated in many different clinical populations, including type 2 diabetes (T2D) (Boudou, Sobngwi, Mauvais-Jarvis, Vexiau, & Gautier, 2003; Little et al., 2011), overweight/obesity (Fisher et al., 2015; Gillen, Percival, Ludzki, Tarnopolsky, & Gibala, 2013; Whyte et al., 2010), cardiovascular disease (Warburton et al., 2005), and metabolic syndrome (Tjønna et al., 2008). Thus, HIIT appears to be a mode of exercise that is well-tolerated, requires minimal time commitment, and provides many of the same cardiometabolic health benefits as MIT in non-disabled individuals.

## High-intensity interval training in individuals with spinal cord injury

Exercise guidelines for non-disabled individuals have been developed based on modes of exercise that require weight bearing activities such as running and walking or lower-body exercise such as cycling. This is uniquely different than exercise programs in individuals with SCI, in which upper body exercise training (arm crank cycling, wheelchair movement, and upper body resistance training) is the primary mode used and consists of upper body muscle mass. When SCI occurs, there is a decline in aerobic capacity, onset of skeletal muscle atrophy, increases in body fat, and impaired cardiometabolic health. Thus, this rapid deconditioning greatly enhances risk of cardiovascular and metabolic diseases then able bodied populations (Bauman & Spungen, 2001). Furthermore, while there have been few long-term studies assessing the benefits of exercise on individuals with SCI a recent study showed improvements in $VO_2$ max and muscular strength compared to no-exercise controls. However, there were no changes in body composition, blood lipids, or blood glucose and insulin following 16 weeks of combined aerobic and resistance training exercise performed twice a week (Totosy de Zepetnek, Pelletier, Hicks, & MacDonald, 2015), suggesting the potential need for greater volume or intensity to modify cardiometabolic risk factors. Thus, HIIT is an attractive form of training to address these potential limitations following lower volume or moderate-intensity exercise. To date, there have been few well-powered studies assessing the potential benefits for improving cardiometabolic health in individuals with SCI. Overall, studies (Brurok, Helgerud, Karlsen, Leivseth, & Hoff, 2011; Harnish, Daniels, & Caruso, 2017; Hasnan, Engkasan, Husain, & Davis, 2013; Tordi et al., 2001) not including a MIT exercise comparison group showed increases in $VO_2$ peak and peak power output, and have also shown improved stroke volume (Brurok et al., 2011), and decrease % body fat (Harnish et al., 2017). Whereas studies that have compared HIIT to MIT training groups have shown similar improvements in $VO_2$ peak (de Groot, Hjeltnes, Heijboer, Stal, & Birkeland, 2003; Graham et al., 2019), insulin sensitivity (de Groot et al., 2003; Graham et al., 2019), and peak power output (de Groot et al., 2003; Mcleod, Diana, & Hicks, 2020). Thus, it appears that HIIT can improve some components of cardiometabolic health, similar to that seen following MIT exercise; however, it remains difficult to compare results between studies due to a lack of standardized exercise guidelines when implementing HIIT.

While there is much work needed to draw conclusions of the efficacy of HIIT for improving health in individuals with SCI, several recent studies have shown greater exercise enjoyment (Astorino & Thum, 2018; Gauthier, Brosseau, Hicks, & Gagnon, 2018) following HIIT compared to MIT (15, 20). Furthermore, HIIT studies to date have reported minimal (Astorino & Thum, 2018; Brurok et al., 2011; Gauthier et al., 2018) or no adverse responses (Graham et al., 2019; Mcleod et al., 2020) to HIIT. Therefore, HIIT appears to be a feasible and tolerable form of training in individuals with SCI. Future well powered and long-term studies are needed to determine if HIIT is a more attractive form of training that can address the major barriers (lack of time, enjoyment, and access to facilities) to exercise participation and exercise adherence in individuals with SCI, and lead to long-term improvements in cardiometabolic health.

## Telehealth strategies to promote exercise adherence to high-intensity interval training

Participation in exercise remains low in individuals with SCI (50% sedentary) (Ginis et al., 2010). Adherence to exercise treatments in disabled individuals is often reported to be less than 50% (Forkan et al., 2006; Kolt & McEvoy, 2003; McLean, Klaber Moffett, Sharp, & Gardiner, 2013). In addition to low adherence rates, the majority of SCI exercise training studies to date have had small sample sizes (mean = 14) and were often underpowered to determine primary treatment effects (Lai, Young, Bickel, Motl, & Rimmer, 2017). Thus, it is critical to identify novel strategies and tools that can increase exercise participation and improve rates of adherence in SCI. Recent advances in technology, such as smartphones and internet streaming, have been used to deliver health care services to individuals at home. Thus, with emerging technology it is possible for the implementation of home-based telehealth exercise, which may enable us to overcome many of the reported barriers, such as transportation, accessibility to exercise facilities, and access to knowledgeable exercise personnel (Scelza et al., 2005), in individuals with SCI.

To date, there have been a few pilot studies testing the efficacy of various telehealth exercise interventions (telephone, video, and web-based approaches) in individuals with SCI (Arbour-Nicitopoulos, Tomasone, Latimer-Cheung, & Martin Ginis, 2014; Coulter et al., 2017; Lai, Rimmer, Barstow, Jovanov, & Bickel, 2016; Sweet, Rocchi, Arbour-Nicitopoulos, Kairy, & Fillion, 2017). Telephone-based exercise counseling has been shown to increase exercise participation in adults with SCI from 35% at baseline to 52% after 6 months (Arbour-Nicitopoulos et al., 2014). A recent pilot study by our group assessed an aerobic exercise training program delivered remotely using an upper body ergometer, tablet, physiological monitor, and custom application that delivered video feed to a remote trainer and monitored and recorded exercise data in real time (Lai et al., 2016). This approach resulted in completion of all 24 sessions during the 8 weeks of training and 100% adherence. Another advantage is that technology-based programs might also increase the likelihood that outcomes are maintained after an exercise trial is completed. In a recent review, we observed that technology-driven interventions sustained 56% of all outcomes measured at follow-up, which was higher than all other exercise modalities and the mean of all studies (32%) (Lai et al., 2018). Collectively, these studies demonstrate that home-based telehealth exercise training can be implemented safely and allows individuals and investigators to overcome many of the common barriers to exercise participation and research in SCI. Given the potent health improvements observed following low-volume HIIT and the ability to implement a home-based telehealth exercise approach, HIIT delivered via telehealth may be an optimal exercise training program that improves cardiometabolic health, requires little time, and reduces environmental barriers that typically prevent exercise in SCI. Current research by our group will assess the potential for home-based telehealth HIIT exercise for improving cardiometabolic health in individuals with SCI. We anticipate greater adherence to exercise and significant health improvements as it has been shown that one-on-one training and attention from the trainers are key factors that will make HIIT enjoyable, as it has been shown that working with personal trainers significantly increase adherence to exercise (Tulloch et al., 2013).

## Applications to other areas of neuroscience

In this chapter, we have reviewed the impact that spinal cord injury has on the risk and prevalence of developing chronic cardiometabolic diseases. We discuss the benefits that exercise training can have for improving or preventing these adverse health outcomes, and highlight the potential role of high-intensity interval training as a mode of exercise that may lead to greater health benefits and adherence in individuals with spinal cord injury. HIIT has gained traction in recent years as it has shown to provide many similar or superior health benefits as prolonged moderate-intensity exercise training despite requiring a significantly reduction in overall time commitment. It remains to be determined if this mode of exercise will provide similar results in SCI as seen in able-bodied individuals, however a number of studies in other neurological diseases have shown promising results. HIIT was shown to be a feasible form of exercise training that improved cardiovascular fitness, vascular hemodynamics, and cognitive function in individuals with Parkinson's disease (Fernandes et al., 2020; Harvey et al., 2019). HIIT has also been shown to reduce pain, fatigue, stiffness, and inflammation in patients with axial spondyloarthritis (Sveaas et al., 2020). The assessment of HIIT in patients with neurological diseases is relatively new, thus future well controlled and statistically powered studies will better determine the long-term health benefits in individuals with SCI and other neurological conditions.

## Mini-dictionary of terms

**Spinal cord injury**: Traumatic damage to the spinal cord causing a certain degree of paralysis and loss of neurological function.
**Paraplegia**: Paralysis of the lower limbs.
**Tetraplegia**: Paralysis of all four limbs.
**Healthcare burden**: The direct (e.g., hospital stay) and indirect cost (e.g., loss of wages) associated with care for the injury and the lasting effects.
**Secondary health conditions**: Conditions associated with a disability, either directly or indirectly, and preventable or manageable through healthy behavior, such as exercise.
**Psychosocial**: The intersection of an individual's psychology and their interaction with their social environment.
**Sample size**: The number of people enrolled in a research study.

**Systematic review**: A review that takes a rigorous approach to aggregating and summarizing the effects of published research studies.
**Health disparity**: Differences in health due to unequal access or resources.
**High-intensity interval training**: A mode of exercise consisting of brief intervals of higher intensity exercise separated by rest intervals of lower intensity exercise or complete rest.
**Moderate-intensity continuous training**: A mode of exercise training that involves steady state continuous exercise without rest intervals.
**Cardiometabolic health**: Risk factors, such as dyslipidemia, elevated blood pressure, impaired glucose intolerance, obesity, reduced cardiovascular fitness, that increase risk of developing diabetes, heart disease, or stroke.

## Key facts of SCI and exercise

- Individuals with SCI have reduced physical functioning due to loss of motor function and loss of muscle mass.
- When SCI occurs there is a decline in aerobic capacity, onset of skeletal muscle atrophy, increases in body fat, and impaired cardiometabolic health.
- The primary cited reason for not participating in regular exercise in non-disabled individuals is perceived "lack of time."
- HIIT is a form of training that consists of brief intervals of higher intensity exercise performed between (~80%–200% of $VO_2$ max) separated by rest intervals of lower intensity exercise or complete rest.
- HIIT appears to be a mode of exercise that is well-tolerated, requires minimal time commitment, and provides many of the same cardiometabolic health benefits as MIT despite requiring significantly less time.

## Summary points

- High rates of secondary medical conditions among individuals with SCI include pressure sores, urinary tract infections, and pain.
- Individuals with SCI have reported lower levels of physical activity compared to individuals without a disability.
- Barriers to exercise for people with SCI include unreliable transportation, inaccessible facilities, and lack of knowledge among exercise trainers.
- Exercise guidelines for SCI recommend a minimum of 20 min of aerobic activity 2 to 3 times per week and strength exercise performed 2 to 3 times.
- More needs to be done in creating effective and enjoyable HIIT exercise interventions for individuals with SCI.
- Telehealth could be used to circumvent the majority of barriers to exercise and provide an enjoyable experience with exercise.

## References

Adriaansen, J., Post, M. W. M., Brinkhof, M. W. G., deGroot, S., van Asbeck, F. W. A., Stolowijk-Swuste, J. M., et al. (2016). Secondary health conditions in persons with spinal cord injury: A longitudinal study from one to five years post-discharge. *Journal of Rehabilitation Medicine*, *45*, 1016–1022.

Arbour-Nicitopoulos, K. P., Tomasone, J. R., Latimer-Cheung, A. E., & Martin Ginis, K. A. (2014). Get in motion: An evaluation of the reach and effectiveness of a physical activity telephone counseling service for Canadians living with spinal cord injury. *PM & R : The Journal of Injury, Function, and Rehabilitation*, *6*(12), 1088–1096. https://doi.org/10.1016/j.pmrj.2014.05.018.

Astorino, T. A., & Thum, J. S. (2018). Interval training elicits higher enjoyment versus moderate exercise in persons with spinal cord injury. *The Journal of Spinal Cord Medicine*, *41*(1), 77–84. https://doi.org/10.1080/10790268.2016.1235754.

Bauman, W. A., & Spungen, A. M. (2001). Carbohydrate and lipid metabolism in chronic spinal cord injury. *The Journal of Spinal Cord Medicine*, *24*(4), 266–277. https://doi.org/10.1080/10790268.2001.11753584.

Boudou, P., Sobngwi, E., Mauvais-Jarvis, F., Vexiau, P., & Gautier, J. F. (2003). Absence of exercise-induced variations in adiponectin levels despite decreased abdominal adiposity and improved insulin sensitivity in type 2 diabetic men. *European Journal of Endocrinology*, *149*(5), 421–424.

Brurok, B., Helgerud, J., Karlsen, T., Leivseth, G., & Hoff, J. (2011). Effect of aerobic high-intensity hybrid training on stroke volume and peak oxygen consumption in men with spinal cord injury. *American Journal of Physical Medicine & Rehabilitation*, *90*(5), 407–414. https://doi.org/10.1097/PHM.0b013e31820f960f.

Buchheit, M., & Laursen, P. B. (2013). High-intensity interval training, solutions to the programming puzzle: Part I: Cardiopulmonary emphasis. *Sports Medicine*, *43*(5), 313–338. https://doi.org/10.1007/s40279-013-0029-x.

Burgomaster, K. A., Howarth, K. R., Phillips, S. M., Rakobowchuk, M., Macdonald, M. J., McGee, S. L., et al. (2008). Similar metabolic adaptations during exercise after low volume sprint interval and traditional endurance training in humans. *The Journal of Physiology*, *586*(1), 151–160. doi: jphysiol.2007.142109 [pii] https://doi.org/10.1113/jphysiol.2007.142109.

Carroll, D. D., Courtney-Long, E. A., Stevens, A. C., Sloan, M. L., Lullo, C., Visser, S. N., et al. (2014). Vital signs: Disability and physical activity—United States, 2009-2012. *MMWR. Morbidity and Mortality Weekly Report*, *63*(18), 407–413.

Coulter, E. H., McLean, A. N., Hasler, J. P., Allan, D. B., McFadyen, A., & Paul, L. (2017). The effectiveness and satisfaction of web-based physiotherapy in people with spinal cord injury: A pilot randomised controlled trial. *Spinal Cord*, *55*(4), 383–389. https://doi.org/10.1038/sc.2016.125.

de Groot, P. C., Hjeltnes, N., Heijboer, A. C., Stal, W., & Birkeland, K. (2003). Effect of training intensity on physical capacity, lipid profile and insulin sensitivity in early rehabilitation of spinal cord injured individuals. *Spinal Cord*, *41*(12), 673–679. https://doi.org/10.1038/sj.sc.3101534.

Eastwood, E. A., Hagglund, K. J., Ragnarsson, K. T., Gordon, W. A., & Marino, R. J. (1999). Medical rehabilitation length of stay and outcomes for persons with traumatic spinal cord injury—1990–1997. *Archives of Physical Medicine and Rehabilitation*, *80*(11), 1457–1463. https://doi.org/10.1016/S0003-9993(99)90258-7.

Ehde, D. M., Jensen, M. P., Engel, J. M., Turner, J. A., Hoffman, A. J., & Cardenas, D. D. (2003). Chronic pain secondary to disability: A review. *The Clinical Journal of Pain*, *19*(1), 3–17.

Fernandes, B., Barbieri, F. A., Arthuso, F. Z., Silva, F. A., Moretto, G. F., Imaizumi, L. F. I., et al. (2020). High-intensity interval versus moderate-intensity continuous training in individuals with Parkinson's disease: Hemodynamic and functional adaptation. *Journal of Physical Activity & Health*, *17*(1), 85–91. https://doi.org/10.1123/jpah.2018-0588.

Fisher, G., Brown, A. W., Bohan Brown, M. M., Alcorn, A., Noles, C., Winwood, L., et al. (2015). High intensity interval- vs moderate intensity-training for improving cardiometabolic health in overweight or obese males: A randomized controlled trial. *PLoS One*, *10*(10), e0138853. https://doi.org/10.1371/journal.pone.0138853.

Forkan, R., Pumper, B., Smyth, N., Wirkkala, H., Ciol, M. A., & Shumway-Cook, A. (2006). Exercise adherence following physical therapy intervention in older adults with impaired balance. *Physical Therapy*, *86*(3), 401–410.

Galea, M. P. (2012). Spinal cord injury and physical activity: Preservation of the body. *Spinal Cord*, *50*(5), 344–351. https://doi.org/10.1038/sc.2011.149.

Gauthier, C., Brosseau, R., Hicks, A. L., & Gagnon, D. H. (2018). Feasibility, safety, and preliminary effectiveness of a home-based self-managed high-intensity interval training program offered to long-term manual wheelchair users. *Rehabilitation Research and Practice*, *2018*, 8209360. https://doi.org/10.1155/2018/8209360.

GBD 2016 Traumatic Brain Injury and Spinal Cord Injury Collaborators. (2016). Global, regional, and national burden of traumatic brain injury and spinal cord injury, 1990–2016: A systematic analysis for the Global Burden of Disease Study. *The Lancet Neurology*, *18*(1), 56–87.

Gibala, M. J., Gillen, J. B., & Percival, M. E. (2014). Physiological and health-related adaptations to low-volume interval training: Influences of nutrition and sex. *Sports Medicine*, *44*(Suppl. 2), S127–S137. https://doi.org/10.1007/s40279-014-0259-6.

Gillen, J. B., Martin, B. J., MacInnis, M. J., Skelly, L. E., Tarnopolsky, M. A., & Gibala, M. J. (2016). Twelve weeks of sprint interval training improves indices of cardiometabolic health similar to traditional endurance training despite a five-fold lower exercise volume and time commitment. *PLoS One*, *11*(4), e0154075. https://doi.org/10.1371/journal.pone.0154075.

Gillen, J. B., Percival, M. E., Ludzki, A., Tarnopolsky, M. A., & Gibala, M. J. (2013). Interval training in the fed or fasted state improves body composition and muscle oxidative capacity in overweight women. *Obesity (Silver Spring)*, *21*(11), 2249–2255. https://doi.org/10.1002/oby.20379.

Ginis, K. A., Latimer, A. E., Arbour-Nicitopoulos, K. P., Buchholz, A. C., Bray, S. R., Craven, B. C., et al. (2010). Leisure time physical activity in a population-based sample of people with spinal cord injury part I: Demographic and injury-related correlates. *Archives of Physical Medicine and Rehabilitation*, *91*(5), 722–728. https://doi.org/10.1016/j.apmr.2009.12.027.

Ginis, K. A. M., van der Scheer, J. W., Latimer-Cheung, A. E., Barrow, A., Bourne, C., Carruthers, P., et al. (2018). Evidence-based scientific exercise guidelines for adults with spinal cord injury: An update and a new guideline. *Spinal Cord*, *56*(4), 308–321.

Gorgey, A., & Gater, D. R. (2007). Prevalence of obesity after spinal cord injury. *Topics in Spinal Cord Injury Rehabilitation*, *12*, 1–7.

Graham, K., Yarar-Fisher, C., Li, J., McCully, K. M., Rimmer, J. H., Powell, D. C., et al. (2019). Effects of high-intensity interval training versus moderate-intensity training on cardiometabolic health markers in individuals with spinal cord injury: A pilot study. *Vol. 25* (pp. 248–259). Topics in Spinal Cord Injury Rehabilitation.

Harnish, C. R., Daniels, J. A., & Caruso, D. (2017). Training response to high-intensity interval training in a 42-year-old man with chronic spinal cord injury. *The Journal of Spinal Cord Medicine*, *40*(2), 246–249. https://doi.org/10.1080/10790268.2015.1136783.

Harvey, M., Weston, K. L., Gray, W. K., O'Callaghan, A., Oates, L. L., Davidson, R., et al. (2019). High-intensity interval training in people with Parkinson's disease: A randomized, controlled feasibility trial. *Clinical Rehabilitation*, *33*(3), 428–438. https://doi.org/10.1177/0269215518815221.

Hasnan, N., Engkasan, J. P., Husain, R., & Davis, G. M. (2013). High-intensity virtual-reality arm plus FES-leg interval training in individuals with spinal cord injury. *Biomedizinische Technik. Biomedical Engineering*. https://doi.org/10.1515/bmt-2013-4028.

Jain, N. B., Ayers, G. D., Peterson, E. N., Harris, M. B., Morse, L., O'Connor, K. C., et al. (2015). Traumatic spinal cord injury in the United States, 1993-2012. *JAMA*, *313*(22), 2236–2243. https://doi.org/10.1001/jama.2015.6250.

Kolt, G. S., & McEvoy, J. F. (2003). Adherence to rehabilitation in patients with low back pain. *Manual Therapy*, *8*(2), 110–116.

Lai, B., Kim, Y., Wilroy, J., Bickel, C. S., Rimmer, J. H., & Motl, R. W. (2018). Sustainability of exercise intervention outcomes among people with disabilities: A secondary review. *Disability and Rehabilitation*, 1–12. https://doi.org/10.1080/09638288.2018.1432704.

Lai, B., Kim, Y., Wilroy, J., Bickel, C. S., Rimmer, J. H., & Motl, R. W. (2019). Sustainability of exercise intervention outcomes among people with disabilities: A secondary review. *Disability and Rehabilitation*, *41*(13), 1584–1595. https://doi.org/10.1080/09638288.2018.1432704.

Lai, B., Rimmer, J., Barstow, B., Jovanov, E., & Bickel, C. S. (2016). Teleexercise for persons with spinal cord injury: A mixed-methods feasibility case series. *JMIR Rehabilitation and Assistive Technologies*, *3*(2), e8. https://doi.org/10.2196/rehab.5524.

Lai, B. W., Rimmer, J. H., Yates, A., Jeter, A., Young, H.-J., Thirumalai, M., et al. (2020). Critical factors influencing the decision to enroll in a physical activity intervention among a predominant group of adults with spinal cord injury: A grounded theory study. *Spinal Cord*. https://doi.org/10.1038/s41393-020-0530-7.

Lai, B. W., Rimmer, J. H., Yates, A., Jeter, A., Young, H. J., Thirumalai, M., et al. (2021). Critical factors influencing the decision to enroll in a physical activity intervention among a predominant group of adults with spinal cord injury: A grounded theory study. *Spinal Cord*, *59*(1), 17–25. https://doi.org/10.1038/s41393-020-0530-7.

Lai, B., Young, H. J., Bickel, C. S., Motl, R. W., & Rimmer, J. H. (2017). Current trends in exercise intervention research, technology, and behavioral change strategies for people with disabilities: A scoping review. *American Journal of Physical Medicine & Rehabilitation*, *96*(10), 748–761. https://doi.org/10.1097/phm.0000000000000743.

Latimer, A., Martin Ginis, K. A., Craven, B., & Hicks, A. (2006). The physical activity recall assessment for people with spinal cord injury: Validity. *Medicine and Science in Sports and Exercise*, *38*, 208–216.

Laughton, G., Buchholz, A. C., Martin Ginis, K. A., & Goy, R. E. (2009). Lowering body mass index cutoffs better identifies obese persons with spinal cord injury. *Spinal Cord*, *47*, 757–762.

Little, J. P., Gillen, J. B., Percival, M. E., Safdar, A., Tarnopolsky, M. A., Punthakee, Z., et al. (2011). Low-volume high-intensity interval training reduces hyperglycemia and increases muscle mitochondrial capacity in patients with type 2 diabetes. *Journal of Applied Physiology (Bethesda, MD: 1985)*, *111*(6), 1554–1560. https://doi.org/10.1152/japplphysiol.00921.2011.

Ma, J. K., & Martin Ginis, K. A. (2018). A meta-analysis of physical activity interventions in people with physical disabilities: Content, characteristics, and effects on behaviour. *Psychology of Sport and Exercise*, *37*, 262–273. https://doi.org/10.1016/j.psychsport.2018.01.006.

MacInnis, M. J., & Gibala, M. J. (2016). Physiological adaptations to interval training and the role of exercise intensity. *The Journal of Physiology*. https://doi.org/10.1113/JP273196.

Mahabaleshwarkar, R., & Khanna, R. (2014). National hospitalization burden associated with spinal cord injuries in the United States. *Spinal Cord*, *52*(2), 139–144. https://doi.org/10.1038/sc.2013.144.

Martin Ginis, K. A., & Hicks, A. L. (2005). Exercise research issues in the spinal cord injured population. *Exercise and Sport Sciences Reviews*, *33*(1), 49–53.

Martin Ginis, K. A., Ma, J. K., Latimer-Cheung, A. E., & Rimmer, J. H. (2016). A systematic review of review articles addressing factors related to physical activity participation among children and adults with physical disabilities. *Health Psychology Review*, *10*(4), 478–494. https://doi.org/10.1080/17437199.2016.1198240.

Martins, C., Kazakova, I., Ludviksen, M., Mehus, I., Wisloff, U., Kulseng, B., et al. (2015). High-intensity interval training and isocaloric moderate-intensity continuous training result in similar improvements in body composition and fitness in obese individuals. *International Journal of Sport Nutrition and Exercise Metabolism*. https://doi.org/10.1123/ijsnem.2015-0078.

McLean, S. M., Klaber Moffett, J. A., Sharp, D. M., & Gardiner, E. (2013). A randomised controlled trial comparing graded exercise treatment and usual physiotherapy for patients with non-specific neck pain (the GET UP neck pain trial). *Manual Therapy*, *18*(3), 199–205. https://doi.org/10.1016/j.math.2012.09.005.

Mcleod, J. C., Diana, H., & Hicks, A. L. (2020). Sprint interval training versus moderate-intensity continuous training during inpatient rehabilitation after spinal cord injury: A randomized trial. *Spinal Cord*, *58*(1), 106–115. https://doi.org/10.1038/s41393-019-0345-6.

National Spinal Cord Injury Statistical Center. (2019). *Spinal cord injury facts and figures at a glance*. National Spinal Cord Injury Statistical Center.

Rajan, S., McNeely, M. J., Warms, C., & Goldstein, B. (2008). Clinical assessment and management of obesity in individuals with spinal cord injury: A review. *The Journal of Spinal Cord Medicine*, *31*, 361–372.

Richards, J. C., Johnson, T. K., Kuzma, J. N., Lonac, M. C., Schweder, M. M., Voyles, W. F., et al. (2010). Short-term sprint interval training increases insulin sensitivity in healthy adults but does not affect the thermogenic response to beta-adrenergic stimulation. *The Journal of Physiology*, *588*(Pt. 15), 2961–2972. https://doi.org/10.1113/jphysiol.2010.189886.

Rimmer, J. H. (2012). Getting beyond the plateau: Bridging the gap between rehabilitation and community-based exercise. *Physical Medicine and Rehabilitation*, *4*, 857–861.

Rimmer, J. H., Schiller, W., & Chen, M. D. (2012). Effects of disability-associated low energy expenditure deconditioning syndrome. *Exercise and Sport Sciences Reviews*, *40*(1), 22–29. https://doi.org/10.1097/JES.0b013e31823b8b82.

Scelza, W. M., Kalpakjian, C. Z., Zemper, E. D., & Tate, D. G. (2005). Perceived barriers to exercise in people with spinal cord injury. *American Journal of Physical Medicine & Rehabilitation*, *84*(8), 576–583.

Skelton, F., Hoffman, J. M., Reyes, M., & Burns, S. P. (2015). Examining health-care utilization in the first year following spinal cord injury. *The Journal of Spinal Cord Medicine*, *38*(6), 690–695. https://doi.org/10.1179/2045772314y.0000000269.

Sveaas, S. H., Bilberg, A., Berg, I. J., Provan, S. A., Rollefstad, S., Semb, A. G., et al. (2020). High intensity exercise for 3 months reduces disease activity in axial spondyloarthritis (axSpA): A multicentre randomised trial of 100 patients. *British Journal of Sports Medicine*, *54*(5), 292–297. https://doi.org/10.1136/bjsports-2018-099943.

Sweet, S. N., Rocchi, M., Arbour-Nicitopoulos, K., Kairy, D., & Fillion, B. (2017). A telerehabilitation approach to enhance quality of life through exercise among adults with paraplegia: Study protocol. *JMIR Research Protocols*, *6*(10), e202. https://doi.org/10.2196/resprot.8047.

The 2018 Annual Statistical Report for The Spinal Cord Injury Model Systems. (2018).

Tian, W., Hsieh, C.-H., DeJong, G., Backus, D., Groah, S., & Ballard, P. H. (2013). Role of body weight in therapy participation and rehabilitation outcomes among individuals with traumatic spinal cord injury. *Archives of Physical Medicine and Rehabilitation*, *94*, S125–S136.

Tjønna, A. E., Lee, S. J., Rognmo, Ø., Stølen, T. O., Bye, A., Haram, P. M., et al. (2008). Aerobic interval training versus continuous moderate exercise as a treatment for the metabolic syndrome: A pilot study. *Circulation*, *118*(4), 346–354. https://doi.org/10.1161/CIRCULATIONAHA.108.772822.

Tordi, N., Dugue, B., Klupzinski, D., Rasseneur, L., Rouillon, J. D., & Lonsdorfer, J. (2001). Interval training program on a wheelchair ergometer for paraplegic subjects. *Spinal Cord*, *39*(10), 532–537. https://doi.org/10.1038/sj.sc.3101206.

Totosy de Zepetnek, J. O., Pelletier, C. A., Hicks, A. L., & MacDonald, M. J. (2015). Following the physical activity guidelines for adults with spinal cord injury for 16 weeks does not improve vascular health: A randomized controlled trial. *Archives of Physical Medicine and Rehabilitation*, *96*(9), 1566–1575. https://doi.org/10.1016/j.apmr.2015.05.019.

Trost, S. G., Owen, N., Bauman, A. E., Sallis, J. F., & Brown, W. (2002). Correlates of adults' participation in physical activity: Review and update. *Medicine and Science in Sports and Exercise*, *34*(12), 1996 2001. https://doi.org/10.1249/01.MSS.0000038974.76900.92.

Tschoepe, R., Benfield, A., Mercer, V., & Posey, R. (2019). Systematic review of community transition programs after acute rehabilitation for adults with traumatic spinal cord injury. *American Journal of Occupational Therapy*, *73*(4_Suppl._1). https://doi.org/10.5014/ajot.2019.73S1-RP102A. 7311510245p7311510241-7311510245p7311510241.

Tulloch, H., Sweet, S. N., Fortier, M., Capstick, G., Kenny, G. P., & Sigal, R. J. (2013). Exercise facilitators and barriers from adoption to maintenance in the diabetes aerobic and resistance exercise trial. *Canadian Journal of Diabetes*, *37*(6), 367–374. https://doi.org/10.1016/j.jcjd.2013.09.002.

Turner, J., Cardenas, D. D., Warms, C. A., & McClellan, C. B. (2001). Chronic pain associated with spinal cord injuries: A community survey. *Archives of Physical Medicine and Rehabilitation*, *82*, 501–508.

U.S. Department of Health and Human Services. (2018). *Physical activity guidelines for Americans* (2nd ed.). Washington, DC: U.S. Department of Health and Human Services.

van der Scheer, J. W., Martin Ginis, K. A., Ditor, D. S., Goosey-Tolfrey, V. L., Hicks, A. L., West, C. R., et al. (2017). Effects of exercise on fitness and health of adults with spinal cord injury: A systematic review. *Neurology*, *89*(7), 736–745. https://doi.org/10.1212/wnl.0000000000004224.

Warburton, D. E., McKenzie, D. C., Haykowsky, M. J., Taylor, A., Shoemaker, P., Ignaszewski, A. P., et al. (2005). Effectiveness of high-intensity interval training for the rehabilitation of patients with coronary artery disease. *The American Journal of Cardiology*, *95*(9), 1080–1084. https://doi.org/10.1016/j.amjcard.2004.12.063.

Whyte, L. J., Gill, J. M., & Cathcart, A. J. (2010). Effect of 2 weeks of sprint interval training on health-related outcomes in sedentary overweight/obese men. *Metabolism*, *59*(10), 1421–1428. https://doi.org/10.1016/j.metabol.2010.01.002.

Chapter 31

# Stem cells and chronic spinal cord injury: Overview

Leonardino A. Digma, Pavan S. Upadhyayula, Joel R. Martin, and Joseph D. Ciacci
*Department of Neurological Surgery, University of California San Diego, San Diego, CA, United States*

## Abbreviations

| | |
|---|---|
| **ALS** | amyotrophic lateral sclerosis |
| **BDNF** | brain-derived neurotrophic factor |
| **CNS** | central nervous system |
| **DNA** | deoxyribonucleic acid |
| **EMG** | electromyography |
| **ESC** | embryonic stem cell |
| **GABA** | gaba-aminobutyric acid |
| **iPSC** | induced-pluripotent stem cell |
| **ISNCSCI** | International standards for neurological classification of spinal cord injury |
| **MSC** | mesenchymal stem cell |
| **ROS** | reactive oxygen species |
| **SCI** | spinal cord injury |

## Introduction

Spinal cord injury (SCI) is a devastating condition that is associated with a poor prognosis and long-lasting disability. Globally, SCI has an incidence of 8 to 246 per million and a prevalence of 236 to 1298 per million (Furlan et al., 2013). The lifetime cost for healthcare and living expenses that are attributable to SCI ranges from $1 to $4 million (*The Journal of Spinal Cord Medicine*, 2014). The destruction and demyelination of neurons that occurs in SCI leads acutely to quadriplegia, paraplegia, quadriparesis, and paraparesis, depending on the degree and location of injury. Furthermore, damage to neurons, glia, and vasculature also triggers a downstream sequence of events mediated primarily by inflammatory cells and cytokines. These events can lead to spreading of injury to adjacent segments. SCI can also lead to chronic complications, including bladder and bowel dysfunction, sexual dysfunction, neurogenic pain, and muscle spasticity. The current available treatments for SCI are limited to those that aim to ameliorate these chronic symptoms and complications (Gwak et al., 2016; Koulousakis & Kuchta, 2007). There are currently no therapies available that produce neurological improvement for SCI patients. Thus, the development of therapies seeking to restore sensory and motor function has been an active area of investigation. One approach, stem cell transplantation, aims to replace the cells and signaling factors that are lost in SCI. In the following chapter, we will discuss stem cells, with a particular emphasis on their potential role as a therapy in SCI.

## Pathophysiology of SCI

The pathophysiology underlying SCI is thought to begin with an initial insult followed by a secondary injury cascade that is characterized by three overlapping phases: acute, sub-acute, and chronic phases (Table 1). The acute phase refers to the period immediately following mechanical insult that leads to compression or transection of the spinal cord. This initial event damages neurons and glial cells, disrupts membrane integrity and electrical conductance, and destroys local vasculature. These processes disturb the blood–spinal cord barrier, leading to ionic imbalance and excitotoxicity (Alizadeh, Dyck, & Karimi-Abdolrezaee, 2019). The sub-acute phase encompasses the inflammatory cascade and resultant cell death by apoptosis, ischemic necrosis, ferroptosis, and demyelination through Wallerian degeneration (Ahuja et al., 2017).

Diagnosis and Treatment of Spinal Cord Injury. https://doi.org/10.1016/B978-0-12-822498-4.00031-2

**TABLE 1** Summary of events in the pathophysiology of spinal cord injury.

| Stage | Key events | Consequences |
|---|---|---|
| Acute | Mechanical compression or transection of spinal cord | –Neuronal and glial injury<br>–Disruption of membrane integrity<br>–Destruction of local vasculature<br>–Excitotoxicity |
| Sub-acute | Triggering of inflammatory and immune responses | –Necrosis<br>–Apoptosis<br>–Demyelination |
| Chronic | Remodeling and healing | –Glial scar<br>–Cystic cavity |

Of note, the key events and consequences can overlap between stages to some degree. However, the above framework is useful for the conceptualization of the cascade of events that follow the initial insult to the spinal cord.

The chronic phase is defined by remodeling and healing which usually produces a glial scar and cystic cavity at the lesion focus (Alizadeh et al., 2019; Pineau & Lacroix, 2007).

This influx of immune cells (such as macrophages, microglia, and neutrophils) is critical for clearing of debris and healing of the initial trauma, but they can also induce further inflammation by cytokine release and spreading of injury to adjacent segments (Kubota et al., 2012; Kumar et al., 2018). In addition to perpetuating further inflammation, immune cells also generate reactive oxygen species (ROS) that can cause oxidative damage to lipids, proteins, and DNA (Su et al., 2019). These oxidative reactions can incite further necrosis and apoptosis (Dizdaroglu et al., 2002; Hausmann, 2003). After the inflammatory process abates, a glial scar is formed from astrocytes. This glial scar serves as a physical and chemical barrier that impedes axonal regeneration and neurite growth (Ahuja, Martin, & Fehlings, 2016).

Ultimately, these processes result in the loss of three components that are critical for sub-serving sensory and motor networks of the spinal cord: neurons, myelin, and neurotrophic factors. Thus, if stem cells could be directed to differentiate into neurons, oligodendrocytes that promote myelination, or neurotrophic factor-producing cells, meaningful functional and clinical recovery may be possible.

## Stem cells: A brief overview

Stem cells are characterized by their capability to undergo both differentiation and self-renewal (Tajbakhsh, Rocheteau, & Le Roux, 2009). Classification of stem cells can be based on their potential differentiation into distinct cell types. Totipotent stem cells can become any cell type in the developing embryo, as well as extra-embryonic cells like the placenta. Pluripotent stem cells have a more restricted differentiation potential and are only able to make cells of the developing embryo. In other words, pluripotent cells can become any cell in the adult body. Multi-potent cells are further restricted and are only able to mature into cell types within a particular germ layer (e.g., any cell type derived from embryonic ectoderm).

## Embryonic stem cells—An overview of pre-clinical findings

During development, embryonic cells can differentiate into a multitude of different cell types that compose the different organs and organ systems of an adult animal. Thus, embryonic stem cells (ESCs) were among the first reservoirs of pluripotent stem cells to be tapped for use in SCI cell-based therapy research (Table 2).

### Animal-derived ESCs

The use of rodent ESCs in animal models of SCI has produced encouraging results at the molecular and functional level. Transplanted ESCs can survive and differentiate into neurons, oligodendrocytes, and astrocytes (Lin et al., 2017; Pajer et al., 2019). Furthermore, these transplanted cells can also form new synaptic connections (Deshpande et al., 2006; Lin et al., 2017). It has been shown that introduction of rodent ESCs with biomaterial scaffolding, may improve survival of transplanted cells, greater retention of cells at the locale of injury, and facilitation of formation of new circuitry (Kourgiantaki et al., 2020; Wang et al., 2018). In addition to scaffolding, survival and maintenance of stem cells may also

**TABLE 2 Summary of different stem cell types that have been investigated in pre-clinical studies or human clinical trials.**

| Stem Cell Type | Source | Pros | Cons |
|---|---|---|---|
| Embryonic stem cells | –Pre-implantation blastocyst | –Very broad potency | –Ethical concerns including destruction of embryo<br>–Potential teratoma formation<br>–Allogeneic transplantation<br>–Potential immune reaction |
| Mesenchymal stem cells | –Bone Marrow<br>–Umbilical Cord<br>–Adipose Tissue | –Large supply of cells<br>–Potential for autologous transplantation | –Limited potency compared to embryonic stem cells<br>–Potential teratoma formation<br>–Potential immune reaction |
| Induced pluripotent stem cells | –Adult somatic (e.g., skin or blood) | –Large supply of cells<br>–Potential for autologous transplantation | –Requires genetic manipulation<br>–Potential teratoma formation<br>–Potential immune reaction |

be promoted with co-transplantation with factors, such as brain-derived neurotrophic factor (BDNF) and vascular endothelial growth factor (Lu et al., 2014a; Robinson & Lu, 2017). Functionally, ESC transplanted cells have been demonstrated to improve locomotion in SCI models (Ramadan et al., 2018; Zhang et al., 2015). Lastly, there is also mounting data from SCI rodent model studies suggesting that transplanted ESCs may be able to reduce some of the long-term complications such as chronic pain and cardiovascular dysfunction (Hou et al., 2020; Jergova et al., 2016).

## Human-derived ESCs

The advent of molecular protocols capable of directing human ESCs, taken from human blastocysts, to neuronal or glial fates facilitated SCI transplantation studies (Lukovic et al., 2012). Demyelination is one of the hallmarks of SCI and, thus, several investigators have pursued transplanting oligodendrocyte progenitor cells to remyelinate descending spinal cord axons. In one study, Keirstead and colleagues transplanted human ESC derived-oligodendrocyte progenitors into the spinal cord of adult rats that were either 7 days or 10 months after thoracic contusion SCI (Keirstead et al., 2005). Among rats that received the transplant 7 days after injury, the investigators observed differentiation of progenitors into mature oligodendrocytes, remyelination of axons, and improved motor functioning. Differentiation into oligodendrocytes was also seen in rats that were transplanted 10 months after injury. However, these rats failed to demonstrate signs of remyelination or locomotor improvement. These data indicate that the value of transplanting stem cells in SCI patients may depend somewhat on the time since injury. In another study, transplanted human ESCs were shown to further express neurotrophic factors such as BDNF and transforming growth factor-B2 (Zhang, Denham, & Thies, 2006). These factors are, among other things, important for axonal regeneration and neurite growth and could further contribute to the utility of human ESCs as a therapeutic agent.

## ESC limitations

Taken together, there is evidence suggesting that transplanted ESCs can survive, mature, reintegrate into existing networks, and promote at least modest motor restoration in vivo. ESC studies, however, have also been accompanied by several limitations. For example, in studying ESCs, they must be collected from the inner cell mass of the blastocyst. This procedure ultimately leads to the demise of the blastocyst (Vazin & Freed, 2010) and has raised ethical concerns (Volarevic et al., 2018). Furthermore, ESC transplantation depends critically on the ability of the stem cell to differentiate specifically into a particular target cell type. When this fails to occur, one of the consequences is the generation of a teratoma, a tumor made up of somatic cells of diverse types (Nussbaum et al., 2007). The final limitation, as is the case for all forms of allogeneic transplantation, is that there always remains a definite concern for inciting an immunologic response against the transplanted cells (Swijnenburg et al., 2008).

Thus, despite the encouraging findings from ESC studies that established stem cells as a promising potential SCI therapy, ESC approaches have suffered from several drawbacks that make clinical translation difficult. For these reasons, alternative sources of stem cells have been investigated.

## Toward autologous transplantation: Mesenchymal stem cells

Among these alternative sources are mesenchymal stem cells (MSC), named for their ability to mature into mesoderm derived tissues, such as bone and cartilage (Table 2). MSCs possess their own benefits and challenges distinct from ESCs. Because they can be derived from adult tissues, MSCs are not shrouded in the same ethical controversies as ESCs. MSCs, like other forms of stem cells, can be transplanted intravenously as they are attracted by inflammatory factors to the site of injury. Lastly, because transplantation can be autologous, MSC interventions would be less likely to elicit a significant immune response. With regards to drawbacks, one of the most critical is the limited potency relative to ESCs. Nevertheless, animal studies suggest that they can differentiate into cell types that have sufficient specificity to integrate into the spinal cord and induce functional gains.

### Bone marrow–derived MSCs

Adult bone marrow harbors MSCs whose usual physiological role is replacing the cellular components of blood. The migratory ability of bone marrow–derived MSCs was demonstrated when murine bone marrow MSCs were injected into the lateral ventricles of neonatal mice and were found to migrate throughout different regions of the brain, such as the hippocampus, cerebellum, and brain stem (Kopen, Prockop, & Phinney, 1999). This was followed by a study in which bone marrow MSCs were retrieved from the lower extremities of rats and centrally injected into rats up to 7 days after thoracic contusion SCI (Hofstetter et al., 2002). Implanted cells differentiated toward astrocytes near the site of injury and, furthermore, gait improvements were found. Notably, more favorable effects were observed when transplantation of MSCs occurred a week after, rather than immediately after injury. Finally, in a step toward assessing the utility of human adult bone marrow MSCs, another study evaluated the effect of intravenous injection of human bone marrow MSC in SCI rat models 7 days after compression injury. The authors demonstrated with immunohistochemical analysis that the intravenously introduced bone marrow MSCs migrated to the site of injury. More specifically, the MSCs integrated within the ventrolateral white matter tracts. With respect to their role within their new home, MSCs were found to differentiate into oligodendrocytes, and increased axonal growth was observed in areas adjacent to injury. Functionally, locomotor improvement was seen at weeks 3 and 4 (Cízková et al., 2006).

### Adipose tissue–derived MSCs

Adipose tissue represents another tissue source with readily available MSCs. Adipose MSCs are unique in that their effect in spinal cord recovery may be mediated by changes in vasculature. One study implanted masses of adipose MSCs from humans into a rat SCI model and found that transplanted cells released angiogenic factors that ultimately led to vascular remodeling (Oh et al., 2012). This was confirmed with another study that intravenously injected adipose MSCs from dorsal fat pads into thoracic contusion SCI model rats 8 days after injury and demonstrated angiogenesis, as well as motor recovery (Ohta et al., 2017). Another interesting feature of adipose tissue is that MSCs isolated from cryopreserved adipose tissue have been shown to exhibit similar biochemical characteristics as adipose MSCs from fresh tissue (Ohta et al., 2018). The implication here is that large stores of adipose can be collected for a tissue bank that can potentially serve as a stem cell reservoir.

### Umbilical cord-derived MSCs

Lastly, MSCs derived from umbilical cord blood may also hold therapeutic potential for SCI. The utility of umbilical cell therapy for CNS injury was first introduced in the context of stroke (Chen et al., 2001) and traumatic brain injury (Lu et al., 2002). Given the promising results in these two contexts, their utility in SCI began to be investigated shortly thereafter. Umbilical MSCs have now been shown to promote improved motor function in thoracic compression and contusion SCI rat models (Saporta et al., 2003). Moreover, the specificity of the umbilical cord MSC's targeting of injured cord, if introduced intravenously, has also been well demonstrated. In the study by Saporta and colleagues, umbilical MSCs were only found in the spinal cord of SCI rats at the level of injury. The introduced cells were absent from non-injured areas of the spinal cord and, moreover, were not found in the spinal cord of non-injured rats. Importantly, similar in spirit to what has been shown for bone marrow, injection of umbilical MSCs 5 days after injury yielded superior results over injection 1 day after injury (Saporta et al., 2003). It is posited that the presence of inflammatory cytokines in the hours immediately after traumatic injury restricts the engraftment of transplanted stem cells into native tissue.

Numerous animal model and basic science studies have illustrated the potential utility of MSCs. From a bench-to-bedside translation standpoint, MSCs offer significant advantages such as mobility that allows for intravenous introduction, autologous transplantation, and relatively large available stores. Although there remains some interest in MSCs for SCI therapy, a few years after the first MSC–SCI studies were reported, advances in molecular techniques ushered in an arguably more promising source for stem cells.

## Cellular reprogramming: Induced pluripotent stem cells

One of the great feats of molecular biology was the discovery of factors that could be applied to adult somatic cells and cause them to dedifferentiate (Takahashi & Yamanaka, 2006). After application of these factors (e.g., SOX2), somatic cells are transformed into pluripotent cells. These cells, termed induced-pluripotent stem cells (iPSCs), evaded two major limitations that had hindered the progress of the development of ESC therapies. First, production of induced-pluripotent stem cells does not necessitate the destruction of a blastocyst, overcoming some of the ethical concerns tied to ESC. Second, because iPSCs can be derived from the adult tissue of an SCI patient, the subsequent transplant would be autologous and would pose lower risk for an immune-mediated transplant reaction. At the outset, iPSCs looked to be a prime candidate for SCI transplant therapy (Table 2). Results from animal model studies have garnered even further advancement.

One of the first steps in attempting to design iPSC-based transplant therapies was the development of protocols that promote iPSC differentiation specifically into neurons, glia, or their precursors. Several studies have now demonstrated this capability with both murine and human iPSCs (Lu et al., 2014b; Tsuji et al., 2010). In addition to showing that neural and glial precursors could be produced from iPSC, there are mounting data that these cells can be integrated into motor networks after transplantation. In one study, murine iPSCs were transplanted into thoracic contusion SCI mice and differentiated into neurons, astrocytes, and oligodendrocytes. The transplanted cells also promoted axonal growth and remyelination, resulting in improvement of motor function (Tsuji et al., 2010). Studies of human iPSCs have likewise produced promising results. In these studies, human iPSCs derived primarily from adult fibroblasts (Lu et al., 2014b) were transformed into neurospheres or neural stem cells and transplanted into animal models of SCI. The transplanted cells were shown to survive in the recipient animal, differentiate into neurons and glia, grow elongated axons, form synapses with the host cells, produce neurotrophic factors, and engage in remyelination. In summary, human iPSCs transplanted into animal models of SCI can replace neurons, myelination, and neurotrophic factors that are lost to injury, adding to the litany of evidence that advocates for the translation of stem cell therapies to human SCI patients.

As discussed above, the use of iPSCs avoids the ethical challenges of ESCs, but iPSCs are not devoid of limitations. As is the case for ESCs, teratoma formation is also a concern in iPSCs (Gutierrez-Aranda et al., 2010). In fact, in one of the studies discussed above, by Tsuji and colleagues, the investigators categorized their iPSC-derived precursors as either safe or unsafe, prior to engraftment (Tsuji et al., 2010). In contrast to the overall positive results observed in mice that received the safe cells, teratoma formation was found in mice that received the unsafe cells. Teratoma formation would serve as a major hurdle to translating iPSCs to clinical use. Thus, alternative protocols for producing iPSC suitable for clinical use that are less likely to produce teratomas are under investigation (Deng et al., 2018). Finally, although the transplant of iPSCs can potentially be autologous rather than allogeneic, they still carry a risk for transplant reaction (Zhao et al., 2011).

## Translation to the bedside

### Development of a more accurate model

Since most animal SCI studies conducted to date have been limited to the setting of acute SCI, there is a paucity of evidence indicating whether stem cell transplantation can be successful in chronic SCI patients. Furthermore, most studies have utilized rodent models, where the pathophysiology and long-term functional deficits do not necessarily recapitulate what is found in humans. For example, even after severe injury, rodents are capable of some level of spontaneous recovery (Fischer & Peduzzi, 2007). These points raise the question of how well findings from such models will translate to human patients with chronic SCI and have motivated the development of animal models that more faithfully reflect the cellular and functional changes found in human patients. In one study, investigators developed a chronic SCI model with Gottingen-Minnesota minipigs (Navarro et al., 2012). Using a compression approach, investigators were able to produce stable long-term neurological deficits as well as histopathological changes like those seen in humans. This model has been recently used in a study where iPSC-derived NPCs were grafted 2.5 months after compression. Grafts were found to be capable of surviving in both autologous and syngeneic settings, and capable of differentiating into neurons several months after transplantation (Strnadel et al., 2018). Further studies like these in large animals that more accurately model human SCI will be critical for identifying the cell-based therapies that have the most potential for success in human trials.

## Human studies and clinical trials

A few clinical trials examining stem cell transplantation in human SCI patients have been completed, with many others ongoing or planned (Table 3). Many of the completed trials have been early stage and were conducted with small sample sizes, and important lessons are to be gleaned. One early report described case studies on eight SCI patients that received bone marrow–derived stem cells (Geffner et al., 2008). The investigators employed multiple routes of administration: spinal cord, spinal canal, and intravenous injections. The procedure was overall safe and well tolerated. At follow-up, patients reported improved quality of life, based on the Barthel Index previously devised for use in patients in the context of stroke. In another study, Mendonca and colleagues conducted a phase I, non-randomized, uncontrolled clinical trial with 14 subjects that had complete thoracic or lumbar level SCI (Mendonça et al., 2014). The study participants received bone marrow–derived stem cell transplants and were evaluated for sensory and motor changes 6 months after the procedure.

**TABLE 3** Examples of human clinical trials for chronic spinal cord injury.

| NCT number | Trial title | Phase(s) | Injury inclusion criteria | Stem cell type studied | Primary outcomes | Location(s) |
|---|---|---|---|---|---|---|
| NCT02352077 | NeuroRegen scaffold with stem cells for chronic spinal cord injury repair | Phase 1 | Complete spinal cord injury at C5-T12 | Bone marrow mononuclear cells or mesenchymal stem cells | Number of patients with adverse effects after transplantation | –Beijing, China<br>–Suzhou, China<br>–Affiliated Hospital of Logistics University of CAPF, Tianjin, China |
| NCT02152657 | Evaluation of autologous mesenchymal stem cell transplantation in chronic spinal cord injury: a pilot study | Phase 1 | Spinal cord injury below T8 | Mesenchymal stem cells | MRI and clinical evaluation for complications | –Salvador, Bahia, Brazil |
| NCT01772810 | Safety study of human spinal cord-derived neural stem cell transplantation for the treatment of chronic SCI | Phase 1 | Spinal cord injury in T2-T12 or C5-C7 | Human spinal cord derived neural stem cells | Adverse events and lab abnormalities | –San Diego, California, United States |
| NCT01325103 | Autologous bone marrow stem cell transplantation in patients with spinal cord injury | Phase 1 | Traumatic spinal cord injury at thoracic or lumbar level | Bone marrow mesenchymal cells | Safety and feasibility of procedure | –Salvador, Bahia, Brazil |
| NCT02688049 | NeuroRegen Scaffold combined with stem cells for chronic spinal cord injury repair | Phase 1 and Phase 2 | Complete spinal cord injury at C5-T12 | Mesenchymal (Phase 1) and neural (Phase 2) stem cells | Improvements in ASIA, SSEP, and MEP | –Tianjin, China |

**TABLE 3** Examples of human clinical trials for chronic spinal cord injury—cont'd

| NCT number | Trial title | Phase(s) | Injury inclusion criteria | Stem cell type studied | Primary outcomes | Location(s) |
|---|---|---|---|---|---|---|
| NCT00816803 | Cell transplant in spinal cord injury patients | Phase 1 and Phase 2 | Traumatic spinal cord injury | Bone marrow mesenchymal cells | Changes in motor, sensory, and impairment scores on ASIA scale; acupuncture score | –Cairo, Egypt |
| NCT01186679 | Safety and efficacy of autologous bone marrow stem cells in treating spinal cord injury | Phase 1 and Phase 2 | Spinal cord injury in C4-T12 | Bone marrow mesenchymal cells | Number of patient with adverse effects; Changes in ASIA impairment scale | –Bangalore, Karnataka, India |
| NCT03505034 | Intrathecal transplantation of UC-MSC in patients with late stage of chronic spinal cord injury | Phase 2 | Traumatic spinal cord injury | Umbilical cord mesenchymal stem cells | Changes in motor and sensory scores on ASIA scale | –Guangzhou, Guangdong, China |
| NCT03521323 | Intrathecal transplantation of UC-MSC in patients with early stage of chronic spinal cord injury | Phase 2 | Traumatic spinal cord injury | Umbilical cord mesenchymal stem cells | Changes in motor and sensory scores on ASIA scale | –Guangzhou, Guangdong, China |
| NCT01393977 | Difference between rehabilitation therapy and stem cells transplantation in patients with spinal cord injury in China | Phase 2 | Traumatic spinal cord injury | Umbilical cord mesenchymal stem cells | Electromyography and brain stem auditory evoked potential | –Beijing, China |
| NCT02163876 | Study of human central nervous system (CNS) stem cell transplantation in cervical spinal cord Injury | Phase 2 | Traumatic spinal cord injury in C5-C7 | Human spinal cord derived neural stem cells | Changes in ISNCSCI motor score | –Various throughout US and Canada |
| NCT01676441 | Safety and efficacy of autologous mesenchymal stem cells in chronic spinal cord injury | Phase 2 and Phase 3 | Cervical spinal cord injury | Mesenchymal stem cells | Changes in motor score on ASIA scale | –Seoul, Korea, Republic of |

*Continued*

**TABLE 3** Examples of human clinical trials for chronic spinal cord injury—cont'd

| NCT number | Trial title | Phase(s) | Injury inclusion criteria | Stem cell type studied | Primary outcomes | Location(s) |
|---|---|---|---|---|---|---|
| NCT01873547 | Different efficacy between rehabilitation therapy and stem cells transplantation in patients with SCI in China | Phase 3 | Traumatic spinal cord injury | Umbilical cord mesenchymal stem cells | Changes in ISNCSCI/ASIA scores | –Beijing, China |
| NCT04213131 | Efficacy and safety of hUC-MSCs and hUCB-MSCs in the treatment of chronic spinal cord injury | Not applicable | Traumatic spinal cord injury | Umbilical cord mesenchymal stem cells | Changes in motor, sensory, and impairment scores on ASIA scale; acupuncture score | –Dalian, Liaoning, China |

Clinical trials were identified from https://www.clinicaltrials.gov/. Information for human clinical trials discussed in the chapter text, but not captured by our search were also included in the table if they are registered on https://www.clinicaltrials.gov/. Abbreviations: *ASIA*: American Spinal Cord Injury Association; *ISNCSCSI*: International Standards for Neurological Classification of Spinal Cord Injury; *MEP*: motor evoked potentials; *MRI*: magnetic resonance imaging; *NCT*: National Clinical Trial; *SSEP*: somatosensory evoked potentials; *UC-MSC*: umbilical cord mesenchymal stem cell.

Apart from one post-operative complication (cerebrospinal fluid leak), no patients experienced adverse events, deeming the transplantation safe. Additionally, several of the study patients demonstrated modest improvements in sensory, motor, and urologic function.

A pair of recent clinical trials made use of human central nervous system derived stem cells rather than autologous bone marrow cells. Curtis and colleagues recruited four patients with thoracic SCI for a phase 1 clinical trial of NSI-566, a neural stem cell line derived from human spinal cord (Curtis et al., 2018). The transplant procedure was well-tolerated by the study participants and, moreover, no serious adverse events were observed during the follow-up period of 18 to 27 months. Two of the four patients demonstrated motor and sensory improvements as measured by ISNCSCI scores. In one additional patient, although ISNCSCI was unchanged during follow-up, new muscle activity was detected via EMG. In another study, Levi and colleagues conducted a phase II trial of HuCNS-SC, a fetal brain-derived human CNS stem cell line, in SCI patients with cervical injury (Levi et al., 2019). The study was completed in two separate steps. First, six patients were transplanted with varying cell doses to define an optimal dose. In the second step, this derived optimal dose was applied to a treatment group of six new patients that was compared to a control group. Again, the procedure was found to be safe and feasible. Hints at motor recovery were shown in an interim analysis.

Although only a few SCI clinical trials of stem cell transplantation have been conducted, the studies carried out to date have documented safety, tolerability, and feasibility. The studies are low-powered and most did not involve a control group. Nevertheless, glimpses of motor, sensory, and urinary improvements were reported across trials that used different stem cell types for engraftment and recruited patients with varying segmental levels and time-delays since injury. These studies suggest that larger, more rigorous clinical trials with long-term follow-up are needed to determine the utility of stem cell transplantation.

## The role of immunity

It is now evident that cells of the immune system become key players in the CNS in the period after SCI. Damage to the spinal cord triggers a cascade of events that recruit several immune and other intrinsic CNS cells to the site of injury. Some macrophage subtypes, for example, communicate with astrocytes to induce reactive astrogliosis (Haan et al., 2015). Macrophages that migrate to the site of injury also can produce nitric oxide that may cause damaged neurons to undergo apoptosis (Satake et al., 2000). Although the activity of immune cells is thought to limit the damage to the spinal cord, the immune response can produce unfavorable consequences and cause detrimental effects on non-damaged cells, such as nitric

**TABLE 4** Areas of neuroscience in which stem cells are being investigated for therapy.

| Neurological disorder | Pathophysiology | Clinical phenotype |
|---|---|---|
| Spinal cord injury | Mechanical, inflammatory, and ischemic insult to neurons, glia, and local vasculature | –Paraplegia, quadriplegia, paraparesis, quadriparesis<br>–Autonomic dysfunction |
| Parkinson's disease | Degeneration of dopaminergic neurons in substantia nigra pars compacta | –Resting tremor<br>–Rigidity<br>–Bradykinesia |
| Huntington's disease | Loss of medium spiny neurons in the striatum | –Chorea<br>–Loss of saccadic eye movements<br>–Neuropsychiatric symptoms |
| Amyotrophic lateral sclerosis | Loss of motor neurons in the spinal cord and brain stem | –Progressive motor dysfunction |

As can be appreciated from the pathophysiology column, all four disorders are characterized by loss of one or more types of cells. Stem cell therapy aims to replace or generate these lost components.

oxide induced apoptosis of healthy cells. Further, myeloperoxidase expressed by neutrophils may exacerbate injury by the generation or upregulation of hypochlorous acid, inflammatory cytokines, and apoptotic-genes (Kubota et al., 2012).

The role of the immune system has always been of interest to stem cell transplant therapeutics due to well-known issues such as allograft rejection. However, with these more contemporary observations that suggest that after SCI, there is a pro-inflammatory and pro-immune milieu that makes conditions potentially sub-optimal for stem cell growth and development, the role of immunity has become of even greater salience. The use of classic immuno-suppressive drugs, such as tacrolimus and cyclosporine, as adjuvants with stem cell transplantation have been evaluated in pre-clinical studies (Nutt et al., 2013). In select human clinical trials of stem cell transplantation, immunosuppressive regimens have also been employed (Curtis et al., 2018; Levi et al., 2019).

## Applications to other areas of neuroscience

Beyond SCI, stem cells have also been explored to regenerate cells in other neurological disorders (Table 4). In the context of Parkinson's disease, which is characterized by loss of cells in the substantia nigra pars compacta, dopaminergic neurons derived from human embryonic stem cells have been transplanted into rat and mouse models of Parkinson's disease, leading to engraftment and improved motor function (Ganat et al., 2012). Clinically, there have been several reports describing outcomes for Parkinson's disease patients that underwent transplantation of stem cell-derived dopaminergic cells (Kefalopoulou et al., 2014; Schweitzer et al., 2020). These small studies have described clinical improvement in the short-term (2 years) (Schweitzer et al., 2020) and long-term (greater than a decade) (Kefalopoulou et al., 2014), but the utility of stem cells in Parkinson's disease remains uncertain without replication of these findings in controlled trials. In Huntington's disease, the underlying pathophysiology involves degeneration of GABA neurons in the basal ganglia. It has been shown in the pre-clinical setting that GABA neurons derived from ESCs can be transplanted into a mouse model of Huntington's and that transplanted cells can form dopaminergic and glutamatergic synapses, leading to motor improvement (Ma et al., 2012). In the clinical setting, however, investigations involving transplantation into human patients have had less success. Multiple studies have reported that long-term survival of grafts is limited, potentially because grafts may begin to assume a degenerative Huntington phenotype after transplant (Barker et al., 2013). Lastly, in amyotrophic lateral sclerosis (ALS), stem cells have also been used to attempt to rescue motor function or delay the widespread degeneration of motor neurons. Neural stem cells, for example, derived from human spinal cords have been used in rat models of familial ALS to delay disease progression and improve survival (Xu et al., 2006). In human patients, early-stage clinical trials involving transplantation of cells from various sources, including neural stem cells and mesenchymal stem cells, have been completed (Mazzini et al., 2015). Studies have consistently demonstrated safety and feasibility. While some transient motor improvements have been described, larger controlled trials are necessary to further assess therapeutic potential in the context of ALS.

## Conclusion

The pathophysiology that underlies SCI can be curtly characterized as loss of neurons, loss of myelination, and loss of neural growth factors. Given stem cells' ability to mature into cells that can restore these lost components, they are an ideal candidate for therapy. The SCI animal model literature is rife with evidence suggesting that transplanted stem cells can survive and integrate into recipient neural networks to promote recovery. Translation of this approach to human patients represents an open frontier that should be further explored, given the encouraging results from early efforts. It is hoped that this foray yields an effective SCI therapy for patients that desperately need one.

## Mini-dictionary of terms

Allogeneic: Cell, tissue, or organ that is of different genetic makeup from a particular recipient.
Angiogenesis: Generation of new blood vessels.
Apoptosis: Genetically programmed cell death.
Astrocyte: Nervous system cell responsible for maintenance of blood–brain barrier and scarring.
Autologous: Cell, tissue, or organ that is from the same recipient.
Blastocyst: Multi-cell structure that forms in the first week after fertilization.
Cytokine: Signaling molecule that allows cell to cell communication.
Excitotoxicity: Pathological overactivation of neurons.
Microglia: Cells in the nervous system that scavenge for debris.
Myelin: Sheaths that surround axons to promote action potential propagation.
Necrosis: Process of cellular death that leads to inflammation.
Neutrophil: Immune system cell responsible for defending against infection.
Neurotrophic factor: Proteins that support growth and reproduction of nervous system cells.
Oligodendrocyte: Nervous system cell responsible for generation of myelin.

## Key facts for stem cells

- Stem cells can self-renew and differentiate into neurons and glia.
- Stem cell reservoirs include embryonic cells, bone marrow, adipose tissue, and umbilical cells.
- Stem cells transplanted into animal models of spinal cord injury can survive and differentiate into neurons and glia.
- Early clinical trials of stem cell transplantation have demonstrated safety and feasibility.

## Key facts for spinal cord injury

- Spinal cord injury is a devastating neurological condition with limited therapeutic options.
- Spinal cord injury is characterized clinically by paraplegia, quadriplegia, paraparesis, quadriparesis, and autonomic dysfunction.
- The pathophysiology of spinal cord injury includes acute, sub-acute, and chronic stages.
- The immune system plays a critical role in the pathophysiology of spinal cord injury.

## Summary points

- Transplanted ESCs can survive, mature, reintegrate into existing networks, and promote at least modest motor restoration in vivo, but there is a concern for inciting an allogeneic immunologic response against the transplanted cells.
- Despite the encouraging findings from ESC studies that established stem cells as a promising potential SCI therapy, ESC approaches have suffered from several drawbacks that make clinical translation difficult. For these reasons, alternative sources of stem cells have been investigated.
- MSCs offer significant advantages such as mobility that allows for intravenous introduction, autologous transplantation, and relatively large available stores.
- Induced-pluripotent stem cells are an attractive candidate for SCI since they do not necessitate the destruction of a blastocyst, and transplant could be autologous posing lower risk for an immune-mediated transplant reaction.
- Only a few SCI clinical trials of stem cell transplantation have been conducted, but have documented safety, tolerability, and feasibility.

- Immunomodulation can ameliorate the unfavorable consequences and cause detrimental effects on non-damaged cells after SCI, such as nitric oxide induced apoptosis of healthy cells.
- In addition to spinal cord injury, the therapeutic utility of stem cells is being explored in other neurological disorders, such as Parkinson's disease and amyotrophic lateral sclerosis.

## Acknowledgements

None.

## Funding

This work was unfunded.

## Disclosure statement

The authors report no conflict of interest.

## References

Ahuja, C. S., Martin, A. R., & Fehlings, M. (2016). Recent advances in managing a spinal cord injury secondary to trauma. *F1000Research*, *5*.

Ahuja, C. S., et al. (2017). Traumatic spinal cord injury. *Nature Reviews. Disease Primers*, *3*, 17018.

Alizadeh, A., Dyck, S. M., & Karimi-Abdolrezaee, S. (2019). Traumatic spinal cord injury: An overview of pathophysiology, models and acute injury mechanisms. *Front. Neurol.*, *10*, 282.

Barker, R. A., et al. (2013). The long-term safety and efficacy of bilateral transplantation of human fetal striatal tissue in patients with mild to moderate Huntington's disease. *Journal of Neurology, Neurosurgery, and Psychiatry*, *84*(6), 657–665.

Chen, J., et al. (2001). Intravenous administration of human umbilical cord blood reduces behavioral deficits after stroke in rats. *Stroke*, *32*(11), 2682–2688.

Cízková, D., et al. (2006). Transplants of human mesenchymal stem cells improve functional recovery after spinal cord injury in the rat. *Cellular and Molecular Neurobiology*, *26*(7-8), 1167–1180.

Curtis, E., et al. (2018). A first-in-human, phase i study of neural stem cell transplantation for chronic spinal cord injury. *Cell Stem Cell*, *22*(6), 941–950.e6.

Deng, J., et al. (2018). Cell transplantation for spinal cord injury: Tumorigenicity of induced pluripotent stem cell-derived neural stem/progenitor cells. *Stem Cells International*, *2018*, 5653787.

Deshpande, D. M., et al. (2006). Recovery from paralysis in adult rats using embryonic stem cells. *Annals of Neurology*, *60*(1), 32–44.

Dizdaroglu, M., et al. (2002). Free radical-induced damage to DNA: Mechanisms and measurement. *Free Radical Biology & Medicine*, *32*(11), 1102–1115.

Fischer, F. R., & Peduzzi, J. D. (2007). Functional recovery in rats with chronic spinal cord injuries after exposure to an enriched environment. *The Journal of Spinal Cord Medicine*, *30*(2), 147–155.

Furlan, J. C., et al. (2013). Global incidence and prevalence of traumatic spinal cord injury. *The Canadian Journal of Neurological Sciences*, *40*(4), 456–464.

Ganat, Y. M., et al. (2012). Identification of embryonic stem cell-derived midbrain dopaminergic neurons for engraftment. *The Journal of Clinical Investigation*, *122*(8), 2928–2939.

Geffner, L. F., et al. (2008). Administration of autologous bone marrow stem cells into spinal cord injury patients via multiple routes is safe and improves their quality of life: Comprehensive case studies. *Cell Transplantation*, *17*(12), 1277–1293.

Gutierrez-Aranda, I., et al. (2010). Human induced pluripotent stem cells develop teratoma more efficiently and faster than human embryonic stem cells regardless the site of injection. *Stem Cells*, *28*(9), 1568–1570.

Gwak, Y. S., et al. (2016). Combined approaches for the relief of spinal cord injury-induced neuropathic pain. *Complementary Therapies in Medicine*, *25*, 27–33.

Haan, N., et al. (2015). Crosstalk between macrophages and astrocytes affects proliferation, reactive phenotype and inflammatory response, suggesting a role during reactive gliosis following spinal cord injury. *Journal of Neuroinflammation*, *12*, 109.

Hausmann, O. N. (2003). Post-traumatic inflammation following spinal cord injury. *Spinal Cord*, *41*(7), 369–378.

Hofstetter, C. P., et al. (2002). Marrow stromal cells form guiding strands in the injured spinal cord and promote recovery. *Proceedings of the National Academy of Sciences of the United States of America*, *99*(4), 2199–2204.

Hou, S., et al. (2020). Grafting embryonic raphe neurons reestablishes serotonergic regulation of sympathetic activity to improve cardiovascular function after spinal cord injury. *The Journal of Neuroscience*, *40*(6), 1248–1264.

Jergova, S., et al. (2016). Recombinant neural progenitor transplants in the spinal dorsal horn alleviate chronic central neuropathic pain. *Pain*, *157*(4), 977–989.

Kefalopoulou, Z., et al. (2014). Long-term clinical outcome of fetal cell transplantation for Parkinson disease: Two case reports. *JAMA Neurology*, *71*(1), 83–87.

Keirstead, H. S., et al. (2005). Human embryonic stem cell-derived oligodendrocyte progenitor cell transplants remyelinate and restore locomotion after spinal cord injury. *The Journal of Neuroscience*, *25*(19), 4694–4705.

Kopen, G. C., Prockop, D. J., & Phinney, D. G. (1999). Marrow stromal cells migrate throughout forebrain and cerebellum, and they differentiate into astrocytes after injection into neonatal mouse brains. *Proceedings of the National Academy of Sciences of the United States of America*, *96*(19), 10711–10716.

Koulousakis, A., & Kuchta, J. (2007). Intrathecal antispastic drug application with implantable pumps: Results of a 10 year follow-up study. *Acta Neurochirurgica. Supplement*, *97*(Pt. 1), 181–184.

Kourgiantaki, A., et al. (2020). Neural stem cell delivery via porous collagen scaffolds promotes neuronal differentiation and locomotion recovery in spinal cord injury. *npj Regenerative Medicine*, *5*, 12.

Kubota, K., et al. (2012). Myeloperoxidase exacerbates secondary injury by generating highly reactive oxygen species and mediating neutrophil recruitment in experimental spinal cord injury. *Spine*, *37*(16), 1363–1369.

Kumar, H., et al. (2018). Neutrophil elastase inhibition effectively rescued angiopoietin-1 decrease and inhibits glial scar after spinal cord injury. *Acta Neuropathologica Communications*, *6*(1), 73.

Levi, A. D., et al. (2019). Clinical outcomes from a multi-center study of human neural stem cell transplantation in chronic cervical spinal cord injury. *Journal of Neurotrauma*, *36*(6), 891–902.

Lin, C.-C., et al. (2017). The therapeutic effectiveness of delayed fetal spinal cord tissue transplantation on respiratory function following mid-cervical spinal cord injury. *Neurotherapeutics*, *14*(3), 792–809.

Lu, D., et al. (2002). Intravenous administration of human umbilical cord blood reduces neurological deficit in the rat after traumatic brain injury. *Cell Transplantation*, *11*(3), 275–281.

Lu, P., et al. (2014a). Promotion of survival and differentiation of neural stem cells with fibrin and growth factor cocktails after severe spinal cord injury. *Journal of Visualized Experiments*, *89*, e50641.

Lu, P., et al. (2014b). Long-distance axonal growth from human induced pluripotent stem cells after spinal cord injury. *Neuron*, *83*(4), 789–796.

Lukovic, D., et al. (2012). Concise review: Human pluripotent stem cells in the treatment of spinal cord injury. *Stem Cells*, *30*(9), 1787–1792.

Ma, L., et al. (2012). Human embryonic stem cell-derived GABA neurons correct locomotion deficits in quinolinic acid-lesioned mice. *Cell Stem Cell*, *10*(4), 455–464.

Mazzini, L., et al. (2015). Human neural stem cell transplantation in ALS: Initial results from a phase I trial. *Journal of Translational Medicine*, *13*, 17.

Mendonça, M. V. P., et al. (2014). Safety and neurological assessments after autologous transplantation of bone marrow mesenchymal stem cells in subjects with chronic spinal cord injury. *Stem Cell Research & Therapy*, *5*(6), 126.

Navarro, R., et al. (2012). Chronic spinal compression model in minipigs: A systematic behavioral, qualitative, and quantitative neuropathological study. *Journal of Neurotrauma*, *29*(3), 499–513.

Nussbaum, J., et al. (2007). Transplantation of undifferentiated murine embryonic stem cells in the heart: Teratoma formation and immune response. *The FASEB Journal*, *21*(7), 1345–1357.

Nutt, S. E., et al. (2013). Caudalized human iPSC-derived neural progenitor cells produce neurons and glia but fail to restore function in an early chronic spinal cord injury model. *Experimental Neurology*, *248*, 491–503.

Oh, J. S., et al. (2012). Transplantation of an adipose stem cell cluster in a spinal cord injury. *Neuroreport*, *23*(5), 277–282.

Ohta, Y., et al. (2017). Intravenous infusion of adipose-derived stem/stromal cells improves functional recovery of rats with spinal cord injury. *Cytotherapy*, *19*(7), 839–848.

Ohta, Y., et al. (2018). Isolation of adipose-derived stem/stromal cells from cryopreserved fat tissue and transplantation into rats with spinal cord injury. *International Journal of Molecular Sciences*, *19*(7).

Pajer, K., et al. (2019). Neuroectodermal stem cells grafted into the injured spinal cord induce both axonal regeneration and morphological restoration via multiple mechanisms. *Journal of Neurotrauma*, *36*(21), 2977–2990.

Pineau, I., & Lacroix, S. (2007). Proinflammatory cytokine synthesis in the injured mouse spinal cord: Multiphasic expression pattern and identification of the cell types involved. *The Journal of Comparative Neurology*, *500*(2), 267–285.

Ramadan, W. S., et al. (2018). Neuroectodermal stem cells: A remyelinating potential in acute compressed spinal cord injury in rat model. *Journal of Biosciences*, *43*(5), 897–909.

Robinson, J., & Lu, P. (2017). Optimization of trophic support for neural stem cell grafts in sites of spinal cord injury. *Experimental Neurology*, *291*, 87–97.

Saporta, S., et al. (2003). Human umbilical cord blood stem cells infusion in spinal cord injury: Engraftment and beneficial influence on behavior. *Journal of Hematotherapy & Stem Cell Research*, *12*(3), 271–278.

Satake, K., et al. (2000). Nitric oxide via macrophage iNOS induces apoptosis following traumatic spinal cord injury. *Brain Research. Molecular Brain Research*, *85*(1–2), 114–122.

Schweitzer, J. S., et al. (2020). Personalized iPSC-derived dopamine progenitor cells for Parkinson's disease. *The New England Journal of Medicine*, *382*(20), 1926–1932.

Strnadel, J., et al. (2018). Survival of syngeneic and allogeneic iPSC-derived neural precursors after spinal grafting in minipigs. *Science Translational Medicine*, *10*(440).

Su, L.-J., et al. (2019). Reactive oxygen species-induced lipid peroxidation in apoptosis, autophagy, and ferroptosis. *Oxidative Medicine and Cellular Longevity*, *2019*, 5080843.

Swijnenburg, R.-J., et al. (2008). Immunosuppressive therapy mitigates immunological rejection of human embryonic stem cell xenografts. *Proceedings of the National Academy of Sciences of the United States of America*, *105*(35), 12991–12996.

Tajbakhsh, S., Rocheteau, P., & Le Roux, I. (2009). Asymmetric cell divisions and asymmetric cell fates. *Annual Review of Cell and Developmental Biology*, *25*, 671–699.

Takahashi, K., & Yamanaka, S. (2006). Induction of pluripotent stem cells from mouse embryonic and adult fibroblast cultures by defined factors. *Cell*, *126*(4), 663–676.

(2014). Spinal cord injury facts and figures at a glance. *The Journal of Spinal Cord Medicine*, *37*(1), 117–118.

Tsuji, O., et al. (2010). Therapeutic potential of appropriately evaluated safe-induced pluripotent stem cells for spinal cord injury. *Proceedings of the National Academy of Sciences of the United States of America*, *107*(28), 12704–12709.

Vazin, T., & Freed, W. J. (2010). Human embryonic stem cells: Derivation, culture, and differentiation: a review. *Restorative Neurology and Neuroscience*, *28*(4), 589–603.

Volarevic, V., et al. (2018). Ethical and safety issues of stem cell-based therapy. *International Journal of Medical Sciences*, *15*(1), 36–45.

Wang, C., et al. (2018). Injectable nanoreinforced shape-memory hydrogel system for regenerating spinal cord tissue from traumatic injury. *ACS Applied Materials & Interfaces*, *10*(35), 29299–29307.

Xu, L., et al. (2006). Human neural stem cell grafts ameliorate motor neuron disease in SOD-1 transgenic rats. *Transplantation*, *82*(7), 865–875.

Zhang, Y. W., Denham, J., & Thies, R. S. (2006). Oligodendrocyte progenitor cells derived from human embryonic stem cells express neurotrophic factors. *Stem Cells and Development*, *15*(6), 943–952.

Zhang, K., et al. (2015). Polarized macrophages have distinct roles in the differentiation and migration of embryonic spinal-cord-derived neural stem cells after grafting to injured sites of spinal cord. *Molecular Therapy*, *23*(6), 1077–1091.

Zhao, T., et al. (2011). Immunogenicity of induced pluripotent stem cells. *Nature*, *474*(7350), 212–215.

Chapter 32

# Viral vector gene therapy approaches for regeneration and repair in spinal cord injury

**Simon J. O'Carroll and Connor A. Clemett**

*Department of Anatomy and Medical Imaging, School of Medical Sciences, Faculty of Medical and Health Sciences, University of Auckland, Auckland, New Zealand*

## List of abbreviations

| | |
|---|---|
| **SCI** | spinal cord injury |
| **CNS** | central nervous system |
| **AdV** | adenovirus |
| **HSV** | herpes simplex virus |
| **LV** | lentivirus |
| **AAV** | adeno-associated virus |
| **IM** | intramuscular |
| **IN** | intraneural |
| **IV** | intravenous |
| **IT** | intrathecal |
| **NGF** | nerve growth factor |
| **BDNF** | brain-derived neurotrophic factor |
| **NT** | neurotrophin |
| **BBB** | blood-brain barrier |
| **CSPG** | chondroitin sulfate proteoglycan |
| **ChABC** | chondroitinase ABC |
| **ADAMTS4** | a disintegrin and metalloproteinase with thrombospondin motifs 4 |

## Introduction

Traumatic spinal cord injury (SCI) results from a physical insult to the spinal cord. This leads to disruption of the blood supply and loss of tissue near the site of impact. Following this, there is ongoing detrimental change to the tissue, including inflammation, formation of an inhibitory scar, and inhibition of axonal regeneration and remyelination (Fig. 1). A number of approaches have been tested in pre-clinical models including increasing the intrinsic regenerative ability of neurons or removing the inhibitory molecules present after injury. Approaches include the delivery of neurotrophic factors (Keefe, Sheikh, & Smith, 2017), strategies to block inhibitory receptor signaling (Lang et al., 2015) or using enzymes to degrade inhibitory molecules (Bradbury et al., 2002). However, conventional methods of delivery of such factors have limitations, such as poor stability in vivo, poor tissue penetration, and the potential for off-target effects (Bo, Wu, Yeh, & Zhang, 2011). Viral vector gene therapy can be applied to a number of these approaches to help address these issues. This chapter describes different viral vectors that have been used to target the spinal cord, different delivery methods, and examples of how viral gene therapy has been used to effectively target therapeutic agents to the spinal cord.

## Viral vector approaches

Viral vectors allow for long-term, stable, and targeted gene expression. A number of viral vectors have been trialed as a gene therapy approach to SCI repair. Each of these approaches has advantages and shortcomings (Fig. 2).

**Diagnosis and Treatment of Spinal Cord Injury. https://doi.org/10.1016/B978-0-12-822498-4.00032-4**

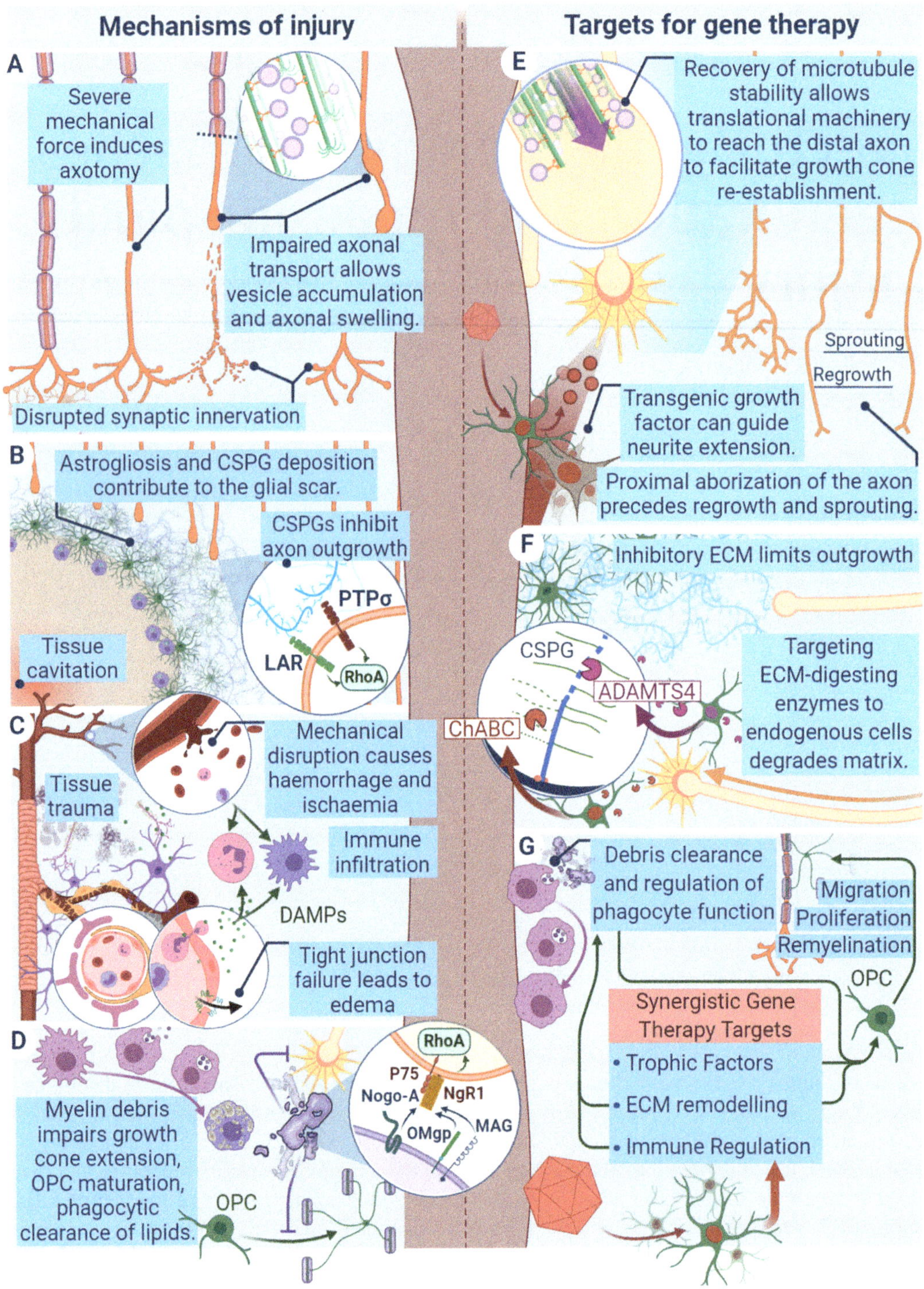


**FIG. 1** Overview of traumatic spinal cord injury mechanisms and potential targets for gene-based therapies. **Mechanism**: (A) Mechanical trauma impairs axonal function leading to synaptic dysfunction. (B) ECM deposition by reactive astrocytes at the lesion border and tissue cavitation significantly impairs axonal regrowth through the injury site. (C) Vascular hemorrhage and leakage disrupts nervous tissue homeostasis whilst a large infiltrating immune cell population responds aggressively to inflammatory signals resulting from tissue trauma. (D) Myelin debris resulting from oligodendrocyte death inhibits remyelination and axon regrowth. Overwhelming of lipid metabolism within phagocytes leads to prolonged inflammatory activation. **Treatment targets**: (E) Growth cone reformation is achieved by the re-establishment of anterograde axonal transport. Neurite extension can be stimulated by the presence of transgenic growth factor cues. (F) The digestion of extracellular matrix (ECM) molecules such as chondroitin sulfate proteoglycans (CSPGs) reduces the inhibition of axonal regrowth exerted by the astroglial scar. (G) The mechanisms of many gene therapy approaches have beneficial effects upon multiple injury processes. For example, stimulation of remyelination using current treatment approaches likely results from the direct effects of transgene products upon endogenous oligodendrocyte precursors (OPCs), i.e., neurotropic factors or removal of CSPGs. However, removal of CSPGs can also regulate phagocytic phenotype and function, which in turn promotes myelin repair. Further work should take consideration the interactions of these factors and how combinational viral gene therapy could be used to optimize this beneficial synergy. *(Created with BioRender.com.)*

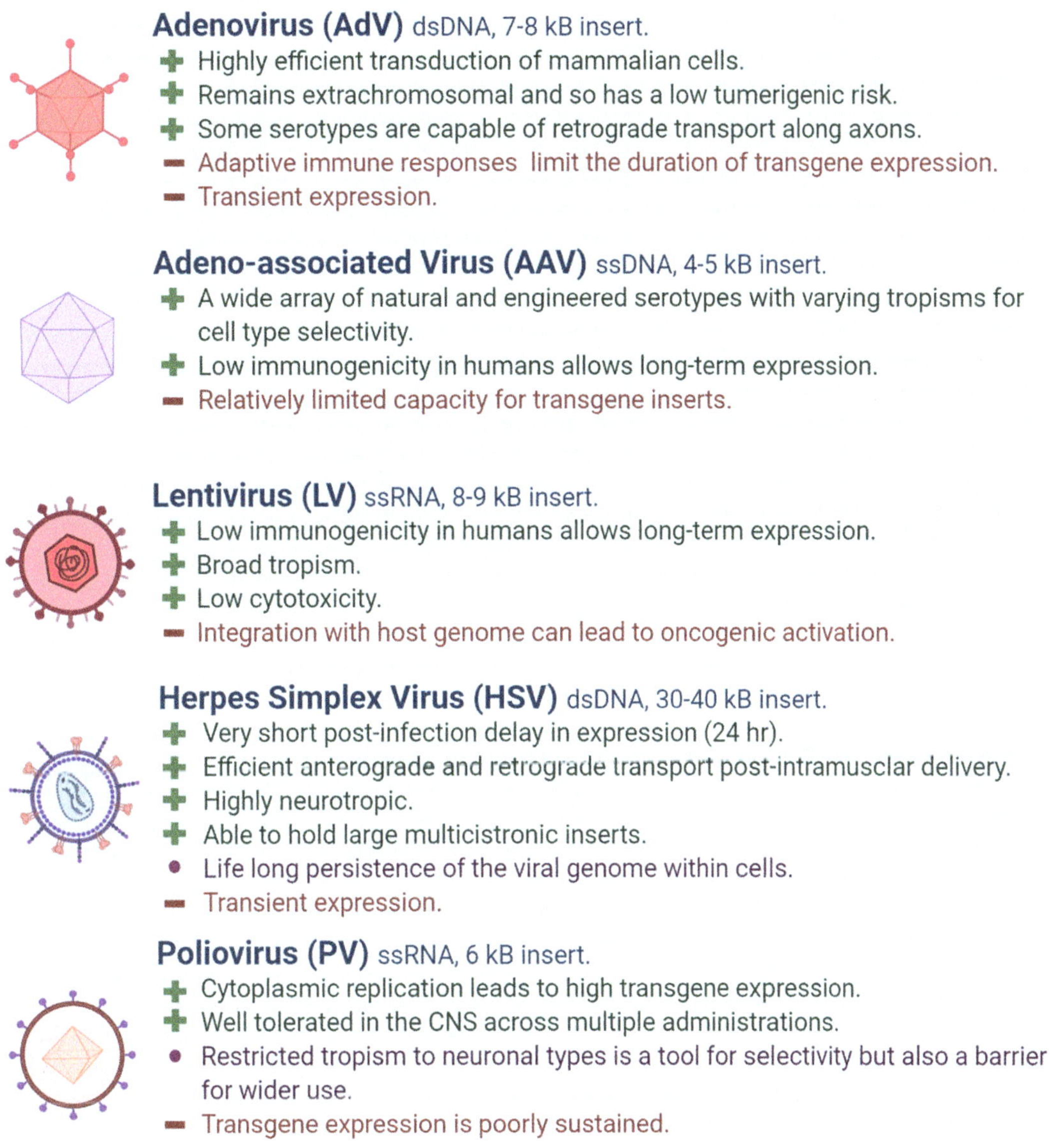


FIG. 2 Overview of viral vector systems commonly used in the treatment and study of spinal cord injury. A number of viral vectors can be used for gene delivery to the spinal cord, each of which has advantages (+) and disadvantages (−). *(Created with BioRender.com.)*

## Adenovirus

Adenoviruses (AdVs) are encapsulated double-stranded DNA viruses with a 36 kb genome and therefore have the capability of holding large transgenes. They can transduce both dividing and non-dividing cells and exhibit high transduction efficiency in a wide variety of cell types. The genome remains extra chromosomal in the nucleus, thereby the transgene is not integrated into the host cell's genome, avoiding risk of tumorigenesis. However, transgene expression is only transient and expression is lost during division as the non-integrated genome is received by only one daughter cell. Also, an immune response is elicited against AdV viral antigens leading to inflammation and limiting the period of transgene expression (Byrnes, MacLaren, & Charlton, 1996).

### Poliovirus

Polioviruses are small, non-enveloped RNA viruses and recombinant poliovirus replicons can be used for gene therapy (Khromykh, 2000). As infection is restricted to motor neurons in the hind brain and spinal cord these have been suggested as a potential vector for use in the spinal cord (Jackson, Cobbs, Peduzzi, Novak, & Morrow, 2001). Advantages of poliovirus include that its RNA genome is replicated in the cytoplasm, leading to very high levels of gene expression (Khromykh, 2000). Multiple inoculations of poliovirus replicons in the spinal cord has been shown to be safe, give good levels of gene expression with no functional deficits or inflammatory response. However, studies show that gene expression is not sustained (Jackson et al., 2001), although this could be of benefit if transient gene expression is required.

### Herpes simplex virus

Herpes simplex virus (HSV) is an enveloped virus that contains a large (150 kb) double-stranded genome, allowing for an almost unlimited packaging capacity for transgenes. They demonstrate an exceptional capability to provide long-lived infections, with the genome being maintained extra-chromosomally in the nucleus of host cells (Glorioso & Fink, 2004). Advantages of HSV are it is neurotrophic and has a tendency to remain latent while permanently transducing target cells. The large size of the vector allows for expression of multiple genes which has been suggested as advantageous in development of effective treatments (Uchida et al., 2014). However, following transduction a number of viral proteins are expressed, leading to cytotoxicity and immune responses against targeted cells (Poulsen, Harrop, & During, 2002).

### Lentivirus

Lentiviral vectors (LV) are a sub-class of retroviruses derived from human immunodeficiency viruses. Due to their ability to transduce dividing and non-dividing cells, LV vectors can transduce post-mitotic neurons, making them a powerful tool for targeting the CNS (Naldini et al., 1996) and as such have been used extensively for gene therapy targeting of the spinal cord (Uchida et al., 2014). Because LV genomes are integrated into the host genome they can provide long-term stable gene expression. While integration poses a risk of oncogenic transformation it has been suggested that this is reduced by the virus integrating into gene ends rather than promoter regions (Baum, Kustikova, Modlich, Li, & Fehse, 2006).

### Adeno-associated virus

One of the most well-characterized vectors for gene therapy is derived from adeno-associated virus (AAV). Wild type AAVs are small, 4.7 kb, linear, single-stranded DNA (ssDNA) viruses in the Parvovirus family. For therapeutic use, the *rep* and *cap* genes, involved in viral replication and capsid production, respectively, are removed and replaced by an expression cassette containing therapeutic transgenes under the control of a promoter and flanked by the AAV ITRs, required for directing genome replication and packaging, forming what is termed a recombinant AAV (rAAV) (During, Young, Baer, Lawlor, & Klugmann, 2003). AAVs are considered ideal for gene therapy approaches as they are small and non-replicative, can transduce dividing and non-dividing cells, are non-pathogenic to humans and can provide long-lasting changes in gene expression. A disadvantage of AAV vectors is their small genome, impacting on the size of transgenes that can be used. A large number of clinical trials using AAV have demonstrated the relative safety of AAV gene therapy (Penaud-Budloo, François, Clément, & Ayuso, 2018).

## Routes of administration

A number of different routes of administration are available for gene therapy delivery to the spinal cord (Fig. 3) and are classified as either remote or direct delivery (Hardcastle, Boulis, & Federici, 2018). Remote delivery involves non-invasive or minimally invasive routes of administration such as intramuscular (IM), intraneural (IN), intravenous (IV), and intrathecal (IT). These are dependent on the capability of specific gene therapy serotypes to travel to the spinal cord following peripheral delivery. Direct delivery involves an invasive surgical approach that directly targets a specific area of the spinal cord tissue via injection of the gene therapy and is referred to as intraparenchymal delivery.

FIG. 3 Overview of common and potential gene therapy delivery routes for the treatment of spinal cord injury. Viral vectors can be delivered to spinal cord via a number of routes allowing for ease of delivery or targeted delivery depending on the desired approach. *(Created with BioRender.com.)*

## Intramuscular and intraneural delivery

Retrograde transport is the transport of viral vector particles to neuronal cell bodies via axonal transport following IM or IN (e.g., sciatic nerve) injection. A number of viral gene therapies can be transported to the spinal cord from peripheral injection sites such as skin, muscle, or peripheral nerves (Franz, Weidner, & Blesch, 2012). Following injection of an HSV gene therapy into the gastrocnemius muscle of the rat, gene expression in neurons of the anterior horn spreads widely through the spinal cord and has been shown to be maintained for up to 182 days post-injection (Keir, Mitchell, Feldman, & Martin, 1995; Yamamura et al., 2000). Adenovirus has been used successfully in a number of studies to target the spinal cord via IM injection and delivery of neuroprotective compounds has been achieved (Nakajima et al., 2007; Nakajima et al., 2010). A number of AAV vector serotypes have been demonstrated to transduce neurons of the spinal cord following both IM and IN injection [reviewed in (Hardcastle et al., 2018) (Boulis et al., 2003; Towne, Schneider, Kieran, Redmond Jr., & Aebischer, 2010)]. However, retrograde transport of AAV vectors can have low efficiency. To overcome this, targeted evolution can be used to improve transport (Sun & Schaffer, 2018; Tervo et al., 2016). When Tet1, a peptide with high affinity for the tetanus toxin GT1b receptor that undergoes retrograde transport in the spinal cord, was grafted onto AAV1, enhanced axonal terminal binding and uptake was seen, with this vector showing a four-fold enhancement in retrograde transport in DRG explants (Davis et al., 2015). While IM injection is relatively easy and non-invasive, this largely restricts transduction to motor neurons or other populations in the anterior horn and so is not able to provide a more diffuse treatment. Also, the large amount of vector that would be required to transduce larger human muscles and the relatively long axons compared to rodents has been raised as challenges to this approach for vector delivery.

## Intravenous delivery

A number of AAV vectors, especially AAV9 and similar serotypes have been found capable of crossing the blood-brain barrier and shown promise for delivering therapeutic agents to the whole spinal cord (Duque et al., 2009). However, systemic delivery using AAV leads to transduction of peripheral organs and there can be reduced levels of transduction due to the presence of antibodies against the virus. Questions also remain as to whether results seen in animal models will translate to humans. A vector, AAV-PHP, was found to have better transduction of the CNS after IV injection than its parent AAV9 but its ability to cross the BBB has been found to be species and even mouse strain dependent due to lack of its receptor (Ly6A) (Batista et al., 2020). No known analogue for this receptor is found in humans and such species differences may impact on the usefulness of such novel BBB-crossing AAV capsids. Also, while whole spinal cord transduction may be useful for developmental disorders, targeted transduction is likely to be of more use to treat SCI.

## Intrathecal delivery

This approach allows for widespread transduction, while avoiding off-target effects (Federici et al., 2012). The vector is delivered into CSF, via the sub-arachnoid space, allowing perfusion into the spinal cord tissue. Gene therapy vectors including AAV (Zhang & Yang, 2020), adenovirus (Mannes, Caudle, O'Connell, & Iadarola, 1998), lentivirus (Thomas, Palma, & Shea, 2015), herpesvirus (Kitamura et al., 2007) and poliovirus ( Jackson et al., 2001) have all been used successfully to target the injured spinal cord and delivery therapeutic agents. While intrathecal injections result in greater spatial distribution, the trade-off is a comparatively lower level of expression compared to delivery into the tissue (Mannes et al., 1998).

## Intraparenchymal delivery

Intraparenchymal injection of viral vectors into the spinal cord of animal models has been widely explored and used successfully by a number of groups (Blits & Bunge, 2006). This allows for focal targeting of a vector but bio-distribution can be increased by multi-level injections to the cord. Intraparenchymal injection allows for much lower volumes of virus to be used compared to remote delivery techniques. However, a major barrier toward clinical translation is the validation of a safe surgical approach for delivery to the spinal cord in large animal models and humans. Pre-clinical studies have shown the safety of delivering AAV to the cervical spinal cord of pigs (Federici, Riley, Park, Bain, & Boulis, 2009), so this does appear to be a feasible approach.

## Viral vector approaches for therapeutic delivery

### Neurotrophins

Following SCI, axons fail to regenerate, at least in part due to low intrinsic regenerative capacity. This has led to a large field of study testing the use of neurotrophic factors, including nerve growth factor (NGF), brain-derived neurotrophic factor (BDNF), and the neurotrophin family (NT-3, NT-4/5), to create a growth permissive environment after an injury (Keefe et al., 2017). However, delivery of neurotrophic factors to the spinal cord has major limitations. Their short half-life and issues with crossing the BBB makes it difficult to deliver biologically relevant amounts to the spinal cord for prolonged periods of time. Also, widespread delivery of neurotrophins can lead to adverse effects (Blesch, Fischer, & Tuszynski, 2012). Viral vector delivery has been used to target neurotrophins to specific neuronal and axonal populations, either directly to the cord for expression from endogenous cells or via viral transduction of cells that are then transplanted into the cord. This has been used successfully to promote axonal regeneration and is extensively reviewed elsewhere (Blesch et al., 2012; Blits & Bunge, 2006; Bo et al., 2011; Franz et al., 2012; Uchida et al., 2014).

While axonal regrowth can be promoted by the presence of neurotrophic factors, growth beyond the spinal cord lesion site is often limited due to continual presence of growth factors and the absence of a gradient of these factors across the injury site (Franz et al., 2012). The use of gene therapy approaches has helped overcome some of these issues through allowing for spatial and temporal regulation of neurotrophin delivery. Viral gene transfer has been used successfully to establish neurotrophin gradients. Viral delivery of NT3 leads to growth of axons through a spinal cord lesion (Alto et al., 2009; Bradbury, Khemani, King, Priestley, & McMahon, 1999) and a gradient established by lentiviral BDNF delivery promoted axonal sprouting toward the highest concentration of BDNF (Bonner, Blesch, Neuhuber, & Fischer, 2010). A study by (Blesch & Tuszynski, 2007) using a tetracycline-regulatable retroviral system to control BDNF expression in transplanted fibroblasts, showed that only transient expression was required to induce axonal growth into the injury site. Viral vectors allow for the targeting of specific cell populations. Targeted evolution of AAV vectors has been successful in promoting specific cell tropism (Deverman, Ravina, Bankiewicz, Paul, & Sah, 2018). As different neurotrophins stimulate different neuronal and axonal populations, the use of engineered AAV viral vectors in conjunction with cell specific promoters would allow for specific targeting of the correct cell type and would help prevent adverse effects caused by widespread neurotrophin delivery.

### CSPG

Another major hurdle preventing axon regeneration and repair after SCI is the release of growth inhibitory molecules; myelin-associated growth inhibitors, chemo-repulsive guidance molecules, and highly sulfated proteoglycans, especially chondroitin sulfate proteoglycans (CSPGs) (Bradbury & McMahon, 2006; He & Koprivica, 2004; Park, Liu, Hu, Kanter, & He, 2010). Following CNS injury, CSPGs are rapidly upregulated at the lesion site by reactive astrocytes within the glial scar (Ohtake & Li, 2015) and form a chemical barrier for axon regeneration via interaction of the sugar GAG side chains with a number of receptor proteins (reviewed in (Sharma, Selzer, & Li, 2012).

Degradation of CSPGs to overcome this inhibition is being trialed as a therapeutic intervention for SCI. Chondroitinase ABC (ChABC) is a bacterial enzyme that degrades the GAG side-chains of CSPGs. In vivo delivery of this enzyme leads to beneficial effects in a number of experimental SCI models (Bradbury & Carter, 2011) although results have been mixed, depending on the type of injury (Bartus et al., 2014). This is thought to be at least in part due to issues with enzyme stability (Bartus et al., 2014). Therefore, gene therapy approaches where cells are transduced with a ChABC viral vector have been tested as a promising tool for achieving stable delivery of the enzyme. Early attempts using gene therapy had limited success in digesting CSPGs (Cafferty, Yang, Duffy, Li, & Strittmatter, 2007; Curinga et al., 2007; Jin, Ketschek, Jiang, Smith, & Fischer, 2011). However, modification of ChABC through mutagenesis of key *N*-glycosylation sites allowed the expression and efficient secretion of active ChABC enzyme by mammalian cells (Muir et al., 2010) and a number of studies have successfully shown that lentiviral delivery of mammalianized ChABC leads to improvement in both cervical and thoracic models of SCI, reduced lesion size, improved sensorimotor function (sensory axon conduction, hindlimb function, forelimb reaching, and grasping), increased axonal conduction, and increased serotonergic and vGlut1innervation (Bartus et al., 2014; Burnside et al., 2018; Didangelos, Iberl, Vinsland, Bartus, & Bradbury, 2014; James et al., 2015; Zhao et al., 2011). While to date gene therapy approaches to deliver ChABC have not been trialed in non-rodent species, promising results in large animal models following delivery of ChABC enzyme to the injured cord have been seen. Injection of ChABC into the primate spinal cord following a C7 transection led to improved hand function (Rosenzweig et al., 2019) and a recent clinical trial using ChABC injection in canines with chronic, naturally occurring spinal cord injuries showed

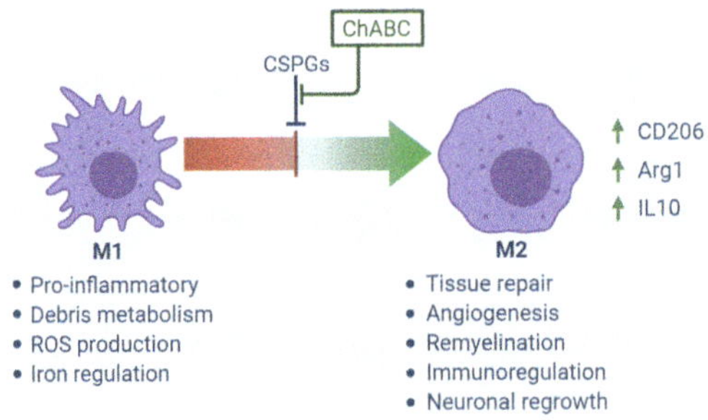


FIG. 4 CSPG digestion by ChABC causes switch to pro-regenerative inflammatory phenotype. Viral vector delivery of chondroitinase ABC modulates the inflammatory response to a pro-regenerative response, promoting tissue repair. *(Created with BioRender.com.)*

improved forelimb-hindlimb coordination (Hu, Granger, Pai, Bellamkonda, & Jeffery, 2018). This suggests that the use of gene delivery of ChABC has potential as a clinical therapy.

It is now well-understood that ECM components, including CSPGs, impact on the inflammatory response following SCI (Gaudet & Popovich, 2014). A number of studies have shown that a lentiviral approach to delivering ChABC, as well as causing extensive remodeling of the extracellular matrix and axonal regeneration, is able to modulate the post-injury immune response (Fig. 4). A study by (Didangelos et al., 2014) using lentiviral delivery of ChABC in a rat thoracic contusion model examined the expression profile of M1/M2 macrophage polarization markers. This study found that ChABC treatment had a distinct effect on the immune signature in the injured tissue, increasing expression of the pro-regenerative cytokine IL-10, with a corresponding reduction in the pro-inflammatory cytokine IL-12B. This demonstrates a distinct, ChABC-mediated anti-inflammatory response following SCI associated with large-scale removal of CSPGs. Lentiviral ChABC delivery leads to an increase in the levels of CD68 and CD206 (Bartus et al., 2014). These are markers of the M2 macrophage phenotype, which promote resolution of inflammation, enhance phagocytosis, angiogenesis, and tissue remodeling following injury (Mantovani, Sozzani, Locati, Allavena, & Sica, 2002).

While there is currently no evidence that long-term expression of ChABC using viral vectors has any detrimental effects, it is known that uncontrolled gene expression can reverse beneficial effects (Fouad, Bennett, Vavrek, & Blesch, 2013) and it has been suggested that long-term, large-scale CSPG digestion and subsequent unintended remodeling could lead to neuropathic pain (Bartus et al., 2014). Therefore, the ability to control the timing and/or duration of treatment administration afforded by the use of viral vector gene therapy would be useful in development of clinically feasible strategies. A study by (Burnside et al., 2018) used a lentivirus to deliver ChABC under the control of the tetracycline-controlled activator. This study included another development by including a glycine-alanine repeat (Hoyng et al., 2014) fused to the transactivator (GARrtTA) to avoid the immune response that can be generated against the system transactivator and which results in cytotoxic T cell-mediated removal of transgene-expressing cells (Latta-Mahieu et al., 2002). This study found short-term ChABC treatment (2.5 weeks) was sufficient to promote improvement in sensory axon conduction and ladder walking performance, however improvement in neuroplasticity of descending motor pathways required for the more complex task of forepaw reaching and grasping required 8 weeks of treatment (Burnside et al., 2018). As well as having clinical potential and allowing for the control of gene expression to avoid detrimental effects, such an approach will be useful in helping determine the most appropriate timing for an intervention, which is also crucial for clinical translation. As lentiviruses integrate into the host genome and so pose a risk of oncogenic transformation, the use of AAV vectors have been trialed for delivery of CSPG-removing enzymes. Our recent study (Griffin et al., 2020) used an AAV5 vector under the control of a GFAP promoter to deliver the human A disintegrin and metalloproteinase with thrombospondin motifs-4 (ADAMTS4) gene, which cause proteolysis of CSPG core proteins, in a rodent model of thoracic contusion SCI. This resulted in significantly decreased lesion size, increased sprouting of hindlimb corticospinal tract axons and serotonergic fiber density and improved hindlimb function when combined with hindlimb exercise rehabilitation. This study used a GFAP promoter to target expression to astrocytes is an example of how viral vector gene therapy can be used to target a specific cell types in the spinal cord to provide therapeutic effect.

## Other approaches

A number of other approaches have been trialed via viral vector delivery, including modulating inhibitory receptors and signaling pathways and cell adhesion and delivery of protective molecules such as calcium sensors and anti-apoptotic proteins that have been successful in improving injury outcomes in animal models (Bo et al., 2011). Delivery of pro-regenerative molecules such as cytokine IL-10 leads to neuronal survival and functional recovery (Zhou, Peng, Insolera, Fink, & Mata, 2009). Using lentivirus to deliver the microRNA miR-133b to the injured mouse spinal cord promotes functional recovery and down regulation of inhibitory pathways and inflammation (Theis et al., 2017). Other recent examples have targeted Hedgehog (Hh) pathway signaling via silencing of the PTC1 and PTC 2 genes (Zhang et al., 2017), apoptotic pathways via overexpression of the peroxisome proliferator activated receptor coactivator 1 alpha which leads to neuronal survival (Hu, Lang, Zhang, Ni, & Lu, 2016) and expression of the mitochondrial protein Prohibitin-1, to protect against mitochondrial dysfunction-induced damage (Li et al., 2015). Viral gene therapy has been used to deliver multiple factors. A study by Wang et al. (Wang, Lu, Chen, Shu, & Qiu, 2015), using lentivirus to deliver the pro-regenerative transcription factor KLF7 and ChABC, enhanced growth of sensory neurons to the injury site following spinal cord transection. Combinational gene therapy approaches have the potential to enhance recovery through targeting of multiple processes. For example, removal of CSPGs using ChABC improves NPC survival, oligodendrocyte differentiation, and remyelination in the presence of growth factors (Karimi-Abdolrezaee, Eftekharpour, Wang, Schut, & Fehlings, 2010; Siebert & Osterhout, 2011). ChABC also regulates macrophage phenotype (Bartus et al., 2014), which in turn promotes myelin repair (Miron & Franklin, 2014). As viral vectors have been used successfully to deliver these therapies, combinational gene therapy to optimize timing of CSPG breakdown and neurotrophic factor delivery could be used to augment the additive beneficial effects of such approaches (Fig. 1G).

In conclusion, viral gene therapy has the ability to deliver a wide-variety of treatments to the cord and with its potential to control the spatiotemporal expression of molecules, has the potential to be a powerful tool in developing effective clinical treatments for SCI.

## Applications to other areas of neuroscience

This chapter provides an overview on the use of viral vectors to target gene therapies for SCI. We describe the different viral vector therapies available, methods of delivery and give an overview of how these have been used to trial potential treatments. Further, we outline how viral vector gene therapy can be used to control the spatiotemporal delivery of therapeutic agents, such as neurotrophic factors and enzymes to remove the glial scar, as a means to improve the efficacy of potential treatments. As is the case with SCI, effective delivery of treatments to the central nervous system to ensure delivery to the correct location or cell type and over the most effective timeframe has potential to improve the efficacy of treatments being developed for a number of neurotraumatic and neurodegenerative conditions. Therefore, the therapeutic potential for viral vector gene delivery is applicable to a number of other conditions including stroke (Craig & Housley, 2016), retinal disease (Campa, Gallenga, Bolletta, & Perri, 2017), Alzheimer's Disease (Sudhakar & Richardson, 2019), Parkinson's Disease (Kabra, Sharma, Kabra, & Baghel, 2018), and Huntington's disease (Estevez-Fraga, Flower, & Tabrizi, 2020).

## Mini-dictionary of terms

**Gene therapy:** Delivery of DNA into cells to produce a therapeutic agent.

**Viral vector:** Tools used to deliver genetic material to cells utilizing mechanisms to efficiently transport their genomes into cells.

**Remote delivery:** Non-invasive or minimally invasive routes of viral vector administration.

**Direct delivery:** Involves an invasive surgical approach that directly targets a specific area of the spinal cord tissue via injection of the gene therapy and is referred to as intraparenchymal delivery.

**Spatio-temporal delivery:** An approach that allows for targeted delivery of a therapeutic agent to a specific region and for a specific period of time.

## Key facts of viral vectors and spinal cord injury

- Viral vectors allow for effective delivery of a number of therapeutic agents to the spinal cord.
- Viral vectors can overcome limitations of drug delivery, such as poor stability in vivo, poor tissue penetration, and the potential for off target effects.
- A number of different viral vectors can be used to target genes to the spinal cord.
- A number of different delivery methods can be used that allow either direct or indirect delivery to spinal cord tissue.
- Viral vectors allow for spatiotemporal regulation of gene expression.

## Summary points

- This chapter focuses on the use of viral vector gene therapy for delivery of therapeutic agents to treat SCI.
- A number of different viral vector gene therapies have been used to deliver therapeutic agents to the spinal cord to overcome a number of issues related to drug delivery to the spinal cord.
- Viral vectors have been used to deliver neurotrophic factors to the spinal cord and have allowed for generation of concentration gradients, which are required for growth of axons through a spinal cord lesion.
- Viral vectors have been used to deliver enzymes to breakdown the glial scar, which overcomes issues of long-term delivery and enzyme stability.
- The use of inducible systems and cell specific promoters allows for spatiotemporal control of gene delivery
- Viral vector gene therapy could be used to optimize deliver of therapeutic agents as part of a combinational approach to treatment.

## References

Alto, L. T., Havton, L. A., Conner, J. M., Hollis, E. R., 2nd, Blesch, A., & Tuszynski, M. H. (2009). Chemotropic guidance facilitates axonal regeneration and synapse formation after spinal cord injury. *Nature Neuroscience*, *12*(9), 1106–1113. https://doi.org/10.1038/nn.2365.

Bartus, K., James, N. D., Didangelos, A., Bosch, K. D., Verhaagen, J., Yanez-Munoz, R. J., … Bradbury, E. J. (2014). Large-scale chondroitin sulfate proteoglycan digestion with chondroitinase gene therapy leads to reduced pathology and modulates macrophage phenotype following spinal cord contusion injury. *The Journal of Neuroscience*, *34*(14), 4822–4836. https://doi.org/10.1523/JNEUROSCI.4369-13.2014.

Batista, A. R., King, O. D., Reardon, C. P., Davis, C., Shankaracharya, P., & V., … Sena-Esteves, M. (2020). Ly6a differential expression in blood-brain barrier is responsible for strain specific central nervous system transduction profile of AAV-PHP.B. *Human Gene Therapy*, *31*(1–2), 90–102. https://doi.org/10.1089/hum.2019.186.

Baum, C., Kustikova, O., Modlich, U., Li, Z., & Fehse, B. (2006). Mutagenesis and oncogenesis by chromosomal insertion of gene transfer vectors. *Human Gene Therapy*, *17*(3), 253–263. https://doi.org/10.1089/hum.2006.17.253.

Blesch, A., Fischer, I., & Tuszynski, M. H. (2012). Gene therapy, neurotrophic factors and spinal cord regeneration. *Handbook of Clinical Neurology*, *109*, 563–574. https://doi.org/10.1016/B978-0-444-52137-8.00035-8.

Blesch, A., & Tuszynski, M. H. (2007). Transient growth factor delivery sustains regenerated axons after spinal cord injury. *The Journal of Neuroscience*, *27*(39), 10535–10545. https://doi.org/10.1523/JNEUROSCI.1903-07.2007.

Blits, B., & Bunge, M. B. (2006). Direct gene therapy for repair of the spinal cord. *Journal of Neurotrauma*, *23*(3–4), 508–520. https://doi.org/10.1089/neu.2006.23.508.

Bo, X., Wu, D., Yeh, J., & Zhang, Y. (2011). Gene therapy approaches for neuroprotection and axonal regeneration after spinal cord and spinal root injury. *Current Gene Therapy*, *11*(2), 101–115. https://doi.org/10.2174/156652311794940773.

Bonner, J. F., Blesch, A., Neuhuber, B., & Fischer, I. (2010). Promoting directional axon growth from neural progenitors grafted into the injured spinal cord. *Journal of Neuroscience Research*, *88*(6), 1182–1192. https://doi.org/10.1002/jnr.22288.

Boulis, N. M., Noordmans, A. J., Song, D. K., Imperiale, M. J., Rubin, A., Leone, P., … Feldman, E. L. (2003). Adeno-associated viral vector gene expression in the adult rat spinal cord following remote vector delivery. *Neurobiology of Disease*, *14*(3), 535–541. https://doi.org/10.1016/j.nbd.2003.08.025.

Bradbury, E. J., & Carter, L. M. (2011). Manipulating the glial scar: Chondroitinase ABC as a therapy for spinal cord injury. *Brain Research Bulletin*, *84*(4–5), 306–316. https://doi.org/10.1016/j.brainresbull.2010.06.015.

Bradbury, E. J., Khemani, S., King, V. R., Priestley, J. V., & McMahon, S. B. (1999). NT-3 promotes growth of lesioned adult rat sensory axons ascending in the dorsal columns of the spinal cord. *The European Journal of Neuroscience*, *11*(11), 3873–3883. https://doi.org/10.1046/j.1460-9568.1999.00809.x.

Bradbury, E. J., & McMahon, S. B. (2006). Spinal cord repair strategies: Why do they work? *Nature Reviews. Neuroscience*, *7*(8), 644–653. https://doi.org/10.1038/nrn1964.

Bradbury, E. J., Moon, L. D., Popat, R. J., King, V. R., Bennett, G. S., Patel, P. N., … McMahon, S. B. (2002). Chondroitinase ABC promotes functional recovery after spinal cord injury. *Nature*, *416*(6881), 636–640. https://doi.org/10.1038/416636a.

Burnside, E. R., De Winter, F., Didangelos, A., James, N. D., Andreica, E. C., Layard-Horsfall, H., … Bradbury, E. J. (2018). Immune-evasive gene switch enables regulated delivery of chondroitinase after spinal cord injury. *Brain*, *141*(8), 2362–2381. https://doi.org/10.1093/brain/awy158.

Byrnes, A. P., MacLaren, R. E., & Charlton, H. M. (1996). Immunological instability of persistent adenovirus vectors in the brain: Peripheral exposure to vector leads to renewed inflammation, reduced gene expression, and demyelination. *The Journal of Neuroscience*, *16*(9), 3045–3055.

Cafferty, W. B., Yang, S. H., Duffy, P. J., Li, S., & Strittmatter, S. M. (2007). Functional axonal regeneration through astrocytic scar genetically modified to digest chondroitin sulfate proteoglycans. *The Journal of Neuroscience*, *27*(9), 2176–2185. https://doi.org/10.1523/JNEUROSCI.5176-06.2007.

Campa, C., Gallenga, C. E., Bolletta, E., & Perri, P. (2017). The role of gene therapy in the treatment of retinal diseases: A review. *Current Gene Therapy*, *17*(3), 194–213. https://doi.org/10.2174/1566523217666171116170040.

Craig, A. J., & Housley, G. D. (2016). Evaluation of gene therapy as an intervention strategy to treat brain injury from stroke. *Frontiers in Molecular Neuroscience*, *9*, 34. https://doi.org/10.3389/fnmol.2016.00034.

Curinga, G. M., Snow, D. M., Mashburn, C., Kohler, K., Thobaben, R., Caggiano, A. O., & Smith, G. M. (2007). Mammalian-produced chondroitinase AC mitigates axon inhibition by chondroitin sulfate proteoglycans. *Journal of Neurochemistry*, *102*(1), 275–288. https://doi.org/10.1111/j.1471-4159.2007.04530.x.

Davis, A. S., Federici, T., Ray, W. C., Boulis, N. M., O'Connor, D., Clark, K. R., & Bartlett, J. S. (2015). Rational design and engineering of a modified adeno-associated virus (AAV1)-based vector system for enhanced retrograde gene delivery. *Neurosurgery*, *76*(2), 216–225. discussion 225 https://doi.org/10.1227/NEU.0000000000000589.

Deverman, B. E., Ravina, B. M., Bankiewicz, K. S., Paul, S. M., & Sah, D. W. Y. (2018). Gene therapy for neurological disorders: Progress and prospects. *Nature Reviews. Drug Discovery*, *17*(10), 767. https://doi.org/10.1038/nrd.2018.158.

Didangelos, A., Iberl, M., Vinsland, E., Bartus, K., & Bradbury, E. J. (2014). Regulation of IL-10 by chondroitinase ABC promotes a distinct immune response following spinal cord injury. *The Journal of Neuroscience*, *34*(49), 16424–16432. https://doi.org/10.1523/JNEUROSCI.2927-14.2014.

Duque, S., Joussemet, B., Riviere, C., Marais, T., Dubreil, L., Douar, A. M., … Barkats, M. (2009). Intravenous administration of self-complementary AAV9 enables transgene delivery to adult motor neurons. *Molecular Therapy*, *17*(7), 1187–1196. https://doi.org/10.1038/mt.2009.71.

During, M. J., Young, D., Baer, K., Lawlor, P., & Klugmann, M. (2003). Development and optimization of adeno-associated virus vector transfer in the central nervous system. *Methods in Molecular Medicine*, *76*, 221–236.

Estevez-Fraga, C., Flower, M. D., & Tabrizi, S. J. (2020). Therapeutic strategies for Huntington's disease. *Current Opinion in Neurology*, *33*(4), 508–518. https://doi.org/10.1097/WCO.0000000000000835.

Federici, T., Riley, J., Park, J., Bain, M., & Boulis, N. (2009). Preclinical safety validation of a stabilized viral vector direct injection approach to the cervical spinal cord. *Clinical and Translational Science*, *2*(2), 165–167. https://doi.org/10.1111/j.1752-8062.2008.00084.x.

Federici, T., Taub, J. S., Baum, G. R., Gray, S. J., Grieger, J. C., Matthews, K. A., … Boulis, N. M. (2012). Robust spinal motor neuron transduction following intrathecal delivery of AAV9 in pigs. *Gene Therapy*, *19*(8), 852–859. https://doi.org/10.1038/gt.2011.130.

Fouad, K., Bennett, D. J., Vavrek, R., & Blesch, A. (2013). Long-term viral brain-derived neurotrophic factor delivery promotes spasticity in rats with a cervical spinal cord hemisection. *Frontiers in Neurology*, *4*, 187. https://doi.org/10.3389/fneur.2013.00187.

Franz, S., Weidner, N., & Blesch, A. (2012). Gene therapy approaches to enhancing plasticity and regeneration after spinal cord injury. *Experimental Neurology*, *235*(1), 62–69. https://doi.org/10.1016/j.expneurol.2011.01.015.

Gaudet, A. D., & Popovich, P. G. (2014). Extracellular matrix regulation of inflammation in the healthy and injured spinal cord. *Experimental Neurology*, *258*, 24–34. https://doi.org/10.1016/j.expneurol.2013.11.020.

Glorioso, J. C., & Fink, D. J. (2004). Herpes vector-mediated gene transfer in treatment of diseases of the nervous system. *Annual Review of Microbiology*, *58*, 253–271. https://doi.org/10.1146/annurev.micro.58.030603.123709.

Griffin, J. M., Fackelmeier, B., Clemett, C. A., Fong, D. M., Mouravlev, A., Young, D., & O'Carroll, S. J. (2020). Astrocyte-selective AAV-ADAMTS4 gene therapy combined with hindlimb rehabilitation promotes functional recovery after spinal cord injury. *Experimental Neurology*, *327*, 113232. https://doi.org/10.1016/j.expneurol.2020.113232.

Hardcastle, N., Boulis, N. M., & Federici, T. (2018). AAV gene delivery to the spinal cord: Serotypes, methods, candidate diseases, and clinical trials. *Expert Opinion on Biological Therapy*, *18*(3), 293–307. https://doi.org/10.1080/14712598.2018.1416089.

He, Z., & Koprivica, V. (2004). The Nogo signaling pathway for regeneration block. *Annual Review of Neuroscience*, *27*, 341–368. https://doi.org/10.1146/annurev.neuro.27.070203.144340.

Hoyng, S. A., Gnavi, S., de Winter, F., Eggers, R., Ozawa, T., Zaldumbide, A., … Verhaagen, J. (2014). Developing a potentially immunologically inert tetracycline-regulatable viral vector for gene therapy in the peripheral nerve. *Gene Therapy*, *21*(6), 549–557. https://doi.org/10.1038/gt.2014.22.

Hu, H. Z., Granger, N., Pai, S. B., Bellamkonda, R. V., & Jeffery, N. D. (2018). Therapeutic efficacy of microtube-embedded chondroitinase ABC in a canine clinical model of spinal cord injury. *Brain*, *141*(4), 1017–1027. https://doi.org/10.1093/brain/awy007.

Hu, J., Lang, Y., Zhang, T., Ni, S., & Lu, H. (2016). Lentivirus-mediated PGC-1alpha overexpression protects against traumatic spinal cord injury in rats. *Neuroscience*, *328*, 40–49. https://doi.org/10.1016/j.neuroscience.2016.04.031.

Jackson, C. A., Cobbs, C., Peduzzi, J. D., Novak, M., & Morrow, C. D. (2001). Repetitive intrathecal injections of poliovirus replicons result in gene expression in neurons of the central nervous system without pathogenesis. *Human Gene Therapy*, *12*(15), 1827–1841. https://doi.org/10.1089/104303401753153893.

James, N. D., Shea, J., Muir, E. M., Verhaagen, J., Schneider, B. L., & Bradbury, E. J. (2015). Chondroitinase gene therapy improves upper limb function following cervical contusion injury. *Experimental Neurology*, *271*, 131–135. https://doi.org/10.1016/j.expneurol.2015.05.022.

Jin, Y., Ketschek, A., Jiang, Z., Smith, G., & Fischer, I. (2011). Chondroitinase activity can be transduced by a lentiviral vector in vitro and in vivo. *Journal of Neuroscience Methods*, *199*(2), 208–213. https://doi.org/10.1016/j.jneumeth.2011.05.007.

Kabra, A., Sharma, R., Kabra, R., & Baghel, U. S. (2018). Emerging and alternative therapies for Parkinson disease: An updated review. *Current Pharmaceutical Design*, *24*(22), 2573–2582. https://doi.org/10.2174/1381612824666180820150150.

Karimi-Abdolrezaee, S., Eftekharpour, E., Wang, J., Schut, D., & Fehlings, M. G. (2010). Synergistic effects of transplanted adult neural stem/progenitor cells, chondroitinase, and growth factors promote functional repair and plasticity of the chronically injured spinal cord. *The Journal of Neuroscience*, *30*(5), 1657–1676. https://doi.org/10.1523/JNEUROSCI.3111-09.2010.

Keefe, K. M., Sheikh, I. S., & Smith, G. M. (2017). Targeting Neurotrophins to specific populations of neurons: NGF, BDNF, and NT-3 and their relevance for treatment of spinal cord injury. *International Journal of Molecular Sciences*, *18*(3). https://doi.org/10.3390/ijms18030548.

Keir, S. D., Mitchell, W. J., Feldman, L. T., & Martin, J. R. (1995). Targeting and gene expression in spinal cord motor neurons following intramuscular inoculation of an HSV-1 vector. *Journal of Neurovirology*, *1*(3–4), 259–267. https://doi.org/10.3109/13550289509114022.

Khromykh, A. A. (2000). Replicon-based vectors of positive strand RNA viruses. *Current Opinion in Molecular Therapeutics*, *2*(5), 555–569.

Kitamura, K., Iwanami, A., Nakamura, M., Yamane, J., Watanabe, K., Suzuki, Y., … Okano, H. (2007). Hepatocyte growth factor promotes endogenous repair and functional recovery after spinal cord injury. *Journal of Neuroscience Research*, *85*(11), 2332–2342. https://doi.org/10.1002/jnr.21372.

Lang, B. T., Cregg, J. M., DePaul, M. A., Tran, A. P., Xu, K., Dyck, S. M., … Silver, J. (2015). Modulation of the proteoglycan receptor PTP sigma promotes recovery after spinal cord injury. *Nature*, *518*(7539), 404–408. https://doi.org/10.1038/nature13974.

Latta-Mahieu, M., Rolland, M., Caillet, C., Wang, M., Kennel, P., Mahfouz, I., … Thuillier, V. (2002). Gene transfer of a chimeric trans-activator is immunogenic and results in short-lived transgene expression. *Human Gene Therapy*, *13*(13), 1611–1620. https://doi.org/10.1089/10430340260201707.

Li, L., Guo, J. D., Wang, H. D., Shi, Y. M., Yuan, Y. L., & Hou, S. X. (2015). Prohibitin 1 gene delivery promotes functional recovery in rats with spinal cord injury. *Neuroscience*, *286*, 27–36. https://doi.org/10.1016/j.neuroscience.2014.11.037.

Mannes, A. J., Caudle, R. M., O'Connell, B. C., & Iadarola, M. J. (1998). Adenoviral gene transfer to spinal-cord neurons: Intrathecal vs. intraparenchymal administration. *Brain Research*, *793*(1–2), 1–6. https://doi.org/10.1016/s0006-8993(97)01422-4.

Mantovani, A., Sozzani, S., Locati, M., Allavena, P., & Sica, A. (2002). Macrophage polarization: Tumor-associated macrophages as a paradigm for polarized M2 mononuclear phagocytes. *Trends in Immunology*, *23*(11), 549–555. https://doi.org/10.1016/s1471-4906(02)02302-5.

Miron, V. E., & Franklin, R. J. (2014). Macrophages and CNS remyelination. *Journal of Neurochemistry*, *130*(2), 165–171. https://doi.org/10.1111/jnc.12705.

Muir, E. M., Fyfe, I., Gardiner, S., Li, L., Warren, P., Fawcett, J. W., … Rogers, J. H. (2010). Modification of N-glycosylation sites allows secretion of bacterial chondroitinase ABC from mammalian cells. *Journal of Biotechnology*, *145*(2), 103–110. https://doi.org/10.1016/j.jbiotec.2009.11.002.

Nakajima, H., Uchida, K., Kobayashi, S., Inukai, T., Horiuchi, Y., Yayama, T., … Baba, H. (2007). Rescue of rat anterior horn neurons after spinal cord injury by retrograde transfection of adenovirus vector carrying brain-derived neurotrophic factor gene. *Journal of Neurotrauma*, *24*(4), 703–712. https://doi.org/10.1089/neu.2006.0004.

Nakajima, H., Uchida, K., Yayama, T., Kobayashi, S., Guerrero, A. R., Furukawa, S., & Baba, H. (2010). Targeted retrograde gene delivery of brain-derived neurotrophic factor suppresses apoptosis of neurons and oligodendroglia after spinal cord injury in rats. *Spine (Phila Pa 1976)*, *35*(5), 497–504. https://doi.org/10.1097/BRS.0b013e3181b8e89b.

Naldini, L., Blomer, U., Gallay, P., Ory, D., Mulligan, R., Gage, F. H., … Trono, D. (1996). In vivo gene delivery and stable transduction of nondividing cells by a lentiviral vector. *Science*, *272*(5259), 263–267. https://doi.org/10.1126/science.272.5259.263.

Ohtake, Y., & Li, S. (2015). Molecular mechanisms of scar-sourced axon growth inhibitors. *Brain Research*, *1619*, 22–35. https://doi.org/10.1016/j.brainres.2014.08.064.

Park, K. K., Liu, K., Hu, Y., Kanter, J. L., & He, Z. (2010). PTEN/mTOR and axon regeneration. *Experimental Neurology*, *223*(1), 45–50. https://doi.org/10.1016/j.expneurol.2009.12.032.

Penaud-Budloo, M., François, A., Clément, N., & Ayuso, E. (2018). Pharmacology of recombinant adeno-associated virus production. *Molecular Therapy - Methods & Clinical Development*, *8*(March), 166–180. https://doi.org/10.1016/j.omtm.2018.01.002.

Poulsen, D. J., Harrop, J. S., & During, M. J. (2002). Gene therapy for spinal cord injury and disease. *The Journal of Spinal Cord Medicine*, *25*(1), 2–9. https://doi.org/10.1080/10790268.2002.11753594.

Rosenzweig, E. S., Salegio, E. A., Liang, J. J., Weber, J. L., Weinholtz, C. A., Brock, J. H., … Tuszynski, M. H. (2019). Chondroitinase improves anatomical and functional outcomes after primate spinal cord injury. *Nature Neuroscience*, *22*(8), 1269–1275. https://doi.org/10.1038/s41593-019-0424-1.

Sharma, K., Selzer, M. E., & Li, S. (2012). Scar-mediated inhibition and CSPG receptors in the CNS. *Experimental Neurology*, *237*(2), 370–378. https://doi.org/10.1016/j.expneurol.2012.07.009.

Siebert, J. R., & Osterhout, D. J. (2011). The inhibitory effects of chondroitin sulfate proteoglycans on oligodendrocytes. *Journal of Neurochemistry*, *119*(1), 176–188. https://doi.org/10.1111/j.1471-4159.2011.07370.x.

Sudhakar, V., & Richardson, R. M. (2019). Gene therapy for neurodegenerative diseases. *Neurotherapeutics*, *16*(1), 166–175. https://doi.org/10.1007/s13311-018-00694-0.

Sun, S., & Schaffer, D. V. (2018). Engineered viral vectors for functional interrogation, deconvolution, and manipulation of neural circuits. *Current Opinion in Neurobiology*, *50*, 163–170. https://doi.org/10.1016/j.conb.2017.12.011.

Tervo, D. G., Hwang, B. Y., Viswanathan, S., Gaj, T., Lavzin, M., Ritola, K. D., … Karpova, A. Y. (2016). A designer AAV variant permits efficient retrograde access to projection neurons. *Neuron*, *92*(2), 372–382. https://doi.org/10.1016/j.neuron.2016.09.021.

Theis, T., Yoo, M., Park, C. S., Chen, J., Kugler, S., Gibbs, K. M., & Schachner, M. (2017). Lentiviral delivery of miR-133b improves functional recovery after spinal cord injury in mice. *Molecular Neurobiology*, *54*(6), 4659–4671. https://doi.org/10.1007/s12035-016-0007-z.

Thomas, A. M., Palma, J. L., & Shea, L. D. (2015). Sponge-mediated lentivirus delivery to acute and chronic spinal cord injuries. *Journal of Controlled Release*, *204*, 1–10. https://doi.org/10.1016/j.jconrel.2015.02.032.

Towne, C., Schneider, B. L., Kieran, D., Redmond, D. E., Jr., & Aebischer, P. (2010). Efficient transduction of non-human primate motor neurons after intramuscular delivery of recombinant AAV serotype 6. *Gene Therapy*, *17*(1), 141–146. https://doi.org/10.1038/gt.2009.119.

Uchida, K., Nakajima, H., Guerrero, A. R., Johnson, W. E., Masri, W. E., & Baba, H. (2014). Gene therapy strategies for the treatment of spinal cord injury. *Therapeutic Delivery*, *5*(5), 591–607. https://doi.org/10.4155/tde.14.20.

Wang, Y. T., Lu, X. M., Chen, K. T., Shu, Y. H., & Qiu, C. H. (2015). Immunotherapy strategies for spinal cord injury. *Current Pharmaceutical Biotechnology*, *16*(6), 492–505. https://doi.org/10.2174/138920101606150407112646.

Yamamura, J., Kageyama, S., Uwano, T., Kurokawa, M., Imakita, M., & Shiraki, K. (2000). Long-term gene expression in the anterior horn motor neurons after intramuscular inoculation of a live herpes simplex virus vector. *Gene Therapy*, *7*(11), 934–941. https://doi.org/10.1038/sj.gt.3301185.

Zhang, H., & Yang, T. (2020). FBXW7alpha promotes the recovery of traumatic spinal cord. *Current Molecular Medicine*, *20*(6), 494–504. https://doi.org/10.2174/1566524020666191223164916.

Zhang, Y. D., Zhu, Z. S., Zhang, D., Zhang, Z., Ma, B., Zhao, S. C., & Xue, F. (2017). Lentivirus-mediated silencing of the PTC1 and PTC2 genes promotes recovery from spinal cord injury by activating the hedgehog signaling pathway in a rat model. *Experimental & Molecular Medicine*, *49*(12). https://doi.org/10.1038/emm.2017.220, e412.

Zhao, R. R., Muir, E. M., Alves, J. N., Rickman, H., Allan, A. Y., Kwok, J. C., … Rogers, J. H. (2011). Lentiviral vectors express chondroitinase ABC in cortical projections and promote sprouting of injured corticospinal axons. *Journal of Neuroscience Methods*, *201*(1), 228–238. https://doi.org/10.1016/j.jneumeth.2011.08.003.

Zhou, Z., Peng, X., Insolera, R., Fink, D. J., & Mata, M. (2009). IL-10 promotes neuronal survival following spinal cord injury. *Experimental Neurology*, *220*(1), 183–190. https://doi.org/10.1016/j.expneurol.2009.08.018.

Chapter 33

# Curcumin usage for inflammation and spinal cord injury

Subum Lee[a] and Kyoung-Tae Kim[b]

[a]Department of Neurosurgery, Korea University Anam Hospital, Seoul, Republic of Korea; [b]Department of Neurosurgery, School of Medicine, Kyungpook National University, Kyungpook National University Hospital, Daegu, Republic of Korea

## List of abbreviations

| | |
|---|---|
| **AGP** | α1-acid glycoprotein |
| **AMPK** | adenosine monophosphate-activated protein kinase |
| **AQP4** | aquaporin 4 |
| **BBB** | Basso, Beattie, Bresnahan |
| **ERKs** | extracellular signal-regulated kinases |
| **GFAP** | glial fibrillary acidic protein |
| **GSK-3β** | glycogen synthase kinase 3 beta |
| **HO-1** | hemeoxygenase 1 |
| **ICE** | interleukin 1β converting enzyme |
| **IL** | interleukin |
| **IP** | intraperitoneal |
| **JAK-STAT** | Janus kinase—signal transducer and activator of transcription |
| **MAPK** | p38 mitogen-activated protein kinase |
| **MDA** | malondialdehyde |
| **MD-2** | myeloid differentiation protein 2 |
| **NF-H** | neurofilament-H |
| **NF-κB** | nuclear factor kappa B |
| **NPC** | neural progenitor cell |
| **Nrf2/ARE** | nuclear erythroid 2-related factor 2/anti-oxidant response element |
| **NSC** | neural stem cells |
| **ROS** | reactive oxygen species |
| **SCI** | spinal cord injury |
| **SC-NPCs** | spinal cord neural progenitor cell |
| **SOD** | serum superoxide dismutase |
| **SOX-9** | sex-determining region Y-box transcription factor 9 |
| **STAT** | signal transducer and activator of transcriptions |
| **TGF-β** | transforming growth factor beta |
| **TNF-α** | tumor necrosis factor alpha |

## Introduction

Traumatic spinal cord injury (SCI) causes necrosis of the central nervous system and subsequent permanent neurological deficit. The mechanism of SCI broadly comprises two stages. First, a primary injury, attributable to the mechanical insult itself and structural damage. A secondary injury is a series of systemic and local neurochemical and physiological changes following the primary injury. The secondary injury occurs via subsequent edema, ischemia, inflammation, cytokine production, free radical damage, glial scar formation, apoptosis, and necrosis (Ormond et al., 2014). Primary injury is immediate and irreversible; in contrast, a secondary injury worsens with time and necessitates therapeutic intervention ( Jo et al., 2018; Kim et al., 2014; Lee et al., 2019). The secondary injury develops within hours or days after SCI, causing neurochemical alterations that lead to neurologic functional impairments. Therapeutic modalities that promote recovery could

Diagnosis and Treatment of Spinal Cord Injury. https://doi.org/10.1016/B978-0-12-822498-4.00033-6

FIG. 1 Curcumin powder and *Curcuma longa* plant. Curcumin is a bright yellow chemical produced by the *C. longa* plants. Curcumin is the principal curcuminoid of turmeric (*C. longa*), a member of the ginger family, Zingiberaceae. It is sold as an herbal supplement, cosmetic ingredient, food-flavoring agent, and food-coloring agent.

be better understood with detailed knowledge of SCI pathophysiology (Ormond et al., 2014). Neurochemical alterations in SCI include increases in the excitatory amino acids, elevation in the calcium influx, stimulation of the calcium-dependent enzymes, generation of reactive oxygen species (ROS), and release of cytokines leading to neuroinflammation (Pouliquen, 2014). These neurochemical alterations affect the neuron and astrocyte activity, induce demyelination, modulate leukocyte infiltration, and activate macrophages (Bramlett & Dietrich, 2004; Jiang et al., 2007). In addition, secreted inflammatory cytokines and growth factors generally up-regulate the pro-survival molecules, such as nuclear factor kappa B (NF-κB) (Ghosh & Hayden, 2008). Thus, the key obstacle in the treatment and recovery process from catastrophic SCI is gliosis caused by the up-regulation of inflammation. Moreover, several studies have demonstrated the potential therapeutic role of curcumin in alleviating the second injury process.

Curcumin [1,7-bis (4-hydroxy-3-methoxyphenyl)-1,6- heptadiene-3, 5-dione] is a yellow extract obtained from *Curcuma longa* and is commonly used in India as a seasoning and food-coloring agent (Fig. 1). It is a polyphenol compound that possesses non-steroidal anti-inflammatory properties. Recent studies indicate that it may have anti-oxidant (Das & Vinayak, 2015), anti-inflammatory (Kim et al., 2014; Lee et al., 2019; Li et al., 2016), neuroprotective (Szczepanowicz et al., 2016) and anti-apoptotic effects (Liu et al., 2015) for SCI. Curcumin has emerged as a promising therapeutic drug in SCI treatment with a tendency to reduce the formation of glial scar and suppress the expression of glial fibrillary acidic protein (GFAP), thus contributing to a more favorable recovery environment (Urdzikova et al., 2016). In a recent study, curcumin inhibited the hypoxia-induced up-regulation of GFAP and neurofilament-H (NF-H) following hypoxia and down-regulated the expression of pro-inflammatory cytokines, such as tumor necrosis factor alpha (TNF-α) and interleukin 1 (IL-1) (Daverey & Agrawal, 2020).

## Structure, pharmacology, and biological targets

Curcumin is a polyphenol substance that has been widely used for medicinal purposes, religious rituals, and local cuisines in the Indian subcontinent. The molecule is symmetric in structure. The keto-enol tautomer in the center, the flexible α,β-unsaturated β-diketo linker, and the terminal o-methoxyphenolic groups comprise the three main components of curcumin molecule (Fig. 2).

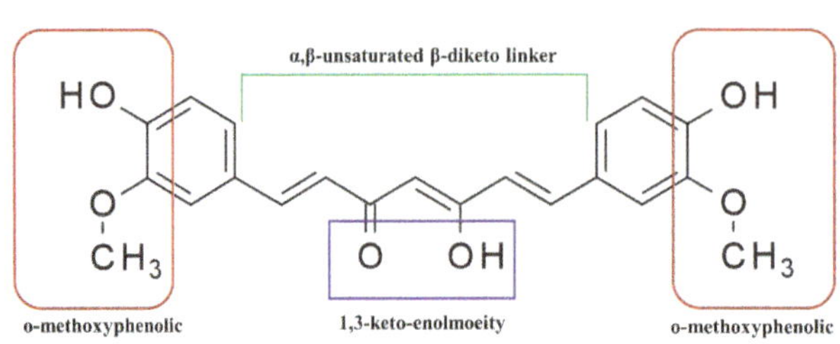


FIG. 2 Structure of curcumin. The curcumin molecule has a symmetric structure and consists of the following three main components: the keto-enol tautomer in the center; the flexible α,β-unsaturated β-diketo linker; and the terminal o-methoxyphenolic groups.

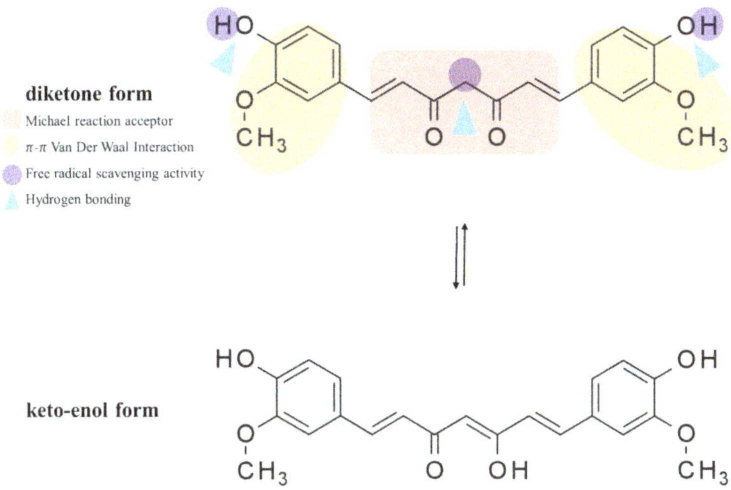


**FIG. 3** The diversity of interactions that curcumin offers. Curcumin has a complex pharmacophore that serves as an anti-oxidant, can chelate metals, and can undergo Michael reaction, hydrogen bonding interactions, π-π van der Waals interactions, and free radical scavenging activity.

Curcumin has a complex pharmacophore that could act as an anti-oxidant, chelate metals, and Michael reaction (Minassi, Sánchez-Duffhues, Collado, Muñoz, & Appendino, 2013). Further, curcumin is a hydrophobic molecule comprising two ferulic acid residues linked by a methylene bridge and has high affinity for cellular membranes (Pérez-Lara, Corbalán-García, & Gómez-Fernández, 2011). Curcumin can participate in hydrogen bonding interactions with its β-diketone moiety and the substituents on the aromatic rings. The aromatic rings can also form π-π Van der Waals interactions (Gupta et al., 2011). Structure-activity relationships demonstrate that the β-diketone (keto-enol) moiety serves as a chelator for cationic metals present in the protein-binding sites and as a Michael reaction acceptor for nucleophilic groups, such as reduced selenocysteine and sulfhydryl, that form covalent bonds with curcumin (Gupta et al., 2011). The phenolic hydroxyl group is essential for the anti-oxidant action of curcumin (Priyadarsini et al., 2003; Selvam, Jachak, Thilagavathi, & Chakraborti, 2005). This group and the methylene hydrogen are crucial for free radical scavenging activity, wherein ROS and nitrogen species are subjected to electron transfer or H-atom abstraction (Priyadarsini et al., 2003). The diversity of the interactions that curcumin offers may explain its binding to multiple proteins (Fig. 3).

Molecular docking studies suggest that curcumin adopts different conformations for the maximization of these interactions, primarily via the α,β-unsaturated β-diketone moiety, and generally favors hydrophilic pockets near the cysteine residues (Ali, Haque, Saleem, & Hsieh, 2013; Gupta et al., 2011). Curcumin shares two important characteristics with the phorbol ester pharmacophore owing to the presence of the hydroxyl and carbonyl groups (Majhi, Rahman, Panchal, & Das, 2010). Removing the methylene group and carbonyl group and cutting off the pharmacophore produces a more potent molecule 1,5-bis (4-hydroxy-3-methoxiphenyl)-1,4-pentadiene-3-1 that maintains all curcumin activity (Appiah-Opong, de Esch, Commandeur, Andarini, & Vermeulen, 2008; Minassi et al., 2013).

Curcumin regulates about 100 biological targets (Das, Pany, Panchal, Majhi, & Rahman, 2011) via various mechanisms, including changing of the activity of cellular proteins via changes in the phosphorylation status (Bill et al., 2012). Generally, curcumin demonstrates its effects at concentrations above the micromolar range. This weak binding affinity has facilitated several attempts to optimize the activity of curcumin using a structural-based drug design.

## Anti-inflammatory effects

After primary SCI, therapeutic strategies and outcomes depend on how we achieve reduction of inflammation and glial scar. Curcumin is a well-known anti-inflammatory molecule that evokes global inhibition of the inflammation network by suppressing transcription factors, such as NF-κB and signal transducer, as well as the activator of transcriptions (STAT) in the upstream signaling pathways of inflammatory mediators, such as prostaglandins, cytokines, and chemokines (Chandran & Goel, 2012). Curcumin may also bind directly to inflammatory mediators and enzymes in downstream inflammation pathways, such as interleukin 1β converting enzyme (ICE), TNF-α, TNF-α converting enzyme (TACE), p38 mitogen-activated protein kinase (MAPK), myeloid differentiation protein 2 (MD-2), α1-acid glycoprotein (AGP), and glycogen synthase kinase 3 beta (GSK-3β) (Elumalai, Muthaiah, & Alf, 2012; Gupta et al., 2011). Curcumin inhibited the expression of pro-inflammatory cytokines and suppressed reactive gliosis. Moreover, curcumin inhibited the generation of transforming growth factor beta (TGF-β)1, TGF-β2, and sex-determining region Y-box transcription factor 9 (SOX-9); decreased the deposition of chondroitin sulfate proteoglycan by inhibiting the transforming growth factors and transcription factor; and improved the microenvironment that enabled nerve growth (Yuan et al., 2015). Occurring concurrently with acute inflammation and preceding fibrosis, spinal cord edema plays a crucial role in neurologic damage and patient symptoms; this is a primary reason behind the clinical usefulness of corticosteroids in SCI patients (Sanivarapu, Vallabhaneni, & Verma, 2016). An experimental rat model that was administered 40 mg/kg curcumin intraperitoneally (IP) showed reduced hemorrhage, edema, and neutrophil infiltration of the traumatic spinal cord. Curcumin also inhibited the SCI-associated aquaporin 4 (AQP4) overexpression and GFAP as well as repressed the unusual activation of the Janus kinase-signal transducer and activator of transcription (JAK-STAT) signaling pathway (Zu et al., 2014). Studies on the effect of curcumin on inflammation, fibrosis, and edema after SCI have been summarized in Table 1.

**TABLE 1** Evidentiary table: anti-inflammatory properties of curcumin.

| Study | Animals/ SCI | Curcumin administration | Description of study | Results |
|---|---|---|---|---|
| (Lee et al., 2019) | SD rats, *N* = 128 or clipping | 200 mg/kg IP daily For 8 weeks | Sham (*n* = 32)<br>SCI only (*n* = 32)<br>SCI & Hyperglycemia (*n* = 32)<br>SCI & Hyperglycemia & Curcumin (*n* = 32) | Curcumin regulate<br>SOD activity increased, MDA level decreased<br>ED-1 macrophage marker level decreased in the SCI-hyperglycemia-curcumin group<br>Reduction in IL-6, IL-8, and TNF-a levels, the phosphorylated-ERKs<br>Lower lesion volume, Higher spared tissue |
| (Ruzicka et al., 2018) | Wistar rats, *N* = 131 or balloon compression | Curcumin, 6 mg/kg; EGCG 17 mg/kg IP daily & Curcumin, 60 mg/kg; EGCG 17 mg/kg) IM weekly For 28 days | Behavioral examinations: saline (*n* = 10), curcumin (*n* = 13), EGCG (*n* = 19), curcumin + EGCG (*n* = 9)<br>Cytokine group studies: Saline (*n* = 20), curcumin (n = 20), EGCG (*n* = 20), curcumin + EGCG (*n* = 20) | Curcumin and EGCG alone or in combination increased axonal sprouting, decreased glial scar formation, and altered the levels of macrophage inflammatory protein 1-alpha, IL-1β, IL-4, and IL-6 |
| (Ruzicka et al., 2018) | Wistar rats, *N* = 135 or balloon compression | 60 mg/kg IT weekly & 6 mg/kg IP daily For 28 days | Saline (*n* = 34),<br>Curcumin (*n* = 27),<br>MSC (*n* = 28),<br>Curcumin + MSC (*n* = 26) | Curcumin + MSC facilitated axonal sprouting, and modulated expression of pro-regenerative factors and inflammatory responses |
| (Yuan et al., 2017) | SD rats, *N* = 280 or blunt dissection and clipping | 100 mg/kg IP daily For 7 days | Sham group (*n* = 70),<br>SCI group (*n* = 70),<br>SCI + curcumin group (*n* = 70),<br>SCI + DMSO group (*n* = 70) | Curcumin regulate both the NF-κb and SOX9 signaling pathways and reduce the expression of intracellular and extracellular glial scar components through dual-target regulating both inflammation and fibrosis after SCI in the rat |
| (Ni et al., 2015) | SD rats, *N* = 48 or clipping | 100 mg/kg IP at 15 min after SCI | Sham group (*n* = 16),<br>SCI group (*n* = 16),<br>SCI + curcumin group (*n* = 16).<br>*N* = 16 per group | Curcumin markedly down regulated the levels of TLR4/NF-κB inflammatory signaling pathway. Significantly ameliorated SCI induced spinal cord edema, and apoptosis |
| (Lin et al., 2015) | Wild-type C57BL/ 6JNarl mice, *N* = 18 or weight drop | 40 mg/kg IP at 30 min after SCI | Sham control (*n* = 6),<br>SCI (*n* = 6),<br>SCI + curcumin (*n* = 6) | Curcumin can attenuate the down-regulation of CISD2 (CISD2 exerts anti-apoptotic and anti-inflammatory effects in neural cells) in SCI and LPS-treated astrocytes |

*Abbreviations:* DMSO, dimethyl sulfoxide; EGCG, epigallocatechin gallate; IM, intramuscular; IP, intraperitoneal; IT, intrathecal; MDA, malondialdehyde; MSC, mesenchymal stem cells; SD, Sprague-Dawley; SOD, superoxide dismutase.

## Anti-oxidant effects

Several trials have assessed the anti-oxidant properties of curcumin for SCI. There is a close association between SCI and free radical production, owing to inflammation (Hausmann, 2003). In an experimental study on the anti-oxidant effect of curcumin on SCI in rats, serum superoxide dismutase (SOD) level in the curcumin group (dose: 200 mg/kg/d orally) was higher than that in the control group and methylprednisolone group (dose: 30 mg/kg intraperitoneally). The malondialdehyde (MDA) level in the curcumin group was lower than in that the control group. Further, curcumin effectively protects the spinal cord tissues against oxidative damage (Sahin Kavaklı, Koca, & Alıcı, 2011). Curcumin acts as a free radical scavenger by donating protons to ROS and nitrogen species and quenches them via electron transfer and H-atom abstraction (Priyadarsini et al., 2003). Consequently, oxidative damage, such as lipid peroxidation is inhibited in the brain and the tissues (Priyadarsini et al., 2003; Scapagnini et al., 2011). Curcumin stimulates the intracellular anti-oxidant defense response via the indication of expression of several anti-oxidants, detoxification, and cytoprotective proteins via the unsilencing of relevant genes on the activation of the nuclear erythroid 2-related factor 2 or anti-oxidant response element (Nrf2/ARE) pathway (Scapagnini et al., 2011). Curcumin can boost the activities of anti-oxidant enzymes, such as plasma catalase, erythrocyte SOD, and plasma glutathione peroxidase (DiSilvestro, Joseph, Zhao, & Bomser, 2012). We have presented the anti-oxidant properties of curcumin as found in SCI studies in Table 2.

**TABLE 2** Evidentiary table: anti-oxidant properties of curcumin.

| Study | Animals or cells/SCI | Curcumin administration | Description of study | Results |
|---|---|---|---|---|
| (Akar et al., 2017) | Wistar rats, $N=40$ or ischemia by clamping the aorta | 100 mg/kg IP at 30 min before ischemia | Sham ($n=10$), Ischemia-reperfusion ($n=10$), Curcumin ($n=10$), Solvent ($n=10$) | Curcumin regulates<br>MDA levels in the spinal cord decreased<br>SOD and GPx levels increased |
| (Daverey & Agrawal, 2016) | HA-sp astrocytes | Cells were treated with curcumin for 24 h | $H_2O_2$ was used to induce oxidative stress in astrocytes<br>50 μM of $H_2O_2$<br>or<br>1 μM of curcumin<br>or<br>50 μM of $H_2O_2$ + 1 μM of curcumin | Curcumin inhibits oxidative stress-induced cytoskeleton disarrangement, and impedes the activation of astrocytes by inhibiting up-regulation of GFAP, vimentin and Prdx6<br>Inhibition of oxidative stress-induced inflammation, apoptosis and mitochondria fragmentation after curcumin treatment |
| (Sanli et al., 2012) | Wistar rats, $N=40$ or weight drop | 300 mg/kg Single IP injection Directly after SCI | Control ($n=8$), SCI alone ($n=8$), Methylprednisolone sodium succinate (30 mg/kg) ($n=8$), Curcumin + DMSO (300 mg/kg) ($n=8$), DMSO alone (0.1 mg/kg) ($n=8$) | Decreased lipid peroxidation and MDA levels in curcumin group |
| (Sahin Kavaklı et al., 2011) | Wistar rats, $N=24$ or weight drop | 200 mg/kg/d orally | SCI + no medication ($n=8$), SCI + curcumin (200 mg/kg/d orally) ($n=8$), SCI + methylprednisolone (30 mg/kg, IP) ($n=8$) | SOD level in the curcumin group was higher than in the control group and methylprednisolone group<br>MDA level in the curcumin group was lower than in the control group |

*Abbreviations*: DMSO, dimethyl sulfoxide, HA-sp, human astrocytes cell line derived from the spinal cord; IP, intraperitoneal; Prdx6, peroxiredoxin.

## Stem cell and progenitor cell proliferation

Although studies using stem cells for SCI are being actively conducted, there are still obstacles in the achievement of remarkable therapeutic results. In particular, it is essential to promote the proliferation of neural stem cells (NSC), and some studies have demonstrated the contribution of curcumin in this subject. As per a recent study, curcumin affects the proliferation of spinal cord neural progenitor cells (SC-NPCs) (Son et al., 2014). This study advocated that the anti-inflammatory effects might promote the proliferation of NPCs and showed that a lower dosage (0.1, 0.5, and 1 μM) of curcumin in the culture medium increased SC-NPC proliferation. However, a higher dosage decreased SC-NPC proliferation. They showed that SC-NPCs are more relevant to extracellular signal-regulated kinases (ERKs) and MAPK. The ERKs protein is reported to be related to cell proliferation, survival, and apoptosis (Gangwal, Bhadauriya, Damre, Dhoke, & Sangamwar, 2013), and the p38 protein played a vital role in cellular responses to external stress signals, such as anti-inflammatory reactions (Dent, 2014). Another trial on the proliferation of fetal rat NSC demonstrated the stem-cell proliferative properties of curcumin. In the cell culture (in vitro) study, curcumin increased the proliferation of fetal rat NSC at low doses (1–6 μM). Consecutive in vivo experiments, curcumin induced NPC proliferation, with the majority of subsequent differentiation into neurons/neuroblasts (Hucklenbroich et al., 2014). Ormond et al. published a study on NSC isolated from the sub-ventricular zone wherein curcumin stimulated NSC proliferation in vitro. In combination with stem-cell therapy, it induced profound recovery from severe SCI as evidenced by improved functional locomotor recovery, increased body weight, and soleus muscle mass. The findings showed that curcumin in conjunction with stem-cell therapy, synergistically improved recovery from severe SCI. Furthermore, their results indicate that the effect of curcumin extends beyond its known anti-inflammatory properties to the regulation of stem-cell proliferation (Ormond et al., 2014).

## Facilitation of neurologic function recovery

In a systematic meta-analysis review, curcumin appeared to significantly improve the neurological function, as assessed using the Basso, Beattie, Bresnahan (BBB) locomotor rating scale in a random-effects model and decreased MDA using a fixed-effects model. Effect size, assessed using the BBB scale, gradually increased as the curcumin dosage was elevated (Yao et al., 2015). The meta-analysis showed that curcumin significantly improved the neurological function, as assessed using the BBB locomotor rating scale (four studies, rats $n = 132$; pooled mean difference = 3.09; 95% confidence interval, 3.40–4.45; $p = 0.04$) (Yao et al., 2015). A strength of many studies that examined the functional recovery is that a uniform scale (BBB score) was used for making inter-study comparisons, thus making extrapolations at least somewhat more reliable than those in studies using different scales (Sanivarapu et al., 2016). In an animal study involving Wistar rats, the rats were randomized into 2 groups, with 30 rats in the curcumin-treatment group and 30 in the control group. Curcumin (applied locally on the surface of the injured spinal cord immediately after an injury and then administered every day intraperitoneally) treatment improved behavioral recovery within the first week after SCI, as indicated by the improved BBB scores and plantar scores sensory performance) (Machova Urdzikova et al., 2015). This functional improvement is attributable to the anti-inflammatory effect described in the previous paragraph. In another animal study on anti-oxidative properties and function recovery, the curcumin-treatment group showed higher neuromotor scores (BBB), increased SOD, decreased MDA, and macrophage markers during seven consecutive days of intraperitoneal injections after SCI (Kim et al., 2014). One study that compared curcumin to methylprednisolone, which is often used to alleviate the secondary injury of SCI in real clinical practice, showed that curcumin was more effective than methylprednisolone after 2 weeks of the SCI, as indicated by the BBB scores. The author group concluded that curcumin has superior therapeutic potential than methylprednisolone with a prolonged treatment time in SCI (Liu et al., 2018). Thus, rat experiments and the use of BBB scores have provided considerable evidence in favor of the effectiveness of curcumin in terms of functional recover following SCI; the data have been summarized in Table 3.

## Conclusion and future direction

Curcumin is a polyphenol compound that is extracted from turmeric rhizomes and is well known for its anti-oxidant and anti-inflammatory properties. These properties are helpful in the protection of neuronal tissue from secondary injury of the spinal cord, such as inflammation, cytokine production, free radical damage, and glial scar formation as well as for superior functional outcomes. Several animal experiments have shown that curcumin enhances active recovery after SCI. There are still several challenges in SCI treatment, and curcumin has several reasonable basic mechanisms as a study subject

**TABLE 3** Evidentiary table: functional recovery outcomes (BBB scores) after curcumin administration.

| Study | Animals/SCI | Curcumin administration | Description of study | Results |
|---|---|---|---|---|
| (Liu et al., 2018) | SD rats<br>$N = 60$/weight drop | 200 mg/kg<br>IP daily<br>For 56 days | SCI-Curcumin ($n = 27$)<br>SCI-MP ($n = 27$)<br>Sham group ($n = 6$) | Paralleled BBB scores between the two treatment groups<br>(MP vs. Curcumin) on Day 56 after SCI |
| (Machova Urdzikova et al., 2015) | Wistar rats<br>$N = 60$<br>or balloon compression | 60 mg/kg<br>Epidural locally<br>Immediately after injury<br>&<br>6 mg/kg<br>IP daily | Curcumin treatment ($n = 30$)<br>Controls ($n = 30$) | Curcumin treatment improved behavioral recovery within the first week following SCI (BBB scores & plantar- sensory performance scores) |
| (Ormond et al., 2012) | SD rats,<br>$N = 14$<br>or weight drop | 60 mg/kg<br>Epidural locally<br>Within 30 min after SCI<br>& Weekly for 6 weeks | Curcumin treatment (n = 7),<br>DMSO treatment ($n = 7$) | In curcumin group, improved functional scores (BBB) after 3 weeks and greater soleus weight |
| (Kim et al., 2014) | SD rats,<br>$N = 36$<br>or clipping | 200 mg/kg<br>IP daily<br>For 7 days | Sham ($n = 12$),<br>SCI/vehicle ($n = 12$),<br>SCI/curcumin ($n = 12$) | Curcumin group shows higher Neuromotor (BBB) scores 7–14 days after surgery |
| (Cemil et al., 2010) | Wistar rats,<br>$N = 40$<br>or clipping | Curcumin:<br>200 mg/kg IP<br>Immediately after injury<br>or<br>MP: 30 mg/kg IP<br>Immediately after injury | Only laminectomy ($n = 8$),<br>Laminectomy + SCI ($n = 8$),<br>Laminectomy + SCI + MP ($n = 8$),<br>Laminectomy + SCI + Curcumin ($n = 8$),<br>Laminectomy + SCI + Vehicle (1 mL of rice bran oil) ($n = 8$) | Curcumin treatment improved neurologic outcome (BBB)<br>Decreased level of tissue MDA and Increased levels of SOD, tissue structure preservation |
| (Lin, Lee, Chiu, & Hung, 2011) | SD rats,<br>$N = 39$<br>or cord hemisection | 40 mg/kg<br>IP daily<br>(1 day before surgery and continuing for 6 days) | Sham ($n = 5$),<br>Vehicle (DMSO) ($n = 17$),<br>Curcumin ($n = 17$) | The BBB scores after hemisection were improved in the curcumin group compared with the vehicle group (on 3–7 days after surgery) |

*Abbreviations*: DMSO, dimethyl sulfoxide; IP, intraperitoneal; MP, methylprednisolone, SD, Sprague-Dawley.

compared to another experimental treatment methods such as stem-cell therapy, electrical nerve stimulation, therapeutic hypothermia that have been attempted.

Future translational research on curcumin is required to enable its clinical application. For the penetration of the blood-brain barrier, oral intake is insufficient, and there is a need to perform a study to establish an effective administration method that would allow the action of curcumin on injured neuronal tissue. Further, safety studies on high-dose administration of curcumin are warranted. In order to improve the bioavailability of curcumin, a method, such as a megadose delivery, may be required, and dosages of previously commercialized curcumin products should be suggested. In most animal studies, continuous or intermittent curcumin administration was administered at the acute stage within 24 h of the primary injury and the chronic stage after the primary injury. Thus, we recommend future studies focus not only the dosage, but also the effective timing and duration of administration.

## Applications to other areas of neuroscience

In this chapter, we have reviewed the effects of curcumin in brain ischemia and brain trauma. These two diseases are active areas in the field of neuroscience for the application of curcumin. First, in an experimental study to evaluate effect of curcumin on focal cerebral ischemic rats, a single intravenous administration of curcumin (1 and 2 mg/kg) alleviated cerebral ischemia and reperfusion injury by preventing nitric oxide-mediated blood-brain barrier disruption. Curcumin reduced the infarct volume, lowered the neurological deficit, and decreased mortality ( Jiang et al., 2007). Curcumin inhibits inducible nitric oxide synthase expression and nitric oxide formation mediated by lipopolysaccharide (LPS)/TNF-α in cultured astrocytes. Based on in vitro and in vivo studies, curcumin prevents peroxynitrite-mediated blood-brain barrier damage, lowers lipid peroxidation-mediated damage, retards apoptotic cell death, and inhibits glial cell activation ( Jiang et al., 2007). Curcumin potently induces the hemeoxygenase 1 (HO-1) expression and activity in rat astrocytes (Scapagnini et al., 2004). HO-1 system is well established as a defensive mechanism for neurons exposed to oxidative stress. The biochemical changes induced by curcumin treatment correlate with its ability to ameliorate changes in the locomotor activity (Wang et al., 2005). Astrocytes that are exposed to curcumin show increased expression of the phase-II detoxification enzymes, quinone reductase and glutathione S-transferase (Scapagnini et al., 2006). Thus, the inhibition of oxidative stress and neuroinflammation with curcumin improves outcomes after brain ischemia. This neuroprotective effect may be attributable to anti-apoptotic mechanisms (Zhao et al., 2010). Collectively, these studies suggest that the neuroprotective activity of curcumin in cerebral ischemia is mediated via its anti-oxidant activity (Thiyagarajan & Sharma, 2004; Zhao et al., 2010).

Studies on the effect of curcumin in traumatic brain injury indicate that this polyphenol substantially lowers oxidative damage; alters the levels of cyclic adenosine monophosphate-response element binding protein; normalizes the levels of brain-derived neurotrophic factor, and reduces the levels of adenosine monophosphate-activated protein kinase (AMPK), ubiquitous mitochondrial creatine kinase, and cytochrome C oxidase II (Sharma, Zhuang, Ying, Wu, & Gomez-Pinilla, 2009; Wu, Ying, & Gomez-Pinilla, 2006). In addition, curcumin and pyrazole curcumin derivatives suppress the increases in calcium-independent phospholipase A2 activity and decreases in the 4-hydroxynonenal levels and increase the fatty acid transport protein. Furthermore, curcumin supplementation counteracts cognitive impairments induced by traumatic brain injury and effectively restores parameters of membrane homeostasis. These results show that oxidative stress acts via the BDNF system to modulate synaptic plasticity membrane homeostasis in the injured brain and that cognition and curcumin can promote neuroprotective mechanisms (Sharma et al., 2009; Sharma, Ying, & Gomez-Pinilla, 2010; Wu et al., 2006) by facilitating endogenous up-regulation of molecules that are important for neural repair and plasticity.

## Mini-dictionary of terms

**Basso, Beattie, and Bresnahan scale:** It is a semi-quantitative scale based on rats' locomotor response that can take on values ranging from 0 to 21. BBB scale is widely used to test behavioral consequences of SCI to the rat.

**Blood-brain barrier:** It is a system that has selective permeability between the blood vessels and the brain, allowing only some substances to pass through and others only through particular pathways such as transport proteins.

**Curcumin:** A major yellow pigment in turmeric, the ground rhizome of *C. longa*. It is a polyphenol compound that pos sesses anti-inflammatory and anti-oxidant properties and has been used for various purposes in Indian medicine for thousands of years.

**Free radical scavenger:** A substance that exhibits anti-oxidant effects by removing free radicals by giving hydrogen atoms to free radicals. Curcumin acts as a free radical scavenger by donating protons to ROS.

**Gliosis:** When the central nervous system is damaged by trauma or ischemia, it refers to the proliferation of glial cells (glia) around it occurs.

**Inflammation:** As one of the defense reactions of injured tissues, it refers to a complex lesion that co-occurs three things: tissue deterioration, circulatory disorders and effusion, and tissue proliferation. It is a critical process in the secondary injury of spinal cord, and curcumin has an anti-inflammatory effect.

**Polyphenol:** It is a substance that has more than two hydroxyls. Anti-oxidant functions are well known and are being applied to food and medical care. The catechins in green tea are typical polyphenol compounds.

**Reactive oxygen species:** A type of unstable molecule that contains oxygen and reacts easily with other molecules in the cell. It accumulates through the secondary injury to the spinal cord, causing DNA, RNA, and proteins damage and causing apoptosis.

**Secondary injury of spinal cord:** A series of systemic and local neurochemical and physiological changes (edema, ischemia, inflammation, cytokine production, free radical damage, glial scar formation, apoptosis, and necrosis) following the primary injury.

**Turmeric:** Ginger family perennial plant. It is about 1 m high. Several trumpet-shaped pink flowers bloom in late spring.

## Key facts of curcumin

Curcumin is a neuroprotective polyphenol compound with pluripotency, oral safety, a long history of use, and reasonable cost.

Many animal experimental studies have showed that curcumin can reduce secondary damage resulting from SCI via anti-inflammatory and anti-oxidant properties as well as stem-cell mobilization.

It decreases the expression of several inflammatory and fibrotic cytokines, viable axonal fibers, GFAP, and MDA after SCI.

Further, curcumin significantly improves the neurological function, as assessed using the Basso, Beattie, and Bresnahan (BBB) locomotor rating scale in a SCI rat model.

Future translational research on curcumin is required to enable its clinical application.

## Summary points

- Curcumin is a key active ingredient extracted from turmeric. Turmeric is a rhizomatous plant of the ginger family.
- Curcumin is well known for its extensive pharmacological activity, such as inflammation relief and anti-oxidant effect.
- The neuroprotective effect has been reported in several experimental studies on the animal brain and neuronal tissue.
- Curcumin is a valid and effective therapeutic agent that could alleviate the catastrophic secondary injury process of the spinal cord, namely inflammation, edema, free radical injury, fibrosis, and glial scar formation.
- This potential is supported by immune pathological results proven in several animal experiments; significantly better outcomes were produced with the evaluation using the Basso, Beattie, and Bresnahan locomotor rating scale for recovery of neurologic function.
- However, the evidence-based effectiveness of curcumin for SCI remains in the experimental stage.

## References

Akar, İ., İnce, İ., Arici, A., Benli, İ., Aslan, C., Şenol, S., … Akbas, A. (2017). The protective effect of curcumin on a spinal cord ischemia-reperfusion injury model. *Annals of Vascular Surgery*, *42*, 285–292. https://doi.org/10.1016/j.avsg.2016.12.016.

Ali, I., Haque, A., Saleem, K., & Hsieh, M. F. (2013). Curcumin-I Knoevenagel's condensates and their Schiff's bases as anticancer agents: Synthesis, pharmacological and simulation studies. *Bioorganic & Medicinal Chemistry*, *21*(13), 3808–3820. https://doi.org/10.1016/j.bmc.2013.04.018.

Appiah-Opong, R., de Esch, I., Commandeur, J. N., Andarini, M., & Vermeulen, N. P. (2008). Structure-activity relationships for the inhibition of recombinant human cytochromes P450 by curcumin analogues. *European Journal of Medicinal Chemistry*, *43*(8), 1621–1631. https://doi.org/10.1016/j.ejmech.2007.10.034.

Bill, M. A., Nicholas, C., Mace, T. A., Etter, J. P., Li, C., Schwartz, E. B., … Lesinski, G. B. (2012). Structurally modified curcumin analogs inhibit STAT3 phosphorylation and promote apoptosis of human renal cell carcinoma and melanoma cell lines. *PLoS One*, *7*(8). https://doi.org/10.1371/journal.pone.0040724, e40724.

Bramlett, H. M., & Dietrich, W. D. (2004). Pathophysiology of cerebral ischemia and brain trauma: Similarities and differences. *Journal of Cerebral Blood Flow and Metabolism*, *24*(2), 133–150. https://doi.org/10.1097/01.Wcb.0000111614.19196.04.

Cemil, B., Topuz, K., Demircan, M. N., Kurt, G., Tun, K., Kutlay, M., … Kucukodaci, Z. (2010). Curcumin improves early functional results after experimental spinal cord injury. *Acta Neurochirurgica*, *152*(9), 1583–1590. discussion 1590 https://doi.org/10.1007/s00701-010-0702-x.

Chandran, B., & Goel, A. (2012). A randomized, pilot study to assess the efficacy and safety of curcumin in patients with active rheumatoid arthritis. *Phytotherapy Research*, *26*(11), 1719–1725. https://doi.org/10.1002/ptr.4639.

Das, J., Pany, S., Panchal, S., Majhi, A., & Rahman, G. M. (2011). Binding of isoxazole and pyrazole derivatives of curcumin with the activator binding domain of novel protein kinase C. *Bioorganic & Medicinal Chemistry*, *19*(21), 6196–6202. https://doi.org/10.1016/j.bmc.2011.09.011.

Das, L., & Vinayak, M. (2015). Long term effect of curcumin in restoration of tumour suppressor p53 and phase-II antioxidant enzymes via activation of Nrf2 signalling and modulation of inflammation in prevention of cancer. *PLoS One*, *10*(4). https://doi.org/10.1371/journal.pone.0124000, e0124000.

Daverey, A., & Agrawal, S. K. (2016). Curcumin alleviates oxidative stress and mitochondrial dysfunction in astrocytes. *Neuroscience*, *333*, 92–103. https://doi.org/10.1016/j.neuroscience.2016.07.012.

Daverey, A., & Agrawal, S. K. (2020). Curcumin protects against white matter injury through NF-κB and Nrf2 cross talk. *Journal of Neurotrauma*, *37*(10), 1255–1265. https://doi.org/10.1089/neu.2019.6749.

Dent, P. (2014). Crosstalk between ERK, AKT, and cell survival. *Cancer Biology & Therapy*, *15*(3), 245–246. https://doi.org/10.4161/cbt.27541.

DiSilvestro, R. A., Joseph, E., Zhao, S., & Bomser, J. (2012). Diverse effects of a low dose supplement of lipidated curcumin in healthy middle aged people. *Nutrition Journal*, *11*, 79. https://doi.org/10.1186/1475-2891-11-79.

Elumalai, M., Muthaiah, R., & Alf, M. A. (2012). Identification of curcumin targets in neuroinflammatory pathways: Molecular docking scores with GSK-3β, p38 MAPK, COX, ICE and TACE enzymes. *Acta Poloniae Pharmaceutica*, *69*(2), 237–245.

Gangwal, R. P., Bhadauriya, A., Damre, M. V., Dhoke, G. V., & Sangamwar, A. T. (2013). p38 mitogen-activated protein kinase inhibitors: A review on pharmacophore mapping and QSAR studies. *Current Topics in Medicinal Chemistry*, *13*(9), 1015–1035. https://doi.org/10.2174/1568026611313090005.

Ghosh, S., & Hayden, M. S. (2008). New regulators of NF-kappaB in inflammation. *Nature Reviews. Immunology*, *8*(11), 837–848. https://doi.org/10.1038/nri2423.

Gupta, S. C., Prasad, S., Kim, J. H., Patchva, S., Webb, L. J., Priyadarsini, I. K., & Aggarwal, B. B. (2011). Multitargeting by curcumin as revealed by molecular interaction studies. *Natural Product Reports*, *28*(12), 1937–1955. https://doi.org/10.1039/c1np00051a.

Hausmann, O. N. (2003). Post-traumatic inflammation following spinal cord injury. *Spinal Cord*, *41*(7), 369–378. https://doi.org/10.1038/sj.sc.3101483.

Hucklenbroich, J., Klein, R., Neumaier, B., Graf, R., Fink, G. R., Schroeter, M., & Rueger, M. A. (2014). Aromatic-turmerone induces neural stem cell proliferation in vitro and in vivo. *Stem Cell Research & Therapy*, *5*(4), 100. https://doi.org/10.1186/scrt500.

Jiang, J., Wang, W., Sun, Y. J., Hu, M., Li, F., & Zhu, D. Y. (2007). Neuroprotective effect of curcumin on focal cerebral ischemic rats by preventing blood-brain barrier damage. *European Journal of Pharmacology*, *561*(1–3), 54–62. https://doi.org/10.1016/j.ejphar.2006.12.028.

Jo, M. J., Kumar, H., Joshi, H. P., Choi, H., Ko, W. K., Kim, J. M., … Han, I. (2018). Oral administration of α-Asarone promotes functional recovery in rats with spinal cord injury. *Frontiers in Pharmacology*, *9*, 445. https://doi.org/10.3389/fphar.2018.00445.

Kim, K. T., Kim, M. J., Cho, D. C., Park, S. H., Hwang, J. H., Sung, J. K., … Jeon, Y. (2014). The neuroprotective effect of treatment with curcumin in acute spinal cord injury: Laboratory investigation. *Neurologia Medico-Chirurgica (Tokyo)*, *54*(5), 387–394. https://doi.org/10.2176/nmc.oa.2013-0251.

Lee, Y. S., Cho, D. C., Kim, C. H., Han, I., Gil, E. Y., & Kim, K. T. (2019). Effect of curcumin on the inflammatory reaction and functional recovery after spinal cord injury in a hyperglycemic rat model. *The Spine Journal*, *19*(12), 2025–2039. https://doi.org/10.1016/j.spinee.2019.07.013.

Li, G., Cao, Y., Shen, F., Wang, Y., Bai, L., Guo, W., … Fan, Z. (2016). Mdivi-1 inhibits astrocyte activation and Astroglial scar formation and enhances axonal regeneration after spinal cord injury in rats. *Frontiers in Cellular Neuroscience*, *10*, 241. https://doi.org/10.3389/fncel.2016.00241.

Lin, C. C., Chiang, T. H., Chen, W. J., Sun, Y. Y., Lee, Y. H., & Lin, M. S. (2015). CISD2 serves a novel role as a suppressor of nitric oxide signalling and curcumin increases CISD2 expression in spinal cord injuries. *Injury*, *46*(12), 2341–2350. https://doi.org/10.1016/j.injury.2015.07.040.

Lin, M. S., Lee, Y. H., Chiu, W. T., & Hung, K. S. (2011). Curcumin provides neuroprotection after spinal cord injury. *The Journal of Surgical Research*, *166*(2), 280–289. https://doi.org/10.1016/j.jss.2009.07.001.

Liu, J., Zhu, P., Song, P., Xiong, W., Chen, H., Peng, W., … Wang, H. (2015). Pretreatment of adipose derived stem cells with curcumin facilitates myocardial recovery via antiapoptosis and angiogenesis. *Stem Cells International*, *2015*, 638153. https://doi.org/10.1155/2015/638153.

Liu, X., Zhang, Y., Yang, Y., Lin, J., Huo, X., Du, X., … Fang, M. (2018). Therapeutic effect of curcumin and methylprednisolone in the rat spinal cord injury. *The Anatomical Record*, *301*(4), 686–696. https://doi.org/10.1002/ar.23729.

Machova Urdzikova, L., Karova, K., Ruzicka, J., Kloudova, A., Shannon, C., Dubisova, J., … Jendelova, P. (2015). The anti-inflammatory compound curcumin enhances locomotor and sensory recovery after spinal cord injury in rats by immunomodulation. *International Journal of Molecular Sciences*, *17*(1). https://doi.org/10.3390/ijms17010049.

Majhi, A., Rahman, G. M., Panchal, S., & Das, J. (2010). Binding of curcumin and its long chain derivatives to the activator binding domain of novel protein kinase C. *Bioorganic & Medicinal Chemistry*, *18*(4), 1591–1598. https://doi.org/10.1016/j.bmc.2009.12.075.

Minassi, A., Sánchez-Duffhues, G., Collado, J. A., Muñoz, E., & Appendino, G. (2013). Dissecting the pharmacophore of curcumin. Which structural element is critical for which action? *Journal of Natural Products*, *76*(6), 1105–1112. https://doi.org/10.1021/np400148e.

Ni, H., Jin, W., Zhu, T., Wang, J., Yuan, B., Jiang, J., … Ma, Z. (2015). Curcumin modulates TLR4/NF-κB inflammatory signaling pathway following traumatic spinal cord injury in rats. *The Journal of Spinal Cord Medicine*, *38*(2), 199–206. https://doi.org/10.1179/2045772313y.0000000179.

Ormond, D. R., Peng, H., Zeman, R., Das, K., Murali, R., & Jhanwar-Uniyal, M. (2012). Recovery from spinal cord injury using naturally occurring antiinflammatory compound curcumin: Laboratory investigation. *Journal of Neurosurgery. Spine*, *16*(5), 497–503. https://doi.org/10.3171/2012.1.Spine11769.

Ormond, D. R., Shannon, C., Oppenheim, J., Zeman, R., Das, K., Murali, R., & Jhanwar-Uniyal, M. (2014). Stem cell therapy and curcumin synergistically enhance recovery from spinal cord injury. *PLoS One*, *9*(2). https://doi.org/10.1371/journal.pone.0088916, e88916.

Pérez-Lara, A., Corbalán-García, S., & Gómez-Fernández, J. C. (2011). Curcumin modulates PKCα activity by a membrane-dependent effect. *Archives of Biochemistry and Biophysics*, *513*(1), 36–41. https://doi.org/10.1016/j.abb.2011.06.010.

Pouliquen, D. L. C. (2014). *Curcumin: Synthesis, emerging role in pain management and health implications*. New York: Nova Science Publishers, Inc.

Priyadarsini, K. I., Maity, D. K., Naik, G. H., Kumar, M. S., Unnikrishnan, M. K., Satav, J. G., & Mohan, H. (2003). Role of phenolic O-H and methylene hydrogen on the free radical reactions and antioxidant activity of curcumin. *Free Radical Biology & Medicine*, *35*(5), 475–484. https://doi.org/10.1016/s0891-5849(03)00325-3.

Ruzicka, J., Urdzikova, L. M., Kloudova, A., Amin, A. G., Vallova, J., Kubinova, S., … Jendelova, P. (2018). Anti-inflammatory compound curcumin and mesenchymal stem cells in the treatment of spinal cord injury in rats. *Acta Neurobiologiae Experimentalis (Wars)*, *78*(4), 358–374.

Ruzicka, J., Urdzikova, L. M., Svobodova, B., Amin, A. G., Karova, K., Dubisova, J., … Jendelova, P. (2018). Does combined therapy of curcumin and epigallocatechin gallate have a synergistic neuroprotective effect against spinal cord injury? *Neural Regeneration Research*, *13*(1), 119–127. https://doi.org/10.4103/1673-5374.224379.

Sahin Kavaklı, H., Koca, C., & Alıcı, O. (2011). Antioxidant effects of curcumin in spinal cord injury in rats. *Ulusal Travma ve Acil Cerrahi Dergisi*, *17*(1), 14–18.

Sanivarapu, R., Vallabhaneni, V., & Verma, V. (2016). The potential of curcumin in treatment of spinal cord injury. *Neurology Research International*, *2016*, 9468193. https://doi.org/10.1155/2016/9468193.

Sanli, A. M., Turkoglu, E., Serbes, G., Sargon, M. F., Besalti, O., Kilinc, K., … Sekerci, Z. (2012). Effect of curcumin on lipid peroxidation, early ultrastructural findings and neurological recovery after experimental spinal cord contusion injury in rats. *Turkish Neurosurgery*, *22*(2), 189–195. https://doi.org/10.5137/1019-5149.Jtn.5193-11.1.

Scapagnini, G., Butterfield, D. A., Colombrita, C., Sultana, R., Pascale, A., & Calabrese, V. (2004). Ethyl ferulate, a lipophilic polyphenol, induces HO-1 and protects rat neurons against oxidative stress. *Antioxidants & Redox Signaling*, *6*(5), 811–818. https://doi.org/10.1089/ars.2004.6.811.

Scapagnini, G., Colombrita, C., Amadio, M., D'Agata, V., Arcelli, E., Sapienza, M., … Calabrese, V. (2006). Curcumin activates defensive genes and protects neurons against oxidative stress. *Antioxidants & Redox Signaling*, *8*(3–4), 395–403. https://doi.org/10.1089/ars.2006.8.395.

Scapagnini, G., Vasto, S., Abraham, N. G., Caruso, C., Zella, D., & Fabio, G. (2011). Modulation of Nrf2/ARE pathway by food polyphenols: A nutritional neuroprotective strategy for cognitive and neurodegenerative disorders. *Molecular Neurobiology*, *44*(2), 192–201. https://doi.org/10.1007/s12035-011-8181-5.

Selvam, C., Jachak, S. M., Thilagavathi, R., & Chakraborti, A. K. (2005). Design, synthesis, biological evaluation and molecular docking of curcumin analogues as antioxidant, cyclooxygenase inhibitory and anti-inflammatory agents. *Bioorganic & Medicinal Chemistry Letters*, *15*(7), 1793–1797. https://doi.org/10.1016/j.bmcl.2005.02.039.

Sharma, S., Ying, Z., & Gomez-Pinilla, F. (2010). A pyrazole curcumin derivative restores membrane homeostasis disrupted after brain trauma. *Experimental Neurology*, *226*(1), 191–199. https://doi.org/10.1016/j.expneurol.2010.08.027.

Sharma, S., Zhuang, Y., Ying, Z., Wu, A., & Gomez-Pinilla, F. (2009). Dietary curcumin supplementation counteracts reduction in levels of molecules involved in energy homeostasis after brain trauma. *Neuroscience*, *161*(4), 1037–1044. https://doi.org/10.1016/j.neuroscience.2009.04.042.

Son, S., Kim, K.-T., Cho, D.-C., Kim, H.-J., Sung, J.-K., & Bae, J.-S. (2014). Curcumin stimulates proliferation of spinal cord neural progenitor cells via a mitogen-activated protein kinase signaling pathway. *Journal of Korean Neurosurgical Society*, *56*(1), 1–4. https://doi.org/10.3340/jkns.2014.56.1.1.

Szczepanowicz, K., Jantas, D., Piotrowski, M., Staroń, J., Leśkiewicz, M., Regulska, M., … Warszyński, P. (2016). Encapsulation of curcumin in polyelectrolyte nanocapsules and their neuroprotective activity. *Nanotechnology*, *27*(35), 355101. https://doi.org/10.1088/0957-4484/27/35/355101.

Thiyagarajan, M., & Sharma, S. S. (2004). Neuroprotective effect of curcumin in middle cerebral artery occlusion induced focal cerebral ischemia in rats. *Life Sciences*, *74*(8), 969–985. https://doi.org/10.1016/j.lfs.2003.06.042.

Urdzikova, L. M., Karova, K., Ruzicka, J., Kloudova, A., Shannon, C., Dubisova, J., … Jendelova, P. (2016). The anti-inflammatory compound curcumin enhances locomotor and sensory recovery after spinal cord injury in rats by immunomodulation. *International Journal of Molecular Sciences*, *17*(1), 15. https://doi.org/10.3390/ijms17010049.

Wang, Q., Sun, A. Y., Simonyi, A., Jensen, M. D., Shelat, P. B., Rottinghaus, G. E., … Sun, G. Y. (2005). Neuroprotective mechanisms of curcumin against cerebral ischemia-induced neuronal apoptosis and behavioral deficits. *Journal of Neuroscience Research*, *82*(1), 138–148. https://doi.org/10.1002/jnr.20610.

Wu, A., Ying, Z., & Gomez-Pinilla, F. (2006). Dietary curcumin counteracts the outcome of traumatic brain injury on oxidative stress, synaptic plasticity, and cognition. *Experimental Neurology*, *197*(2), 309–317. https://doi.org/10.1016/j.expneurol.2005.09.004.

Yao, M., Yang, L., Wang, J., Sun, Y. L., Dun, R. L., Wang, Y. J., & Cui, X. J. (2015). Neurological recovery and antioxidant effects of curcumin for spinal cord injury in the rat: A network meta-analysis and systematic review. *Journal of Neurotrauma*, *32*(6), 381–391. https://doi.org/10.1089/neu.2014.3520.

Yuan, J., Liu, W., Zhu, H., Chen, Y., Zhang, X., Li, L., … Lin, J. (2017). Curcumin inhibits glial scar formation by suppressing astrocyte-induced inflammation and fibrosis in vitro and in vivo. *Brain Research*, *1655*, 90–103. https://doi.org/10.1016/j.brainres.2016.11.002.

Yuan, J., Zou, M., Xiang, X., Zhu, H., Chu, W., Liu, W., … Lin, J. (2015). Curcumin improves neural function after spinal cord injury by the joint inhibition of the intracellular and extracellular components of glial scar. *The Journal of Surgical Research*, *195*(1), 235–245. https://doi.org/10.1016/j.jss.2014.12.055.

Zhao, J., Yu, S., Zheng, W., Feng, G., Luo, G., Wang, L., & Zhao, Y. (2010). Curcumin improves outcomes and attenuates focal cerebral ischemic injury via antiapoptotic mechanisms in rats. *Neurochemical Research*, *35*(3), 374–379. https://doi.org/10.1007/s11064-009-0065-y.

Zu, J., Wang, Y., Xu, G., Zhuang, J., Gong, H., & Yan, J. (2014). Curcumin improves the recovery of motor function and reduces spinal cord edema in a rat acute spinal cord injury model by inhibiting the JAK/STAT signaling pathway. *Acta Histochemica*, *116*(8), 1331–1336. https://doi.org/10.1016/j.acthis.2014.08.004.

Chapter 34

# Use of (−)-epigallocatechin-3-gallate on spinal cord injury

Ali Reza Khalatbary
*Molecular and Cell Biology Research Center, Faculty of Medicine, Mazandaran University of Medical Sciences, Sari, Iran*

## List of abbreviations

| | |
|---|---|
| **AGEs** | advanced glycation end products |
| **AD** | Alzheimer's disease |
| **ACR** | acrylamide |
| **AKT** | protein kinase B |
| **APP** | amyloid precursor protein |
| **AQP4** | astrocyte expressing aquaporin-4 |
| **ARE** | antioxidant response element |
| **α7nAChR** | alpha-7 nicotinic acetylcholine receptor |
| **BBB** | blood-brain barrier |
| **Bax** | Bcl-2-associated X protein |
| **Bcl-2** | B-cell lymphoma 2 |
| **BDNF** | brain-derived neurotrophic factor |
| **CAT** | catalase |
| **COX-2** | cyclooxygenase-2 |
| **EGCG** | (−)-epigallocatechin-3-gallate |
| **eNOS** | endothelial nitric oxide synthase |
| **FASN** | fatty acid synthase |
| **FGF2** | fibroblast growth factor 2 |
| **GFAP** | glial fibrillary acidic protein |
| **GDNF** | glial cell line-derived neurotrophic factor |
| **GSH-Px** | glutathione peroxidase |
| **GSK-3** | glycogen synthase kinase-3 |
| **HMGB1** | high-mobility group box 1 protein |
| **IRI** | ischemia-reperfusion injury |
| **iNOS** | inducible nitric oxide synthase |
| **IL-1β** | interleukin-1 beta |
| **IL-4** | interleukin-4 |
| **IL-6** | interleukin-6 |
| **67LR** | 67 kDa laminin receptor |
| **LPS** | lipopolysaccharide |
| **MDA** | malodialdehyde |
| **MDI** | motor deficit index |
| **MMP-9** | matrix metalloproinase-9 |
| **MIP-1α** | macrophage inflammatory protein 1-alpha |
| **MPO** | myeloperoxidase |
| **NF-κB** | nuclear factor Kappa B |
| **NGF** | nerve growth factor |
| **Nrf2** | nuclear erythroid 2-related factor 2 |
| **NF-L** | neurofilament triplet L |
| **NADPH-d** | nicotinamide adenine dinucleotide phosphate-diaphorase |
| **NO** | nitric oxide |

Diagnosis and Treatment of Spinal Cord Injury. https://doi.org/10.1016/B978-0-12-822498-4.00034-8

| | |
|---|---|
| **NOS** | neuronal nitric oxide synthase |
| **nNOS** | neuronal nitric oxide synthase |
| **NT3** | neurotrophin-3 |
| **NGFR-p75** | nerve growth factor receptor p75 |
| **PARP** | poly(ADP-ribose) polymerase |
| **PD** | Parkinson's disease |
| **PKC** | protein kinase C |
| **PI3K** | phosphatidylinositol 3-kinase |
| **ROS** | reactive oxygen species |
| **RhoA** | ras homologue gene family member A |
| **SCI** | spinal cord injury |
| **SOD** | superoxide dismutase |
| **TNFα** | tumor necrosis factor alpha |
| **TUNEL** | terminal deoxynucleotidyl transferase dUTP nick end labeling |
| **TLR4** | Toll-like receptor 4 |
| **Trk-B** | tropomyosin receptor kinase B |
| **Trk-C** | tropomyosin receptor kinase C |
| **VEGF** | vascular endothelial growth factor |

## Introduction

Over the past decades, a large number of polyphenolic compounds with neuroprotective effects have been described. One of the main sources of these molecules is green tea, the most widely consumed beverage next to water. The chemical composition of green tea contains a number of bioactive components, mainly including polyphenols, caffeine, and amino acids (Dulloo et al., 1999). Green tea polyphenols, generally known as catechins which make up 30% of the dry weight of green tea leaves, are the main bioactive constituents of green tea with a wide variety of beneficial health effects (Graham, 1992). The main catechins in green tea include (+)-catechin, (−)-epicatechin, (+)-gallocatechin, (−)-epigallocatechin, (+)-catechin gallate, (−)-epicatechin gallate, (+)-gallocatechin gallate, and (−)-epigallocatechin-3-gallate (EGCG) (Sutherland, Rahman, & Appleton, 2006). EGCG, chemically (2*R*,3*R*)-5,7-dihydroxy-2-(3,4,5trihydroxyphenyl) chroman-3-yl 3,4,5-trihydroxybenzoate (Fig. 1), is the most abundant composition of the tea catechins, account for 65% of the total catechin content and is thought to be responsible for the majority of biological activity of green tea extracts (Zaveri, 2006). In terms of bioavailability, EGCG is predominantly absorbed in the intestine and presented more than 77% in a free form in plasma as well as its metabolites excreted in urine after oral administration (Lee et al., 2002; Meng et al., 2002;

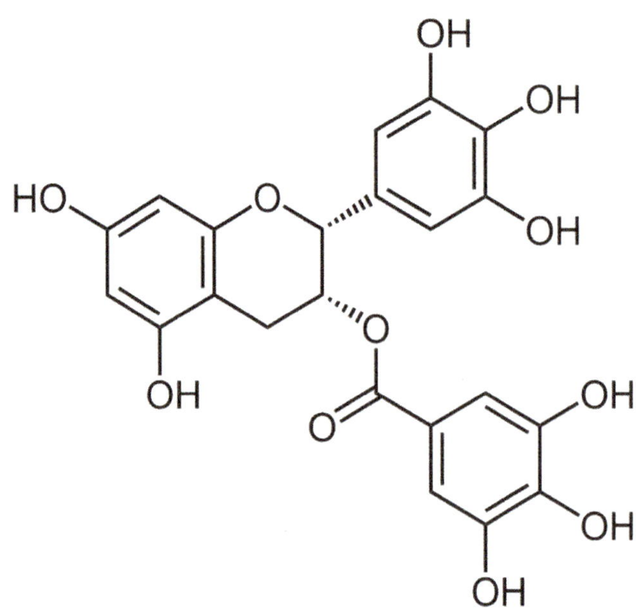

FIG. 1 Chemical structure of epigallocatechin-3-gallate. Chemical structure of epigallocatechin-3-gallate with the molar mass of 458.372 g/mol obtained from Sigma-Aldrich website. The hydroxyl groups on the B- and D-rings can bind to the free radicals and neutralized those.

Moore, Jackson, & Minihane, 2009). The half-life of EGCG in a purified form is around 3 h (Williamson, Dionisi, & Renouf, 2011). Meanwhile, it is well documented that EGCG is able to cross the blood-brain barrier (BBB) and can reach to nervous tissue even at a very low concentration, the first requirement of a dietary compound to apply neuroprotective effects (Lin, Wang, Tseng, Sung, & Tsai, 2007; Pervin et al., 2017). Several experimental studies have shown that EGCG can afford neuroprotection against brain (Itoh et al., 2011) and spinal cord (Urdzikova et al., 2017) injuries, neurodegenerative diseases (Xu, Langley, Kanthasamy, & Reddy, 2017), and peripheral nerve injuries (Renno, Benov, & Khan, 2017). These beneficial effects have been mainly attributed to free radical scavenging or anti-oxidant, anti-inflammatory, and anti-apoptotic properties of EGCG (Ohishi, Goto, Monira, Isemura, & Nakamura, 2016; Oliveira, Nabavi, Daglia, Rastrelli, & Nabavi, 2016; Weinreb, Amit, Mandel, & Youdim, 2009). In this regard, it is well known that green tea catechins due to the hydroxyl groups on the B- and D-rings of the galloylated catechins can bind to the free radicals and neutralized those (Salah et al., 1995). On the other hand, scavenging effects of EGCG lead to attenuation of nuclear factor kappa B (NF-κB) activity (Ohishi et al., 2016), which regulates genes involved in inflammatory processes such as tumor necrosis factor alpha (TNFα), cyclooxygenase-2 (COX-2), and interleukin-1 beta (IL-1β), beside modulation of nitric oxide synthase isoforms (Lawrence & Fong, 2010). Also, catechins have proven to modulate apoptosis at various points in the sequence and messengers, and so regulate the mitochondrial membrane permeabilization (Kalfon, Youdim, & Mandel, 2007; Mandel, Maor, & Youdim, 2004).

In this chapter, we have reviewed the neuroprotective effects of EGCG and its molecular mechanisms responsible for the neuroprotection following spinal cord injury (SCI). Meanwhile, we also compared the neuroprotective effects of EGCG in SCI to other neurological disorders.

## In vivo studies

SCI is a complex multifactorial process caused first by ischemia or mechanical trauma and then by various mechanisms of secondary injury (Amar & Levy, 1999). The neurological outcome of SCI depends on the extent of secondary cellular, molecular, and biochemical cascades such as oxygen free radical-induced lipid peroxidation, inflammatory reaction, and apoptosis (Hall, 1993; Liu et al., 1997; Popovich, Wei, & Stokes, 1997). In recent years, much attention has been paid to secondary injuries, as they appear to be prone to therapeutic interventions that may include the use of natural anti-oxidants, anti-inflammatory, and anti-apoptotic agents (Khalatbary, 2014) such as EGCG (Table 1). In this regard, our previous studies have shown that intraperitoneal injection of EGCG (50 mg/kg, i.p., immediately and 1 h after SCI)

**TABLE 1** Neuroprotective effects of epigallocatechin-3-gallate on spinal cord injury.

| Model of injury | Treatment schedule | Finding | Possible mechanism | Author | Year |
|---|---|---|---|---|---|
| Weight-drop method | 50 mg/kg, intraperitoneal, immediately and 1 h after spinal cord injury | Decreased bcl-2-associated X protein and increased b-cell lymphoma 2 expression | Anti-apoptotic activity | Khalatbary et al. | 2010 |
| Weight-drop method | 50 mg/kg, intraperitoneal, immediately and 1 h after spinal cord injury | Decreased tissue lipid peroxide level | Anti-oxidant activity | Khalatbary and Ahmadvand | 2010 |
| Weight-drop method | 50 mg/kg, intraperitoneal, immediately and 1 h after spinal cord injury | Attenuated tumor necrosis factor alpha, interleukin-1 beta, Nitrotyrosine, inducible nitric oxide synthase, cyclooxygenase-2 and poly (ADP-ribose) polymerase expression, and decreased myeloperoxidase activity | Anti-inflammatory activity | Khalatbary and Ahmadvand | 2011 |
| Vascular clip method | 100 mg/kg, between 24 and 72 h after spinal cord injury | Down-regulated astrocyte expressing aquaporin-4 and glial fibrillary acidic protein | Anti-edema effect | Ge et al. | 2013 |

*Continued*

**TABLE 1 Neuroprotective effects of epigallocatechin-3-gallate on spinal cord injury—cont'd**

| Model of injury | Treatment schedule | Finding | Possible mechanism | Author | Year |
|---|---|---|---|---|---|
| Vascular clip method | 100 mg/kg, intraperitoneal, immediately after spinal cord injury | Reduced the releasing of tumor necrosis factor alpha and interleukin-1 beta, decreased the water content | Regulating p38MAPK\NF -κB \AQP4 signaling pathway | Tao and Zhu | 2013 |
| Weight-drop method | Intrathecal administration of 10 or 20 mg/kg, immediately after spinal cord injury | Improved locomotor function | Up-regulation of brain-derived neurotrophic factor and glial cell line-derived neurotrophic factor | Tian et al. | 2013 |
| Weight-drop method | 20 mg/kg/hour for continuous 36 h, intravenous, in acutely (initiated within 4 h post-trauma 0 time) and chronically (initiated after 12 months of spinal cord injury onset | Improved motor and sensory functions | Reduced lesion size area | Renno et al. | 2014 |
| Weight-drop method | 30 mg/kg, intraperitoneal, 30 min after and daily during the first week post-spinal cord injury | Reduced thermal hyperalgesia and gliosis | Via RhoA and FASN pathway | Álvarez-Pérez et al. | 2016 |
| Balloon compression method | Intra-spinal injection of 50 mg/kg, immediately and then weekly for up to 28 days | Attenuated nuclear Factor Kappa B pathway and increased gene expression of fibroblast growth factor 2 and vascular endothelial growth factor | Enhanced neuroregeneration | Urdzikova et al. | 2017 |
| Weight-drop method | 50-mg/kg, intraperitoneal, before and after ischemia-reperfusion injury | Decreased malodialdehyde, attenuated caspase-3, tumor necrosis factor alpha, and inducible nitric oxide synthase, reduction of motor deficit index | Anti-inflammatory and anti-apoptotic activity | Ahadi et al. | 2019 |

A summary of in vivo studies on the neuroprotective effects of epigallocatechin-3-gallate against spinal cord injury, along with treatment schedules, findings, and related molecular mechanisms, is presented in this table.

can protect spinal cord tissue and improve behavioral function after traumatic and ischemia-reperfusion injury (IRI) in rats (Ahadi, Zargari, & Khalatbary, 2019; Khalatbary & Ahmadvand, 2010; Khalatbary et al., 2010). In regard to mechanisms of neuroprotective effects of EGCG, our results revealed that malodialdehyde (MDA) levels as an index of tissue lipid peroxidation, an important pathologic event in post-traumatic neuronal degeneration, were significantly decreased in traumatized spinal cord tissue after EGCG treatment (Khalatbary & Ahmadvand, 2010). It is well documented that EGCG due to hydroxyl groups can bind to the free radicals and neutralize those, while can indirectly increase the body's endogenous antioxidants (Salah et al., 1995). We detected immunohistochemically anti-apoptotic properties of EGCG with decreasing pro-apoptotic protein Bax and increasing anti-apoptotic protein Bcl-2 staining in EGCG-treated rats, which was correlated with terminal deoxynucleotidyl transferase dUTP nick end labeling (TUNEL) staining method (Khalatbary et al., 2010). Also, our other study showed that EGCG treatment (50 mg/kg, i.p., immediately and 1 h after SCI) attenuated pro-inflammatory cytokines such as TNFα and IL-1β in traumatized spinal cord tissue (Khalatbary & Ahmadvand, 2011). In line with the anti-inflammatory properties of EGCG, Ge et al., 2013 documented that EGCG treatment (100 mg/kg, between 24 and 72 h after SCI) decreased spinal tissue edema through down-regulation of astrocyte expressing aquaporin-4 (AQP4), which plays a critical role in the transport of water from blood/CSF to spinal cord parenchyma, and through down-regulation of glial fibrillary acidic protein (GFAP) as a specific marker of astrocytes. Also, another study found that attenuated NF-κB pathway is a pivotal anti-inflammatory affect of EGCG (intra-spinal injection of 50 mg/kg EGCG immediately and then weekly for up to 28 days) after SCI (Urdzikova et al., 2017). Meanwhile, this study showed a significant increase in the gene expression of fibroblast growth factor 2 (FGF2) and vascular endothelial growth factor (VEGF) after SCI. Tao and Zhu (2013) documented that EGCG (100 mg/kg, i.p., immediately after SCI) can protect secondary SCI by potential mechanism of regulating p38MAPK\NF -κB\AQP4 signaling pathway and thus reduce edema after SCI in rats. Also, another study documented that intrathecal administration of EGCG (10 or 20 mg/kg immediately after SCI) can significantly improve locomotors recovery, which may be related to the inhibition of Bcl-2-associated X protein (Bax) and to the up-regulation of brain-derived neurotrophic factor (BDNF) and glial cell line-derived neurotrophic factor (GDNF) (Tian et al., 2013). Studies have shown that EGCG has beneficial effects not only on the tissue destruction but also on behavioral and functional outcome, so that intravenous infusion of EGCG (20 mg/kg/hour for continuous 36 h) in acutely (initiated within 4 h post-trauma 0 time) and chronically (initiated after 12 months of SCI onset) spinal cord injured rats improved significantly motor and sensory functions measured using Basso-Beattie-Bresnehan behavioral score test, Louisville forced swim test, and pain behavior assessment tests (Renno et al., 2014). One of the common complications of SCI is neuropathic pain. About the effect of EGCG on neuropathic pain, Álvarez-Pérez et al. (2016) documented that short time EGCG treatment (30 mg/kg i.p., 30 min after and daily during the first week post-SCI) reduced thermal hyperalgesia and gliosis via ras homologue gene family member A (RhoA) and fatty acid synthase (FASN) pathway. In this regard, it is mentioned that some biological properties of EGCG are attributed to its inhibitory action on FASN (Singh, Shankar, & Srivastava, 2011). In another study, the synergistic effects of curcumin and EGCG in an animal model of acute SCI were investigated (Ruzicka et al., 2018). Results of this study revealed that combination of curcumin and EGCG reduced glial scar formation, increased axonal sprouting, and changed the amount of macrophage inflammatory protein 1-alpha (MIP-1α), IL-1β, interleukin-4 (IL-4), and interleukin-6 (IL-6). Recently, our laboratory assessed the neuroprotective effects of EGCG on spinal cord IRI in rats (Ahadi et al., 2019). The level of MDA was significantly reduced in EGCG-treated rats. Attenuated caspase-3 (Fig. 2), TNFα (Fig. 3), and inducible nitric oxide synthase (iNOS) expression could be significantly detected in the EGCG-treated rats. Also, EGCG reduced the extent of degeneration of the spinal cord neurons, in addition to a significant reduction of motor deficit index (MDI). Overall, the behavioral, biochemical, and histopathological evidences demonstrated that pre- (50 mg/kg, i.p., before IRI) and post-ischemic (50 mg/kg, i.p., after IRI) treatment with EGCG had protective effects against spinal cord IRI in rats.

## In vitro studies

A number of in vitro studies regarding EGCG's neuroprotection and its underlying molecular mechanisms in neural cells are starting to accumulate which summarized in Table 2.

In regard to anti-apoptotic effects of EGCG, some studies indicated that EGCG led to significant inhibition of apoptosis in PC12 cells by scavenging reactive oxygen species (ROS), modulating cytochrome *c* and caspases, activation of phosphoinositide 3-kinase (PI3K), activation of survival-promoting pathways, and inhibition of glycogen synthase kinase-3 (GSK-3) (Hou et al., 2008; Jung, Han, et al., 2007; Jung, Mo, et al., 2007; Koh et al., 2003; Mandel, Reznichenko, Amit, & Youdim, 2003; Nie, Cao, & Zhao, 2002; Nie, Jin, Cao, Shen, & Zhao, 2002; Srividhya & Kalaiselvi, 2013). Also, studies showed that EGCG has a protective effect on cell death in SH-SY5Y (or SHSY-5Y) cell line through inhibiting ROS production and caspase activity along with activation of protein kinase C (PKC) (Avramovich-Tirosh et al., 2007; Jeong et al., 2004; Levites, Amit, Youdim, & Mandel, 2002; Tai & Truong, 2010). EGCG protects N18D3 neural cells from quinolinic acid-induced apoptosis by regulation of phosphatidyl inositol 3-kinase (PI3K) and modulation of pro- and anti-apoptotic genes due to decreased intra-cellular calcium levels and nitric oxide (NO) production (Zhang et al., 2014). Results of a study showed that EGCG protects the neuronal cells against human prion protein-induced damage through inhibiting Bax and cytochrome *c* translocation and autophagic pathways in primary neuron cells (Lee et al., 2015). Recently, it was documented that EGCG reduced amyloid beta-induced neurotoxicity via inhibiting endoplasmic reticulum stress-mediated apoptosis in SH-SY5Y cells in a dose-dependent manner which is contributed to the down-regulation of cleaved-caspase-3 and -12 (Du et al., 2018).

In regard to microglial activation, a study indicated that EGCG exerts significant protection against microglial activation-induced neuronal injury through the down-regulation of iNOS and TNFα expression (Li, Huang, Fang, & Le, 2004).

In regard to antioxidant properties of EGCG, in vitro study revealed that EGCG treatment decreased MDA level, up-regulated superoxide dismutase (SOD) and catalase (CAT) levels against advanced glycation end products (AGEs)-induced

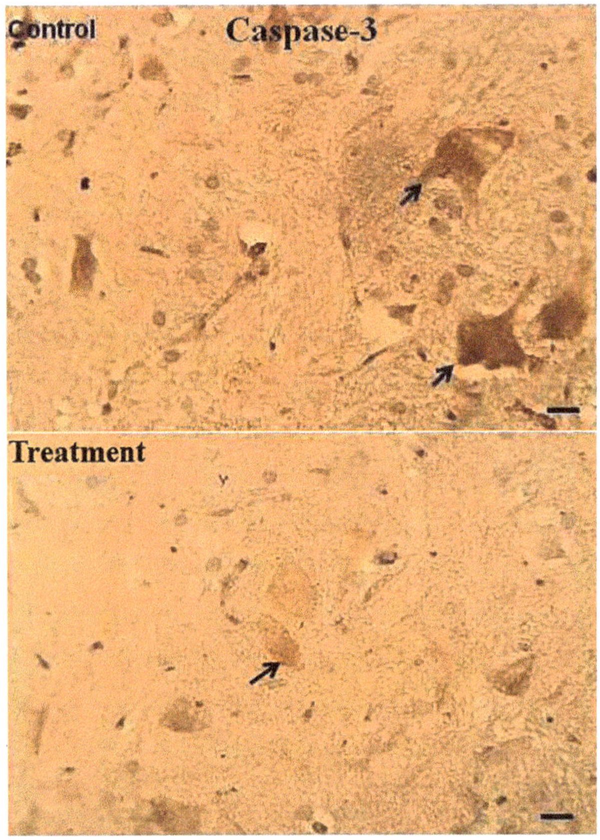


FIG. 2 Immunohistochemical staining of casepase-3. Light photomicrographs show immunohistochemical expression of caspase-3 as an index of apoptosis in control and epigallocatechin-3-gallate treatment groups after spinal cord injury (magnification, ×200). The positive staining of caspase-3 is presented by a brown color of cytoplasm (arrows). This figure prepared in our laboratory.

injury in neuronal cell culture (Lee & Lee, 2007). Also, it was documented that EGCG decreased markedly ROS production via the SIRT1/PGC-1α signaling pathway in 1methyl4-phenyl-pyridine (MMP+)-induced PC12 cells damage (Ye et al., 2012). Evaluation of EGCG's effects on acrylamide-induced cytotoxicity in PC12 cells showed that EGCG inhibited the cytotoxicity of acrylamide and increased the cell viability (Esmaeelpanah, Razavi, Vahdati Hasani, & Hosseinzadeh, 2018).

In regard to anti-inflammatory properties of EGCG, in vitro studies also revealed that EGCG treatment potently inhibited lipopolysaccharide (LPS)-activated microglial secretion of TNFα and significantly exerted anti-inflammatory effects partly as a suppressor of TNFα signaling via the NF-κB pathway in SH-SY5Y (Li et al., 2004) and RPE cells (Fu & Koo, 2006).

It is well known that anti-oxidants can also affect the stem cells differentiation and potency. In this regard, using the PC12 cell model showed for the first time that low concentration of EGCG and unfractionated green tea polyphenols potentiate nerve growth factor (NGF)-induced and BDNF-induced neuritogenesis likely through induction sub-lethal levels of ROS or dependent on 67 kDa laminin receptor (67LR) and $H_2O_2$ (Gundimeda et al., 2014; Gundimeda, McNeill, Schiffman, Hinton, & Gopalakrishna, 2010).

## Applications to other areas of neuroscience

It is well known that EGCG has neuroprotective effects in another neurological disorder such as brain injury, nerve injury, and neurodegenerative diseases (Table 3).

In regard to brain ischemia, it was documented that EGCG through deoxidizing peroxynitrate/peroxynitrite attenuates NO-3/NO-2 concentration without affecting blood flow (Nagai et al., 2002). Another study demonstrated that EGCG due to attenuation of the malondialdehyde level and oxidized/total glutathione ratio protects brain tissue against transient middle cerebral artery occlusion (Choi, Kim, Lee, Kim, & Kim, 2004). A possible mechanism involved in the neuroprotective

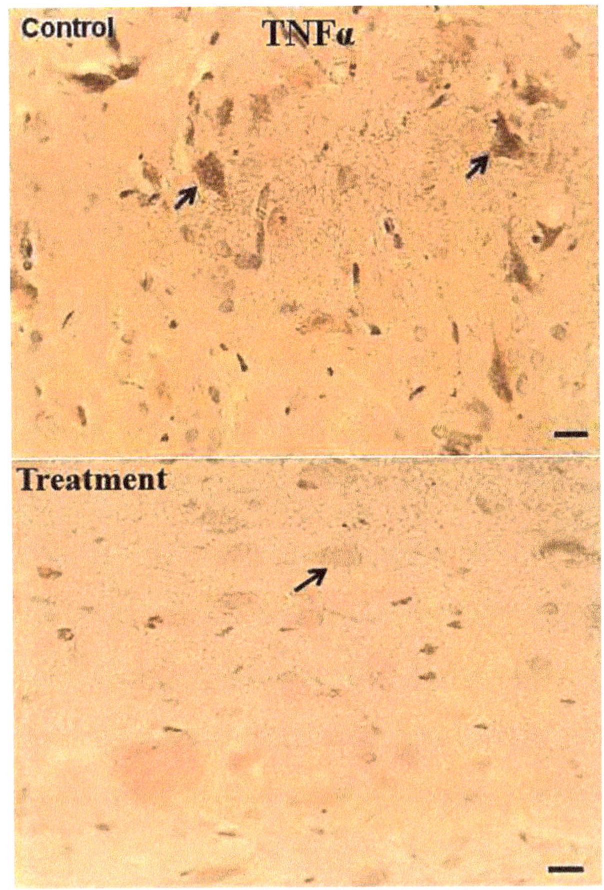


FIG. 3 Immunohistochemical staining of tumor necrosis factor alpha. Light photomicrographs show immunohistochemical expression of tumor necrosis factor alpha (TNFα) as an index of inflammation in control and epigallocatechin-3-gallate treatment groups after spinal cord injury (magnification, ×200). The positive staining of TNFα is presented by a brown color of cytoplasm (arrows). This figure prepared in our laboratory.

effect of EGCG is inhibition of matrix metalloproinase-9 (MMP-9) activity which plays an important role in the pathophysiology of brain ischemia (Park et al., 2010). Also, EGCG exerts protective effect via the activation of the nuclear erythroid 2-related factor 2 (Nrf2) against cerebral ischemia (Han et al., 2014). A further study showed that EGCG promotes angiogenesis via up-regulation of the Nrf2 signaling pathway in the early stage of ischemic stroke (Bai et al., 2017). Modulation of the inflammation-related molecules such as TNFα, IL-1β, IL-6, NF-κB /p65, COX-2, and iNOS is another possible mechanism of action of EGCG following focal cerebral ischemia/reperfusion injury (Zhang, Li, Jiang, Chen, & Huang, 2015). It was found that EGCG also preserves mitochondrial energetics (complex I—V) and citrate synthase activity after brain ischemia (Sutherland et al., 2005). While, it was documented that EGCG through eliminating free radical generation, inhibiting apoptosis, and increasing neural stem cells protects brain against traumatic injury in rats (Itoh et al., 2011, 2012, 2013).

In regard to nerve injury, results of a study showed significantly higher neurofilament triplet L (NF-L) protein expression in EGCG-treated rats compared to non-treated rats following optic nerve crush injury (Xie et al., 2010). Also, EGCG through attenuation of nicotinamide adenine dinucleotide phosphate-diaphorase (NADPH-d)/neuronal nitric oxide synthase (nNOS) increased motor neuron survival time following peripheral nerve crush injury in rats (Wei et al., 2011). Intrathecal injection of EGCG improved pain-like behaviors through decreasing NF-κB, Toll-like receptor 4 (TLR4), TNFα, high-mobility group box 1 protein (HMGB1), and IL-1β following chronic constriction injury (Kuang et al., 2012). EGCG due to the down-regulation of apoptosis related genes, up-regulation of glutathione reductase, and up-regulation of survivin gene accelerates nerve regeneration and improves functional recovery following rat sciatic nerve crush injury (Renno et al., 2017; Renno, Al-Maghrebi, Rao, & Khraishah, 2015). Also, Renno et al. documented that the EGCG neuroprotective effects are mediated through the modulation of BDNF, GDNF and neurotrophin-3 (NT3) neurotrophic factor and their receptors tropomyosin receptor kinase B (Trk-B), tropomyosin receptor kinase C (Trk-C), and nerve growth factor receptor p75 (NGFR-p75) following sciatic nerve crush injury (Renno, Khan, & Benov, 2016).

**TABLE 2** In vitro neuroprotective effects of epigallocatechin-3-gallate.

| Model of induction | Cell type | Finding | Possible mechanism | Author | Year |
|---|---|---|---|---|---|
| Apoptosis | PC12 | Inhibited apoptosis | Modulation of cytochrome *c* and caspases, activation of phosphatidylinositol 3-kinase and survival-promoting pathways, inhibition of glycogen synthase kinase-3 | Nie et al., Koh et al., Mandel et al., Jung et al., Hou et al., Strividhya and Kalaisali | 2002a, 2002b, 2003, 2003, 2007a, 2007b, 2008, 2013 |
| Apoptosis | SH-SY5Y | Inhibited apoptosis | Inhibition of reactive oxygen species production and caspase activity, activation of protein kinase C | Levites et al., Jeong et al., Avramovich-Tirosh et al., Tia and Truong | 2002, 2004, 2007, 2010 |
| Apoptosis | N18D3 | Attenuated apoptosis | Regulation of phosphatidylinositol 3-kinase and modulation of pro- and anti-apoptotic genes | Zhang et al. | 2014 |
| Apoptosis | Primary neuron cells | Attenuated apoptosis | Inhibition of Bcl-2-associated X protein and cytochrome *c* translocation and autophagic pathways | Lee et al. | 2015 |
| Apoptosis | SH-SY5Y | Reduced apoptosis | Inhibiting endoplasmic reticulum stress-mediated apoptosis | Du et al. | 2018 |
| Microglial activation | SH-SY5Y | Inhibited LPS-activated microglial secretion of nitric oxide and tumor necrosis factor alpha | Down-regulation of inducible nitric oxide synthase and tumor necrosis factor alpha expression | Lee et al. | 2004 |
| Oxidative stress | Neuronal cells culture | Up-regulated superoxide dismutase and catalase levels, decreased malodialdehyde and carbonyl levels, and advanced glycation end products formation | Antioxidative properties and interfering with AGEs-RAGE interaction mediated pathways | Lee and Lee | 2007 |
| Oxidative stress | PC12 | Decreased reactive oxygen species production | via the SIRT1/PGC-1$\alpha$ signaling pathway | Ye et al. | 2012 |
| Oxidative stress | PC12 | Inhibited the cytotoxicity of acrylamide and increased the cell viability | Without providing possible mechanisms | Esmaeelpanah et al. | 2018 |
| Inflammation | SH-SY5Y and ARPE-19 | Anti-inflammatory effects | Suppression of tumor necrosis factor alpha signaling via the nuclear Factor Kappa B pathway | Li et al., Fu and Koo | 2004, 2006 |
| Neurogenesis | PC12 | Potentiated nerve growth factor- and brain derived neurotrophic factor-induced neuritogenesis | Involvement of sub-lethal levels of reactive oxygen species or dependent on 67 LR and H2O2 | Gundimeda et al. | 2010, 2014 |

A summary of in vitro studies on the neuroprotective effects of epigallocatechin-3-gallate with related molecular mechanisms is presented in this table.

**TABLE 3** Neuroprotective effects of epigallocatechin-3-gallate against other neurological disorders.

| Model of injury | Treatment schedule | Finding | Possible mechanism | Author | Year |
|---|---|---|---|---|---|
| Brain injury | 50 mg/kg, intraperitoneal, immediately after ischemia | Deoxidizing peroxynitrate/ peroxynitrite without affecting blood flow | Attenuation of $NO_3^-$/ $NO_2^-$ | Nagai et al. | 2002 |
| Brain injury | 25 mg and 50 mg/kg, intraperitoneal, immediately after ischemia | Attenuated malodialdehyde level and oxidized/total glutathione ratio | Antioxidant effects | Choi et al. | 2004 |
| Brain injury | 50 mg/kg, intraperitoneal, immediately after ischemia | Reduced neuronal damage | Reduction of matrix metalloproinase-9 activity | Park et al. | 2010 |
| Brain injury | 40 mg/kg, intraperitoneal, once daily for three consecutive days prior to surgery | Reduced neuronal damage | Activation of nuclear erythroid 2-related factor 2 /antioxidant response element signaling | Han et al. | 2014 |
| Brain injury | 50 mg/kg, intraperitoneal, immediately after surgery and once daily till day 7 | Promoted angiogenesis | Up-regulation of the nuclear erythroid 2-related factor 2 signaling pathway | Bai et al. | 2017 |
| Brain injury | 50 mg/kg, intraperitoneal, immediately after ischemia | Modulated inflammatory responses | Amelioration of tumor necrosis factor alpha, interleukin-1 beta, interleukin-6, nuclear factor kappa B /p65, cyclooxygenase-2, inducible nitric oxide synthase | Zhang et al. | 2015 |
| Brain injury | 50 mg/kg, daily for 1 day and 1 h before ischemia | Reduced nitric oxide synthase and inducible nitric oxide synthase, increased endothelial nitric oxide synthase, preserved complex I–V | Complex signal transduction mechanisms | Sutherland et al. | 2005 |
| Brain injury | Consuming 0.1% (*w/v*) epigallocatechin-3-gallate drinking water ad libitum before and after traumatic injury | Increased neural stem cells number around the damaged area | Antioxidant and antiapoptotic activities | Itoh et al. | 2011, 2013, 2012 |
| Brain injury | 100 mg/kg, intraperitoneal, after traumatic injury | Reduced interleukin-1 beta and tumor necrosis factor alpha mRNA expression, inhibited astrocyte expressing aquaporin-4 and glial fibrillary acidic protein, elevated superoxide dismutase and glutathione peroxidase activities, decreased nicotinamide adenine dinucleotide phosphate oxidase activation | Inhibition of inflammation and oxidative stress | Zhang et al. | 2015 |

*Continued*

**TABLE 3** Neuroprotective effects of epigallocatechin-3-gallate against other neurological disorders—cont'd

| Model of injury | Treatment schedule | Finding | Possible mechanism | Author | Year |
|---|---|---|---|---|---|
| Nerve injury | 25 mg/kg, injected daily for 5 days and 2 mg/kg orally daily afterwards | Increased density of retinal ganglion cells | Up-regulation of neurofilament triplet L protein expression | Ahadi et al. | Xie 2010 |
| Nerve injury | 25 mg/kg, intraperitoneal or orally, daily for 5 days and 2 mg/kg orally daily afterwards | Enhanced motor neuron survival time | Attenuation of nicotinamide adenine dinucleotide phosphate-diaphorase-d/neuronal nitric oxide synthase | Wei et al. | 2011 |
| Nerve injury | 1 mg/kg, intrathecal, once daily from 1 day before to 3 days after chronic constriction injury | Decreased chronic neuropathic pain | Down-regulation of Toll-like receptor 4, nuclear factor kappa B, high-mobility group box 1 protein, tumor necrosis factor alpha and interleukin-1 beta protein expression | Kuang et al. | 2012 |
| Nerve injury | 50 mg/kg, intraperitoneal, following rat sciatic nerve crush injury | Enhanced functional recovery and accelerated nerve regeneration | Down-regulation of apoptosis related genes, up-regulation of survivin gene, and up-regulation of glutathione reductase | Renno et al. | 2017, 2015 |
| Nerve injury | 50 mg/kg, intraperitoneal, starting 1 h after sciatic nerve crush injury | Modulated brain-derived neurotrophic factor, glial cell line-derived neurotrophic factor and neurotrophin-3 and their receptors tropomyosin receptor kinase B, tropomyosin receptor kinase C, and nerve growth factor receptor p75 | Up-regulation of neurotrophic factors and their receptors | Renno et al. | 2016 |
| Nerve injury | 50 mg/kg, intraperitoneal, 1 h after nerve transaction and followed for 3 days | Reduced neural apoptosis | Modulation of malodialdehyde levels, catalase and superoxide dismutase activities, and Caspase3 and cyclooxygenase-2 expression | Kian et al. | 2019 |
| Alzheimer's disease | 20 mg/kg, intraperitoneal, daily for 60 days | Effective prophylaxis for Alzheimer's disease | Modulation of amyloid precursor protein cleavage and cerebral amyloidosis | Rezai-Zadeh et al. | 2005 |
| Neurodegenerative disease | 2 or 6 mg/day for 4 weeks, intragastrically | Improved degenerative changes of the brain | Modulation of beta-amyloid and amyloid precursor protein | He et al. | 2012 |

**TABLE 3** Neuroprotective effects of epigallocatechin-3-gallate against other neurological disorders—cont'd

| Model of injury | Treatment schedule | Finding | Possible mechanism | Author | Year |
|---|---|---|---|---|---|
| Neurodegenerative disease | 2 mg/kg/day for 4 weeks, intragastrically | Ameliorated learning and memory deficits | Modulation of TrkA/ $p75^{NTR}$ signaling and apoptosis | Liu et al. | 2014 |
| Parkinson's disease | No available | Alleviated dopaminergic neuronal injury in Parkinson's disease | Inhibition of nitric oxide and tumor necrosis factor alpha | Li et al. | 2004 |
| Parkinson's disease | 25 mg/kg, oral administration for 7 days | Neurorescue effects in Parkinson's disease | Regulation of the iron-export protein ferroportin and reduction of oxidative stress | Xu et al. | 2017 |
| Neurodegenerative disease | 300 µg, oral gavage twice daily | Delayed disease onset, induced regeneration of hippocampal axons | Reduction of the inflammatory infiltrates | Herges et al. | 2011 |

A summary of in vivo studies on the neuroprotective effects of epigallocatechin-3-gallate against other neurological disorders, along with treatment schedules, findings, and related molecular mechanisms, is presented in this table.

Recently, we found that EGCG protects sensory and motor neurons through attenuation of COX-2 and caspase-3, and elevation of SOD and CAT activities in rat sciatic nerve transaction model (Kian, Khalatbary, Ahmadvand, Karimpour Malekshah, & Shams, 2019).

In regard to neurodegenerative diseases, research on Alzheimer's disease (AD) revealed that EGCG treatment in Alzheimer transgenic mice reduces cerebral amyloidosis and modulates amyloid precursor protein (APP) cleavage (Rezai-Zadeh et al., 2005). Some existing evidence suggests that the EGCG's neuroprotection against Aβ-induced cell death is somewhat due to phosphatidylinositol 3-kinase (PI3K)/protein kinase B (AKT) transduction signaling and alpha-7 nicotinic acetylcholine receptor (α7nAChR) activity (Zhang et al., 2014). Liu et al. (2014) documented that EGCG through modulating the balance of TrkA/p75NTR signaling in APP/PSI transgenic mice ameliorated memory and learning deficits. Results of a study supported that EGCG in Parkinson's disease (PD) can be a therapeutic approach to alleviate microglia-mediated dopaminergic neuronal injury (Li et al., 2004). Recently the neurorescue effects of EGCG have been proven in a mouse model of PD assessed with behavioral and neurotransmitter analysis (Xu et al., 2017). Also, results of another study indicated that combination therapy of glatiramer acetate and EGCG can be a safe and promising approach for multiple sclerosis (Herges et al., 2011).

## Conclusion

In recent years, dietary polyphenols have attracted a lot of attention due to their interesting biological activities. An expanding body of results obtained by pre-clinical in vitro and in vivo studies on the use of EGCG as a therapeutic option for SCI confirmed its neuroprotective properties. These beneficial effects have been mainly attributed to free radical scavenging or anti-oxidant, anti-inflammatory, and anti-apoptotic properties of EGCG. The evidence presented in this review suggests the potential of EGCG in both preventive and therapeutic usages for a variety of SCI models. At the same time, there is a lot of evidence that this natural polyphenol has proper absorption and crosses the BBB with minimal side effects. Of course, considering that no study has been conducted on the neuroprotective effects of EGCG following SCI in human, the design of clinical trials are required regarding EGCG's neuroprotection and its possible mechanisms as well as determining the dose and course of treatment remain to be explored following SCI.

## Mini-dictionary of terms

- Catechins: They are natural polyphenol compounds with potent anti-oxidant activity harvested from a variety of sources.
- Free radical scavenging: Inactivation of free radicals in a biological system.

- Endogenous anti-oxidants: They are products of the body's metabolism which act as free radical scavengers.
- Synergistic effects: An effect arising between two or more substances that create a greater effect than the sum of their individual effects.
- Motor deficit index: Evaluation of neurological function with scoring of the placing/stepping reflex and ambulation.
- Intrathecal administration: Injection of therapeutic agents into the spinal sub-arachnoid space.
- Neuritogenesis: A complex phenomenon involving multiple interactions between the extracellular matrix and growing neurite.

## Key facts of EGCG

- EGCG has anti-oxidant, anti-inflammatory, and anti-apoptotic properties.
- EGCG is able to cross the BBB and can reach to injured nervous tissue even at a very low concentration.
- Several experimental studies have shown that EGCG can afford neuroprotection against brain and SCIs, neurodegenerative diseases, and peripheral nerve injuries.
- The behavioral, biochemical, and histopathological evidences demonstrated EGCG treatment has protective effects against experimental model of SCI.
- The neuroprotective effects of EGCG are dose-dependent and therefore it is necessary to determine the appropriate dose.

## Summary points

- Epigallocatechin-3-gallate is thought to be responsible for the most biological activities of green tea catechins, including anti-oxidant, anti-inflammatory, and anti-apoptotic properties.
- Epigallocatechin-3-gallate has been shown to be some of the neuroprotective effects against injury to the peripheral and central nervous system.
- Treatment of SCI with epigallocatechin-3-gallate improves neurological function and decreases spinal tissue loss.
- Neuroprotective effects of epigallocatechin-3-gallate against SCI are largely due to its anti-inflammatory, anti-apoptotic, and anti-oxidant properties.
- This paper summarizes current knowledge on neuroprotective effects of epigallocatechin-3-gallate and their molecular mechanisms responsible for the neuroprotection following SCI.

## References


Ahadi, S., Zargari, M., & Khalatbary, A. R. (2019). Assessment of the neuroprotective effects of (−)epigallocatechin-3-gallate on spinal cord ischemiareperfusion injury in rats. *The Journal of Spinal Cord Medicine*, 1–8.

Álvarez-Pérez, B., Homs, J., Bosch-Mola, M., Puig, T., Reina, F., Verdú, E., et al. (2016). Epigallocatechin-3-gallate treatment reduces thermal hyperalgesia after spinal cord injury by down-regulating RhoA expression in mice. *European Journal of Pain*, *20*, 341–352.

Amar, A. P., & Levy, M. L. (1999). Pathogenesis and pharmacological strategies for mitigating secondary damage in acute spinal cord injury. *Neurosurgery*, *44*, 1027–1039.

Avramovich-Tirosh, Y., Reznichenko, L., Mit, T., Zheng, H., Fridkin, M., Weinreb, O., et al. (2007). Neurorescue activity, APP regulation and amyloid-beta peptide reduction by novel multi-functional brain permeable iron-chelating-antioxidants, M-30 and green tea polyphenol, EGCG. *Current Alzheimer Research*, *4*, 403–411.

Bai, Q., Lyu, Z., Yang, X., Pan, Z., Lou, J., & Dong, T. (2017). Epigallocatechin-3-gallate promotes angiogenesis via up-regulation of Nfr2 signaling pathway in a mouse model of ischemic stroke. *Behavioral Brain Research*, *321*, 79–86.

Choi, Y. B., Kim, Y. I., Lee, K. S., Kim, B. S., & Kim, D. J. (2004). Protective effect of epigallocatechin gallate on brain damage after transient middle cerebral artery occlusion in rats. *Brain Research*, *1019*, 47–54.

Du, K., Liu, M., Zhong, X., Yao, W., Xiao, Q., Wen, Q., et al. (2018). Epigallocatechin gallate reduces amyloid β-induced neurotoxicity via inhibiting endoplasmic reticulum stress-mediated apoptosis. *Molecular Nutrition & Food Research*, *62*, e1700890.

Dulloo, A. G., Duret, C., Rohrer, D., Girardier, L., Mensi, N., Fathi, M., et al. (1999). Efficacy of a green tea extract rich in catechin polyphenols and caffeine in increasing 24-h energy expenditure and fat oxidation in humans. *The American Journal of Clinical Nutrition*, *70*, 1040–1045.

Esmaeelpanah, E., Razavi, B. M., Vahdati Hasani, F., & Hosseinzadeh, H. (2018). Evaluation of epigallocatechin Gallate and epicatechin gallate effects on acrylamide-induced neurotoxicity in rats and cytotoxicity in PC 12 cells. *Drug and Chemical Toxicology*, *41*, 441–448.

Fu, Y., & Koo, M. W. (2006). EGCG protects HT-22 cells against glutamateinduced oxidative stress. *Neurotoxicity Research*, *10*, 23–30.

Ge, R., Zhu, Y., Diao, Y., Tao, L., Yuan, W., & Xiong, X. C. (2013). Anti-edema effect of epigallocatechin gallate on spinal cord injury in rats. *Brain Research*, *1527*, 40–46.

Graham, H. N. (1992). Green tea composition, consumption, and polyphenol chemistry. *Preventive Medicine*, *21*, 334–350.

Gundimeda, U., McNeill, T. H., Fan, T. K., Deng, R., Rayudu, D., Chen, Z., et al. (2014). Green tea catechins potentiate the neuritogenic action of brain-derived neurotrophic factor: Role of 67-kDa laminin receptor and hydrogen peroxide. *Biochemical and Biophysical Research Communications, 445*, 218–224.

Gundimeda, U., McNeill, T. H., Schiffman, J. E., Hinton, D. R., & Gopalakrishna, R. (2010). Green tea polyphenols potentiate the action of nerve growth factor to induce neuritogenesis: Possible role of reactive oxygen species. *Journal of Neuroscience Research, 88*, 3644–3655.

Hall, E. D. (1993). Lipid antioxidants in acute central nervous system injury. *Annals of Emergency Medicine, 22*, 1022–1027.

Han, J., Wang, M., Jing, X., Shi, H., Ren, M., & Lou, H. (2014). (−)-epigallocatechin gallate protects against cerebral ischemiainduced oxidative stress via Nrf2/ARE signaling. *Neurochemical Research, 39*, 1292–1299.

Herges, K., Millward, J. M., Hentschel, N., Infante-Duarte, C., Aktas, O., & Zipp, F. (2011). Neuroprotective effect of combination therapy of glatiramer acetate and epigallocatechin-3-gallate in neuroinflammation. *PLoS One, 6*, e25456.

Hou, R. R., Chen, J. Z., Chen, H., Kang, X. G., Li, M. G., & Wang, B. R. (2008). Neuroprotective effects of (−)-epigallocatechin-3-gallate (EGCG) on paraquat-induced apoptosis in PC12 cells. *Cell Biology International, 32*, 22–30.

Itoh, T., Imano, M., Nishida, S., Tsubaki, M., Hashimoto, S., Ito, A., et al. (2011). (−)-Epigallocatechin-3-gallate protects against neuronal cell death and improves cerebral function after traumatic brain injury in rats. *Neuromolecular Medicine, 13*, 300–309.

Itoh, T., Imano, M., Nishida, S., Tsubaki, M., Mizuguchi, N., Hashimoto, S., et al. (2012). (−)-Epigallocatechin-3-gallate increases the number of neural stem cells around the damaged area after rat traumatic brain injury. *Journal of Neural Transmission (Vienna), 119*, 877–890.

Itoh, T., Tabuchi, M., Mizuguchi, N., Imano, M., Tsubaki, M., Nishida, S., et al. (2013). Neuroprotective effect of (−)-epigallocatechin3-gallate in rats when administered pre- or post-traumatic brain injury. *Journal of Neural Transmission (Vienna), 120*, 767–783.

Jeong, J. H., Kim, H. J., Lee, T. J., Kim, M. K., Park, E. S., & Choi, B. S. (2004). Epigallocatechin 3-gallate attenuates neuronal damage induced by 3-hydroxykynurenine. *Toxicology, 195*, 53–60.

Jung, J. Y., Han, C. R., Jeong, Y. J., Kim, H. J., Lim, H. S., Lee, K. H., et al. (2007). Epigallocatechin gallate inhibits nitric oxide-induced apoptosis in rat PC12 cells. *Neuroscience Letters, 411*, 222–227.

Jung, J. Y., Mo, H. C., Yang, K. H., Jeong, Y. J., Yoo, H. G., Choi, N. K., et al. (2007). Inhibition by epigallocatechin gallate of CoCl2-induced apoptosis in rat PC12 cells. *Life Sciences, 80*, 1355–1363.

Kalfon, L., Youdim, M. B., & Mandel, S. A. (2007). Green tea polyphenol (−)-epigallocatechin-3-gallate promotes the rapid protein kinase C- and proteasome-mediated degradation of bad: Implications for neuroprotection. *Journal of Neurochemistry, 100*, 992–1002.

Khalatbary, A. R. (2014). Natural polyphenols and spinal cord injury. *Iranian Biomedical Journal, 18*, 120–129.

Khalatbary, A. R., & Ahmadvand, H. (2010). Effects of epigallocatechin gallate on tissue lipid peroxide levels in traumatized spinal cord of rat. *Iranian Journal of Basic Medical Sciences, 13*, 139–142.

Khalatbary, A. R., & Ahmadvand, H. (2011). Anti-inflammatory effect of the epigallocatechin gallate following spinal cord trauma in rat. *Iranian Biomedical Journal, 15*, 31–37.

Khalatbary, A. R., Tiraihi, T., Boroujeni, M. B., Ahmadvand, H., Tavafi, M., & Tamjidipoor, A. (2010). Effects of epigallocatechin gallate on tissue protection and functional recovery after contusive spinal cord injury in rats. *Brain Research, 1306*, 168–175.

Kian, K., Khalatbary, A. R., Ahmadvand, H., Karimpour Malekshah, A., & Shams, Z. (2019). Neuroprotective effects of (−)-epigallocatechin-3-gallate (EGCG) against peripheral nerve transection-induced apoptosis. *Nutritional Neuroscience, 22*, 578–586.

Koh, S. H., Kim, S. H., Kwon, H., Park, Y., Kim, K. S., Song, C. W., et al. (2003). Epigallocatechin gallate protects nerve growth factor differentiated PC12 cells from oxidative-radical-stress-induced apoptosis through its effect on phosphoinositide 3-kinase/Akt and glycogen synthase kinase-3. *Molecular Brain Research, 118*, 72–81.

Kuang, X., Huang, Y., Gu, H. F., Zu, X. Y., Zou, W. Y., Song, Z. B., et al. (2012). Effects of intrathecal epigallocatechin gallate, an inhibitor of toll-like receptor 4, on chronic neuropathic pain in rats. *European Journal of Pharmacology, 676*, 51–56.

Lawrence, T., & Fong, C. (2010). The resolution of inflammation: Antiinflammatory roles for NF-kappaB. *The International Journal of Biochemistry & Cell Biology, 42*, 519–523.

Lee, S. J., & Lee, K. W. (2007). Protective effect of (−)-epigallocatechin gallate against advanced glycation end products-induced injury in neuronal cells. *Biological and Pharmaceutical Bulletin, 30*, 1369–1373.

Lee, M. J., Maliakal, P., Chen, L., Meng, X., Bondoc, F. Y., Prabhu, S., et al. (2002). Pharmacokinetics of tea catechins after ingestion of green tea and (−)-epigallocatechin-3-gallate by humans: Formation of different metabolites and individual variability. *Cancer Epidemiology, Biomarkers & Prevention, 11*, 1025–1032.

Lee, J. H., Moon, J. H., Kim, S. W., Jeong, J. K., Nazim, U. M., Lee, Y. J., et al. (2015). EGCG-mediated autophagy flux has a neuroprotection effect via a class III histone deacetylase in primary neuron cells. *Oncotarget, 6*, 9701–9717.

Levites, Y., Amit, T., Youdim, M. B., & Mandel, S. (2002). Involvement of protein kinase C activation and cell survival/ cell cycle genes in green tea polyphenol (−)-epigallocatechin 3-gallate neuroprotective action. *Journal of Biological Chemistry, 277*, 30574–30580.

Li, R., Huang, Y. G., Fang, D., & Le, W. D. (2004). (−)-epigallocatechin gallate inhibits lipopolysaccharide-induced microglial activation and protects against inflammation-mediated dopaminergic neuronal injury. *Journal of Neuroscience Research, 78*, 723–731.

Lin, L. C., Wang, M. N., Tseng, T. Y., Sung, J. S., & Tsai, T. H. (2007). Pharmacokinetics of (−)-epigallocatechin-3-gallate in conscious and freely moving rats and its brain regional distribution. *Journal of Agricultural and Food Chemistry, 55*, 1517–1524.

Liu, M., Chen, F., Sha, L., Wang, S., Tao, L., Yao, L., et al. (2014). (−)-Epigallocatechin-3-gallate ameliorates learning and memory deficits by adjusting the balance of TrkA/p75NTR signaling in APP/PS1 transgenic mice. *Molecular Neurobiology, 49*, 1350–1363.

Liu, X. Z., Xu, X. M., Hu, R., Du, C., Zhang, S. X., McDonald, J. W., et al. (1997). Neuronal and glial apoptosis after traumatic spinal cord injury. *The Journal of Neuroscience*, *17*, 5395–5406.

Mandel, S., Maor, G., & Youdim, M. B. (2004). Iron and alpha-synuclein in the substantia nigra of MPTP-treated mice: Effect of neuroprotective drugs R-apomorphine and green tea polyphenol (−)-epigallocatechin-3-gallate. *Journal of Molecular Neuroscience*, *24*, 401–416.

Mandel, S., Reznichenko, L., Amit, T., & Youdim, M. B. (2003). Green tea polyphenol (−)-epigallocatechin-3-gallate protects rat PC12 cells from apoptosis induced by serum withdrawal independent of P13-Akt pathway. *Neurotoxicity Research*, *5*, 419–424.

Meng, X., Sang, S., Zhu, N., Lu, H., Sheng, S., Lee, M. J., et al. (2002). Identification and characterization of methylated and ringfission metabolites of tea catechins formed in humans, mice, and rats. *Chemical Research in Toxicology*, *15*, 1042–1050.

Moore, R. J., Jackson, K. G., & Minihane, A. M. (2009). Green tea (*Camellia sinensis*) catechins and vascular function. *British Journal of Nutrition*, *102*, 1790–1802.

Nagai, K., Jiang, M. H., Hada, J., Nagata, T., Yajima, Y., Yamamoto, S., et al. (2002). (−)-epigallocatechin gallate protects against NO stressinduced neuronal damage after ischemia by acting as an antioxidant. *Brain Research*, *956*, 319–322.

Nie, G., Cao, Y., & Zhao, B. (2002). Protective effects of green tea polyphenols and their major component. (−)-epigallocatechin-3-gallate (EGCG), on 6-hydroxydopamine-induced apoptosis in PC12 cells. *Redox Report*, *7*, 171–177.

Nie, G., Jin, C., Cao, Y., Shen, S., & Zhao, B. (2002). Distinct effects of tea catechins on 6-hydroxydopamine-induced apoptosis in PC12 cells. *Archives of Biochemistry and Biophysics*, *397*, 84–90.

Ohishi, T., Goto, S., Monira, P., Isemura, M., & Nakamura, Y. (2016). Anti-inflammatory action of green tea. *Anti-Inflammatory & Anti-Allergy Agents in Medicinal Chemistry*, *15*, 74–90.

Oliveira, M. R., Nabavi, S. F., Daglia, M., Rastrelli, L., & Nabavi, S. M. (2016). Epigallocatechin gallate and mitochondria—A story of life and death. *Pharmacological Research*, *104*, 70–85.

Park, J. W., Hong, J. S., Lee, K. S., Kim, H. Y., Lee, J. J., & Lee, S. R. (2010). Green tea polyphenol (−)-epigallocatechin gallate reduces matrix metalloproteinase-9 activity following transient focal cerebral ischemia. *The Journal of Nutritional Biochemistry*, *21*, 1038–1044.

Pervin, M., Unno, K., Nakagawa, A., Takahashi, Y., Iguchi, K., Yamamoto, H., et al. (2017). Blood brain barrier permeability of (−)-epigallocatechin gallate, its proliferation-enhancing activity of human neuroblastoma SH-SY5Y cells, and its preventive effect on age-related cognitive dysfunction in mice. *Biochemistry and Biophysics Reports*, *9*, 180–186.

Popovich, P. G., Wei, P., & Stokes, B. T. (1997). Cellular inflammatory response after spinal cord injury in Sprague-Dawley and Lewis rats. *Journal of Comparative Neurology*, *377*, 443–464.

Renno, W. M., Al-Khaledi, G., Mousa, A., Karam, S. M., Abul, H., & Asfar, S. (2014). (−)-Epigallocatechin-3-gallate (EGCG) modulates neurological function when intravenously infused in acute and, chronically injured spinal cord of adult rats. *Neuropharmacology*, *77*, 100–119.

Renno, W. M., Al-Maghrebi, M., Rao, M. S., & Khraishah, H. (2015). (−)-Epigallocatechin-3-gallate modulates spinal cord neuronal degeneration by enhancing growth-associated protein 43. Bcell lymphoma; 2, and decreasing B-cell lymphoma 2-associated x protein expression after sciatic nerve crush injury. *Journal of Neurotrauma*, *32*, 170–184.

Renno, W. M., Benov, L., & Khan, K. M. (2017). Possible role of antioxidative capacity of (−)-epigallocatechin-3-gallate treatment in morphological and neurobehavioral recovery after sciatic nerve crush injury. *Journal of Neurosurgery. Spine*, *27*, 593–613.

Renno, W. M., Khan, K. M., & Benov, L. (2016). Is there a role for neurotrophic factors and their receptors in augmenting the neuroprotective effect of (−)-epigallocatechin-3-gallate treatment of sciatic nerve crush injury? *Neuropharmacology*, *102*, 1–20.

Rezai-Zadeh, K., Shytle, D., Sun, N., Mori, T., Hou, H., Jeanniton, D., et al. (2005). Green tea epigallocatechin-3-gallate (EGCG) modulates amyloid precursor protein cleavage and reduces cerebral amyloidosis in Alzheimer transgenic mice. *The Journal of Neuroscience*, *25*, 8807–8814.

Ruzicka, J., Urdzikova, L. M., Svobodova, B., Amin, A. G., Karova, K., Dubisova, J., et al. (2018). Does combined therapy of curcumin and epigallocatechin Gallate have a synergistic neuroprotective effect against spinal cord injury? *Neural Regeneration Research*, *13*, 119–127.

Salah, N., Miller, N. J., Paganga, G., Tijburg, L., Bolwell, G. P., & Rice-Evans, C. (1995). Polyphenolic flavanols as scavengers of aqueous phase radicals and as chain-breaking antioxidants. *Archives of Biochemistry and Biophysics*, *322*, 339–346.

Singh, B. N., Shankar, S., & Srivastava, R. K. (2011). Green tea catechin, epigallocatechin-3-gallate (EGCG): Mechanisms, perspectives and clinical applications. *Biochemical Pharmacology*, *82*, 1807–1821.

Srividhya, R., & Kalaiselvi, P. (2013). Neuroprotective potential of epigallo catechin-3-gallate in PC-12 cells. *Neurochemical Research*, *38*, 486–493.

Sutherland, B. A., Rahman, R. M., & Appleton, I. (2006). Mechanisms of action of green tea catechins, with a focus on ischemia-induced neurodegeneration. *The Journal of Nutritional Biochemistry*, *17*, 291–306.

Sutherland, B. A., Shaw, O. M., Clarkson, A. N., Jackson, D. N., Sammut, I. A., & Appleton, I. (2005). Neuroprotective effects of (−)-epigallocatechin gallate following hypoxia-ischemia-induced brain damage: Novel mechanisms of action. *FASEB Journal*, *19*, 258–260.

Tai, K. K., & Truong, D. D. (2010). (−)-Epigallocatechin-3-gallate (EGCG), a green tea polyphenol, reduces dichlorodiphenyl-trichloroethane (DDT)-induced cell death in dopaminergic SHSY-5Y cells. *Neuroscience Letters*, *482*, 183–187.

Tao, L., & Zhu, Y. (2013). Anti-edema effect of epigallocatechin against acute spinal cord injury in rats and its correlation with the P38MAPK/NF-κb signaling pathway. *Journal of Spine*, *S4*.

Tian, W., Han, X. G., Liu, Y. J., Tang, G. Q., Liu, B., Wang, Y. Q., et al. (2013). Intrathecal epigallocatechin gallate treatment improves functional recovery after spinal cord injury by upregulating the expression of BDNF and GDNF. *Neurochemical Research*, *38*, 772–779.

Urdzikova, L. M., Ruzicka, J., Karova, K., Kloudova, A., Svobodova, B., Amin, A., et al. (2017). A green tea polyphenol epigallocatechin-3-gallate enhances neuroregeneration after spinal cord injury by altering levels of inflammatory cytokines. *Neuropharmacology*, *126*, 213–223.

Wei, I. H., Tu, H. C., Huang, C. C., Tsai, M. H., Tseng, C. Y., & Shieh, J. Y. (2011). (−)-Epigallocatechin gallate attenuates NADPH-d/nNOS expression in motor neurons of rats following peripheral nerve injury. *BMC Neuroscience*, *12*, 52.

Weinreb, O., Amit, T., Mandel, S., & Youdim, M. B. (2009). Neuroprotective molecular mechanisms of (−)-epigallocatechin-3-gallate: A reflective outcome of its antioxidant, iron chelating and neuritogenic properties. *Genes & Nutrition*, *4*, 283–296.

Williamson, G., Dionisi, F., & Renouf, M. (2011). Flavanols from green tea and phenolic acids from coffee: Critical quantitative evaluation of the pharmacokinetic data in humans after consumption of single doses of beverages. *Molecular Nutrition & Food Research*, *55*, 864–873.

Xie, J., Jiang, L., Zhang, T., Jin, Y., Yang, D., & Chen, F. (2010). Neuroprotective effects of epigallocatechin-3-gallate (EGCG) in optic nerve crush model in rats. *Neuroscience Letters*, *479*, 26–30.

Xu, Q., Langley, M., Kanthasamy, A. G., & Reddy, M. B. (2017). Epigallocatechin gallate has a neurorescue effect in a mouse model of Parkinson disease. *The Journal of Nutrition*, *147*, 1926–1931.

Ye, Q., Ye, L., Xu, X., Huang, B., Zhang, X., Zhu, Y., et al. (2012). Epigallocatechin-3-gallate suppresses 1-methyl-4-phenyl-pyridine-induced oxidative stress in PC12 cells via the SIRT1/ PGC-1α signaling pathway. *BMC Complementary and Alternative Medicine*, *12*, 82.

Zaveri, N. T. (2006). Green tea and its polyphenolic catechins: Medicinal uses in cancer and noncancer applications. *Life Sciences*, *78*, 2073–2080.

Zhang, F., Li, N., Jiang, L., Chen, L., & Huang, M. (2015). Neuroprotective effects of (−)-epigallocatechin-3-gallate against focal cerebral ischemia/reperfusion injury in rats through attenuation of inflammation. *Neurochemical Research*, *40*, 1691–1698.

Zhang, X., Wu, M., Lu, F., Luo, N., He, Z. P., & Yang, H. (2014). Involvement of α7 nAChR signaling cascade in epigallocatechin gallate suppression of β-amyloid-induced apoptotic cortical neuronal insults. *Molecular Neurobiology*, *49*, 66–77.

Chapter 35

# Vitamin D and spinal cord injury

**Kelly Pritchett, Brianna Rice, and Sophia Berg**

*Department of Health Sciences, College of Education and Professional Studies, Central Washington University, Ellensburg, WA, United States*

## Abbreviations

| | |
|---|---|
| **25(OH)D** | 25-hydroxyvitamin D, also known as calcidiol |
| **1,25(OH)$_2$D** | 1,25-dihydroxyvitamin D, also known as calcitriol |
| **IOM** | Institute of Medicine |
| **UVB** | ultraviolet B |
| **SCI** | spinal cord injury |
| **VDR** | vitamin D receptor |

## Introduction

Based on data collected from the National Health and Nutrition Examination Survey (NHANES), the prevalence of vitamin D deficiency (<50 nmol/L) in the United States is estimated to be 39.9% (Parva et al., 2018). Historically, vitamin D is known for its role in supporting bone health and preventing osteoporosis (Jin, 2018; Laird, Ward, McSorley, Strain, & Wallace, 2010). However, vitamin D is also important for various functions in the body including immunity, cell growth, and neuromuscular function (Flueck, Hartmann, Strupler, & Perret, 2016). The discovery of vitamin D receptors in skeletal muscle tissue has led to an increase in research examining vitamin D and its potential effects on exercise performance. Active individuals with spinal cord injury (SCI) have been suggested to be at an increased risk for vitamin D insufficiency and deficiency due to low dietary intake of foods high in vitamin D (Gerrish et al., 2017), anti-convulsant medications which interfere with vitamin D metabolism (Gröber & Kisters, 2012), and lifestyle factors such as reduced exposure to sunlight (Flueck & Perret, 2017; Pritchett, Pritchett, Stark, Broad, & LaCroix, 2019). Thus, a sufficient vitamin D (>75 nmol/L) status is dependent on diet, location of residence, skin pigmentation, time spent outdoors and whether an individual wears sunscreen or sun protective clothing (Flueck & Perret, 2017; Pritchett et al., 2016). Research has suggested that dietary intake for both micronutrients and macronutrients are insufficient in the para-athlete population (Scaramella, Kirihennedige, & Broad, 2018).

## Vitamin D recommendations: Intake, sources, and status

Vitamin D, also referred to as calciferol, is a fat-soluble vitamin that acts like a prohormone. Vitamin D can be obtained from sun exposure, supplements, and foods like fatty fish, beef liver, eggs, mushrooms, and fortified foods (see Table 1), which are the main source of dietary vitamin D in American diets (National Institutes of Health [NIH], 2020).

While there are several forms (vitamers) of vitamin D, two major forms include cholecalciferol ($D_3$) and ergosterol ($D_2$). Ergosterol is added to foods to fortification purposes (i.e., cereal), and it naturally present in fungi like mushrooms. Vitamin $D_3$ is synthesized in the skin as well as consumed in the diet through animal sources (Ross, Taylor, Yaktine, & Valle, 2011). When exposed to ultraviolet B (UVB) radiation, 7-dehydrocholesterol (7-DHC) is converted into previtamin $D_3$, and later converted vitamin D (Hossein-nezhad & Holick, 2013) (Fig. 1).

This type of synthesis is influenced by variables like skin exposure, season, and location of residence. There are seasonal variations in vitamin D synthesis for individuals greater than 35 to 37 degrees N latitude during the winter due to the unavailability of UVB radiation during winter months in these locations. Excessive pollution and cloud cover are also variables that can impact the ability to synthesize vitamin D through UVB radiation (Cannell, Hollis, Sorenson, Taft, & Anderson, 2009). Other variables that affect the ability to synthesize vitamin D via sunlight are sunscreen and clothing choice. Recent research suggests that sunscreen use may not play a large role in low 25(OH)D status when using moderate sun protection

Diagnosis and Treatment of Spinal Cord Injury. https://doi.org/10.1016/B978-0-12-822498-4.00035-X

**TABLE 1** Dietary sources of vitamin.

| Food | Standard Portion | Approximate Vitamin D Content (IU) |
|---|---|---|
| **Fortified foods** | | |
| Milk, low fat | 1 cup | 117 |
| Soy beverage (soy milk) | 1 cup | 119 |
| Yogurt, low fat | 8 oz | 116 |
| Cheese, American | 1.5 oz | 85 |
| Orange juice | 1 cup | 100 |
| **Nonfortified foods** | | |
| Salmon (various) | 3 oz | 383–570 |
| Rainbow trout, freshwater | 3 oz | 645 |
| Light tuna, canned | 3 oz | 231 |
| Mushrooms (various | 1 cup | 114–110 |
| Eggs | 1 egg | 37 |
| Beef liver | 3 oz | 42 |

Source: U.S. Department of Agriculture, Agricultural Research service. Food Data Central, 2019. fdc.nal.usda.gov

factor (SPF). Currently, research has not examined the use of high SPF sunscreens and the effect they may have on 25(OH) D concentrations (Neale et al., 2019). Protective clothing as well as increased indoor activity does decrease skin exposure to the sun which may negatively impact 25(OH)D status. Individuals that have dark pigmented skin are at higher risk for vitamin D deficiency as well since darker skin pigmentation requires more sun exposure to absorb the same amount of UVB radiation as someone with lighter skin pigmentation (Holick & Chen, 2008). Fig. 2 displays risk factors for low 25(OH)D status.

Whether synthesized in the skin or ingested through foods and supplements, vitamin D is hydroxylated by the liver into the major circulating form of vitamin D, 25(OH)D, which is the form of vitamin D measured to assess vitamin D status (Hossein-nezhad & Holick, 2013). 25(OH)D is further hydroxylated in the kidneys to form 1,25(OH)$_2$D Vitamin D, which is the biologically active form of vitamin D, performing many functions in the body related to its interactions with vitamin D receptors (VDR) (Hossein-nezhad & Holick, 2013). Vitamin D that has been synthesized by the sun stays in circulation two to three times longer than vitamin D that is taken in supplement form. This is because vitamin D taken in supplement form is partially bound to lipoproteins which is cleared from the system before being absorbed (Wacker & Holick, 2013). Since vitamin D is a fat-soluble vitamin, there is research showing an inverse relationship between adiposity and low vitamin D status, which puts those who are obese at greater risk for vitamin D deficiency (Larson-Meyer & Willis, 2010; Ogan & Pritchett, 2013).

The recommended intakes of vitamin D from the Institutes of Medicine (IOM) and Endocrine Society are noted in Table 2. The IOM guidelines are created from a population model aiming to prevent vitamin D deficiency in most of the population (Hossein-nezhad & Holick, 2013). IOM notes that most of the population intakes enough vitamin D to reach "sufficient" serum 25(OH)D levels (20 ng/mL) (Hossein-nezhad & Holick, 2013). Guidelines from the Endocrine Society recommend higher intakes of vitamin D, as it is based from a medical model that incorporates the wide array of vitamin D functions, and suggests 30 ng/mL 25(OH)D levels are beneficial, and that there is a higher upper limit than IOM suggests (Hossein-nezhad & Holick, 2013).

25(OH)D status can be measured in either nanograms (ng/mL) or in nanomoles (nmol/L). The World Health Organization and Endocrine Society considers a vitamin D deficiency as <50 nmol/L (<20 ng/ml), insufficiency as 50 to 75 nmol/L (25–30 ng/mL) and a sufficient status would be >75 nmol/L (>30 ng/mL) (Holick et al., 2011). The best indicator of vitamin D status is a serum concentration of 25(OH)D according to the National Institute of Health due to its 15-day half-life (National Institutes of Health, 2020). 1,25(OH)$_2$D concentration, on the other hand, is not a good indicator of vitamin D status since concentrations are constantly changing related to metabolic processes (National Institutes of Health, 2020).

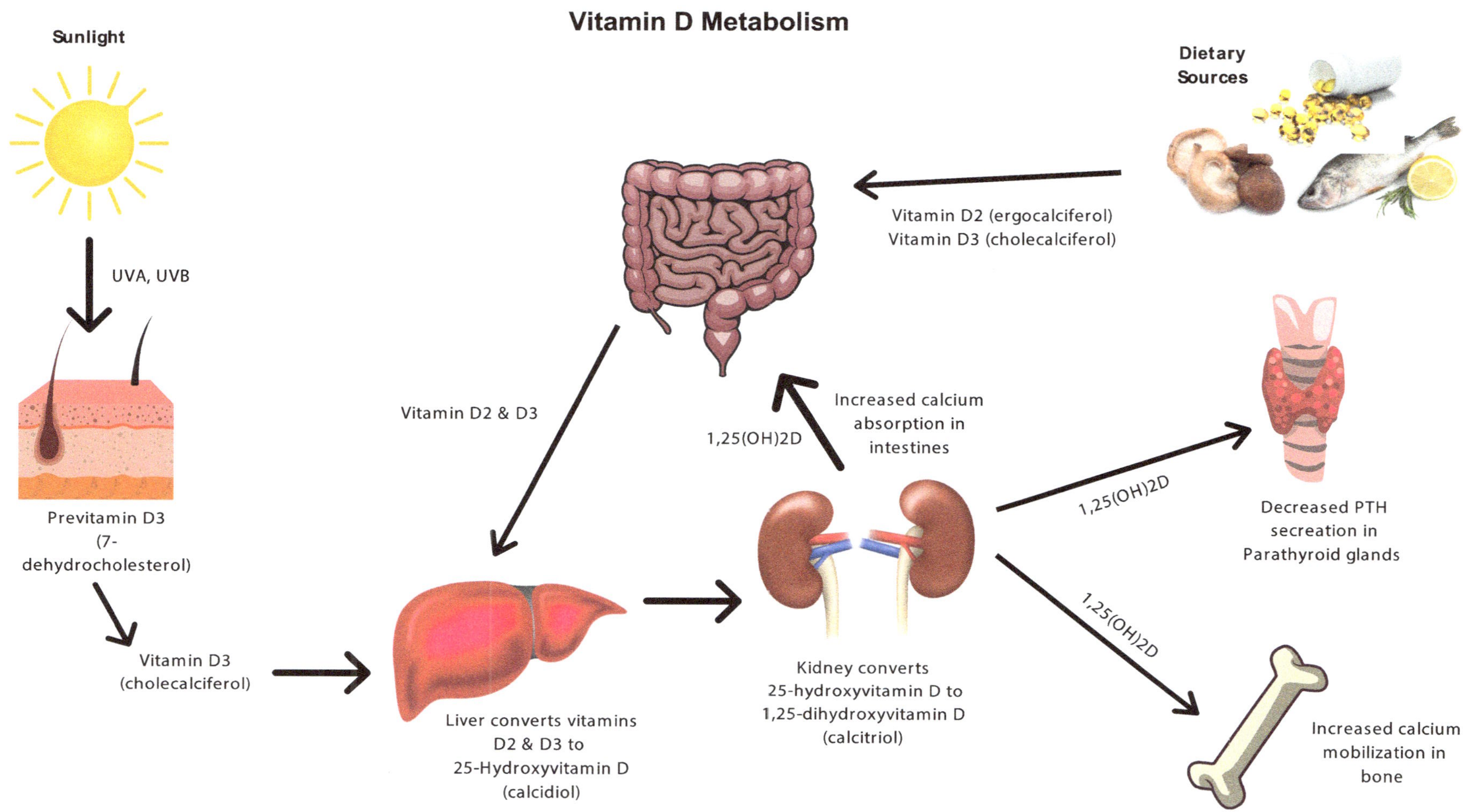


FIG. 1 Vitamin D metabolism.

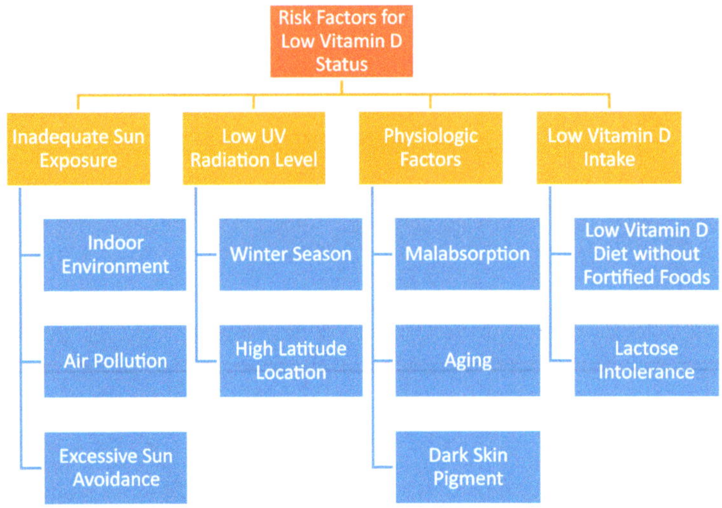


FIG. 2 Risk factors for insufficient/deficient 25(OH)D status.

**TABLE 2 Recommended Intake and Upper Limit for Vitamin D.**

| Age | Recommended Intake (IU/day) | Upper Limit (IU/day) |
|---|---|---|
| National Institute of Medicine | | |
| Children (0–18 years) | 400–600 | 2500 (1–3 years)<br>3000 (4–8 years)<br>4000 (13–18 years) |
| Adults (19–70 years) | 600 | 4000 |
| Older Adults (>70 years) | 800 | 4000 |
| Pregnancy/Lactation | 600 | 4000 |

Table from Ogan, D., & Pritchett, K. (2013). Vitamin D and the athlete: Risks, recommendations, and benefits. *Nutrients*, *5* (6), 1856–1868. doi:10.3390/nu5061856.

## Function and physiology

Vitamin D plays an important role in many different functions of the body, including immunity, bone health, muscle and neuromuscular function, hormone synthesis, cell development, and gene expression. Vitamin D has also been linked to a reduction in cancer risk, autoimmune disorders, and chronic disease (Ross et al., 2011).

### Bone health

Vitamin D is known for its role in bone health and prevention of osteoporosis. Vitamin D is vital to the transcription of calbindin, a carrier protein present in enterocytes which aids in the absorption of calcium, the most abundant mineral in bones. Vitamin D is also involved in the regulation of serum calcium and phosphorus, which is important to the processes of bone mineralization and contraction of muscles (National Institutes of Health, 2020). In athletes, vitamin D status is important to consider for bone health and preventing fractures (Ogan & Pritchett, 2013). Lappe et al. (2008) provided 2000 mg of calcium and 800 IU of vitamin D to a group of female Navy recruits while the other group received a placebo for 8 weeks. The supplement group had a 20% lower prevalence of stress fractures than the placebo group (Lappe et al., 2008).

## Muscle function

Vitamin D receptors (VDR) have been identified in nearly every type of cell in the body, including muscle cells (Ogan & Pritchett, 2013). The VDR has either a positive or negative response after being bound with vitamin D which targets gene expression, and therefore may play a large role in regulating skeletal muscle tissue (Ogan & Pritchett, 2013; Tanner & Harwell, 2015). Vitamin D may also affect muscle contractions by influencing calcium levels in the sarcoplasmic reticulum of muscles through alteration in the quantity and/or efficiency of calcium binding sites (Ogan & Pritchett, 2013). In addition to associations found between vitamin D status and muscle weakness, pain, and balance, research has also suggested that vitamin D plays a role in fast twitch muscle fiber morphology. It has been suggested that with an increase in vitamin D status from deficient to sufficient, fast twitch muscle fiber size and number will increase (Cannell et al., 2009; Girgis, Clifton-Bligh, Hamrick, Holick, & Gunton, 2013). However, further research is needed to determine the role of vitamin D in muscle fiber morphology.

## Extraskeletal functions

Adequate vitamin D status has been associated with decreased risk of certain cancers, autoimmune disorders, infectious disease, mental disorders, cardiovascular disease, and Type 2 diabetes (Hossein-nezhad & Holick, 2013). Vitamin D can work as an immune modulator, mainly through interactions with receptors and affecting T-cell differentiation (Hossein-nezhad & Holick, 2013). Vitamin D may decrease risk of cardiovascular disease through action in the regulation of the renin-angiotensin-aldosterone system and inflammatory pathways, although clinical trials in this area have not been conclusive (National Institutes of Health, 2020). Vitamin D is involved in neurological function, as VDRs are present on neurons and glia (National Institutes of Health, 2020). VDRs are also present on pancreatic beta cells, and may be involved in insulin signaling, glucose control, and reducing inflammation, thereby possibly decreasing risk of Type 2 diabetes, although more research is needed in this area (National Institutes of Health, 2020).

# Vitamin D status of active individuals with SCI

The vitamin D status of active individuals has become a popular area of study over the past decade. Athletes are potentially at an increased risk for vitamin D deficiency due to the increased enzymatic activity that occurs during exercise (Dahlquist, Dieter, & Koehle, 2015). Many studies have found that athletes are at an increased risk for 25(OH)D deficiency during the winter months and that athletes participating in an indoor sport have a higher rate of than those that play an outdoor sport (Cannell et al., 2009; Close et al., 2013; Fields, Payne, Gallo, Busteed, & Jones, 2019; Halliday et al., 2011; Krzywanski et al., 2016; Larson-Meyer & Willis, 2010; Ogan & Pritchett, 2013; Owens, Allison, & Close, 2018; Wiciński et al., 2019).

Recently, research has examined the vitamin D status of athletes with a physical impairment. This includes athletes with SCI, amputations, cerebral palsy, and other impairments as well. Athletes with SCI may be at higher risk for vitamin D deficiency due to a lack of mobility and thermoregulatory dysfunction. Athletes with SCI may also have additional malabsorption concerns due to taking anticonvulsant medications (Flueck, Schlaepfer, & Perret, 2016; Krempien & Barr, 2011; Oleson, Patel, & Wuermser, 2010; Pritchett et al., 2016). These aspects decrease the amount of time that individuals with SCI spend outside, thus decreasing the ability to synthesize vitamin D by sunlight (Flueck & Perret, 2017). It is important for active individuals with SCI to get adequate vitamin D as it is important for bone health.

Krempien and Barr (2011) examined the nutrient intake of elite Canadian athletes with SCI (Krempien & Barr, 2011). This study used self-reported food diaries to assess diet and found energy and micronutrient intake, including vitamin D, to be lower than recommended. Scaramella and colleagues further highlighted that para-athletes are at an increased for carbohydrate, iron, protein, and vitamin D deficiencies (Scaramella et al., 2018).

Studies have also focused on 25(OH)D status related to lifestyle factors in athletes with SCI. These lifestyle factors include diet, place of residence, and skin color. Other factors that are important for determining 25(OH)D status in athletes with SCI are whether they play an indoor or outdoor sport, medication use that may inhibit vitamin D absorption and limited time outdoors due to impaired thermoregulation (Pritchett et al., 2016).

Multiple studies have found that athletes with SCI are at greater risk for vitamin D status. Oleson et al. (2010) found that 96.4% of participants with chronic SCI had an insufficient 25(OH)D status during the winter months, and 81% were insufficient during the summer months (Oleson et al., 2010). This was further supported by another study in Swiss wheelchair athletes that compared 25(OH)D status between winter and summer months, and as indoor versus outdoor athletes. Flueck et al. found that vitamin D insufficiency was more prevalent in athletes during winter months and in athletes that played indoor sports (Flueck, Hartmann, et al., 2016; Flueck, Schlaepfer, & Perret, 2016).

An observational study in athletes from the Canadian Wheelchair Sports Association and the US Olympic Committee Paralympic program was conducted to examine the effects of lifestyle factors on 25(OH)D status during the autumn and winter months (Pritchett et al., 2016). There was no significant difference found in vitamin D status between autumn and winter, with 41% to 51% of athletes having an insufficient and 15.4% having a deficient vitamin D status. Time spent outside was significantly correlated with vitamin D status in the autumn, but not during the winter. Furthermore, a seasonal decline in vitamin D status was not observed regardless that athlete sun exposure decreased from autumn to winter. In this study, a greater percentage of athletes that competed in indoor sports were deficient compared to those that played outdoor sports (Pritchett et al., 2016). Thus, research is needed to identify the optimal 25(OH)D levels for active individuals with SCI.

### Vitamin D and athletic performance in athletes with SCI

There have been conflicting results in the literature as to whether vitamin D provides an ergogenic benefit for exercise performance and recovery in active individuals. Flueck et al. examined the effects of vitamin D supplementation in athletes with a chronic SCI on 25(OH)D status and whether it would subsequently impact performance (Flueck, Schlaepfer, & Perret, 2016). Twenty indoor elite wheelchair athletes received 6000 IU of vitamin $D_3$ over a 12-week period. After 6 weeks of supplementation, vitamin D status increased significantly and reached an optimal level in all athletes. However, there was no significant difference found for any performance measurements except for isokinetic strength in the nondominant hand, although this result may have been influenced by training. In order to truly test whether the neuromuscular changes were due to vitamin D supplementation or a change in 25(OH)D status, a placebo group is needed (Flueck, Schlaepfer, & Perret, 2016).

Another study examined the efficacy of a 12- to 16-week sliding scale vitamin D supplementation protocol on 25(OH)D status and performance measures (handgrip strength, and 20-m wheelchair spring) in elite Paralympic athletes (Pritchett et Al., 2019). Participants with sufficient 25(OH)D status received a maintenance dose (15,000 IU/week of vitamin D), insufficient participants received 35,000 IU/week for 4 weeks and then received the maintenance dose, and participants that were deficient received 50,000 IU/week of vitamin D for 8 weeks and then received the maintenance dose. Mean 25(OH)D status significantly increased after supplementation ($66.3 \pm 24.3$ nmol/mL presupplementation; $111.3 \pm 30.8$ nmol/mL postsupplementation). There were no significant differences in 20 m wheelchair sprints or handgrip strength; however, 62% of participants did improve handgrip strength.

## Conclusion

Further research is warranted in a number of areas surrounding athletic populations, especially athletes with SCI. Recommendations for optimal vitamin D status for the para-populations need to be addressed as this population can be more susceptible to deficiency. Based on previous findings, athletes with SCI should be screened biannually for 25(OH)D status to determine whether supplementation is needed. A 12- to 16-week sliding scale supplementation protocol based on the individual's vitamin D status can be applied for individuals with an insufficient or deficient status. Furthermore, the relationship between athletic performance and vitamin D status remains unclear and there is a need for more research in this area.

## Application to other areas of neuroscience (200–750 words)

This chapter reviewed vitamin D status, functions, and physiology. Vitamin D is important for various functions in the body including immunity, cell growth, and neuromuscular function. The discovery of vitamin D receptors in skeletal muscle tissue has led to an increase in research examining vitamin D and its potential effects on exercise performance. Current research suggests that active individuals with SCI are vulnerable populations with an increased risk for vitamin D deficiency. Active individuals with SCI present unique challenges in terms of nutritional needs, including vitamin D status. Thus, it is important that active individuals with SCI to obtain adequate vitamin D as it is important for bone health. It is suggested that active individuals with SCI may be a at higher risk for vitamin D deficiency due to a lack of mobility, thermoregulatory challenges, and malabsorption concerns due to taking anticonvulsant medications. These aspects decrease the amount of time that individuals with SCI spend outside, thus decreasing the ability to synthesize vitamin D by sunlight. Given that the diet does not supply enough vitamin D, active individuals with SCI may need vitamin D supplementation to obtain optimal vitamin D levels. Active individuals with SCI should receive specific and appropriate nutrition assessments and interventions, according to their needs.

## Mini-dictionary of terms

Vitamin D—the fat-soluble vitamin available in food and from the sun.

Cholecalciferol—or vitamin $D_3$, the form of vitamin D found in the skin when exposed to sunlight, and the preferred form found in supplements.

Ergocalciferol—or vitamin $D_2$ found in supplements and in mushrooms and fungi in the diet.

Vitamers—various forms of vitamins that have a similar physiological activity in the body (i.e., cholecalciferol, and ergocalciferol).

Calcitriol—(1,25 hydroxyvitamin $D_3$) the active from of vitamin by made by the kidney but used to treat low calcium in individuals with chronic kidney disease or decreased function of the parathyroid glands.

## Key facts of vitamin D and spinal cord injury

- Vitamin D is known for its role in bone health and prevention of osteoporosis.
- Vitamin D deficiency (<50 nmol/L) prevalence in the United States is estimated to be 39.9%
- Vitamin D status is assessed by serum 25(OH)D concentration
- Vitamin D can be obtained from sun exposure, foods, and supplements.
- Rickets was one of the first discovered conditions associated with a vitamin D deficiency.

## Summary points

- Vitamin D is important for skeletal, muscle, immune, and neurological health.
- Athletes are at higher risk for vitamin D deficiency, related to increased enzymatic activity during exercise and indoor sports.
- Athletes with SCI may be a at higher risk for vitamin D deficiency due to a lack of mobility, thermoregulatory dysfunction, and malabsorption concerns due to taking anticonvulsant medications.
- Active individuals with SCI should be screened biannually for 25(OH)D status to determine whether supplementation is needed.

## References


Cannell, J. J., Hollis, B. W., Sorenson, M. B., Taft, T. N., & Anderson, J. J. B. (2009). Athletic performance and vitamin D. *Medicine & Science in Sports & Exercise*, *41*(5), 1102–1110. https://doi.org/10.1249/MSS.0b013e3181930c2b.

Close, G. L., Russell, J., Cobley, J. N., Owens, D. J., Wilson, G., Gregson, W., et al. (2013). Assessment of vitamin D concentration in non-supplemented professional athletes and healthy adults during the winter months in the UK: Implications for skeletal muscle function. *Journal of Sports Sciences*, *31* (4), 344–353. https://doi.org/10.1080/02640414.2012.733822.

Dahlquist, D. T., Dieter, B. P., & Koehle, M. S. (2015). Plausible ergogenic effects of vitamin D on athletic performance and recovery. *Journal of the International Society of Sports Nutrition*, *12*(1), 33. https://doi.org/10.1186/s12970-015-0093-8.

Fields, J. B., Payne, D. C., Gallo, S., Busteed, D. R., & Jones, M. T. (2019). Vitamin D status differs by sex, sport-season, and skin pigmentation among elite collegiate basketball players. *Sports*, *7*(11). https://doi.org/10.3390/sports7110239.

Flueck, J. L., Hartmann, K., Strupler, M., & Perret, C. (2016). Vitamin D deficiency in Swiss elite wheelchair athletes. *Spinal Cord*, 1–5. https://doi.org/10.1038/sc.2016.33.

Flueck, J. L., & Perret, C. (2017). Vitamin D deficiency in individuals with a spinal cord injury: A literature review. *Spinal Cord*, *55*(5), 428–434. https://doi.org/10.1038/sc.2016.155.

Flueck, J. L., Schlaepfer, M. W., & Perret, C. (2016). Effect of 12-week vitamin D supplementation on 25[OH]D status and performance in athletes with a spinal cord injury. *Nutrients*, *8*(10). https://doi.org/10.3390/nu8100586.

Gerrish, H., Pritchett, R., Broad, E., Ogan, D., LaCroix, M., & Pritchett, K. (2017). Nutrient intake of elite athletes with spinal cord injury. *International Journal of Exercise Science*, *10*(7), 1018–1028.

Girgis, C. M., Clifton-Bligh, R. J., Hamrick, M. W., Holick, M. F., & Gunton, J. E. (2013). The roles of vitamin D in skeletal muscle: Form, function, and metabolism. *Endocrine Reviews*, *34*(1), 33–83. https://doi.org/10.1210/er.2012-1012.

Gröber, U., & Kisters, K. (2012). Influence of drugs on vitamin D and calcium metabolism. *Dermato-Endocrinology*, *4*(2), 158–166. PubMed https://doi.org/10.4161/derm.20731.

Halliday, T. M., Peterson, N. J., Thomas, J. J., Kleppinger, K., Hollis, B. W., & Larson-Meyer, D. E. (2011). Vitamin D status relative to diet, lifestyle, injury, and illness in college athletes. *Medicine and Science in Sports and Exercise*, *43*(2), 335–343. https://doi.org/10.1249/MSS.0b013e3181eb9d4d.

Holick, M., & Chen, T. (2008). Vitamin D deficiency: A worldwide problem with health consequences. *The American Journal of Clinical Nutrition*, *87*(4), 1080S–1086S.

Holick, M. F., Binkley, N. C., Bischoff-Ferrari, H. A., Gordon, C. M., Hanley, D. A., Heaney, R. P., et al. (2011). Evaluation, treatment, and prevention of vitamin D deficiency: An Endocrine Society clinical practice guideline. *The Journal of Clinical Endocrinology & Metabolism*, *96*(7), 1911–1930. https://doi.org/10.1210/jc.2011-0385.

Hossein-nezhad, A., & Holick, M. F. (2013). Vitamin D for health: A global perspective. *Mayo Clinic Proceedings*, *88*(7), 720–755. https://doi.org/10.1016/j.mayocp.2013.05.011.

Jin, J. (2018). Vitamin D and calcium supplements for preventing fractures. *JAMA*, *319*(15), 1630. https://doi.org/10.1001/jama.2018.3892.

Krempien, J. L., & Barr, S. I. (2011). Risk of nutrient inadequacies in elite Canadian athletes with spinal cord injury. *International Journal of Sport Nutrition and Exercise Metabolism*, *21*(5), 417–425. https://doi.org/10.1123/ijsnem.21.5.417.

Krzywanski, J., Mikulski, T., Krysztofiak, H., Mlynczak, M., Gaczynska, E., & Ziemba, A. (2016). Seasonal vitamin D status in polish elite athletes in relation to sun exposure and Oral supplementation. *PLoS One*, *11*(10). https://doi.org/10.1371/journal.pone.0164395.

Laird, E., Ward, M., McSorley, E., Strain, J. J., & Wallace, J. (2010). Vitamin D and bone health; potential mechanisms. *Nutrients*, *2*(7), 693–724. https://doi.org/10.3390/nu2070693.

Lappe, J., Cullen, D., Haynatzki, G., Recker, R., Ahlf, R., & Thompson, K. (2008). Calcium and vitamin D supplementation decreases incidence of stress fractures in female navy recruits. *Journal of Bone and Mineral Research*, *23*(5), 741–749. https://doi.org/10.1359/jbmr.080102.

Larson-Meyer, D. E., & Willis, K. S. (2010). Vitamin D and athletes. *Current Sports Medicine Reports*, *9*(4), 220. https://doi.org/10.1249/JSR.0b013e3181e7dd45.

National Institutes of Health. (2020, October 9). *Vitamin D—Health Professional Fact Sheet*. Office of Dietary Supplements. https://ods.od.nih.gov/factsheets/VitaminD-HealthProfessional/#en29.

Neale, R. E., Khan, S. R., Lucas, R. M., Waterhouse, M., Whiteman, D. C., & Olsen, C. M. (2019). The effect of sunscreen on vitamin D: A review. *The British Journal of Dermatology*, *181*(5), 907–915. https://doi.org/10.1111/bjd.17980.

Ogan, D., & Pritchett, K. (2013). Vitamin D and the athlete: Risks, recommendations, and benefits. *Nutrients*, *5*(6), 1856–1868. https://doi.org/10.3390/nu5061856.

Oleson, C. V., Patel, P. H., & Wuermser, L.-A. (2010). Influence of season, ethnicity, and chronicity on vitamin D deficiency in traumatic spinal cord injury. *The Journal of Spinal Cord Medicine*, *33*(3), 202–213.

Owens, D. J., Allison, R., & Close, G. L. (2018). Vitamin D and the athlete: Current perspectives and new challenges. *Sports Medicine (Auckland, N.Z.)*, *48* (Suppl 1), 3–16. PubMed https://doi.org/10.1007/s40279-017-0841-9.

Parva, N. R., Tadepalli, S., Singh, P., Qian, A., Joshi, R., Kandala, H., et al. (2018). Prevalence of vitamin D deficiency and associated risk factors in the US population (2011 – 2012). *Cureus*.. https://doi.org/10.7759/cureus.2741.

Pritchett, K., Pritchett, R., Ogan, D., Bishop, P., Broad, E., & LaCroix, M. (2016). 25(OH)D status of elite athletes with spinal cord injury relative to lifestyle factors. *Nutrients*, *8*(6), 374. https://doi.org/10.3390/nu8060374.

Pritchett, K., Pritchett, R. C., Stark, L., Broad, E., & LaCroix, M. (2019). Effect of vitamin D supplementation on 25(OH)D status in elite athletes with spinal cord injury. *International Journal of Sport Nutrition and Exercise Metabolism*, *29*(1), 18–23. https://doi.org/10.1123/ijsnem.2017-0233.

Ross, A. C., Taylor, C. L., Yaktine, A. L., & Valle, H. B. D. (2011). *Overview of vitamin D*. National Academies Press (US). https://www.ncbi.nlm.nih.gov/books/NBK56061/.

Scaramella, J., Kirihennedige, N., & Broad, E. (2018). Key nutritional strategies to optimize performance in Para athletes. *Physical Medicine and Rehabilitation Clinics of North America*, *29*(2), 283–298. https://doi.org/10.1016/j.pmr.2018.01.005.

Tanner, S. B., & Harwell, S. A. (2015). More than healthy bones: A review of vitamin D in muscle health. *Therapeutic Advances in Musculoskeletal Disease*, *7*(4), 152–159. https://doi.org/10.1177/1759720X15588521.

Wacker, M., & Holick, M. F. (2013). Sunlight and vitamin D. *Dermato-Endocrinology*, *5*(1), 51–108. https://doi.org/10.4161/derm.24494.

Wiciński, M., Adamkiewicz, D., Adamkiewicz, M., Śniegocki, M., Podhorecka, M., Szychta, P., et al. (2019). Impact of vitamin D on physical efficiency and exercise performance—A review. *Nutrients*, *11*(11). https://doi.org/10.3390/nu11112826.

Chapter 36

# Corticospinal tract regeneration after spinal cord injury: Implications for treatment and recovery

Yuanyuan Li[a] and Paul Lu[a,b]

[a] *Department of Neurosciences, University of California San Diego, La Jolla, CA, United States;* [b] *Veterans Administration San Diego Healthcare System, San Diego, CA, United States*

## Abbreviations

| | |
|---|---|
| **A** | adult |
| **C** | cervical |
| **ChABC** | chondroitinase ABC |
| **CNS** | central nervous system |
| **CSPG** | chondroitin sulfate proteoglycans |
| **CST** | corticospinal tract |
| **E** | embryonic |
| **Htt** | huntingtin gene |
| **KLF** | Krüppel-like factor |
| **MSE** | motor synergy encoder |
| **mTOR** | mammalian target of rapamycin |
| **NgR** | Nogo-A receptor |
| **NPC** | neural progenitor cell |
| **NT-3** | neurotrophine-3 |
| **P** | post-natal |
| **PTEN** | tumor suppressor phosphatase and tensin homolog |
| **SCI** | spinal cord injury |
| **SOCS3** | suppressor of cytokine signaling 3 |
| **T** | thoracic |
| **YA** | young adult |

## Introduction

The corticospinal tract (CST) is an axonal bundle that starts from cortical layer V pyramid neurons and travels through brain and brainstem to terminate in spinal cord. One major characteristics of CST is that the majority of the CST axons cross the midline at the pyramidal decussation, resulting the right side of brain to control the left side of the body and vice versa. The location of CST is different among different species, from ventral dorsal column in rodent to dorsal lateral funiculus in cat, non-human primate, and human (Welniarz, Dusart, & Roze, 2017). In addition, the CST termination is different among different species, from dorsal to ventral gray matter.

The main function of the CST is to control primary motor activity of the body, especially the voluntary movements. Disruption of CST along with other tracts after spinal cord injury (SCI) definitely interferes with communication between brain and the spinal cord below the injury, resulting in loss of motor, sensory, and autonomic function. Although studies show that CST sprouts after SCI from uninjured or injured axons, especially after genetically manipulation and therapeutic treatments (Blackmore et al., 2012; García-Alías, Barkhuysen, Buckle, & Fawcett, 2009; Liu et al., 2010; Rosenzweig et al., 2019), true regeneration from transected CST is rarely reported until recently from our lab (Kadoya et al., 2016).

Diagnosis and Treatment of Spinal Cord Injury. https://doi.org/10.1016/B978-0-12-822498-4.00036-1

Regeneration of CST, including sprouting and regenerative sprouting of CST, is vitally important to restore motor function, especially the voluntary skilled motor function, after SCI.

In this chapter, we will review development of CST and its innervation of spinal cord neurons. In addition, we will review injury-induced CST sprouting, and genetical manipulation and therapeutic treatment to promote CST sprouting and regenerative sprouting. Finally, we will discuss our own work for promotion of CST regeneration by transplantation of caudalized neural progenitor cells (NPCs) after SCI.

# CST development

## CST outgrowth

The corticospinal tract (CST) is a motor pathway in central nervous system (CNS) starting from the cerebral cortex, traveling through internal capsule and brainstem, most of them crossing the midline at the pyramidal decussation, and ending at the spinal cord (Welniarz et al., 2017). In the popular rodent model of CNS development, CST axons reach brainstem at embryonic day 17 (E17), caudal medulla at E19, and cross the middle line and enter the spinal cord around post-natal day 0 (P0). CST axons continue to extend into entire spinal cord, including cervical cord around P1–2, thoracic around P3, and lumbar around P5–7 (Fig. 1) ( Joosten et al., 1989; Welniarz et al., 2017). The development of CST in primate most occurs before birth. For example, CST reaches the entire spinal cord at birth in non-human primate (Welniarz et al., 2017) and reach the entire cervical spinal cord at 24 weeks post-conceptional age in human (Eyre, 2003). The growth of CST is leaded by a small number of "pioneer" axons followed by other axons that eventually populated the CST tract ( Joosten et al., 1989). Following the outgrowth of CST, myelination starts around P10 and finishes in the entire spinal cord around 4 weeks in rodent spinal cord (Fig. 1), indicating readiness for electro impulse transmission ( Joosten et al., 1989).

## CST location and termination pattern in spinal cord

CST travels in the spinal cord white matter, which just likes other descending systems in the spinal cord. However, the location where CST travels in the spinal cord varies among species (Welniarz et al., 2017). While the crossed main CST locates in the most ventral portion of dorsal column in rodent, it shifts to dorsal lateral funiculus in cat and primates, including human (Fig. 2). The small proportion of uncrossed lateral and ventral CST are in similar location among different species. In addition, the termination pattern of CST into spinal cord gray matter is different among species. In rodent, the CST terminates most in dorsal and intermediate zone, but not directly into motor neurons in the ventral horn (Fig. 2A). However, in primates, both crossed main CST tract and uncrossed CST shift their termination ventrally into intermediate zone and ventral horn (Fig. 2C and D), indicating direct cortico-motoneuronal connections (Welniarz et al., 2017).

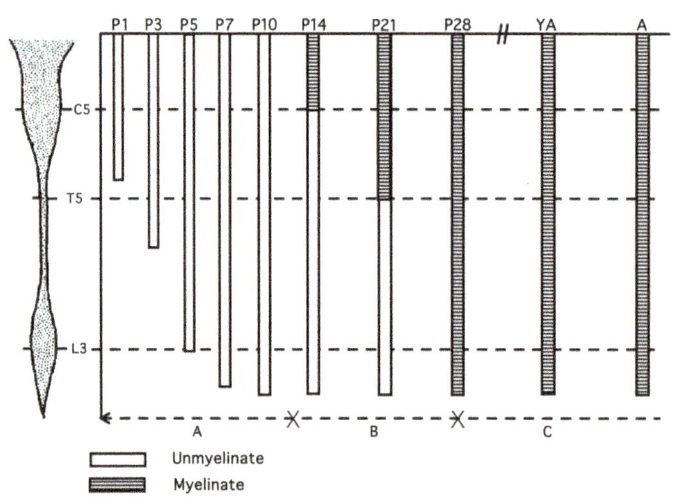


FIG. 1 Corticospinal tract outgrowth, myelination, and maturation. CST outgrowth (A) from post-natal day 1 (P1) to P10, myelination (B) from P10 to P24, and maturation (C) in young adult (YA) and adult (A) spinal cord in rats. *Adopted and redrawn from Joosten, E. A., Gribnau, A. A., & Dederen, P. J. (1989). Postnatal development of the corticospinal tract in the rat. An ultrastructural anterograde HRP study.* Anatomy and Embryology (Berlin), 179, *449–456 with permission.*

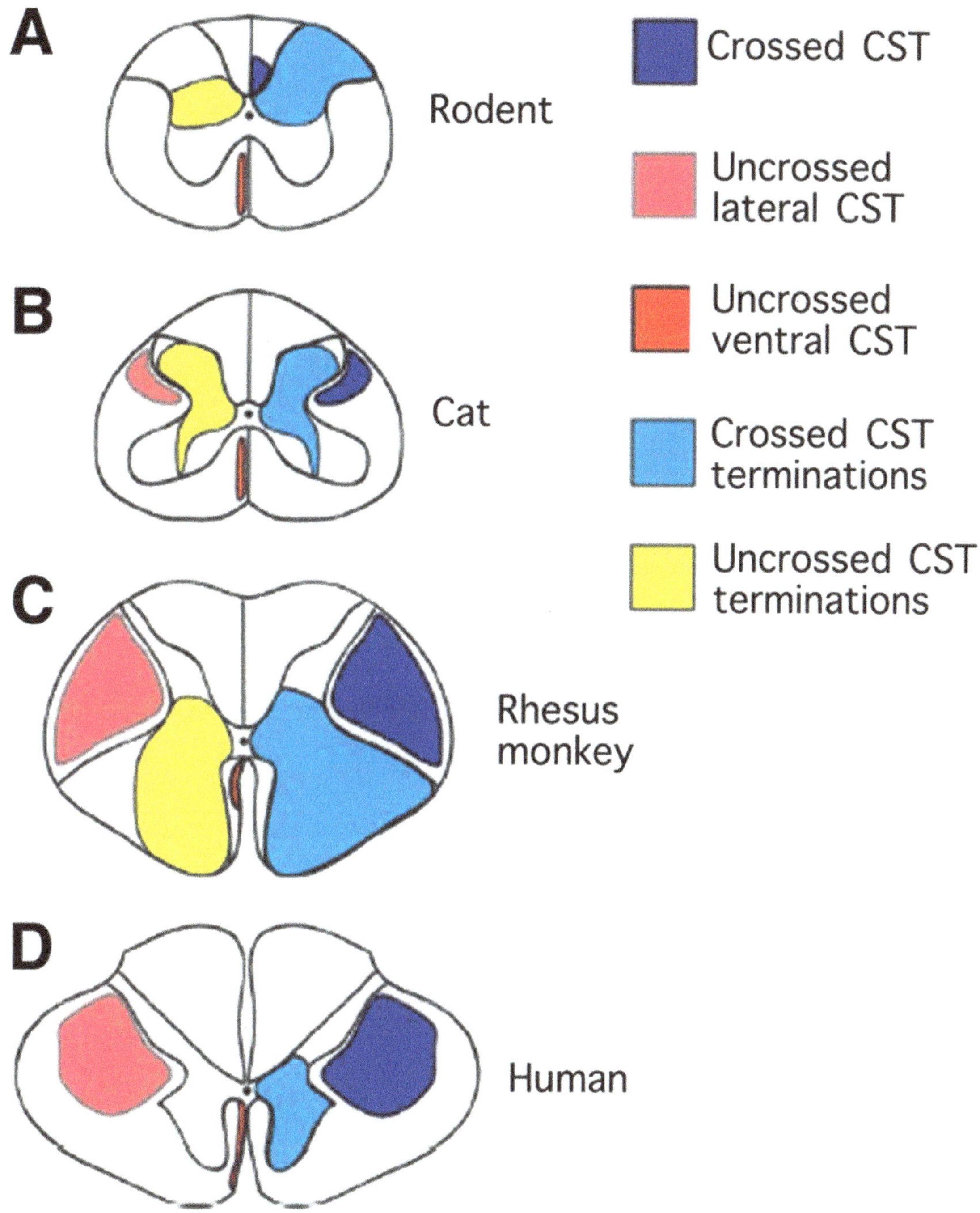


FIG. 2 Location and termination pattern of crossed and uncrossed CST in the spinal cord in mammals. (A) Rodent; (B) cat; (C) monkey; and (D) human. *Reproduced from Welniarz, Q., Dusart, I., & Roze, E. (2017). The corticospinal tract: Evolution, development, and human disorders.* Developmental Neurobiology, *77, 810–829 with permission.*

Furthermore, our own work demonstrates additional extensive decussation and bilateral termination of CST in cervical region in monkey, nearly twice as many CST axons decussating in the cervical midline (Rosenzweig et al., 2009).

## Precise control of skilled motor functions by defined CST populations

A recent study identifies spatially defined CST populations with distinct spinal projections that control different musculature groups and function in skilled forelimb motor function in mice (Wang et al., 2017). They reveal a sequential activation of topographically organized CST neurons during skilled forelimb performance using in vivo calcium imaging with intersectional approach to specifically label and monitor different regions of CST neurons. Manipulation of region-specific CST neurons by ablation of specific population with diphtheria toxin identifies that caudal forelimb area controls reaching while rostral forelimb area controls grasping. The identification of these spatially defined groups of CST neurons that controls different skilled forelimb movements enables us to understand how CST system precisely works and how to repair it after SCI.

### Identification of cellular node for CST control of motor function in mice

Since CST terminate mostly in dorsal and intermediate zone in rodent, it is important to identify those segmental interneurons and propriospinal neurons that relay CST signal into motor neurons. Levine et al. (2014) identify a neuronal population term "motor synergy encoder (MSE)" in the mouse spinal cord that may function as relay neurons from CST to motor neurons for voluntary motor movement. The MSE directly receive CST inputs in addition to sensory pathways, and have monosynaptic connection to spinal motor neurons. Molecular study reveals three candidate genes, *Tfap2b*, *Satb1*, and *Satb2*, that specifically identify MSE. Identification of MSE could direct neural stem cell study to generate these MSE progenitor cells as graft to attract CST regeneration to mimic their relays for CST input into spinal cord motor neurons below SCI.

## Spinal cord injury induced spontaneous sprouting of CST

### Classical work

Spontaneous sprouting of spared axons may innervate denervated targets and has been regarded as an underlying factor for functional recovery after SCI. In a classical work to investigate anatomical plasticity of CST, Kuang and Kalil (1990) unilaterally lesioned CST tracted at medullary pyramids in young hamster from age 5 to 23 days. They injected vulgaris leucoagglutinin (PHA-L) into intact side to label uninjured CST for 2 weeks. Immunohistochemical study reveals that PHA-L labeled CST sprouts into contralaterally denervated spinal cord side 4 weeks post-injury. CST sprouting is maximal in very young age of 5 days and declines with increased ages. Interestingly, the sprouted CST axons keep the same topographical specificity as normal CST arborization. Their results suggest that sprouting occurs in response to local signals which govern the specificity of denervation and induce sprouting similar to those normal development of corticospinal connections.

### Dorsal injury: Ventral CST sprouting

Collateral sprouting after CNS injuries has been proposed as a key mechanism for spontaneous functional recovery. Several studies examined the mechanisms underlying recovery after lesions of motor systems in rodents, including rats and mice. After complete lesions of the main dorsal CST that comprises more than 95% of all CST axons at C3, the uninjured ventral CST spontaneously sprouts into medial motoneuron pools within cervical spinal cord, which correlates with functional recovery (Weidner, Ner, Salimi, & Tuszynski, 2001). When both dorsal and ventral CST axons are lesioned to remove almost all CST axons, functional recovery abolishes. These results indicate that extensive spontaneous structural plasticity of CST is a mechanism underneath the motor functional recovery after SCI.

### Dorsal column injury: Dorsolateral CST spouting

The dorsolateral CST is dispersed throughout the dorsolateral columns. The pathways underlying dorsolateral CST spouting remain poorly understood. By performing optogenetic mapping of motor cortex after a mid-cervical dorsal column SCI that interrupts most corticospinal transmission in mice, Hilton et al. (2016) demonstrate that the motor cortex can reestablish output to the limbs longitudinally. Importantly, they dissociated for the first time the role of spared dorsolaterally projecting corticospinal neurons in mediating spontaneous recovery after SCI. They found that silencing of dorsolateral corticospinal neurons projecting caudal to the injury site by inhibitory DREADD (designer receptor exclusively activated by designer drug) was sufficient to result in the reappearance of deficits on the horizontal ladder task early after injury (Hilton et al., 2016). Their results demonstrate that spared dorsolateral CST plays an important role for spontaneous recovery of skilled locomotion in adult mice despite the limited reconstitution of corticospinal density after injury.

### Unilateral hemisection: CST cross into denervated side in monkeys

Unilateral hemisection is an important SCI model to assess extensive spontaneous regeneration of adult CST axons that contribute to limited recovery after SCI. This model is especially valid in non-human primates since the main crossed CST locates in dorsolateral funiculus in each spinal cord hemisphere (Fig. 2C). To investigate mechanisms underlying spontaneous recovery after incomplete SCI, Rosenzweig et al. (2010) performed unilateral C7 spinal cord hemisections to adult rhesus monkeys and analyzed behavioral, electrophysiological and anatomical outcomes. They found enhanced spontaneous plasticity of CST axons that cross the midline from intact CST and into denervated injury site below hemisection site with 60% reconstitution of pre-lesion axon density. This extensive CST reconstitution is associated with injured forelimb functional recovery, including coordinated muscle recruitment, hand function and locomotion. This study demonstrates a very extensive CST sprouting to reconstitute injured spinal cord, which highlights the importance of primate model of SCI.

To further dissect the mechanism underlying association of CST sprouting and the forelimb functional recovery, Nakagawa, Ninomiya, Yamashita, and Takada (2015) performed C7–8 unilateral SCI in adult macaque and assessed extent of spontaneous recovery of manual dexterity using a reaching/grasping task. They found gradual recovery of the impaired dexterous manual movements. CST from uninjured hemisphere sprouts into the denervated site below injury. However, they found changes of the laminar distribution of CST axons. The sprouting CST axons preferentially innervate into spinal cord lamia IX containing the spinal motor neuron pool, indicating direct innervation of the motor neurons. On the other hand, sprouting CST axons rarely innervate the dorsal laminae. This study indicates that the reorganization of CST axons below the SCI site is an important mechanism associated with recovery of forelimb movements (Nakagawa et al., 2015).

### Rewiring hindlimb CST

Large thoracic spinal cord injuries in adult rats result in disruption of hindlimb function. To assess whether the hindlimb CST can rewire after thoracic spinal cord injury, Bareyre et al. (2004) demonstrate that transected hindlimb CST axons sprout into the cervical gray matter to contact short and long propriospinal neurons (PSNs). Interestingly, long PSNs naturally innervate lumbar motor neurons, which forms a new intraspinal circuit relay that transmits the hindlimb CST input to its original lumbar spinal targets. In a follow-up study, the same group found additional CST plasticity (Ghosh et al., 2010). They found that injured hindlimb CST sprouts into cervical spinal cord after a thoracic SCI, which forms a new forelimb CST projection in the rostral part of the original hindlimb cortex. They confirm this expansion of forelimb CST in sacrifice of injured hindlimb CST by voltage-sensitive dye (VSD) imaging and blood-oxygen-level-dependent functional magnetic resonance imaging (BOLD fMRI). These studies indicate the existence of great CST plasticity and rewiring after SCI.

## Promotion of CST regenerative sprouting

### Promotion of CST regenerative sprouting by NT-3

Besides above reports about spontaneous sprouting of CST axons after SCI, many researchers attempt to promote further CST sprouting and regenerative sprouting that could support functional recovery using various treatments. In an early prototype study to search whether neurotrophins can enhance SCT sprouting, Schnell, Schneider, Kolbeck, Barde, and Schwab (1994) found that neurotrophine-3 (NT-3) can enhance CST sprouting first in intact post-natal rats (P2–5) just after CST extends into thoracic cord and next in young adult rats (4–7 weeks) that received low thoracic dorsal hemisection injury. Notably, the main CST is completely transected after injury, and there is no regeneration from cut axon tips into lesion site that is an empty cavity. However, CST axons sprout into ventral cord above injury site and continue to grow along the intact ventral portion of cord to reach caudal cord below injury. Combination of NT-3 single injection with an antibody that can block inhibition greatly enhance long distance of CST growth (Schnell et al., 1994).

While Schnell et al. (1994) delivered NT-3 for a single protein injection into spinal cord above injury site, which lasts short period of time, Grill, Murai, Blesch, Gage, and Tuszynski (1997) used an ex vivo gene therapy approach to deliver NT-3 by grafting of genetically modified fibroblasts expressing NT-3 into the same SCI model, low-thoracic dorsal hemisection, for continuous local supply of NT-3 in the rat SCI site. Immunohistochemistry demonstrates significant growth of CST axons at and caudal to the injury site comparing to control rats 3 months post-graft. In addition, a significant partial functional recovery on grid walking is reported, which is associated with CST regenerative sprouting. Similarly, delivery of NT-3 by this ex vivo gene therapy promotes chronically (3 months) injured CST regenerative sprouting up to 15 mm distal to SCI site (Tuszynski et al., 2003). In another study using gene therapy, a long-term NT-3 in vivo delivery to the lumbar enlargement area through retrograde transport of NT-3 by sciatic injection of adenoviral vector expressing NT-3 promotes CST axonal growth from intact CST into injured side where injury is far rostral at medulla level (Zhou, Baumgartner, Hill-Felberg, McGowen, & Shine, 2003). These studies indicate that NT-3, especially for local and sustained delivery, support CST axonal re-growth after CST injury.

In vivo NT-3 gene delivery may not be easily translated to clinical application. A recent study implants a one-centimeter long biomaterial chitosan scaffold to slowly release NT-3 into a T8 lateral hemisection site in adult rhesus monkey (Rao et al., 2018). Antegrade labeled CST axons regenerate into scaffold and reach the caudal end while there is no regeneration of CST in lesion only control monkey. This is a true regeneration since lateral hemisection completely transects the main CST located in dorsal lateral funiculus and the CST axons grow from their cut ends (Rao et al., 2018). However, there is no empty chitosan scaffold control and no quantification of proportion of CST axon regenerated. In addition, this study shows motor and sensory functional recovery. These studies indicate that NT-3, especially for local and sustained delivery, support CST axonal re-growth after CST injury.

## Promotion of CST regenerative sprouting by overcome of myelin inhibitors

While certain neurotrophins can function as chemoattractant to attract CST axon to grow, overcoming of inhibitory environment in the injured adult spinal cord could promote further growth of CST. One of the major inhibitions is myelin-associated inhibitors, such as a 35- and a 250-kd membrane proteins named NI-35 and NI-250, initially discovered by Martin Schwab's group (Sartori, Hofer, & Schwab, 2020). Afterwards they raised a monoclonal antibody, IN-1, against these two inhibitory proteins (Schnell & Schwab, 1990). CST sprouts into caudal cord through spared ventral tissue after mid-thoracic dorsal hemisection in young rats (2–6 weeks) after administration of IN-1 produced by tumors intracerebrally. Following this seminar study, various studies were conducted to demonstrate growth of CST axons after similar incomplete SCI model in adult rodent and monkey using IN-1 or its related antibodies termed anti-Nogo-A with or without other treatments, although the number of sprouting axons is very modest and most injury models are hemisection that is rare in human (Chen et al., 2017; Sartori et al., 2020; Wang et al., 2020). Nevertheless, clinical trial using Nogo-A antibody intrathecally infused continuously for 1 month for acute and severe SCI patients started as early in 2006 and finished in 2011 with certain safety data (Sartori et al., 2020). A phase II trial started 8 years later in 2019 and arrowed down to acute cervical SCI.

In addition to various IN-1 or Nogo-A antibody studies, the cloning of Nogo gene opens a new door to study its receptors for inhibitory signal transduction mechanism as well as CST regeneration in Nogo and its receptor deficient mice (Sartori et al., 2020). Although early studies report variable amount of CST regeneration in incomplete low-thoracic dorsal hemisection in Nogo mutant mice, a later study re-assessed the early findings and found no enhancement of CST regeneration (Lee et al., 2009). Similarly, no enhanced CST regeneration in mice that are deficient of Nogo-A receptor (NgR) that is supposed to mediate inhibition of all three myelin-associated proteins, Nogo, myelin-associated glycoprotein, and oligodendrocyte myelin glycoprotein (Zheng et al., 2005). Inhibition of NgR with their antagonist peptide NEP1–40 also yields controversy results of CST regeneration (Steward, Sharp, Yee, & Hofstadter, 2008). Nevertheless, a soluble function-blocking NgR termed NgR1(310)ecto-Fc, or AXER-204 is tested in rodent, monkey and is in clinical trial at present (Wang et al., 2020).

## Promotion of CST regenerative sprouting by overcome of CSPG

Besides myelin-associated inhibitors, another major inhibitor is extracellular matrix molecules associated with glia scar around SCI site, such as chondroitin sulfate proteoglycans (CSPGs) (Tran, Warren, & Silver, 2018). Digestion of CSPG by chondroitinase ABC (ChABC) around C4 dorsal column injury site in rats promotes CST regeneration beyond the injury site, which is associated with functional recovery (Bradbury et al., 2002). Later, the same group delivers ChABC specifically into CST neurons by gene therapy using lentiviral vectors, which reduces CST axon die-back and promotes CST sprouting, but only in short range (Zhao et al., 2011). Following these exciting results in rodents, Rosenzweig et al. (2019) delivered ChABC below C7 lateral hemisection by multiple intraparenchymal injections in rhesus monkeys and found that ChABC delivery promotes significant amount of spared CST axonal sprouting from intact side into gray matter below the injury site where it controls hand function. Delivery of ChABC significantly improves hand function comparing with vehicle-injected controls using non-linear principal component analysis. This large animal primate study may pave a path for clinical translation of this strategy to promote CST regenerative sprouting after SCI.

### Promotion of CST regenerative sprouting by genetic manipulation of certain genes

The extrinsic inhibitory environment after SCI is not the only factor to prevent CST axonal regeneration. The down-regulation of intrinsic regeneration-associated gene program in the adult neurons after SCI is another important factor for the failure of CST regeneration (Fagoe, Heest, & Verhaagen, 2014). A seminal study from Zhigang He's group shows that the tumor suppressor phosphatase and tensin homolog (PTEN) is one these genes that are responsible to down-regulate growth-associated genes, such as mammalian target of rapamycin (mTOR) (Liu et al., 2010). Genetic deletion of PTEN in mice reactivates mTOR expression and promotes substantial amount CST axon growth after SCI. Subsequently, the same group finds another cortical suppressor of cytokine signaling 3 (SOCS3) and deletion of SOCS3 can promote CST sprouting from uninjured CST axons to the denervated spinal cord in a unilateral pyramidotomy model ( Jin et al., 2015). Co-deletion of SOCS3 and PTEN significantly enhances CST axon sprouting, which results in significant recovery of forelimb skilled locomotion. The deletion of PTEN and SOCS3, however, is at development stage, post-natal day 1, in these studies. A later study demonstrates that diminishing effect of PTEN deletion on CST regeneration in a low-thoracic dorsal hemisection model as the age of mice increases (Geoffroy, Hilton, Tetzlaff, & Zheng, 2016).

Besides deletion of PTEN and SOCS3, directly over-expression of developmentally regulated transcription factors, such as Krüppel-like factor (KLF) family and Sox11, in adult CST neurons by gene therapy improves their regenerative ability after incomplete C4–5 dorsal quadrant injury (Blackmore et al., 2012; Wang, Reynolds, Kirry, Nienhaus, & Blackmore, 2015). In addition, a recent study finds that a subunit of voltage-gated calcium channels alpha2delta2 involving synapse assembly negatively regulates CST axon growth and regeneration (Sun et al., 2019). Pharmacological suppression of alpha2delta2 with gabapentin promotes CST regenerative sprouting from the injury site into the intact side in a C5 lateral hemisection model in mice. The CST reorganization is associated with forelimb motor functional recovery (Sun et al., 2019). Notably, the up-regulation of alpha2delta2 parallels the maturation of CST in the development and the start of synaptogenesis. This indicates that synaptic activity may suppress axonal regeneration (Meve & Zheng, 2016). If this new mechanism for failure of axonal regeneration is proved to be true, it opens a new door to enhance axon regeneration by suppressing synaptic activity.

## Robust CST regeneration into caudalized NPC graft after SCI

### Transplantation of non-neural cells or fetal neural tissue

Although various above studies show sprouting and regenerative sprouting of CST axons after SCI, very few shows that transected CST can regenerate into a large clinically relevant injury site. A large injury site usually contains cavities where no axons can grow without a cellular substrate. Thus, various tissues or cell types are transplanted into SCI sites in combination with neurotrophin or other therapeutic interventions. However, transplantation of non-neural cells, such as fibroblasts, bone marrow stromal cells into the spinal cord lesion site, does not elicit any CST regeneration (Grill et al., 1997; Kadoya et al., 2016; Tuszynski et al., 2003). Transplantation of fetal (embryonic day 14) spinal cord tissues in a cervical over-hemisection site, however, does elicit very moderate regeneration of CST axons into lesion/transplant sites (Bregman et al., 2002). Sparse regenerated CST axons are usually restricted to the rostral a few hundred micrometers. Delivery of BDNF and NT-3 increases regeneration of CST axons, but the effect is very moderate (Bregman et al., 2002). This study indicates that early stage neural cells could serve as permissive cellular substrates for CST regeneration.

### Transplantation of caudalized neural progenitor cells

Based on these previous studies, our group recently modified a transplantation protocol and demonstrated the robust regeneration of CST axons into the caudalized neural progenitor cell (NPC) graft (Kadoya et al., 2016) (Fig. 3). We transplanted freshly dissociated NPCs in single cell suspensions rather than embryonic spinal cord tissues, supported by growth factor cocktail embedding in fibrin matrixes. Transplanted NPCs survive, completely fill in the lesion site, and integrate very well with host tissue, resulting seamless and continuous connections from the transplant to the host tissue (Lu et al., 2012). The good integration of NPC transplant with host is critical since regeneration of CST requires direct contact of transected axon stumps with graft (Kadoya et al., 2016). In addition, CST regeneration requires caudalized and homotypic NPC grafts since CST fail to regenerate into telencephalic grafts, less into hindbrain grafts, and mostly into spinal cord grafts (Kadoya et al., 2016).

Injured CST axons not only robustly regenerate into the NPC graft, they also form synaptic connections with graft-derived neurons (Fig. 4) (Kadoya et al., 2016). Furthermore, our recent studies show that regenerated CST axons make appropriate connections with graft-derived spinal cord neurons (Dulin et al., 2018; Kumamaru, Lu, Rosenzweig,

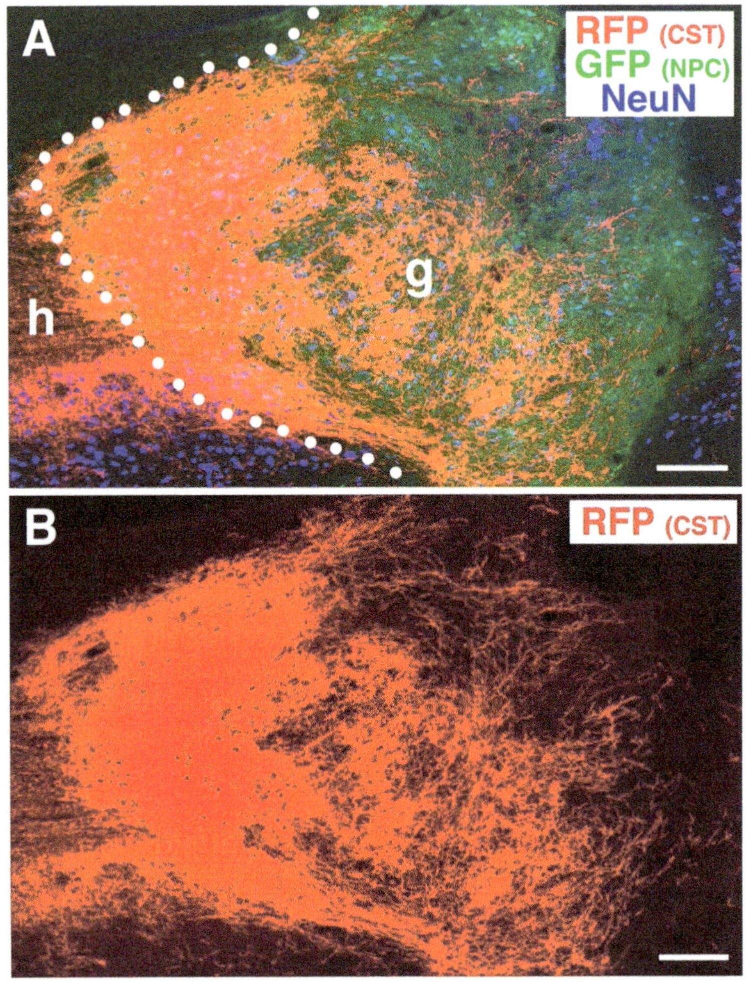


FIG. 3 Robust CST regeneration into NPC graft. (A and B) Triple immunolabeling demonstrates robust regeneration of RFP-labeled CST axons into GFP-labeled NPC graft with differentiation of NeuN-labeled neurons in a C3 dorsal column lesion site that completely transects the main CST. Dashed lines indicate host (h) and graft (g) in the rostral interface in a sagittal section view. Scale bar = 200 μm.

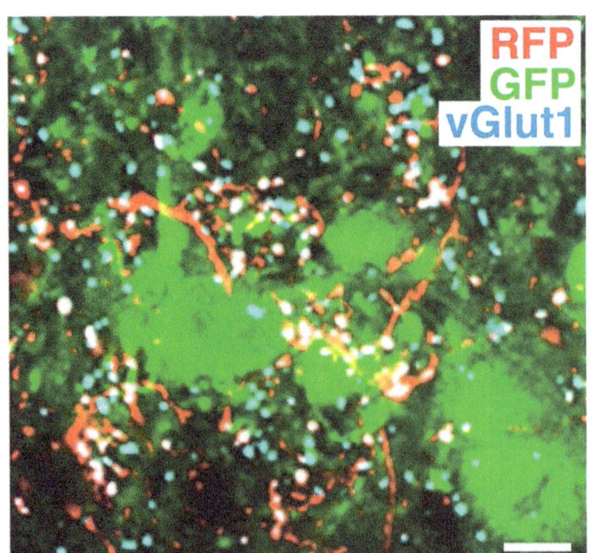


FIG. 4 Connectivity of regenerated CST axons with graft-derived neurons. Triple labeling for RFP, vGlut1, and GFP demonstrates that RFP labeled regenerated CST axons within GFP-labeled NPC graft exhibit bouton-like structures that co-localized with pre-synaptic marker vGlut1, indicating synaptic connections with grafted neurons. Scale bar = 4 μm.

Kadoya, & Tuszynski, 2019). *trans*-Synaptic tracing with herpes simplex virus reveals innervation of CST axons with pre-motor interneurons, such as V2a and Chx10+ excitatory interneurons and motor synergy encoder neurons positive for Satb1 and Ap2b immunolabeling (Levine et al., 2014), which are appropriate target of CST axons. On the other hand, CST axons avoid inappropriate targets of sensory cluster neurons derived from the spinal cord NPC graft. These results indicate that injured CST retains their intrinsic ability to recognize appropriate targets, which is important to achieve functional recovery after SCI.

For translational purpose, our group tests whether CST can regenerate into human NPC graft placed into a low-cervical hemisection site in non-human primate (Rosenzweig et al., 2018). As described above, the main CST in primate is located in the dorso-lateral funiculus that can be completely transected after lateral hemisection. CST axons regenerate by penetration into human fetal spinal cord-derived NPC grafts. Similarly, regenerated CST axons project toward Prdm8 positive pre-motor interneurons and avoid those Tlx3 positive sensory neuron clusters, indicating an appropriate connection with grafted human neurons (Kumamaru et al., 2019). However, it appears qualitatively that the density and distance of primate CST regeneration are less than rodent CST. This could be due to sub-optimal conditions of prolong cultured human neural progenitor cells in vitro.

### Mechanisms of CST regeneration

In a follow-up mechanism study, our group discovers that transplanted NPCs are enable CST neurons to reverse to an embryonic transcriptional state for regeneration by transcriptome analysis (Poplawski et al., 2020). We find that the huntingtin gene (Htt) is a central hub in our analysis since conditional deletion of Htt in mice significantly reduces CST regeneration. The discovery of plasticity of injured adult CST neurons that can be reversed back to embryonic transcriptional state is significant since CST neurons and other injured adult neurons can be further manipulated to enhance their regeneration after SCI.

## Reconnect injured spinal cord through neuronal relays

Although CST axons can regenerate into NPC graft, regenerated CST axons can only reach certain distance within the graft, usually about 2 mm (Kadoya et al., 2016). In a large severe and clinically relevant cervical contusive injury, regenerated CST can only reach the first half rostral NPC graft and leave the caudal half graft un-occupied. Unlike long-distance growth along the white matter in development, adult CST regeneration is similar to CST termination into gray matter neurons since NPC graft virtually re-constitutes spinal cord gray matter that contains large number of neurons (Lu et al., 2012), some of which are innervated by regenerated CST axons (Kadoya et al., 2016; Kumamaru et al., 2019). The limited distance of CST regeneration within the NPC graft is a challenge since CST signals rarely reach directly into spinal cord neurons caudal (below) to the injury to restore their motor functions.

On the other hand, graft-derived neurons can extend large number of axons into host, up to 29,000 axons counted at 0.5 mm caudal from graft in a rat upper-thoracic transection model (Lu et al., 2012). Graft-derived axons, just like developing neurons, travel along host white matter for long distance crossing several segments in both rostral and caudal directions (Lu et al., 2012). Indeed, our recent study demonstrates that host white matter myelin actually supports growth of NPC graft-derived axons (Poplawski et al., 2018). Grafted human NPCs extend their axons almost into the entire rat CNS axis (Fig. 5) (Lu et al., 2014, 2017). The large number and long-distance growth of graft-derived axons in the host are confirmed in a primate model of SCI where up to 150,000 graft-derived human axons are counted at 2 mm caudal to the lesion and reach up to 50 mm from the graft in caudal direction (Rosenzweig et al., 2018).

The ingrowth of regenerated CST axons and other supraspinal descending axons into the graft and the outgrowth of graft-derived axons into caudal host could constitute neuronal relays to re-connect injured CST neurons with their target spinal cord neurons below the injury (Fig. 6). Indeed, our electrophysiological study demonstrates partial recovery of evoked responses below a T3 complete transection site at T6 when stimulation is above the transection at C7 (Lu et al., 2012). This evoke response recovery depends on the presence of synapses in the graft since the application of kynurenic acid, the *N*-methyl-D-aspartic acid (NMDA) receptor blocker, abolishes this recovery. These results indicate excitatory synaptic transmission across the graft functions as neuronal relays. In addition, a recent study from our lab shows that stimulation of host CST by optogenetic method produces distinct and segregated neuronal network responses in the entire NPC graft in mouse SCI models. Furthermore, stimulation of graft-derived axons also elicits spinal cord neuronal responses below injury, indicating potential neuronal relay formation by grafted neurons (Ceto, Sekiguchi, Takashima, Nimmerjahn, & Tuszynski, 2020). These neuronal relays, in theory, could re-connect injured CSTs with spinal cord

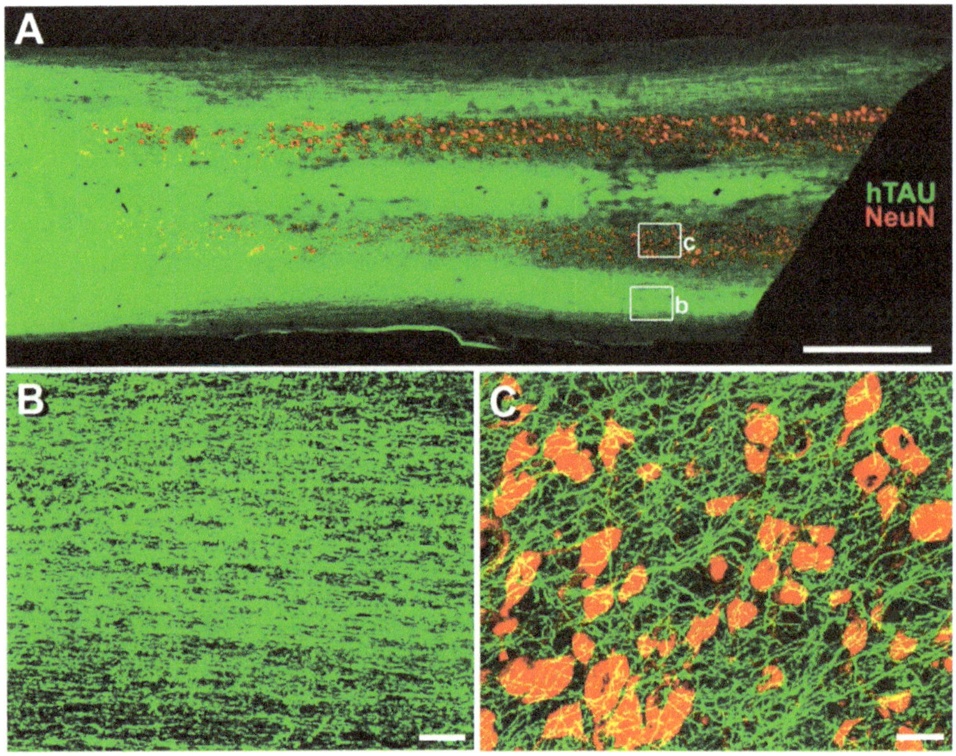


FIG. 5 NPC graft-derived axonal extension. (A) A low magnification horizontal image showing that human Tau labeled human neural progenitor cells (NPCs) extend their axons caudally into host rat spinal cord after NPC graft into the T10 bilateral contusion site for 3 months. NeuN labels both host and graft-derived neurons. Rostral, left, and caudal, right. (B and C) Higher magnification confocal *z*-stack images from the boxed area in panel (A) showing robust and extensive human graft-derived axonal growth in both host (B) white matter and (C) gray matter. Scale bar = 1 mm (A); 32 μm (B and C).

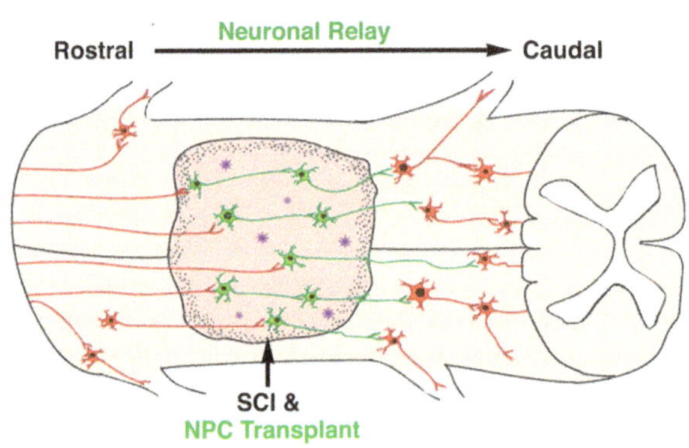


FIG. 6 A cartoon diagram of a neuronal relay formation by transplanted NPCs after SCI. Descending motor axons, including CST (*red*), regenerate into neural progenitor cell (NPC) transplant in spinal cord injury (SCI) site and make synaptic connections with graft-derived neurons (*green*). Graft-derived neurons extend their axons (*green*) into caudal host spinal cord and make synaptic connection with host spinal cord neurons (*red*). The reciprocal connection of host neurons with grafted neurons constitutes a neuronal relay formation. *Drawn by Audelia Arasheben.*

neurons below the injury by either mono-synaptic or poly-synaptic connections between grafted neurons that directly receive regenerated CST inputs and those grafted neurons that directly project into caudal spinal cord (Lu et al., 2019).

## Future prospective

Since the major function of CST is cortical control of spinal cord activity, regeneration of CST is vitally important to restore motor function after SCI. Fortunately, studies demonstrate plasticity of CST system that is enable CST axon to sprout from uninjured axons, regenerative sprouting and regeneration either spontaneously or with therapeutic manipulations.

However, the major challenges remain, such as how to enhance CST sprouting in an incomplete SCI since there are always some spared axons where sprouting originates. For a large severe and complete SCI, promotion of CST regeneration crossing the injury site or through the neuronal relay is the only option. Although regenerated CST form functional synapses (Ceto et al., 2020; Jayaprakash et al., 2016; Kadoya et al., 2016), this does not spontaneously translate to motor behavioral recovery. The new circuit from regenerated CST and relay neurons need to be trained to be functional, especially for skilled and fine motor activity. In addition, whether regenerated CST axons are from the appropriate or inappropriate cortical area (forelimb vs hindlimb cortical area) in a defined cervical SCI is unknown. Promotion of CST regeneration from the appropriate cortical area is another potential challenge in the future.

## Applications to other areas of neuroscience

Injury to the CST occurs not only in SCI, but also in stroke and brain injuries. The mechanism of spontaneous CST sprouting or therapeutic treatments to enhance CST sprouting and regeneration in SCI can be used for promotion of CST regeneration after stroke and brain injuries. However, the injury mechanism and the location of SCI are different from stroke and brain injuries. For example, a stroke results interruption of the blood supply to certain part of the brain, such as cortical area that contains CST neurons, and affects neuronal survival immediately. Restoration of blood supply at early stage of stroke along with therapeutic strategies discussed in this chapter could save CST neurons and restore their functional connectivity. Once a severe stroke occurs in the cortex and damages a large number of CST soma, a neuronal replacement strategy can be applied by transplantation of cortical NPCs. Our studies demonstrate early stage NPCs have great potential to extend their axons for long-distance growth (Lu et al., 2012). A recent study demonstrates that transplanted human embryonic stem cell-derived cerebral organoids into mouse cerebral cortices extend human axons along the corticospinal tract (Kitahara et al., 2020). These results indicate a great feasibility of a neuronal replacement therapy to restore CST connectivity for stroke and brain injuries.

## Mini-dictionary of terms

***Corticospinal tract***: An axonal bundle that starts from cortical layer V pyramid neurons and travels through brain and brainstem to terminate in spinal cord for control of voluntary motor function of the body.

***Spinal cord injury***: A traumatic damage to the spinal cord that disrupts the connection between brain and the spinal cord below injury and results paralysis below injury.

***Neural progenitor cells***: Are cells that are capable of dividing a limited number of times and have the capacity to differentiate into a restricted repertoire of neuronal and glial cell types.

***Central nervous system***: Consists of the brain and spinal cord. It integrates the received information and coordinates and influences the activity of all parts of the bodies of bilaterally symmetric animals.

***Motor synergy encoder***: Represent a central node in neural pathways for voluntary and reflexive movement.

***CSPG, chondroitin sulfate proteoglycans***: Are proteoglycans consisting of a protein core and a chondroitin sulfate side chain.

***PTEN, tumor suppressor phosphatase and tensin homolog***: Is encoded by the PTEN gene and possesses lipid and protein phosphatase-dependent as well as phosphatase-independent activities.

***Axon regeneration***: A new growth occurs from tip of a transected axon, which may lead to reinnervation of its normal target.

***Regenerative sprouting***: Wherein new growth arises not from tip of transected axon, but from region of an axon close to injury sites.

***Axonal sprouting***: Injury disrupts a pathway that induces compensatory growth of new connections from nearby undamaged axons.

## Key facts of corticospinal tract regeneration after spinal cord injury

*Spinal cord injury* irreversibly damages CST and leads to loss of voluntary motor function below injury.

Uninjured corticospinal tract can spontaneously sprout into denervated injury sites.

Injury can induce regenerative sprouting from injured corticospinal tract axons.

Therapeutic manipulation of inhibitory environment or enhancement of intrinsic growth capacity can promote CST sprouting and regenerative sprouting.

Transplanted neural progenitor cells can enhance robust regeneration of CST.

## Summary points

- The corticospinal tract (CST) is the most important voluntary motor control system in humans.
- The regeneration of CST is vitally important for the restoration of the voluntary motor function after spinal cord injury.
- The CST has plasticity since they can spontaneously sprout from uninjured axons into denervated injury sites.
- The CST, however, cannot spontaneously regenerate after injury.
- Therapeutic manipulation is necessary in order to enhance CST regeneration or regenerative sprouting.

## References


Bareyre, F. M., Kerschensteine, M., Raineteau, O., Mettenleiter, T. C., Weinmann, O., & Schwab, M. E. (2004). The injured spinal cord spontaneously forms a new intraspinal circuit in adult rats. *Nature Neuroscience*, *7*, 269–277.

Blackmore, M. G., Wang, Z., Lerch, J. K., Motti, D., Zhang, Y. P., Shields, C. B., et al. (2012). Krüppel-like factor 7 engineered for transcriptional activation promotes axon regeneration in the adult corticospinal tract. *Proceedings of the National Academy of Sciences of the United States of America*, *109*, 7517–7522.

Bradbury, E. J., Moon, L. D., Popat, R. J., King, V. R., Bennett, G. S., Patel, P. N., et al. (2002). Chondroitinase ABC promotes functional recovery after spinal cord injury. *Nature*, *416*, 636–640.

Bregman, B. S., Coumans, J. V., Dai, H. N., Kuhn, P. L., Lynskey, J., McAtee, M., et al. (2002). Transplants and neurotrophic factors increase regeneration and recovery of function after spinal cord injury. *Progress in Brain Research*, *137*, 257–273.

Ceto, S., Sekiguchi, K. J., Takashima, Y., Nimmerjahn, A., & Tuszynski, M. H. (2020). Neural stem cell grafts form extensive synaptic networks that integrate with host circuits after spinal cord injury. *Cell Stem Cell*, *27*, 430–440.

Chen, K., Marsh, B. C., Cowan, M., Al'Joboori, Y. D., Gigout, S., Smith, C. C., et al. (2017). Sequential therapy of anti-Nogo-A antibody treatment and treadmill training leads to cumulative improvements after spinal cord injury in rats. *Experimental Neurology*, *292*, 135–144.

Dulin, J. N., Adler, A. F., Kumamaru, H., Poplawsik, G. H. D., Lee-kubli, C., Strobl, H., et al. (2018). Injured adult motor and sensory axons regenerate into appropriate organotypic domains of neural progenitor grafts. *Nature Communications*, *9*, 1–13.

Eyre, J. A. (2003). Development and plasticity of the corticospinal system in man. *Neural Plasticity*, *10*, 93–106.

Fagoe, N. D., Heest, J. V., & Verhaagen, J. (2014). Spinal cord injury and the neuron-intrinsic regeneration-associated gene program. *Neuromolecular Medicine*, *16*, 799–813.

García-Alías, G., Barkhuysen, S., Buckle, M., & Fawcett, J. W. (2009). Chondroitinase ABC treatment opens a window of opportunity for task-specific rehabilitation. *Nature Neuroscience*, *12*, 1145–1151.

Geoffroy, C. G., Hilton, B. J., Tetzlaff, W., & Zheng, B. (2016). Evidence for an age-dependent decline in axon regeneration in the adult mammalian central nervous system. *Cell Reports*, *15*, 238–246.

Ghosh, A., Haiss, F., Sydekum, E., Schneider, R., Gullo, M., Wyss, M. T., et al. (2010). Rewiring of hindlimb corticospinal neurons after spinal cord injury. *Nature Neuroscience*, *13*, 97–104.

Grill, R., Murai, K., Blesch, A., Gage, F. H., & Tuszynski, M. H. (1997). Cellular delivery of neurotrophin-3 promotes corticospinal axonal growth and partial functional recovery after spinal cord injury. *Journal of Neuroscience*, *17*, 5560–5572.

Hilton, B. J., Anenberg, E., Harrison, T. C., Boyd, J. D., Murphy, T. H., & Tetzlaff, W. (2016). Re-establishment of cortical motor output maps and spontaneous functional recovery via spared dorsolaterally projecting corticospinal neurons after dorsal column spinal cord injury in adult mice. *Journal of Neuroscience*, *36*, 4080–4092.

Jayaprakash, N., Wang, Z., Hoeynck, B., Krueger, N., Kramer, A., Balle, E., et al. (2016). Optogenetic interrogation of functional synapse formation by corticospinal tract axons in the injured spinal cord. *Journal of Neuroscience*, *36*, 5877–5890.

Jin, D., Liu, Y., Sun, F., Wang, X., Liu, X., & He, Z. (2015). Restoration of skilled locomotion by sprouting corticospinal axons induced by co-deletion of PTEN and SOCS3. *Nature Communications*, *6*, 8074–8085.

Joosten, E. A., Gribnau, A. A., & Dederen, P. J. (1989). Postnatal development of the corticospinal tract in the rat. An ultrastructural anterograde HRP study. *Anatomy and Embryology (Berlin)*, *179*, 449–456.

Kadoya, K., Lu, P., Nguyen, K., Lee-Kubli, C., Kumamaru, H., Yao, L., et al. (2016). Spinal cord reconstitution with homologous neural grafts enables robust corticospinal regeneration. *Nature Medicine*, *22*, 479–487.

Kitahara, T., Sakaguchi, H., Morizane, A., Kikuchi, T., Miyamoto, S., & Takahashi, J. (2020). Axonal extensions along corticospinal tracts from transplanted human cerebral organoids. *Stem Cell Reports*, *15*, 467–481.

Kuang, R. Z., & Kalil, K. (1990). Specificity of corticospinal axon arbors sprouting into denervated contralateral spinal cord. *Journal of Comparative Neurology*, *302*, 461–472.

Kumamaru, H., Lu, P., Rosenzweig, E. S., Kadoya, K., & Tuszynski, M. H. (2019). Regenerating corticospinal axons innervate phenotypically appropriate neurons within neural stem cell grafts. *Cell Reports*, *26*, 2329–2339.

Lee, J. K., Chan, A. F., Luu, S. M., Zhu, Y., Ho, C., Tessier-Lavigne, M., et al. (2009). Reassessment of corticospinal tract regeneration in Nogo-deficient mice. *Journal of Neuroscience*, *29*, 8649–8654.

Levine, A. J., Hinckley, C. A., Hilde, K. L., Driscoll, S. P., Poon, T. H., Montgomery, J. M., et al. (2014). Identification of a cellular node for motor control pathways. *Nature Neuroscience*, *17*, 586–593.

Liu, K., Lu, Y., Lee, J. K., Samara, R., Willenberg, R., Sears-Kraxberger, I., et al. (2010). PTEN deletion enhances the regenerative ability of adult corticospinal neurons. *Nature Neuroscience*, *13*, 1075–1081.

Lu, P., Ceto, S., Wang, Y., Graham, L., Wu, D., Kumamaru, H., et al. (2017). Prolonged human neural stem cell maturation supports recovery in injured rodent CNS. *Journal of Clinical Investigation*, *127*, 3287–3299.

Lu, P., Gomes-Leal, W., Anil, S., Dobkins, G., Huie, J. R., Ferguson, A. R., et al. (2019). Origins of neural progenitor cell-derived axons projecting caudally after spinal cord injury. *Stem Cell Reports*, *13*, 105–114.

Lu, P., Wang, Y., Graham, L., McHale, A. K., Gao, M., Wu, D., et al. (2012). Long-distance growth and connectivity of neural stem cells after severe spinal cord injury. *Cell*, *150*, 1264–1273.

Lu, P., Woodruff, G., Wang, Y., Graham, L., Wu, D., Poplawski, G., et al. (2014). Long-distance axonal growth of neural stem cells derived from human induced pluripotent stem cells after spinal cord injury. *Neuron*, *83*, 789–796.

Meve, J. M., & Zheng, B. (2016). Synaptic suppression of axon regeneration. *Neuron*, *92*, 267–269.

Nakagawa, H., Ninomiya, T., Yamashita, T., & Takada, M. (2015). Reorganization of corticospinal tract fibers after spinal cord injury in adult macaques. *Scientific Reports*, *5*, 11986–11994.

Poplawski, G. H. D., Kawaguchi, R., Van Niekerk, E., Lu, P., Mehta, N., Canete, P., et al. (2020). Injured adult neurons regress to an embryonic transcriptional growth state. *Nature*, *581*, 77–82.

Poplawski, G. H. D., Lie, R., Hunt, M., Kumamaru, H., Kawaguchi, R., Lu, P., et al. (2018). Adult rat myelin enhances axonal outgrowth from neural stem cells. *Science Translational Medicine*, *10*, 2563–2575.

Rao, J. S., Zhao, C., Zhang, A., Duan, H., Hao, P., Wei, R. H., et al. (2018). NT3-chitosan enables de novo regeneration and functional recovery in monkeys after spinal cord injury. *Proceedings of the National Academy of Sciences of the United States of America*, *115*, 5595–5604.

Rosenzweig, E. S., Brock, J. H., Culbertson, M. D., Lu, P., Moseanko, R., Edgerton, V. R., et al. (2009). Extensive spinal decussation and bilateral termination of cervical corticospinalprojections in rhesus monkeys. *Journal of Comparative Neurology*, *513*, 151–163.

Rosenzweig, E. S., Brock, J. H., Lu, P., Kumamaru, H., Salegio, E. A., Kadoya, K., et al. (2018). Restorative effects of human neural stem cell grafts on the primate spinal cord. *Nature Medicine*, *24*, 484–490.

Rosenzweig, E. S., Courtine, G., Jindrich, D. L., Brock, J. H., Ferguson, A. R., Strand, S. C., et al. (2010). Extensive spontaneous plasticity of corticospinal projections after primate spinal cord injury. *Nature Neuroscience*, *13*, 1505–1510.

Rosenzweig, E. S., Salegio, E. A., Liang, J. J., Weber, J. L., Weinholtz, C. A., Brock, J. H., et al. (2019). Chondroitinase improves anatomical and functional outcomes after primate spinal cord injury. *Nature Neuroscience*, *22*, 1269–1275.

Sartori, A. M., Hofer, A. S., & Schwab, M. E. (2020). Recovery after spinal cord injury is enhanced by anti-Nogo-A antibody therapy from animal models to clinical trials. *Current Opinion in Physiology*, *14*, 1–6.

Schnell, L., Schneider, R., Kolbeck, R., Barde, Y. A., & Schwab, M. E. (1994). Neurotrophin-3 enhances sprouting of corticospinal tract during development and after adult spinal cord lesion. *Nature*, *367*, 170–173.

Schnell, L., & Schwab, M. E. (1990). Axonal regeneration in the rat spinal cord produced by an antibody against myelin-associated neurite growth inhibitors. *Nature*, *343*, 269–272.

Steward, O., Sharp, K., Yee, K. M., & Hofstadter, M. (2008). A re-assessment of the effects of a Nogo-66 receptor antagonist on regenerative growth of axons and locomotor recovery after spinal cord injury in mice. *Experimental Neurology*, *209*, 446–468.

Sun, W., Larson, M. J., Kiyoshi, C. M., Annett, A. J., Stalker, W. A., Peng, J., et al. (2019). Gabapentinoid treatment promotes corticospinal plasticity and regeneration following murine spinal cord injury. *Journal of Clinical Investigation*, *130*, 345–358.

Tran, A. P., Warren, P. M., & Silver, J. (2018). The biology of regeneration failure and success after spinal cord injury. *Physiological Reviews*, *98*, 881–917.

Tuszynski, M., Grill, R., Jones, L., Brant, A., Blesch, A., Low, K., et al. (2003). NT-3 gene delivery elicits growth of chronically injured corticospinal axons and modestly improves functional deficits after chronic scar resection. *Experimental Neurology*, *181*, 47–56.

Wang, X., Liu, Y., Li, X., Zhang, Z., Yang, H., Zhang, Y., et al. (2017). Deconstruction of corticospinal circuits for goal-directed motor skills. *Cell*, *171*, 440–455.

Wang, Z., Reynolds, A., Kirry, A., Nienhaus, C., & Blackmore, M. G. (2015). Overexpression of Sox11 promotes corticospinal tract regeneration after spinal injury while interfering with functional recovery. *Journal of Neuroscience*, *35*, 3139–3145.

Wang, X., Zhou, T., Maynard, G. D., Terse, P. S., Cafferty, W. B., Kocsis, J. D., et al. (2020). Nogo receptor decoy promotes recovery and corticospinal growth in non-human primate spinal cord injury. *Brain*, *143*, 1697–1713.

Weidner, N., Ner, A., Salimi, N., & Tuszynski, M. H. (2001). Spontaneous corticospinal axonal plasticity and functional recovery after adult central nervous system injury. *Proceedings of the National Academy of Sciences of the United States of America*, *98*, 3513–3518.

Welniarz, Q., Dusart, I., & Roze, E. (2017). The corticospinal tract: Evolution, development, and human disorders. *Developmental Neurobiology*, *77*, 810–829.

Zhao, R. R., Muir, E. M., Alves, J. N., Rickman, H., Allan, A. Y., Kwok, J. C., et al. (2011). Lentiviral vectors express chondroitinase ABC in cortical projections and promote sprouting of injured corticospinal axons. *Journal of Neuroscience Methods*, *201*, 228–238.

Zheng, B., Atwal, J., Ho, C., Case, L., He, X. L., Garcia, K. C., et al. (2005). Genetic deletion of the Nogo receptor does not reduce neurite inhibition in vitro or promote corticospinal tract regeneration in vivo. *Proceedings of the National Academy of Sciences of the United States of America*, *102*, 1205–1210.

Zhou, L., Baumgartner, B. J., Hill-Felberg, S. J., McGowen, L. R., & Shine, H. D. (2003). Neurotrophin-3 expressed in situ induces axonal plasticity in the adult injured spinal cord. *Journal of Neuroscience*, *23*, 1424–1431.

Chapter 37

# Trophic factors in patients with spinal cord injury

J. Luis Quintanar[a], María del Carmen Díaz-Galindo[a], Denisse Calderón-Vallejo[a,b], and Francisco Jaramillo-González[b]

[a]*Department of Physiology and Pharmacology, Center for Basic Sciences, Autonomous University of Aguascalientes, Aguascalientes, Ags, Mexico;* [b]*Department of Morphology, Center for Basic Sciences, Autonomous University of Aguascalientes, Aguascalientes, Ags, Mexico*

## Abbreviations

| | |
|---|---|
| **BDNF** | brain-derived neurotrophic factor |
| **CNTF** | ciliary neurotrophic factor |
| **ES** | electrical stimulations |
| **FES** | functional electrical stimulation |
| **FGF** | fibroblast growth factors |
| **GDNF** | glial-derived neurotrophic factor |
| **GH** | growth hormone |
| **GnRH** | gonadotropin-releasing hormone |
| **IGF-1** | insulin-like growth factor-1 |
| **iPSCs** | induced pluripotent stem cells |
| **MSCs** | mesenchymal stem cells |
| **NGF** | nerve growth factor |
| **NPCs** | neural progenitor cells |
| **NT3** | neurotrophin 3 |
| **NT4/5** | neurotrophin 4/5 |
| **OECs** | olfactory cells |
| **OPCs** | oligodendrocyte progenitor cells |
| **SCI** | spinal cord injury |
| **TRH** | thyrotropin-releasing hormone |

## Introduction

Spinal cord injury (SCI) causes temporary or permanent changes in the motor, sensory, and autonomic functions of the spinal cord.

The main symptoms of SCI vary according to where the spinal cord damage occurred. Cervical injuries produce partial or total tetraplegia, shown by paralysis of the four extremities, while injuries in the lower areas induce paraplegia, in which paralysis of the lower body occurs and, in both cases urogenital and digestive alterations.

SCI can directly induce the death of different cell types such as neurons, oligodendrocytes, and astrocytes. Once damage has been generated in the spinal cord, secondary alterations such as demyelination of axons and apoptosis processes in oligodendrocytes occur (see review, Fakhoury, 2015) (Fig. 1).

Thus, in this chapter, we will review the main trophic factors that are released spontaneously in patients when SCI occurs. Likewise, to those trophic factors that are secreted when patients undergo rehabilitation and finally, the perspectives of its use in clinical trials after SCI.

Currently, the concept of trophic factor is applied to substances of a known or unknown nature that have specific effects on cells or tissues through specific receptors called Trk and p75. Particularly when the trophic factor act, on neurons and glia, they are called neurotrophic factors and its actions include support the growth, survival, neurotransmitters release, and differentiation in different stages of development, in maturity or in regenerative processes.

Diagnosis and Treatment of Spinal Cord Injury. https://doi.org/10.1016/B978-0-12-822498-4.00037-3

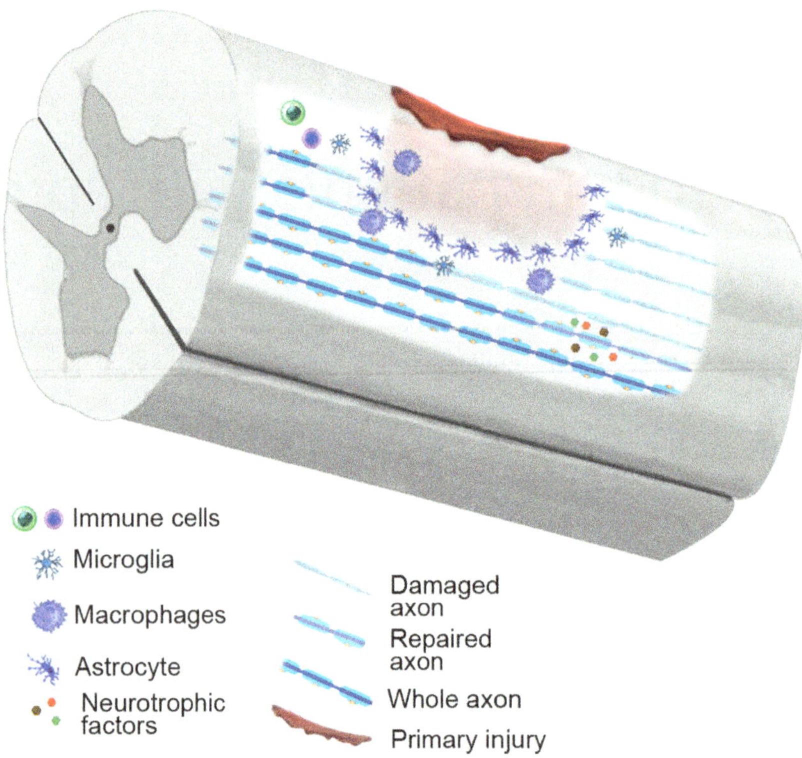


FIG. 1 Spinal cord injury. The image represents the main factors involved in the mechanisms of spinal cord injury. Initially, the primary lesion can produce a hemorrhagic site and edema, as well as cell death due to necrosis and apoptosis. On the other hand, the arrival of immune response cells to the site of injury occurs. At a later stage, a glial scar is formed that incorporates astrocytes and microglia. Therefore, many ascending and descending axons are disrupted and do not regenerate over long distances. Despite this, due to intrinsic neuronal plasticity and the presence of neurotrophic factors, it is possible that some axons grow to form new neuronal circuits.

Although there is evidence of spontaneous recovery, they will largely depend on the magnitude of the injury, spinal level, age of the injury, patient's condition, among others. However, spontaneous repair processes are initiated simultaneously in motor, sensory, and autonomous functions. Thus, the ability of intact corticospinal axons to sprout after SCI could present, as well as remyelination of injured axons.

Another approach to restoring the SCI and perhaps the most common strategy is rehabilitation and that produces environmental conditions that favor neurological recovery. Physical rehabilitation (exercise), electrical stimulation both directly to the muscle and to the spinal cord, produces motor and sensory improvements after SCI.

Cell therapies are used in order to create a suitable environment for restoration after SCI. The therapies include transplantation of stem cells, glial cells, and cells of the olfactory nervous system.

The treatment with trophic factor has been shown to improve conditions in different experimental animal models of SCI. The potential use of trophic factors in patients with SCI is evident, considering the possibility of improving the quality of life. However, a large number of clinical trials for its pharmacological use have to be carried out.

## Role of neurotrophic factors in human neural plasticity after SCI

Spontaneous functional recovery in patients with spinal cord injury is limited, it occurs especially in patients in whom at least part of the sensory and motor functions has been preserved. This functional recovery stabilizes between 12 and 18 months and is mainly due to the conservation and sprouting of axons within the conserved medullary tissue adjacent to the injury site (Fawcett, Curt, Steeves, et al., 2006). After SCI, plasticity processes involving axonal regeneration can occur, this reconnection of axons involves the regrowth of transected axons through a site of injury toward their original synaptic targets, while other forms of axonal sprouting result in the reorganization of the circuit; all these processes can lead to restoration of function (Onifer, Smith, & Fouad, 2011). However, many of the axonal sprouts are usually not beneficial, and even cause adverse functional effects such as neuropathic pain, spasticity, and dysreflexia (Kathe, Hutson, McMahon, & Moon, 2016).

Neurotrophic factors can play an important role in structural and functional recovery after spinal cord injury and they are up or down regulation after the damage. These molecules are involved in axonal growth, remyelination, and modulation of the glial response after injury, among others (Afshari, Kappagantula, & Fawcett, 2009). This group of proteins is involved in the processes of natural plasticity after a spinal cord injury, which in patients has been described a relevant functional recovery without any type of treatment (Waters, Adkins, Sie, & Yakura, 1996). Not all neurons or spinal tracts express the same types of receptors, so they may preferentially be sensitive to a particular neurotrophic factor (Blesch, Fischer, & Tuszynski, 2012). The importance of the main trophic factors involved in the intrinsic recovery of neural tissue after the spinal injury is described below.

## Nerve growth factor (NGF)

It is known that NGF activity increases after injury and that this factor is partly responsible for the initial regenerative response of the central nervous system, such as the stimulation of axonal sprouting and synaptic plasticity (Mocchetti & Wrathall, 1995). Besides, in the spinal cord, the NGF can influence neural responses to injury on cell types that display specific receptors to this factor, such as nociceptive sensory neurons and α motor neurons (Tuszynski et al., 1996; Verge, Richardson, Benoit, & Riopelle, 1989). The expression of NGF within the spinal cord induces a strong sprouting of nociceptive axons and hyperalgesia (Romero et al., 2000). Because the major trophic effects of NGF following SCI are seen in small-diameter sensory neurons, clinical trials have focused on investigating improvements in sensory components of neurological examinations and in pain assessments (McArthur, Yiannoutsos, Simpson, et al., 2000).

## Brain-derived neurotrophic factor (BDNF)

BDNF has been shown to have neuroprotective and growth-promoting effects on different neuronal types after injury. It has particularly been observed in the rubrospinal, reticulospinal, and vestibulospinal tracts, as well as in the proprioceptive neurons of the nucleus of Clarke of the lumbar spinal cord. The neuroprotective effects can be attributed to the signaling cascades activated when BDNF binds to TrkB receptors. There is also evidence showing that BDNF can decrease glutamate-induced apoptosis-mediated cell death in the site of the injury (Almeida, Manadas, et al., 2005; Keefe, Sheikh, & Smith, 2017). However, in addition to the positive effects on the regeneration of motor tracts, the over-secretion of BDNF would only add to an environment that is favorable to not providing adequate plasticity and pain hypersensitivity after SCI (Coull, Beggs, Boudreau, et al., 2005).

## Neurotrophin NT3 and NT4/5

These proteins are expressed in low amounts under normal conditions; however, after injury to the nervous system, inflammatory factors stimulate their expression and protect uninjured neurons by promoting the growth and repair of injured nervous tissue (Fang, Liu, et al., 2017). Also, the intrinsic secretion of NT-4/5 attracted Schwann cells to the injury site, causing myelination of affected axons at the injury site. Likewise, neurotrophins induced long-distance axonal growth of propriospinal and supraspinal axons (Engele, Schubert, & Bohn, 1991).

## Glial-derived neurotrophic factor (GDNF)

Astrocytes, oligodendrocytes, and Schwann cells produce GDNF (Walker & Xu, 2018). After a spinal cord injury, a dense astrocytic glial scar occurs that surrounds the injury and this astrogliosis process is known to be positive, since it limits both the spread of excitotoxic molecules and the area of injury. In this sense, GDNF has been shown to positively modulate astrogliosis by improving the conditions of the lesion (Ansorena, De Berdt, Ucakar, et al., 2013). On the other hand, one of the mechanisms by which GDNF achieves functional improvement is by reducing secondary damage after SCI. GDNF has been shown to reduce the permeability of the blood–spinal cord barrier (BSCB) and down-regulate nitric oxide synthase (NOS). In this way, edema can be reduced and the survival of different neurons could be favored. In addition, neural apoptosis in the spinal cord after injury is also reduced and this also favors a decrease in secondary damage and greater recovery (Rosich, Hanna, Ibrahim, Hellenbrand, & Hanna, 2017).

## Fibroblast growth factors (FGFs)

FGFs are present in the central and peripheral nervous system during development and throughout life, stimulating neuronal differentiation and migration. Its function in the context of spinal cord injuries focuses on the following aspects: stimulating the regeneration and regrowth of axons, guiding and accelerating the regeneration of the axon, carrying out the chemotactic attraction of the cells of the immune response, as well as the stimulation of the formation of blood vessels; attenuate inflammation, astrocyte activation, and glial scar formation; finally, reducing the death of injured neurons and mediating the gene expression of injured neurons (Zhou, Wang, Li, Li, & Xiao, 2018).

Considering that spinal cord injuries in humans are generally incomplete, any therapeutic strategy that improves the cellular and synaptic function of surviving neurons is potentially useful. In this context, neurotrophic factors become in an important target for the clinical approach, which is discussed in the next sections.

## Different environments enhance trophic factors delivery

Microenvironment consequences after primary injury in SCI involves hypoxia, ischemia, cellular swelling, free radical production, immune response, and several other factors that leads an injury tissue. Anti-regenerative profile promoted by damage, glial scar conformation, cellular death, and limited cellular infiltration conform chronic characteristics where axonal regeneration is unable and determinates clinical permanent disability in patients (Yang et al., 2020). Nerve disruptions avoid voluntary muscle contraction control and entails paralyzed muscles. Muscle atrophy determinate fiber lost, diameter decrease and metabolic conversions (Fu, Wang, Deng, & Li, 2016).

Currently, physical exercise is a fundamental therapeutic approach in the management of SCI, due, in part, to the fact that it increases the release of neurotrophic factors. This chapter describes its importance in the next section.

### Exercise

Despite great efforts for new SCI treatments so far acute handling is based on high steroid doses and acute rehabilitation. Exercise enhances functional prognosis through neuroprotection actions beyond strength muscle improving.

Experimental studies demonstrate that exercise promotes blood–brain barrier strengthening, this integrity conserves an adequate filtration, and less edema occurs. During exercise, metabolic demand increase linked to further angiogenesis; regular physical practices enhance immune function, anti-inflammatory profile, and neuroprotection (Kohut et al., 2006). Neuroprotective actions also include enhancement of neurotrophins production, these polypeptides induce neuronal preservation, migration, and proliferation (Cobianchi, Arbat-Plana, Lopez-Alvarez, & Navarro, 2017).

Even when in acute injury neurotrophins concentration increase, trophic support decline over the time (Gold, Schulz, Hartmann, et al., 2003). Up-regulation of BDNF due exercise it considers the most important and assessed factor related to neural repair and brain circuit construction in humans. It is primarily produced in central nervous system (75% of circulating BDNF) but also platelets keep a large amount of BDNF protein. There is another neurotrophic factor extensively studied as NGF, NT3 in animal models related to exercise and neuroregeneration. Until now only BDNF has easy access for detection by enzyme-linked immunosorbent assay (ELISA) in humans (Gold et al., 2003; Leech & Hornby, 2017; Rosenfeld et al., 1995; Zeller, Abel, Rojas-Vega, Foitschik, & Strueder, 2015).

Different ways to exercise determinates neurotrophic factors release; variables like intensity, duration, and frequency impact on the expression of these factors, moderate to intermediate intensity are related to a better improve that observed in exhaustive exercise, daily no longer session induces an increase in plasmatic levels of BDNF. Several studies at different circumstances like adult healthy human male and female, SCI athletes, and multiple sclerosis patients, exhibit an elevates concentration of BNDF against control group when exercise develops (Gold et al., 2003; Leech & Hornby, 2017; Zeller et al., 2015).

BDNF is implied at multiple metabolic complex processes like neuroplasticity; up-regulation allows respond to situation as neural network reconstruction; added release is very susceptible to physical activity, and early time after injury exhibits a greater improve.

Furthermore, do not dismiss exercise is a primarily part of the SCI treatment conserving muscle and joins tone and flexibility, reduces contractures, and enhances internal organ functions (Rayegani et al., 2011). Likewise, functional magnetic resonance imaging (fMRI) and electroencephalogram (EEG) records show modifications in cortical sensorimotor area in SCI patients against healthy subjects when exercise perform, functional remodeling and provide functional recovery (Fu et al., 2016).

Injury level determinates the ability to perform physical rehabilitation, below the lesion no voluntary contraction exist. Electrical stimulation supplies neuronal connection, trough electric impulse promoting an action potential in a single muscle or group of muscles.

## Electrical stimulation

Electrical stimulations (ES) appear as clinical application in 20th century, since the beginning stimulating electrodes were used to activate motor system and produce muscle contraction, although in some of the patient integrity of peripheral nervous system pathways it is intact, neuronal connection interruption between central nervous system and these tracts impair action potential generation, ES supply as an alternative to conserve muscle contraction, diminish atrophy and delay join alterations.

ES improves axon growth, myelin sheath formation and impact directly in reconstruction pattern independent of voluntary action. Transcutaneous and subcutaneous stimulation system can be applied directly in motor nerve or sensory nerve activation through spinal reflex activation (Yang et al., 2020).

The aim of stimulation determinates electrodes localization; muscle mass stimulation distributes depolarization to fibers around the application site, while nerve trunk stimulation depolarizes entire muscle. Orthodromic and antidromic activation generates a bidirectional stimulus and impacts to cortical brain reorganization (Milosevic et al., 2020).

Currently, complex patterns of stimulations make possible limbs movements control and gait reproduction consequently of transcutaneous implants electrode stimulation, this technique is referred like Functional Electrical Stimulation (FES). Rehabilitation was the first objective of FES, considering some experiments that exhibit TrkB receptor expression and BDNF serum levels enhance after electrical stimulation and brains network reorganization. FES became a treatment strategy (FEST) while patients could stand up and develop gait movements supporting by team assistances moving around short distances (Hogan, Hamilton, & Horner, 2020).

Recently brain–computer interface (BCI) an emerging technology, potency bidirectional stimulation utilizing FES movements control and brain outputs signals that synchronize FES and cortical instructions. Simultaneous activations enhance networks reconstruction (neuroplasticity). Few SCI patients have participant on this technology.

Electrical stimulation similar to physical activities, increase neurotrophic factor release, brain and spinal circuit reorganization, neuroplasticity and improve motor function (Rojas Vega et al., 2008).

Reinforce of neuronal reorganization could be supported by cellular implanted, another way to increase neuroplasticity and neurotrophic factor release.

## Cell transplantation

Replace of injury neuronal cells is one of the objectives of cell transplantation; additionally, they promote axon regeneration, neurotrophic factor release and myelination (Griffin & Bradke, 2020).

Clinical application of these cells started since 2010, a relative new strategy, Geron Corporation explored the introduction of oligodendrocyte progenitor cells (OPCs) at injured place, ethical component was very discussed field about origin of the cells, preservation place and growth methods, all these components were determined with a strict limitation ( Jin, Medress, Azad, Doulames, & Veeravagu, 2019).

Clinical assays implement different cells for SCI transplantation including neural progenitor cells (NPCs), Schwann cells, olfactory cells (OECs), mesenchymal stem cells (MSCs) and induced pluripotent stem cells (iPSCs) ( Jin et al., 2019). Fig. 2 shows the trophic factors released in the different therapeutic approaches.

Additionally, route of administration is a critical point, and determinates potency of treatment most of them are invasive: intramedullary, intrathecal, intravenous, and percutaneous. All administrations represent a certainty risk but also allows introduce cells by infusion or graft containing cells.

Furthermore, after cell transplantation, the rate of cell differentiation and survival is low. This makes it necessary to administer more cells in a second application.

Currently, clinical trial in phase 1 or 2 are a few (<30); they are evaluating safety adverse effects and viability of management (Silvestro, Bramanti, Trubiani, & Mazzon, 2020).

Mesenchymal stem cells obtain from: bone marrow, umbilical cord blood, adipose tissue, is one of the experimental transplanted cells, promotes conditions against apoptosis and substitutes neuronal cell death. Side effects after administration consist in fever, headache, abdominal distention, and backache (Dasari, Veeravalli, & Dinh, 2014). These patients are evaluated through MRI for long periods (more than a year) regarding adverse effects as tumor (ectopic tissue), death or immune reactions. A main risk of stem cell transplantation involves tumors and phenotype conversion in transplanted cells.

Regarding of these treatment options, different techniques exist to over expression and release of neurotrophic factors, exercise and electrical stimulations are safety than cell transplantation although all of them represent an alternative option to improve motor activation in SCI patients.

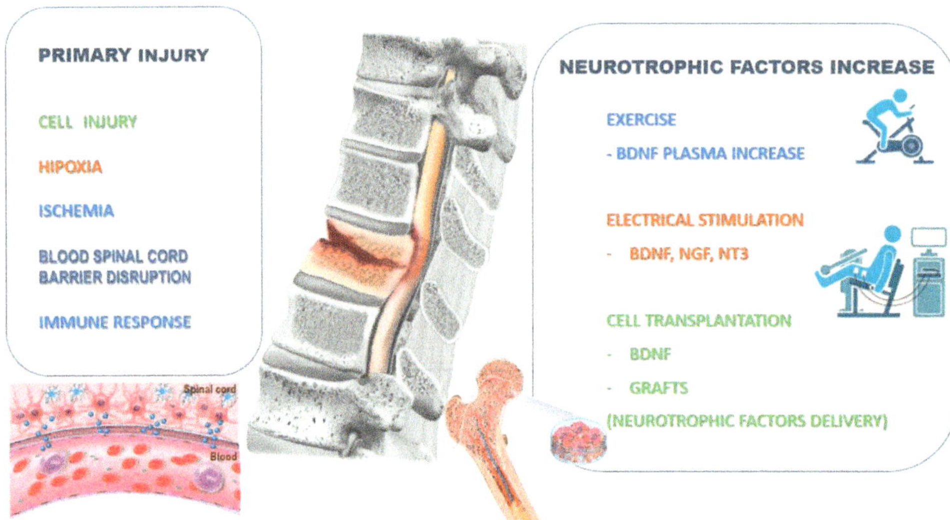


FIG. 2 Representative image of mechanisms presents during primary injury against possible interventions that enhance neurotrophic factor release as a possible management for SCI patient.

## Therapeutic potential of trophic factors

Trophic factors mainly participate in the regulation of neuron survival or repair, in the generation of synaptic plasticity, in the secretion of neurotransmitters and in the induction of neuritic changes such as axonal growth, as well as in glial recovery (Harvey et al., 2015; Keefe et al., 2017) (Fig. 3).

Most of them have been tested in animal models with SCI, where their neurotrophic capacity is shown, improving motor conditions, sensitivity, and autonomic functions.

The main neurotrophic factors and their effects and, where appropriate, clinical trials in patients with spinal cord injury are described below.

The first trophic factor to be characterized was the NGF, which regulates nerve regeneration by promoting growth, the formation of new synaptic contacts, inducing differentiation in certain neuronal types, and the recovery of constitutive structures of neural networks. These facts have been described in experimental animals (Xu, Wu, Qin, et al., 2019). However, there is no clinical trial in patients that supports its administration.

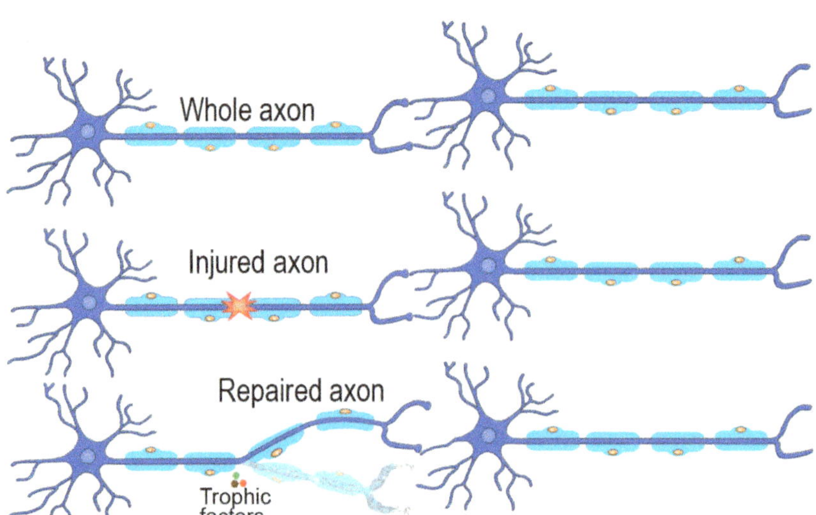


FIG. 3 Schematic representation of trophic factor-induced neuronal plasticity after spinal cord injury. Integral neurons with synaptic contact with the whole axons. After injury, the axon remains incomplete but still viable. Finally, a possible mechanism is shown by which an axon can regenerate induced by trophic factors and return to make synaptic contact with the next neuron.

The BDNF participates in neuroprotective functions in both neurons and glial cells. It induces neuronal growth, particularly of axons and their regeneration in the event of neurological lesions, as well as axonal remyelination (Keefe et al., 2017). One of the most important risks in the management of these neurotrophic factors is their side effects. For this reason, treatment with these factors requires more studies that allow it to be used in patients with SCI.

In the case of neurotrophins, NT-3, NT-4, and NT-5, their role within the nervous system is to repair damaged neurons and promote growth, as well as induce neuronal differentiation (Veneruso et al., 2019). Currently, there is no treatment protocol for patients with SCI, only experimental trials in animal models.

The GDNF promotes axonal recovery and regulates the proliferation of glial cells such as astrocytes. In addition, it has anti-inflammatory effects, which favors the reduction in the size of the lesion, thereby improving the environmental conditions for its recovery (Rosich et al., 2017; Walker & Xu, 2018). Treatment with this neurotrophic factor has been tested in experimental animals, but there are no reports that propose a treatment scheme for patients with SCI.

The ciliary neurotrophic factor (CNTF) plays an important role in the survival of injured nervous tissue, either directly in neurons in regenerating axons or in damaged oligodendrocytes (Ong, Pinese, & Chew, 2019). It is perhaps the trophic factor that could have the greatest impact on a possible treatment for neuronal regeneration in the case of patients with SCI. Its properties are very specific and with few side effects.

Other neurotrophic factor is FGF which induce axonal regeneration and recovery of damaged neurons. Likewise, they induce the activation of astrocytes and the reduction of inflammation (Zhou et al., 2018). A clinical study in patients with SCI, FGF administration after 24 months of treatment, partially improved sensory and motor functions. The most important inconvenient is that an invasive method is required for application of FGF since it is administered at the site of injury (Wu, Huang, Chen, et al., 2011).

Insulin-like growth factor 1 (IGF-1) is involved in axon growth, oligodendrocyte survival, and myelin recovery. In addition, it has antioxidant effects (Allahdadi, de Santana, Santos, et al., 2019). Unfortunately, there are no reports of studies performed in clinical trials with SCI patients treated with IGF-1, only in experimental models in animals with lesion of spinal cord.

## Molecules with properties of trophic factor in clinical trials

It has been described that Gonadotropin-releasing hormone (GnRH) presents neurotrophic effects in experimental animal models with traumatic SCI. Recently has been demonstrated the presence of GnRH receptor in the spinal cord of humans (Díaz-Galindo, Calderón-Vallejo, Hernández-Jasso, et al., 2020). The use of synthetic analogues of GnRH such as leuprolide acetate, also improve the conditions of SCI. Leuprolide acetate administered via intramuscular injection, cross the blood-spinal cord barrier (Barrera, Kastin, Fasold, & Banks, 1991). We have reported that in patients with chronic SCI, intra-muscular administration of leuprolide acetate each month during 6 months induces recovery in ASIA (American Spinal Injury Association) scale scores of motors, sensory and functional independence measure (Quintanar, Díaz-Galindo, Calderón-Vallejo, et al., 2018). The main advantage of this treatment is of easy and safe application and that it is a non-invasive method (Fig. 4).

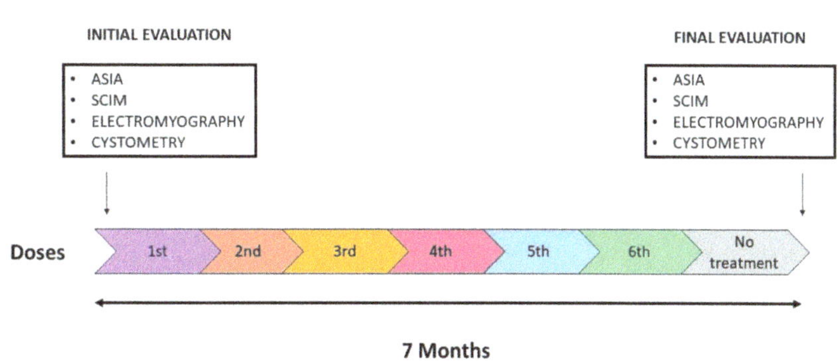


FIG. 4 Scheme of a possible treatment with a GnRH agonist (leuprolide acetate) as a trophic factor administered intramuscularly every month for 6 months in patients with SCI (Quintanar et al., 2018). ASIA score, American Spinal Cord Injury Association; SCIM, Spinal Cord Independence Measure.

Thyrotropin-releasing hormone (TRH) stimulates the release of the thyroid stimulating hormone produced by pituitary gland, which stimulates the thyroid gland for the secretion of thyroid hormones (Fröhlich & Wahl, 2019). Studies have described that patients with SCI treated with TRH, improves of motor or sensitive functions when they were treated within 12 h after trauma. They were evaluated up to 4 months after lesion (Pitts, Ross, Chase, & Faden, 1995). It is possible that the neuroprotective effect of TRH and its analogues is through the stabilization of neuronal membrane, its anti-inflammatory properties and its antioxidant capacity (Faden, Vink, & McIntosh, 1989).

Growth hormone (GH) is a peptide produced by pituitary gland. Has been described its role in neuroregeneration processes, promotion of innervation and neurogenesis (Arámburo, Alba-Betancourt, Luna, & Harvey, 2014; Martínez-Moreno, Calderón-Vallejo, Harvey, Arámburo, & Quintanar, 2018). A clinical case has been published in a patient with incomplete spinal cord development and impaired limb motor activity in which was administered with GH. Treatment induced an improvement in motor function (Devesa, Alonso, López, et al., 2017). This neurotrophic effect could be due to the fact that GH participates in the expression of neurotrophins such as BDNF and IGF-1 (Sharma, 2007).

The hormone melatonin is a peptide produced by the pineal gland (Aygün, Kaplan, Odaci, Onger, & Altunkaynak, 2012). In SCI rats, systemic melatonin administration during 4 weeks induces a reduction of free radical, and therefore oxidative stress, giving its recovery (Erşahin, Özdemir, Özsavcı, et al., 2012).

The estrogens are hormones produced mainly in the ovaries and they have neuroprotective effects. In SCI animal models, treatment of estrogen 48 h after the lesion induces a reduction of calpain and caspase-3 expression which have anti-apoptotic effects (Samantaray, Sribnick, Das, et al., 2010). Likewise, in SCI rats, locomotion was improved and neurons of the spinal cord, the survival was higher after estrogen treatment (Table 1) (Kachadroka, Hall, Niedzielko, Chongthammakun, & Floyd, 2010).

The most important challenges that have to be considered in clinical use of trophic factors in patients are: (a) its half-life is very short, (b) in most cases they do not cross the blood-spinal barrier, (c) the administration would be topical, so they would use invasive techniques (surgical), (d) possible non-specific effects on uninjured tissue, (e) adverse side effects, and (f) the economic costs of a possible treatment.

However, perhaps one of the most important and pending challenges is that there are preclinical trials that have been performed in experimental models using trophic factors and that have not yet been tested in patients with SCI.

It is possible that with the conjunction of several simultaneous therapeutic approaches, such as rehabilitation and the administration of trophic factors, it may give a better result in terms of neurological recovery, thus giving a better quality of life for patients with SCI.

## Applications to other areas of neuroscience

Because of the complex nature of the nervous system, when injuries occur to it, possible treatment approaches are also complex. To understand the pathophysiological phenomena that occur with SCI, experimental animal models are initially used. Later the translational consideration would be its applicability in patients with SCI. In the specific case of the application of neurotrophic factors such as nerve growth factor, brain-derived neurotrophic factor, neurotrophin 3 and 4/5, glial-derived neurotrophic factor, fibroblast growth factors, ciliary neurotrophic factor, and insulin-like growth factor-1, have been widely studied. However, not all these factors have been clinical trials, either due to the routes of administration, their chemical characteristics or the adverse effects that they may produce. On the other hand, there are molecules with trophic properties such as gonadotropin-releasing hormone, thyrotropin-releasing hormone, growth hormone, melatonin, and estrogens. These molecules have among their functional properties, neuronal plasticity. Due to these properties, these hormones have been used for the treatment of spinal cord injuries both in experimental models and in some clinical trials. The results observed include improvements in motor, sensory, and autonomic function.

It is possible that these trophic factors can also be used in neurological lesions in other sites of the nervous system such as peripheral nerves and in the brain.

**TABLE 1** Summary of trophic factors and their possible therapeutic effects in spinal cord injury patients.

| TROPHIC FACTOR | POTENTIAL TERAPHEUTIC EFFECTS IN SCI | REFERENCES |
|---|---|---|
| BDNF | ✓Stimulate neuroplasticity and neural network reconstruction as well as modifications in cortical sensorimotor area.<br>✓Conserving muscle and joins tone and flexibility, reduces contractures, and enhance internal organ functions.<br>✓Develop gait movements supporting by team assistances moving around short distances. | Rayegani et al., 2011; Fu et al., 2016; Hogan et al., 2020. |
| NT-3, NT-4 and NT-5 | ✓Repair damaged neurons and promote growth.<br>✓Attracted Schwann cells to the injury site, causing myelination of affected axons at the injury site.<br>✓Induced long-distance axonal growth of propriospinal and supraspinal axons. | *Engele et al., 1991; Veneruso et al., 2019.* |
| GDNF | ✓ Promotes axonal recovery.<br>✓ Regulates the proliferation of glial cells such as astrocytes.<br>✓It has anti-inflammatory effects, which favors the reduction in the size of the lesion. | Rosich et al., 2017; Walker et al., 2018. |
| CNTF | ✓ Plays an important role in the survival of injured nervous tissue, either directly in neurons in regenerating axons or in damaged oligodendrocytes.<br>✓ In SCI patients, its properties are very specific and with few side effects. | Ong et al., 2019. |
| FGF | ✓ Induce axonal regeneration and recovery of damaged neurons.<br>✓ induce the activation of astrocytes and the reduction of inflammation.<br>✓ In SCI patients, partially improved sensory and motor functions. | Zhou et al., 2018; Wu et al. 2011. |
| IGF-1 | ✓ Involved in axon growth, oligodendrocyte survival, and myelin recovery.<br>✓ It has antioxidant effects. | Allahdadi et al., 2019. |
| GnRH and analogues | ✓ In SCI patients, a partial recovery of motor and sensorial functions.<br>✓ Increase in functional independence measure in SCI patients. | Quintanar et al, 2018. |
| TRH and analogues | ✓ In SCI patients, a partial recovery of motor and sensory functions.<br>✓ Anti-inflammatory properties and antioxidant capacity. | Fade et al, 1989. |
| GH and analogues | ✓ In SCI patients, an improvement in motor function.<br>✓ Stimulate the release of BDNF and IGF-1. | Devesa et al. 2017; Sharma, 2007. |
| Melatonin | ✓ Reduction of oxidative stress, thereby reducing the deterioration of the spinal cord lesion. | Ersahin et al. 2012. |
| Oestrogens | ✓ Reduction of apoptosis.<br>✓ Improve in locomotion.<br>✓ The survival of spinal neurons was higher. | Samantaray et al., 2010; Kachadroka et al., 2010. |

The column on the left side indicates the name of the different trophic factors. The potential therapeutic effects of each factor are listed in the middle column. The right column contains the information of the references that can be consulted for more details.

## Mini-dictionary of terms

***Spinal cord injury***: An alteration of the spinal cord, caused by trauma or disease. A spinal cord injury often causes permanent loss of strength, sensation, and mobility below the site of injury.

***Trophic factors***: A group of substances, most of them of a protein nature that, together with hormones and neurotransmitters, induce morphofunctional changes in tissues.

***Neural plasticity***: Is the ability of the nervous system to change its structure and function as a reaction to the diversity of the environment.

***Electrical stimulation***: It is the technique that uses electrical current to cause a muscle contraction. Using a device called an electrostimulator, an electrical current is generated to induce muscle contractions or tingling sensations that serve to prevent injuries, train or treat muscles, obtain a therapeutic effect and improve performance.

***Physical rehabilitation***: A group of methodologies that aims to recover and improve the functional capacity and quality of life of those people who suffer from disability due to illness or injury.

***Neurotrophins***: A family of proteins which regulate the generation, survival, proliferation, differentiation, and death of neurons in the peripheral and central nervous system. They include NGF, BDNF, NT3, and NT4/5.

***Glial-derived neurotrophic factor***: A polypeptide factor that binds and activates the TrkB receptor, it has an important role in the physiological processes underlying the plasticity and development of the nervous system.

***Ciliary neurotrophic factor***: A member of the interleukin-6 family of cytokines. This factor is a differentiating cytokine that drives cells toward a predominantly astrocytic fate.

***Fibroblast growth factors***: A family of 18-member proteins grouped in 6 subfamilies that have biological activity during early embryo development regulating positive and negative cellular growth, differentiation and survival and during adulthood, mediating metabolic functions, tissue repair and regeneration.

***Insulin-like growth factor-1***: A polypeptide hormone secreted in multiple tissues by the effect of growth hormone. Most of circulating IGF-I is of hepatic origin and it has been seen to be important in modulating of neurogenesis.

## Key facts of spinal cord injury

In the world, more than a million people have affectations caused by SCI. Currently there are no treatment proposals that can give a full recovery of SCI. Depending on the degree, age, level of the injury, and treatment, the degree of neurological recovery and therefore functional improvement of the patient with SCI can be predicted.

Different therapeutic strategies have been developed to improve the conditions of patients with SCI.

## Key facts therapeutic approaches

Among the strategies is rehabilitation, which include physical exercise, electrical stimulation, and cell transplantation. These strategies have in common that among their nerve regeneration mechanisms is the induction of trophic factors.

The trophic factor acts on neurons and glia, they are called neurotrophic factors and its actions include support the growth, survival, neurotransmitters release and differentiation in several stages of development, in maturity or in regenerative processes.

Another possible strategy for recovery from spinal cord injury is the administration of neurotrophic factors or molecules that have neurotrophic properties.

Either in preclinical protocols or in clinical trials, different trophic factors have been tested, including nerve growth factor, brain-derived neurotrophic factor, neurotrophin 3 and 4/5, glial-derived neurotrophic factor, fibroblast growth factors, ciliary neurotrophic factor, insulin-like growth factor-1, gonadotropin-releasing hormone, thyrotropin-releasing hormone, growth hormone, melatonin, and estrogens.

The results obtained after carrying out any of the strategies, be it rehabilitation or the administration of trophic factors, have been improvements in motor activity, sensory, and autonomic function with the perspective of giving a possible better quality of life for patients with SCI.

## Summary points

- Spinal cord injury in patients causes motor, sensory and autonomic dysfunction.
- There are therapeutic approaches to induce neurological recovery in different experimental models or in clinical trials.
- Rehabilitation induces neurological recovery through the release of neurotrophic factors.

- Therapeutic treatment with directly administered trophic factors induces neurological recovery in SCI.
- There are preclinical trials that have been performed in experimental models using neurotrophic factors and that have not yet been tested in patients with SCI.

## References


Afshari, F. T., Kappagantula, S., & Fawcett, J. H. (2009). Extrinsic and intrinsic factors controlling axonal regeneration after spinal cord injury. *Expert Reviews in Molecular Medicine*, *8*(11), e37.

Allahdadi, K. J., de Santana, T. A., Santos, G. C., et al. (2019). IGF-1 overexpression improves mesenchymal stem cell survival and promotes neurological recovery after spinal cord injury. *Stem Cell Research & Therapy*, *10*, 146.

Almeida, R. D., Manadas, B. J., Melo, et al. (2005). Neuroprotection by BDNF against glutamate-induced apoptotic cell death is mediated by ERK and PI3-kinase pathways. *Cell Death and Differentiation*, *12*, 1329–1343.

Ansorena, E., De Berdt, P., Ucakar, B., et al. (2013). Injectable alginate hydrogel loaded with GDNF promotes functional recovery in a hemisection model of spinal cord injury. *International Journal of Pharmaceutics*, *455*(1–2), 148–158.

Arámburo, C., Alba-Betancourt, C., Luna, M., & Harvey, S. (2014). Expression and function of growth hormone in the nervous system: A brief review. *General and Comparative Endocrinology*, *203*, 35–42.

Aygün, D., Kaplan, S., Odaci, E., Onger, M. E., & Altunkaynak, M. E. (2012). Toxicity of non-steroidal anti-inflammatory drugs: A review of melatonin and diclofenac sodium association. *Histology and Histopathology*, *27*, 417–436.

Barrera, C. M., Kastin, A. J., Fasold, M. B., & Banks, W. A. (1991). Bidirectional saturable transport of LHRH across the blood-brain barrier. *American Journal of Physiology. Endocrinology and Metabolism*, *261*(3), E312–E318.

Blesch, A., Fischer, I., & Tuszynski, M. H. (2012). Gene therapy, neurotrophic factors and spinal cord regeneration. *Handbook of Clinical Neurology*, *109*, 563–574.

Cobianchi, S., Arbat-Plana, A., Lopez-Alvarez, V. M., & Navarro, X. (2017). Neuroprotective effects of exercise treatments after injury: The dual role of neurotrophic factors. *Current Neuropharmacology*, *15*(4), 495–518.

Coull, J. A. M., Beggs, S., Boudreau, D., et al. (2005). BDNF from microglia causes the shift in neuronal anion gradient underlying neuropathic pain. *Nature*, *438*(7070), 1017–1021.

Dasari, V. R., Veeravalli, K. K., & Dinh, D. Z. (2014). Mesenchymal stem cells in the treatment of spinal cord injuries: A review. *World Journal of Stem Cells*, *6*(2), 120.

Devesa, J., Alonso, A., López, N., et al. (2017). Growth hormone (GH) and rehabilitation promoted distal innervation in a child affected by caudal regression syndrome. *International Journal of Molecular Sciences*, *18*(1), 230.

Díaz-Galindo, C., Calderón-Vallejo, D., Hernández-Jasso, I., et al. (2020). Gonadotropin-releasing hormone receptor expression in human spinal cord. *Neurochemical Research*. https://doi.org/10.1007/s11064-020-03178-w (Online ahead of print).

Engele, J., Schubert, D., & Bohn, M. C. (1991). Conditioned media derived from glial cell lines promote survival and differentiation of dopaminergic neurons in vitro: Role of mesencephalic glia. *Journal of Neuroscience Research*, *30*(2), 359–371.

Erşahin, M., Özdemir, Z., Özsavcı, D., et al. (2012). Melatonin treatment protects against spinal cord injury induced functional and biochemical changes in rat urinary bladder. *Journal of Pineal Research*, *52*(3), 340–348.

Faden, A. I., Vink, R., & McIntosh, T. K. (1989). Thyrotropin-releasing hormone and central nervous system trauma. *Annals of the New York Academy of Sciences*, *553*, 380–384.

Fakhoury, M. (2015). Spinal cord injury: Overview of experimental approaches used to restore locomotor activity. *Reviews in the Neurosciences*, *26*(4), 397–405.

Fang, H., Liu, C., Yang, et al. (2017). Neurotrophic factor and Trk signaling mechanisms underlying the promotion of motor recovery after acute spinal cord injury in rats. *Experimental and Therapeutic Medicine*, *14*(1), 652–656.

Fawcett, J. W., Curt, A., Steeves, J. D., et al. (2006). Guidelines for the conduct of clinical trials for spinal cord injury as developed by the ICCP panel: Spontaneous recovery after spinal cord injury and statistical power needed for therapeutic clinical trials. *Spinal Cord*, *45*, 190–205.

Fröhlich, E., & Wahl, R. (2019). The forgotten effects of thyrotropin-releasing hormone: Metabolic functions and medical applications. *Frontiers in Neuroendocrinology*, *52*, 29–43.

Fu, J., Wang, H., Deng, L., & Li, J. (2016). Exercise training promotes functional recovery after spinal cord injury. *Neural Plasticity*, *2016*, 1–7.

Gold, S. M., Schulz, K.-H., Hartmann, S., et al. (2003). Basal serum levels and reactivity of nerve growth factor and brain-derived neurotrophic factor to standardized acute exercise in multiple sclerosis and controls. *Journal of Neuroimmunology*, *138*(1–2), 99–105.

Griffin, J. M., & Bradke, F. (2020). Therapeutic repair for spinal cord injury: Combinatory approaches to address a multifaceted problem. *EMBO Molecular Medicine*, *12*(3), e11505.

Harvey, A. R., Lovett, S. J., Majda, B. T., Yoon, J. H., Wheeler, L. P., & Hodgetts, S. I. (2015). Neurotrophic factors for spinal cord repair: Which, where, how and when to apply, and for what period of time? *Brain Research*, *1619*, 36–71.

Hogan, M. K., Hamilton, G. F., & Horner, P. J. (2020). Neural stimulation and molecular mechanisms of plasticity and regeneration: A review. *Frontiers in Cellular Neuroscience*, *14*, 271.

Jin, M. C., Medress, Z. A., Azad, T. D., Doulames, V. M., & Veeravagu, A. (2019). Stem cell therapies for acute spinal cord injury in humans: A review. *Neurosurgical Focus*, *46*(3), E10.

Kachadroka, S., Hall, A. M., Niedzielko, T. L., Chongthammakun, S., & Floyd, C. L. (2010). Effect of endogenous androgens on 17beta-estradiol-mediated protection after spinal cord injury in male rats. *Journal of Neurotrauma*, *27*(3), 611–626.

Kathe, C., Hutson, T. H., McMahon, S. B., & Moon, L. D. F. (2016). Intramuscular neurotrophin-3 normalizes low threshold spinal reflexes, reduces spasms and improves mobility after bilateral corticospinal tract injury in rats. *eLife*, *5*, e18146.

Keefe, K. M., Sheikh, I. S., & Smith, G. M. (2017). Targeting neurotrophins to specific populations of neurons: NGF, BDNF, and NT-3 and their relevance for treatment of spinal cord injury. *International Journal of Molecular Sciences*, *18*(3), 548.

Kohut, M. L., McCann, D. A., Russell, D. W., Konopka, D. N., Cunnick, J. E., Franke, W. D., et al. (2006). Aerobic exercise, but not flexibility/resistance exercise, reduces serum IL-18, CRP, and IL-6 independent of β-blockers, BMI, and psychosocial factors in older adults. *Brain, Behavior, and Immunity*, *20*(3), 201–209.

Leech, K. A., & Hornby, T. G. (2017). High-intensity locomotor exercise increases brain-derived neurotrophic factor in individuals with incomplete spinal cord injury. *Journal of Neurotrauma*, *34*(6), 1240–1248.

Martínez-Moreno, C. G., Calderón-Vallejo, D., Harvey, S., Arámburo, C., & Quintanar, J. L. (2018). Growth hormone (GH) and gonadotropin-releasing hormone (GnRH) in the central nervous system: A potential neurological combinatory therapy? *International Journal of Molecular Sciences*, *19* (2), 375.

McArthur, J. C., Yiannoutsos, C., Simpson, D. M., et al. (2000). A phase II trial of nerve growth factor for sensory neuropathy associated with HIV infection. AIDS Clinical Trials Group team 291. *Neurology*, *54*, 1080–1088.

Milosevic, M., Marquez-Chin, C., Masani, K., Hirata, M., Nomura, T., Popovic, M. R., et al. (2020). Why brain-controlled neuroprosthetics matter: Mechanisms underlying electrical stimulation of muscles and nerves in rehabilitation. *Biomedical Engineering Online*, *19*(1), 81.

Mocchetti, I., & Wrathall, J. R. (1995). Neurotrophic factors in central nervous system trauma. *Journal of Neurotrauma*, *12*, 853–870.

Ong, W., Pinese, C., & Chew, S. Y. (2019). Scaffold-mediated sequential drug/gene delivery to promote nerve regeneration and remyelination following traumatic nerve injuries. *Advanced Drug Delivery Reviews*, *149–150*, 19–48.

Onifer, S. M., Smith, G. M., & Fouad, K. (2011). Plasticity after spinal cord injury: Relevance to recovery and approaches to facilitate it. *Neurotherapeutics*, *8*, 283–293.

Pitts, L. H., Ross, A., Chase, G. A., & Faden, A. I. (1995). Treatment with thyrotropin-releasing hormone (TRH) in patients with traumatic spinal cord injuries. *Journal of Neurotrauma*, *12*(3), 235–243.

Quintanar, J. L., Díaz-Galindo, C., Calderón-Vallejo, D., et al. (2018). Neurological improvement in patients with chronic spinal cord injury treated with leuprolide acetate, an agonist of GnRH. *Acta Neurobiologiae Experimentalis*, *78*(4), 352–357.

Rayegani, S. M., Shojaee, H., Sedighipour, L., Soroush, M. R., Baghbani, M., & Amirani, O. B. (2011). The effect of electrical passive cycling on spasticity in war veterans with spinal cord injury. *Frontiers in Neurology*, *2*, 39.

Rojas Vega, S., Abel, T., Lindschulten, R., Hollmann, W., Bloch, W., & Strüder, H. K. (2008). Impact of exercise on neuroplasticity-related proteins in spinal cord injured humans. *Neuroscience*, *153*(4), 1064–1070.

Romero, M. I., Rangappa, N., Li, L., Lightfoot, E., Garry, M. G., & Smith, G. M. (2000). Extensive sprouting of sensory afferents and hyperalgesia induced by conditional expression of nerve growth factor in the adult spinal cord. *Journal of Neuroscience*, *20*, 4435–4445.

Rosenfeld, R. D., Zeni, L., Haniu, M., Talvenheimo, J., Bennett, L., Miller, J. A., et al. (1995). Purification and identification of brain-derived neurotrophic factor from human serum. *Protein Expression and Purification*, *6*, 465–471.

Rosich, K., Hanna, B. F., Ibrahim, R. K., Hellenbrand, D. J., & Hanna, A. (2017). The effects of glial cell line-derived neurotrophic factor after spinal cord injury. *Journal of Neurotrauma*, *34*, 3311–3325.

Samantaray, S., Sribnick, E. A., Das, A., et al. (2010). Neuroprotective efficacy of estrogen in experimental spinal cord injury in rats. *Annals of the New York Academy of Sciences*, *1199*, 90–94.

Sharma, H. S. (2007). A select combination of neurotrophins enhances neuroprotection and functional recovery following spinal cord injury. *Annals of the New York Academy of Sciences*, *1122*, 95–111.

Silvestro, S., Bramanti, P., Trubiani, O., & Mazzon, E. (2020). Stem cells therapy for spinal cord injury: An overview of clinical trials. *International Journal of Molecular Sciences*, *21*(2), 659.

Tuszynski, M. H., Gabriel, K., Gage, F. H., Suhr, S., Meyer, S., & Rosetti, A. (1996). Nerve growth factor delivery by gene transfer induces differential outgrowth of sensory, motor, and noradrenergic neurites after adult spinal cord injury. *Experimental Neurology*, *137*, 157–713.

Veneruso, V., et al. (2019). Stem cell paracrine effect and delivery strategies for spinal cord injury regeneration. *Journal of Controlled Release*, *300*, 141–153.

Verge, V. M., Richardson, P. M., Benoit, R., & Riopelle, R. J. (1989). Histochemical characterization of sensory neurons with high-affinity receptors for nerve growth factor. *Journal of Neurocytology*, *18*, 583–591.

Walker, M. J., & Xu, X. M. (2018). History of glial cell line-derived neurotrophic factor (GDNF) and its use for spinal cord injury repair. *Brain Sciences*, *8* (6), 109.

Waters, R. L., Adkins, R. H., Sie, I. H., & Yakura, J. S. (1996). Motor recovery following spinal cord injury associated with cervical spondylosis: A collaborative study. *Spinal Cord*, *34*, 711.

Wu, J. C., Huang, W. C., Chen, Y. C., et al. (2011). Acidic fibroblast growth factor for repair of human spinal cord injury: A clinical trial. *Journal of Neurosurgery: Spine*, *15*(3), 216–227.

Xu, D., Wu, D., Qin, M., et al. (2019). Efficient delivery of nerve growth factors to the central nervous system for neural regeneration. *Advanced Materials*, *31*(e), 1900727.

Yang, B., Zhang, F., Cheng, F., Ying, L., Wang, C., Shi, K., et al. (2020). Strategies and prospects of effective neural circuits reconstruction after spinal cord injury. *Cell Death & Disease*, *11*(6), 439.

Zeller, S., Abel, T., Rojas-Vega, S., Foitschik, T., & Strueder, H. K. (2015). Brain-derived neurotrophic factor concentrations in tetraplegic athletes. *Spinal Cord*, *53*(11), 791–794.

Zhou, Y., Wang, Z., Li, J., Li, X., & Xiao, J. (2018). Fibroblast growth factors in the management of spinal cord injury. *Journal of Cellular and Molecular Medicine*, *22*(1), 25–37.

Section E

# Rehabilitation in spinal injury

Chapter 38

# Spinal cord injury: Multiple family group (MFG) education and support

Dennis G. Dyck[a], Crystal Lederhos Smith[b], and Bruce E. Becker[c]

[a]*Psychology Department, Washington State University Spokane, Spokane, WA, United States;* [b]*Elson S. Floyd College of Medicine, Washington State University, Spokane, WA, United States;* [c]*University of Washington School of Medicine, Seattle, WA, United States*

## Introduction

Damage to the spinal cord can be a life-changing event, usually occurring suddenly as a result of an accident, and presenting both acute and long-term medical effects to individuals and their families. Because these events usually happen unexpectedly, suddenly the individual is thrown into the chaos of the health care system. Rarely do people have any idea of the implications on future health, lifestyle, or economic consequences of this kind of event. The impact on the affected individual is profound, producing both paralysis and medical complications to many body systems. Medical prognostication is often uncertain, and many myths surround spinal cord injury (SCI; sciprogress.com, 2020).

## Epidemiology of spinal cord injury

Over the past 50 years, the age at time of acute spinal injury has increased by a decade or more, but still, the event affects individuals in the prime of their lives, currently averaging 43 years of age (National Spinal Cord Injury Statistical Center, 2020a). Nearly 80% of all SCI individuals are male. As a result of the age at onset, consequent effects have major socioeconomic impacts on work-lives, family income, and life roles. Despite many improvements in motor vehicles, motor vehicle accidents represent approximately 40% of all SCI cases, with falls accounting for another 30%, gunshots 13.5%, and sports-related causes about 8% (National Spinal Cord Injury Statistical Center, 2020a). Medical issues such as transverse myelitis and intraspinal hemorrhages account for the remainder. Regardless of cause, the effect is hugely traumatizing.

There are approximately 17,700 new cases of SCI each year, representing an annual incidence of 54 cases per 1 million population. Because the pathology is often lasting, it is estimated that around 300,000 persons with SCI are currently living in the United States (National Spinal Cord Injury Statistical Center, 2020b). SCI related disability may affect the cervical, thoracic, lumber and sacral portions of the spinal cord structures, with higher level injuries producing more disability (National Spinal Cord Injury Statistical Center, 2020b). The physical impacts on the spinal cord may produce an incomplete injury with some functions retained or may result in a near-complete loss of function distally. Nearly 60% of all SCIs are at the cervical level producing tetraplegia, and 2/3rds of those are incomplete. Of the 40% remaining injuries producing paraplegia, half are incomplete (National Spinal Cord Injury Statistical Center, 2020b). Less than 1% of affected individuals experience complete recovery by the time of hospital discharge (National Spinal Cord Injury Statistical Center, 2020b).

## Life impact and medical support

Everyday activities, from prior independence of basic life activities through vocational function undergo massive change. SCI greatly affects a person's level of independence in mobility, self-care and economic well-being. A little over 30% of affected individuals are married at the time of injury, with 60% single or divorced at injury (National Spinal Cord Injury Statistical Center, 2020b). While nearly 60% are employed at the time of injury, by the end of year one post-injury only about 12% are (National Spinal Cord Injury Statistical Center, 2020b). As a consequence, reliance upon social support systems greatly increases. The level of such support systems varies greatly across the United States, rarely covering all the injured person's needs, especially immediately following injury.

Diagnosis and Treatment of Spinal Cord Injury. https://doi.org/10.1016/B978-0-12-822498-4.00038-5

Medical care for acute spinal cord injury has changed significantly over the years, with improvements in surgical and medical management resulting in improved survival. But as the physiologic impacts of SCI are usually long-term, this presents a need for comprehensive rehabilitation in order to strengthen the areas affected and instruct the injured person in self-management skills including dressing, hygiene, toileting and skin care, in addition to teaching mobility skills such as wheelchair use and other supported mobility needs. During the 1970s, most SCI-affected individuals spent months (average 142 days) in inpatient rehabilitation facilities learning these skills, gaining competence in self-management, and educating families and immediate caregivers on appropriate ways of providing needed support. By discharge, these skills had been practiced, needs anticipated and support structures provided. During those years, the average length of stay in an inpatient rehabilitation facility for a cervical cord injury was around 5 months and during this period the individual often had opportunities for weekend passes to practice skills in their home environment. Persons with paraplegia stayed for 2 months or longer, with similar opportunities to practice home skills. During those years, the rehabilitation inpatient staff team frequently included physical therapists, occupational therapists, social workers, recreation therapists, vocational counselors and psychologists to teach, guide and support individuals with SCI in the areas affected by their injury. Health care systems have changed dramatically since then, with the overall average length of stay currently just over 1 month (National Spinal Cord Injury Statistical Center, 2020b). Rehabilitation staffs have shrunk as well, with only basic therapy needs provided in many American inpatient rehabilitation facilities. Specialty SCI rehabilitation hospitals may have much larger staffs and longer lengths of stay, but funding for such care is not available for many acute SCI patients. As a consequence of these systematic changes in care provision, there are larger burdens placed upon caregivers and patients.

The dramatic compression of lengths of stay result in a much busier hospital day, especially given that nearly 40% of traumatic SCI patients have other injuries in addition to their spinal cord and 80% will have undergone spinal surgeries (National Spinal Cord Injury Statistical Center, 2020b). There are common co-morbidities including traumatic brain injury and thoracic and abdominal trauma. Focus is on physical recovery, wound healing, pain management and other medical needs, as well as rehabilitation therapies and education. Individuals with significant pain issues have consequent learning difficulties that interfere with the teaching process. The burden of healing, coping, adaptation, and multi-tasking overwhelms even emotionally stable individuals. The cliché of drinking from a firehose is not an exaggeration. The inpatient stay period is intensely stressful for all, patients, spouses, family, and caregivers.

## Spinal cord injury after-effects

There are common medical changes resultant from the injury. Bladder management poses issues for about 75% of individuals with SCI, with a variety of urologic adaptations needed, including intermittent catheterization, indwelling catheters, or some form of urinary diversion (National Spinal Cord Injury Statistical Center, 2020b). Medical complications of these alternative solutions may include subsequent urinary tract infections (National Spinal Cord Injury Statistical Center, 2020b). Bowel management may also pose issues, requiring modifications of typical routines, medications and physical techniques which may necessitate caregiver aid. Education regarding skin care is an essential part of the inpatient training process, but follow-through is critical post-discharge and this too may pose caregiver burden, especially if pressure sores occur (National Spinal Cord Injury Statistical Center, 2020b). The need for family and caregiver education is critical to support the person with an acute SCI both medically and with mobility needs.

The preponderance of SCI survivors (87.4%) are discharged to a private home (National Spinal Cord Injury Statistical Center, 2020b). Architectural accessibility of the home to mobility limitations more often than not presents significant problems, especially for those with higher level injuries. Thus, architectural barriers are routinely problematic, as 60% of individuals with SCI rely on a wheelchair for mobility, including both manual and powered chairs (National Spinal Cord Injury Statistical Center, 2020b). In many homes, bedrooms are upstairs, bathrooms are small and inaccessible to wheelchairs and entry and exit present their own challenges. Steps and stairs, narrow doorways, cupboards, and toilet areas cause difficulties for most at discharge. Only rarely is funding for architectural modification available. Transportation poses a common problem, as many vehicles are difficult to enter and exit for someone in a wheelchair and public transportation is often unavailable, especially for rural residents.

The changes in spinal cord injury care over the past decades are many, with some good news in reduction of medical complications, more sophisticated spinal surgery resulting in enhanced neuro recovery and better understanding of medical needs. But there is also less positive news in terms of the reduction in lengths of stay resulting in less opportunity for patient and family education, less medical oversight in the months immediately following the injury and less emphasis on community re-entry education. Community support systems are insufficient. As a consequence of these latter issues, the Multi-family Group Education (MFG) process poses a real opportunity in the long-term approach to the many problems posed by SCI.

## Impacts of SCI on family caregivers

There is nearly always a need for caregiver support of people with SCI. For married individuals, this usually falls to the spouse, placing both physical and substantial emotional demands. For unmarried persons, often this role falls to the family and obviously such support varies greatly in time, interest and ability. The caregiving position encompasses more than physical assistance and often includes educating, advising, financially supporting, advocating for and becoming the prevention / management specialist for the injured family member (Guilcher et al., 2012). Family caregivers are thrust into learning much of this on the fly and often don't know how to accomplish all the necessary tasks. As a result, they often become as, or more stressed and overwhelmed as the person with SCI by the added responsibility and adjustments needed to their daily lives. They can experience frustration, isolation and feelings of guilt, as well as resentment toward the injured family member they care for (Kester, Rothblum, Lobato, & Milhous, 1988). Qualitative research on the experience of family caregivers for people with SCI indicates that caregivers experience deterioration in their relationship with the person who has SCI and that they require intentional re-building and maintaining of their relationships ( Jeyathevan, Cameron, Craven, Munce, & Jaglal, 2019). Spouses who become caretakers of an SCI injured partner have been found to have significantly more depressive affect than their spouse with SCI and have significantly more physical and emotional stress, burnout, fatigue, anger and resentment, depressive affect and somatic depression than spouses who are not caregivers (Weitzenkamp, Gerhart, Charlifue, Whiteneck, & Savic, 1997). Negative psychological and physical health impacts on caregivers have been indicated across countries and cultures (Ebrahimzadeh et al., 2013).

In an examination of the unique contribution of social support to SCI caregiver burden researchers have found that social support, social interactions, and social integration are significant predictors of caregiver burden, suggesting that interventions such as MFG that promote social support, interaction, and integration hold the potential to decrease caregiver burden (Rodakowski, Skidmore, Rogers, & Schulz, 2012). There is a pressing need for individualized education, support and coaching not only for individuals with SCI, but also for their caregivers, particularly when those caregivers are family members (North, 1999). This need begins immediately and should be in place upon discharge from the hospital.

## Impact of multiple family groups (MFG) on psychiatric outcomes

Chronic health conditions require multiple adjustments for both those afflicted and their family caregivers. They are invariably extremely isolating conditions that require coping skills both by patients and their family caregivers. There is now accumulated evidence over a number of decades which supports the idea that, for a number of chronic conditions, a combination of education, support and problem-solving services improves rehabilitative outcomes, community supports and in some cases reduces caregiver burden (Glynn, Dixon, Cohen, & Murray-Swank, 2008; Mueser, 2003; Pilling et al., 2002). Chronic conditions such as severe mental disorders (Dyck, Hendryx, Short, Voss, & McFarlane, 2002; Hazel et al., 2004; McFarlane, 2002), brain injury (Perlick et al., 2013; Rodgers et al., 2007; Straits-Troster et al., 2013), mild cognitive impairment (Schmitter-Edgecombe & Dyck, 2014) and chronic pain (Smith et al., 2017) have all been studied. A pioneering approach which served as a model for our intervention for SCI was developed by McFarlane in the late 70s for the management of schizophrenia (McFarlane, 2002). The intervention called multiple family group (MFG), provides education, support and problem-solving activities to six to eight persons with schizophrenia and their respective family members. The treatment process is a structured interactive format consisting of two mental health clinicians to assist families and patients to improve coping, social support and illness management skills. McFarlane initially enrolled in-patients with schizophrenia and their family caregivers in a 2-year long intervention. The intervention focused initially on preventing relapse and then gradually shifted focus to rehabilitative and occupational goals with community supports. MFG prevented relapse and re-hospitalization and improved rehabilitation outcomes (cf. McFarlane, 2002), relative to single family groups, and MFG became an evidence based best practice in mental health. In the 1990s, psychiatric hospitalization stays became increasingly shorter and more infrequent as treatment services for persons with severe mental illness shifted to the community. We therefore adapted the MFG intervention to a community mental health care setting (Dyck et al., 2000). We enrolled persons with schizophrenia receiving outpatient services and their family members into a two arm randomized clinical trial comparing a 2-year MFG intervention to Standard Care and demonstrated that relative to standard care, relapse, negative symptoms and hospitalization care and costs were all reduced (Dyck et al., 2000; McDonell, Short, Hazel, Berry, & Dyck, 2006). We also found reductions in family caregiver burden and distress (Hazel et al., 2004).

## Adapting multiple family group treatment for brain and spinal cord injury

Approximately 15 years ago Becker, Dyck and colleagues sought to adapt MFG for persons with traumatic brain injury (TBI) and SCI. Below we summarize that treatment development process and the quantitative and qualitative outcome evidence supporting these efforts (Rodgers et al., 2007). We reasoned that although the medical conditions and social and behavioral features of schizophrenia are unique, there are also common themes such as chronicity, the need for illness management skills and community supports and extensive impacts on the family support system. It is also the case that these conditions occur most often in younger age groups, resulting in significant functional disability and restriction of vocational options. These conditions also share the increased risk of social isolation and altered family roles to support the affected individuals. We therefore initiated a field development project to adapted MFG for TBI and SCI. The MFG protocol followed the original model (McFarlane, 1994). The stages included (1) joining with patients and the families through several individual sessions, (2) conducting an educational workshop, and (3) teaching management skills for physical and cognitive challenges as well as improving psychosocial functioning and supports through the use of structured problem-solving group sessions. The group sessions lasted approximately 75–90 min and each consisted of: socialization, go-around, problem identification, and problem solving. We retained the structure and format of the original model but modified the content to accommodate the unique issues of TBI and SCI. For example, during the joining sessions we gathered information on the injury event, its impacts on the patient and the family, how and how successfully they were coping with the adjustments and setbacks. The educational workshop covered neuroanatomy basics, aspects of physical rehabilitation and the support and coping responses to the injury needed by patients and family members. In the problem-solving session, the go-arounds often focused on medical complications, equipment concerns and needs, challenges to travel and caregiver issues. The complete intervention lasted between 12 and 18 months and included initially bi-monthly and then monthly meetings. Two clinicians led each group. The clinicians were experienced physical, speech, and occupational therapists and mental health clinicians (social workers and psychologists). Overall, the results were encouraging. For patients, life satisfaction for patients improved over 18 months (Fig. 1). Patients also reported a decrease in depressive symptoms and anger expression. Among family caregivers, burden was reduced (Fig. 2). Focus groups and qualitative analysis further supported the idea that the MFG experience led to a normalization of the caregiving experience, provided important informational and social support and improved the coping and management skills of both injured persons and their caregivers. Several limitations of this initial effort were that, being a treatment development study, the sample was small, there was no control group and the drop-out rate was approximately 20%. In retrospect, the drop-out rate could be the result of a combination of factors: not enrolling participants early enough after the injury, the length of the intervention, and transportation and related logistic issues.

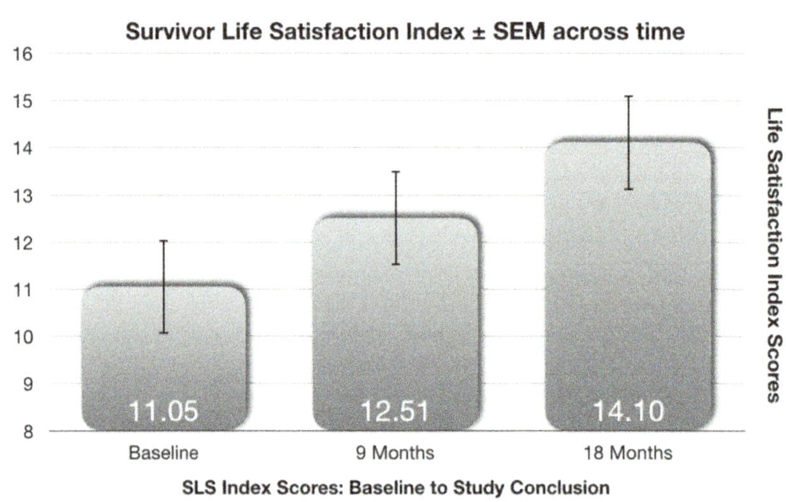


FIG. 1 Mean survivor life satisfaction index scores ± standard error of measure. *Figures adapted from Rodgers, M. L., Strode, A. D., Norell, D. M., Short, R. A., Dyck, D. G., & Becker, B. (2007). Adapting multiple-family group treatment for brain and spinal cord injury intervention development and preliminary outcomes. American Journal of Physical Medicine & Rehabilitation, 86(6), 482–492. https://doi.org/10.1097/Phm.0b013e31805c00a1.*

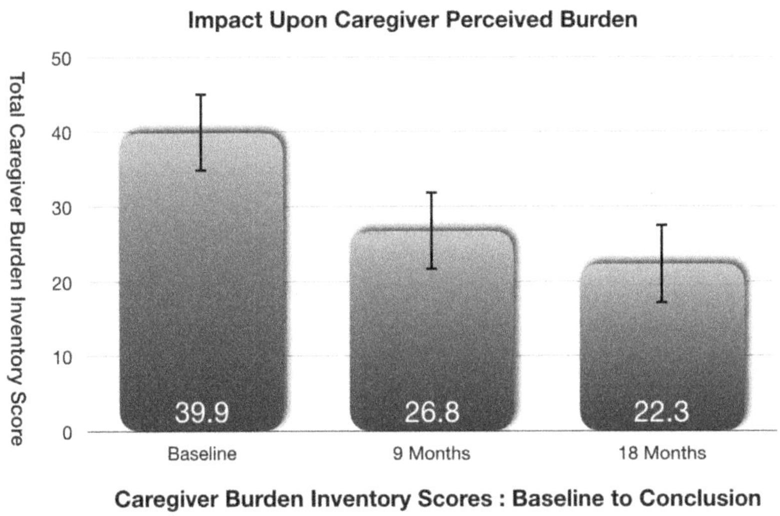


FIG. 2 Mean impact upon caregiver perceived burden ± standard error measurement. *Figures adapted from Rodgers, M. L., Strode, A. D., Norell, D. M., Short, R. A., Dyck, D. G., & Becker, B. (2007). Adapting multiple-family group treatment for brain and spinal cord injury intervention development and preliminary outcomes. American Journal of Physical Medicine & Rehabilitation, 86(6), 482–492. https://doi.org/10.1097/Phm.0b013e31805c00a1.*

## Comparison of MFG to educational control (EC)

We have recently completed a small randomized clinical trial comparing MFG to an active Education Control (Dyck, Weeks, Smith, & Shaw, 2020). In an effort to improve treatment retention, we shortened the MFG group to 9 months and recruited participants within 3 years of their injury. All participants were provided a copy of the 4th Edition of the book titled: Yes you can: a guide to self-care for persons with spinal cord injury (Hammond, 2000). We block randomized participants (SCI outpatients and their caregivers), in groups of three or four dyads, to either MFG or EC. Participants with SCI and their caregivers were assessed both with psychometrically validated quantitative measures at baseline, at the conclusion of the 9-month intervention and 6 months following the interventions. Focus groups were conducted with all participants at the end of the intervention to determine participants' perceptions of the value of the overall experience, how helpful it was and recommendations for improvement. SCI participants were eligible if they had a paraplegic or quadriplegic injury with complete or incomplete lesion; were discharged from inpatient rehabilitation within 3 years, were living in the community in a non-group setting, at least 16 years old, had a mobility impairment and planned to remain in the geographic area for at least 12 months. SCI caregivers were eligible if they were the primary instrumental and/or emotional support person for a spouse, partner, relative, or friend with SCI, spent at least 2 h of face to face contact per week with the participant, lived with or near the participant with SCI, were at least 18 years old, and planned to remain in the geographical area for at least 12 months. Exclusion criteria included terminal illness with life expectancy of less than 12 months, being in active cancer treatment and cognitive impairment. An overview and global comparison of MFG and the EC interventions is provided in Table 1. The primary difference was that MFG was an interactive experience that provided education, support, and individualized problem solving to participants, while the EC condition was a typical didactic classroom experience with limited opportunity for interaction or individual problem solving. The EC condition covered the topics in the "Yes, you can" book. While content for the MFG condition was loosely based on the book, participants had considerable voice in selecting what problems were identified for problem solving in the sessions. The content comparison between the MFG and EC interventions is shown in Table 2. Treatment integrity in both conditions was supported by videotaping, weekly supervision, and treatment manuals. The results generally supported what we observed in the earlier treatment development study.

Quantitative analyses indicated that following treatment, MFG participants with SCI more frequently disagreed with passive coping statements from the Family Crisis Personal Evaluation Scales (F-COPES; McCubbin, Larsen, & Olson, 1985) compared to EC participants. Thus, MFG group participants more frequently rejected passive coping strategies such as "believing if we wait long enough the problem will go away" that are counter-productive beliefs for rehabilitation. It is important to note that not only did MFG participants with SCI reject passive coping strategies, so too did MFG caregivers. We also found that MFG participants with SCI also reported higher levels of subjective and overall social support on the Abbreviated Duke Social Support Index (ADSSI; Koenig et al., 1993). A number of other quantitative measures such as

**TABLE 1** Global comparison of multiple family group (MFG) and education control interventions (EC).

| Treatment component | MFG | EC |
|---|---|---|
| Therapeutic strategy | Skills training, problem solving, support | Information only |
| Contents | SCI effects on the body, maximizing function, coping, living and staying healthy with SCI | SCI effects on the body, maximizing function, coping, living and staying healthy with SCI |
| Target Group | Persons with SCI and caregivers | All persons with SCI and caregivers |
| Use of group dynamics/ cohesion | *Social support promoted*: Entire group participates in problem-solving for each dyad and gives support and encouragement | *Social support minimized*: Individual health issues not discussed, education is general, group interaction minimized |
| Therapeutic stance | Educator stance is collaborative | Educator stance is didactic |
| Room set-up | Round table | Lecture style (all chairs face forward) |
| Source of material | Drawn from everyday problems brought in by group members | Supplied by educator |
| Homework | Assigned and reviewed at the start of the following session | Handouts but not homework provided |

Tables reprinted from Dyck, D. G., Weeks, D. L., Smith, C. L., & Shaw, M. (2020). Multiple family group intervention for spinal cord injury: Quantitative and qualitative comparison with standard education. *The Journal of Spinal Cord Medicine, 44*, 1–11 with publisher permission.

**TABLE 2** Content comparison between multiple family group (MFG) and education control (EC) interventions.

| Treatment component | MFG-SCI | # sessions | SCIEC | # sessions |
|---|---|---|---|---|
| Joining | *Dyad-tailored education*[a]: SWOT analysis, SCI problems identified and corrected Formulation of management problems and coping. Recommend one or more strategies and adjustments (individual and dyad) | 2(3)[b] | *Standard dyad intake*: History of person with SCI and caregiver focusing on current health, skin care, bladder management, bowel management. No skills training, interventions, or formulation of management problems and needed adjustments | 2(3)[b] |
| Group introductory sessions | *Educational workshop*: ASIA classification, clinical syndromes, rehab therapy, medications, health lifestyle, the family and adjustment, family guidelines. Structure and function of multiple family group, how it can help | 2 | *Education introduction*: Structure and rationale for intervention. Rules of conduct. Overview of topics to be covered | 1 |
| Ongoing group sessions | *Problem-solving and skills training sessions*: Problem-solving designed to address specific problems associated with SCI. Compensatory strategies for SCI problems, planning ahead | 12 | *SCIEC education*: General information provided to promote healthy living in areas relevant for persons with SCI and caregivers (bladder/bowel management, nutrition, use of alcohol, drugs, safe exercise) Personal health concerns not discussed; however, discuss referral to provider | 13 |
| Total | | 16 | | 16 |

[a]*The default is two sessions, an optional third session may be used to maintain contact with group members recruited early, or where the dyads are uncertain about continued participation.*
[b]*In addition to basic intake.*

Tables reprinted from Dyck, D. G., Weeks, D. L., Smith, C. L., & Shaw, M. (2020). Multiple family group intervention for spinal cord injury: Quantitative and qualitative comparison with standard education. *The Journal of Spinal Cord Medicine, 44*, 1–11 with publisher permission.

patient activation (Hibbard, Mahoney, Stockard, & Tusler, 2005) and reduced anger expression (Spielberger et al., 1985) approached but did not reach statistical significance, potentially due to a small sample size (Tables 3 and 4).

Qualitative analysis of the focus groups indicated that MFG participants experienced an enhanced sense of belonging and teamwork. They expressed that, the MFG group experience provided them with opportunities to engage with others experiencing similar challenges, which decreased their feelings of isolation. Both participants with SCI and their caregivers requested more and earlier opportunities for this engagement and suggested that the intervention begin while patients with SCI were still inpatients. Participants valued the knowledge and support of the group leaders and spoke at length about their personal ongoing quests for knowledge related to SCI.

Although there were study limitations, such as a small self-selected sample, lack of racial and ethnic diversity and lack of a treatment-as-usual comparison, the results support the earlier treatment development study with a more methodologically rigorous comparison of MFG to an active control intervention.

**TABLE 3 Six-step formal problem-solving process utilized in the multiple family group intervention.**

| Step | Process |
|---|---|
| 1. | Define the problem or goal (MFG members and facilitators) |
| 2. | List all possible solutions (MFG members) |
| 3. | Discuss advantages and disadvantages of each in turn (MFG members and facilitators) |
| 4. | Choose the solution that best fits the situation (MFG members) |
| 5. | Plan how to carry out this solution (facilitators) |
| 6. | Review implementation (facilitators) |

Steps are based on brainstorming methods from organizational and business practices.
Tables reprinted from Dyck, D. G., Weeks, D. L., Smith, C. L., & Shaw, M. (2020). Multiple family group intervention for spinal cord injury: Quantitative and qualitative comparison with standard education. *The Journal of Spinal Cord Medicine, 44,* 1–11 with publisher permission.

**TABLE 4 Means (standard deviations) per group for significant group main effects in the outcome measures ($P < 0.05$).**

| | | Group | | |
|---|---|---|---|---|
| | Analysis | Multiple family group | Education control | Effect size |
| *Participants with SCI* | | | | |
| F-COPES passive appraisal scale[a] | Pooled imputed | 17.7(2.7) | 14.9(3.1) | 0.9 |
| | Complete case | 18.6(2.1) | 15.5(2.2) | 1.4 |
| ADSSI subjective social support scale[b] | Pooled imputed | 19.3(1.5) | 15.8(1.7) | 2.1 |
| | Complete case | 19.4(1.9) | 15.4(2.0) | 2.0 |
| ADSSI total score | Pooled imputed | 27.1(1.8) | 24.0(2.1) | 1.5 |
| | Complete case | 28.1(2.2) | 23.2(2.3) | 2.1 |
| *Caregivers* | | | | |
| F-COPES passive appraisal scale | Pooled imputed | 18.3(2.4) | 13.9(2.7) | 1.6 |
| | Complete case | 18.0(2.5) | 14.7(2.6) | 1.3 |

[a]*F-COPES, family crisis oriented personal evaluation scales. Higher scores represent lower passive appraisal.*
[b]*ADSSI, Abbreviated Duke social support index.*

Tables reprinted from Dyck, D. G., Weeks, D. L., Smith, C. L., & Shaw, M. (2020). Multiple family group intervention for spinal cord injury: Quantitative and qualitative comparison with standard education. *The Journal of Spinal Cord Medicine, 44,* 1–11 with publisher permission.

## Implementation and dissemination opportunities and barriers

While clinicians and researchers consistently request interventions such as MFG to support individuals with SCI (Ebrahimzadeh et al., 2013; Jeyathevan et al., 2019), and the evidence base for MFG SCI and similar group approaches as an important component of outpatient rehabilitation is growing, it is not yet an evidence based best practice. It is, however, important to consider the barriers and opportunities to bridge the chasm between research and practice. In addition to systemic barriers (e.g., cost, effort and reimbursement issues), one of the biggest challenges to recruitment and retention that we have encountered is the high cost of commitment to an intensive long-term intervention with time and travel demands. The time and travel commitments are especially burdensome in the winter for those who live in rural communities impacted by inclement weather. This not only impacts vehicle travel, but often times sidewalks and other paths are not cleared sufficiently for wheelchair use, which makes it even more difficult during an already challenging time. We believe that in this age of telehealth, MFG could be implemented much more readily and likely as effectively through the use of telehealth technology. Choi et al. (2014) recently reported both improved depression and disability outcomes following telehealth delivered (via Skype video call) problem-solving therapy (PST) for depressed, low income homebound older adults. They compared the efficacy of telehealth PST, to in person PST and telephone care calls to homebound older depressed adults. Both telehealth and in person PST were effective treatment, relative to telephone care; however, the effect of telehealth PST on both depression and disability outcomes were sustained longer than those of in-person PST. Although their study was not a group-based intervention for SCI, it is nevertheless a promising example of improving access to care among low income, underserved, home bound adults by providing them with a laptop and wireless card. A systematic review by Gentry, Lapid, Clark, and Rummans (2019) found that videoconference groups were feasible and resulted in similar outcomes to in-person groups, with high participant satisfaction despite some initial technical learning challenges. Additional research was recommended to identify the optimal method of videoconference group delivery to maximize outcomes and benefit.

### Application to other areas of neuroscience

MFG is being adapted to other areas of applied neuroscience. We will summarize two areas that we have had direct involvement in: TBI and mild cognitive impairment (MCI).

TBI: TBI survivors and their families experience many of the same challenges that persons with SCI experience. The injury immediately changes the lives of injured persons and their families, especially relationships with spouses. Our initial field development work (Rodgers et al., 2007) adapted MFG for persons with TBI. Given that in the United States TBI became the signature injury of the Gulf wars, there was a pressing need to provide education and support to Veterans with TBI (often with posttraumatic stress disorder (PTSD) as a co-morbid condition). Taking our initial development work with civilian TBI as a starting point, several investigators (Perlick, Troster, and colleagues) sought our collaboration to adapt MFG for military Veterans with TBI (Straits-Troster et al., 2013). Straits-Troster et al. (2013) evaluated the feasibility, acceptability and helpfulness of MFG for veterans of the wars Iraq and Afghanistan who had sustained a TBI, and their family members. Both veterans and family members reported that MFG helped them explore common struggles, reduced their isolation and helped restore and improve relationships, particularly with their spouses. They also reported a better understanding of the connection between TBI and PTSD. The focus group results were similar to those in our work with SCI and TBI and supported the feasibility, acceptability and need for the MFG support program among veterans with TBI. Perlick et al. (2013) reported that treatment was associated with decreased anger expression and increased social support and occupational activity in veterans, while caregivers reported decreased burden and increased empowerment.

Mild Cognitive Impairment (MCI): Schmitter-Edgecombe and Dyck (2014) integrated the MFG intervention with cognitive rehabilitation (CR) for persons recently diagnosed with MCI and their spousal caregivers. The combined intervention was designed to support the adoption of newly learned cognitive strategies including the use of a memory notebook into everyday life. The combined CR-MFG intervention was associated with improved pre-post group differences on measures of everyday functioning and neuropsychological tests of memory. There was also suggestion that care partners noticed the improvements in the everyday functioning of their spouses with MCI. While preliminary, these results suggest that MFG may produce modest functional benefit for persons with MCI. The study also shows that MFG can be successfully combined with other therapeutic strategies. Thus, MFG would potentially have benefit for other neurological conditions such as stroke and multiple sclerosis.

## Conclusion

Although significant surgical and related medical advances have been made in improving clinical and functioning outcomes in persons with SCI, for most individuals SCI is a life-changing injury. This is true not only for the person with SCI but also for their caregivers and family members. We are encouraged that functional outcomes for persons with SCI and caregiver support systems can be strengthened by providing education and support through multiple family groups. However, there remain a number of future opportunities and challenges in implementing and refining our systems of care for persons with SCI and their families. The use of telehealth to conduct education and support should be explored to help overcome logistic and access challenges faced by persons with SCI. In our experience, the biggest challenges to successful implementation of innovative practices are often systemic and financial. In this regard, while it can be argued that MFG has been shown to be cost effective in mental health settings by reducing relapse and re-hospitalization, this has yet to be documented in rehabilitation settings. We recommend this as the next primary goal for future research.

Given the substantial social, quality of life and medical costs of SCI to affected individuals and their families, it is critical to explore methods to reduce these burdens. These costs are unlikely to diminish in the future through existing technology. The multiple family group education process has been shown to reduce the impact of some of these burdens humanely, safely, and with reasonable efficiency. It deserves front-of-mind consideration in the management of spinal cord injury care.

## Summary points

a. Spinal cord injury is an unexpected event that immediately and permanently changes the lives of those who experience it and the lives of their loved ones.
b. The majority of persons with spinal cord injury are discharged to home under the care of family members after increasingly abbreviated hospitalization.
c. Family caregivers are often as stressed and overwhelmed as persons with spinal cord injury by the adjustments needed. There is a pressing need for individualized education and support for caregivers.
d. Evidence from research on other chronic medical and psychiatric conditions indicates that individualized education, problem solving, and support provided in a group format improves clinical outcomes.
e. Multiple family group is a group intervention that provides clinical care, support and individualized education to persons with spinal cord injury and their caregivers.
f. In persons with SCI, multiple family group was found to decrease depressive symptoms, anger expression, and passive coping and to increase life satisfaction and levels of subjective and overall social support.
g. Multiple family group participants experience an enhanced sense of belonging and teamwork, reduced isolation, increased knowledge and support.
h. Multiple family group intervention has relevant applications to other areas of Neuroscience such as Traumatic Brain Injury and Mild Cognitive Impairment
i. Implementation and dissemination opportunities and challenges, including telehealth and system and financial barriers should be considered for future dissemination of the multiple family group intervention.

## References

Choi, N. G., Hegel, M. T., Marti, C. N., Marinucci, M. L., Sirrianni, L., & Bruce, M. L. (2014). Telehealth problem-solving therapy for depressed low-income homebound older adults. *The American Journal of Geriatric Psychiatry*, *22*(3), 263–271.

Dyck, D. G., Hendryx, M. S., Short, R. A., Voss, W. D., & McFarlane, W. R. (2002). Service use among patients with schizophrenia in nychoeducational multiple-family group treatment. *Psychiatric Services*, *53*(6), 749–754.

Dyck, D. G., Short, R. A., Hendryx, M. S., Norell, D., Myers, M., Patterson, T., et al. (2000). Management of negative symptoms among patients with schizophrenia attending multiple-family groups. *Psychiatric Services*, *51*(4), 513–519.

Dyck, D. G., Weeks, D. L., Smith, C. L., & Shaw, M. (2020). Multiple family group intervention for spinal cord injury: Quantitative and qualitative comparison with standard education. *The Journal of Spinal Cord Medicine*, *44*(4), 572–582.

Ebrahimzadeh, M. H., Shojaei, B.-S., Golhasani-Keshtan, F., Soltani-Moghaddas, S. H., Fattahi, A. S., & Mazloumi, S. M. (2013). Quality of life and the related factors in spouses of veterans with chronic spinal cord injury. *Health and Quality of Life Outcomes*, *11*(1), 48.

Gentry, M. T., Lapid, M. I., Clark, M. M., & Rummans, T. A. (2019). Evidence for telehealth group-based treatment: A systematic review. *Journal of Telemedicine and Telecare*, *25*(6), 327–342.

Glynn, S. M., Dixon, L. B., Cohen, A., & Murray-Swank, A. (2008). *The family member provider outreach program*. Psychiatric Services.

Guilcher, S. J., Casciaro, T., Lemieux-Charles, L., Craven, C., McColl, M. A., & Jaglal, S. B. (2012). Social networks and secondary health conditions: The critical secondary team for individuals with spinal cord injury. *The Journal of Spinal Cord Medicine*, *35*(5), 330–342.

Hammond, M. C. (2000). *Yes, you can!: A guide to self-care for persons with spinal cord injury*. Paralyzed Veterans of America.

Hazel, N., McDonell, M. G., Short, R., Berry, C. M., Voss, W., Rodgers, M., et al. (2004). Impact of multiple-family groups for outpatients with schizophrenia on caregivers distress and resources. *Psychiatric Services*, *55*(1), 35–41.

Hibbard, J. H., Mahoney, E. R., Stockard, J., & Tusler, M. (2005). Development and testing of a short form of the patient activation measure. *Health Services Research*, *40*(6p1), 1918–1930.

Jeyathevan, G., Cameron, J. I., Craven, B. C., Munce, S. E., & Jaglal, S. B. (2019). Re-building relationships after a spinal cord injury: Experiences of family caregivers and care recipients. *BMC Neurology*, *19*(1), 1–13.

Kester, B. L., Rothblum, E. D., Lobato, D., & Milhous, R. L. (1988). Spouse adjustment to spinal cord injury: Long-term medical and psychosocial factors. *Rehabilitation Counseling Bulletin*, *32*, 4–21.

Koenig, H. G., Westlund, R. E., George, L. K., Hughes, D. C., Blazer, D. G., & Hybels, C. (1993). Abbreviating the Duke social support index for use in chronically ill elderly individuals. *Psychosomatics*, *34*(1), 61–69.

McCubbin, H. I., Larsen, A., & Olson, D. (1985). F-COPES: Family crisis oriented personal evaluation scales. In *Family inventories: Inventories used in a national survey of families across the family life cycle* Science Open, Inc (David H. Olson…[et al.]).

McDonell, M. G., Short, R. A., Hazel, N. A., Berry, C. M., & Dyck, D. G. (2006). Multiple-family group treatment of outpatients with schizophrenia: Impact on service utilization. *Family Process*, *45*(3), 359–373.

McFarlane, W. R. (1994). Multiple-family groups and psychoeducation in the treatment of schizophrenia. *New Directions for Mental Health Services*, (62), 13.

McFarlane, W. R. (2002). *Multiple family groups in the treatment of severe psychiatric disorders*. New York, NY: Guilford Press.

Mueser, K. T. (2003). *Integrated treatment for dual disorders: A guide to effective practice*. Guilford Press.

National Spinal Cord Injury Statistical Center. (2020a). *Spinal cord injury facts and figures at a glance*. Retrieved from https://www.nscisc.uab.edu/Public/Facts%20and%20Figures%202020.pdf.

National Spinal Cord Injury Statistical Center. (2020b). *Spinal cord injury model systems 2019 annual report-complete public version*. 2019. Retrieved from https://www.nscisc.uab.edu/public/2019%20Annual%20Report%20-%20Complete%20Public%20Version.pdf.

North, N. (1999). The psychological effects of spinal cord injury: A review. *Spinal Cord*, *37*(10), 671–679.

Perlick, D. A., Straits-Troster, K., Strauss, J. L., Norell, D., Tupler, L. A., Levine, B., et al. (2013). Implementation of multifamily group treatment for veterans with traumatic brain injury. *Psychiatric Services*, *64*(6), 534–540.

Pilling, S., Bebbington, P., Kuipers, E., Garety, P., Geddes, J., Orbach, G., et al. (2002). Psychological treatments in schizophrenia: I. Meta-analysis of family intervention and cognitive behaviour therapy. *Psychological Medicine*, *32*(05), 763–782.

Rodakowski, J., Skidmore, E. R., Rogers, J. C., & Schulz, R. (2012). Role of social support in predicting caregiver burden. *Archives of Physical Medicine and Rehabilitation*, *93*(12), 2229–2236.

Rodgers, M. L., Strode, A. D., Norell, D. M., Short, R. A., Dyck, D. G., & Becker, B. (2007). Adapting multiple-family group treatment for brain and spinal cord injury intervention development and preliminary outcomes. *American Journal of Physical Medicine & Rehabilitation*, *86*(6), 482–492. https://doi.org/10.1097/Phm.0b013e31805c00a1.

Schmitter-Edgecombe, M., & Dyck, D. G. (2014). Cognitive rehabilitation multi-family group intervention for individuals with mild cognitive impairment and their care-partners. *Journal of the International Neuropsychological Society*, *20*(9), 897–908.

sciprogress.com. (2020). *Common myths and misconceptions about spinal cord injury*. Retrieved from https://sciprogress.com/common-myths-misconceptions-spinal-cord-injuries/.

Smith, C. L., Skaer, T. L., Orr, M. F., Barbosa-Leiker, C., Howell, D. N., & Dyck, D. G. (2017). Development of multi-family group and mindfulness treatment for chronic pain: Preliminary outcomes. *Journal of Musculoskeletal Disorders and Treatment*, *3*(4). https://doi.org/10.23937/2572-3243.1510042.

Spielberger, C. D., Johnson, E. H., Russell, S. F., Crane, R. J., Jacobs, G. A., & Worden, T. J. (1985). The experience and expression of anger: Construction and validation of an anger expression scale. In *Anger and hostility in cardiovascular and behavioral disorders* (pp. 5–30). New York: Hemisphere.

Straits-Troster, K., Gierisch, J. M., Strauss, J. L., Dyck, D. G., Dixon, L. B., Norell, D., et al. (2013). Multifamily group treatment for veterans with traumatic brain injury: What is the value to participants? *Psychiatric Services*, *64*(6), 541–546. https://doi.org/10.1176/appi.ps.001632012.

Weitzenkamp, D. A., Gerhart, K. A., Charlifue, S. W., Whiteneck, G. G., & Savic, G. (1997). Spouses of spinal cord injury survivors: The added impact of caregiving. *Archives of Physical Medicine and Rehabilitation*, *78*(8), 822–827.

Chapter 39

# Spinal cord injury rehabilitation: Linking service delivery and community integration

Linda Barclay and Gillean Hilton
*Department of Occupational Therapy, Monash University, Frankston, VIC, Australia*

## Abbreviations

| | |
|---|---|
| **AR** | active rehabilitation |
| **ICF** | International Classification of Functioning Disability and Health |
| **SCI** | spinal cord injury |
| **TR** | transitional rehabilitation |
| **UNCRPD** | United Nations Convention on the Rights of Persons with Disabilities |
| **VR** | vocational rehabilitation |

## Introduction

Community integration includes having access to appropriate housing, being able to mobilize in the community, being able to participate in work, leisure or educational activities of choice, and engaging in satisfying social relationships (Barclay, Lalor, Migliorini, & Robins, 2020; Chang, Coster, & Helfrich, 2013). It is synonymous with the term *participation* in the International Classification of Functioning Disability and Health (ICF) (World Health Organization, 2002), which refers to involvement in life situations, and includes active engagement at the community level and social connectedness with other people (Hammel et al., 2008). Participating in meaningful roles and interests in the community, such as leisure and employment, can enable a person to reframe their views on acquired disability (Kratz, Chadd, Jensen, Kehn, & Kroll, 2015), but poor community integration and lack of social participation is one of the biggest challenges for someone following SCI (Craig, Nicholson Perry, Guest, Tran, & Middleton, 2015). Services and models for SCI rehabilitation need to prioritize community integration in order to maximize outcomes and quality of life.

## Prioritizing community integration

Following SCI a person will usually spend several months undergoing inpatient rehabilitation in a specialized spinal unit (Dwyer & Mulligan, 2015). Following this, the injured person will be discharged to live in either their own home, or some other form of temporary or alternate accommodation. This period of transition from hospital to home is a very challenging time for people, as they are removed from the safety of the hospital environment with the associated supports, both physical and social, and plunged into environments that may be inaccessible and even hostile toward people with disabilities (Nunnerley, Hay-Smith, & Dean, 2013). This process is made more challenging for people in developing countries, where community infrastructure such as roads, and transport is even less accessible for people with disabilities (Liu et al., 2020). The funding environment, policies and infrastructure in the jurisdiction in which the injured person lives, impacts the availability of the resources they have access to, and may facilitate or hinder this transition process. People with SCI have described feeling unprepared for the difficulties and barriers they are going to face in the real world (Nunnerley et al., 2013). Therefore, in the context of shrinking health budgets, maximizing the efficacy of inpatient rehabilitation to prepare people to reintegrate into their community must be one of the primary goals of rehabilitation (Kratz et al., 2015).

**Diagnosis and Treatment of Spinal Cord Injury. https://doi.org/10.1016/B978-0-12-822498-4.00039-7**

## Interplay between health condition, person factors, and environmental factors

As depicted in the ICF (Fig. 1), activity and participation are influenced by the health condition and body functions and structures (in this case the injury and associated levels of neurological impairment and secondary health conditions), in addition to person factors such as age, gender, other health conditions, weight, coping style, education and other life experiences, and environmental factors. Environmental factors are frequently not modifiable (e.g., culture, climate, physical environment such as hills). Some environmental factors occur at a societal level and require policy or government involvement (e.g., availability of rehabilitation services, funding for equipment, public transport, discriminatory attitudes, and access to housing) (Carr et al., 2017), while others occur at a family or local community level (for example, social supports, and private transport), and are potentially modifiable (Barclay, McDonald, Lentin, & Bourke-Taylor, 2016). Environmental barriers are often more challenging in developing countries (Sekaran et al., 2010). Personal and environmental factors have a significant role in facilitating or hindering the participation or community integration of people following SCI.

## Models and approaches to community integration

Most commonly, service models or programs that aim to support the community integration of people with SCI in the immediate period following inpatient rehabilitation are conducted by specialist SCI units (Kendall, Ungerer, & Dorsett, 2003). Such programs typically include a core multidisciplinary team consisting of physiotherapists, occupational therapists, community nurses and social workers (Momsen, Rasmussen, Nielsen, Iversen, & Lund, 2012). Additional staff may include vocational consultants, psychologists, peer mentors, leisure specialists and exercise physiologists. Community integration programs usually address a large range of issues, including community-based wheelchair skills, physical health education, strategies to increase self-efficacy and self-management skills, goal setting, training or managing carers, and working on improving independence in self-care (Barclay, Lalor, et al., 2020).

There are a range of community integration and/or transitional living models or approaches being used by specialist SCI units internationally, and most can be categorized into one of four models (Barclay, Lalor, et al., 2020). Each of these models is outlined in the next section. It is also important to consider goal setting as an integral outcome tool within these models. In a systematic review of the qualitative literature exploring their experience of, and their perspectives on, goal setting in rehabilitation, people with SCI recognized the importance of being self-directed (Maribo, Jensen, Madsen, & Handberg, 2020). Furthermore, they aspired to take an active role, and had a preference for goal setting oriented around their daily life. This was at times at odds with the health professional, who tended to use the hospital as their point of reference (Maribo et al., 2020). A shared process with goal setting relevant to persons with SCI and their everyday life is needed (Maribo et al., 2020) and formally or informally integrated in community integration models (Fig. 2).

### Community integration embedded within inpatient rehabilitation

The first approach to community integration involves preparing people for discharge as part of the inpatient rehabilitation service, that is, there is no separate or specific service for community integration (Fig. 3). This model tends to be in place in countries where length of hospital rehabilitation is longer, and there is less pressure to discharge people quickly. Some services are able to offer patients the opportunity to return home for a short period of time (for example, 1 month) prior

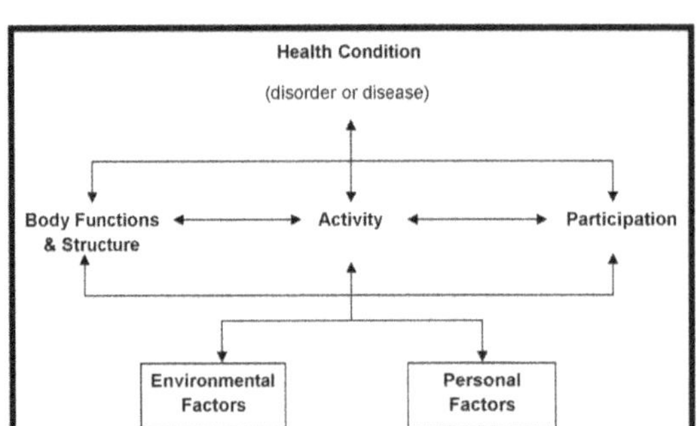


FIG. 1 International classification of functioning disability and health (ICF) (World Health Organization, 2002). Depicts International Classification of Functioning Disability and Health framework. Permission from WHO is not required for the use of WHO materials issued under the Creative Commons Attribution-Non Commercial-ShareAlike 3.0 Intergovernmental Organization (CC BY-NC-SA 3.0 IGO) licence.

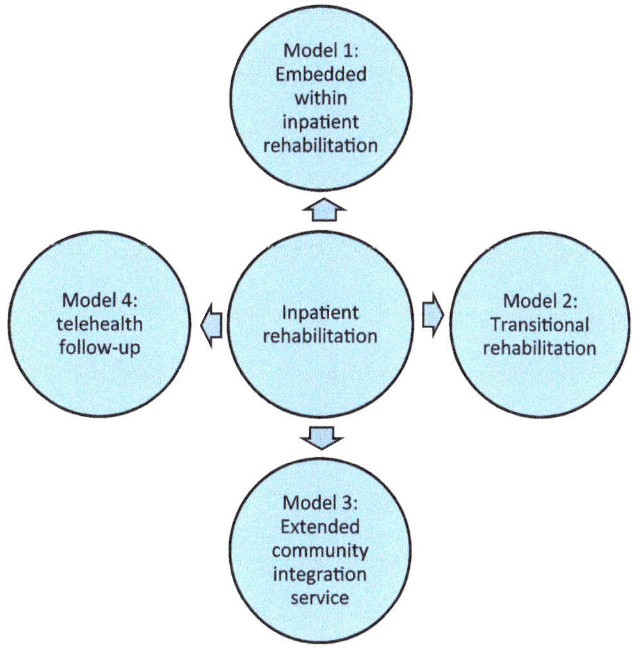


FIG. 2 Models to support community integration. Depicts four models or approaches to community integration.

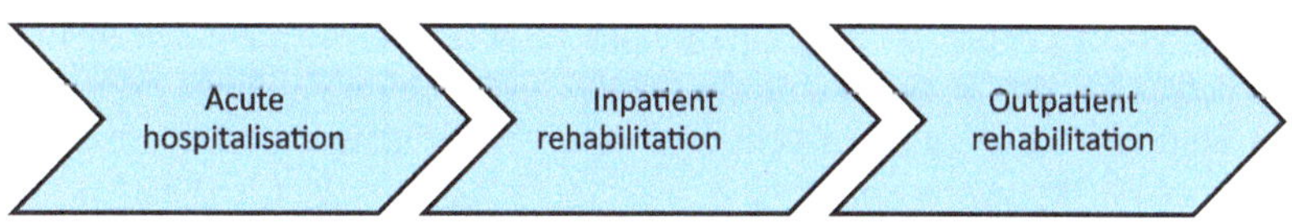


FIG. 3 Community integration embedded in inpatient rehabilitation. Illustrates one model where community integration is embedded within inpatient rehabilitation.

to discharge, and then return to the facility work to on goals that were identified while at home. Other SCI units offer ongoing outpatient therapy on a regular basis for up to 3–4 months, provided the patient lives within driving distance of the SCI unit. Following this, patients are referred onto community-based providers. This approach has advantages and disadvantages. While it allows the injured person and their family time to adjust and prepare for the potential challenges that may be faced when the patient returns home, particularly in relation to organizing assistive technology and an accessible home environment, it is also likely that long periods of hospitalization impact negatively on the patient's self-determination and self-efficacy. The longer the hospital stay, the more likely there are to be entrenched patterns of institutionalization, and the development of greater anxiety when the time comes to return to living in the community (Craig & Nicholson Perry, 2013; Kendall et al., 2003). For example, in hospital, routines are imposed and there are few opportunities to assume responsibility and take an active role in rehabilitation decisions (Craig & Nicholson Perry, 2013). Therefore, the possible benefits of a longer hospital stay need to be weighed against the likely negative impacts.

## Transitional rehabilitation program

A transitional rehabilitation (TR) program typically consists of a short "stay" of between 4 and 8 weeks (Fig. 4). In some instances, the patient lives in a self-contained unit during this period, either on or close to the SCI unit or hospital site. In this model, specific goals for community integration are identified collaboratively between the injured person and the

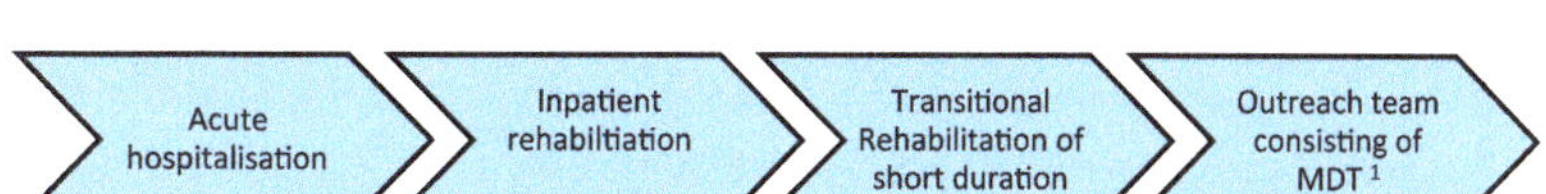


FIG. 4 Transitional Rehabilitation model. [1]MDT = multi-disciplinary team. Depicts the transitional rehabilitation model.

rehabilitation team, using a goal-setting tool. Examples of such tools include Goal Attainment Scaling (GAS) (SRALAB, 2020a), the Needs Assessment Checklist (SRALAB, 2020b), and the Multi-disciplinary Goal Attainment Measure (MGAM) (Kendall & Wallace, 2016). Goals may include, for example, to return to study, to learn to use public transport, or to drive (Kendall & Wallace, 2016). The identified goals are addressed during the 4–8 weeks of the program, are reviewed regularly, and again at the completion of the program. Following completion of the TR program, the patient may either be referred to an SCI outreach service for a short period of time, or referred to community-based services to continue to work on identified goals or set new goals. This approach has the benefit of enabling the patient to address the community integration goals they wish to address, without the negative impacts of institutionalization described above.

### Extended community integration service

A third approach provides patient support for a longer period that TR, and can be for up to 12 months. This enables longer term goals to be addressed, allowing the injured person to make use of the expertise of the specialist SCI for longer, and providing more time to address and resolve issues that come up once they have returned home. While this may be helpful for patients that have had a shorter length of inpatient rehabilitation, it is also possible that this approach fosters a greater dependency on the SCI unit (Kendall et al., 2003). It is important when undertaking this approach that patients are actively engaged in decision-making, thereby minimizing dependency and enhancing the patient's confidence and independence skills in preparation for returning home (Craig et al., 2015).

### Telehealth follow-up

The final approach involves following up people using telehealth (by either telephone or video conferencing), for a fixed period of time (e.g., up to 60 days). While this approach is becoming more common, it could be more widely utilized to facilitate community integration of people with SCI, particularly for health maintenance or management of secondary health conditions, and to build capacity of community-based therapists (Woo, Guihan, Frick, Gill, & Ho, 2011). While there has been some work done assessing the effectiveness of using telehealth specifically for health self-management following SCI (Houlihan et al., 2017; Phillips, Vesmarovich, Hauber, Wiggers, & Egner, 2001), less information is available regarding using tele-based services for capacity building of community providers such as therapists, general practitioners and community nurses. Utilizing telephone follow-up is particularly useful in countries where people live in rural locations or there is poor infrastructure (Hossain et al., 2017).

## Assistive technology, transport, housing

Access to suitably accessible housing, appropriate assistive technology and transport are key in facilitating community integration after SCI (Wright, Colley, & Kendall, 2020).

### Assistive technology

In order to facilitate the process of returning home, there is a variety of assistive technologies required for someone to ensure safety and maximize independence (Barclay et al., 2016). Such technologies range from wheelchairs, pressure cushions, and bathing aids (Friesen, Theodoros, & Russell, 2016 ), through to environmental control systems and Smart Home technology (Hooper, Verdonck, Amsters, Myburg, & Allan, 2018). A detailed discussion of this area is beyond the scope of this chapter; however, it is an important element that needs to be addressed as part of the plan for transitioning the injured person to the community, taking into account funding available and policies within the person's jurisdiction.

### Transport

Having access to transport is a key requirement to enable people with SCI to shop, to use the bank, to get to work and education, and to engage in social and leisure activities in the community (Barclay et al., 2016). *The United Nations Convention on the Rights of Persons with Disabilities (UNCRPD) [Article 9]* (n.d.), specifically refers to the importance of the rights of people with disabilities to be able to access transportation on an equal basis with others (United Nations). Yet, transportation issues are consistently identified by people with disabilities as a major challenge and barrier to community participation. This is even more problematic in low and middle income countries (Kett, Cole, & Turner, 2020).

Common issues raised by people with disabilities, such as SCI, regarding using public transport include concerns about accessibility, safety, reliability, affordability, limited services especially in rural areas, and the attitude of the public (Kett et al., 2020). Community integration programs may address these concerns prior to discharge from hospital, by accompanying their patients into the local community where possible to practice using public transport and problem-solve issues that arise (Price, Stephenson, Krantz, & Ward, 2011).

It is common for people in high income countries to prefer the use of private vehicles, with high rates of return to self-driving being associated with high community integration and better life satisfaction (Norweg, Jette, Houlihan, Ni, & Boninger, 2011). However, for someone with an SCI to be able to self-drive a car, or travel as a passenger in a wheelchair, expensive modifications such as hand controls, ramps and modified vehicles are usually needed. Availability of such modifications will depend on the funding and policy environment of the country in which the person lives, but are also influenced by socioeconomic factors as education, income, and employment status (Norweg et al., 2011). Rehabilitation programs need to have a strong focus on working toward driving as a goal, which may require follow-up once the injured person has returned home and for some time after.

## Housing

Challenges for people with SCI relate to both housing availability and housing suitability (Wright et al., 2020). An important part of the process of preparing someone with an SCI to transition to live back in the community is ensuring they have somewhere suitable to live. Quality of life and successful community integration after SCI have been associated with discharge residence (Bergmark, Winograd, & Koopman, 2008).

After inpatient rehabilitation, there are a number of possible discharge scenarios for someone with SCI. They may be discharged back to their own or to a previous family home, which will likely require modifications, they may move into physically accessible social housing, or they may need to move into some form of supported accommodation or residential care facility (Wright et al., 2020) (Fig. 5). In addition, access to appropriate formal and informal supports are integral to an individual's experience of optimal social participation (Braaf, Lennox, Nunn, & Gabbe, 2018; Murphy, O'Hare, & Wallis, 2010).

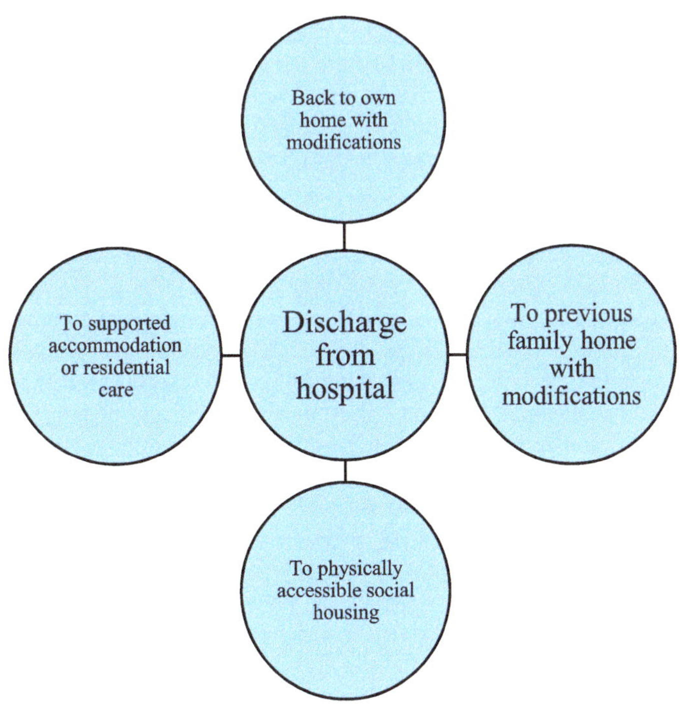


FIG. 5 Discharge destinations. Highlights possible destinations for discharge following inpatient rehabilitation.

## Leisure and social participation

### Participating in leisure assists adjustment following SCI

Following SCI, people experience a disruption to their leisure identity and a changed leisure repertoire (Lundström, Lilja, Petersson, Lexell, & Isaksson, 2014). How people use their time also changes. A large reduction in time spent in paid employment, potentially makes more time available for avocational pursuits (Barclay et al., 2011). With demonstrated links between engagement in leisure and subjective wellbeing (Kuykendall, Tay, & Ng, 2015; Lloyd & Auld, 2002), exploring leisure identity as part of rehabilitation/community integration can assist people to achieve meaningful adaptation to changed circumstances. Engagement in leisure has a direct influence on the adjustment of individuals following SCI, therefore developing skills in managing leisure becomes critical in the process of community reintegration (Loy, Dattilo, & Kleiber, 2003). In addition, for those unable to reintegrate into paid work, participation in non-vocational activities should also be considered a goal of rehabilitation (Schonherr, Groothoff, Mulder, & Eisma, 2005).

### Health benefits of physical leisure participation

Physical leisure activities have an important role in addressing some of the negative health effects associated with lack of physical activity following SCI, including loss of muscle mass, obesity, cardiovascular disease, type-2 diabetes, depression, pain and fatigue (Bombardier et al., 2020). Sports participation has also been associated positively with community integration and quality of life (McVeigh, Hitzig, & Craven, 2009).

## The role of peer mentors in community integration

Peer mentors are typically people who have faced similar experiences (such as SCI), and are therefore well positioned to provide practical, emotional and informational support (Barclay & Hilton, 2019). Peer mentors have a vital contribution in assisting people with SCI to have hope and visualize possibilities for the future (Barclay et al., 2016; Veith, Sherman, Pellino, & Yasui, 2006). Peers often use language that can open people's minds. They help create ability in others because of their own critical reflection, and they do this not as a provider of service—but as a role model (Mead & MacNeil, 2006). These reflections can be extrapolated to hospital and community-based settings for people with SCI.

### Peer mentors assist in transition from hospital to home

Peer mentors can play an important role in supporting people during the transition from hospital to home, and often fill a service gap between formal hospital-based services and community-based services (Barclay & Hilton, 2019). Peers can promote physical activity (Latimer-Cheung et al., 2013), assist with wheelchair skills training (Best, Miller, Huston, Routhier, & Eng, 2016), and provide effective support in managing ongoing health conditions (Houlihan et al., 2017). Specialist SCI units utilizing emerging best practice approaches to rehabilitation and community integration, include people with lived experience as an employed member of the rehabilitation team (Barclay, Robins, Migliorini, & Lalor, 2020).

### Active rehabilitation

Once people with SCI are discharged from inpatient rehabilitation, peer mentors can continue to have a positive influence on development of self-efficacy, through role modelling success and demonstrating how to overcome failures (Divanoglou & Georgiou, 2017). Active Rehabilitation (AR) is a community peer-based approach that typically involves people with SCI attending camps that offer intensive goal-oriented group-based training and peer support activities in a community setting (Divanoglou, Tasiemski, Augutis, & Trok, 2017). Typical activities addressed during AR are wheelchair skills training, activities of daily living, sports and recreational activities, education and setting goals. These are delivered through a combination of SCI peer mentors, and non-disabled assistants. AR offers a unique learning environment through a mixture of resources, activities and a "can-do" attitude, where the peer mentors are viewed as credible role models who have "lived it all" (Divanoglou et al., 2017).

## Return to work

Varied meanings of work for a person following SCI have been described as re-developing a sense of self, re-establishing place in the community and regaining economic self-sufficiency, (Ullah, Fossey, & Stuckey, 2018). Since the development of the first SCI rehabilitation centers, gaining paid employment has been considered an indicator of achieving optimal community integration (Guttmann, 1954) and a measure of rehabilitation success (Britell, 1991; Guttmann, 1954; Post, Reinhardt, & Escorpizo, 2020).

### Benefits of employment

Seeking, gaining and maintaining employment after SCI is valued by individuals and the process contributes to people's sense and strength of identity (Krause, Saunders, & Acuna, 2012; Hilton, Unsworth, Stuckey, & Murphy, 2018). Employment has been described as a source of mental stimulation and an opportunity for personal growth and independence (Chan & Man, 2005), and also as part of one's identity and representative of "living a normal life" (Hay-Smith, Dickson, Nunnerley, & Sinnott, 2013). The health and social benefits of employment, including having a place within the community and economic self-sufficiency, are being increasingly recognized (Holmes et al., 2016; Ullah et al., 2018).

### Vocational rehabilitation

Vocational rehabilitation (VR) to restore or maintain an individual's vocational participation after major injury may be more effective if actioned sooner (Chan & Man, 2005; Fadyl & McPherson, 2010; Ferdiana et al., 2014), and should be a major focus of community integration (Hilton, Unsworth, & Murphy, 2018). Current evidence supports the practice of intervention as early as when the patient is in the inpatient rehabilitation phase (Hilton, Unsworth, Browne, Murphy, & Olver, 2017; Middleton et al., 2015), and can address modifiable factors associated with barriers to employment (Trenaman, Miller, Querée, & Escorpizo, 2015). In addition to vocational rehabilitation, these modifiable factors include education, functional independence, social support and financial disincentives (Trenaman et al., 2015). The community integration models described earlier in this chapter may also provide appropriate settings for delivery of this early intervention vocational rehabilitation (Hilton et al., 2017; Middleton et al., 2015).

Pre and post injury education consistently appears to facilitate post injury employment, (Ferdiana et al., 2014; Hilton, Unsworth, Stuckey, et al., 2018). Similarly, further education for skill development strengthens opportunities for post injury employment (Hilton, Unsworth, Stuckey, et al., 2018). Vocational rehabilitation within the community integration setting can assist to explore further education and training, facilitating employment trajectories (Hilton, Unsworth, Stuckey, et al., 2018).

## Self-management, education and health literacy

### Importance of education in maintaining physical and mental health

An important part of being successfully integrated back into the community after SCI is maintaining physical and mental health, and preventing the occurrence of secondary health conditions that may require hospital readmission. People with SCI are at higher risk of hospital readmission (Gassaway et al., 2017), or significant disruption to their social and employment participation due to health problems such as skin injuries, urinary tract infections, chronic pain, spasm and spasticity, and circulatory problems (Callaway, Barclay, McDonald, Farnworth, & Casey, 2015). Rehabilitation has the role of educating people with SCI and their families as to how to stay physically and mentally healthy after discharge. However, people with SCI report that they often leave hospital feeling unprepared to assume responsibility for their health needs (Gassaway et al., 2017). A variety of approaches used by specialist spinal units to educate patients and their families are utilized. Approaches include using hard copy educational handouts, conducting education sessions in hospital, and developing "How To" videos for people to watch at a convenient time (Barclay, Lalor, et al., 2020). After discharge home, ongoing support of patients and their families in relation to maintaining health is vital. Telephone counselling has the potential to reduce medical complications and health care utilization, and improve psychosocial outcomes in the early stages following discharge (Mackelprang & Salsgiver, 2016).

Whatever education is provided should be presented in a form that is suitable for people with low levels of health literacy. It is also recognized that people with SCI often have issues with fatigue and mild cognitive impairment, and therefore may need information repeated frequently and at different time points to maximize impact (Craig, Guest, Tran, & Middleton, 2017). Improvements in knowledge do not necessarily translate into improvements in problem-solving. Therefore, incorporating more active learning strategies, such as role plays within patient education programs may facilitate better transfer of knowledge within life situations (May, Day, & Warren, 2006).

### Role of peer educators in facilitating self-management

Facilitating self-management of people with SCI, can help them to take a more active role in managing their health care, and potentially minimize or prevent the development of secondary health conditions listed above. Chronic disease self-management programs focus on increasing perceived self-efficacy to solve problems and make decisions to manage day-to-day health (Hirsche, Williams, Jones, & Manns, 2011). People with SCI report that barriers to self-management include funding and funding policies, and lack of physical access to medical clinics and buildings, while facilitators include peer support (Munce et al., 2014). With appropriate training peers can play an active role in education as they are well placed to form empathic relationships, self-reflect and build understanding through practical life examples (Barclay & Hilton, 2019). One example of integration of lived experience in a peer educator role, which demonstrates benefits in self-management, is the peer-led, telephone based, health self-management intervention called 'My Care My Call' for adults with chronic SCI (Houlihan et al., 2017). Such interventions, however, need to be integrated with, rather than separate to, any other SCI-specific services the patient is receiving. In addition, it is important that specific goals are set with the patient and their family in order for interventions to be targeted and effective (Mackelprang & Salsgiver, 2016).

## Summary

This chapter presented an overview of the important process of preparing people with SCI to transition from inpatient rehabilitation to living in the community. The interplay between modifiable and non-modifiable environmental factors, the injury and person factors was highlighted. A range of models and approaches to community integration were presented. Benefits of employment and vocational rehabilitation were described, while for those unable to reintegrate into paid work, the importance of participation in non-vocational activities as a goal of rehabilitation was discussed. The important role of peer mentors in supporting people during the transition from hospital to home, by filling a service gap between formal hospital-based services and community-based services was outlined. Finally, maintenance of physical and mental health, and prevention of secondary health conditions, including ways to facilitate self-management was discussed as a key to maintaining ongoing successful community integration.

## Applications to other areas of neuroscience

In this chapter we have discussed the importance of community integration as a major goal of rehabilitation following spinal cord injury. The importance of the link between service delivery during inpatient rehabilitation and the transition to community based services, and the ways in which environmental factors and person based factors influence community integration outcomes is very relevant to other areas of neurological rehabilitation, such as acquired brain injury, or stroke. Rehabilitation services for people following acquired brain injury and stroke function in similar ways to those for spinal cord injury, particularly in high income countries (Cott, Wiles, & Devitt, 2007; Turner et al., 2007). Inpatient rehabilitation usually occurs within a specialist rehabilitation unit, with follow up services occurring in the community (Turner-Stokes, 2008). Similar issues face people with acquired brain injury and stroke as for those with spinal cord injury as they transition to community living. These challenges include low rates of employment (Van Velzen, Van Bennekom, Edelaar, Sluiter, & Frings-Dresen, 2009), challenges with social participation (McLean, Jarus, Hubley, & Jongbloed, 2014), issues with housing and transport (Roessler et al., 2013), and barriers regarding assistive technology (de Joode, van Heugten, Verhey, & van Boxtel, 2010). As with spinal cord injury, there is evidence of the positive influence of peer support for traumatic brain injury survivors and their caregivers in areas of social support, coping, behavioral control and physical quality of life (Wobma, Nijland, Ket, & Kwakkel, 2016). Self-management of ongoing health and secondary conditions is also an issue following stroke, acquired brain injury and multiple sclerosis. Chronic disease health management programs such as those used in spinal cord injury can be equally beneficial in the case of these other conditions (Hirsche et al., 2011).

## Mini-dictionary of terms

***Community integration***: Having access to appropriate housing, being able to mobilize in the community, being able to participate in work, leisure or educational activities of choice, and being able to engage in satisfying social relationships.
***Vocational rehabilitation***: Aims to restore or maintain an individual's vocational participation after major injury.
***Participation***: Involvement in life situations.
***Secondary health conditions***: Common ongoing health concerns experienced after spinal cord injury.
***Telehealth***: Health care provided remotely by telecommunications technology.

***Peer mentors***: People with lived experience of spinal cord injury.

***Active rehabilitation***: A community peer-based approach that usually involves people with SCI attending camps that offer intensive goal-oriented group based training and peer support activities in a community setting.

## Key facts of spinal cord injury community integration

- Community integration services and models for spinal cord injury rehabilitation need to prioritize community integration in order to maximize outcomes and quality of life.
- Access to suitably accessible housing, appropriate assistive technology and transport are key facilitators of community integration.
- Participation in leisure activities, and employment after spinal cord injury provides benefits to counter the difficulties of post-injury adjustment.
- People with lived experience working as paid peer mentors support people during the transition home.
- Education to stay physically and mentally healthy after discharge is important.

## Summary points

- This chapter focuses on the link between rehabilitation services and community integration following spinal cord injury.
- Community integration is similar in meaning to the term participation in the International Classification of Functioning Disability and Health (ICF).
- There are a range of community integration and/or transitional living models or approaches.
- Access to suitably accessible housing, appropriate assistive technology and transport are key in facilitating community integration.
- Participation in leisure activities provides benefits to counter the difficulties of adjusting to spinal cord injury.
- Peer mentors provide support to people transitioning home from hospital.
- Seeking, gaining, and maintaining employment after spinal cord injury contributes to people's identity.
- Vocational rehabilitation may be more effective if actioned sooner.
- Facilitating self-management of people with spinal cord injury, can help them to take a more active role in managing their health care.

## References


Barclay, L., Callaway, L., McDonald, R., Farnworth, L., Brown, T., & Broom, L. (2011). Time use following spinal cord injury: An examination of the literature. *British Journal of Occupational Therapy*, *74*(12), 573–580.

Barclay, L., & Hilton, G. M. (2019). A scoping review of peer-led interventions following spinal cord injury. *Spinal Cord*, *57*(8), 626–635.

Barclay, L., Lalor, A., Migliorini, C., & Robins, L. (2020). A comparative examination of models of service delivery intended to support community integration in the immediate period following inpatient rehabilitation for spinal cord injury. *Spinal Cord*, *58*(5), 528–536.

Barclay, L., McDonald, R., Lentin, P., & Bourke-Taylor, H. (2016). Facilitators and barriers to social and community participation following spinal cord injury. *Australian Occupational Therapy Journal*, *63*, 19–28. https://doi.org/10.1111/1440-1630.12241.

Barclay, L., Robins, L., Migliorini, C., & Lalor, A. (2020). Community integration programs and interventions for people with spinal cord injury: A scoping review. *Disability and Rehabilitation*, *1–11*. https://doi.org/10.1080/09638288.2020.1749889.

Bergmark, B., Winograd, C., & Koopman, C. (2008). Residence and quality of life determinants for adults with tetraplegia of traumatic spinal cord injury etiology. *Spinal Cord*, *46*(10), 684–689.

Best, K. L., Miller, W. C., Huston, G., Routhier, F., & Eng, J. J. (2016). Pilot study of a peer-led wheelchair training program to improve self-efficacy using a manual wheelchair: A randomized controlled trial. *Archives of Physical Medicine and Rehabilitation*, *97*(1), 37–44. https://doi.org/10.1016/j.apmr.2015.08.425.

Bombardier, C. H., Dyer, J. R., Burns, P., Crane, D. A., Takahashi, M. M., Barber, J., et al. (2020). A tele-health intervention to increase physical fitness in people with spinal cord injury and cardiometabolic disease or risk factors: A pilot randomized controlled trial. *Spinal Cord*, *59*, 1–11.

Braaf, S. C., Lennox, A., Nunn, A., & Gabbe, B. J. (2018). Experiences of hospital readmission and receiving formal carer services following spinal cord injury: A qualitative study to identify needs. *Disability and Rehabilitation*, *40*(16), 1893–1899.

Britell, C. W. (1991). Why aren't they working? *The Journal of the American Paraplegia Society*, *14*(1), 1–2.

Callaway, L., Barclay, L., McDonald, R., Farnworth, L., & Casey, J. (2015). Secondary health conditions experienced by people with spinal cord injury within community living: Implications for a National Disability Insurance Scheme. *Australian Occupational Therapy Journal*, *62*(4), 246–254.

Carr, J., Kendall, M., Amsters, D., Pershouse, K., Kuipers, P., Buettner, P., et al. (2017). Community participation for individuals with spinal cord injury living in Queensland, Australia. *Spinal Cord*, *55*(2), 192–197.

Chan, S. K., & Man, D. W. (2005). Barriers to returning to work for people with spinal cord injuries: A focus group study. *Work*, *25*(4), 325–332.

Chang, F. H., Coster, W. J., & Helfrich, C. A. (2013). Community participation measures for people with disabilities: A systematic review of content from an international classification of functioning, disability and health perspective. *Archives of Physical Medicine and Rehabilitation*, *94*(4), 771–781.

Cott, C. A., Wiles, R., & Devitt, R. (2007). Continuity, transition and participation: Preparing clients for life in the community post-stroke. *Disability and Rehabilitation*, *29*(20 – 21), 1566–1574.

Craig, A., Guest, R., Tran, Y., & Middleton, J. W. (2017). Cognitive impairment and mood states after spinal cord injury. *Journal of Neurotrauma*, *34*(6), 1156–1163. https://www.liebertpub.com/doi/10.1089/neu.2016.4632.

Craig, A., & Nicholson Perry, K. (2013). *Guide for health professionals on the psychosocial care of adults with spinal cord injury*. Retrieved from Agency for Clinical Innovation.

Craig, A., Nicholson Perry, K., Guest, R., Tran, Y., & Middleton, J. W. (2015). Adjustment following chronic spinal cord injury: Determining factors that contribute to social participation. *British Journal of Health Psychology*, *20*(4), 807–823. https://doi.org/10.1111/bjhp.12143.

de Joode, E., van Heugten, C., Verhey, F., & van Boxtel, M. (2010). Efficacy and usability of assistive technology for patients with cognitive deficits: A systematic review. *Clinical Rehabilitation*, *24*(8), 701–714.

Divanoglou, A., & Georgiou, M. (2017). Perceived effectiveness and mechanisms of community peer-based programmes for spinal cord injuries—A systematic review of qualitative findings. *Spinal Cord*, *55*(3), 225–234. https://doi.org/10.1038/sc.2016.147.

Divanoglou, A., Tasiemski, T., Augutis, M., & Trok, K. (2017). Active rehabilitation—A community peer-based approach for persons with spinal cord injury: International utilisation of key elements. *Spinal Cord*, *55*(6), 545–552.

Dwyer, K. J., & Mulligan, H. (2015). Community reintegration following spinal cord injury: Insights for health professionals in community rehabilitation services in New Zealand. *New Zealand Journal of Physiotherapy*, *43*(3), 75–85.

Fadyl, J., & McPherson, K. (2010). Understanding decisions about work after spinal cord injury. *Journal of Occupational Rehabilitation*, *20*(1), 69–80. https://doi.org/10.1007/s10926-009-9204-1.

Ferdiana, A., Post, M. W. M., Hoekstra, T., van der Woude, L. H., van der Klink, J. J., & Bultmann, U. (2014). Employment trajectories after spinal cord injury: Results from a 5-year prospective cohort study. *Archives of Physical Medicine and Rehabilitation*, *95*(11), 2040–2046. https://doi.org/10.1016/j.apmr.2014.04.021.

Friesen, E. L., Theodoros, D., & Russell, T. G. (2016). *Assistive technology devices for toileting and showering used in spinal cord injury rehabilitation—A comment on terminology*. Taylor & Francis.

Gassaway, J., Jones, M. L., Sweatman, M., Hong, M., Anziano, P., & DeVault, K. (2017). Effects of peer mentoring on self-efficacy and hospital readmission after inpatient rehabilitation of individuals with spinal cord injury: A randomized controlled trial. *Archives of Physical Medicine and Rehabilitation*, *98*, 1526–1534.

Guttmann, L. (1954). Statistical survey on one thousand paraplegics. *Proceedings of the Royal Society of Medicine*, *47*(12), 1099.

Hammel, J., Magasi, S., Heinemann, A., Whiteneck, G., Bogner, J., & Rodriguez, E. (2008). What does participation mean? An insider perspective from people with disabilities. *Disability and Rehabilitation*, *30*(19), 1445–1460.

Hay-Smith, E. J., Dickson, B., Nunnerley, J., & Sinnott, A. K. (2013). The final piece of the puzzle to fit in: An interpretative phenomenological analysis of the return to employment in New Zealand after spinal cord injury. *Disability and Rehabilitation*, *35*(17), 1436–1446. https://doi.org/10.3109/09638288.2012.737079.

Hilton, G. M., Unsworth, C. A., Browne, M., Murphy, G. C., & Olver, J. (2017). Longitudinal employment outcomes of an early-intervention vocational rehabilitation service for people admitted to rehabilitation with a traumatic spinal cord injury. *Spinal Cord*, *55*, 743–752. https://doi.org/10.1038/sc.2017.24.

Hilton, G. M., Unsworth, C. A., & Murphy, G. (2018). The experience of attempting to return to work following spinal cord injury: A systematic review of the qualitative literature. *Disability and Rehabilitation*, *40*(15), 1745–1753. https://doi.org/10.1080/09638288.2017.1312566.

Hilton, G., Unsworth, C. A., Stuckey, R., & Murphy, G. C. (2018). The experience of seeking, gaining and maintaining employment after traumatic spinal cord injury and the vocational pathways involved. *Work*, *59*(1), 67–84.

Hirsche, R. C., Williams, B., Jones, A., & Manns, P. (2011). Chronic disease self-management for individuals with stroke, multiple sclerosis and spinal cord injury. *Disability and Rehabilitation*, *33*(13–14), 1136–1146.

Holmes, J., Phillips, J., Morris, R., Bedekar, Y., Tyerman, R., & Radford, K. (2016). Development and evaluation of an early specialised traumatic brain injury vocational rehabilitation training package. *British Journal of Occupational Therapy*, *79*(11), 693–702.

Hooper, B., Verdonck, M., Amsters, D., Myburg, M., & Allan, E. (2018). Smart-device environmental control systems: Experiences of people with cervical spinal cord injuries. *Disability and Rehabilitation. Assistive Technology*, *13*(8), 724–730.

Hossain, M., Harvey, L., Rahman, M., Bowden, J., Islam, M., Taylor, V., et al. (2017). A pilot randomised trial of community-based care following discharge from hospital with a recent spinal cord injury in Bangladesh. *Clinical Rehabilitation*, *31*(6), 781–789.

Houlihan, B. V., Brody, M., Everhart-Skeels, S., Pernigotti, D., Burnett, S., Zazula, J., et al. (2017). Randomized trial of a peer-led, telephone-based empowerment intervention for persons with chronic spinal cord injury improves health self-management. *Archives of Physical Medicine and Rehabilitation*, *98*, 1067–1076. https://doi.org/10.1016/j.apmr.2017.02.005.

Kendall, M. B., Ungerer, G., & Dorsett, P. (2003). Bridging the gap: Transitional rehabilitation services for people with spinal cord injury. *Disability and Rehabilitation*, *25*(17), 1008–1015.

Kendall, M. B., & Wallace, M. A. (2016). Measuring goal attainment within a community-based multidisciplinary rehabilitation setting for people with spinal cord injury. *Edorium Journal of Disability and Rehabilitation*, *2*, 43–52.

Kett, M., Cole, E., & Turner, J. (2020). Disability, mobility and transport in low-and middle-income countries: A thematic review. *Sustainability*, *12*(2), 589.

Kratz, A. L., Chadd, E., Jensen, M. P., Kehn, M., & Kroll, T. (2015). An examination of the psychometric properties of the community integration questionnaire (CIQ) in spinal cord injury. *The Journal of Spinal Cord Medicine*, *38*(4), 446–455.

Krause, J. S., Saunders, L. L., & Acuna, J. (2012). Gainful employment and risk of mortality after spinal cord injury: Effects beyond that of demographic, injury and socioeconomic factors. *Spinal Cord*, *50*(10), 784–788. https://doi.org/10.1038/sc.2012.49.

Kuykendall, L., Tay, L., & Ng, V. (2015). Leisure engagement and subjective well-being: A meta-analysis. *Psychological Bulletin*, *141*(2), 364.

Latimer-Cheung, A., Arbour-Nicitopoulos, K., Brawley, L., Gray, C., Wilson, A., Prapavessis, H., et al. (2013). Developing physical activity interventions for adults with spinal cord, injury. Part 2: Motivational counseling and peer-mediated interventions, for people intending to be active. *Rehabilitation Psychology*, *58*(3), 307–315.

Liu, H., Hossain, M. S., Islam, M. S., Rahman, M. A., Costa, P. D., Herbert, R. D., et al. (2020). Understanding how a community-based intervention for people with spinal cord injury in Bangladesh was delivered as part of a randomised controlled trial: A process evaluation. *Spinal Cord*, *58*, 1–10.

Lloyd, K. M., & Auld, C. J. (2002). The role of leisure in determining quality of life: Issues of content and measurement. *Social Indicators Research*, *57*(1), 43–71.

Loy, D. P., Dattilo, J., & Kleiber, D. A. (2003). Exploring the influence of leisure on adjustment: Development of the leisure and spinal cord injury adjustment model. *Leisure Sciences*, *25*(2–3), 231–255.

Lundström, U., Lilja, M., Petersson, I., Lexell, J., & Isaksson, G. (2014). Leisure repertoire among persons with a spinal cord injury: Interests, performance, and well-being. *The Journal of Spinal Cord Medicine*, *37*(2), 186–192.

Mackelprang, R. W., & Salsgiver, R. O. (2016). *Disability: A diversity model approach in human service practice*. Oxford University Press.

Maribo, T., Jensen, C. M., Madsen, L. S., & Handberg, C. (2020). Experiences with and perspectives on goal setting in spinal cord injury rehabilitation: A systematic review of qualitative studies. *Spinal Cord*, *58*, 1–10.

May, L., Day, R., & Warren, S. (2006). Evaluation of patient education in spinal cord injury rehabilitation: Knowledge, problem-solving and perceived importance. *Disability and Rehabilitation*, *28*(7), 405–413.

McLean, A. M., Jarus, T., Hubley, A. M., & Jongbloed, L. (2014). Associations between social participation and subjective quality of life for adults with moderate to severe traumatic brain injury. *Disability and Rehabilitation*, *36*(17), 1409–1418.

McVeigh, S. A., Hitzig, S. L., & Craven, B. C. (2009). Influence of sport participation on community integration and quality of life: A comparison between sport participants and non-sport participants with spinal cord injury. *The Journal of Spinal Cord Medicine*, *32*(2), 115–124.

Mead, S., & MacNeil, C. (2006). Peer support: What makes it unique. *International Journal of Psychosocial Rehabilitation*, *10*(2), 29–37.

Middleton, J. W., Murphy, G. C., Ramakrishnan, K., Savage, N., Harper, R., Compton, J., et al. (2015). Early access to vocational rehabilitation for spinal cord injury inpatients. *Journal of Rehabilitation Medicine*, *47*, 626–631.

Momsen, A.-M., Rasmussen, J. O., Nielsen, C. V., Iversen, M. D., & Lund, H. (2012). Multidisciplinary team care in rehabilitation: An overview of reviews. *Journal of Rehabilitation Medicine*, *44*(11), 901–912.

Munce, S. E., Webster, F., Fehlings, M. G., Straus, S. E., Jang, E., & Jaglal, S. B. (2014). Perceived facilitators and barriers to self-management in individuals with traumatic spinal cord injury: A qualitative descriptive study. *BMC Neurology*, *14*(1), 48.

Murphy, G. C., O'Hare, M. A., & Wallis, L. (2010). The relationships among different types of social support accessed by traumatic spinal cord injury survivors. *International Journal of Disability Management*, *5*(2), 59–66.

Norweg, A., Jette, A. M., Houlihan, B., Ni, P., & Boninger, M. L. (2011). Patterns, predictors, and associated benefits of driving a modified vehicle after spinal cord injury: Findings from the National Spinal Cord Injury Model Systems. *Archives of Physical Medicine and Rehabilitation*, *92*(3), 477–483.

Nunnerley, J. L., Hay-Smith, E., & Dean, S. (2013). Leaving a spinal unit and returning to the wider community: An interpretative phenomenological analysis. *Disability and Rehabilitation*, *35*(14), 1164–1173.

Phillips, V., Vesmarovich, S., Hauber, R., Wiggers, E., & Egner, A. (2001). Telehealth: Reaching out to newly injured spinal cord patients. *Public Health Reports*, *116*(Supp 1), 94–102.

Post, M. W., Reinhardt, J. D., & Escorpizo, R. (2020). Return to work after spinal cord injury. In *Handbook of disability, work and health* (pp. 1–13). Springer.

Price, P., Stephenson, S., Krantz, L., & Ward, K. (2011). Beyond my front door: The occupational and social participation of adults with spinal cord injury. *OTJR: Occupation, Participation and Health*, *31*(2), 81–88.

Roessler, R. T., Bishop, M., Rumrill, P. D., Sheppard-Jones, K., Waletich, B., Umeasiegbu, V., et al. (2013). Specialized housing and transportation needs of adults with multiple sclerosis. *Work*, *45*(2), 223–235.

Schonherr, M., Groothoff, J., Mulder, G., & Eisma, W. (2005). Vocational perspectives after spinal cord injury. *Clinical Rehabilitation*, *19*(2), 200–208. https://doi.org/10.1191/0269215505cr845oa.

Sekaran, P., Vijayakumari, F., Hariharan, R., Zachariah, K., Joseph, S. E., & Kumar, R. S. (2010). Community reintegration of spinal cord-injured patients in rural South India. *Spinal Cord*, *48*(8), 628–632.

SRALAB. (2020a). *Goal attainment scale*. Retrieved from https://www.sralab.org/rehabilitation-measures/goal-attainment-scale.

SRALAB. (2020b). *Needs assessment checklist*. Retrieved from https://www.sralab.org/rehabilitation-measures/needs-assessment-checklist.

Wright, C. J., Colley, J., & Kendall, E. (2020). Exploring the efficacy of housing alternatives for adults with an acquired brain or spinal injury: A systematic review. *Brain Impairment*, *21*(2), 125–153. https://doi.org/10.1017/BrImp.2019.33.

*The United Nations Convention on the Rights of Persons with Disabilities (UNCRPD) [Article 9]*. Retrieved from: https://www.un.org/development/desa/disabilities/convention-on-the-rights-of-persons-with-disabilities/article-9-accessibility.html.

Trenaman, L. M., Miller, W. C., Querée, M., & Escorpizo, R. (2015). Modifiable and non-modifiable factors associated with employment outcomes following spinal cord injury: A systematic review. *The Journal of Spinal Cord Medicine*, *38*(4), 422–431.

Turner, B., Fleming, J., Cornwell, P., Worrall, L., Ownsworth, T., Haines, T., et al. (2007). A qualitative study of the transition from hospital to home for individuals with acquired brain injury and their family caregivers. *Brain Injury*, *21*(11), 1119–1130.

Turner-Stokes, L. (2008). Evidence for the effectiveness of multi-disciplinary rehabilitation following acquired brain injury: A synthesis of two systematic approaches. *Journal of Rehabilitation Medicine*, *40*(9), 691–701.

Ullah, M. M., Fossey, E., & Stuckey, R. (2018). The meaning of work after spinal cord injury: A scoping review. *Spinal Cord*, *56*(2), 92.

Van Velzen, J., Van Bennekom, C., Edelaar, M., Sluiter, J. K., & Frings-Dresen, M. (2009). How many people return to work after acquired brain injury?: A systematic review. *Brain Injury*, *23*(6), 473–488.

Veith, E. M., Sherman, J. E., Pellino, T. A., & Yasui, N. Y. (2006). Qualitative analysis of the peer-mentoring relationship among individuals with spinal cord injury. *Rehabilitation Psychology*, *51*(4), 289–298.

Wobma, R., Nijland, R. H., Ket, J. C., & Kwakkel, G. (2016). Evidence for peer support in rehabilitation for individuals with acquired brain injury: A systematic review. *Journal of Rehabilitation Medicine*, *48*(10), 837–840.

Woo, C., Guihan, M., Frick, C., Gill, C. M., & Ho, C. H. (2011). What's happening now! Telehealth management of spinal cord injury/disorders. *The Journal of Spinal Cord Medicine*, *34*(3), 322–331.

World Health Organization. (2002). *Toward a common language for functioning, disability and health*. Geneva: World Health Organization.

Chapter 40

# Rehabilitation in spinal cord injury: Exercise and testing for cardiorespiratory endurance and musculoskeletal fitness

Kerri A. Morgan[a], Kelly L. Taylor[b], and Susan Tucker[a]
[a]*Washington University School of Medicine, St. Louis, MO, United States;* [b]*Murray State University, Paducah, KY, United States*

## Abbreviations

| | |
|---|---|
| **SCI** | spinal cord injury |
| **PwSCI** | persons with a spinal cord injury |
| **$VO_{2peak}$** | peak volume of oxygen consumption |
| **$VO_{2max}$** | maximal volume of oxygen consumption |
| **ACE** | arm crank ergometer |
| **1-RM** | one repetition maximum |
| **WMS** | WheelMill System |
| **6-MAT** | 6-minute arm test |
| **6MPT** | 6-minute push test |
| **RPE** | rate of perceived exertion |
| **ASIA** | The American Spinal Injury Association Impairment Scale |
| **PO** | power output |
| **W** | watts |
| **RM** | repetition maximum |
| **ROM** | range of motion |
| **RER** | respiratory exchange ratio |

## Introduction

Persons with a spinal cord injury (PwSCI) are at increased risk of leading less active lifestyles as compared to the general population. Spinal cord injury (SCI) results in declines in physical capacity and an increased risk of secondary health conditions. Specifically, PwSCI are three times more likely to have one or more chronic diseases such as heart disease, stroke, diabetes, cancer, or obesity compared to persons without disabilities ( Jacobs & Beekhuizen, 2005; Keyser, Rasch, Finley, & Rodgers, 2003). SCIs often result in various physiological alterations affecting multiple body systems, including blunted hemodynamics, altered sympathetic innervation, altered glycemic control, reduced active muscle mass, deteriorated circulatory vessels below level of injury, and impaired arterial compliance and function (West, AlYahya, Laher, & Krassioukov, 2013). These pathophysiological alterations amplify the risk and prevalence of cardiovascular and metabolic diseases in PwSCI. PwSCI are also at an increased risk of negative mood states including depression and anxiety (Post & van Leeuwen, 2012). An approach to managing the potential physiological and psychological implications of an SCI is engaging in physical activity or exercise.

## Exercise and persons with a spinal cord injury

The typical daily routine of PwSCI does not stress the cardiorespiratory system enough to produce positive health-related changes; therefore, participation in regular exercise is needed to enhance physical capacity and to reduce the likelihood of secondary complications (Hjeltnes & Vokac, 1979; Jacobs & Nash, 2004). Similar to the general population, regular exercise is critical to reducing the risk of developing chronic health conditions and has positive effects on the physiological

**Diagnosis and Treatment of Spinal Cord Injury.** https://doi.org/10.1016/B978-0-12-822498-4.00040-3

health and psychological well-being of PwSCI (Carroll et al., 2014; Hicks et al., 2011; Martin Ginis et al., 2010). Participation in exercise is also connected to more established social networks and greater participation in life activities (Crawford, Hollingsworth, Morgan, & Gray, 2008; Rimmer, Chen, McCubbin, Drum, & Peterson, 2010). Increased cardiorespiratory fitness in PwSCI promotes not only improved functional independence, but also improved cardiovascular and metabolic health (van der Scheer et al., 2017).

## Exercise and spinal cord injury considerations

PwSCI may experience secondary health conditions that are important to take into consideration when engaging in exercise. Bone loss and osteoporosis are complications that may occur following an SCI and can lead to an increased risk of fractures, which can have further negative consequences on health and functioning ( Jiang, Dai, & Jiang, 2006). Upper extremity pain and chronic overuse injuries, particularly in the shoulder joints of PwSCI, are commonly associated with wheelchair self-propulsion and transfers (Lal, 1998). Due to the direct effects of an SCI, PwSCI often lack normal protective sensation below the level of their injury. This lack of sensation combined with the reduced mobility associated with an SCI can result in the development of pressure ulcers (Vidal & Sarrias, 1991). Having an understanding of these possible secondary conditions is important when exercising. Proper biomechanical positioning when exercising, reducing the number of transfers onto exercise equipment, and implementing a regular shoulder strengthening exercise program targeted at the rotator cuff are among the actions that can be taken to prevent or delay these secondary conditions from impacting the exercise routine of a PwSCI.

## Exercise recommendations for persons with spinal cord injury

Basic programming guidelines for exercise and physical activity are set forth by organizations including the American College of Sports Medicine's Guidelines for Exercise Testing and Prescription (2020) and the Physical Activity Guidelines for Adults with Chronic Health Conditions and Adults with Disabilities from the U.S. Department of Health and Human Services (2018). The U.S. Department of Health and Human Services recommends at least 150 minutes per week of moderate-intensity aerobic activity or 75 minutes per week of vigorous-intensity aerobic activity for adults with chronic conditions or disabilities, as well as muscle strengthening activities of moderate or greater intensity, 2 or more days a week. Diagnosis-specific guidelines exist for PwSCI and recommend shorter durations of weekly exercise time. For example, to gain cardiometabolic health and fitness and muscle strength benefits, the International Exercise Guidelines for PwSCI recommend that a PwSCI should engage in at least 20 to 30 min of moderate-to-vigorous-intensity aerobic exercise two to three times per week and three sets of strength exercises for each major functioning muscle group, at a moderate-to-vigorous intensity, two times per week (Martin Ginis et al., 2018).

## Barriers to exercise

Despite the widely recognized health benefits of regular exercise, PwSCI report lower levels of exercise compared to the general population (Carroll et al., 2014; Martin Ginis et al., 2010; Sullivan, 2009). Because of decreased sympathetic innervation, muscle atrophy, and decreased cardiac reserves, it is difficult for PwSCI to achieve intensity levels necessary to make health-related changes without proper support (Cowell, Squires, & Raven, 1986; Hicks et al., 2011). PwSCI face numerous challenges to participating in exercise, including personal (e.g., lack of motivation and self-efficacy) and environmental (e.g., transportation, access to proper equipment) barriers (Adam & Morgan, 2018). Unlike the surplus of fitness centers and equipment available for the general population, there is a considerable lack of accessible equipment, support, and professional guidance for PwSCI (Cardinal & Spaziani, 2003; Johnson, Stoelzle, Finco, Foss, & Carstens, 2012). In addition, PwSCI do not always have access to the information needed to monitor their progress. Accurate and reliable measures are important so that physical activity levels of PwSCI can be assessed, recommendations can be made to better inform development of exercise programs, and PwSCI can independently track their exercise and take ownership of their health.

## Importance of exercise testing

An important aspect of supporting PwSCI to successfully engage, initiate, and maintain exercise is the ability to have reliable information to inform and develop individualized programs and track change. Whether a person is an elite athlete training for the Paralympics or a newly injured PwSCI, having information to guide programs and help people achieve their goals is crucial. The following will discuss different types of testing, testing protocols, considerations for PwSCI, and equipment and adaptations.

## Exercise testing for persons with a spinal cord injury

Physical fitness is the ability to perform activities of daily living, leisure activities, and other occupations without excessive fatigue. Physical fitness testing, also known as exercise testing, is important to conduct to determine a baseline fitness level, identify strengths and areas of improvement, and establish realistic goals for individuals. Comprehensive testing best informs exercise professionals to design and implement appropriate, client-centered exercise prescriptions. Follow-up testing results can also be performed and compared to baseline data to evaluate progress during and/or following a prescribed exercise program (Gibson, Wagner, & Heyward, 2019a). Accurate and reliable assessments are important so that exercise and fitness levels can be evaluated, recommendations can be made to better inform development of exercise programs, and PwSCI can independently track their exercise and take ownership of their health.

Exercise testing can include a number of different components; however, the two most common are cardiorespiratory endurance, also known as functional capacity, and musculoskeletal fitness testing. Musculoskeletal fitness is comprised of two components, muscle strength and muscle endurance (Arena et al., 2007; Gibson, Wagner, & Heyward, 2019b).

### General considerations for exercise testing for people with spinal cord injury

PwSCI may experience a myriad of multi-system physiological alterations as a result of their injury (West et al., 2013). For these reasons, specific considerations and precautions are recommended when conducting exercise testing with PwSCI. Exercise testing participants should obtain physician approval and signed release prior to initiation of testing. Testers should be well trained in the testing protocol, procedures, and equipment, and also familiar with potentially dangerous conditions associated with SCI, such as autonomic dysreflexia. Testing participants should be supervised and closely monitored throughout the duration of the testing session. Ambient temperature of the testing environment should be comfortable for the participant, keeping in mind persons with a cervical level injury may not be able to control their temperature by sweating and may need a fan blowing on them or cooler room temperatures. Devices such as heart rate monitors and/or other wearable sensors should be worn by testing participants for accurate monitoring of participant status. Vitals and pain level should be taken prior to initiation of testing to ensure that the session can safely proceed. The testing session should be terminated if any of the following distressful issues occur: abnormal heart rate or blood pressure, presence of a bladder infection, presence of a pressure ulcer, unusual spasticity, or autonomic dysreflexia. Testers should also determine a pain threshold at which to terminate testing; recommended thresholds may vary; however, terminating testing if reported pain is $\geq 6$ out of 10 has been acceptable (Morgan, Taylor, Tucker, Cade, & Klaesner, 2018).

## Cardiorespiratory endurance testing

Cardiorespiratory testing can be categorized into laboratory or field tests. While laboratory tests provide greater accuracy and reliability, they also require advanced technology and training as well as increased cost to perform. The gold standard for assessing cardiorespiratory fitness is the measurement of peak volume of oxygen consumption ($VO_{2peak}$) during a graded exercise test (Vanhees et al., 2005). Field tests, while less sensitive than$VO_{2peak}$, are typically more cost-effective, more readily available to administer in non-laboratory settings, and still provide a submaximal estimation of cardiorespiratory endurance. Examples of cardiorespiratory field tests that will be discussed include the 6-Minute Arm Test (6-MAT) and the 6-Minute Push Test (6MPT).

### $VO_{2peak}$ testing in people with spinal cord injury

While the maximal volume of oxygen consumption ($VO_{2max}$) is widely accepted as the benchmark for measuring cardiorespiratory fitness, the $VO_2$ plateau and maximal heart rate criteria required to reliably establish $VO_{2max}$ may be difficult for PwSCI to achieve given the various pathophysiological changes with SCI, specifically altered sympathetic innervation and reduced muscle activation. Therefore, $VO_{2peak}$ is considered the gold standard to assess cardiorespiratory fitness and aerobic metabolism for this population. Breath-by-breath cardiometabolic and physiological measures can be obtained during testing using a computer-integrated metabolic measurement system; these measures include respiratory exchange ratio (RER), heart rate, pulmonary ventilation, energy expenditure, and ventilatory threshold (Crouter, Antczak, Hudak, DellaValle, & Haas, 2006; Lovell, Shields, Beck, Cuneo, & McLellan, 2012). $VO_{2peak}$ testing is typically completed in 8 to 12 minutes, depending on the individual's fitness level ( Jelleyman et al., 2015). Assessing the participant's rating of perceived exertion (RPE) throughout the graded exercise test is also highly recommended (Lewis, Nash, Hamm, Martins, & Groah, 2007). $VO_{2peak}$ determination typically includes achievement of RER $\geq 1.1$ and RPE $\geq 15$–17 as standard indicators for test termination; however, other criteria may be incorporated, including blood lactate and heart rate

(Morgan et al., 2018; Tørhaug, Brurok, Hoff, Helgerud, & Leivseth, 2016). Heart rate is only reliable for people with neurological levels of injury below T7, as autonomic dysfunction and blunted sympathetic innervation occur in individuals with higher injury levels (West et al., 2013). Evidence of norm-reference values for $VO_{2peak}$ in adults with SCI is sparse; however, it is accepted that $VO_{2peak}$ may be impacted by sex and neurological level of injury ( Janssen, Dallmeijer, Veeger, & van der Woude, 2002; Simmons, Kressler, & Nash, 2014).

## Equipment options for $VO_{2peak}$ fitness testing in people with spinal cord injury

For non-wheelchair users, $VO_{2peak}$ is commonly measured on treadmills or stationary bikes, both common forms of exercise for this population. For individuals with SCI using wheelchairs, the most common method of measuring $VO_{2peak}$ is during a graded exercise test using an arm crank ergometer (Ilias, Xian, Inman, & Martin, 2009), wheelchair ergometer, or a roller-based wheelchair dynamometer (Ilias et al., 2009; Molik et al., 2017).

### *Arm crank ergometers*

An arm crank ergometer (ACE) is a low-impact exercise system for both lower and upper extremity conditioning that has hand cranks and foot pedals and a removable seat for wheelchair access. For PwSCI who use a wheelchair for functional mobility, the hand cranks are used during testing. It is recommended to use an ACE that includes adjustable arm cranks and a removable seat for wheelchair access. Arm cranks should be adjusted where the axes of rotation are positioned just below shoulder level of the participant to ensure proper alignment of the glenohumeral joint to the ACE (Fig. 1A). Proper biomechanics also include ensuring that participants' arms are slightly flexed at the elbows at the moment of furthest reach (Fig. 1B). Participants are able to use their personal wheelchairs (manual or power) during testing; however, if trunk stability safely allows, participants may opt to transfer to the ACE seat.

While ACEs are the most widely used and commercially available devices for cardiorespiratory testing (Ilias et al., 2009; Janssen et al., 2002; Simmons et al., 2014; Tørhaug et al., 2016), have shown to be effective in testing fitness, and provide health-related changes in PwSCI, they present with limitations (Myers, Lee, & Kiratli, 2007). The operation of an ACE requires a movement pattern not typically employed during functional mobility or activities of daily living (Morgan et al., 2018). Functional impairments in hand strength and trunk control can yield further limitations of ACEs. Ace bandage wraps or specialized grip-assist gloves (Fig. 1C) can be used when hand grip impairments are present. Abdominal binders may be used to assist with trunk stabilization during testing.

### *Wheelchair-based systems*

Propulsion is the primary functional movement pattern for PwSCI who use a manual wheelchair. Testing devices such as wheelchair ergometers and roller-based systems provide a simulated over-ground propulsion experience, allowing the individual to remain stationary while engaging in a functional movement pattern during testing. Each of these devices allows manual wheelchair users to use functional propulsion to complete the stepwise $VO_{2peak}$ protocols. Evidence suggests that each device has strengths and weaknesses with no consensus on which type is best for exercise testing (Stephens & Engsberg, 2010).

Wheelchair ergometers, sometimes referred as dynamometers, utilize treadmill and bicycle components to allow participants to propel a manual wheelchair; however, these testing devices have limitations. Wheelchair ergometers often require participants to sit in and propel a designated laboratory wheelchair that has been pre-fit for the device. Participants report that the glide differs from typical wheelchair propulsion, potentially impacting body mechanics and performance

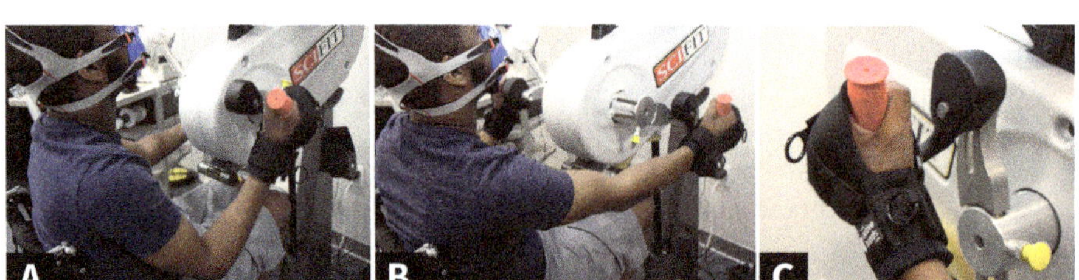


FIG. 1 Setup for $VO_{2peak}$ testing using an arm crank ergometer. (A) Proper alignment of glenohumeral joint in relation to ACE axes of rotation; (B) proper setup for slightly flexed elbows at the moment of furthest reach during cranking motion; (C) example of specialized grip-assist gloves known as Active Hands; $VO_{2peak}$, peak volume of oxygen consumption. *Photo credit: Michele Berhorst/Washington University.*

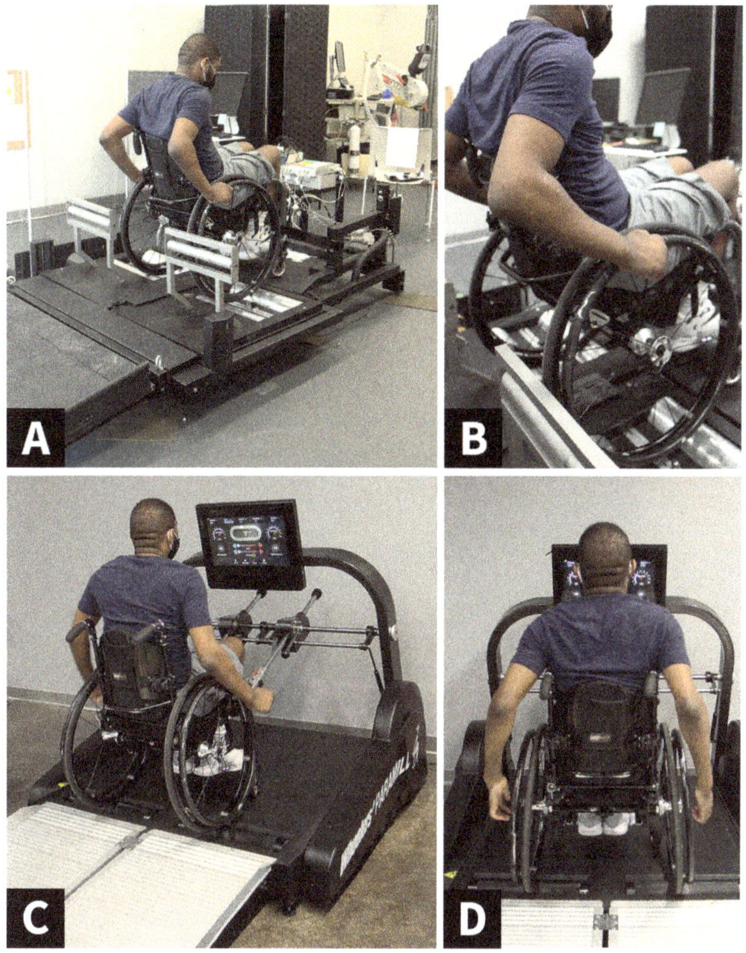


FIG. 2 Examples of roller-based systems allowing for use of personal manual wheelchairs. (A and B) WheelMill System; (C and D) The Paramill. *Photo credit: Michele Berhorst/Washington University.*

during testing. Wheelchair ergometers typically have limited customization or fine tuning that is highly beneficial when working with a diagnosis such as SCI (Rodgers, Keyser, Rasch, Gorman, & Russell, 2001).

Roller-based systems, also referred to as wheelchair dynamometers, consist of one or two parallel rollers and may allow for customizable manipulation of variables such as resistance (Klaesner, Morgan, & Gray, 2014; Morgan, Engsberg, & Klaesner, 2015). Roller-based systems (Fig. 2) further promote the functional movement pattern of propulsion by allowing for participants to use their personal wheelchairs, including sports wheelchairs, optimizing body mechanics, propulsion patterns, and performance during testing. An example of a roller-based system that has been used for $VO_{2peak}$ testing for PwSCI is the WheelMill System (WMS). The WMS is a computer-controlled, motor-driven wheelchair dynamometer that has been used for $VO_{2peak}$ testing (Klaesner et al., 2014; Morgan et al., 2015). Morgan et al. (2018) found that despite similarities in outcomes between ACEs and the WMS during $VO_{2peak}$ testing, participants consistently rated their RPE higher on the ACE compared to the WMS despite the same workload (Morgan et al., 2018). At this time, roller-based systems are not as widely available as ACE, and testers require enhanced training to operate and execute protocols using these devices. Participants should be directly supervised or spotted when transitioning their wheelchair to and from the rollers to prevent falls. Falls and wheelchair tipping should also be prevented during testing using safety straps to secure the wheelchair to the roller-based system.

## Submaximal field tests

### *6-Minute arm test*

The 6-MAT is a submaximal cardiorespiratory fitness test performed at a constant power output (PO), which is selected based on the individual's score on the American Spinal Injury Association (ASIA) Scale, sex, mobility device (manual vs

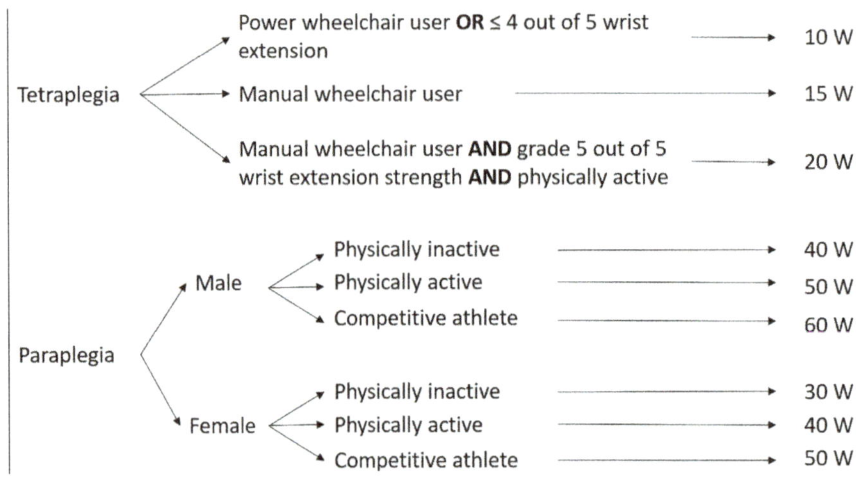


FIG. 3 Power output selection for the 6-minute arm test. The constant power output is based on level of injury, sex, mobility device, manual muscle test, and current physical activity level; W, Watts. *Figure adapted from Hol, A. T., Eng, J. J., Miller, W. C., Sproule, S., & Krassioukov, A. V. (2007). Reliability and validity of the six-minute arm test for the evaluation of cardiovascular fitness in people with spinal cord injury. Archives of Physical Medicine and Rehabilitation, 88(4), 489–495, Table 5.*

power), manual muscle testing, and current physical activity level (inactive, active, or competitive; Fig. 3). A standard, adjustable ACE is used for the 6-MAT. Individuals are instructed to maintain a consistent cadence of 60 rpm at the selected PO for the 6-minute duration of the test. Heart rate should be measured in 1-minute intervals, while RPE may be assessed either at 1-minute intervals and/or immediately following completion of the test (Hol, Eng, Miller, Sproule, & Krassioukov, 2007; Totosy de Zepetnek, Au, Hol, Eng, & MacDonald, 2016). Normative values have not been established for this submaximal field test; however, testers can use wearable sensors to gather data such as heart rate, RPE, total distance cranked, and estimated energy expenditure (Hiremath, Intille, Kelleher, Cooper, & Ding, 2016; Nightingale, Rouse, Thompson, & Bilzon, 2017) to compare pre-post for purposes of determining exercise programming effectiveness. The 6-MAT is appropriate to administer for PwSCI using either a manual or power wheelchair. Considerations for the 6-MAT are related to the aforementioned limitations and considerations of using an ACE.

### *6-Minute push test*

The 6MPT is a physical fitness and functional mobility test involving manual wheelchair users propelling in a designated 30-meter loop. Participants are asked to propel as far as possible on the course for 6 minutes. Total distance propelled is determined by calculating the number of completed 30-meter loops and measuring the partially completed final lap using a tape measure. While the 6MPT employs a familiar, functional movement pattern, the primary limitation of this test is that it is only meaningful for persons able to independently propel a manual wheelchair (Cowan, Callahan, & Nash, 2012).

## Musculoskeletal fitness testing

Muscle strength and endurance are key components related to fitness (Liguori & American College of Sports Medicine, 2020). Strength testing for PwSCI may be tested in the same manner as testing done in persons without disability. Consideration should be given to levels of SCI that may require stabilization and/or adaptive equipment to compensate for lack of grip strength and/or trunk control, as well as accessible equipment (Liguori & American College of Sports Medicine, 2020; Martin Ginis et al., 2018). Static strength testing, such as hand dynamometry and other isometric testing methods, are somewhat limited. These testing methods usually test isolated muscle groups at one joint angle, which is not a typical strength demand used for functional activities. The one repetition maximum (1-RM) has been considered the standard to assess dynamic muscular strength for many years (Liguori, & American College of Sports Medicine, 2020; Neto, Guanais, Dornelas, Coutinho, & Costa, 2017), and more recently multiple repetition maximum (RM) tests have been used (Gail, Rodefeld, & Künzell, 2015; Liguori & American College of Sports Medicine, 2020). A 1-RM is the maximal amount of resistance a person can move through their full range of motion (ROM), for a given exercise using proper form, one time. A multiple RM involves using multiple repetitions, usually 5 to 10, having a person move through their full ROM for a given exercise to the point of muscle failure, and then calculating their 1-RM from the multiple RM achieved (Reynolds, Gordon, & Robergs, 2006). 1-RM predictive equations commonly used in people without disability are accurate for use with adults with SCI who complete multiple RM testing of 4 to 12 repetitions (Baechle & Groves, 1998; Brzycki,

1993; Lombardi, 1989; Neto et al., 2017). More recently, some have suggested that, rather than using 1-RM, methods such as determining a number of repetitions completed within a specific time frame, that can be completed with proper form to fatigue, can help determine a baseline level of strength and that level of strength can then be used as a point of comparison for musculoskeletal fitness testing (Carpinelli, 2011; Liguori & American College of Sports Medicine, 2020).

## Considerations for strength testing in persons with spinal cord injury

Most PwSCI use a wheelchair for functional mobility. Prior to testing strength with PwSCI, consideration should be given to the accessibility of the equipment used, transfer skills required to use the equipment and the ability of the person to complete the movement required. Consideration should be given to testing the person in their wheelchair vs on equipment that would require a person to transfer. If a transfer is required to use equipment, testers should first determine the individual's level of comfort with transfers as well as their usual method of transfers to determine whether it is safe and reasonable for the PwSCI to transfer to the equipment. Testers should ensure proper positioning of the individual, the test equipment, and spotter(s) prior to beginning the strength testing. Generally, PwSCI above level C8 will have a range of hand and upper extremity weakness from minimal to total paralysis of the muscles of the hand and wrist, and may require grip aids.

Persons with an SCI above L2 may have impairment in the abdominal and core regions ranging from minimal weakness to total paralysis of core musculature, and may require stabilization during strength testing via a human spotter and/or positioning straps at the hip, abdominal, and/or chest area. Additionally, consideration should be given for positioning and stabilization needs related to the wheelchair if the person being tested will remain in their wheelchair. For example, if testing will use a machine similar to the Equalizer (Fig. 4) to perform a vertical bench press, which allows a person to remain in their wheelchair, consideration should be taken to ensure that the back pad is placed at a height and position that will prevent rearward tipping of the wheelchair. Wheelchair locks should be used to assist in keeping the wheelchair stationary during testing, and additional methods such as tie-down straps may be necessary to prevent the wheelchair from sliding on floor surfaces. Persons who perform strength testing out of their wheelchair, such as a standard bench press in the supine position, should be spotted during transfers, may need straps or additional assistance to achieve and maintain the positioning required to complete the test in a safe manner.

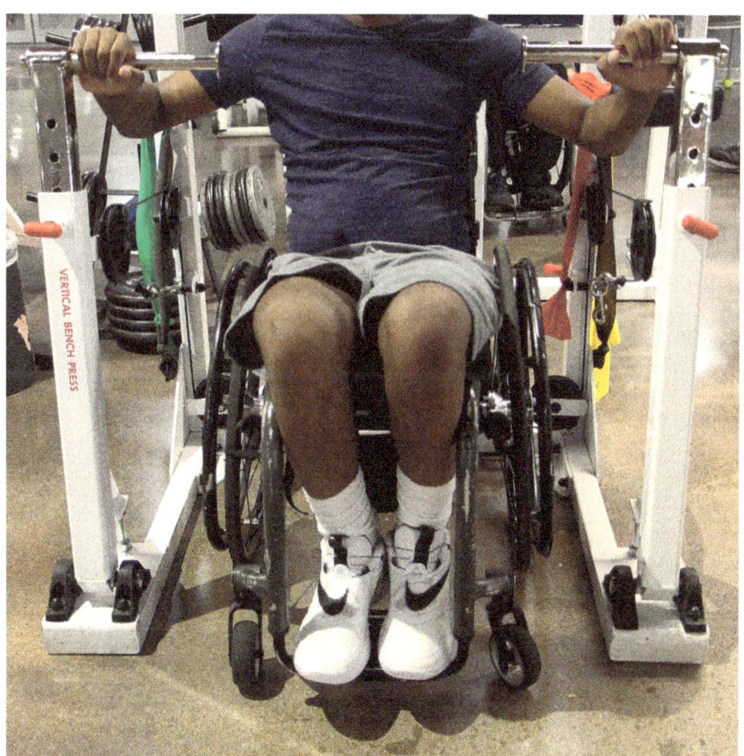


FIG. 4 Equalizer bench press equipment. Example of vertical bench press strength testing setup using the Equalizer for manual wheelchair user with a spinal cord injury. *Photo credit: Michele Berhorst/Washington University.*

## Strength testing equipment

Dynamic strength testing may be performed on several types of equipment. Equipment used for strength testing with PwSCI should be adjustable to allow for proper positioning, particularly if the person will remain seated in their wheelchair or requires any adaptations to compensate for muscle weakness in order to safely complete the required lift. Standard resistance training equipment, often found in most commercial fitness centers, can be used if the PwSCI being tested is able to transfer to the equipment and remain in a safe and stable position to complete the specific lift. Universally designed resistance training equipment, such as Equalizer and/or Uppertone, allows both wheelchair users and non-wheelchair users to access the equipment. The Uppertone is a wheelchair accessible device that targets upper limb strengthening by using stackable weights for adjusting resistance. The Equalizer is an accessible weight lifting machine that can be used by people who are ambulatory and people who utilize wheelchairs for mobility. For example, the Equalizer 1005VB machines and attachments allow the following exercises to be done: bench press, lat pull down, upright rowing, overhead press, vertical butterfly, biceps curl, triceps extension, leg curl, and leg extension. On the vertical bench press station, each arm works independently so different weights can be used if a participant's upper extremity strength varies between their right and left side. The vertical bench press height, backrest, and hand grips are adjustable to accommodate various users.

Rehabilitation testing equipment such as the BTE™ PrimusRS (Fig. 5) or Biodex dynamometer systems may be used to test isokinetic strength in PwSCI (Ambrosio et al., 2005; Freitas, Serenza, Santana, Manoel, & Riberto, 2017; Shechtman, MacKinnon, & Locklear, 2001; Souza et al., 2005). Typically, strength measures using rehabilitation testing equipment for isokinetic testing gives results in peak torque/weight, work, and power vs a weight resistance value for 1-RM that would be discovered from weight resistance training equipment, and may not readily translate to the standard 1-RM.

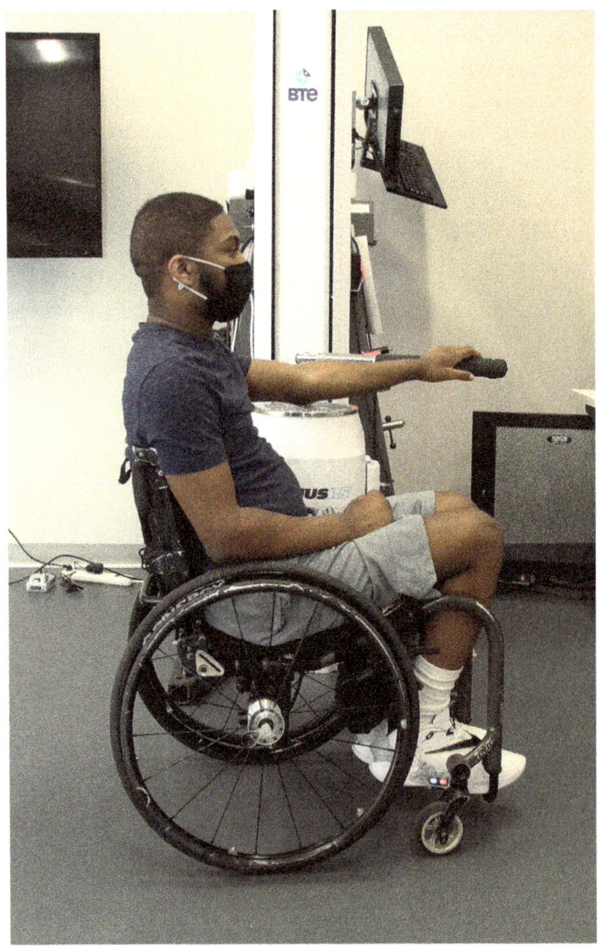


FIG. 5 Isokinetic testing equipment set up for a bench press motion using a BTE™ PrimusRS. Example setup of vertical bench press isokinetic strength testing for a manual wheelchair user with a spinal cord injury. *Photo credit: Michele Berhorst/Washington University.*

### Testing procedures

The procedures for conducting 1-RM and multiple RM testing have been well established (Gail et al., 2015; Liguori and American College of Sports Medicine, 2020; Neto et al., 2017). Additional steps may need to be taken for PwSCI, particularly those who are wheelchair users. PwSCI should be instructed to complete the motion of each exercise prior to adding weight resistance; those demonstrating sufficient ROM to complete the exercise with proper form can proceed with testing. The process to conduct 1-RM or multiple RM testing with PwSCI is shown in Table 1. As noted previously, it is important to ensure proper positioning of the individual and equipment prior to beginning the strength testing. Also, it is important to ensure that the individual being tested is stabilized to ensure safety and minimize compensatory movement patterns that would make the lift unsafe.

## Application to other areas in neuroscience

In this chapter, we have reviewed the importance of exercise, barriers to exercise, recommendations for exercising, and methods for exercise testing for persons with a spinal cord injury. Many of the themes discussed specific to spinal cord injury are seen in persons with other neurological conditions. This includes the impact exercise can have on physical and emotional health and the personal and environmental barriers to exercise experienced. Other medical conditions that also involve impairment of the corticospinal tracts of the central nervous system include multiple sclerosis and cerebral palsy. The benefits and focus of exercise often emphasize walking and balance for persons with multiple sclerosis and balance for persons with cerebral palsy, whereas, for persons with a spinal cord injury, function and aerobic fitness are where improvements were seen in the literature. Persons with a spinal cord injury also experience physiological alterations that impact organ systems that are different than other neurological conditions such as autonomic dysreflexia and temperature control. Many of the methods reviewed for exercise testing could also be adapted to persons with other neurological conditions. The methods were written and adapted for this chapter with the main assumption that persons had the cognitive ability to process instructions. For persons with a stroke or traumatic brain injury, the instructions may need to be modified. Also, for persons with hemiplegia versus paraplegia or tetraplegia, the adaptations may be different.

## Mini-dictionary of terms

*Cardiorespiratory endurance:* integration of the cardiovascular and pulmonary systems to efficiently provide adequate oxygen and nutrients to working skeletal muscles in order to perform tasks and activities requiring sustained aerobic metabolism.
*Musculoskeletal fitness:* ability of the musculoskeletal system to exert adequate force in order to perform work.
*Muscle strength:* maximal force producible by a muscle group.

**TABLE 1 Strength testing procedures for 1-RM for PwSCI.**

| | |
|---|---|
| 1. | Identify any grip aids, core stabilization aids, wheelchair positioning, and wheelchair stabilization needs specific to the equipment being used to complete testing |
| 2. | Complete 1-RM or multiple repetition maximum test after individual has been familiarized/completed practice sessions of the movement to be tested, all necessary adaptations and wheelchair setup and stabilization needs have been trialed, and stabilization and adaptation needs have been established |
| 3. | Warm up with completion of submaximal repetitions of the exercise being tested; following warm-up, adjust any positioning supports and/or stabilizations as needed |
| 4. | Determine 1-RM (or multiple repetition maximum) within four trials; rest for 3–5 min between trials |
| 5. | Select starting weight within perceived capacity |
| 6. | Progressively increase resistance 5–10% from previous attempt for upper body exercise until selected repetitions cannot be completed. Make sure speed of all lift movements is consistent |
| 7. | Record final weight lifted with correct form through full ROM as the 1-RM or multiple repetition maximum |

Recommended procedures for conducting upper extremity 1-RM testing for PwSCI. 1-RM, one-repetition maximum; PwSCI, persons with spinal cord injury. Adapted from Liguori, G., & American College of Sports Medicine. (2020). ACSM's guidelines for exercise testing and prescription. Lippincott Williams & Wilkins., p. 99 Box 3.9. Adapted from Ligouri & American College of Sports Medicine, 2020, p. 99, Box 3.9.

*Muscle endurance:* ability of a muscle group to sustain submaximal force for a prolonged time.
*Autonomic dysreflexia:* life-threatening condition involving a sudden increase in blood pressure from a reflexive overreaction of the autonomic nervous system in response to noxious stimulation.
*Dynamometery:* measurement of force produced and used, typically musculoskeletal strength.
*1-rep max:* maximal amount of resistance that can be moved, one time, through full range of motion for a given exercise using proper technique.
*Multiple rep max:* maximal amount of resistance that can be moved 5 to 10 repetitions through full range of motion for a given exercise to the point of muscle failure.
*Isokinetic:* muscle action that maintains a constant velocity.

## Key facts of rehabilitation in spinal cord injury: Exercise and testing for cardiorespiratory endurance and musculoskeletal fitness

### Key facts of spinal cord injury and exercise

- With less than 25% of adults with disabilities participating in sufficient physical activity to achieve health benefits, persons with spinal cord injury have the lowest engagement in physical activity and exercise
- Persons with spinal cord injury are three times more likely to have one or more chronic diseases such as heart disease, stroke, diabetes, cancer, or obesity compared to persons without disabilities
- Exercise guidelines have been developed specific to persons with a spinal cord injury and recommend at least 90 minutes of moderate-to-vigorous-intensity exercise each week to achieve cardiometabolic health benefits.

### Key facts of exercise testing for persons with spinal cord injury

- Peak volume of oxygen consumption testing is the gold standard for assessing cardiorespiratory endurance in persons with a spinal cord injury.
- Exercise test administrators must be well familiar with potentially dangerous conditions associated with spinal cord injury included but not limited to autonomic dysreflexia, temperature dysregulation, abnormal heart rhythms, decubitus ulcers, and bladder infections.
- People with spinal cord injury may require adaptations and/or extra stabilization or use of accessible strength testing equipment to complete dynamic strength testing in a similar manner as persons without disabilities.

## Summary points

- Habitual exercise is critical for persons with spinal cord injury (PwSCI) as they are at a greater risk for major health conditions and poorer health outcomes than persons without spinal cord injury.
- Significant barriers exist for PwSCI who desire to participate in exercise, including personal and environmental barriers.
- An important aspect of supporting PwSCI to successfully engage in physical activity is the ability to have reliable information to inform and develop individualized programs and track change.
- Specific considerations and precautions are recommended when conducting exercise testing with PwSCI.
- Cardiorespiratory endurance testing can be conducted in a controlled laboratory setting or during submaximal field tests
- Common equipment used for cardiorespiratory endurance testing for PwSCI include arm crank ergometers, wheelchair ergometers, and roller-based dynamometers.
- PwSCI may participate in strength testing with a variety of resistance equipment when appropriate precautions are taken to ensure safety and proper lifting technique.

## References

Adam, S., & Morgan, K. A. (2018). Meaningful components of a community-based exercise program for individuals with disabilities: A qualitative study. *Disability and Health Journal*, *11*(2), 301–305.

Ambrosio, F., Boninger, M. L., Souza, A. L., Fitzgerald, S. G., Koontz, A., & Cooper, R. A. (2005). Biomechanics and strength of manual wheelchair users. *The Journal of Spinal Cord Medicine*, *28*(5), 407–414.

Arena, R., Myers, J., Williams, M. A., Gulati, M., Kligfield, P., Balady, G. J., … Fletcher, G. (2007). Assessment of functional capacity in clinical and research settings. *Circulation*, *116*(3), 329–343.

Baechle, T. R., & Groves, B. R. (1998). *Weight training: Steps to success* (2nd ed.). Champaign, IL: Human Kinetics Publishers.

Brzycki, M. (1993). Strength testing—Predicting a one-rep max from reps-to-fatigue. *Journal of Physical Education, Recreation, & Dance*, *64*, 88–90.

Cardinal, B. J., & Spaziani, M. D. (2003). ADA compliance and the accessibility of physical activity facilities in Western Oregon. *American Journal of Health Promotion*, *17*(3), 197–201.

Carpinelli, R. N. (2011). Assessment of one repetition maximum (1RM) and 1RM prediction equations: Are they really necessary. *Medicina dello Sport*, *15* (2), 91–102.

Carroll, D. D., Courtney-Long, E. A., Stevens, A. C., Sloan, M. L., Lullo, C., Visser, S. N., … Dorn, J. M. (2014). Vital signs: Disability and physical activity—United States, 2009-2012. *Morbidity and Mortality Weekly Report*, *63*(18), 407–413.

Cowan, R. E., Callahan, M. K., & Nash, M. S. (2012). The 6-min push test is reliable and predicts low fitness in spinal cord injury. *Medicine and Science in Sports and Exercise*, *44*(10), 1993–2000.

Cowell, L. L., Squires, W. G., & Raven, P. B. (1986). Benefits of aerobic exercise for the paraplegic: A brief review. *Medicine and Science in Sports and Exercise*, *18*(5), 501–508.

Crawford, A., Hollingsworth, H. H., Morgan, K., & Gray, D. B. (2008). People with mobility impairments: Physical activity and quality of participation. *Disability and Health Journal*, *1*(1), 7–13.

Crouter, S. E., Antczak, A., Hudak, J. R., DellaValle, D. M., & Haas, J. D. (2006). Accuracy and reliability of the ParvoMedics TrueOne 2400 and MedGraphics VO2000 metabolic systems. *European Journal of Applied Physiology*, *98*(2), 139–151.

Freitas, P. S., Serenza, F. D., Santana, T. S., Manoel, L. S., & Riberto, M. (2017). Isokinetic analysis of rotator cuff musculature of sedentary wheelchair users and basketball athletes with spinal cord injury. *Journal of Physical Medicine, Rehabilitation & Disabilities*, *3*(3), 017.

Gail, S., Rodefeld, S., & Künzell, S. (2015). Reproducibility of a 5-repetition maximum strength test in older adults. *Isokinetics and Exercise Science*, *23* (4), 291–295.

Gibson, A. L., Wagner, D. R., & Heyward, V. H. (2019a). Principles of assessment, prescription, and exercise program adherence. In A. L. Gibson, D. R. Wagner, & V. H. Heyward (Eds.), *Advanced fitness assessment and exercise prescription* (pp. 55–78). Champaign IL: Human Kinetics.

Gibson, A. L., Wagner, D. R., & Heyward, V. H. (2019b). Assessing cardiorespiratory fitness. In A. L. Gibson, D. R. Wagner, & V. H. Heyward (Eds.), *Advanced fitness assessment and exercise prescription* (pp. 79–124). Champaign IL: Human Kinetics.

Hicks, A. L., Martin-Ginis, K. A., Pelletier, C. A., Ditor, D. S., Foulon, B., & Wolfe, D. L. (2011). The effects of exercise training on physical capacity, strength, body composition and functional performance among adults with spinal cord injury: A systemic review. *Spinal Cord*, *49*, 1103–1127.

Hiremath, S. V., Intille, S. S., Kelleher, A., Cooper, R. A., & Ding, D. (2016). Estimation of energy expenditure for manual wheelchair users using a physical activity monitoring system. *Archives of Physical Medicine and Rehabilitation*, *97*(7), 1146–1153.

Hjeltnes, N., & Vokac, Z. (1979). Circulatory strain in everyday life of paraplegics. *Scandinavian Journal of Rehabilitation Medicine*, *11*(2), 67–73.

Hol, A. T., Eng, J. J., Miller, W. C., Sproule, S., & Krassioukov, A. V. (2007). Reliability and validity of the six-minute arm test for the evaluation of cardiovascular fitness in people with spinal cord injury. *Archives of Physical Medicine and Rehabilitation*, *88*(4), 489–495.

Ilias, N. A., Xian, H., Inman, C., & Martin, W. H. (2009). Arm exercise testing predicts clinical outcome. *American Heart Journal*, *157*, 69–76.

Jacobs, P. L., & Beekhuizen, K. S. (2005). Appraisal of physiological fitness in persons with spinal cord injury. *Topics in Spinal Cord Injury Rehabilitation*, *10*(4), 32–50.

Jacobs, P. L., & Nash, M. S. (2004). Exercise recommendations for individuals with Spinal Cord Injury. *Sports Medicine*, *34*, 727–751.

Janssen, T. W., Dallmeijer, A. J., Veeger, D. J., & van der Woude, L. H. (2002). Normative values and determinants of physical capacity in individuals with spinal cord injury. *Journal of Rehabilitation Research and Development*, *39*(1), 29–39.

Jelleyman, C., Yates, T., O'Donovan, G., Gray, G., King, J. A., Khunti, K., & Davies, M. J. (2015). The effects of high-intensity interval training on glucose regulation and insulin resistance: A meta-analysis. *Obesity Reviews*, *16*, 942–961.

Jiang, S.-D., Dai, L. Y., & Jiang, L. S. (2006). Osteoporosis after spinal cord injury. *Osteoporosis International*, *17*(2), 180–192.

Johnson, M. J., Stoelzle, H. Y., Finco, K. L., Foss, S. E., & Carstens, K. (2012). ADA compliance and accessibility of fitness facilities in western Wisconsin. *Topics in Spinal Cord Injury Rehabilitation*, *18*(4), 340–353.

Keyser, R. E., Rasch, E. K., Finley, M., & Rodgers, M. M. (2003). Improved upper-body endurance following a 12-week home exercise program for manual wheelchair users. *Journal of Rehabilitation Research and Development*, *40*(6), 501.

Klaesner, J., Morgan, K. A., & Gray, D. B. (2014). The development of an instrumented wheelchair propulsion testing and training device. *Assistive Technology: The Official Journal of RESNA*, *26*(1), 24–32.

Lal, S. (1998). Premature degenerative shoulder changes in spinal cord injury patients. *Spinal Cord*, *36*(3), 186–189.

Lewis, J. E., Nash, M. S., Hamm, L. F., Martins, S. C., & Groah, S. L. (2007). The relationship between perceived exertion and physiologic indicators of stress during graded arm exercise in persons with spinal cord injuries. *Archives of Physical Medicine and Rehabilitation*, *88*(9), 1205–1211.

Liguori, G., & American College of Sports Medicine. (2020). *ACSM's guidelines for exercise testing and prescription*. Lippincott Williams & Wilkins.

Lombardi, V. P. (1989). *Beginning weight training: The safe and effective way*. Dubuque, IA: WC Brown.

Lovell, D., Shields, D., Beck, B., Cuneo, R., & McLellan, C. (2012). The aerobic performance of trained and untrained handcyclists with spinal cord injury. *European Journal of Applied Physiology*, *112*(9), 3431–3437.

Martin Ginis, K. A., Arbour-Nicitopoulos, K. P., Latimer, A. E., Buchholz, A. C., Bray, S. R., Craven, C., … Wolfe, D. L. (2010). Leisure time physical activity in a population-based sample of people with spinal cord injury part II: Activity types, intensities, and durations. *Archives of Physical Medicine and Rehabilitation*, *91*, 729–733.

Martin Ginis, K. A., van der Scheer, J. W., Latimer-Cheung, A. E., Barrow, A., Bourne, C., Carruthers, P., … Goosey-Tolfrey, V. L. (2018). Evidence-based scientific exercise guidelines for adults with spinal cord injury: An update and a new guideline. *Spinal Cord*, *56*, 308–321.

Molik, B., Kosmol, A., Morgulec-Adamowicz, N., Lencse-Mucha, J., Mroz, A., Gryko, K., et al. (2017). Comparison of aerobic performance testing protocols in elite male wheelchair basketball players. *Journal of Human Kinetics*, *60*, 243–254. https://doi.org/10.1515/hukin-2017-0140.

Morgan, K. A., Engsberg, J. R., & Klaesner, J. (2015). The testing of an instrumented wheelchair propulsion testing and training device. *Journal of Physical Medicine, Rehabilitation, & Disabilities*, *1*, 003.

Morgan, K. A., Taylor, K. L., Tucker, S. M., Cade, W. T., & Klaesner, J. W. (2018). Exercise testing protocol using a roller system for manual wheelchair users with spinal cord injury. *Journal of Spinal Cord Medicine*, *42*(3), 288–297.

Myers, J., Lee, M., & Kiratli, J. (2007). Cardiovascular disease in spinal cord injury: An overview of prevalence, risk, evaluation, and management. *American Journal of Physical Medicine and Rehabilitation*, *86*(2), 614–625.

Neto, F. R., Guanais, P., Dornelas, E., Coutinho, A. C. B., & Costa, R. R. G. (2017). Validity of one-repetition maximum predictive equations in men with spinal cord injury. *Spinal Cord*, *55*(10), 950–956.

Nightingale, T. E., Rouse, P. C., Thompson, D., & Bilzon, J. L. J. (2017). Measurement of physical activity and energy expenditure in wheelchair users: Methods, considerations, and future directions. *Sports Medicine—Open*, *3*(1), 10.

Post, M. W., & van Leeuwen, C. M. (2012). Psychosocial issues in spinal cord injury: A review. *Spinal Cord*, *50*, 382–389.

Reynolds, J. M., Gordon, T. J., & Robergs, R. A. (2006). Prediction of one repetition maximum strength from multiple repetition maximum testing and anthropometry. *The Journal of Strength & Conditioning Research*, *20*(3), 584–592.

Rimmer, J. H., Chen, M. D., McCubbin, J. A., Drum, C., & Peterson, J. (2010). Exercise intervention research on persons with disabilities: What we know and where we need to go. *American Journal of Physical Medicine and Rehabilitation*, *89*, 249–263.

Rodgers, M. M., Keyser, R. E., Rasch, E. K., Gorman, P. H., & Russell, P. J. (2001). Influence of training on biomechanics of wheelchair propulsion. *Journal of Rehabilitation Research and Development*, *38*(5), 505–511.

Shechtman, O., MacKinnon, L., & Locklear, C. (2001). Using the BTE Primus® to measure grip and wrist flexion strength in physically active wheelchair users: An exploratory study. *American Journal of Occupational Therapy*, *55*(4), 393–400.

Simmons, O. L., Kressler, J., & Nash, M. S. (2014). Reference fitness values in the untrained spinal cord injury population. *Archives of Physical Medicine and Rehabilitation*, *95*, 2272–2278.

Souza, A. L., Boninger, M. L., Fitzgerald, S., Shimada, S. D., Cooper, R. A., & Ambrosio, F. (2005). Upper limb strength in individuals with spinal cord injury who use manual wheelchairs. *The Journal of Spinal Cord Medicine*, *28*(1), 26–32.

Stephens, C. L., & Engsberg, J. R. (2010). Comparison of overground and treadmill propulsion patterns of manual wheelchair users with tetraplegia. *Disability and Rehabilitation. Assistive Technology*, *5*(6), 420–427.

Sullivan, K. J. (2009). Healthcare reform: A call to action for our patients. *Journal of Neurologic Physical Therapy*, *33*(3), 171–172.

Tørhaug, T., Brurok, B., Hoff, J., Helgerud, J., & Leivseth, G. (2016). Arm crank and wheelchair ergometry produce similar peak oxygen uptake but different work economy values in individuals with spinal cord injury. *BioMed Research International*, *2016*, 5481843. https://doi.org/10.1155/2016/5481843.

Totosy de Zepetnek, J. O., Au, J. S., Hol, A. T., Eng, J. J., & MacDonald, M. J. (2016). Predicting peak oxygen uptake from submaximal exercise after spinal cord injury. *Applied Physiology, Nutrition, and Metabolism*, *41*(7), 775–781.

U.S. Department of Health and Human Services. (2018). *2018 physical activity guidelines for Americans*. Washington, DC: U.S. Department of Health and Human Services.

van der Scheer, J. W., Martin Ginis, K. A., Ditor, D. S., Goosey-Tolfrey, V. L., Hicks, A. L., West, C. R., & Wolfe, D. L. (2017). Effects of exercise on fitness and health of adults with spinal cord injury. *Neurology*, *89*(7), 736–745.

Vanhees, L., Lefevre, J., Philippaerts, R., Martens, M., Huygens, W., Troosters, T., & Beunen, G. (2005). How to assess physical activity? How to assess physical fitness? *European Journal of Cardiovascular Prevention & Rehabilitation*, *12*(2), 102–114.

Vidal, J., & Sarrias, M. (1991). An analysis of the diverse factors concerned with the development of pressure sores in spinal cord injured patients. *Paraplegia*, *29*(4), 261–267.

West, C. R., AlYahya, A., Laher, I., & Krassioukov, A. (2013). Peripheral vascular function in spinal cord injury: A systematic review. *Spinal Cord*, *51*, 10–19.

Chapter 41

# Community-based activity-based therapy for spinal cord injuries rehabilitation

Natalia Padula[a] and Camila Quel de Oliveira[b,]*

[a]*Acreditando Center of Neuromotor Recovery, Health and Wellness, Sao Paulo, Brazil;* [b]*Discipline of Physiotherapy, Graduate School of Health, University of Technology Sydney, Ultimo, NSW, Australia*

## Abbreviations

| | |
|---|---|
| **SCI** | spinal cord injury |
| **ABT** | activity-based therapy |
| **CPGs** | central pattern generators |
| **LT** | locomotor training |
| **ISNCSCI** | International Standards for Neurological Classification of Spinal Cord Injuries |
| **ICF** | International Classification of Functioning, Disability and Health |
| **FES** | functional electrical stimulation |
| **NMES** | neuromuscular electrical stimulation |
| **BWSTT** | body weight support treadmill training |
| **WBV** | whole body vibration |

## Introduction

Injuries to the spinal cord result in profound disability and involve many body systems, resulting in severe physiological changes. The most obvious consequence is paralysis, which occurs immediately after injury and may persist throughout life (Harvey, Glinsky, & Bowden, 2016). Paralysis and deconditioning are responsible for reduced mobility and independence and play a significant role in secondary health complications, such as cardiovascular and respiratory dysfunction, metabolic syndrome, bone demineralization, immunosuppression, and musculoskeletal deterioration (Galea, 2012).

Despite all rehabilitation efforts to promote recovery of the injured spinal cord, more than 50% of individuals face long-term disability (Ditunno, 1999). Traditionally, rehabilitation strategies have been successful in improving independence mainly by providing compensatory strategies focused on the use of the preserved muscles above the level of injury and using leverage and momentum to move weak or paralyzed body parts. However, most individuals only have access to physical interventions during the acute and subacute phases of injury through inpatient and outpatient rehabilitation programs (Behrman & Harkema, 2007). After discharge from hospital, most individuals with a spinal cord injury (SCI) face several barriers to continue to exercise and be active in the community, such as lack of specialized professionals, programs, and facilities. Inactivity and sedentary behavior during the chronic phases of injury can lead to decline in mobility and independence and impose a greater risk of developing co-morbidities (Ginis et al., 2011).

The reduction in physical activity levels and the systemic impairments after injury result in a sedentary lifestyle (Wyndaele & Wyndaele, 2006). Sedentary behavior and inactivity are the main risk factors for the development of metabolic and cardiovascular disorders and reduction in muscle strength and aerobic capacity. Therefore, it is necessary to offer specialized physical therapies, exercise programs, sports, and recreation to individuals with SCI to increase their physical activity levels and improve health (Barker et al., 2009).

Over the past years, there has been growing evidence related to the central nervous system plasticity, repair, and regeneration. However, the concept of an irreparable, "hard-wired" nervous system still guides rehabilitation after SCI in most

*Dr. Camila Quel de Oliveira is the first author of this chapter.

Diagnosis and Treatment of Spinal Cord Injury. https://doi.org/10.1016/B978-0-12-822498-4.00041-5

countries. In general, physical rehabilitation still focuses solely on teaching patients how to achieve day-to-day function by using compensatory strategies and assistive devices, despite significant neurological deficits (Behrman, Bowden, & Nair, 2006).

Activity, defined as both function-specific motor tasks and exercise appears to be a necessity for optimization of functional, metabolic, and neurological status in chronic paralysis (Sadowsky & McDonald, 2009). Exercises that target systems below the level of injury (sensory, musculoskeletal, vascular) are essential to promote recovery after SCI and prevent secondary complications related to paralysis, such as pressure injuries, joint contractures, and pneumonia, promoting a better life (Galea, 2012). And even more important, as the life expectancy after SCI approaches that of the able-bodied population and when the worldwide number of survivors continues to grow to over 2 million people (Fawcett et al., 2007).

The evidence regarding spinal cord plasticity that emerged from animal (Battistuzzo, Callister, Callister, & Galea, 2012) and human studies (Fouad, Krajacic, & Tetzlaff, 2011) and the knowledge that augmented sensory stimulation and repeated active movement can lead to functional improvements (Harkema, Schmidt-Read, Lorenz, Edgerton, & Behrman, 2012) led to the necessity of developing more comprehensive rehabilitative interventions. New interventions focused on functional improvement through neurological recovery aroused. Those interventions were termed Activity-Based Therapies (ABT), and numerous programs emerged globally with the goal of assisting people with SCI at any level or severity to achieve their maximal potential for recovery (Backus, Apple, & Hudson, 2011; Harkema, Behrman, & Barbeau, 2012).

## Recovery after spinal cord injuries

The advances in the acute medical management of people with a spinal cord injury (SCI), such as early spinal decompression, pharmacological interventions, and intra-operative temperature management (Kraft, Karpenko, & Rincon, 2016) have been successful in reducing secondary neuronal damage and the extension of the injury, resulting in greater clinical recovery and better outcomes. Neuroprotective strategies have reduced the incidence of complete injuries and have increased the chances for functional recovery.

However, the intrinsic potential for recovery of the spinal cord after injury seems to be limited and not responsive to the conventional treatment strategies. It is well known that during the first weeks, and even months after injury, spontaneous recovery may occur due to the resolution of the swelling and return of the neural activity below the lesion site in response to the acute injury (Marino, Ditunno, Donovan, & Maynard, 1999). Spontaneous motor and sensory recovery have been reported during the first 2 years, with major recovery occurring between 3 and 6 months post-injury (Kirshblum & O'Connor, 1998). Critical factors in determining recovery in the first year after a traumatic SCI include the initial neurological level of injury, initial motor strength and, most importantly, the severity of injury. Kirshblum, Millis, McKinley, and Tulsky (2004) investigated the rate of neurological recovery of 987 individuals with a traumatic SCI during their first 5 years and found small degrees of neurologic recovery between 1 and 5 years, with 5.6% of cases presenting late conversion from a sensorimotor complete to an incomplete lesion status, demonstrating that spinal cord plasticity, similarly to the brain, can persist for many years post-injury.

Substantial functional recovery was seen after SCI in animals in response to training (Fouad & Tse, 2008). The degree of sensorimotor recovery was defined largely by the type and intensity of motor training and sensory stimulation provided following the injury (Edgerton, Tillakaratne, Bigbee, De Leon, & Roy, 2004). Recovery after SCI has been related to the spinal circuits' capacity for automaticity and plasticity. Automaticity is the ability of the spinal cord to perform complex motor tasks without supraspinal inputs and plasticity refers to the ability to recover or change, potentially stimulated by biochemical changes at the cellular level in the spinal cord. There is evidence that plasticity can be induced by physical activity, thus leading to sensorimotor recovery (Lynskey, Belanger, & Jung, 2008). The physiological mechanisms behind spinal cord plasticity are possibly related to changes in the cerebral cortex, growth of new axonal branches, remodeling of synapses, changes in neuronal excitability, and modulation of neurotransmitters in the spinal cord (Edgerton & Roy, 2009).

Training and exercise seem to play an important role in enabling spinal plasticity by inducing changes at the cellular and molecular levels. It is suggested that the spinal cord has the ability to change when the sensory circuits are repeatedly activated (Roy, Harkema, & Edgerton, 2012). This capacity can be due to the activation of complex neuron networks, often termed central pattern generators (CPGs), which are responsible for organizing complex motor movements within the spinal cord.

In humans, similar responses to activity and training have been found, demonstrating the capacity of the human spinal cord to respond to sensory inputs and adapt after injury. Repetitive training is critical for promoting changes to the spinal networks resulting in functional recovery. Therefore, physical interventions aimed to optimize the neuroplastic mechanisms of the spinal cord are potential tools to promote functional recovery (Edgerton, Kim, Ichiyama, Gerasimenko, & Roy, 2006).

Furthermore, recent studies have shown that approximately 50% of the injuries functionally classified as sensorimotor complete (International Standards for Neurological Classification of Spinal Cord Injury—ISNCSCI A) are actually anatomically incomplete with some degree of preservation of the nervous fibers. This new knowledge unveils the potential for recovery of injuries previously considered immutable, especially in response to interventions aimed at increasing the excitability and activation of the preserved spinal circuitries (Wrigley, Siddall, & Gustin, 2018).

In summary, the scientific findings that support the capacity of the spinal cord to change and recover, even years after injury, warrant that clinicians should employ interventions aimed at the preservation of the function of spinal neuronal circuits under the level of injury as an important condition for regeneration (Dietz & Curt, 2006). The dysfunctional central nervous system (CNS) reorganization and repair are dependent on maintaining an optimal level of neurological activity above and below the site of injury (Sadowsky & McDonald, 2009).

## Conventional rehabilitation after spinal cord injuries

Traditionally, management of a person with SCI involves the determination of the person's level of injury and functional capacity, and then the generation of a problem list of activity limitations and prescription for therapies, with the overall goal of having the patient achieve their maximum functional capacity, according to the overall level and severity of neurological injury (Stampas & Tansey, 2014).

Often, clinicians are guided by the World Health Organization International Classification of Functioning, Disability and Health (ICF) framework to assess the rehabilitation needs of people with SCI. The ICF is an integrative and biopsychosocial approach that comprehensively describes the impact of a health condition and health interventions on an individual's functioning. It consists of assessing three domains: (1) Body structure and functioning; (2) Activity limitations; and (3) Participation restrictions (Escorpizo & Bemis-Dougherty, 2015). The ultimate goal of rehabilitation after SCI is to enable the person to return to a productive and satisfying life (Harvey, 2016). Therefore, rehabilitative interventions delivered in the community should focus on strategies to increase activity levels in order to decrease participation restrictions.

Conventional physical rehabilitation aims to restore independence, physical capacity, and community participation, focusing on mobility and self-care (Van Langeveld et al., 2011). Exercise therapy is often employed to the preserved or partially preserved muscles to improve strength, regulate tone and augment muscle length and joint mobility. Moreover, increased fitness, reduced pain and prevention of cardio-metabolic conditions are priorities. Retraining of motor tasks, such as hand and arm function, bed mobility, transfers and wheelchair use, standing, and walking are frequent goals during inpatient rehabilitation (Harvey, Lin, Glinsky, & De Wolf, 2009).

To achieve the rehabilitation goals, therapists rely substantially on providing compensatory strategies focusing on the use of the preserved muscles above the level of injury and using leverage and momentum to teach patients how to move weak or paralyzed body parts. Orthotic devices and modifications to the environment are provided to compensate the loss of function, and the use of the upper limbs is encouraged to substitute for lacking movements of the trunk and lower limbs (Behrman et al., 2006).

While these strategies are effective in increasing mobility, independence and participation (Karimi, Omar, & Fatoye, 2014), they do not target the recovery of the affected areas below the site of injury and do not consider that further functional gains may be possible. They also neglect the need of exercise and activity to the paralyzed areas in order to preserve the body systems below the site of injury and to prevent the deleterious effects of non-use (Galea, 2012). Hence, in the last decades, new physical therapies based on the advances in spinal cord neuroplasticity have emerged, leading to a change in the focus of rehabilitation after SCI from the compensatory model to a recovery-based model (Fig. 1).

## Activity-based therapies

### Definition

Activity-based therapies (ABTs) are a group of interventions that aim to maximize recovery by focusing on areas above and below the site of injury and this is known as the activity-dependent recovery model (Edgerton & Roy, 2009). ABT employs repetitive task-specific training using weight bearing and external facilitation of neuromuscular activation (Dolbow et al., 2015). In contrast to the traditional approach, ABT interventions aim to drive changes to the nervous and muscular systems

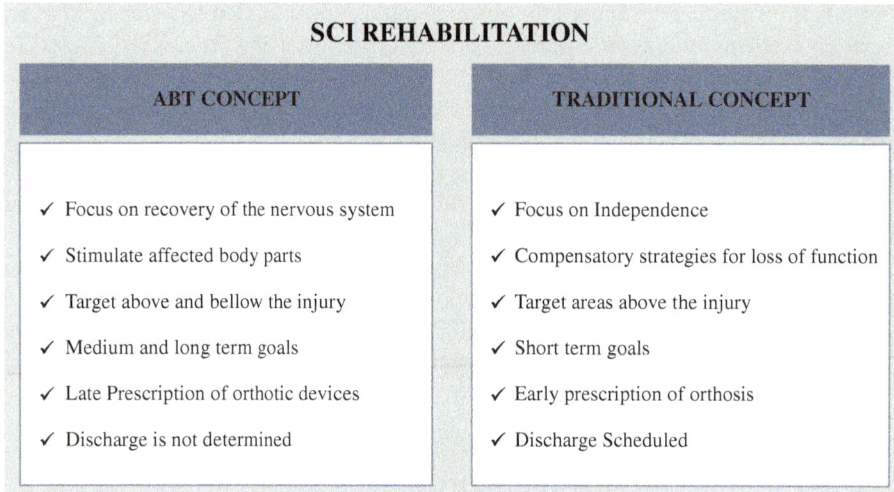


FIG. 1 Differences between the two concepts of rehabilitation in spinal cord injuries.

via repetitive activation of the neuromuscular system above and below the level of injury (Behrman et al., 2006). The key components of ABT interventions that elicit plasticity to the spinal cord are the sensory input, the specificity of the task and the repetitive practice of the task.

Exercise targeted at regions and anatomical structures below the spinal cord injury promoted neuroplastic changes at cellular and biochemical levels in animal models (Roy, Harkema, & Edgerton, 2012) and led to cortical changes in humans with incomplete SCI (Chisholm, Peters, Borich, Boyd, & Lam, 2015). Besides preserving muscle mass and restoring motor and sensory function, exercise can induce synaptic plasticity by increasing the concentration of neurotrophic factors and the number of regenerating neurons (Sandrow-Feinberg & Houle, 2015). This exercise-induced plasticity and the emerging conceptions that the spinal cord has the capacity to undergo plastic changes and control complex movements provide foundation for ABT interventions.

Backus, Apple, and Hudson (2009) defined ABT as "any intervention that specifically uses tools and to improve muscle activation or sensory function below the level of injury in the spinal cord, and does not rely on compensatory mechanisms for improving function after SCI. Such an approach includes interventions that combine intensive active movement with one or more of the following: facilitation techniques (use of tactile or vibratory stimulation; electrical stimulation applied to muscles or nerves (surface or indwelling); body-weight supported locomotor training (manual or robotic); use of upper extremity robotics; or massed-practice training)."

## Goals

The general goals of ABT are to optimize the neurological system function to offset the rapid aging, physical deterioration and secondary complications associated with SCI, while promoting functional recovery, increasing autonomy and participation. The ICF framework is used for the decision-making and planning of the rehabilitation program.

## Multimodal activity-based therapy

The Reeve Foundation's Neurorecovery Network (NRN) was the pioneer in the research of ABT modalities and its clinical implementation across a network of rehabilitation hospitals and community-based clinics in North America. Other examples of ABT clinics around the world are the NeuroMoves Exercise Programs in Australia, the Acreditando Centre of Neuromotor Recovery in Brazil, and Neurokinex in the United Kingdom.

There is growing evidence on the effects of multimodal ABT delivered in community settings on body structures and function after SCI. Significant improvements in neurological function and walking was reported in chronic individuals with SCI after 6 months of training (Harness, Yozbatiran, & Cramer, 2008). Improvements in sitting balance, general mobility and independence were also found after 3–9 months of ABT delivered in the community to individuals with chronic SCI (Quel de Oliveira, Refshauge, Middleton, de Jong, & Davis, 2017). Furthermore, the various components of ABT (i.e., FES, activities in load bearing and muscle strengthening above and below the level of injury) reversed negative body composition changes after SCI, resulting in the reduction of risk of cardiovascular and other metabolic diseases (Dolbow et al., 2015).

Community-based ABT clinics and gyms deliver multimodal exercise programs individually tailored to the clients' goals and functional abilities by an exercise physiologist or physiotherapist. Exercises target whole-body strengthening and are performed out of the wheelchair and in positions where load bearing is applied to the paralyzed extremities (e.g., standing, 4-point kneeling). The exercise programs usually involve a combination of (1) activation of motor patterns (locomotor training, and functional electrical stimulation (FES) associated with ergometry); (2) non-standardized motor activation (recruitment, strengthening, and training of specific tasks); and (3) sensory stimulation (sensorimotor training, and body vibration on vibrating platforms) (Sadowsky & McDonald, 2009). Those interventions are discussed in more detail next.

## Locomotor and gait training

Locomotor training (LT) aims to improve sensory, motor and autonomic functioning, health and quality of life. The main focus is to promote recovery of postural control and walking after SCI. LT is based on the principles derived from studies using animal models that demonstrated recovery of the ability to weight bear on the hind limbs and walk after exposure to repetitive step training (i.e., by training for the specific motor task) (Harkema, Hillyer, et al., 2012).

The LT protocol developed by the NeuroRecovery Network involves the use of an overhead body-weight support system attached to a harness where the individual is suspended over a treadmill. Trained therapists facilitate control of balance and manually assist trunk and leg movements during stepping and standing to generate sensory information that is consistent with locomotion (Fig. 2A). In conjunction with the body-weight supported treadmill training, LT programs usually include an overground training component, where activity limitations are assessed and task-specific training incorporating sensory feedback is performed outside the body-weight support system with the aim to contribute to functional recovery (Behrman et al., 2005; Harkema, Schmidt-Read, Behrman, et al., 2012).

LT is the most investigated ABT intervention to date and there is mounting evidence of its benefits after acute and chronic SCI. A large number of studies have investigated the effects of LT on walking ability in individuals with incomplete SCI and have demonstrated improvements in walking speed, distance and coordination (Adams & Hicks, 2011; Dobkin et al., 2006; Field-Fote & Roach, 2011). Moreover, neurological recovery has been reported through studies that showed increments in lower extremity motor score and ISNCSCI classification after LT (Anwer et al., 2014). Gains in balance, functional mobility and independence were also reported (Alexeeva et al., 2011).

LT can also be deployed as an exercise modality to improve general health after complete and incomplete SCI. Improvements in cardiorespiratory health (Jeffries et al., 2015), reversal of muscle atrophy (Giangregorio et al., 2006), and enhanced autonomic function related to blood pressure control have been reported after LT (Harkema, Ferreira, Van Den Brand, & Krassioukov, 2008).

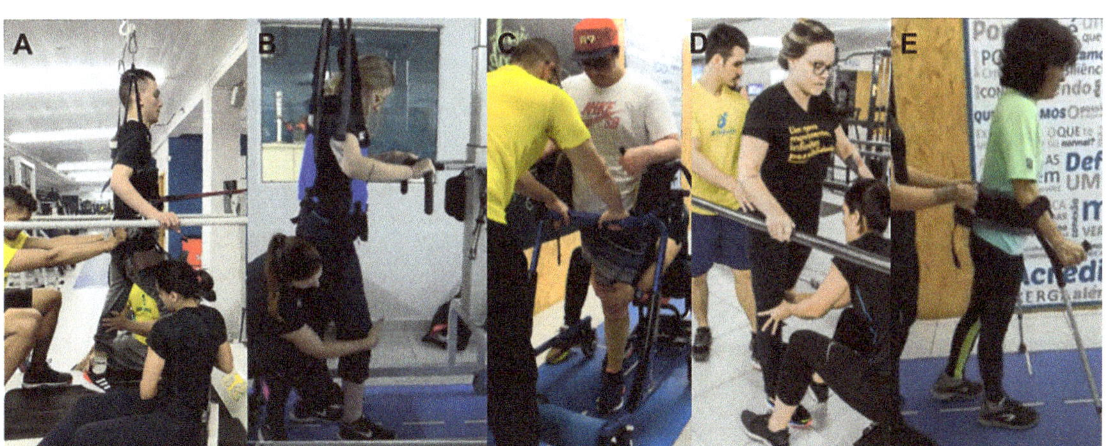


FIG. 2 Examples of locomotor and gait training. (A) Body-weight support treadmill locomotor training, (B) body-weight support overground gait training, (C) overground gait training with walker and assistance, (D) overground gait training in parallel bars and assistance, (E) overground gait training with crutches and supervision.

Although LT has demonstrated to be beneficial for individuals with SCI, the training volume required to achieve significant gains is high, ranging from 4 to 5 times per week for 1.5 h per session. Hence, its application in community rehabilitation settings is not always feasible due to the challenges in maintaining adherence and the financial burden to the participants. Therefore, community-based clinics often offer multimodal ABT programs, where LT can be a component of the rehabilitation program. It is worth noting that LT is also employed as a training strategy with the overall aim is to promote neurological recovery and health benefits. The goal may not always be related to walking.

Task-specific gait training strategies are also employed in ABT programs with the aim to promote functional recovery of walking. Similarly to traditional rehabilitation, ABT programs may use external assistance, walking aids and orthotic devices to retrain walking according to the client's abilities and functional goals, as illustrated in Fig. 2B–E. However, in ABT, the prescription of orthotic devices and walking aids is delayed, when compared to traditional rehabilitation, and will occur only when the strategies for recovery are exhausted and the recovery of motor function is not achieved. LT and other modalities of gait training can be used in ABT programs as a form of whole-body exercise with the aim of improving gross motor function (when walking is not a goal), and cardiovascular fitness due to the physiological benefits of moving in the upright position.

## Neuromuscular electrical stimulation (NMES) and functional electrical stimulation (FES)

When applied to certain body systems, electrical currents have the property of eliciting an electrical response in different cells, such as muscle and neurons. Over the years, novel currents and devices have been developed to safely elicit a neurological response, such as muscle contraction, in paralyzed or weakened muscles to assist individuals with different medical conditions, including SCI, to gain muscle strength and reverse the deleterious effects of paralysis.

Neuromuscular stimulation (NMES) is the technical name used when electric current is used to reduce deficits resulting from clinical conditions, such as non-use. NMES can increase the individuals' participation in voluntary activities producing contractions of the paralyzed muscles that are still innervated, and as a result can improve muscle trophism, strength, and tone (Marquez-Chin & Popovic, 2020).

Functional electrical stimulation (FES) is a subtype of NMES, in which the stimulation of a nerve or a paralyzed muscle promotes assistance to specific and functional movement such as cycling as illustrated in Fig. 3A, assisted walking (Fig. 3B) or grasping (Fig. 3C and D). FES and other types of neuromuscular electrical stimulation improve blood circulation, range of motion, muscle strength and endurance, and reduce spasticity, atrophy and loss of bone mineral density, while increasing metabolism and cardiorespiratory function (Luo, Xu, Zuo, Liu, & All, 2020; Martin, Sadowsky, Obst, Meyer, & McDonald, 2012).

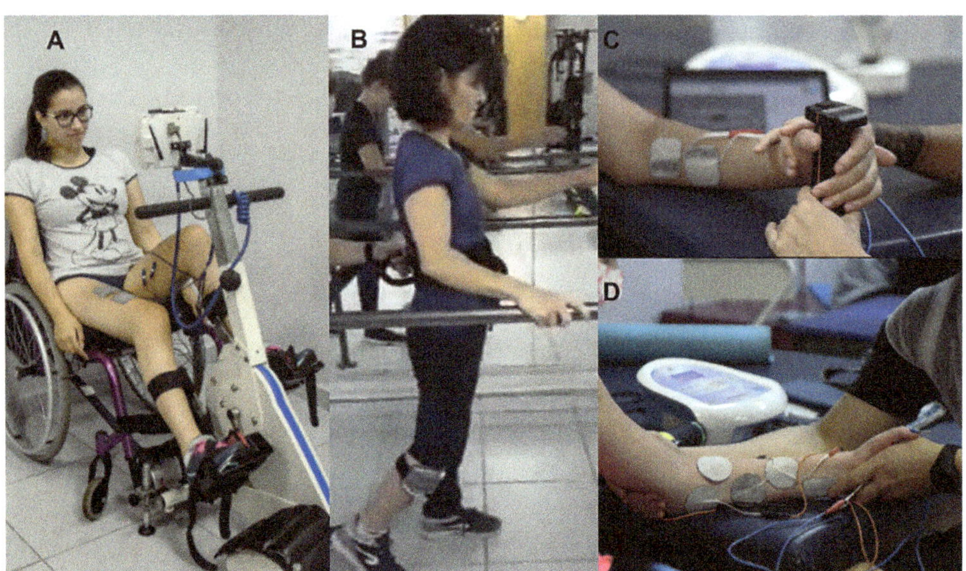


FIG. 3 Applications of functional electrical stimulation (FES). (A) FES associated with cycling. (B) FES associated with gait training. (C and D) FES associated with grasping.

Due to the evidence on the benefits of the use of electrical stimulation in people with complete and incomplete SCI, this modality is frequently used in ABT programs to increase the activation and strength of the paralyzed or partially paralyzed muscles to promote recovery of function and mobility. Furthermore, when simultaneously applied to multiple large paralyzed muscles, it increases energy consumption and cardiac output and therefore, can be used as a cardiovascular fitness modality in people with severe paralysis.

## Developmental postures: Weight-bearing activities

Developmental sequencing is focused on strengthening the primary stabilizing muscles of the trunk and pelvis because of their role in axial stability; this approach involves training in various positions and contributes to better posture, increased trunk control, general mobility (including upright function and walking) and independence. These include activities performed in quadruped (on all fours), kneeling, sitting, and standing ( Jones et al., 2014), as illustrated in Fig. 4. Movements and positions are designed to mimic the human developmental process and to allow weight bearing to the paralyzed or partially paralyzed limbs. Weight bearing, especially in standing can improve bowel and bladder function, bone mineral density and joint range of motion; prevent pressure injuries and muscle contractures; enhance autonomic and cardiovascular function (Can, Dosoglu, Karacan, & Karamehmetoglu, 2007); reduce spasticity (Adams & Hicks, 2011); and improve motor function and quality of life.

## Vibration

Whole-body vibration (WBV), using vibratory platforms, is commonly used as an exercise or therapeutic tool with numerous benefits for the musculoskeletal system in both able-bodied and clinical populations. In ABT, it is commonly associated with developmental postures and standing (Fig. 5) with the goal to improve spasticity and muscle activation ( Ji et al., 2017), proprioception, balance and functional activities such as standing and walking (In, Jung, Lee, & Cho, 2018).

## Strength and aerobic training

Strength and aerobic training are essential to prevent clinical complications and promote health in general populations as well as in SCI (Valent, Dallmeijer, Houdijk, Talsma, & van der Woude, 2007). The physical activity guidelines for people with SCI recommends that individuals should engage in aerobic training activities at least 2 days per week (minimum of 20 min continuously) and perform strengthening exercises with every functioning muscle group; both modalities should be exerted at moderate-to-vigorous intensity. In order to exercise safely and effectively, it is important to adapt body positions and equipment according to the individual's level of functionality as demonstrated in Figs. 6 and 7 (Tweedy et al., 2017). Regarding the type of aerobic fitness exercises, arm/leg cycling and wheelchair propulsion training with/without functional electrical stimulation (FES) have been recommended. Also, progressive resistance training, alone or in combination with FES, has shown to increase muscle strength in individuals with SCI (Eitivipart, de Oliveira, Arora, Middleton, & Davis, 2019).

FIG. 4 Developmental Postures. Exercises in load bearing positions. (A) Prone. (B) Quadruped. (C) Kneeling. (D) Standing.

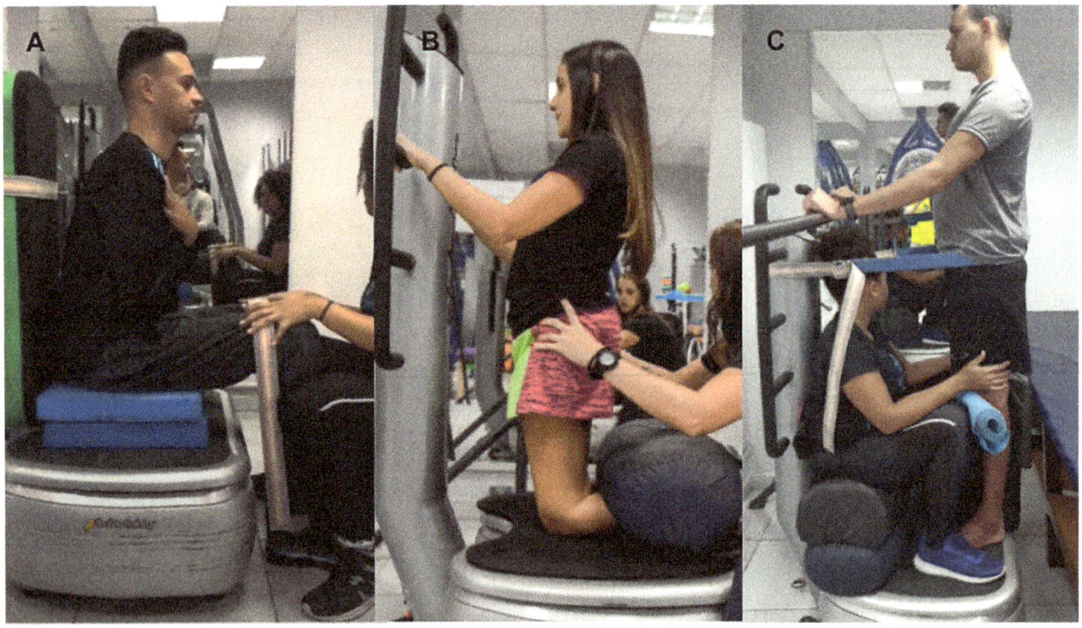

FIG. 5 Examples of whole body vibration (WBV) exercises. (A) WBV for trunk control in sitting. (B) WBV in kneeling. (C) WBV in standing.

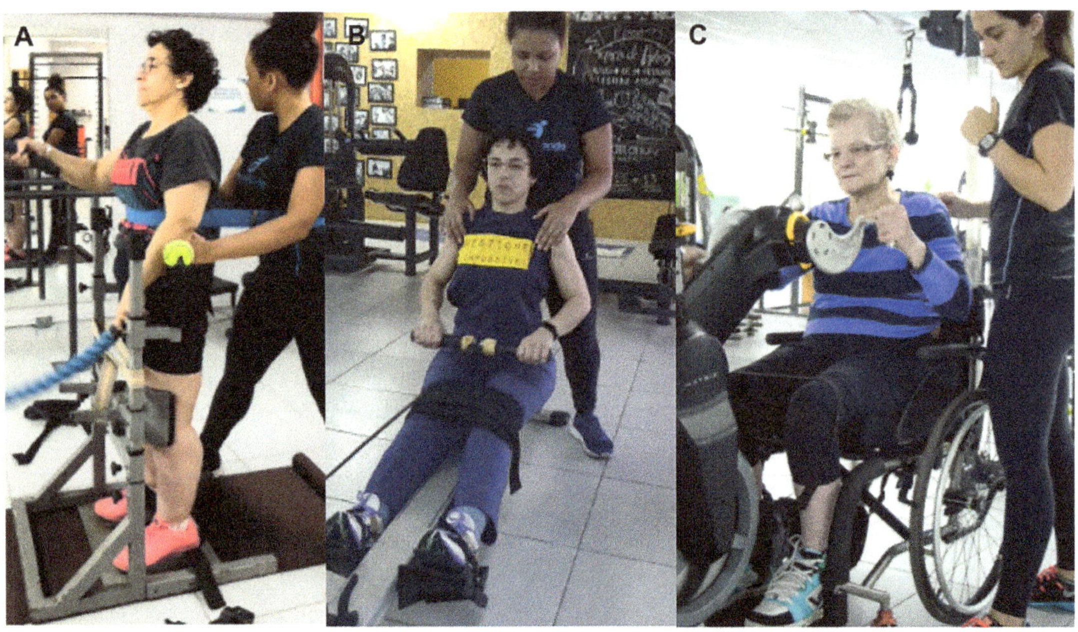

FIG. 6 Examples of aerobic exercises. (A) Battle rope in standing. (B) Adapted rowing. (C) Upper limb cycle ergometer.

For this reason, most ABT programs in the community include progressive resistance training and aerobic activities as part of the exercise routine. The exercises are performed mainly out of the wheelchair with engagement of the body parts below the site of injury via load bearing (Fig. 6A), FES (Fig. 3A) or body-weight supported treadmill training (Fig. 2A).

## Active and active-assisted exercises

In the body parts where clients have little to no voluntary movement, ABT therapists employ active or active-assisted exercises using facilitation strategies such as: manual guidance, gravity eliminated positions, suspension of the limb, reduction of friction and lever. Therapists help the client through different ranges of motion while assisting or providing resistance

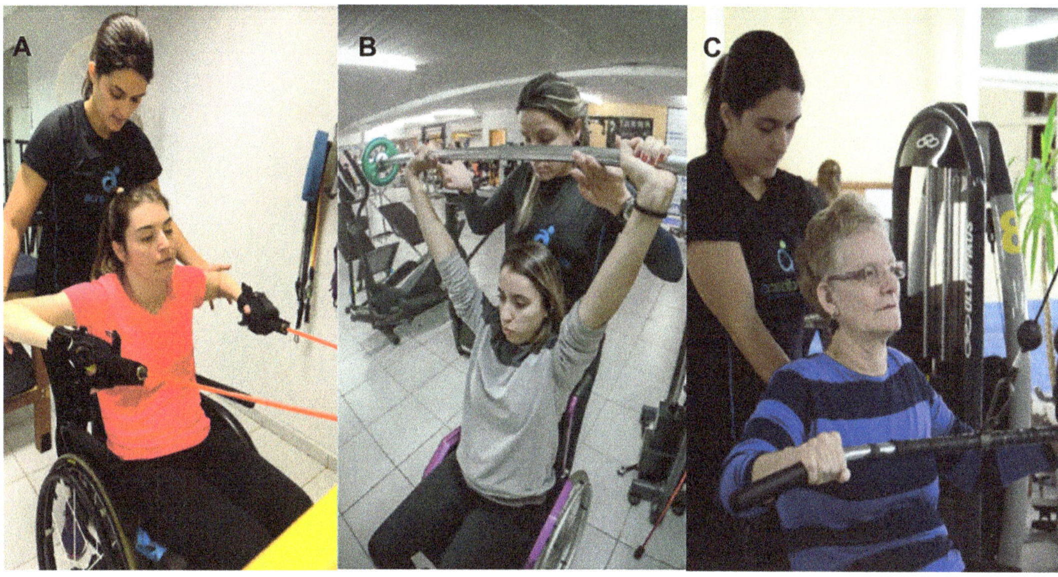


FIG. 7 Examples of strengthening exercise. (A) Exercise using elastic resistance. (B) Exercise using free weights. (C) Exercise using weight training machine.

less than gravity, according to the client's control of the body segment. Clients are instructed to attempt or visualize actively assisting or resisting the movement performed, even when voluntary movement is not possible. Active-assisted exercises attempt to provide sensory stimulus and elicit a motor response, even if minimal, with the goal of generating neural activity, and ultimately, using a high number of repetitions, achieving long-term neural potentiation ( Jones et al., 2012).

## Dosage

High volumes of training (i.e., frequency and duration) have shown to promote motor learning and skill acquisition for people with neurological conditions (Waddell, Birkenmeier, Moore, Hornby, & Lang, 2014), and therefore facilitate neuroplastic mechanisms. Unlike conventional rehabilitation, ABT interventions require high-volume practice, involving a high number of repetitions and high frequency (Behrman et al., 2006; Edgerton, Kim, Ichiyama, Gerasimenko, & Roy, 2006). ABT performed for 2 to 5 h per day at a frequency of three to five times per week produced benefits in participants with chronic SCI that included improved lower limb muscle strength, balance, mobility, increased gait speed, symmetry, and endurance in participants with chronic SCI (Behrman & Harkema, 2007).

## Assessment and outcome measures

SCI has a vast range of clinical features leading to high variability in the classification, which reduces the ability to test the therapeutic efficacy of an intervention accurately (Behrman et al., 2012). There is a critical need to develop valid, reliable, and sensitive scales and assessments to be used in many aspects of rehabilitation of people with SCI, as well as for carrying out patient-centered clinical trials. The lack of more specific scales to assess mobility, functional independence, and quality of life may prevent the detection of changes after an intervention. There is a need to create new tools that are more sensitive to functional gains and changes in movements to assess the effects of programs that are aimed at recovery of function, such as ABT, especially in the chronic phase (Padula et al., 2015; Wilson & Fehlings, 2021).

Community ABT programs often employ multiple tools to assess the individual holistically, using the ICF framework to monitor the effects of the program overtime. Some examples of commonly measures to assess body function and structure in ABT programs are: muscle strength (manual muscle testing, 10 repetition maximum); muscle tone (Modified Ashworth and Tardieu Scales), range of motion, sensation (proprioception, pin prick, light touch). It is essential to also include measures of activity limitations and participation restrictions that are valid to the SCI population (Gaspar, Padula, Freitas, de Oliveira, & Torriani-Pasin, 2019). Some examples of measures of activity limitations are: Spinal Cord Injury Independence Measure (SCIM) and the Walking Index for Spinal Cord Injuries (WISCI). The Community Integration Questionnaire and the Quality of life Index can be used to assess participation outcomes in people with SCI engaging in ABT programs.

The NeuroRecovery Network advises the use of the NeuroRecovery Scale (NRS) to assess recovery and monitor the effects of ABT programs, especially locomotor training. It is an 11-item scale that assesses the quality of movement without compensatory movement patterns by comparing sitting, walking, standing and transfers to typical movement (Harkema, Shogren, Ardolino, & Lorenz, 2016). Development of measures, such as the NRS, is particularly relevant because rehabilitation strategies are encompassing ABT and will soon partner with regenerative medicine to advance neuromuscular recovery post-injury (Behrman et al., 2012; Tester et al., 2016).

### Practical aspects of ABT in the community

Most of the ABT training programs are staffed by physiotherapists and exercise specialists. A client entering the program is evaluated by a clinician who designs and implements an ABT program based on their level of injury, strength, skills, personal goals, preferences ( Jones et al., 2014), and segments to be stimulated (tetraplegia—exercises for the upper limbs, trunk, and lower limbs and paraplegia—exercises for the trunk and lower limbs) (Padula et al., 2015).

Jones et al. (2014) developed algorithms to guide decision-making and progression of ABT interventions, especially locomotor training. The algorithms include muscle tone, trunk control and gait ability, taking into consideration the client's functional status to prescribe the type, duration and combination of interventions. Thus, individuals with greater deficits receive activities with greater assistance, such as active-assisted exercises, and spend more hours in developmental postures (i.e., standing in a standing frame or with manual guidance from a therapist), while individuals with better functional status, spend most of the time performing activities with less external assistance and less aids.

In conclusion, considering the literature about the needs and objectives of rehabilitation and physical monitoring of people with SCI living in the community, as well as the evidence of the benefits of ABT interventions, a multi-modal ABT program seems to be an appropriate exercise program for this population. Studies that evaluated the effects of a multimodal program within the approach, demonstrated positive results in promoting functional recovery and activity as well as in the prevention and treatment of common comorbidities in this population. Since ABT is a novel approach, with a constantly evolving literature, more studies and protocols are needed to improve assessments, program designs and dosage, while considering the heterogeneity of the population with SCI.

## Applications to other areas of neuroscience

In this chapter, we addressed the principles of activity-based therapy and its applicability in the rehabilitation of the people with chronic spinal cord injuries living in the community. However, the same principles apply to in-hospital rehabilitation for individuals in the acute and subacute phases of injury. It's worth noting that activity-based therapy is currently applied to people with other neurological injuries such as traumatic brain injury, stroke, cerebral palsy and multiple sclerosis. Activity-based therapy principles and practices may have similar benefits to conditions such as spina bifida and transverse myelitis, although research in other conditions is still scarce.

Fundamentals of neuroplasticity (intensity, repetition and specificity) form the basis for activity-based therapies. Neuroscientists and neurophysiologists should further investigate the mechanisms of plasticity and adaptability of the spinal cord after injury in order to further develop activity-based therapies, explore its optimal dosage and maximize recovery after spinal cord injuries. Moreover, activity-based therapy seems to be crucial to optimize the effects of upcoming regenerative interventions and treatments, such as stem cells (Belegu, Oudega, Gary, & McDonald, 2007) and epidural stimulation (Rejc, Angeli, Atkinson, & Harkema, 2017) in reversing paralysis and other deleterious consequences from an SCI. It is important to highlight that maintaining a healthy lifestyle and a healthy muscular, nervous and vascular system through exercising below the injury may be a key factor for eligibility and success of those novel regenerative strategies.

Overall, we have gathered information on clinical practice to guide rehabilitation professionals (physicians, physiotherapists, occupational therapists, and exercise physiologists) and academics in their decision-making and understanding of this novel intervention.

## Mini dictionary of terms:

**Metabolic syndrome:** Group of risk factors that raises one's risk for heart disease and other health problems, such as diabetes and stroke.

**Compensatory strategies:** Environmental modifications or behavioral strategies designed to bypass a persistent impairment. I.e.: use of the arms to transfer in the presence of lower limb paralysis.

**Plasticity (neuroplasticity):** The ability of the nervous system to form and reorganize synaptic connections, especially in response to learning or experience or following injury.
**Orthotic devices:** Braces that help support weakened muscles during a functional task such as walking.
**Body function and structure:** According to the World Health Organization's International Classification of Functioning, Disability and Health, body structure and function refer to impairments related to the physiological functions of body systems and/or to the anatomical parts of the body such as organs, limbs and their components.
**Activity limitations:** According to the World Health Organization's International Classification of Functioning, Disability and Health, activity limitations refer to difficulties an individual may have in executing activities (task or action).
**Participation restrictions:** According to the World Health Organization's International Classification of Functioning, Disability and Health, participation restrictions refer to the problems an individual may experience in involvement in life situations.
**Transcutaneous electrical stimulation (TENS):** Therapy that involves the use of low-voltage electric currents to treat pain.
**Task-specific training:** Treatment approach focused on function that evolved out of the movement science and motor learning literature. The basic premise is goal-directed practice of functional tasks instead of focusing on impairment reduction exercises.
**Neurotrophic factors:** Endogenous substances that control cell proliferation and differentiation in the nervous system.

## Key facts of activity-based therapy in the community

- Activity-based therapy is a novel concept of rehabilitation of spinal cord injury.
- Activity-based therapy is focused on the recovery of function to the areas affected by the injury.
- Locomotor training, exercises in weight bearing, electrical stimulation and whole-body strengthening exercises, are modalities that compose an activity-based therapy program.
- Activity based therapies have shown to improve motor and sensory function for acute and chronic spinal cord injury population.
- Activity-based therapies also seem to have a role in reduction of co-morbidities related to paralysis and sedentary behavior.

## Summary points

- Spinal cord injury is a condition that results in motor, sensory and systemic dysfunction that reduces mobility, independence, quality of life, and general health status.
- The main objectives of conventional rehabilitation are independence and functionality, targeting areas above the level of injury through compensatory strategies for loss of function.
- In recent years, there has been an increase in scientific evidence on the spinal cord neuroplasticity, showing that interventions that stimulate below the level of injury, such as locomotor training, are effective strategies for functional recovery.
- Activity-based therapy is a novel concept of rehabilitation with a focus on recovery of the damaged nervous system, using activities that stimulate the segments affected by the injury.
- Activity-based therapy employs different modalities in an intensive, holistic, and individualized program.
- It includes multiple interventions for the spinal cord injury population, such as Locomotor Training, Functional Electrical Stimulation, Whole-Body Vibration, developmental postures, strength, and aerobic exercises.
- There is scientific evidence to support the effects of activity-based therapies on functional recovery, participation, quality of life and reduction of secondary complications.
- The activity-based therapy concept can be used in hospital-based or community-based settings and has the potential to optimize the effects of new regenerative interventions such as stem cells and epidural stimulation.

## References

Adams, M. M., & Hicks, A. L. (2011). Comparison of the effects of body-weight-supported treadmill training and tilt-table standing on spasticity in individuals with chronic spinal cord injury. *The Journal of Spinal Cord Medicine*, *34*(5), 488–494.

Alexeeva, N., Sames, C., Jacobs, P. L., Hobday, L., di Stasio, M. M., Mitchell, S. A., et al. (2011). Comparison of training methods to improve walking in persons with chronic spinal cord injury: A randomized clinical trial. *The Journal of Spinal Cord Medicine*, *34*(4), 362–379.

Anwer, S., Equebal, A., Palekar, T. J., Nezamuddin, M., Neyaz, O., & Alghadir, A. (2014). Effect of locomotor training on motor recovery and walking ability in patients with incomplete spinal cord injury: A case series. *Journal of Physical Therapy Science*, *26*(6), 951–953.

Backus, D., Apple, D., & Hudson, L. (2009). Systematic review of activity-based interventions to improve neurological outcomes after SCI January 1998—March 2009. In *Disability Research Right to Know*. http://www.bu.edu/drrk/research-syntheses/spinal-cord-injuries/activity-based-interventions.

Backus, D., Apple, D., & Hudson, L. (2011). Neural and functional outcomes after lower extremity and walking activity-based interventions for persons with spinal cord injury: A research synthesis. *Topics in Spinal Cord Injury Rehabilitation*, *16*(65).

Barker, R. N., Kendall, M. D., Amsters, D. I., Pershouse, K. J., Haines, T. P., & Kuipers, P. (2009). The relationship between quality of life and disability across the lifespan for people with spinal cord injury. *Spinal Cord*, *47*(2), 149–155.

Battistuzzo, C. R., Callister, R. J., Callister, R., & Galea, M. P. (2012). A systematic review of exercise training to promote locomotor recovery in animal models of spinal cord injury. *Journal of Neurotrauma*, *29*(8), 1600–1613.

Behrman, A. L., Ardolino, E., Vanhiel, L. R., Kern, M., Atkinson, D., Lorenz, D. J., & Harkema, S. J. (2012). Assessment of functional improvement without compensation reduces variability of outcome measures after human spinal cord injury. *Archives of Physical Medicine and Rehabilitation*, *93*(9), 1518–1529.

Behrman, A. L., Bowden, M. G., & Nair, P. M. (2006). Neuroplasticity after spinal cord injury and training: An emerging paradigm shift in rehabilitation and walking recovery. *Physical Therapy*, *86*(10), 1406–1425.

Behrman, A. L., & Harkema, S. J. (2007). Physical rehabilitation as an agent for recovery after spinal cord injury. *Physical Medicine and Rehabilitation Clinics of North America*, *18*(2), 183–202.

Behrman, A. L., Lawless-Dixon, A. R., Davis, S. B., Bowden, M. G., Nair, P., Phadke, C., et al. (2005). Locomotor training progression and outcomes after incomplete spinal cord injury. *Physical Therapy*, *85*(12), 1356–1371.

Belegu, V., Oudega, M., Gary, D. S., & McDonald, J. W. (2007). Restoring function after spinal cord injury: Promoting spontaneous regeneration with stem cells and activity-based therapies. *Neurosurgery Clinics of North America*, *18*(1), 143–168 (xi).

Can, A., Dosoglu, M. S., Karacan, I., & Karamehmetoglu, S. (2007). Effect of axial loading on bone mineral density in patients with traumatic spinal cord injury. *Ulusal Travma ve Acil Cerrahi Dergisi*, *13*(2), 101–105.

Chisholm, A. E., Peters, S., Borich, M. R., Boyd, L. A., & Lam, T. (2015). Short-term cortical plasticity associated with feedback-error learning after locomotor training in a patient with incomplete spinal cord injury. *Physical Therapy*, *95*(2), 257–266.

Dietz, V., & Curt, A. (2006 Aug). Neurological aspects of spinal-cord repair: Promises and challenges. *Lancet Neurology*, *5*(8), 688–694.

Ditunno, J. F., Jr. (1999). Predicting recovery after spinal cord injury: A rehabilitation imperative. *Archives of Physical Medicine and Rehabilitation*, *80*(4), 361–364.

Dobkin, B., Apple, D., Barbeau, H., Basso, M., Behrman, A., Deforge, D., et al. (2006). Weight-supported treadmill vs over-ground training for walking after acute incomplete SCI. *Neurology*, *66*(4), 484–493.

Dolbow, D. R., Gorgey, A. S., Recio, A. C., Stiens, S. A., Curry, A. C., Sadowsky, C. L., et al. (2015). Activity-based restorative therapies after spinal cord injury: Inter-institutional conceptions and perceptions. *Aging and Disease*, *6*(4), 254–261.

Edgerton, V. R., Kim, S. J., Ichiyama, R. M., Gerasimenko, Y. P., & Roy, R. R. (2006). Rehabilitative therapies after spinal cord injury. *Journal of Neurotrauma*, *23*(3–4), 560–570.

Edgerton, V. R., & Roy, R. R. (2009). Activity-dependent plasticity of spinal locomotion: Implications for sensory processing. *Exercise and Sport Sciences Reviews*, *37*(4), 171–178.

Edgerton, V. R., Tillakaratne, N. J. K., Bigbee, A. J., De Leon, R. D., & Roy, R. R. (2004). Plasticity of the spinal neural circuitry after injury. *Annual Review of Neuroscience*, 145–167.

Eitivipart, A. C., de Oliveira, C. Q., Arora, M., Middleton, J., & Davis, G. M. (2019). Overview of systematic reviews of aerobic fitness and muscle strength training after spinal cord injury. *Journal of Neurotrauma*, *36*(21), 2943–2963.

Escorpizo, R., & Bemis-Dougherty, A. (2015). Introduction to special issue: A review of the international classification of functioning, disability and health and physical therapy over the years. *Physiotherapy Research International*, *20*(4), 200–209.

Fawcett, J. W., Curt, A., Steeves, J. D., Coleman, W. P., Tuszynski, M. H., Lammertse, D., ... Short, D. (2007). Guidelines for the conduct of clinical trials for spinal cord injury as developed by the ICCP panel: Spontaneous recovery after spinal cord injury and statistical power needed for therapeutic clinical trials. *Spinal Cord*, *45*(3), 190–205.

Field-Fote, E. C., & Roach, K. E. (2011). Influence of a locomotor training approach on walking speed and distance in people with chronic spinal cord injury: A randomized clinical trial. *Physical Therapy*, *91*(1), 48–60.

Fouad, K., Krajacic, A., & Tetzlaff, W. (2011). Spinal cord injury and plasticity: Opportunities and challenges. *Brain Research Bulletin*, *84*(4–5), 337–342.

Fouad, K., & Tse, A. (2008). Adaptive changes in the injured spinal cord and their role in promoting functional recovery. *Neurological Research*, *30*(1), 17–27.

Galea, M. P. (2012). Spinal cord injury and physical activity: Preservation of the body. *Spinal Cord*, *50*(5), 344–351.

Gaspar, R., Padula, N., Freitas, T. B., de Oliveira, J. P. J., & Torriani-Pasin, C. (2019). Physical exercise for individuals with spinal cord injury: Systematic review based on the international classification of functioning, disability, and health. *Journal of Sport Rehabilitation*, *28*(5), 505–516.

Giangregorio, L. M., Webber, C. E., Phillips, S. M., Hicks, A. L., Craven, B. C., Bugaresti, J. M., et al. (2006). Can body weight supported treadmill training increase bone mass and reverse muscle atrophy in individuals with chronic incomplete spinal cord injury? *Applied Physiology, Nutrition, and Metabolism*, *31*(3), 283–291.

Ginis, K. A., Hicks, A. L., Latimer, A. E., Warburton, D. E., Bourne, C., Ditor, D. S., … Wolfe, D. L. (2011). The development of evidence informed physical activity guidelines for adults with spinal cord injury. *Spinal Cord*, *49*, 1088–1096.

Harkema, S., Behrman, A., & Barbeau, H. (2012). Evidence-based therapy for recovery of function after spinal cord injury. *The Hand*, *109*, 259–274.

Harkema, S. J., Ferreira, C. K., Van Den Brand, R. J., & Krassioukov, A. V. (2008). Improvements in orthostatic instability with stand locomotor training in individuals with spinal cord injury. *Journal of Neurotrauma*, *25*(12), 1467–1475.

Harkema, S. J., Hillyer, J., Schmidt-Read, M., Ardolino, E., Sisto, S. A., & Behrman, A. L. (2012). Locomotor training: As a treatment of spinal cord injury and in the progression of neurologic rehabilitation. *Archives of Physical Medicine and Rehabilitation*, *93*(9), 1588–1597.

Harkema, S. J., Schmidt-Read, M., Behrman, A. L., Bratta, A., Sisto, S. A., & Edgerton, V. R. (2012). Establishing the NeuroRecovery network: Multisite rehabilitation centers that provide activity-based therapies and assessments for neurologic disorders. *Archives of Physical Medicine and Rehabilitation*, *93*(9), 1498–1507.

Harkema, S. J., Schmidt-Read, M., Lorenz, D. J., Edgerton, V. R., & Behrman, A. L. (2012). Balance and ambulation improvements in individuals with chronic incomplete spinal cord injury using locomotor training-based rehabilitation. *Archives of Physical Medicine and Rehabilitation*, *93*(9), 1508–1517.

Harkema, S. J., Shogren, C., Ardolino, E., & Lorenz, D. J. (2016). Assessment of functional improvement without compensation for human spinal cord injury: Extending the neuromuscular recovery scale to the upper extremities. *Journal of Neurotrauma*, *33*(24), 2181–2190.

Harness, E. T., Yozbatiran, N., & Cramer, S. C. (2008). Effects of intense exercise in chronic spinal cord injury. *Spinal Cord*, *46*(11), 733–737.

Harvey, L. A. (2016). Physiotherapy rehabilitation for people with spinal cord injuries. *Journal of Physiotherapy*, *62*(1), 4–11.

Harvey, L. A., Glinsky, J. V., & Bowden, J. L. (2016). The effectiveness of 22 commonly administered physiotherapy interventions for people with spinal cord injury: A systematic review. *Spinal Cord*, *54*(11), 914–923.

Harvey, L. A., Lin, C. W. C., Glinsky, J. V., & De Wolf, A. (2009). The effectiveness of physical interventions for people with spinal cord injuries: A systematic review. *Spinal Cord*, *47*(3), 184–195.

In, T., Jung, K., Lee, M. G., & Cho, H. Y. (2018). Whole-body vibration improves ankle spasticity, balance, and walking ability in individuals with incomplete cervical spinal cord injury. *NeuroRehabilitation*, *42*(4), 491–497.

Jeffries, E. C., Hoffman, S. M., de Leon, R., Dominguez, J. F., Semerjian, T. Z., Melgar, I. A., et al. (2015). Energy expenditure and heart rate responses to increased loading in individuals with motor complete spinal cord injury performing body weight-supported exercises. *Archives of Physical Medicine and Rehabilitation*, *96*(8), 1467–1473.

Ji, Q., He, H., Zhang, C., Lu, C., Zheng, Y., Luo, X. T., & He, C. (2017). Effects of whole-body vibration on neuromuscular performance in individuals with spinal cord injury: A systematic review. *Clinical Rehabilitation*, *31*(10), 1279–1291.

Jones, M. L., Evans, N., Tefertiller, C., Backus, D., Sweatman, M., Tansey, K., & Morrison, S. (2014). Activity-based therapy for recovery of walking in individuals with chronic spinal cord injury: Results from a randomized clinical trial. *Archives of Physical Medicine and Rehabilitation*, *95*(12), 2239.

Jones, M. L., Harness, E., Denison, P., Tefertiller, C., Evans, N., & Larson, C. A. (2012). Activity-based therapies in spinal cord injury: Clinical focus and empirical evidence in three independent programs. *Topics in Spinal Cord Injury Rehabilitation*, *18*(1), 34–42.

Karimi, M., Omar, A. H., & Fatoye, F. (2014). Spinal cord injury rehabilitation: Which way forward? *NeuroRehabilitation*, *35*(2), 325–340.

Kirshblum, S., Millis, S., McKinley, W., & Tulsky, D. (2004). Late neurologic recovery after traumatic spinal cord injury. *Archives of Physical Medicine and Rehabilitation*, *85*(11), 1811–1817.

Kirshblum, S. C., & O'Connor, K. C. (1998). Predicting neurologic recovery in traumatic cervical spinal cord injury. *Archives of Physical Medicine and Rehabilitation*, *79*(11), 1456–1466.

Kraft, J., Karpenko, A., & Rincon, F. (2016). Intraoperative targeted temperature management in acute brain and spinal cord injury. *Current Neurology and Neuroscience Reports*, *16*(2), 18.

Luo, S., Xu, H., Zuo, Y., Liu, X., & All, A. H. (2020). A review of functional electrical stimulation treatment in spinal cord injury. *Neuromolecular Medicine*, *22*(4), 447–463.

Lynskey, J. V., Belanger, A., & Jung, R. (2008). Activity-dependent plasticity in spinal cord injury. *Journal of Rehabilitation Research and Development*, *45*(2), 229–240.

Marino, R. J., Ditunno, J. F., Jr., Donovan, W. H., & Maynard, F., Jr. (1999). Neurologic recovery after traumatic spinal cord injury: Data from the model spinal cord injury systems. *Archives of Physical Medicine and Rehabilitation*, *80*(11), 1391–1396.

Marquez-Chin, C., & Popovic, M. R. (2020). Functional electrical stimulation therapy for restoration of motor function after spinal cord injury and stroke: A review. *Biomedicine Engineering Online*, *19*(1), 34.

Martin, R., Sadowsky, C., Obst, K., Meyer, B., & McDonald, J. (2012). Functional electrical stimulation in spinal cord injury: From theory to practice. *Topics in Spinal Cord Injury Rehabilitation*, *18*(1), 28–33.

Padula, N., Costa, M., Batista, A., Gaspar, R., Motta, C., Palma, G., et al. (2015). Long-term effects of an intensive interventional training program based on activities for individuals with spinal cord injury: A pilot study. *Physiotherapy Theory and Practice*, *31*(8), 568–574.

Quel de Oliveira, C., Refshauge, K., Middleton, J., de Jong, L., & Davis, G. M. (2017). Effects of activity-based therapy interventions on mobility, independence, and quality of life for people with spinal cord injuries: A systematic review and meta-analysis. *Journal of Neurotrauma*, *34*(9), 1726–1743.

Rejc, E., Angeli, C. A., Atkinson, D., & Harkema, S. J. (2017). Motor recovery after activity-based training with spinal cord epidural stimulation in a chronic motor complete paraplegic. *Scientific Reports*, *7*(1), 13476.

Roy, R. R., Harkema, S. J., & Edgerton, R. (2012). Basic concepts of activity-based interventions for improved recovery of motor function after spinal cord injury. *Archives of Physical Medicine and Rehabilitation*, *93*(9), 1487–1497.

Sadowsky, C. L., & McDonald, J. W. (2009). Activity-based restorative therapies: Concepts and applications in spinal cord injury-related neurorehabilitation. *Developmental Disabilities Research Reviews*, *15*(2), 112–116.

Sandrow-Feinberg, H. R., & Houle, J. D. (2015). Exercise after spinal cord injury as an agent for neuroprotection, regeneration and rehabilitation. *Brain Research*, *1619*, 12–21.

Stampas, A., & Tansey, K. E. (2014). Spinal cord injury medicine and rehabilitation. *Seminars in Neurology*, *34*(5), 524–533.

Tester, N. J., Lorenz, D. J., Suter, S. P., Buehner, J. J., Falanga, D., Watson, E., … Michele Basso, D. (2016). Responsiveness of the neuromuscular recovery scale during outpatient activity-dependent rehabilitation for spinal cord injury. *Neurorehabilitation and Neural Repair*, *30*(6), 528–538.

Tweedy, S. M., Beckman, E. M., Geraghty, T. J., Theisen, D., Perret, C., Harvey, L. A., & Vanlandewijck, Y. C. (2017). Exercise and sports science Australia (ESSA) position statement on exercise and spinal cord injury. *Journal of Science and Medicine in Sport*, *20*(2), 108–115.

Valent, L., Dallmeijer, A., Houdijk, H., Talsma, E., & van der Woude, L. (2007). The effects of upper body exercise on the physical capacity of people with a spinal cord injury: A systematic review. *Clinical Rehabilitation*, *21*(4), 315–330.

Van Langeveld, S. A., Post, M. W., van Asbeck, F. W., Gregory, M., Halvorsen, A., Rijken, H., et al. (2011). Comparing content of therapy for people with a spinal cord injury in postacute inpatient rehabilitation in Australia, Norway, and the Netherlands. *Physical Therapy*, *91*(2), 210–224.

Waddell, K. J., Birkenmeier, R. L., Moore, J. L., Hornby, T. G., & Lang, C. E. (2014). Feasibility of high-repetition, task-specific training for individuals with upper-extremity paresis. *American Journal of Occupational Therapy*, *68*(4), 444–453.

Wilson, J. R., & Fehlings, M. G. (2021). Management strategies to optimize clinical outcomes after acute traumatic spinal cord injury: Integration of medical and surgical approaches. *Journal of Neurosurgical Sciences*, *56*(1), 1–11.

Wrigley, P. J., Siddall, P. J., & Gustin, S. M. (2018). New evidence for preserved somatosensory pathways in complete spinal cord injury: A fMRI study. *Human Brain Mapping*, *39*(1), 588–598.

Wyndaele, M., & Wyndaele, J. J. (2006). Incidence, prevalence and epidemiology of spinal cord injury: What learns a worldwide literature survey? *Spinal Cord*, *44*(9), 523–529.

Chapter 42

# Mobile health apps and self-management for spinal cord injury rehabilitation

Gurkaran Singh[a,b,c], Megan K. MacGillivray[d], Ethan Simpson[a,b,c], Bonita J. Sawatzky[c,e], and W. Ben Mortenson[a,b,c]

[a]*Rehabilitation Research Program, G.F. Strong Rehabilitation Centre, Vancouver, BC, Canada;* [b]*Department of Occupational Science and Occupational Therapy, University of British Columbia, Vancouver, BC, Canada;* [c]*International Collaboration on Repair Discoveries, Blusson Spinal Cord Centre (3rd Floor), Vancouver, BC, Canada;* [d]*Human Kinetics, St. Francis Xavier University, Antigonish, NS, Canada;* [e]*Department of Orthopaedics, University of British Columbia, Vancouver, BC, Canada*

## Abbreviations

| | |
|---|---|
| **SCI** | spinal cord injury |
| **mHealth** | mobile health |
| **eHealth** | electronic health |
| **PA** | physical activity |
| **SUS** | system usability scale |

## Introduction

For individuals living with spinal cord injury (SCI), inadequate health management can result in adverse psychological and physical health outcomes including autonomic dysreflexia, pressure sores, bowel and bladder dysfunction, depression, and spasticity (Adriaansen et al., 2016). Learning to self-manage health, i.e., symptoms, treatment, and lifestyle changes inherent to living with a chronic condition, has been shown to reduce the severity and prevalence of secondary conditions among those with chronic health condition (Barlow, Wright, Sheasby, Turner, & Hainsworth, 2002; Mortenson et al., 2019). Recent advancements in smartphone technology have resulted in consumers becoming more actively engaged with using mobile health (mHealth) apps to manage their day-to-day health (Gill, Kamath, & Gill, 2012). Because of the widespread adoption of smartphones and tablets, mHealth has the potential to become an accessible, low-cost approach of self-management intervention delivery (Lewis, Boissaud-Cooke, Aungst, & Eysenbach, 2014). This chapter provides insight into the implementation of mHealth apps for the self-management and prevention of secondary conditions related to SCI. Specifically, this chapter will (1) provide background on chronic disease management, (2) describe traditional approaches to SCI-related health self-management, and (3) highlight the latest research and advancements in mHealth apps for individuals living with SCI.

## Chronic disease management

Chronic conditions are long lasting and generally incurable conditions that require varying degrees of management (Shiel Jr., 2020). Examples of conditions meeting these criteria include diabetes, arthritis, dementia, and SCI (Government of Canada, 2020). Given the varied nature of the aforementioned conditions, their risk factors, prevalence, and management vary considerably. The presence of one chronic condition or disease is often associated or accompanied by subsequent chronic diseases resulting in multimorbidity (Sakib, Shooshtari, John, & Menec, 2019). Demographic characteristics have also been shown to be related to outcomes associated with chronic conditions. For example, increased age and longer time with injury are both associated with higher rates of secondary conditions among people with SCI ( Jensen et al., 2013). Furthermore, the age at which SCI is acquired displays a bimodal distribution in North America (Noonan et al., 2012). SCIs earlier in life occur primarily among males, who tend to be injured from high velocity events such as motor vehicle accidents, while SCIs in later life tend to occur from falls equally for both men and women (Chen, Tang, Vogel, & Devivo, 2013).

Diagnosis and Treatment of Spinal Cord Injury. https://doi.org/10.1016/B978-0-12-822498-4.00042-7

**Common Spinal Cord Injury Secondary Conditions**

- Respiratory (or breathing-related) problems
- Genitourinary (or bowel and bladder) dysfunction
- Cardiovascular problems
- Pressure sores (or pressure ulcers)
- Sexual dysfunction
- Blood clots
- Chronic pain (or fatigue)
- Spasticity (or impaired muscle contraction)
- Pneumonia
- Autonomic dysreflexia (or hyperreflexia),
- Depression
- Increased likelihood of certain cancers, including bladder cancer

FIG. 1 List of common SCI secondary conditions. A list of common spinal cord injury-related secondary conditions. *This previously unpublished figure was created by the authors of this book chapter.

Given the prevalence of chronic conditions, interventions that improve self-management have a substantial potential impact both personally and societally. Chronic conditions are thought to be responsible for the deaths of 41 million people each year with a disproportionate percentage of deaths occurring in less resourced countries (World Health Organization, 2019). Given the global significance of this problem, the World Health Organization Global Action Plan set targets to reduce the prevalence of chronic conditions by 2025 (World Health Organization, 2013). In Canada age-standardized chronic condition prevalence has been decreasing over time (Hamm et al., 2019). However, chronic conditions still represent a significant burden to the health care system with 44% of Canadian adults having at least 1 of 10 common chronic conditions and direct health care costs to the Canadian economy of $68 billion annually (Chronic Disease Prevention Alliance of Canada, 2017; Government of Canada, 2019). In England, more than 40% of people aged 65 years and over report having at least one chronic condition and 15% reported having at least two (Organization for Economic Cooperation and Development, 2019). Similarly, in the United States, chronic conditions equate for 75% of health care costs with 133 million residents living with at least one chronic condition and almost a third living with multiple chronic conditions (National Health Council, 2014). Improving management of these chronic conditions would result in considerable personal and societal benefits.

Similar to other chronic conditions, SCI requires complex management as a result of a constellation of both physiological and psychological challenges. Common secondary conditions (Fig. 1) include pressure sores, problematic spasticity, genitourinary dysfunction, autonomic dysreflexia, and depression (Adriaansen et al., 2016). Pressure sores occur when an area of skin, often over bony areas, is placed under constant pressure, such as prolonged sitting, resulting in the breakdown of the skin and underlying tissue (Bhattacharya & Mishra, 2015). Spasticity can be characterized by increased muscle tone, increased tendon reflexes, and increased reflex responses from external stimuli (Skold, Levi, & Seiger, 1999). Genitourinary dysfunction relates to disturbance to the urinary tract and sexual function (Beck & Fowler, 1994). Injuries at thoracic level T6 or above may disrupt the autonomic nervous system, which can result in autonomic dysreflexia characterized by high blood pressure along with headaches, sweating, flushing, bradycardia, nasal congestion, feelings of apprehension, and piloerection (Ekland, Krassioukov, McBride, & Elliott, 2008). Poor management of these secondary conditions, among others, can limit social participation, reduce quality of life, and may prove fatal in extreme cases (Adriaansen et al., 2016). Leading causes of death for individuals with SCI include respiratory disease, cardiovascular disease, and sepsis, thus prevention of these through self-management is essential (Savic et al., 2017).

## Traditional SCI self-management interventions

Following SCI, self-management is encouraged early in the rehabilitation process. Rehabilitation hospitals often provide client education sessions and resources related to self-management techniques. Key self-management topics identified in the client education sessions may involve pressure sores, urinary tract infections, sexual health, pain management, and others; however, motivation to participate in optional sessions and length of stay in inpatient rehabilitation may minimize opportunities to develop self-management skills. Several interpersonal barriers may relate to engagement in self-management classes within inpatient rehabilitation including pain, fatigue, emotional status, denial, low self-efficacy (Jerant, von Friederichs-Fitzater, & Moore, 2005).

A study by Munce et al. (2016) revealed the different perspectives patients, caregivers, and managers have about SCI-related self-management interventions. Patients with traumatic SCI emphasized that self-management requires them to take responsibility for their health by incorporating "wellness awareness" and "monitoring for secondary complications" (Munce et al., 2016, p. 7). Furthermore, caregivers and patients with SCI felt that an understanding of self-management required "ownership of one's own care/empowerment in care" (Munce et al., 2016, p. 7). In contrast, managers indicated a belief in shared responsibility between patients and health professionals with implementing self-management interventions (Munce et al., 2016). These multiple perspectives are relevant for the development and administration of self-management programs for individuals with SCI.

Historically, self-management programs for chronic conditions have focused on providing information, drug management, symptom management, management of psychological consequences, lifestyle, social support, communication, and others (e.g., goal setting, problem solving, spirituality) (Barlow et al., 2002). A recent scoping review of self-management interventions following SCI identified that the primary component of programming has been the delivery of general or topic-specific SCI information (McIntyre et al., 2020), which is consistent with the historical focus of self-management programs for chronic conditions.

Although individuals with SCI have identified self-management needs in different areas, programming options are limited. For example, individuals with SCI would like to improve self-management related to exercise nutrition, pain management, aging, communication with healthcare professionals, problem solving, community reintegration, and confidence (Munce et al., 2014). Despite the variety of topics believed to be important for SCI self-management interventions, many of the 102 unique self-management programs related to SCI, identified in the aforementioned scoping review, have focused on pain (McIntyre et al., 2020). This suggests a large unmet need.

The self-management needs of people with SCI are diverse and complex. Stakeholders (i.e., individuals with SCI, and their formal and informal caregivers) involved in the development of a self-management app for people with SCI emphasized that it needed to be "individualized and user friendly" (Mortenson et al., 2019). To address the diverse nature of SCI, 54% of interventions included in the aforementioned scoping review incorporated "self-tailoring" (McIntyre et al., 2020). Self-tailoring has the potential to address the unique needs among individuals with SCI.

Many rehabilitation hospitals and experimental interventions have implemented components of Bandura's social cognitive theory, which "posits that people learn from one another, via observation, imitation, and modeling" (Nabavi, 2012). Strategies congruent with social cognitive theory include the incorporation of a social context in self-management delivery that allows for dynamic and reciprocal interactions (Hirsche, Williams, Jones, & Manns, 2011; Kim & Cho, 2017). Not surprisingly, upwards of 45% of SCI self-management programs have adopted a group or mixed (group and individual) approach (McIntyre et al., 2020). Peer-based inpatient programs have been evaluated in major rehabilitation hospitals and have identified benefits such as improved self-efficacy and reduced unplanned readmissions (Gassaway et al., 2017; Jones, Gassaway, & Sweatman, 2019).

Nearly 63% of individuals with SCI identified that the best mode of delivery for a self-management program would be "internet-based" (Munce et al., 2014). With increasing technology development, the delivery mode of self-management education and resources has been becoming increasingly digital. Digital resources may improve accessibility for individuals with SCI who are based in rural locations, as they are often wide-reaching, cost effective, and accessible for people with SCI using personal computers they have already adapted to using, provided they have access to the internet. In a scoping review of studies published prior to April 2018, McIntyre et al. (2020) identified 17 SCI self-management programs that incorporated online elements, 10 of which exclusively used an online format.

## eHealth technology and mHealth apps

First coined by Mitchell and expanded on by Eysenbach, eHealth can be described as information or a health service delivered or enhanced by the internet or related technologies (Eysenbach, 2001; Mitchell, 1999). Instead of being considered simply an adjunctive electronic health technology that supplements traditional face-to-face services, as technology becomes more embedded into healthcare over time, "eHealth" may become an integral part of healthcare (Cunningham, Wake, Waller, & Morris, 2014). Since the turn of the 21st century, there has been a dramatic shift toward the use of eHealth technologies for health-related self-management and information seeking (Gill et al., 2012). Common examples of eHealth technologies include electronic health records, telemedicine, and mHealth apps.

With advancements in smartphones and tablets, mHealth apps have emerged as means to seek health information and better health service delivery and accessibility (Lewis et al., 2014). Furthermore, the introduction of mHealth apps has made it easier to provide cost-effective, population-based, self-management interventions for individuals living with chronic conditions such as SCI. For example, a study examining the views of individuals with traumatic SCI on the use of mHealth apps for health self-management, found that 63% of participants felt it was the best means of self-management delivery (Munce et al., 2014). Popular examples of mHealth apps that promote mainstream self-management include Apple Health, Google Fit, and Samsung Health. However, for individuals with chronic conditions such as SCI, these "one-size-fits-all" mHealth apps may not meet their specific health management needs. As a result, multiple mHealth apps have been developed to promote health self-management training among individuals with SCI.

## mHealth app needs among individuals with SCI

Multiple studies have evaluated the mHealth app-specific needs among individuals with SCI. For example, a consumer led study conducted by The Hopkins Centre (Queensland, Australia) explored the perspectives of 138 participants with SCI on a self-management mobile app. It found that a majority of participants (58%) reported that they would use an app to manage at least one SCI-related secondary condition (Colley, Atresh, Parekh, & Salehi, 2017). Furthermore, participants felt that if an app were developed, it should include three key features: (1) communication with healthcare providers, (2) effective self-management advice, and (3) resources and videos on relevant health topics (Colley et al., 2017). As part of a needs assessment to guide the development of a healthy lifestyle mobile app for wheelchair users with SCI, goal attainment and social support were identified as critical personal and environment determinants that promote physical activity and healthy dietary behaviors (Holla et al., 2020; Van den Akker et al., 2020). To identify preferred potential features for a web-based physical activity self-management app, a participatory study explored key stakeholders' perspectives including individuals with SCI and health care professionals (Pancer et al., 2019). The study identified five key features: (1) guidance management (i.e., exercise tutorials), (2) achievements system (i.e., tool monitoring achievements and milestones), (3) self-regulation strategies (i.e., goal setting and reminders), (4) interactivity (i.e., peers and health professional integration), and (5) format (i.e., appearance and ease of use) (Pancer et al., 2019). These studies speak to the range in health and lifestyle needs across diverse SCI populations.

## Development and pilot implementation of mHealth apps for SCI self-management

Over the past decade, several studies have been conducted to develop and pilot test mHealth apps for SCI self-management (Fig. 2). Apps range from those designed to manage specific secondary conditions to those that serve a variety of self-care-related purposes. In terms of secondary condition-specific apps, for individuals experiencing chronic bladder retention, a web-based mobile app was developed to self-monitor and manage intermittent catheterization, fluid intake, urinary output, and symptoms of urinary tract infections (Wilde et al., 2017). A study conducted on the feasibility of this app revealed that in a sample of 30 participants, self-management of bladder dysfunction improved, and participants' self-management scale scores increased significantly ($P = 0.032$) (Wilde et al., 2017). Similarly, to assist with self-managing secondary conditions such as pain, weight gain, and fatigue, a small-scale study was conducted to document the impact of a physical activity (PA) tracking self-management mobile app on overall PA levels among individuals with SCI (Hiremath et al., 2019). By providing real-time feedback on PA levels, in a sample of 20 participants, the majority (69%) demonstrated greater light to

**Mobile/Web Apps for Spinal Cord Injury Self-Management**

- **Managing an Intermittent Catheter app (Wilde et al., 2016)**
- **Personal Health Informatics and Rehabilitation Engineering (PHIRE) app (Hiremath et al., 2019)**
- **Amman et al.'s mobile app for pressure injuries (Amann et al., 2019)**
- **SCI Health Storylines app (Mortenson et al., 2018)**
- **iMHere mobile app (Parmanto et al., 2013)**

FIG. 2 List of mobile and web apps for SCI self-management and rehabilitation. A list of mobile and web apps for spinal cord injury self-management and rehabilitation. *This previously unpublished figure was created by the authors of this book chapter.

moderate intensity PA compared to baseline (Hiremath et al., 2019). In order to develop a mHealth app for pressure injury prevention among individuals with SCI, a study was conducted to develop a working app prototype and conduct usability testing (Amann et al., 2020). Using a co-design approach, a working prototype was developed with features that included a pressure injury diary, smart camera, expert consultation service, and knowledge repository. Results from usability testing demonstrated above average system usability scale (SUS) score (78.5/100) with a majority of participants being able to fluently navigate the app (Amann et al., 2020; Lewis, 2018).

In order to provide individuals living with SCI with an app that can address both general and SCI-specific health self-management needs, we developed a mHealth app funded by the Craig Nielsen Foundation called *SCI Health Storylines* (Mortenson et al., 2019). *SCI Health Storylines* (Fig. 3) is a mHealth app that includes various general health tools (e.g., medication tracker, health journal), SCI-specific tools (e.g., SCI goal-setting tool, bowel and bladder tracker, pain tracker), and lifestyle tools (e.g., physical activity tracker, recreational resources tool, nutritional planner). Our app was developed using a user-centered design process with input from rehabilitation inpatients, clinicians, and informal caregivers (Mortenson et al., 2019). During the app's development, participants identified three core themes about the functionality of the app which needed to (1) be tailor-made for individuals with SCI, (2) target goals to promote self-management, and (3) increase social participation in the community (Mortenson et al., 2019). Furthermore, small-scale

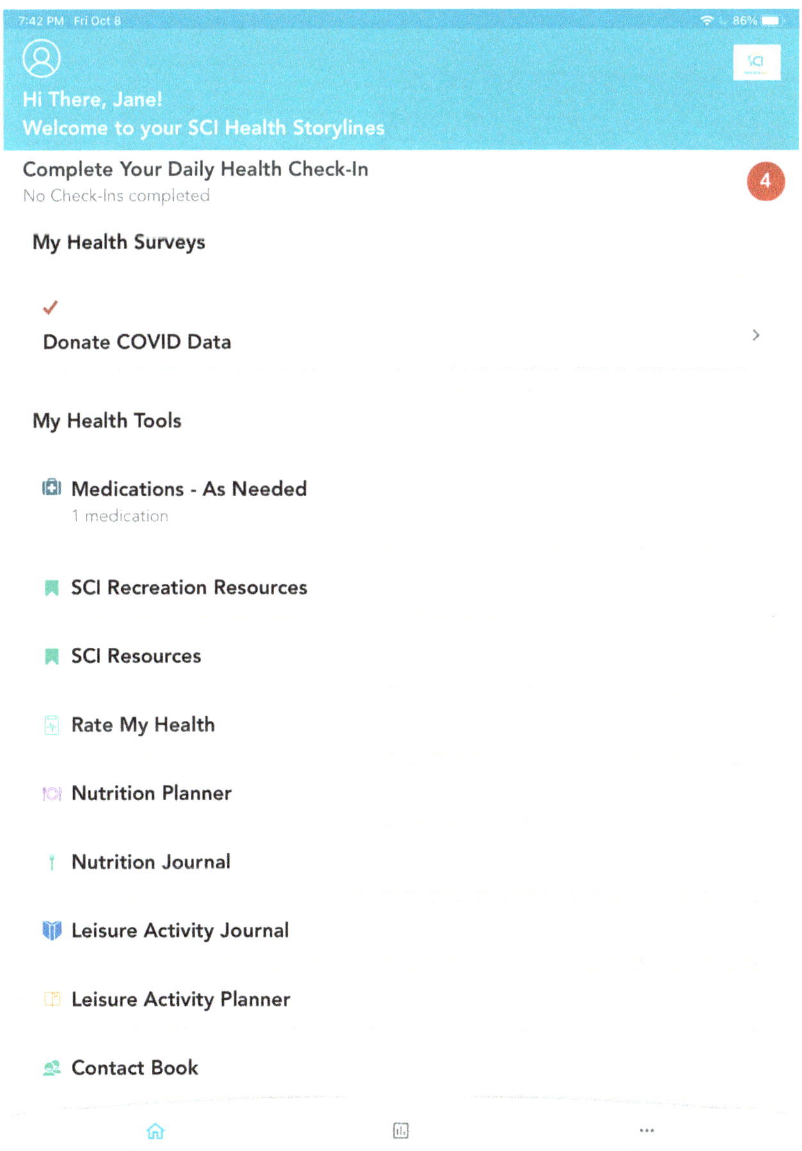


FIG. 3 *SCI Health Storylines* mHealth App. A screenshot of the splash page of the SCI Health Storylines mHealth app. *The source of this previously unpublished figure is the *SCI Storylines App* on the iOS Platform. Copyright permission has been obtained from Self Care Catalysts (Toronto, Canada) to use this figure in this book chapter.

intervention feasibility and app usability studies were also conducted among 20 rehabilitation inpatients on the implementation of our app during inpatient rehabilitation and following community discharge (MacGillivray et al., 2019; Singh et al., 2019). In terms of feasibility, our research revealed retention was 85% at inpatient discharge and 70% 3 months post-discharge, participants' bowel self-management confidence improved between admission and discharge ($P < 0.01$), and indicators of strong overall feasibility to support a larger clinical trial (MacGillivray et al., 2019). Furthermore, our mobile app was found to have above average SUS scores (Lewis, 2018) both at discharge (78.1/100) and 3 months post-discharge (71.6/100).

## Clinical interventions on mHealth apps for SCI self-management

With the recent emergence of mHealth apps designed to promote self-management and SCI rehabilitation, only two large-scale clinical interventions have been or are in the process of being conducted at the time of writing this chapter. Kryger and colleagues completed a randomized controlled trial (RCT) to determine if their mHealth app, *iMHere* (Fig. 4), could improve overall health outcomes and promote the self-management of secondary conditions such as urinary tract infection, depression, and pressure sores over a 9-month period (Kryger et al., 2019; Parmanto et al., 2013). Key findings from the RCT included statistically significant reductions in urinary tract infections ($P = .03$), and a non-significant trend toward

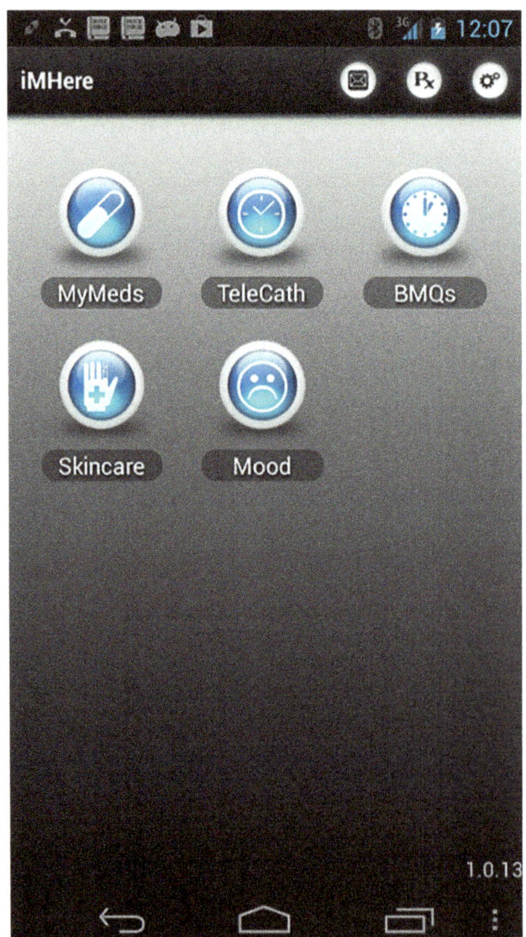


FIG. 4 *iMHere* mHealth App. A screenshot of the splash page of the iMHere mHealth app (Parmanto et al., 2013). *This figure was obtained from an open-access article distributed under the terms of the Creative Commons Attribution License (http://creativecommons.org/licenses/by/2.0/), which permits unrestricted use, distribution, and reproduction in any medium, provided the original work, first published in JMIR mHealth and uHealth, is properly cited. The complete bibliographic information, a link to the original publication on http://mhealth.jmir.org/, as well as this copyright and license information must be included. © Bambang Parmanto, Gede Pramana, Daihua Xie Yu, Andrea D Fairman, Brad E Dicianno, Michael P McCue. Originally published in JMIR mHealth and uHealth (http://mhealth.jmir.org), 11.07.2013.

reduced depressive symptoms (Kryger et al., 2019), among 38 participants randomized to two groups (intervention: mHealth app and standard care; control: standard care only). The chapter authors are in the process of completing a mixed-methods rater-blinded RCT, with a stepped wedge design (Hemming, Haines, Chilton, Girling, & Lilford, 2015), evaluating the efficacy of our self-management mHealth app, *SCI Health Storylines*, in the community setting (Mortenson et al., 2018). The principle outcome measures of this RCT include successful attainment of self-selected self-management goals, and reductions in both self-reported SCI-related secondary conditions and adverse health events. With the study nearing completion, findings on our mHealth app's efficacy in promoting self-management delivery are not currently available. However, three key themes emerged from a qualitative study exploring user's expectations of our mHealth app intervention, including participants' desire to improve their overall health outcomes, learn about the mHealth app's potential and functionality, and improve their sense of personal and social autonomy (Mortenson et al., 2019). If the results of our clinical intervention are positive, this study will provide new knowledge on the implementation of mHealth apps for self-management delivery in clinical practice.

## Applications to other areas of neuroscience

In this chapter, we reviewed chronic disease management, with a particular emphasis on SCI-related secondary condition management. We also highlighted the traditional approaches to SCI health self-management and the emerging transition toward the use of mHealth apps to provide population-based self-management delivery. The interconnection between mHealth apps and chronic disease self-management can be applied to disability groups more broadly in other areas of neuroscience, including stroke (Kang et al., 2019; Li et al., 2020; Sarfo & Ovbiagele, 2017), multiple sclerosis (Ruzic, Mahajan, & Sanford, 2018), traumatic brain injury (Christopher, Alsaffarini, & Jamjoom, 2019), and Parkinson's disease (Bot et al., 2016). Similar to the various SCI-related apps reviewed in this chapter, many apps designed for other neurological diseases and conditions feature similar core tools to promoting health self-management including medication tracking, symptom tracking, and daily journaling tools. Whereas similar general health tools can be seen in mHealth apps across many areas of neuroscience, studies in other areas of neuroscience also highlight the benefits of app tools relevant to their targeted chronic disease population. For example, an app for SCI may include tools related to monitoring various involuntary physiological processes (e.g., blood pressure, digestion, and respiration) and apps designed for traumatic brain injury self-management often include tools for sports-related concussion monitoring (Christopher et al., 2019; Kryger et al., 2019; Mortenson et al., 2019). As novel mHealth apps continue to be introduced to individuals living with other neurological conditions, it is important they provide individualized tools that meet the needs of those specific populations and they undergo independent evaluation to determine their efficacy.

## Mini-dictionary of terms

**Traumatic Spinal Cord Injury:** An injury to the spinal cord that is a result of physical trauma including during falls, motor vehicle accidents, and sports injuries (Pickett, Campos-Benitez, Keller, & Duggal, 2006).

**Non-Traumatic Spinal Cord Injury:** An injury to the spinal cord as a result of factors unrelated to an external force, including tumors, infections, post-surgical complications, and neurogenerative diseases (New et al., 2015).

**Secondary Condition:** An additional physical, physiological or psychological condition resulting from a primary disabling condition (Rowland, 2020).

**Spasticity:** A common secondary condition characterized with increased muscle tone, sustained involuntary somatic reflexes, and/or muscle spasms (Rabchevsky & Kitzman, 2011).

**Pressure Sores:** An area of localized soft tissue necrosis resulting from prolonged pressure to skin (Agrawal & Chauhan, 2012).

**Autonomic Dysreflexia:** An abnormal stimulation of the autonomic nervous stimulation than can result in symptoms including changes in heart rate, high blood pressure, and excessive sweating (Allen & Leslie, 2020).

**Self-Management:** A person's ability to manage the symptoms, treatment, and lifestyle changes inherent to living with a chronic condition or disease (Barlow et al., 2002).

**Mixed-Methods Study:** The collection or analysis of quantitative and qualitative data within a single study (Creswell, 2015).

**Telemedicine:** The delivery of clinical health services remotely (Sood et al., 2007).

**eHealth:** Information or a health service being delivered or enhanced by the Internet and related technologies (Eysenbach, 2001).

**Mobile Health (mHealth):** The application of mobile and wireless devices (e.g., cell phones and tablets) to improve health care delivery and health outcomes (Tucker, 2015).

**Mobile App:** A software program that operates on a mobile or wireless device (Harrison, Flood, & Duce, 2013).

**System Usability Scale:** A short questionnaire designed to determine the usability of website, software program, or application (Brooke, 1996).

## Key facts of spinal cord injury

Spinal cord injury is a neurological disorder.
It occurs from damage to the spinal cord from contusion, rotation, compression, or distraction.
Based on location and severity of damage, the injury can result in a variety of secondary conditions.
It can be classified as traumatic or non-traumatic.
Traumatic spinal cord injury is caused by physical trauma from factors like falls, sports injuries, and motor vehicle accidents.
Non-traumatic spinal cord injury is caused by factors unrelated to an external force such as infections, diseases, tumors, and post-surgical complications.

## Key facts of eHealth

eHealth was first coined in 1999 by John Mitchell.
It is described as information or a health service delivered or enhanced by the Internet or related technologies.
Advancements in Internet technologies have resulted in increased popularity of the eHealth field.
Mobile health falls under the eHealth umbrella.
An individual's ability to seek, find, and understand health information using Internet technologies is called eHealth literacy.

## Summary points

- Spinal cord injury is a chronic condition that may lead to various physical, and psychological sequelae
- In-person, health self-management training has been found to reduce the prevalence and reoccurrence of spinal cord injury-related secondary conditions
- Recent advancements in smartphone technology and access, mobile health apps have emerged as a cost-effective, population-based means to facilitate behavioral change related to self-management
- Multiple mobile health apps have been developed to promote self-care and assist with secondary condition management among people with spinal cord injury
- Despite their emerging prevalence and success in small-scale implementation studies, further clinical trials and large-scale interventions are needed to determine the effectiveness of mHealth apps for spinal cord injury populations

## References

Adriaansen, J., Ruijs, L., van Koppenhagen, C., van Asbeck, F., Snoek, G., van Kuppevelt, D., et al. (2016). Secondary health conditions and quality of life in persons living with spinal cord injury for at least then years. *Journal of Rehabilitation Medicine*, *48*, 853–860.

Agrawal, K., & Chauhan, N. (2012). Pressure ulcers: Back to the basics. *Indian Journal of Plastic Surgery*, *45*(2), 244–254.

Allen, K. J., & Leslie, S. W. (2020). *Autonomic dysreflexia*. Treasure Island, FL: StatsPearls Publishing.

Amann, J., Fiordelli, M., Brach, M., Bertschy, S., Scheel-Sailer, A., & Rubinelli, S. (2020). Co-designing a self-management app prototype to support people with spinal cord injury in the prevention of pressure injuries: Mixed methods study. *JMIR mHealth and uHealth*, *8*(7), e18018.

Barlow, J., Wright, C., Sheasby, J., Turner, A., & Hainsworth, J. (2002). Self-management approaches for people with chronic conditions: A review. *Patient Education and Counselling*, *48*(2), 177–187.

Beck, R., & Fowler, C. (1994). Clinical neurophysiology in the investigation of genitourinary tract dysfunction. In D. N. Rushton (Ed.), *Handbook of neuro-urology* (pp. 151–181). New York: Marcel Dekker Inc.

Bhattacharya, S., & Mishra, R. (2015). Pressure ulcers: Current understanding and newer modalities of treatment. *Indian Journal of Plastic Surgery*, *48*, 4–16.

Bot, B. M., Suver, C., Neto, E. C., Kellen, M., Klein, A., Bare, C., et al. (2016). The mPower study, Parkinson disease mobile data collected using ResearchKit. *Scientific Data*, *6*, 160011.

Brooke, J. (1996). *A "quick and dirty" usability scale*. London: Taylor & Francis.

Chen, Y., Tang, Y., Vogel, L., & Devivo, M. (2013). Causes of spinal cord injury. *Topics in Spinal Cord Injury Rehabilitation*, *19*, 1–8.

Christopher, E., Alsaffarini, K. W., & Jamjoom, A. A. (2019). Mobile health for traumatic brain injury: A systematic review of the literature and mobile application market. *Cureus*, *11*(7), e5120.

Chronic Disease Prevention Alliance of Canada. (2017, August 4). *2018 pre-budget submission to the House of Commons Standing Committee on Finance*. Retrieved from https://www.ourcommons.ca/content/committee/421/fina/brief/br9073636/br-external/chronicdiseasepreventionallianceofcanada-e.pdf.

Colley, J., Atresh, S., Parekh, S., & Salehi, N. (2017). *Potential of smartphone apps to assist people with spinal cord injury*. Retrieved from https://www.hopkinscentre.edu.au/project/potential-of-smartphone-apps-to-assist-people-31.

Creswell, J. W. (2015). *A concise introduction to mixed methods research*. Thousand Oaks, CA: Sage.

Cunningham, S., Wake, D., Waller, A., & Morris, A. (2014). Definitions of eHealth. In *eHealth, care and quality of life*. Milan: Springer.

Ekland, M. B., Krassioukov, A. V., McBride, K. E., & Elliott, S. L. (2008). Incidence of autonomic dysreflexia and silent autonomic dysreflexia in men with spinal cord injury undergoing sperm retrieval: Implications for clinical practice. *Journal of Spinal Cord Medicine*, *31*(1), 33–39.

Eysenbach, G. (2001). What is e-health? *Journal of Medical Internet Research*, *3*(2), e20.

Gassaway, J., Jones, M. L., Sweatman, W. M., Hong, M., Anziano, P., & DeVault, K. (2017). Effects of peer mentoring on self-efficacy and hospital readmission after inpatient rehabilitation of individuals with spinal cord injury: A randomized controlled trial. *Physical Medicine and Rehabilitation*, *98*(8), 1526–1534.

Gill, P. S., Kamath, A., & Gill, T. S. (2012). Distraction: An assessment of smartphone usage in health care work settings. *Risk Management and Healthcare Policy*, *5*, 105–114.

Government of Canada. (2019, February 5). *Prevalence of chronic diseases among Canadian adults*. Retrieved from https://www.canada.ca/en/public-health/services/chronic-diseases/prevalence-canadian-adults-infographic-2019.html.

Government of Canada. (2020, August 31). *Chronic Diseases*. Retrieved from https://www.canada.ca/en/public-health/services/chronic-diseases.html.

Hamm, N., Pelletier, L., Ellison, J., Tennenhouse, L., Reimer, K., Paterson, J., et al. (2019). Original quantitative research—Trends in chronic disease incidence rates from the Canadian Chronic Disease Surveillance System. *Health Promotion and Chronic Disease Prevention in Canada*, *39*, 216–224.

Harrison, R., Flood, D., & Duce, D. (2013). Usability of mobile applications: Literature review and rationale for a new usability model. *Journal of Interaction Science*, 1–16.

Hemming, K., Haines, T. P., Chilton, P. J., Girling, A. J., & Lilford, R. J. (2015). The stepped wedge cluster randomised trial: Rational, design, analysis, and reporting. *BMJ*, *350*, h391.

Hiremath, S. V., Amiri, A. M., Thapa-Chhetry, B., Snethen, G., Schmidt-Read, M., Ramos-Lamboy, M., et al. (2019). Mobile health-based physical activity intervention for individuals with spinal cord injury in the community: A pilot study. *PLoS One*, *14*(10), e0223762.

Hirsche, R. C., Williams, B., Jones, A., & Manns, P. (2011). Chronic disease self-management for individuals with stroke, multiple sclerosis and spinal cord injury. *Disability and Rehabilitation*, *33*(13–14), 1136–1146.

Holla, J. F. M., van den Akker, L. E., Dadema, T., de Groot, S., Tieland, M., Weijs, P. J. M., et al. (2020). Determinants of dietary behaviour in wheelchair users with spinal cord injury or lower limb amputation: Perspectives of rehabilitation professionals and wheelchair users. *PLoS One*, *15*(1), e0228465.

Jensen, M. P., Truitt, A. R., Schomer, K. G., Yorkston, K. M., Baylor, C., & Molton, I. R. (2013). Frequency and age effects of secondary health conditions in individuals with spinal cord injury: A scoping review. *Spinal Cord*, *51*(12), 882–892.

Jerant, A. F., von Friederichs-Fitzater, M. M., & Moore, M. (2005). Patients' perceived barriers to active self-management of chronic conditions. *Patient Education and Counselling*, *57*(3), 300–307.

Jones, M. L., Gassaway, J., & Sweatman, M. W. (2019). Peer mentoring reduces unplanned readmissions and improves self-efficacy following inpatient rehabilitation for individuals with spinal cord injury. *The Journal of Spinal Cord Medicine*, 1–9.

Kang, Y. N., Shen, H. N., Lin, C. Y., Elwyn, G., Huang, S. C., Wu, T. F., et al. (2019). Does a mobile app improve patients' knowledge of stroke risk factors and health-related quality of life in patients with stroke? A randomized controlled trial. *BMC Medical Informatics and Decision Making*, *19*(282), 1–9.

Kim, J. Y., & Cho, E. (2017). Evaluation of a self-efficacy enhancement program to prevent pressure ulcers in patients with a spinal cord injury. *Japan Journal of Nursing Science*, *14*, 76–86.

Kryger, M. A., Crytzer, T. M., Fairman, A., Quinby, E. J., Karavolis, M., Pramana, G., et al. (2019). The effect of the interactive mobile health and rehabilitation system on health and psychosocial outcomes in spinal cord injury: Randomized Controlled Trial. *Journal of Medical Internet Research*, *21* (7), e14305.

Lewis, J. R. (2018). The system usability scale: Past, present, and future. *International Journal of Human Computer Interaction*, *34*(7), 577–590.

Lewis, T. L., Boissaud-Cooke, M. A., Aungst, T. D., & Eysenbach, G. (2014). Consensus on use of the term "app" versus "application" for reporting of mHealth research. *Journal of Medical Internet Research*, *16*(7), e174.

Li, L., Huang, J., Wu, J., Jiang, C., Chen, S., Xie, G., et al. (2020). A mobile health app for the collection of functional outcomes after inpatient stroke rehabilitation: Pilot randomized controlled trial. *JMIR mHealth and uHealth*, *8*(5), e17219.

MacGillivray, M. K., Sadeghi, M., Mills, P. B., Adams, J., Sawatzky, B. J., & Mortenson, W. B. (2019). Implementing a self-management app for spinal cord injury during inpatient rehabilitation and following community discharge: A feasibility study. *The Journal of Spinal Cord Medicine*, *43*(5), 676–684.

McIntyre, A., Marrocco, S. L., McRae, S. A., Sleeth, L., Hitzig, S., Jaglal, S., et al. (2020). A scoping review of self-management interventions following spinal cord injury. *Topics in Spinal Cord Injury Rehabilitation*, *26*(1), 36–63.

Mitchell, J. (1999). *From telehealth to e-health: The unstoppable rise of e-health*. Canberra, Australia: National Office for the Information Technology.

Mortenson, W. B., Mills, P. B., Adams, J., Singh, G., MacGillivray, M., & Sawatzky, B. (2018). Improving self-management skills among people with spinal cord injury: Protocol for a mixed-methods study. *JMIR Research Protocols*, *7*(11), e11069.

Mortenson, W. B., Singh, G., MacGillivray, M., Sadeghi, M., Mills, P., Adams, J., et al. (2019). Development of a self-management app for people with spinal cord injury. *Journal of Medical Systems*, *43*(6), 145.

Munce, S. E., Fehlings, M. G., Straus, S. E., Nugaeva, N., Jang, E., Webster, F., et al. (2014). Views of people with traumatic spinal cord injury about the components of self-management programs and program delivery: A Canadian pilot study. *BMC Neurology*, *14*, 209.

Munce, S. E. P., Webster, F., Fehlings, M. G., Straus, S. E., Jang, E., & Jaglal, S. B. (2016). Meaning of self-management from the perspective of individuals with traumatic spinal cord injury, their caregivers, and acute care and rehabilitation managers: An opportunity for improved care delivery. *BMC Neurology*, *16*(11), 1–9.

Nabavi, R. T. (2012). *Bandura's social learning theory & social cognitive learning theory*. Retrieved from https://www.researchgate.net/publication/267750204_Bandura%27s_Social_Learning_Theory_Social_Cognitive_Learning_Theory.

National Health Council. (2014). *About chronic disease*. Retrieved from https://nationalhealthcouncil.org/wp-content/uploads/2019/12/AboutChronicDisease.pdf.

New, P., Reeves, R., Smith, E., Townson, A., Eriks-Hoogland, I., Gupta, A., et al. (2015). International retrospective comparison of inpatient rehabilitation for patients with spinal cord dysfunction epidemiology and clinical outcomes. *Archives of Physical Medicine and Rehabilitation*, *96*, 1080–1087.

Noonan, V., Fingas, M., Farry, A., Baxter, D., Singh, A., Fehlings, M., et al. (2012). Incidence and prevalence of spinal cord injury in Canada: A national perspective. *Neuroepidemiology*, *38*, 219–226.

Organization for Economic Cooperation and Development. (2019). *State of health in the EU: United Kingdom country health profile 2019*. Retrieved from https://www.oecd-ilibrary.org/docserver/744df2e3-en.pdf?expires=1605149953&id=id&accname=guest&checksum=38BD5378D6BB08537E80336C84B485EC.

Pancer, M., Manganaro, M., Pace, I., Marion, P., Gagnon, D. H., Laramee, M. T., et al. (2019). A web-based physical activity portal for individuals living with a spinal cord injury: Qualitative study. *JMIR Formative Research*, *3*(3), e12507.

Parmanto, B., Pramana, G., Yu, D. X., Fairman, A. D., Dicianno, B. E., & McCue, M. P. (2013). iMHere: A novel mHealth system for supporting self-care in management of complex and chronic conditions. *JMIR mHealth and uHealth*, *1*(2), e10.

Pickett, G. E., Campos-Benitez, M., Keller, J. L., & Duggal, N. (2006). Epidemiology of traumatic spinal cord injury in Canada. *Spine*, *31*(7), 799–805.

Rabchevsky, A. G., & Kitzman, P. H. (2011). Latest approaches for the treatment of spasticity and autonomic dysreflexia in chronic spinal cord injury. *Neurotherapeutics*, *8*(2), 274–282.

Rowland, J. (2020). *Defining secondary conditions for people with disabilities*. Retrieved from https://www.nchpad.org/360/2050/Defining~Secondary~Conditions~for~People~with~Disabilities.

Ruzic, L., Mahajan, H. P., & Sanford, J. A. (2018). Evaluating an evidence-based mHealth application, MS assistant, individuals again with multiple sclerosis. *International Journal on Advances in Life Sciences*, *10*(3), 156–168.

Sakib, M., Shooshtari, S., John, P., & Menec, V. (2019). The prevalence of multimorbidity and associations with lifestyle factors among middle-aged Canadians: an analysis of Canadian Longitudinal Study on Aging data. *BMC Public Health*, *19*.

Sarfo, F. S., & Ovbiagele, B. (2017). Mobile health for stroke: A promising concept for research and practice. *mHealth*, *3*, 4.

Savic, G., DeVivo, M. J., Frankel, H. L., Jamous, M. A., Soni, B. M., & Charlifue, S. (2017). Causes of death after traumatic spinal cord injury—A 70-year British study. *Spinal Cord*, *55*, 891–897.

Shiel, W. C., Jr. (2020). *Medical definition of chronic disease*. Retrieved from https://www.medicinenet.com/script/main/art.asp?articlekey=33490.

Singh, G., MacGillivray, M., Mills, P., Adams, J., Sawatzky, B., & Mortenson, W. B. (2019). Patients' perspectives on the usability of a mobile app for self-management following spinal cord injury. *Journal of Medical Systems*, *44*(1), 26.

Skold, C., Levi, R., & Seiger, A. (1999). Spasticity after traumatic spinal cord injury: Nature, severity and location. *Archives of Physical Medicine and Rehabilitation*, *80*, 1548–1557.

Sood, S., Mbarika, V. W. A., Jugoo, S., Doohky, R., Doarn, C. R., Prakash, N., et al. (2007). What is telemedicine? A collection of 104 peer-reviewed perspectives and theoretical underpinnings. *Telemedicine Journal and eHealth*, *13*(5), 573–590.

Tucker, S. (2015). Welcome to the world of mHealth. *mHealth*, *1*, 1.

Van den Akker, L., Holla, J. F. M., Dadema, T., Visser, B., Valent, L. J., de Groot, S., et al. (2020). Determinants of physical activity in wheelechair users with spinal cord or lower limb amputation: Perspectives of rehabilitation professionals and wheelchair users. *Disability and Rehabilitation*, *42*(14), 1934–1941.

Wilde, M. H., McMohan, J. M., Fairbanks, E., Brasch, J., Parshall, R., Zhang, F., et al. (2017). Feasibility of a web-based self-management intervention for intermittent urinary catheter users with spinal cord injury. *Journal of Wound, Ostomy, and Continence Nursing*, *43*(5), 529–538.

World Health Organization. (2013). *Global action plan for the prevention and control of noncommunicable diseases 2013–2020*. Retrieved from https://apps.who.int/iris/bitstream/handle/10665/94384/9789241506236_eng.pdf;jsessionid=B7659CDABAE694BE369DAD0E42B777B4?sequence=1.

World Health Organization. (2019). *Noncommunicable diseases*. Retrieved from https://www.who.int/en/news-room/fact-sheets/detail/noncommunicable-diseases.

Chapter 43

# Biomaterials, spinal cord injury, and rehabilitation: A new narrative

Elisa Lacroce[a], Giuseppe Perale[b,c], and Filippo Rossi[a]

[a]*Department of Chemistry, Materials and Chemical Engineering "Giulio Natta", Politecnico di Milano, Milan, Italy;* [b]*Faculty of Biomedical Sciences, University of Southern Switzerland (USI), Lugano, Switzerland;* [c]*Ludwig Boltzmann Institute for Experimental and Clinical Traumatology, Vienna, Austria*

## List of abbreviations

| | |
|---|---|
| **AEMA** | 2-aminoethyl methacrylate |
| **BBB** | Basso, Beattie, Bresnahan score |
| **BDNF** | brain-derived neutrophic factor |
| **BSCB** | blood–spinal cord barrier |
| **CNS** | central nervous system |
| **ECM** | extracellular matrix |
| **HA** | hyaluronic acid |
| **HP** | hydroxyphenyl |
| **MOEAA** | [2-(methacryloyloxy)ethoxy]acetic acid |
| **MOETA+** | [2-(methacryloyloxy)ethyl]trimethylammonium chloride |
| **MSCs** | mesenchymal stem cells |
| **NFs** | nanofibers |
| **NPs** | nanoparticles |
| **NTs** | nanotubes |
| **NT-3** | neurotrophin-3 |
| **NWs** | nanowires |
| **PCL** | poly(ε-caprolactone) |
| **PEG** | polyethylene glycol |
| **PEI** | polyethyleneimine |
| **PHEMA** | poly(2-hydroxyethyl methacrylate) |
| **PHPMA** | poly(N-(2-hydroxypropyl)-methacrylamide) |
| **PLA** | polylactic acid |
| **PLGA** | poly(lactic-*co*-glycolic acid) |
| **RGD** | Arg-Gly-Asp |
| **SC** | spinal cord |
| **SCI** | spinal cord injury |
| **SC-ECM** | spinal cord extracellular matrix |
| **SEM** | scanning electron microscopy |
| **SIKVAV** | ser-Ile-Lys-Val-Ala-Val |
| **UB-ECM** | urinary bladder extracellular matrix |
| **UC-ECM** | umbilical cord extracellular matrix |

## Introduction

Spinal cord injury (SCI) is a lesion of the spinal cord which leads to the permanent loss of sensory and motor functions below the injury site. The traumatic effect of SCI is due to the low regenerative capability of the tissue, which is in contrast

Diagnosis and Treatment of Spinal Cord Injury. https://doi.org/10.1016/B978-0-12-822498-4.00043-9

with the fast inflammatory response that occurs in the first minutes after the injury and starts the secondary injury in the following hours. The latter is characterized by demyelinization and the formation of a glial scar tissue which represents a physical barrier to the growth of axons. Nowadays, there are no effective clinical treatments able to regenerate the nervous tissue and restore the motor functions, so new strategies are being developed by researchers in order to overcome these limitations. One strategy is represented by scaffolds able to provide a structure that mimics the extracellular matrix (ECM) and at the same time supports the cellular attachment, growth and differentiation. The scaffolds must be biocompatible, non-toxic and have mechanical and morphological properties suitable for the tissue regeneration. In addition, they can chemically bind or physically entrap one or more drugs and release them in a controlled manner. The materials used for scaffold development can be synthetic or natural. Examples of the most used chemically synthetized materials are aliphatic polyesters such as polylactide, polyglycolide, and polycaprolactone (Pires & Pêgo, 2015). Synthetic materials present the advantage of being able to be controlled and modified from the point of view of chemistry, mechanical and structural properties in order to mimic as much as possible the ECM. In contrast, natural scaffolds are more similar in composition to ECM because some of the molecules are already present in the ECM, such as e.g., collagen, fibronectin, and hyaluronic acid but they have some differences in composition depending on their origin and previous treatments. Other natural materials such as alginate, agarose, and chitosan are widely used too. In addition, also the mechanical properties of the scaffold should be similar to the target biological tissue in order to avoid adverse effects. For the treatment the SCI, a stiff scaffold is not suitable because it is not able to support flexible movements of spinal cord without further lesioning other surrounding tissues. In addition, it was demonstrated that a stiff material promotes astrocyte growth (Georges, Miller, Meaney, Sawyer, & Janmey, 2006) and causes glial cell activation which leads to inflammation response and formation of a fibrotic tissue. Hence, soft scaffold, such as hydrogel, better matches the mechanical properties of the nervous tissue. Hydrogels are suitable for this purpose not only because of their flexibility, but also because of their bioadhesive and swelling properties, which confer the ability to stay localized in situ and to exchange metabolites with the surrounding tissue fluids. Other factors can influence the tissue regeneration, such as the pore sized distribution of the scaffold which has to guarantee the possibility for cells and fluids to enter inside the scaffold. The topography, the charge, the composition of the surface and the orientation of the fibers influence actin cytoskeleton and hence cell adhesion, spreading and differentiation.

In the case of nerve tissue regeneration, fibers arranged in a longitudinal way and pore size of 50 to 100 μm enhance nerve regeneration ( Jurga et al., 2011; Yuan et al., 2014). Moreover, scaffold can be made of non-degradable or degradable materials. In the first case, the scaffold remains inside the body so the tissue can partially regenerate occupying the space between the fibers, whereas in the second case, it is necessary that the rate of degradation of the scaffold matches tissue regeneration speed.

# Hydrogels and scaffolds

## Synthetic-based hydrogels

Among synthetic non-biodegradable hydrogel used for SCI repair poly(N-(2-hydroxypropyl)-methacrylamide) (PHPMA) is very promising. The research group of Woerly et al. developed a hydrogel made of PHPMA, obtained by a radical polymerization of the monomer HPMA with the use of a divinyl cross-linking agent (Woerly et al., 1999), functionalized by a synthetic peptide which includes RGD sequence (Woerly, Pinet, De Robertis, Van Diep, & Bousmina, 2001). The implantation of the hydrogel into the neonatal and adult spinal cord reveals a good infiltration of cells and blood vessels and the following implantation of the hydrogel seeded with rat mesenchymal stem cells (MSCs) in rats results in better Basso, Beattie, Bresnahan (BBB) score than the control group without the implantation (Hejčl et al., 2010). Another investigated polymer in nerve tissue regeneration is the biocompatible and hydrophilic poly(2-hydroxyethyl methacrylate) (PHEMA). As previously said, the charge and the structure of the hydrogel can influence the behavior of cells and the ingrowth of the new tissue. A study by Hejcl et al. (Hejčl et al., 2010) was conducted to compare the different effects of the surface charge and structure of HPMA and HEMA hydrogels on tissue regeneration. Specifically, four different hydrogels were prepared and seeded with rat MSCs: one HPMA-RGD hydrogel by using heterophase separation (HPMA-HS-RGD), which resulted in a structure characterized by microparticles, two HPMA hydrogels by using a solid porogen, one functionalized by RGD peptide (HPMA-SP-RGD), and one without functionalization (HPMA-SP), which resulted in network structures, and the last hydrogel made by positively charged copolymers of HEMA with [2-(methacryloyloxy)ethyl]trimethylammonium chloride (MOETA+). After successful in vitro studies, hydrogels were implanted into rat SCI hemisection model. The best results in terms of in vitro adhesiveness and in vivo survival of MSC was found in the positively charged HEMA-MOETA+ hydrogel, whereas the best results in terms of axonal ingrowth and vascularization was found in the HPMA-SP-RGD hydrogel demonstrating the higher efficacy of the network architecture respect to the globular ones. With respect to the influence of the RGD peptide, it increases the vascularization but has no effect the growth of axons.

## Hydrogel functionalization with cell-adhesive peptides

The presence of cell-adhesive peptides on hydrogels, such as the laminin-derived peptide sequence SIKVAV [Ser-Ile-Lys-Val-Ala-Val] and fibronectin-derived peptide RGD [Arg-Gly-Asp], can enhance cell adhesion, migration on the scaffold, proliferation, and differentiation (Rossi & van Griensven, 2014). The research group of Kubinova et al. functionalized a copolymer-based hydrogel of HEMA and 2-aminoethyl methacrylate (AEMA) with the laminin-derived Ac-CGGASIKVAVS-OH peptide by disulfide bridges (Kubinová et al., 2010). The functionalization with SIKVAV and RGD (Macková et al., 2016) was made on the same type of hydrogel also through the maleimide-thiol coupling reaction. All the functionalized hydrogels guarantee the adhesion and proliferation of rat MSCs maintaining their multi-lineage potential. In addition, in vivo studies results shown a higher connective tissue and vascularization on fibronectin-modified HEMA hydrogel compared to the non-functionalized one (Hejčl et al., 2018). Other molecules such as serotonin can be used as neurotransmitter and can improve neuronal differentiation of implanted or endogenous neuronal progenitor precursors. Despite promising in vitro results, in vivo model studies on implanted PHEMA functionalized with serotonin showed a migration of seeded neural progenitor out from the polymer leading to a fail in proving a long-term effect on nerve tissue reconstruction (Růžička et al., 2013).

## Porosity orientation

The orientation of fibers is important for tissue regeneration because it provides preferential lines along with the cells growth and proliferation. In addition, adequate porosity and mechanical properties have to support the movements of the organ and the regeneration of the new tissue. Hence, in the case of scaffolds for nervous tissue regeneration, the best one has to be characterized by parallel guiding channels and pores. The group of Kubinová et al. (2015) developed SIKVAV-modified PHEMA hydrogels with parallel oriented pores prepared by a salt-leaching method with ammonium oxalate needle-like crystals, and added 8%, 4%, and 0% (wt%) of [2-(methacryloyloxy)ethoxy]acetic acid (MOEAA) obtaining three hydrogels with 57%–77% porosity, pore diameter of $\sim$60 mm, and an elastic modulus of 6.7, 27.4, and 45.3 kPa along the pore axis and 2.9, 3.6, and 11 kPa in a perpendicular direction. After 2 months of implantation the results showed that the softest hydrogel collapses because of the thinness of walls causing a sparse axonal growth inside the hydrogel, whereas the stiffest hydrogel supported axonal ingrowth into the pore guides but cyst formed at the tissue-scaffold interface because of difference in mechanical properties between the two components. The best results in terms of axonal ingrowth, presence of blood vessels and Schwann cells are obtained using the hydrogel with the moderate elasticity modulus of 27.4 kPa along the pores. Unfortunately, the use of the moderate scaffold seeded with MSCs was not able to promote a sufficient axonal growth.

Indeed, the rate of axonal growth resulted very slow and after 6 months from the implantation only few axons were be able to cross the hydrogel and infiltrate the caudal stump (Hejčl et al., 2018). Therefore, other factors are necessary in order to promote axonal regeneration. For example, the presence of MSCs overexpressing of an NT-3 receptor (Zeng et al., 2015) or brain-derived neutrophic factor (BDNF) (Gao et al., 2013) on the scaffold can be added in order to enhance axonal growth and recovery of motor functions.

## Natural-based hydrogels

This type of hydrogels may be made by ECM derived components such as collagen or hyaluronic acid. Hyaluronic acid (HA) is a natural biocompatible polymer, biodegradable and non-toxic, but it is does not favor the attachment of cells. A possible overcoming solution is represented by the use the hydroxyphenyl derivative of HA which is able to covalently crosslink in situ, forming a hydrogel in presence of horseradish peroxidase enzyme and hydrogen peroxidase (Kučera et al., 2015). Moreover, the RGD peptide can be linked to the HA-PH derivative (Zaviskova et al., 2018) in order to favor the attachment of cells on (HP-HA) hydrogel. Human Wharton's jelly derived mesenchymal stem cells (hWJ-MSCs) were encapsulated in the hydrogel, which then was injected in the sub-acute spinal cord hemisection. In situ crosslinking had no cytotoxic effect or negative effect on cells. HA-PH-RGD hydrogel was able to favor axonal ingrowth and the presence of hWJ-MSCs increases the effect. However, there were no improvements of motor function probably due to the low quantity of cells incapsulated inside the gel.

### Extracellular matrix-based hydrogels

Another type of natural-based hydrogel is represented by decellularized ECM: it is suitable for tissue regeneration because of its biocompatibility, biomolecular and complex chemical composition which characterize it and distinguish it from other scaffolds. Decellularization is performed by different chemical, physical or enzymatic method and then the decellularized ECM is transformed in a liquid phase using pepsin solubilization at pH $< 2$ in order to be injected into the site of injury. The physiological temperature and pH favor its crosslinking in situ, leading to its original structure. The research group of Kubinova tried to use ECM-based hydrogels derived from CNS, such as porcine spinal cord (SC-ECM), and non-CNS derived, such as human umbilical cord tissue (UB-ECM) and porcine urinary bladder (UB-ECM). After implantation into injured spinal cord they stimulated nerve tissue regeneration and no differences on biological response were seen between the use of CNS and non-CNS–derived ECM (Kočí et al., 2017; Medberry et al., 2013). However, a critical problem was represented by the fast degradation rate of the scaffold, which was due to the infiltration of resident cells present in the site of lesion.

Therefore, inadequate structure was provided to the new tissue and a correct regeneration of the tissue was compromise. In order to decrease the rate of degradation it was necessary to increase the number of crosslinks. This can be done using crosslinking agents such as genipin, which is able to bridge free amino groups present in the ECM. Its use on the UC-ECM hydrogel did not increase in vivo inflammatory response (Výborný et al., 2019), moreover the lack of ethical problems and the allogeneic source leads to consider promising the use of umbilical cord in neural tissue regeneration.

## Nanomaterials

### Nanotechnology and nanomedicine

Nanotechnology is the synthesis and characterization of nanosystems and their application in different fields, from the research to the industrial practice. When nanotechnology is applied in medicine and healthcare, it is called nanomedicine. Nanomedicine covers different medical fields such as prevention, diagnosis and treatment. It uses nanomaterials in the range of 10–1000 nm for interacting with biological systems at the molecular level. In addition, the resulting high surface area per unit volume favors a higher number of interactions with biological systems. Thanks to the binding with specific cellular receptors nanosystems can also deliver drugs and molecules in specific site without damaging the surrounding healthy tissue. Nanomedicine developed a lot of structures such as nanoparticles, nanotubes, nanorods, nanogels, quantum dots, etc., but the most used in SCI field are nanoparticles, nanogels, and nanotubes.

### Properties of nanomaterials

The treatment of SCI with drugs administrated by oral, intravenous or intra-arterial ways is not effective due to the filtrating action of the blood–spinal cord barrier (BSCB) which prevents the passage of foreign and immunological substance from bloodstream to the SC parenchyma. Nanomaterials can be developed by top-down, bottom-up or hybrid methods. The first method consists in transforming a bulk material to a nanosized material, the second one consists in forming a nanomaterial starting from molecular arrangements and interactions whereas the last method is based on mixing the previous two. Nano-carriers have to correspond to specific size in order to favor their migration across the biological barrier of spinal cord and the target the desired tissue. Smaller particles are more suitable for this purpose and the possible presence of ligands on their surface can bind to receptor molecule of neural cells favoring the activation of specific cellular response. The drawback of using small particles is the limited control on modification in a batch-to-batch synthesis approach (Saraiva et al., 2016). Furthermore, the shape of the nanovectors influences their behavior inside biological environments and their cellular uptake too.

Specifically, in the case of nervous system, nanorods characterized by targeting peptides are considered more able than nanoparticles in accumulating in specific vascular environment without activating immune clearance (Kolhar et al., 2013) or, for example, biconcave nanoparticles enhance the release of drug respect to spherical or tubular particles (Zuidema, Gilbert, & Osterhout, 2016). Finally, surface charge has a key role with respect to the final aim of particles. In general, positively charged nanoparticles are better internalized by cells (Xiao et al., 2011; Yue et al., 2011) but the modification with chemical groups or peptides can change the surface charge leading to a different aim such as to target a specific area or avoid activation of immune systems. As for the composition of hydrogels, also nanomaterials can be made by synthetic or natural materials. Natural nanomaterials are in general biocompatible, non-toxic and very similar in composition and chemical features with the biological environment allowing a weak immune response, but it is difficult to achieve a good reproducibility during their development and production. On the other hand, synthetic nanomaterials guarantee a high reproducibility and possibility to modify their chemical, physical and morphological properties adapting them to the final purpose, but their immunogenicity is higher compared to the natural ones. Natural materials used for developing nanoparticles are collagen, lipids, albumin, fibrin, silicone, alginate, agarose, hyaluronic acid, chitosan, cellulose, heparin and chondroitin sulfate, whereas synthetic materials used are polyethylene glycol (PEG), polyethyleneimine (PEI), polylactic acid (PLA), poly(lactic-co-glycolic acid) (PLGA), polyglycolic derivatives, polymethacrylate, polyacrylates, polycyanoacrylates, and poly(ε-caprolactone) (PCL). Fig. 1 represents a summary of natural and synthetic polymers.

## Nanoparticles

Nanoparticles are colloidal systems made of polymer chains from which is possible to obtain nanospheres or nanocapsules (Fig. 2). Nanosphere are systems with a size of 100–200 nm composed of a solid matrix with physically or chemically entrapped drug (Liu, Xiao, & Allen, 2004). They can be covered at their surface with surfactants or hydrophilic polymers which avoid opsonization and subsequent internalization from immune cells. Nanocapsules are nanosystems composed by an external polymeric layer which surround a lipophilic core.

These systems are very useful in encapsulating hydrophobic drugs in the core of nanoparticle and releasing them in situ. Generally, they are made of PLA, PLGA or PCL surrounded by hydrophilic PEG in order to avoid the activation of immune system. Finally, polymersomes are nanocapsules made of an aqueous core able to encapsulate hydrophilic drugs. In this case the aqueous core is enclosed by amphiphilic copolymers which expose hydrophilic segments in the core and external surface, whereas hydrophobic segment in the middle.

There are different techniques used to synthetize nanoparticles:

- Emulsion: it consists in emulsifying an oil phase containing hydrophobic monomers with a water phase containing surfactants. The polymerization of hydrophobic monomers starts after the addition of oil-soluble initiators forming polymeric particles inside an aqueous phase. In order to avoid opsonization and aggregation between nanoparticles, surfactants bind to the surface of particles by a polycondensation reaction between the two monomers present in the oil and in the aqueous phase or by the presence of initiators. The drug can be encapsulated during the polymerization process or absorbed at the end of the polymerization.
- Nanoprecipitation: it consists in desolvation of polymers dissolved in the solvent solution after the addition of it to the non-solvent solution.
- Solvent evaporation: emulsifying agents dissolved in water phase are added to an organic phase containing drug and polymer dissolved. The formation of oil/water emulsions is followed by solvent evaporation by using temperature or low pressure obtaining the nanoparticles.
- Salting out: the organic solution containing the polymer is added to an aqueous phase containing an emulsifier and a high concentration of salts. Then, pure water is added to promote the diffusion of organic solvent into water phase forming nanoparticles.
- Controlled gelification: gel nanospheres can be formed by using sodium alginate and calcium chloride.
- Desolvation: this method can be used only on natural polymers which are dissolved in aqueous environment. The following drip of a desolvating agent, such as ethanol or acetone, containing active molecule and the addition of cross-linking molecules in polymeric solution allow to obtain nanoparticles.
- Coacervation: coacervates are formed by electrostatic interactions between cargo aqueous phase and polymer.

As regard to nanocapsules, they are formed by mixing an oil-containing lipophilic surfactants with an aqueous phase miscible with organic solvent containing polymeric chains and therapeutic molecules. Under stirring, oil droplets are forming in the aqueous phase and polymers interact with the two phases exposing the hydrophobic chain toward the oil component and the hydrophilic one toward the aqueous phase. A method recently developed is based on following addition of water to the system favoring the passage of solvent from the center of nanoparticles to the external phase. Finally, polymersomes are formed starting from a copolymers dissolved in an organic solvent. Then the solvent evaporates leading to the formation of a polymeric layer and water is added to rehydrate polymers.

The following sonication and extrusion of solution lead to the formation of polymersomes. In the case the therapeutic molecule is a protein, some steps such as the use of organic solvent or sonication can denature the protein leading to its inactivation. In this case, the addition of natural salts or alcohols can affect its 3D structure favoring aggregates with polymer chains. At last, the use of glutaraldehyde can stabilize the nanoparticles.

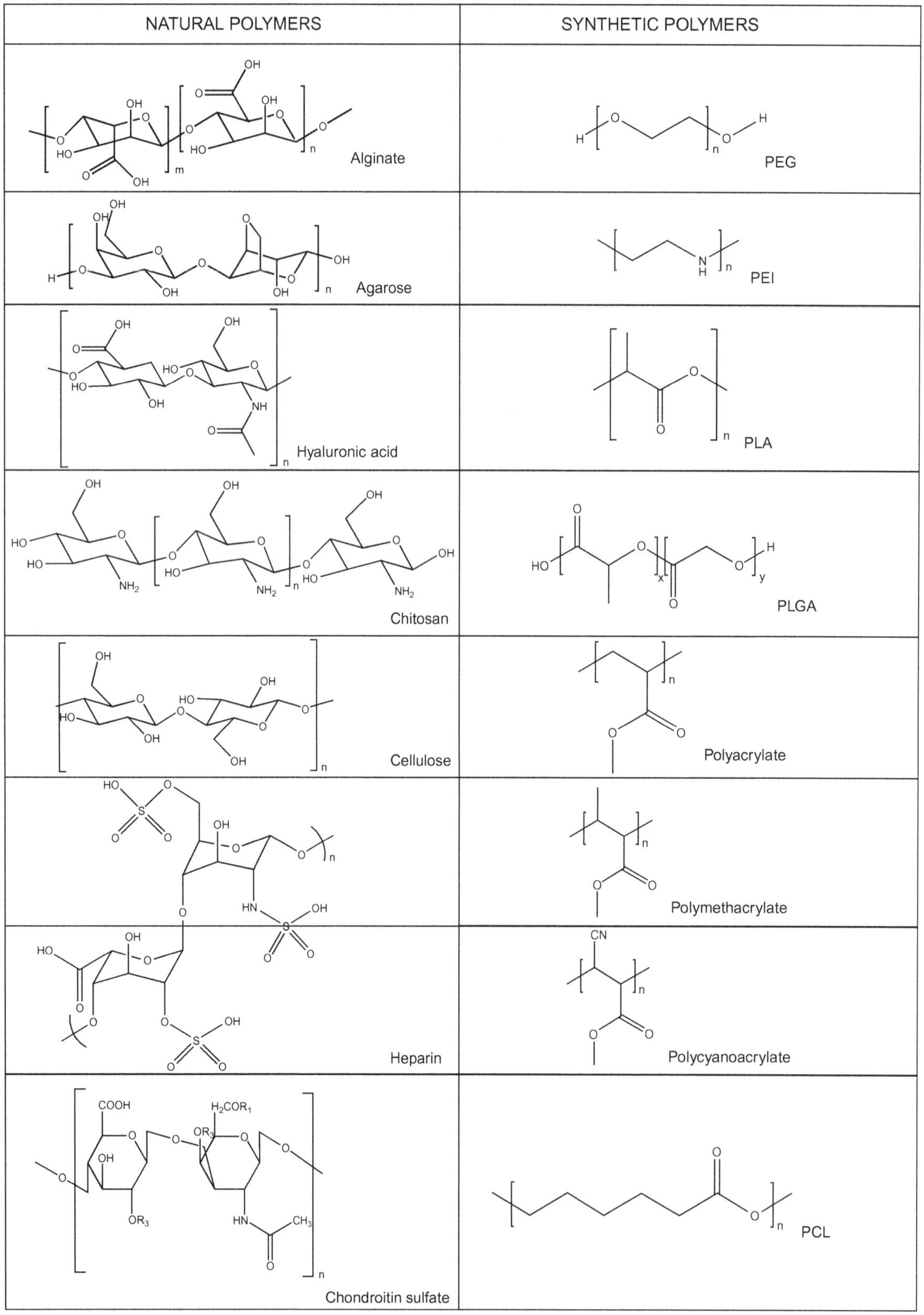


FIG. 1 Polymers for NPs. Chemical structure of natural and synthetic polymers used for developing NPs.

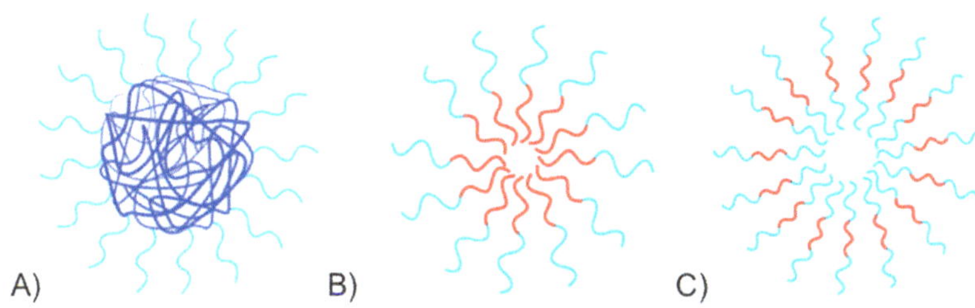


FIG. 2 Different NPs. Structure of (A) nanospheres, (B) nanocapsules, and (C) polymersomes.

## Functionalization of nanoparticles

Nanoparticle functionalization is needed in order to increase its half-life and favor its interaction with the targeted cells. In case of SCI it is important that nanoparticles are able to pass across the BSCB and reach the site of injury providing neuroprotective and/or neuro-regenerative effect. The functionalization can consist in addition of surfactants, biomolecule, dyes (for in vitro and in vivo tracking) and peptides by linking them to functional polymer groups such as hydroxyl, amine, carboxylic or alkyl groups. For example the presence of antioxidant enzyme superoxide dismutase on the surface of nanoparticles induces neuroprotective action (Varma et al., 2013) or the release of encapsulated fibroblast growth factor-2 inside PLGA nanoparticles reduces vasoconstriction in SCI during primary injury and favors angiogenesis (Kang, Baumann, Tator, & Shoichet, 2012). Reactions with functional groups such as the carboxyl group can link chitosan to a PEG grafting biotin able to attract monoclonal antibody OX26 leading to a decrement of neuronal cell death in injured spinal cord (Aktaş et al., 2005). The functionalization can be performed also on cation polymers such as PEI or chitosan. The most common methods to functionalize amine groups are making a reaction with thiol or maleimide group forming a disulfide bond.

## The effects of encapsulated neurotrophin

Neurotrophin is a protein that induces axonal regeneration and can be used as therapeutic molecule in case of SCI. The research group of Elliot Donaghue et al. was able to encapsulate neurotrophin-3 (NT-3) inside PLGA 220 nm nanoparticles through a double emulsion-solvent evaporation method, and then to entrap nanoparticles inside a hyaluronan/methyl cellulose matrix in order to have a more confined and controlled release of NT-3 (Elliott Donaghue, Tator, & Shoichet, 2015). In particular, the diffusion of NT-3 outside the matrix lasted 50 days in in vitro studies and 28 days in in vivo studies, leading to consider this system as a possible solution in order to limit the number of dosing. In addition, a higher locomotor recovery and axonal growth was seen in mouse model study after the treatment with PLGA NPS loaded with NT-3, compared to the controls.

## Nanogels

Nanogels are innovative nanoparticles with hydrophilic properties and high colloidal stability. They are characterized by swelling behavior which gives them the unique ability to exchange ions and biological molecules with the surrounding environment maintaining an equilibrium of metabolites between the internal and external parts. In addition, their deformability allows an easy passage through biological barrier and this characteristic, together with the swelling behavior, allows considering nanogels soft materials because of their similar properties to hydrogels. Nanogels can be developed starting from monomers of low molecular weight or from polymeric precursors. In the first case, monomers polymerize thanks to a controlled living radical polymerization using an initiator molecule from which the polymerization starts and propagates forming the nanoparticle, whereas in the second case the process is characterized by an inter-polymer interactions. Particularly, functional groups of polymer precursors chemically interact each other forming covalent bonding between polymer chains. Another possible technique is based on physical interactions between polymer chains such as hydrogen bonding, electrostatic interactions, Van der Walls forces or hydrophilic/hydrophobic interactions. In this case, the final nanogel has a low stability and its structure can be easily compromised by temperature, pH or external forces. A new strategy that can be used to produce nanogels is represented by the non-wetting templates (PRINT) technology. It is a lithographic technique that uses non-wetting elastomeric templates inside which nanogels are formed allowing a high reproducibility. Another new approach is represented by molecular imprinting, although its use is very difficult in nanogel developing. It consists in a linkage of a chosen protein on a functional monomer in order to form a template molecule. Then the polymerization starts using cross-linking agent and the protein is detached. The remaining polymer will present a cavity complementary to the protein and it will be used for selective cell targeting during the treatment of SCI.

## Functionalization of nanogels

The functionalization of nanogels with drug, peptides, proteins, enzymes, dyes, etc. can be performed using different types of reactions (Fig. 3): formation of amide bond from an ester bond, esterification, ring opening and Schiff base reactions,

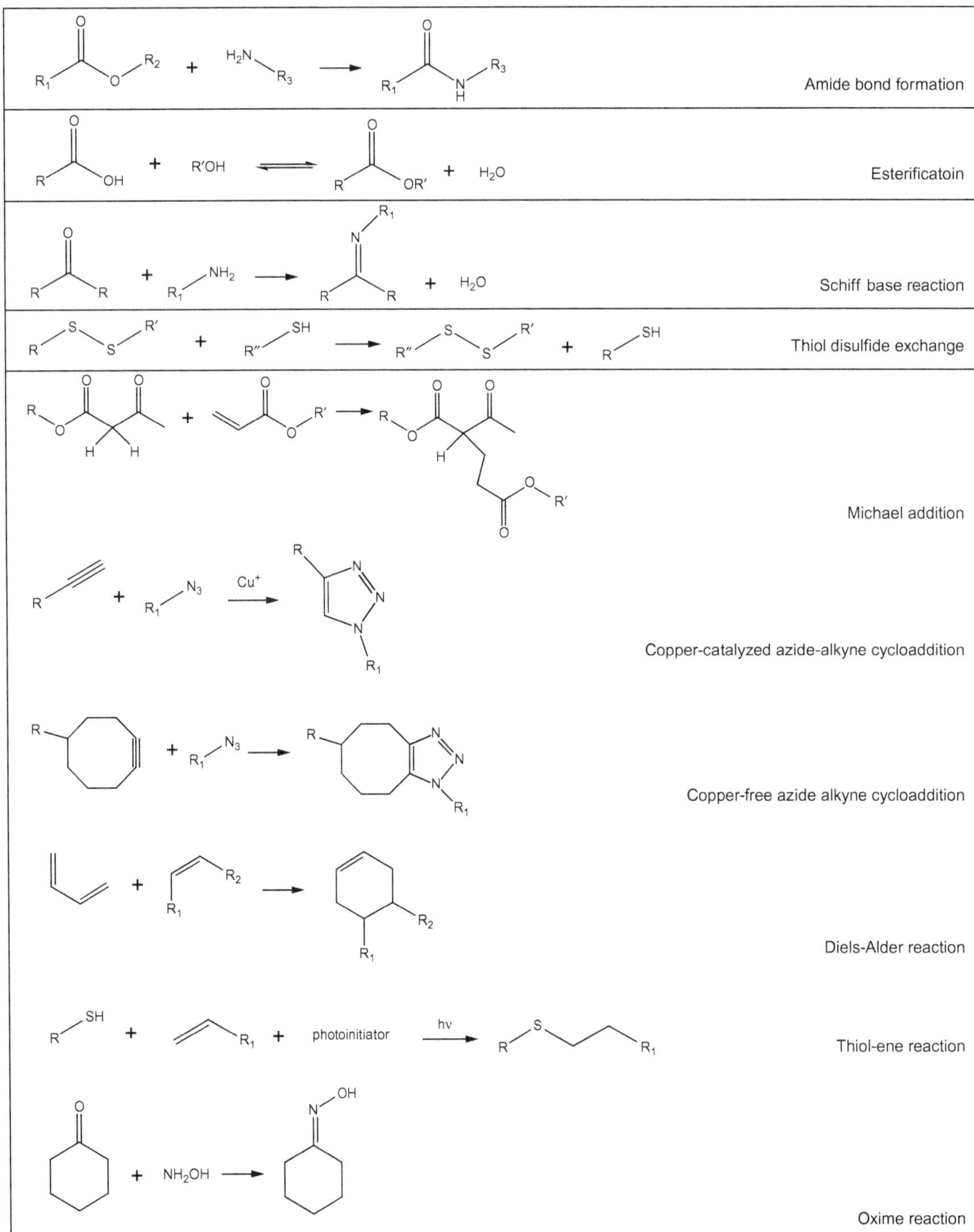


FIG. 3 Polymer functionalization. Different strategies of polymer functionalization.

thiol disulfide exchange and finally click chemistry which includes Michael type addition, copper-catalyzed and copper-free azide-alkyne cycloaddition, Diels–Alder reaction, thiol-ene reaction or oxime reaction. Depending on the type of chemical bond, the *cargo* can be released according to a change of temperature, pH, the presence of enzymes or other external stimuli able to break the bond. For example, redox-responsive NGs are able to accumulate in the target tissue and release the cargo only when redox stimulus is applied (Ghorbani & Hamishehkar, 2019).

In another study, the research group of Mauri et al. (2017) developed nanogels able to be internalized by microglia cells and release the therapeutic *cargo* only in the cytosol. In this case, the fast release in biological fluids that characterizes the hydrophilic drugs and the following rapid clearance from the body is avoided. The drug mimetic, rhodamine, was linked to PEG using the thiol chemistry forming a disulfide bond that can be broken in the cytosol by glutathione or cysteines, whereas the Cy5 dye was linked to the PEI by copper-catalyzed alkyne-azide cycloaddition. Finally, carbamate bonds formed the final nanogel. In vitro studies showed florescent signals inside the cytosol demonstrating the internalization of nanogels from microglia and the release of drug in the cytosol after 4 days. The functionalization can also be non-covalent if other interactions occur. This is the case of PEI functionalization in which electrostatic attraction between genes, peptide or growth factor and protonated amine on polymer chains is used to functionalize the nanogel.

### Nanochannels, nanotubes, nanowires, and conduits

It is worth mentioning other nanostructures that are used in SCI repair: tubular particles such as nanotubes (NTs), nanowires (NWs), and nanofibers (NFs) can guide axonal regeneration and limit the local inflammation at the same time.

Nanotubes are cylinders with a diameter in the order of nanometers, made of graphene. These structures are similar to cytoskeletal elements in neurons, signaling proteins and ion channels, so their presence does not activate immune or inflammatory responses. Their flexibility, electrical conductivity, and durability allow their implantation in the spinal cord for a long time. In addition, they can be functionalized with active molecules, such as 4-hydroxynonenal which promotes neurons spatial orientation and interconnections (Mattson, Haddon, & Rao, 2000), neurotrophic factors which provide neuroprotective effect or chemical groups which confers superficial charge to stimulate axonal growth. In vivo studies about the injection of NTs during the secondary injury phase showed a reduction of injured site, an increment of neurofilament-positive fibers and a partial recovery of locomotor functions (Roman, Niedzielko, Haddon, Parpura, & Floyd, 2011).

Nanowires are structures similar in shape with that on nanotubes but in this case the length is much longer than the diameter. In addition, they are generally made with metals, semiconductors, insulators or polymers with electric properties. They can be used for SCI repair in order to provide cell adhesion, proliferation, and electric stimulation as made by Bechara and coworkers (Bechara, Wadman, & Popat, 2011) with PCL NWs linked with polypyrrole, an electroconductive polymer.

Nanofibers (NFs) are the third most common nanosystems used for nerve regeneration. Their spatial orientation and diameter are able to positively influence cell behavior and differentiation. In particular, cell aggregation decreases and cell proliferation increases as the diameter of NFs decreases, whereas aligned NFs results in higher rate of neural stem cell differentiation than NFs oriented in random way (Xie et al., 2009; Yang, Xu, Kotaki, Wang, & Ramakrishna, 2004). Also in this case, NFs can be functionalized at their surface with specific chemical groups or molecules which confers surface charges, such as Rolipram in PLGA NFs (Zhu et al., 2010), able to improve axonal growth and reduce inflammatory response in the site of injury.

Conduits are cylindrical systems used to cover nerve gap and provide a guide for nerve regeneration. Promising conduits are made of PLGA-chitosan or PCL because of their effect in promoting remyelination of axons. Even if their use has positive effects on nerve repair, their implantation is invasive causing infections, inflammation and other permanent damages, making the risk/benefit balance unfavorable.

## Case study: Agarose–carbomer-based hydrogels

Even if the polymers used for biomaterials preparation are biocompatible, this does not ensure they promote a correct cell viability; functional compounds are hence needed in order to provide an interaction between the cells and the polymeric scaffold. RGD peptide, for example, can be used for scaffold functionalization because of its ability to bind to the receptors present on the surface of cell membrane and activate a cell adhesion response. The research group of Perale and Rossi (Caron et al., 2016; Papa et al., 2018) proved the extremely promising results obtained with functionalized poly-acrylic acid (PAA) and polyethylene glycol (PEG) with RGD peptide by using a click chemistry strategy. In particular, PAA polymer was previously functionalized with an alkyne group (Fig. 4) (Mauri et al., 2018). Then, CuAAC click reaction between RGD azide and alkyne polymers was conducted at 50°C–60°C forming the triazole. Finally, the hydrogel was synthesized by microwave-assisted polycondensation between mixed RGD-functionalized polymers and agarose.

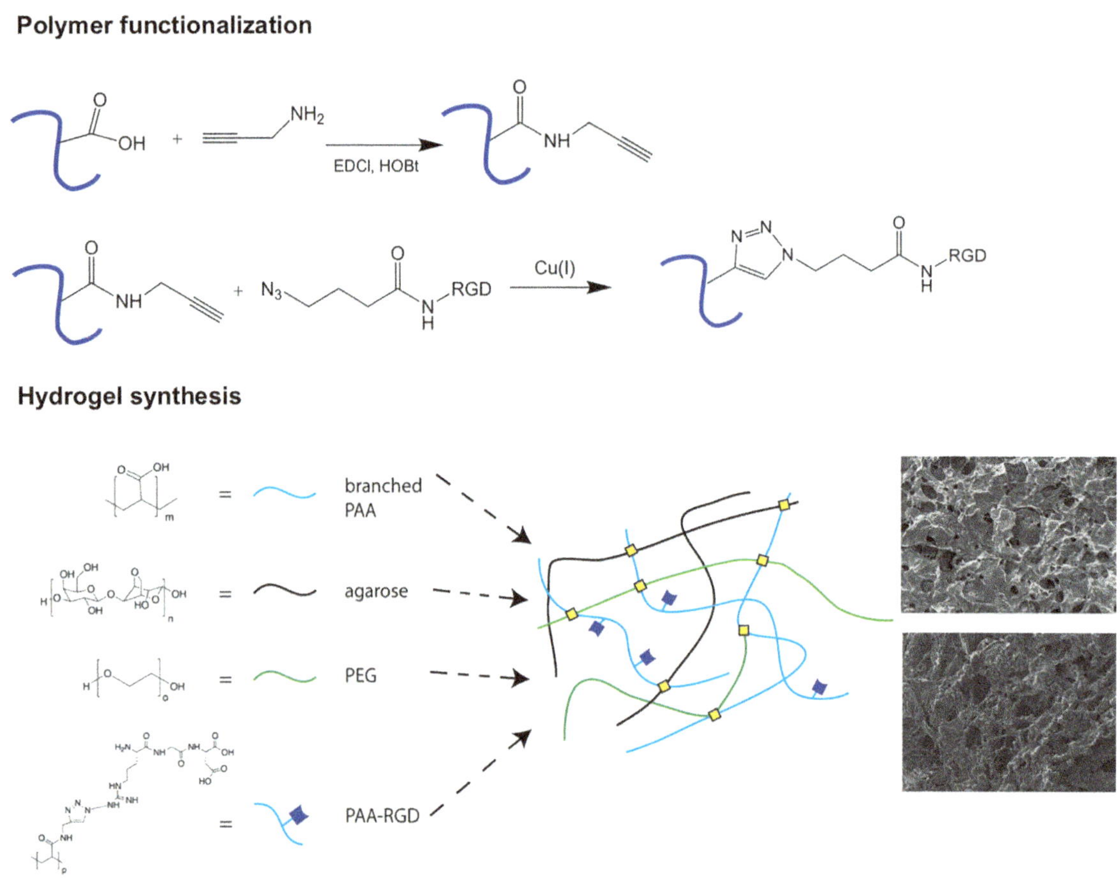


FIG. 4 Agarose–Carbomer-based hydrogels. Schematic representation of RGD-functionalized (HG-RGD) hydrogels. *(Adapted from Mauri, E., Sacchetti, A., Vicario, N., Peruzzotti-Jametti, L., Rossi, F., & Pluchino, S. (2018). Evaluation of RGD functionalization in hybrid hydrogels as 3D neural stem cells culture systems.* Biomaterials Science, *6(3), 501–510. doi:10.1039/c7bm01056g, with permission from the Royal Society of Chemistry.)*

From SEM analysis of the final hydrogel, there were not differences in polymer network and porous structure compared to the non-functionalized one, and adequate mechanical and morphological properties result from physical and chemical characterization. The system is highly biocompatible, can remain localized in the lesion site, can maintain the stemness of the loaded cells (Fig. 5) in vitro and in vivo improving the locomotor performances of mice (Fig. 6).

## Applications to other areas of neuroscience

SCI remains one of the most devastating conditions in neurological diseases. Most of the post traumatic degeneration of the tissue is caused by a multifactorial secondary injury including several interconnected processes. Relevant is the involvement of acute and chronic inflammation, represented mostly by inflammation that contributes to the cascade of harmful events during the secondary injury, in the end leading to spreading and chronicity of SCI. An unresolved inflammation is a pathological hallmark of many neuropathologies and microglial cells can play a relevant role in these scenarios. The current view suggests that under normal physiological condition the acute inflammatory response is a transitory process, aiming at eliminating many potential toxic stimuli, which is followed by resolution of the inflammation and a return to homeostasis. Hence, an acute neuro-inflammatory response is considered generally beneficial to the CNS, since it tends to limits the damages and contributes to the repair of injured tissue. However, in many neuropathologies, an over-activation and an accumulation of microglial cells occurs, due to persistent insults or triggered by factors released in the damaged environment by dead cell. Indeed, a sustained release of pro-inflammatory mediators can propagate the inflammatory reaction, promoting microglia proliferation and further releasing pro-inflammatory factors that fed an uncontrolled response. Therefore, a prolonged and unresolved chronic inflammation due to over-activation of microglial cells can have neurotoxic consequences that could lead to the exacerbation of the pathology.

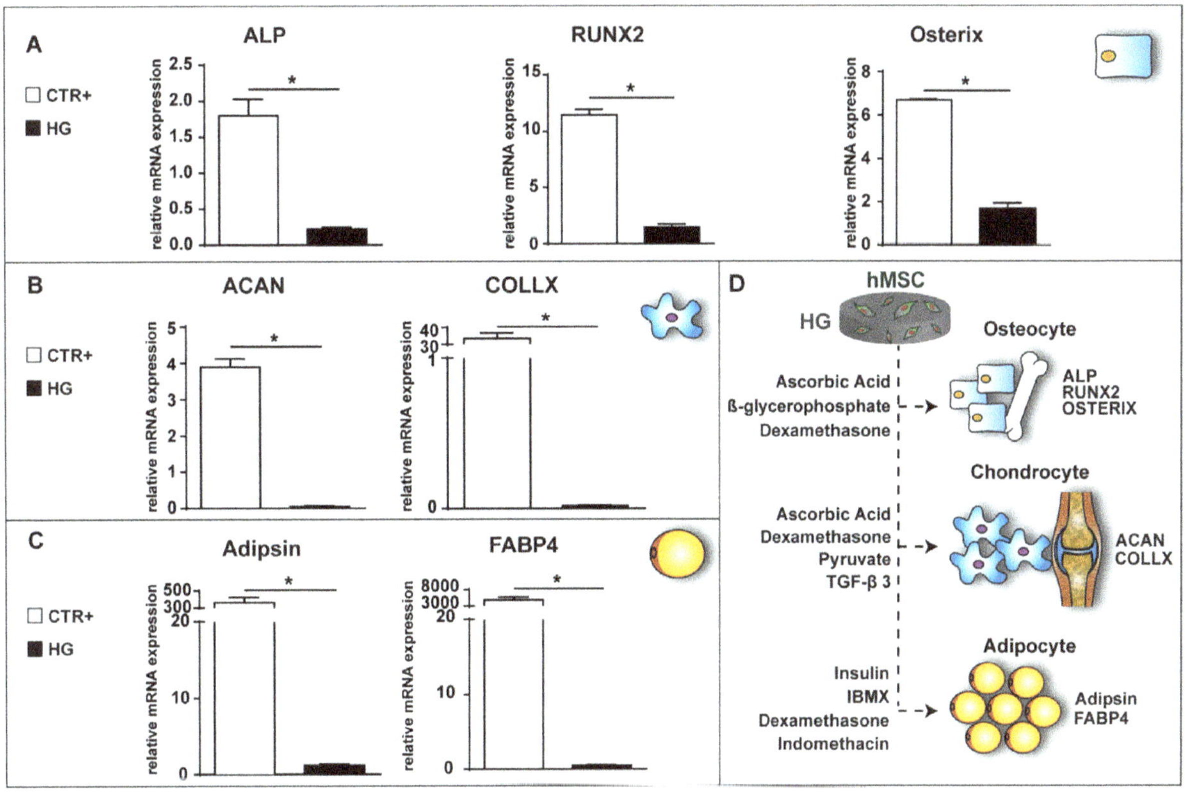


FIG. 5 mRNA analysis of hMSCs encapsulated within biomimetic scaffold. Graphs representing the expression of specific genes related to three differentiation lineages: alkaline phosphatase (ALP), runt-related transcription factor 2 (RUNX2) and osterix for osteogenic differentiation; aggrecan (ACAN) and collagen type X (COLLX) for chondrogenic differentiation and adipsin and fatty acid binding-protein 4 (FABP4) for adipogenic differentiation. hMSCs encapsulated within HG for 21 days are compared to the positive control represented by hMSCs loaded in HG and treated with specific differentiating media for 21 days. *(Reprinted with permission from Caron, I., Rossi, F., Papa S., Aloe, R., Sculco, M., Mauri, E., Sacchetti, A., et al. (2016). A new three dimensional biomimetic hydrogel to deliver factors secreted by human mesenchymal stem cells in spinal cord injury.* Biomaterials, *75(1), 135–147. doi:10.1016/j.biomaterials.2015.10.024, Elsevier.)*

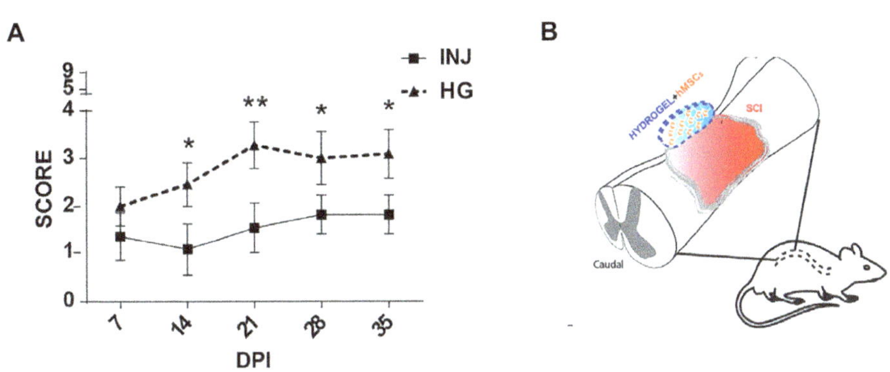


FIG. 6 HG ability to improve locomotor performance in SCI mice. In vitro HG ability to improve locomotor performance in SCI mice: (A) untreated SCI mice (INJ) or treated (HG) 1 DPI examined weekly starting 7 days post treatment, using the Basso Mouse Scale-BMS (score 0, complete paralysis, 9 complete mobility, referred to healthy mice). (C) Positioning of the hydrogel + cells in the SCI mouse model. *(Reprinted with permission from Papa, S., Vismara, I., Mariani, A., Barilani, M., Rimondo, S., De Paola, M., et al. (2018). Mesenchymal stem cells encapsulated into biomimetic hydrogel scaffold gradually release CCL2 chemokine in situ preserving cytoarchitecture and promoting functional recovery in spinal cord injury.* Journal of Controlled Release, 278*(10), 49–56. doi:10.1016/j.jconrel.2018.03.034, Elsevier.)*

Several neurodegenerative CNS disorders, including traumatic brain injury, spinal cord injury, stroke, amyotrophic lateral sclerosis, Alzheimer's disease, Parkinson's disease, epilepsy and multiple sclerosis are associated with chronic neuro-inflammation and high levels of several cytokines. For these reasons, new therapeutic approaches able to modulate activated microglial cells are needed. In recent years, new evidences, from both in vitro and in vivo studies, suggest that nanoparticles can be selectively internalized by a specific phagocytic activity of macrophages, exploiting them as Trojan horses to selectively treat these cells. This delivery approach represents a promising strategy to develop tailored treatment during the inflammatory response.

## Mini-dictionary of terms

**Colloids:** system composed by a disperse phase with a size range between 1 and 1000 nm dissolved in an incompatible continuous phase.

**Drug delivery system:** engineered materials able to load and deliver drugs with a controlled kinetics, maintaining the pharmacological activity during time.

**Emulsion:** thermodynamically unstable colloid constituted by two immiscible liquids.

**Functionalization:** chemical reaction between two reactive sites with consequent formation of covalent chemical bond.

**Hydrogel:** network of cross-linked hydrophilic polymeric chains able to absorb an extremely large amount of water (dispersion medium).

**Nanogels:** nanoparticle, usually in the tens to hundreds of nanometers in diameter, composed of a cross-linked hydrophilic polymer network composed of synthetic polymers or biopolymers chemically or physically cross-linked.

**Nanomedicine:** medical applications of nanomaterials that range from biological devices, to nanoelectronics biosensors, molecular nanotechnology such as biological machines.

**Nanoparticles:** particles of matter with size range between 1 and 100 nm (nm) in terms of diameter.

**Polymer:** substance constituted by very large molecules, or macromolecules, composed of many repeating subunits and that can be synthesized by step-growth or chain-growth mechanisms.

**Tissue engineering:** biomedical engineering discipline that uses a combination of cells, engineering, materials methods, and suitable factors to maintain, restore or replace different types of biological tissues.

## Key facts of "Biomaterials, spinal cord injury, and rehabilitation: A new narrative"

- SCI is the most frequent disabling spinal injury, estimated 2.5 million people worldwide live with SCI
- SCI is a multifactorial where most of the medical problems are caused by cascade of events (secondary injury)
- A winning therapeutic strategy is represented by the possibility to work against different pathological mechanisms.
- Hydrogels, three-dimensional polymeric networks thanks to their water affinity to maintain cells viable and able to restore the damaged tissue.
- Nanoparticles, thanks to their ability to be cell selective, can carry and deliver drugs into specific cells working as Trojan horses.

## Summary points

- SCI is a debilitating condition caused by damage to the spinal cord.
- More than 130,000 new spinal cord injuries are reported every year.
- Hydrogel can restore the tissue carrying cells within the damage site.
- Scaffold can drive axonal growth across their ordered pores.
- Nanoparticles can selectively deliver drugs within cells reducing secondary injury issues.

## References

Aktaş, Y., Yemisci, M., Andrieux, K., Gürsoy, R. N., Alonso, M. J., Fernandez-Megia, E., et al. (2005). Development and brain delivery of chitosan-PEG nanoparticles fnnctionalized with the monoclonal antibody OX26. *Bioconjugate Chemistry*, *16*(6), 1503–1511. https://doi.org/10.1021/bc050217o.

Bechara, S., Wadman, L., & Popat, K. C. (2011). Electroconductive polymeric nanowire templates facilitates in vitro C17.2 neural stem cell line adhesion, proliferation and differentiation. *Acta Biomaterialia*, *7*(7), 2892–2901. https://doi.org/10.1016/j.actbio.2011.04.009.

Caron, I., Rossi, F., Papa, S., Aloe, R., Sculco, M., Mauri, E., et al. (2016). A new three dimensional biomimetic hydrogel to deliver factors secreted by human mesenchymal stem cells in spinal cord injury. *Biomaterials*, *75*(1), 135–147. https://doi.org/10.1016/j.biomaterials.2015.10.024.

Elliott Donaghue, I., Tator, C. H., & Shoichet, M. S. (2015). Sustained delivery of bioactive neurotrophin-3 to the injured spinal cord. *Biomaterials Science*, *3*(1), 65–72. https://doi.org/10.1039/c4bm00311j.

Gao, M., Lu, P., Bednark, B., Lynam, D., Conner, J. M., Sakamoto, J., et al. (2013). Templated agarose scaffolds for the support of motor axon regeneration into sites of complete spinal cord transection. *Biomaterials*, *34*(5), 1529–1536. https://doi.org/10.1016/j.biomaterials.2012.10.070.

Georges, P. C., Miller, W. J., Meaney, D. F., Sawyer, E. S., & Janmey, P. A. (2006). Matrices with compliance comparable to that of brain tissue select neuronal over glial growth in mixed cortical cultures. *Biophysical Journal*, *90*(8), 3012–3018. https://doi.org/10.1529/biophysj.105.073114.

Ghorbani, M., & Hamishehkar, H. (2019). Redox-responsive smart nanogels for intracellular targeting of therapeutic agents: Applications and recent advances. *Journal of Drug Targeting*, *27*(4), 408–422. Taylor and Francis Ltd https://doi.org/10.1080/1061186X.2018.1514041.

Hejčl, A., Kekulová, K., Svobodová, B., Proks, V., Macková, H., Jiránková, K., et al. (2018). Modified methacrylate hydrogels improve tissue repair after spinal cord injury. *International Journal of Molecular Sciences*, *19*, 2481. https://doi.org/10.3390/ijms19092481.

Hejčl, A., Růžička, J., Proks, V., Macková, H., Kubinová, Š., Tukmachev, D., et al. (2018). Dynamics of tissue ingrowth in SIKVAV-modified highly superporous PHEMA scaffolds with oriented pores after bridging a spinal cord transection. *Journal of Materials Science: Materials in Medicine*, *29* (7), 89. https://doi.org/10.1007/s10856-018-6100-2.

Hejčl, A., Šedý, J., Kapcalová, M., Toro, D. A., Amemori, T., Lesný, P., et al. (2010). HPMA-RGD hydrogels seeded with mesenchymal stem cells improve functional outcome in chronic spinal cord injury. *Stem Cells and Development*, *19*(10), 1535–1546. https://doi.org/10.1089/scd.2009.0378.

Jurga, M., Dainiak, M. B., Sarnowska, A., Jablonska, A., Tripathi, A., Plieva, F. M., et al. (2011). The performance of laminin-containing cryogel scaffolds in neural tissue regeneration. *Biomaterials*, *32*(13), 3423–3434. https://doi.org/10.1016/j.biomaterials.2011.01.049.

Kang, C. E., Baumann, M. D., Tator, C. H., & Shoichet, M. S. (2012). Localized and sustained delivery of fibroblast growth factor-2 from a nanoparticle-hydrogel composite for treatment of spinal cord injury. *Cells, Tissues, Organs*, *197*(1), 55–63. https://doi.org/10.1159/000339589.

Kočí, Z., Výborný, K., Dubišová, J., Vacková, I., Jäger, A., Lunov, O., et al. (2017). Extracellular matrix hydrogel derived from human umbilical cord as a scaffold for neural tissue repair and its comparison with extracellular matrix from porcine tissues. *Tissue Engineering—Part C: Methods*, *23*(6), 333–345. https://doi.org/10.1089/ten.tec.2017.0089.

Kolhar, P., Anselmo, A. C., Gupta, V., Pant, K., Prabhakarpandian, B., Ruoslahti, E., et al. (2013). Using shape effects to target antibody-coated nanoparticles to lung and brain endothelium. *Proceedings of the National Academy of Sciences of the United States of America*, *110*(26), 10753–10758. https://doi.org/10.1073/pnas.1308345110.

Kubinová, Š., Horák, D., Hejčl, A., Plichta, Z., Kotek, J., Proks, V., et al. (2015). SIKVAV-modified highly superporous PHEMA scaffolds with oriented pores for spinal cord injury repair. *Journal of Tissue Engineering and Regenerative Medicine*, *9*(11), 1298–1309. https://doi.org/10.1002/term.1694.

Kubinová, Š., Horák, D., Kozubenko, N., Vaněček, V., Proks, V., Price, J., et al. (2010). The use of superporous Ac-CGGASIKVAVS-OH-modified PHEMA scaffolds to promote cell adhesion and the differentiation of human fetal neural precursors. *Biomaterials*, *31*(23), 5966–5975. https://doi.org/10.1016/j.biomaterials.2010.04.040.

Kučera, L., Weinfurterová, R., Dvořákova, J., Kučera, J., Pravda, M., Foglarová, M., et al. (2015). Chondrocyte cultivation in hyaluronan-tyramine cross-linked hydrogel. *International Journal of Polymeric Materials and Polymeric Biomaterials*, *64*(13), 661–674. https://doi.org/10.1080/00914037.2014.996715.

Liu, J., Xiao, Y., & Allen, C. (2004). Polymer-drug compatibility: A guide to the development of delivery systems for the anticancer agent, ellipticine. *Journal of Pharmaceutical Sciences*, *93*(1), 132–143. https://doi.org/10.1002/jps.10533.

Macková, H., Plichta, Z., Proks, V., Kotelnikov, I., Kučka, J., Hlídková, H., et al. (2016). RGDS- and SIKVAVS-modified superporous poly(2-hydroxyethyl methacrylate) scaffolds for tissue engineering applications. *Macromolecular Bioscience*, *16*(11), 1621–1631. https://doi.org/10.1002/mabi.201600159.

Mattson, M. P., Haddon, R. C., & Rao, A. M. (2000). Molecular functionalization of carbon nanotubes and use as substrates for neuronal growth. *Journal of Molecular Neuroscience*, *14*(3), 175–182. https://doi.org/10.1385/JMN:14:3:175.

Mauri, E., Sacchetti, A., Vicario, N., Peruzzotti-Jametti, L., Rossi, F., & Pluchino, S. (2018). Evaluation of RGD functionalization in hybrid hydrogels as 3D neural stem cells culture systems. *Biomaterials Science*, *6*(3), 501–510. https://doi.org/10.1039/c7bm01056g.

Mauri, E., Veglianese, P., Papa, S., Mariani, A., De Paola, M., Rigamonti, R., et al. (2017). Double conjugated nanogels for selective intracellular drug delivery. *RSC Advances*, *7*(48), 30345–30356. https://doi.org/10.1039/c7ra04584k.

Medberry, C. J., Crapo, P. M., Siu, B. F., Carruthers, C. A., Wolf, M. T., Nagarkar, S. P., et al. (2013). Hydrogels derived from central nervous system extracellular matrix. *Biomaterials*, *34*(4), 1033–1040. https://doi.org/10.1016/j.biomaterials.2012.10.062.

Papa, S., Vismara, I., Mariani, A., Barilani, M., Rimondo, S., De Paola, M., et al. (2018). Mesenchymal stem cells encapsulated into biomimetic hydrogel scaffold gradually release CCL2 chemokine in situ preserving cytoarchitecture and promoting functional recovery in spinal cord injury. *Journal of Controlled Release*, *278*(10), 49–56. https://doi.org/10.1016/j.jconrel.2018.03.034.

Pires, L. R., & Pêgo, A. P. (2015). Bridging the lesion-engineering a permissive substrate for nerve regeneration. *Regenerative Biomaterials*, *2*(3), 203–214. https://doi.org/10.1093/rb/rbv012.

Roman, J. A., Niedzielko, T. L., Haddon, R. C., Parpura, V., & Floyd, C. L. (2011). Single-walled carbon nanotubes chemically functionalized with polyethylene glycol promote tissue repair in a rat model of spinal cord injury. *Journal of Neurotrauma*, *28*(11), 2349–2362. https://doi.org/10.1089/neu.2010.1409.

Rossi, F., & van Griensven, M. (2014). Polymer functionalization as a powerful tool to improve scaffold performances. *Tissue Engineering Part A*, *20*(15), 2043–2051. https://doi.org/10.1089/ten.tea.2013.0367.

Růžička, J., Romanyuk, N., Hejčl, A., Vetrík, M., Hrubý, M., Cocks, G., … Jendelova, P. (2013). Treating spinal cord injury in rats with a combination of human fetal neural stem cells and hydrogels modified with serotonin. *Acta Neurobiologiae Experimentalis (Wars)*, *73*, 102–115.

Saraiva, C., Praça, C., Ferreira, R., Santos, T., Ferreira, L., & Bernardino, L. (2016). Nanoparticle-mediated brain drug delivery: Overcoming blood-brain barrier to treat neurodegenerative diseases. *Journal of Controlled Release*, *235*, 34–47. Elsevier B.V https://doi.org/10.1016/j.jconrel.2016.05.044.

Varma, A. K., Das, A., Wallace, G., Barry, J., Vertegel, A. A., Ray, S. K., et al. (2013). Spinal cord injury: A review of current therapy, future treatments, and basic science frontiers. *Neurochemical Research*, *38*(5), 895–905. Springer New York LLC https://doi.org/10.1007/s11064-013-0991-6.

Výborný, K., Vallová, J., Kočí, Z., Kekulová, K., Jiráková, K., Jendelová, P., et al. (2019). Genipin and EDC crosslinking of extracellular matrix hydrogel derived from human umbilical cord for neural tissue repair. *Scientific Reports*, *9*(1). https://doi.org/10.1038/s41598-019-47059-x.

Woerly, S., Petrov, P., Syková, E., Roitbak, T., Simonová, Z., & Harvey, A. R. (1999). Neural tissue formation within porous hydrogels implanted in brain and spinal cord lesions: Ultrastructural, immunohistochemical, and diffusion studies. *Tissue Engineering*, *5*(5), 467–488. https://doi.org/10.1089/ten.1999.5.467.

Woerly, S., Pinet, E., De Robertis, L., Van Diep, D., & Bousmina, M. (2001). Spinal cord repair with PHPMA hydrogel containing RGD peptides (NeuroGel™). *Biomaterials*, *22*(10), 1095–1111. https://doi.org/10.1016/S0142-9612(00)00354-9.

Xiao, K., Li, Y., Luo, J., Lee, J. S., Xiao, W., Gonik, A. M., et al. (2011). The effect of surface charge on in vivo biodistribution of PEG-oligocholic acid based micellar nanoparticles. *Biomaterials*, *32*(13), 3435–3446. https://doi.org/10.1016/j.biomaterials.2011.01.021.

Xie, J., MacEwcm, M. R., Willerth, S. M., Li, X., Moran, D. W., Sakiyama-Elbert, S. E., et al. (2009). Conductive core-sheath nanofibers and their potential application in neural tissue engineering. *Advanced Functional Materials*, *19*(14), 2312–2318. https://doi.org/10.1002/adfm.200801904.

Yang, F., Xu, C. Y., Kotaki, M., Wang, S., & Ramakrishna, S. (2004). Characterization of neural stem cells on electrospun poly(L-lactic acid) nanofibrous scaffold. *Journal of Biomaterials Science, Polymer Edition*, *15*(12), 1483–1497. https://doi.org/10.1163/1568562042459733.

Yuan, N., Tian, W., Sun, L., Yuan, R., Tao, J., & Chen, D. (2014). Neural stem cell transplantation in a double-layer collagen membrane with unequal pore sizes for spinal cord injury repair. *Neural Regeneration Research*, *9*(10), 1014–1019. https://doi.org/10.4103/1673-5374.133160.

Yue, Z. G., Wei, W., Lv, P. P., Yue, H., Wang, L. Y., Su, Z. G., et al. (2011). Surface charge affects cellular uptake and intracellular trafficking of chitosan-based nanoparticles. *Biomacromolecules*, *12*(7), 2440–2446. https://doi.org/10.1021/bm101482r.

Zaviskova, K., Tukmachev, D., Dubisova, J., Vackova, I., Hejcl, A., Bystronova, J., et al. (2018). Injectable hydroxyphenyl derivative of hyaluronic acid hydrogel modified with RGD as scaffold for spinal cord injury repair. *Journal of Biomedical Materials Research Part A*, *106*(4), 1129–1140. https://doi.org/10.1002/jbm.a.36311.

Zeng, X., Qiu, X. C., Ma, Y. H., Duan, J. J., Chen, Y. F., Gu, H. Y., et al. (2015). Integration of donor mesenchymal stem cell-derived neuron-like cells into host neural network after rat spinal cord transection. *Biomaterials*, *53*, 184–201. https://doi.org/10.1016/j.biomaterials.2015.02.073.

Zhu, Y., Wang, A., Shen, W., Patel, S., Zhang, R., Young, W. L., et al. (2010). Nanofibrous patches for spinal cord regeneration. *Advanced Functional Materials*, *20*(9), 1433–1440. https://doi.org/10.1002/adfm.200901889.

Zuidema, J. M., Gilbert, R. J., & Osterhout, D. J. (2016). Nanoparticle technologies in the spinal cord. *Cells, Tissues, Organs*, *202*(1–2), 102–115. S. Karger AG https://doi.org/10.1159/000446647.

Chapter 44

# Support in spinal cord injury: A focus on robotics

Angel Gil-Agudo[a,b] and Guillermo Asín-Prieto Eng[b,c]

[a]*Neurorehabilitation and Biomechanics Unit (HNP-SESCAM), Associate Unit CSIC, Physical Medicine and Rehabilitation Department, National Hospital for Paraplegics, SESCAM, Toledo, Spain;* [b]*Neurorehabilitation and Biomechanics Unit (HNP-SESCAM), Associate Unit CSIC, Biomechanics and Technical Aids Department, National Hospital for Paraplegics, SESCAM, Toledo, Spain;* [c]*Gogoa Mobility Robots, S.L., Abadiño, Vizcaya, Spain*

## List of abbreviations

| | |
|---|---|
| **ADL** | activity of daily living |
| **AIS** | ASIA impairment scale |
| **ARTIC** | advance robotic therapy integrated centers |
| **ASIA** | American Spinal Injury Association |
| **CNS** | central nervous system |
| **CPG** | central pattern generator |
| **FES** | functional electrical stimulation |
| **SCI** | spinal cord injury |

## Introduction

The incorporation of robotics in the field of neurorehabilitation is taking place rapidly, both in research and in its clinical applications, and is presented as a very promising tool that is changing therapeutic paradigms. In the late 1980s and early 1990s, basic research findings constituted a major change in therapeutic intervention in neurorehabilitation. One of most relevant was that in experimental models with cats subjected to a spinal cord injury (SCI), the subsequent training of the locomotor function applied to them offered good results. In fact, it was shown that these cats with SCI walked effectively when placed on a treadmill with partial weight support (Barbeau & Rossignol, 1987). The suggested mechanism is the activation of the basic neuronal circuitries sufficient to generate efficient stepping patterns and independent standing. Indeed, the operations underlying the elaboration of motor patterns for walking and standing are essentially achieved by the neuronal networks embedded within the lumbosacral segments of the spinal cord (Grillner & Zangger, 1984). These findings led to the concept of spinal learning via activity-dependent plasticity. Following this concept, it was found that locomotor activity can be activated in patients with severe SCI via passive activation of the legs on a treadmill (Barbeau, Danakas, & Arsenault, 1993). Synchronous reciprocal movements of both legs, simulating normal walking are required to activate the locomotor centers in the spinal cord. The repetitive and simultaneous activation of certain sensory and motor pathways with task-specific training can select and reinforce those spinal circuits improving the ability to perform the practiced movement successfully. Thus, functional rehabilitation (i.e., walking) had to be intensive and task-oriented. Intensive and task-oriented are the two are the pillars of the motor learning neuroplasticity-based neurorehabilitation concepts that also justify the development of robotic therapy (Cai et al., 2006; Edgerton, Courtine, Gerasimenko, et al., 2008).

Although these interventions appear promising, in order to translate them into clinical practice in humans, a great effort is needed to standardize the assessments of the therapies applied (Curt, Schwab, & Dietz, 2004). Gait training using partial weight bearing systems on treadmills in patients who had suffered a stroke or SCI was extended in the early 1990s following motor learning principles. This therapy initially presented high costs in terms of personnel and effort, as it required the participation of at least two physiotherapists to mobilize the paralyzed lower extremities of the patient with the intention of reproducing the treadmill walking cycle (Dietz & Harkema, 2004). The great effort that this activity demanded from the physiotherapists limited the duration of the treatment sessions. This limitation led to the idea that a robotic device could

**Diagnosis and Treatment of Spinal Cord Injury. https://doi.org/10.1016/B978-0-12-822498-4.00044-0**

serve as an alternative to manual treatment and that such a device could cover the demands of functional training (Colombo, Joerg, Schreier, & Dietz, 2000). This led to the first robotic systems for walking training with weight suspension on treadmills.

These robotic assistive devices enable to start a functional and task-oriented training as soon as possible after the injury and allow an intensive application of adequate afferent feedback and a high number of repetitions of functional movements (Wirz & Rupp, 2012).

Furthermore, the outcome of rehabilitation is better if the patient is more motivated and involved in the treatment (Weber & Stein, 2018). All these without forgetting one of the most evident shortcomings of conventional systems, which is the need to incorporate sensors that provide objective variables of the patient's condition or of the execution of the task, need to be trained. These issues are satisfactorily addressed by robotic devices. This therapy can be applied alone or in combination with other new technologies such as functional electrical stimulation (FES) or virtual reality.

Robotic therapy has experienced a huge boom in the last 15 years. In fact, different clinical guidelines approved its use as a complementary element to conventional therapy in the rehabilitation of patients with upper limb deficits after suffering a stroke (Department of Veterans Affairs et al., 2010). Robotic devices are appropriately adapted to the need to assist limb movements based on their ability to perform simple, repetitive tasks in a consistent manner that facilitates functional recovery and adaptive plasticity (Edgerton & Roy, 2009). There are two main categories: distal end effector devices and exoskeleton-type devices. Distal end effectors were the first to appear and are characterized by the fact that they use a single distal point of contact to guide the movement of the entire limb. In the upper extremity, it can make contact in the hand or forearm, facilitating the movements of the elbow and shoulder. They produce combined movements being difficult to isolate pure simple movements. The operation of exoskeletons is different. They are structures located in parallel to the different parts of the extremities with more than one point of interaction with the person. They provide direct control over each segment of the limb by incorporating individualized motors, also called actuators, which coincide with the anatomical axis of each joint. Thus, each actuator triggers the movement of each joint on which it is located. The design of exoskeletons seems to be more suitable than that of distal effector systems to achieve large joint paths (Krebs, Conroy, Bever, & Hogan, 2012).

In this chapter, we will focus on upper limb robots, stationary and ambulatory lower limb exoskeletons.

## Upper limb robots

Cervical SCI can result in partial or complete tetraplegia. Each small improvement in motor control of the upper extremity can translate to significant ameliorations in function and increases independence for the individual. As mentioned above, this type of therapy offers new possibilities in the rehabilitation not only for the lower limbs but also for the upper limbs. The robotic devices allow the application of high-intensity sessions during longer periods of time, remaining invariable certain physical parameters such as speed, strength, or precision (Page, Hill, & White, 2013; Takahashi, Der-Yeghiaian, Le, Motiwala, & Cramer, 2008). There is evidence that suggests task-based therapy specifically designed to deal with lost abilities produce better results than resistance strengthening exercises (Teasell & Kaira, 2004). This task should be performed by the patient as far as possible. That's why the devices should be equipped with a controller that provides the least assistance needed to accomplish the movement (assist as needed) and reproducible treatment protocols.

Some studies point out that by focusing the improvement of robotic therapy more on the proximal recovery of the upper limb (shoulder and elbow), it does not translate into improvement of the functional ability that depends on hand control. However, the best results seem to be found by adding the application of both types of therapies (Bayona, Bitensky, Salter, & Teasell, 2005). Despite the low number of studies, results from these studies suggest that robotic training protocols are feasible and well tolerated and have a positive impact on improving arm and hand functions in selected patients with cervical SCI, but the results must be interpreted with caution (Mehrholz, Platz, Kugler, & Pohl, 2009). In any case, studies with larger samples are needed, especially those that analyze the distal region of the upper limb, in order to have solid conclusions about the effectiveness of these devices.

Most of the current devices include a virtual reality module with visual or haptic feedback to improve sensory feedback, as well as patient motivation and engagement. They also have the capability to obtain movement kinematics that can provide precise information about movement quality that otherwise is not included in functional assessments (Esquenazi & Talaty, 2019).

Although there is a number of different robotic devices currently used for neurorehabilitation of the upper extremities following SCI (Fig. 1), we will now focus on the most commonly used:

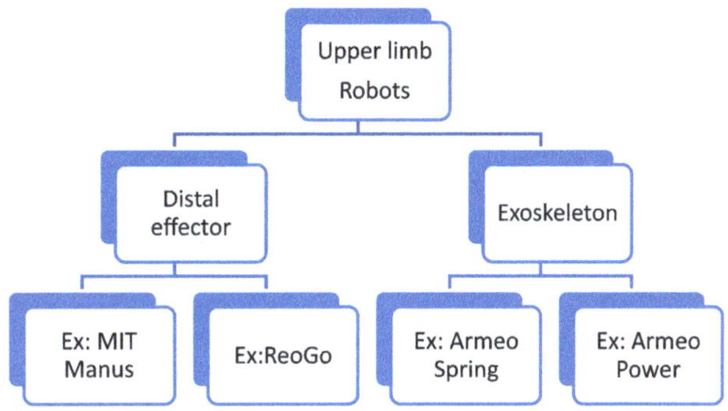


FIG. 1 Types of upper limb robots.

## MIT MANUS

It was designed to provide high-intensity and reproducible upper limb rehabilitation in adults and older children. This modular distal effector system consists of a series of proximal and distal components that can be used individually or together for upper extremity training. It comprises two modules and 5 degrees of freedom, two for elbow and forearm motion, and three for wrist motion and allows patients to perform reaching movements in horizontal plane. The robot can move, guide, or perturb the movement of a patient's upper limb and record quantities, such as position, velocity, and force. The operating paradigm is the so-called "assist as needed." Thanks to motion sensors, the mobility of the joint segments can always be monitored. The patient–robot interface consists of video games for elbow, shoulder, and wrist exercises that can be used to increase the quality of therapy sessions as well as keep the user engaged (Krebs, Hogan, Aisen, & Volpe, 1998). It has initially been used in the rehabilitation of the upper limb of stroke patients, proving effective in the sub-acute and chronic phases by reducing motor deficits, improving function and bringing about a lasting change (Bayón-Calatayud et al., 2014; Fasoli et al., 2004). The commercialized version of MIT-MANUS, INMOTION (Bionik Laboratories Corp., Toronto, Canada), has been used in patients with SCI to a limited extent although one study demonstrated that after a training protocol, significant improvements in quality of movement were found with no changes in upper extremity strength, pain, or spasticity (Cortés et al., 2013).

## ReoGo

The ReoGo system (Motorika Medical, Caesarea, Israel) is a stationary fixed based end-effector arm rehabilitation robot, which facilitates the mobilization of the upper limb on a support that allows a wide range of movements in the 3 dimensions of space. The Reo-Go allows for movements at the shoulder, elbow, and wrist. It also uses a real-time visual feedback monitor to display games for the subject to perform. Although it has primarily been used for stroke patients, it has also been applied in SCI. Reo-Go was incorporated into an acute incomplete SCI patient therapy protocol. The subject demonstrated remarkable improvements in muscle strength, active range of motion and functional assessment (Siedziewski, Schaaf, & Mount, 2012).

## Armeo

Armeo devices (Hocoma AG, Volketswil, Switzerland) were the first unilateral upper extremity exoskeletons marketed for upper limb rehabilitation. This range of devices includes the Armeo Power for the most affected patients, the Armeo Spring, the Armeo Spring for children and the Armeo Senso for those less affected.

The Armeo system is a well-documented device and is the only device shown to offer better functional results after stroke compared to traditional therapy. Two studies showed the utility of this device for upper limbs in SCI subjects focusing on the potential of these devices in performing upper limb assessment (Rudhe, Albisser, Starkey, Curt, & Bolliger, 2012; Zariffa et al., 2012). The Armeo Power is one of the most advanced active exoskeletons for upper extremity rehabilitation. It is based on the ARMin device, which consists of an exoskeleton covering the upper limb and allowing anthropometric adaptations. It provides support for the weight of the patient's upper limb and features different modes of use, such as mobilization mode, 2D and 3D games and functional training of daily life activities. ARMin provides three actuated degrees of freedom for the shoulder and one for the elbow joint. It offers three different therapy modes: the movement therapy, the game therapy and the ADL (activity of daily living) training mode. Like the MIT-MANUS, Armeo

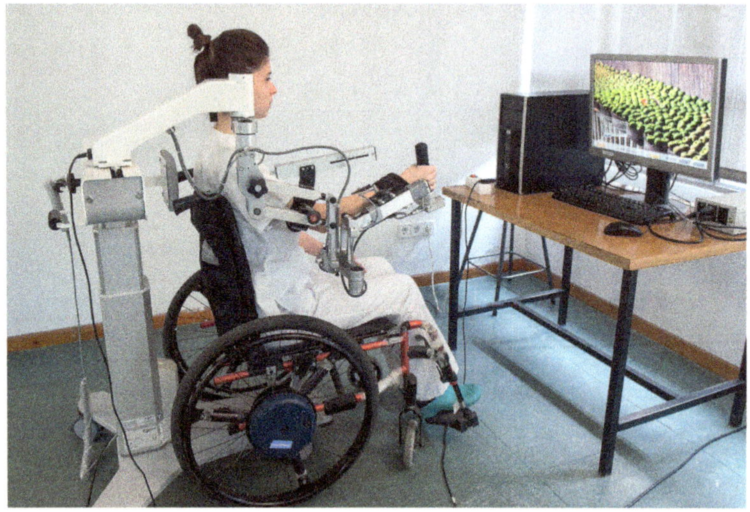

FIG. 2 Armeo Spring. Upper limb device that works through a system of springs that eliminate the weight of the body as an enabler.

Power uses an "assist as needed" mode of operation, allowing the clinician to adapt the difficulty of the task to the degree of recovery. There are studies that demonstrate its usefulness in patients with SCI (Rudhe et al., 2012).An earlier version is the Armeo Spring that manages to cover the shoulder and elbow, and also works the wrist flexo-extension and manual gripping. It is a passive exoskeleton (Fig. 2). It works through a system of springs that eliminate the weight of the body as an enabler instead of using motors to assist movement as the Armeo Power does. Both feature a monitor with motivational games to encourage repetitive movements. The software allows the clinician to select the task and its degree of difficulty by defining the required joint path and the rhythm of the selected game.

### Other devices

There are other devices on the market such as the DIEGO (Tyromotion, Graz, Austria) which uses a wiring system to support and mobilize the limb, the Bi-Manu-Track (RehaStim, Germany) which facilitates the treatment of both upper limbs simultaneously. There are devices that focus on the individual mobility of each finger but allow practice in gripping by controlling the performance of each finger. This would be the case of AMADEO (Tyromotion, Graz, Austria), HAND-CARE 2 and RUTGERS-MASTER II (Rutgers University, USA), although the latter excludes the treatment of the fifth finger. It is not common but in some cases two robotic devices have been used in combination, such as the Armeo Power and the Amadeo, using the first for the shoulder, elbow, and carpal and the second for the shoulder, elbow, and carpal.

## Stationary lower limb robots

Thoracic and lower SCIs can result in partial to complete paralysis of the lower extremities. Independent mobility for many can only be achieved at a wheelchair level, although walking oftentimes remains a priority (Dittuno, Patrick, Stineman, & Dittuno, 2008). Lower limb robots have emerged as potential upright mobility devices for those with lower limb paralysis. Locomotor training focuses on retraining the motor function via plastic change (Morawietz & Moffat, 2013; Nam et al., 2017), and the neurophysiological mechanism underlying the restoration of human restoration after SCI involve enhancing the afferent input to the spinal cord and activating CPG (central pattern generator) embedded within the lumbosacral spinal cord (Dietz, Wirz, & Curt, 1998). Plastic changes can be induced in both the spinal cord level and sensory motor cortex via intensive locomotor training, mainly in incomplete SCI subjects (Hubli & Dietz, 2013).

As it has been previously referred, to manually replicate a normal walking pattern with the patient in body weight-supported on a treadmill two or three therapists are needed to control and move lower limbs. This is a strenuous and exhausting task for therapist, so sophisticated automated electromechanical devices have been developed (Tefertiller, Pharo, Evans, & Winchester, 2011) that offers several advantages, including the ability to increase the intensity and total duration of training while maintaining a physiological gait pattern.

**TABLE 1** Types of lower limb robots.

| Lower limb robots | | |
|---|---|---|
| Stationary robots | | Ambulatory exoskeletons |
| Distal effector | Stationary exoskeletons | Ex: Ekso Bionics, Rewalk, Indego, etc. |
| Ex: Gait Trainer | Ex: Lokomat | |

As in the case of the upper limbs, in the lower limb robots we also find, depending on their structure, distal effectors and exoskeletons (Table 1). Among lower limb robotic exoskeletons, we can distinguish the stationary and the ambulatory ones. In this section we will discuss the stationary ones, and the ambulatory will be analyzed in the following section.

## End effector devices

End-effector-based systems work like conventional elliptical trainers: the subject's feet are strapped to two footplates moving along a gait-like trajectory, as in an elliptical trainer, moving the entire lower limb. They work based on a constraint at the distal end of the kinetic chain that specifies the trajectory there and the proximal joints can simply move as the body geometry and articulations dictate. The footplates generate the stance and swing phases in most instances with symmetric motion. The main difference compared with exoskeletons with a treadmill is that the feet are always in contact with the moving platform, simulating the gait phases but not necessarily generating true swing and stance phases. The trajectories of the footplates, as well as the vertical and horizontal movements of the center of mass, are programmable. The end-effector design lends itself to gait retraining and star climbing (Hesse, Waldner, & Tomelleri, 2010). Examples of end-effector devices include Gait Trainer GT1 (Reha-Stim, Berlin, Germany), G EO (Reha Technologies, Switzerland) and Lokohelp. In relation to SCI patients, 3-dimensional data were obtained with Lokomat and G-EO. Their kinematic data were compared when devices were used by SCI or traumatic brain injury patients. The results confirmed a more controlled and repetitive gait pattern when using Lokomat and the G-EO system provided a gait pattern that had more variability of motion for the hips and knees, with slightly reduced knee motion, and the gait pattern differed slightly from that observed during overground walking (Esquenazi & Talaty, 2019).

## Stationary exoskeletons

Stationary exoskeletons have a device that surrounds the patient's legs, which may be suspended from an overhead guide rail, supported by a metal frame on wheels, or the exoskeleton can even be directly supported by a mobile robot. They are usually connected directly to the ground through a rigid frame or bolted to a wall, enhancing and ensuring total safety. Stationary exoskeletons can have a large and powerful motors and controllers. They often involve walking on a treadmill. These devices are less complex in their engineering requirements and more stable and safer than ambulatory exoskeletons that allow overground walking due to the elimination of fall risk. They are less accommodating of individual gait variations, such as changes of speed or direction. This group of stationary exoskeletons includes the Lokomat, Walk-Trainer, LOPES or ReoAmbulator.

The Lokomat (Hocoma AG, Volketswil, Switzerland) is the most clinically implanted and studied robot on the market. The Lokomat is a bilaterally driven gait orthosis that is used in conjunction with a body support system (Colombo et al., 2000). It is essentially a robotic implementation of the treadmill walking training system with partial weight support and manual mobilization of the patient by physiotherapists. This system consists of a treadmill, a partial weight support system and a bilateral exoskeleton that provides action on the hips and knees with the ankle being passively supported by a spring to facilitate dorsiflexion of the swing phase of walking (Riener, 2012). The Lokomat moves the patient legs through the gait cycle mainly in the sagittal plane (Fig. 3). The device's hip and knee are actuated by linear drives integrated into an exoskeleton structure. There is no actuator on the ankle and dorsal flexion during the swing phase is achieved passively by means of springs. The lower limb motion can be controlled with highly repeatable predefined hip and knee joint trajectories on the basis of a conventional position control strategy. The exoskeleton is fixed to the rigid frame of the body weight support system and the patient is fixed to the exoskeleton with straps around the waist, thighs and shanks.

The hip and knee joint trajectories can be manually adjusted to the individual patient by changing amplitude and offsets. Signals obtained from force sensors may be used to determine the interactions torques between the patient and the device,

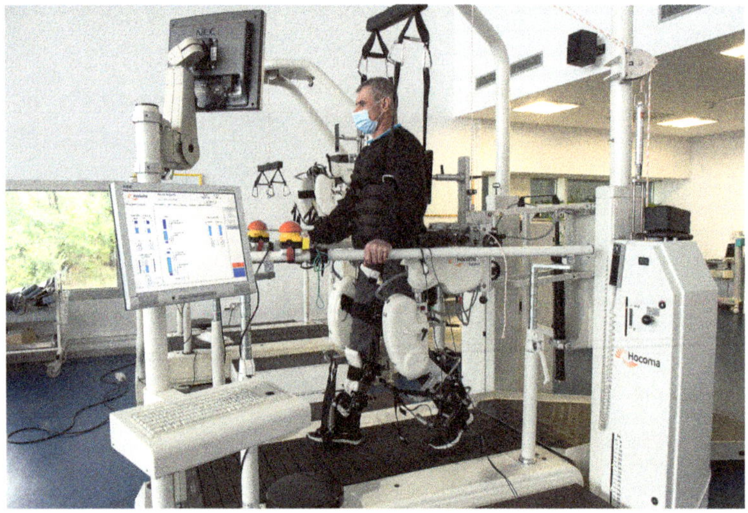

FIG. 3 Lokomat. Lokomat system with a spinal cord-injured patient.

which inform about the voluntary muscle effort produced by the patient (Riener, 2012). The device allows some anthropometrical adaptation to the lower limb segments size via telescopic bars so that the exoskeleton can be used by subjects with different shank and thigh lengths. The width of the hip exoskeleton may also be adjusted by changing the distance between the two lower limbs.

The body weight support system consists of a harness worn by the patient, ropes and pulleys and a counterweight used to partial unload the patient. A patient-cooperative control strategy has been developed that recognize the patient's movement intention and motor ability by monitoring muscular efforts and adapt the robotic assistance to the patient's contribution (Riener et al., 2005). It is recommended that the control and strategies should do the same as a human therapist assisting the patient's movement only as much as needed and informing the patient how to optimize voluntary muscle efforts.

The largest body of scientific is for Lokomat when used by individuals with SCI or stroke. However, there is no consensus of whether and how it affects outcomes in comparison with conventional therapies (Alcobendas-Maestro et al., 2012; Ucar, Parker, & Bugdayci, 2014; Westlake & Patten, 2009) although a recent review provide evidence that acute SCI patients treated with Lokomat showed significantly greater improvement in gait distance and functional level of mobility and independence, and chronic SCI patients a significantly greater improvement in speed and balance were observed than in the group with no intervention (Nam et al., 2017). The Advance Robotic Therapy Integrated Centers (ARTIC) network has recently been set up to collect a large amount of data in order to obtain results with statistical significance. The database includes almost 600 patients not only with SCI but with other neurological conditions with gait deficits who used the Lokomat as part of their rehabilitation (Van Hedel et al., 2018). Other devices, such as the ReoAmbulator (Motorika, New Jersey, USA) have very limited published reports with inconclusive results (Mantone, 2006). A report on LOPES (University of Twente, the Netherlands) showed improved walking ability, as well as gait quality, in subjects with incomplete SCI after an 8 weeks treatment program, with slower walking subjects showing greater benefits (Fleerkotte et al., 2014).

## Ambulatory exoskeletons

Ambulatory exoskeletons are used as a powerful tool in the clinical environment and promoting gait training. Both patients with complete and incomplete SCI can use these exoskeletons but with different aim. Patients with incomplete injuries present an improvement prognosis considering the exoskeletons as a rehabilitation tool. In those cases of complete SCI in which recovery is not foreseeable, its use is intended with the aim of permitting the patient to gain a standing up position, walking short distances and replacing the wheelchair as a means of movement in the community in the future.

They adapt to the lower limbs and have electric motors or other kind of powered actuators that mobilize the joints to produce an automatic overground gait. Furthermore, they offer different approaches on the intelligence of the system, from merely healthy normal gait pattern repetition to EMG-based actuation, passing through error augmentation (Marchal-Crespo & Reinkensmeyer, 2009). These robotic systems make it possible for subjects with SCI to perform the action of walking over ground without the need of partial weight support, harnesses, or the treadmill.

Probably the most popular of these robotic exoskeletons for ambulatory walking is the ReWalk (ReWalk Robotics, Inc., Marlborough, MA, USA). It is a lower limb exoskeleton with two active joints (knee and hip), intended to be used with patients with SCI from T4 to L5 and allowing standing up, sitting down, walking, climbing and descending stairs. There are two versions: for personal use and for rehabilitation. Both exoskeletons are composed of a metallic structure that is adjusted by means of tapes or straps, a pelvic support and motors at the hip and knee joints. The difference between both versions is that the exoskeleton for personal use is customized to the dimensions of the user, whereas, in the case of rehabilitation version, the hip and lateral components are replaceable. It offers several levels of assistance and starts ambulation thanks to a sensor that detects the forwards-leaning of the trunk as a signal to start walking.

The different modes of action (walking forward, going from sitting to standing, stopping, going from standing to sitting) are controlled by a control unit located on the patient's wrist. The device has also a "manual" mode (only in sitting position), where the user can control each joint from its local control system (interface at each joint), useful mode for hazards such as spasticity.

Another lower limb exoskeleton is the Vanderbilt exoskeleton, marketed as Indego (Parker Hanninfin Corp., Cleveland, OH, USA), with a modular design that facilitates its adaptation. This product is intended to be used with patients with SCI or stroke. As other commercial exoskeletons, the hip and knee are motorized. On the other hand, the knee joints consist of an electromechanical brake that blocks the motor in the event of a power failure, to avoid the fall of the patient. This exoskeleton allows gaits at a speed of up to 0.8 km/h (with a battery life of up to an hour). The control of this product is based on postural information and is composed of three detachable elements facilitating the donning and doffing of the user. It can be used in patients with level of injury C7 to L5 in rehabilitation facilities.

Like the previous two, the Ekso (Ekso Bionics, Richmond, CA, USA) has actuators on the hips and knees and has a backpack that contains the batteries and controllers. Ekso Bionics mainly commercializes two lower limb exoskeletons: Hulc (Human Universal Load Carrier) and Ekso (eLEGS at the start). The first one is a hydraulic exoskeleton intended to be used for the transport and handling of loads, not a medical device, thus falling outside of the focus of this short review. Ekso, on the other hand, was introduced by the company to allow paraplegics to stand up and walk using crutches or a walker. This exoskeleton is made up of force and movement sensors, which collect information and transfer it to movement. The approximate weight of this product is 20 kg and it can reach a speed of 3.2 km/h with a battery life of up to 6 h. Its software allows the clinician to adjust the amount of assistance provided at each limb and. The control of the device is performed by the therapist accompanying the patient. Patients with level of injury from T4 to L5 can use this exoskeleton or even from C7 if AIS (ASIA Impairment Scale) D (Mekki, Delgado, Fry, Putrino, & Huang, 2018).

HAL (Hybrid Assitive Limb) developed by the Japanese company Cyberdine was initially developed to assist older adults with muscle weakness in walking (Kawamoto & Sankai, 2002) although it is also used for gait rehabilitation in patients with SCI. It consists of a modular design that provides uni-or bilateral actuation at the hip and/or knee joints. The system allows automatic and voluntary control thanks to the activation of certain muscles whose signal is collected by EMG electrodes.

Currently marketed HAL exoskeleton version is intended to be used for different applications, namely rehabilitation, work that requires force, rescue work and even entertainment. There are currently different versions but HAL-5 is full-body exoskeleton for paraplegic users. Both the hip and the knee function actively; however, the ankle is a passive joint.

These four exoskeletons are approved by the FDA, ReWalk and Indego for use in clinical centers and in the community, while Ekso is only for clinical use with medical supervision.

Unlike already described exoskeletons, Hank (Gogoa Mobility Robots, Guipúzcoa, Spain) has six actuated joints including the two ankles to avoid the effect of foot drop during gait (Asín-Prieto, Intxaurburu Sarasua, Fernández Seco, & Fernández Isoird, 2020). It is based on Exo-H2 (Technaid S.L., Madrid, Spain) (Bortole et al., 2015). Hank is intended for patients with incomplete SCI. Its operating system is also based on the "assist as needed" mode and allows a certain deviation from the ideal gait pattern before applying the correcting force (Fig. 4). It presents an open control architecture to be able to make it compatible with other neural interfaces such as brain computer interfaces (BCI) or brain–machine interface systems, or other technologies that facilitate the recovery process such as functional electrical stimulation (FES). The control modes range from rigid trajectory tracking, to transparent mode, passing thru adjustable assistance per joint. The trajectory is tuned depending on the selected speed and can also be adjusted to user constraints. The device can also perform sit to stand and stand to sit actions.

The first exoskeleton marketed that has its own balance system is the REX (REX Bionics, New Zealand), freeing the patient from using crutches for use, as is the case with other devices. In this way, its use is preferably reserved to treat alterations in postural balance although it also allows walking. It is also the first device intended to be used without any help, as a substitute of the wheelchair, in a daily environment.

As is to be expected due to the novelty of its appearance, the experiences registered with exoskeletons that have been carried out so far present small samples that make it difficult to obtain significant results. Several studies on specific

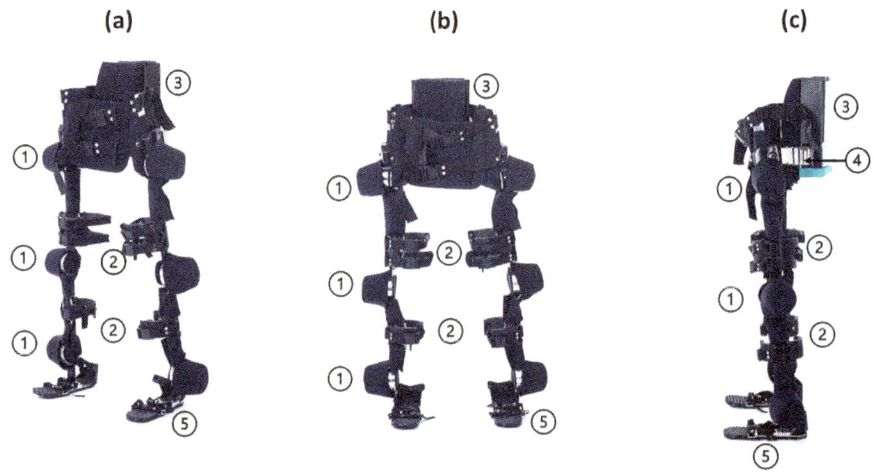


FIG. 4 Hank exoskeleton. HANK exoskeleton; (A) oblique view, (B) frontal view, (C) lateral view; (1) actuators, (2) attachment and fitting, (3) backpack tat containing battery, main microprocessor and the communication electronics connects, (4) flexible arm for adaptation to pelvis, (5) force-sensing-resistors for measuring foot-floor contact.

exoskeletons and feasibility have been conducted and they have found to be practical for use (Bach Baunsgaard et al., 2018; Benson, Hart, Tussler, & van Middentrop, 2016; Esquenazi, Talaty, Packel, & Saulino, 2012; Tefertiller et al., 2018). There are some studies that compare different exoskeletons systems as tools for rehabilitation in the chronic SCI population (Contreras-Vidal et al., 2016). The benefits that have been reported to date include strengthening the muscles, increasing speed and gait efficiency, as well as improvements in aspects of SCI such as spasticity, pain, cardiovascular and metabolism, in the control of intestinal rhythm, in osteoporosis and in quality of life (Winchester et al., 2005) and benefits also in the budget for the recovery (Pinto et al., 2020).

## Applications to other areas of neuroscience

In this chapter we have presented the new features of robotic-based treatments from the point of view of neuroplasticity and their application in therapy. There is an increasing evidence to support the concept for reorganization and plasticity of the injured central nervous system (CNS). The potential for reorganization is particularly high after CNS injury but also possible at later stages. Reorganization in a functionally meaningful way seems to depend on motor activity as executed during rehabilitative training and followed by functional improvements. The science behind exercise in CNS disorders is supported by the therapy concept of increased dosage effect. Task oriented, high repetition movements based on the principles of motor learning can improve muscle strength, motor control, and movement coordination in patients with neurological impairments. All these findings are also applied not only after a SCI but also after brain damage (ictus, traumatic brain injury, cerebral palsy). Robots enhance the rehabilitation process and may improve therapeutic outcomes and have the potential to support clinical evaluation by allowing instrumented measurement of physiological and performance parameters, precisely control and measure the therapeutic interventions, implement novel forms of mechanical manipulation impossible for therapists to provide and supply different forms of feedback, thereby increasing patient's motivating and improving outcomes.

## Mini-dictionary of terms

***Neuroplasticity:*** Ability of the Central Nervous System to make functional changes after injury and adapt to new situation.
***Task-oriented training:*** Training focused on recovering a specific task such as walking.
***Robotic device:*** Device that use robotic technology in rehabilitation programs.
***Functional electrical stimulation (FES):*** Type of electrotherapy aimed at achieving a functional improvement (such as walking) and not the analytical stimulation of a muscle group without having a functional objective.
***Virtual reality:*** It is an environment of scenes or objects of real appearance. The most common meaning refers to an environment generated by computer technology, which creates in the user the sensation of being immersed in it.

***Distal end effector devices:*** Distal end effectors are characterized by the fact that they use a single distal point of contact to guide the movement of the entire limb.
***Exoskeleton-type devices:*** They are structures located in parallel to the different parts of the extremities with more than one point of interaction with the person. They provide direct control over each segment of the limb by incorporating individualized motors.
***Actuators:*** This term is synonymous with motors.
***Haptic:*** Haptic perception is based on the forces experienced during contact with the robotic device, this has allowed the creation of virtual haptic sensations with different qualities of perception.
***Degrees of freedom:*** It refers to the number of planes in which a joint can be moved.
***Assist as needed:*** This term refers to the robotic control strategy in which the actuators act to complete a certain joint path that the patient cannot perform.
***Swing phase:*** This term refers to the gait cycle phase in which the foot is not in contact with the ground and allows the limb to move forward.
***Stance phase***: This term refers to the phase of the walking cycle in which the foot is in contact with the ground giving stability to the limb.

## Key facts of functional recovery

- It is based on the concept of spinal learning via activity-dependent plasticity.
- The training effects of any motor task depend on the provision of sufficient and appropriate stimuli.
- Locomotor activity can be activated in patients with severe SCI via passive activation of the legs on a treadmill.
- Functional rehabilitation (i.e., walking) had to be intensive and task-oriented.
- Gait training using partial weight support systems on treadmills is based on the principles of functional recovery.
- Robotic therapy allows task-oriented treatments and intensive.

## Summary points

- New technologies in neurorehabilitation represents a huge change in treatment protocols for spinal cord injuries
- The training effects depend on the provision of sufficient and appropriate stimuli
- Training must be task-oriented
- Training must be intensive
- Partial body weight support systems on treadmills is based on the principles of functional recovery
- Robotic therapy allows task-oriented and intensive treatment
- Robots offers objective data of patient performance
- There are robots for upper limb and lower limbs
- Robots be classified in distal end effector devices, stationary exoskeletons and ambulatory exoskeletons

## References

Alcobendas-Maestro, M., Esclarín-Ruz, Casado-López, R. M., Muñoz-González, A., Pérez-Mateos, G., González-Valdizán, E., et al. (2012). Lokomat robotic-assisted versus overground training within 3 to 6 months of incomplete spinal cord lesion: Randomized controlled trial. *Neurorehabilitation and Neural Repair*, *26*(9), 1058–1063.

Asín-Prieto, G., Intxaurburu Sarasua, A., Fernández Seco, M., & Fernández Isoird, C. (2020). *HANK_case_study.pdf.* https://doi.org/10.6084/m9.figshare.13041983.v1.

Bach Baunsgaard, C., Vig Nissen, U., Katrin Brust, A., Frotzler, A., Ribeill, C., Kalke, Y. B., et al. (2018). Gait training after spinal cord injury: Safety, feasibility and gait function following 8 weeks of training with the exoskeleton form Ekso bionics. *Spinal Cord*, *56*(2), 106–116.

Barbeau, H., Danakas, M., & Arsenault, B. (1993). The effects of locomotor training in spinal cord injured subjects: A preliminary study. *Restorative Neurology and Neuroscience*, *5*, 81–84.

Barbeau, H., & Rossignol, S. (1987). Recovery of locomotion after chronic spinalization in the adult cat. *Brain Research*, *412*, 84–95.

Bayona, N. A., Bitensky, J., Salter, K., & Teasell, R. (2005). The role of task-specific training rehabilitation therapies. *Topics in Stroke Rehabilitation*, *12* (3), 58–65.

Bayón-Calatayud, M., Gil-Agudo, A., Benavente-Valdepeñas, A., Drozdowskyj-Palacios, O., Sánchez-Martín, G., & del Alamo-Rodríguez, M. J. (2014). Efficacy of new therapies in neurorehabilitation of the upper limb in stroke. *Rehabilitación (Madr)*, *48*, 232–240.

Benson, I., Hart, K., Tussler, D., & van Middentrop, J. J. (2016). Lower-limb exoskeletons for indivuduals with chronic spinal cord injury: Findings from a feasibility study. *Clinical Rehabilitation*, *30*(1), 73–84.

Bortole, M., Venkatakrishna, A., Zhu, F., Moreno, J. C., Francisco, G. E., Pons, J. L., et al. (2015). The H2 robotic exoskeleton for gait rehabilitation after stroke: Early findings form a clinical study. *Journal of Neuroengineering and Rehabilitation*, *12*, 54.

Cai, L. L., Courtine, G., Fong, A. J., Burdick, J. W., Roy, R. R., & Edgerton, V. R. (2006). Plasticity of functional connectivity on the adult spinal cord. *Philosophical Transactions of the Royal Society of London. Series B, Biological Sciences*, *361*(1473), 1635–1646.

Colombo, G., Joerg, M., Schreier, R., & Dietz, V. (2000). Treadmill training of paraplegic patients using a robotic orthosis. *Journal of Rehabilitation Research and Development*, *37*, 693–700.

Contreras-Vidal, J. L., Bhagat, N. A., Brantley, J., Cruz-Garza, J. G., He, Y., Manley, Q., et al. (2016). Powered exoskeletons for bipedal locomotion after spinal cord injury. *Journal of Neural Engineering*, *13*(3), 031001.

Cortés, M., Elder, J., Rykman, A., Murray, L., Avedissian, M., & Stampas, A. (2013). Improved motor performance in chronic spinal cord injury following upper limb robotic training. *NeuroRehabilitation*, *33*(1), 57–65.

Curt, A., Schwab, M. E., & Dietz, V. (2004). Providing the clinical basis for new international therapies: Refined diagnosis and assessment of recovery after spinal cord injury. *Spinal Cord*, *42*, 1–6.

Department of Veterans Affairs. Department of Defense, American Heart Association, American Stroke Association. (2010). *VA/DoD Clinical Practice Guideline for the management of stroke rehabilitation*. The Office of Quality and Performance, VA, Washington, DC and Quality Management Division, United States Army MEDCOM. (Version 3.0). Available form http://www.healthquality.va.gov. or http://www.qmo.amedd.army.mil.

Dietz, V., & Harkema, S. J. (2004). Locomotor activity in spinal cord injured persons. *Journal of Applied Physiology*, *96*, 1954–1960.

Dietz, V., Wirz, M., & Curt, A. (1998). Locomotor pattern in paraplegic patients: Training effects and recovery of spinal cord function. *Spinal Cord*, *36*, 380–390.

Dittuno, P. L., Patrick, M., Stineman, M., & Dittuno, J. F. (2008). Who wants to walk? Preferences for recovery after SCI: A longitudinal and cross-sectional study. *Spinal Cord*, *46*(7), 500–506.

Edgerton, V. R., Courtine, G., Gerasimenko, Y. P., et al. (2008). Training locomotor networks. *Brain Research Reviews*, *57*(1), 241–254.

Edgerton, V. R., & Roy, R. R. (2009). Robotic training and spinal cord injury plasticity. *Brain Research Bulletin*, *78*(1), 4–12.

Esquenazi, A., & Talaty, M. (2019). Robotics for lower limb rehabilitation. *Physical Medicine and Rehabilitation Clinics of North America*, *30*, 385–397.

Esquenazi, A., Talaty, M., Packel, A., & Saulino, M. (2012). The ReWalk powered exoskeleton to restore ambulatory function to individuals with thoracic-level motor-complete spinal cord injury. *American Journal of Physical Medicine & Rehabilitation*, *91*(11), 911–921.

Fasoli, S. E., Krebs, H. I., Stein, J., Frontera, W., Hughes, R., & Hogan, N. (2004). Robotic therapy for chronic motor impairments after stroke: Follow-up results. *Archives of Physical Medicine and Rehabilitation*, *85*(7), 1106–1111.

Fleerkotte, B. M., Koopman, B., Buurke, J. H., van Asseldonk, E. H. F., van der Kooij, H., & Rietman, J. S. (2014). The effect of impedance-controlled robotic gait training on walking ability and quality in individuals with chronic incomplete spinal cord injury: An explorative study. *Journal of Neuroengineering and Rehabilitation*, *11*, 26.

Grillner, S., & Zangger, P. (1984). The effect of dorsal root transection on the efferent motor pattern in the cat's hindlimb during locomotion. *Acta Physiologica Scandinavica*, *120*(3), 393–405.

Hesse, S., Waldner, A., & Tomelleri, C. (2010). Innovative gait robot for the repetitive practice of floor walking and stair climbing up and down in stroke patients. *Journal of Neuroengineering and Rehabilitation*, *7*, 30.

Hubli, M., & Dietz, V. (2013). The physiological basis of neurorehabilitation-locomotor training after spinal cord injury. *Journal of Neuroengineering and Rehabilitation*, *105*.

Kawamoto, H., & Sankai, Y. (2002). Power assist system HAL-3 for gait disorder person. In *International conference on computers for handicapped persons* (pp. 196–203). Berlin, Heidelberg: Springer.

Krebs, H. I., Conroy, S. S., Bever, C. T., & Hogan, N. (2012). Forging Mens et Manus: The MIT experience in upper extremity robotic therapy. In V. Dietz, T. Nef, & W. Zev Rymer (Eds.), *Neurorehabilitation Technology* (pp. 125–140). London: Springer-Verlag.

Krebs, H. I., Hogan, N., Aisen, M. L., & Volpe, B. T. (1998). Robot aided neurorehabilitation. *IEEE Transactions on Rehabilitation Engineering*, *6*(1), 75–87.

Mantone, J. (2006). Getting a leg up? Rehab patients get an assist from devices such as Health South's Auto Ambulator, but the robot's clinical benefits are still in doubt. *Modern Healthcare*, *36*(7), 58–60.

Marchal-Crespo, L., & Reinkensmeyer, D. J. (2009). Review of control strategies for robotic movement training after neurologic injury. *Journal of Neuroengineering and Rehabilitation*, *6*(1), 1–15.

Mehrholz, J., Platz, T., Kugler, J., & Pohl, M. (2009). Electromechanical and robot-assisted arm training for improvement arm function and activities of daily living after stroke. *Stroke*, *40*(5), E392–E393.

Mekki, M., Delgado, A. D., Fry, A., Putrino, D., & Huang, V. (2018). Robotic rehabilitation and spinal cord injury: A narrative review. *Neurotherapeutics*, *15*, 604–617.

Morawietz, C., & Moffat, F. (2013). Effects of locomotor training after incomplete spinal cord injury: A systematic review. *Archives of Physical Medicine and Rehabilitation*, *94*, 2297–2308.

Nam, K. Y., Kim, H. J., Kwon, B. S., Park, J. W., Lee, H. J., & Yoo, A. (2017). Robot-assisted gait training (Lokomat) improves walking function and activity in people with spinal cord injury: A systematic review. *Journal of Neuroengineering and Rehabilitation*, *14*, 24.

Page, S. J., Hill, V., & White, S. (2013). Portable upper extremity robotics is a efficacious as upper extremity rehabilitation therapy: A randomized controlled pilot trial. *Clinical Rehabilitation*, *27*(6), 494–503.

Pinto, D., Garnier, M., Barbas, J., Chang, S. H., Charlifue, S., Field-Fote, E., et al. (2020). Budget impact analysis of robotic exoskeleton use for locomotor training following spinal cord injury in four SCI Model Systems. *Journal of Neuroengineering and Rehabilitation*, *17*(1), 1–13.

Riener, R. (2012). Technology of the robotic gait orthosis Lokomat. In V. Dietz, T. Nef, & W. Zev Rymer (Eds.), *Neurorehabilitation technology* (pp. 221–232). London: Springer-Verlag.

Riener, R., Lunenburger, L., Jezernik, S., Anderschitz, M., Colombo, G., & Dietz, V. (2005). Patient-cooperative strategies for robot-aided treadmill training: First experimental results. *IEEE Transactions on Neural Systems and Rehabilitation Engineering*, *13*, 380–394.

Rudhe, C., Albisser, U., Starkey, M. L., Curt, A., & Bolliger, M. (2012). Reliability of movement workspace measurements in a passive arm orthosis used in spinal cord injury rehabilitation. *Journal of Neuroengineering and Rehabilitation*, *9*, 37.

Siedziewski, L., Schaaf, R. C., & Mount, J. (2012). Use of robotics in spinal cord injury: A case report. *The American Journal of Occupational Therapy*, *66* (1), 51–58.

Takahashi, C. D., Der-Yeghiaian, L., Le, V. H., Motiwala, R., & Cramer, S. C. (2008). Robot-based hand motor therapy after stroke. *Brain*, *131*(Pt 2), 425–437.

Teasell, R. W., & Kaira, L. (2004). What's new in stroke rehabilitation. *Stroke*, *35*, 383–385.

Tefertiller, C., Hays, K., Jones, J., Jayaraman, A., Hartigan, C., Bushnik, T., et al. (2018). Initial outcomes from a multicenter study utilizing the Indego powered exoskeleton in spinal cord injury. *Topics in Spinal Cord Injury Rehabilitation*, *24*(1), 78–85.

Tefertiller, C., Pharo, B., Evans, N., & Winchester, P. (2011). Efficacy of rehabilitation robotics for walking training in neurological disorders: A review. *Journal of Rehabilitation Research and Development*, *48*, 387–416.

Ucar, D. E., Parker, N., & Bugdayci, D. (2014). Lokomat: A therapeutic chance for patients with chronic hemiplegia. *NeuroRehabilitation*, *34*(3), 47–453.

Van Hedel, H. J. A., Severini, G., Scarton, A., O'Brien, A., Reed, T., Gaebler-Spira, D., et al. (2018). Advanced robotic therapy integrated centers (ARTIC): An international collaboration facilitating the application of rehabilitation technologies. *Journal of Neuroengineering and Rehabilitation*, *15*, 30.

Weber, L. M., & Stein, J. (2018). The use of robots in stroke rehabilitation. A narrative review. *NeuroRehabilitation*, *43*, 99–110.

Westlake, K. P., & Patten, C. (2009). Pilot study of Lokomat versus manual-assisted treadmill training for locomotor recovery post-stroke. *Journal of Neuroengineering and Rehabilitation*, *6*, 18.

Winchester, P., McColl, R., Querry, R., Foreman, N., Mosby, J., Tansey, K., et al. (2005). Changes in supraspinal activation patterns following robotic locomotor therapy in motor-incomplete spinal cord injury. *Neurorehabilitation and Neural Repair*, *19*(4), 313–324.

Wirz, M., & Rupp, R. (2012). Application issues for robotic. In V. Dietz, T. Nef, & W. Zev Rymer (Eds.), *Neurorehabilitation technology* (pp. 23–38). London: Springer Verlag.

Zariffa, J., Kapadia, N., Kramer, J. L., Taylor, P., Alizadeh-Meghrazi, M., Zivanovic, V., et al. (2012). Relationship between clinical assessment of function and measurements from an upper-limb robotic rehabilitation device in cervical spinal cord injury. *IEEE Transactions on Neural Systems and Rehabilitation Engineering*, *20*(3), 341–350.

Section F

# Resources

Chapter 45

# Recommended resources and sites for the neuroscience of spinal cord injury

Rajkumar Rajendram[a,b,c], Vinood B. Patel[d], and Victor R. Preedy[e]

[a]College of Medicine, King Saud bin Abdulaziz University for Health Sciences, Riyadh, Saudi Arabia; [b]Department of Medicine, King Abdulaziz Medical City, King Abdullah International Medical Research Center, Ministry of National Guard Health Affairs, Riyadh, Saudi Arabia; [c]Stoke Mandeville Hospital, Buckinghamshire Healthcare NHS Trust, Aylesbury, United Kingdom; [d]University of Westminster, School of Life Sciences, Department of Biomedical Science, London, United Kingdom; [e]Diabetes and Nutritional Sciences Research Division, Faculty of Life Science and Medicine, King's College London, London, United Kingdom

## List of Abbreviation

**SCI** Spinal cord injury

## Introduction

Any insult to the spinal cord temporarily or permanently affecting its function can be defined as a spinal cord injury (SCI). Motor vehicle incidents are currently the most common reason for SCI (Chen, Tang, Vogel, & Devivo, 2013). Approximately, a third of all new SCI is attributable to this single preventable cause (Chen et al., 2013). This is particularly upsetting because, depending on the location (i.e., level) and severity of insult, SCI may significantly impair autonomic, sensory, and/or motor function. As such, SCI often afflicts young people and results in permanent, life-changing, and devastating disabilities.

The first documented reports of patients with SCI are contained in the Edwin Smith Papyrus which arises from around 2500 years BC (Hughes, 1988). Indeed, it is important to note that this seminal document states that SCI is "an ailment not to be treated" (Donovan, 2007; Hughes, 1988). Nearly 5000 years ago, most SCI was probably related to injuries sustained in combat (Donovan, 2007). In that setting, it was probably appropriate to triage the scare resources available on the battlefield to those patients with injuries which would not prevent a return to active military service (Donovan, 2007). Yet, regrettably, in the 21st century, besides those few specialists in neurorehabilitation, many clinicians still approach SCI with a significant degree of therapeutic nihilism.

Until relatively recently, the limited clinical literature on SCI focused purely on the feasibility and appropriateness of surgical intervention (Donovan, 2007). This was in part because developments in the field of anesthesia facilitated surgery for SCI. Regardless, technological advances such as advanced orthotic devices (To, Kirsh, Kobetic, & Triolo, 2005) and powered wheelchairs (Algood, Cooper, Fitzgerald, Cooper, & Boninger, 2005) allow those who are managed conservatively (i.e., without surgery) to have a good quality of life.

Perhaps the most internationally renowned clinician for the rehabilitation of patients with SCI was Sir Ludwig Guttmann (Donovan, 2007). He is most widely recognized as the founder of the Stoke-Mandeville Games which subsequently became the Paralympics (Donovan, 2007). Yet his contribution to improving the outcomes of SCI is equally important. A neurosurgeon appointed to lead the SCI unit at Stoke-Mandeville Hospital, Buckinghamshire, England in 1944; he advocated a holistic approach to this cohort and highlighted the importance of their physicians focusing on rehabilitation rather than acting as single organ "ologists" (Donovan, 2007; Guttmann, 1976). The National Spinal Injuries Centre (NSIC) at Stoke-Mandeville Hospital became a role-model for the handful of centers which subsequently blossomed worldwide.

The NSIC continues to advocate for this complex cohort. It is important to prevent insidious neglect from the misconception that the outcomes of patients with SCI are poor. Indeed, a recent series of patients with SCI admitted to the intensive care unit at Stoke-Mandeville Hospital found that survival to hospital discharge is very good (78%; Adam, Rouse, Ali, & Rajendram, 2019). Thus, although, as yet, there is no cure for SCI, therapeutic nihilism is unwarranted.

The inability of victims of SCI to regain neurological function has been thought (for over 100 years) to be due to the failure of the neurons of the central nervous system to regenerate (Cajal, 1928). Thus, considerable resources have focused

Diagnosis and Treatment of Spinal Cord Injury. https://doi.org/10.1016/B978-0-12-822498-4.00045-2

on attempts to stimulate neuronal regeneration. As a consequence, novel tools for the study of SCI have recently become available. Our understanding of the neuroscience of SCI has advanced, although more slowly than desired. Importantly, the neurons of the central nervous system have been shown to have greater plasticity and greater capacity to regenerate than originally thought (Barnabe-Heider & Frisen, 2008).

Although the promise of being able to initiate neuronal regeneration looms elusively on the horizon, extensive further research is required for SCI to become an ailment that can be cured. Regardless, it even experienced scientists struggle to remain up to date. To assist colleagues who are interested in understanding more about the neuroscience of spinal cord injury, we have therefore produced tables containing up-to-date resources in this chapter. The experts who assisted with the compilation of these tables of resources are acknowledged below.

## Resources

Tables 1–5 list the most up-to-date information on the regulatory bodies (Table 1), journals (Table 2), books (Table 3), professional societies (Table 4), research groups, and centers emerging technologies, platforms, and other resources (Table 5) that are relevant to an evidence-based approach to the neuroscience of spinal cord injury. Some organizations are listed in more than one table as they occasional fulfill more than one role.

**TABLE 1** Regulatory bodies and relevant organizations.

| | |
|---|---|
| American Spinal Injury Association (ASIA) | https://asia-spinalinjury.org/ |
| American Society for Surgery of the Hand (ASSH) | https://www.assh.org/hande/s/tetraplegia |
| Asociación de personas con lesión medular y otras discapacidades física (ASPAYM) | https://www.aspaym.org |
| Associação Brasileira de Fisioterapia Neurofuncional | https://abrafin.org.br/ |
| Associazione Aspal Paratetraplegici Liguria | www.associazione-paratetraplegici-liguria.it |
| Associazione Gruppo Animazione Lesionati Midollari (GALM) | www.galm.it |
| Associazione Il Melograno Organizzazine di Volontariato | http://www.ilmelogranoodv.org/ |
| Associazione Medullolesi Siciliana | http://www.ass-medullolesi.org/ |
| Associazione Paraplegici di Roma e del Lazio | www.apromaelazio.it |
| Associazione Paraplegici Lombardia - Onlus | www.apl-onlus.it |
| Associazione Paraplegici Marche | https://www.apmarche.org/ |
| Associazione Paraplegici Toscana | http://www.atponlus.org |
| Associazione Paratetraplegici Nord Est | http://www.paratetraplegicinordest.it |
| Associazione Spina Bifida Italia | www.spinabifidaitalia.it; |
| Associazione Tetra-Paraplegici Friuli Venezia Giulia Onlus | http://www.paraplegicifvg.it/ |
| Associazione Voglia di Vivere | https://www.vdvpistoia.org/ |
| Australian Spinal Injury Alliance | https://spinalinjuryalliance.com.au/ |
| Canadian Spinal Cord Injury Rehabilitation Association | https://cscira.ca/ |
| Canadian Spinal Research Organization | https://www.csro.com/ |
| Centre for the Rehabilitation of the Paralysed (CRP) Bangladesh | https://www.crp-bangladesh.org/ |
| Christopher and Dana Reeve Foundation | https://www.christopherreeve.org/ |
| Comitato Paralimpico Italiano | http://www.comitatoparalimpico.it/ |
| Craig H Neilsen Foundation | https://chnfoundation.org/ |
| elearnSCI | http://www.elearnsci.org/ |
| European Commission | https://ec.europa.eu/info/index_en |

**TABLE 1 Regulatory bodies and relevant organizations—cont'd**

| | |
|---|---|
| European Network on Independent Living | www.enil.it |
| European Paralimpic Committee | https://www.europaralympic.org/ |
| European Spinal Cord Injury Federation | http://www.escif.org/ |
| Federation of European Societies for Surgery of the Hand (FESSH) | https://fessh.com/ |
| Federazione Associazioni Italiane Paratetraplegici | http://www.faiponline.it/drupal/ |
| Fundación Lesionado Medular | www.medular.org |
| International Group for Research into Spinal Cord Injury (SCI-Research Group) | https://sites.hss.univr.it/npsy-labvr/spinal-cord-injury-research-center/ |
| International Spinal Cord Society | https://www.iscos.org.uk/ |
| International Spinal Research Trust | https://spinal-research.org/ |
| Japan Spinal Cord Foundation | http://www.jscf.org/ |
| Life Rolls On | Liferollson.org |
| Ministério da Saúde Ministry of Health of Brazil | https://www.gov.br/saude/pt-br |
| National Council on Independent Living (NCIL) | https://ncil.org |
| National Institute for Health and Care Excellence | https://www.nice.org.uk/guidance/ng41 |
| National Institute of Neurological Disorders and Stroke | https://www.commondataelements.ninds.nih.gov/Spinal%20Cord%20Injury |
| National Spinal Cord Injury Association of Illinois | https://sci-illinois.org/ |
| Paralyzed Veterans of America | https://www.pva.org/ |
| Reeve Foundation (also known as Christopher & Dana Reeve Foundation) | https://www.christopherreeve.org/ |
| Rick Hansen Foundation | https://www.rickhansen.com/ |
| Sarah Network Rehabilitation Hospitals | www.sarah.br |
| Sheperd Center. Rehabilitation Hospital | https://www.shepherd.org/ |
| Spinal Cord Injuries Australia | https://scia.org.au/ |
| Spinal Cord Injuries: Clinical Trials | https://stemcellsportal.com/clinical_trials_spinal_cord_injuries |
| Spinal Cord Injury Alberta | https://sci-ab.ca/ |
| Spinal Cord Injury British Columbia | https://sci-bc.ca/ |
| Spinal Cord Injury Research Program- Mayo Clinic | https://www.mayo.edu/research/centers-programs/spinal-cord-injury-research-program |
| Spinal injuries association | https://www.spinal.co.uk/ |
| SpinalCord-ItalianLab | https://spinalcord-italianlab.it/ |
| The Community Research and Development Information Service (CORDIS) | https://cordis.europa.eu/article/id/89878-spinal-cord-injury-treatment-and-rehabilitation |
| The International Spinal Cord Society | https://www.iscos.org.uk/ |
| The National Spinal Cord Injury Foundation | www.spinalcord.org |
| Unite2fight paralysis | https://u2fp.org |
| United Spinal Association | https://unitedspinal.org/ |
| United States Food and Drug Administration (FDA) | https://www.fda.gov/home |
| World Health Organization | https://www.who.int |

This table lists the regulatory bodies and organizations involved with the neuroscience of spinal cord injury and associated specialties or interests. The links were accurate at the time of going to press but may move or alter. In these cases, the use of the "Search" tabs should be explored at the parent address or site. See also Table 4.

**TABLE 2 Relevant journals publishing original research and review articles related to the neuroscience of spinal cord injury.**

| |
|---|
| Spinal Cord |
| Journal of Spinal Cord Medicine |
| World Neurosurgery |
| Journal of Neurotrauma |
| Neural Regeneration Research |
| Archives of Physical Medicine and Rehabilitation |
| Experimental Neurology |
| Scientific Reports |
| Spinal Cord Series and Cases |
| PLoS One |
| Topics in Spinal Cord Injury Rehabilitation |
| Chinese Journal of Tissue Engineering Research |
| Spine |
| Neuroscience Letters |
| International Journal of Molecular Sciences |
| Journal of Neurosurgery Spine |
| Disability and Rehabilitation |
| European Spine Journal |
| Neurourology and Urodynamics |
| Journal of Neuroscience |
| Spine Journal |
| Molecular Neurobiology |
| Neuroscience |
| Frontiers in Neuroscience |
| Journal of Neuroinflammation |
| Frontiers in Neurology |
| Global Spine Journal |
| Frontiers in Cellular Neuroscience |
| American Journal of Physical Medicine and Rehabilitation |
| Medicine United States |

Journals publishing original research and review articles related to the neuroscience of spinal cord injury. Included in this list are the top 30 journals which have published the most number of articles on spinal cord injury over the past 5 years. Data derived from Scopus.

## Application to other areas of neuroscience

The pathophysiology and management of spinal cord injuries are similar to traumatic injuries to the other components of the nervous system. These include the brain and the peripheral nervous system. Thus, the contents of this chapter are also relevant to the understanding of traumatic brain injuries and peripheral neuropathies.

**TABLE 3** Relevant books.

| Book title | Authors or editors | Publisher | Year |
|---|---|---|---|
| AACD Reabilitação | Fernandes AC, Ramos ACR, Morais Filho MC, Ares MJJ | Manole | 2014 |
| AOSpine Masters Series, Volume 7: Spinal Cord Injury and Regeneration | Vialle LR, Fehlings M, Weidner N | Thieme | 2016 |
| CNS Regeneration: Basic Science and Clinical Advances | Tuszynski MH, Kordower J | Elsevier | 1999; 2007 |
| Critical Care in Spinal Cord Injury | Fehlings M | Future Medicine LTD | 2013 |
| Delisa's Physical Medicine and Rehabilitation: Principles and Practice | Delisa JA | Lippincott Williams & Wilkings | 2010 |
| Diagnostic Imaging: Spine | Ross JS, Moore R | Elsevier | 2020 |
| Diseases of the Spinal Cord | Hattingen, E | Springer | 2015 |
| Diseases of the Spinal Cord: Novel Imaging, Diagnosis and Treatment | Hattingen, EHattingen E, Weidauer S, Setzer M, Klein JC, Vrionis F | Springer | 2013 |
| Essentials of Spinal Cord Injury. Basic Research to Clinical Practice | Fehlings MG, Vaccaro AR, Boakye M, Rossignol S, Dituno JF Jr., Burns AS | Thieme Medical Publishers | 2012 |
| Essentials of the Adult Neurogenic Bladder | Corcos J, Karsenty G, Kessler T, Ginsberg D, | Taylor & Francis Group | 2015 |
| Functional Electrical Stimulation: Standing and Walking After Spinal Cord Injury | Kralj AR, Bajd T | CRC Press | 1989 |
| Hand Function A Practical Guide to Assessment | Duruöz MT | Springer | 2019 |
| Incontinence 6th edition | Abrams P, Cardozo L, Wagg A, Wein A | International Consultation on Urological Diseases | 2017 |
| Ischemic and Traumatic Brain and Spinal Cord Injuries Mechanisms and Potential Therapies | Farooqui A | Academic Press | 2018 |
| Lesión medular. Enfoque multidisciplinar | Esclarin de Ruz A | Panamericana | 2020 |
| Living with Spinal Cord Injury: A Wellness Approach | Cristian A | Demos Medical Publishing | 2010 |
| Management and Rehabilitation of Spinal Cord Injuries | Ko HY | Springer | 2019 |
| Management of Spinal Cord Injuries: A Guide for Physiotherapists | Harvey L | Elsevier | 2008 |
| Manual de Medicina Fisica y Rehabilitación | Frontera WR, Silver JK, Rizzo T | Elsevier España | 2020 |
| Medicina e Reabilitação: Princípios e Práticas | Fernandes AC, Ramos ACR, Casilis MEP, Herbert SK | Artes Médicas | 2013 |
| Neurological Aspects of Spinal Cord Injury | Weidner N, Rupp R, Tansey K | Springer | 2017 |
| Neurotrauma: A Comprehensive Textbook on Traumatic Brain Injury and Spinal Cord Injury | Wang K | Oxford University Press | 2018 |
| Programa de Atualização em Fisioterapia Neurofuncional (PROFISIO) | Faria C, Leite H | Artmed | 2021 |
| Recovery of Motor Function Following Spinal Cord Injury | Fuller, H | IntechOpen | 2016 |
| Rehabilitation in Spinal Cord Injuries | Reznik J, Simmons J | Elsevier | 2020 |

*Continued*

**TABLE 3 Relevant books—cont'd**

| Book title | Authors or editors | Publisher | Year |
|---|---|---|---|
| Spinal Cord Injuries Management and Rehabilitation | Sisto SA, Druin E, Sliwinski MM | Elsevier | 2008 |
| Spinal cord injury | Holtz A, Levi R | Oxford University Press. | 2010 |
| Spinal Cord Injury (SCI) Repair Strategies | Giuseppe P, Filippo R | Elsevier Health Sciences | 2019 |
| Spinal Cord Injury: A guide for patient and families (American Academy of Neurology) | Selzer M, Dobkin B | Demos Health | 2008 |
| Spinal Cord Injury: Management and Rehabilitation | Sisto SA, Druin E, Sliwinski MM | Mosby-Elsevier | 2009 |
| Spinal Cord Medicine | Kirshblum S, Campagnolo DI | Lippincott Williams & Wilkins | 2011 |
| Spinal Cord Medicine, 3rd Edition | Kirshblum S, Vernon WL | Springer | 2018 |
| Spinal Trauma: Imaging, Diagnosis and Management | Schwaartz ED, Flanders AE | Lippincott Williams & Wilkins | 2006 |
| Spine and Spinal Cord Trauma | Vaccaro A, Fehlings M, Dvorak M | Thieme Medical Publishing | 2011 |
| Textbook of neurogenic bladder Third edition | Corcos J, Ginsberg D, Karsenty G | Taylor & Francis Group | 2015 |
| The art of healthy living with physical impairments | Lagerstrom A-C, Wahman K | Spinalis | 2014 |
| The Physiology of Exercise in Spinal Cord Injury | Taylor JA | Springer | 2017 |
| Therapeutic Strategies to Spinal Cord Injury | Jendelova P | MDPI | 2018 |
| Traumatic Brain and Spinal Cord Injury. Challenges and Developments | Morganti-Kossmann C, Raghupathi R, Maas A | Cambridge | 2012 |
| Urologic management of the spinal cord injured patient | Elliott S, Gomez R | SIU Academy | 2017 |
| Vascular Anatomy of the Spinal Cord | Thron AK | Springer | 2016 |

This table lists books on the neuroscience of spinal cord injury.

**TABLE 4 Professional societies and other organizations.**

| Society name | Web address |
|---|---|
| Academy of Spinal Cord Injury Professionals | https://www.academyscipro.org/ |
| Academy of Spinal Cord Injury Professionals (ASCIP) | https://www.academyscipro.org/ |
| American Academy of Physical Medicine and Rehabilitation (AAPM&R) | https://www.aapmr.org/ |
| American Congress of Rehabilitation Medicine (ACRM) | https://acrm.org/ |
| American Spinal Cord Injury Association (ASIA) | https://asia-spinalinjury.org/ |
| AO Spine Knowledge Forum Spinal Cord Injury | https://aospine.aofoundation.org |
| Asian Spinal Cord Network (ASCoN) | https://ascon.info/ |
| Asociación Española de Enfermería especializada en Lesión Medular | www.aselme.com |
| ASSH (American Society for Surgery of the Hand) | https://www.assh.org/hande/s/tetraplegia |
| Association of Academic Physiatrists (AAP) | https://www.physiatry.org/ |

**TABLE 4 Professional societies and other organizations—cont'd**

| Society name | Web address |
|---|---|
| Australian and New Zealand Spinal Cord Society | https://anzscos.org/ |
| Canadian Spinal Cord Injury Rehabilitation Association | https://cscira.ca/ |
| Christopher & Dana Reeve Foundation | www.christopherreeve.org |
| European Spinal Cord Injury Federation (ESCIF) | http://www.escif.org/ |
| Federation of European Societies for Surgery of the Hand (FESSH) | https://fessh.com/ |
| Fehlings Lab Twitter | www.twitter.com/DrFehlings |
| Fehlings Lab Website | www.drfehlings.ca |
| International Continence Society | https://www.ics.org |
| International Neuro-Urology Society | https://www.neuro-uro.org |
| International Society of Physical Medicine and Rehabilitation (ISPRM) | https://www.isprm.org/ |
| International Spinal Cord Society (ISCoS) | https://www.iscos.org.uk/ |
| Korean Spinal Cord Society | http://www.koscos.kr/ |
| National Organization For Rare Disorders (NORD) | https://rarediseases.org/organizations/national-spinal-cord-injury-association/ |
| North American Spine Society | https://www.spine.org/ |
| Praxis Spinal Cord Institute | www.praxisinstitute.org |
| Protection Center of Spinal Cord Disables of Iran | www.irannokhaa.ir (not viable at the time of going to press) |
| Sheperd Center. Rehabilitation Hospital | https://www.shepherd.org/ |
| Shirley Ryan Hability Lab | https://www.sralab.org/conditions/spinal-cord-injury |
| Sociedad Española de Paraplejia | www.sociedaddeparaplejia.com |
| Società Italiana Chirurgia della Mano (SICM) | https://www.sicm.it/ |
| Society for neuroscience | https://www.sfn.org |
| Spinal Cord Injury Associations & Organizations (USA) | https://www.sci-info-pages.com/spinal-cord-injury-organizations/ |
| Spinal Cord Injury Canada | https://sci-can.ca/about-us |
| Spinal Cord Injury Ontario | https://sciontario.org/ |
| Spinal Cord Injury Research Evidence (SCIRE) | https://scireproject.com/ |
| Spinal Cord Society-Indian Chapter | http://www.scs-isic.com/ |
| Spinal Injuries Association | https://www.spinal.co.uk/ |
| The Academy of Spinal Cord Injury Professionals | https://www.academyscipro.org/ |
| The Asian Spinal Cord Network | https://ascon.info/ |
| The Canadian Spinal Research Organization (CSRO)/American Spinal Research Organization (ASRO) | https://www.csro.com/ |
| The International Spinal Cord Society | https://www.iscos.org.uk/ |
| The Nordic Spinal Cord Society | http://noscos.org/ |
| United Spinal Association | https://unitedspinal.org/ |

This table lists some societies and organizations devoted to understanding the neuroscience of spinal cord injury. Please note, occasionally the location of the websites or web address changes. Not viable at the time of going to press indicates that the site has changed or is the process of being changed. See also Table 1.

**TABLE 5** **Emerging techniques, platforms, and other sites of interest relevant to the neuroscience of spinal cord injury.**

| Organization or company or society name | Web address |
|---|---|
| Acreditando | https://www.acreditando.com.br/ |
| American Society for Surgery of the Hand (ASSH) | https://www.assh.org/hande/s/tetraplegia |
| American Spinal Injury Association (ASIA)-e Learning Center | https://asia-spinalinjury.org/learning/ |
| Association for Assistance of Disabled Children (AACD) | https://aacd.org.br/centro-de-reabilitacao |
| Avery biomedical devices | https://www.averybiomedical.com/spinal-cord-injury-treatments/ |
| California Institute for Regenerative Medicine (CIRM) | https://www.cirm.ca.gov |
| Christopher & Dana Reeve Foundation | https://www.christopherreeve.org/ |
| Clinical Trials. Gov | https://clinicaltrials.gov |
| Dalhousie University Faculty of Medicine Wheelchair Skills Program (WSP) | https://wheelchairskillsprogram.ca/en/ |
| Elearning-SCI | http://www.elearnsci.org |
| European Multicenter Study about Spinal Cord Injury (EMSCI) | https://www.emsci.org |
| facing disability | https://facingdisability.com/resources/assistive-technology |
| Federation of European Societies for Surgery of the Hand (FESSH) | https://fessh.com/ |
| Gaylord Speciality Healthcare | https://www.gaylord.org/Patients-Families/Conditions-Services/Spinal-Cord-Injury-Program |
| inspire neurocare | https://www.inspireneurocare.co.uk/ |
| Instituto de Medicina Física e Reabilitação (Rede Lucy Montoro) | https://www.redelucymontoro.org.br/site/programa-de-reabilitacao.html |
| Instituto Novo Ser | http://www.novoser.org.br/index.html |
| International Collaboration on Repair Discoveries (ICORD) | https://icord.org/ |
| International Group for Research into Spinal Cord Injury (SCI-Research Group) | https://sites.hss.univr.it/npsy-labvr/spinal-cord-injury-research-center/ |
| International Society of Spinal Cord Injury | www.iscos.org.uk |
| Kentucky Spinal Cord Injury Research Center | https://louisville.edu/kscirc |
| Kessler Foundation | https://kesslerfoundation.org/ |
| Mayo Foundation for Medical Education and Research | https://www.mayo.edu/research/centers-programs/spinal-cord-injury-research-program |
| Miami Project to Cure Paralysis | https://www.themiamiproject.org/ |
| Model System Knowledge Translation Center (MSKTC) | https://msktc.org/sci |
| National Institute of Neurological Disorders and Stroke: SCI | https://www.ninds.nih.gov/Disorders/Patient-Caregiver-Education/Hope-Through-Research/Spinal-Cord-Injury-Hope-Through-Research |
| Ontario Neurotrauma Foundation | https://spauldingrehab.org/ |
| Physiopedia (Assistive Devices for Spinal Cord Injury) | https://www.physio-pedia.com/Assistive_Devices_for_Spinal_Cord_Injury |
| Praxis Spinal Cord Institute | https://praxisinstitute.org/ |

**TABLE 5** Emerging techniques, platforms, and other sites of interest relevant to the neuroscience of spinal cord injury—cont'd

| Organization or company or society name | Web address |
|---|---|
| ProBed | https://www.pro-bed.com/blog/info/top-10-spinal-cord-research-organizations |
| PropelPysiotherapy | https://propelphysiotherapy.com/spinal-cord-injury/assistive-devices-sci-rehabilitation |
| Rede SARAH (Specialized assistance in rehabilitation) | https://www.sarah.br/especialidades/neurorreabilitacao-em-lesao-medular/ |
| ReWalk | https://rewalk.com/ |
| Shepherd Center | https://www.shepherd.org/resources-healthcare-professionals/research/spinal-cord-injury/current |
| Società Italiana Chirurgia Della Mano (SICM) | https://www.sicm.it/ |
| Spaulding Rehabilitation Network | https://onf.org/knowledge-mobilization/spinal-cord-injury-2/knowledgemobilization-sci-networkssummits/ |
| Spinal Cord Injury Adaptive Equipment And Assistive Technology | https://www.sci-info-pages.com/adaptive-equipment/ |
| Spinal Cord Injury and You (SCI-U) | http://sci-u.ca/ |
| Spinal Cord Injury Research Evidence | https://scireproject.com/ |
| Spinal Cord Injury Trials Finder | https://scitrialsfinder.net |
| Spinal Cord Outcomes Partnership Endeavor | https://scope-sci.org/ |
| Spinal Cord Research Centre | https://scrc.umanitoba.ca/wp/ |
| Spinal Research | https://spinal-research.org/ |
| Spinal Research Institute | https://www.thesri.org/spinal-cord-research-hub/ |
| SpineUniverse | https://www.spineuniverse.com/ |
| The American Trauma Society | https://www.amtrauma.org/ |
| The Big Idea | https://reevebigidea.org/ |
| Transforming Research and Clinical Knowledge in Spinal Cord Injury (TRACK-SCI) | https://spinalcordinjury.ucsf.edu |
| Unite 2 Fight Paralysis | https://u2fp.org/ |
| United Spinal Association | https://askus-resource-center.unitedspinal.org/index.php?pg=kb.book&id=32 |

This table lists some emerging technologies and platforms relevant to the neuroscience of spinal cord injury. Please note, occasionally the location of the websites or web address changes.

## Mini-dictionary of terms

**Orthotic device:** A support/brace for the spine or limbs.
**Neuronal plasticity:** The ability of neural networks to adapt and/or change by reorganization and/or growth.
**Neuronal regeneration:** The repair/regrowth of neurons by the formation of new axons, synapses neurons, or glia.
**Neurorehabilitation:** The process which aims to restore function to patients who have sustained a neurological insult such as stroke or spinal cord injury.
**Therapeutic nihilism:** The perception that it is impossible to improve the outcome of a patient with a specific condition.

## Key facts of spinal cord injury

- Any insult to the spinal cord temporarily or permanently affecting, its function can be defined as a spinal cord injury.
- Spinal cord injury often afflicts young people and results in permanent, life-changing, and devastating disabilities.

- Following spinal cord injury, therapeutic nihilism is unwarranted as survival to discharge home is good and technological advances have greatly improved quality of life.
- Lack of functional recovery post spinal cord injury is thought to be due to failure of central neurons to regenerate.
- Despite great advances, vast amounts must still be learned about the neuroscience of spinal cord injury before this devastating condition can be cured.

## Summary points

- Patients with spinal cord injury often survive to be discharged at home, and their quality of life has been improved by technological advances.
- Although there is currently no cure for spinal cord injury, prognostic pessimism is unwarranted.
- There is significant interest in stimulating neuronal regeneration to promote functional recovery after spinal cord injury.
- Recent advances have suggested that central neurons have greater plasticity than previously thought. This seed plants the hope that the ability to control neuronal regeneration is on the horizon.
- The expansion of the knowledge and understanding of the neuroscience of spinal cord injury has been slow but steady. It is becoming increasingly difficult for those interested in this field to remain up to date.

## Acknowledgements (in alphabetical order)

We would like to thank the following authors for contributing to the development of this resource. We apologize if some suggested material was not included in this chapter or has been moved to different sections.
Abou L, Aftab K, Alam M, Bernabéu-Sanz A, Breshnan J, Calderón Vallejo D, Champs A, Cunha do Espírito Santo, C da Silver F, Desneves K, Díaz Galindo M, Enam SA, Fehlings M, Ferreiro Velasco E, Fisher G, Ftouni L, Fusini F, Hong Y, Khalatbary AR, Lacerda de Araújo AV, Lu P, Mortenson B, Moussalem C, Mujtaba SB, Palazon R, Park K, Pavel J, Scandola M, Sekido N, Siraj MA, Sumsuzzman D, Yohann A.

## References

Adam, M., Rouse, A., Ali, T., & Rajendram, R. (2019). Intensive care unit resource use by patients with spinal cord injury. *Journal of the Intensive Care Society*, *20*(Suppl. 1 EP.240), 182.

Algood, S. D., Cooper, R. A., Fitzgerald, S. G., Cooper, R., & Boninger, M. L. (2005). Effect of a pushrim-activated power-assist wheelchair on the functional capabilities of persons with tetraplegia. *Archives of Physical Medicine and Rehabilitation*, *86*, 380–386.

Barnabe-Heider, F., & Frisen, J. (2008). Stem cells for spinal cord repair. *Cell Stem Cell*, *3*, 16–24.

Cajal, R. (1928). *Degeneration and regeneration of the nervous system*. London: Oxford University Press.

Chen, Y., Tang, Y., Vogel, L. C., & Devivo, M. J. (2013). Causes of spinal cord injury. *Topics in Spinal Cord Injury Rehabilitation*, *19*, 1–8.

Donovan, W. H. (2007). Donald Munro lecture. Spinal cord injury- -past, present, and future. *The Journal of Spinal Cord Medicine*, *30*, 85–100.

Guttmann, L. (1976). Management of spinal fractions. In L. Guttmann (Ed.), *Spinal cord injuries, comprehensive management and research* (pp. 7–21). London: Blackwell Scientific Publications, Oxford Press.

Hughes, J. T. (1988). The Edwin Smith Papyrus; an analysis of the first case reports of spinal cord injuries. *Paraplegia*, *26*, 71–82.

To, C. S., Kirsh, R. F., Kobetic, R., & Triolo, R. J. (2005). Simulation of a functional neuromuscular stimulation powered mechanical gait orthosis with coordinated joint locking. *IEEE Transactions on Neural Systems and Rehabilitation Engineering*, *13*, 227–235.

# Index

Note: Page numbers followed by *f* indicate figures and *t* indicate tables.

## D

## G



## H



## I

## O



## P

## Q



## R

## T